Mechanisms and Machine Science

Volume 125

This book series establishes a well-defined forum for monographs, edited Books, and proceedings on mechanical engineering with particular emphasis on MMS (Mechanism and Machine Science). The final goal is the publication of research that shows the development of mechanical engineering and particularly MMS in all technical aspects, even in very recent assessments. Published works share an approach by which technical details and formulation are discussed, and discuss modern formalisms with the aim to circulate research and technical achievements for use in professional, research, academic, and teaching activities.

This technical approach is an essential characteristic of the series. By discussing technical details and formulations in terms of modern formalisms, the possibility is created not only to show technical developments but also to explain achievements for technical teaching and research activity today and for the future.

The book series is intended to collect technical views on developments of the broad field of MMS in a unique frame that can be seen in its totality as an Encyclopaedia of MMS but with the additional purpose of archiving and teaching MMS achievements. Therefore, the book series will be of use not only for researchers and teachers in Mechanical Engineering but also for professionals and students for their formation and future work.

The series is promoted under the auspices of International Federation for the Promotion of Mechanism and Machine Science (IFToMM).

Prospective authors and editors can contact Mr. Pierpaolo Riva (publishing editor, Springer) at: pierpaolo.riva@springer.com

Indexed by SCOPUS and Google Scholar.

More information about this series at https://link.springer.com/bookseries/8779

Zuzana Dimitrovová · Paritosh Biswas ·
Rodrigo Gonçalves · Tiago Silva
Editors

Recent Trends in Wave Mechanics and Vibrations

Proceedings of WMVC 2022

Springer

Editors
Zuzana Dimitrovová
Departamento de Engenharia Civil
Universidade Nova de Lisboa
Caparica, Portugal

IDMEC, Instituto Superior Técnico
Universidade de Lisboa
Lisbon, Portugal

Rodrigo Gonçalves
Departamento de Engenharia Civil
Universidade Nova de Lisboa
Caparica, Portugal

Paritosh Biswas
University of North Bengal
Jalpaiguri, West Bengal, India

Tiago Silva
Depart. de Eng Mecânica e Industrial
Universidade Nova de Lisboa
Caparica, Portugal

ISSN 2211-0984 ISSN 2211-0992 (electronic)
Mechanisms and Machine Science
ISBN 978-3-031-15757-8 ISBN 978-3-031-15758-5 (eBook)
https://doi.org/10.1007/978-3-031-15758-5

This Springer imprint is published by the registered company Springer Nature Switzerland AG
The registered company address is: Gewerbestrasse 11, 6330 Cham, Switzerland

Preface

The 10th International Conference on Wave Mechanics and Vibrations (10th WMVC) was announced by the Von Karman Society for Advanced Study and Research in Mathematical Sciences and held at Lisbon's city centre, in Portugal, at VIP Executive Zurique Hotel, on 4–6 July 2022.

The 10th WMVC was jointly organized by the Department of Civil Engineering of the NOVA School of Science and Technology, NOVA University of Lisbon, (DEC/FCT/UNL) and IDMEC, Institute of Mechanical Engineering of the Instituto Superior Técnico of the University of Lisbon (IDMEC/IST/UL).

The WMVC series has its roots in India; it was founded by Prof. Paritosh Biswas from University of North Bengal, India, as an annual national conference. Due to the pandemic situation, the 9th edition was cancelled. The 10th edition, held in Lisbon, elevated the WMVC series to the international level. Given the current situation, the organizers chose a hybrid in-person/remote format, which proved to be the most suitable.

The scientific programme consisted of four plenary lectures given in-person, three keynote lectures given remotely, 105 in-person regular presentations and 96 remote ones, grouped according to 25 Mini-Symposia and a General Conference Topic. The conference was attended by 88 remote participants including three keynote speakers and 127 in-person participants. The total number of represented countries was 36.

The organization was closely linked to the Journal of Vibration Engineering & Technologies (JVET, Springer Nature), with the support of Editor-in-Chief Prof. C. Nataraj, Director of Villanova Center for Analytics of Dynamic Systems, Villanova University, Pennsylvania, USA. Springer Nature issued certificates for the Best Paper Award for Ph.D. students, handed to the best theee students by Associate Editor Mrs. Sonal Choudhary. A special issue in JVET will be organized from selected contributions.

The main objective of the conference was to bring together academicians, engineers, scientists, researchers, mathematicians and technologists devoting their work to wave mechanics and vibration-related problems in different areas of scientific and engineering applications on a common platform. Wave mechanics and vibration problems are frequently encountered in various areas of mechanical, civil, aeronautical, robotics, marine, nuclear, biology and earthquake engineering, science and technology. Mathematical theories, numerical simulations, machine intelligence techniques, physical experiments and their engineering and science applications have been presented in this event, providing an outstanding opportunity for university teachers, researchers and industry experts for sharing new ideas and experience. We strongly believe that the 10th WMVC had a significant impact on the development of contemporary analytical, numerical and experimental methods in wave mechanics and vibration technology and created an opportunity for opening a forum for discussion and collaboration amongst the participants.

The submission of full-length papers was not mandatory, but several authors who presented their work at this conference chose to prepare them. These submissions were subjected to peer review. We are very pleased to announce that this resulted in a total of 125 accepted papers linked to regular presentations and one to a keynote lecture. It is sincerely hoped that the readers will find in this volume of **Recent Trends in Wave Mechanics and Vibrations** purposeful, challenging and stimulating new ideas related to all areas of science and engineering in general and to wave mechanics and vibration in particular.

We would like to thank the organizing institutions DEC/FCT/UNL and IDMEC/IST/UL for their support, to the VIP Executive Zurique Hotel staff for the excellent service and to the travel Agencies TOP Atlântico and Abreu for managing the reservations and registrations.

We wish to thank all participants for contributing to a diverse set of topics covered by their research presentations and full-length articles and for recreating the spirit of a conference by their presentations and active discussions inside as well as outside the presentation rooms.

We extend our thanks to the supporting staff responsible for registration and clarifying any participants' doubts and questions. The valuable help provided by our undergraduate students and research fellows responsible for maintaining the programme schedule and ensuring a smooth switch between remote and in-person presentations is also gratefully acknowledged.

Acknowledgments are also extended to the Ph.D. students who responded to our appeal and participated in the Best Paper Award competition, and to the judging panel, for their assessment.

We are grateful for the contributions of the plenary and keynote speakers for their rich and high scientific quality presentations.

We wish to thank to the Mini-Symposia organizers for attracting participants to their topics and chairing the respective sessions. Our thanks are extended to the reviewers who, together with the Mini-Symposia organizers, contributed to the scientific quality of this volume. The papers in this volume are ordered according the Mini-Symposia listed in alphabetical order of their abbreviation. Within each Mini-Symposium, the papers order is also alphabetical, according to the paper title.

Last but not least, the editors express profuse thanks to Springer Nature for taking up the responsibility to publish this volume.

Zuzana Dimitrovová
Conference Chairperson

Rodrigo Gonçalves
Tiago Silva
Local Organizing Committee

Contents

Keynote Lecture

Vibration of Flexible Robots: A Theoretical Perspective

Debanik Roy(✉)

Division of Remote Handling and Robotics, Department of Atomic Energy, Bhabha Atomic Research Centre and Homi Bhabha National Institute, Govt. of India, Mumbai 400085, India
deroy@barc.gov.in

Abstract. Flexible Robotic Systems, by and large, are prone to inherent vibration that recreates itself in several modal frequencies. This in-situ vibration in flexible robots or in any such complaint robotic units becomes tricky so far as the control system architecture is concerned. Thus, customization of the design and firmware of higher-order flexible robots is highly challenging due to its inherent parameters related to real-time vibration. Subsuming technological challenges, the field of *Flexible Robotics* has come out as a niche ensemble of harnessing non-linearity in dynamics of the robotic system(s).

Keywords: Flexible robot · Compliance · Vibration · Modal frequency · Non-linearity

1 Introduction

Modeling of 3D highly-flexible frames (structures) play a crucial role in characterizing the inherent random vibration of Flexible Robotic System (FRS). In fact, a near-perfect analytical model involving flexible/complaint dynamics of jointed frames is the foundation for successful real-time control of multi-degrees-of-freedom flexible robot (e.g., applications in space industry, highly flexible antenna, radio telescope, solar panel etc.) as well as Compact Compliant Robot (C2R). Many of the features and physics of motion of FRS & C2R finally get translated into a modified variety of *Assistive Robotics*, which is the emerging domain of Compliant Bio-inspired Robot (CBR) (e.g., rolling-type, crawling-type, swaging-tail type etc.). Such robotic structures necessarily have low mass and very high flexibility, so that we can consider large displacement behavior of such systems during real-time modeling. This modeling of ultra-high flexible multi-body systems is worth studying for the purpose of successful prototyping of various pertinent robotic system that undergo medium to large trembling under run-time conditions, leading to instability as well as modal vibrations with high eigen vectors.

A first-hand approach to imbibe 'flexibility' and/or 'compliance' effect in the analysis of super-slender jointed mechanism is to adopt a hypothesis of 'small displacements' and 'rotations' in an element-attached frame. This 'frame' can be selected with respect to an infinitesimal segment of the original flexible structure (ideally, the flexible link) and all the analyses will proceed thereof. These analyses are extremely instrumental

Z. Dimitrovová et al. (Eds.): WMVC 2022, MMS 125, pp. 3–16, 2023.
https://doi.org/10.1007/978-3-031-15758-5_1

in judging the penetration of the vibration and can be categorized into two metrics, namely: i] rheology-based analysis & ii] tribology-based analysis. It may so happen that we may have to perform either of these two types of analysis based on the real-time characteristics of the in-situ trembling or modal vibrations of the flexible robotic gadgets. While rheological studies are prudent for FRS; tribological studies are must for CBR. Under such conditions for all the three members of the triad, namely, FRS, C2R & CBR, linear finite element analysis can be extended to handle problems in a non-linear domain. However, this hypothesis of breaking the flexible body into infinitesimal segments is not very well suited to model ultra-flexible members and/or 'soft' members.

In spite of carrying an engineering boon of possessing very low tare-weight the prime shortcoming for the practical usability of FRS is essentially linked to the perpetual trembling of its constituent members as well as the end-of-arm tooling (gripper). The source of this inherent vibration can be attributed to the internal stress/strain; however, this very vibration is structure- independent as well as design-invariant. The signature of natural vibration of FRS is quantitatively ascertained through two facets, viz. Modal frequency & eigen value. We do observe chronological built-up of trembling of the slender links of FRS and at times, shaking of the link-joint interface zone. Due to multi-dimensionality of the workspace. The development of working prototype of FRS is instrumental in bringing out the issues pertaining to its ready deployment in several social & non-manufacturing sectors. However, indigenous firmware of FRS is often customized, in order to suit the requirement for indoor assistive services, e.g., patients and elderly persons. The facets of rheology (stress-strain paradigms), randomized vibration, sensory instrumentation and non-linear coupled dynamics are some of the paradigms that need extended research in a synergistic way.

In this paper, we will focus on two representative varieties of flexible robot, one each from FRS & C2R family, designed and developed with complete indigenous technology. In course of the detailing, focused references will be made on the novel prior arts of prototypes of our earlier flexible robots, developed chronologically in past six years. We will refer to the eight indigenously-designed & developed FRS, with specific reference to the novel vibration harnessing mechanisms. Chronologically, the prototypes are titled as: 1] DDFR (Direct Drive Flexible Robot); 2] FSFR-I (Flexible Shaft-driven Flexible Robot: Type I); 3] FSFR-II (Flexible Shaft-driven Flexible Robot: Type II); 4] PAR (Patient Assistant Robot); 5] FUMoR (Flexible Universal Modular Robot); 6] Flex2R (Flexible Rotary-joint Robot); 7] SAIPAR (Sensor Augmented Intelligent Patient Assistant Robot) and 8] FlexFAR (Flexible Feeding Assistant Robot). While Flex2R is completely manually-operated experimental robotic system, dedicated for the study on run-time vibration and its measure via force closure of grasp using a novel sensor-augmented miniaturized robotic gripper, DDFR &FUMoR are manufactured with direct-drive transmission. The rest of the prototypes are actuated with loop-through drive mechanism, namely via flexible shaft(s). The last two prototypes (SAIPAR &FlexFAR) are at advanced stages of manufacturing. Out of these varieties of FRS, we will focus on the characteristics of FUMoR in this paper.

In-line with the development of the FRS-family hardware, we will delineate the working prototype of a C2R and design insight for a miniature CBR. While the prototype

C2R is aimed at grasping semi-soft objects within a very compact workspace (e.g. table-top), the design of CBR is conceptualized with the fundamentals of perpetually of the rolling motion over planar surfaces. The perpetually of motion has been ensured through the novel design of a cam-like structure, called, *retainer ring,* that utilizes the kinetic energy generated out of dynamic friction with the ground surface. The designed CBR has gradual decrement of external envelope with a final aim of deployment for internal inspection of co-planar pipelines of smaller orifices (~ 35–50 mm.). The study of inherent vibration for the tiny CBR as well as small-sized C2R need subtle dynamic modeling in order to establish a robust control system algorithm, either under tethered or wireless communication.

Prior to design of multi-degrees-of-freedom FRS, we must have a foolproof dynamic model that is commensurate to the expected work-output of the system [1, 2]. Thus, we need to consider quantitative metrics for vibration attenuation with the help of dynamic model before the onset of the design for manufacturing of FRS [3]. Extended modeling of the spring-mass tuple has been found to be very effective in establishing new dynamic model of FRS [4]. Several other poignant methodologies have been reported for the reduction of vibration and subsequent real-time control of the FRS [5–8].We may appreciate that non-field trials on the real-time performance of mono-link flexible manipulator (mostly without gripper) have accurized a considerable momentum in recent past. A majority of these flexible manipulators have been used extensively to validate a variety of novel control strategies [9–11]. Nonetheless, grasp-based design of the multi-link FRS having near-compliant sub-assemblies as well as miniature gripper at the free-end, remains an open research domain till date.

We have investigated interesting scenarios of control dynamics for a multi-degrees-of-freedom FRS fitted with mini-gripper [12], aided by novel spring-structural model (for vibration signature of both manipulator-links & gripper) and strain gauge-induced model (for dynamic strain signature). Firmware of a multi-link serial-chain FRS using flexible shafts for drive motion of the joints have been developed by the author's group [13]. The complete design set-up with modeling for real-time vibration as well as beta version of the hardware for a slender serial-chain three degrees-of-freedom flexible robot, meant for patient assistance is described by the author [14].

It may be stated that experiment-based case-study of the performance of the FRS-gripper amidst in-situ vibration demands a synergistic coupling of two metrics: a] optimization of design parameters of the FRS-gripper and b] joint-space redundancy of a general serial-chain slender manipulator, such as FRS. Topology optimization-based modeling & novel hypothesis towards decision matrix for robotic grasp have been delineated [15]. On the other hand, novel theory and analytical model for joint-space redundancy of the serial-chain slender manipulator is reported [16]. Based on these earlier accomplishments, a new theology on deformation & deflection for the time-specific adjudgment of natural frequency of vibration will be presented here.

The paper has been organized in six sections. An overview of the fundamental modeling semantics of the perpetual vibration of serial-chain FRS is presented in the next section. Section 3 reports the paradigms of run-time vibration with inherent flexibility, with illustration and physical prototypes built (FRS & C2R). The characteristics of the run-time data on vibration analysis are presented in Sect. 4. Section 5 addresses the preludes to the experimental investigation on run-time vibration of prototype flexible robots and finally, Sect. 6 concludes the paper.

2 Perpetual Vibration of Flexible Robots: Modeling Fundamentals

The most fundamental aspect of real-time vibration of flexible robot is instantaneous deflection of its link(s) that can be quantified through displacement vectors. This deflection is assumed to be uniform and wide-spreaded inside the body of the link. In other words, the deflection of the link is assumed to be homogeneous and uniformly distributed over the surface as well as inside the link-body. So, fundamentally, this run-time deflection of FRS or C2R is very much an intrinsic property and it is highly dependent on the material of construction of the link. Of course, stability of the link(s) of FRS or C2R post-vibration is always a technological issue that needs careful measurement of the stable deflection values. Hence, at the nucleus of the analytical model supporting the vibration of FRS we will study the vector analysis of the micro-displacements of the '*elements*' of the FRS-link. This basic novel conceptual framework is represented schematically in Figs. 1a,b.

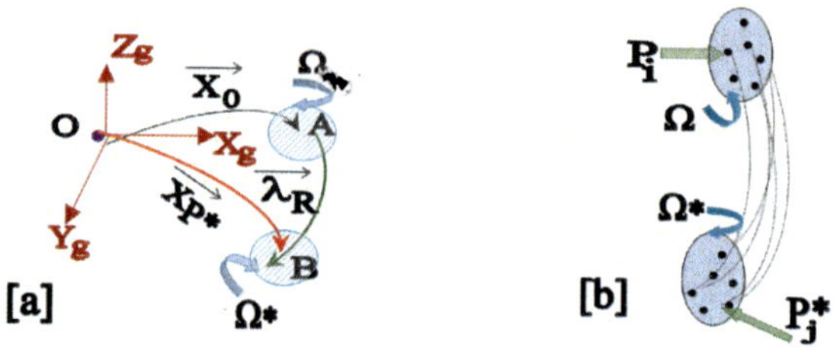

Fig. 1. Conceptual framework of the in-situ deflection of the link of flexible robot: [a] general view; [b] intra-link assessment

The basic nature of the run-time deflection of the FRS-link is highlighted in Fig. 1a, wherein the link OA is getting displaced instantaneously to the position OB, with respect to the global/home reference frame [X_g Y_g Z_g]. The position vectors for points 'A' & 'B' are respectively $\vec{X}_0$ & $\vec{X}_{p*}$ in accordance to the zones of in-situ deflection of the link, namely, 'Ω' & 'Ω*'. It is important to note here that there is no pre-assumed trajectory of this deflection; rather we can't predict this trajectory using mathematical tools a-priori. This is, in fact, the biggest irony for the design & run-time control of the flexible robots. Thus, the very nature of the said trajectory, indexed by $\vec{\lambda}_R$ needs to be ascertained only through repeated experimentations, using a range of 'excitations' (external force-functions). It is true that these three vectors ($\vec{X}_0$, $\vec{X}_{p*}$ & $\vec{\lambda}_R$) may form a triad, which, at times, can be solved geometrically. But physical assessment of the deflection can be only be made via experimental data. Large set of testings were reported on this facet by several research-groups and those do give a clear picture of the nature of this deflection in quantitative terms. The schematic of Fig. 1b essentially brings out this concept of experimental variation. We can observe that a representative cross-section of the FRS-link (at zone Ω) is zoomed in to highlight the internal fixed locations (the black dots) that undergo excitation by an external forcing function, 'P_i'. In other words, all the elements inside a fixed cross-section/zone of the FRS-link will undergo in-situ micro-deformation but the extents of such deformations are not similar all the times. The deformed state of the said cross-section has been demarcated by the zone Ω^* that contains the same internal elements but in deformed states. The curved line-segments of Fig. 1b represent those unequal deformations of FRS-link at element-level. The modeling paradigms of the perpetual vibration of flexible robots are highly dependent on the external excitation and the corresponding forcing function(s). Thus, we can assume that a separate forcing function, 'P^*_j' is acting over the other cross-section. This vibrational forcing function is the key for the detailed vibration study of the FRS subsequently.

We shall now investigate another crucial aspect of the vibration model, namely, the Spring-Structural Model (SSM). The model has been conceptualized and expanded indigenously, with specific reference to serial-chain flexible manipulators. Figure 2 illustrates the so-called *longitudinal syntax* of the in-situ deflection of the FRS-link. Here, the modeling of flexibility has been proposed through a tri-junction, namely coupling of spring constant (K_N), viscous damping coefficient (C_N) and transient amplitude factor (T_N). In other words, the quantum of final deflection of the FRS-link will be governed by the numerical values of these three members.

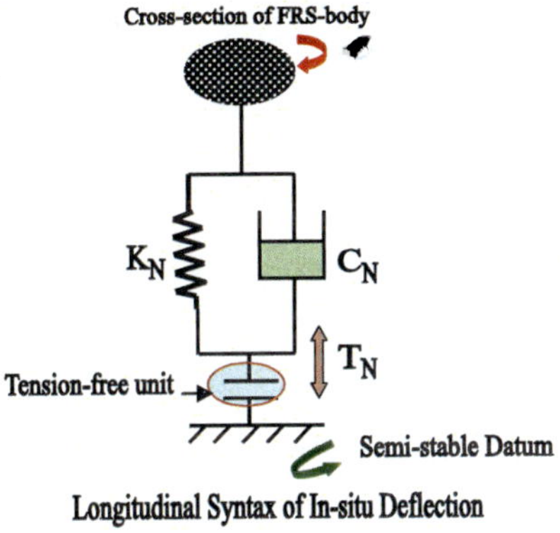

Fig. 2. Illustrative schematic of the longitudinal syntax of in-situ deflection of a serial-chain flexible robot

It is also important to state that the concept of semi-stable datum in Fig. 2 is largely imaginary, unless the flexible robot is mounted over a fixed frame. For the sake of generality, we have visualized the SSM to be in slow rotation mode, i.e. the link(s) of the flexible manipulator will be subjected to some twist or slow jerking (external excitation).

3 Run-time Vibration Under Inherent Flexibility

Customization of the design and firmware of higher-order flexible robots is highly challenging due to its inherent parameters related to run-time vibration. Although design ensemble of serial-chain FRS has been refined over the years with novel research paradigms, it is still crucial to study the niche for perpetual vibration especially in flexible robots that has highly-compliant links. In our endeavor of designing & prototyping various kinds of serial-chain FRS, we have observed that flexible robot with gooseneck-styled slender link is dynamically more challenging than the same having standard cross-section hollow cylindrical link. We made experimental investigation towards ascertaining the most optimal cross-section of the FRS-link out of three possible alternatives, viz. Circular, rectangular & square. Likewise, another crucial factor in judging run-time vibration is the ensemble morphology of the FRS-link: be it external or internal. While external metrics can be with varieties of links having either straight or tapered or stepped cross-section, we can realize the physical hardware of the flexible robot with two internal facets, viz., solid or hollow cross-section. Naturally, the dynamics of FRS will alter significantly as per the finalized design ensemble as discussed above. The other crucial parameter for vibration characterization is material of fabrication of the FRS-link(s), which by & large, gets manifested through type of material (preferably non-metallic), density & Young's Modulus.

We have found over the chronological development of seven hardware of FRS that run-time vibration differs significantly based on the joint paradigms. This facet will not be a parent for mono-link flexible robots because joint dynamics is the least in such cases, except the rotational motion of the base joint. In our first prototype, christened as, 'Direct-Drive Flexible Robot' (*DDFR*), we have observed the influence of the mating pairs of the three revolute joints therein. Unlike traditional way of fabricating revolute joints around a vertical axis, in DDFR we opted for revolute joints between two adjacent links in side-by-side alignment. However, in our subsequent prototypes, namely, 'Flexible Shaft-Driven Flexible Robots' (*FSFR-I & II*) we used vertical alignment of the revolute joints (refer: https://youtu.be/P9-FrIKBBSM). The run-time vibration gets heavily influenced by the joint sub-assembly that partakes link-joint interface as well. The concept of high-efficiency revolute joint sub-assembly was used in a well-acclaimed prototype FRS of ours, namely, 'Patient Assistant Robot–*Beta version*' (*PARv1.0*): its performance including vibration harnessing techniques can be viewed at: https://www.youtube.com/watch?v=Nwbt94E8Tds&t=98s. In nutshell, the design features of a typical revolute joint of FRS can be grouped on the basis of its type, mating pairs, alignment and sub-assembly so far as vibration abetment is concerned. The other ancillary parameters that might be having some influence over the harnessing of in-situ perpetual vibration of FRS are drive mechanism and end-of-arm attachment. We have tried both types of drive mechanisms that are possible in FRS, viz. Direct-to-joint (e.g. DDFR) and via flexible shaft (e.g. FSFR-I, FSFR-II, PARv1.0). Flexible shaft-driven FRS is slightly robust so far as vibration attenuation is concerned and on the basis of this experience, we have finalized the design for manufacturing for our next two FRS-hardware, namely, Sensor-Augmented Intelligent Patient assistant Robot:*SAIPAR* and Flexible Feeding Assistance Robot:*FlexFAR*, with multiple indigenously-manufactured flexible shafts. Like drive mechanism, we have also noticed moderate to heavy variation in vibration signature for FRS with different kinds of end-of-arm tooling, like tip-mass, miniature gripper & end-effector.

The conceptual framework of single-degree-of-freedom flexible robot with slender gooseneck-styled flexible link is illustrated in Fig. 3a. The corresponding solid model through Computer Aided Design (CAD) is shown in Fig. 3b. The mapping of the legends between the two figures may be noted.

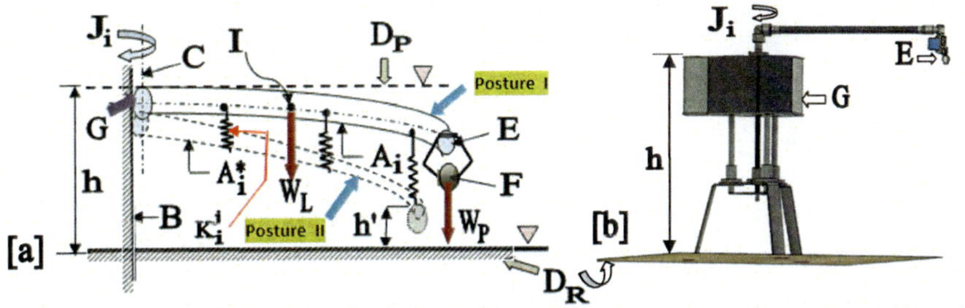

Index: $\mathbf{A_i}$: i[th] Gooseneck Link; $\mathbf{A^*_i}$: Deflected Posture of the ith Link; **B**: Vertical Support (imaginary); C: Axis of the Base Joint (Revolute); $\mathbf{D_P}$: Pesudo Datum; $\mathbf{D_R}$: Real Datum; **E**: Gripper; **F**: Object being Grasped; **G**: Revolute Joint Assembly; **h**: Height of the Flexible Robot (undeformed posture); **h'**: Height of the Flexible Robot (after deflection); **I**: Centre of Gravity of the Flexible Link; $\mathbf{J_i}$: Rotation of the Base Joint (for i[th] Link); $\mathbf{K_i^j}$: j[th] Imaginary Spring for the ith Link; $\mathbf{W_L}$: Self-Weight of the Link; $\mathbf{W_P}$: Payload of the Gripper; *Possture I*: Original Stable Disposition of the Link; *Posture II*: Deflected Disposition of the Link (after attaining stability, post-vibration)

Fig. 3. Layout of single degree-of-freedom flexible robot: [a] schematic view; [b] 3D solid model (CAD) view

It is crucial to note that the vibration and/or rheological features of the FRS do alter under two situations of end-of-arm attachment, viz. End-/tip-mass vis-à-vis a mini-gripper (at the distal link of the FRS). With the advent of our expertise in prototyping multi-link FRS, we have brought in the concept of modularity in the hardware. The prototype development of the flexible robot having CAD view as that shown above (refer Fig. 3b) has been accomplished with the coherence of prior arts and allied exposure through troubleshooting on earlier prototypes. Figure 4 presents the final working-level prototype of the said flexible robot, entitled, Flexible Universal Modular Robot (*FUMoR*). FUMoR is the first on this kind of mono-link flexible modular robot that uses a non-standard geometry of the link, namely, gooseneck-styled. The prototype FUMoR has one gooseneck link, one custom-built revolute joint, end-connectors & adapters for

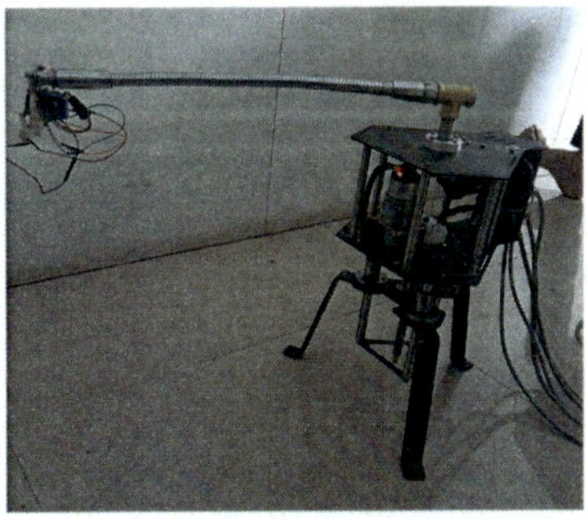

Fig. 4. Photographic view of the working prototype of flexible universal modular robot ["Namy-omeet"]

fitting with the tripod base at one end and miniature gripper at the other end. The functionalities and grasp performance of FUMoR can be visualized at: https://youtu.be/liKWRwHfHVA.

We will now study the prototype hardware of a novel C2R, which has been designed with two focused intentions, viz.: a] highly slender & low tare-weight links and b] adaption to multiple grippers having design variations in the jaws. Both of these objectives have been successfully accomplished through our novel hardware. Like FUMoR or for that matter any other FRS, perpetual vibration is a concern here too, which was studied in depth before the grasp. Figure 5 shows the photographic views of the prototype C2R in working postures using two different grippers for the grasp of semi-soft object.

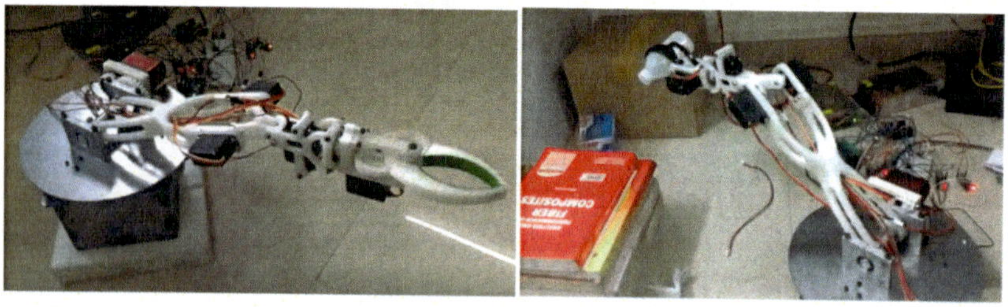

Fig. 5. Photographic views of the prototypes of compact compliant robot fitted with two different grippers

It is important to note that in all experiments with the prototype C2R selection of the rotational speed of the joint-motor(s) becomes instrumental so far as the proper functioning of the control system is concerned, amidst slow jerking/vibration of the system. Object manipulation (by the gripper-jaws) was carried out with a slow speed at the wrist, which is approximately 25% of the base rotation speed (~3–4 rpm). We have also incorporated a time delay between every incremental angle of rotation so as to ensure jerk-free movement of the wrist & gripper assembly during task execution. The customized program module has been made interactive in order to systemize multiple grippers. In order to combat the run-time vibration of the prototype C2R, two posture-driven strategies have been implemented in order to evolve an effective control strategy. This novel control semantics has become a boon for the prototype C2R, wherein objects are being grasped through two modes, namely: a] in-plane grasping and b] off-plane lifting. Both of these two strategies have been found to be very effective in minimizing the system trembling or run-time vibration. The real-time performance of this prototype C2R can be viewed at: https://www.youtube.com/watch?v=lpkQ8JM1-rY and https://www.youtube.com/watch?v=K6OZYSr-8pk.

4 Source Term and Characteristics of Run-time Data on Vibration

Contrary to the traditional proposition of non-linear analysis of in-situ deflection of FRS or CBR, the alternate approximations towards solving the said transient Finite Element (FE) analysis involve the definition of a set of generalized deformation parameters, degenerated at the element level or the formulation of 3D degenerated beam elements. Another formulation in this context consists of deriving the '*beam equation*' for the in-situ vibration directly from a 3D non-linear theory that accounts for ensemble finite rotations of the joints of the FRS or CBR. Nonetheless, solution of any 3D non-linear model necessarily calls for assumptions and therefore, we need to introduce approximate beam kinematics in the said dynamics model of vibration as assumptions. This transformed 3D non-linear theory with approximate beam kinematics leads to the so-called concept of 'Geometrically Exact Beam Theory (GEBT)'. New formulations of run-time dynamics for our prototype FRS & CBR have been achieved by successfully adapting the basic lemma of GEBT.

It is prudent and computationally or scientifically perfect to employ an updated Lagrangian formulation for the said novel modeling using GEBT, wherein vibration-induced rotations of the flexible/compliant systems are described as increments with respect to previous configuration. The resulting mathematical formulation becomes a set of Eulerian transformation with regard to the rotational degrees-of-freedom of the specific flexible system. Our novel mathematical formulae allow, on the other hand, a relatively simple derivation of fully linearized static and dynamic operators, but, on the other hand, these operators are non-symmetric in a general case of FRS or CBR. Nonetheless, 'symmetry' is shown to be recovered at equilibrium post-vibration so far as statics of the vibrating bodies are concerned, provided appropriate conditions of conservatively of external loads to these flexible robots are respected.

We have used celebrated Euler-Bernoulli beam theory for the FE-based dynamic analysis of the prototype FUMoR. We have assumed that the end-displacement of the FRS, namely, that of the pair of jaws of the miniature gripper can be separated into two parts, respectively dependent on position and time as shown below:

$$w(x, t) = \Lambda(x)\Psi(t) \tag{1}$$

where, w(x,t): overall displacement of the gripper-jaws; Λ(x): component of the displacement as a function of position and Ψ(t): component of the displacement as a function of time. The governing equation that was used for estimating these two components of displacement under FEA to ensure simple harmonic motion is presented below:

$$\frac{EI}{kA(x)}\frac{\partial^4\Lambda(x)}{\partial x^4} = \frac{1}{\Psi(t)}\frac{\partial^2\Psi(t)}{\partial t^2} = -\omega^2 \tag{2}$$

where, E: Young's modulus of the material; I: Moment of inertia of the '*beam*'(gripper-jaw/body); A: Cross-sectional area of the '*beam*'; k: Linear mass density of the '*beam*'(= ρA; ρ: material density); ω: Frequency of vibration of the '*beam*'. Now, segregation of the position variable of Eq. (2) will lead to the following FE-equation:

$$\Lambda(x) = C_1 Sinh(\delta x) + C_2 Cosh(\delta x) + C_3 Sin(\delta x) + C_4 Cos(\delta x) \tag{3a}$$

where, $\{C_1,..C_4\}$ are constants and

$$\delta = \left(\frac{\rho A\omega^2}{EI}\right)^{1/4} \tag{3b}$$

Likewise, we can get the following extraction of Eq. (2) for time variable, as used in FE-analysis:

$$\frac{\partial^2 \Psi(t)}{\partial t^2} + \omega^2 \Psi(t) = 0 \tag{4}$$

The FE-analysis-based solution for Eq. (4) will be:

$$\Psi(t) = C_5 Sin(\omega t) + C_6 Cos(\omega t) \tag{5}$$

where, C_5 and C_6 are constants.

Finally, the combined FE-equation for the displacement of the gripper-jaw/body becomes:

$$\text{w(x,t)} = (C_1 Sinh\delta x + C_2 Cosh\delta x + C_3 Sin\delta x + C_4 Cos\delta x)X\,(C_5 sin\omega t + C_6 Cos\omega t) \tag{6}$$

where, the constants $\{C_1, C_2, C_3, C_4\}$ can be obtained from the boundary conditions and $\{C_5, C_6\}$ are obtainable from the initial conditions under the FE-analysis.

The above formulation proves that appropriate description of the flexible members in many cases requires the use of '*beam*' formalism that incorporates inherent non-linear effects of the run-time vibration of such flexible/complaint robotic gadgets in a systematic manner. 'Geometric Stiffening' is one such source of non-linearity that can be studied extensively in case of FRS. Effective modeling of our prototype FRS was carried out by incorporating the theory of geometric stiffening. It is, therefore, essential to rely upon a tailor-made GEBT, which should have important kinematic assumptions, like: i] the flexible beam will be straight initially; ii] cross-sections of the beam remain in plane and do not deform during elastic deformation of the flexible frame (unless it is a soft flexible member); iii] the neutral axis of the flexible beam undergoes shear deformation and iv] the rotational kinetic energy of the cross-sections is taken into account (especially true for CBR). In nutshell, we need to consider the episodes of turn, twist and deflection of the segmented infinitesimal cross-section(s) of the flexible link/body/member, by defining the co-ordinates of any arbitrary point inside the said infinitesimal cross-section(s). It is important to note that the positional paradigm of any point inside the cross-section will have an additional vector component due to the inherent vibration of the FRS or at times, CBR. However, this additional vector will be manifested in those two Cartesian axes that are in quadrature to the third axis, representing the direction of in-plane as well as in-situ vibration.

5 Preludes to the Experimental Investigation of Vibration of Flexible Robots

There are two fundamental pathways for the experimental manifestation of the tailor-made deflection model of a flexible robot, namely: a] via angular measurement & b] via Cartesian frame measurement. Both of these methods do rely on the global Cartesian frame of reference, which may or may not coincide with the first link of the multi-link

FRS/C2R or first body-segment of the CBR. The commonality between the experimental study for FRS and CBR is the evaluation of the rotational momemtum of either the flexible link (for FRS) or the body-segment (for CBR). Figure 6 schematically illustrates the representative CBR, having two body-segments and a pair of grippers. This specific design of the CBR is a direct adaptation of the prototype C2R and can be viewed as the alteration of design morphology (*'links'* are transformed to *'chambers'* etc.). As the schematic unfolds, we can be sure of angular measurements so as to calculate the ensemble deflection of the system due to run-time vibration.

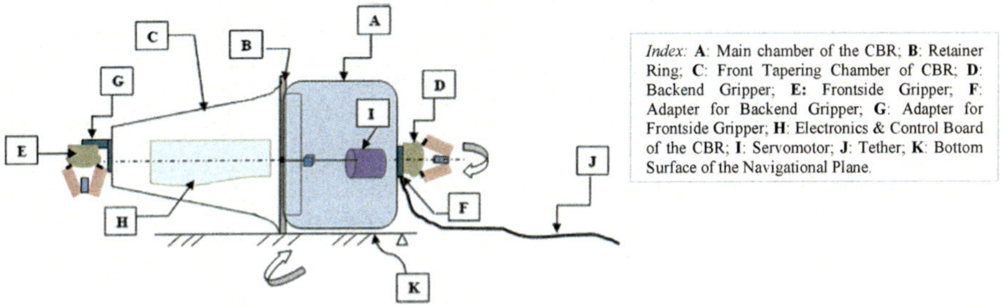

Fig. 6. Schematic disposition of a representative compliant bio-robot with two body-segments

The commonality of both methods is the evaluation of the displacement of the infinitesimal cross-section of the flexible link that will have direct functional relationship with the rotation vector of the flexible link using Euler angle parenthesis. Cartesian co-ordinates of each & every points inside these infinitesimal cross-sections of the flexible link(s) need to be defined a-priori, which can be treated as the primary pre-requisite for building up the deflection model of the system. Accordingly, all such co-ordinates must have a component, signifying the effect of vibration. But, this effect due to in-situ vibration will be attributed only in two quadrature axes, leaving apart the axis around which rotation is taking place. For example, if the rotation occurs around X-axis, then deflection due to vibration will be apparent in Y & Z axes. It may be stated that all points inside the said infinitesimal cross-sections will be twisted as well as slightly deformed to the corresponding points under the deflected postures of those infinitesimal segments. In fact, twisting effect of vibration/trembling should be inspected with due priority as the effect can be infused to the adjacent links as well. The basic ingredient for experiments on twisting due to vibration is to design and fabricate slender plate-type link. This is unique in many aspects, be it the extended zone of twist or procrastination of vibration. Accordingly, the design thought-process for FRS needs a paradigm shift and conceptually, the usual hollow cylindrical shaped links are to be altered to flat-type slender links with fixtures in between.

Figure 7 pictorially describes such a typical experimental test-bed that is equipped with two flat-type slender links. The set-up has been fabricated in the form of a Flexible Rotary-Joint Robot (Flex2R) that has been used for the evaluation of force closure of grasp under different conditions of run-time vibration: both natural and under external excitation of the flexible link.

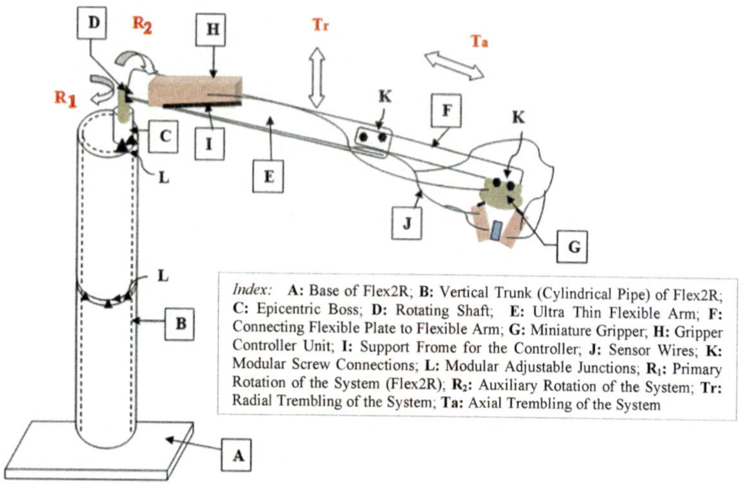

Fig. 7. Schematic of the flexible rotary-joint robot as the experimental test-bed for vibration study

The combined effect of turn, twist & deformation of these infinitesimal segments of a specific flexible link will be apparent in the quadrature axes to the direction of the principal stress. Mathematically, we can compute the positional vector due to the said combined effect through the vector cross-product of unit vectors along the Cartesian co-ordinates of the original state and the deflected state of the flexible link(s). The global co-ordinate system for this computation need not be always at the end-point of the flexible link and if we assume that the global co-ordinate frame is located at a finite vector-distance with respect to the flexible link then we have to factor in another positional vector. Nonetheless, the rotation vector, symbolizing the vibration, can be applied to all co-ordinate reference systems that are part & parcel of the said FRS or CBR. Likewise, Cartesian reference frames can be put up at any location inside the flexible link(s), depending upon the modeling requirement. The said position vectors can be computed accordingly, i.e. based on the specific fixations of the co-ordinate frames. Figure 8 postulates this lemma on co-ordinate frames, as a precursor for experimental investigation on the run-time vibration of FRS or CBR. The successive position vectors, P_i & P_{i*} are being mapped via the series of revolute joints of FRS, viz. J_i, $J_{i*}, \ldots J_n$ & J_{n+1}.

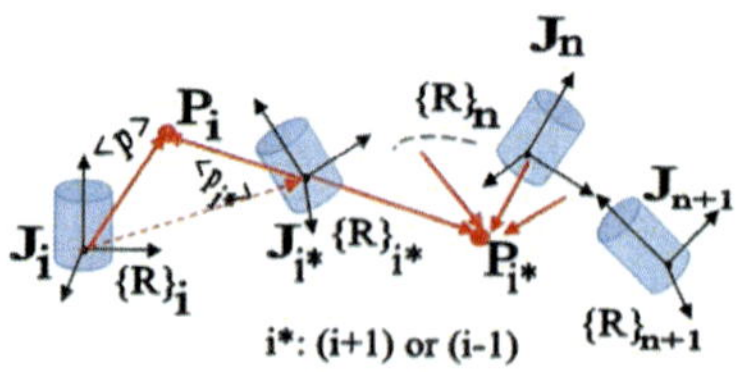

Fig. 8. Analytical postulation on co-ordinate frames for FRS or CBR for vibration study

We may note the angular postures of the joints and thereby, variations in the position vectors. These are important theoretical framework for the evaluation of the vibration through frequency analysis method. The turn & twist of the joints and more importantly, the link-joint interface zone are crucial for enhancement of deflection and thereby natural frequency of vibration and finally, modal frequencies.

The other important paradigm of the vibration analysis of flexible robots is related to the evaluation of the dynamic 'distance function' of the arbitrary position of a point inside the infinitesimal cross-section. The dynamicity of the said distance function is mainly attributed to the run-time shear deformation of the flexible link, wherein cross-section(s) do not remain orthogonal to the neutral axis of the said 'beam'. The micro-rotations, leading to trembling and subsequently, vibrations of the flexible link, do occur along the quadrature axes with respect to the basic axis of shear deformation. Analytically, this leads to equations of 'curvature' of the neutral axis of the flexible 'beam' that can happen in two postures, viz. 'single-curve' & 'double-bulge'. This paradigm is illustrated through the analytical postulation of Fig. 9 below, which depicts the spring-effect of deflection and subsequent vibration.

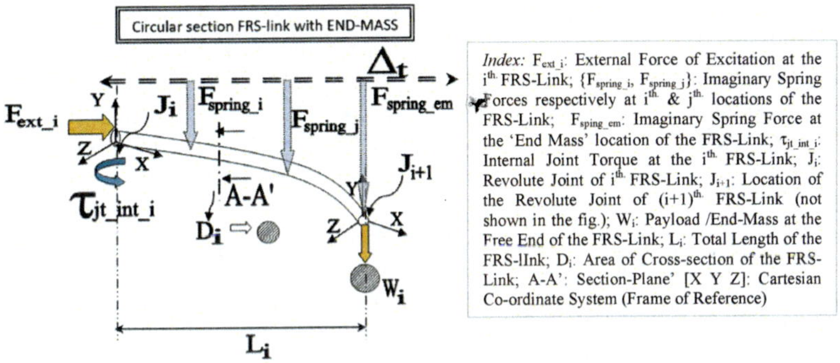

Fig. 9. Schematics showing spring-effect of deflection and subsequent vibration of FRS-link

6 Conclusions

The field of *Assistive Flexible Robotics* is a niche ensemble of harnessing non-linearity in dynamics of the robotic system(s). The most challenging episode of real-time control of in-situ vibration of flexible robots is harnessing its perpetually. As soon as one set of vibration retreats or subsides down to a reasonable extent, another set of vibration stages in, which finally transforms the flexible robot into a perpetually vibrating system. This perpetual vibration gives enough jitter to the run-time control of the FRS & CBR. The theoretical perspective of the sourcing of this vibration is, thus, very important for the application technologists because final fine-tuning of the system controller will depend on the formulation as well as experimental investigation. The test-data are generated mostly from the sensors, mounted on the FRS or CBR, which need to be processed through analytical model to arrive at the frequency of oscillation.

References

1. Chen, W.: Dynamic modeling of multi-link flexible robotic manipulators. Comput. Struct. **79**(2), 183–195 (2001)
2. Feliu, V., Somolinos, J.A., Garcia, A.: Inverse dynamics-based control system for a three degrees-of-freedom flexible arms. IEEE Trans. Robot. Autom. **19**(6), 1007–1014 (2003)
3. Singer, N.C., Seering, W.C.: Preshaping command inputs to reduce system vibration. J. Dyn. Syst., Meas. Control-Trans. ASME **112**, 76–82 (1990)
4. Zhang, J., Tian, Y., Zhang, M.: Dynamic model and simulation of flexible manipulator based on spring & rigid bodies. In: Proceedings of the 2014 IEEE International Conference on Robotics & Biomimetics ('ROBIO-2014'), pp. 2460–2464 (2014)
5. Chen, Y.P., Hsu, H.T.: Regulation & vibration control of an FEM-based single-link flexible arm using sliding-mode theory. J. Vibr. Control **7**(5), 741–752 (2001)
6. Tjahyadi, H., Sammut, K.: Multi-mode vibration control of a flexible cantilever beam using adaptive resonant control. Smart Mater. Struct. **15**, 270–278 (2006)
7. Trapero-Arenas, J.R., Mboup, M., Pereira-Gonalez, E., Feliu, V.: Online frequency and damping estimation in a single-link flexible manipulator based on algebraic identification. In: Proceedings of the 16th Mediterranean Conference on Control & Automation (IEEE), pp. 338–343 (2008)
8. Pereirea, E., Aphale, S.S., Feliu, V., Moheimani, S.O.R.: Integral resonant control for vibration damping and precise tip-positioning of a single-link flexible manipulator. IEEE/ASME Trans. Mechatron. **16**(2), 232–240 (2011)
9. Feliu, J., Feliu, V., Cerrada, C.: Load adaptive control of single-link flexible arms based on a new modeling technique. IEEE Trans. Robot. Automat. **15**(5), 793–804 (1999)
10. Canon, R.H., Schmitz, E.: Initial experiments on the end-point control of a flexible robot. The Int. J. Robot. Res. **3**(3), 62–75 (1984)
11. Kotnick, T., Yurkovich, S., Ozguner, U.: Acceleration feedback control for a flexible manipulator arm. J. Robot. Syst. **5**(3), 181–196 (1998)
12. Roy, D.: Control of inherent vibration of flexible robotic systems and associated dynamics. In: Chakraverty, S., Biswas, Paritosh (eds.) Recent Trends in Wave Mechanics and Vibrations: Select Proceedings of WMVC 2018, pp. 201–222. Springer Singapore, Singapore (2020). https://doi.org/10.1007/978-981-15-0287-3_16
13. Warude P., et al.: On the design and vibration analysis of a three-link flexible robot interfaced with a mini-gripper", Springer Book, "Lecture Notes in Mechanical Engineering: Recent Trends in Wave Mechanics and Vibrations", ISBN: 978-981-15-0286, 2019, pp. 29–46; Selected proceedings of the 8th National Conference on Wave Mechanics and Vibrations ("WMVC-2018"), Rourkela, India, Nov. 2018. https://doi.org/10.1007/978-981-15-0287-3 (2019)
14. Roy, D.: Design, modeling and indigenous firmware of patient assistance flexible robotic system-type I: beta version. Adv. Rob. Mech. Engineering **2**(3), 148–159 (2020). https://doi.org/10.32474/ARME.2020.02.000140
15. Bhelsaikar, A., Atpadkar, V., Roy, D.: Design optimization of a curvilinear-jaw robotic gripper aided by finite element analysis. Int. J. Mech. Prod. Eng. (IJMPE) **8**(10), 9–14 (2020). http://iraj.doionline.org/dx/IJMPE-IRAJ-DOIONLINE-17606
16. Jain, T., Jain, J.K., Roy, D.: Joint space redundancy resolution of serial link manipulator: an inverse kinematics and continuum structure numerical approach. Mater. Today: Proc. **38**, 423–431 (2021). https://doi.org/10.1016/j.matpr.2020.07.608

ABM: Advanced Beam Models

GBT-based Vibration Analysis of Cracked Steel-Concrete Composite Beams

David Henriques[1], Rodrigo Gonçalves[2(✉)], Carlos Sousa[3], and Dinar Camotim[4]

[1] PPE Structural Engineering, R. Jaime Batalha Reis 1C, 1500-679 Lisbon, Portugal
[2] CERIS and Departamento de Engenharia Civil, Faculdade de Ciências e Tecnologia, Universidade NOVA de Lisboa, 2829-516 Caparica, Portugal
rodrigo.goncalves@fct.unl.pt
[3] CONSTRUCT, Departamento de Engenharia Civil, Faculdade de Engenharia, Universidade do Porto, R. Dr. Roberto Frias, 4200-465 Porto, Portugal
cfsousa@fe.up.pt
[4] CERIS and DECivil, Instituto Superior Técnico, Universidade de Lisboa, Av. Rovisco Pais 1, 1049-001 Lisbon, Portugal
dcamotim@civil.ist.utl.pt

Abstract. This paper presents a computationally efficient finite element, based on Generalised Beam Theory (GBT), that enables assessing the vibration serviceability limit state (the calculation of undamped natural frequencies and vibration mode shapes) of steel-concrete composite beams, accounting for concrete cracking and cross-section in-plane and out-of-plane deformation. It is shown that the modal decomposition features of GBT enable an in-depth characterisation of the vibration modes. A numerical example is presented to illustrate the accuracy and efficiency of the proposed element, through comparison with refined shell finite element model results.

Keywords: Steel-concrete composite beams · Generalised Beam Theory (GBT) · Natural frequencies · Vibration modes · Concrete cracking · Cross-section deformation

1 Introduction

The serviceability limit state (SLS) design checks for conventional steel-concrete composite floors require ensuring that the fundamental (undamped natural) frequency is higher than a specified value. Although simplified calculation methods are available [27], accurate values can only be generally obtained using shell/solid finite element models [16,19], since cracking and cross-section deformation can play a relevant role.

Generalised Beam Theory (GBT) is a thin-walled beam theory incorporating cross-section in-plane and out-of-plane deformation through the inclusion of

Z. Dimitrovová et al. (Eds.): WMVC 2022, MMS 125, pp. 19–29, 2023.
https://doi.org/10.1007/978-3-031-15758-5_2

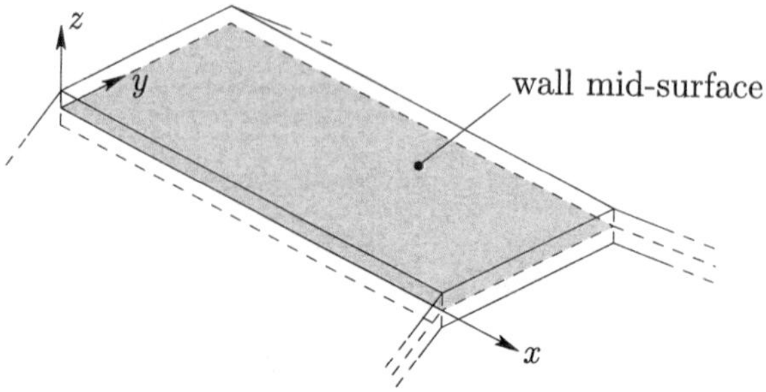

Fig. 1. Wall mid-surface local axes (x, y, z).

hierarchic and structurally meaningful DOFs—the so-called cross-section "deformation modes" (DMs) [25]. Consequently, GBT-based finite elements generally lead to accurate, computational efficient and insightful solutions in a wide range of problems (see, e.g., [3,4,7,20] and the bibliography at `gbt.info` and `www.civil.ist.utl.pt/gbt`). Although previous research concerning the application of GBT to the dynamic case is quite significant (e.g., [2,10,17,21,24,26]), only the linear elastic case has been dealt with so far. Quite recently, the authors have been applying GBT to steel-concrete composite beams [8,11–14,28]. This paper extends this previous work to enable the calculation of undamped natural frequencies and associated vibration modes (VMs), accounting for concrete cracking and arbitrary cross-section deformation. An illustrative example is presented, to show that the proposed element leads to very accurate solutions with a small computational cost and that the GBT modal decomposition features provides an unique structural insight into the characteristic of each VM.

2 The Proposed Finite Element

Using the wall mid-surface local axes (x, y, z) shown in Fig. 1, where x is parallel to the beam axis and z defines the through-thickness direction, the displacement vector can be written as

$$\boldsymbol{U}(x,y,z) = \begin{bmatrix} U_x \\ U_y \\ U_z \end{bmatrix} = \boldsymbol{\Xi}_U(y) \begin{bmatrix} \boldsymbol{\phi}(x) \\ \boldsymbol{\phi}_{,x}(x) \end{bmatrix}, \quad \boldsymbol{\Xi}_U(y) = \bar{\boldsymbol{\Xi}}_U^M(y) + z\bar{\boldsymbol{\Xi}}_U^B(y), \tag{1}$$

with the auxiliary matrices

$$\bar{\boldsymbol{\Xi}}_U^M(y) = \begin{bmatrix} \mathbf{0} & \bar{\boldsymbol{u}}^T \\ \bar{\boldsymbol{v}}^T & \mathbf{0} \\ \bar{\boldsymbol{w}}^T & \mathbf{0} \end{bmatrix}, \quad \bar{\boldsymbol{\Xi}}_U^B(y) = -\begin{bmatrix} \mathbf{0} & \bar{\boldsymbol{w}}^T \\ \bar{\boldsymbol{w}}_{,y}^T & \mathbf{0} \\ \mathbf{0} & \mathbf{0} \end{bmatrix}, \tag{2}$$

where (i) Kirchhoff's thin-plate assumption was employed, (ii) superscripts M and B indicate membrane and bending terms, respectively, (iii) $\bar{\boldsymbol{u}}(y)$, $\bar{\boldsymbol{v}}(y)$, $\bar{\boldsymbol{w}}(y)$ are column vectors that collect the mid-line displacement components of each cross-section DM along the local axes (x, y, z, respectively), (iv) column vector $\boldsymbol{\phi}(x)$ stores the corresponding amplitude functions (unknowns) and (iv) the subscript commas denote a derivative with respect to the indicated local axis (the derivative in $\boldsymbol{\phi}$ affecting $\bar{\boldsymbol{u}}$ is required to allow accommodating Vlasov's null membrane shear strain assumption [25]. The strains follow directly from the displacement field, reading

$$\boldsymbol{\varepsilon} = \begin{bmatrix} \varepsilon_{xx} \\ \varepsilon_{yy} \\ \gamma_{xy} \end{bmatrix} = \boldsymbol{\Xi}_{\varepsilon}(y) \begin{bmatrix} \boldsymbol{\phi} \\ \boldsymbol{\phi}_{,x} \\ \boldsymbol{\phi}_{,xx} \end{bmatrix}, \quad \boldsymbol{\Xi}_{\varepsilon}(y) = \boldsymbol{\Xi}_{\varepsilon}^{M}(y) + z\boldsymbol{\Xi}_{\varepsilon}^{B}(y), \tag{3}$$

$$\boldsymbol{\Xi}_{\varepsilon}^{M}(y) = \begin{bmatrix} \mathbf{0} & \mathbf{0} & \bar{\boldsymbol{u}}^{T} \\ \bar{\boldsymbol{v}}_{,y}^{T} & \mathbf{0} & \mathbf{0} \\ \mathbf{0} & (\bar{\boldsymbol{u}}_{,y} + \bar{\boldsymbol{v}})^{T} & \mathbf{0} \end{bmatrix}, \quad \boldsymbol{\Xi}_{\varepsilon}^{B}(y) = -\begin{bmatrix} \mathbf{0} & \mathbf{0} & \bar{\boldsymbol{w}}^{T} \\ \bar{\boldsymbol{w}}_{,yy}^{T} & \mathbf{0} & \mathbf{0} \\ \mathbf{0} & 2\bar{\boldsymbol{w}}_{,y}^{T} & \mathbf{0} \end{bmatrix}. \tag{4}$$

The cross-section DMs for a steel-concrete composite beam are obtained as described in the authors' previous works (see, e.g., [14]) and therefore only a short overview is provided here. The steel rebars in the concrete slab behave uniaxially and no slip is allowed. As shown in Fig. 2(a), the cross-section mid-line is discretised using 4-DOF nodes and several simplifying assumptions are adopted to reduce the DOF number while ensuring no significant loss of accuracy, namely: (i) $\varepsilon_{yy}^{B} = \gamma_{xy}^{M} = 0$ in the steel flanges, (ii) $\varepsilon_{yy}^{M} = 0$ in the steel beam and (iii) no slip is allowed at the steel-concrete interface (hence the rigid link in the figure). Furthermore, to model cracking properly, quadratic warping and transverse displacement functions in the concrete walls are added, leading to the 31 DM shown in Fig. 2(b): (i) the first four correspond to axial extension, major/minor-axis bending and torsion about the shear center, (ii) mode 5 corresponds to a constant membrane shear strain in the web, (iii) modes 6–15 are obtained assuming $\gamma_{xy}^{M} = 0$ in all walls (the Vlasov assumption) and $\varepsilon_{yy}^{M} = 0$, (iv) modes 16–23 are linear and quadratic warping modes (18/19/22/23 are reflections of 17/16/21/20, respectively) and (v) modes 24–31 are linear and quadratic transverse displacement modes (26/27/30/31 are reflections of 25/24/29/28, respectively).

In a first step, a geometrically static analysis is carried out, increasing the loads up to the desired level, including cracking and cross-section deformation. This element is described in detail in [14] and therefore only its main features are explained. The steel beam and rebars follow a St. Venant-Kirchhoff hyperelastic material law (plane stress and uniaxial laws, respectively). A strain decomposition approach is adopted for concrete [1,5,15,22,23], where a fixed smeared crack model is implemented, with either one or two orthogonal cracks. The plain concrete follows the plane stress St. Venant-Kirchhoff law, whereas the uniaxial crack law in the normal crack direction n is displayed in Fig. 3 and the shear stresses are calculated from the shear strains using a reduced shear modulus $\beta G_c/(1-\beta)$ [23], where G_c is the concrete elastic shear modulus and β is the so-called shear retention factor.

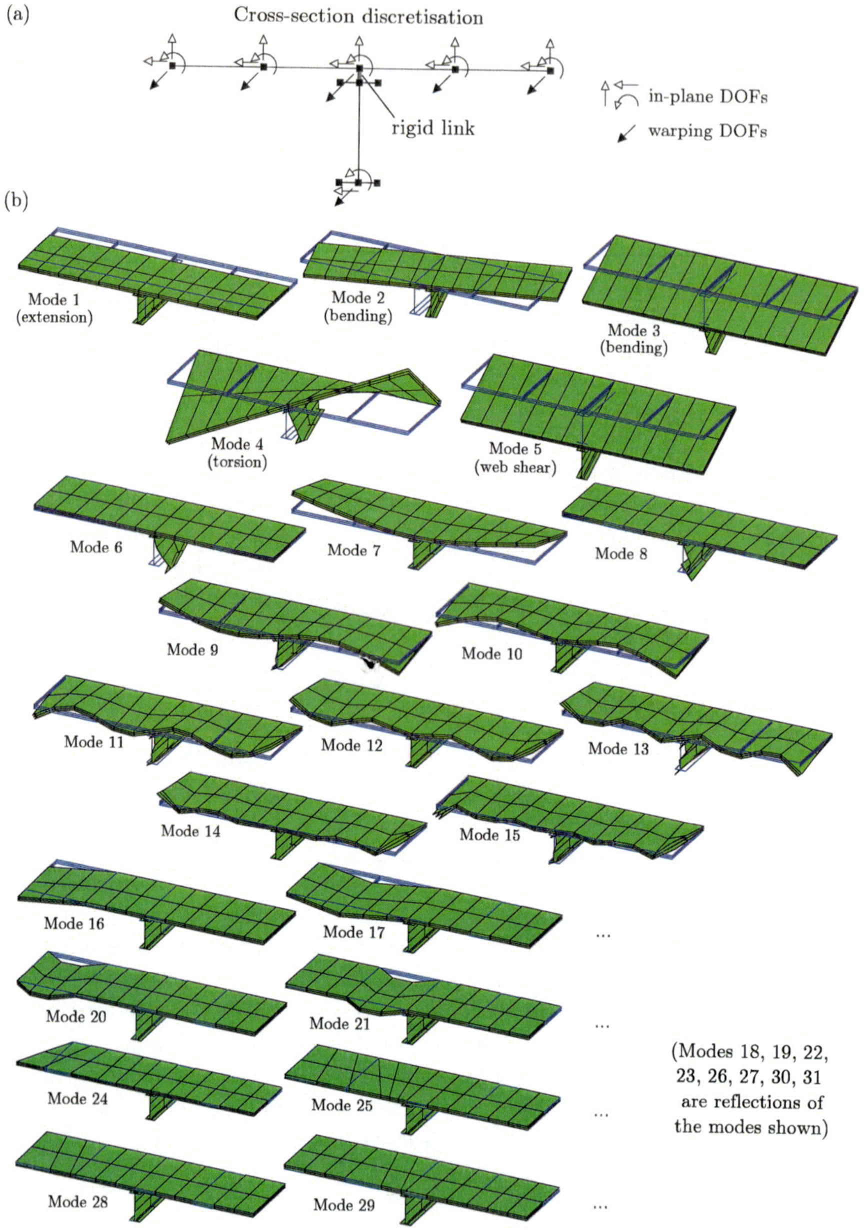

Fig. 2. Steel-concrete composite beam GBT (a) discretisation and (b) cross-section deformation modes.

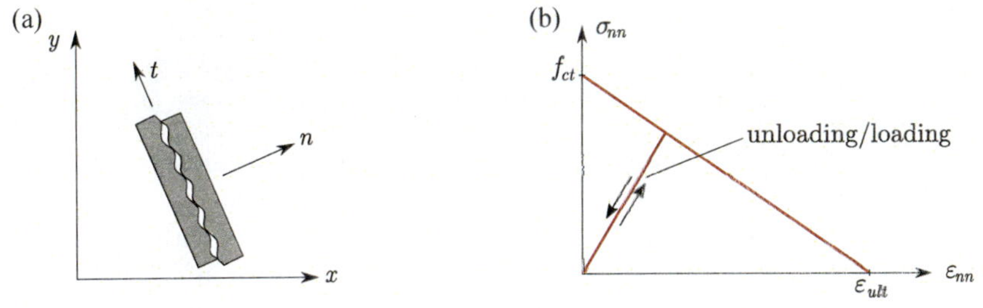

Fig. 3. (a) Wall mid-surface (x, y) and crack local (n, t) axes, and (b) crack law for normal stresses.

The finite element is obtained by approximating functions ϕ_k using Hermite cubic polynomials except for the DMs involving only warping, for which Lagrange quadratic hierarchical functions are employed for $\phi_{k,x}$, as in [9]. If the approximation is written as $\boldsymbol{\phi} = \boldsymbol{\Psi d}_e$, where matrix $\boldsymbol{\Psi}$ collects the interpolation functions and vector $\boldsymbol{d}_e$ collects the element DOFs, the element tangent stiffness matrix and the internal/external force vectors read

$$\boldsymbol{K}_e = \int_{V_e} \begin{bmatrix} \boldsymbol{\Psi} \\ \boldsymbol{\Psi}_{,x} \\ \boldsymbol{\Psi}_{,xx} \end{bmatrix}^T \boldsymbol{\Xi}_\varepsilon^T \boldsymbol{C}_t \boldsymbol{\Xi}_\varepsilon \begin{bmatrix} \boldsymbol{\Psi} \\ \boldsymbol{\Psi}_{,x} \\ \boldsymbol{\Psi}_{,xx} \end{bmatrix} dV, \tag{5}$$

$$(\boldsymbol{f}_{int})_e = \int_{V_e} \begin{bmatrix} \boldsymbol{\Psi} \\ \boldsymbol{\Psi}_{,x} \\ \boldsymbol{\Psi}_{,xx} \end{bmatrix}^T \boldsymbol{\Xi}_\varepsilon^T \boldsymbol{\sigma} \, dV, \tag{6}$$

$$(\boldsymbol{f}_{ext})_e = \int_{V_e} \rho \begin{bmatrix} \boldsymbol{\Psi} \\ \boldsymbol{\Psi}_{,x} \end{bmatrix}^T \boldsymbol{\Xi}_U^T \boldsymbol{g} \, dV + \int_{\Omega_e} \begin{bmatrix} \boldsymbol{\Psi} \\ \boldsymbol{\Psi}_{,x} \end{bmatrix}^T \left(\bar{\boldsymbol{\Xi}}_U^M\right)^T \bar{\boldsymbol{q}} \, d\Omega, \tag{7}$$

where V_e and Ω_e are the element volume and mid-surface, respectively, $\boldsymbol{C}_t$ is the tangent constitutive matrix, ρ is the mass density, vector $\boldsymbol{g}$ defines the acceleration of gravity along wall local axes and $\bar{\boldsymbol{q}}^T = [\bar{q}_x \ \bar{q}_y \ \bar{q}_z]$ are mid-surface forces. A standard incremental/iterative procedure was implemented in MATLAB [18], using load/displacement control. The steel contribution (I-beam and rebars) is elastic and is calculated only once, using analytical integration, but the concrete contribution must be evaluated at each iteration, due to cracking. In the latter case a 3×3 Gauss point grid in each wall mid-surface and Simpson's rule along the thickness are employed for the integrations.

When the incremental/iterative procedure achieves the desired load, the mass matrix $\boldsymbol{M}$ is assembled from its element counterpart

$$\boldsymbol{M}_e = \int_{V_e} \rho \begin{bmatrix} \boldsymbol{\Psi} \\ \boldsymbol{\Psi}_{,x} \end{bmatrix}^T \boldsymbol{\Xi}_U^T \boldsymbol{\Xi}_U \begin{bmatrix} \boldsymbol{\Psi} \\ \boldsymbol{\Psi}_{,x} \end{bmatrix} dV, \tag{8}$$

and the vibration eigenvalue problem

$$\left(\boldsymbol{K}_t - \omega^2 \boldsymbol{M}\right) \boldsymbol{d} = \boldsymbol{0}, \tag{9}$$

is solved, where $\boldsymbol{K}_t$ is the global tangent stiffness matrix from the last iteration, the eigenvalues ω correspond to the natural angular frequencies (in the example presented next, only the ordinary frequencies $f = \omega/2\pi$ are provided) and the eigenvectors $\boldsymbol{d}$ are the associated VMs. The solution procedure in MATLAB is quite fast using the `eigs(k,'smallestabs')` function, which calculates only the first k frequencies (it takes less than one second in the example presented next).

3 Illustrative Example

The simply supported beam shown in Fig. 4(a) is analysed. A SLS load is applied, comprising self-weight and an uniformly distributed frequent load $p = 2.4$ kN/m^2. The supports prevent the vertical displacements of all cross-section points (including the slab). Both uncracked (elastic and not accounting for the steel rebars) and cracked analyses are carried out.

The GBT cross-section discretization adopted and the resulting DMs are shown in Fig. 2, leading to a 115 DOF element. A convergence analysis for the first seven frequencies showed that 10 equal-length elements suffice, which amounts to a model with 673 DOFs (before eliminating the restrained DOFs at the supports), well below the shell model value reported next.

For comparison purposes, results obtained with the Reissner-Mindlin 8-node CQ40S shell finite element model shown in Fig. 4(b) are provided, which involves approximately 15000 DOFs and was analysed using DIANA [6]. For the concrete material law, a multi-directional fixed crack model with a large threshold angle was adopted, which is quite similar to the model implemented in the GBT-based element (in this example only one crack opens at each integration point). The connection between the steel top flange and concrete mid-lines is ensured using rigid links, as in the GBT model.

The static step for the cracked case was carried out in 20 equal increments (for the uncracked case a single step obviously suffices). The resulting crack patterns are displayed in Fig. 5, which shows one quarter of the slab (the static steps has double symmetry). Even though the current release of DIANA only shows the cracks at the most cracked integration point in each shell element layer, the crack patterns are in quite good agreement.

Figure 6 displays the first seven frequencies (Hz) and associated VMs obtained with both models. An excellent agreement is observed, being worth noting that the differences in the frequency values are less than 4% (generally well below). Although the mode shapes match quite accurately, modes 5 and 6 are swapped in the uncracked case (their frequencies virtually coincide). It is interesting to note that the nature of the first four modes does not change for the uncracked/cracked cases, even though the frequencies obviously drop when cracking is considered.

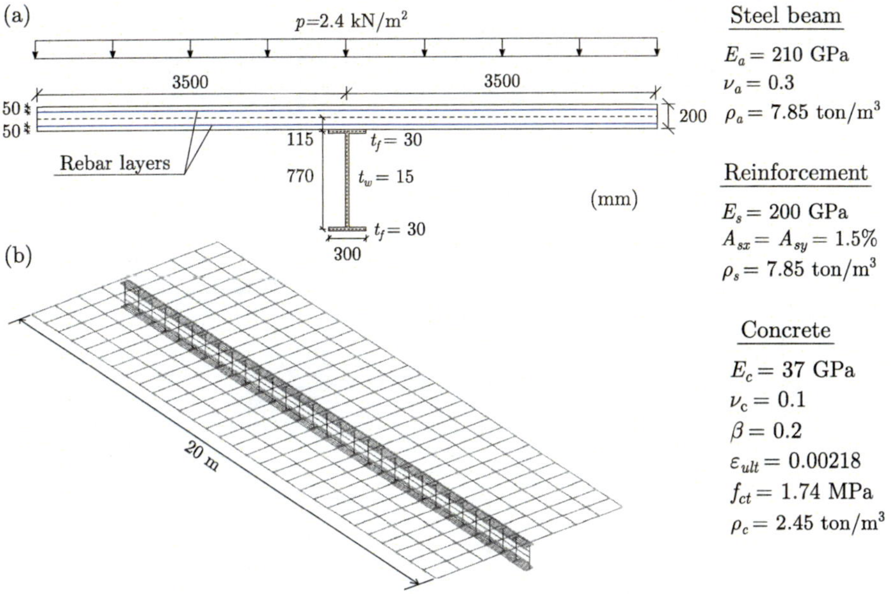

Fig. 4. Simply supported beam subjected to SLS loading: (a) cross-section geometry, loading, material parameters and (c) shell finite element model.

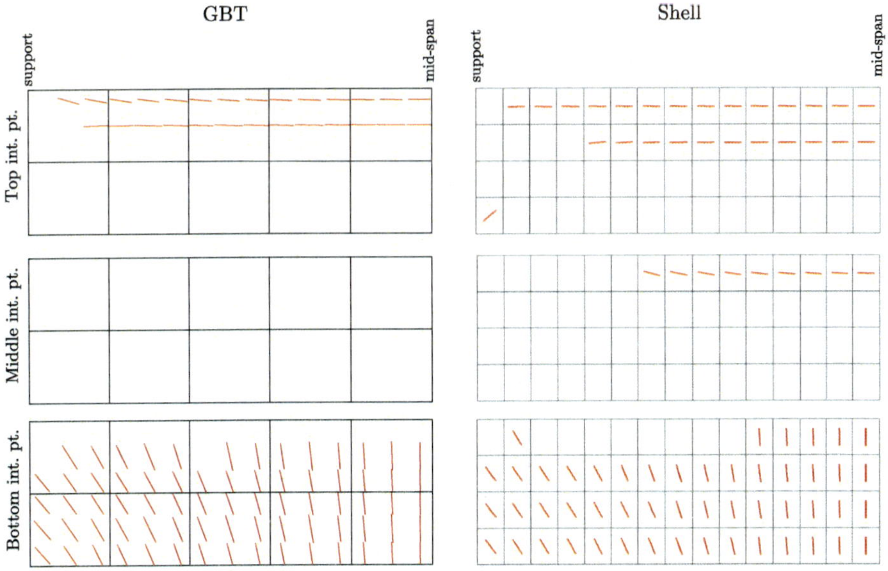

Fig. 5. Crack patterns (only one quarter of the slab is shown).

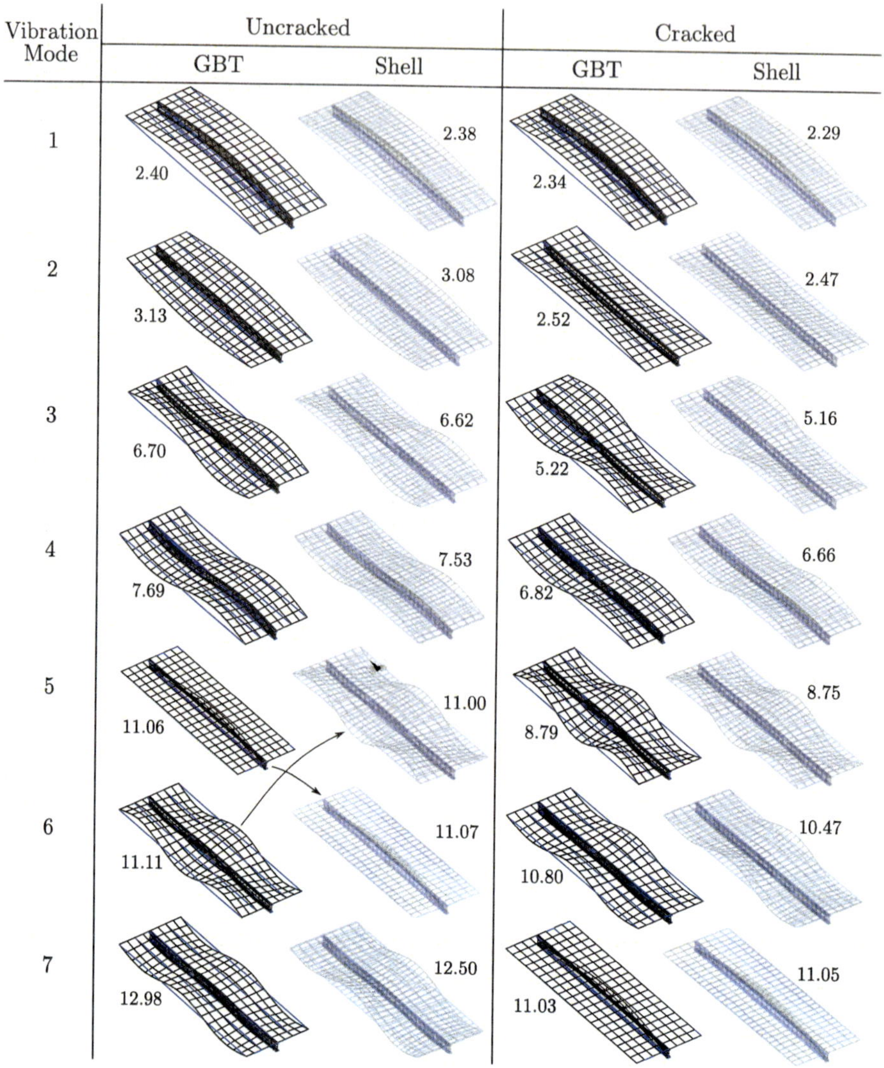

Fig. 6. First seven natural frequencies (in Hz) and vibration modes shapes.

Further insight can be obtained from the DM amplitude function graphs which are displayed in Fig. 7. Note that only a few DMs (out of the total 31) participate in each VM, due to their remarkable hierarchical properties and structural meaning. Three VM types can be identified with an increasing number of half-waves along the length, in agreement with the shapes in Fig. 6: (i) VMs 1, 4 and 7/6 (uncracked/cracked cases), which involve vertical bending (3), web shear (5) and single-curvature symmetric bending of the slab (7), the latter becoming predominant as the number of half-waves increase, (ii) VMs 2,

3 and 6/5 (uncracked/cracked cases), combining torsion (4), steel beam distortion, mode 9 and, for VM 6, also mode 8, and (iii) VMs 5/7 (uncracked/cracked cases), with a single half-wave distortional DM (6).

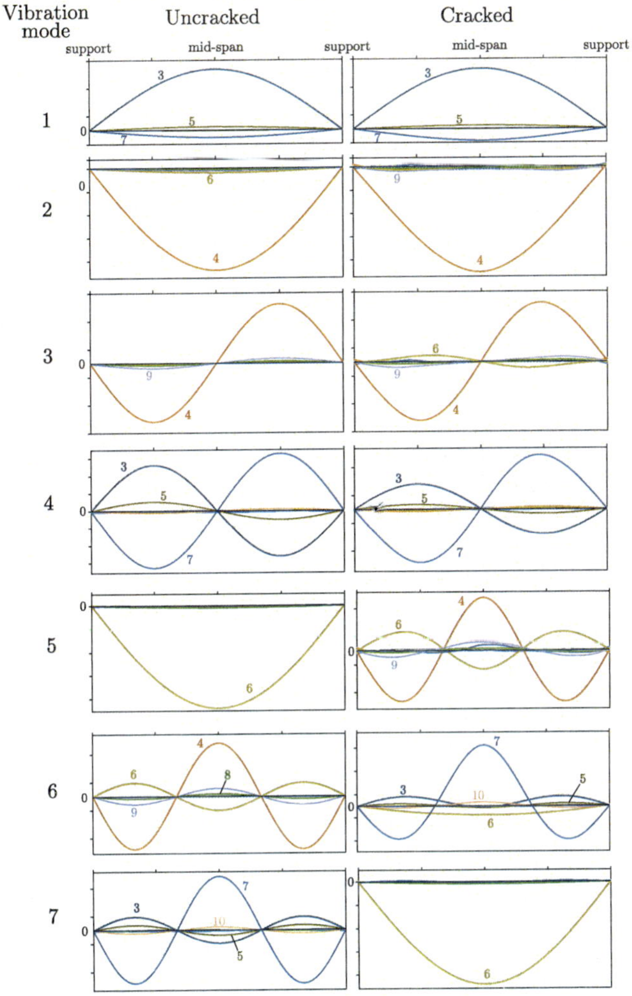

Fig. 7. GBT deformation mode amplitude functions.

4 Concluding Remarks

This paper presented an accurate and efficient GBT-based beam finite element that is able to calculate undamped natural frequencies and vibration mode shapes of steel-concrete composite beams undergoing cross-section deformation and concrete cracking, for SLS design checks. In a first step an incremental static and geometrically linear analysis is carried out, up to the target load, to obtain the global tangent stiffness matrix. In a second step, the mass matrix is assembled and the vibration eigenvalue problem is solved. An illustrative numerical example was presented, showing that (i) the results obtained with the proposed element are very close to those obtained with a refined shell finite element model, but with much less DOFs, and (ii) the GBT unique modal decomposition features enable an in-depth characterisation of the vibration modes.

Acknowledgements. The authors are grateful for the Foundation for Science and Technology's support through funding UIDB/04625/2020 from the research unit CERIS.

References

1. Bazant, Z.P., Gambarova, P.: Rough cracks in reinforced concrete. J. Struct. Div. (ASCE) **106**(4), 819–842 (1980)
2. Bebiano, R., Camotim, D., Silvestre, N.: Dynamic analysis of thin-walled members using Generalised Beam Theory (GBT). Thin-Walled Struct. **72**, 188–205 (2013)
3. Camotim, D., Gonçalves, R., Basaglia, C.: Analysis and Design of plated Structures. Volume 1: Stability, Chap. Developments on the GBT-Based Stability Analysis of Thin-Walled Members and Structural Systems, pp. 133–175. Woodhead Publishing (2022)
4. Casafont, M., Bonada, J., Pastor, M.M., Roure, F., Susín, A.: Linear buckling analysis of perforated cold-formed steel storage rack columns by means of the Generalised Beam Theory. Int. J. Struct. Stab. Dyn,. **18**(1), 1850004 (2018)
5. de Borst, R., Nauta, P.: Non-orthogonal cracks in a smeared finite element model. Eng. Comput. **2**(1), 35–46 (1985)
6. DIANA-FEA: DIANA FEM software release 10.5 (2021)
7. Duan, L., Zhao, J.: GBT deformation modes for thin-walled cross-sections with circular rounded corners. Thin-Walled Struct. **136**, 64–89 (2019)
8. Gonçalves, R., Camotim, D.: Steel-concrete composite bridge analysis using Generalised Beam Theory. Steel Compos. Struct. **10**(3), 223–243 (2010)
9. Gonçalves, R., Camotim, D.: Geometrically non-linear Generalised Beam Theory for elastoplastic thin-walled metal members. Thin-Walled Struct. **51**, 121–129 (2012)
10. Habtemariam, A.K., Tartaglione, F., Zabel, V., Könke, C., Bianco, M.J.: Vibration analysis of thin-walled pipes with circular axis using the Generalized Beam Theory. Thin-Walled Struct. **163**, 107628 (2021)
11. Henriques, D., Gonçalves, R., Camotim, D.: GBT-based finite element to assess the buckling behaviour of steel-concrete composite beams. Thin-Walled Struct. **107**, 207–220 (2016)
12. Henriques, D., Gonçalves, R., Camotim, D.: A physically non-linear GBT-based finite element for steel and steel-concrete beams including shear lag effects. Thin-Walled Struct. **90**, 202–215 (2015)
13. Henriques, D., Gonçalves, R., Camotim, D.: A visco-elastic GBT-based finite element for steel-concrete composite beams. Thin-Walled Struct. **145**, 106440 (2019)
14. Henriques, D., Gonçalves, R., Sousa, C., Camotim, D.: GBT-based time-dependent analysis of steel-concrete composite beams including shear lag and concrete cracking effects. Thin-Walled Struct. **150**, 106706 (2020)

15. Litton, R.: A contribution to the analysis of concrete structures under cyclic loading. Ph.D. thesis, University of California, Berkeley (1974)
16. Malveiro, J., Ribeiro, D., Sousa, C., Calçada, R.: Model updating of a dynamic model of a composite steel-concrete railway viaduct based on experimental tests. Eng. Struct. **164**, 40–52 (2018)
17. Manta, D., Gonçalves, R., Camotim, D.: Combining shell and GBT-based finite elements: vibration and dynamic analysis. Thin-Walled Struct. **167**, 108187 (2021)
18. MATLAB: version 7.10.0 (R2010a). The MathWorks Inc., Massachusetts (2010)
19. Matsuoka, K., Collina, A., Somaschini, C., Sogabe, M.: Influence of local deck vibrations on the evaluation of the maximum acceleration of a steel-concrete composite bridge for a high-speed railway. Eng. Struct. **200**, 109736 (2019)
20. de Miranda, S., Madeo, A., Melchionda, D., Ubertini, F.: A high performance flexibility-based GBT finite element. Comput. Struct. **158**, 285–307 (2015)
21. Peres, N., Gonçalves, R., Camotim, D.: GBT-based dynamic analysis of thin-walled members with circular axis. Thin-Walled Struct. **170**, 108533 (2022)
22. Riggs, H.R., Powell, G.H.: Rough crack model for analysis of concrete. J. Eng. Mech. **112**(5), 448–464 (1986)
23. Rots, J., Nauta, P., Kuster, G., Blaauwendraad, J.: Smeared crack approach and fracture localization in concrete. Heron **30**(1–48), 629–638 (1985)
24. Schardt, R., Heinz, D.: Vibrations of thin-walled prismatic structures under simultaneous static load using Generalized Beam Theory. In: European Conference on Structural Dynamics, EURODYN 1990, pp. 921–927 (1991)
25. Schardt, R.: Verallgemeinerte Technische Biegetheorie. Springer Verlag, Berlin (1989). https://doi.org/10.1007/978-3-642-52330-4(German)
26. Silvestre, N., Camotim, D.: Vibration behaviour of axially compressed cold-formed steel members. Steel Compos. Struct. **6**(3), 221 (2006)
27. Smith, A., Hicks, S., Devine, P.: Design of floors for vibration: a new approach (revised edn.). No. P354, Steel Construction Institute, Ascot, UK (2009)
28. Vieira, L., Gonçalves, R., Camotim, D., Pedro, J.O.: Generalized Beam Theory deformation modes for steel-concrete composite bridge decks including shear connection flexibility. Thin-Walled Struct. **169**, 108408 (2021)

Projection Approach to Spectral Analysis of Thin Axially Symmetric Elastic Solids

Georgy Kostin(✉)

Ishlinsky Institute for Problems in Mechanics RAS, Prospekt Vernadskogo 101-1, Moscow 119526, Russia
kostin@ipmnet.ru
https://ipmnet.ru/en/labs/cms/

Abstract. Natural vibrations of a homogeneous isotropic elastic solid of rotation are studied. A modification of the Petrov–Galerkin method is applied to find numerically the eigenfrequencies and eigenforms of the rod when its surface is free of loads. According to this approach, local stress-strain and momentum-velocity relations are replaced by an integral equality. Approximate solution to the eigeproblem is based on polynomial semi-discretization of displacements and stresses: a finite-dimensional expansion in powers of two lateral coordinates is used. The rod's motions are decomposed into independent groups of eigenmodes due to axial symmetry. Four of these groups are considered, namely, torsional, longitudinal, and two type of bending vibrations. At the lowest approximation order, the governing equations are reduced to two differential systems of fourth order for bending and two systems of second order for longitudinal and torsional motions. Spectrum features related to the variability of the cross-sectional radius are discussed.

Keywords: Linear elasticity · Eigenvalue problem · Petrov–Galerkin method · Decoupling system

1 Introduction

For decades such conventional mechanical systems as thin elastic rods and related problems have been attracting attention of scientists [3]. From the practical point of view, elastic structures often consist of thin elements: prismatic bars, shafts, pipes, ans so on [9]. To describe the dynamic behavior of these elements, variational or projection formulations can be applied resulting in reliable models governed by ordinary differential equations (ODEs). The ODE models suppose specific spatial distribution of displacement and stress fields [7]. Eigenvalue problems for specific elastic bodies may have symmetry properties that give one possibility to decrease the dimension of the original system of partial differential

This work was partially supported by the Ministry of Science and Higher Education within the framework of the Russian State Assignment under contract No. AAAA-A20-120011690138-6 and partially supported by RFBR Grant No. 21-51-12004.

Z. Dimitrovová et al. (Eds.): WMVC 2022, MMS 125, pp. 30–39, 2023.
https://doi.org/10.1007/978-3-031-15758-5_3

equations (PDEs) and apply more effectively order-reduction approaches. Certain difficulties, of course, arise in spectral analysis of rod systems if geometric and mechanical characteristics are variable in length [1].

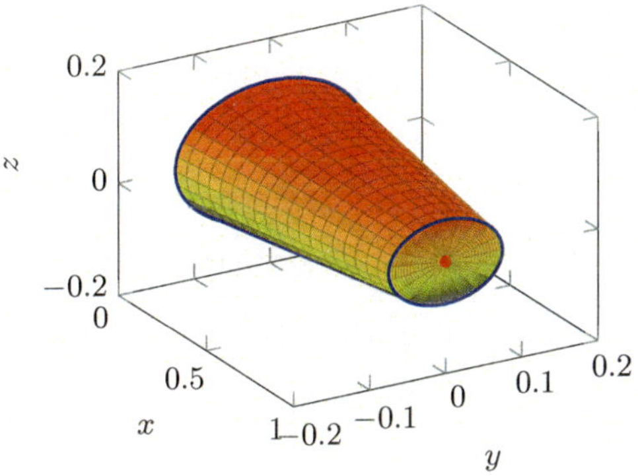

Fig. 1. Thin elastic solid of revolution with length $L = 1$.

In Sect. 2 of this paper, a projection procedure is used in the frame of linear elasticity to obtain finite-dimensional systems of differential-algebraic equations (DAEs), which describe natural vibrations of rectilinear elastic rods. The approach is based on the method of integrodifferential relations (MIDR) utilizing special integral projections of constitutive equations to relate the stress and strain tensors as well as the vectors of velocities and momentum density [6]. As compared with [5,8], rods with the variable circular cross section are studied. In Sect. 3, an example of longitudinal vibrations of a rod with conical frustum shape is analyzed.

2 Natural Vibrations of an Elastic Rod

Let us consider a thin elastic solid (rod) of revolution with length $L = 1$ in the dimensionless form as shown in Fig. 1. A Cartesian reference frame $Oxyz$, $\boldsymbol{x} = (x, y, z)$, has the origin O in the center of a boundary cross section, and the x-axis is directed towards the center of the opposite boundary face. The radius of cross section $0 < r_0(x) \ll 1$ can change and is a rather smooth function of x. The domain occupied by the rod is defined as

$$\Omega = \left\{ \boldsymbol{x} : x \in (0, 1), y^2 + z^2 < r_0^2(x) \right\}.$$

We analyze natural vibrations of the rod of homogeneous isotropic material [4]. The vibrations are described by the governing equations

$$\boldsymbol{p}_t(t, \boldsymbol{x}) = \nabla \cdot \sigma(t, \boldsymbol{x}), \quad \varepsilon := \tfrac{1}{2} \left(\nabla \boldsymbol{w} + \nabla \boldsymbol{w}^{\mathrm{T}} \right), \tag{1}$$

$$\boldsymbol{v}(t,\boldsymbol{x})=0,\quad \xi(t,\boldsymbol{x})=0,\quad \boldsymbol{v}:=\boldsymbol{w}_t-\rho^{-1}\boldsymbol{p},\quad \xi:=\varepsilon-\mathbf{C}^{-1}:\sigma, \tag{2}$$

where $\mathbf{C}$ is the elastic modulus tensor, $\boldsymbol{x}\in\Omega$. The volume density ρ and Young's modulus E of such material can be always chosen as $\rho=E=1$ in dimensionless units.

The first equation in (1) describes the change of momentum density vector $\boldsymbol{p}(t,\boldsymbol{x})$ in time t due to the divergence of the stress tensor field $\sigma(t,\boldsymbol{x})$ (Newton's second law). The second kinematic relation in (1) defines Cauchy's strain tensor $\varepsilon(t,\boldsymbol{x})$ via the displacement vector $\boldsymbol{w}(t,\boldsymbol{x})$. The constitutive law relating the velocity $\boldsymbol{w}_t$ and the momentum density $\boldsymbol{p}$ as well as the strains ε and the stresses σ (Hooke's law) is represented in (2) via the constitutive vector $\boldsymbol{v}$ and tensor ξ. The boundary of the rod is free of stresses according to

$$\boldsymbol{x}\in\partial\Omega:\quad \sigma(t,\boldsymbol{x})\cdot\boldsymbol{n}(\boldsymbol{x})=0, \tag{3}$$

where $\boldsymbol{n}$ is the outward unit normal to the boundary.

In what follows, the unknown variables are expressed in the form

$$\boldsymbol{w}=\tilde{\boldsymbol{w}}(\boldsymbol{x})\sin\omega t,\quad \boldsymbol{p}=\tilde{\boldsymbol{p}}(\boldsymbol{x})\cos\omega t,\quad \sigma=\tilde{\sigma}(\boldsymbol{x})\sin\omega t,$$

where ω is the eigenfrequency. The tilde is further omitted. Then the momentum can be found from (1) as $\boldsymbol{p}=-\omega^{-1}\nabla\cdot\sigma$.

In accordance with the MIDR, the constitutive equations (2) are represented in an integral form. This means that the unknown displacements $\boldsymbol{w}$ and stresses σ must satisfy the boundary condition (3) and the projection relation

$$\int_\Omega(\boldsymbol{v}\cdot\boldsymbol{u}+\xi:\tau)\mathrm{d}\Omega=0\quad \text{for } \forall\boldsymbol{u},\tau. \tag{4}$$

The vector $\boldsymbol{u}(\boldsymbol{x})$ and tensor $\tau(\boldsymbol{x})$ are test functions square integrable in Ω.

Taking into account the smallness of the cross-sectional radius $r_0(x)\ll 1$, we will try to construct an approximation of a thin rod in the framework of the linear theory of elasticity. To the components w_i, $\sigma_{ij}=\sigma_{ji}$ $(i,j=1,2,3)$ of the displacement vector $\boldsymbol{w}$ and the stress tensor σ, the approximation, which is finite-dimensional (polynomial) in the coordinates y and z, is chosen as

$$\begin{aligned} w_i &= \sum_{m+n\le 1} y^m z^n w_i^{(mn)}(x),\quad \sigma_{11}=\sum_{m+n\le 1} y^m z^n \sigma_{11}^{(mn)}(x),\\ \sigma_{ij} &= \sum_{m+n\le 2} y^m z^n \sigma_{ij}^{(mn)}(x),\quad i=1,2,3,\quad j=2,3, \end{aligned} \tag{5}$$

where $w_i^{(mn)}$, $\sigma_{ij}^{(mn)}$ are unknown trial functions of x.

Due to the symmetry of the rod, at least four decoupled rod motions can be specified. Namely, there are lateral bending around the y-axis and the z-axis as well as longitudinal tension-compression along the x-axis and torsion around this axis. The last two eigenforms (longitudinal and torsional) are independent of the angular component ϕ of cylindrical coordinates (x,r,ϕ). The bending forms depend only on two trigonometric functions $\cos\phi$, $\sin\phi$. In the Cartesian coordinates, the displacements and stresses of these four types of vibrations are approximated by polynomials with monomials $y^m z^n$ of only certain degrees m and n. Evenness or oddness of these degrees corresponding to bending, tension-compression, and torsion is indicated in Table 1 for all components w_i, σ_{ij}.

Table 1. Symmetry property of solution.

Variables	Bending y		Bending z		Extension		Torsion	
	m	n	m	n	m	n	m	n
w_1, σ_{jj}	Even	Odd	Odd	Even	Even	Even	Odd	Odd
w_2, σ_{12}	Odd	Odd	Even	Even	Odd	Even	Even	Odd
w_3, σ_{13}	Even	Even	Odd	Odd	Even	Odd	Odd	Even
σ_{23}	Odd	Even	Even	Odd	Odd	Odd	Even	Even

The proposed approximation (5) is enough to satisfy the boundary condition (3) on the lateral surface of the rod ($r = r_0(x)$). To do so, we need to express the cross-sectional coordinates in the polar form: $y = r\cos\phi$, $z = r\sin\phi$. In the Cartesian coordinates, the normal $\boldsymbol{n}$ in (3) is then represented on this surface as

$$\boldsymbol{n} = \cos\alpha(x)\cdot\big(-r_0'(x), \cos\phi, \sin\phi\big), \quad \alpha(x) = \arctan r_0'(x). \tag{6}$$

After expansion into a trigonometric series in ϕ, the Eq. (3) at $r = r_0(x)$ is reduced to a linear algebraic system with respect to the variables $\sigma_{ij}^{(mn)}$.

To model the vibrations of a thin rod in this study, we will confine ourselves to linear terms of bending, longitudinal, and torsional displacements. Thus, so-called 'breathing' eigenforms, when the cross section (not the central line) is only deformed, should be excluded as discussed in [5]. For that we assume that some variables are related: $w_3^{(01)} = w_2^{(10)}$ and $w_3^{(10)} = -w_2^{(01)}$. The number of remaining displacement functions is then equal to seven.

The constitutive relation (4) allows us to compose a consistent finite-dimensional system of DAEs with respect to the unknown displacements $w_i^{(mn)}$ and free stresses $\sigma_{ij}^{(mn)}$. To guarantee the system consistency, the test functions are chosen as the variation of trial ones according to $\boldsymbol{u} = \delta\boldsymbol{w}$, $\tau = \delta\sigma$. Considering the variety of the Petrov–Galerkin method [2], this is not a unique way to compose projections in (4). It should be noted that in the case of boundary conditions specified in stresses and the selected form of test functions, formula (1) corresponds to the necessary stationarity conditions of the Hellinger–Reissner principle up to signs [10]. The relationship of the various principles in the linear theory of elasticity with the proposed MIDR is discussed in [6].

What is provided with the approach proposed is non-degeneracy of the resulting DAE system in terms of first derivatives of $w_j^{(n,m)}(x)$ and $\sigma_{1j}^{(nm)}(x)$, which appear respectively in the first and second terms of the integrand $\boldsymbol{v}\cdot\boldsymbol{u} + \xi : \tau$

in (4). After satisfying the lateral boundary condition, the total number of the unknown stress functions $\sigma_{1j}^{(n,m)}$ contributing to the differential order of the DAEs is eight. This order can be reduced by equating two quadratic coefficients to zero: $\sigma_{12}^{(02)} = \sigma_{13}^{(20)} = 0$.

By taking into account (3) and (6), only four trial variables in the components $\sigma_{22}, \sigma_{23}, \sigma_{33}$ of the stress tensor σ remain independent. These variables together with the displacement function $w_2^{(10)}$ appear in the system only algebraically (without derivatives) and can be directly expressed through the other variables. By virtue of the DAE system, the function of shear stresses of torsional vibrations is equal to zero: $\sigma_{23}^{(00)} = 0$. It follows from the same equations that the functions of transversal normal stresses in tension-compression are equal to each other: $\sigma_{33}^{(00)} = \sigma_{22}^{(00)}$. Finally, we arrive at a linear system of 14 DAEs of 12th order with polynomial coefficients depending on ω and the longitudinal coordinate x.

This system is split into four independent subsystems. The bending around the y-axis is described by four ODEs with respect to the displacements $w_3^{(00)}$, $w_1^{(01)}$ and the stresses $\sigma_{13}^{(00)}$, $\sigma_{11}^{(01)}$ in the form

$$\begin{aligned}
9\tfrac{d}{dx}w_3^{(00)} &= \big(16(\nu+1)+6(5+\nu)r_0'^2\big)\sigma_{13}^{(00)} \\
&\quad -9w_1^{(01)} + 2\big((7+\nu)r_0r_0' - 6(1-\nu)r_0r_0'^3\big)\sigma_{11}^{(01)}, \\
r_0\tfrac{d}{dx}w_1^{(01)} &= \big(\tfrac{13\nu-5}{9}r_0' - \tfrac{33-15\nu}{9}r_0'^3\big)\sigma_{13}^{(00)} \\
&\quad + \tfrac{1}{9}\big(9r_0 - (37\nu-5)r_0r_0'^2 + 24(1-\nu)r_0r_0'^4\big)\sigma_{11}^{(01)}, \\
3r_0\tfrac{d}{dx}\sigma_{13}^{(00)} &= -4\omega^2r_0w_3^{(00)} + \omega^2r_0^2r_0'w_1^{(01)} - 9r_0'\sigma_{13}^{(00)} - r_0^2r_0''\sigma_{11}^{(01)}, \\
r_0^2\tfrac{d}{dx}\sigma_{11}^{(01)} &= -\omega^2r_0^2w_1^{(01)} + 3\sigma_{13}^{(00)} - 3r_0r_0'\sigma_{11}^{(01)}.
\end{aligned} \tag{7}$$

The boundary conditions for this motions are

$$\sigma_{13}^{(00)}(0) = \sigma_{11}^{(01)}(0) = \sigma_{13}^{(00)}(1) = \sigma_{11}^{(01)}(1) = 0. \tag{8}$$

Similarly, 4 ODEs for the bending around the z-axis include the displacements $w_2^{(00)}$, $w_1^{(10)}$ and the stresses $\sigma_{12}^{(00)}$, $\sigma_{11}^{(10)}$ according to

$$\begin{aligned}
9\tfrac{d}{dx}w_2^{(00)} &= \big(16(\nu+1)+6(5+\nu)r_0'^2\big)\sigma_{12}^{(00)} \\
&\quad -9w_1^{(10)} + 2\big((7+\nu)r_0r_0' - 6(1-\nu)r_0r_0'^3\big)\sigma_{11}^{(10)}, \\
r_0\tfrac{d}{dx}w_1^{(10)} &= \big(\tfrac{13\nu-5}{9}r_0' - \tfrac{33-15\nu}{9}r_0'^3\big)\sigma_{12}^{(00)} \\
&\quad + \tfrac{1}{9}\big(9r_0 - (37\nu-5)r_0r_0'^2 + 24(1-\nu)r_0r_0'^4\big)\sigma_{11}^{(10)}, \\
3r_0\tfrac{d}{dx}\sigma_{12}^{(00)} &= -4\omega^2r_0w_2^{(00)} + \omega^2r_0^2r_0'w_1^{(10)} - 9r_0'\sigma_{12}^{(00)} - r_0^2r_0''\sigma_{11}^{(10)}, \\
r_0^2\tfrac{d}{dx}\sigma_{11}^{(10)} &= -\omega^2r_0^2w_1^{(10)} + 3\sigma_{12}^{(00)} - 3r_0r_0'\sigma_{11}^{(10)}.
\end{aligned} \tag{9}$$

The corresponding boundary conditions are

$$\sigma_{12}^{(00)}(0) = \sigma_{11}^{(10)}(0) = \sigma_{12}^{(00)}(1) = \sigma_{11}^{(10)}(1) = 0. \tag{10}$$

The torsion around the x-axis is described by the unknowns $w_2^{(01)}$ and $\sigma_{12}^{(01)}$ satisfying two ODEs of order two

$$\begin{aligned} \tfrac{d}{dx} w_2^{(01)} &= 2(1+\nu)(r_0'^2+1)\sigma_{12}^{(01)}, \\ r_0 \tfrac{d}{dx}\sigma_{12}^{(01)} &= -\omega^2 r_0 w_2^{(01)} - 4r_0'\sigma_{12}^{(01)} \end{aligned} \tag{11}$$

and the boundary conditions

$$\sigma_{12}^{(01)}(0) = \sigma_{12}^{(01)}(1) = 0. \tag{12}$$

Four DAEs of second differential order approximate tension-compression of the elastic rod and depend on the variables $w_1^{(00)}$, $\sigma_{11}^{(00)}$ as well as their first derivatives and only algebraically on $w_2^{(10)}$, $\sigma_{22}^{(00)}$. Depending parametrically on Poisson's ratio ν and the radius of cross section $r_0(x)$, this system has the form

$$\begin{aligned} &24\tfrac{d}{dx} w_1^{(00)} + 6r_0 r_0' \tfrac{d}{dx} w_2^{(10)} = -12 r_0'^2 w_2^{(10)} \\ &\qquad + \big(24 + 12(1-\nu)r_0'^2 + (7-\nu)r_0'^4\big)\sigma_{11}^{(00)} - \big(36\nu - (5-11\nu)r_0'^2\big)\sigma_{22}^{(00)}, \\ &r_0 \tfrac{d}{dx}\sigma_{11}^{(00)} = -\omega^2 r_0 w_1^{(00)} - 2r_0'\sigma_{11}^{(00)}, \\ &\omega^2 r_0 r_0' w_1^{(00)} - \omega^2 r_0^2 w_2^{(10)} - r_0 r_0''\sigma_{11}^{(00)} + 3\sigma_{22}^{(00)} = 0, \\ &36 w_2^{(10)} + \big(36\nu + (11\nu-5)r_0'^2\big)\sigma_{11}^{(00)} + (25\nu - 31)\sigma_{22}^{(00)} = 0. \end{aligned} \tag{13}$$

The conditions at the end cross sections are given by

$$\sigma_{11}^{(00)}(0) = \sigma_{11}^{(00)}(1) = 0. \tag{14}$$

Although the integral constitutive relation (4) with the proposed trial and test functions gives only approximate solution to the eigenvalue problem, the MIDR allows us to estimate the relative error by using the ratio of the functionals

$$\begin{aligned} \Delta &= \frac{\Phi}{\Psi} \geq 0, \quad \Phi = \tfrac{1}{4}\int_\Omega (\rho \boldsymbol{v}\cdot\boldsymbol{v} + \xi : \mathbf{C} : \xi)\, d\Omega, \\ \Psi &= \tfrac{1}{2}\int_\Omega \big(\rho\omega^2 \boldsymbol{w}\cdot\boldsymbol{w} + \sigma : \mathbf{C}^{-1} : \sigma\big)\, d\Omega = 1. \end{aligned} \tag{15}$$

For the chosen degree of approximation, smallness of the value $\Delta \ll 1$ can be regarded as a criterion of solution quality, see [6]. Here, $\Phi[\boldsymbol{w}, \sigma, \omega]$ is the constitutive residual functional, $\Psi[\boldsymbol{w}, \sigma, \omega]$ is the total mechanical energy stored in the rod.

3 Numerical Analysis of Vibrations

Poisson's ratio is equal to $\nu = 0.3$ in the example. The simplest representative of solids of revolution with variable cross section is a conical frustum, in which the radius of cross section is an affine function $r_0(x) = ax + b$. For numerical analysis, the mean thickness of the rod and the slope of its lateral surface are chosen so that $a = -\frac{1}{20}$, $b = \frac{1}{8}$ (see Fig. 1).

Since Poisson's ratio is not equal to zero $\nu \neq 0$, longitudinal vibrations, which are governed by the DAE system (13) and (14), is accompanied by axial deformation of the cross section. This corresponds to the Rayleigh correction to the natural frequencies of the rod. With increasing the mode number n, the frequencies ω_n become lower in comparison with those ω_n^0 for the classical model. This fact is illustrated in Table 2 for the selected parameters. The conventional eigenproblem for longitudinal vibrations of a thin rectilinear elastic rod is formulated as follows.

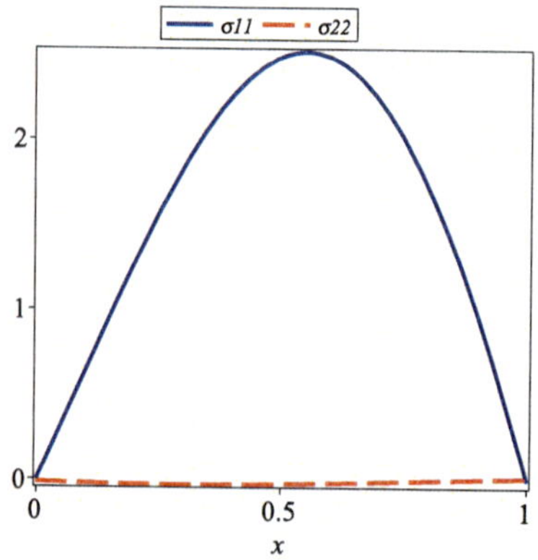

Fig. 2. Stresses $\sigma_{11}(x, 0, 0)$ and $\sigma_{22}(x, 0, 0)$ for the first mode.

$$\frac{d}{dx}\left(EA_0(x)\frac{d}{dx}w(x)\right) + \omega^2 \rho A_0(x) w(x) = 0, \quad w'(0) = w'(1) = 0. \tag{16}$$

Here, $w(x)$ denotes the amplitude of longitudinal displacements, $A_0 = \pi r_0^2(x)$ is the area of the circular cross section, $E = \rho = 1$. The pairs ω_n and ω_n^0 in Table 2 are quite close for the lowest modes, but they noticeably diverge with increasing the number n.

Table 2. Eigenfrequencies and error for tension-compression.

n	1	2	3	4	5	6	7	8
ω_n	3.21	6.26	9.22	12.0	14.4	16.1	17.0	29.4
ω_n^0	3.22	6.33	9.45	12.6	15.7	18.9	22.0	29.1
Δ_n	0.10	0.28	0.72	1.81	4.60	11.8	32.2	10.2

For the first mode, the amplitudes of longitudinal stresses $\sigma_{11}(x, 0, 0)$ and transversal stresses $\sigma_{22}(x, 0, 0)$ are presented in Fig. 2 by solid and dashed curves, respectively. As expected for the lowest vibration mode, the stress functions vanish at the ends of the interval $x \in (0, 1)$ and have no nodal points inside. The transversal normal stresses are much less intensive than the longitudinal ones and have the opposite sign. In its turn, the corresponding displacements $w_1(x, 0, 0)$ and $w_2(x, r_0, 0)$ are shown in Fig. 3 respectively by solid and dash curves. If the rod is stretched at the ends, what is always possible to achieve by choosing a free scaling factor of the eigenform amplitudes, the cross section is slightly compressed in the middle.

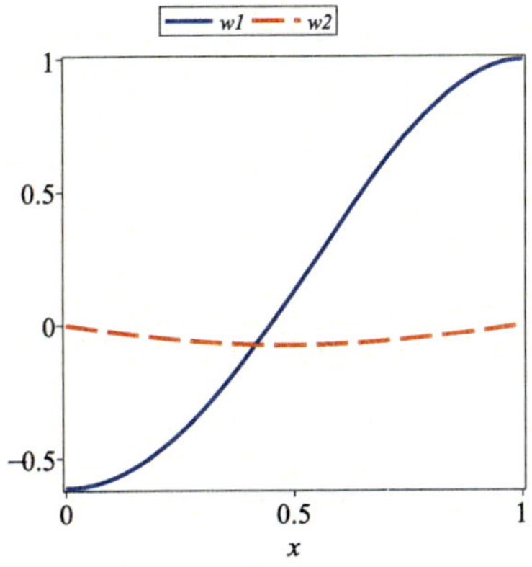

Fig. 3. Displacements $w_1(x, 0, 0)$ and $w_2(x, r_0, 0)$ for the first mode.

If we return to Table 2, it is worth noting that the frequencies ω_n and ω_n^0 diverge significantly at $n = 7$, and then the value of ω_n at $n = 8$ jumps up suddenly. This anomalous behavior of the spectrum is due to the fact that the proposed model (13), (14) has a frequency interval $\omega \in I_d \approx (17.1, 28.6)$, at which the degeneration of the system occurs at two closely spaced points $x_1, x_2 \in (0, 1)$. Thus, any solution to the boundary value problem does not exist. The model qualitatively reflects the zone of eigenform rearrangement at wavelengths comparable to the cross-sectional diameter.

Such rearrangement is also revealed by the value of the relative error $\Delta_n = \Delta(n)$ calculated in accordance with (15). The error Δ_n increases greatly, when approaching the lower end of the degeneracy interval I_d, and decreases notably just after this interval. In this frequency area, a more accurate model for the spatial distributions of displacements and stresses (higher orders of polynomials) is required.

The components $\sigma_{11}(x, 0, 0)$ and $\sigma_{22}(x, 0, 0)$ of the stress tensor σ are presented in Fig. 4 for the eighth mode (with $\omega_8 > I_d$) again by solid and dash. For comparison, the same figure shows the distribution of the normal stress $s(x) = Ew'(x)$ for the eighth mode of the classical model (16) (the dot curve). As follows from the Sturm–Liouville theorem, this function has seven nodal points inside the domain $x \in (0, 1)$. In contrast to the classical form of vibration, normal longitudinal and lateral stresses have only 6 nodes for $n = 8$. The maximum of the amplitude of the stresses $\sigma_{22}(x, 0, 0)$ is almost twice the maximum of the longitudinal stress, and both functions are close in phase (cf. the first mode in Fig. 2).

The displacements $w_1(x, 0, 0)$ and $w_2(x, r_0, 0)$ for ω_8 are presented in Fig. 5 by solid and dash. The classical form $w(x)$ is given by dotted curve. The functions $w_1(x, 0, 0)$ and $w_2(x, r_0, 0)$ are almost in antiphase and the transverse displacements w_2 have a maximum closer to the thinner end of the rod ($x = 1$).

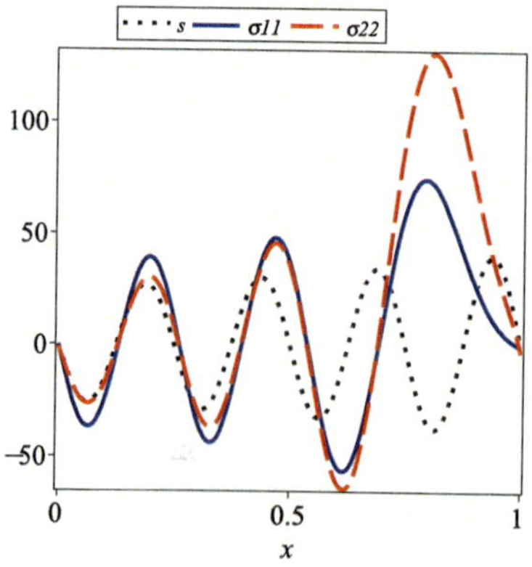

Fig. 4. Stresses $\sigma_{11}(x, 0, 0)$ and $\sigma_{22}(x, 0, 0)$ for the eighth mode.

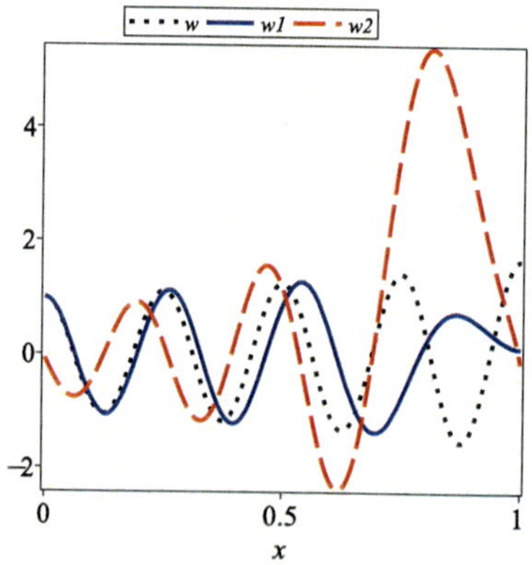

Fig. 5. Displacements $w_1(x, 0, 0)$ and $w_2(x, r_0, 0)$ for the eighth mode.

4 Conclusions

In the frame of the MIDR, a model of natural vibrations of elastic rods with the circular cross section has been developed. Approximation of torsional, longitudinal, and bending vibrations is based on a polynomial representation of displacement and stress fields with respect to the lateral (cross-sectional) Cartesian coordinates. The original 3D elastic system is reduced to a DAE system. As an example, a thin homogeneous isotropic rod with the geometric shape of a conical frustum is studied. Natural vibrations of the rod are decomposed due to axial geometrical and physical symmetry. Spectrum properties of the rod have been compares for the proposed projection and classical beam models.

References

1. Akulenko, L.D., Nesterov, S.V.: High-Precision Methods in Eigenvalue Problems and Their Applications. Charman and Hall/CRC, Boca Raton (2005)
2. Atluri, S.N., Zhu, T.: A new meshless local Petrov-Galerkin (MLPG) approach in computational mechanics. Comput. Mech. **22**, 117–127 (1998). https://doi.org/10.1007/s004660050346
3. Donnell, L.H.: Beams, Plates and Shells. McGraw-Hill, New York (1976)
4. Haan, H.G.: Elastizitätstheorie. B. G. Teubner, Stuttgart (1985)
5. Kostin, G.V., Saurin, V.V.: Modelling and analysis of the natural vibrations of a prismatic elastic beam based on a projection approach. J. Appl. Math. Mech. **75**(6), 700–710 (2011)
6. Kostin, G.V., Saurin, V.V.: Dynamics of Solid Structures. Methods Using Integrodifferential Relations. De Gruyter, Berlin (2018)
7. Levinson, M.: On Bickford's consistent higher order beam theory. Mech. Res. Commun. **12**(1), 1–9 (1985)
8. Saurin, V., Kostin, G.: A projection approach to analysis of natural vibrations for beams with non-symmetric cross sections. In: Neittaanmäki, P., Repin, S., Tuovinen, T. (eds.) Mathematical Modeling and Optimization of Complex Structures. CMAS, vol. 40, pp. 153–173. Springer, Cham (2016). https://doi.org/10.1007/978-3-319-23564-6_10
9. Timoshenko, S.: Strength of Materials, Elementary Theory and Problems, vol. 1. D. Van Nostrand Reinhold, Princenton (1956)
10. Washizu, K.: Variational Methods in Elasticity and Plasticity. Pergamon Press, Oxford (1982)

AIM: Advances in Impact Mechanics and Computational Sciences

Dynamic Response of a Reinforced Concrete Column Under Axial Shock Impact

Sergey Savin[1,3](✉), Vitaly Kolchunov[1,2], and Nataliya Fedorova[1,3]

[1] Moscow State University of Civil Engineering, Moscow, Russian Federation
savinsyu@mgsu.ru

[2] South-West State University, Kursk, Russian Federation

[3] Research Institute of Building Physics of Russian Academy of Architecture and Construction Sciences, Moscow, Russian Federation

Abstract. Analysis of the disproportionate collapses of buildings and structures that have occurred in recent decades, as well as the results of experimental studies on this problem, show dynamic effects in structures adjacent to the site of initial local destruction such as sudden column removal. In this case, the columns adjacent to the bay of the initial local destruction can be subjected to the simultaneous action of the axial shock impact. In this regard, the subject of this research is the dynamic response of reinforced concrete columns exposed to shock impact. The paper proposes an analytical solution to the problem based on Bessel function. It considers structural damping, physical nonlinearity and geometric nonlinearity of the first order (P - delta effect). To take into account the physical nonlinearity of the material and the P - delta effect, the solution is carried out step by step by dividing the time into separate steps, within which the load is considered as a linear function of time. The provided parametric analysis shows that the maximum dynamic deflection of a compressed reinforced concrete column, subjected to a shock impact, increases almost linearly with an increase in slenderness ratio at a load value in the range from 0.6 to 8 of ultimate bearing capacity. With an increase in the slenderness ratio of the column up to 20, dynamic deflections begin to grow more intensively with increasing of load.

Keywords: Dynamic response · Shock impact · Bessel function · Reinforced concrete column · Slenderness ratio · Reinforcement percentage

1 Introduction

During the service life, buildings and structures are exposed to loads and actions of various nature. Some loads act throughout the entire life cycle of the structure. Other one act for a short time only. The third group, which is associated with emergency situations and sudden structural transformation of the load bearing system, have a low probability of occurrence. However, neglect of the final group of loads can lead to catastrophic consequences: a progressive collapse of a part or the entire facility. Examples of this are the collapses of the Ronan Point residential building in London [1], the Alfred P. Murrah federal building in Oklahoma City [2–4], the Champlain Tower in Florida and etc.

Z. Dimitrovová et al. (Eds.): WMVC 2022, MMS 125, pp. 43–52, 2023.
https://doi.org/10.1007/978-3-031-15758-5_4

Experimental studies of the progressive collapse behavior of the San Diego Hotel [5, 6] reinforced concrete frame, prepared for demolition, showed a rapid strains increase in the columns, when the corner column was destructed due to the blast impact. The strain increment was observed during tenths of a second. At the same time, the strains increase along the opposite faces of the column was uneven. This indicated a change in the eccentricity of the force in the column.

Similar results were revealed when testing scale models of reinforced concrete frames [7–12]. However, these studies were focused on the behavior of slabs or beams [13–16]. But the influence of the parameters of vertical load-bearing structural members and the features of its loading in emergency situation was not studied separately. In this regard, the subject of this research is the influence of the geometrical parameters of a reinforced concrete column and the features of its loading on the dynamic response under axial shock impact.

2 Models and Methods

2.1 Modeling the Object of Study, Loads and Actions

A compressed reinforced concrete column with a square cross section with dimensions of 400 × 400 mm was chosen as the object of study in the work. Construction materials: concrete of class B30 (compressive strength of concrete prisms $R_{bn} = 22$ MPa, initial modulus of elasticity of concrete $E_c = 32500$ MPa), longitudinal reinforcement of class A500C (tensile strength of steel reinforcement $R_{sn} = 500$ MPa, compressive strength of steel reinforcement $R_{scn} = 400$ MPa, elastic modulus of steel $E_s = 200000$ MPa). The distance from the surface of the column to the center of gravity of the longitudinal reinforcement in tensioned (or less compressed) and compressed zone are a = 0.05 m and a′ = 0.05 m respectively. Three options for the reinforcement percentage μ_s were adopted for the study: 1, 2, 3%. Three variants of the effective length of the column l_0 are considered: 4 m, 6 m, 8 m, which correspond to the slenderness ratio $\lambda_h = (l_0/h)$: 10, 15, 20 respectively.

Before the shock impact, the column was statically preloaded with axial force $P_0 = 0.5\ N_{ult}$ applied with a random eccentricity $e_0 = h/30 = 0.013$ m. N_{ult} is the ultimate force for the compressed reinforced concrete column, determined according to Building Code of RF SP 63.13330.2018 by formula (1):

$$N_{ult} = \varphi \cdot (R_{bn} \cdot A_b + R_{scn} \cdot A_{s,\ tot}), \tag{1}$$

where φ is buckling coefficient for long-term load action according to SP 63.13330.2018;

A_b is concrete cross section area, [m^2];
$A_{s,tot}$ is total cross section area of longitudinal steel reinforcement, [m^2].

The paper considers three options for loading a column with a shock impact $P = (P_0 + P_d)$: 0.6 N_{ult}, 0.8 N_{ult}, and N_{ult}, where P_d is the value of the dynamic additional loading of the column. The time of dynamic loading was the same for all three options $t_d = 0.01$ s.

The model of an eccentrically compressed physically and geometrically non-linear bar was used (Fig. 1) to simulate the dynamic response of reinforced concrete columns. Accounting for physical and geometric non-linearity is performed by dividing the solution into time steps, within which the load and stiffness are assumed to be constant. In the first approximation, piecewise linear diagrams of the deformation of concrete (Fig. 2a) and reinforcement (Fig. 2b) and the nonlinear elastic law of deformation are used for concrete and steel reinforcement, i.e., unloading occurs in accordance with the same deformation diagram as during loading.

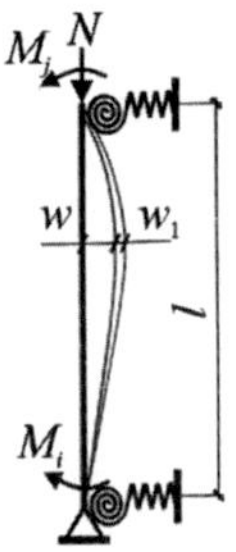

Fig. 1. Column design scheme

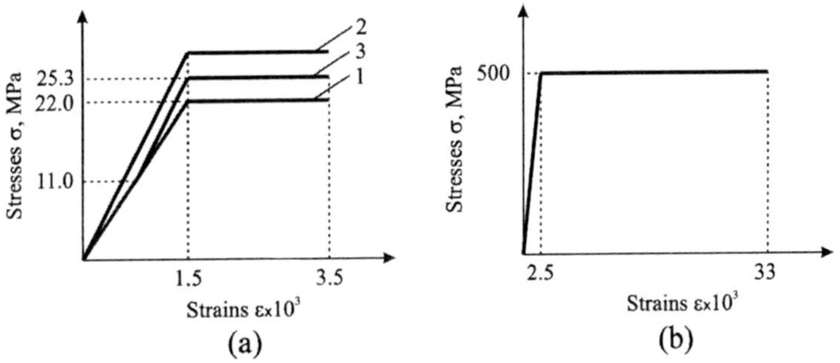

Fig. 2. Piece-wise linear Stress vs. Strain curves for concrete (a) and reinforcing steel (b) under uniaxial compression. 1 - quasi-static loading, 2 - dynamic loading, 3 - dynamic loading of statically pre-load concrete

The bearing capacity of members after axial impact was evaluated as for eccentrically compressed elements with small eccentricities according to SP 63.13330.2018:

$$M_{ult} = \varphi_b \cdot R_{bn} \cdot b \cdot x(d - 0.5x) + (R_{scn} \cdot A_s' - N/2)(d - a'), \tag{2}$$

where b is a width of the column cross section, x is the depth of compressed zone, $d = h - a$ is the effective depth of a cross section, a and a' are the distance from the surface of the column to the center of gravity of the longitudinal reinforcement in tensioned (or less compressed) and compressed zone respectively; $\varphi_b = 1.15$ is the dynamic hardening factor for concrete [17].

2.2 Analytical Solution of the Dynamic Buckling Problem

Deformation of the bar member (Fig. 1a) during its dynamic buckling caused by impact load, considering damping, can be described by the differential Eq. (3) [18]:

$$EJ\frac{\partial^4(w_1 - w_0)}{\partial x^4} + P\frac{\partial^2 w_1}{\partial x^2} + \frac{\gamma \cdot A}{g}\frac{\partial^2 w}{\partial t^2} + k\frac{\partial w}{\partial t} = 0, \tag{3}$$

where $w_0 = w_0(x)$ is the initial deflection of the elastic axis of the bar from the vertical axis passing through the center of gravity of one of the end sections. The appearance of the initial deflection is due to geometric and(or) physical imperfections in the sections of the column acquired during manufacture and operation, as well as buckling from a statically applied operating load.

$w_1 = w_1\ (x, t)$ is the total lateral deflection of the elastic axis of the rod at time t;
$w = w_1 - w_0$ is the additional lateral deflection of the bar member at the moment of time t;
E is the initial modulus of elasticity of concrete;
J is the moment of inertia of the reduced section of a reinforced concrete column;
$P = P(t)$ is the relation for the change in time of the axial force, adopted in the form of a piecewise linear dependence (4):

$$P = \begin{cases} P_0 + \Omega \cdot t, \ for \ \ 0 \le t \le t_1; \\ P_0 + \Omega \cdot t_1, \quad for \ \ t > t_1, \end{cases} \tag{4}$$

where P_0 is the axial force of the static preloading;
Ω is the rate of dynamic axial force, N/s;
t_1 is the time of the forced impact;
γ is the specific gravity of the material;
A is the area of reduced cross section;
g is the gravity acceleration;
k is the coefficient considering the damping properties, which is proportional to the natural frequency of the bar member.

The mode of oscillations can be described by a half-wave sinusoid (5) with sufficient accuracy for practical purposes:

$$w_0 = f_0 \cdot sin\left(\frac{m \cdot \pi \cdot x}{l}\right); \quad w_1 = f_1(t) \cdot sin\left(\frac{m \cdot \pi \cdot x}{l}\right), \tag{5}$$

Then we obtain the following solution for non-stationary oscillations of the bar element (6):

$$f = A_1 \cdot e^{\frac{\alpha \cdot t_\omega}{2}} \cdot z^{\frac{1}{2}} \cdot J_{1/3}\left(\frac{2}{3} z^{\frac{3}{2}}\right) + A_2 \cdot e^{\frac{\alpha \cdot t_\omega}{2}} \cdot z^{\frac{1}{2}} \cdot J_{-1/3}\left(\frac{2}{3} z^{\frac{3}{2}}\right), \tag{6}$$

where the following notation is adopted
$J_{1/3}\left(\frac{2}{3} z^{\frac{3}{2}}\right), \ J_{-1/3}\left(\frac{2}{3} z^{\frac{3}{2}}\right)$ are Bessel functions;
A_1, A_2 are the arbitrary constants determined from the initial conditions of the problem.

$$\alpha = \frac{k \cdot \omega \cdot l^4}{\pi^4 \cdot E \cdot J}, \quad t_\omega = t \cdot \omega = b - a^{\frac{1}{6}} \cdot z, \quad \omega = \frac{\pi^2}{l^2} \sqrt{\frac{E \cdot g}{\gamma} \cdot \frac{J}{A}}, \quad P_E = \frac{\pi^2 \cdot E \cdot J}{l^2};$$

$$b = \frac{\omega}{\Omega}\left(P_E \cdot n^2 - P_0 - \frac{\alpha^2 \cdot P_E}{4 \cdot n^4}\right), \quad a = \left(\frac{\omega}{\Omega} \frac{P_E}{n^2}\right)^2;$$

For steady oscillations of the rod element, we obtain the Eq. (7):

$$f_1 = e^{-\frac{\beta \cdot t_2}{2}} \cdot \left(D_1 \cdot \cos\left(\frac{\lambda \cdot t_2}{2}\right) + D_2 \cdot \sin\left(\frac{\lambda \cdot t_2}{2}\right)\right) + \frac{f_0}{1 - \frac{P}{m^2 \cdot P_E}}, \tag{7}$$

where D_1, D_2 are the arbitrary constants determined from the initial conditions, which are the lateral deflection and speed of the bar movement at the time $t = t_1$, obtained from expression (4) and its first derivative, respectively;

$$\beta = \frac{k \cdot l^2 \cdot \omega^2}{P_E \cdot \pi^2}, \quad t_2 = t - t_1 \quad for \quad t > t_1, \quad \lambda^2 = \beta^2 - 4\omega^2 \cdot m^4\left(1 - \frac{P}{m^2 \cdot P_E}\right).$$

It should be noted that the accepted assumption $P < P_E$ corresponds to the case $\lambda^2 < 0$. Analysis of the expression for the parameter λ^2 shows that if the value of the axial force approaches P_E, then λ^2 changes its sign [19, 20], and the first term of expression (7) is written in hyperbolic functions.

3 Results and Discussion

Using PTC Mathcad Prime 4.0 and the computational models presented in Sects. 2.1 and 2.2, the calculation of the maximum dynamic response f_{max} (maximum value of f_1 for the time range from 0 to t) of the considered reinforced concrete columns has been performed. On the basis of this, the design bending moments relative to the center of gravity of the tensioned (least compressed) longitudinal reinforcement have been determined using the formula (8):

$$M = P \cdot (f_{max} + 0.5h_0 - - 0.5a'), \tag{8}$$

and the bearing capacity for a special limit state has been checked from the condition $M \le M_{ult}$. The calculation results are presented in Table 1.

Table 1. Calculation results

$\lambda_h = l_0/h$	μ_s, %	P/N_{ult}	N_{ult}, [kN] Eq. (1)	$N_{ult,sh}$, [kN] Eq. (1)	f_{max}, [m] Eq. (7)	M, [kN·m] Eq. (8)	M_{ult}, [kN·m] Eq. (2)	Special limiting state
10	1	0.6	3888	3888	0.0149	384.7	602.0	+
10	1	0.8	3888	3888	0.0154	514.5	631.2	+
10	1	1	3888	3888	0.0159	645.0	630.4	X
10	2	0.6	4608	4608	0.015	456.2	713.1	+
10	2	0.8	4608	4608	0.0156	610.5	729.6	+
10	2	1	4608	4608	0.0161	765.4	722.5	X
10	3	0.6	5328	5328	0.0151	527.8	815.5	+
10	3	0.8	5328	5328	0.0157	706.3	825.5	+
10	3	1	5328	5328	0.0163	886.0	815.4	X

(continued)

Table 1. *(continued)*

$\lambda_h = l_0/h$	μ_s, %	P/N_{ult}	N_{ult}, [kN] Eq. (1)	$N_{ult,sh}$, [kN] Eq. (1)	f_{max}, [m] Eq. (7)	M, [kN·m] Eq. (8)	M_{ult}, [kN·m] Eq. (2)	Special limiting state
15	1	0.6	3585.6	3780	0.0167	358.6	590.9	+
15	1	0.8	3585.6	3780	0.018	481.9	625.4	+
15	1	1	3585.6	3780	0.0194	607.4	634.3	+
15	2	0.6	4249.6	4480	0.017	425.8	705.9	+
15	2	0.8	4249.6	4480	0.0185	572.8	727.0	+
15	2	1	4249.6	4480	0.0201	722.9	728.1	+
15	3	0.6	4913.6	5180	0.0173	493.2	810.3	+
15	3	0.8	4913.6	5180	0.0189	663.9	824.5	+
15	3	1	4913.6	5180	0.0207	838.8	821.7	**X**
20	1	0.6	3024	3672	0.0188	306.3	565.9	+
20	1	0.8	3024	3672	0.0211	413.9	606.8	+
20	1	1	3024	3672	0.0242	526.8	629.5	+
20	2	0.6	3584	4352	0.0193	364.1	689.0	+
20	2	0.8	3584	4352	0.022	493.2	716.1	+
20	2	1	3584	4352	0.0257	629.7	728.9	+
20	3	0.6	4144	5032	0.0198	422.2	797.7	+
20	3	0.8	4144	5032	0.0228	572.9	817.6	+
20	3	1	4144	5032	0.027	733.5	825.3	+

Notes:
$N_{ult,sh}$ has been calculated by formula (1) with a buckling coefficient φ for the case of short-term loading.
\+ corresponds the case when the bearing capacity of the eccentrically compressed element is ensured;
X corresponds the case when the bearing capacity of the eccentrically compressed element is not ensured.

Analysis of the data presented in Table 1 shows that at $P = N_{ult,l}$, the bearing capacity of structures from the condition $M \leq M_{ult}$ is provided. This is due to the fact that the bearing capacity of compressed reinforced concrete column according to formula (1) was calculated at values of the buckling coefficient φ corresponding to a long-term load, while the considered columns were subjected to dynamic additional loading. Formula (2) does not take into account the duration of the load, so its use to assess the bearing capacity of eccentrically compressed reinforced concrete column according to the criteria of a special limit state is justified. However, the operation of the structure after such an impact is not safe, since over time, static deflections will increase due to the development of plastic deformations and creep, and the column will lose stability.

Figures 3, 4, 5 show graphs of the dependence of the maximum dynamic deflection of the column on slenderness ratio, the percentage of reinforcement and the level of loading. Here f_{max} is the maximum value of f_1 for the time range from 0 to t.

The graphs show that the dynamic deflection of a compressed reinforced concrete column with additionally loaded impact load increases almost linearly with an increase in slenderness ratio at axial force in the range from $0.6\,N_{ult}$ to $0.8N_{ult}$. When the axial force reaches the ultimate value for long-term loading, the relationship between slenderness ratio and deflection becomes non-linear. Deflections begin to increase more intensively as slenderness ratio increases.

A similar picture is observed when varying the axial force. With slenderness ratio in the range from 10 to 15, an almost linear relationship between the deflection and the magnitude of the applied load remains. However, at a slenderness ratio of 20, the dependence becomes non-linear, and the deflections begin to grow more intensively with increasing load.

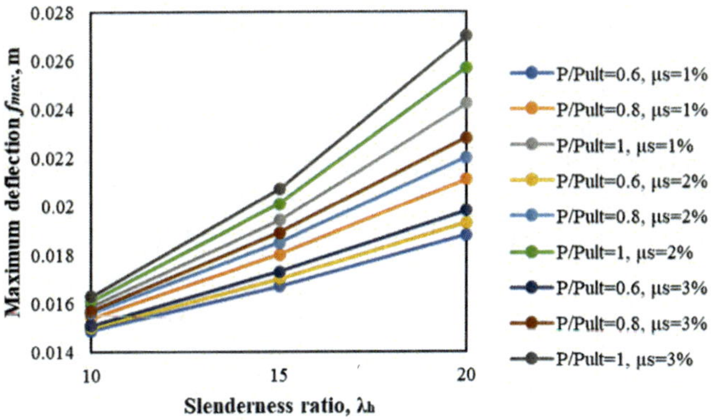

Fig. 3. Slenderness ratio vs. Maximum deflection diagram

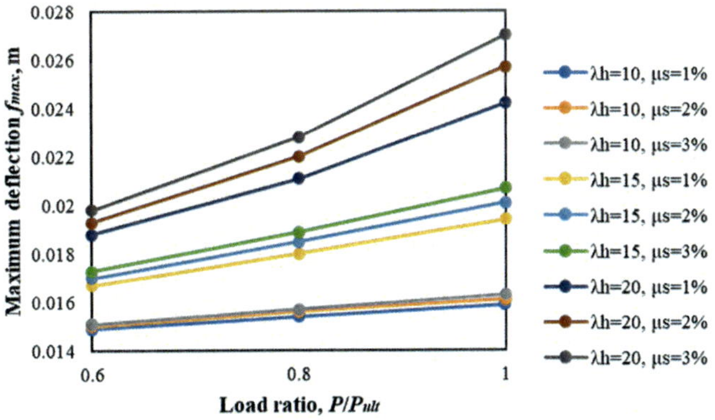

Fig. 4. Load ratio vs. Maximum deflection diagram

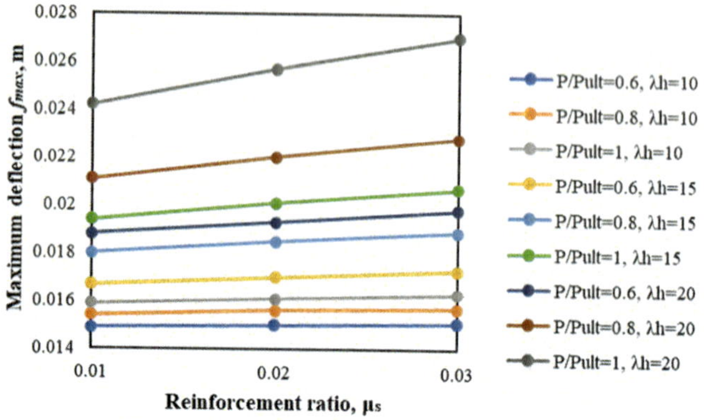

Fig. 5. Reinforcement ratio vs. Maximum deflection diagram

Figure 5 shows that an increase of the calculated reinforcement percentage led to an increase in deflections due to a decrease in the relative stiffness of the section per unit of applied axial force, if all reinforcement in cross section is calculated. The relationship between the dynamic deflection and the calculated reinforcement percentage remains linear in the entire range of values of the variable parameters. For columns with a slenderness ratio of 10 reinforcement percent has practically no effect on the magnitude of the dynamic deflection. With an increase in the slenderness ratio of the column, an increase of the calculated reinforcement percentage leads to a negative effect. Thus, with a slenderness ratio of 20 and axial force $P = N_{ult,l}$, an increase in the percentage of calculated reinforcement by 2% leads to an increase in the dynamic deflection by 11.6%. From the point of view the protection eccentrically compressed reinforced concrete columns against progressive collapse under special emergency actions, attention should be paid not only to strength, but also to its bending stiffness.

4 Conclusion

The article presents an analytical solution to the problem of the dynamic buckling of a compressed reinforced concrete column of a multi-storey building frame. The analysis of the influence of its slenderness ratio, the percentage of reinforcement and the magnitude of the axial force on the dynamic response of a reinforced concrete column is carried out. When varying these parameters, the bearing capacity of the columns was evaluated according to the criteria of a special limiting state.

Based on the results of the study, the following conclusions can be formulated:

(a) *Slenderness ratio of the column.* The dynamic deflection of a compressed reinforced concrete column loaded with an shock load increases almost linearly with an increase in slenderness ratio at a loading level in the range from 0.6 $N_{ult,l}$ to $0.8N_{ult,l}$. When the axial force reaches the ultimate value for long-term loading, a more intensive increase in deflections with an increase in slenderness ratio is observed.

(b) *The magnitude of the axial force.* With an increase in the slenderness ratio of the column up to 20, dynamic deflections begin to grow more intensively with increasing load. When the dynamic axial force in the column reaches the value of its bearing capacity under prolonged loading, the structure is able to resist destruction for a short time. However, over time, static deflections in the structure will increase due to the development of plastic deformations and creep, and the column will lose stability.

(c) *Calculated reinforcement percentage.* An increase of the calculated reinforcement percentage for column with a slenderness ratio exceeding 15 leads to a negative effect - an increase in dynamic deflections. In this regard, to ensure the bearing capacity of eccentrically compressed reinforced concrete column, it is advisable to increase their bending stiffness.

References

1. Pearson, C., Delatte, N.: Ronan point apartment tower collapse and its effect on building codes. J. Perform. Constr. Facil. **19**(2), 172–177 (2005)
2. Tagel-Din, H., Rahman, N.A.: Simulation of the Alfred P. Murrah federal building collapse due to blast loads. In: AEI 2006 Build. Integr. Solut. - Proc. 2006 Archit. Eng. Natl. Conf., vol. 32 (2006)
3. Byfield, M., Paramasivam, S.: Murrah building collapse: reassessment of the transfer girder. J. Perform. Constr. Facil. **26**(4), 371–376 (2012)
4. He, X.-H.-C., Yi, W.-J., Yuan, X.-X.: A non-iterative progressive collapse design method for RC structures based on virtual thermal pushdown analysis. Eng. Struct. **189**, 484–496 (2019)
5. Sasani, M., Sagiroglu, S.: Progressive collapse resistance of hotel San Diego. J. Struct. Eng. **134**(3), 478–488 (2008)
6. Sasani, M., Werner, A., Kazemi, A.: Bar fracture modeling in progressive collapse analysis of reinforced concrete structures. Eng. Struct. **33**(2), 401–409 (2011)
7. Kolcunov, V.I., Tuyen, V.N., Korenkov, P.A.: Deformation and failure of a monolithic reinforced concrete frame under accidental actions. IOP Conf. Series: Mater. Sci. Eng. **753**, 032037 (2020)
8. Fedorova, N.V., Korenkov, P.A.: Static and dynamic deformation of monolithic reinforced concrete frame building in ultimate limit and beyond limits states. Build. Reconstr. **6**(68), 90–100 (2016)
9. Adam, J.M., et al.: Research and practice on progressive collapse and robustness of building structures in the 21st century. Eng. Struct. **173**, 122–149 (2018)
10. Fedorova, N.V., Savin, S.Y.: Progressive collapse resistance of facilities experienced to localized structural damage - an analytical review. Build. Reconstr. **95**(3), 76–108 (2021)
11. Tavakoli, H.R., Kiakojouri, F.: Progressive collapse of framed structures: suggestions for robustness assessment. Sci. Iran. **21**(2), 329–338 (2014)
12. Wang, H., et al.: A review on progressive collapse of building structures. Open Civ. Eng. J. **8**(1), 183–192 (2014)
13. Shan, L., et al.: Robustness of RC buildings to progressive collapse: influence of building height. Eng. Struct. **183**, 690–701 (2019)
14. Du, K., et al.: Experimental investigation of asymmetrical reinforced concrete spatial frame substructures against progressive collapse under different column removal scenarios. Struct. Des. Tall Spec. Build. **29**, e1717 (2020)

15. Lim, N.S., Tan, K.H., Lee, C.K.: Experimental studies of 3D RC substructures under exterior and corner column removal scenarios. Eng. Struct. **150**, 409–427 (2017). https://doi.org/10.1016/j.engstruct.2017.07.041
16. Kai, Q., Li, B.: Dynamic performance of RC beam-column substructures under the scenario of the loss of a corner column—Experimental results. Eng. Struct. **42**, 154–167 (2012)
17. Building Code of Russian Federation SP 385.1325800.2018: Protection of Buildings and Structures against Progressive Collapse. Design code. Basic Statements, vol. 26. Ministry of Russia, Moscow (2018)
18. Savin, S., Kolchunov, V.: Dynamic behavior of reinforced concrete column under accidental impact. Int. J. Comput. Civ. Struct. Eng. **17**(3), 120–131 (2021)
19. Volmir, A.S.: Stability of Deformable Systems, vol. 984. Nauka, Moscow (1967)
20. Kamke, E.: Handbook of Ordinary Differential Equations, vol. 576. Nauka, Moscow (1971)

Energy Absorption Characteristics of Aluminium Alloy Tubes Subjected to Quasi-static Axial Load

M. Kulkarni Sudhanwa(✉), Vemu Priyal, D. Mali Kiran, and M. Kulkarni Dhananjay

Department of Mechanical Engineering, Birla Institute of Technology and Sciences - Pilani, K. K. Birla Goa Campus, Goa, India
{p20190071,f20170819,kiranm,dmk}@goa.bits-pilani.ac.in

Abstract. In the current numerical study, axial quasi-static crashworthiness performance of circular crash-boxes having the same geometry but made up of various frequently used wrought aluminium alloys is investigated using ABAQUS/Explicit. All the considered crash-boxes showed a stable progressive axi-symmetric collapse. Technique for Order of Preference by Similarity to Ideal Solution (TOPSIS), one of the most widely used Multi-Criteria Decision-Making (MCDM) techniques, is then employed to determine the most suitable aluminium alloy for the application of a crash-box in an automobile. It is found that, out of the considered aluminium alloys AA 6060 T4 is the most suitable aluminium alloy to be used as a crash-box in an automobile.

Keywords: Aluminium alloy tubes · Crash-box · Crashworthiness · Crush force efficiency · Energy absorption · TOPSIS

1 Introduction

To improve safety of driver and passengers it is necessary to enhance crashworthiness of an automobile. So, various active and passive safety devices are employed in automobiles. Thin-walled collapsible energy absorber also known as crash-box is one of the essential passive safety devices, located in between the bumper and the front side rails of an automobile [1]. Crash-box is a thin walled curved shell tubular structure. Crash-boxes of various cross-sectional profiles have been tested and implemented in the automobiles. They undergo stable progressive collapse to absorb the energy of external impact during crush events and protects the occupants by reducing the forces transferred to them [2]. Various materials including different grades of steel, aluminium alloys, magnesium alloys etc. are used to manufacture crash-boxes.

Gupta [3] concluded that the crashworthiness response for both mild steel and aluminium alloy circular tubes changed considerably after annealing as the material properties for both the tubes are altered due to annealing heat treatment. Moreover, it is also noticed that energy absorption (*EA*) of mild steel tubes is higher than that of aluminium tubes. Dindar et al. [4] reported that *EA* and initial peak force (F_{peak}) of spring steel tubes

Z. Dimitrová et al. (Eds.): WMVC 2022, MMS 125, pp. 53–64, 2023.
https://doi.org/10.1007/978-3-031-15758-5_5

subjected to axial impact is increased after austenitizing because austenitizing makes the material more brittle. Bambatch et al. [5] observed that the crashworthiness performance of hexagonal tubes of same dimensions but made up of dual phase steel and heat-treated high manganese steel subjected to axial impact loads is different.

Niu et al. [6] showed that the mean crushing force (F_{mean}) and *EA* of spot-welded top hat energy absorbers fabricated from high strength low alloy steel is lower than that fabricated from dual phase steel. Yamashita et al. [7] conducted low velocity impact tests on top and double hat structures made up of aluminium alloys namely, AA 1050 H24 and AA 5052 H34 and concluded that both the structures made up of AA 5052 H34 collapsed more regularly showing about 1.3 times higher crush strength than that made up of AA 1050 H24.

Gronostajski et al. [8] concluded that the crash-boxes made up of magnesium alloy AZ31 showed catastrophic failure when subjected to low velocity impact as against the steel crash-box that showed regular progressive collapse. However, the specific energy absorption (*SEA*) of magnesium alloy tubes is the highest and can be further increased by using aluminium foam filling. Deng et al. [9] compared the crashworthiness performance of star shaped energy absorbers made up of different grades of aluminium alloys and observed that the crashworthiness response of the tubes change substantially as the material of the crash-box is changed. Rai et al. [10] came to the conclusion that even though the tubes made up of steel (E235) and aluminium alloy (AA 6063 T6) collapse in diamond mode, aluminium tubes are superior to steel tubes if stroke efficiency, *CFE* and *SEA* are considered under same impact conditions.

It can be seen from the above discussion that the crashworthiness performance of thin-walled energy absorbing tubes can be tailored either by the use of different materials or by changing their material properties through heat treatment. Also, to bring down production cost, to improve fuel efficiency [10] and to lower down pollution from an automobile to meet extremely stringent environmental regulations [11], automobile manufacturers prefer lightweight materials in place of heavier materials like steel in an automobile structure. For the same reasons, wrought aluminium alloys having low density and high strength are replacing thin-walled crash-boxes manufactured from various grades of steel in an automobile.

A comparative analysis of crashworthiness performance of crash-boxes fabricated from commonly used aluminium alloys is rarely found in the literature. Therefore, this numerical study focuses on determining the crashworthiness performance of circular crash-boxes made up of different aluminium alloys. Crashworthiness performance of all the aluminium alloy tubes is then compared based on different crashworthiness performance indicators and their mode of collapse. Technique for Order of Preference by Similarity to Ideal Solution (TOPSIS) - a multi-criteria decision-making (MCDM) technique is then used to determine the most suitable aluminium alloy for the application of a crash-box.

2 Materials and Methods

2.1 Geometry

In this work, a seamless circular tube as shown in Fig. 1 is considered. The undeformed length (l), outer diameter (D) and wall thickness (t) of the tube are 90 mm, 44.5 mm and 1.4 mm, respectively. Tube having the same dimensions but made up of AA 6xxx and AA 7xxx was studied by Bade et al. [12].

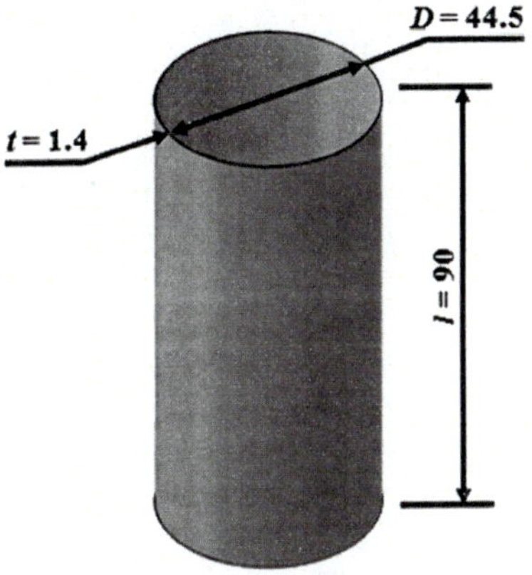

Fig. 1. Overall dimensions of crash-box (in mm)

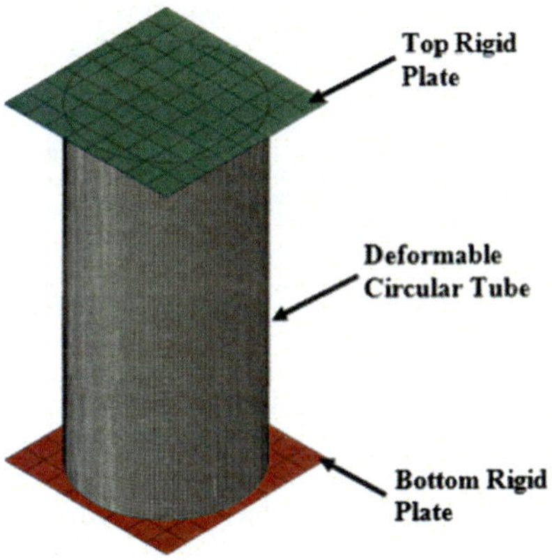

Fig. 2. Finite element model

2.2 Material Properties

The circular crash-boxes of above dimensions but made up of different aluminium alloys as listed in Table 1 are considered. Mechanical and material properties of these aluminium alloys are derived from the literature as mentioned in Table 1.

2.3 Crashworthiness Performance Indicators

To quantify and compare the crashworthiness performance of the thin-walled collapsible energy absorbers, following crashworthiness performance indicators have been defined.

Initial Peak Force ($\mathbf{F_{peak}}$). Initial peak force (F_{peak}) is that maximum force which is required to generate first permanent deformation i.e. first fold in the tube [13]. High value of F_{peak} indicates that the forces transferred to the occupants will be high which will cause damage to them [14]. F_{peak} thus, should have as minimum value as possible.

Energy Absorption (EA). Energy absorption (EA) is the work done by the external compressive load (F_S) on the crush tube in the longitudinal direction up to given structural deformation (S_{max}) [15]. Mathematically, EA can be defined as Eq. (1)

$$EA = \int_0^{S_{\max}} F_{(S)} dS \tag{1}$$

Table 1. References for material properties of aluminium alloys.

Simulation no.	Aluminium alloy grade	Reference no.
S1	AA 2024 T3	[15]
S2	AA 3003 H12	[18]
S3	AA 6060 T4	[19]
S4	AA 6061 T4	[16]
S5	AA 6061 T5	[20]
S6	AA 6061 T6	[12]
S7	AA 6063 T5	[21]
S8	AA 6063 T6	[10]
S9	AA 7005 T6	[12]
S10	AA 7003 T6	[12]
S11	AA 7075 T6	[22]

For an energy absorber to be crashworthy, *EA* should be as high as possible.

Mean Crushing Force ($\mathbf{F}_{mean}$). Mean crushing force (F_{mean}) is defined as the energy absorbed by the crash-box when it is deformed by unit distance [16]. If a tube has absorbed *EA* amount of energy while getting deformed by S_{max}, then F_{mean} is given as Eq. (2)

$$F_{mean} = \frac{EA}{S_{\max}} \tag{2}$$

The high value of F_{mean} all over the deformed length is desirable for a crash-box to be crashworthy [14].

Crush Force Efficiency (CFE). A ratio of F_{mean} to F_{peak} is termed as crush force efficiency (*CFE*) given by Eq. (3) [13]. The value of this non-dimensional quantity ranges between zero and one.

$$CFE = \frac{F_{mean}}{F_{peak}} \tag{3}$$

The value of *CFE* for a tube should be nearer to unity, which indicates that the jerks felt by occupants during an accident will be low as the tube is being crushed at a load nearer to F_{peak} [13]. At the same time, near unity value of *CFE* also signifies uniformity in load-displacement characteristic of the crash-box [17]. Thus, the maximum value of *CFE* is beneficial for an energy absorber to be crashworthy.

Specific Energy Absorption (SEA). Energy absorbed (*EA*) per unit mass of the tube (m) during the crushing process is defined as specific energy absorption (*SEA*) given by Eq. (4) [14].

$$SEA = \frac{EA}{m} \tag{4}$$

A high value of SEA for a tube indicates either more value of energy absorbed by the tube or low mass of the tube. Both the conditions are favorable for a tube to be crashworthy. Thus, a crash-box to be crashworthy should have a high value of *SEA*.

2.4 Finite Element Model

The commercially available finite element (FE) software ABAQUS/Explicit (6.14) is used to simulate axial quasi-static collapse of the thin-walled energy absorber. Figure 2 displays the FE model that consists of a deformable circular tube and two flat rigid plates of size 100 mm × 100 mm, one at bottom while other at top of the tube. As the ratio of diameter of tube (D) to its thickness (t) is greater than 10, shell elements (S4R) are used to model the geometry of thin-walled energy absorbers [11, 12]. After conducting mesh sensitivity analysis it was observed that shell elements of size 1 mm with 5 integration points through thickness could appropriately represent the geometry of the tube. For this configuration, artificial energy of the model is found to be less than 5% of internal energy, which shows insignificant hourglass effect. Top and bottom plates are meshed with rigid quadrilateral elements (R3D4).

Bottom plate is fixed completely whereas the top plate is moved only along the length of the tube with a velocity of 1000 mm/s. To minimize the inertia effects in the quasi-static simulation velocity is ramped up smoothly, using SMOOTH STEP sub-option, from zero to its full value within 0.12 s. As the kinetic energy of the system is found out to be less than 5% of the internal energy, inertia effects are negligible and thus the applied velocity is appropriate. All the tubes are crushed by 60 mm [12]. To model the interactions between rigid plates and the tube, and interactions between the lobes that are formed during collapse of the tube, general contact algorithm with a friction coefficient of 0.2 is found out to be suitable [12].

2.5 Technique for Order of Preference by Similarity to Ideal Solution (TOPSIS)

TOPSIS is a MCDM technique first developed by Hwang and Yoon [23]. It tries to find the alternative, which is nearer to the hypothetical ideal solution and at the same time far away from the hypothetical worst solution. Various steps involved in TOPSIS technique are as follows.

Deciding the objective and identifying the relevant attributes for the problem.
Developing the decision matrix 'D', in such a way that every entry m_{ij} in the matrix represents the value of jth attribute for the ith alternative.
Obtaining the normalized decision matrix R_{ij} by using formula given in Eq. (5)

$$R_{ij} = \frac{m_{ij}}{\left[\sum_{j=1}^{M} m_{ij}^2\right]^{1/2}} \tag{5}$$

Deciding the relative weights (w_j) for each attribute such that summation of all weights must be equal to 1.

Obtaining weighted normalized matrix V_{ij} using Eq. (6)

$$V_{ij} = w_j R_{ij} \tag{6}$$

Obtaining the best (V^+) and the worst (V^-) solution for each attribute from the weighted normalized matrix (V).
Obtaining the separation measures (S^+ and S^-) of each alternative from the best and the worst solution.
Determining the degree of closeness of each alternative from the best solution by using Eq. (7)

$$P_i = \frac{S_i^-}{(S_i^+ + S_i^-)} \tag{7}$$

Ranking the alternatives depending on its degree of closeness to the best solution i.e. P value.

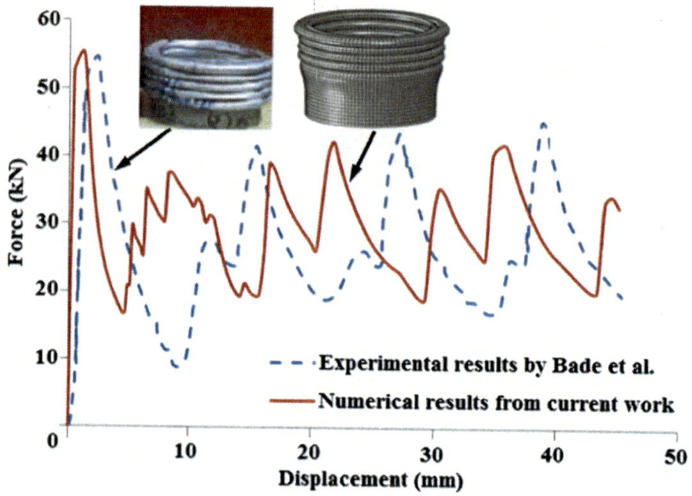

Fig. 3. Force - displacement characteristics of AA 7005 T6 tube (snippet: mode of collapse of tube)

3 Results and Discussion

3.1 Validation of the Finite Element Model with Previous Work

Table 2 and Fig. 3 compare the results obtained from the numerical model presented above with the results obtained by Bade et al. [12]. Here, the numerical results are compared with the experimental results obtained by Bade et al. [12]. In the case of experimental analysis, when crash-box is fabricated by any of the manufacturing processes, imperfections in the form of residual stresses, waviness of the sidewalls, non-parallel ends, wall thickness variations, and so on are introduced into it. During numerical simulations, these imperfections induced in the crash-boxes are not taken into account. Moreover, material properties assigned to the crash-box, and experimental conditions simulated in numerical analysis are assumed to be ideal. These assumptions simplify the simulations to a great extent, but at the same time, they affect the simulation results. Therefore, the results obtained from the simulations deviate from the results obtained from experiments. However, in the current study, this deviation is not more than 12% in any case as can be seen from Table 2. In addition, Fig. 3 shows close correlation between the mode of collapse and force-displacement curves obtained from both the studies.

Table 2. Comparison of FE results with previous work.

	AA 6061 T6		AA 7003 T6		AA 7005 T6	
	Bade et al. [12]	Present study	Bade et al. [12]	Present Study	Bade et al. [12]	Present study
F_{peak} (kN)	49.85	53.30	61.00	65.70	51.90	55.50
EA (kJ)	1.10	1.22	1.34	1.54	1.21	1.34
F_{mean} (kN)	24.40	27.11	29.78	34.22	26.89	29.78
SEA (kJ/kg)	47.94	53.17	56.10	64.94	54.54	56.50
CFE	0.49	0.51	0.49	0.52	0.52	0.54

Because the numerical results differ by less than 12% from the experimental data, they have little impact on the conclusions. The provided FE model is robust enough to accommodate changes in material properties of the crash-boxes, and hence can be used for further study.

3.2 Crashworthiness Performance of Circular Aluminium Alloy Tubes

A series of quasi-static crushing simulations are carried out as per the numerical procedure described above. The results obtained from different simulations are discussed in this section.

Table 3 compares the crashworthiness of the various aluminium alloy tubes under consideration. Table 3 shows how changes in the material properties of the tube affect the crashworthiness performance indicators. Despite the fact that the material properties of the crash-boxes were altered, all of the crash-boxes studied here collapsed in an axi-symmetric concertina mode. Changes in the material properties of the tube cause a modest difference in the mechanism of collapse. Except for the AA 6061 T4 tube, which started collapsing in concertina mode from the bottom fixed end, all other tubes collapse from the top end where the load is applied. A representative force-displacement characteristic and mode of collapse for AA 7005 T6 circular tube is shown in Fig. 3.

Table 3. Comparison of crashworthiness performance indicators of the tubes.

Simulation no	Material of tube	F_{peak} (kN)	*SEA* (kJ/kg)	*CFE* (%)
S1	AA 2024 T3	65.9	48.2	57.61
S2	AA 3003 H12	31.6	21.2	52.83
S3	AA 6060 T4	22.3	18.2	62.71
S4	AA 6061 T4	28.6	22.2	59.35
S5	AA 6061 T5	37.5	26.9	54.81
S6	AA 6061 T6	53.3	36.3	52.06
S7	AA 6063 T5	41.3	26.6	49.39
S8	AA 6063 T6	42.1	29.6	53.94
S9	AA 7005 T6	55.5	36.9	53.72
S10	AA 7003 T6	65.7	43.6	52.49
S11	AA 7075 T6	105.1	72.3	54.51

3.3 Identifying the Suitable Aluminium Alloy to be Used for Crash-box

The crash-box having minimum value of F_{peak}, and maximum values of *SEA*, *CFE* is said to be crashworthy and that is to be used in an automobile. It can be seen from Table 3 that the tube made up of AA 7075 T6 is having the highest *SEA* and moderate *CFE*. Nevertheless, it has a maximum value of F_{peak}, which is undesirable. Similarly, AA 6060 T4 tube is having the highest value for *CFE*. However, it has the smallest value of *SEA*. Due to these conflicting situations, it becomes difficult to select the suitable material for the crash-box from the list. In such cases, where it is required to select the best alternative from a number of alternatives by comparing more than one attribute (criteria), MCDM techniques are used.

TOPSIS, one of the widely accepted MCDM techniques, is employed here to determine the most suitable aluminium alloy for the crash-box out of 11 alternative alloys based on three attributes i.e. criteria, namely F_{peak}, *CFE* and *SEA*. All these 11 alternatives and three attributes along with their values are tabulated in Table 3. The same crashworthiness performance indicators are used as attributes by Baaskaran et al. [15]. Table 4 shows the representative calculations for the current MCDM problem using TOPSIS technique considering equal relative weights (w_j) for all of the three attributes i.e. $w_1 = w_2 = w_3 = (1/3)$. For a crash-box, F_{peak} is non-beneficial while *CFE* and *SEA* are beneficial. This means that, the lowest value of F_{peak} and the highest values of *CFE* and *SEA* are considered as the best alternative solution (V^+) while the highest value of F_{peak} and the lowest value of *CFE* and *SEA* are termed for the worst alternative (V^-) [15].

Table 4. Calculations of current MCDM problem using TOPSIS.

Step 2: $D =$			Step 3: $R =$			Step 5: $V =$		
65.9	48.2	57.61	0.3637	0.3848	0.3160	0.1212	0.1283	0.1053
31.6	21.2	52.83	0.1742	0.1690	0.2898	0.0581	0.0563	0.0966
22.3	18.2	62.71	0.1230	0.1446	0.3440	0.0410	0.0482	0.1147
28.6	22.2	59.35	0.1580	0.1766	0.3255	0.0527	0.0589	0.1085
37.5	26.9	54.81	0.2072	0.2148	0.3007	0.0691	0.0716	0.1002
53.3	36.3	52.06	0.2941	0.2895	0.2855	0.0980	0.0965	0.0952
41.3	26.6	49.39	0.2280	0.2122	0.2709	0.0760	0.0707	0.0903
42.1	29.6	53.94	0.2324	0.2362	0.2958	0.0775	0.0787	0.0986
55.5	36.9	53.72	0.3062	0.2946	0.2947	0.1021	0.0982	0.0982
65.7	43.6	52.49	0.3626	0.3483	0.2879	0.1209	0.1161	0.0960
10.5	72.3	54.51	0.58020	0.5766	0.2990	0.1934	0.1922	0.0997

Step 6:	Step 7: $S^+ =$	Step 7: $S^- =$	Step 8: $P =$	Step 9: *Rank* =
$V^+ =$ [0.0410 0.1922 0.1147]	0.1030	0.1088	0.5137	3
	0.1381	0.1357	0.4956	6
	0.1440	0.1543	0.5174	1
	0.1340	0.1423	0.5151	2
$V^- =$ [0.1934 0.0482 0.0903]	0.1247	0.1269	0.5044	4
	0.1131	0.1070	0.4861	7
	0.1287	0.1195	0.4815	9
	0.1203	0.1202	0.4998	5
	0.1133	0.1044	0.4795	10
	0.1120	0.0100	0.4707	11
	0.1531	0.1443	0.4851	8

Assigning weights to different attributes involved in the MCDM problem is a crucial and the most significant task. One may think that the crash-box is a component designed to absorb the kinetic energy of external impact during an accident by progressively getting collapse. Therefore, *SEA* is a more critical factor than F_{peak} and *CFE*. As a result, *SEA* should be given more weight than F_{peak} and *CFE* from a practical standpoint. Other may think that all the attributes are equally important and thus equal weights should be assigned to them all. To avoid this confusion, the most acceptable and the least desired aluminium alloys for fabricating a crash-box, are determined by TOPSIS using various combinations of weights for the attributes (w_1; w_2; w_3) and the same are represented in Table 5.

Table 5. Effect of different attribute weights on material selection

Sr. no.	Attribute weights			Best alternative	Worst alternative
	F_{peak} (w_1)	*SEA* (w_2)	*CFE* (w_3)		
01	1/3	1/3	1/3	AA 6060 T4	AA 7003 T6
02	0.5	0.25	0.25	AA 6060 T4	AA 7075 T6
03	0.5	0.4	0.1	AA 6060 T4	AA 7075 T6
04	0.5	0.1	0.4	AA 6060 T4	AA 7075 T6
05	0.2	0.4	0.4	AA 7075 T6	AA 3003 H12
06	0.2	0.7	0.1	AA 7075 T6	AA 3003 H12
07	0.2	0.1	0.7	AA 6060 T4	AA 7075 T6
08	0.25	0.5	0.25	AA 7075 T6	AA 3003 H12
09	0.4	0.5	0.1	AA 7075 T6	AA 6063 T5
10	0.1	0.5	0.4	AA 7075 T6	AA 3003 H12
11	0.4	0.2	0.4	AA 6060 T4	AA 7075 T6
12	0.7	0.2	0.1	AA 6060 T4	AA 7075 T6
13	0.1	0.2	0.7	AA 7075 T6	AA 6063 T5
14	0.25	0.25	0.5	AA 6060 T4	AA 7003 T6
15	0.1	0.4	0.5	AA 7075 T6	AA 3003 H12
16	0.4	0.1	0.5	AA 6060 T4	AA 7075 T6
17	0.4	0.4	0.2	AA 6060 T4	AA 7003 T6
18	0.1	0.7	0.2	AA 7075 T6	AA 3003 H12
19	0.7	0.1	0.2	AA 6060 T4	AA 7075 T6

Table 5 shows that the selection of the most suitable aluminium alloy for fabrication of the crash-box is dependent on the relative attribute weights (w_j). In eleven of the nineteen scenarios studied above, AA 6060 T4 is the best material to fabricate circular crash-boxes, while AA 7075 T6 is the best material in only eight cases. As aluminium alloy AA 6060 T4 appears more times, it can be considered the most suitable material for manufacturing a crash-box. Similarly, eight cases reveal that AA 7075 T6 is the least preferred material for crash-box construction. Aluminium alloys AA 3003 H12, AA 7003 T6, and AA 6063 T5 appear six, three and two times in the worst alternative column, respectively.

4 Conclusions

Numerical simulations of axial quasi-static collapse of circular thin-walled energy absorbers having same dimensions but made up of different aluminium alloys is carried out. Crashworthiness performance of these circular tubes made up of different aluminium alloys is then compared based on various crashworthiness performance indicators. Following conclusions are drawn from the current study.

- All the tubes collapsed in axi-symmetric concertina mode. All the tubes started collapsing from the top end where load is applied except the tube made of AA 6061 T4, which started collapsing from fixed bottom end.
- F_{peak} for the tubes ranges between 22.3 kN for AA 6060 T4 to 105.3 kN for AA 7075 T6. Tube made up of AA 6060 T4 has the lowest *SEA* of 18.2 kJ/kg but it has the highest *CFE* of 62.72% when compared with the other tubes.
- To determine the most suitable aluminium alloy to be used for the application of crash-box in an automobile a MCDM technique known as TOPSIS is employed. F_{peak}, CFE, and *SEA* are considered as attributes (criteria) in TOPSIS technique. Out of the eleven aluminium alloys considered in this study, it is revealed that AA 6060 T4 is the most suitable aluminium alloy to be used for crash-box. However, AA 7075 T6 is the last choice to be used for crash-box as per TOPSIS technique.

References

1. Yang, K.: Design optimization of energy absorption structures with origami patterns. Ph.D. Thesis (2018)
2. Kathiresan, M.: Studies on crashworthiness characteristics of thin-walled conical frusta under axial loading. Ph.D. Thesis, Anna University (2016)
3. Gupta, N.K.: Some aspects of axial collapse of cylindrical thin-walled tubes. Thin-Walled Struct. **32**, 111–126 (1998). https://doi.org/10.1016/S0263-8231(98)00029-9
4. Dindar, Ç., Beytüt, H., Karagöz, S.: 51CrV4 yay çeliğinde isil işlem ve kesit geometrisinin çarpişma dayanikliliği üzerine etkisi. Uludağ Univ. J. Fac. Eng. **24**(2), 477–488 (2019). https://doi.org/10.17482/uumfd.477569
5. Bambach, M., Conrads, L., Daamen, M., Güvenç, O., Hirt, G.: Enhancing the crashworthiness of high-manganese steel by strain-hardening engineering, and tailored folding by local heat-treatment. Mater. Des. **110**, 157–168 (2016). https://doi.org/10.1016/j.matdes.2016.07.065
6. Niu, J., Zhu, P., Guo, Y.: Crush performance of top-hat tubular structures considering different forming conditions. Adv. Mater. Res. **139–141**, 571–575 (2010). https://doi.org/10.4028/www.scientific.net/AMR.139-141.571
7. Yamashita, M., Kenmotsu, H., Hattori, T.: Dynamic crush behavior of adhesive-bonded aluminum tubular structure - experiment and numerical simulation. Thin-Walled Struct. **69**, 45–53 (2013). https://doi.org/10.1016/j.tws.2013.04.005
8. Gronostajski, Z., Kaczyński, P., Polak, S., Bartczak, B.: Energy absorption of thin-walled profiles made of AZ31 magnesium alloy. Thin-Walled Struct. **122**, 491–500 (2018). https://doi.org/10.1016/j.tws.2017.10.035
9. Deng, X., Liu, W., Lin, Z.: Experimental and theoretical study on crashworthiness of star-shaped tubes under axial compression. Thin-Walled Struct. **130**, 321–331 (2018). https://doi.org/10.1016/j.tws.2018.06.002
10. Rai, V., Ghasemnejad, H., Watson, J.W., Gonzalez-Domingo, J.A., Webb, P.F.: Developed trigger mechanisms to improve crush force efficiency of aluminium tubes. Eng. Struct. **199**, 109620 (2019). https://doi.org/10.1016/j.engstruct.2019.109620
11. Simhachalam, B., Rao, C.L., Srinivas, K.: Compression behavior and energy absorption of aluminum alloy AA6061 tubes with multiple holes. Int. J. Comput. Methods Eng. Sci. Mech. **15**, 232–241 (2014). https://doi.org/10.1080/15502287.2014.882433

12. Simhachalam, B., Srinivas, K., Rao, C.L.: Energy absorption characteristics of aluminium alloy AA7XXX and AA6061 tubes subjected to static and dynamic axial load. Int. J. Crashworthiness. **19**, 139–152 (2014). https://doi.org/10.1080/13588265.2013.878974
13. Tarlochan, F., Samer, F., Hamouda, A.M.S., Ramesh, S., Khalid, K.: Design of thin wall structures for energy absorption applications: enhancement of crashworthiness due to axial and oblique impact forces. Thin-Walled Struct. **71**, 7–17 (2013). https://doi.org/10.1016/j.tws.2013.04.003
14. Mohamed, A.S., Laban, O., Tarlochan, F., Al, S.E., Khatib, M.S., Matar, E.M.: Experimental analysis of additively manufactured thin-walled heat-treated circular tubes with slits using AlSi10Mg alloy by quasi-static axial crushing test. Thin-Walled Struct. **138**, 404–414 (2019). https://doi.org/10.1016/j.tws.2019.02.022
15. Baaskaran, N., Ponappa, K., Shankar, S.: Study of the effect of varying shapes of holes in energy absorption characteristics on aluminium circular windowed tubes under quasi-static loading. Struct. Eng. Mech. **70**, 153–168 (2019). https://doi.org/10.12989/sem.2019.70.2.153
16. Liu, W., Jin, L., Luo, Y., Deng, X.: Multi-objective crashworthiness optimisation of tapered star-shaped tubes under oblique impact. Int. J. Crashworthiness. **26**, 328–342 (2020). https://doi.org/10.1080/13588265.2020.1717917
17. Fu, J., Liu, Q., Liufu, K., Deng, Y., Fang, J., Li, Q.: Design of bionic-bamboo thin-walled structures for energy absorption. Thin-Walled Struct. **135**, 400–413 (2019). https://doi.org/10.1016/j.tws.2018.10.003
18. Zhang, X., Zhang, H., Ren, W.: Axial crushing of tubes fabricated by metal sheet bending. Thin-Walled Struct. **122**, 252–263 (2018). https://doi.org/10.1016/j.tws.2017.09.023
19. Zheng, G., Wu, S., Sun, G., Li, G., Li, Q.: Crushing analysis of foam-filled single and bitubal polygonal thin-walled tubes. Int. J. Mech. Sci. **87**, 226–240 (2014). https://doi.org/10.1016/j.ijmecsci.2014.06.002
20. Hu, D., Wang, Y.Y., Song, B., Wang, Y.Y.: Energy absorption characteristics of a foam-filled tri-tube under axial quasi-static loading: experiment and numerical simulation. Int. J. Crashworthiness. **23**, 417–432 (2018). https://doi.org/10.1080/13588265.2017.1331494
21. Estrada, Q., et al.: Effect of radial clearance and holes as crush initiators on the crashworthiness performance of bi-tubular profiles. Thin-Walled Struct. **140**, 43–59 (2019). https://doi.org/10.1016/j.tws.2019.02.039
22. Sampath, V.: Numerical investigation of empty and foam filled AA7075 tubes under impact loading (2018)
23. Rao, R.V.: Decision Making in the Manufacturing Environment Using Graph Theory and Fuzzy Multiple Attribute Decision Making Methods. Springer-Verlag, London (2007)

Estimation on Accuracy of Compressive and Tensile Damage Parameters of Concrete Damage Plasticity Model

K. Senthil(✉) and Rachit Sharma

Dr. B. R. Ambedkar National Institute of Technology, Jalandhar 144011, India
urssenthil@yahoo.co.in

Abstract. The behavior of reinforced concrete (RC) structures under extreme demands, as in strong ground motions or impact loading, is quite intricate because of the joint operation of concrete and steel coupled with respective failure models. In addition to that, new advances are made in the field of dynamics engineering due to the complexity and importance for seismic and impact activities worldwide. However, numerical models based on finite element codes had occurrence on several developments because of their simplicity and cost effectiveness. Therefore, the manuscript focused on the estimation of accuracy of compressive and tensile damage parameters (d_c and d_t) of the Concrete Damage Plasticity (CDP) model based on continuum mechanics to simulate the concrete behavior in ABAQUS finite element software. Within the context, it has been summarized analytical equations describing behavior of concrete for stress-inelastic strain in uniaxial compression and stress-displacement in uniaxial tension and their respective damage variables in uniaxial compression and tension, presented by several researchers over the past decade. The methodology provides a basis for the evaluation of analytical models which will be applicable to CDP model. The accuracy and reliability are validated with the experimental results available for uniaxial tension and compression tests. The parameters associated with the biaxial and triaxial behavior of concrete for CDP model will be kept same as originally proposed in Lubliner/ Lee/ Fenves formulation and were given in ABAQUS manual. The numerical results thus obtained were compared with the experimental results and it was postulated that each model predicted static results very accurately. For impact problem, the model by Alfarah et al. [16] best predicted crack pattern and peak load with a 6% difference.

Keywords: Concrete · Analytical models · ABAQUS · Concrete damage plasticity model · Reinforced concrete slab

1 Introduction

Concrete is innately heterogeneous in nature on different scales ranging from nano to macro which makes it quite unique and renders its intricate non-linear inelastic behavior [1]. Owning to the complex microstructure, the mechanical behavior of concrete differs

Z. Dimitrovová et al. (Eds.): WMVC 2022, MMS 125, pp. 65–76, 2023.
https://doi.org/10.1007/978-3-031-15758-5_6

due to the strength and mixing composition. Since the compressive and tensile strength of concrete is an elementary property, the understanding of stress-strain relationships governing its behavior is critical to simulate the structural response and design concrete structures under varying loading rates [2].

The character of concrete has changed with advancement in material technology. With new advancements, the strength of concrete in range of 100–120 MPa has become common more recently. In addition, a more precise mathematical exemplification of concrete behavior is sought to employ intense computing power of digital computers. It may lead to economic design rationalities in the design and analysis of complex structures [3]. There are several researchers that had published experimental and analytical results on defining the strength of concrete over a wide range of strengths ranging from 10 to 120 MPa [4–6]. Among which several material models were analytically developed and considered in standard codes [7, 8]. The use of analytical models may enable one to skip the tedious procedures of experiments and is useful for finite element-based engineers that have an absence of proper instrumentation to characterize the behavior of concrete.

Reinforced concrete structure may get subjected to accidental loading during their life span. Typical examples of accidental loadings include low velocity impacts, hail storm impacts and ballistic impacts among others [9]. Recently, the studies on the performance and vulnerability of RC structures under impact loadings is getting limelight [10, 11] Finite element software such as ABAQUS, LS Dyna provides a necessary platform for the critical evaluations of such structures. Therefore, engineers are tending to numerical models to carry out design, evaluation, and critical checks for safeguarding these structures.

In the evaluation of structural response, the moment-curvature relationship is of prime importance. In such case, the stress-strain variation of concrete and its sectional properties are required over a wide range of compressive strength. The material models are evolved to consider the size of aggregate effect on fracture strength [7] and further the size effect of structural element tested on the mechanical properties of concrete [12]. Therefore, for non-linear analysis and design of structures the knowledge of their stress-strain behavior is required. Earlier analytical models of concrete are based on parabolic function to represent the stress-strain behavior of concrete, however they failed to generate the full curve of stress-strain relationships [13]. To overcome this, some models uses two different formulas for hardening and softening behavior [8]. Therefore, there is still a gap on which model works better under finite element simulation of structural elements.

This paper is an attempt to investigate the effect of material models on the static and dynamic response of reinforced concrete slab. Under this context, the five widely accepted stress-strain models were adopted and incorporated into concrete damage plasticity model available in ABAQUS/CAE. Section 2 highlights the analytical equations for stress-strain relationship in compression and tension. Section 3 highlights the numerical model parameters built using analytical models and experimental test setup of Said and Mabrook Mouwainea [10]. Section 4 highlights the comparison of different models with the experimental results under static and dynamic loading. Section 5 concludes with the major outcomes of the study.

2 Stress-Strain Relationship Under Compression and Tension

For practical applications the design codes simplify stress-strain relationships to parabola-rectangle approximation. The main parameters which govern the formation of these curve are initial tangent modulus, strain at maximum stress, and the shape factor for governing the softening branch. Various stress-strain models are proposed among which some are summarized in Table 1. More details about the analytical models can be found in their respective studies.

Table 1. Complete Stress-strain relationship proposed by several researchers.

Study	Stress-strain relationship of concrete
Under compression	
Carreira and Chu [14]	$f_c = f_c'\left[\frac{\beta\left(\frac{\varepsilon}{\varepsilon_0}\right)}{\beta - 1 + \left(\frac{\varepsilon}{\varepsilon_0}\right)^{\beta}}\right]; \beta = \frac{1}{1-\left(\frac{f_c'}{\varepsilon_0 E_{it}}\right)}$
CEB Model Code [7]	For $0 \leq \varepsilon = \varepsilon_{max}$; $\varepsilon_{max} =$ *concrete strain* at $0.5 f_c'$ on ascending part $f_c = f_c'\left[\frac{\left(\frac{E_{it}}{E_0}\right)\left(\frac{\varepsilon}{\varepsilon_0}\right)-\left(\frac{\varepsilon}{\varepsilon_0}\right)^2}{1+\left(\frac{E_{it}}{E_0}-2\right)\left(\frac{\varepsilon}{\varepsilon_0}\right)}\right]$; $E_0 =$ *secant modulus at peak stress* For $\varepsilon > \varepsilon_{max}$; $f_C = \frac{f_c'}{\left(\frac{1}{\left(\varepsilon_{max}/\varepsilon_0\right)}\zeta - \frac{2}{\left(\varepsilon_{max}/\varepsilon_0\right)^2}\right)\left(\frac{\varepsilon}{\varepsilon_0}\right)^2 + \left[\frac{4}{\left(\varepsilon_{max}/\varepsilon_0\right)} - \zeta\right]\frac{\varepsilon}{\varepsilon_0}}$; where $\zeta = -\frac{4\left[\left(\varepsilon_{max}/\varepsilon_0\right)^2\left(\frac{E_{it}}{E_0}-2\right)+2\left(\varepsilon_{max}/\varepsilon_0\right)-\frac{E_{it}}{E_0}\right]}{\left[\left(\varepsilon_{max}/\varepsilon_0\right)\left(\frac{E_{it}}{E_0}-2\right)+1\right]^2}$; and $E_{it} = 21500\left(\frac{f_c'}{10}\right)^{0.33}$
Chinese Code, GB-50010 [8]	$\sigma = \begin{cases} f_c[\alpha_a(\varepsilon/\varepsilon_c)] + (3-2\alpha)(\varepsilon/\varepsilon_c)^2 + (\alpha_a - 2)(\varepsilon/\varepsilon_c)^2, & \varepsilon/\varepsilon_c \leq 1 \\ \frac{(f_c\varepsilon/\varepsilon_c)}{\left[\alpha_d(\varepsilon/\varepsilon_c - 1)^2\right]} + \varepsilon/\varepsilon_c, & \varepsilon/\varepsilon_c > 1 \end{cases}$

(*continued*)

Table 1. (*continued*)

Study	Stress-strain relationship of concrete
Aslani and Jowarmamendi [15]	$s_c(e_c) = \frac{n\left(\frac{\varepsilon_c}{\varepsilon_{cp}}\right)f_c'}{n-1+\left(\frac{\varepsilon_c}{\varepsilon_{cp}}\right)^n}$; where $n = n_1 = \left(1.02 - 1.17\frac{fc'}{E_c\varepsilon_{cp}}\right)^{-0.74}$ for $e_c < \varepsilon_{cp}$ and $a = 3.5(12.4 - 0.0166f_c')^{-0.46}$ and $b = 0.83e^{-911/f_c'}$
Alfarah et al. [16]	For $0 \le \varepsilon = \varepsilon_{max}$; $\varepsilon_{max} =$ *concrete strain* at $0.5f_c'$ on ascending part $f_c = f_c'\left[\frac{\left(\frac{E_{it}}{E_0}\right)\left(\frac{\varepsilon}{\varepsilon_0}\right)-\left(\frac{\varepsilon}{\varepsilon_0}\right)^2}{1+\left(\frac{E_{it}}{E_0}-2\right)\left(\frac{\varepsilon}{\varepsilon_0}\right)}\right]$; $E_0 =$ *secant modulus at peak stress* For $\varepsilon > \varepsilon_{max}$, $\sigma_c = \left(\frac{2+\gamma_c f_{cm}\varepsilon_{cm}}{2f_{cm}} - \gamma_c\varepsilon_c + \frac{\varepsilon_c^2\gamma_c}{2\varepsilon_{cm}}\right)^{-1}$ where $\gamma_c = 2\frac{\pi^2 f_{cm}\varepsilon_{cm}}{\left(\frac{G_{ch}}{l_{eq}}-0.5f_{cm}\left(\varepsilon_{cm}(1-b)+b\frac{f_{cm}}{E_0}\right)\right)^2}$ and $b = \frac{\varepsilon_c^{pl}}{\varepsilon_c^{ch}}$
Under tension	
Reinhardt et al. [17]	$\frac{\sigma_t}{f_t} = \left(1 + \left(c_1\frac{w}{w_c}\right)^3\right)\exp\left(-c_2\frac{w}{w_c}\right) - \frac{w}{w_c}\left(1 + c_1^3\right)\exp(-c_2)$ where $c_1 = 3$; $c_2 = 6.93$ and $w_c = 5.136G_F/f_t$
Chinese Code, GB-50010 [8]	$\sigma = \begin{cases} f_t[1.2(\varepsilon/\varepsilon_t) - 0.2(\varepsilon/\varepsilon_c)^6], & \varepsilon/\varepsilon_t \le 1 \\ f_t\varepsilon/\varepsilon_c[\alpha_t(\varepsilon/\varepsilon_t)^{1.7} + \varepsilon/\varepsilon_c], & \varepsilon/\varepsilon_c > 1 \end{cases}$

The stress-strain equations for $f_c = 29$ MPa was constructed for each model to validate the experimental results of Said and Mabrook Mouwainea [10].

3 Numerical Modelling and Experimental Setup

3.1 Experimental Setup

The experimental data was based on a series of experiments performed by Said and Mabrook Mouwainea [10]. The size of reinforced concrete slab is 1800 × 1800 × 100 mm and reinforced with Ø8 @82 mm c/c on both faces and the details a shown in Fig. 1. The data acquisition system consists of load cell recording at 96 kHz and laser sensor and LVDT recording at 2.4 kHz. The static tests were conducted using similar boundary conditions. It was not given the dimensions of the loading arrangement, therefore dimensions like the impactor for impact loading was used in the static testing in numerical simulations.

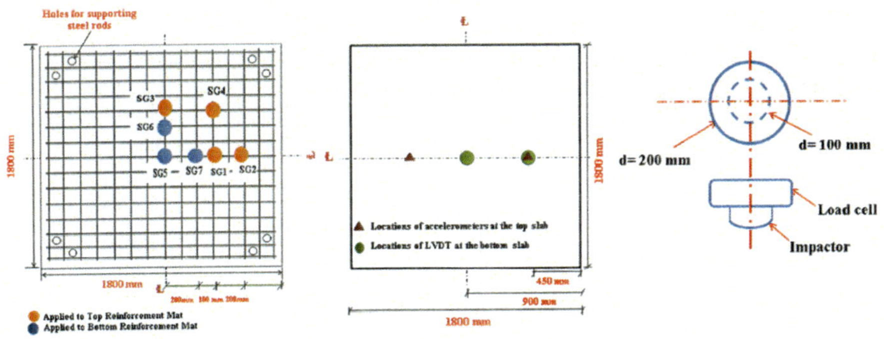

Fig. 1. Location of data acquisition instruments used in the experiments [10].

3.2 Constitutive Modelling

The concrete damage plasticity (CDP) model was used for concrete material. The concrete damage plasticity model uses the concept of isotropic damaged elasticity combined with isotropic tensile and compressive plasticity to represent the inelastic behaviour of concrete. This model allows the definition of strain hardening in compression and can be defined sensitive to strain rate, thus resembling the impact phenomenon more realistically. The model was developed by Lubliner et al. [18] and, later modified by Lee and Fenves [19] for dynamic and cyclic loading, and adopted in *ABAQUS/EXPLICIT*. The stress-strain behavior using analytical models are shown in (Fig. 1(a)) for compression and in (Fig. 1(b)) for tension behavior (Fig. 2).

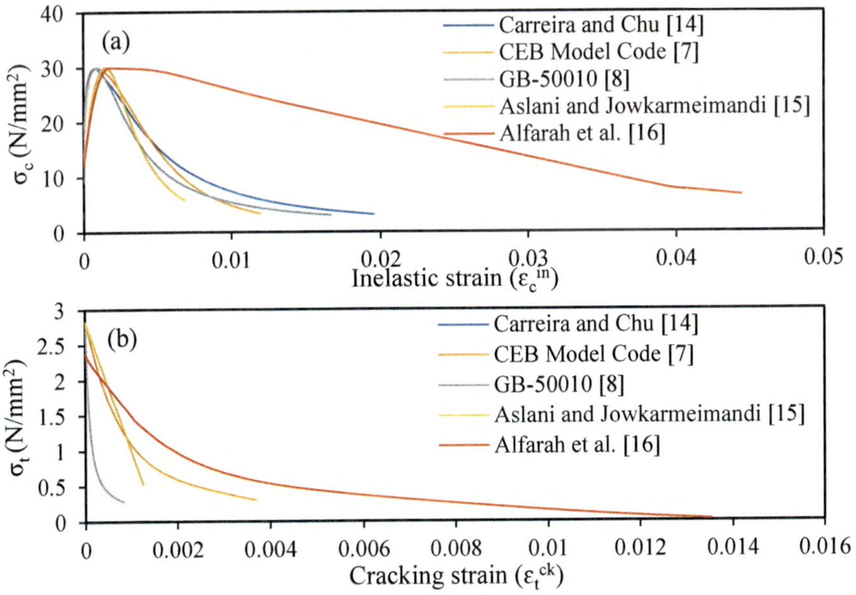

Fig. 2. Stress-strain relationship used in CDP model for (a) compression and (b) tension.

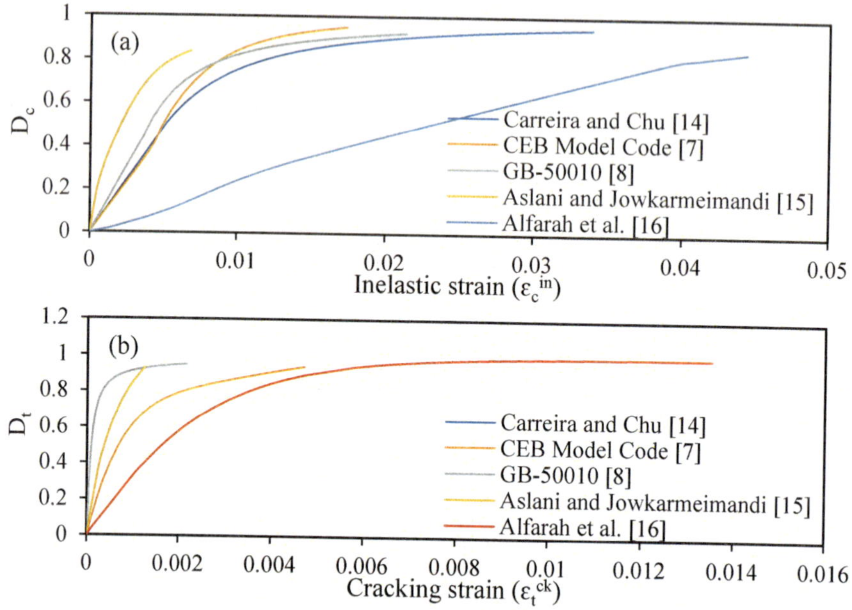

Fig. 3. Damage relationship used in CDP model for (a) compression and (b) tension.

The isotropic damage characterized by different evolutions in compression and tension were shown in (Fig. 3(a), (b)). Other parameters of CDP model were kept similar as originally proposed in Lubliner/ Lee/ Fenves formulation. The parameter was defined as dilation angle, $\Psi = 31°; {f_{b0}}/{f_{c0}} = 1.16, \ \in \ = 0.1, K_c = 0.67$. The parameter controls the multi axial behavior of concrete under different confining pressures and loading conditions. For rebar, Johnson-cook model was used. The model parameters are taken from Cadoni et al. [20]. The damage parameters were not included in the model as no damage was observed in the experiments.

3.3 Numerical Modelling

The numerical model has Lagrangian description for the motion. The concrete slab was discretized with eight-node hexahedron elements with reduced integration. The elements for concrete slab were kept at an aspect ratio of 1 with a mesh se of 20 mm. One-point gauss integration and relax stiffness hourglass modes were used for the element of concrete slab. The reinforcement was discretised as two node 3-D truss elements. The embedded constraint was enforced to couple the beam elements with the solid elements of host, see (Fig. 4).

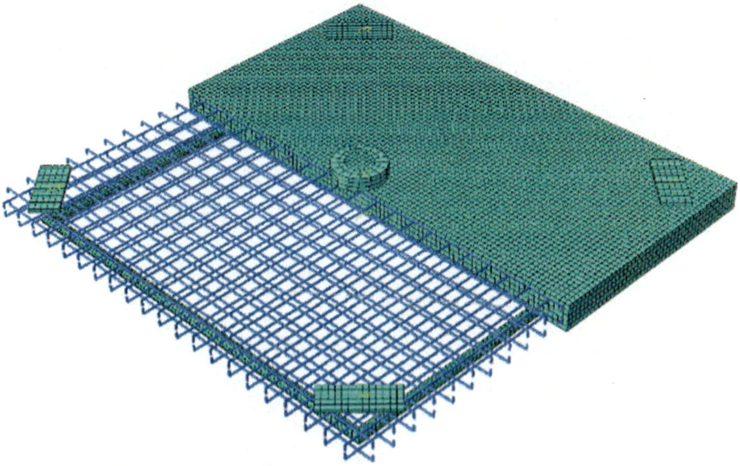

Fig. 4. Numerical model of reinforced concrete slab under low velocity impact.

The supports and impactor were modelled as discrete rigid parts with mass inertia applied to them. The impactor was given a mass of 50 kg and a velocity of 5.4 m/s was applied to it. Kinematic surface to surface contact with finite sliding was used to model the interaction between the impactor and the deformable concrete slab. The tangential behaviour was defined using Coulomb's law with a coefficient of friction as 0.2 and normal HARD behaviour was used. The supports were tied to the concrete slab such that movement was restricted in all six axes at the supports. ABAQUS/EXPLICIT solution scheme was used for both static as well as dynamic simulation.

4 Results and Discussions

4.1 Static Loading

The load-deflection curve of static test on reinforced concrete slab is shown in (Fig. 5). The peak load in experiment was 138 kN. On comparing analytical models, the prediction performance of most of the model are identical. The GB-50010 model was diverging from other curves and may have less stiffness as compared to other models however the final load was observed at the desired displacement. The peak load observed for static test for different models are 144 kN, 149, 139, 146 and 145 kN for [7, 8, 14, 16] and [15] respectively. The overall prediction capability considering computational cost is quite satisfactory for all the models. The minimum percentage deviation was observed for GB-50100 [8] as 0.3% with maximum deviation of 7.8% for Carreira and Chu [14].

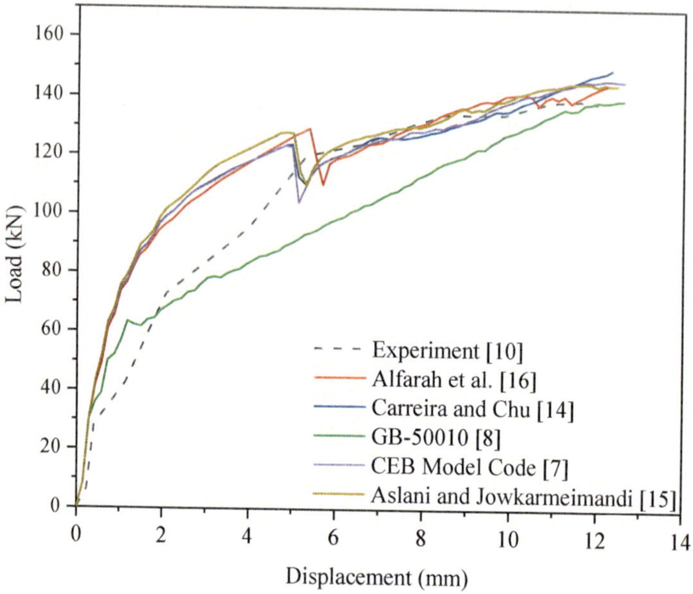

Fig. 5. Comparison of load-displacement response of RC slab using analytical models.

4.2 Dynamic Loading

Under low velocity impact, the impact load-time profile was found to be like experimental results, see (Fig. 6). However, on closer look the analytical model of Alfarah et al. [16] has better prediction accuracy as compared to other models which also capture the plateau of load. For peak load, the experimental value was 148 kN. Only [16] model overpredicted the load at 157 kN. All other analytical models predicted impact load less than experimental impact load maybe due to the absence of fracture energy criterion in defining the stress-strain behavior of concrete under compression. The peak impact load predicted by different analytical models are 126 kN, 117 kN, 130 kN and 113 kN for [7, 8, 14] and [15] respectively. It can be concluded that stress-strain models based on fracture energy criterion and mesh objectively in defining the parameters should be preferred for impact simulations.

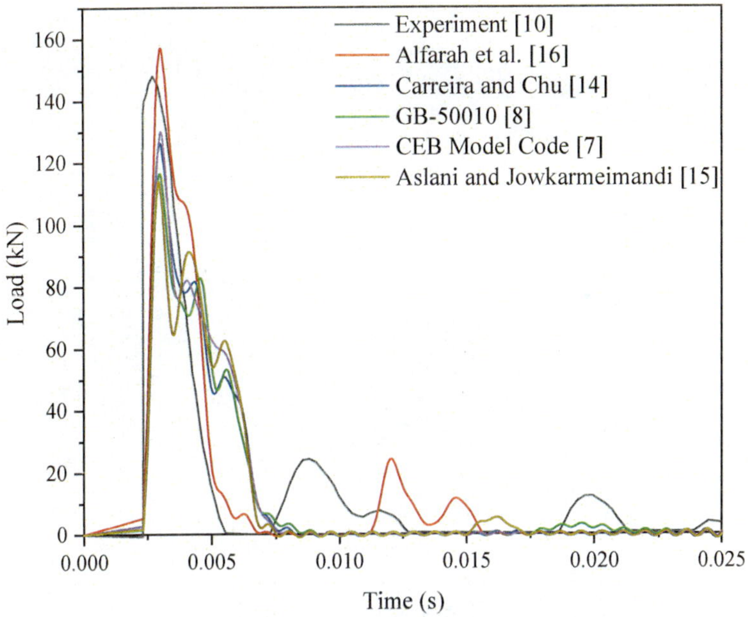

Fig. 6. Comparison of Load-time response of RC slab using analytical models.

The crack pattern predicted by numerical analysis using various analytical model definition is shown in (Fig. 7). It was observed that the crack pattern in experimental results in major scabbing the rear face as shown in (Fig. 7(a)). The crack pattern predicted by [16] and [15] seems the most appropriate for describing the failure pattern under impact, see (Fig. 7(b) & (f)). The crack pattern for [8] shows very high damage to the reinforced concrete slab indicating that this analytical model is suitable for impact loading when crack pattern is concerned. The scabbing phenomenon was successfully captured by the CDP model incorporating analytical models. The tensile damage to the rear face of the slab seems to be under predicted using [14] and [7], see (Fig. 7(c) & (e)).

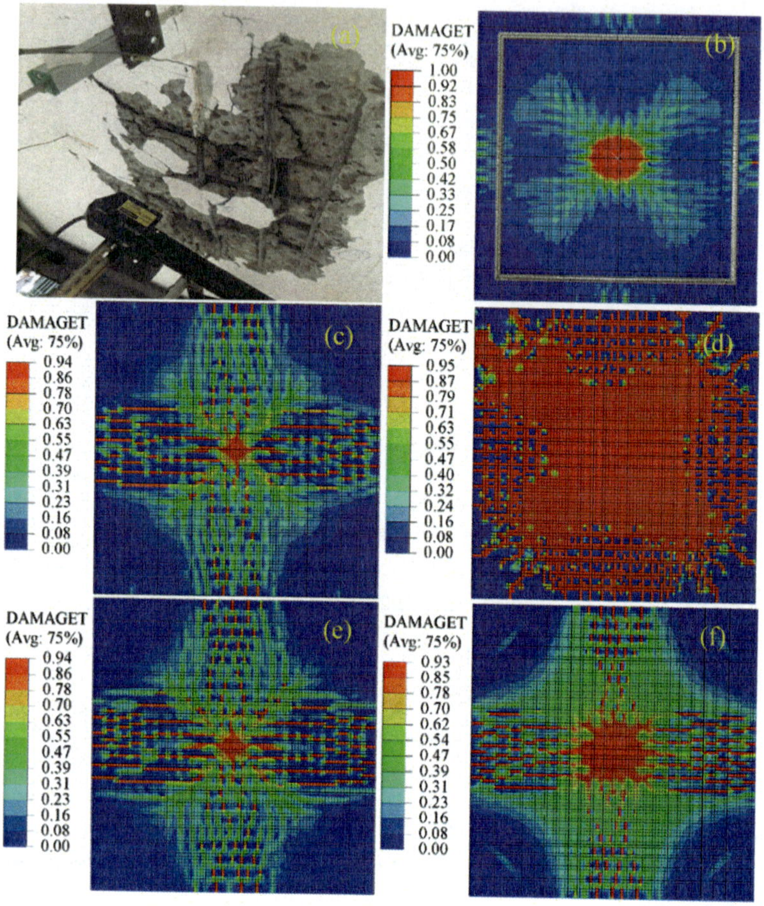

Fig. 7. Crack pattern comparison for impact simulation using analytical model formulation with (a) experiment for (b) [16], (c) [14], (d) [8], (e) [7] and (f) [15].

5 Conclusions

A numerical study was conducted to evaluate the analytical models available for predicting the stress-strain behavior of concrete under compression and tension. Different analytical models were tabulated, and the corresponding stress-strain values were made. The analytical models were evaluated for study of reinforced concrete slab under static and impact loading. Based on the numerical study the following conclusion are drawn:

- For static analysis, the CDP model incorporating analytical models performed satisfactory with a maximum percentage difference of 7.8% using Carreira and Chu model [14]. The GB-50010 [8] was the best among other with only a percentage difference of 0.3% from experimental results.
- Under dynamic loading, Alfarah et al. [16] was the best suited model with the percentage difference of 6% from the experimental results. The Carreira and Chu [14]

model is not suitable for dynamic loading when static definitions are used to define stress-strain characteristics as the concrete part shows erroneous damage.

References

1. Cui, J., Hao, H., Shi, Y.: Discussion on the suitability of concrete constitutive models for high-rate response predictions of RC structures. Int. J. Impact Eng. **106**, 202–216 (2017). https://doi.org/10.1016/j.ijimpeng.2017.04.003
2. Hanif, M.U., Ibrahim, Z., Ghaedi, K., Hashim, H., Javanmardi, A.: Damage assessment of reinforced concrete structures using a model-based nonlinear approach – A comprehensive review. Constr. Build. Mater. **192**, 846–865 (2018). https://doi.org/10.1016/j.conbuildmat.2018.10.115
3. Kumar, P.: A compact analytical material model for unconfined concrete under uni-axial compression. Mater. Struct. Constr. **37**, 585–590 (2004). https://doi.org/10.1617/13974
4. Yang, K.H., Mun, J.H., Cho, M.S., Kang, T.H.K.: Stress-strain model for various unconfined concretes in compression. ACI Struct. J. **111**, 819–826 (2014). https://doi.org/10.14359/51686631
5. Lu, Z.-H., Zhao, Y.-G.: Empirical stress-strain model for unconfined high-strength concrete under uniaxial compression. J. Mater. Civ. Eng. **22**, 1181–1186 (2010). https://doi.org/10.1061/(asce)mt.1943-5533.0000095
6. Samani, A.K., Attard, M.M.: A stress-strain model for uniaxial and confined concrete under compression. Eng. Struct. **41**, 335–349 (2012). https://doi.org/10.1016/j.engstruct.2012.03.027
7. CEB-FIP MODEL CODE 1990. Thomas Telford Publishing (1993)
8. GB: 50010–2002: Chinese standard: code for design of concrete structures (2002)
9. Chen, Y., May, I.M.: Reinforced concrete members under drop-weight impacts. Proc. Instit. Civil Eng. – Struct. Buildings **162**(1), 45–56 (2009). https://doi.org/10.1680/stbu.2009.162.1.45
10. Said, A.M.I., Mabrook Mouwainea, E.: Experimental investigation on reinforced concrete slabs under high-mass low velocity repeated impact loads. Structures **35**, 314–324 (2022). https://doi.org/10.1016/j.istruc.2021.11.016
11. Said, A.I., Mouwainea, E.M.: Behaviours of reinforced concrete slabs under static loads and high-mass low velocity impact loads. IOP Conf. Ser. Mater. Sci. Eng. **1067**, 012036 (2021). https://doi.org/10.1088/1757-899x/1067/1/012036
12. Bažant, Z.P.: Size effect aspects of measurement of fracture characteristics of quasibrittle material. Adv. Cem. Based Mater. **4**, 128–137 (1996). https://doi.org/10.1016/s1065-7355(96)90081-4
13. Hognestad, E.: Confirmation of inelastic stress distribution in concrete. J. Struct. Div. **83**, (1957). https://doi.org/10.1061/JSDEAG.0000093
14. Carreira and Chu: Stress-strain relatonship for reinforced concrete in compression. ACI Struct. J. 797–804 (1985)
15. Aslani, F., Jowkarmeimandi, R.: Stress-strain model for concrete under cyclic loading. Mag. Concr. Res. **64**, 673–685 (2012). https://doi.org/10.1680/macr.11.00120
16. Alfarah, B., López-Almansa, F., Oller, S.: New methodology for calculating damage variables evolution in Plastic Damage Model for RC structures. Eng. Struct. **132**, 70–86 (2017). https://doi.org/10.1016/j.engstruct.2016.11.022
17. Reinhardt, H.W., Cornelissen, H.A.W., Hordijk, D.A.: Tensile tests and failure analysis of concrete. J. Struct. Eng. **112**, 2462–2477 (1986). https://doi.org/10.1061/(ASCE)0733-9445(1986)112:11(2462)

18. Lubliner, J., Oliver, J., Oller, S., Oñate, E.: A plastic-damage model for concrete. Int. J. Solids Struct. **25**, 299–326 (1989). https://doi.org/10.1016/0020-7683(89)90050-4
19. Lee, J., Fenves, G.L.: Plastic-damage model for cyclic loading of concrete structures. J. Eng. Mech. **124**, 892–900 (1998). https://doi.org/10.1061/(ASCE)0733-9399(1998)124:8(892)
20. Cadoni, E., Dotta, M., Forni, D., Tesio, N.: Dynamic behaviour of reinforcing steel bars in tension. Appl. Mech. Mater. **82**, 86–91 (2011). https://doi.org/10.4028/www.scientific.net/AMM.82.86

Influence of Constitutive Models on the Behaviour of Clay Brick Masonry Walls Against Multi Hit Impact Loading

Ankush Thakur, Senthil Kasilingam(✉), and Amrit Pal Singh

Department of Civil Engineering, National Institute of Technology Jalandhar, Jalandhar 144011, India

urssenthil85@yahoo.co.in, kasilingams@nitj.ac.in

Abstract. Masonry structures are sensibly vulnerable against low velocity out of plane impacts caused due to vehicular collision, rockfalls, debris impact. Sometimes, these impacts cause severe consequences hence methods are developed based upon experimental investigations to assess the response of wall systems. However, in experiments, physical limitations are present accompanied by the increased cost of the study hence computational analysis of masonry structural systems are being used widely. In this manuscript, an attempt has been made to exploit the constitutive models available in ABAQUS in-built library for the modelling of masonry walls. The response of walls subjected to multiple hits were studied experimentally and validated numerically using Drucker Prager (DP), Mohr-coulomb (MC) and concrete damage plasticity model (CDP). It was observed that among chosen models, MC model overpredicts the structural response of walls measured in terms of contact force whereas DP and CDP model predicts the contact force under 10% deviation from experimental results. Further, the CDP model was able to simulate the crack within brick units whereas DP and MC model were able to simulate joint failure only. It was concluded that DP and CDP models can be adopted for simulating the structural response of masonry walls subjected to multiple hits.

Keywords: Brick masonry wall · Experiment and simulations · Multiple hit · Large mass

1 Introduction

The masonry structures are often widely used as protective structures and commercial applications due to its ease of construction and suitability to withstand gravitational loads. However, masonry structures are sensibly vulnerable against low velocity out of plane impacts caused due to vehicular collision, rockfalls, debris impact. In order to understand the effect of impact loads, methods have been developed in the past based on the experimental findings. Other than that, computational tools are also widely used to predict the response of masonry structures to impact loads and, therefore, assess the safety of a structure. However, numerical modelling of masonry walls subjected to impact

Z. Dimitrovová et al. (Eds.): WMVC 2022, MMS 125, pp. 77–86, 2023.
https://doi.org/10.1007/978-3-031-15758-5_7

conditions is a complex phenomenon. Various authors in the past have proposed different constitutive relations to accurately simulate the response of masonry structures under various in-plane and out-of-plane loading conditions. Lourenco (1996) [1] developed a constitutive model to simulate the in-plane response of masonry walls. The main feature of the model is it includes a compression cap, tension cut-off, and a composite yield criterion. Rafsanjani et al. (2015) [2] extended the same model and incorporated the effect of dynamic increase factors. Extended model was a strain rate dependent anisotropic continuum model for impact loadings and suitable for simulating masonry structures under impact loadings. Mehrabi and Shing (1997) [3] developed a constitutive model to simulate behaviour of masonry joints interface. Further the models include, compressive hardening behaviour, reversal of shear dilatancy, and normal contraction of interfaces. Compressive hardening was simulated using a $\sqrt{J_2}$ plasticity-based model which is combined with Rankine tension cut-off criterion to simulate cracks in masonry infilled reinforced concrete frames. It was observed that the model can simulate the nonlinear behaviour of infilled frames. Burnett et al. (2007) [4] simulated impact response of masonry walls using FEM solver LS DYNA. The constitutive models used to simulate hardening, mortar joint interface and residual strength was Mohr-coulomb, interface model, and dilatant friction model, respectively. It was concluded that with appropriate calibration of material parameters, response of walls can be simulated in proximity of experimental results. Wei and Hao (2009) [5] developed a dynamic plastic damage model with strain rate effects to simulate the response of masonry walls subjected to blast loadings. Model is divided into three parts namely (i) Strength criterion (ii) Damage definition, and (iii) Strain rate effects. It was concluded that the model can predict the damage levels in a reasonable agreement with experimental observations. Bolhassani et al. (2015) [6] used concrete damage plasticity model (CDP) to simulate the response of solid and hollow concrete brick masonry assemblages. It was concluded that the CDP model fairly predicted the response of assemblages under static loading conditions and can be used to model masonry structures. Similarly, Daltri et al. (2019) [7] used the CDP model to define the elastic-plastic behaviour of masonry walls. Also, a cohesive contact approach with frictional contact was used to simulate the joints between brick units. It was concluded that model properties can be characterized from small-scale experimental tests and are suitable to analyse the cyclic response of masonry structures. Recently, Asad et al. (2021) [8] proposed a multi surface plasticity model to simulate the low velocity impact response of masonry walls. Rankine and Hill type yield surfaces were used to define failure surfaces in tensions and compression, respectively. It was concluded that the model fairly predicted the response of walls in terms of residual displacement with a variation of 6.5% from experimental results.

Based on the detailed literature survey, it was concluded that the prediction on the response of masonry structures depends on the appropriate constitutive model. Most of the constitutive models discussed above are incorporated using user defined computer codes into general purpose finite element solvers such as ANYSYS, LSDYNA and ABAQUS. The drawback of such codes is that they are not available for general use. Hence, an attempt has been made to exploit the constitutive models available in ABAQUS/explicit in-built library to study the behavior of masonry walls against low velocity impact load. The response of walls subjected to multiple hits were studied experimentally, see Sect. 2. The Drucker Prager (DP), Mohr-coulomb (MC) and concrete damage plasticity model (CDP) are discussed in Sect. 3. Further, brief numerical modelling strategy along with mesh sensitivity study is discussed in Sect. 4. The results are compared with the experimental observation in terms of peak force and residual displacements are discussed in Sect. 5.

2 Experimental Investigation

The experiments were performed on a masonry wall of size 1.2 × 1.2 × 0.110 m composed of clay bricks and 1:4 Cement: Sand mortar. The impactor used is hemispherical in shape and comprises hard steel with a length of 630 mm and an impact head radius of 80 mm. The impactor has a mass of 60 kg and a velocity of 7 km/h. The specimen was studied repeatedly until the specimen failed ultimately. The test was performed with the same mass and velocity at the midpoint of the masonry wall and the pendulum angle was between 28°–30°, see Fig. 1.

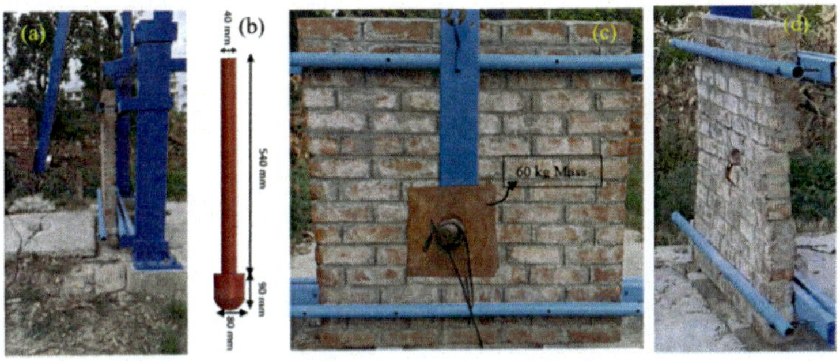

Fig. 1. Experimental facility of (a) pendulum impact testing frame (b) impactor and masonry wall (c) before and (d) after test

The response history was measured using a dynamic load cell of 250 kN capacity connected to a high-speed data acquisition system. The test data were recorded at 50 kHz sampling rate. For clarity purposes, Hits 1–5 were not presented, considering that the hits 1–7 were similar in nature and magnitude. Resistance offered by the wall up to 7 hits were in the range of 16–18 kN as peak force, see Fig. 2. The resistance offered is almost the same however, damage increases with every hit. Lastly at 11th hit, horizontal crack throughout the wall width, near impact region of masonry wall was observed.

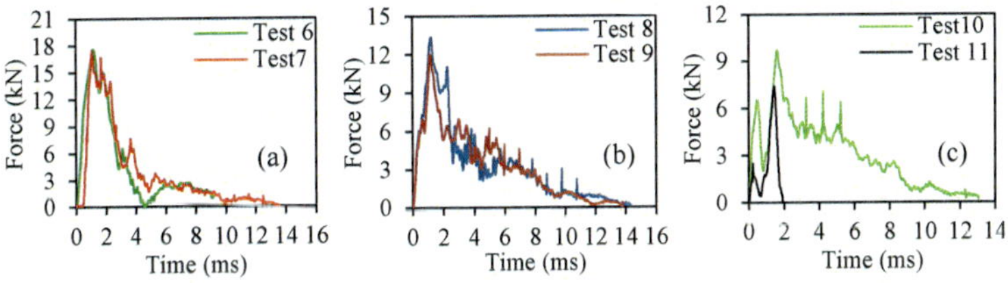

Fig. 2. Response from (a) hit 6 and 7 (c) hit 8 and 9 (d) hit 10 and 11

3 Constitutive Behaviour and Numerical Modelling

Three different constitutive models (i) Drucker Prager (DP) (ii) Mohr-coulomb (MC) and (iii) concrete damage plasticity model (CDP) was used to predict the response of masonry walls which is readily available in ABAQUS materials library. Further, the interface between brick and mortar was simulated using traction- separation law was presented in this Section.

3.1 Drucker Prager Constitutive Behaviour

Drucker Prager (DP) model, a modified version of Mohr- Coulomb model is used for simulation of granular materials. Linear yield criterion of DP model in combination with non-associated flow rule given as $F = t - p\ tan\beta - d = 0$, is applied in present study. Where, β is friction angle, d is cohesion value and p is hydrostatic pressure. Damage is incorporated using uniaxial compressive strength (σc) using hardening option in terms of yield stress versus plastic strain. The input curve for the model is shown in Fig. 3. Further, general material properties to simulate elastic plastic response of bricks are given Table 1.

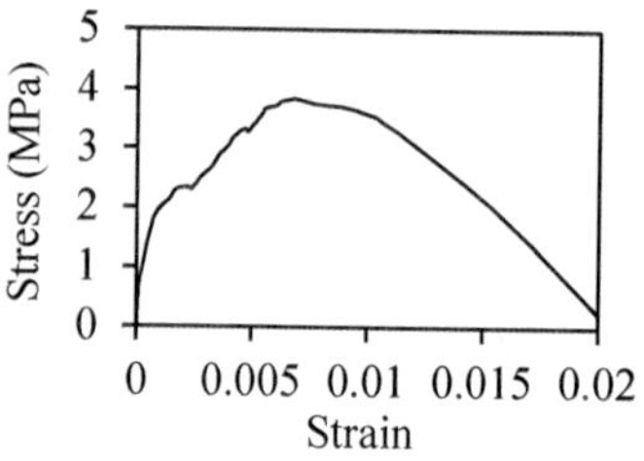

Fig. 3. Stress strain response of 3-D FE wall

3.2 Mohr Coulomb Constitutive Behaviour

The Mohr-Coulomb (MC) elasto-plastic model is perhaps the most widely used model in discrete element simulations. It allows tensile and shear failure in the brick material. Table 1 shows the properties of MC model for 3-D FE bricks considered in the present study. In the MC model, governing Equation is described as $\tau = d + \tan\beta$, where τ is shear stress, d is the cohesion and β is the friction angle. Further, cohesion input was given in the form of isotropic hardening and softening in terms of stress and equivalent plastic strains. More information regarding MC model available in [9]. Lastly, general material properties to simulate elastic plastic response of bricks are given Table 1.

3.3 Concrete Damage Plasticity Constitutive Behaviour

The CDP model considers the non-associated plastic flow rule, was developed by [10]. Two independent damage variables, namely, tension and compression, are used whose values lie between $0 \leq dt$ and $d_c \leq 1$. Also, the stress-strain relationship under tension (σt) and compression (σc) are defined as: $\sigma_t = (1 - d_t)E_0(\varepsilon_t - \varepsilon_t^p)$, and $\sigma_c = (1 - d_c)E_0(\varepsilon_c - \varepsilon_{tc}^p)$, respectively. Where E_0 is the initial young's modulus of the material, ε_t and ε_c are the uniaxial tensile and compressive strains, and ε_t^p and ε_c^p t are uniaxial tensile and compressive plastic strains. Particularly, the curves depicted in Fig. 4 represent the primary input data of the model. For the yielding surface, a multiple-hardening Drucker-Prager type surface is assumed. It is characterized by equibiaxial and initial uniaxial compressive strength (f_{bo}/f_{co}) ratio and a constant K_c, which is the ratio of the second stress invariant on the tensile to that of the compressive meridian. In addition to that, the general parameters for the CDP model are shown in Table 2.

Table 1. Properties of masonry walls

Particulars	Value	Remarks
Brick units		
Density (γ), kg/m^3	1980	Determined from 5 tests
Compressive strength, (f_b), MPa	24	Determined from 5 tests
Elastic modulus (E_u), MPa	7200	$E_u = 300 \times f_b$ [11]
Mortar		
Compressive strength, (f_j), MPa	4.04	Determined from 5 tests
Elastic modulus (E_m), MPa	1200	$E_m = 300 \times f_j$ [11]
Brick- mortar interface		
Coefficient of friction	0.78	Assumed
Normal behaviour	"HARD" Contact	To avoid penetration of nodes
K_{nn} (N/mm^3)	96	Calculated from Eq. (1)
K_{ss} (N/mm^3)	46	Calculated from Eq. (2)
K_{tt} (N/mm^3)	46	Same as above
Shear strength, MPa	0.24	Determined from 9 triplet test
Tensile strength, MPa	4	Determined from mortar compressive tests
Mode-I fracture energy, (G_{IC})	12	Lourenco, 1994 [1]
Mode-II fracture energy, (G_{IIC})	40	Angelillo et al. 2014 [12]
Impactor-wall interface		
Coefficient of friction	0.03	Assumed due to low speed
Penalty stiffness, N/mm	10^7	Burnett et al. 2007 [4]
Masonry walls properties for DP and MC model		
Density (γ), kg/m^3	1980	Determined from lab test

(*continued*)

Table 1. (*continued*)

Particulars	Value	Remarks
Compressive strength, (f_p), MPa	3.83	Determined from 3 masonry prism test
Elastic modulus (E_u), MPa	2075	Same as above
Poisson ratio, (υ)	0.15	Assumed
Flow stress ratio	1	Simulia 2014 [9]
Friction angle (ϕ)	41	Singhal and Rai 2013 [13]
Dilatation angle φ	11	Mosalam et al. 2009 [14]

Table 2. General parameters for concrete damage plasticity model [9]

Dilatation angle (φ)	Eccentricity, ϵ	f_{bo}/f_{co}	K_c
31°	0.1	1.16	0.67

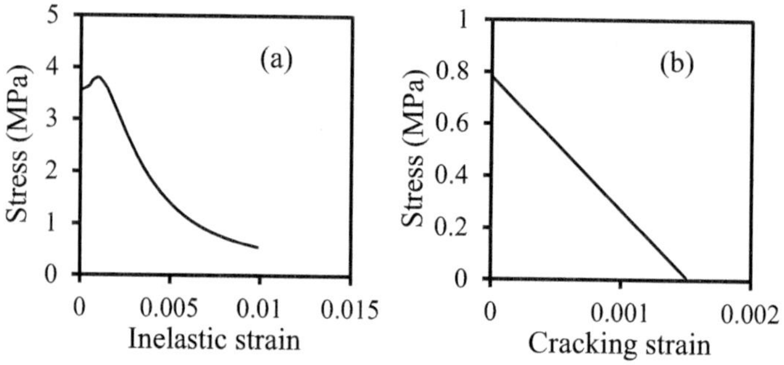

Fig. 4. Behaviour of concrete bricks under uniaxial (a) compression and (b) tension

3.4 Brick Mortar Interface Constitutive Behaviour

Modelling of contact between mortar and bricks is achieved by introducing Cohesive zone modelling (CZM) between brick units. In the present study, a cohesive contact approach was used as cracks confined to the formation along with the layers of the bed and head joints. The mechanical behaviour of cohesive contact is introduced by traction-separation laws, including damage initiation and damage evolution. Before the damage, it is assumed that cohesive behaviour follows a linear traction separation law see, Fig. 5.

The stiffness coefficients shown in Fig. 5 (K_{nn}, K_{ss} and K_{tt}) are calculated based on elastic modulus of brick (E_u), mortar (E_m) and thickness of joint (t_m) using Eq. 1 and Eq. 2.

$$K_{nn} = \frac{E_u E_m}{t_m(E_u - E_m)} \tag{1}$$

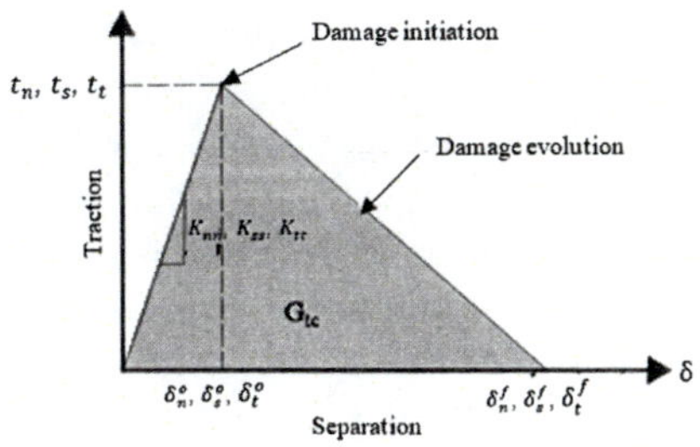

Fig. 5. Traction separation response of brick interface

$$K_{ss} = \frac{G_u G_m}{t_m(G_u - G_m)} \tag{2}$$

For damage, Quadratic traction criteria, see Eq. 1, which initiates damage up on user-defined parameters, was used. The normal (t_n, ,) and shear stress (t_s, and t_t) vectors represents; Mode-I, Mode-II and Mode-III failure mode. Other properties, such as cohesion, shear strength, tensile strength, friction coefficient and fracture energy of joints is shown in Table 1.

$$\left(\frac{t_n}{t_n^{max}}\right)^2 + \left(\frac{t_s}{t_s^{max}}\right)^2 + \left(\frac{t_t}{t_t^{max}}\right)^2 = 1 \tag{3}$$

4 Numerical Modelling and Mesh Convergence

The numerical modelling technique along with mesh convergence study has been discussed in [15] however main points are recalled here. To visualize the crack patterns in the brick mortar interfaces, simplified micro modelling (SMM) technique was used. SMM was successively used in the past studies [6, 7, 16]. The masonry walls elastic-plastic properties were incorporated by means of various constitutive models discussed above. For simulation, the size of brick considered was 230 mm, 110 mm, 80 mm (length, breadth, height). To simulate interface between brick-and-mortar interface, cohesive behaviour was introduced between head and bed joints of brick units. The cohesive interface properties were discussed shown in Table 1. The velocity 1.98 ms^{-1} was incorporated using a predefined option. The contact between impactor head and wall face is employed using surface-to-node (explicit). For simulating tangential behaviour, the coefficient of friction between these bodies was assumed to be 0.03 due to the low velocity of the impactor. Normal behaviour is simulated using the penalty contact option with a stiffness of 10^7 N/mm [4]. In all the simulations, the impactor was considered as master surface and wall as node-based slave surface. The hourglass control is kept default for all the 3D brick elements. Lastly, based on the mesh convergence study, it was concluded that a mesh size of 32 mm is suitable to carry out simulation as it matched well with the experimental force time history curves [15].

5 Comparison of Experiment and FE Simulation Results

The numerical simulations were carried out to understand the influence of constitutive material models against low velocity impact loads. The results thus presented in terms of peak force and residual displacement through detailed experimental and numerical investigation, see Table 3. The walls simulated using a simplified micro modelling approach were presented herein, by considering rigid elements at back and front face of walls to act as boundary conditions similar to experimental setup. The simulations were identified as first impact, second impact and third impact against hit 6–7, hit 8–9 and hit 10–11, respectively, see Fig. 6. It was observed that experimental peak force was found to be 17.62, 13.07, and 9.76 for first impact, second impact and third impact, respectively. For first impact peak force was found to be 19.47, 28.45 and 17.11 kN for DP, MC, and CDP models, respectively. Also, local residual displacements were recorded at the end of the test as 45 mm, see Fig. 7, whereas from simulation, it was found to be 50.68, 307, and 49.88 mm for DP, MC, and CDP models, respectively. It was observed that among chosen models, MC model overpredicts the wall resistance measured in terms of peak force whereas DP and CDP model predicts the contact force under 10% deviation from experimental results. Further, the CDP model was able to simulate the crack within brick units whereas DP and MC model were able to simulate joint failure only. It was concluded that DP and CDP models can be adopted for simulating the structural response of masonry walls subjected to multiple hits.

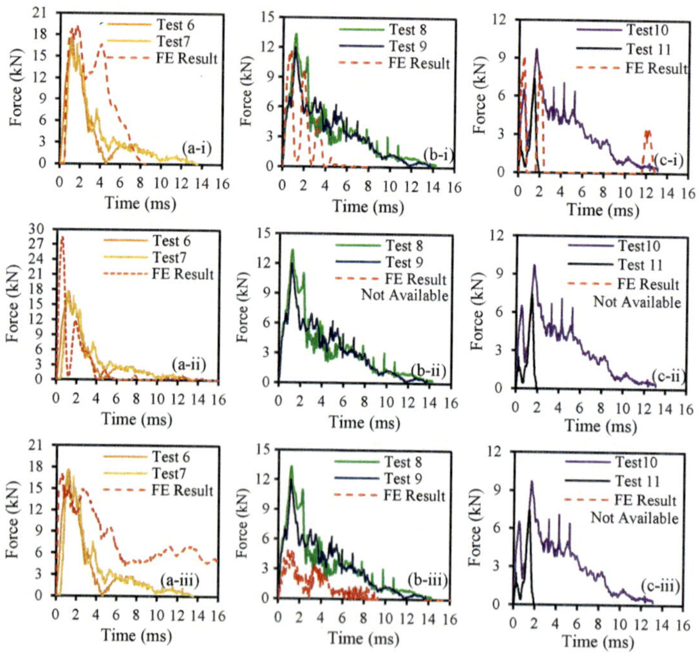

Fig. 6. Wall resistance in terms of peak force for (a) first (b) second (c) third impact with (i) DP (ii) MC and (iii) CDP model

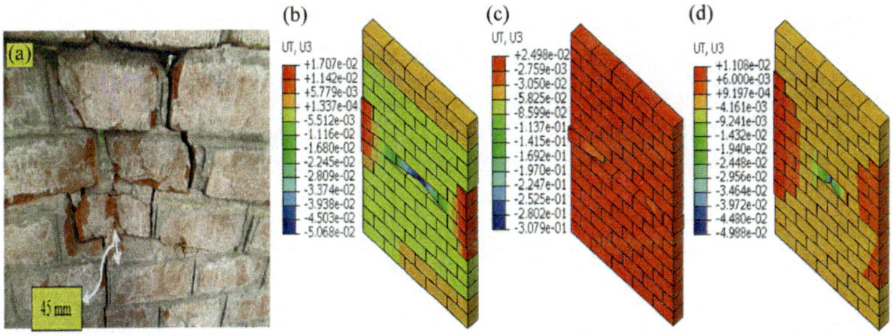

Fig. 7. Local displacement on (a) experiment and FE simulation using (b) DP (C) MC and (d) CDP material model

Table 3. Comparison of experiment and FE simulations with various material models

	Peak force				Residual displacement			
Impact No	Experiment	DP	MC	CDP	Experiment	DP	MC	CDP
First	17.62	19.47	28.45	17.11	–	11.56	131	9.06
Second	13.07	13.07	–	4.89	–	25.62	234	29.77
Third	9.76	9.17	–	–	45	50.68	307	49.88

6 Conclusion

The experiment as well as FE simulation were performed to study the response of masonry walls under repeated loading. The FE simulations were performed using various material models available in ABAQUS inbuilt library. The influence of various constitutive models on the response of clay brick masonry walls is studied and the results thus predicted were compared with the experimental results. Based on the study, following conclusion were drawn:

- It was observed that the response of masonry walls changes significantly with change in the constitutive model. Hence, it was concluded that the material model should be chosen carefully such that accurate results can be obtained from the simulation.
- It was concluded that DP and CDP models were able to predict the peak force with deviation of under 10% from experimental value whereas MC model overpredicts the peak force by 61% as compared to experiment.

References

1. Lourenço, P.B.: Computational strategies for masonry structures (1996)
2. Rafsanjani, S.H., Lourenço, P.B., Peixinho, N.: Implementation and validation of a strain rate dependent anisotropic continuum model for masonry. Int. J. Mech. Sci. **104**, 24–43 (2015)
3. Mehrabi, A.B., Benson Shing, P., Schuller, M.P., Noland, J.L.: Experimental evaluation of masonry-infilled RC frames. J. Struct. Eng. **122**, 228–237 (1996)
4. Burnett, S., Gilbert, M., Molyneaux, T., Beattie, G., Hobbs, B.: The performance of unreinforced masonry walls subjected to low-velocity impacts: Finite element analysis. Int. J. Impact Eng. **34**, 1433–1450 (2007)
5. Wei, X., Hao, H.: Numerical derivation of homogenized dynamic masonry material properties with strain rate effects. Int. J. Impact Eng. **36**, 522–536 (2009)
6. Bolhassani, M., Hamid, A.A., Lau, A.C.W., Moon, F.: Simplified micro modeling of partially grouted masonry assemblages. Constr. Build. Mater. **83**, 159–173 (2015)
7. D'Altri, A.M., Messali, F., Rots, J., Castellazzi, G., de Miranda, S.: A damaging block-based model for the analysis of the cyclic behaviour of full-scale masonry structures. Eng. Fract. Mech. **209**, 423–448 (2019)
8. Asad, M., Dhanasekar, M., Zahra, T., Thambiratnam, D.: Failure analysis of masonry walls subjected to low velocity impacts. Eng. Fail. Anal. **116**, 1–24 (2020)
9. Simulia: ABAQUS User Guide (2019)
10. Lubliner, J., Oliver, J., Oller, S., Onate, E.: Damage model for concrete. Int. J. Solids Struct. **25**, 299–326 (1989)
11. Kaushik, H.B., Rai, D.C., Jain, S.K.: Stress-Strain characteristics of clay brick masonry under uniaxial compression. J. Mater. Civ. Eng. **19**, 728–739 (2007)
12. Angelillo, M., Lourenço, P.B., Milani, G.: Masonry behaviour and modelling. In: Angelillo, M. (ed.) Mechanics of Masonry Structures. CICMS, vol. 551, pp. 1–26. Springer, Vienna (2014). https://doi.org/10.1007/978-3-7091-1774-3_1
13. Singhal, V., Rai, D.C.: Suitability of half-scale burnt clay bricks for shake table tests on masonry walls. J. Mater. Civ. Eng. **26**, 644–657 (2013)
14. Mosalam, K., Glascoe, L., Bernier, J.: Mechanical properties of unreinforced brick masonry. Section **1**, 1–26 (2009)
15. Senthil, K., Thakur, A., Singh, A.P.: Multihit impact response of masonry wall. In: Maiti, D.K., Maity, D., Patra, P.K., Jana, P., Mistry, C.S., Ghoshal, R., Afzal, M.S. (eds.) Recent Advances in Computational and Experimental Mechanics, vol. II. Springer Nature, IIT Kharagpur (2021)
16. Abdulla, K., Cunningham, L., Gillie, M.: Simulating masonry wall behaviour using a simplified micro-model approach. Eng. Struct. **151**, 349–365 (2017). https://doi.org/10.1016/j.engstruct.2017.08.021

CEW: Computational Efficiency in Wave Propagation and Structural Dynamics Analyses

Effect of Solid Dust Particles on the Propagation of Magnetogasdynamical Shock Waves in a Non-ideal Gas with Monochromatic Radiation

P. K. Sahu(✉)

Department of Mathematics, Government Shyama Prasad Mukharjee College, Sitapur 497111, Chhattisgarh, India
praveensahu173@gmail.com

Abstract. The expansion of either plane or spherical or cylindrical shock waves, in this analysis, is examined by the influence of either axial or azimuthal magnetic field; monochromatic radiation in a dust-pervade gas. Equilibrium circumstances are expected to be retained for the flow and the advancing piston provides the varying energy input constantly. With a view to getting the self-similar solutions, the density of the uninterrupted media is considered to remain unchanged. The flow framework of the parameters was calculated numerically. Detailed data is provided on the impacts of the parameter of gas non-idealness as well as existence of magnetic field. It's noticeable that the pressure and density at the expansive region wear down in the presence of an azimuthal magnetic field, resulting in the creation of a vacuum at the symmetry's centre, it is perfectly consistent with the settings in the lab that created the shock wave.

Keywords: Magnetic field · Two-phase flow · Shock wave · Monochromatic radiation · Non-ideal gas

1 Introduction

In traditional gas-dynamics, energy transmission via radiation has been usually ignored. But now, the gas is acceptable to be partly ionized at high temperatures in many phenomena, hence a great interest has generated to study the outcomes of radiative-gas-dynamics. Therefore, a complete analysis of such problems is carried out through the observation of gas-dynamic flow, radiation, and electromagnetic fields simultaneously. The perusal of the radiative magnetogasdynamics is also useful on plasma heating for guided explosive systems and the fluctuations in the density in interstellar clouds. Many academics have researched on the issue in recent years in an attempt to perceive the mechanics of the physical processes occurring (see, [1–12] and references therein).

Z. Dimitrovová et al. (Eds.): WMVC 2022, MMS 125, pp. 89–98, 2023.
https://doi.org/10.1007/978-3-031-15758-5_8

The expansion of shock to conductive gas under the influence of magnetic field should be taken in to account in experiments including the pinch effect, bursting wires, etc. Magnetic fields dominate the cosmos and play an essential part in many celestial circumstances. Perusers may get involved in the exciting study that has served as the major motivation for the ongoing study, which involves shock waves at the magnetic field interaction, in Refs. [13–21] and references therein.

Recently, interaction of waves in reacting real gases has obtained attention because of its appliance to an ample number of physical processes relating to managed nuclear fusion, spacecraft propulsion, burners, combustion industries, furnaces, fire extinguishers, etc. Lately, the fluid model flow in a dust-pervade gas has recently gained popularity due to its vast applications in both municipal and biological domains. Many academics have researched on the issue in recent years in an attempt to perceive the shock waves expansion in a dust-pervade gas (see Refs. [22–28] and the references cited therein).

Nonetheless, dust-pervade gases have been extensively discussed in the literature, however, they have received little attention when considering the magnetic field. The possibility of a magnetic field and monochromatic radiation in dust-pervade gases in plane/cylindrical/spherical shock still hasn't been explored in any of the research. The study extends Nath and Takhar's [4] approach by enabling the results of an magnetic field in dust-pervade gas.

2 Ruling Equations

The governing equations describing adiabatic, unsteady, planar or cylindrically or spherically symmetric, non-perfect dust-pervade gas flow accompanied by monochromatic radiation and magnetic field may be written like (c.f. Khudyakov [1], Nath and Takhar [4], Nath [9], Sahu [18], Pai et al. [22], Sahu [26])

$$\frac{\partial \rho}{\partial t}+v\frac{\partial \rho}{\partial x}+\rho\frac{\partial v}{\partial x}+\frac{iv\rho}{x}=0, \tag{1}$$

$$\rho\frac{\partial v}{\partial t}+\rho v\frac{\partial v}{\partial x}+\frac{\partial h}{\partial x}+\frac{\partial p}{\partial x}+\frac{2mh}{x}=0, \tag{2}$$

$$\frac{\partial h}{\partial t}+v\frac{\partial h}{\partial x}+2h\frac{\partial v}{\partial x}+\frac{2jhv}{x}=0, \tag{3}$$

$$\frac{\partial E_m}{\partial t}+v\frac{\partial E_m}{\partial x}+p\left(\frac{\partial}{\partial t}+v\frac{\partial}{\partial x}\right)\frac{1}{\rho}=\frac{1}{\rho x^i}\frac{\partial}{\partial x}\left(Jx^i\right), \tag{4}$$

$$\frac{\partial J}{\partial x}=\alpha_{ab}J, \tag{5}$$

in which x and t are space and time coordinates; v, p, ρ, and E_m depute dust-pervade gas's velocity, pressure, density, and internal energy; $h=\frac{\kappa H^2}{2}$ deputes magnetic pressure, H deputes the intensity of the magnetic field; κ deputes magnetic permeability; J deputes flux of monochromatic radiation; and α_{ab} deputes absorption coefficient. Then we accept the afterward cases:-

Case A: $m=0$, $i=0$, $j=0$ coincides with planar symmetry.

Case B: $m = 1$, $i = 1$, $j = 0$ coincides with cylindrically symmetric flow interpenetrating azimuthal magnetic field.

Case C: $m = 0$, $i = 1$, $j = 1$ coincides with cylindrically symmetric flow interpenetrating axial magnetic field.

Case D: $m = 1$, $i = 2$, $j = 1$ coincides with spherically symmetric flow interpenetrating azimuthal magnetic field.

Case E: $m = 0$, $i = 2$, $j = 2$ coincides with spherically symmetric flow interpenetrating axial magnetic field.

The α_{ab} (absorption coefficient) must be altered as (c.f. Khudyakov [1], Nath and Takhar [3], Nath and Takhar [4], Nath [9], Sahu [10])

$$\alpha_{ab} = \alpha_{ab_a}\, \rho^{\Psi}\, p^{\Xi}\, J^{q}\, x^{s}\, t^{l}, \tag{6}$$

in which Ψ, Ξ, q, s and l be rational numbers; α_{ab_a} is a dimensional constant.

The dust-prevade gas in its reference condition is shown in Fig. 1. The reader is referred to Sahu [28], Steiner and Hirschler [29] for details. The governing Eqs. (1)–(5) ought to be enclosed by an equation of state (see, Sahu [28]).

$$p = \frac{(1-\mu_p)}{(1-Z)}\,[1 + (1-\mu_p)\, b\,\rho]\; R^{*}\rho T, \quad E_m = \frac{(1-Z)\,p}{\rho\,(\Gamma-1)\;[1+(1-\mu_p)b\rho]}. \tag{7}$$

Flow inputs shortly before the shock are recognised. as

$$v = v_a = 0, \tag{8}$$

$$\rho = \rho_a = constant, \tag{9}$$

$$H = H_a = H^{*}\; x_s^{-\epsilon} \tag{10}$$

$$p = p_a = \frac{(m-\epsilon)}{2\epsilon}\kappa {H^{*}}^{2}\; x_s^{-2\epsilon}. \tag{11}$$

in which H^{*} and ϵ depute dimensional constants; x_s deputes shock radius and the subscript a deputes conditions forthwith preceding the shock.

The Rankine-Hugoniot conditions are (c.f. (Khudyakov [1], Nath and Takhar [3], Nath and Takhar [4], Nath [9], Sahu [10])) namely,

$$v_{x=x_s} = (1-\beta)W_s, \qquad Z_{x=x_s} = \frac{Z_a}{\beta}, \qquad j_{x=x_s} = p_a^{3/2}\;\rho_a^{-1/2},$$

$$\rho_{x=x_s} = \frac{\rho_a}{\beta}, \qquad h_{x=x_s} = \frac{h_a}{\beta^2},$$

$$p_{x=x_s} = \left[(1-\beta) + C_a\left(1 - \frac{1}{\beta^2}\right) + \frac{1}{\gamma M^{*2}}\right]\rho_a\, W_s^2,$$

in which $W_s \left(= \frac{dx_s}{dt}\right)$ deputes shock front velocity, $M^{*} = \left(\frac{\rho_a W_s^2}{\gamma p_a}\right)^{\frac{1}{2}}$ and $C_a = \left(\frac{h_a}{\rho_a W_s^2}\right)$ depute shock-Mach and Cowling number. The density ratio β is specified during the shock using following relationship.

$$\frac{(\Gamma+1)}{2}\beta^3 - \left[\frac{\Gamma}{\gamma M^{*2}} + \Gamma\, C_a + Z_a + \frac{(\Gamma-1)}{2}\;\{1 - \bar{b}\,(1-\mu_p\}\right]\beta^2 \tag{12}$$

$$-\beta\left[(\Gamma-1)\,\overline{b}\,(1-\mu_p)\left\{\frac{1}{2}+C_a\right\}-2C_a\,(\Gamma-1)+C_a(\Gamma-Z_a)+\frac{\Gamma\overline{b}\,(1-\mu_p)}{\gamma\,M^{*2}}-\frac{\left\{Z_a+\overline{b}\,(1-\mu_p)\right\}\overline{b}\,(1-\mu_p)}{\left\{1+\overline{b}\,(1-\mu_p)\right\}\gamma\,M^{*2}}\right]$$

$$+\Gamma\,C_a\overline{b}\,(1-\mu_p)+\left\{Z_a-\overline{b}\,(1-\mu_p)\right\}\,C_a=0.$$

Here $\overline{b}=b\rho_a$ deputes gas non-idealness parameter.

3 Self-Similarity Transformations

The inner range of flow is said to be extensive behind the shock. According to Sedov [30], the expanded region's velocity is advised to obey as follows in the forming of self-similarity (Steiner and Hirschler [29], Zel'Dovich & Raizer [31])

$$W_p=\frac{dx_p}{dt}=V_0\,t^n, \tag{13}$$

where x_p deputes expansive region's radius, n deputes constant and V_0 deputes dimensional constant.

According to the extension specification, the shock velocity is comparable to the expanding area velocity, as illustrated below

$$W_s=CV_0\,t^n, \tag{14}$$

in which C deputes dimensionless constant. Also,

$$\eta\,(self-similarity\;\;variable)=\frac{x}{x_s}=\frac{(n+1)\,x}{CV_0\,t^{n+1}}. \tag{15}$$

Without a doubt, onto the shock $\eta=1$ and $\eta_p\left(=\frac{x_p}{x_s}\right)$ upon the expanded region. To to get similarity solutions, the unknowable factors should be remarked as (Nath [9], Sahu [10], Sahu [18], Sahu [26], Steiner and Hirschler [29])

$$v=\frac{x}{t}\,U(\eta),\qquad p=\rho_a\frac{x^2}{t^2}P(\eta),\qquad \rho=\rho_aR(\eta), \tag{16}$$

$$h=\rho_a\,\frac{x^2}{t^2}\,B(\eta),\qquad J=\rho_a\,\frac{x^3}{t^3}\,F(\eta),\qquad Z=Z_aR(\eta),$$

here U, P, R, B and F are function of η only.

M^* and C_a are supposed to remain unchangeable for the similarity of the results.

$$\epsilon=-\frac{n}{n+1}. \tag{17}$$

The governing mechanism of Eqs. (1–5) morphs into the following mechanism of ODEs when the Eq. (16) is used:

$$\frac{dR}{d\eta}=\frac{1}{L}\left[\frac{(1-Z_a\,R)R}{\eta}\left\{1+\overline{b}\,R(1-\mu_p)\right\}[RU(U-1)+2P+2(1+m)B-RU(1+i)\,\{U-(n+1)\}]\right. \tag{18}$$

$$+ R(1-Z_a R)\left[2B\{U(i-j-1)+1\} - 2P(U-1)\right]\frac{\left\{1+\bar{b}R(1-\mu_p)\right\}}{l}\eta\left\{U-(n+1)\right\}$$

$$\left. + \left\{1+\bar{b}R(1-\mu_p)\right\}^2\left\{i+\xi_R R^{\Psi}P^{\Xi}F^{q}\right\}\frac{(\Gamma-1)RF}{l}\eta\left\{U-(n+1)\right\}\right],$$

$$\frac{dU}{d\eta} = -\frac{(1+i)U}{\eta} - \frac{[U-(n+1)]}{R}\frac{dR}{d\eta}, \tag{19}$$

$$\frac{dB}{d\eta} = \frac{2B}{R}\frac{dR}{d\eta} + \frac{2B}{\eta[U-(n+1)]}\left[1+U(i-j-1)\right], \tag{20}$$

$$\frac{dP}{d\eta} = -\frac{dB}{d\eta} - \frac{2}{\eta}(m+1)B - \frac{2P}{\eta} - \frac{RU(U-1)}{\eta} - \frac{dU}{d\eta}R\left[U-(n+1)\right] \tag{21}$$

$$\frac{dF}{d\eta} = \frac{1}{\eta}\left[\xi_R R^{\Psi} P^{\Xi} F^{q+1} - 3F\right], \tag{22}$$

where

$$L = (1-Z_a R)\left\{\{U-(n+1)\}^2 R - 2B\right\}\left\{1+\bar{b}R(1-\mu_p)\right\} - \Gamma P\left\{1+\bar{b}(1-\mu_p R)\right\}^2 \tag{23}$$

$$+\bar{b}PR^2(1-\mu_p)\left\{\bar{b}(1-\mu_p+Z_a)\right\}.$$

Here $\xi_R = \alpha_{ab_a}\ \rho_a^{\Psi+\Xi+q}$. Moreover, for acquiring the similarity solution, essential to contemplate $2\Xi + 3q - l = 0$. Further ξ_R is a non-dimensional constant, that is contracted as the factor defining the relationship seen into gas and monochromatic radiation (c.f. Khudyakov [1], Nath and Takhar [3], Nath and Takhar [4], Nath [9], Sahu [10]).

The modified shock conditions are calculated using an Eq. (16)

$$P(1) = \left[1 + C_a\left(1-\frac{1}{\beta^2}\right) - \beta + \frac{1}{\gamma M^{*^2}}\right](n+1)^2,$$

$$R(1) = \frac{1}{\beta},\ \ B(1) = \frac{C_a}{\beta^2}(n+1)^2,\ \ F(1) = \frac{(n+1)^3}{\gamma^{\frac{3}{2}} M^{*^3}},\ \ U(1) = (n+1)(1-\beta). \tag{24}$$

Therewith, the prerequisite at the expansive region is such the fluid velocity identical to expansive region velocity. Aforementioned kinematic condition inscribed as

$$U(\eta_p) = (n+1). \tag{25}$$

Also, the disturbance's total energy is stated as

$$E = 2\pi i\int_{x_p}^{x_s}\rho\left[E_m + \frac{1}{2}(v^2) + \frac{h}{\rho}\right]x^i dx.$$

Using (7) and (16), equation (3) becomes

$$E = 2\pi\, i\, \rho_a \left[\frac{CV_0}{(n+1)}\right]^{\frac{2}{n+1}} x_s^{i+3-\frac{2}{n+1}} \int_{\eta_p}^{1} \left[\frac{P(1 - Z_a R)}{(\Gamma - 1)\left\{1 + \bar{b}\, R\,(1 - \mu_p)\right\}} + \frac{RU^2}{2} + B\right] \eta^{i+2} d\eta. \tag{26}$$

Consequently, the shock wave's total energy is non-stationary and alter as $x_s^{i+3-\frac{2}{n+1}}$ (see, Freeman and Craggs [32], Sahu [33], Sahu [34]).

It is preferable to assemble variables v, p, ρ, h and J in the following order to acquire numerical results

$$\frac{v}{v_n} = \frac{U(\eta)}{U(1)}\,\eta, \quad \frac{p}{p_n} = \frac{P(\eta)}{P(1)}\,\eta^2, \quad \frac{\rho}{\rho_n} = \frac{Z}{Z_n} = \frac{R(\eta)}{R(1)}, \quad \frac{h}{h_n} = \frac{B(\eta)}{B(1)}\,\eta^2 \quad \frac{J}{J_n} = \frac{F(\eta)}{F(1)}\,\eta^3;$$

in which the subscript n notify the situations instantaneously at the back of the shock front.

4 Results and Discussion

The dispersion of flow characteristics at $\eta = 1$ to $\eta = \eta_p$ is determined by computationally integrate equations (18–22) with (24–25) that use the fourth-order Runge-Kutta technique. The study extends Nath and Takhar's [4] approach by enabling the results of an magnetic field in dust-pervade gas. Figure 2 shows that curve-1 relates to Nath and Takhar's [4] solution. This curve illustrates that the obtained resolution agrees well with the existing solution of Nath and Takhar [4].

4.1 Upshot of the Existence of the Magnetic Field:

By intensifying the value of C_a; the extent between expansive region and shock as well as β intensify (see Table 1). Physically it approach that the gas at the back of the shock contracted wane, i.e. the shock strength wanes in the existence of an azimuthal magnetic field. $\frac{v}{v_n}$ and $\frac{h}{h_n}$ decreases; however the density $\frac{\rho}{\rho_n}$ as well as the radiation flux $\frac{J}{J_n}$ increases while moving inside towards expansive region through shock (see Figs. 2(a, c-e)). $\frac{p}{p_n}$ increases for ideal dust-pervade gas, but decreases in-general for non-ideal dust-pervade gas (see Fig. 2(b)). $(C_{adi})\ p_a$ intensify adjoining shock and wanes adjoining expansive region (see Fig. 2(f)).

4.2 Upshot of the Intensifying Non-idealness Parameter of the Gas ($\bar{b}$)

The intensify in $\bar{b}$ affects to intensify η_p and wane β (see Table 1) (i.e. to intensify the shock strength. $\frac{v}{v_n}$ intensify for non-magnetic dust-pervade gas, but it intensify adjoining shock and wanes adjoining expansive region in-general for non-ideal non-magnetic dust-pervade gas. $\frac{p}{p_n}$ and $\frac{\rho}{\rho_n}$ wanes in the existence of the magnetic field, however, intensify adjoining shock and wanes adjoining

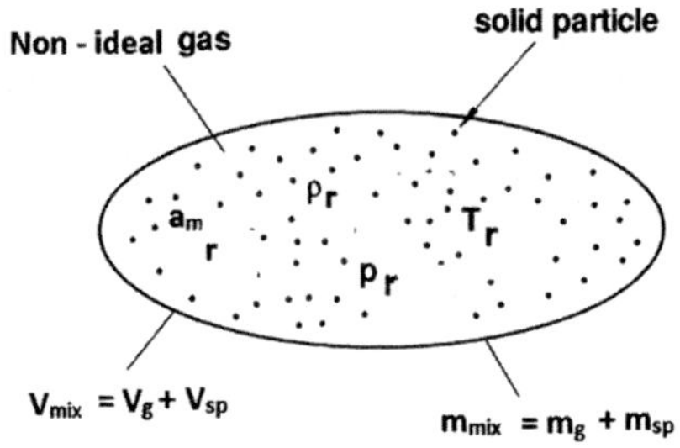

Fig. 1. The dust-prevade gas in its reference condition [The reader is referred to Sahu [28], Steiner and Hirschler [29] for details].

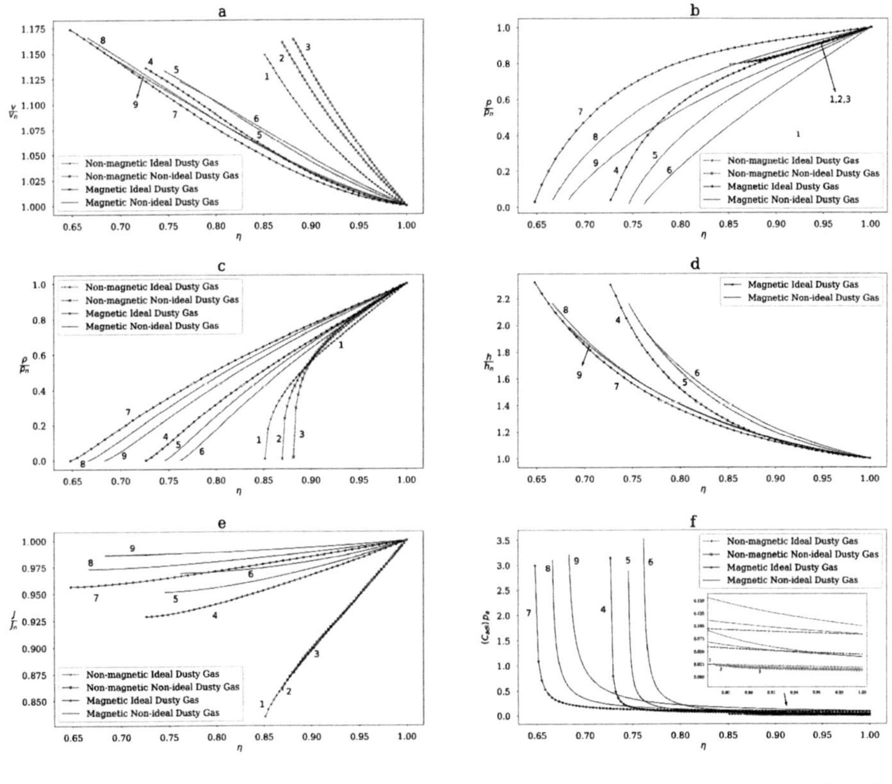

Fig. 2. Variability of dust-prevade gas parameters at back of the shock (a) $\frac{v}{v_n}$, (b) $\frac{p}{p_n}$, (c) $\frac{\rho}{\rho_n}$, (d) $\frac{h}{h_n}$,(e) $\frac{J}{J_n}$, (f) $(C_{adi})\ p_a$: (Reference Table 1 for further information on input variables).

expansive region in the absence of magnetic field; though, inverted conduct is executed for the radiation flux. $\frac{h}{h_n}$ intensify for ideal to non-ideal dust-pervade gas. $(C_{adi})\ p_a$ wanes in the non-existence of a magnetic field but intensify in the existence of the magnetic field. (see Figs. 2(a–f)).

Table 1. Taking into account different aspects of C_a and $\bar{b}$ for $\gamma = \frac{5}{3}$, $m = 1$, $i = 2$, $j = 1$, $\beta' = 1$, $M^* = 5$, $\mu_p = 0.2$, $G_a = 10$, $n = -0.4$, $\Psi = -\frac{1}{2}$, $\Xi = \frac{3}{2}$, $q = -\frac{5}{3}$, $l = -2$, $\xi_R = 10^{-4}$.

C_a	$\bar{b}$	β	η_p	Case No. in Fig. 2
0.0	0.0	0.258834	0.851284	1
	0.1	0.250680	0.869739	2
	0.2	0.242343	0.881605	3
0.05	0.0	0.360391	0.726426	4
	0.1	0.342217	0.745421	5
	0.2	0.321513	0.761809	6
0.1	0.0	0.448205	0.647747	7
	0.1	0.428630	0.666123	8
	0.2	0.406499	0.683107	9

5 Conclusion

On the basis of the above work, the findings might be drawn:

(i) The study of shock expansion in the existence of magnetic field with monochromatic radiation in a non-ideal dust-pervade gas for planar or spherical or cylindrical shock, has not been made previously.
(ii) It's noticeable that the pressure and density at the expansive region wear down in the presence of an azimuthal magnetic field, resulting in the creation of a vacuum at the symmetry's centre, it is perfectly consistent with the settings in the lab that created the shock wave.
(iii) The physical deliberation sets a limit on n as $\epsilon = -\frac{n}{n+1}$.
(iv) The shock strength intensify with intensifying the non-idealness of the gas in the existence of magnetic field. The non-idealness contemplation produces outstanding distinction in the flow variables distribution.

Acknowledgments. "The author is thankful to **Prof. M. K. Verma**, Department of Physics, Indian Institute of Technology Kanpur, Kanpur–208016, India for fruitful discussions. This work was supported by research grant no. **TAR/2018/000150** under **Teachers Associateship for Research Excellence (TARE)** scheme from the **Science and Engineering Research Board (SERB)**, India. The author gratefully acknowledges financial support from **SERB**."

References

1. Khudyakov, V.M.: The self-similar problem of the motion of a gas under the action of monochromatic radiation. Soviet Phys. Doklady **28**, 853 (1983)
2. Zheltukhin, A.: A family of exact solutions of the equations of the one-dimensional motion of a gas under the influence of monochromatic radiation. J. Appl. Math. Mech. **52**(2), 262–263 (1988)

3. Nath, O., Takhar, H.S.: Propagation of cylindrical shock waves under the action of monochromatic radiation. Astrophys. Space Sci. **166**(1), 35–39 (1990). https://doi.org/10.1007/BF00655604
4. Nath, O., Takhar, H.S.: Spherical MHD shock waves under the action of monochromatic radiation. Astrophys. Space Sci. **202**(2), 355–362 (1993). https://doi.org/10.1007/BF00626888
5. Nath, G., Sahu, P.K., Dutta, M.: Magnetohydrodynamic cylindrical shock in a rotational axisymmetric non-ideal gas under the action of monochromatic radiation. Procedia Eng, **127**, 1126–1133 (2015)
6. Nath, G., Sahu, P.K.: Unsteady adiabatic flow behind a cylindrical shock in a rotational axisymmetric non-ideal gas under the action of monochromatic radiation. Procedia Eng. **144**, 1226–1233 (2016)
7. Nath, G., Sahu, P.K.: Flow behind an exponential shock wave in a rotational axisymmetric non-ideal gas with conduction and radiation heat flux. Int. J. Appl. Comput. Math. **3**(4), 2785–2801 (2017). https://doi.org/10.1007/s40819-016-0260-x
8. Nath, G., Sahu, P.K.: Similarity solution for the flow behind a cylindrical shock wave in a rotational axisymmetric gas with magnetic field and monochromatic radiation. Ain Shams Eng. J. **9**(4), 1151–1159 (2018)
9. Nath, G.: Shock wave driven out by a piston in a mixture of a non-ideal gas and small solid particles under the influence of the gravitation field with monochromatic radiation. Chin. J. Phys. **56**(6), 2741–2752 (2018)
10. Sahu, P.K.: Similarity solution for a spherical shock wave in a non-ideal gas under the influence of gravitational field and monochromatic radiation with increasing energy. Math. Methods Appl. Sci. **42**(14), 4734–4746 (2019)
11. Nath, G.: Spherical shock generated by a moving piston in a nonideal gas under gravitation field with monochromatic radiation and magnetic field. J. Eng. Phys. Thermophys. **93**(4), 911–923 (2020). https://doi.org/10.1007/s10891-020-02193-6
12. Sahu, P.K.: Similarity solution for one dimensional motion of a magnetized self-gravitating gas with variable density under the absorption of monochromatic radiation. Zeitschrift für Naturforschung A. (2022). https://doi.org/10.1515/zna-2021-0254
13. Hartmann, L.: Accretion Processes in Star Formation. Cambridge University Press, Cambridge (1998)
14. Balick, B., Adam, F.: Shapes and shaping of planetary nebulae. Annu. Rev. Astron. Astrophys. **40**, 439 (2002)
15. Nath, G., Sahu, P.K.: Flow behind an exponential shock wave in a rotational axisymmetric perfect gas with magnetic field and variable density. SpringerPlus **5**(1), 1–18 (2016). https://doi.org/10.1186/s40064-016-3119-z
16. Nath, G., Sahu, P.K., Chaurasia, S.: An exact solution for the propagation of cylindrical shock waves in a rotational axisymmetric nonideal gas with axial magnetic field and radiative heat flux. Model. Meas. Control B **87**(4), 236–243 (2018)
17. Nath, G., Sahu, P.K., Chaurasia, S.: Self-similar solution for the flow behind an exponential shock wave in a rotational axisymmetric non-ideal gas with magnetic field. Chin. J. Phys. **58**, 280–293 (2019)
18. Sahu, P.K.: Shock wave driven out by a piston in a mixture of a non-ideal gas and small solid particles under the influence of azimuthal or axial magnetic field. Braz. J. Phys. **50**(5), 548–565 (2020). https://doi.org/10.1007/s13538-020-00762-x

19. Sahu, P.K.: Magnetogasdynamic exponential shock wave in a self-gravitating, rotational axisymmetric non-ideal gas under the influence of heat-conduction and radiation heat-flux. Ricerche di mat **1**, 1–37 (2021). https://doi.org/10.1007/s11587-021-00563-7
20. Sahu, P.K.: Shock wave propagation in perfectly conducting rotational axisymmetric two-phase medium with increasing energy under the action of heat conduction and radiation heat flux. Chin. J. Phys. **72**, 176–190 (2021)
21. Sahu, P.K.: Flow behind the magnetogasdynamical cylindrical shock wave in rotating non-ideal dusty gas with monochromatic radiation. Plasma Res. Express **3**(4), 045004 (2021)
22. Pai, S. I.: Two-Phase Flows, Vol. 3. Springer-Verlag (2013)
23. Nath, G., Sahu, P.K.: Self-similar solution of a cylindrical shock wave under the action of monochromatic radiation in a rotational axisymmetric dusty gas. Commun. Theor. Phys. **67**(3), 327 (2017)
24. Nath, G., Sahu, P.K.: Propagation of a cylindrical shock wave in a mixture of a non-ideal gas and small solid particles under the action of monochromatic radiation. Combust. Explos. Shock Waves **53**(3), 298–308 (2017). https://doi.org/10.1134/S0010508217030078
25. Sahu, P.K.: Self-similar solution of spherical shock wave propagation in a mixture of a gas and small solid particles with increasing energy under the influence of gravitational field and monochromatic radiation. Commun. Theor. Phys. **70**(2), 197 (2018)
26. Sahu, P.K.: Propagation of an exponential shock wave in a rotational axisymmetric isothermal or adiabatic flow of a self-gravitating non-ideal gas under the influence of axial or azimuthal magnetic field. Chaos, Solitons Fractals **135**, 109739 (2020)
27. Sahu, P.K.: Analysis of magnetogasdynamic spherical shock wave in dusty real gas with gravitational field and monochromatic radiation. Eur. Phys. J. Plus **136**(4), 1–19 (2021). https://doi.org/10.1140/epjp/s13360-021-01282-6
28. Sahu, P.K.: The influence of magnetic and gravitational fields in a non-ideal dusty gas with heat conduction and radiation heat flux. Indian J. Phys. **4**, 1–15 (2022). https://doi.org/10.1007/s12648-021-02269-w
29. Steiner, H., Hirschler, T.: A self-similar solution of a shock propagation in a dusty gas. Eur. J. Mech.-B/Fluids **21**(3), 371–380 (2002)
30. Sedov, L.I.: Similarity and dimensional methods in mechanics. Academic Press, New York (1959)
31. Zel'Dovich, Y. B., Raizer, Y. P.: Physics of shock waves and high-temperature hydrodynamic phenomena. Courier Corporation (2002)
32. Freeman, R.A., Craggs, J.D.: Shock waves from spark discharges. J. Phys. D Appl. Phys. **2**(3), 421 (1969)
33. Sahu, P.K.: Similarity solution for the flow behind an exponential shock wave in a rotational axisymmetric non-ideal gas under the influence of gravitational field with conductive and radiative heat fluxes. In: Dawn, S., Balas, V.E., Esposito, A., Gope, S. (eds.) ICIMSAT 2019. LAIS, vol. 12, pp. 1060–1070. Springer, Cham (2020). https://doi.org/10.1007/978-3-030-42363-6_122
34. Sahu, P.K.: Unsteady flow behind an MHD exponential shock wave in a rotational axisymmetric non-ideal gas with conductive and radiative heat fluxes. In: Dawn, S., Balas, V.E., Esposito, A., Gope, S. (eds.) ICIMSAT 2019. LAIS, vol. 12, pp. 1049–1059. Springer, Cham (2020). https://doi.org/10.1007/978-3-030-42363-6_121

Seismic Vulnerability Assessment of Old Brick Masonry Buildings: A Case Study of Dhulikhel

Subarna Pandey[1](✉) and Shyam Sundar Khadka[2]

[1] Department of Civil Engineering, Kathmandu University, Dhulikhel, Kavre, Nepal
subaranap097@gmail.com

[2] Department of Civil Engineering, Kathmandu University, Dhulikhel, Nepal

Abstract. Unreinforced brick masonry (URM) in mud mortar is the most common building typology in Dhulikhel. Most of these buildings were constructed more than 100 years ago with traditional construction techniques and locally available materials. These old buildings are made up of thick walls and consist timber floors /roofs representing the main structural system of the typology. Dhulikhel is considered one of the oldest and historical cities in Nepal and buildings constructed here with old Newari architecture exhibit high historical and cultural value. These buildings experienced devastating earthquakes and most of them are still serviceable after traditional maintenance. However, due to the limited intrinsic capacity of unreinforced masonry against lateral load, construction without seismic provisions, and construction deficiencies such as inadequate wall-to-wall or wall-to-floor connections, these buildings are categorized as the most vulnerable constructions. Past earthquakes showed severe damage to URM buildings in Nepal with frequent out-of-plane failure. In the last decades, masonry structures have gained increasing attention after the realization of their historical and cultural importance and the great risk that they suffer in seismic areas. In this study, linear dynamic analysis is performed to develop fragility curves for a typical representative building. The building is selected based on a rapid visual screening survey carried out in the city. The result of this study highlights that the URM buildings in Dhulikhel are highly susceptible to out of plane failure because of the long unsupported length of a masonry wall and flexible floor.

Keywords: Unreinforced brick masonry · Structural system · Flexible floor · Out-of-plane damage · Seismic vulnerability assessment

1 Introduction

Unreinforced brick Masonry (URM) has been used as principal construction material for buildings in Nepal from the early age of modern civilization [6]. In Nepal, URM houses comprise a majority of building stock. ~62% of houses in Nepal are of masonry either in the form of mud mortar or cement mortar [3]. The present study focuses on the vulnerability assessment of existing URM buildings of Dhulikhel. The city is located southeast of Kathmandu, the capital city of Nepal. It is a historical city and buildings constructed here with indigenous Newari architecture exhibit high historical and cultural

Z. Dimitrovová et al. (Eds.): WMVC 2022, MMS 125, pp. 99–106, 2023.
https://doi.org/10.1007/978-3-031-15758-5_9

value. URM in mud mortar is the most common building typology in Dhulikhel. Most of these buildings were constructed more than a hundred years ago with traditional construction techniques and locally available materials. Generally, these buildings are low rise, and most of them have been constructed for residential purposes.

Nepal lies in a highly seismically active zone and the frequent occurrence of disastrous earthquakes verifies the seismicity of the country. Past studies on the aftermath of the earthquakes show a higher vulnerability of URM houses [4]. Out-of-plane failure and delamination of masonry walls are predominant failure modes [4]. Due to the limited intrinsic capacity of unreinforced masonry against lateral load, construction without seismic provisions, and construction deficiencies such as inadequate wall-to-wall or wall-to-floor connections, these buildings are categorized as the most vulnerable constructions [5, 6]. However, most of the traditional buildings of Dhulikhel are still serviceable after traditional maintenance. In the last decades, masonry structures have gained increasing attention after the realization of their archeological importance and the great risk that they suffer in seismic areas. It is necessary to check strength against disastrous earthquakes and strengthen the structural system to downscale the severity of seismic effects. Hence, this study is aimed to estimate the seismic fragility of existing URM buildings of Dhulikhel.

2 Traditional Buildings in Dhulikhel: General Features and Structural System

The study is focused on the Narayansthan area of Dhulikhel city. Most of the buildings in this area are coalesced in a row-housing form. These are private buildings with traditional architecture and were constructed with primordial technology. URM buildings in the locality are three to five-storied and mostly four-storied. Floor height is limited between 2.0 to 2.5 m. Generally, these buildings are characterized by regular configuration and opening patterns. Sufficient openings are provided in the façade and rear walls. The openings made of timber are usually double framed, small-sized, square-shaped, and placed symmetrically in the walls. The most prominent feature of the traditional buildings is the latticed window typically called *San Jhya* (Fig. 1e), a richly decorated window covering most of the façade at the third story. On the ground floor, a certain portion of the façade is built with a timber frame. It comprises a series of twin posts, capital for each, and a double beam running over each capital. The ground floor space associated with the timber frame is typically called *dalan* (Fig. 1b) and recently, it is used for commercial purposes. The structural system of these traditional buildings is mainly based on thick superstructure walls, timber floor, roof, and shallow wall foundation [6]. Load-bearing walls are constructed in Wythe system with locally available burnt clay bricks and mud-mortar. Mostly an exterior thin layer of burnt brick (*ma apa*) and the inner thick layer of unburnt bricks (*kachi apa*) or use of burnt bricks only can be seen in load-bearing walls. Wall width varies from 0.45 to 0.6 m. Generally, two external walls with a spine wall at the center are usually seen in a long direction. At the upper story, the spine wall is sometimes replaced by a timber frame system (Fig. 1c) that is similar to the frame provided at *dalan*. Along short direction, cross walls are usually less in number and sometimes not provided. In most cases, the only interior face of the wall is plastered with a mixture of mud, rice husk, and dung. On floors (Fig. 1d), closely spaced timber joists are used as a structural member over which timber planks with a thick layer of mud topping or sometimes bricklayer is applied. This conventional technique of floor construction represents a flexible floor diaphragm [5]. Most of the masonry houses have sloped roofs with roofing tiles made of burnt clay (*jhingati*). For vertical access, in many cases, two ladders are present in different locations. Timber pins, pegs sometimes shear keys are used for connections.

3 Case Study: Representative Building in Dhulikhel

A private building (Fig. 1a) was selected as a case study based on a rapid visual assessment survey which represents most features of the traditional buildings of the locality. The building is four-storied, 12.82 m long in length and narrow 6.2 m in width, the aspect ratio is around two. The floor height of the ground story is 2.1 m and the overall height of the building is about 8.77 m from the ground to the ridge of the roof. The Ground floor plan of the building is shown in Fig. 2. Typical features and explanations of the structural system have been presented in Sect. 2. The existing condition of the building looks different from its original construction by appearance, possibly due to additional requirements of the owner. Initially, the building was built with three floors, but, after several years from construction time, a story was added. During the addition of the floor, a double-pitched purlin roof with local roofing tiles (*jhingati*) was modified to a single pitched roof with Corrugated Galvanized Iron (CGI) covering. Generally, the condition of the building is good and fully occupied. The ground floor of the building is used for commercial purposes and the rest floors are used for residential purposes. The building experienced minor damage of non-structural components during the Gorkha earthquake 2015 and it was renovated immediately.

4 Numerical Modelling

A linear elastic numerical model, based on a macro modeling strategy, is developed in the commercially available software ETABS 19 [1, 7]. The masonry wall is modeled using homogenized isotropic surface/shell elements and all timber members are represented with line/frame elements. The model of the building is shown schematically in Fig. 3. The foundation of the building is considered fully restrained. Since most of the buildings in the area are low-rise light structures and founded on stiff soil, neglecting Soil-Structure Interaction (SSI) can be reasonable. Only considering SSI with the addition of computational burden cannot make model realistic, since several assumptions were made and the study is undertaken within a scope. It was difficult to identify the fixity of timber joist into the wall, thus, the connection of the timber floor/roof with the masonry wall is assumed to be pinned. Subsequently, floors and roofs are defined with a flexible diaphragm. The mechanical properties of the materials used during modeling are tabulated in Table 1.

Fig. 1. (a) Typical URM house (b) *Dalan* at ground floor (c) twin posts in the third story with shear key (red encircled) (d) closely spaced timber joists (e) latticed window (*san jhya*) at the third story

5 Analysis and Seismic Vulnerability Assessment

The model is analyzed by the Linear Time History Analysis (LTHA) using Newmark's Direct Integration method. For the definition of variation of acceleration over a time step, average acceleration method is used. Parameters Gamma(Υ) and Betta (β) are assigned with values $\Upsilon = 0.5$ and $\beta = 0.25$. Three orthogonal pairs of real recorded time history data of three different earthquakes: El Centro 1940, Kobe 1995, and Gorkha 2015 are used as dynamic load. Ground motion selection is based on criteria recommended by ASCE/SEI 41–13. Damping is 5% [8], and Rayleigh damping coefficients (mass proportional coefficient and stiffness proportional coefficient) are calculated based on the first and second mode vibration periods. As lateral load resisting horizontal structural system is considered flexible, the mass of the structure is distributed throughout the height and thus, lateral force is distributed uniformly [6]. Ground motions are scaled within the range of peak ground acceleration between 0.1 g to 1.0 g. Seismic analysis is performed for both orthogonal directions (X and Y) of the building. The result of Finite Element Analysis (FEA) in the form of element stress (normal and shear) is further evaluated to identify the failure elements using the failure criterion proposed by Asteris and Syrmakezis (2009). The failure results obtained are used for the calculation of a damage index (DI) and damage states are defined by the values of DI. A value of DI less than 10% is interpreted as insignificant damage; from 10% to less than 20%, as moderate damage; and larger or equal than 20% as heavy damage [1]. Evaluation of fragility is performed in a probabilistic approach taking account of the uncertainties associated with the structural system and seismic demand (record-to-record variability).

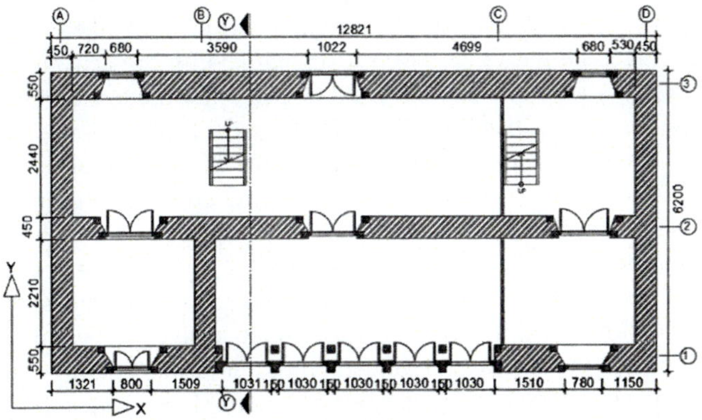

Fig. 2. Ground floor plan of case study building

Fig. 3. Numerical model in ETABS

Table 1. Mechanical properties of the materials (Mean values) (Rits-DMUCH 2012)

Material	Unit weight (kN/m^3)	Compressive strength(N/mm^2)	Tensile strength(N/mm^2)	Young's modulus(N/mm^2)	Poisson's ration
Wall	17.68	1.82	Nil	794	0.25
Timber	4.47	22	42	8100	0.3

6 Results and Discussions

Analysis result of the existing building shows maximum out-of-plane displacement of long walls and higher values of DI due to seismic load in the Y direction (Fig. 4). Along this direction, the threshold limit of 10% and 20% is attained at 0.044 g and 0.091 g respectively, with very low levels of seismicity. On the contrary, the intensity levels are comparatively high and the values are 0.24 g and 0.345 g respectively (Fig. 4) in the other direction. This indicates that the building can resist moderate to heavy shakings with moderate damages when the direction of lateral load is along a long wall. To determine the resistance offered by the double frame of openings and timber frames, a model was created replacing the timber frames with fictitious walls and removing the openings' frames. The analysis results in terms of damage index show the significant addition of resistance due to timber members (Fig. 5). The result can be interpreted compared with the fictitious condition of the building having large openings without these provisions. Another comparison is made by truncating the building to only three stories to determine the increase in vulnerability due to the addition of a story over the then building. The comparative study shows a remarkable increase in vulnerability i.e., 61% and 15% increments of DI are observed at intensity levels of 0.1 g and 0.36 g respectively (Fig. 6). This shows the greater increment of vulnerability at lower intensity levels. Certain interventions are applied and results are checked with the previous results of the existing building. Cross walls are provided along the short wall direction against the long walls. After applying interventions, there is a 68% reduction in out-of-plane displacement value (Fig. 7) and a 51% reduction in DI at an intensity level of 0.36 g (Fig. 8).

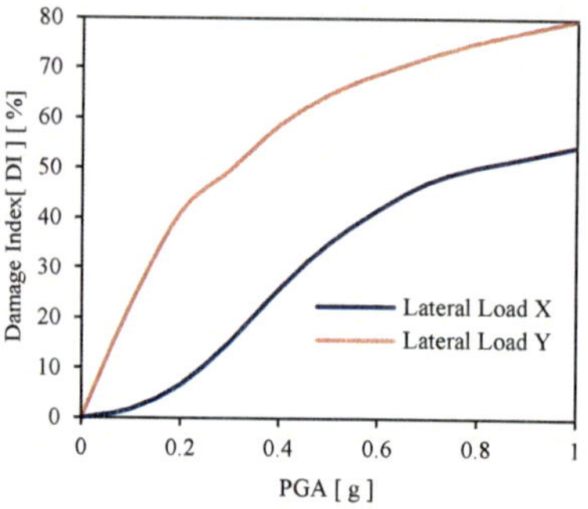

Fig. 4. Vulnerability curves for existing building

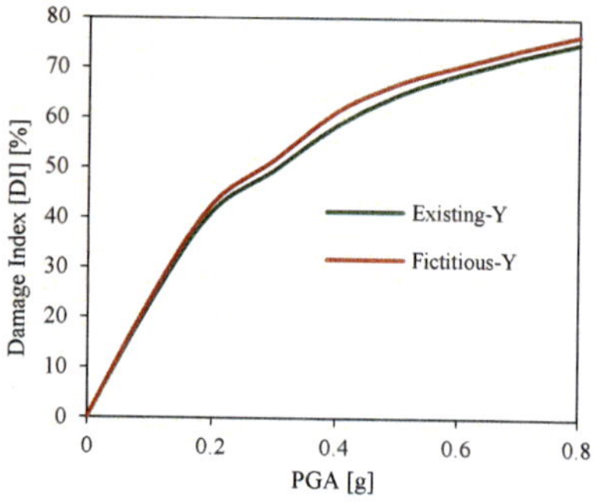

Fig. 5. Vulnerability curves (Y-direction)

As URM buildings are non-engineered traditional constructions, a high probability of heavy damage is expected in case of moderate to high peak ground acceleration. As shown in Fig. 9, the probability of getting heavily damaged is 50% at the PGA of 0.415 g, when the load direction is along longitudinal walls. On the contrary, the value of the intensity measure is very low, that is, 0.138 g for the other direction of lateral load. Similarly, 50% probability of exceeding moderate damage state for longitudinal shaking is at 0.28 g and for transverse shaking the PGA value is 0.078 g. Other significant values can be obtained from Fig. 9.

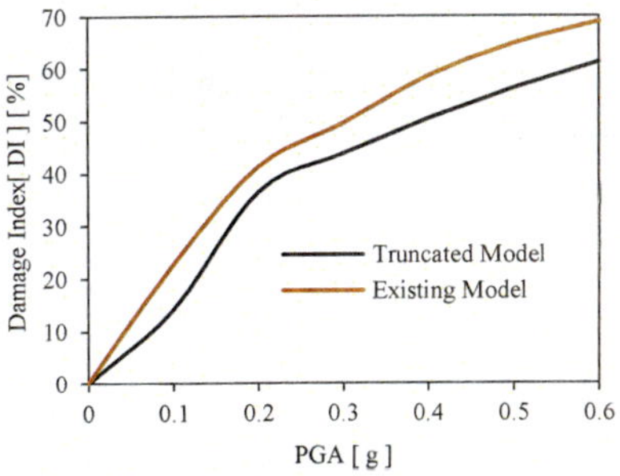

Fig. 6. Vulnerability curves (Y-direction)

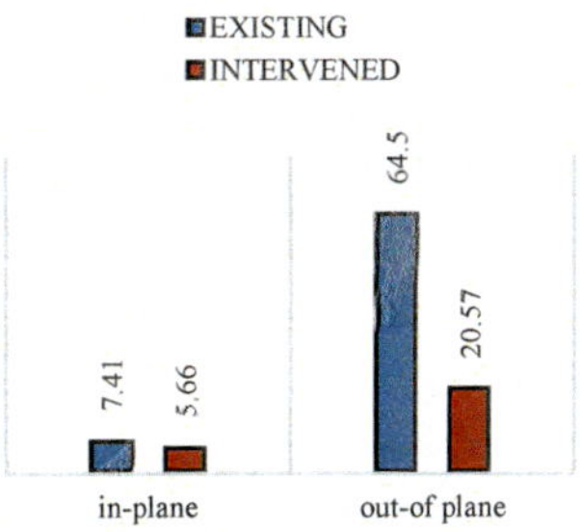

Fig. 7. Maximum displacement (mm) (Hazard level: 0.36 g)

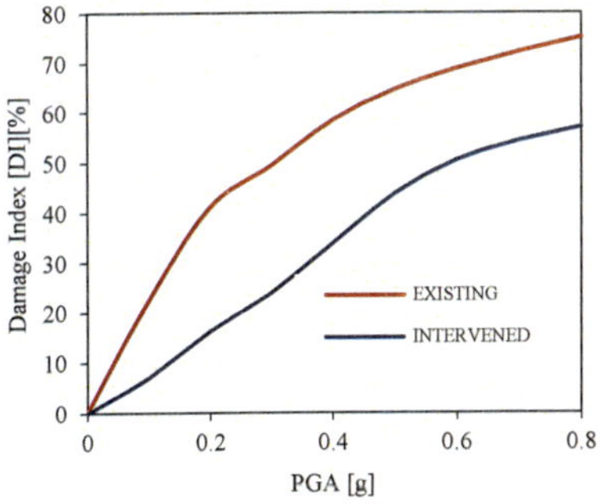

Fig. 8. Vulnerability curves (Y-direction)

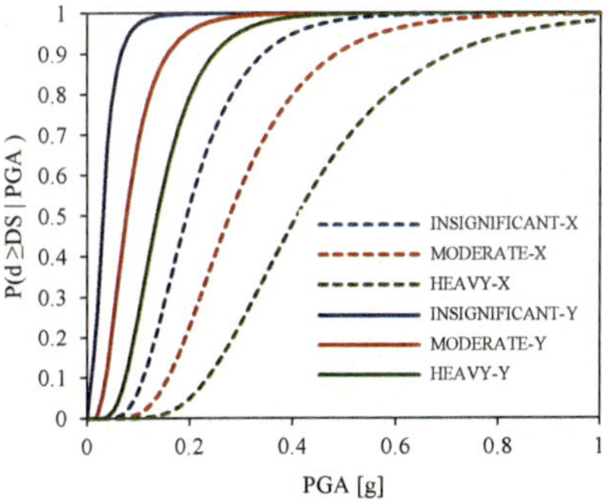

Fig. 9. Seismic fragility curves for existing building

7 Conclusions

To assess the seismic vulnerability of URM buildings in mud mortar, a typical representative of the traditional building is taken for a case study. Based on this study, certain conclusions are drawn. The building is highly vulnerable to earthquakes in the short direction and shows good resistance in the long direction. The seismic resistance is due to the thickness of the wall and the low height of the wall on each floor. The long wall is highly susceptible to out-of-plane failure because of its long unsupported length with usually less number of cross walls and flexibility of floor diaphragm. The timber frames have added significant strength to the total resistance of the building. A significant increase in seismic vulnerability is noted due to the addition of a story over the original form of the building. The increment values are large at lower intensities. The provision of interventions has shown a significant reduction in the out-of-plane displacement and damage index. The building is at seismic risk not only because of its structural system, but the direction of ground motion is also an important aspect of its risk.

Acknowledgments. The author would like to sincerely acknowledge the Department of Civil Engineering, Kathmandu University for continuous support, Buddha Shrestha for building drawings and a great help during fieldwork, and the local people of Dhulikhel for the kind cooperation during the field survey.

References

1. Asteris, P.G., et al.: Seismic vulnerability assessment of historical masonry structural systems. Eng. Struct. **62–63**, 118–134 (2014). https://doi.org/10.1016/j.engstruct.2014.01.031
2. Asteris, P.G., Syrmakezis C.A.: Non-dimensional masonry failure criterion under biaxial stress state. In: 11th Canadian Masonry Symposium, Toronto, Ontario, May 31–June 3 (2009)
3. Central Bureau of Statistics (CBS): National Population and Housing Census 2011 (National Report), vol. 01NPHC. Government of Nepal (2011)
4. EERI.:M7.8 Gorkha: Nepal Earthquake on April 25, 2015 and its Aftershocks. Earthquake Engineering Research Institute (EERI), Oakland, CA, USA (2016)
5. Gautam, D., Rodrigues, H., Bhetwal, K.K., Neupane, P., Sanada, Y.: Common structural and construction deficiencies of Nepalese buildings. Innov. Infrastruct. Solut. **1**(1), 1–18 (2016). https://doi.org/10.1007/s41062-016-0001-3
6. Khadka, S.S.: Seismic performance of traditional unreinforced masonry building in Nepal. Kathmandu University Journal of Science, Engineering and Technology **9**(I), 15–28 (2013)
7. Lourenço, P.B.: Computational Strategies for Masonry Structures. Ph.D., Delft University of Technology, Netherlands (1996)
8. Rits-DMUCH.: Disaster risk management for the historic city of Patan, Nepal. Research Center for Disaster Mitigation of Urban Cultural Heritage, Kyoto, Japan (2012)

Test of an Idea for Improving the Efficiency of Nonlinear Time History Analyses When Implemented in Seismic Analysis According to NZS 1170.5:2004

Aram Soroushian[1(✉)] and Peter Wriggers[2]

[1] International Institute of Earthquake Engineering and Seismology, 19537 Tehran, Iran
a.soroushian@iiees.ac.ir

[2] Institut für Kontinuumsmechanik, Leibniz Universität Hannover, 30167 Hannover, Germany
wriggers@ikm.uni-hannover.de

Abstract. For nonlinear time history analysis, employing a time integration method and some nonlinearity iterative method (for implicit analyses) is a broadly accepted practice. In 2015, the authors proposed a change in the analysis, according to which, when the nonlinearity iterations do not converge, the analysis proceeds to the next integration step. In this paper, this change, and no other change, is applied to seismic analysis according to the seismic code of New Zealand, NZS 1170.5:2004. The purpose is to clarify whether the change can increase the analysis efficiency. The possibility of improving the efficiency is shown via analysis of a tall building's structural model. The observations are explained, discussed, and generalized. As a main result, the change proposed in 2015 may considerably improve the efficiency when the tolerance is sufficiently small and is ineffective when the nonlinearity tolerance is sufficiently large.

Keywords: Nonlinear structural dynamics · Time integration · Nonlinearity iterations · Iterations convergence · Analysis efficiency · NZS 1170.5:2004

1 Introduction

Structural systems' actual behavior is nonlinear and dynamic. For analyzing this behavior, a versatile approach is to discretize the system in space, and to solve the resulting initial value problem, typically expressed as [1]:

$$\begin{gathered} \mathbf{M}\,\ddot{\mathbf{u}}(t\,) + \mathbf{f}_{\text{int}}(t) = \mathbf{f}(t) \qquad\qquad 0 \le t \le t_{end} \\ \text{Initial Conditions}: \left| \begin{array}{l} \mathbf{u}(t=0) = \mathbf{u}_0 \\ \dot{\mathbf{u}}(t=0) = \dot{\mathbf{u}}_0 \\ \mathbf{f}_{\text{int}}(t=0) = \mathbf{f}_{\text{int}_0} \end{array} \right. \\ \text{Additional Constraints}: \mathbf{Q} \end{gathered} \tag{1}$$

In Eq. (1), t implies the time; t_{end} stands for the duration of the dynamic behavior; $\mathbf{M}$ is the mass matrix; $\mathbf{f}_{\text{int}}$ and $\mathbf{f}(t)$ indicate the vectors of internal force and excitation;

Z. Dimitrovová et al. (Eds.): WMVC 2022, MMS 125, pp. 107–114, 2023.
https://doi.org/10.1007/978-3-031-15758-5_10

$\mathbf{u}(t)$, $\dot{\mathbf{u}}(t)$, and $\ddot{\mathbf{u}}(t)$, denote the vectors of displacement, velocity, and acceleration; $\mathbf{u}_0$, $\dot{\mathbf{u}}_0$, and $\mathbf{f}_{\text{int}_0}$, define the initial status of the model ($\mathbf{f}_{\text{int}_0}$ may be essential in presence of material nonlinearity); and $\mathbf{Q}$ represents restricting conditions because of nonlinearity, e.g., additional constraints in problems involved in impact or elastic-plastic behavior. A broadly accepted method for analyzing Eq. (1) is direct time integration, briefly reviewed in Fig. 1. The step-by-step computation, though important and common between linear and nonlinear analyses [1, 2], is not the focus of the discussion here. The main attention is to nonlinearity iteration, which is an essential section of many nonlinear analyses [1, 2]. Regardless of the iterative method [4], the iterations continuation/stop (or stop of the analysis) generally is according to [1]:

$$\begin{aligned} &\|\delta_k\| > \bar{\delta} \text{ and } k < \bar{k} : \text{Continue the iterations} \\ &\|\delta_k\| \le \bar{\delta} \text{ and } k \le \bar{k} : \text{Stop the iterations and continue the analysis} \\ &\|\delta_k\| > \bar{\delta} \text{ and } k = \bar{k} : \text{Stop the iterations and stop the analysis} \\ &\qquad\qquad k = 1, 2, 3, \ldots k' \quad , \quad k' \le \bar{k} \quad , \quad \bar{k} < \infty \end{aligned} \tag{2}$$

In Eq. (2), $\|\delta_k\|$ denotes the inaccuracy in modeling the nonlinearity after k iterations, $\bar{\delta}$ implies the tolerance (maximum acceptable inaccuracy in modeling the nonlinearity), and $\bar{k}$ is the maximum acceptable number of the iterations representing the available computational facility. In 2015, the authors proposed replacing Eq. (2) with [3]:

$$\begin{aligned} &\|\delta_k\| > \bar{\delta} \text{ and } k < \bar{k} : \text{Continue the iterations} \\ &\|\delta_k\| \le \bar{\delta} \text{ or } k = \bar{k} : \text{Stop the iterations and continue the analysis} \\ &\qquad\qquad k = 1, 2, 3, \ldots k' \quad , \quad k' \le \bar{k} \quad , \quad \bar{k} < \infty \end{aligned} \tag{3}$$

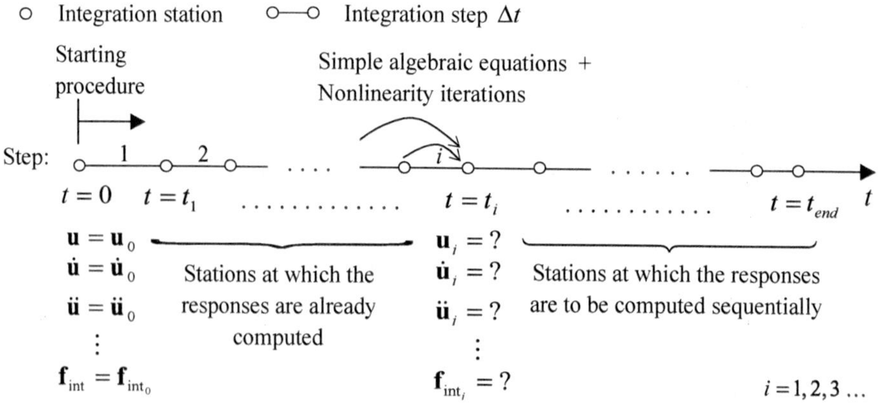

Fig. 1. A pictorial review of direct time integration analysis of Eq. (1) [1].

To study the resulting effect on computational efficiency, a clear procedure for arriving at the history of the target response is needed. Such a procedure, that considers Eq. (2) for ending the non-linearity iterations, is stated in [5, 6], as follows:

1. Make some initial selections, e.g. the target response, the integration method, the method and parameters needed for nonlinearity iteration, and set $i = 1$.
2. Carry out a first time integration analysis, using Eq. (2) for nonlinearity iterations, and an integration step Δt obtained from:

$$\Delta t = \mathrm{Min}\left(\frac{T_1}{\chi},\ T_n,\ \Delta t_{cr},\ \Delta t_{CFL},\ {}_f\Delta t,\ 0.01\right) \tag{4}$$

   where, T_1 is the period associated with the first mode of vibration, χ equals 100, unless when the behavior is involved in impact and $\chi = 1000$, T_n is the period of the highest mode, in the same direction as T_1, required to achieve the 90% mass as described in the modal response spectrum method [5], Δt_{cr} is the upper-bound because of the linear theory of numerical stability [2], Δt_{CFL} is the upper-bound because of the CFL condition [7], and ${}_f\Delta t$ is the step at which the excitation is digitized [1, 5, 6],
3. If the analysis is not stopped, by the last *Stop* in Eq. (2), continue to Step 5.
4. Set $i = i + 1$, repeat the last analysis with half steps, and return to Step 3.
5. Determine the peak target response.
6. If i is larger than one, continue to Step 8.
7. Return to Step 4.
8. If the peaks of the target response in the last two analyses are in a relative difference not more than 5%, accept the last response as final and stop.
9. Return to Step 4.

Evidently, in replacing Eq. (2) with Eq. (3), further changes are needed in the above procedure. Accordingly, it is unclear how the replacement will affect the response accuracy and specifically the analysis efficiency. Considering these, the objective of this paper is to compare Eqs. (2) and (3), when used in nonlinear time history analysis according to NZS 1170.5:2004. First, brief theoretical efforts are presented to clarify the ambiguities, including a new analysis procedure. Next, the possibility of increasing the analysis efficiency is displayed. The observations are then explained and generalized, and finally, the paper is concluded with a set of the achievements.

2 Theory in Brief

Replacing Eq. (2) with Eq. (3) in time history analysis according to NZS 1170.5:2004 [5, 6] changes the analysis procedure (stated in Sect. 1) to the following simpler procedure:

1. Make some initial selections, e.g. the target response, the integration method, the method and parameters needed for nonlinearity iteration, and set $i = 1$.
2. Carry out a first time integration analysis, using Eq. (3) for nonlinearity iterations, and an integration step Δt obtained from Eq. (4),
3. Determine the peak target response.
4. If i is larger than one, continue to Step 6.
5. Set $i = i + 1$, repeat the last analysis with half steps, and return to Step 3.

6. If the peaks of the target response in the last two analyses are in a relative difference not more than 5%, accept the last response as final and stop.
7. Return to Step 5.

In view of the differences between the two procedures, and the following facts:

1. The nonlinearity tolerance is not discussed in NZS 1170.5:2004 [5, 6].
2. The possibility of the last *Stop* in Eq. (2) is less, for larger values of $\overline{\delta}$.
3. The integration steps of the first analyses of the two procedures are equal.
4. Comparing the relative difference of the last two peak target responses with 5% is part of both procedures.

the change proposed in 2015, and the new analysis procedure, seem capable of providing accurate target responses more efficiently. At least, for some of the analyses, using the procedure introduced in this section, instead of that stated in the previous section, may reduce the analysis run-time, without notable inaccuracy. With regard to the run-time reduction, the reason is the less number of time history analyses, and the final analysis' larger integration steps, when implementing the new procedure. And the good accuracy can be explained by the procedures' common ending criterion.

3 Numerical Investigation

3.1 Preliminary Notes

In view of the key role of NZS 1170.5:2004 [5, 6] in the discussion in this paper, the structural systems are considered subjected to ground motions, and (see [8]):

$$\mathbf{f}(t) = -\mathbf{M}\,\Gamma\,\ddot{u}_g \tag{5}$$

where, $\ddot{u}_g$ is the ground acceleration, and Γ is a vector, needed for matrix multiplication and considering spatial changes of $\ddot{u}_g$.

Material nonlinearity is inherent in earthquake engineering studies [1, 8]. Besides, χ has its intermediate value, when the nonlinearity is because of material [1]. Considering these, material nonlinearity is the nonlinearity considered in this section.

Evidently, the nonlinearity tolerance may affect the influence of replacing Eq. (2) with Eq. (3) on the analysis run-time and efficiency. Accordingly, in this study, the analyses are carried out with different values of nonlinearity tolerance, including the conventional value, i.e. 10^{-4} (for relative error) [9]. The maximum number of nonlinearity iterations will be set equal to the conventional values. The iteration method is set to fractional time stepping [10], considering the simplicity of fractional time stepping, the objective of this paper, and that by using fractional time stepping the run-time can be represented by the total number of integration steps [1]. Finally, the integration method is set to the average acceleration method of Newmark [2, 8, 9].

3.2 The Example

Figure 2(a) displays a preliminary model of a tall building, subjected to the ground acceleration displayed in Fig. 2(b), where g stands for the acceleration of gravity. The model's main properties are reviewed in Table 1 (the units are in S.I.).

Specifically, the stiffness is linear-elastic/perfectly-plastic (u_y is the yield displacement), and unloading is considered, and hence the behavior is potentially nonlinear. For the corresponding linear system, the natural frequencies are computed, leading to:

$$T_1 \cong 2.4477\,\text{sec}, \quad T_n \cong 0.1302\,\text{sec} \tag{6}$$

The rest of the unknowns in the right hand side of Eq. (4) are determined as:

$$\chi = 100, \quad \Delta t_{cr} = \Delta t_{CFL} = \infty, \quad {}_f\Delta t = 0.01\,\text{sec} \tag{7}$$

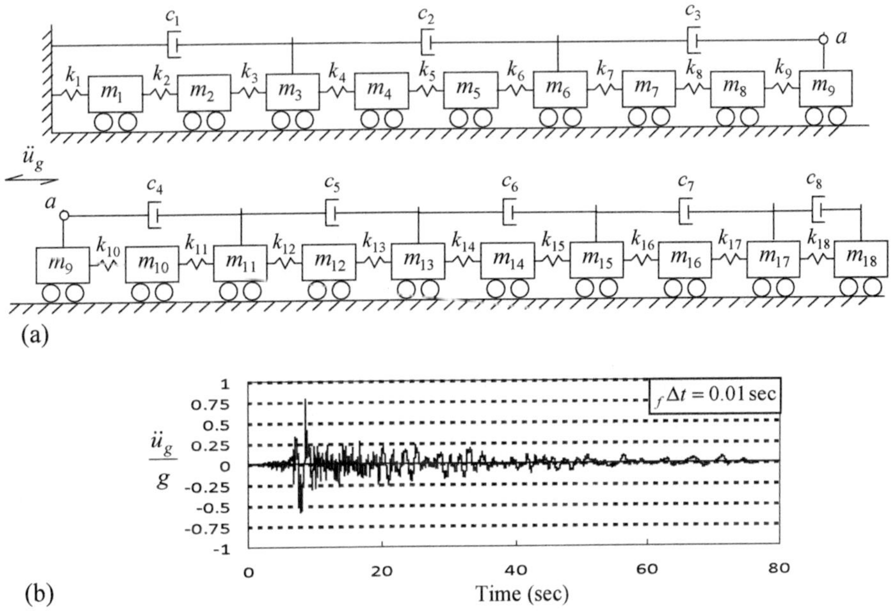

Fig. 2. The example under study: (a) structural model, (b) ground acceleration.

Table 1. Main properties of the structural system.

Property	i																	
	1	2	3	4	5	6	7	8	9	10	11	12	13	14	15	16	17	18
$10^{-9} \times m_i$	3	3	3	3	3	3	3	3	1.5	1.5	1.5	1.5	1.5	1.5	0.5	0.5	0.5	0.5
$10^{-12} \times k_i$	2	2	2	2	1.2	1.2	1.2	1.2	0.6	0.6	0.6	0.6	0.6	0.6	0.1	0.1	0.1	0.1
$10^{-8} \times c_i$	12	8	6	2.5	2.5	1.5	0.5	0.2										
$10^{2} \times u_{y_i}$	8	8	8	8	8	8	5	5	5	5	5	5	5	5	3	3	3	3

Consequently, in the time history analysis according to NZS 1170.5:2004 [5, 6], for the first analysis, the integration step equals 0.01 s. The top floor's lateral displacement is considered as the target response, for which, the exact history is displayed in Fig. 3. Figure 3 also displays that the behavior is considerably nonlinear and $\chi = 100$ (see Eq. (7)). The model is analyzed once with attention to the original procedure, stated in Sect. 1, and then, using the procedure stated in Sect. 2. The maximum number of nonlinearity iterations is set to 3 [10], i.e.

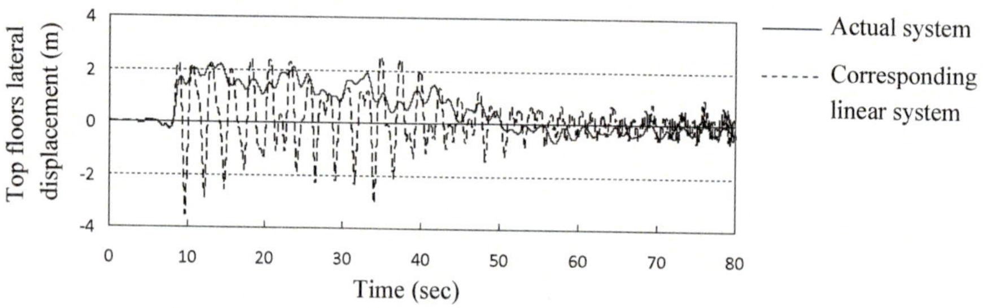

Fig. 3. History of the exact target response for the example under study.

$$\bar{k} = 3 \tag{8}$$

Each of the two analyses is carried out thrice, with the following three values of the nonlinearity tolerance:

$$\bar{\delta} = 10^{-2},\ 10^{-4},\ 10^{-6} \tag{9}$$

The study has led to Table 2 and Fig. 4. Evidently, in this problem, the change proposed in [3] can positively affect the efficiency of nonlinear time history analysis according to NZS 1170.5:2004 [5, 6]. The positive effect is especially notable when the nonlinearity tolerance is sufficiently small. This is in agreement with the claim stated in Sect. 2. An effort to explain and generalize the observations is presented next.

Table 2. Total numbers of integration steps in analysis of the example according to NZS 1170.5:2004 using the original/new procedures (with/without the change proposed in [3]).

Analysis	$\bar{\delta} = 10^{-2}$	$\bar{\delta} = 10^{-4}$	$\bar{\delta} = 10^{-6}$
Original procedure (without the change)	59128	111328	> 6.4E6
New procedure (with the change)	59128	38522	44046

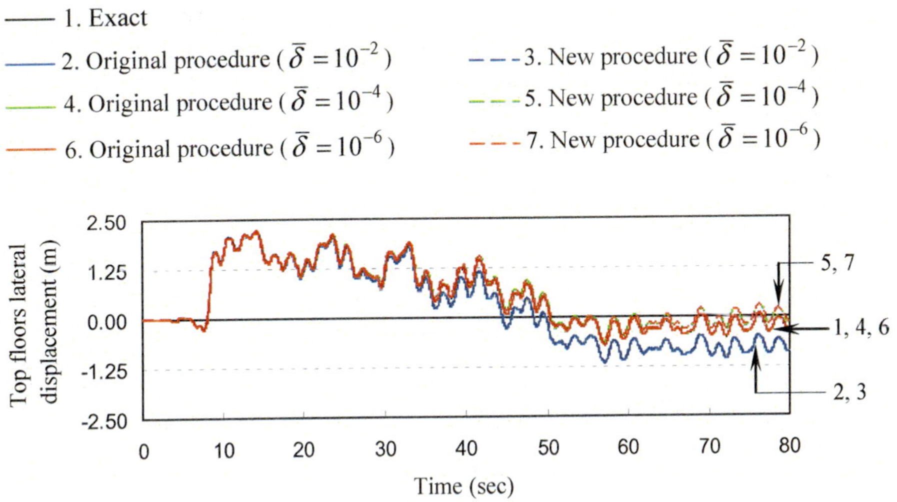

Fig. 4. The responses obtained from analysis according to NZS 1170.5:2004 using the original/new procedures (with/without the change proposed in [3]).

4 Discussion

Consider several nonlinear time history analyses of a structural system, each with a different value of nonlinearity tolerance. Evidently, for sufficiently large nonlinearity tolerances, the nonlinearity iteration in the original procedure of analysis ends with satisfying the second line of Eq. (2). In this case, the effect of the change proposed in [3] disappears. This is always valid, but the notion of sufficiently large nonlinearity tolerance depends on the problem, the integration method, and the iterative method. The result, for the example in Sect. 3, is the identical analyses according to the procedures in Sects. 1 and 2, when $\overline{\delta} = 10^{-2}$, as observed. For sufficiently small nonlinearity tolerances, the case is completely different. In this case, by implementing the proposed change, the response accuracy will be the accuracy that, depending on the available computational facilities (implied in $\overline{k}$), the integration step can provide. In this case, without implementing the change, the analysis stops before $t = t_{end}$ and an additional analysis starting from $t = 0$ is essential. This implies considerable additional run-time, without necessarily more accuracy, considering the two independent errors contributing the analysis, e.g. see the cases $\overline{\delta} = 10^{-4}$ and $\overline{\delta} = 10^{-6}$ in Sect. 3. To sum up, while for sufficiently large tolerances, using Eq. (3) instead of Eq. (2) does not change the analysis, for sufficiently small tolerances, the replacement can considerably improve the efficiency without notable inaccuracy.

5 Conclusions

The effect on computational efficiency of a change proposed in nonlinearity iterations [3] is tested in nonlinear time history analysis according to NZS 1170.5:2004 [5, 6]. The study is based on a numerical test and brief theoretical discussion. Consequently,

1. The change proposed in [3], may increase the efficiency of time history analysis according to NZS 1170.5:2004 [5, 6], without notable inaccuracy.
2. The increase in efficiency may be considerable, when $\overline{\delta}$ is sufficiently small.
3. The observations can be explained and generalized.

Finally, it is recommended to study more examples, specifically with different nonlinear behaviors, different integration and iterative methods, and different values of $\overline{k}$.

References

1. Soroushian, A.: Integration step size and its adequate selection in analysis of structural systems against earthquakes. In: Papadrakakis, M., Plevris, V., Lagaros, N.D. (eds.) Computational Methods in Earthquake Engineering. CMAS, vol. 44, pp. 285–328. Springer, Cham (2017). https://doi.org/10.1007/978-3-319-47798-5_10
2. Belytschko, T., Hughes, T.J.R.: Computational Methods for Transient Analysis. Elsevier, The Netherlands (1983)
3. Soroushian, A., Wriggers, P., Farjoodi, J.: From the notions of nonlinearity tolerances towards a deficiency in commercial Transient Analysis softwares and its solution. In: Papadrakakis, M., Papadopoulos, Plevris, V. (eds.) 5th International Conference On Computational Methods In Structural Dynamics and Earthquake Engineering 2015, Compdyn 2015, Crete, Greece, May 25–27 2015, vol. 1, pp. 1899–1907. National Technical University of Athens, Athens (2015)
4. Allgower, E.L., Georg, K.: Numerical Continuation Methods, an Introduction. Springer, USA (1980)
5. NZS 1170.5: Structural Design Actions Part 5: Earthquake Actions-New Zealand. Standards New Zealand, New Zealand (2004)
6. NZS 1170.5 Supp 1: Structural Design Actions Part 5: Earthquake Actions-New Zealand-Commentary. Standards New Zealand, New Zealand (2004)
7. Courant, R., Friedrichs, K., Lewy, H.: Über die partiellen differenzengleichungen der mathematischen physik. Math. Ann. **100**(1), 32–74 (1928)
8. Clough, R.W., Penzien, J.: Dynamics of Structures. McGraw-Hill, Singapore (1993)
9. SAP2000: User's Guide: A Structural Analysis Program for Static of Linear Systems (Educational Version). Computers and Structures Inc., USA (1999)
10. Nau, J.M.: Computation of inelastic spectra. ASCE J. Eng. Mech. **109**(1), 279–288 (1983)

DBM: Dynamics of Bridge Structures – Mathematical Modelling and Monitoring

Development of a Remote and Low-Cost Bridge Monitoring System

Airton B. S. Júnior, Gabriel E. Lage, Natália C. Caruso, Epaminondas Antonine, and Pedro H. C. Lyra(✉)

Instituto Mauá de Tecnologia, São Caetano do Sul SP, Brasil
{gabriel.lage,pedro.lyra}@maua.br

Abstract. Structures made from concrete, metal or any other material, are designed to have a long service life. However, to ensure a durable structure, it must have constant and efficient maintenance. Structural Health Monitoring (SHM) can be described as the method of monitoring and evaluating the structural health through the collection and analysis of data extracted from sensors, which are connected or not to the structure in the matter. In this way, the difficulties are in how to collect that data, analyze it and have information about it to give a final response about the structure status. This research focus at the development of a low-cost bridge management system, which will allow remote monitoring of special structures and shows their real-time status. Subjects as Finite Element Method (FEM), dynamic test and Internet of Things (IoT) are studied to create a viable system. The first step is modelling a case-study bridge on SAP2000 to analyze the theoretical behavior of the bridge. The second step is using low-cost devices to extract data with low noise, process it and send a response about the structure status, involving Wireless Sensor Networks. Then, experimental results are analyzed.

Keywords: SHM · Finite element analysis · IoT · Dynamic test

1 Introduction

To ensure a durable structure, it must have constant and efficient maintenance. However, due to the high cost of local inspection that involves sophisticated equipment and qualified professionals, there is a lack of maintenance of special structures, as bridges and viaducts, in Brazil.

Bolina and Pacheco [1] presents that the problems caused by vibration have become more perceptive in the last few years, which is a consequence of slimmer and more flexible structures under dynamic load. Therefore, it is necessary to carry out a dynamic analysis, considering the structure's natural frequency, which can be a good indicator of its status. When a structural system is loaded with an extreme force, disturbing its static balance, and then unloaded, it is noticeable that the system vibrates around its natural balance condition [2]. This kind of vibration is known as natural vibration, and every structure has one, which means, it has vibration modes and their corresponding

Z. Dimitrovová et al. (Eds.): WMVC 2022, MMS 125, pp. 117–124, 2023.
https://doi.org/10.1007/978-3-031-15758-5_11

frequencies that depend on the system parameters, as mass distribution, stiffness and support constraints [3].

As bridges are concerned, the vibration comes from the live loads that can be the wind or the car traffic on them. The live load on a special structure can cause resonance phenomenon which is very significant and might lead to a very high level of vibration, if the introduced frequency matches with the structure's natural frequency, causing serious problems or even dangerous accidents [4].

Certainly, the monitoring of bridges would have avoided the waste of time and money from governments or private companies that must fix damages in the structures due to the lack of inspection. Furthermore, it would have saved lives lost in accidents caused by decrepit structures [5].

For these reasons, the necessity of monitoring special structures has been increasing throughout the years. This research aims at the development of a low-cost bridge management system, which will allow a remote monitoring of the structure's real time status. Subjects as Finite Element Method (FEM), dynamic test for modelling calibration and Internet of Things (IoT) have been studied to develop the system proposed.

2 Methodology

2.1 Object of Study

In order to create the structure monitoring system, a case-study bridge was selected to be used as a reference for analysis, tests and validations of what is proposed. Therefore, it is necessary to understand its main characteristics. The bridge is located on the 13th quilometer of the Anchieta highway, São Paulo, Brazil. With the project provided by responsible dealership of the bridge, it was possible to identify its elements, such as girders, cross girders, crossbeams, foundation and elastomeric support devices. Figure 1 shows the cross-section of the studied bridge. The bridge consists of 4 longitudinal central girders, 2 longitudinal lateral girders, 2 transversal lateral beams, 1 slab and 6 foundation piles connected to the super-structure by 12 elastomeric pads.

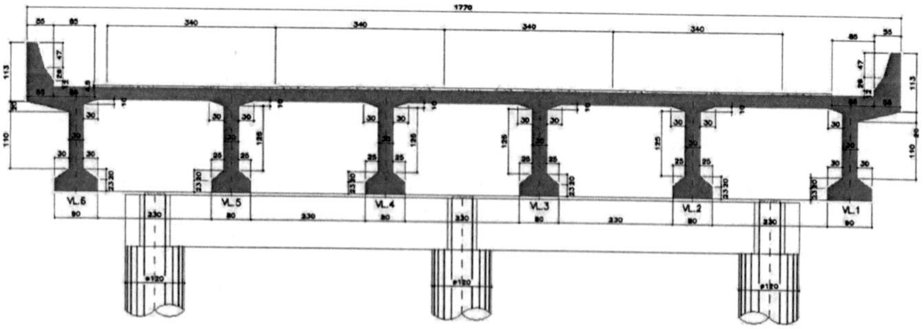

Fig. 1. Cross-section of the analysed bridge (dimensions in centimeters).

2.2 Finite Element Model

The first step was modelling the bridge on SAP2000, which is a software used to perform the structural and dynamic analysis of the bridge. Also, it is a Finite Element program, and the structure was introduced through its 3D interface, where the bridge elements are represented by bars and shell elements. These bars and shells receive the physical and geometric properties of the structural element they correspond to in the model, such as area and inertia. The bridge is made from C30 concrete (compression strength of 30 MPa [6]), apart for the support device, an elastomer that behaves similarly to rubber. Material properties were also considered in the model. Girder self-weight (38,1 kN/m), slab self-weight (25 kN/m^3), pavement (24 kN/m^3), and wheel guards (7,85 kN/m^3 applied for a width of 1 m) are placed on the shell element that represents the slab. The self-weight load is inserted per element to avoid interference. When the model is completed, it must be processed to analyze the results and thus proceed to the next steps of the work.

2.3 BIM Model

The bridge was also modeled in BIM, using Revit at first, an Autodesk® software. For the time being, the main elements of the structure were modeled, such as foundation, girders, cross girders, support devices, deck and the wheel guard. As the angles do not form exactly 90°, it was necessary to create specific families for the bridge beams and for the wheel guard. With the bridge already modeled, InfraWorks software was used to implement the model in its surroundings, importing the area from Google Earth, selecting the desired area and inserting the bridge in its corresponding geographical position. Figure 2 represents the BIM model of the bridge from both Autodesk® softwares Revit (Fig. 2a) and InfraWorks (Fig. 2b).

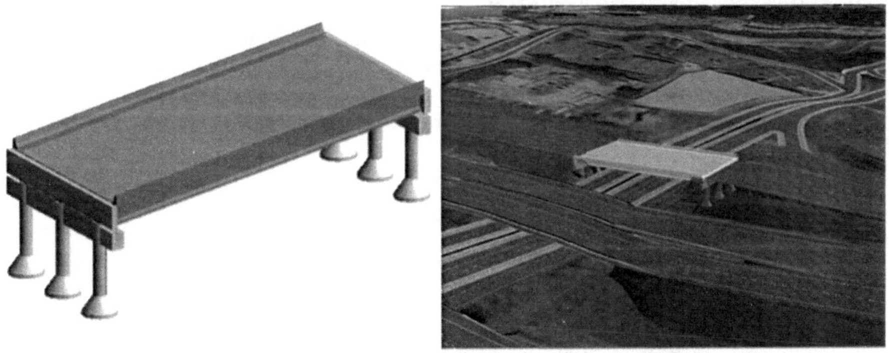

(a) (b)

Fig. 2. BIM models of the bridge using Autodesk® softwares. (a) Revit (b) Infraworks.

2.4 Proposed Monitoring System

Briefly, the challenge of this analysis was the development of a structure monitoring system that allows the company responsible for it to identify possible damage to the structure in real time, allowing actions to be taken to prevent further damage, seeking the lowest cost possible. In this way, the idealized system works by verifying if there has been a significant change in the dynamic behavior of the structure under analysis and issuing an alert if positive.

The use of a structure monitoring system is important to manage the bridge. With the correct parameters inserted in the virtual BIM model, it is possible to extract fundamental specifications that, when linked to the information obtained by the monitoring system, allow to identify the structure's behavior in certain situations. In addition, forecasting and scheduling dates for visual inspection and possible maintenance becomes more assertive.

Regarding the identification of the dynamic behavior of the structure, accelerometers were used to capture vibrations at critical points previously stipulated according to the geometric characteristics of the bridge. In this case, since it is a bi-supported bridge, it is understood that the largest deflections occur in the middle of the span, which makes this region more susceptible to vibrations. For the identification of the vibration modes, if desired, three devices capable of obtaining acceleration data in the middle of the longitudinal span are installed at three different points of the cross section: one device more to the left, one central and one more to the right.

As the proposal is to install three different devices to capture acceleration data, it is necessary to have a fourth device capable of receiving information from the other three, processing it and then sending it to the monitoring center. To send the information from the acceleration devices to the central device, it was decided to add a radio frequency module to the devices, allowing the sending of signals between them.

There will be a monitoring center and for it to receive data from the central device, it was determined that communication would be via 4G mobile network, since it allows a high rate of data transmission over long distances, which provides real-time monitoring getting all the desired information about the state of the devices.

Therefore, there is a final dashboard in the monitoring center, which allows those responsible for the bridge to analyze the data and results obtained, defining actions to be taken in case of alerts. Figure 3 represents the position of the installed devices on the bridge and the system's communication flow, from the accelerometers to the final dashboard.

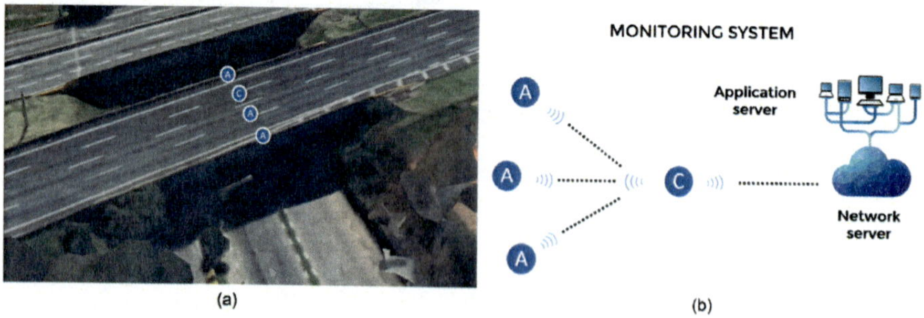

Fig. 3. Scheme of the monitoring system. (a) position of devices on the bridge (b) system's communication flow.

3 Results and Discussion

Up to the present time, it has been possible to advance in parts in the development of the proposed monitoring system, as well as in the computational modeling of the object of study, both in BIM modeling and Finite Element modeling. BIM model used softwares such as Revit and Infraworks, as presented at the methodology section. Figure 4 shows the FEM analysis developed through SAP2000. In addition, part of the algorithms that will be used in the final monitoring system has also been developed, as well as field experiments for collecting and analyzing bridge data, in order to obtain results that could demonstrate relationships with the numerical model and generate insights for other approaches to the problem.

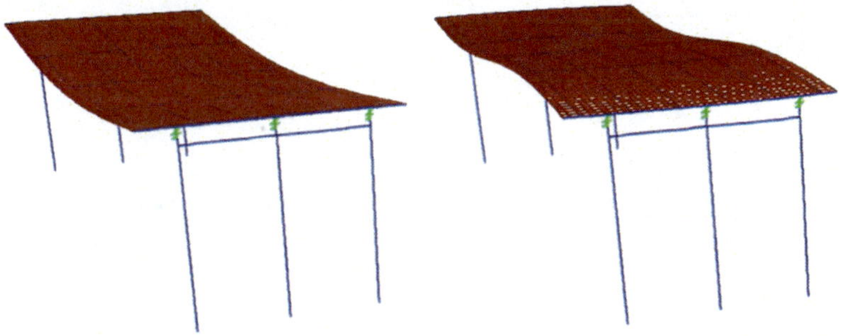

Fig. 4. Example of bridge's modes of vibration through SAP2000.

Two initial prototypes were developed, one being the acceleration measuring device (A) and the other the central device (C) as shown in Fig. 3. Device A contains a microcontroller Arduino NANO and device C contains both an Arduino UNO and a microcomputer Raspberry Pi 4B. At the beginning of the prototype's development, the basic idea was to start developing the programs of the devices for capturing accelerations, analyzing the data and sending data to the cloud.

To obtain data and understand the usual behavior of the studied bridge, an in-loco test was performed, using the device developed by Oliveira [7]. Oliveira's device works in a similar way to device A that is being developed in the present paper. There is an accelerometer (MMA8451) connected to a microcontroller (ESP32). A real-time clock (RTC), a SD card read/write module and a lithium battery are also connected to the microcontroller. Thus, it is a remote device that allows the collection of acceleration data and recording of that data on the SD card for later analysis with the aid of a computer, presenting extreme ease for its use. The test lasted one hour and three minutes and was carried out under normal traffic conditions on the bridge. Two devices were used to capture acceleration, which were positioned in the middle of the bridge span, close to the wheel guard.

While the sensors measured the accelerations on the bridge, the exact times of the moments when trucks produced perceptible vibrations were noted for later checking together with the data recorded by the accelerometers. During the test period, eighteen sections of truck passing were noted. A program was developed to calculate the Fast

Fourier Transform (FFT) of the test analysis sections and save a general report with graphs and results obtained.

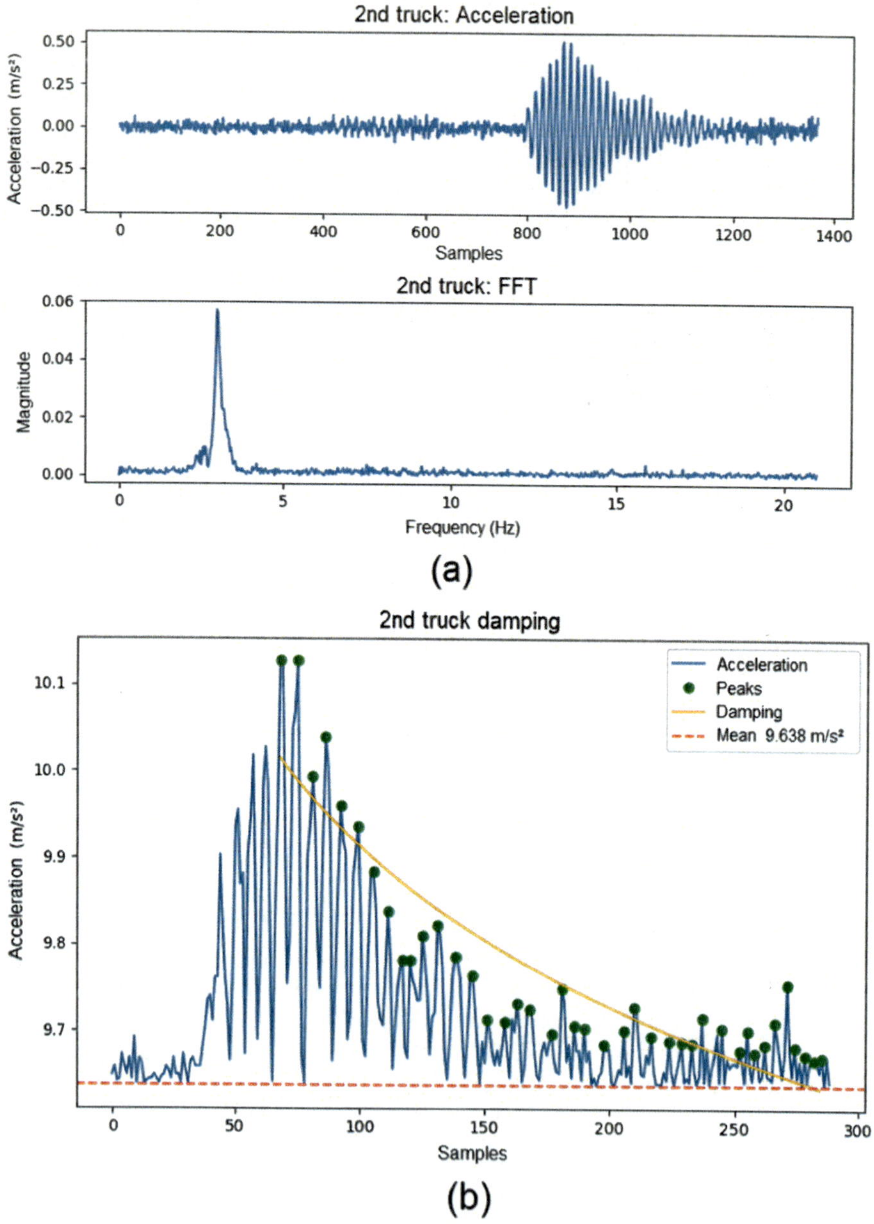

Fig. 5. Results obtained for the second truck. (a) accelerations and frequency (b) damping analysis.

It is noted that the vibration was visibly identified by the devices. Thus, through the FFT, a peak frequency of 2.903 Hz was found, which represents the natural frequency of the bridge for the vibration mode at the respective moment when the truck passed it. In this way, it is intended to adjust the numerical model in order to find a natural frequency close to that obtained experimentally for the vibration mode in question.

Another point of analysis was the possibility of visualizing the bridge's damping in this same case of truck. Therefore, this case is still being analyzed individually in order to obtain other relevant information, such as the bridge's damping rate. Figure 5a shows the measured acceleration data and Fig. 5b shows a linear regression model that was performed to adjust the damping curve.

4 Conclusion

As the research has demonstrated, great part of the computational modelling and the monitoring system of bridges have been developed. Some of the algorithms that will be used in the final system have already been created, also, a partial test using a device to collect accelerations took place on the bridge in order to guide the further steps of the research. The FE model in SAP2000 needs to be adjusted to be as precise as possible comparing to the real bridge case. To do so, a dynamic test will take place on the bridge, thus, the necessary adjustments will be done in the model to achieve the experimental frequency.

The results so far, from the partial test, show visibly the vibration of the bridge, from which the structure's natural frequency was obtained through the Fast Fourier Transformation (FFT). Therefore, the device that will be used in the monitoring system is able to capture a clear signal of the vibration, allowing a good analysis of the frequencies. On top of that, the results also showed that the bridge damping can be detected by the devices, permitting the use of damping rates in the criteria that will be used to decide whether there were changes on the dynamic behavior of the bridge.

Finally, even though there are only partial tests and results yet, they are very promising in leading the research to its objective, the development of a low-cost bridge monitoring system that is precise and efficient.

References

1. Bolina, C.C., Pacheco, E.U.L.: Vibrações: As Frequências Naturais Estimada e Experimental de uma Estrutura. Pontifica Universidade Católica de Goiás, Goiás (2014)
2. Clough, R.W., Penzien, J.: Dynamics of Structures, 3rd edn. Computers & Structures Inc, Berkeley, Ca (2003)
3. Soriano, H.L.: Introdução à dinâmica das estruturas, 1st edn. Elsevier, Rio de Janeiro (2014)
4. Moutinho, C.M.R.: Controlo de vibrações em estruturas em engenharia civil. Porto (2007)
5. Colombo, A.B.: Applications of Structural Health Monitoring and Field Testing Techniques to Probabilistic Based Llife-Cycle Evaluation of Reinforced Concrete Bridges. In: 2016. 127 f. Tese (Doutorado) - Curso de Engenharia Civil. Universidade de São Paulo, São Paulo (2016)

6. Junior, S., et al.: Desenvolvimento de um sistema de baixo custo para monitoramento de Obras de Arte Especiais. In: 2020. 121 f. TCC (Graduação) - Curso de Engenharia Civil. Instituto Mauá de Tecnologia, São Caetano do Sul (2020)
7. Oliveira, C.C., et al.: Desenvolvimento e validação de instrumento medidor de vibrações com ênfase em estruturas da construção civil. In: 2019. 197 f. TCC (Graduação) - Curso de Engenharia Civil. Instituto Mauá de Tecnologia, São Caetano do Sul (2019)

Fractional Mass-Spring-Damper System Described by Conformable Fractional Differential Transform Method

Basem Ajarmah(✉)

Al-Istiqlal University, Jericho, Palestine
basem@pass.ps

Abstract. This paper proposes analytical solutions of the mass-spring-damper systems described by conformable fractional differential transform method. Conformable fractional transform method is based on new relation between fractional calculus and calculus based on basic limit definition were used. The behaviors of the analytical solutions of the mass-spring-damper systems described by conformable fractional transform method were represented analytically and graphically. Then the effect of the orders of the fractional derivative on the system was analyzed. The results obtained by this approach provide new explanation on the importance of fractional calculus on mechanical systems and showed clearly the amplitude of steady state solution depends on time; with contrary to what is known.

Keywords: Mass-spring-damper · Conformable fractional differential transform calculus · Fractional derivative · Analytical fractional · Numerical solutions · Steady state

1 Introduction

In recent years, there has been an increase in attempts to establish a definition or method agreed between mathematicians for determining the general definition of fractional derivative operator [1–3]. Two trends have emerged in the consideration of fractional calculus operators. Firstly, there exists a desire to explore and create new definitions and models for fractional integral and differential operators. Secondly, there exists a desire to impose criteria and strict definitions for what we call a "fractional derivative" or "fractional integral": which operators between functions should be named as such and which should not [4, 5]. Therefore, you will find many published definitions and methods scattered throughout various articles [4] and book [5] all of them are mainly depend on the basics of the fractional order definition of the derivative. These differences in definitions makes it important in diversify studies of the same model or problem and compare more results. Therefore, researchers and mathematicians find that they re-examine the same model multiple times with different definitions, methods, or algorithms. If look closely at the results, you will find that they all lead to the same idea which is the importance of the fractional derivative for finding or discovering new aspects in any model applied to it.

Z. Dimitrovová et al. (Eds.): WMVC 2022, MMS 125, pp. 125–132, 2023.
https://doi.org/10.1007/978-3-031-15758-5_12

Because these new approaches make them easier to understand the results of experiments and their success convergence to mathematical models.

In this study, we revisited the most important physical-mechanical vibration model the Mass-Spring-Damper [6], that is very widely used in all engineering aspects, physics, mechanics, science and engineering, bio-mathematics, biology, finance and economics. The interest in vibrations stems from our interest in predicting the response of structures or mechanical systems to external forces; using fractional derivative in an attempt to understand other trends in this model. we used a relatively new and recent definition in mathematics called conformable fractional derivative (CFD) [7–10] which is based on the basic limit definition of the derivative.

The CFD definitions of fractional derivative appear that they satisfy the linear property. However, Khalil et al. [9], have shown that there are some drawbacks in the characteristics of the definitions of the Riemann-Liouville and Caputo. They [9] use the chain rule definition in proving all the properties of fractional derivation which was applied to the base of the series. This definition in Eq. (1) is a natural extension of the usual derivatives, and it achieves the general characteristics of the fractional derivation for $f : [0, \infty) \rightarrow \Re$. The CDF of f with a fractional degree γ is given as follows [8]:

$$f^{\gamma}(t) = \lim_{\tau \to 0} \frac{f\left(t + \tau t^{1-\gamma}\right) - f(t)}{\tau}, \forall t > 0, 0 < \gamma < 1 \tag{1}$$

that implies

$$f^{\gamma}(t) = D^{\gamma} f(t) = t^{1-\gamma} \frac{d}{dt}(f(t)) \tag{2}$$

The paper is organized as follows. Section 2 presents the adopted definition and problem formulation. Sections 3 proposed model results and discussion and Sect. 4 outlines some brief remarks conclusion.

2 Problem Formulation for Mass-Spring Damper System

The mass-spring-damper model consists of discrete mass nodes distributed throughout an object and interconnected via a network of springs and dampers. This model is well-suited for modelling object with complex material properties such as nonlinearity and viscoelasticity. We obtain the analytical and numerical solutions for the forced viscoelastic fractional mass-spring-damper equation described by the CFD. We consider the case of forced viscoelastic fractional mass-spring damper Eqs. (1), (2) represented by CFD [8, 11, 12]

$$D^2 x(t) + 2\xi \omega_n D^{\gamma} x(t) + \omega_n^2 x(t) = F_{ext}/m \tag{3}$$

where $\omega_n = \sqrt{\frac{k}{m}}$; $\xi = \frac{c}{2m\omega_n}$; $\mathrm{U} = \frac{1}{2}kx^2$ Such that m is the mass connected to the spring-dumber, ω_n is the un-damped natural frequency and ξ represents the damping ratio, and k represents the spring coefficient. The mass m is displaced from its equilibrium position, then it vibrates by the orthogonal external harmonic force. Also, assume that the external force is in the form of sinusoidal which means it is changing at certain periodicity w, and amplitude $F_0 =$ constant and $F_{ext} = F_0 \cos(\omega t)$.

Assume the solution of Eq. (3) is

$$x(t) = Acos(\omega t - \phi) \tag{4}$$

so

$$x(t) = A_s cos(\omega t) + B_s sin(\omega t) \tag{5}$$

where A is the amplitude of the motion, ϕ is called a phase shift, $A_s = Acos(\phi)$, $B_s = Asin(\phi)$, so $A = \sqrt{A_s^2 + B_s^2}$, and $\phi = tan^{-1}\left(\frac{B_s}{A_s}\right)$. Substitute Eq. (5) in Eq. (3) using

$$Dx(t) = -A_s \omega sin(\omega t) + \omega B_s cos(\omega t) \tag{6}$$

$$D^2 x(t) = -A_s \omega^2 cos(\omega t) - \omega^2 B_s sin(\omega t) \tag{7}$$

And from CDF using Eq. (2) we find $D^\gamma x(t)$ as

$$D^\gamma x(t) = -A_s \omega t^{1-\gamma} sin(\omega t) + \omega B_s t^{1-\gamma} cos(\omega t) \tag{8}$$

we get

$$\begin{aligned} &-\omega^2 A_s cos(\omega t) - \omega^2 B_s sin(\omega t) + 2\xi \omega_n \left(-A_s \omega t^{1-\gamma} sin(\omega t) + \omega B_s t^{1-\gamma} cos(\omega t)\right) + \\ &\omega_n^2 (A_s cos(\omega t) + B_s sin(\omega t)) = \frac{F_0}{m} \cos(\omega t) \end{aligned} \tag{9}$$

From Eq. (9) we find

$$-\omega^2 A_s + 2\xi \omega_n \omega B_s t^{1-\gamma} + \omega_n^2 A_s = \frac{F_0}{m} \tag{10}$$

$$-\omega^2 B_s - 2\xi \omega_n \omega A_s t^{1-\gamma} + \omega_n^2 B_s = 0 \tag{11}$$

Equation (10) and Eq. (11) is indeed represent the solution for classical spring-damper system (simple harmonic motion). That to say using ordinary derivative when $\gamma = 1$.

From Eq. (10) and Eq. (11) we conclude that

$$A_s = \frac{\omega_n^2 - \omega^2}{(\omega_n^2 - \omega^2)^2 + \left(2\xi \omega_n \omega t^{1-\gamma}\right)^2} \frac{F_0}{m} \tag{12}$$

$$B_s = \frac{2\xi \omega_n \omega t^{1-\gamma}}{(\omega_n^2 - \omega^2)^2 + \left(2\xi \omega_n \omega t^{1-\gamma}\right)^2} \frac{F_0}{m} \tag{13}$$

so

$$A = \frac{F_0}{m}\sqrt{\frac{1}{(\omega_n^2 - \omega^2)^2 + \left(2\xi\omega_n\omega t^{1-\gamma}\right)^2}} \tag{14}$$

$$\phi = tan^{-1}\left(\frac{2\xi\omega_n\omega t^{1-\gamma}}{\omega_n^2 - \omega^2}\right) \tag{15}$$

this implies

$$x(t) = \frac{F_0}{m}\sqrt{\frac{1}{(\omega_n^2 - \omega^2)^2 + \left(2\xi\omega_n\omega t^{1-\gamma}\right)^2}}cos\left(\omega t - tan^{-1}\left(\frac{2\xi\omega_n\omega t^{1-\gamma}}{\omega_n^2 - \omega^2}\right)\right) \tag{16}$$

The steady state solution of Eq. (14) is indeed depends on time and fractional power ($t^{1-\gamma}$) as in Eq. (17)

$$A = \frac{F_0}{m\omega_n}\sqrt{\frac{1}{\left(1 - \left(\frac{\omega}{\omega_n}\right)^2\right)^2 + \left(2\xi\frac{\omega}{\omega_n}t^{1-\gamma}\right)^2}} \tag{17}$$

When fractional derivative order $\gamma \approx 1$ the amplitude A and phase ϕ did not depend on time as in the classical analytic ordinary solution *x(t)*. so the classical amplitude A are a function with respect to frequency ratio $\beta = \frac{\omega}{\omega_n}$ and damping ratio ξ. Therefore, $A\ (\beta, \zeta)$ depends in two parameter only but in fractional spring–mass system of order derivative γ the amplitude $A\ (\beta, \zeta, t, \gamma)$ depends on four parameters β, ζ, t, and γ which provide more ranges and aspects of solutions that may fit better the experimental data.

The importance of CFD provides analytical solutions is that its simpler and lesser calculations than other definitions approaches of fractional derivative like Riemann-Liouville, Caputo, Grünwald-Letnikov, Weyl, Marchaud, Hadamard, Chen, Davidson-Essex, Coimbra, Canavati, Riesz, Cossar, Yang, Osler, and Hilfer [6]. Because These methods require more complex calculations and huge computational powers than conformable fractional derivative.

3 Results and Analysis

The particular numerical solutions of fractional spring–mass system *x(t)* Eq. (16) and amplitude A Eq. (17) that depends on β, ζ, t, γ are given in Fig. 1 below, we assume different parameters $\zeta = \{0.05, 0.125, 0.25, 0.50, 1.0, 1.772\}$, $\gamma = [0.1, 0.5, 0.99]$, $m = 10$, $k = 10$, $F_0 = 10$ and different t = {0.9, 4.5, 8.1}.

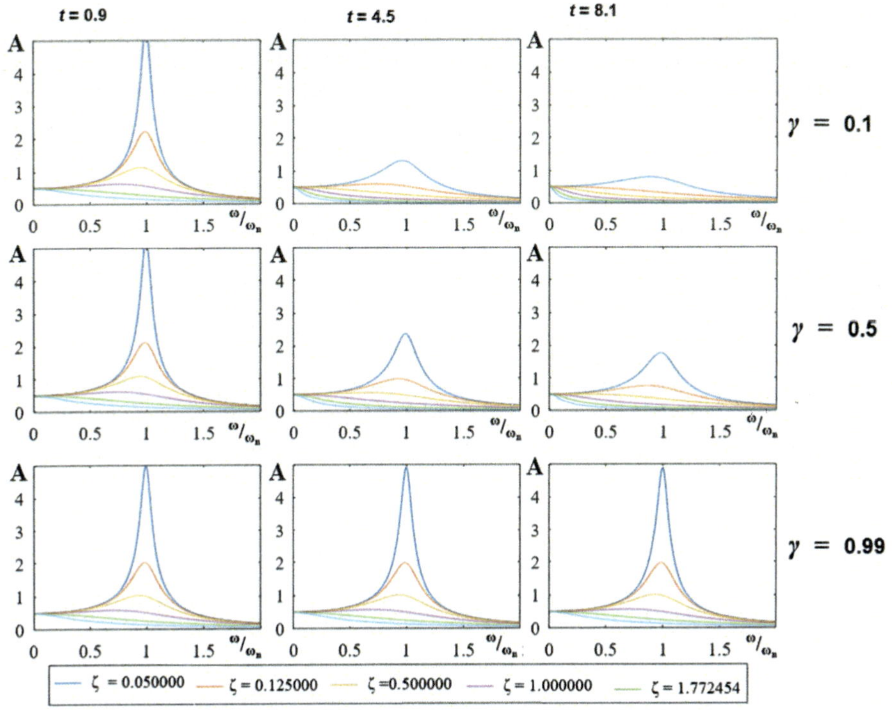

Fig. 1. Amplitude vs frequency ratio $\frac{\omega}{\omega_n}$ of fractional damped vibration of a force spring-mass with different parameters $\zeta = \{0.05, 0.125, 0.25, 0.50, 1.0, 1.772\}$, $\gamma = [0.1, 0.5, 0.99]$, m = 10, k = 10, $F_0 = 10$ and different time t = {0.9, 4.5, 8.1}.

In Fig. 1 its clearly shown the most important observation does not agree with what is known about the steady state amplitude, because the solution of the Eq. (17) in our model depends inversely proportional on time to the power of (1-γ); which means the system reacts in a changing manner with time. Also, a stationary state amplitude classically occurs by Eq. (17) only if the time is one, even if the value of the fractional derivative γ in the region which has changed from zero to one; or when the fractional parameter is one. Therefore, it is necessary to take into consideration the initial conditions when designing any model. Otherwise, they design system when the time equal unity.

In Fig. 2. We observe the following behaviors of system when using CFD fractional model on mass spring damping vibrations.

- If time $t < 1$ the maximum amplitude vs frequency ratio will decrease when fractional parameter γ increased from 0 to 1 until reach steady state solutions.
- If time $t > 1$ the maximum amplitude vs frequency ratio will increase when fractional parameter γ increased from 0 to 1 until reach steady state solutions.
- The maximum amplitude vs frequency ratio always decreases with time.
- When $t = 1$ the maximum amplitude vs frequency reach the steady state solution.

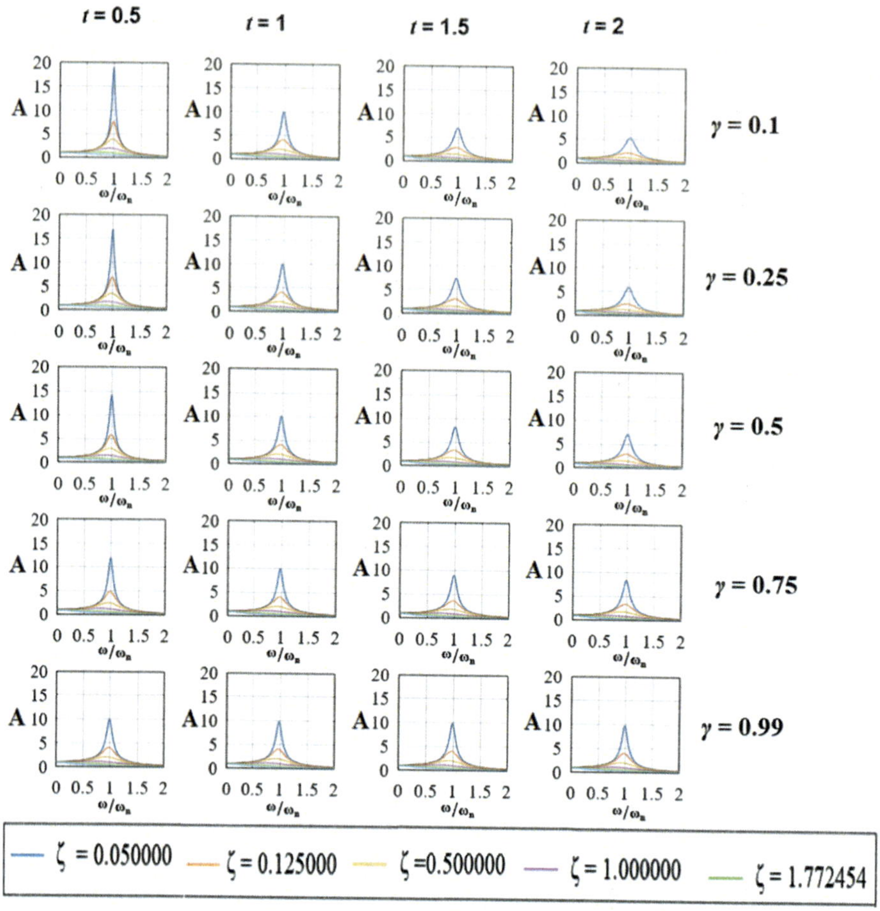

Fig. 2. Amplitude vs frequency ratio $\frac{\omega}{\omega_n}$ of fractional damped vibration of a force spring-mass with different parameters $\zeta = \{0.05, 0.125, 0.25, 0.50, 1.0, 1.772\}$, $\gamma = [0.1, 0.25, 0.5, 0.75, 0.99]$, m = 1kg, k = 1N/m, F_0 = 1600N and different time t = {0.5, 1, 1.5, 2}.

As in Fig. 2. The second column (when time = 1) and the last row (when $\gamma = 0.99 \approx 1$) are represents the steady state solution of the system; actually equal to each other.

Another observation from the Fig. 3. Above in this study shown that each time effect the response and resonance of mass spring fractional damper [12]. Actually where the resonance will start under different conditions of fractional derivative and damping ratio.

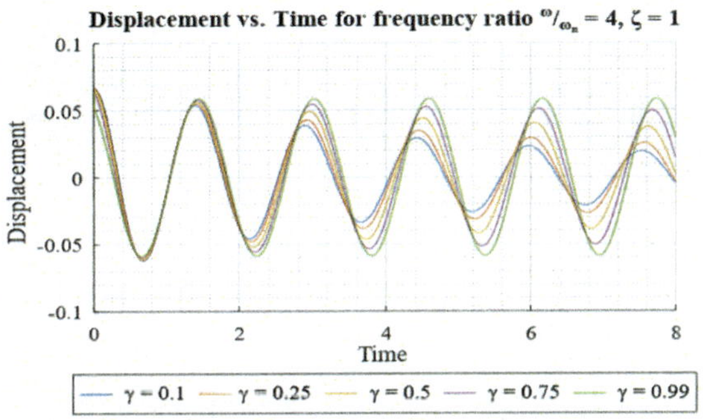

Fig. 3. Displacement vs. Time with parameters frequency ratio $\frac{\omega}{\omega_n} = 4$, $\zeta = 1$, $\gamma = [0.1, 0.25, 0.5, 0.75, 0.99]$, m = 10, k = 10, and $F_0 = 10$.

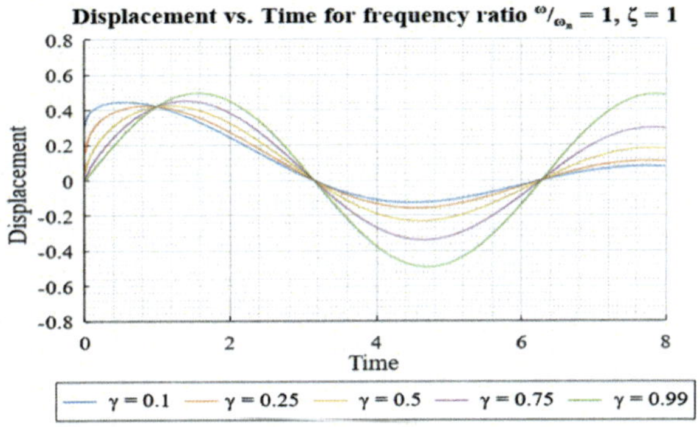

Fig. 4. Displacement vs. Time with parameters frequency ratio $\frac{\omega}{\omega_n} = 1$, $\zeta = 1$, $\gamma = [0.1, 0.25, 0.5, 0.75, 0.99]$, m = 10, k = 10, and F0 = 10.

These results in Fig. 4. Agree with [6, 11] which stat that the different γ values are modifies the time constant of the systems and exhibit fractional structures (components that show an intermediate behavior between a system that is conservative and dissipative).

4 Conclusion

The problem of mass-spring damper vibrations described by Conformable Fractional Differential transform method in general is one of the most important topics in the field of mechanical engineering, because it is prevalent in several areas, including bridges, buildings, cars, aircraft, kinetic machines, electron microscopes, etc. Conformable fractional transform method is based on new relation between fractional calculus and calculus based on basic limit definition were used. The behaviors of the analytical solutions of the mass-spring-damper systems described by conformable fractional transform method were represented analytically and graphically. Then the effect of the orders of the fractional derivative on the system shows the importance of initial condition on the model. The results obtained by this study provide that the amplitude of steady state solution is dependent on time when use different fractional parameter.

Acknowledgments. The research was supported by the AL-Istiqlal university. No external funding.

References

1. Valério, D., Trujillo, J.J., Rivero, M., Machado, J.A.T., Baleanu, D.: Fractional calculus: A survey of useful formulas. The European Physical Journal Special Topics **222**(8), 1827–1846 (2013). https://doi.org/10.1140/epjst/e2013-01967-y
2. de Oliveira, E.C., Machado, J.T.: A review of definitions for fractional derivatives and integral. Mathematical Problems in Engineering, 6 (2014). Article ID 238459
3. Miller, K.S., Ross, B.: An Introduction to the Fractional Calculus and Fractional Differential Equations. John Wiley & Sons, New York, NY, USA (1993)
4. Grigoletto, E.C., de Oliveira, E.C.: Fractional versions of the fundamental theorem of calculus. Appl. Math. **4**, 23–33 (2013)
5. Baleanu, D., Diethelm, K., Scalas, E., Trujillo J.: Fractional Calculus: Models and Numerical Methods, vol 3 of Series on Complexity, Nonlinearity and Chaos. World Scientific, Singapore (2012)
6. Gómez-Aguilar, J.F., Yépez-Martínez, H., Calderón-Ramón, C., Cruz-Orduña, I., Eecobar-Jiménez, R.F., Olivares-Peregrino, V.H.: Modeling of a mass-spring-damper system by fractional derivatives with and without a singular kernel. Entropy **17**, 6289–6303 (2015)
7. Ünal, E., Gökdoğan, A.: Solution of conformable fractional ordinary differential equations via differential transform method. Optik **128**, 264–273 (2017)
8. Mamat, M., Syouri, S, Alghrouz, I.M., Sulaiman, I.M., Sufahani, S.F.: Conformable fractional differential transform method for solving fractional derivatives. Int. J. Adv. Sci. Technol. **29**(7), 1734–1743 (2020)
9. Roshdi, K., Yousef, H.M., Sababheh, M.: A new Definition of Fractional Derivative. Comput. Appl. Math. **246**, 65–70 (2014)
10. Syouri, S., Mamat, M., Alghrouz, I.M., Sulaiman, I.M.: Conformable Fractional Differintegral. Int. J. Sci. Technol. Res. **9**(3), 292–295 (2020)
11. Sene, N., Gómez-Aguilar, J.F.: Fractional mass-spring-damper system described by generalized fractional order derivatives. Fractal and Fractional **3**(39) (2019)
12. Ray, S.S., Sahoo, S., Das, S.K.: Formulation and solutions of fractional continuously variable order mass–spring–damper systems controlled by viscoelastic and viscous–viscoelastic dampers. Advances in Mechanical Engineering **8** (2015)

Pushover Analysis Accounting for Torsional Dynamic Amplifications for Pile-Supported Wharves

Enrico Zacchei[1,2](✉), Pedro H. C. Lyra[3,4], and Fernando R. Stucchi[4,5]

[1] Itecons, Coimbra, Portugal
enricozacchei@gmail.com
[2] University of Coimbra, CERIS, Coimbra, Portugal
[3] Mauá Institute of Technology (MIT), 1 Mauá Square, Mauá, São Caetano Do Sul, São Paulo, Brazil
[4] Polytechnic School of São Paulo, University of São Paulo (USP), 380 Prof. Luciano Gualberto Avenue, São Paulo-SP 05508-010, Brazil
[5] EGT Engineering Ltda, Fábia Avenue, São Paulo 05051-030, Brazil

Abstract. In this paper the pushover analysis for pile-supported wharf structures (PSWSs) is carried out. This nonlinear static analysis is a common method to estimate the optimum displacement demand for seismic design of constructions. The main aspect of this study is to account for the dynamic magnification factors due to torsional effects. 2D and 3D models by finite element method (FEM) were computed considering the seismic loadings, modal analysis, and the lateral stiffnesses imposed to consider the soil-piles interactions. The seismic design of the piles consists in accounting for the formation of plastic hinges. The purpose is to determine, in a deterministic way, the performance points of the structure in terms of base shears and displacements, then, in a stochastic analysis, to estimate the probability of exceedance of horizontal displacements of a semi-flexible model with respect to a flexible model. The case study is a real pile-supported wharf in ports placed in a high seismicity area, in this sense, this paper should provide some useful technical data. The advantage of using the nonlinear pushover analysis in terms of displacements was estimated about 30%.

Keywords: PSWS · Pushover analysis · Reliability analysis · Numerical analysis

1 Introduction

Pile-supported wharf structures (PSWSs) are large-scale constructions placed in ports. Their design regards different area, as structural, geotechnical, hydraulic engineering, for this, these constructions can attract great interest for researchers and engineers.

However, in literature there are not many published studies probably because the construction of ports is not frequent than other structures since they are more expensive, and they need large periods for buildings.

In this paper the pushover analyses applied on a real PSWS have been carried out in a deterministic and stochastic way.

Z. Dimitrovová et al. (Eds.): WMVC 2022, MMS 125, pp. 133–141, 2023.
https://doi.org/10.1007/978-3-031-15758-5_13

Regarding this specific analysis, in literature there are some studies. In [1], it is identified the damage on piles and the bound limits of seismic demands for estimation the fragility curves. In [2], a parametric study to carried out the performance of PSWSs under liquefaction and lateral spreading soil conditions was carried out.

In [3, 4] they are estimated seismic fragility curves, the limits states by considering the displacement ductility factors, and the effects of geotechnical uncertainties by probabilistic methods. In [5] new damping equations for PSWSs based on the total energy dissipation of soil and structure are proposed. The contribution of soil-pile interactions on system damping are also studied. In [6] seismic performances for transient and full liquefaction conditions were carried out. For this, new displacement demand models were proposed. Finally, in [7] new methods to analyse seismic performances of structures beyond the limit of force balances were proposed.

All these studies treated important aspects highlighting two ones: (i) the soil-pile interaction considering liquefaction effects [2, 6] and (ii) the estimation of fragility curves to evaluate the structural vulnerability [1, 3, 4].

From previous studies [8, 9] on the pushover analyses for PSWSs, in this paper some data have been collected to carry out a probabilistic analysis.

The focus is to quantify the non-linear response of a semi-flexible model with respect a flexible model. This is to reduce the difficulty in estimating soil effects and torsional dynamic amplifications.

2 Case Study

A real PSWS of a port placed at northern of Venezuela is the case study in this paper. This area has as very high seismic hazard [10]. The structure has an area of 6520.0 m^2, i.e., 163.0 (parallel to the sea) × 40.0 m. The deck of the wharf is formed by reinforced transversal beams of 1.95 m high, and a concrete slab with a transversal slope of 0.5% and an average thickness of 0.85 m. The deck average thickness is 1.254 m. The wharf structure is supported on 6 rows of vertical piles with 36 total piles of length of 8.0 m/pile connected to the concrete deck.

Each pile has a metal jacket to prevent the corrosion of steel bars and chemical attacks in marine/tidal environmental [11]. All piles have the same the cross section with an external diameter of 130.0 mm and include longitudinal steel bars with diameter between 12.50–32.0 mm and a metal jacket of thickness of 25.0 mm. The great thickness of the reinforced deck plus the reinforced concrete of the piles provides a high rigidity in the horizontal plane.

Figure 1(a) shows the PSWS drawing and the port rendering, whereas Fig. 1(b) shows the 3D whole model and a cross section of the pile with metal jacket (the blue area represents the jacket thickness) by finite element method (FEM) using SAP2000 software [12].

The FEM model is basically formed by vertical elements (piles), longitudinal/transversal elements (beams and decks) and 630.0 small elements (elastic springs) of 1.0 m on the pile to simulate the pile-soil interactions in two directions x and y by using Winkler's springs [7].

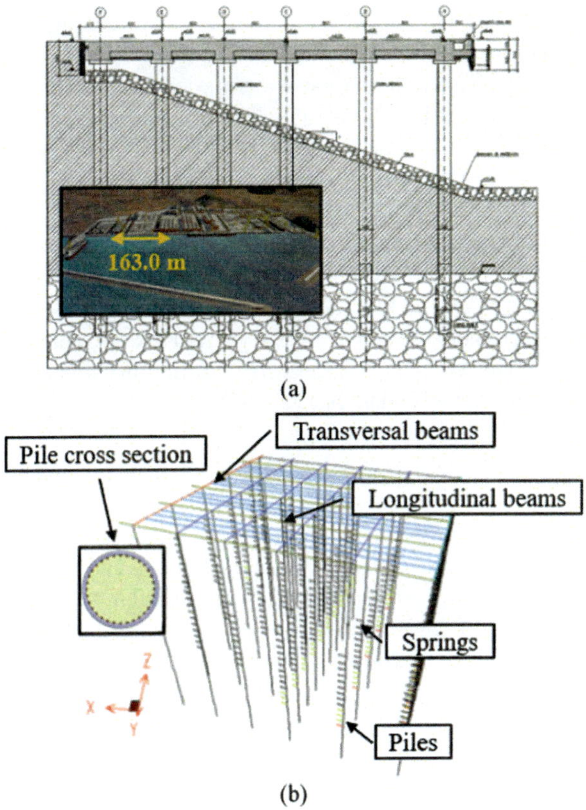

Fig. 1. Studied PSWS: (a) drawing and rendering of the whole port; (b) FEM model.

The model is constructed trying to reproduce in a better way the real structure, for this, the axis of each bar coincides with its centre of gravity, the nodes of the slab coincide with the nodes of the beams, and the dead weight of each element has been manually inserted to avoid the double calculation in the overlapped sections.

3 Materials and Methods

3.1 Materials

In this section the main characteristics of the materials (i.e., concrete and steel), soil, seismic data, and seismic masses are described.

The concrete characteristics are: density of 25.0 kN/m^3, elastic modulus of 30.0 GPa, compressive strength of 50.0 MPa, and structural damping ratio of 5.0%

The used steel has an elastic modulus of 200.0 GPa and yield strengths between 250.0–500.0 MPa (the metal jacket has a lower yield strength).

The confined concrete was estimated by the Mander's model [13] with values of the compressive strengths between 45.0 – 75.0 MPa. The estimated length of the 44 plastic hinges defined by [14] has a mean value of 1.0 m. These values are necessary to analyse the material nonlinear behaviour.

The soil with a 3:1 superficial slope is composed by clay, sand and rock with a stiffness that ranges between k = 3.25–195.0 MN/m^3. This range includes two possible limits, i.e., 0.5 k and 1.5 k, in order to quantify a flexible and semi-flexible behaviour of the soil-pile model, respectively. As mentioned, the studied structure is placed in an area with high seismic hazard with a peak ground acceleration of 0.48 g [15, 16].

Finally, in accordance with the literature [17–19], the seismic masses applied in the structure are: (1) superstructure dead loading of the wharf = 8778.0 kN; (2) 10% of the accidental loading = 2800.0 kN; (3) 1/3 of the pile loading between the deck and 5 times the pile diameter under the soil = 1283.0 kN; (4) hydrodynamic force, considered equivalent to the pile volume, placed at the top of the piles in a transversal plane = 683.0 kN. Therefore, the total seismic mass is 13.54 MN.

3.2 Methodology

The methodology of this study consists in obtaining the pushover curves of the studied PSWS and then, by a stochastic approach, in estimating the no-exceedance probabilities of the flexible model with respect to the semi-flexible model.

Former methodology follows the classical approach described in [8, 9, 20], thus here it is not shown, whereas the latter methodology is described below.

In the reliability analysis, the failure probability of a system with basic random variable $x \in \mathbb{R}^n$ can be expressed by [8, 21]:

$$P_f = \int_{\mathbb{R}^n} I_F(x) f(x) dx \tag{1}$$

where f(x) is the probability density function (PDF) of x, and $I_F(\cdot)$ is a binary function (= 1.0 if the point x coincides with the failure domain F, and 0 otherwise).

For a random variable x = T(u) to be expressed in terms of independent standard Gaussian random variables u, Eq. (1) can be written as:

$$P_f = \int_{\mathbb{R}^n} I_F[T(u)] \varphi(u) du \tag{2}$$

where φ(u) denotes the multivariate Gaussian PDF.

The solutions of Eq. (2) involve the construction of a sequence of failure domains, thus the failure domain of interest, F, is expressed by:

$$F = \bigcap_{j=1}^{M} F_j \tag{3}$$

whit $F_1 \supset F_2 \supset \ldots \supset F_M$, and $F = F_M$. The failure probability $P_f = P_r(u \in F)$ can be written as:

$$P_r(u \in F) = \prod_{j=1}^{M} P_r(u \in F_j | u \in F_{j-1}) \tag{4}$$

whit $F_0 = \mathbb{R}^n$, thus $P_r(u \in F_0) = 1$ (the range is $F > 0$ and $F < 0$). Each $P_r(u \in F_j|u \in F_{j-1})$ in Eq. (4) can be compute using:

$$P_r(u \in F_j|u \in F_{j-1}) = \int_{\mathbb{R}^n} I_{F_j}(u)\varphi(u|F_{j-1})du \tag{5}$$

where $\varphi(u|F_{j-1})$ is the truncated $\varphi(u)$. Finally, Eq. (5) can be approximated by Monte Carlo simulation (MCS) with:

$$P_r(u \in F_j|u \in F_{j-1}) \cong \frac{1}{N}\sum_{i=1}^{N} I_{F_j}(u_i) = \frac{N_f}{N} \tag{6}$$

where u_i are the sample generated from conditional PDF $\varphi(u|F_{j-1})$, and N_f is the number of simulations with $I_{Fj}(u_i) \leq 0$.

4 Analyses and Results

Figure 2 shows the results of the pushover analysis for 0.5 k and 1.5 k in terms of base shear and elasto-plastic displacements. It is shown the elastic spectrum for 5% and 10% (i.e., total damping ratio [8]), the linear-elastic capacity curve, bi-linear elastic curve, and pushover curve (in x-direction). The fundamental structural period was estimated as 1.65 s.

The pushover curve can be described by 4 main points: point 1 where the bi-liner curve intersects the pushover curve; point 2 where the pushover curve intersects the 10% elastic spectrum; point 3 where the pushover curve intersects the 5% elastic spectrum; point 4 where the pushover curve reaches its ultimate value.

The point of interest in this analysis is called "performance point", d* (i.e., point 2), which represents the maximum allowable degradation (elastic + inelastic).

The pushover analysis demonstrates the benefit of the non-linearity of the material. The results show a d* = 13.74 cm (base shear = 6.63 MN) and d* = 12.54 cm (base shear = 7.27 MN) for 0.5 k and 1.5 k model, respectively. The ductility for two cases was estimated as: 13.74/8.66 = 1.59 and 12.54/6.67 = 1.88; both < 5 [14].

These values have been amplified by considering the dynamic magnification factor due to torsional effects studied in [8, 18] providing a $d^*_{maj} = 17.86$ cm and $d^*_{maj} = 18.18$ cm (see Table 1) for 0.5 k and 1.5 k model, respectively.

The reliability analysis should analyse the following aspects: (1) the differences of the pushover curves of the two models in a probabilistic way; (2) the no-exceedance probability of the flexible model with respect the semi-flexible model; (3) the probability that the displacement differences, i.e., $d^*_{maj}-d'$, assumes a value less than or equal to 0: $I_{Fj}(u_i) = d^*_{maj}-d' \leq 0$ (Eq. (6)).

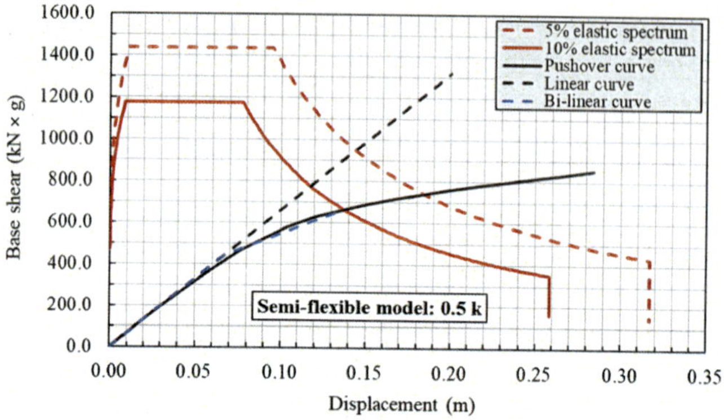

(a)

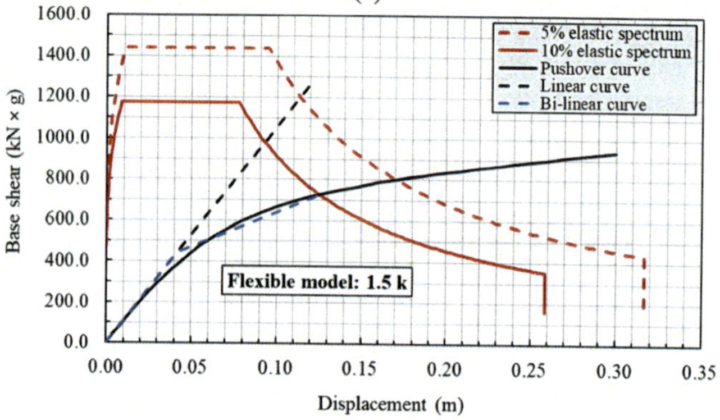

(b)

Fig. 2. Pushover curves in x-direction for (a) 0.5 k and (b) 1.5 k.

Rigorously, the results of P_r are accurate if the samples N tends to ∞; in practice the number of N should be 1.0×10^k, where the choice of exponent k depends on the order of magnitude of P_r to be reached.

Table 1. Generated parameters for the displacements by MCS for 1.5 k model.

	Point 1	Point 2 [a]	Point 3	Point 4
μ (cm)	9.68	18.18	24.90	43.50
$\pm\,\sigma$ (cm)	2.78	5.22	7.16	12.54
CV (%)	29.0	29.0	29.0	29.0

[a] Amplified performance point.

Fir generating stochastic data, random variables (RVs) with a probability distribution in the range RV_{min} to RV_{max} up to a list of N random values must be defined. The considered range is $-0.5 \leq RV \leq +0.5$ with $N = 1.0 \times 10^6$ [22]. A very large range of RV could provide a very large data dispersion.

Table 1 lists the generated data indicating the mean value, μ, its standard deviation, $\pm$ σ, and the coefficient of variation (CV = σ/μ). Figure 3 shows the PDFs for the 4 points.

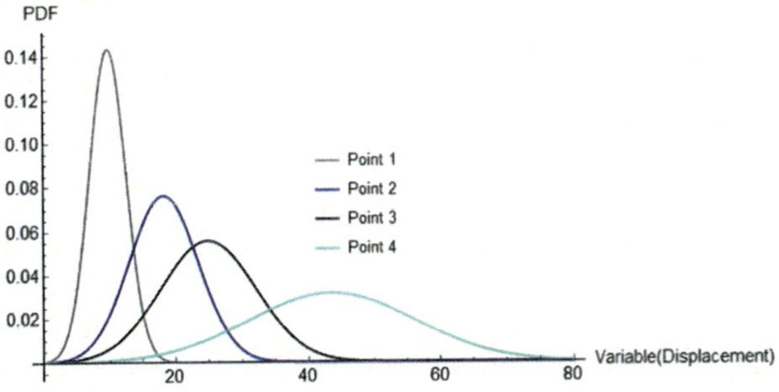

Fig. 3. PDFs of the generated 4 points by MCS.

Figure 4 shows the probability of no-exceedance in function of a certain displacement d' estimated by Eq. (6). These scenarios, generated by 16.0 $\times$ 10^6 analyses, regard the 4 points of the pushover curve. It was estimated the probability that the points of the pushover curve for a semi-flexible model (i.e., 0.5 k) assumes a value $\leq$ the points of the pushover curve for a flexible model (i.e., 1.5 k) accounting for the dynamic magnification factor due to the torsional effects.

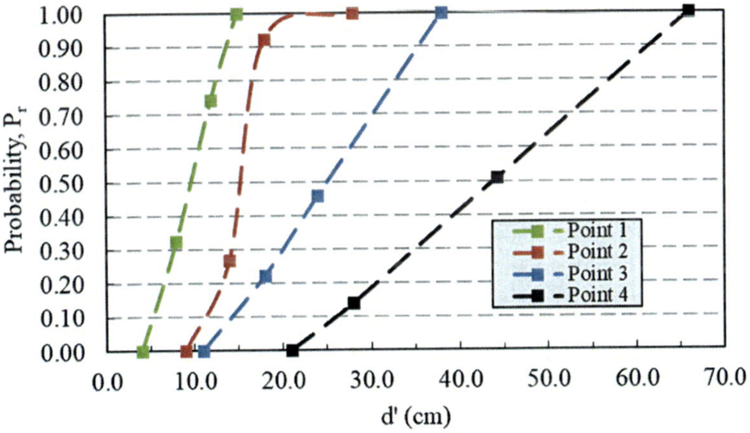

Fig. 4. Probability of no-exceedance for the points 1–4.

In Fig. 4, when a P_r curve increases the slope, is more probable that a d^*_{maj} value does not exceeds a d' value. In this sense, when d' value increases the structure is more flexible, otherwise the structure is more rigid. Point 4 refers to the ultimate value therefore its P_r curve provides a high probability with high d'.

5 Conclusions

In this paper a pushover analysis for a real PSWS placed in an area with very high seismic hazard was treated. The analysis has been carried out in a deterministic and probabilistic way.

Pushover curves were obtained in x-direction where performance displacements between 13.74–12.54 cm were registered. They correspond to the base shears of 6.63–7.72 MN, respectively. This non-linear analysis quantified an advantage, in terms of displacements, due to the material ductility of 18–38%.

The reliability analysis considering the torsional effects provided (i) the gap of the pushover curves for two models by a probabilistic approach (d^*_{maj}–d' ≤ 0), and (ii) the probability of non-exceedance of the flexible structure with respect the semi-flexible structure (P_r ranges between 0 and 1 for d' which ranges between 4.0 and 66.0 cm).

References

1. Su, L., Wan, H.P., Dong, Y., Frangopol, D.M., Ling, X.Z.: Seismic fragility assessment of large-scale pile-supported wharf structures considering soil-pile interaction. Eng. Struct. **186**, 270–281 (2019)
2. Khazi, M.F.R., Vazeer, M.: Behaviour of pile supported wharf in liquefied soils. International Journal of Geomate **13**, 186–193 (2017)
3. Heidary-Torkamani, H., Bargi, K., Amirabadi, R., McCllough, N.J.: Fragility estimation and sensitivity analysis of an idealized pile-supported wharf batter piles. Soil Dyn. Earthq. Eng. **61–62**, 92–106 (2014)
4. Chiou, J.S., Chiang, C.H., Yang, H.H., Hsu, S.Y.: Developing fragility curves for a pile-supported wharf. Soil Dyn. Earthq. Eng. **31**, 830–840 (2011)
5. Gao, S., Gong, J., Feng, Y.: Equivalent damping for displacement-based seismic design of pile-supported wharves with soil-pile interaction. Ocean Eng. **125**, 12–25 (2016)
6. Lombardi, D., Bhattacharya, S.: Evaluation of seismic performance of pile-supported models in liquefiable soils. Earthquake Engineering & Structural Analysis **45**, 1019–1038 (2016)
7. Nozu, A., Ichii, K., Sugano, T.: Seismic design of port structures. Journal of Japan Association for Earthquake Engineering **4**(3), 195–208 (2004)
8. Zacchei, E., Lyra, P.H.C., Stucchi, F.: Pushover analysis for flexible and semi-flexible pile-supported wharf structures accounting the dynamic magnification factors due to torsional effects. Struct. Concr. **2020**, 1–20 (2020)
9. Zacchei, E., Lyra, P.H.C., Stucchi, F.R.: Nonlinear static analysis of a pile-supported wharf. Ibracon Struct Mater J. **12**, 998–1009 (2019)
10. Global earthquake model (GEM): database. Accessed in 02 2022. https://downloads.openquake.org/countryprofiles/VEN.pdf
11. Zacchei, E., Nogueira, C.G.: Calibration of boundary conditions correlated to the diffusivity of chloride ions: An accurate study for random diffusivity. Cement Concr. Compos. **126**, 1–14 (2022)
12. Sap2000: Version 16.0.0 Plus. Computers and Structures Inc, California/New York (2013)
13. Mander, J.B., Priestley, M.J.N., Park, R.: Theoretical stress-strain model for confined concrete. J. Struct. Eng. **114**(8), 1804–1825 (1988)
14. European Committee for Standardization (CEN): Eurocode 8: Design of structures for earthquake resistance - part 2: Bridges, EN 1998–2:2005. Brussels, Belgium (2005)

15. Fundación Venezolana de Investigaciones Sismológicas Covenin 1756–1:2001. Funvisis. Edificaciones Sismorresistentes, Parte 1: Requisitos. Caracas: Fondonorma (2001)
16. European Committee for Standardization (CEN): Eurocode 8: Design of structures for earthquake resistance - part 1: General rules, seismic actions and rules for buildings, EN 1998–1:2004 (2004)
17. Permanent International Association for Navigation Congresses (PIANC): Seismic Design Guidelines for Port Structures. Working group n.34 of the Maritime Navigation Commission, International Navigation Association, 1[a] ed., A.A. Balkema Publishers, p. 474 (2001)
18. Port of Long Beach (POLB): Wharf Design Criteria. Version 2.0 (2009)
19. American Society of Civil Engineers (ASCE): Seismic Design of Piers and Wharves. ASCE/COPRI 61–14, p. 90 (2014)
20. Applied Technology Council (ATC-40): Seismic Evaluation and Retrofit of Concrete Buildings. California Seismic Safety Commission, Redwood City, California, v.1, report n. SSC 96–01 (1996)
21. Wang, Z., Broccardo, M., Song, J.: Hamiltonian monte carlo methods for subset simulation in reliability. Struct. Saf. **76**, 51–67 (2019)
22. Zacchei, E., Nogueira, C.G.: 2D/3D numerical analyses of corrosion initiation in RC structures accounting fluctuations of chloride ions by external actions. KSCE J. Civ. Eng. **25**, 1–23 (2021)

Service Life Assessment of Steel Girder Bridge Under Actual Truck Traffic

Sahan Chanuka Bandara and Panon Latcharote(✉)

Faculty of Engineering, Mahidol University, Nakhon Pathom 73170, Thailand
panon.lat@mahidol.ac.th

Abstract. Fatigue behavior of a bridge is considered as a main concern for structural engineers especially for bridge construction due to the rapid growth of overloaded trucks. Truck traffic has increased significantly on bridge structures in Bangkok. Moving loads due to overloaded trucks on a steel girder bridge would cause fracture formation and propagation. The study is conducted to assess the safety of a steel girder bridge under actual truck traffic data. The study addresses the fatigue behavior of the bridge structure using FEM analysis. This will help to evaluate the remaining service life under the operational performance of several truck classes. The service life of this steel girder bridge can be evaluated by S-N curve application based on Miner's damage rule which is recommended by AASHTO LRFD standard specifications. The bridge consists of 19 spans with 1.75 m transverse width of each five longitudinal steel girders including reinforced concrete slabs. The dynamic behavior of the steel girders was validated by field instrumentation. The fatigue truck model was constructed using data of 102,546 truck movements in 2020. The data included gross weight, axle weight, and axle spacing. The number of stress cycles can be measured by analytical investigation at the midspan of each girder. The result obtained from analytical investigation revealed that the steel girder bridge would be affected fatigue failure with the expected growth of truck traffic and the classified fatigue trucks based on the actual truck data.

Keywords: Steel girder bridge · Overloaded truck · Fatigue behavior · Service life · FEM

1 Introduction

Increases in problems on road networks around the world have occurred due to continuous increases in freight demand. Truck loads, load configurations, and numbers of trucks lead to pavement deterioration, necessitating load limitations and early replacement. Fatigue is an essential parameter consideration in the design and analysis of any bridge structures. Continuous use by heavy trucks can lead to cracking and premature failure (Khaleel and Anditani, 1993). This is considered as generative deterioration by forming crack growth due to the increase in overloaded truck traffic crossing on the bridge which produce a reduction in bridge service life. Many existing bridges show major effects of fatigue damage due to exceeding the designed load limits on the bridge

Z. Dimitrovová et al. (Eds.): WMVC 2022, MMS 125, pp. 142–151, 2023.
https://doi.org/10.1007/978-3-031-15758-5_14

(Frangopol et al., 2008). The stress history is caused by the fatigue truck passage can be used to determine the design stress range for fatigue and the number of cycles depending on the bridge span. For shorter spans, the design shows two distinct peaks, whereas, for longer spans there was one overall peak (Schilling, 1978). The practice for bridge fatigue design may have underestimated the effect of dynamic vehicle loading and truck overloading on the fatigue life of steel bridges (Wang et al., 2016). The passage of trucks on a bridge deck can cause vertical tensile stresses in the welded connections between cross-frame connection plates and girder bottom flanges. The tensile stresses in the connection plates depend on the magnitudes and positions of the wheel loads of crossing vehicles. Zhao et al. (2018) suggested that dynamic loads during peak time periods may cause the highest tensile stress in the connections of concern, so that any increase in truck weight limits would affect the safety of existing bridges. Increasing the weight limits would shorten the repair and replacement time of many bridges.

Kwad et al. (2017) studied the fatigue behavior using FEM of the bridge calibrated by measured responses for an ambient vibration test. The highest condition of structural damage was identified from a single 5-axle vehicle and the overall damage was calculated over 42 years. Zhao et al. (2018) conducted different truck axle loads on a steel girder bridge in Maryland for assessment of fatigue behavior using FEM analysis. The fatigue behavior was studied based on AASHTO LRFD standard specifications and collected truck moving load data at the girder using strain gauges It was determined that the bridge structure would perform for more than its design life of 75 years. Habeeba et al. (2015) also estimated the remaining service life of a composite steel girder bridge in Kerala, India. Palmgren's Miner theory was applied for evaluating fatigue damage. The results obtained from analytical study found that the bridge could provide the service life between 50 and 100 years. Aggarwal and Parameswaran (2015) presented the effect of overloaded trucks on fatigue damage of a bridge to study the relationship between overloaded truck and fatigue damage accumulation on the bridge structure in Bangalore. In Bangkok, Suparp and Joyklad (2014) revealed that the overall of average maximum bridge responses were approximately 25% and 18% greater than those obtained from HS20–44, so the problem of overloaded trucks is a serious issue. According to the statistics in 1996, overloading occurred on ten-wheel trucks, and it involved around 25% of all trucks, which carry 78% of shipments by weight (Chan, 2008). Based on the statistics collected by WIM, they found that 33% of ten-wheel trucks are overloaded. 94% of these weighed between 21 tons and 30 tons, while 21 tons is the legal weight for a ten-wheel truck. Chan (2008) indicated that 81% of total damage to the highway is caused by 33% overloaded trucks.

In this study, the fatigue performance of an existing steel girder bridge in Bangkok was investigated for evaluating the remaining service life using 3D FEM analysis. The main objectives for this study are to evaluate the service life cycle of a steel bridge under current traffic conditions and to examine the maximum service life of the structure under operational performance of different truck classes to enhance service life improvement.

2 Bridge Deck and Girder Geometry

The selected steel girder bridge was built with approximately 17 m span and 8.8 m width as shown in Fig. 1. It has two traffic lanes each of 4.4 m in width and two sidewalks each of 0.9 m in width. The bridge has 19 spans and total length of the bridge is 327.3 m. The superstructure was constructed using reinforced concrete slabs. The distance between each of girders is 1.75 m. The thickness of a reinforced concrete slab with asphaltic concrete is 0.2 m. The steel sections used welded wide flanges made with yield stress steel grade of 253 MPa with ultimate strength of 380 MPa. The steel girders were adapted with young's modulus of 1.96×10^5 MPa. The concrete slab has specified compression strength of 40 MPa. Figure 1 shows the cross-sectional view of a steel I girder containing the height of the girder, top and bottom width of flanges, and web thickness. A standard drawing obtained from Department of Highways in Thailand provides information for bridge geometry for FEM analysis.

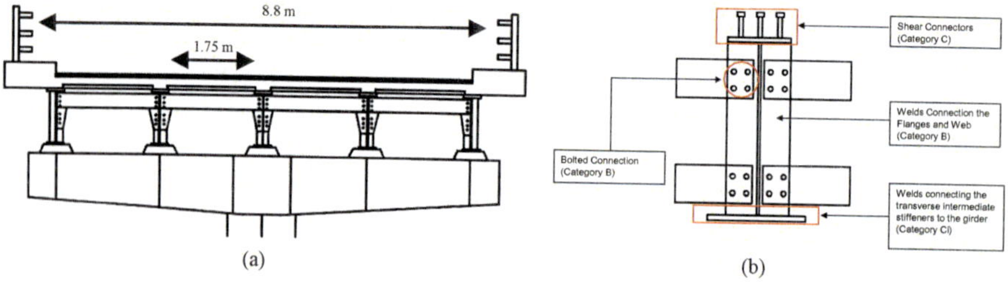

Fig. 1. The standard drawing of a steel girder bridge

3 Fatigue Truck Model

A fatigue truck model is developed using available data collected at site containing gross weight, axle load, and axle spacing. AASHTO LRFD standard specifications have delivered development of a fatigue truck model among researchers to observe the fatigue behavior on the bridge structure. The fatigue truck model with gross weight of 240 kN is represented in AASHTO LRFD standard specifications for determination of fatigue strength (AASHTO, 2017). The gross weight is calculated from more than 27,000 actual truck traffic data from 30 sites nationwide collected by WIM (Snyder et al., 1985). Its configuration was approximated based on the axle weight ratios and axle spacing of four- and five-axle trucks, which cause a high percentage of the fatigue damage in typical bridges. The fatigue truck in AASHTO is developed with axle truck spacing of 4.27 m and 9.14 m for front and rear with axle width of 1.83 m. In this study, Root Mean Cube (RMC) method was used to determine the effective gross weight for AASHTO fatigue truck modification when the gross weight data has been collected at a site in Bangkok.

4 Actual Truck Traffic Database

It is challenging to quantify variety of truck configurations, weights, and volumes. Truck configurations affect bridges differently depending on the span length and girder spacing.

Bridge Weigh-In-Motion Technology (BWIM) system is used to determine the bridge deformation and responses under different truck traffic loading. BWIM system contains strain gauge sensors to detect axle loading and axle spacing in different time- period from truck loads travelling on the bridge structure at the normal traffic speed. The actual truck traffic database collected from this system is important to verify with FEM analysis model. The Closed-Circuit Television (CCTV) connected with the system is used for remote visualization to observe the truck population including lane, axle weight, axle spacing, speed, time of arrival, classification, and number of axles. Regardless of the sensor involved, the calibration is needed before these devices can produce actual data without interference.

Table 1. Recorded truck data in 2020 (BMA, 2020)

Truck type	Axle nos	Jan	Feb	Mar	Apr	May	Jun	Jul	Aug	Sep	Oct	Nov	Dec	Total	**Overload**
No.2	2	2888	5211	10845	5013	8919	9123	8770	5842	6646	4461	5463	5526	78707	17.4%
No.3	3	566	1313	3377	2295	2948	2233	1744	1668	1754	1404	1651	1781	22734	41.1%
No.4	4	8	24	30	27	51	42	33	53	65	64	40	64	501	66.7%
No.5	4	1	6	4	-	1	-	-	-	7	-	-	-	19	10.5%
No.6	5	1	-	-	-	-	-	-	-	-	-	-	-	1	0%
No.7	5	15	25	20	-	1	-	-	-	40	1	-	-	102	3.9%
No.9.1	6	11	8	4	-	-	-	-	-	27	1	-	-	51	23.5%
No.11	4	15	58	13	3	-	-	-	-	5	-	-	-	94	8.5%
No.12	5	2	1	-	-	-	-	-	-	1	-	-	-	4	50%
No.14	6	35	24	13	23	52	31	14	33	35	29	33	11	333	82.9%

Actual truck traffic data was collected from 102,546 trucks in 2020 by the Bangkok Metropolitan Administration (BMA). Number of trucks in each type was provided in Table 1 with the percentage distribution of overloaded trucks. From Table 1, Truck Type No.2 has the highest distribution and accounted for 78,707 trucks, while 67% is the maximum overloaded truck weight distribution found on the database. From Table 1, the effective gross weight for 2020 was 238 kN.

5 Finite Element Modelling

Using FEM analysis for a steel girder bridge, dimensional and material properties of the bridge need to be found including the moments of inertia estimated for each main girder along the cross section. The standard drawing was used to estimate the width of reinforced concrete slabs that can contribute to bending resistance resulting in the effective flange depended on the slab thickness. To create the bridge model, a combination of solid and frame elements was used to create a deck system and truss members including dimensions, sizes, and properties. The bridge, as presented in Fig. 2a, was modelled using solid elements for the reinforced concrete slab and frame elements for the steel girder, stiffeners, diaphragm, and connectors. Figure 2b shows 3D FEM rendering of the steel girder bridge. All solid elements were assigned with reinforced concrete properties to simulate the actual condition of the slab on the bridge. Element sizes were minimized to create a more refined mesh, maintaining a low aspect ratio. The aspect ratio is defined as the ratio of the longest dimension to the shortest dimension of the element shape. All frame elements were assigned with steel properties as mentioned above to model the 3D frame including biaxial bending, torsion, axial deformation, and biaxial shear deformations. The frame element was modelled with the connection of two points as a straight line with respect to the local coordination system for section property definition of the bridge structure. The bridge girders were created using the given frame section. Material properties of steel and concrete were defined as elasticity modulus and compressive strength. The FEM analysis was conducted by comparing observed strain between the field experiment and the analytical model for load paths in a single lane and in both lanes using three-axle trucks.

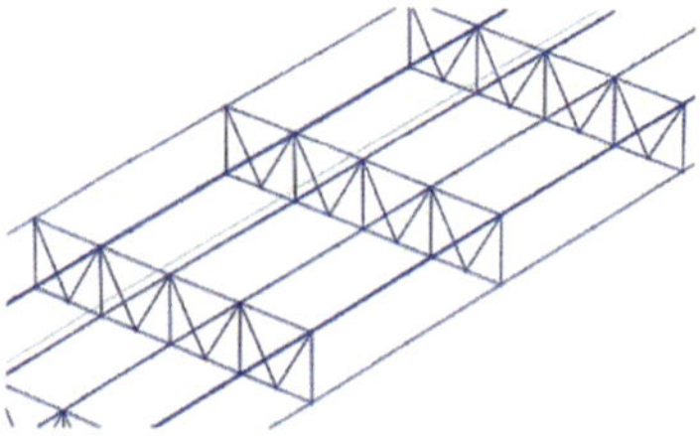

(a)

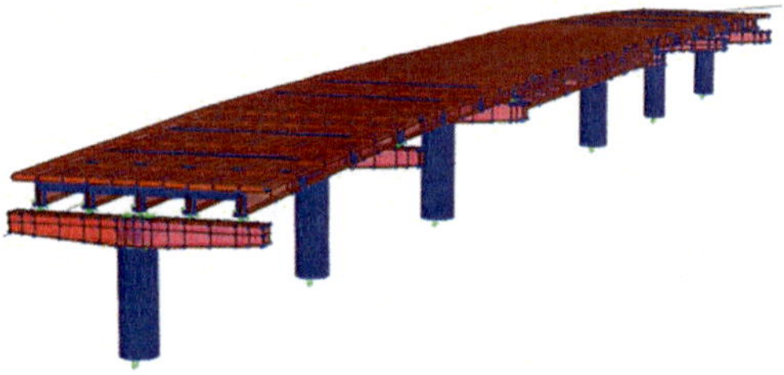

(b)

Fig. 2. Model of steel girder bridge

6 Model Validation

From the 19 spans in the steel girder bridge, two span types including main span and typical span were selected for model validation between field investigation and FEM analysis. A dynamic load test was conducted on the bridge after completion of construction. The results were used to calibrate and verify the analytical model to establish the behavior of the bridge prior to service. Measurement devices were prepared to record strain on the bridge structure. The installation was done in each girder at the main span and typical span for calibration compared with FEM analysis.

No.1 and No.2 Thai trucks with different loading were selected to validate the bridge structural model. Both trucks have three axles but have different vehicle dimensions. Total loading weight of No.1 and No.2 trucks is 238.9 kN and 226.1 kN, respectively. Three loading cases were investigated, including No.1 truck passage in the left-side lane, No.2 truck passage in the right-side lane, and passage of both No.1 and No.2 trucks in both side lanes, as shown in Fig. 3.

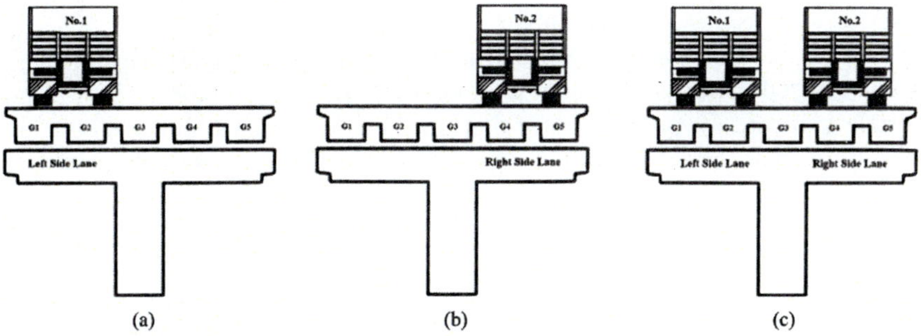

Fig. 3. FEM loading case of truck passage on (a) Left side lane (b) Right side lane (c) Both (left and right) lane

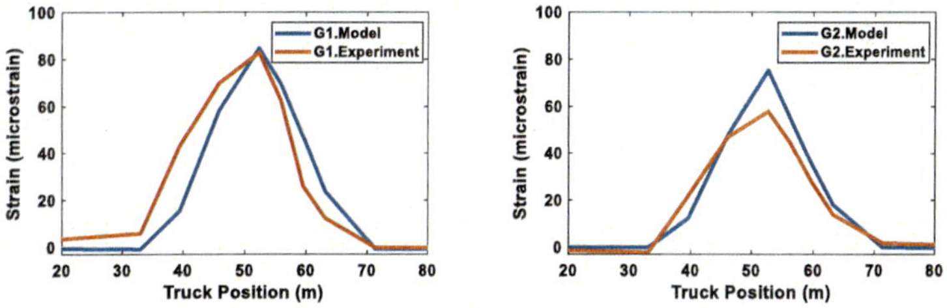

(a) Main Span

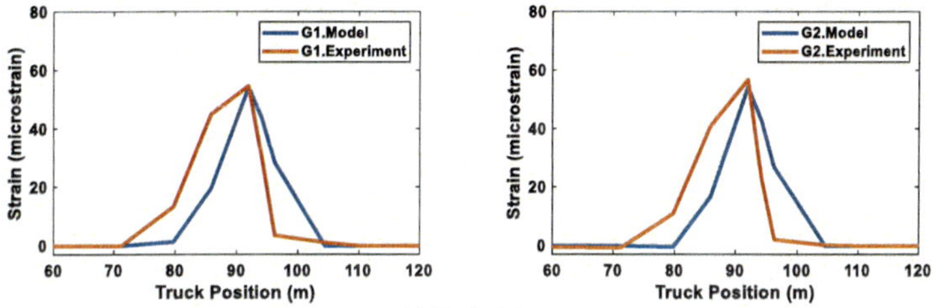

(b) Typical Span

Fig. 4. Dynamic responses for stage I model validation at main span and typical span

Figure 4 shows the maximum strain value related to G1 girder, which was able to receive when No.1 truck was moving along the left-side lane. The strain value at G5 girder represents the lowest value at stage I compared to other girders. However, the highest strain value was recorded at G5 girder when the truck was moving along the right-side lane. Maximum strain value occurred at G2 and G3 girders for stage III and this implies that when both trucks move on the bridge structure, the girders show higher strain values due to heavy dynamic loadings. Moving load analysis for No.1 and No.2 trucks was conducted using CSiBridge (CSI, 2018) to compute the responses from each girder due to truck load. Both trucks on the bridge were modelled as point loads from each of the axles. Any number of vehicle axle loads in gravity direction can be defined for each lane of the bridge structure. The vehicle axle load of No.1 and No.2 Thai trucks were applied in FEM analysis. The stress time history can be generated after they have completed running simultaneously in the selected lane. It provides accurate results of responses, such as maximum and minimum stresses, due to effect of truck loading for discussion of the structural behavior using influence lines. The results from influence line can be delivered by various sensitivity responses to the axle loads. The stress response was able to record for observation of dynamic behavior of each girder on the bridge and the stress results were converted into micro strain with young's modulus of steel as presented in Fig. 4.

7 Evaluation of Bridge Service Life

Fatigue is defined as a process that causes premature failure or damage to a component subject to repetitive loading (Leitão, 2012). Unlike, the strength limit state that is defined by maximum load for the entire service life of a structure, the fatigue damage has been accumulated over time in several years. The truck data in Table 1 cover a period of 12 months from January 2020 to December 2020. The effective gross weights computed from the WIM data were assigned for the gross weight of the modified AASHTO fatigue truck. The maximum stress can be obtained at each girder of the bridge from the simulation of the modified AASHTO fatigue truck over the analytical bridge model. A rain-flow cycle counting method was implemented for determining stress peaks and the number of cycles in a time history graph.

AASHTO Description	Drawing Detail	Detail Category
Welds connection the Flanges and Web (I Girder)	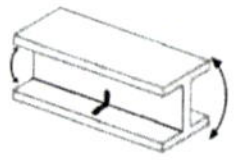	B
Welds connecting the transverse intermediate stiffeners to the girder	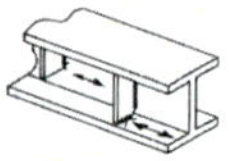	C'
Bolts connecting the transverse intermediate stiffeners to channel	* **High strength bolted connection**	B
Shear Connectors (Welds)	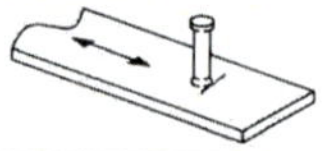	C
Longitudinal continuous welds in buildup plates and shape	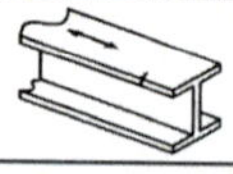	B

Fig. 5. Selected detail category

The S-N curve is used to determine fatigue damage for each detail category designated by Category A to Category E' (AASHTO, 2017). The stress range is defined as the change in stress level during its variation of one loading cycle. The fatigue damage of a bridge member then depends on three factors, i.e., stress range, number of loading cycles, and member classification. The design life, which refers to the time-period during which a structure must perform safely with a reasonable probability of avoiding failure due to fatigue cracking, is set at 75 years. (AASHTO, 2017). The service life of each structural element and fatigue damage were estimated by the given S-N curve. The detail categories were selected based on the standard specifications and bridge drawings presented in Fig. 5.

The service life was estimated using the limit state of Fatigue II according to the AASHTO design specification. The *A* value is the design category constant taken from the reference of AASHTO. The *n* value is number of cycles per truck passage of the fatigue truck. The *A* value provided by AASHTO specification for detail category B and detail category C or C' are 120x10^8 (ksi^3) and 44 x10^8 (ksi^3), respectively. The value of fatigue strength $(\Delta F)_n$ is taken from the fatigue truck simulation. The fatigue truck was developed using RMC method with the collected gross weight database.

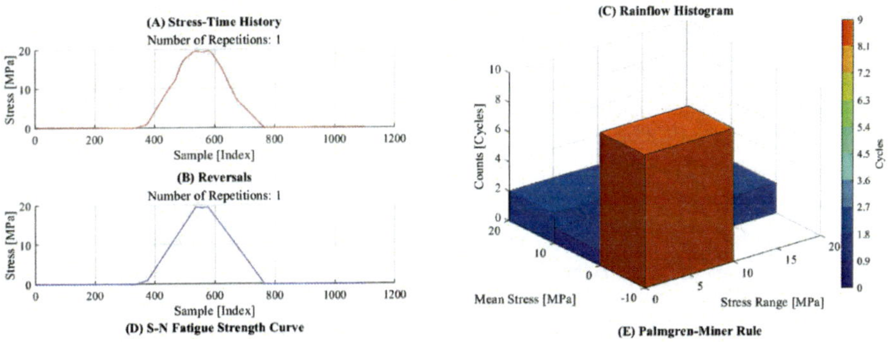

Fig. 6. Calculated stress cycle and fatigue damage (Carlos, 2021)

The worst case of service life estimation can be generated by using the truck data in March, which is the highest traffic over 12 months. According to Table 1, the data from 14,306 truck was collected with different truck types in March 2020. The AASHTO fatigue truck was modified using the RMC method and it was assigned with three axle loads. However, the stress cycle obtained from FEM analysis resulted in complexity with many stress peaks and included diversified values. Since it is difficult to calculate stress cycles manually, the rain-flow cycle counting method was used for this study and the number of cycles per truck passage was obtained for service life estimation. The histogram of calculated stress cycle and estimated fatigue damage using Palmgren's miner rule was presented in Fig. 6 and the number of cycles per truck passage (*n*) was 12 counts. The amount of stress reached at maximum was 19.58 MPa at G1 girder. Then, the service life was calculated as 259 years for detail category B and 95 years for detail category C.

Table 2. Estimated service life at G1 girder

2020 database	Total truck	W (kN)	AADT	Max stress (MPa)	Service life (years)	
					Detail category B	Detail category C and C'
All	102,546	238.17	280.18	18.38	516	189
Growth factor 4%			280.18	18.38	79	55
Growth factor with classified type		*Depend on classified truck			92	67

Table 2 illustrates outcome of the service life of the bridge structure for each detail categories B and C/C' by using the modified AASHTO fatigue truck from all actual truck data collected in 2020. The result in Table 2 shows the bridge structure would perform under a secure situation more than 75 years for both detail categories without considering the growth factor. However, when the growth factor of 4% in Bangkok was considered, the service life would be under 75 years for detail category C and C'. According to Table 1, the number of stress cycles and the maximum stress can be determined from average gross weight, axle weight, and axle spacing for each truck type. Then, the service life from all classified trucks can be considered based on ten truck models, which was over the service life calculated from all trucks.

8 Summary and Conclusions

The steel girder bridge model was developed with given material and sectional properties to estimate the service life. The actual traffic data with several truck types was used to calculate effective gross weights and develop a fatigue truck model based on actual truck database in 2020. With the guidance of AASHTO LRFD standard specifications, the fatigue strength and service life were computed using different fatigue truck models from effective gross weight for AASHTO fatigue truck modification with the gross weight data collected at the site. Based on the analytical investigation, it was found that the service life can be varied compared with AASHTO fatigue truck gross weight. The results show that different gross weights applied to the fatigue truck can cause considerable overestimation or underestimation of actual fatigue damage. This study was conducted considering the load-induced dynamic effect of the bridge to estimate the fatigue strength in terms of service life. It can be concluded that the bridge structure performs in a secure condition for more than 75 years without considering the growth factor of 4%. However, the service life for selected detail categories would cause the reduction of fatigue strength with the growth factor of future traffic demand. In addition, the fatigue strength could be affected from other factors, such as corrosion and surface roughness.

Acknowledgment. The authors would like to thank to Prof. Tospol Pinkaew, Department of Civil Engineering, Chulalongkorn University, and Bangkok Metropolitan Administration (BMA) for providing relevant data throughout the research.

References

AASHTO: LRFD Bridge Design Specifications. American Society of State Highway and Transportation Officials, Washington DC (2017)

Aggarwal, V., Parameswaran, L.: Effect of overweight trucks on fatigue damage of a birdge. In: Matsagar, V. (ed.) Advances in Structural Engineering 2015. Springer, India (2015)

Carlos, S.: x. https://github.com/carlosouto/Fatigue-Damage-Accumulation GitHub. last accessed 14 June 2021

Chan, Y.: Truck Overloading Study in Developing Countries and Strategies to minimize its Impact. Master's Thesis. Queensland University of Technology, Brisbane (2008)

CSI: CsiBridge User Manual, 21st edn. Berkeley, California (2018)

Frangopol, D.M., Strauss, A., Kim, S.: Bridge reliability assessment based on monitoring. J. Bridg. Eng. **13**, 258–270 (2008)

Habeeba, A., Sabeena, M.V., Anjusha, R.: Fatigue evaluation of reinforced concrete highway bridge. Int. J. Innov. Res. Sci. Eng. Technol. **4**(4), (2015)

Khaleel, M.A., Anditani, R.Y.: Effect of Alternative Truck Configurations and Weights on the Fatigue Life of Bridges. Transportation Research Record 1393. National Academy Press. Washington DC (1993)

Kwad, J., Alencar, A., Correia, J., Jesus, A.: Fatigue assessment of an existing steel bridge by finite element modelling and field measurements. Journal of Physics Conference Series **843**(1), (2017)

Leitão, F.N.: Fatigue analysis and life prediction of composite highway bridge decks under traffic loading. Latin American Journal of Solids and Structures **10**, 505–522 (2012)

Snyder, R.E., Likins, G.E., Moses, F.: Loading spectrum experienced by bridge structures in the United States. Bridge Weighing Systems Inc, Ohio (1985)

Suparp, S., Joyklad, P.: Effects of increasing truck weight limits on highway bridges in Thailand. IABSE Madrid Symposium Report. **102**, 660–667 (2014)

Wang, W., Deng, L., Shao, X.: Fatigue design of steel bridges considering the effect of dynamic vehicle loading and overloaded truck. Journal of Bridge Engineering **21**(9), (2016)

Zhao, G., Fu, C., Lu, Y., Saad, T.: Fatigue assessment of highway bridges under traffic loading using microscopic traffic simulation. In: Zhou, Y.L. (ed.) Bridge Optimization. IntechOpen (2018)

DHM: Dynamics and Control in Human-Machine Interactive Systems

A Two-Dimensional Model to Simulate the Effects of Ankle Joint Misalignments in Ankle-Foot Orthoses

Vishal K. Badari and Ganesh M. Bapat(✉)

Department of Mechanical Engineering, BITS Pilani - K.K.Birla Goa Campus, Zuarinagar, Goa 403726, India
ganeshm@goa.bits-pilani.ac.in

Abstract. Misalignment of an orthotic ankle joint with respect to the anatomical ankle joint leads to "pistoning motion" which is the relative sliding motion between a limb and its externally wearable orthosis. This pistoning motion and resulting pressure points are the cause of skin problems due to which an orthosis user experiences pain and discomfort. This work quantifies the effects of sagittal plane ankle joint misalignments in terms of relative motion between the limb and the orthosis. A 2D link segment model was developed using MATLAB software to simulate relative motion between the limb and the orthosis for a functional range of ankle motion. The orthotic ankle joint was methodically misaligned with respect to the anatomical ankle joint in the Anterior-Posterior (A-P), Proximal-Distal (P-D) directions, and their combinations to simulate orthosis sliding and locate pressure points on the limb. Simulation results showed that A-P misalignments caused a significantly greater pistoning motion than P-D misalignments, which agrees with previous studies. Combined misalignments (Anterior-Proximal, Anterior-Distal, Posterior-Proximal, and Posterior-Distal) were found to have a greater effect on overall relative motion between the limb and the orthosis as a result of the superposition of relative motions from the A-P and P-D directions. The 2D model also predicts pressure point locations due to joint misalignment, which supports the results from previous studies. Simple 2D simulations presented in this work can be used to interpret the consequences of orthotic ankle joint misalignments. It further emphasizes the importance of accurate alignment of orthotic and anatomical ankle joints and provides insights for modifications in orthosis for improved user comfort. Such a simulation-based model can be used to guide anatomical and external joint alignments in ankle-foot orthoses and lower-limb exoskeleton devices.

Keywords: Orthoses · Ankle joint · Biomechanics · Exoskeleton · Pistoning · Misalignment

1 Introduction

An orthosis is an externally applied device that assists in correcting the alignment and deformities of movable body segments. An Ankle-Foot Orthosis (AFO) is a type of

Z. Dimitrovová et al. (Eds.): WMVC 2022, MMS 125, pp. 155–165, 2023.
https://doi.org/10.1007/978-3-031-15758-5_15

lower-limb orthosis that extends from the calf region to the foot and aims to provide stability along with improvement in gait and physical functioning of the lower limb. The ankle joint is the key functional component in AFO. The most common AFO designs consist of a single-axis ankle joint that allows sagittal plane motions only.

The congruency between anatomical and orthotic ankle joint axes, which can be affected due to the human factor (depending upon the orthotist's level of training and experience) or anatomical ankle joint approximation, is important since misalignments result in relative motion between the limb and the orthosis and generate undesirable forces (both compressive and shear) for a range of ankle joint motion. This relative sliding motion is also known as "pistoning motion," and the resulting pressure points on the leg are the consequences of these undesirable forces. The empirical diagrams presented in the New York University (NYU) lower-limb orthotics manual [1] about linear misalignments between ankle joints suggest the same. This pistoning motion and resulting pressure points are the cause of skin problems such as rashes, calluses and friction blisters [2], due to which an orthosis user experiences pain and discomfort. A simple 2D model consisting of an inversion of the 4-bar slider-crank mechanism was proposed by Fatone et al., (2007) [3] that determines the degree and direction of calf band travel over a functional range of ankle joint motion. Later studies [4, 5] have studied the effects of ankle joint misalignments using 3D simulation models.

The present work investigates the effects of sagittal plane ankle joint misalignments by methodically misaligning the orthotic ankle joint about the anatomical ankle joint (with both the ankle-joint and the orthosis allowing 1 degree of freedom in the sagittal plane). Simple 2D simulations presented here help determine relative motions between the limb and orthosis and locate resulting pressure points on the leg. The analysis done in this study is applicable to articulated AFO designs.

2 Methodology

2.1 Model Description

The anatomical ankle joint was located at the origin of the Anatomical Coordinate System (ACS), which acts as the reference frame for the analysis. The orthotic ankle joint, which was systematically misaligned with respect to the ACS, was considered at the origin of the Orthotic Coordinate System (OCS). The x-axis and y-axis are considered to be along the horizontal and vertical directions respectively. Both ACS and OCS would be coincident in an ideal case of accurate joint alignment. Leg segments rotate about ACS while the orthosis rotation occurs about OCS. The idea is to simulate the motions of the leg and orthosis about their respective axes of rotation in their corresponding reference frames and finally transform back the orthosis motion about OCS to ACS to determine the relative motion between the leg and the orthosis. All the simulations were carried out using MATLAB (R2021b; The MathWorks Inc, Natick, MA, USA) software.

Figure 1A shows the link segment model superimposed on an AFO and illustrates the terminologies used in the model with respect to the AFO. The 2D link-segment diagram of the proposed model used for the analysis is shown in Fig. 1B. SP_0 and SP_1 are the strap points where SP_1 is a moving point that is coincident with the shin region (i.e. anterior shank portion) and the strap, while SP_0 is a fixed point on the orthosis shell.

TP_1 and TP_2 are the touch points that are coincident with the orthosis shell and the calf region at 0° flexion of the shank. As the shank rotates from −20° (plantar flexion) to 20° (dorsiflexion), the line connecting the strap points (SP_0 and SP_1) and the line connecting the touch points (TP_1 and TP_2) rotate about ACS and OCS. Affine transformations were used for translation/shift and rotation about an axis. The positions of the lines after respective transformations are indicative of the relative motions between the leg and orthosis. The relative displacements of the moving strap point (SP_1) and the line joining the touch points were analyzed. Pistoning motion was calculated as the displacement of SP_1 while the gap between the TP lines was used to predict the resulting pressure on the leg.

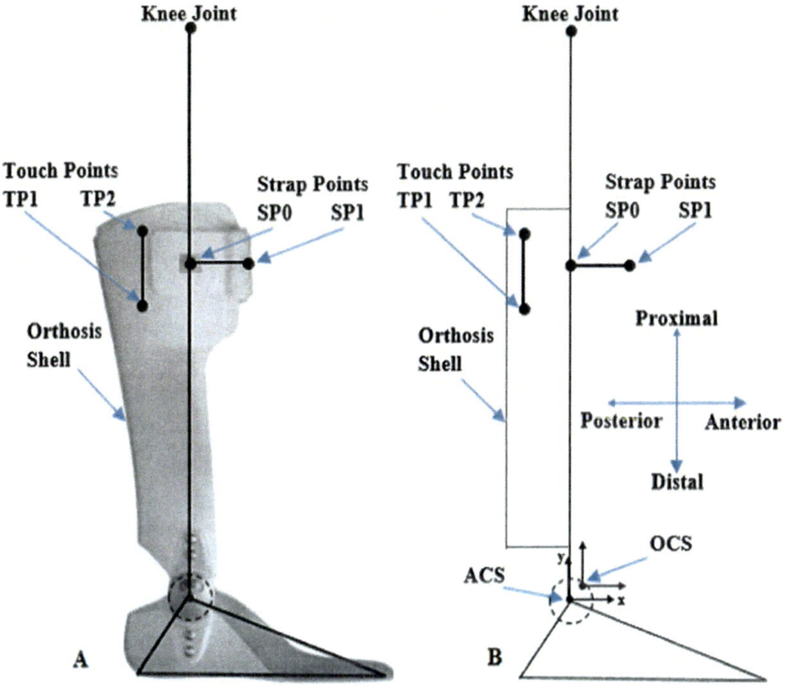

Fig. 1. Illustration of the terminologies used in the simulation model. **A**, Link segment diagram of the model superimposed on the AFO. **B**, Link segment diagram of the proposed 2D simulation model with the conventions used for analysis.

2.2 Mathematical Formulation

The mathematical formulation used for the simulation is similar to the previous study by Bapat et al., (2021) [6]. The Anterior-Posterior (A-P) and Proximal-Distal (P-D) relative motions were calculated between the TP and the SP lines for various ankle joint misalignments by systematically introducing OCS offset with respect to the ACS.

2.3 Ankle Joint Misalignment Cases

As the shank rotates from $-20°$ to $20°$, based on the practical feasibility, the OCS was misaligned in the following four cases with a 5 mm step size for each misalignment:

- *A-P misalignments*: OCS was moved from 20 mm posterior to 20 mm anterior of the ACS ($\delta x = -20{:}5{:}20$ mm, $\delta y = 0$ mm).
- *P-D misalignments*: OCS was moved from 20 mm distal to 20 mm proximal of the ACS ($\delta x = 0$ mm, $\delta y = -20{:}5{:}20$ mm).
- *Posterior-Distal to Anterior-Proximal misalignments*: OCS was moved from 20 mm posterior and 20 mm distal to 20 mm anterior and 20 mm proximal of the ACS in a diagonal manner ($\delta x = -20{:}5{:}20$ mm, $\delta y = -20{:}5{:}20$ mm).
- *Posterior-Proximal to Anterior-Distal misalignments*: OCS was moved from 20 mm posterior and 20 mm proximal to 20 mm anterior and 20 mm distal of the ACS in a diagonal manner ($\delta x = -20{:}5{:}20$ mm, $\delta y = 20{:}-5{:}-20$ mm).

3 Results

Relative motions simulated by the model were quantified and visualized in the A-P and P-D directions. Displacements in the P-D direction indicate slippage or shearing between the leg and the orthosis, and those in A-P direction indicate resulting pressure on the leg. For the A-P relative motion, the orthosis TP line segment leading the limb TP line segment was considered as positive displacement (interference), while the opposite was considered a negative displacement (clearance). For P-D relative motion, the orthosis SP line segment sliding up with respect to the limb SP line was considered as positive displacement, while the opposite was considered a negative displacement.

Figure 2 and Fig. 3 illustrate the simulations of relative motion measured between the TP lines for the four misalignment cases. Similar results are obtained for the simulations of relative motion measured between the SP lines. The effects of OCS misalignments for all the misalignment cases are summarized in Table 1.

- *A-P misalignments*: The A-P relative motion increases as the shank flexion increases on either side of 0° (Fig. 2A). The maximum A-P misalignment of $\pm$ 20 mm results in a maximum gap of $\pm$ 1.21 mm between the limb and orthosis at $\pm$ 20° shank angle. The P-D relative motion increases as the shank flexion increases on either side of 0°, but takes a reversal in its direction of motion (Fig. 2B). The maximum P-D misalignment of $\pm$ 20 mm results in a maximum pistoning of $\pm$ 6.84 mm at $\pm$ 20° shank angle.
- *P-D misalignments*: The A-P relative motion increases as the shank flexion increases on either side of 0°, but takes a reversal in its direction of motion (Fig. 2C). The maximum A-P misalignment of $\pm$ 20 mm results in a maximum gap of $\pm$ 6.84 mm between the limb and orthosis at $\pm$ 20° shank angle. The P-D relative motion increases as the shank flexion increases on either side of 0° (Fig. 2D). The maximum P-D misalignment of $\pm$ 20 mm results in a maximum pistoning of $\pm$ 1.21 mm at $\pm$ 20° shank angle.

- *Posterior-Distal to Anterior-Proximal misalignments*: Both A-P and P-D relative motions decrease as the shank rotates from −20° to 0°, reach a minimum at 0°, and then increase with a reversal in the direction of their motions (Fig. 3A and 3B). The maximum misalignments (20 mm posterior and 20 mm distal, and 20 mm anterior and 20 mm proximal) result in a maximum gap of ± 8.05 mm between the limb and orthosis at 20° shank angle and a maximum pistoning of ± 8.05 mm at −20° shank angle.
- *Posterior-Proximal to Anterior-Distal misalignments*: Both A-P and P-D relative motions decrease as the shank rotates from −20° to 0°, reach a minimum at 0°, and then increase with a reversal in the direction of their motions (Fig. 3C and 3D). The maximum misalignments (20 mm posterior and 20 mm proximal, and 20 mm anterior and 20 mm distal) result in a maximum gap of ± 8.05 mm between the limb and orthosis at −20° shank angle and a maximum pistoning of ± 8.05 mm at 20° of shank flexion.

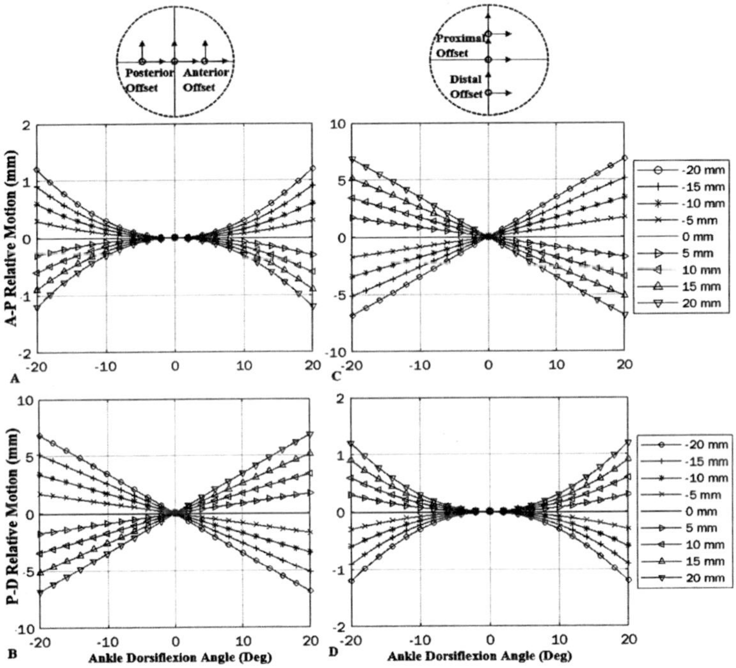

Fig. 2. A-P and P-D relative motion measured between the TP lines for orthotic ankle joint misalignments in A-P and P-D directions.

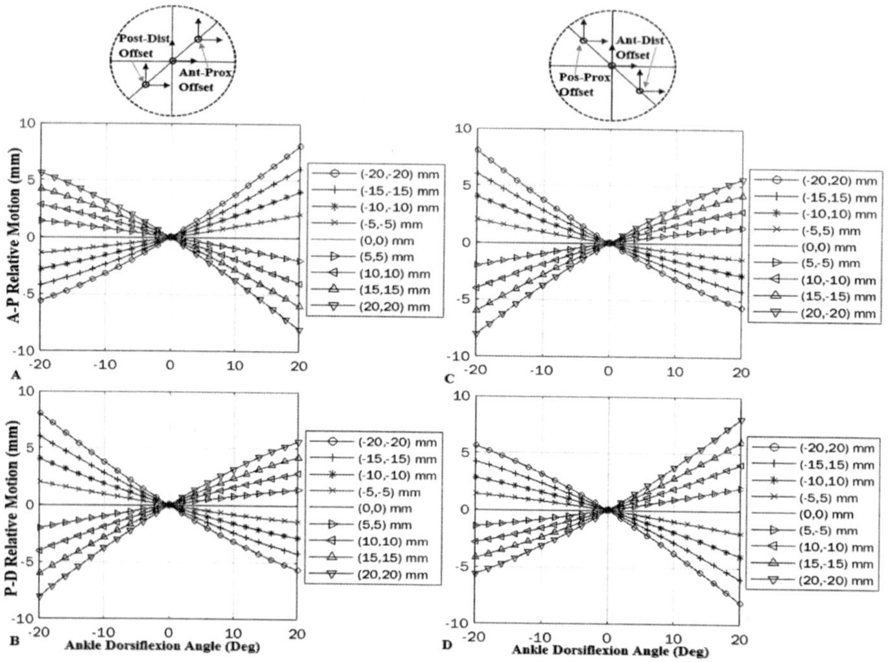

Fig. 3. A-P and P-D relative motion measured between the TP lines for combined misalignments in A-P and P-D directions of orthotic ankle joint.

Table 1. Summary of the effects of orthotic ankle joint misalignments predicted by the model

Direction of misalignment	Direction of strap/orthosis movement with respect to leg (determined through SP lines)		Resultant pressure point location on leg (determined through TP lines)	
	Plantar flexion	Dorsiflexion	Plantar flexion	Dorsiflexion
Posterior	Anterior & Proximal	Anterior & Distal	Posterior	Posterior
Anterior	Posterior & Distal	Posterior & Proximal	Anterior	Anterior
Proximal	Anterior & Distal	Posterior & Distal	Posterior	Anterior
Distal	Posterior & Proximal	Anterior & Proximal	Anterior	Posterior
Posterior-Distal	Posterior & Proximal	Anterior & Distal	Anterior	Posterior

(*continued*)

Table 1. (*continued*)

Direction of misalignment	Direction of strap/orthosis movement with respect to leg (determined through SP lines)		Resultant pressure point location on leg (determined through TP lines)	
	Plantar flexion	Dorsiflexion	Plantar flexion	Dorsiflexion
Anterior-Proximal	Anterior & Distal	Posterior & Proximal	Posterior	Anterior
Posterior-Proximal	Anterior & Proximal	Posterior & Distal	Posterior	Anterior
Anterior-Distal	Posterior & Distal	Anterior & Proximal	Anterior	Posterior

Figure 4A shows the simulation of the shank movement from −20° to 20° with a posteriorly offset OCS, and Fig. 4B shows the effects of an anteriorly offset OCS in terms of resulting relative motion and pressure points. Similar sets of simulations with OCS offsets along the P-D direction, and for combined misalignments (Posterior-Distal to Anterior-Proximal and Posterior-Proximal to Anterior-Distal) are shown in Fig. 5A and Fig. 5B, Fig. 6A and Fig. 6B, and Fig. 7A and Fig. 7B respectively.

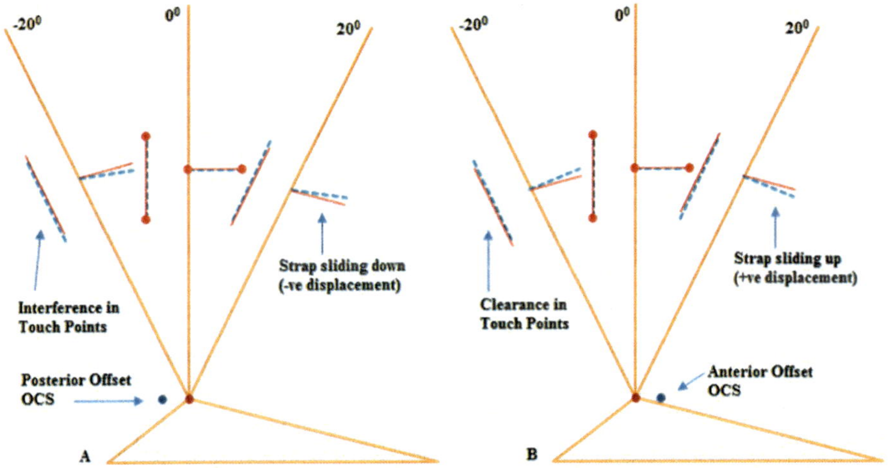

Fig. 4. Simulation of shank movement with OCS misalignments in A-P direction; solid red line with circular end points represents coincident touch and strap points at 0° flexion, solid red line (without end points) represents touch and strap points moving with the orthosis, and the dashed blue line represents touch and strap points moving with the limb. A, Posterior offset. B, Anterior offset.

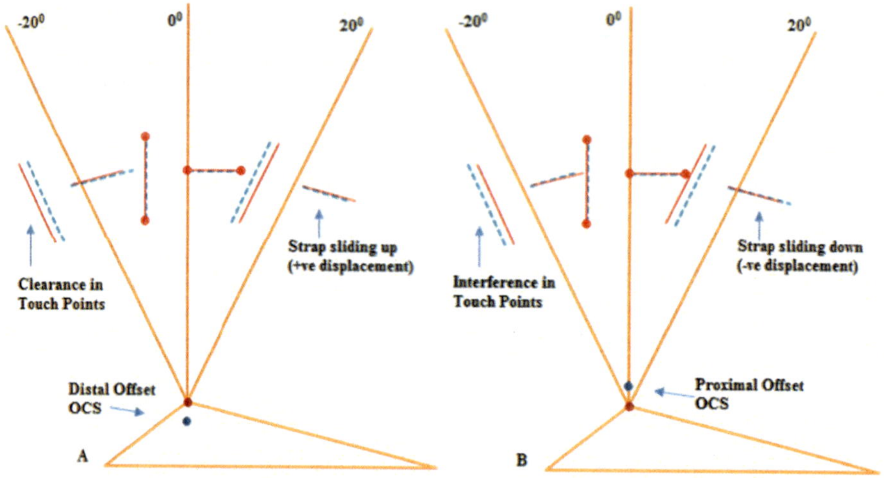

Fig. 5. Simulation of shank movement with OCS misalignments in P-D direction; solid red line with circular end points represents coincident touch and strap points at 0° flexion, solid red line (without end points) represents touch and strap points moving with the orthosis, and the dashed blue line represents touch and strap points moving with the limb. A, Distal offset. B, Proximal offset.

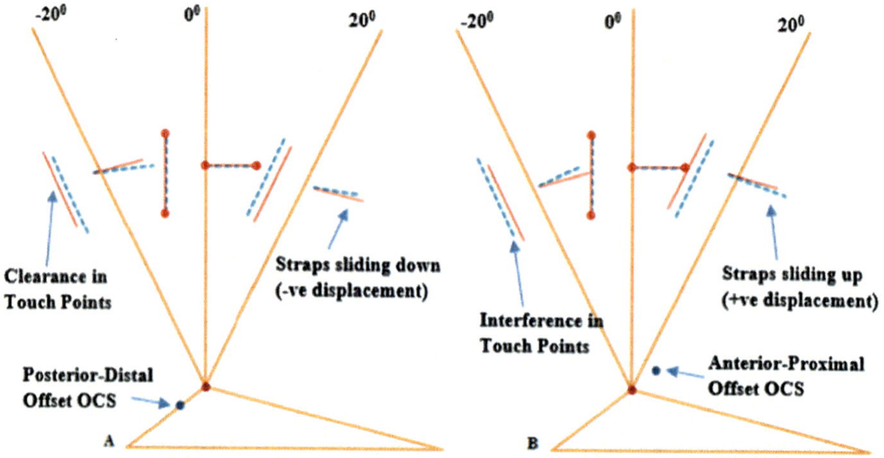

Fig. 6. Simulation of shank movement with combined OCS misalignments; solid red line with circular end points represents coincident touch and strap points at 0° flexion, solid red line (without end points) represents touch and strap points moving with the orthosis, and the dashed blue line represents touch and strap points moving with the limb. A, Posterior-Distal Offset. B, Anterior-Proximal Offset.

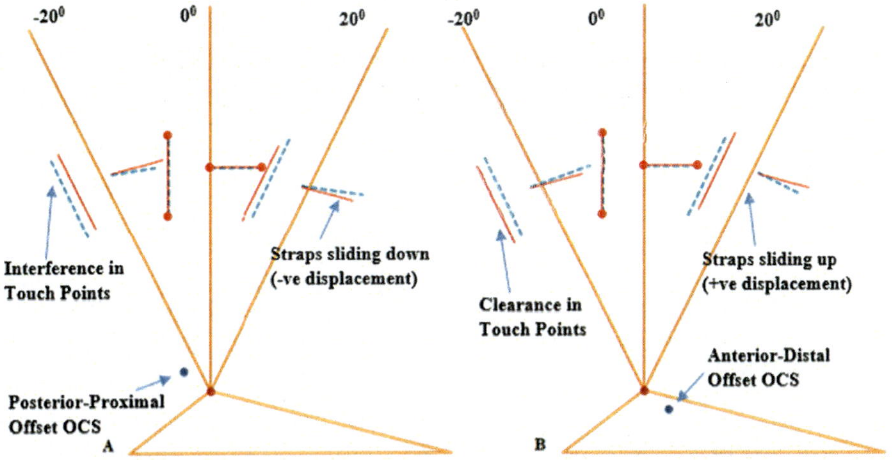

Fig. 7. Simulation of shank movement with combined OCS misalignments; solid red line with circular end points represents coincident touch and strap points at 0° flexion, solid red line (without end points) represents touch and strap points moving with the orthosis, and the dashed blue line represents touch and strap points moving with the limb. A, Posterior-Proximal offset. B, Anterior-Distal offset.

4 Discussion

In this work, a 2D simulation-based model was proposed to study the effects of orthotic ankle joint misalignments in an articulated AFO for the functional range of ankle joint motion. The OCS was systematically misaligned about ACS in the A-P and P-D directions and their combinations to analyze the relative motion between the limb and orthosis and locate pressure points on the leg. The model was superimposed on one of the articulated AFO designs that assist in both dorsiflexion and plantar flexion. While some articulated AFOs (hinged, adjustable, or spring-assisted AFOs, etc.) take into account the stiffness factors and allow controlled plantar flexion [7, 8], for the sake of simulation purposes, the plantar flexion range has been considered to be till 20° (as considered in previous studies [3–5]).

Strap movements and resulting pressures on the leg predicted by the model (shown in Table 1) for A-P and P-D misalignments agree with the NYU diagrams [1]. Simulation results showed that misalignments in the A-P direction have a greater pistoning motion than those in P-D direction (Fig. 2B and Fig. 2D) which corroborates with a previous study in the literature [3]. The simulation results also show that the resulting pressure on the leg is greater for P-D misalignments than A-P misalignments. This is indicative from Figs. 2A and 2C where A-P relative motion (representing the gap between TP lines) is greater for P-D misalignments than A-P misalignments. Additionally, the proposed model also simulates the combinations of A-P and P-D misalignments. Results show that the combined misalignments lead to an even greater pistoning motion and resulting pressures than the standalone A-P and P-D misalignments (Fig. 2 and Fig. 3). This result implies that there is a superposition of relative motions from A-P and P-D directions for combined misalignments. It has also been observed from simulations that the relative sliding motion and resulting pressure points are of repetitive and of reversing nature. Based on the findings in this work, it can be concluded that orthotic ankle joint misalignments have a significant impact on the AFO user comfort. The simulation-based model presented here can be used to guide anatomical and external joint alignment in lower-limb orthoses and exoskeleton devices.

The results from the proposed model are limited to linear misalignments in the sagittal plane and do not predict angular misalignment effects. The model assumes that the link segment is rigid and there is no relative motion between the foot and foot shell of the AFO. The future scope of work includes more sophisticated models that consider stiffness, friction, and other physical parameters between the limb and the orthosis. Furthermore, actual gait data measurements of orthosis users during walking, stand-to-sit activity, etc., can be collected to validate the model.

5 Conclusions

Simple 2D simulation-based model to study the consequences of sagittal plane ankle joint misalignments was presented and validated with the previous literature. The mathematical formulation helps in quantifying and visualizing the relative motions between the leg and orthosis in the A-P and P-D directions. The model showed that A-P misalignments caused a greater pistoning motion than P-D misalignments supporting previous findings in the literature [1, 3]. The model also showed that P-D misalignments caused greater resulting pressure on the leg than A-P misalignments. The proposed model additionally showed that combined misalignments caused greater pistoning motion and pressure due to the superposition of relative motions from A-P and P-D directions. Overall, this work emphasizes the importance of accurate alignment and fitting of orthosis for improved patient comfort and care in a rehabilitation setting.

Acknowledgments. This research work was supported through the Research Initiation Grant (2021–2023) from BITS Pilani K K Birla Goa Campus, Goa, India. This report of findings does not represent the official views of the funding agencies.

References

1. New York University: Lower-limb orthotics: including orthotists' supplement. New York: Prosthetics and Orthotics, New York University Post-Graduate Medical School (1974)
2. Roniger, L.: Skin-care issues related to orthotic device wear. LER, May (2016). https://lermagazine.com/article/skin-care-issues-related-to-orthotic-device-wear
3. Fatone, S., Hansen, A.H.: A model to predict the effect of ankle joint misalignment on calf band movement in ankle-foot orthoses. Prosthet Orthot Int **31**(1), 76–87 (2007)
4. Fatone, S., Johnson, W., Kwak, S.: Using a three-dimensional model of the ankle-foot orthosis/leg to explore the effects of combinations of axis misalignments. Prosthet Orthot Int **40**(2), 247–520 (2014)
5. Fatone, S., Johnson, W.B., Tucker, K.: A three-dimensional model to assess the effect of ankle joint axis misalignments in ankle-foot orthoses. Prosthet Orthot Int **40**(2), 240–246 (2016)

6. Bapat, G.M., Sujatha, S.: A two-dimensional mathematical model to simulate the effects of knee center misalignment in lower-limb orthoses. Journal of Prosthetics and Orthotics **33**(1), (2021)
7. Yamamoto, M., Shimatani, K., Hasegawa, M., Kurita, Y.: Effect of ankle-foot orthosis on gait kinematics and kinetics: case study of post-stroke gait using a musculoskeletal model and an orthosis model. ROBOMECH Journal **6**(1), 6:9,(2019)
8. Asa, F., Mattias, M., Anton, A.: The effect of ankle foot orthosis' design and degree of dorsiflexion on achilles tendon biomechanics - tendon displacement, lower leg muscle activation, and plantar pressure during walking. Front Sports Act Living **2**, 16 (2020)

DIM: Direct and Inverse Methods for Wave Propagation Prediction

Theorical Modelling of Longitudinal Wave Propagation Emitted by a Tunnel Boring Machine in a Finite Domain

Antoine Rallu(✉) and Denis Branque

University of Lyon, ENTPE LTDS - UMR CNRS 5513 - CeLyA, Vaulx-en-Velin, France
antoine.rallu@entpe.fr

Abstract. Boring shallow tunnels in urban areas generates ground-borne vibrations with energy distributed over a wide range of frequencies, up 100 Hz during excavation. As a multiple vibration source, the TBM emits waves until existing building foundations and the free surface. They do not depend only on the source and its surrounding environment but also on the propagation environment. Based on several dynamic in-situ campaigns with synchronized measurements inside the TBM and on the ground, a numerical modelling of the propagation of waves emitted by a tunnel boring machine is proposed with an original considering of the source term.

Keywords: Ground-borne vibrations · Tunnel boring machine · In-situ measurements

1 Introduction

This paper addresses the wave propagation in a layered finite medium with an intern dynamic source of vibration. This problem reflects the modelisation of vibrations emitted by a tunnel boring machine, excavating a shallow tunnel in a free field, that is whithout inclusions inside the ground. The closure of the domain is a need for solving the problem with Finite Difference Method or Finite Element Method. For this purpose, absorbing boundary conditions will be applied.

The key part of this work is the modelisation of the dynamic sollicitation. For achieving it, we propose an empirical model combining (i) strength and pressure measurements inside the chamber of the TBM, and (ii) dynamic velocity measurements inside the manlock of the TBM. These measurements were carried out during the excavation of a tunnel in Lyon (France) in a granite hill.

2 Tunnel Excavation in Lyon, France

2.1 Presentation of the Site

The measurement campaign has been performed on the project of extension of the line B of the Lyon subway in september 2020 in a granite hill, see Fig. 1. The

Z. Dimitrovová et al. (Eds.): WMVC 2022, MMS 125, pp. 169–177, 2023.
https://doi.org/10.1007/978-3-031-15758-5_16

tunnel boring machine was a variable density pressure TBM (Herrenknecht), whose the main features are presented on Table 1.

Table 1. TBM average parameters during the campaign.

Parameter	TBM
Chamber	Full of bentonite
Excavation diameter	9.8 m
Penetration pitch	7 mm/revolution
Bentonite face pressure in the tunnel axis (P_{chamber})	180 kPa
Wheel/ground contact force (F_{contact})	15 000 kN
Area cutter head ($A_{\text{cutter head}}$)	75 m^2

According to Fig. 1, the cover of the tunnel consists approximatively in 20 m of good granite (moderate to high uniaxial compressive strength $\sigma_c = 90 \pm 70$ MPa and rock quality designation (RQD) between 50% and 75%, that is moderately fractured), 5 m of altered granite and 2 m of sandy clayey limes (alluvium).

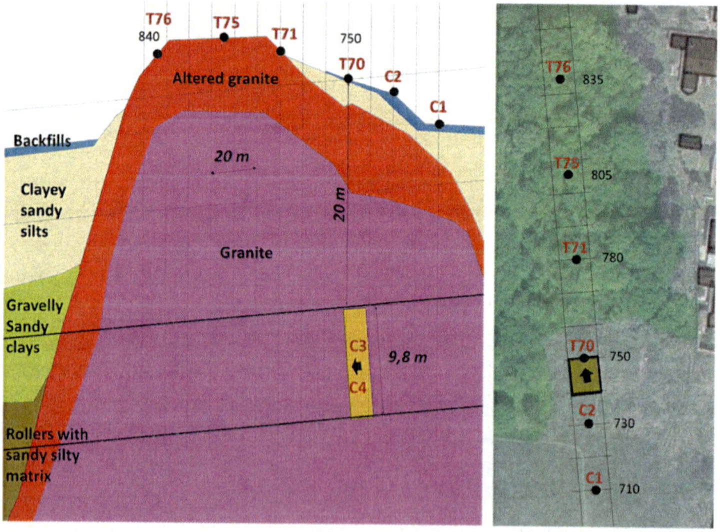

Fig. 1. Plan and longitudinal section of the Lyon-granites measurements (source [4]).

2.2 Instrumentation and Acquisition

The experimental set-up is made of ©TROMINO sensors (speed acquisition) of the Moho brand, denoted C1, C2, C3, C4, T70, T71, T75 and T76. These sensors get three orthogonal acquisition channels (X, Y, Z) of high sensitivity (sensors designed to measure ambient noise). Note that the X-direction is oriented in the tunnel axis in the excavation sens, the Y-direction in the transverse direction, and Z is oriented in the vertical direction.

Before each acquisition, all the sensors are synchronized by GPS at the same location, then moved according to the experimental set-up (see Fig. 1) on the surface and inside the two manlocks of the TBM. This allow to sensors inside he TBM to be synchronized, even if the connexion to GPS is impossible. Moreover, because of the low vibration levels and the heavy weight of the sensors non sealement is apply to connect them to their support. All measurements were acquired at a sampling rate 512 Hz.

A measurement campaign is carried out in two steps:

Ambient noise. Only for sensors on the surface, a measurement phase (about 30 min) under ambient mechanical noise (mainly road and pedestrian traffic), allowing to estimate (i) the characteristic ambient noise level and (ii) frequency content of the ground, that is the frequency ranges where the response of the ground is the highest.

Excavation. For sensors on the surface and inside the TBM, a synchronized acquisition phase during the excavation phase.

Each acquisition is pre-processed in order to clean the signals: (i) restriction of the time histories to a relevant (quasi)-stationary signal, (ii) suppression of the time-history offset, and (iii) band-pass filtering in the range [0.05–40]Hz corresponding to the usual frequencies to investigate ground borne vibrations. All the raw data are available in [3] and in the dataset https://mycore.core-cloud.net/index.php/s/ybajakmrNwcRSqw.

3 Geometry of the Domain

The goal of this model is not to be representative of the topography of the site. In particular, the hill is not modelized and a infinite two-layers stratified flat domain is considered. The idea is to take into account only sensors T71–T75 and T76 (Fig. 1) which are quite on the same altitude and are over only altered granite and good granite layers.

Because of the symmetry of the problem, only one half of the field is modelized as a box ($\ell \times L \times H$ respectively according to $(\underline{\mathbf{e}}_x, \underline{\mathbf{e}}_y, \underline{\mathbf{e}}_z)$, see Fig. 2) defining the set

$$\mathcal{V} =]0, \ell[\times]0, L[\times] - H, 0[$$

whose the six boundaries are denoted respectively:

$$\partial\mathcal{V} = \partial\mathcal{V}_{top} \cup \partial\mathcal{V}_{bottom} \cup \partial\mathcal{V}_{b,x} \cup \partial\mathcal{V}_{e,x} \cup \partial\mathcal{V}_{b,y} \cup \partial\mathcal{V}_{b,y}$$

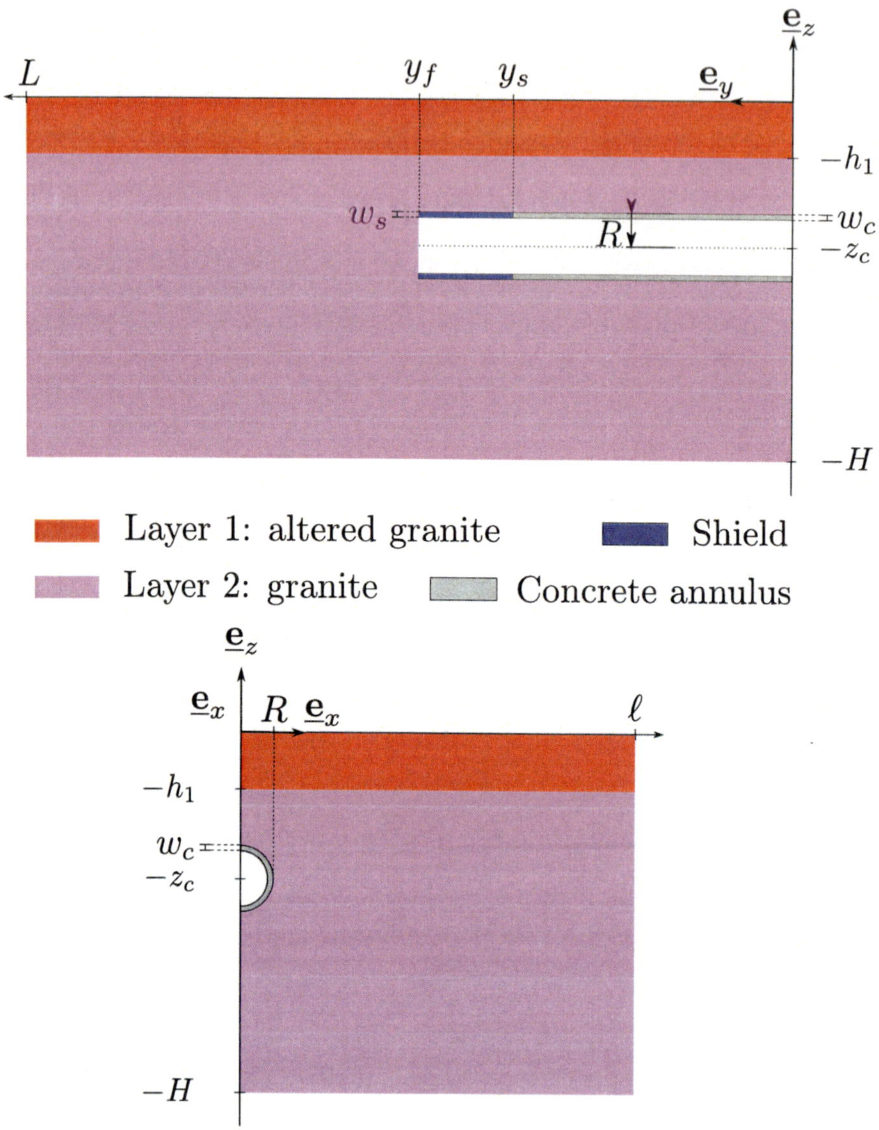

Fig. 2. Geometry of the problem illustrated according two plane views. Top: longitudinal view $\partial\mathcal{V}_{b,x}$ (plane $(\underline{\mathbf{e}}_y, \underline{\mathbf{e}}_z)$ along the symmetry plane $x = 0$). Bottom: cross section $\partial\mathcal{V}_{b,y}$ (plane $(\underline{\mathbf{e}}_x, \underline{\mathbf{e}}_z)$ at $y = 0$).

with

$$\partial\mathcal{V}_{bottom} = [0, \ell] \times [0, L] \times \{z = -H\} \quad ; \quad \partial\mathcal{V}_{top} = [0, \ell] \times [0, L] \times \{z = 0\}$$
$$\partial\mathcal{V}_{b,y} = [0, \ell] \times \{y = 0\} \times [-H, 0] \quad ; \quad \partial\mathcal{V}_{e,y} = [0, \ell] \times \{y = L\} \times [-H, 0]$$
$$\partial\mathcal{V}_{b,x} = \{x = 0\} \times [0, L] \times [-H, 0] \quad ; \quad \partial\mathcal{V}_{e,x} = \{x = \ell\} \times [0, L] \times [-H, 0]$$

The geology of the ground is simplified as a two layers, each of height ($\{h_1, h_2\}$ as $h_1 + h_2 = H$) defining two sets:

$$\overset{1}{\mathcal{V}} =]0, \ell[\times]0, L[\times] - h_1, 0[\quad ; \quad \overset{2}{\mathcal{V}} =]0, \ell[\times]0, L[\times] - H, -h_1[$$

The excavated tunnel is a half-pipe (radius R_t) whose the center is at the depth $-z_c$ on the symmetry plane ($x = 0$) alongside the direction $y \in]0, y_f[$, as:

$$\mathcal{T} = \{x^2 + (z + z_c)^2 < R_t^2, \, x > 0\} \cup \{y \in]0, y_f[\}$$

At the intrados of the excavated tunnel, concrete annulus (width w_c) are located in the built part of the tunnel ($0 < y \leq y_s$) whereas the shield of the TBM is also an annulus, of width w_s located in ($y_s < y < y_f$).

$$\begin{aligned}
\mathcal{T}_c &= \{(R_t - w_c)^2 < x^2 + (z + z_c)^2 < R_t^2, \, x > 0\} \cup \{y \in]0, y_s[\} \\
\mathcal{T}_s &= \{(R_t - w_s)^2 < x^2 + (z + z_c)^2 < R_t^2, \, x > 0\} \cup \{y \in]y_s, y_f[\}
\end{aligned}$$

Consequently, the empty volume of air $\mathcal{E}$ inside the excavation is defined as ($\bar{\cdot}$ denotes the closure of a set):

$$\begin{aligned}
\mathcal{E} &= \overline{\mathcal{T}}|_{\overline{\mathcal{T}_c \cup \mathcal{T}_s}} \\
&= \{(x^2 + (z + z_c)^2 < (R_t - w_c)^2, \, x > 0\} \cup \{y \in]0, y_s[\} \\
&\quad \bigcup \{(x^2 + (z + z_c)^2 < (R_t - w_s)^2, \, x > 0\} \cup \{y \in]y_s, y_f[\}
\end{aligned}$$

Therefore, the studied domain Ω is defined as the union of the four homogeneous domains ($\mathcal{D} = [\![1; 4]\!]$):

$$\Omega = \mathscr{V}|_{\overline{\mathscr{E}}} = \bigcup_{j \in \mathcal{D}} \Omega_j = \overset{1}{\Omega} \cup \overset{2}{\Omega} \cup \mathcal{T}_c \cup \mathcal{T}_s \quad \text{with} \quad \overset{1}{\Omega} = \overset{1}{\mathcal{V}}; \, \overset{2}{\Omega} = \overset{2}{\mathcal{V}}|_{\mathcal{T}} \tag{1}$$

whose the boundaries are:

$$\partial\Omega = \partial\Omega_{bottom} \cup \partial\Omega_{top} \cup \partial\Omega_{b,y} \cup \partial\Omega_{e,y} \cup \partial\Omega_{b,x} \cup \partial\Omega_{e,x} \cup \partial\Omega_f \cup \partial\Omega_c \cup \partial\Omega_s$$

with

$$\begin{aligned}
&\partial\Omega_{bottom} = \partial\mathcal{V}_{bottom} && \partial\Omega_{top} = \partial\mathcal{V}_{top} \\
&\partial\Omega_{b,y} = \partial\mathcal{V}_{b,y}|_{\{x^2+(z+z_c)^2<(R_t-w_c)^2, \, x>0\}} && \partial\Omega_{e,y} = \partial\mathcal{V}_{e,y} \\
&\partial\Omega_{b,x} = \partial\mathcal{V}_{b,x}|_{\{0\leq y\leq y_s, |z+z_c|\leq R_t-w_c\}\cup\{y_s\leq y\leq y_f, |z+z_c|\leq R_t-w_s\}} && \partial\Omega_{e,x} = \partial\mathcal{V}_{e,x} \\
&\partial\Omega_f = \{x^2 + (z + z_c)^2 \leq (R_t - w_s)^2, x \geq 0\} \cup \{y = y_f\} \\
&\partial\Omega_c = \{x^2 + (z + z_c)^2 = (R_t - w_c)^2, \, x > 0\} \cup \{0 \leq y \leq y_s\} \\
&\partial\Omega_s = \{x^2 + (z + z_c)^2 = (R_t - w_s)^2, \, x > 0\} \cup \{y_s \leq y \leq y_f\}
\end{aligned}$$

Finally, the interfaces between domains $\{\Omega_j\}_{j \in \mathcal{D}^2}$ are denoted $\{\partial\Omega_{ij}\}_{(i,j) \in \mathcal{D}}$.

4 Formulation of the Problem

4.1 Materials

In this study, all materials (in particular granites) are assumed to work in a small strain rate and to behave as linear isotropic elastic media, whose the elastic parameters are listed in Table 2. That is, for each domain Ω_j (tr() is the trace operator and $\underline{\underline{I}}$ the unity matrix)

Table 2. Mechanical parameters

Material	j	ρ [kg/m^3]	E [GPa]	ν [−]	K_0 [−]
Layer 1: Altered granite	1	2400	7.5	0.25	0.6
Layer 2: Granite	2	2700	60	0.25	0.4
Concrete	3	2500	30	0.2	
Shield	4	7800	200	0.3	

$$\overset{j}{\underline{\underline{\sigma}}}(\underline{x},t) = \frac{E_j}{1+\nu_j}\left(\overset{j}{\underline{\underline{\varepsilon}}}(\underline{x},t) + \frac{\nu_j}{1-2\nu_j}\mathrm{tr}\left(\overset{j}{\underline{\underline{\varepsilon}}}(\underline{x},t)\right)\underline{\underline{I}}\right) \qquad \forall \underline{x} \in \Omega_j \tag{2}$$

The coefficient K_0 is the pressure coefficient at rest, allowing to initialize the horizontal stresses in the ground according the vertical stress (geostatic stresses). The value of this coefficient varies according to the nature of the ground and of its geological history. Values in Table 2 were defined in the geotechnical synthesis report of the project.

4.2 Modelling the Dynamic Sollicitation

In the in-situ experimental campaign, two sensors were located in the manlocks of the TBM. However, as shown in [4], the vibration level measured inside a manlock is not consistent with the vibration levels measured on surface; in other words, only one part of the wave field measured on the manlock propagates into the ground. Nevertheless, the two frequency contents are consistent.

On the other hand, we get the driving parameters of the TBM, allowing to estimate the different stresses applied on the ground. In this section, in order to generate a relevant longitudinal dynamic stress, we will show how combine (i) frequency content of from dynamic sensors and (ii) "static" TBM stress and force parameters. As a first step, in this study we only focus on a dynamic sollicition in the longitudinal (alongside y) direction.

The force applied on the area of the cutter head, denoted F_{face}, is made of two terms:

$$F_{\text{face}} = F_{\text{contact}} + F_{\text{ground pressure}}$$

with F_{contact}, the force applied on the ground by the cutter head of TBM measured by a sensor in the jack, and $F_{\text{ground pressure}} = P_{\text{chamber}} \times A_{\text{cutter head}}$ the force resulting from the confinement pressure measured by sensors in the center of chamber coming from ground, and applied on $A_{\text{cutter head}}$, the area of the cutter head.

Considering the values of Table 1, for an area $A_{\text{cutter head}} = 75\,\text{m}^2$ the value of $F_{\text{face}} = 28500$ kN. Moreover, the mean pressure applied on the TBM surface to the ground is then defined as:

$$P_{\text{mean}} = \frac{F_{\text{face}}}{A_{\text{cutter head}}} = 380\,\text{kPa}$$

It should be noted that the lateral friction between the TBM and the ground is weak according to the total thrust force in the thrust jacks. In addition, the annular gap between the rings of segments and the ground is filled with mortar, but this is not modelled as the injection of mortar does not generate significant vibrations in the ground.

The goal is to generate a dynamic stress oscillating around P_{mean}, the characteristic longitudinal pushing stress measured by the sensors inside the TBM. For this purpose, from the measured velocities inside the manlock of the TBM we select a small ($T = 1$ s) relevant part (i.e. a stationary part without shock or special event) of the longitudinal channel. Moreover this signal is filtered in the frequency range [0–40] Hz characterizing ground-borne vibrations. This signal is denoted $v(t)$. In order to only keep the fluctuations (i.e. frequency content) of the signal without its amplitude, this latter is normalized by its maximum value $v_{\max} = \max_{t \in [0,T]} |v(t)|$. Finally, the following dynamic normal stress is proposed to modelize the time-varying pressure applied on the front:

$$P(t) = P_{\text{mean}} \left(1 + \alpha \frac{v(t)}{v_{\max}}\right) \tag{3}$$

with $\alpha \leq 1$ a small parameter to calibrate.

4.3 Boundary Conditions

All the boundary conditions are written in (4), the two first one being carried by the kinematic variable $\underline{u}$, the others by the associated stress $\underline{\underline{\sigma}}$. Vectors $\underline{n}$ and $\underline{t}$ respectively represent the outgoing normal and tangential vectors of a boundary. $\forall t \in]0, T]$,

$$\underline{u}(\underline{x}, t) = \underline{0} \qquad \forall \underline{x} \in \partial\Omega_{bottom} \tag{4a}$$

$$\underline{u}(\underline{x}, t).\underline{n} = 0 \qquad \forall \underline{x} \in \partial\Omega_{b,x} \tag{4b}$$

$$\underline{\underline{\sigma}}(\underline{x}, t).\underline{n} = \underline{0} \qquad \forall \underline{x} \in \partial\Omega_{top} \cup \partial\Omega_c \cup \partial\Omega_s \tag{4c}$$

$$\begin{cases} \underline{\underline{\sigma}}(\underline{x}, t).\underline{n} = -\rho(\underline{x}) C_P \partial_t \underline{u}(\underline{x}, t).\underline{n} \\ \underline{\underline{\sigma}}(\underline{x}, t).\underline{t} = -\rho(\underline{x}) C_S \partial_t \underline{u}(\underline{x}, t).\underline{t} \end{cases} \qquad \forall \underline{x} \in \partial\Omega_{b,y} \cup \partial\Omega_{e,y} \cup \partial\Omega_{e,x} \tag{4d}$$

$$\underline{\underline{\sigma}}(\underline{x}, t).\underline{n} = P(t) \qquad \forall \underline{x} \in \partial\Omega_f \tag{4e}$$

The homogeneous Dirichlet condition (4a) expresses that at the bottom on the domain stands a bedrock, in other words Ω is embedded in $z = -H$. Equation (4b) naturally translates the symmetry of the kinematics field $\underline{u}$. The homogeneous conditions (4c) classically express that normal stresses vanish at the boundary between a medium and the air. In order to close the domain, the two conditions (4d) refer to the modelisation of an Absorbing Boundary Condition, see [2], with C_P and C_S the velocities of P-waves and S-waves, as:

$$C_P = \sqrt{\frac{E}{2\rho\,(1+\nu)\,(1-2\nu)}} \quad ; \quad C_S = \sqrt{\frac{E}{2\rho\,(1+\nu)}}$$

Finally, the Eq. (4e) imposes a dynamic normal stress to the medium, see Sect. 4.2.

4.4 Interfaces

At the interfaces $\{\partial\Omega_{ij}\}_{(i,j)\in\mathcal{D}^2}$ between material domains $\{\Omega_j\}_{j\in\mathcal{D}}$ we classically impose the continuity of the kinematic variable and of the normal stresses:

$$\forall (i,j) \in \mathcal{D}^2,\ \forall (\underline{x},t) \in \partial\Omega_j \times [0,T], \quad \begin{cases} \underline{u}|_{\Omega_i}(\underline{x},t) = \underline{u}|_{\Omega_j}(\underline{x},t) \\ \underline{\underline{\sigma}}(\underline{x},t)|_{\Omega_i}.\underline{n} = \underline{\underline{\sigma}}(\underline{x},t)|_{\Omega_j}.\underline{n} \end{cases} \tag{5}$$

4.5 Initial Conditions

The initial conditions are defined in (6), that are nullity of the displacement and velocity fields. These initial values were chosen because the measured ambient noise was negligible compared to the velocities during excavation.

$$\forall \underline{x} \in \Omega, \begin{cases} \underline{u}(\underline{x},0) = \underline{0} \\ \partial_t \underline{u}(\underline{x},0) = \underline{0} \end{cases} \tag{6}$$

4.6 Balance of the Problem

Finally, the theorical formulation of the problem can be written:

$$\begin{cases} \rho(\underline{x})\partial_t^2 \underline{u}(\underline{x},t) = \underline{\mathrm{div}}\left(\underline{\underline{\sigma}}(\underline{x},t)\right) + \rho(\underline{x})\underline{g} & \forall (\underline{x},t) \in \Omega \times]0,T] \\ \text{Boundary conditions (4)} & \forall (\underline{x},t) \in \partial\Omega \times]0,T] \\ \text{Interfaces conditions (5)} & \forall (\underline{x},t) \in \partial\Omega \times]0,T] \\ \text{Initial conditions (6)} \ \forall \underline{x} \in \Omega & \end{cases} \tag{7}$$

This second-order, linear and hyperbolic partial differential equation with two initial conditions and mixed boundary conditions has a unique solution. The proof of existence and uniqueness of this solution is out of the topic of this paper, but we can refer for example to [1,5,6].

5 Conclusion

In this paper we have built a theorical framework for the resolution of the problem of 3D wave propagation generated by a TBM in a free field through a stratified ground. This simple model expresses some phenomena occurring during the excavation of a tunnel in Lyon (France) when the TBM had crossed a granite hill, whose the material behaviors are reasonably considered linear elastic.

An original empirical model for the longitudinal source of vibration is proposed, combining (i) pressure and force measurements inside the TBM and (ii) dynamic measurement inside the manlock of the TBM.

This model has been written as a boundary condition problem with absorbing boundary conditions to close the domain, in order to be numerically solved by a Finite Difference Method or a Finite Element Method. The next step is the resolution of this problem and testing the sensibility of parameter α (3) on the amplitudes of waves on the surface. To complete this work, a dynamic torque on the cutter wheel must be taken into account in the same spirit as (3).

References

1. Duvaut, G., Lions, J.L.: Les Inequations en Mecanique et en Physique. Dunod, Paris (1972)
2. Lysmer, J., Kuhlemeyer, R.L.: Finite dynamic model for infinite media. J. Eng. Mech. Div. **95**(4), 859–877 (1969)
3. Rallu, A., Berthoz, N.: Vibrations induced by tunnel boring machines in urban areas: Dataset of synchronized in-situ measurements inside the shield and on the surface. Data Brief **41**, 107826 (2022)
4. Rallu, A., Berthoz, N., Branque, D., Charlemagne, S.: Vibrations induced by tunnel boring machines in urban areas: in-situ measurements and methodology of analysis. J. Rock Mech. Geotech. Eng. (2022). https://doi.org/10.1016/j.jrmge.2022.02.014
5. Trefethen, L.N., Halpern, L.: Well-Posedness of One-way wave equations and absorbing boundary conditions. Math. Comput. **47**(176), 421 (1986). https://doi.org/10.2307/2008165
6. Vacus, O.: Mathematical analysis of absorbing boundary conditions for the wave equation: the corner problem. Math. Comput. **74**(249), 177–200 (2004)

Ultrasonic Wave Propagation in Imperfect Concrete Structures: XFEM Simulation and Experiments

Long Nguyen-Tuan[1(✉)], Matthias Müller[2], Horst-Michael Ludwig[2], and Tom Lahmer[1]

[1] Institute of Structural Mechanics, Bauhaus-University Weimar, Weimar, Germany
long.nguyen.tuan@uni-weimar.de

[2] Institute of Material Science, Bauhaus-University Weimar, Weimar, Germany

Abstract. In this paper, we introduce an Extended Finite Element Method (XFEM) to simulate the wave propagation in an imperfect concrete structure. The XFEM allows simulating a small discontinuity in the structure and the change of this discontinuity without re-meshing the whole structure. Therefore, the meshed structure is kept consistent while the structural discontinuity changes. Simultaneously, we improve the model with XFEM mass lumping technique and viscous boundary conditions in order to simulate efficiently the damping of signals' amplitudes. Furthermore, ultrasonic experiments were carried out in our lab to validate the numerical simulation. The simulations and experiments with and without cracks are presented. The effects of the crack on received signal, e.g., time-of-flight, wavelength, and spectrum are also discussed in the paper.

Keywords: XFEM · Ultrasonic wave · Concrete

1 Introduction

Ultrasonic based crack detection by computing time-of-flight of the signal in structures is widely used in the detection of damages and imperfection in the structures. However, when the damage is complex and experimental data is limited, the reconstructed damages are no longer accurate. Model based inverse analysis emerges as an alternative method, where the full waveform recorded is inverse-analysed in order to re-construct the imperfection in the structure. Such problems are nonlinear and generally ill-posed, so they require particular treatment. The method requires an accurate forward numerical model, allowing to reproduce the natural behaviour of the real structure. In this paper, we introduce an XFEM-based formulation to simulate the wave propagation in imperfect concrete structures. The XFEM has been used to simulate longitudinal wave propagation in composite materials [3,12,14]. However, no quantitative comparison is presented to prove the accuracy of the simulation.

Z. Dimitrovová et al. (Eds.): WMVC 2022, MMS 125, pp. 178–188, 2023.
https://doi.org/10.1007/978-3-031-15758-5_17

In order to simulate the propagation of waves in structures, Menouillar et al. [14] have introduced an XFEM technique based on the modified Newmark scheme [15]. This work was carried out based on the dynamical mechanic formulation. This method allows avoiding critical time step reaching to zero when the crack tip is close to element nodes. However, their proposed lumped mass matrix fails to reproduce the exact kinetic energies of all possible rigid body translations [1]. Consequently, partial energy of the wave transmits through the cracked surface. Herein, Asareh et al. [1] propose a technique, in which the mass of each side of the interface is lumped into the finite element nodes and the time integration is performed by adopting an explicit-implicit technique [4].

Damping is an important phenomenon to be considered during the simulations. Most of the previous works on XFEM do not consider damping in their models [3,12]. In many cases, concrete is considered as perfectly elastic at the stress values smaller than 100 kPa in the analysis of dynamic structures. However, there is a fact that damping occurs during analysis of an acoustic wave. The source of damping can be the loss of energy at the interfaces of the specimen, especially at the solid to solid contacts, and damping in the material itself. We assume that the energy loss at the boundary is much greater than the energy dissipation in the material. Many efforts have been made to describe the absorbing boundary conditions [11,18]. Motivated from the model of unbounded half-space simulated by bounded finite element method such as earthquake, sound, and explosion propagations, the boundary value problem must absorb all the incident waves as if it goes to infinity. The researchers sought a technique to absorb perfectly the incident wave onto the boundary. Such technique is call artificial boundary conditions, which are classified to two types [9] : Global absorbing boundary conditions [8] and local absorbing boundary conditions [13]. In this paper, we simulate the absorbing boundary conditions using local viscous boundary method [13].

Additionally, ultrasonic experiments were performed to validate the XFEM model. The concrete specimens with cracks and without crack were prepared. The results show that XFEM is a promising method to simulate accurately the ultrasonic wave propagation in concrete materials; although, the model must be improved quantitatively in waveform comparison.

2 Wave Propagation Using Dynamic XFEM Formulations

2.1 Constitutive Mechanical Equations

We aim in solving the equation of motion in finite element forms, presented in [12]

$$\mathbb{M}\ddot{\mathbf{u}} + \mathbb{C}\dot{\mathbf{u}} + \mathbb{K}\mathbf{u} = \mathbf{f}, \tag{1}$$

where $\mathbb{M}$ is the mass matrix, described in equation Eq. 3; $\mathbb{C}$ is the is the damping matrix expressed as Eq. 5, $\mathbb{K}$ is the stiffness matrix, $\mathbf{F}$ is the external force vector; $\mathbf{u}$, $\dot{\mathbf{u}}$, and $\ddot{\mathbf{u}}$ denote the displacement, velocity, and acceleration vectors discretized on the complete basis of shape functions. The shape function at the enriched elements is a result of voids and cracks in the structure. Each constituent of the coupled matrix in Eq. 1 is extended to include both the crack tip and heaviside enrichment as

$$\begin{bmatrix} \mathbb{M}_{uu}\mathbb{M}_{ua}\mathbb{M}_{ub} \\ \mathbb{M}_{au}\mathbb{M}_{aa}\mathbb{M}_{ab} \\ \mathbb{M}_{bu}\mathbb{M}_{ba}\mathbb{M}_{bb} \end{bmatrix} \begin{Bmatrix} \ddot{\mathrm{u}} \\ \ddot{\mathrm{a}} \\ \ddot{\mathrm{b}} \end{Bmatrix} + \begin{bmatrix} \mathbb{C}_{uu}\mathbb{C}_{ua}\mathbb{C}_{ub} \\ \mathbb{C}_{au}\mathbb{C}_{aa}\mathbb{C}_{ab} \\ \mathbb{C}_{bu}\mathbb{C}_{ba}\mathbb{C}_{bb} \end{bmatrix} \begin{Bmatrix} \dot{\mathrm{u}} \\ \dot{\mathrm{a}} \\ \dot{\mathrm{b}} \end{Bmatrix} + \begin{bmatrix} \mathbb{K}_{uu}\mathbb{K}_{ua}\mathbb{K}_{ub} \\ \mathbb{K}_{au}\mathbb{K}_{aa}\mathbb{K}_{ab} \\ \mathbb{K}_{bu}\mathbb{K}_{ba}\mathbb{K}_{bb} \end{bmatrix} \begin{Bmatrix} \mathrm{u} \\ \mathrm{a} \\ \mathrm{b} \end{Bmatrix} = \begin{Bmatrix} \mathrm{f}_u \\ \mathrm{f}_{\mathrm{a}} \\ \mathrm{f}_b \end{Bmatrix} \tag{2}$$

where $\mathbb{M}_{\alpha\beta}$ is the mass sub-matrix corresponding to the original mesh of FEM and the additional degree of freedom (DoF) in the enrichment α and β; the subscript is defined here as $\alpha\beta = (uu, ua, ub, aa, aa, ab...)$. It is similar for the stiffness sub-matrix $\mathbb{K}_{\alpha\beta}$ and the damping sub-matrix $\mathbb{C}_{\alpha\beta}$. These enrichments are implemented for elements that contain crack tips and also for elements completely cut through by a crack. The individual components of the XFEM mass matrix, stiffness matrix, are defined as follows

$$\mathbb{M}_{\alpha\beta} = \int_{\Omega} (\boldsymbol{N}_u^{\alpha})^T \rho \boldsymbol{N}_u^{\beta} \mathrm{d}\Omega, \tag{3}$$

where $\boldsymbol{N}_u$ is the displacement shape function, $\mathbb{K}_{\alpha\beta}$ is the usual standard stiffness matrix, given as

$$\mathbb{K}_{\alpha\beta} = \int_{\Omega} (\boldsymbol{B}_u^{\alpha})^T \boldsymbol{D} \boldsymbol{B}_u^{\beta} \mathrm{d}\Omega, \tag{4}$$

where $\boldsymbol{D}$ is the stress-strain matrix, which is defined by Young modulus and Poison's ratio; $\boldsymbol{B}_u$ is the derivative matrix of the shape function, see [16] for further details. Additionally, the damping matris is composed as

$$\mathbb{C}_{\alpha\beta} = \lambda \mathbb{M}_{\alpha\beta} + \mu \mathbb{K}_{\alpha\beta}, \tag{5}$$

where λ and μ are Rayleigh damping coefficients [2]. Assuming that cyclic loads are applied only on the original FEM nodes, the load vector reads as:

$$\mathbf{f}_u = \int_{\Gamma_t} (\boldsymbol{N}_u)^T \, \bar{\boldsymbol{t}} \, \mathrm{d}\Gamma, \tag{6}$$

where $\bar{\boldsymbol{t}}$ is the traction force prescribed on the boundaries Γ.

2.2 Enrichment Scheme

To solve the equation of motion, XFEM is used to obtain the values of $u(\boldsymbol{x})$. The displacement field is approximated by locally enriching the finite element space to account for the discontinuity in the displacement derivatives across the domain. Sukumar et al. [17] modified [5] using a level set function ϕ in XFEM to define the interface. One can see Nguyen Tuan et al. [16] for further definition of ψ and Φ functions at the crack zone.

2.3 Mass Lumping Technique

Asareh et al. [1] propose a new mass lumping of enriched elements for explicit transient analysis, which satisfies the kinetic energy conservation. By this method, the mass of each side of the interface is lumped to the finite element nodes and the mass of enriched DoFs are assigned to zero. To overcome the singularity of zero mass matrix, an explicit-implicit algorithm is applied, where the enriched DOFs are treated implicitly. Consequently, the critical time step size is independent from the location of the discontinuity within the elements. This technique, according to the authors, allows conserving the kinetic energy across the crack interface. Based on Asareh et al. [1], for a four-node quadrilateral element with a crack passing through, the mass matrix is written as follows:

$$\mathbb{M} = \begin{bmatrix} \mathbb{M}^E & 0 \\ 0 & M^I \end{bmatrix} = \begin{bmatrix} \mathbb{M}^E & 0 \\ 0 & 0 \end{bmatrix}, \tag{7}$$

where $\mathbb{M}^E$ is the standard FE mass matrix, in which the equations are solved explicitly; $\mathbb{M}^I$ denotes enriched terms, in which the equations are solved implicitly. By this method, $\mathbb{M}^I$ is assigned to zero. The element m_k in $\mathbb{M}^E$ mass matrix is computed as

$$m_k = \frac{m_e}{\mathrm{mes}(\Omega^e)} \int_{\Omega^e} \mathcal{H}(\phi) \mathrm{d}\Omega, \tag{8}$$

where m_e is the mass of element cut by a crack; $\mathrm{mes}(\Omega^e)$ is the area of element Ω^e, $\mathcal{H}(\phi)$ is Heaviside function; k is the node number.

2.4 Absorbed Boundary Conditions

Local absorbing boundary conditions include: Viscous boundary [13], perfectly matched layer [7,10], infinite elements [6].

Our problem is the simulation of acoustic waves propagating in construction elements. The elements are decomposed into finite elements, the contact between the elements or elements and foundation can cause of energy loss, thus, the damping of vibration at the measurement points is to be considered. Therefore, instead of seeking a perfectly absorbing boundary technique, we seek a technique, which can reproduce a partial energy loss at the boundary. The viscous boundary technique [13] is a method suitable for this aim. The force at the boundary is computed as

$$\mathbf{f}^E = \mathbf{f}^{ext} - \int \boldsymbol{\sigma}_d \dot{\mathbf{u}} \mathrm{d}\Omega \tag{9}$$

where $\mathbf{f}^{ext}$ is the external forces, $\boldsymbol{\sigma}_d = \{\sigma, \tau\}$ is the damping stress, which is composed by a normal stress and shear stress τ at boundary.

$$\sigma = \alpha_\sigma \rho c_p, \tag{10}$$

$$\tau = \alpha_\tau \rho c_s, \tag{11}$$

where α_σ and α_τ are the dimensionless coefficients, ρ is the mass density, c_p and c_s is the velocities of P-wave and S-wave respectively

$$c_p = \sqrt{\frac{\lambda + 2G}{\rho}} \text{ and } c_s = \sqrt{\frac{G}{\rho}}, \tag{12}$$

$$\lambda = \frac{E\nu}{(1+\nu)(1-2\nu)} \text{ and } G = \frac{E}{2(1+\nu)}, \tag{13}$$

where E and ν are elastic modulus und Poisson ratio, respectively.

3 Experiments

The ultrasonic wave experiments were performed with apparatus in our lab. The experimental specimens were prepared by mixing cement paste with sand (<2 mm). The specimens with the size of $20 \times 10 \times 10$ cm, (Fig. 1a, b) were cured in water 28 days before the ultrasonic experiments. The ultrasonic apparatus consists of the ultrasonic sender at one side and the receiver on the opposite side of the specimen. Both sender and receiver sensors were mounted on a steel frame (Fig. 1c). The ultrasonic generator can send a signal band about 200 to 250 kHz. The signal is measured directly by a receiver by contacting the sender to the receiver. This signal is used as a boundary condition for numerical simulation after cutting the echo tail.

The specimen without crack was tested first. After that, the specimen was sawed to generate artificial cracks as shown in Fig. 1 (a, b). So far, only 2.5 cm and 5 cm cracks are produced to calibrate the signals. The results measured were calibrated considering travelling time through cable, the effects of glue layer on the transmission speed, and the effect of basement where the specimen lying on, were calibrated.

4 Results and Discussion

Experimental results show the degree of damping in the signals observed. The causes of this damping are the transformation of potential and kinetic energies to thermal energy, and the energy dispersed from the system through the boundaries, especially though the solid-to-solid contacts.However, determination of damping fractions i.e. material damping (energy transformation) and boundary damping cannot be done. Therefore, we consider that concrete is perfectly elastic ($\mathbb{C} = 0$) at a small amplitude of vibration. The damping comes from the energy loss at the boundaries (Sect. 2.4).

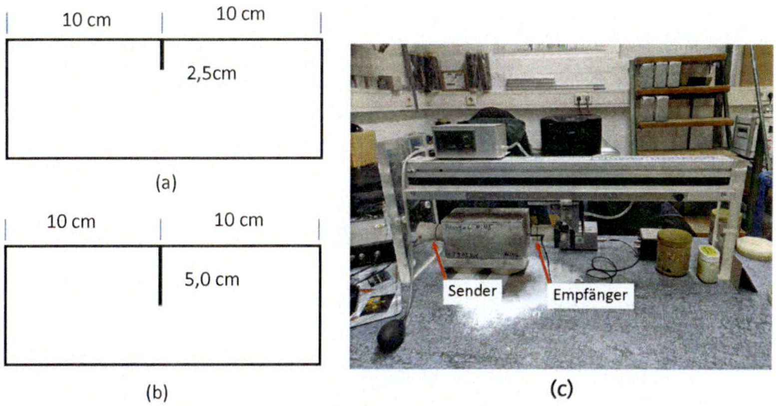

Fig. 1. (a) Artificial cracks of the specimen generated by sawing 2.5 cm, (b) extending the crack by sawing 5.0 cm, (c) ultrasonic apparatus to test the quality of the specimen.

Table 1. Model parameters

E (GPa)	ν (-)	ρ(kg/m^3)	α_σ	α_τ
39.5	0.18	2260	10.0	5.0

The measured signals of the specimens with and without crack are shown in Fig. 2 (top). It is clearly seen that the time-of-flight (t_L) in two specimens are the same, however, the wave forms are different. It confirms that when the line between sender and receiver does not cut across the crack, the time-of-flights t_L does not change. Consequently, the crack will not be found with the conventional method. By observing frequency analysis using Fast Fourier Transform (FFT) (Fig. 2 bottom), it can be noticed the difference in eigenfrequency between 2 tests, an additional peak in the test of the specimen with crack appears.

In simulation, we attempt to simulate all the aspects determined in experiments such as damping, boundary conditions, material properties. The model mesh with 80×40 elements is generated to satisfy the critical conditions about wavelength . Time discretization has to satisfy the critical conditions. The model parameters are shown in Table 1. Figure 4 (above) shows the signals measured and simulation output at the same location. It can be noticed that the time-of-flights in measurement and simulation are identical, whereas their amplitudes of the signals remain different. The discrepancy can be interpreted for the following reasons. First and foremost are the boundary conditions. We generate the excitation using a piezoelectric head. The excitation load is harmonic, consisting both compressing and tension forces. However, the clay glue is only compressible, it transmits only compressive signals to the boundary. Consequently, the excitation signal is not always identical with the electrical signal sent to the piezoelectric head. How to measure the real excitation at the specimen boundary? so far, we measure directly by contacting the sender to the receiver. Secondly, simplifying 3D boundary to 2D boundary attributes to an effective error, this can be improved by using a 3D model.

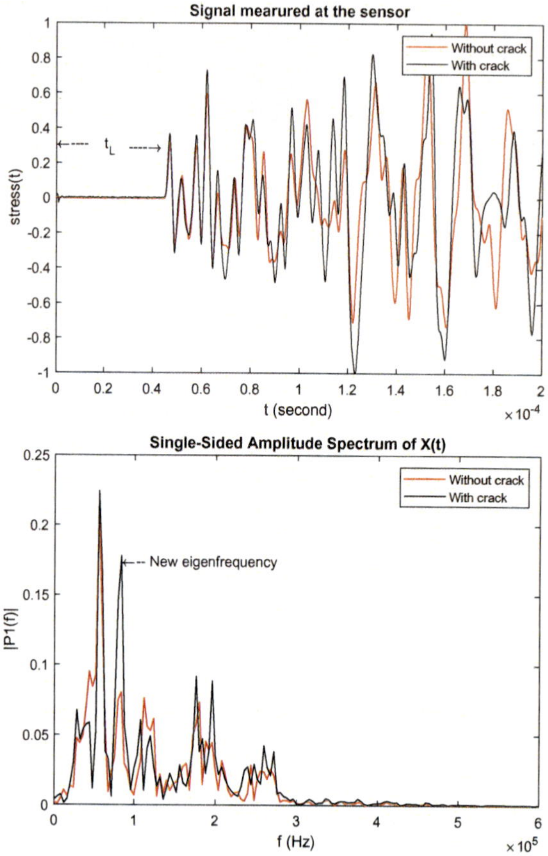

Fig. 2. Experimental results between specimens with crack 2.5 cm and without crack: (top) time domain measurement, (bottom) frequency domain.

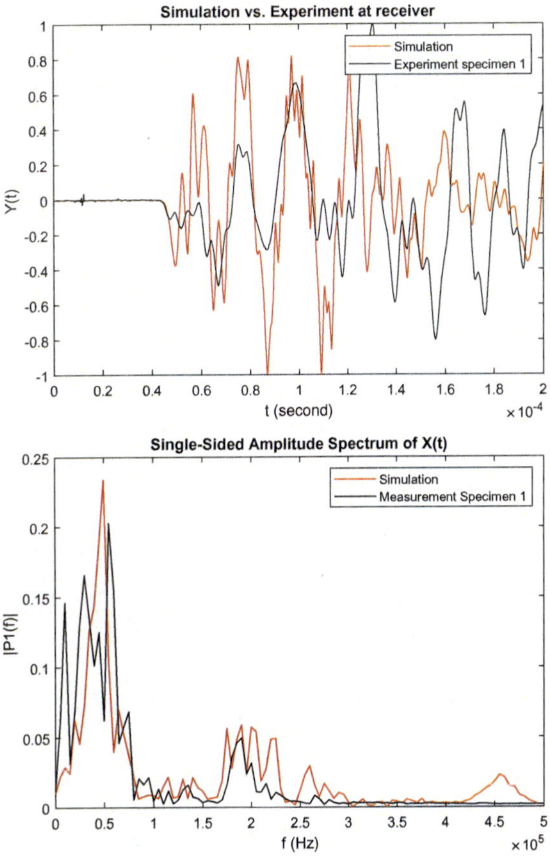

Fig. 3. Numerical simulation vs. experiment of the specimen without crack: (top) received signal and the measurement at the same location, (bottom) FFT analysis.

The difficulty in time domain analysis is favourably resolved by using frequency domain analysis. By applying an FFT, time domain results are transformed into frequency domain graphs (Fig. 4 (bottom)). One can see in this plot the similarity between two signals, namely in the regions 200 to 800 kHz, and 1600 to 2300 kHz.

Additionally, we show the visualization of signals transmitting across the crack (Fig. 5). Clear differences between the model with crack and without crack are noted by distribution of longitudinal stress at the crack. It confirms that the XFEM does not allow elastic wave transmitting across the crack, with a new mass lumping technique [1].

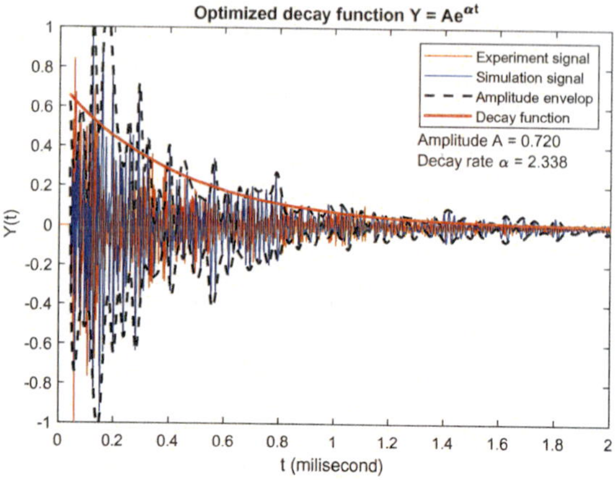

Fig. 4. Experiment and simulation signal damping

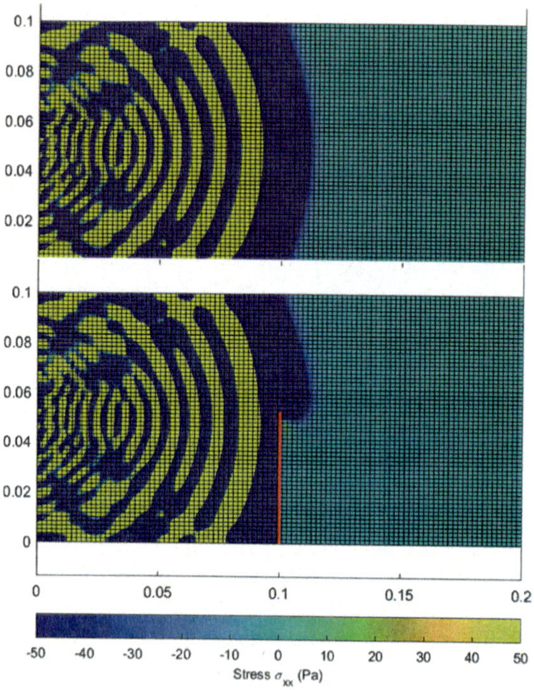

Fig. 5. Visualization example of stress distribution at time 0.86 µ sec: (top) specimen without crack, (bottom) specimen with crack (5 cm)

5 Conclusion

Ultrasonic experiments found no difference between time-of-flight of the signal measured in the specimen with crack and without crack using frequency band of maximum 250 kHz. However, the waveforms of these signal can be noticed. It is clearly seen in the frequency analysis, where a new eigenfrequency appears with the crack. The comparison between measured and simulated signals is still different in time domain graph. Nevertheless, the similarity between the two can be improved by using FFT analysis. The XFEM with new mass lumping technique [1] shows that the method is effective to hinder the energy leaking across the crack surface.

References

1. Asareh, I., Song, J.H., Mullen, R.L., Qian, Y.: A general mass lumping scheme for the variants of the extended finite element method. Int. J. Numer. Methods Eng. **121**(10), 2262–2284 (2020)
2. Bathe, K.J.: Finite Element Procedures. Prentice-Hall (1996)
3. Belytschko, T., Chen, H., Xu, J., Zi, G.: Dynamic crack propagation based on loss of hyperbolicity and a new discontinuous enrichment. Int. J. Numer. Methods Eng. **58**(12), 1873–1905 (2003)
4. Belytschko, T., Mullen, R.: Stability of explicit-implicit mesh partitions in time integration. Int. J. Numer. Methods Eng. **12**(10), 1575–1586 (1978)
5. Belytschko, T., Parimi, C., Moës, N.S, N., Sukumar, N., Usui, S.: Structured extended finite element methods for solids defined by implicit surfaces. Int. J. Numer. Methods Eng. **56**(4), 609–635 (2003)
6. Bettess, P., Zienkiewicz, O.C.: Diffraction and refraction of surface waves using finite and infinite elements. Int. J. Numer. Methods Eng. **11**(8), 1271–1290 (1977)
7. Chew, W., Liu, Q.: Perfectly matched layers for elastodynamics: a new absorbing boundary condition. J. Comput. Acoust. **04**(4) (1997). https://doi.org/10.1142/S0218396X9800020X
8. Givoli, D.: A spatially exact non-reflecting boundary condition for time dependent problems. Comput. Methods Appl. Mech. Eng. **95**(1), 97–113 (1992)
9. Hamdan, N.: Two-Dimensional Numerical Modelling of Wave Propagation in Soil Media, Ph.D. thesis, Heriot-Watt University, Edinburgh, UK (2013)
10. Kaltenbacher, B., Kaltenbacher, M., Sim, I.: A modified and stable version of a perfectly matched layer technique for the 3-d second order wave equation in time domain with an application to aeroacoustics. J. Comput. Phys. **235**(100), 407–422 (2013)
11. Liu, G., Quek Jerry, S.: A non-reflecting boundary for analyzing wave propagation using the finite element method. Finite Elem. Anal. Des. **39**(5), 403–417 (2003)
12. Liu, Z., Oswald, J., Belytschko, T.: XFEM modeling of ultrasonic wave propagation in polymer matrix particulate/fibrous composites. Wave Motion **50**(3), 389–401 (2013)
13. Lysmer, J., Kuhlemeyer, R.L.: Finite dynamic model for infinite media. J .Eng. Mech. Div. **95**(4), 859–877 (1969)
14. Menouillard, T., Rethore, J., Combescure, A., Bung, H.: Explicit time stepping for the eXtended finite element method (X-FEM). Int. J. Numer. Methods Eng. **68**(9), 911–939 (2006)

15. Newmark, N.: A method of computation for structural dynamics. J. Eng. Mech. Div. **85**(EM), 67–94 (1959)
16. Nguyen-Tuan, L., Nanthakumar, S.S., Lahmer, T.: Identification of multiple flaws in dams using inverse analysis based on hydro-mechanical XFEM and level sets. Comput. Geotech. **110**, 211–221 (2019)
17. Sukumar, N., Chopp, D., Moës, N., Belytschko, T.: Modeling holes and inclusions by level sets in the extended finite-element method. Comput. Methods Appl. Mech. Eng. **190**, 6183–6200 (2001)
18. Zhao, M., Li, H., Du, X., Wang, P.: Time-domain stability of artificial boundary condition coupled with finite element for dynamic and wave problems in unbounded media. Int. J. Comput. Methods **16**(04), 1850099 (2019)

Vibration Analysis of Pressurized and Rotating Cylindrical Shells by Rayleigh-Ritz Method

Ivo Senjanović, Damjan Čakmak, Ivan Ćatipović, Neven Alujević(✉), and Nikola Vladimir

Faculty of Mechanical Engineering and Naval Architecture, University of Zagreb, Ivana Lučića 5, 10 000 Zagreb, Croatia
neven.alujevic@fsb.hr

Abstract. A numerical procedure for vibration analysis of pressurised and rotating cylindrical shells by the Rayleigh Ritz method is presented. Potential and kinetic energies are specified. Stiffness matrix, geometrical stiffness matrix and three mass matrices due to centrifugal forces, Coriolis forces and the inertia forces are derived. Three basic sets of boundary conditions are considered, i.e. free, clamped, and simply supported in all the three directions. Appropriate sets of orthogonal trigonometric functions are used for the approximation of the displacement fields. The accuracy of the results obtained is verified analytically, by the finite strip method (FSM), and by the Finite Element Method (FEM).

Keywords: Cylindrical shell · Vibration · Rotation · Pressure · Rayleigh-Ritz

1 Introduction

An instructive approach to the shell theory, adopted to the engineering level, is presented in [1]. Most papers on the vibration of cylindrical shells, either rotating or stationary, are related to the simply supported shell ends. The reason for this boundary conditions is a possibility of getting relatively simple analytical solution, [2, 3]. Another complicate analytical solution of pressurized and rotating cylindrical shells with arbitrary boundary conditions has been presented in [4]. Such a problem can be relatively easily solved numerically by the Rayleigh-Ritz method, [5]. Recently, a finite strip for modelling pre-stressed rotating cylindrical and toroidal shells has been developed, [6, 7]. In this paper, the Rayleigh-Ritz method is applied for vibration analysis of pressurized and rotating cylindrical shells.

2 Strain and Kinetic Energy

The general thin shell theory is presented in [1] using Lamé parameters, A_1, A_2, the main radii of curvature, R_1, R_2, the in-plane displacements, u_1, u_2, and normal displacement (deflection), u_3, Fig. 1. The strain-displacement relationships are specified, i.e. in-plain

Z. Dimitrovová et al. (Eds.): WMVC 2022, MMS 125, pp. 189–198, 2023.
https://doi.org/10.1007/978-3-031-15758-5_18

strains, ε_x, ε_φ, $\varepsilon_{x\varphi}$, curvature changes, κ_x, κ_φ, $\kappa_{x\varphi}$, and rotation angles, β_x, β_φ. The ordinary strain energy due to tension and bending, according to [1], reads

$$E_s = \frac{1}{2}\int_0^{2\pi}\int_0^l \left\{ K\left[\varepsilon_x^2 + \varepsilon_\varphi^2 + 2\nu\varepsilon_x\varepsilon_\varphi + \frac{1}{2}(1-\nu)\varepsilon_{x\varphi}^2\right] + \right.$$
$$\left. +D\left[\kappa_x^2 + \kappa_\varphi^2 + 2\nu\kappa_x\kappa_\varphi + \frac{1}{2}(1-\nu)\kappa_{x\varphi}^2\right]\right\} \mathrm{d}x\, r\, \mathrm{d}\varphi, \quad (1)$$

where x and φ are axial and circumferential coordinate, r is shell radius, and K and D are the tensional and bending stiffness, respectively.

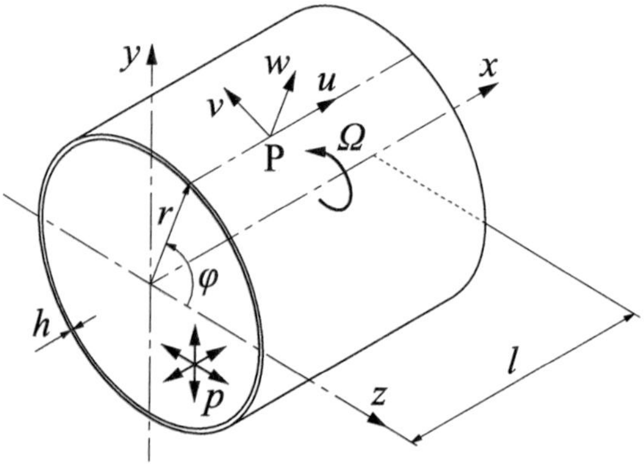

Fig. 1. Rotating cylindrical shell

Referring to [1], the strain energy due to pre-stressing by tension forces N_x and N_φ is presented in the form

$$E_G = \int_0^{2\pi}\int_0^l \left(\varepsilon_x^* N_x + \varepsilon_\varphi^* N_\varphi\right) \mathrm{d}x\, r\, \mathrm{d}\varphi, \quad (2)$$

where ε_x^* and ε_φ^* are the second order strains based on the Green-Lagrange tensor [8]

$$\varepsilon_x^* = \frac{1}{2}\left[\varepsilon_x^2 + \left(\varepsilon_\varphi^{(2)}\right)^2 + \beta_x^2\right]$$
$$\varepsilon_\varphi^* = \frac{1}{2}\left[\varepsilon_\varphi^2 + \left(\varepsilon_\varphi^{(1)}\right)^2 + \beta_\varphi^2\right]. \quad (3)$$

where $\varepsilon_\varphi^{(1)}$ and $\varepsilon_\varphi^{(2)}$ are the first and second terms of $\varepsilon_{x\varphi}$, [1].

According to [1], the kinetic energy of the rotating cylindrical shell reads

$$E_k = \frac{1}{2}\rho h \int_0^{2\pi}\int_0^l \left[\left(\frac{\partial u}{\partial t}\right)^2 + \left(\frac{\partial v}{\partial t} + \Omega w\right)^2 + \left(\frac{\partial w}{\partial t} - \Omega\, v\right)^2\right] \mathrm{d}x\, r\, \mathrm{d}\varphi. \quad (4)$$

where Ω is rotation speed.

3 Strain and Kinetic Energy Condensed to Shell Cross-Section

For a closed cylindrical shell, one can assume harmonic variation of displacements in circumferential direction, i.e.

$$\begin{aligned} u(x,\varphi,t) &= U(x)\cos(n\varphi+\omega t) \\ v(x,\varphi,t) &= V(x)\sin(n\varphi+\omega t) \\ w(x,\varphi,t) &= W(x)\cos(n\varphi+\omega t), \end{aligned} \tag{5}$$

where $U(x)$, $V(x)$ and $W(x)$ are amplitudes of the cross-sectional mode profiles, n is circumferential mode number, and ω is natural frequency. The unified argument of trigonometric functions, $n\varphi+\omega t$, is used in order to enable the description of the travelling modes commonly appearing in rotating shells. Substituting Eq. (5) into strain-displacement relationship, [6], and then into the strain energies (1) and (2), one obtains products of two displacement amplitudes or their derivatives, with squares of sine and cosine functions (5). Their integrals over the circumferential angle φ within the domain $0-2\pi$ equals π. Thus, the temporal variation vanishes, and the strain and kinetic energies become time-invariant. This is due to the fact that the natural modes rotate while keeping a fixed cross-sectional profile. As a result, Eq. (1) is reduced to the following form:

$$\begin{aligned} E_{\mathrm{s}} = \int_0^l \Big[& \frac{1}{2}a_1(U')^2 + \frac{1}{2}a_2U^2 + \frac{1}{2}a_3(V')^2 + \frac{1}{2}a_4V^2 + a_5U'V + a_6UV' + \\ & + \frac{1}{2}b_1(W'')^2 + \frac{1}{2}b_2(W')^2 + \frac{1}{2}b_3W^2 + b_4W''W + b_5WU' + b_6W'V' + b_7WV + \\ & b_8W''V \Big]\mathrm{d}x, \end{aligned} \tag{6}$$

where coefficients a_i and b_i are specified in [6].

Furthermore, substituting Eq. (5) into the strain-displacement relationship, [6], and further into Eqs. (3) and (2), yields

$$E_{\mathrm{G}} = \int_0^l \left[\frac{1}{2}c_1(U')^2 + \frac{1}{2}c_2(V')^2 + \frac{1}{2}c_3(W')^2 + \frac{1}{2}c_4U^2 + \frac{1}{2}c_5V^2 + \frac{1}{2}c_6W^2 + c_7WV \right]\mathrm{d}x, \tag{7}$$

where coefficients c_i are specified in [6].

In similar way, by substituting Eq. (5) into Eq. (4) one obtains for the condensed kinetic energy

$$E_k = \frac{1}{2}\pi\rho hr \int_0^i [\omega^2U^2 + (\omega^2+\Omega^2)V^2(\omega^2+\Omega^2)W^2 + 4\omega\Omega VW]dx. \tag{8}$$

4 Stiffness and Mass Matrices

4.1 Stiffness Matrix

Vibrations of pressurized and rotating cylindrical shells can be analyzed by the Rayleigh-Ritz method. In order to enable different boundary conditions let us assume the displacement fields in the form of two series

$$U(x) = \left\langle \langle f_m \rangle \; \langle g_m \rangle \right\rangle \begin{Bmatrix} \{A_m\} \\ \{B_m\} \end{Bmatrix}$$
$$V(x) = \left\langle \langle f_m \rangle \; \langle g_m \rangle \right\rangle \begin{Bmatrix} \{C_m\} \\ \{D_m\} \end{Bmatrix} \tag{9}$$
$$W(x) = \left\langle \langle f_m \rangle \; \langle g_m \rangle \right\rangle \begin{Bmatrix} \{E_m\} \\ \{F_m\} \end{Bmatrix},$$

where f_m and g_m are the coordinate functions and A_m, B_m, C_m, D_m, E_m and F_m are accompanying unknown coefficients. The coefficients are determined from condition of minimum total shell energy. For this purpose let us substitute Eqs. (9) into the strain energy, Eq. (6), and differentiate it with respect to corresponding coefficients. As a result system of three matrix equations are obtained which can be presented in the condensed form

$$\left\{ \frac{\partial E_s}{\partial \{\delta\}} \right\} = [K]\{\delta\}, \tag{10}$$

where

$$\{\delta\}^{\mathrm{T}} = \langle \delta \rangle = \left\langle \langle A_m \rangle \; \langle B_m \rangle \; \langle C_m \rangle \; \langle D_m \rangle \; \langle E_m \rangle \; \langle F_m \rangle \right\rangle \tag{11}$$

is the vector of unknown coefficients and

$$[K] = \begin{bmatrix} [K]_{11} & [K]_{12} & [K]_{13} \\ [K]_{21} & [K]_{22} & [K]_{23} \\ [K]_{31} & [K]_{32} & [K]_{33} \end{bmatrix} \tag{12}$$

is the stiffness matrix. Submatrices $[K]_{ij}$, $i, j = 1, 2, 3$ encompass the integrals from Eq. (6), [6].

4.2 Geometric Stiffness Matrix

In a similar way one can determine the geometric stiffness matrix. Substituting Eq. (9) for displacements into the energy equation due to pre-stressing (7) and differentiating it with respect to coefficients one obtains

$$\left\{ \frac{\partial E_G}{\partial \{\delta\}} \right\} = [G]\{\delta\}, \tag{13}$$

where

$$[G] = \begin{bmatrix} [G]_{11} & [0] & [0] \\ [0] & [G]_{22} & [G]_{23} \\ [0] & [G]_{32} & [G]_{33} \end{bmatrix} \tag{14}$$

is the geometric stiffness matrix. Submatrices $[\mathrm{G}]_{ij}$, $i, j = 1, 2, 3$ represent the integrals in Eq. (7), [6].

Tension forces N_x and N_φ, due to internal pressure and shell rotation, which are present in coefficients c_i, [6], read

$$N_x = \frac{1}{2} rp \; , \; N_\varphi = rp + \rho h r^2 \Omega^2. \tag{15}$$

Hence, the geometric stiffness matrix can be decomposed into two matrices, i.e. one due to internal pressure and the other related to the centrifugal force

$$[G] = p[G]_p + \Omega^2 [G]_\Omega. \tag{16}$$

4.3 Mass Matrices

Mass matrices are derived from the kinetic energy, Eq. (8). By substituting expressions (9) into (8), and differentiating the kinetic energy with respect to coefficients, one obtains the following system of algebraic equations:

$$\left\{ \frac{\partial E_{\mathrm{k}}}{\partial \{\delta\}} \right\} = \left(\Omega^2 [B] + \omega \Omega [C] + \omega^2 [M] \right) \{\delta\}, \tag{17}$$

where

$$[B] = \alpha \begin{bmatrix} [0] & [0] & [0] \\ [0] & [K]_2 & [0] \\ [0] & [0] & [K]_2 \end{bmatrix}$$
$$[C] = 2\alpha \begin{bmatrix} [0] & [0] & [0] \\ [0] & [0] & [K]_2 \\ [0] & [K]_2 & [0] \end{bmatrix} \tag{18}$$
$$[M] = \alpha \begin{bmatrix} [K]_2 & [0] & [0] \\ [0] & [K]_2 & [0] \\ [0] & [0] & [K]_2 \end{bmatrix}$$

and $\alpha = \pi \rho h r$. Mass matrices $[B]$, $[C]$ and $[M]$ are related to the centrifugal force, the Coriolis force, and the inertia force respectively, [6].

5 Matrix Equation of Motion

In the considered case of a rotating cylindrical shell, the total energy reads

$$\Pi = E_\mathrm{s} + E_\mathrm{G} - E_\mathrm{k}, \tag{19}$$

where E_s, E_G and E_k is ordinary strain energy, strain energy due to pre-stressing and kinetic energy, respectively. All these energies are time-invariant due to rotation of fixed mode profile around the axis of symmetry. Natural frequency is in fact the speed of this rotation. Nevertheless, each particle on the shell still undergoes motions where minima and maxima of the displacement and velocity are interchanged. Total energy for an exact vibration mode $\Pi = 0$, Eq. (19). If the modes are determined approximately with truncated series, the governing equation of motion can be obtained from the minimum total energy principle [9]

$$\left\{\frac{\partial \Pi}{\partial\{\delta\}}\right\} = \left\{\frac{\partial E_\mathrm{s}}{\partial\{\delta\}}\right\} + \left\{\frac{\partial E_\mathrm{G}}{\partial\{\delta\}}\right\} - \left\{\frac{\partial E_\mathrm{k}}{\partial\{\delta\}}\right\} = \{0\}. \tag{20}$$

Taking into account relations (10), (13) and (17) respectively, the following matrix equation for natural vibrations is obtained

$$\Big([K] + p[G]_p + \Omega^2([G]_\Omega - [B]) - \omega\Omega[C] - \omega^2[M]\Big)\{\delta\} = \{0\}. \tag{21}$$

Matrix $[C]$ multiplying the mixed $\omega\Omega$ term, resulting from the Coriolis force, is the only term which causes a bifurcation of the natural frequencies. The geometric stiffness matrix $[G]_\Omega$ and the mass matrix $[B]$ are related to the centrifugal force with stiffening and softening effects respectively.

6 Boundary Conditions and Coordinate Functions

In the Rayleigh-Ritz method, it is well known that each coordinate function in the assumed series has to satisfy boundary conditions and the functions have to be mutually orthogonal. Hence, the choice of the coordinate functions for approximation of displacement fields in Eqs. (9) depends on boundary conditions.

6.1 Free Cylindrical Shell

In this case, boundary forces and moments equal zero, while displacements are free. Free displacement fields can be described by the Fourier series and the coordinate functions in Eqs. (9) are

$$\begin{aligned} f_0 &= 1, \quad f_m = \cos\frac{m\pi x}{l}, \quad m = 1, 2\ldots \\ g_0 &= \frac{2x}{l} - 1, \quad g_m = \sin\frac{m\pi x}{l}, \quad m = 1, 2\ldots \end{aligned} \tag{22}$$

Functions f_0 and g_0 enable translation and rotation of the shell generatrix, respectively.

6.2 Clamped Cylindrical Shell

In-plane displacements U and V at the shell boundaries are zero, as well as deflection W and rotation $\Phi = \mathrm{d}W/\mathrm{d}x$ in radial plane. Displacements U and V are approximate with sine functions, while deflection is approximate with cosine functions

$$
\begin{aligned}
g_m &= \sin \frac{m\pi x}{l}, \quad m = 1, 2 \ldots \\
f_m &= \cos \frac{(m+1)\pi x}{l} - 1, \quad m = 1, 3, 5 \ldots \\
f_m &= \cos \frac{(m+1)\pi x}{l} - \cos \frac{\pi x}{l}, \quad m = 2, 4, 6 \ldots
\end{aligned}
\tag{23}
$$

In this case Eqs. (9) for displacement fields are reduced to

$$
U = \langle g_m \rangle \{B_m\}, V = \langle g_m \rangle \{D_m\}, W \langle f_m \rangle \{E_m\}. \tag{24}
$$

6.3 Simply-Supported Cylindrical Shell

All displacement fields are assumed in the form of sine functions

$$
U = \langle g_m \rangle \{B_m\}, V = \langle g_m \rangle \{D_m\}, W \langle g_m \rangle \{F_m\}. \tag{25}
$$

where

$$
g_m = \sin \frac{m\pi x}{l}, \quad m = 1, 2 \ldots \tag{26}
$$

7 Illustrative Numerical Examples

7.1 Free Cylindrical Shell

In order to illustrate the application of the presented theory and evaluate its accuracy, three numerical examples considering various boundary conditions are presented. First, natural vibration analysis of a free cylindrical shell with the following geometric particulars and material properties of rubber is carried out by using the example from [4]: $l = 0.2$ m, $r = 0.1$ m, $h = 0.002$ m, $E = 0.45$ GPa, $\nu = 0.45$, $\rho = 1452$ kg/m^3.

Natural frequencies for case $p = 0$ and $\Omega = 0$ are calculated for the circumferential wave number $n = 0,1,2$ by utilizing 20 coordinate functions, Eqs. (22). The obtained values are listed in Tables 1, 2 and 3, where M is the ordinary number of the axial mode profile which is identical to the number of modes. The calculated natural frequencies are compared with those determined by the finite strip method (FSM), [6], and the analytical method according to the theory presented in [4]. Very good agreement between all results is achieved. Some characteristic natural modes determined by the Rayleigh-Ritz method for $M = 0$, 1, 2 and $n = 0$, 1, 2 have the same shape as those determined analytically, [4], and by FSM, [6].

Table 1. Natural frequencies of free cylindrical shell, ω (Hz), $p = 0$, $\Omega = 0$, $n = 0$

Mode M	0	1	2	3	4	5
RRM	0	817.3	839.0	887.9	900.4	929.5
FSM [6]	0	818.1	839.4	890.2	904.9	935.9
Analytical [5]	0	817.3	839.0	887.9	900.4	929.5

Table 2. Natural frequencies of free cylindrical shell, ω (Hz), $p = 0$, $\Omega = 0$, $n = 1$

Mode M	0	1	2	3	4	5
RRM	0	661.0	772.3	847.3	879.2	917.4
FSM [6]	0	661.4	773.5	849.9	883.6	924.0
Analytical [5]	0	661.0	772.3	847.4	879.2	917.4

Table 3. Natural frequencies of free cylindrical shell, ω (Hz), $p = 0$, $\Omega = 0$, $n = 2$

Mode M	0	1	2	3	4	5
RRM	15.2	20.7	547.7	716.4	818.4	879.4
FSM [6]	15.2	20.7	549.1	718.9	822.9	885.4
Analytical [5]	15.2	20.6	547.8	716.4	818.4	879.5

7.2 Clamped Cylindrical Shell

A clamped steel cylindrical shell with the following geometric particulars and physical properties, and $p = 0$ and $\Omega = 0$, is considered next: $l = 2$ m, $r = 1$ m, $h = 0.01$ m, $E = 2.1 \cdot 10^{11}$ N/m^2, $\nu = 0.3$, $\rho = 7850$ kg/m^3. The corresponding coordinate functions for this case of boundary conditions are given in Sect. 6.2. The first twelve natural frequencies are listed in Table 4 and compared with those determined by FSM and FEM, [6]. The obtained results for all three approaches agree very well. The corresponding three natural modes generated by NASTRAN [10], are shown in Fig. 2.

7.3 Simply-Supported Cylindrical Shell

Natural vibration analysis of the cylindrical shell having the same geometric particulars and physical properties as in the previous example are performed. The coordinate functions for simply-supported boundary conditions are specified in Sect. 6.3. The first twelve natural frequencies are given in Table 5 and compared with those calculated by FSM and FEM. The agreement of results is very good. The corresponding three natural modes determined by Abaqus, [11], are shown in Fig. 3.

Table 4. Natural frequencies of clamped cylindrical shell, ω (Hz), $p = 0$, $\Omega = 0$

Mode No *N*	1 6	2 5	3 7	4 4	5 8	6 9	7 7	8 3	9 8	10 6	11 9	12 10
RRM	130.5	136.1	144.5	165.5	172.5	210.2	218.8	223.4	226.8	231.1	251.0	255.1
FSM	133.0	139.1	146.3	168.6	173.9	211.3	226.1	226.2	233.1	238.9	256.3	256.0
FEM	132.1	138.2	145.5	167.8	173.0	210.2	223.0	225.6	229.7	236.4	254.4	255.3

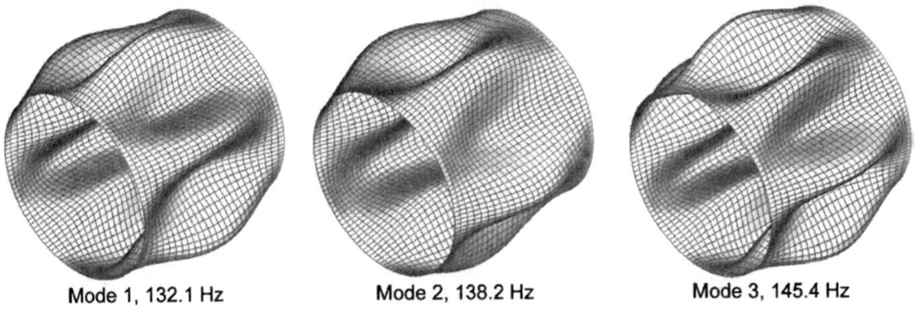

Fig. 2. Natural modes of clamped cylindrical shell, $p = 0$, $\Omega = 0$

Table 5. Natural frequencies of simply-supported cylindrical shell, ω(Hz), $p = 0$, $\Omega = 0$

Mode No *N*	1 6	2 5	3 7	4 4	5 8	6 9	7 7	8 3	9 8	10 6	11 9	12 10
RRM	130.5	136.2	144.5	165.5	172.6	210.2	218.9	223.4	226.8	231.2	251.1	255.1
FSM	131.0	136.7	144.8	166.1	172.7	210.4	220.5	223.9	228.2	233.1	252.1	255.2
FEM	130.4	136.0	144.3	165.5	172.3	209.8	218.6	223.4	226.4	230.9	250.5	254.6

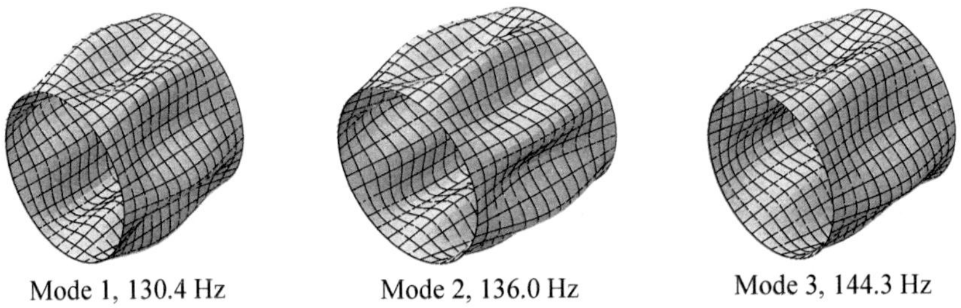

Fig. 3. Natural modes of simply-supported cylindrical shell, $p = 0$, $\Omega = 0$

8 Conclusion

Cylindrical shell is the basic configuration of the most shell structures. Internal pressure and rotation make shell dynamic behaviour significantly more complex. In this paper problem is solved numerically by the Rayleigh-Ritz method. Three sets of elementary boundary conditions are considered, i.e. free, clamped and simple-supported edges. Displacements are assumed in form of appropriate trigonometric series. The ordinary stiffness matrix, geometric stiffness matrix and mass matrices due to centrifugal force, Coriolis force, and the inertia force are derived. The presented numerical procedure is checked by performing free vibration analysis for a cylindrical shell with different edge support. The obtained results are checked analytically, by the FSM and FEM, and very good agreement is achieved.

Acknowledgement. This work has been supported by the Croatian Science Foundation under the project IP-2019-04-5402.

References

1. Soedel, W.: Vibrations of Shells and Plates, 3rd edn. Marcel Dekker, Inc, New York (2004)
2. Kim, Y.-J., Bolton, J.S.: Effects of rotation on the dynamics of a circular cylindrical shell with application to tire vibration. J. Sound Vib. **275**, 605–621 (2004)
3. Huang, S.C., Soedel, W.: On the forced vibration of simply supported rotating cylindrical shells. J. Acoust. Soc. Am. **84**(1), 275–285 (1988)
4. Alujević, N., Campillo-Davo, N., Kindt, P., Desmet, W., Pluymers, B., Vercammen, S.: Analytical solution for free vibrations of rotating cylindrical shells having free boundary conditions. Eng. Struct. **132**, 152–171 (2017)
5. Sun, S., Cao, D., Han, Q.: Vibration studies of rotating cylindrical shells with arbitrary edges using characteristic orthogonal polynomials in the Rayleigh-Ritz method. Int. J. Mech. Sci. **68**, 180–189 (2013)
6. Senjanović, I., Ćatipović, I., Alujević, N., Vladimir, N., Čakmak, D.: A finite strip for the vibration analysis of rotating cylindrical shells. Thin-Walled Struct. **122**, 158–172 (2018)
7. Senjanović, I., Ćatipović, I., Alujević, N., Čakmak, D., Vladimir, N.: A finite strip for the vibration analysis of rotating toroidal shell under internal pressure. J. Vib. Acoust. **141 and 2**(021013), 1–17 (2019)
8. Crisfield, M.A.: Non-linear finite element analysis of solids and structures, Volume 1: essentials. John Wiley & Sons, Chichester, West Sussex, England (1991)
9. Szilard, R.: Theories and Applications of Plate Analysis, Classical, Numerical and Engineering Methods. John Wiley & Sons Inc, Hoboken, New Jersey (2004)
10. MSC.MD NASTRAN 2010 Dynamic Analysis User's Guide, MSC Software (2010)
11. Dassault Systèmes: Abaqus 6.9 User's guide and theoretical manual, Hibbitt, Karlsson & Sorensen, Providence, RI, Inc. (2009)

DSP: Dynamic Stability, Deterministic, Chaotic and Random Post-critical States

Analytical Solution of the Problem of Free Vibrations of a Plate Lying on a Variable Elastic Foundation

Mykola Surianinov(✉), Yurii Krutii, Vladimir Osadchiy, and Oleksii Shyliaiev

Odessa State Academy of Civil Engineering and Architecture,
Didrikson str., 4, Odessa 65029, Ukraine
sng@ogasa.org.ua

Abstract. An analytical solution of the problem of free bending vibrations of rectangular plates with Levy boundary conditions lying on a continuous variable elastic foundation, which is described by the Winkler model, is given. An exact solution of the differential equation of free vibrations of plates when the bedding coefficient is an arbitrary continuous function of one variable is found. The quadratures for numerical realization of the found solutions are derived. The formulas for dynamic state parameters, which allow one to investigate free vibrations of the plates under any boundary conditions at two parallel edges, are obtained. The dependence of the frequency of free vibrations of the system under consideration on its other parameters is established in the analytical form. Computational formulas for determining the spectrum of frequencies of free vibrations of the plates are obtained. The general form of frequency equation and formulas for the main forms of vibrations corresponding to the three cases of boundary conditions are established. Frequency spectra of free vibrations of hinged plates resting on a variable elastic base for four different laws of bedding coefficient changes are determined. It is shown that in the case of constant bedding coefficient the frequencies calculated by the author's method practically coincide with the frequencies calculated by the known exact formula.

Keywords: Plate · Variable foundation · Winkler model · Analytical solution · Vibrations · Frequency

1 Introduction

One of the many problems arising in the design of structures with variable geometric and (or) mechanical parameters is the problem of bending vibrations of plates on a variable elastic base. Such a plate is a common computational model of structural elements of objects of construction, mechanical engineering, shipbuilding, etc. In calculations of such mechanical systems one has to face differential equations (or systems of differential equations) with variable coefficients [1–5]. To date, only some cases of construction of exact analytical solutions are known [6–11]. And mainly approximate and numerical methods are used to solve differential equations with variable coefficients [12–15]. Of

Z. Dimitrovová et al. (Eds.): WMVC 2022, MMS 125, pp. 201–210, 2023.
https://doi.org/10.1007/978-3-031-15758-5_19

course, it is possible to obtain a result with almost any given accuracy. However, this is only a quantitative result that requires verification. The construction of analytical solutions is still relevant.

2 Review of Research and Publications

The development of analytical methods for calculating plates of variable thickness is devoted to the publications of E.B. Koreneva [16, 17]. Here we deal with circular and annular plates. An analytical solution is also the basis of the study [9] on the bending of a rectangular plate, which lies on a two-parameter elastic base. In [18] a generalized differential quadrature method for transverse free vibrations of rectangular plates of variable thickness on an elastic base is used. The vibrations of an arbitrarily shaped plate on an elastic Winkler-type base with arbitrary boundary conditions are considered in [19]. Note also the articles [20–23]. In all these works, the base remains constant. The purpose of our study is to analytically solve the problem of free vibrations of a plate lying on a variable elastic base, the reaction of which is accounted for by the Winkler model.

3 Research Results

Consider the free bending vibrations of a rectangular plate of constant cylindrical stiffness D lying on a variable elastic base, which is described by the Winkler model. As we know, in such a model, the base reaction $R(x, y, t)$ and the deflection $w(x, y, t)$ are related by equality $R(x, y, t) = -kw(x, y, t)$, where $k-$ is the bedding coefficient. In the general case, the bed coefficient is a function of x and y. Here we will assume that it is given by an arbitrary continuous function which depends only on x, remaining constant along the y axis, i.e. $k = k(x)$.

Let's represent the bed coefficient as $k(x) = k_0 A(x)$, where k_0- is a constant dimensional parameter (bed coefficient at some characteristic point, for example, at point $x = 0$), and $A(x)-$ is a dimensionless continuous function that determines the law of variation of the bed coefficient on segment $0 \leq x \leq a$. Any boundary conditions can be set on the sides of the plate $x = 0,\ \ x = a$ and the sides $y = 0,\ \ y = b$ are assumed to be hinged. The solution of the problem is reduced to the solution of the partial differential equation [3]

$$D\Delta\Delta w + \rho h \frac{\partial^2 w}{\partial t^2} + k_0 A(x) w = 0, \tag{1}$$

where E – material modulus of elasticity, h – plate thickness, μ – Poisson's ratio; Δ – Laplace operator; $w = w(x, y, t)$ – dynamic deflection; ρh – mass of unit area.

We find the solution of "Eq. (1)" in the form

$$w = W(x, y)T(t), \tag{2}$$

where $W(x, y)$ – the unknown main form of vibration, and $T(t)$ – the time function, which has the form

$$T(t) = T(0)\cos\omega t + \frac{dT(0)/dt}{\omega}\sin\omega t.$$

Here $\omega-$ is the unknown frequency of free vibrations of the plate.

Then for the main form of vibrations we obtain

$$\frac{\partial^4 W}{\partial x^4}+2\frac{\partial^4 W}{\partial x^2\partial y^2}+\frac{\partial^4 W}{\partial y^4}+\frac{1}{D}\left(k_0A(x)-\rho h\omega^2\right)W=0. \tag{3}$$

At the edges $y=0$ and $y=b$ the deflection and the bending moment must be zero:

$$w=0,\ M_y=-D\left(\frac{\partial^2 w}{\partial y^2}+\mu\frac{\partial^2 w}{\partial x^2}\right)=0.$$

We find the solution "Eq. (3)" in the form

$$W(x,y)=X(x)\sin\frac{m\pi y}{b}. \tag{4}$$

The problem is reduced to finding a function $X(x)$ from the differential equation.

$$X''''(x)-2\left(\frac{m\pi}{b}\right)^2X''(x)+\left(\frac{k_0}{D}A(x)-\left(\frac{m\pi}{b}\right)^4\lambda\right)X(x)=0, \tag{5}$$

where λ – unknown dimensionless parameter,

$$\lambda=\left(\frac{b}{m\pi}\right)^4\frac{\rho h\omega^2}{D}-1. \tag{6}$$

In addition to deflection w, the state of the plate is also characterized by dynamic parameters – angles of rotation $\varphi_x,\ \varphi_y$, moments $M_x,\ M_y,\ M_{xy}$, and forces $V_x,\ V_y$.

$$w=X(x)\sin\frac{m\pi y}{b}T(t); \tag{7}$$

The formulas for the state parameters, expressed through $X(x)$, are well known. Along with "Eq. (5)" we will consider an equivalent system of differential equations

$$\frac{d\Phi(x)}{dx}=P(x)\Phi(x), \tag{8}$$

where the vector of unknowns and the matrix of coefficients are of the form:

$$\Phi(x)=\begin{pmatrix}X(x)\\X'(x)\\X''(x)\\X'''(x)\end{pmatrix};\ P(x)=\begin{pmatrix}0&1&0&0\\0&0&1&0\\0&0&0&1\\-(k_0/D)A(x)+(m\pi/b)^4\lambda&0&2(m\pi/b)^2&0\end{pmatrix} \tag{9}$$

Let us construct the solution of equation "Eq. (5)" by the method of direct integration, the essence of which was described in [12]. The goal is to express the desired solution $X(x)$ through dimensionless fundamental functions. Consider four infinite systems of functions $\alpha_{n,0}(x),\ \alpha_{n,k}(x)$ $(n=1,\ 2,\ 3,\ 4)$ $(k=1,\ 2,\ 3,\ \ldots)$. Using these functions and their derivatives we form series:

$$X_n(x)=\alpha_{n,0}(x)+\alpha_{n,1}(x)+\alpha_{n,2}(x)+\alpha_{n,3}(x)+\ldots; \tag{10}$$

$$X_n^{(\nu)}(x) = \alpha_{n,0}^{(\nu)}(x) + \alpha_{n,1}^{(\nu)}(x) + \alpha_{n,2}^{(\nu)}(x) + \alpha_{n,3}^{(\nu)}(x) + \ldots \; (\nu = 1,\ 2,\ 3,\ 4). \tag{11}$$

We assume that series "Eqs. (10), (11)" are uniformly convergent, which means that the operation of their sequential differentiation is possible.

The functions $\alpha_{n,0}(x)$ $(n = 1,\ 2,\ 3,\ 4)$ will be called initial, the functions $\alpha_{n,k}(x)$ $(n = 1,\ 2,\ 3,\ 4)$ $(k = 1,\ 2,\ 3,\ \ldots)$ – forming $X_n(x)$ $(n = 1,\ 2,\ 3,\ 4)$ satisfy "Eq. (5)".

$$X_n''''(x) - 2\left(\frac{m\pi}{b}\right)^2 X_n''(x) + \left(\frac{k_0}{D}A(x) - \lambda\left(\frac{m\pi}{b}\right)^4\right)X_n(x) = 0 \; (n = 1,\ 2,\ 3,\ 4). \tag{12}$$

Substituting their values instead of $X_n'''(x)$, $X_n''(x)$, $X_n(x)$ we will have.

$$\alpha_{n,0}''''(x) + \sum_{k=1}^{\infty}\left(\alpha_{n,k}''''(x) - 2\left(\frac{m\pi}{b}\right)^2 \alpha_{n,k-1}''(x) + \left(\frac{k_0}{D}A(x) - \lambda\left(\frac{m\pi}{b}\right)^4\right)\alpha_{n,k-1}(x)\right) = 0.$$

The last equality will be satisfied when the conditions are satisfied:

$$\alpha_{n,0}''''(x) = 0; \tag{13}$$

$$\alpha_{n,k}''''(x) = 2\left(\frac{m\pi}{b}\right)^2 \alpha_{n,k-1}''(x) - \left(\frac{k_0}{D}A(x) - \lambda\left(\frac{m\pi}{b}\right)^4\right)\alpha_{n,k-1}(x) \; (k = 1,\ 2,\ 3,\ \ldots). \tag{14}$$

The fundamental system of solutions of equation "Eq. (13)" is chosen in the form

$$\alpha_{n,0}(x) = \frac{1}{(n-1)!}\left(\frac{x}{a}\right)^{n-1} \; (n = 1,\ 2,\ 3,\ 4). \tag{15}$$

Four times integrating "Eq. (14)", we express $\alpha_{n,k}(x)$ through $\alpha_{n,k-1}(x)$ and set appropriate boundary conditions for all:

$$\alpha_{n,k}(0) = \alpha_{n,k}'(0) = 0 \; (k = 1,\ 2,\ 3,\ \ldots); \tag{16}$$

$$\alpha_{n,k}''(0) = 2\left(\frac{m\pi}{b}\right)^2 \alpha_{n,k-1}(0); \; \alpha_{n,k}'''(0) = 2\left(\frac{m\pi}{b}\right)^2 \alpha_{n,k-1}'(0) \; (k = 1,\ 2,\ 3,\ \ldots). \tag{17}$$

As a result, for values of $k = 1,\ 2,\ 3,\ \ldots$ we get

$$\begin{aligned}\alpha_{n,k}(x) = {} & 2\left(\frac{m\pi}{b}\right)^2 \int_0^X \int_0^X \alpha_{n,k-1}(x)dxdx + \lambda\left(\frac{m\pi}{b}\right)^4 \int_0^X \int_0^X \int_0^X \int_0^X \alpha_{n,k-1}(x)dxdxdxdx - \\ & - \frac{k_0}{D} \int_0^X \int_0^X \int_0^X \int_0^X A(x)\alpha_{n,k-1}(x)dxdxdxdx\end{aligned} \tag{18}$$

As a result, we have a recurrence formula according to which each initial function $\alpha_{n,0}(x)$ will have its own infinite set of generating functions $\alpha_{n,k}(x)$. Thus, four solutions

$X_n(x)$ $(n = 1,\ 2,\ 3,\ 4)$ are defined. Each of these solutions corresponds to a vector, the solution of system "Eq. (8)". Then the matrix composed of these vectors

$$\Omega(x) = \begin{pmatrix} X_1(x) & X_2(x) & X_3(x) & X_4(x) \\ X_1'(x) & X_2'(x) & X_3'(x) & X_4'(x) \\ X_1''(x) & X_2''(x) & X_3''(x) & X_4''(x) \\ X_1'''(x) & X_2'''(x) & X_3'''(x) & X_4'''(x) \end{pmatrix}, \tag{19}$$

also satisfies the system "Eq. (8)".

Let's calculate the values of $\Omega(0)$. Counting $x = 0$ and considering the boundary conditions "Eqs. (16), (17)", we get.

$$\Phi_n(0) = \begin{pmatrix} X_n(0) \\ X_n'(0) \\ X_n''(0) \\ X_n'''(0) \end{pmatrix} = \begin{pmatrix} \alpha_{n,0}(0) \\ \alpha_{n,0}'(0) \\ \alpha_{n,0}''(0) + 2(m\pi/b)^2\alpha_{n,0}(0) \\ \alpha_{n,0}'''(0) + 2(m\pi/b)^2\alpha_{n,0}'(0) \end{pmatrix}.$$

Herefrom

$$\Phi_1(0) = \begin{pmatrix} 1 \\ 0 \\ 2(m\pi/b)^2 \\ 0 \end{pmatrix};\ \Phi_2(0) = \frac{1}{a}\begin{pmatrix} 0 \\ 1 \\ 0 \\ 2(m\pi/b)^2 \end{pmatrix};\ \Phi_3(0) = \frac{1}{a^2}\begin{pmatrix} 0 \\ 0 \\ 1 \\ 0 \end{pmatrix};\ \Phi_4(0) = \frac{1}{a^3}\begin{pmatrix} 0 \\ 0 \\ 0 \\ 1 \end{pmatrix}.$$

So,

$$\Omega(0) = \begin{pmatrix} 1 & 0 & 0 & 0 \\ 0 & 1/a & 0 & 0 \\ 2\,(m\pi/b)^2 & 0 & 1/a^2 & 0 \\ 0 & 2\,(m\pi/b)^2/a & 0 & 1/a^3 \end{pmatrix}. \tag{20}$$

Obviously, the determinant of the matrix "Eq. (19)" is a Wronskian [6, 13], for which we find $|\Omega(0)| = \frac{1}{a^4} \neq 0$ taking into account "Eq. (20)". It follows [13] that the vectors "Eq. (19)" are linearly independent. Thus, matrix $\Omega(x)$ is the fundamental matrix of the system "Eq. (8)". By multiplying $\Omega(x)$ by a constant matrix $\Omega^{-1}(0)$, we have a new fundamental matrix $\Lambda(x) = \Omega(x)\Omega^{-1}(0)$, for which

$$\Lambda(x) = \begin{pmatrix} 1\,0\,0\,0 \\ 0\,1\,0\,0 \\ 0\,0\,1\,0 \\ 0\,0\,0\,1 \end{pmatrix}. \tag{21}$$

As we know [6, 13], a fundamental matrix that has property "Eq. (21)" is uniquely defined and is called a matrixant. Thus, assuming in the beginning the uniform convergence of series "Eqs. (10), (11)", we conclude that matrix $\Lambda(x)$, which is formed from the sums $X_n(x),\ X_n^{(\upsilon)}(x)$ $(\upsilon = 1,\ 2,\ 3)$ $(n = 1,\ 2,\ 3,\ 4)$ of these series, is the matrixant for the system "Eq. (8)". On the other hand, it has been proved in the theory of differential equations that the matrixant of a system of equations with continuous coefficients is always an absolutely and uniformly convergent matrix series [6]. Hence, absolute and uniform convergence of series "Eqs. (10), (11)" for values $\upsilon = 1, 2, 3$. The same convergence of series "Eq. (11)" for values $\upsilon = 4$ follows directly from "Eq. (12)". The solution of the system "Eq. (8)" is expressed by the matrixant with the well-known formula $\Phi(x) = \Lambda(x)\Phi(0)$. From this we have:

$$X(x) = X(0)\left(X_1(x) - 2\left(\frac{m\pi a}{b}\right)^2 X_3(x)\right) + aX'(0)\left(X_2(x) - 2\left(\frac{m\pi a}{b}\right)^2 X_4(x)\right) + a^2X''(0)X_3(x) + a^3X'''(0)X_4(x) \tag{22}$$

$$X^{(\nu)}(x) = X(0)\left(X_1^{(\nu)}(x) - 2\left(\frac{m\pi a}{b}\right)^2 X_3^{(\nu)}(x)\right) + aX'(0)\left(X_2^{(\nu)}(x) - 2\left(\frac{m\pi a}{b}\right)^2 X_4^{(\nu)}(x)\right) + a^2X''(0)X_3^{(\nu)}(x) + a^3X'''(0)X_4^{(\nu)}(x)\ (\nu = 1, 2, 3) \tag{23}$$

Thus, the general solution of equation "Eq. (5)" is found. The constants of integration in this solution are expressed in terms of initial parameters $X(0),\ X'(0),\ X''(0),\ X'''(0)$. The realization of boundary conditions at the edges of plate $x = 0$, $x = a$ will lead to a system of linear equations with respect to the unknown initial parameters $X(0)$, $X'(0)$, $X''(0)$, $X'''(0)$. From this we obtain the frequency equation. The unknown in the frequency equation will be the dimensionless numerical parameter λ. Therefore, our next goal is to ensure that we can write the frequency equation as a numerical series in powers of parameter λ. For this purpose, the fundamental solutions of $X_n(x)$, as well as functions $\tilde{X}_n(x)$, $\underset{n}{\frown{X}}(x)$, $\hat{X}_n(x)$ must be represented as series in powers of λ.

After finding the dimensionless parameter λ, for the frequency of free vibrations of the plate on the basis of formula "Eq. (6)" we will have

$$\omega = \sqrt{(1+\lambda)}\left(\frac{m\pi}{b}\right)^2\sqrt{\frac{D}{\rho h}}. \tag{24}$$

It is clear that by successively substituting expressions for the $\alpha_{n,0}(x)$, $\alpha_{n,1}(x)$, ..., $\alpha_{n,k-1}(x)$, functions into formula "Eq. (18)", we obtain a polynomial of degree k with respect to parameter λ with variable coefficients. In other words, for the generating functions the representation

$$\alpha_{n,k}(x) = \beta_{n,k,0}(x) + \lambda\beta_{n,k,1}(x) + \ldots + \lambda^k\beta_{n,k,k}(x)\ (k = 1, 2, 3, \ldots), \tag{25}$$

where functions $\beta_{n,k,l}(x)$ $(k = 1, 2, 3, \ldots)$ $(l = 0, 1, 2, \ldots, k)$ are to be defined.

Substituting in formula "Eqs. (16), (17)" instead of functions $\alpha_{n,k}(x)$ their values "Eq. (25)" and equating in obtained equations coefficients at equal degrees λ, we obtain boundary conditions for functions $\beta_{n,k,l}(x)$.

Due to the found representations, the frequency equation can be represented as

$$\eta_0 + \eta_1\lambda + \eta_2\lambda^2 + \eta_3\lambda^3 + \ldots = 0, \tag{26}$$

where η_k $(k = 0, 1, 2, \ldots)$ – dimensionless coefficients, the calculation of which will depend on the boundary conditions at the edges $x = 0,\ x = a$. Also these coefficients will depend on the parameter m. Let us denote by $\lambda_{1,m},\ \lambda_{2,m},\ \lambda_{3,m},\ \ldots$ the roots of the frequency equation "Eq. (26)", which correspond to the given parameter m and are written in ascending order. Then by formula "Eq. (24)" we obtain the spectrum of frequencies of free vibrations of the plate

$$\omega_{jm} = \sqrt{\left(1+\lambda_{jm}\right)}\left(\frac{m\pi}{b}\right)^2\sqrt{\frac{D}{\rho h}} \quad (m = 1, 2, 3, \ldots)\ (j = 1, 2, 3, \ldots). \tag{27}$$

The following solution of "Eq. (5)" corresponds to the found roots λ_{jm}:

$$X_{jm}(x) = X_{jm}(0)\left(X_1\left(x, \lambda_{jm}\right) - 2\left(\frac{m\pi a}{b}\right)^2 X_3\left(x, \lambda_{jm}\right)\right)$$
$$+ aX'_{jm}(0)\left(X_2\left(x, \lambda_{jm}\right) - 2\left(\frac{m\pi a}{b}\right)^2 X_4\left(x, \lambda_{jm}\right)\right)$$
$$+ a^2X''_{jm}(0)X_3\left(x, \lambda_{jm}\right) + a^3X'''_{jm}(0)X_4\left(x, \lambda_{jm}\right)\ \ (m = 1, 2, 3, \ldots)\ \ (j = 1, 2, 3, \ldots)$$

As a result, according to "Eq. (4)", the main waveforms of the plate will be written.

$$W_{jm}(x, y) = X_{jm}(x)\sin\frac{m\pi y}{b} \quad (m = 1, 2, 3, \ldots)\ \ (j = 1, 2, 3, \ldots).$$

Writing down the formulas for the main forms of vibrations in each case of boundary conditions at the $x = 0,\ x = a$, we further conditionally represent them as

$W_{jm}(x, y) = D_{jm}W^*_{jm}\left(\frac{x}{a}, \frac{y}{b}\right)$, where D_{jm} – is a constant dimensional multiplier, and $W^*_{jm}\left(\frac{x}{a}, \frac{y}{b}\right)$ – is a dimensionless function that determines the law of the main form of vibrations.

Numerical realization of the proposed analytical solution of the problem is performed for three cases of boundary conditions: hinged support along the whole contour; plate sides $x = 0,\ x = a$ are rigidly fixed, while the other two have hinged support; side $x = 0$ is rigidly fixed, while the other three have hinged support. The frequencies and forms of natural vibrations are determined for four laws of variation of the bedding coefficient – constant; according to the linear law; according to the concave parabola law; according to the convex parabola law. In view of the limited volume of the article, here we present only the results of calculation of a square plate ($a = 1,\ b = 1$) with articulated support along the contour, when the bedding coefficient varies according to the law of a concave parabola, with the following initial data:

$$E = 2\cdot 10^8\ kPa,\ \ \rho = 7800\ kg/m^3,\ \ h = 0,05\ m,\ \ \mu = 0,3.$$

The results of the calculation are presented in Table 1. The roots $\lambda_{q,\,m}$ $(q = 1, 2)$ $(m = 1, 2)$ of the frequency equation, the frequencies corresponding to these roots, and the graphs of the dimensionless laws of the main forms of vibrations are shown here. For comparison, Table 2 presents the first four frequencies calculated at constant bedding coefficient by the author's method and using the exact formula known for articulated contour support.

Table 1. Calculation results.

Elastic base type	Initial parameters
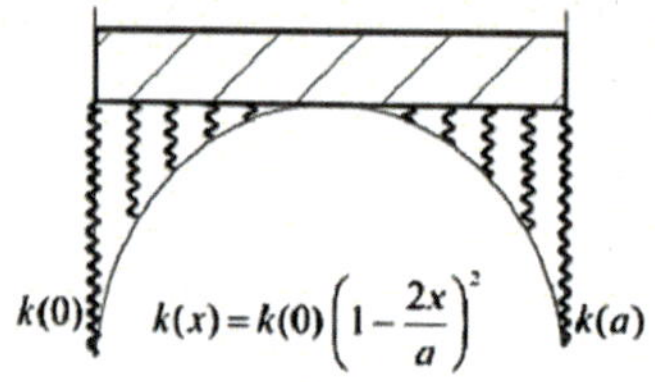	$k(0)=k(a)=4\cdot10^6 kN/m^3$; $s=2,\ A_0=1,\ A_1=-4,\ A_2=4$. Frequencies $\omega_{q,m}$

	$m=1$	$m=2$
$q=1$	1893,3363	3950,5305
$q=2$	4143,8796	6283,089

$\lambda_{1,1}=5,2691$

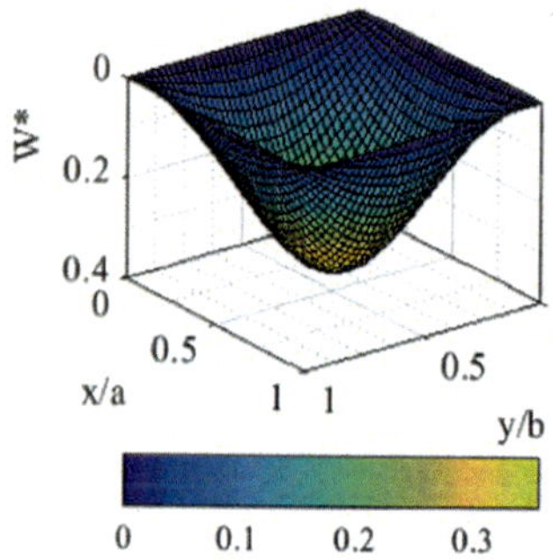

$\lambda_{1,2}=0,7058$

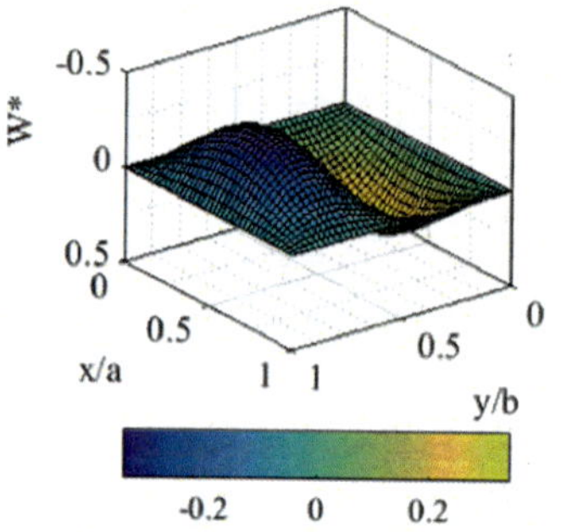

$\lambda_{2,1}=29,0305$

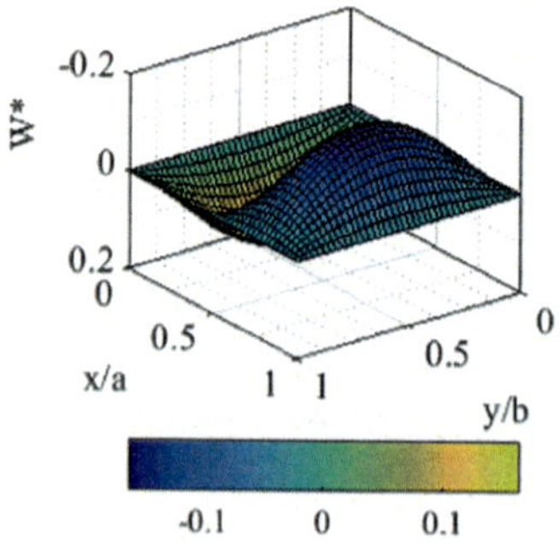

$\lambda_{2,2}=3,3149$

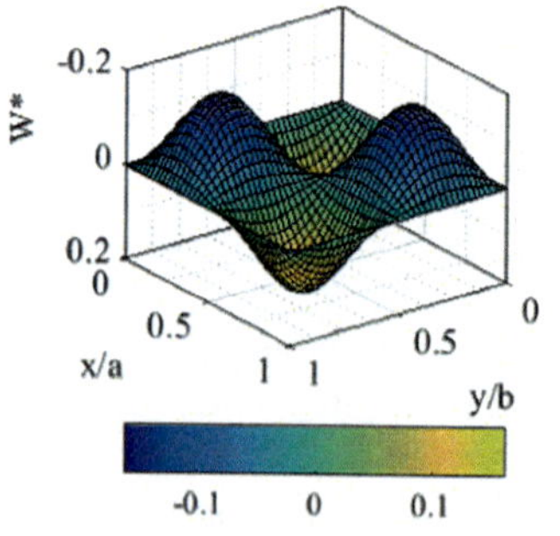

Table 2. Comparison of frequencies at a constant bedding factor.

	Frequencies $\omega_{q,\,m}$				Relative error, %	
	Author's method		The exact formula			
	$m = 1$	$m = 2$	$m = 1$	$m = 2$	$m = 1$	$m = 2$
$q = 1$	1516,5958	3782,6031	1516,5959	3782,6032	0,000011	0,000004
$q = 2$	3782,6026	6050,5120	3782,6032	6050,5127	0,000017	0,000011

4 Conclusions

An analytical solution of the problem of free bending vibrations of rectangular plates with Levy boundary conditions lying on a continuous variable elastic base, which is described by the Winkler model, is obtained. An exact solution of the differential equation of free vibrations of plates when the bedding coefficient is an arbitrary continuous function of one variable is found. The quadratures for numerical realization of the found solutions are derived. The formulas for dynamic state parameters, which allow one to investigate free vibrations of the plates under any boundary conditions at two parallel edges, are obtained. The dependence of the frequency of free vibrations of the system under consideration on its other parameters is established in the analytical form. Computational formulas for determining the spectrum of frequencies of free vibrations of the plates are obtained. The general form of frequency equation, frequency equations and formulas for the main forms of vibrations corresponding to the three cases of boundary conditions are established. Frequency spectra of free vibrations of hinged plates resting on variable elastic bases for four different laws of bedding coefficient changes are determined. It is shown that in the case of constant bedding coefficient the frequencies calculated by the author's method practically coincide with the frequencies calculated by the known exact formula.

References

1. Atay, M.T.: Determination of buckling loads of tilted buckled column with varying flexural rigidity using variational iteration method. Int. J. Nonlin. Sci. Numer. Simul. **11**(2), 97–103 (2010)
2. Balachandran, B., Magrab, E.B.: Vibrations. Cengage Learning (2008)
3. Buchacz, A., Żółkiewski, S.: Longitudinal vibrations of mechanical systems with the transportation effect. J. Achiev. Mater. Manuf. Eng. **32**(1), 29–36 (2009)
4. Coskun, S.B.: Advances in Computational Stability Analysis. InTech (2012). https://doi.org/10.5772/3085
5. Leissa, A.W., Qatu, M.S.: Vibrations of Continuous Systems. McGraw-Hill (2011)
6. Arbabi, F., Li, F.: Buckling of variable cross-section columns – integral equation approach. J. Struct. Eng. ASCE **117**(8), 2426–2441 (1991)
7. Valchenko, A.I., Metelev, V.A., Khalypa, V.M.: Calculation of the strength of a cylindrical tank for storing toxic liquids. Probl. Supranatl. Situat. **12**, 50–55 (2010)
8. Vasylenko, M.V., Alekseychuk, O.M.: Theory of Oscillations and Stability of Motion. Higher school, Kyiv (2004)

9. Bolshakov, A.A.: Rectangular plate on a two-parameter elastic foundation: an analytical solution. Bull. SamGU Nat. Sci. Ser. **8**(89), 128–133 (2011)
10. Doronin, A.M., Soboleva, V.A.: Natural vibrations of a round plate lying on a Winkler-type variable elastic foundation. Bull. Nizhny Novgorod Univ. Lobachevsky **4**(1), 254–258 (2014)
11. Schwabyuk, V.I., Rotko, S.V., Guda, O.V.: Oscillations of a transtropic plate on an elastic basis under fluid pressure. Modern problems of mechanics and mathematics. J.S. Hairdresser of the National Academy of Sciences of Ukraine, pp. 182–184 (2013)
12. Gabbasov, R.F., Moussa, S.: Generalized equations of the finite difference method and their application to the calculation of flexible plates with variable rigidity. Izvestiya vuzov. Building **5**, 17–22 (2004)
13. Gabbasov, R.F., Uvarov, N.B.: Application of generalized equations of the finite difference method to the calculation of plates on an elastic foundation. Vestnik MGSU **4**, 102–107 (2012)
14. Andreev, V.I., Barmenkova, E.V., Matveeva, A.V.: Calculation of slabs of variable stiffness on an elastic foundation by the finite difference method. Vestnik MGSU **4**, 30–38 (2014)
15. Zaporozhets, E.V., Krasovsky, V.L.: Finite element calculation of axisymmetric bending of plates on an elastic foundation. Bull. Prydniprovska State Acad. Life Archit. **5**, 16–23 (2002)
16. Koreneva, E.B.: Development of analytical methods for calculating plates of variable thickness and their practical applications. Abstract of a Doctoral Dissertation. MGSU, Moscow (1999)
17. Koreneva, E.B.: Analytical Methods for Calculating Plates of Variable Thickness and Their Practical Applications. ASV, Moscow (2009)
18. Teng, Z., Ding, S., Zheng, P.: Free vibration analysis of rectangular plates with variable thickness on elastic foundation by using GDQ method. Chin. J. Appl. Mech. **31**, 236–241 (2014)
19. Kanev, N.G.: The natural frequencies of the plate oscillations on an elastic base of the Winkler type. Noise Theory Pract. **2**(6), 28–35 (2020)
20. Kägo, E., Lellep, J.: Free vibrations of plates on elastic foundation. Procedia Eng. **57**, 489–496 (2013). https://doi.org/10.1016/j.proeng.2013.04.063
21. Teng, Z.-C., Wang, W.-B., Zheng, W.: Free vibration analyses of porous Fgm rectangular plates on a Winkler-Pasternak elastic foundation considering the temperature effect. Eng. Mech. https://doi.org/10.6052/j.issn.1000-4750.2021.02.0152
22. Minh, P.P., Manh, D.T., Duc, N.D.: Free vibration of cracked FGM plates with variable thickness resting on elastic foundations. Thin-Walled Struct. **161**, 107425 (2021). https://doi.org/10.1016/j.tws.2020.107425
23. Khetib, M., Abbad, H., Elmeiche, N., Mechab, I.: Effect of the viscoelastic foundations on the free vibration of functionally graded plates. Int. J. Struct. Stab. Dyn. (2019). https://doi.org/10.1142/s0219455419501360

Dynamic Mixed Problem of Elasticity for a Rectangular Domain

Pozhylenkov Oleksii(✉) and Vaysfeld Nataly

Department of Mathematics, Physics and IT,
Odesa Mechnikov University, Odesa, Ukraine
leshiy12345678@gmail.com
http://onu.edu.ua/en/

Abstract. The dynamic elasticity problem for rectangular domain is presented in this paper. The conditions of the second main elasticity problem are given at the lateral sides. It is necessary to find the wave field of the rectangular domain under steady state loading. The integral Fourier transform was applied to the formulated boundary value problem and reduced it to a one dimensional vector boundary problem. The last one was solved with the method based on matrix differential calculations which was successfully applied earlier to solve the analogous static problem [1,2]. As a result, the non-homogeneous one-dimensional boundary problem was derived and solved with the help of Green's matrix-function apparatus [3]. The wave field is constructed as a superposition of the homogeneous and non-homogeneous solutions and contains the unknown derivative of the displacements. To find this unknown displacement the singular integral equation is derived and solved with the help of the orthogonal polynomial method. Application of the inverse integral Fourier transform finalized the construction of the stated problem solution. The analysis of wave field of the rectangular domain depending on different loading types, frequency values and domain's size was done.

Keywords: Vector boundary problem · Rectangular domain · Dynamic loading · Singular integral equation

1 Introduction

The problem of the rectangular domain is not a new one, but unfortunately a lot of unsolved issues remain. This problem was considered and solved in the different statements important to engineering applications with the help of analytical methods and numerical ones.

The last approach is demonstrated in the paper [4], where the boundary element-free method was applied to two the dimensional problem of elasticity. This method is a numerical which combines the boundary integral equation method and an improved moving least-square approximation that gives higher computational accuracy. At [5] the finite element method is applied and discussion of the condition's type necessary for the penalty methods to provide a basis.

Z. Dimitrovová et al. (Eds.): WMVC 2022, MMS 125, pp. 211–218, 2023.
https://doi.org/10.1007/978-3-031-15758-5_20

For finite and convergent elements, schemes are proposed for stable calculations. The mixed finite element approximation of a stress-displacement system derived from the Hellinger-Reissner variation principle for the linear elasticity problem was considered at [6]. Many benefits of the numerical methods can attributed by their existence in many numerical software applications, for easy use by engineers.

But if one needs to provide the calculation of the stress at the rectangular domain in the neighborhood of the angular points (points where stress increases significant at the vertices of the domain), the numerical methods lose their efficiency as it is known. These points of the changing boundary condition cause the stress with a special order of a singularity. To take these singularities into consideration one must use analytical approaches [7], methods solving such problems with regard to the singularities existence. One of such analytical methods was proposed at [8,9] for a two dimensional problem for the rectangular domain. Then it was developed and applied simultaneously with the matrix differential calculations in papers [10,11]. Such methodology allows derivation of the exact solution for some particular cases of the boundary conditions. Also, the well-known paper [12] was one of the pioneer papers in this direction. The solution of the plane thermoelasticity problem for a rectangular domain was constructed with the help of a new solving method. This method permits the construction of an analytical solution, corresponding to the Saint-Venan principle in the form of trigonometric series expansion using an orthogonal set of the eigen and associated functions. These investigations were successfully continued in [13].

The novelty of the presented paper is in the application of the new approach represented in [14] to solve the dynamic elasticity problem for a rectangular domain. The solution is studied by the authors in case of different loading, domains size and frequency.

2 Statement of the Problem

The elastic rectangular domain $-a \leq x \leq a$, $0 \leq y \leq b$ (G is a shear modulus, μ is a Poison's coefficient, E is a Young's modulus, c_1, c_2 are the wave velocities, ω is a frequency) meets a load at the upper face of the domain,

$$\sigma_y(x,b,t) = -p(x,t), \quad \tau_{xy}(x,b,t) = 0 \tag{1}$$

The lower side is under the condition of ideal contact

$$u_y(x,0,t) = 0, \quad \tau_{xy}(x,0,t) = 0 \tag{2}$$

Displacements at the left and the right side of the domain are equal to zero

$$u_x(\pm a,y,t) = 0, \quad u_y(\pm a,y,t) = 0 \tag{3}$$

It is required to estimate the stress state of the rectangular domain satisfying boundary conditions (1)–(3) and the Lame's equations

$$U''(x,y,t) + U^{**}(x,y,t) + \mu_0(U''(x,y,t) + V'^{*}(x,y,t)) = \frac{1}{c_1^2}\frac{\partial^2 U(x,y,t)}{\partial t^2}$$
$$U''(x,y,t) + U^{**}(x,y,t) + \mu_0(U''(x,y,t) + V'^{*}(x,y,t)) = \frac{1}{c_2^2}\frac{\partial^2 V(x,y,t)}{\partial t^2} \quad (4)$$

Here we use the following notations $U(x,y,t) = u_x(x,y,t)$, $V(x,y,t) = u_y(x,y,t)$, $f'(x,y,t) = \frac{\partial f(x,y,t)}{\partial x}$, $f^{*}(x,y,t) = \frac{\partial f(x,y,t)}{\partial y}$, $\mu_0 = \frac{1}{1-2\mu}$

To solve the stated problem only half of the domain is considered $0 \le x \le a$, $0 \le y \le b$ and with the usage of the symmetric properties, and the boundary conditions are written in the form:

$$u_x(0,y,t) = 0, \quad \tau_{xy}(0,y,t) = 0 \quad (5)$$
$$u_x(a,y,t) = 0, \quad u_y(a,y,t) = 0 \quad (6)$$

$p(x,t)$ should be symmetric function on the given interval.

The case of the harmonic vibrations is considered $U(x,y,t) = e^{i\omega t}U(x,y)$, $V(x,y,t) = e^{i\omega t}V(x,y)$, $p(x,t) = e^{i\omega t}p(x)$.

Lame's equations are transformed

$$U''(x,y) + U^{**}(x,y) + \mu_0(U''(x,y) + V'^{*}(x,y)) = -\frac{\omega^2}{c_1^2}U(x,y)$$
$$U''(x,y) + U^{**}(x,y) + \mu_0(U''(x,y) + V'^{*}(x,y)) = -\frac{\omega^2}{c_2^2}V(x,y) \quad (7)$$

3 Solving the Problem

The Fourier's integral transforms are applied to the Eqs. (7) with the following scheme

$$\begin{pmatrix} U_n(y) \\ V_n(y) \end{pmatrix} = \int_0^a \begin{pmatrix} U(x,y) * \sin(\alpha_n x) \\ V(x,y) * \cos(\alpha_n x) \end{pmatrix} dx, \quad \alpha_n = \frac{\pi}{a}(n - \frac{1}{2}) \quad (8)$$

It leads to a non-homogeneous system of ordinary differential equations in the transform's domain

$$U_n''(y) - \mu_0\alpha_n V_n'(y) + \frac{\omega^2}{c_1^2}U_n(y) - \alpha_n^2 U_n(y) - \mu_0\alpha_n^2 U_n(y) = f_n(y)$$
$$V_n''(y) + \mu_0 V_n''(y) + \mu_0\alpha_n U_n'(y) + \frac{\omega^2}{c_2^2}V_n(y) - \alpha_n^2 V_n(y) = 0 \quad (9)$$

where $f_n(y) = (-1-\mu_0)sin(\alpha_n x)\frac{\partial U(x,y)}{\partial x}|_{x=a}$

Boundary conditions (1)–(2), (5)–(6) are reformulated in terms of displacements transforms

$$(2G+1)V_n'(b) + \lambda\alpha_n U_n(b) = -p_n \tag{10}$$

$$V_n(0) = 0 \tag{11}$$

$$U_n(b) - \alpha_n V_n(b) = 0 \tag{12}$$

$$U_n(0) - \alpha_n V_n(0) = 0 \tag{13}$$

The initial problem in the transform's domain (9)–(13) is reformulated into a vector boundary problem with the help of vector differential operator of the second order:

$$\begin{gathered} L_2[Z_n(y)] = AZ_n''(y) + BZ_n''(y) + CZ_n(y), \\ L_2[Z_n(y)] = F_n(y), \end{gathered} \tag{14}$$

Boundary conditions:

$$\begin{gathered} U_i[Z_n(y)] = p_i, \\ U_i[Z_n(y)] = D_i Z_n'(b_i) + E_i Z_n(b_i), \\ b_0 = 0, b_1 = b \end{gathered} \tag{15}$$

The vectors and matrices of the vector boundary problem are input

$$\begin{gathered} D_0 = \begin{pmatrix} 1 & 0 \\ 0 & (2G+\lambda) \end{pmatrix}, \quad E_0 = \begin{pmatrix} 0 & -\alpha_n \\ \alpha_n\lambda & 0 \end{pmatrix} \\ D_1 = \begin{pmatrix} 1 & 0 \\ 0 & 0 \end{pmatrix}, \quad E_1 = \begin{pmatrix} 0 & -\alpha_n \\ 0 & 1 \end{pmatrix} \\ p_1 = \begin{pmatrix} 0 \\ -p_n \end{pmatrix}, \quad p_2 = \begin{pmatrix} 0 \\ 0 \end{pmatrix} \end{gathered} \tag{16}$$

To solve this vector boundary problem (14)–(15), the fundamental solution matrices are constructed [3]. Firstly the matrix $e^{\xi}I$ (where I is the unit matrix) must be substituted into the Eq. (14). From the equality $L_2[e^{\xi}I] = M(\xi)e^{\xi y}$, one can derive the $M(\xi)$ matrix

$$M(\xi) = \begin{pmatrix} \xi^2 - \alpha_n^2 - \mu_0\alpha_n^2 + \frac{\omega^2}{c_1^2} & -\xi\alpha_n\mu_0 \\ \xi\alpha_n\mu_0 & \xi^2 + \xi^2\mu_0 - \alpha_n^2 + \frac{\omega^2}{c_2^2} \end{pmatrix} \tag{17}$$

The fundamental solution matrices are found in the following form [15]

$$Y(y) = \frac{1}{2\pi i}\oint_C e^{\xi y} M^{-1}(\xi) d\xi \tag{18}$$

The calculation of the integral requires knowing all poles of the integral function below. To do it the determinant of the matrix $M(\xi)$ was derived

$$detM(\xi) = (\xi - a_1)(\xi + a_1)(\xi - a_2)(\xi + a_2) \tag{19}$$

Roots $a_1, -a_1, a_2, -a_2$ of the $detM(\xi)$ were found exactly. After cthe ontour integration procedure the four linear independent solutions of the matrix equation were derived

$$\frac{1}{2\pi i}\oint_C e^{\xi y}M^{-1}(\xi)d\xi = \sum_{i=0}^{4} Res[e^{\xi y}M^{-1}(\xi)] = \frac{1}{1+\mu_0}(Y_0(y)+Y_1(y)+Y_2(y)+Y_3(y)),$$

$$Y_i(y) = (-1)^i \frac{e^{(-1)^i y a_1}}{2a_1(a_1^2-a_2^2)} \begin{pmatrix} a_1^2 + a_1^2\mu_0 - \alpha_n^2 + \frac{\omega^2}{c_2^2} & (-1)^i a_1\alpha_n\mu_0 \\ (-1)^{i+1}a_1\alpha_n\mu_0 & a_1^2 - \alpha_n^2 - \mu_0\alpha_n^2 + \frac{\omega^2}{c_1^2} \end{pmatrix} \quad (20)$$

where $Y_i(y), i = 0,1,2,3$ - are fundamental solution matrices.

The obtained fundamental solution matrices allow the derivation of the fundamental solution in the following form

$$\Psi_n^i(y) = \frac{1}{1+\mu_0}(Y_0(y)+Y_1(y))C_1^i + \frac{1}{1+\mu_0}(Y_2(y)+Y_3(y))C_2^i, \quad i=1,2 \quad (21)$$

Applying the boundary conditions $U_i[\Psi_n^j(y)] = k_{i,j}, k_{i,j} = \begin{cases} 0, & i=j, \\ I, & i\neq j, \end{cases}$ one obtained the linear algebraic system from which matrices C_j^i, $i=1,2$, $j=1,2$, were found.

With the help of Green's matrix the solution of the non-homogeneous vector boundary problem was constructed [3]

$$Z_n(y) = \int_0^b G(y,\xi)F(\xi)d\xi + \Psi_n^1(y)p_1 + \Psi_n^2(y)p_2 \quad (22)$$

Green's matrix has been found in the form

$$G_n(y,\xi) = \begin{cases} \Psi_n^1(y)\Psi_n^2(\xi), & 0\le y<\xi, \\ \Psi_n^2(y)\Psi_n^1(\xi), & \xi<y\le b, \end{cases} \quad (23)$$

Solution of the stated problem in the transforms domain is derived

$$\begin{cases} U_n(y) = \int_0^b g_{1,1}(y,\xi)f_n(\xi)d\xi - \Psi_{1,2}^1(y)p_n \\ V_n(y) = \int_0^b g_{2,1}(y,\xi)f_n(\xi)d\xi - \Psi_{2,2}^1(y)p_n \end{cases} \quad (24)$$

Here $g_{1,1}(y,\xi), g_{1,2}(y,\xi)$ elements of the Green's matrix, $\Psi_{1,2}^1(y), \Psi_{2,2}^1(y)$ elements of $\Psi_n^1(y)$ matrix, $f_n(\xi) = (-1-\mu_0)sin(\alpha_n x)\frac{\partial U(x,\xi)}{\partial x}|_{x=a}$. With the help of the inversion Fourier's transform the solution of stated problem was found

$$\begin{cases} U(x,y) = \frac{2}{a}\sum_{n=1}^{\infty} U_n(y)sin(\alpha_n x) \\ V(x,y) = \frac{2}{a}\sum_{n=1}^{\infty} V_n(y)cos(\alpha_n x), \quad \alpha_n = \frac{\pi}{a}(n-\frac{1}{2}) \end{cases} \quad (25)$$

This solution will define the wave field if function $f(\xi)$ is found. To find it one must demand the satisfying of the boundary condition (10), it leads to the integral singular equation relatively to the unknown function. The last part which has to be found it is $f(\xi) = \frac{\partial U(x,\xi)}{\partial x}|_{x=a}$ function. To find it, the boundary condition $\sigma_y(x,b) = -p(x)$ was applied and integral equation for the unknown function was obtained.

$$\frac{2(1+\mu_0)}{a} \int_0^b \sum_{n=1}^{\infty} (-1)^n K_n(x-\xi) f(\xi) d\xi = H(x) \tag{26}$$

To find the solution of the integral equation, the orthogonal polynomial's method was applied. It leads to the solution of the singular integral equation to the infinite algebraic system, which can be solved applying the reduction method that has been shown in [16].

4 Conclusions

New method was applied for the dynamic problem of elasticity for a rectangular domain with the fixed lateral sides. The solving leads to the integral equation which was effectively solved by the orthogonal-polynomial method. Investigation of the wave field depending on different domain sizes, loading types and frequencies was done.

5 Numerical Results

Some numerical results are presented for different types of loading and domain sizes.

Here stress values $\sigma_x(x,y,t)$, $\sigma_y(x,y,t)$ are shown at the Fig. 1, 2, 3 and 4 for the external loading $p(x) = (x-2.5)^2$, $0 < x < 5$, $0 < y < 6$, $t = 0$.

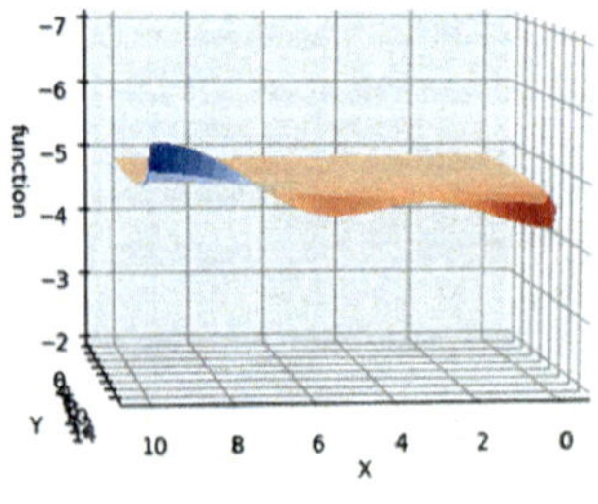

Fig. 1. $\sigma_x(x,y,t)$, $\omega = 0.75$

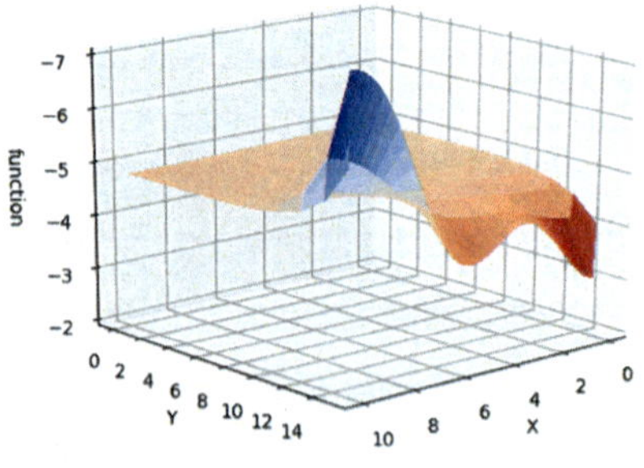

Fig. 2. $\sigma_x(x,y,t)$, $\omega = 1.5$

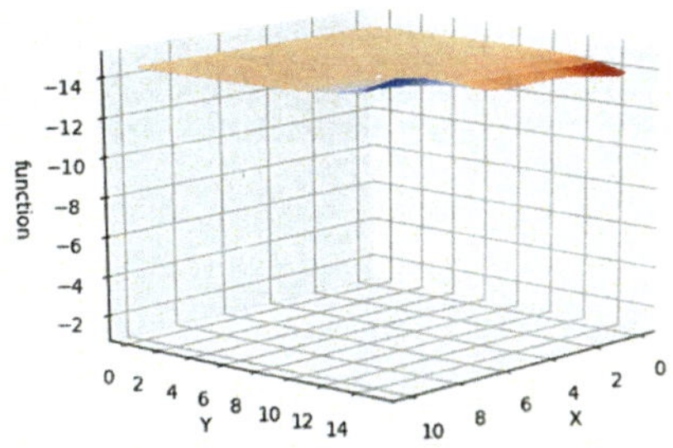

Fig. 3. $\sigma_y(x, y, t)$, $\omega = 0.75$

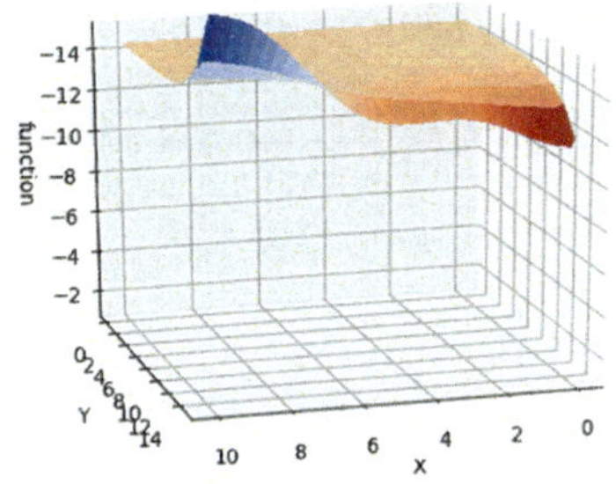

Fig. 4. $\sigma_y(x, y, t)$, $\omega = 1.5$

References

1. Pozhylenkov, O.V.: The stress state of the rectangular elastic domain. Odesa I.I. Mechnikov National University, Odesa (2019), vol. 24, no. 2(34) (2019)
2. Pozhyenkov, O., Vaysfeld, N.: Stress state of a rectangular domain with the mixed boundary conditions. Procedia Struct. Integrity **28**, 458–463 (2020)
3. Popov, G., Abdimanapov, S., Efimov, V.: Green's Functions and Matrixes of the One-Dimensional Problems. Student Manual for the Physical and Mathematical Faculties, Almaty (1999)
4. Liew, K.M., Cheng, Y., Kitipornchai, S.: Boundary element-free method (BEFM) and application to two dimensional elasticity problems. Int. J. Numer. Methods Eng. **65**(8), 1310–1332 (2005)
5. Oden, J.T., Kikuchi, N.: Finite element methods for constrained problems in elasticity. Int. J. Numer. Methods Eng. **18**(5), 701–725 (1982)
6. Shi, D., Li, M.: Superconvergence analysis of the stable conforming rectangular mixed nite elements for the linear elasticity problem. J. Comput. Math. **32**(2), 205–214 (2014)
7. Prasad, S.N., Chatterjee, S.N.: Some mixed boundary value problems of elasticity in a rectangular domain. Int. J. Solids Struct. **9**(10), 1193–1210 (1973)
8. Popov, G.: The Elastic Stress Concentration Around Dies, Cuts, Thin, Inclusions and Reinforcements. Nauka, Moscow (1982). (in Russian)
9. Popov, G.Y., Protserov, Y.S.: Axisymmetric problem for an elastic cylinder of infinite length with fixed lateral surface with regard for its weight. J. Math. Sci. **212**(1), 67–82 (2016)
10. Popov, G., Vaysfeld, N., Zozulevich, B.: The exact solution of elasticity mixed plain boundary value problem in a rectangular domain. In: 20-th International Conference Engineering Mechanics, Srvatka, Czech Republic (2014)
11. Zhuravlova, Z.Y.: The plane mixed problem for an elastic semi-strip under different load types at its short edge. Int. J. Mech. Sci. **144**, 526–530 (2018)
12. Vihak, V.M., Yuzvyak, N.Y., Yasinskij, A.V.: The solution of the plane thermoelasticity problem for a rectangular domain. J. Thermal Stresses **21**(5), 545–561 (1998)
13. Vihak, V.M., Tokovyy, Yu.: Construction elementary solutions of the elasticity plane problem for the rectangular domain. Int. Appl. Mech. **32**(7), 79–87 (2002)
14. Popov, G., Vaysfeld, N.: The steady-one oscillations of the elastic infinite come loaded at a vertex by a concentrated force. Acta Mechanica **221**(3–4), 261–270 (2011)

15. Gantmakher, F.R.: The Theory of Matrices. AMS Chelsea Publishing, Providence (1998)
16. Popov, G.: On the method of orthogonal polynomials in contact problems of the theory of elasticity. J. Appl. Math. Mech. **33**(3), 503–517 (1969)

Free Flexural Axisymmetric Vibrations of Generalized Circular Sandwich Plate

Krzysztof Magnucki[1] and Ewa Magnucka-Blandzi[2](✉)

[1] Łukasiewicz Research Network, Poznan Institute of Technology, Poznan, Poland
[2] Institute of Mathematics, Poznan University of Technology, Poznan, Poland
ewa.magnucka-blandzi@put.poznan.pl

Abstract. The work is devoted to the mathematical modeling of a sandwich circular plate. Free flexural axisymmetric vibrations of generalized three-layer plate is studied. The mechanical properties of the plate's core vary through its thickness. The mechanical properties of faces are constant. The individual nonlinear theory of deformation of the straight normal line to the middle plane of the plate is elaborated taking into account the shear effect. Two cases of the supported edge of the plate are considered: simply supported, and clamped. Based on the Hamilton's principle two differential equations of motion are obtained. This system of equations is approximately solved and the fundamental natural frequencies are obtained for two circular sandwich plates.

Keywords: Mathematical modelling · Generalized sandwich plate · Free flexural vibrations

1 Introduction

The sandwich constructions initiated in the mid-twentieth century are the subject of the contemporary studies. Ventsel and Krauthammer [1] presented theories of thin plates and shells, emphasizing analytical and numerical methods for solving linear and nonlinear problems, including variational and numerical methods. Reddy [2] provided complete, detailed coverage of the various theories, analytical solutions, and finite element models of laminated composite plates and shells. The monograph reflects materials modeling in general and composite materials and structures in particular. It also includes a chapter devoted to the theory and analysis of laminated shells, discussions on smart structures and functionally graded materials. Yang and Qiao [3] presented a higher-order impact model to simulate the response of a soft-core sandwich beam subjected to a foreign object impact. A free vibration problem of sandwich beams was solved, and the results were validated by comparing with numerical finite element modeling results of ABAQUS. Then a foreign object impact process was incorporated in the higher-order model. The calculated stresses caused by a foreign object impact were then used to assess failure locations, failure time, and failure modes in sandwich beams, which were shown to compare well with the available experimental results. The higher-order impact model of sandwich beams developed in the study provides accurate predictions of the generated

Z. Dimitrovová et al. (Eds.): WMVC 2022, MMS 125, pp. 219–229, 2023.
https://doi.org/10.1007/978-3-031-15758-5_21

stresses and impact process and can be used effectively in design analysis of anti-impact structures made of sandwich materials. Zenkour [4] presented the static response for a simply supported functionally graded rectangular plate subjected to a transverse uniform load. The generalized shear deformation theory obtained by the author in other papers was used. This theory was simplified by enforcing traction-free boundary conditions at the plate faces. No transversal shear correction factors were needed because a correct representation of the transversal shearing strain was given. Material properties of the plate were assumed to be graded in the thickness direction according to a simple power-law distribution in terms of the volume fractions of the constituents. The equilibrium equations of a functionally graded plate were given based on a generalized shear deformation plate theory. The results were verified with the known results in the literature. Carrera and Brischetto [5] described and assessed a large variety of plate theories to evaluate the bending and vibration of sandwich structures. The work includes significant review papers and developments on sandwich structure modelings. The kinematics of classical, higher order, zigzag, layerwise, and mixed theories were described. An exhaustive numerical assessment of the whole theories was provided in the case of closed form solutions of simply supported panels made of orthotropic layers. Based on the three-dimensional theory, Wang et al. [6] presented a direct displacement method to investigate the free axisymmetric vibration of transversely isotropic circular plates, whose material is functionally graded and properties obey the exponential law along the thickness direction of the plate. Yas and Tahouneh [7] investigated the free vibration of functionally graded annular plates on elastic foundations, based on the three-dimensional theory of elasticity, using the differential quadrature method for different boundary conditions including simply supported–clamped, clamped–clamped and free–clamped ends. The material properties change continuously through the thickness of the plate, which can vary according to power law, exponentially or any other formulations in this direction. A semi-analytical approach composed of differential quadrature method (DQM) and series solution were adopted to solve the equations of motions. Sayyad and Ghugal [8] reviewed the recent research done on the free vibration analysis of multilayered laminated composite and sandwich plates using various methods available for the analysis of plates. Displacement fields of various displacement based shear deformation theories were presented and compared. Also, some numerical results related to fundamental flexural mode frequencies of laminated composite and sandwich plates were presented using a trigonometric shear and normal deformation theory. This article cites 391 references. Sarangan and Singh [9] presented new shear deformation theories (algebraic, exponential, hyperbolic, logarithmic and trigonometric) were developed to analyze the static, buckling and free vibration responses of laminated composite and sandwich plates using Navier closed form solution technique. Sekkal et al. [10] developed a new higher shear deformation theory (HSDT) for the free vibration and buckling of functionally graded (FG) sandwich plates. The proposed theory includes a new displacement field by using undetermined integral terms. Equations of motion were obtained via Hamilton's principle. The analytical solutions of FG sandwich plates were determined by employing the Navier method. Nasihatgozar and Khalili [11] presented the effect of different boundary conditions on the free vibration analysis response of a sandwich plate using the higher order shear deformation theory. The face sheets were orthotropic laminated composites. The motion equations were derived considering the continuity boundary conditions between the layers based on the Hamilton's principle. Arshid and Khorshidvand [12] studied the free vibration of a circular plate made up of a porous material integrated by piezoelectric actuator patches. The plate was assumed to be thin and its shear deformations were neglected. It was assumed that the porous material properties vary through the plate thickness according to some given functions. Using Hamilton's variational principle and the classical plate theory (CPT) the governing motion equations were obtained. Simple and clamped supports were considered. The results were compared with the similar ones in the literature. Beni and Dehkordi [13] extended the Carrera Unified Formulation (CUF) in the polar coordinates for analyzing the sandwich circular plate with the functionally graded material core. In order to apply variations in the properties of the functionally graded material, the variable kinematic method was used in the frame work of CUF. The functionally graded material was modeled as a mixture of ceramics and metal, whose properties change according to a power distribution in the direction of thickness. Magnucka-Blandzi et al. [14] studied thin-walled clamped symmetrical three-layer circular plate which consists of two facings, and a metal foam core. The mechanical properties of the core plate vary along its radius, remaining constant in the facings. The main goal of the study was to elaborate a mathematical model of the compressed circular plate in its middle plane, analytical description and solution of the global buckling problem. The analytical model was verified numerically with the use of Finite Element Analysis. On the basis of the classical plate theory, Żur [15] presented the analysis and numerical results regarding free vibrations of functionally graded circular plates elastically supported on a concentric ring. The quasi-Green's function was employed to solve the boundary value problem of the free vibrations of functionally graded circular plates with clamped, free and simply supported edges. Meksi et al. [16] introduced a new shear deformation plate theory to illustrate the bending, buckling and free vibration responses of functionally graded material sandwich plates. A new displacement field containing integrals was proposed which involves only four variables. Based on the suggested theory, the equations of motion were derived from Hamilton's principle. The Navier solution technique was adopted to derive analytical solution for simply supported rectangular sandwich plates. Magnucki et al. [17] presented a rectangular plate with symmetrically varying mechanical properties in the thickness direction. The nonlinear hypothesis of deformation of the straight line normal to the plate neutral surface was assumed, and based on the Hamilton's principle differential equations of motion were obtained. The system of equations was analytically solved. The critical loads and fundamental natural frequencies were derived. The FEM model of the plate in the ABAQUS system was developed, and the calculation results of these two methods were compared. In another paper, Magnucki et al. [18] studied a beam with symmetrically varying mechanical properties in the depth direction. The proposed formulation of the functions of the properties makes a certain generalization in the research of functionally graded materials and allows to describe homogeneous, nonlinearly variable and sandwich structures with the use of only one, consistent analytical model. The individual nonlinear hypothesis for planar cross section was assumed. Basing on the Hamilton's principle equations of motion were obtained and then analytically solved. The results were compared with numerical solutions obtained with FEM. Hadji and Avcar [19] performed free vibration analysis of the square sandwich plate with functionally graded (FG) porous face sheets and isotropic homogenous core under various boundary conditions, i.e. combinations of clamped (C), simply supported (SS), and free (F) edges. It was supposed to the material properties of the sandwich plate vary continuously through the thickness direction. A hyperbolic shear displacement theory was used in the kinematic relation of the FG porous sandwich plate, and the equations of motion were derived based on the Hamilton's principle. Then analytical solutions were achieved. A circular plate with clamped edge supported on elastic foundation was also studied by Magnucki [20]. It was assumed that mechanical properties of the plate symmetrically vary in its thickness direction. Free axisymmetric flexural vibration problem of the plate with consideration of the shear effect was analytically studied. The equations of motion based on the Hamilton principle were obtained, and then analytically solved. Magnucki et al. [21] studied simply supported beams with bisymmetrical cross-sections under a generalized load. Based on the Zhuravsky shear stress formula, the shear deformation theory of a planar beam cross-section was formulated. The deflections and the shear stresses were determined analytically, and the numerical-FEM computations of these beams were carried out. Magnucki et al. [22] presented a simply supported circular plate with symmetrically varying mechanical properties in the thickness direction. In extreme cases of this variability, the plate becomes a single-layer or three-layer structure. The main goal of the study was to develop a mathematical description of both a single-layer and three-layer structure using one formula. The mathematical model also includes all intermediate states of the plate in the axisymmetric bending and buckling problems with consideration of the shear effect. Magnucki and Magnucka-Blandzi [23] generalized the analytical model of sandwich structures. The continuous variation of mechanical properties in thickness direction of the structure wall was proposed. The individual nonlinear theory of deformation of the straight line normal to the neutral surface was developed. This analytical model of sandwich structures was presented in detail for the example rectangular plate. Javed and Al Mukahal [24], based on the theory of higher order shear deformation, analysed the free vibration of composite annular circular plates using the spline approximation technique.

The subject of the studies is a circular sandwich plate with symmetrically thickness-wise varying mechanical properties of radius R and total thickness h with simply supported edge (Fig. 1) or with clamped edge.

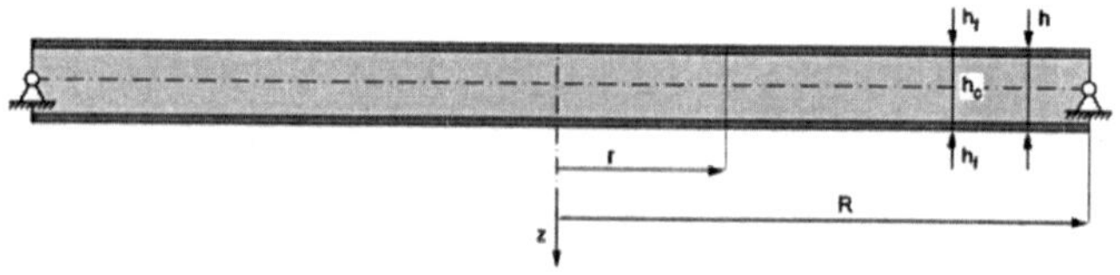

Fig. 1. Schemes of the circular sandwich plate with simply supported edge or clamped edge.

The mechanical properties of the successive layers are as follows:

- the upper face ($-1/2 \le \zeta \le -\chi_c/2$) and the lower face ($\chi_c/2 \le \zeta \le 1/2$)

$$E_f(\zeta) = E_f = const.,\ \rho_f(\zeta) = \rho_f = const. \tag{1}$$

where: E_f – the Young's modulus of faces, ρ_f – mass density of faces, $\chi_c = h_c/h$ – dimensionless thickness of the core, $\zeta = z/h$ – dimensionless coordinate.

- the core ($-\chi_c/2 \le \zeta \le \chi_c/2$)

$$E_c(\zeta) = E_f f_e(\zeta),\ \rho_c(\zeta) = \rho_f f_\rho(\zeta), \tag{2}$$

where dimensionless functions, with consideration of the paper [23], are in the form

$$f_e(\zeta) = e_0 + (1 - e_0)\left(\frac{2}{\chi_c}\zeta\right)^{k_e},\ f_\rho(\zeta) = \sqrt{e_0} + \left(1 - \sqrt{e_0}\right)\left(\frac{2}{\chi_c}\zeta\right)^{k_e}, \tag{3}$$

and k_e – even exponent, $e_0 = E_0/E_f$ – dimensionless coefficient, $E_0 = E_c(0)$.

2 Analytical Model of the Circular Sandwich Plate

The deformation of the straight line normal to the neutral surface of the plate is shown in Fig. 2.

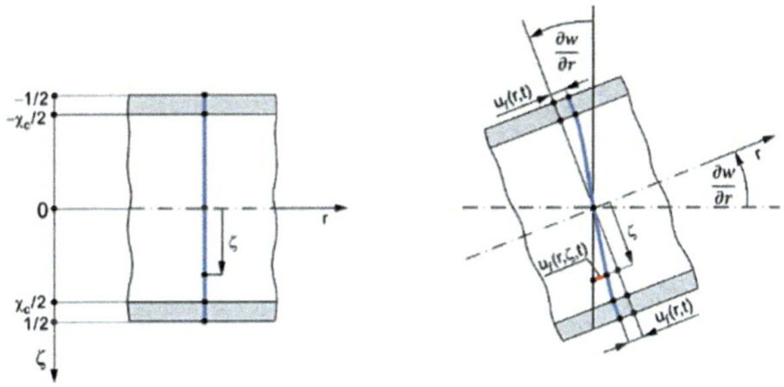

Fig. 2. The scheme of the deformation of the straight normal line of the plate.

The longitudinal displacements, strains and stresses are as follows:

- the upper face ($-1/2 \le \zeta \le -\chi_c/2$) and the lower face ($\chi_c/2 \le \zeta \le 1/2$)

$$u(r, \zeta, t) = -h\left[\zeta\frac{\partial w}{\partial r} \pm \psi(r, t)\right], \tag{4}$$

$$\varepsilon_r^{(uf/lf)}(r, \zeta, t) = \frac{\partial u}{\partial r} = -h\left[\zeta\frac{\partial^2 w}{\partial r^2} \pm \frac{\partial \psi}{\partial r}\right], \tag{5}$$

$$\varepsilon_{\varphi}^{(uf/lf)}(r,\zeta,t)=\frac{u(r,\zeta,t)}{r}=-h\left[\zeta\frac{\partial w}{r\partial r}\pm\frac{\psi(r,t)}{r}\right],\tag{6}$$

$$\gamma_{rz}^{(uf/lf)}(r,\zeta,t)=\frac{\partial w}{\partial r}+\frac{\partial u}{h\partial\zeta}=0,\ \tau_{rz}^{(uf)}(r,\zeta,t)=0\tag{7}$$

$$\sigma_{r}^{(uf/lf)}(r,\zeta,t)=\frac{E_f}{1-\nu^2}\left[\varepsilon_{r}^{(uf/lf)}(r,\zeta,t)+\nu\varepsilon_{\varphi}^{(uf/lf)}(r,\zeta,t)\right],\tag{8}$$

$$\sigma_{\varphi}^{(uf/lf)}(r,\zeta,t)=\frac{E_f}{1-\nu^2}\left[\varepsilon_{\varphi}^{(uf/lf)}(r,\zeta,t)+\nu\varepsilon_{r}^{(uf/lf)}(r,\zeta,t)\right],\tag{9}$$

where: $w(r,t)$ – defection, $\psi(r,t)=u_f(x,t)/h$ – dimensionless displacement function.

- the core $(-\chi_c/2\leq\zeta\leq\chi_c/2)$

$$u(r,\zeta,t)=-h\left[\zeta\frac{\partial w}{\partial r}-f_d(\zeta)\psi(r,t)\right],\tag{10}$$

$$\varepsilon_{r}^{(c)}(r,\zeta,t)=\frac{\partial u}{\partial r}=-h\left[\zeta\frac{\partial^2 w}{\partial r^2}-f_d(\zeta)\frac{\partial\psi}{\partial r}\right],\tag{11}$$

$$\varepsilon_{\varphi}^{(c)}(r,\zeta,t)=\frac{u(r,\zeta,t)}{r}=-h\left[\zeta\frac{\partial w}{r\partial r}-f_d(\zeta)\frac{\psi(r,t)}{r}\right],\tag{12}$$

$$\gamma_{rz}^{(c)}(r,\zeta,t)=\frac{\partial w}{\partial r}+\frac{\partial u}{h\partial\zeta}=\frac{df_d}{d\zeta}\psi(r,t),\tag{13}$$

$$\sigma_{r}^{(c)}(r,\zeta,t)=\frac{E_f}{1-\nu^2}\left[\varepsilon_{r}^{(c)}(r,\zeta,t)+\nu\varepsilon_{\varphi}^{(c)}(r,\zeta,t)\right]f_e(\zeta),\tag{14}$$

$$\sigma_{\varphi}^{(c)}(r,\zeta,t)=\frac{E_f}{1-\nu^2}\left[\varepsilon_{\varphi}^{(c)}(r,\zeta,t)+\nu\varepsilon_{r}^{(c)}(r,\zeta,t)\right]f_e(\zeta),\tag{15}$$

$$\tau_{rz}^{(uf)}(r,\zeta,t)=\frac{E_f}{2(1+\nu)}\gamma_{rz}^{(c)}(r,\zeta,t)f_e(\zeta),\tag{16}$$

where $f_d(\zeta)$ – dimensionless function of deformation of the straight line of the core (Fig. 2). The dimensionless function of deformation of the straight line of the core, with consideration of the paper [21], is elaborated in the following form

$$f_d(\zeta)=\frac{1}{C_0}\int\frac{1-\chi_c^2+\left(\chi_c^2-4\zeta^2\right)e_0+8(1-e_0)J_c(\zeta)}{f_e(\zeta)}d\zeta,\tag{17}$$

where: $C_0=\int_0^{\chi_c/2}\frac{1-\chi_c^2+\left(\chi_c^2-4\zeta^2\right)e_0+8(1-e_0)J_c(\zeta)}{f_e(\zeta)}d\zeta$ – dimensionless coefficient, $J_c(\zeta)=\frac{\chi_c^2}{4(k_e+2)}\left[1-\left(\frac{2}{\chi_c}\zeta\right)^{k_e+2}\right]$ – dimensionless function, and $f_d\left(\mp\frac{\chi_c}{2}\right)=\mp 1$.

The Hamilton's principle

$$\delta\int_{t_1}^{t_2}(U_k-U_\varepsilon)dt=0,\tag{18}$$

where the kinetic energy of the plate

$$U_k = \pi \rho_p h \int_0^R \left(\frac{\partial w}{\partial t}\right)^2 r dr, \tag{19}$$

the mass density of the plate, with consideration of the expression (3b), is as follows

$$\rho_p = (1-\chi_c)\rho_f + \rho_f \int_{-\frac{\chi_c}{2}}^{\frac{\chi_c}{2}} f_\rho(\zeta) d\zeta = \rho_f C_\rho, \tag{20}$$

dimensionless coefficient $C_\rho = 1 - \frac{(1-\sqrt{e_0})k_e}{1+k_e}\chi_c$, the elastic strain energy of the plate

$$U_\varepsilon = \pi h \int_0^R \left[F_\varepsilon^{(uf)}(r,t) + F_\varepsilon^{(c)}(r,t) + F_\varepsilon^{(lf)}(r,t)\right] dr, \tag{21}$$

where: $F_\varepsilon^{(uf)}(r,t) = \int_{-\frac{1}{2}}^{\frac{-\chi_c}{2}} \left[\sigma_r^{(uf)}\varepsilon_r^{(uf)} + \sigma_\varphi^{(uf)}\varepsilon_\varphi^{(uf)}\right] d\zeta$, $F_\varepsilon^{(c)}(r,t) = \int_{-\chi_c/2}^{\chi_c/2} \left[\sigma_r^{(c)}\varepsilon_r^{(c)} + \sigma_\varphi^{(c)}\varepsilon_\varphi^{(c)} + \tau_{rz}^{(c)}\gamma_{rz}^{(c)}\right] d\zeta$, $F_\varepsilon^{(lf)}(r,t) = \int_{\chi_c/2}^{1/2} \left[\sigma_r^{(lf)}\varepsilon_r^{(lf)} + \sigma_\varphi^{(lf)}\varepsilon_\varphi^{(lf)}\right] d\zeta$.

Therefore, the elastic strain energy (21) after integration in thickness direction of the plate and after simply transformation is as follows

$$U_\varepsilon = \pi \frac{E_f h^3}{1-\nu^2} \int_{R_0}^{R_1} \left[C_{ww} f_{ww} - 2C_{w\psi} f_{w\psi} + C_{\psi\psi} f_{\psi\psi} + \frac{1}{2}(1-\nu)\frac{J_3}{h^2}\psi^2\right] r dr, \tag{22}$$

where: $J_1 = \int_{-\chi_c/2}^{\chi_c/2} f_d(\zeta) f_e(\zeta) \zeta d\zeta$, $J_2 = \int_{-\chi_c/2}^{\chi_c/2} f_d^2(\zeta) f_e(\zeta) d\zeta$, $f_{ww}(r,t) = \left(\frac{\partial^2 w}{\partial r^2}\right)^2 + 2\nu \frac{\partial^2 w}{\partial r^2} \cdot \frac{\partial w}{r\partial r} + \left(\frac{\partial w}{r \partial r}\right)^2$, $f_{\psi\psi}(r,t) = \left(\frac{\partial \psi}{\partial r}\right)^2 + 2\nu \frac{\partial \psi}{\partial r} \cdot \frac{\psi(r,t)}{r} + \left(\frac{\psi(r,t)}{r}\right)^2$, $f_{w\psi}(r,t) = \frac{\partial^2 w}{\partial r^2} \cdot \frac{\partial \psi}{\partial r} + \nu\left(\frac{\partial^2 w}{\partial r^2} \cdot \frac{\psi(r,t)}{r} + \frac{\partial w}{r\partial r} \cdot \frac{\partial \psi}{\partial r}\right) + \frac{\partial w}{r\partial r} \cdot \frac{\psi(r,t)}{r}$, $C_{ww} = \frac{1}{12}\left[1 - (1-e_0)\frac{k_e}{k_e+3}\chi_c^3\right]$, $C_{w\psi} = \frac{1}{4}\left(1-\chi_c^2\right) + J_1$, $C_{\psi\psi} = 1 - \chi_c + J_2$, $J_3 = \frac{1}{C_0^2} \int_{-\chi_c/2}^{\chi_c/2} \frac{\left[1-\chi_c^2 + (\chi_c^2 - 4\zeta^2)e_0 + 8(1-e_0)J_c(\zeta)\right]^2}{f_e(\zeta)} d\zeta$.

Thus, based on the Hamilton's principle (18), the system of two differential equations of motion of this circular plate is obtained in the following form

$$\rho_p h r \frac{\partial^2 w}{\partial t^2} + \frac{E_f h^3}{1-\nu^2} \cdot \frac{\partial}{\partial r}\left\{ r \frac{\partial}{\partial r}\left\{ \frac{1}{r}\frac{\partial}{\partial r}\left[r\left(C_{ww}\frac{\partial w}{\partial r} - C_{w\psi}\psi(r,t)\right)\right]\right\}\right\} = 0, \tag{23}$$

$$\frac{\partial}{\partial r}\left\{\frac{1}{r}\frac{\partial}{\partial r}\left[r\left(C_{w\psi}\frac{\partial w}{\partial r} - C_{\psi\psi}\psi(r,t)\right)\right]\right\} + \frac{1-\nu}{2} \cdot \frac{J_3}{h^2}\psi(r,t) = 0. \tag{24}$$

Moreover, the radial bending moment

$$M_r(r,t) = h^2 \left[\int_{-1/2}^{-\chi_c/2} \sigma_r^{(uf)} \zeta d\zeta + \int_{-\chi_c/2}^{\chi_c/2} \sigma_r^{(c)} \zeta d\zeta + \int_{\chi_c/2}^{1/2} \sigma_r^{(lf)} \zeta d\zeta\right]. \tag{25}$$

Therefore, after integration, with consideration of the expressions (8), (14), and after simply transformation, one obtains

$$M_r(r,t) = -\frac{E_f h^3}{1-\nu^2}\left[C_{ww}\left(\frac{\partial^2 w}{\partial r^2} + \nu\frac{\partial w}{r\partial r}\right) - C_{w\psi}\left(\frac{\partial \psi}{\partial r} + \nu\frac{\psi(r,t)}{r}\right)\right]. \tag{26}$$

3 Analytical Solution of the Free Flexural Axisymmetric Vibration of the Plate

3.1 Circular Sandwich Plate with Simply Supported Edge

Taking into account the bending of the circular sandwich plate under uniformly distributed load with simply supported edge, the deflection and dimensionless displacement functions are assumed in the following form

$$w(\xi,t) = \left[2c_\nu\left(1-\xi^2\right) - 1 + \xi^4\right]w_a(t),\ \psi(\xi,t) = -\left(c_\nu - \xi^2\right)\xi\,\psi_a(t), \tag{27}$$

where: $c_\nu = (3+\nu)/(1+\nu)$ – dimensionless coefficient, $w_a(t)$, $\psi_a(t)$ – functions of the time, $\xi = r/R$ – dimensionless coordinate.

These functions satisfy boundary conditions: $w(1,t) = 0$, $\psi(0,t) = 0$. Substituting these functions into the Eq. (21) and using the Galerkin method, after simply transformation one obtains

$$\psi_a(t) = 4\frac{C_{w\psi}}{C_{\psi\psi} + \frac{1-\nu}{32}\cdot\frac{6c_\nu^2-8c_\nu+3}{3c_\nu-2}J_3\left(\frac{R}{h}\right)^2}\cdot\frac{w_a(t)}{R}. \tag{28}$$

Then, substituting the functions (27) into the Eq. (20) and using the Galerkin method, after simply transformation one obtains the following differential equation

$$\frac{d^2 w}{dt^2} + k_\nu\left(C_{ww} - C_{s,s}\right)\frac{E_f h^2}{\rho_p R^4}w_a(t) = 0, \tag{29}$$

where $k_\nu = \frac{264}{1-\nu^2}\cdot\frac{7c_\nu-5}{11(3c_\nu-4)c_\nu+15}$, $C_{s,s} = \frac{C_{w\psi}^2}{C_{\psi\psi}+\frac{1-\nu}{32}\cdot\frac{6c_\nu^2-8c_\nu+3}{3c_\nu-2}J_3\left(\frac{R}{h}\right)^2}$.

This Eq. (29) is solved with the use of the function

$$w_a(t) = w_a\sin(\omega t) \tag{30}$$

where: w_a [mm] – amplitude of the flexural vibration, ω [rad/s] – fundamental natural frequency. Substituting this function into the Eq. (29) one obtains

$$\omega = \frac{10^6 h}{R^2}\sqrt{k_\nu\left(C_{ww} - C_{s,s}\right)\frac{E_f}{\rho_p}}. \tag{31}$$

Exemplary calculations are carried out for the following data of the plates (Fig. 1a): Young's modulus of faces $E_f = 72000$ MPa, mass density of faces $\rho_f = 2710$ kg/m^3, Young's modulus in the middle of the core $E_c(0) = E_0 = 2400$ MPa, Poisson's ratio $\nu = 0.3$, radius $R = 500$ mm and thicknesses $h = 24$ mm, $h_c = 21$ mm, $h_f = 1.5$ mm. Thus: $\chi_c = 7/8$ and $e_0 = 1/30$. The results of the calculations are specified in Table 1.

Table 1. The results of the analytical study of the circular plate with simply supported edge

k_e	2	10	50
C_{ww}	0.061747	0.041821	0.032422
$C_{s,s}$	0.0010842	0.0011364	0.00081684
ω[rad/s]	952.40	953.89	909.6664

3.2 Circular Sandwich Plate with Clamped Edge

Taking into account the bending of the circular sandwich plate under uniformly distributed load with clamped edge, the deflection and dimensionless displacement functions are assumed in the following form

$$w(\xi, t) = \frac{1}{4}\left(1 - \xi^2\right)^2 w_a(t),\ \psi(\xi, t) = -\left(1 - \xi^2\right)\xi\,\psi_a(t). \tag{32}$$

These functions satisfy boundary conditions: $w(1, t) = 0$, $\partial w/\partial\xi\,|_1 = 0$, $\psi(0, t) = 0$, $\psi(1, t) = 0$. Substituting these functions into the Eq. (21) and using the Galerkin method, after simply transformation one obtains

$$\psi_a(t) = 4\frac{C_{w\psi}}{C_{\psi\psi} + \frac{1-\nu}{32}J_3\left(\frac{R}{h}\right)^2} \cdot \frac{w_a(t)}{R}. \tag{33}$$

Then, substituting the functions (32) into the Eq. (20) and using the Galerkin method, after simply transformation one obtains the following differential equation

$$\frac{d^2w}{dt^2} + \frac{132}{1-\nu^2}\left(C_{ww} - C_{s,c}\right)\frac{E_f h^2}{\rho_p R^4} w_a(t) = 0, \tag{34}$$

where the dimensionless shear coefficient $C_{s,c} = \frac{C_{w\psi}^2}{C_{\psi\psi} + \frac{1-\nu}{32}J_3\left(\frac{R}{h}\right)^2}$.

This Eq. (34) is analogous to the Eq. (29), therefore is solved with the use of the function (30). Thus, the fundamental natural frequency of the circular sandwich plate with clamped edge is as follows

$$\omega = 2\frac{10^6 h}{R^2}\sqrt{\frac{33}{1-\nu^2}\left(C_{ww} - C_{s,c}\right)\frac{E_f}{\rho_p}}. \tag{35}$$

Exemplary calculations are carried out for the following data of the plates (Fig. 1a): Young's modulus of faces $E_f = 72000$ MPa, mass density of faces $\rho_f = 2710$ kg/m^3, Young's modulus in the middle of the core $E_c(0) = E_0 = 2400$ MPa, Poisson's ratio $\nu = 0.3$, radius $R = 500$ mm and thicknesses $h = 24$ mm, $h_c = 21$ mm, $h_f = 1.5$ mm. Thus: $\chi_c = 7/8$ and $e_0 = 1/30$. The results of the calculations are specified in Table 2.

Table 2. The results of the analytical study of the circular plate with clamped edge

k_e	2	10	50
C_{ww}	0.061747	0.041821	0.032422
$C_{s,c}$	0.0039211	0.0040129	0.0029009
ω[rad/s]	1981.34	1959.37	1873.3208

4 Conclusions

The presented studies of the circular sandwich plates allow to formulate the following conclusions: 1) The proposed hypothesis describes the vibrational behavior of the analyzed structures well, 2) The Young's modulus variability enables to consider the plates from the homogeneous to the sandwich ones, 3) This theory may be applicated to the analytical modelling of the plates.

References

1. Ventsel, E., Krauthammer, T.: Thin plates and shells, theory, analysis, and applications. Marcel Dekker, Inc., New York, Basel (2001)
2. Reddy, J.N.: Mechanics of laminated composite plates and shells: Theory and analysis. CRC Press, Boca Raton, London, New York, Washington (2004)
3. Yang, M., Qiao, P.: Higher-order impact modeling of sandwich structures with flexible core. Int. J. Solids Struct. **42**(20), 5460–5490 (2005)
4. Zenkour, A.M.: Generalized shear deformation theory for bending analysis of functionally graded plates. Appl. Math. Model. **30**(1), 67–84 (2006)
5. Carrera, E., Brischetto, S.: A survey with numerical assessment of classical and refined theories for the analysis of sandwich plates. Appl. Mech. Rev. **62**(1), 010803 (2009)
6. Wang, Y., Xu, R., Ding, H.: Free axisymmetric vibration FGM circular plate. Appl. Math. Mech. **30**, 1077–1082 (2009)
7. Yas, M.H., Tahouneh, V.: 3-D Free vibration analysis of thick functionally graded annular plates on Pasternak elastic foundation via differential quadrature method (DQM). Acta Mech. **223**, 43–62 (2012)
8. Sayyad, A.S., Ghugal, Y.M.: On the free vibration analysis of laminated composite and sandwich plates: A review of recent literature with same numerical results. Compos. Struct. **129**, 177–201 (2015)
9. Sarangan, S., Singh, B.N.: Higher-order closed-form solution for the analysis of laminated composite and sandwich plates based on new shear deformation theories. Compos. Struct. **138**, 391–403 (2016)
10. Sekkal, M., Fahsi, B., Tounsi, A., Mahmoud, S.R.: A novel and simple higher order shear deformation theory for stability and vibration of functionally graded sandwich plate. Steel Compos. Struct. **25**(4), 389–401 (2017)
11. Nasihatgozar, M., Khalili, S.: Free vibration of a thick sandwich plate using higher order shear deformation theory and DQM for different boundary conditions. Journal of Applied and Computational Mechanics **3**(1), 16–24 (2017)
12. Arshid, E., Khorshidvand, R.: Free vibration analysis of saturated porous FG circular plates integrated with piezoelectric actuators via differential quadrature method. Thin-Walled Structures **125**, 220–233 (2018)

13. Beni, N.N., Dehkordi, M.B.: An extension of Carrera unified formulation in polar coordinate for analysis of circular sandwich plate with FGM core using GDQ method. Compos. Struct. **185**, 421–434 (2018)
14. Magnucka-Blandzi, E., Wiśniewska-Mleczko, K., Smyczyński, M.J.: Buckling of symmetrical circular sandwich plate with variable mechanical properties of the core in the radial direction. Compos. Struct. **204**, 88–94 (2018)
15. Żur, K.K.: Quasi-Green's function approach to free vibration analysis of elastically supported functionally graded circular plates. Compos. Struct. **183**, 600–610 (2018)
16. Meksi, R., Benyoucef, S., Mahmoudi, A., Tounsi, A., Bedia, E.A.A.: Mahmoud SR. An analytical solution for bending, buckling and vibration responses of FGM sandwich plates. Journal of Sandwich Structures and Materials **21**(2), 727–757 (2019)
17. Magnucki, K., Witkowski, D., Magnucka-Blandzi, E.: Buckling and free vibrations of rectangular plates with symmetrically varying mechanical properties – Analytical and FEM studies. Compos. Struct. **220**, 355–361 (2019)
18. Magnucki, K., Witkowski, D., Lewinski, J.: Bending and free vibrations of porous beams with symmetrically varying mechanical properties – Shear effect. Mech. Adv. Mater. Struct. **27**(4), 325–332 (2020)
19. Hadji, L., Avcar, M.: Free vibration analysis of FG porous sandwich plate under various boundary conditions. Journal of Applied and Computational Mechanics (2020)
20. Magnucki, K.: Free axisymmetric flexural vibrations of circular plate with symmetrically varying mechanical properties supported on elastic foundation. Vibrations in Physical Systems **31**(2), 2020217 (2020)
21. Magnucki, K., Lewinski, J., Magnucka-Blandzi, E.: A shear deformation theory of beams of bisymmetrical cross sections based on the Zhuravsky shear stress formula. Eng. Trans. **68**(4), 353–370 (2020)
22. Magnucka-Blandzi, E., Magnucki, K., Stawecki, W.: Bending and buckling of a circular plate with symmetrically varying mechanical properties. Appl. Math. Model. **89**(2), 1198–1205 (2021)
23. Magnucki, K., Magnucka-Blandzi, E.: Generalization of a sandwich structure model: Analytical studies of bending and buckling problems of rectangular plates. Compos. Struct. **255**, 112944 (2021)
24. Javed, S., Al Mukahal, F.H.H.: Free vibration of annular circular plates based on higher-order shear deformation theory: a spline approximation technique. International Journal of Aerospace Engineering (2021)

FVF: Forced Vibrations in Structures and Vibration Fatigue

Estimation of Fatigue Crack Growth at Transverse Vibrations of a Steam Turbine Shaft

A. Bovsunovsky(✉) and Wu Yi Zhao

National Technical University of Ukraine "Igor Sikorsky Kyiv Polytechnic Institute", 37 Peremohy Ave., Kyiv 03056, Ukraine
apbovsunovsky@gmail.com

Abstract. During operation, a steam turbine shaft is subjected to a wide range of thermomechanical and thermochemical loading. Despite significant reserve of static and dynamic strength, laid down at the stage of turbine design, fatigue cracks still appear in its structural elements, which lead to catastrophic failures. Potential reasons of damage in turbine shafts are all technological operations used in the process of manufacture (forging, turning, and milling, heat treatment), since they are accompanied with plastic deformation of material. Damage accumulates during long-term cyclic deformation and turns into local damage of a fatigue crack type. In addition, cracking in turbine shafts is caused by the presence of stress concentrators. The analytical model of high-pressure rotor of the K-200-130 steam turbine has been developed to study the transverse vibrations when rotor passes through the first critical speed. The growth of crack is predicted based on fracture mechanics approaches through the determined maximal stresses in the cracked section and on the experimental dependences of the crack growth rate on the stress intensity factor range for the rotor steel.

Keywords: Steam turbine shaft · Fatigue damage · Transverse vibration · Crack growth

1 Introduction

Steam turbines are complex mechanical system that operates in conditions of high temperature and high static and dynamic loads. Significant reserves of static strength laid down at the design stage cannot fully prevent the occurrence of fatigue damage in structural elements of turbines [1]. Such damage is especially intense during transient modes of turbine operation, that is, during start-ups and shutdowns (lateral vibrations) [2] and when a turbine generator is connected to the power network (torsional vibrations) [3]. Accumulating for a long period of time, fatigue damage is localized in the form of a fatigue crack, which quickly reaches critical size and causes catastrophic fractures [4, 5].

There is a huge number of works devoted to investigations of vibrations of cracked turbines shaft [6]. The main goal of these works is to understand the features of influence

Z. Dimitrovová et al. (Eds.): WMVC 2022, MMS 125, pp. 233–240, 2023.
https://doi.org/10.1007/978-3-031-15758-5_22

of a crack on the transverse and torsional vibrations of rotating shafts and to develop damage detection methods based on the analysis of vibrations.

However, these studies do not make possible to predict the process of crack growth in real turbine operating conditions, that is, during transient vibrations arising in the process of turbines start-ups and shutdowns, when there is a transition through the resonance [7]. The stress amplitudes at resonance are usually below the endurance limit of rotor steels. However, in the presence of even small surface crack, these loading conditions may be sufficient for the further growth of crack.

Potential causes of initial damage in turbines structural elements are practically all technological operations used in manufacture (forging, turning, and milling, heat treatment), since they are accompanied by plastic deformation. In addition, damage in turbine shafts is due to the complex geometry, that is, by the presence of fillets and grooves, which are stress concentrators and, therefore, potential sites for the initiation and growth of fatigue cracks [8].

Inspections of steam turbines during scheduled repairs reveal cracks of various depths and lengths on the surface of shafts [8]. The turbine operating instruction [9] admits the presence of cracks up to 1 mm deep. During major repair, cracks of greater depth are removed by turning the surface to the depth of crack.

The appearance of even small crack completely changes the approach to the assessment of structure's bearing capacity. Analytical methods based on classical mechanics do not allow the assessment of stress-strain state in the vicinity of crack. The process of crack growth can be described with the sufficient for practice accuracy only using the approaches of fracture mechanics [10]. At this it is necessary to consider a lot of factors, the influence of which can be determined exclusively experimentally. These include material fracture toughness, crack growth rate dependence on the SIF range, the effect of temperature, scale factor, chemical aggressiveness of the medium, etc.

The purpose of this study is a comparative assessment of the effect of different operation factors on the crack growth in high-pressure rotor of K-200-130 turbine during transverse vibrations in the process of turbine's start-ups and shutdowns.

2 Simulation of the Crack Growth Process

Estimation of crack growth in a rotating rotor is based on the solution of forced vibrations problem with account for damping. The high-pressure rotor of K-200-130 steam turbine of capacity 200 MW was chosen as a model object for the research (Fig. 1, where S1 and S2 are the supports). Since the diameters of rotor stages are slightly different from each other, the solution is built for a homogeneous shaft with a diameter D and a length L.

If the harmonic disturbing force is represented in a complex form, then the equation of forced vibrations of the rotor caused by the eccentricity because of deflection under its own weight have the form

$$\mu \frac{\partial^2 y}{\partial t^2} + \left(1 + \frac{i\delta}{\pi}\right) EI \frac{\partial^4 y}{\partial x^4} = F(x) e^{i\omega t}, \tag{1}$$

where μ is the mass of a rotor length unit; δ is the logarithmic decrement of vibration; E is the modulus of elasticity; I is the axial moment of inertia of the cross section; $F(x)$

is the driving force; ω is the frequency of driving force; $i = \sqrt{-1}$. In Eq. (1), the second term considers the inelastic resistance proportional to the elastic restoring force [11], which is typical for metallic materials.

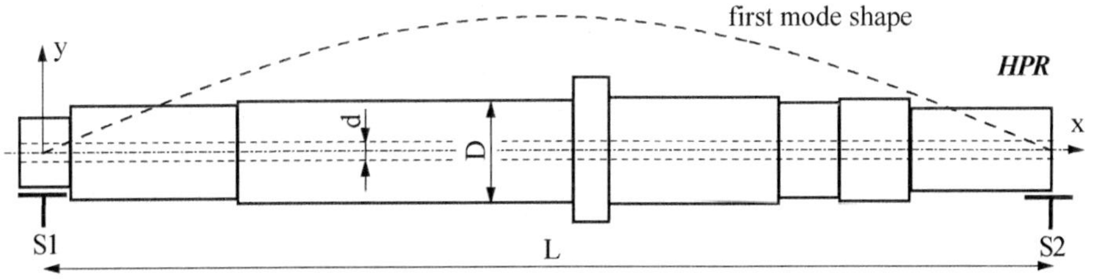

Fig. 1. High pressure rotor of K-200-130 turbine.

The solution of Eq. (1) can be found in the form

$$y(x, t) = \sum_{m=1}^{\infty} a_m w_m(x) e^{i\omega t}, \tag{2}$$

where w_m are the mode shapes of the rotor, satisfying the condition

$$EI \frac{\partial^4 w_m(x)}{\partial x^4} = \mu \omega_m^2 w_m(x). \tag{3}$$

In Eq. (3) ω_m are the natural frequencies of the rotor, that is $\omega_m = \frac{m^2\pi^2}{L^2}\sqrt{\frac{EI}{\mu}}$.

The distribution function of the force *F(x)* along the length of the rotor is represented as an expansion in terms of natural mode shapes

$$F(x) = \sum_{m=1}^{\infty} b_m w_m(x), \tag{4}$$

where

$$b_m = \frac{\int_0^L F(x) w_m(x) dx}{\int_0^L w_m^2(x) dx}. \tag{5}$$

The substitution of Eq. (2) into Eq. (1) gives

$$\mu \sum_{m=1}^{\infty} a_m \left[\left(1 + \frac{i\delta}{\pi} \right) \omega_m^2 - \omega^2 \right] w_m(x) = \sum_{m=1}^{\infty} b_m w_m(x), \tag{6}$$

from where

$$a_m = \frac{b_m}{\mu} \frac{e^{i\psi_m}}{\sqrt{(\omega_m^2 - \omega^2)^2 + \frac{\delta^2 \omega_m^4}{\pi^2}}}, \quad \text{where } tg\,\psi_m = -\frac{\delta \omega_m^2}{\pi \left(\omega_m^2 - \omega^2 \right)}. \tag{7}$$

Then the sought-for solution of Eq. (1), that is, its real part, takes the form

$$y(x, t) = \sum_{m=1}^{\infty} \frac{b_m}{\mu} \frac{w_m(x)}{\sqrt{(\omega_m^2 - \omega^2)^2 + \frac{\delta^2 \omega_m^4}{\pi^2}}} \cos(\omega t + \psi_m). \tag{8}$$

Since the resonance is considered when passing through the first critical speed of the rotor, the problem is limited by considering only the first mode of vibration (m = 1). Thus, as it follows from Eqs. (4) and (5) that $F(x) = \mu\omega_1{}^2 w_1(x)$ and $b_1 = \mu\omega_1{}^2$. At the exact resonance, that is, at $\omega = \omega_1$, the phase shift is $\psi_1 = -\pi/2$ and

$$y(x, t) = \frac{\pi}{\delta} w_1(x) sin\omega_1 t. \tag{9}$$

With the solution of the static rotor deflection problem under the action of its own weight, the equation of the first mode shape has the appearance

$$w_1(x) = \frac{g\mu L^4}{\pi^4 EI} sin\frac{\pi x}{L}. \tag{10}$$

One of the reasons for the initiation of cracks on the surface of rotor is the action of thermal stresses, especially in the areas of stress concentration, during the start-ups of turbines [12]. When passing through the critical speed of rotation, the amplitude of rotor's vibrations increases to values at which crack can grow. To predict this grow the Paris equation is used [10]

$$\frac{da}{dN} = C(\Delta K_I)^n, \tag{11}$$

where a is the depth of crack; N is the number of loading cycles; C and n are empirical parameters; ΔK_I is the stress intensity factor (SIF) range, which changes from the value ΔK_{Ith} (threshold value) to the value ΔK_{Ic} (critical value).

The orientation of crack relatively to the phase of vibration can be arbitrary. The extreme cases when crack is subjected to tensile or compressive stresses are demonstrated in Figs. 2a and 2b, respectively. There are many possible intermediate positions of the crack, in which its behavior is extremely difficult to predict.

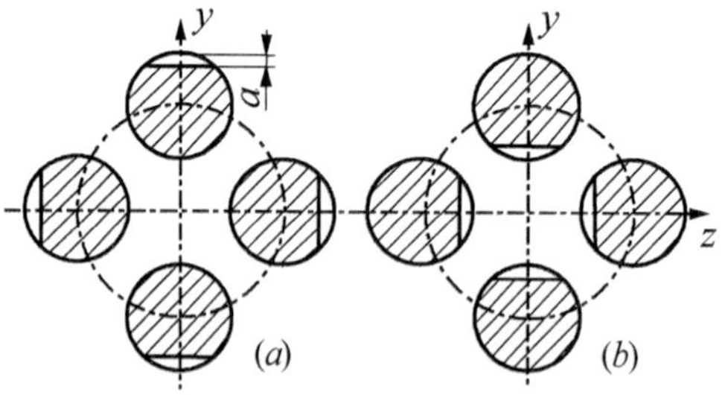

Fig. 2. Possible orientations of crack in rotation of rotor.

Here the worst-case scenario of crack orientation during rotor vibration is considered, namely, shown in Fig. 2a. In this case, the SIF for the edge crack with a straight front is determined by the equation [13]

$$K_I = \sigma_c\sqrt{\pi a}\left(1.1105 - 2.6475\gamma + 5.6875\gamma^2\right), \tag{12}$$

where σ_c is the stress in the cross-section with crack; $\gamma = a/D$.

Based on Eqs. (8) and (10), the distribution of stress amplitudes along the length of rotor at first mode vibration takes the form

$$\sigma_1(x,t) = \frac{g\mu L^2}{\pi^2\kappa}\frac{\omega_1^2}{\sqrt{(\omega_1^2-\omega^2)^2+\frac{\delta^2\omega_1^4}{\pi^2}}}sin\frac{\pi x}{L}. \tag{13}$$

where g is the acceleration of gravity; κ is the axial moment of resistance of the cross section.

Since the worst scenario of crack growth is considered, crack was considered in the middle of the rotor ($L_c = L/2$). In this case Eq. (13) is simplified to the form

$$\sigma_c(L/2,t) = \frac{g\mu L^2}{\pi^2\kappa}(\frac{\omega_1^2}{\sqrt{(\omega_1^2-\omega^2)^2+\frac{\delta^2\omega_1^4}{\pi^2}}}+1). \tag{14}$$

In Eq. (14), the static component of stresses caused by the weight of the rotor in the lower phase of vibrations was added to the stress amplitude.

One revolution of the rotor is considered as one loading cycle. The angular speed of rotor, accelerating or decelerating with angular acceleration A, at the N-th revolution (more precisely, at the end of the N-th revolution) is determined by the formula $\omega_N = \sqrt{\beta NA}$, where $\beta = 4\pi$ if the acceleration is set in rad/s^2, and $\beta = 2$ – if in r/s^2 (accordingly, in the first case, the angular velocity is measured in rad/s, and in the second - in r/s).

The algorithm of calculations is as follows. It is assumed that a crack with a depth of 1 mm has formed on the surface of rotor, which is the maximal permissible according to the instructions for the safe operation of turbine [9]. Let's call it the initial crack (a_0). The angular acceleration is assigned and the angular speed of the rotor at each of N revolutions is determined. Then the stress amplitude in the cracked section and the corresponding value of SIF are calculated according to Eqs. (14) and (12).

When approaching the first critical speed of rotation, the stress amplitude and SIF increases. At this, the K_I is compared with ΔK_{Ith}. The fulfillment of the condition $K_I \geq \Delta K_{Ith}$ means that the crack in one loading cycle grows by the value da_{N_1} and becomes equal to $a_{N_1} = a_0 + da_{N_1}$. At the next loading cycle, the calculation is carried out taking into account the changed size of the crack, namely $a_{N_2} = a_{N_1} + da_{N_2}$, and so on ($N_1$, N_2 … N_d are the damaging loading cycles). The process continues until either the number of start-ups and shutdowns of the turbine in total reaches 4000, or the condition $K_I \geq \Delta K_{Ic}$ is fulfilled, which, in fact, means a loss of load-bearing capacity of the structure. It is considered that the set number of turbine start-ups is $S = 2000$. Since the rotor passes through the resonance both at start-ups and shutdowns, the number of passes through the resonance is equal to twice the number of start-ups.

3 Effect of Operating Factors on Crack Growth

Considering the unavailability or inconsistency of some properties of rotor steel, the calculation was carried out in a certain plausible range of their values.

The calculations were carried out based on the dependence of the crack growth rate on the SIF range for the rotor steel, presented in [14]. It is approximated by the Paris Eq. (11) with parameters $C = 2.75 \cdot 10^{-10}$, $n = 3.86$ (with these parameters, the crack is measured in mm). The damping characteristic, which significantly affects the amplitude of resonant vibrations, varied in the range $\delta = 0.03 \ldots 0.07$. The lower limit of this range is close to the property of rotor steels [15], and the upper one hypothetically considers, in addition, the effect of structural damping. The diameter of rotor varied in the range $D = 0.44 \ldots 0.84$ m (at this $d = 0.112$ m). The length of rotor was $L = 4.5$ m and the density of the steel was $\rho = 7800$ kg/m^3. In the process of the start-up of turbine, the angular acceleration varies over a wide range. In the model the angular acceleration was fixed at one of the values from the range $A = 0.01 \ldots 1$ r/s^2. This idealization makes it possible to reveal the effect of angular acceleration on the rate of crack growth solely.

Figure 3 illustrates the process of crack propagation caused by multiple passes of the rotor through the first critical frequency (in this case, $f_1 = 29.9$ Hz, which is close to the real one) with different angular acceleration. Obviously, the higher the angular acceleration, the fewer damaging cycles when passing through the resonance and the more start-ups the rotor can stand before the crack reaches a critical size. The rate of crack growth increases significantly at the last stage of its development. Therefore, there is a small window of opportunity for the detection of crack to prevent catastrophic failures.

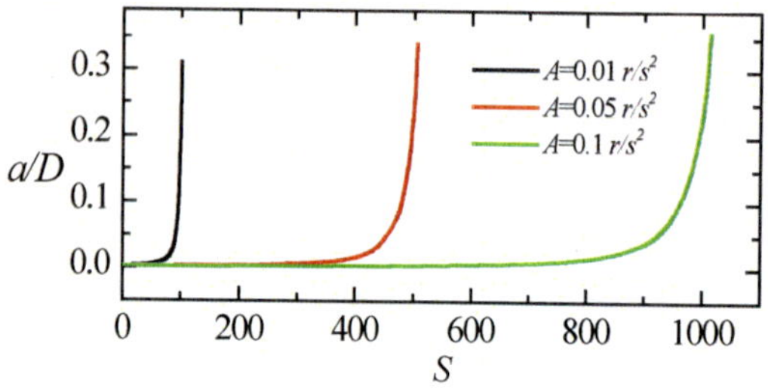

Fig. 3. Crack growth in rotor as a function of start-ups number at different angular acceleration ($D = 0.44$; $\delta = 0.03$; $f_1 = 29.9$ Hz).

Since the data on the influence of operating factors (in particular, temperature) on the characteristics K_{Ith} and K_{Ic} are contradictory or unavailable, the calculation of crack growth was performed in a wide range of their variation (Fig. 4). As can be seen, the influence of K_{Ith} on the crack growth manifests itself starting from a certain value, which practically does not depend on the value K_{Ic} in the range $K_{Ic} = 100 \ldots 500$ MPa m$^{1/2}$. At the same time, the influence of K_{Ic} on the start-ups number is practically insignificant.

The increase of damping essentially reduces the amplitude of resonant vibrations and, consequently, the number of damaging cycles. As can be seen from Fig. 5, the effect of damping on the number of rotor start-ups is significantly manifested only at relatively small angular accelerations. In general, there is a damping level (in our case, it close to $\delta \approx 0.053$), when crack has no possibility to grow. The increase of the critical speed f_1 unambiguously leads to the intensification of crack growth and, accordingly, to the decrease of start-ups number (here the dependance of crack growth rate on the frequency of loading [10] was not considered). To preserve the mass and geometric characteristics of the rotor, the change of critical frequency was achieved by varying the elastic modulus.

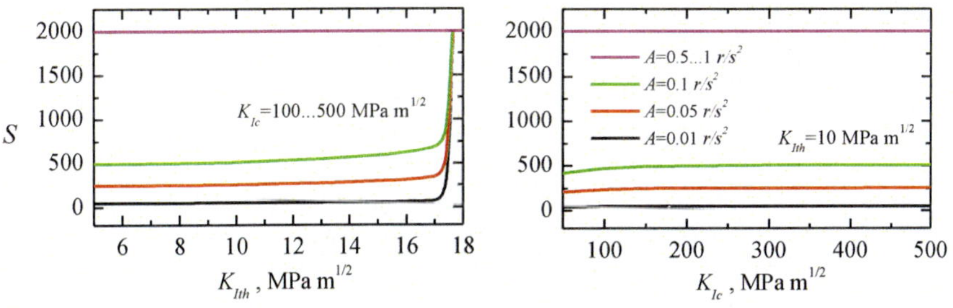

Fig. 4. Number of start-ups as a function of K_{Ith} and K_{Ic} ($D = 0.44$; $\delta = 0.03$; $f_1 = 29.9$ Hz).

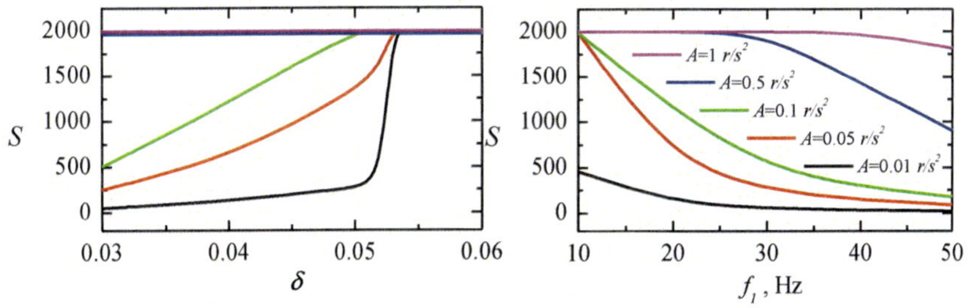

Fig. 5. Number of start-ups as a function of δ and f_1 ($D = 0.44$; $\delta = 0.03$).

The increase of the rotor diameter, all other factors being equal, reduces the stress level in the cracked cross section and thereby increases the rotor durability (Fig. 6). The constancy of critical frequency was also achieved by varying the elastic modulus.

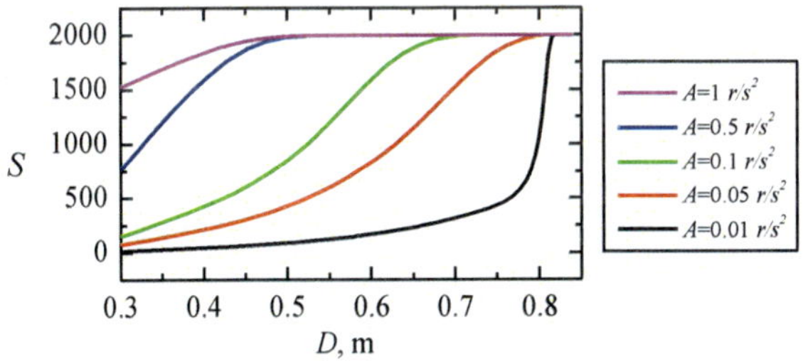

Fig. 6. Number of start-ups as a function of rotor's diameter D ($\delta = 0.03$; $f_1 = 29.9$ Hz).

4 Conclusions

The proposed model makes it possible to carry out a qualitative comparative analysis of the influence of different factors on the process of crack growth in a rotating rotor during transient processes. Some of these factors, considered in the article, can be controlled in a relatively wide range (angular acceleration, frequency, geometrical characteristics), and some of them are practically not amenable to such control, since they are a property of steel (δ, K_{Ith}, K_{Ic}). The first category of factors has the greatest influence on the crack growth. It is possible to reduce the possibility of crack propagation by increasing the angular acceleration and/or rotor diameter and by decreasing the resonant frequency. Of the second category of factors, the damping characteristic is most influential. The development of rotor steels with high damping characteristics and the use of effective dampers are the promising ways to prevent crack growth in turbine rotors.

A further work supposes the creation of a more complete turbine shafting model for more accurate prediction of stresses during transient processes, including both transverse and torsional vibrations, as well as other types of cracks (slant, longitudinal).

References

1. Troshchenko, V.T., Khamaza, L.A.: Fatigue fracture stages of metals and alloys and stage-to-stage transition criteria. Strength Mater. **50**(3), 529–539 (2018)
2. Zhou, T., Xu, J., Sun, Z.: Dynamic analysis and diagnosis of a cracked rotor. Trans. ASME. J. Vib. Acoust. **123**(4), 539–543 (2001)
3. Bovsunovsky, A.P.: Fatigue damage of steam turbine shaft at asynchronous connections of turbine generator to electrical network. J. Phys. Confe. Ser. **628**(conference 1), 012001 (2015)
4. Kramer, L.D., Randolph, D.D.: Analysis of the Tennessee valley authority, Gallatin unit N2 turbine rotor burst. In: ASME-MPC Symposium on Creep-Fatigue Interaction, p. 1 (1976)
5. Zagretdinov, I., Kostyuk, A.G., Trukhnii, A.D., Dolzhanskii, P.R.: Failure of the 300 MW turbine unit of the state district power station at Kashira: causes, consequences and conclusions. Therm. Eng. **5**, 5–15 (2004)
6. Bovsunovsky, A., Surace, C.: Non-linearities in the vibrations of elastic structures with a closing crack: a state of the art review. Mech. Syst. Signal Process. **62–63**, 129–148 (2015)
7. Gunter, E.J.: Unbalance response and field balancing of an 1150-MW turbine-generator with generator bow. In: 7th IFToMM-Conference on Rotor Dynamics, pp. 25–28 (2006)
8. Chernousenko, O., Ryndiuk, D., Peshko, V.: Estimation of residual life and extension of operation of high-power steam turbines. Part 3. NTUU KPI, Kyiv (2020)
9. Typical instructions for metal control and extension of the service life of the main elements of boilers, turbines and pipelines of thermal power plants (RD 10-577-03). Scientific and Technical Center for Industrial Safety, Moscow (2003)
10. Broek, D.: The Practical Use of Fracture Mechanics. Kluwer Academic Publishers, Dordrecht/Boston/London (1988)
11. Sorokin, E.: On the Issue of Inelastic Resistance of Building Materials During Vibrations. State Publishing House of Literature on Construction, Moscow (1954)
12. Shul'zhenko, M.G., Gontarovskyi, P.P., Garmash, N.G.: Thermostressed state and crack growth resistance of rotors of the NPP turbine K-1000-60/1500. Strength Mater. **42**(1), 114–119 (2010)
13. Murakami, Y. (Editor-in-Chief): Stress Intensity Factors Handbook, 3 vols., vol. 1. The Society of Materials Science, Japan and Pergamon Press (1987)
14. Shlyannikov, V., Kosov, D., Fedorenkov, D., Zhang, X.C., Tu, S.T.: Size effect in creep–fatigue crack growth interaction in P2M steel. Fatigue Fract. Eng. Mater. Struct. **44**(12), 3301–3319 (2021)
15. Bovsunovskii, A.P.: Experimental studies on high-cycle fatigue and damping properties of R2MA rotor steel in torsion. Strength Mater. **43**(4), 455–463 (2011)

Forced Vibration of Bus Bodyworks and Estimates of Their Fatigue Damage

Miloslav Kepka(✉) and Miloslav Kepka Jr.

University of West Bohemia, Pilsen, Czech Republic
kepkam@rti.zcu.cz

Abstract. Driving a road vehicle on uneven roads causes its bodywork to vibrate. Vibration of the bodywork can result in fatigue damage of critical bodywork nodes. In the case of buses (trolleybuses, battery electric buses), the critical structural nodes are welds of thin-walled profiles, most often corners of openings for doors and windows. The fatigue properties of critical structural nodes are expressed by S-N lines. S-N lines can be determined in advance by performing laboratory fatigue tests. The time series data of stresses in the bodywork have the character of a random process. These stresses can only be determined with sufficient accuracy by strain gauge measurements, either while the vehicle is running on real tracks or on a proving ground. Fatigue damage can then be estimated using the appropriate fatigue damage accumulation hypothesis. The paper demonstrates a case study that use data from the authors' practical cooperation with bus producers.

Keywords: Bus bodywork · Model testing track · Real road · Special test track · Stress measurement · Stress-time history · Rain-flow counting · Stress spectra · S-N line · Fatigue life calculation

1 Introduction

The summary of the methodology of computational and experimental investigation of strength and fatigue life of bodyworks of road vehicles for mass passenger transport - which was used for designing many Skoda trolleybuses and buses – has already been presented to the public [1]. It continues to be developed by the Regional Technological Institute, which is a research center of the Faculty of Mechanical Engineering of University of West Bohemia [2]. The procedure of the development of a new bus includes computational and experimental activities: CAD - MBS - FEM - stand tests. In the final phase, the measurement of stresses at various types of forced bodywork vibrations and fatigue life calculations play an important role. This is shown by the data of a case study.

2 Load Cases

The measurement is usually performed both with an empty bus and with a load that represents a fully occupied vehicle. For both of these load cases, the stresses in the critical bodywork nodes can be determined at various excitation types (see Fig. 1) [3].

Z. Dimitrovová et al. (Eds.): WMVC 2022, MMS 125, pp. 241–248, 2023.
https://doi.org/10.1007/978-3-031-15758-5_23

Fig. 1. Types of forced body vibration: crossing artificial standardized obstacles; driving on real roads with different surface quality; driving on special tracks on the test circuit.

The following Fig. 2, Fig. 3 and Fig. 4 illustrate examples of measured stresses that were used for the parametric case study.

Maximum service stress amplitudes (or stress ranges) are estimated by strain gauge measurements on model testing track. The model testing track is usually made from artificial obstacles. The cylinder segments are laid on an even asphalt road in such composition, that overruns of them occurred gradually by right-hand wheels, by both wheels simultaneously and by left-hand wheels.

The most important are the measurement while the vehicle prototype is driving in real traffic on various unevenness´s of real roads. The measured random stress-time histories are converted to fatigue damage using cumulative damage rules which have been proposed by various authors, for example Palmgren [4] and Miner [5], Haibach [6] and others.

Some customers also require measuring of the bodywork during tests of the vehicle on the proving ground that contains special tracks with defined parameters.

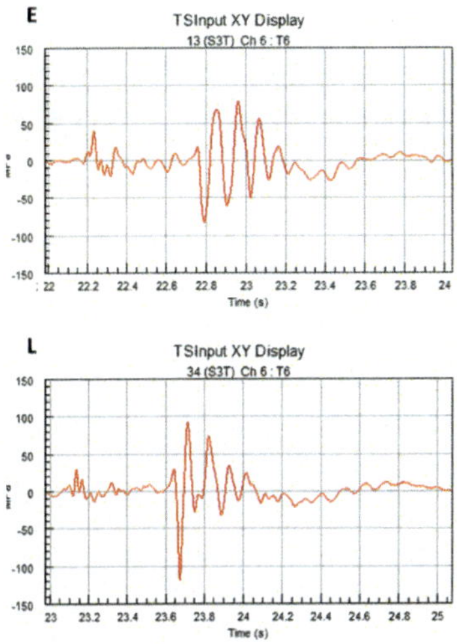

Fig. 2. Crossing of artificial obstacles by both wheels at the same time (E-empty bus, L-loaded bus).

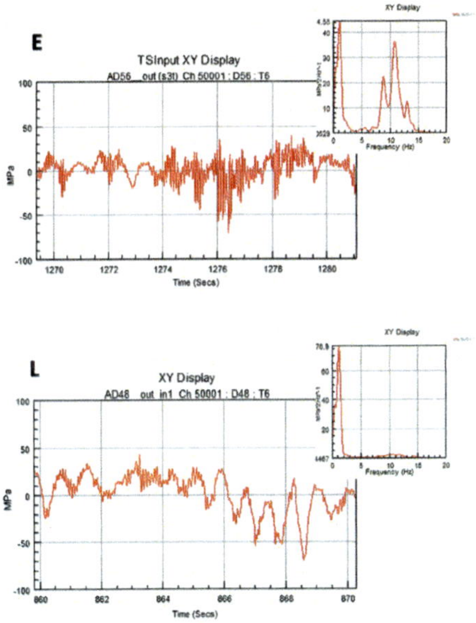

Fig. 3. Driving on real roads - short cuts from the measured signals. (E-empty bus, L-loaded bus).

3 Case Study

3.1 S-N Line

The detail of interest was a severely stressed beam welded joint in the top corner of the door opening in the bus bodywork shown in Fig. 5. These critical cross section was

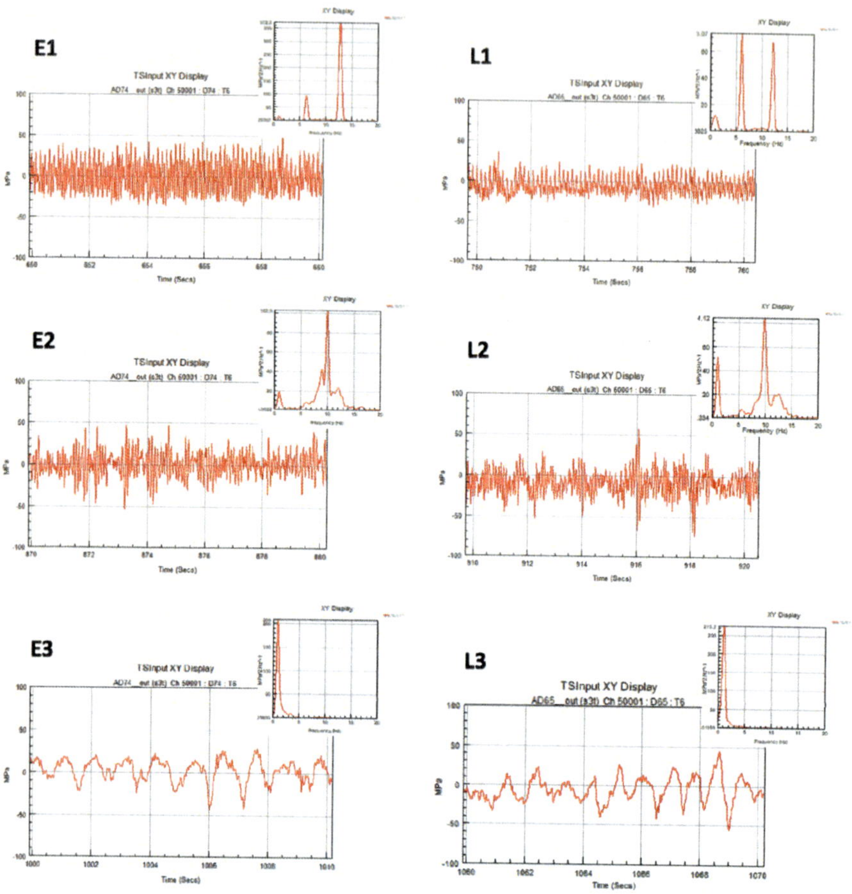

Fig. 4. Driving on special tracks on the proving ground - short cuts from the measured signals (E-empty bus, L-loaded bus, 1–sinus resonance road, 2–paved road, 3-Belgian paving).

monitored by strain gauge T6. Also, brief examples of stress-time courses in the previous figures are from this sensor.

In order to determine the fatigue strength of the evaluated structural detail, laboratory fatigue testing was carried out. Test specimens were equivalent to the considered the structural node in terms of basic material, shape and manufacturing technology (welding). The test specimens were made from thin-walled welded closed profiles, which had 70 × 50 mm cross-section and 2 mm wall thickness and were made of S235JR steel. The critical cross-section of the joint was subjected to reverse bending load (the cycle stress ratio was $R = -1$). The limit state was defined by the instant at which a macroscopic fatigue crack was formed (1 to 2 mm). In all cases, fatigue cracks initiated in the transition zone of the fillet weld.

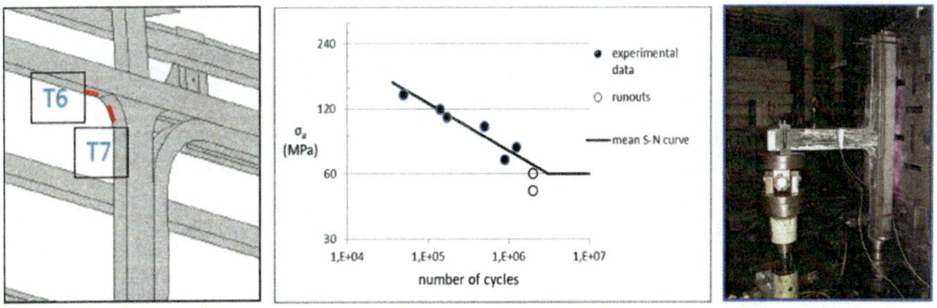

Fig. 5. Detail of interest (T6): schematic illustration; S-N line; test stand.

Statistical evaluation of the fatigue test data yielded the parameters of the mean *S-N* line for the structural detail in the form:

$$log\,(N) = 14.54 - 4.53 \cdot log\,(\sigma_a);\;\; \sigma_c = 60\,\text{MPa} \tag{1}$$

3.2 Stress Spectra

The service stress-time histories were measured for a city bus riding on an irregular surface along a city route whose total length was $L_m \approx 40$ km.

In a similar manner to representative urban traffic, the stress spectra on the test polygon were evaluated. The test polygon offers tracks with various longitudinal road profile and different surface quality. Table 1 specifies the composition of the proposed test route. The measurement was repeated tree times with empty vehicle and three times with fully loaded vehicle. Total length of the measured route was $L_m \approx 35.5$ km.

The measured stress-time histories were transformed into stress spectra via rain-flow counting method. In high-cycle fatigue of welded structures, the mean stress of the cycle does not play a major role. Therefore, only the one-parameter stress spectra $\sigma_{ai} - n_i$ were used for subsequent calculations.

3.3 Fatigue Life Calculations

The mostly the fatigue damage D is calculated using the linear cumulative damage rule. According to this rule, the limit state with respect to fatigue is reached (i.e. the fatigue life of the structural part is exhausted) when the following condition is met:

$$D = \sum n_i / N_i = D_{lim} \tag{2}$$

D - fatigue damage caused by the stress spectrum imposed,
n_i - number of cycles applied at the i-th level of stress with the amplitude σ_{ai},
N_i - limit life under identical loading, the number of cycles derived from S-N line of a considered component at the amplitude σ_{ai},
D_{lim} - limit value of fatigue damage.

Table 1. Composition of tracks on test polygon.

Section	Length (km)
Slope circuit	3.80
Speed circuit	2.80
Arrival to special roads	0.07
Panel road	0.45
Exit/arrival	0.13
Sine resonance road	0.40
Exit/arrival	0.15
Paved road	0.40
Exit/arrival	0.15
Paved road	0.40
Exit/arrival	0.15
Paved road	0.40
Exit/arrival	0.16
Belgian paving	0.40
Exit from special roads	0.08
Speed circuit	1.90
TOTAL	11.84

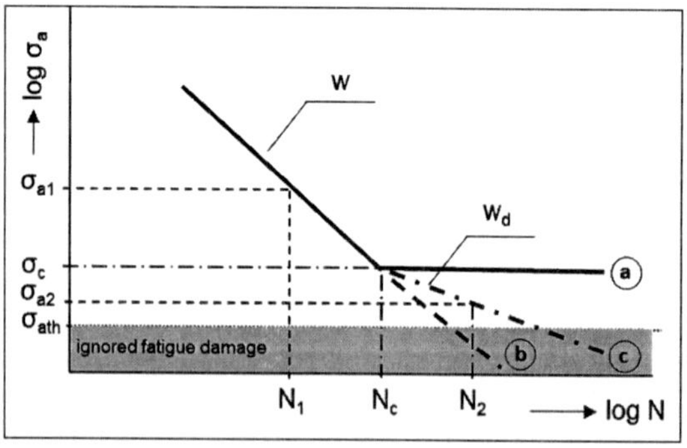

Fig. 6. Boundary conditions for calculating cumulative fatigue damage: (a - Palmgren-Miner original, b - Palmgren-Miner elementar, c – Haibach).

Various rules apply different boundary conditions to the fatigue damage calculation. A schematic representation of these boundary conditions is shown in Fig. 6.

Account is taken of the damage caused by cycles with small amplitudes, which occur very frequently $(\sigma_{ath} < \sigma_{ai} < \sigma_c)$ [7]. A threshold value σ_{ath} is applied to the conversion of stress to damage, and therefore the damage caused by cycles with amplitudes of is neglected ($\sigma_{ai} < \sigma_{ath} < \sigma_c$). In the present case, the Haibach-modified version of the Palmgren-Miner rule was chosen for calculating fatigue damage. The limit number of cycles Ni was determined as follows:

$$\sigma_{ai} \geq \sigma_c: \qquad N_i = N_c \cdot (\sigma_c/\sigma_{ai})^w \tag{3}$$

$$\sigma_c \geq \sigma_{ai} > \sigma_{ath}: \; N_i = N_c \cdot (\sigma_c/\sigma_{ai})^{wd} \tag{4}$$

Haibach recommends the exponent for the lower part of the *S-N* line to be set as $w_d = 2 \cdot w - 1$. In this study, the value was $w_d = 8$. The threshold stress amplitude for taking the fatigue damaging into account was given as $\sigma_{ath} = 0.5 \cdot \sigma_c$ in the present case. The limit value of fatigue damage was taken as $D_{lim} = 0.5$.

The computational estimation of service fatigue life (in kilometre run) is obtained from equation:

$$L = (D \,/\, D_{lim}) \cdot L_m \tag{5}$$

Table 2 provides predicted fatigue life for the service stress spectra and the spectra measured on the test polygon. The ratio of these predicted fatigue life is an estimation of the acceleration (shortening) of the driving fatigue test that could be achieved on the test polygon compared to normal vehicle operation [8].

Table 2. Table captions should be placed above the tables.

Life in kilometers	Real operation	Proving ground	Acceleration
Empty bus	3,178,000	203,000	15,7
Loaded bus	4,309,000	337,000	12,8

In theory, an even more aggressive composition of the test tracks could be designed. It would involve a larger proportion of those sections of the proving ground which produce the most severe damage. However, suspension elements would have to be protected from degradation. The sequence of the test track sections should enable them to "relax"; particularly the shock absorbers would need to cool down, because they might overheat during riding on some types of road surface.

4 Conclusions

The response of bus bodyworks to operating loads has the character of random processes. Their statistical characteristics and time courses can be precisely analysed only on the basis of long-term measurements.

Fatigue analysis is fundamentally different from dynamic analysis. Dynamic analyses are performed in the frequency domain; fatigue analyses must be performed in the time domain.

An important phenomenon can be the occurrence of extraordinary overloads, which occur due to extraordinary operating situations as for example driving over high road roughness or driving over a pothole.

The case study indicated how the experimentally obtained data from strain gauge measurement and laboratory fatigue tests can be used for estimation of fatigue life of bodywork welded nodes.

The authors prepared this paper with the support of the Ministry of Industry and Trade of the Czech Republic, the project FV40260 "On-line measurement and analysis of the operational loading of structures with adaptive virtual models".

References

1. Kepka, M.: Durability and fatigue life investigation of bus and trolleybus structures: review of SKODA VYZKUM methodology. In: 2nd International Conference on Material and Component Performance under Variable Amplitude Loading, pp. 1231–1240. DVM, Darmstadt (2009). ISBN 978-3-00-027049-9
2. Kepka, Jr., M.: Degradation of mechanical properties of cyclically loaded materials and structural nodes. Dissertation, University of West Bohemia, Pilsen (2021). (in Czech)
3. Chmelko, V., Garan, M.: Long-term monitoring of strains in a real operation of structures. In: Proceedings of the 14th IMEKO TC10 Workshop on Technical Diagnostics, Milano, Italy, pp. 333–336 (2016)
4. Palmgren, A.Z.: Die Lebensdauer von Kugellagern. Z. Ver. Deutsch. Ing. **68**, 339 (1924)
5. Miner, M.A.: Cumulative damage in fatigue. Trans. ASME Ser. E. J. Appl. Mech. **12**, 159–164 (1945)
6. Haibach, E.: Modified linear damage accumulation hypothesis accounting for a decreasing fatigue strength during the increasing fatigue damage. TM Nr. 50. LBF, Darmstadt (1970)
7. Chmelko, V.: Cyclic anelasticity of metals. Metal. Mater. **52**, 353–359 (2014)
8. Kepka, M., Kepka, Jr., M.: Design, service and testing grounds stress spectra and their using to fatigue life assessment of bus bodyworks. In: 12th International Fatigue Congress. MATEC Web of Conferences, vol. 165, p. 17007 (2018)

Integrated Force Shaping and Optimized Mechanical Design in Underactuated Linear Vibratory Feeders

Dario Richiedei, Iacopo Tamellin(✉), and Alberto Trevisani

Department of Management and Engineering, University of Padova, Vicenza, Italy
{dario.richiedei,iacopo.tamellin,alberto.trevisani}@unipd.it

Abstract. This study proposes a method to improve the feeding speed in an underactuated and flexible linear vibratory feeder through the simultaneous synthesis of the optimal excitation forces and the tuning of the mechanical design. The target is to increase the flow of the parts by ensuring uniform displacements along the tray with the prescribed throw angle, despite the relevant tray flexibility. First, a multi-DOF (degrees of freedom) analytical model of the underactuated feeder is formulated in term of actuated and unactuated coordinates. Hence, the subspace of allowable motion is computed, to highlight the relation between the achievable displacements and the physical parameters of the system (mass, stiffness and force distribution matrices and excitation frequency). Then, such subspace is optimized through dynamic structural modification where parametric sensitivities are adopted to select the structural modifications that enable to reduce the actuation effort. Once the mechanical design has been optimized, the optimal harmonic forces are computed through an ad-hoc inverse dynamics algorithm, defined "force shaping". Finally, a multibody model developed in MSC Adams is used to evaluate the flow of the parts over the feeder. The improved mechanical design of the system together with the shaped forces evidence the improved system performances with reduced actuation forces.

Keywords: Vibratory feeders · Underactuated systems · Feedforward control · Dynamic structural modification · Multibody simulation

1 Introduction

Vibratory feeders are devices used in the industry to convey products in manufacturing plants by exploiting mechanical vibrations [1]. Linear vibratory feeders convey the products along a linear path [2] and the flexibility of these devices is exploited to achieve a large amplitude of vibration through a low actuation effort [3]. In general, the number of actuators is smaller than the number of degrees of freedom (DOFs) arising in the system due to its flexibility, leading to an underactuated system. It is well known in the literature that the control of underactuated systems is not trivial. In particular, in the case of vibratory feeders these devices work in open-loop. Hence, it is fundamental to carefully plan a-priori the actuation forces to achieve high performances [4]. In this

Z. Dimitrovová et al. (Eds.): WMVC 2022, MMS 125, pp. 249–258, 2023.
https://doi.org/10.1007/978-3-031-15758-5_24

light, this paper proposes the application of a force shaping technique proposed in a previous paper by the Authors [5] to boost the performances of linear vibratory feeders. The mathematical analysis provided in the paper highlights that the achievable performances are constrained by the subspace of the allowable motion of the system, hence a dynamic structural modification approach is adopted to redesign the vibratory feeder in such a way that it is optimized to obtain the desired tray displacements. The structural modification variables, i.e., the stiffnesses and masses to be modified, are chosen through an ad-hoc sensitivity index, developed in this work, that enables to reduce the actuation effort. The proposed method is applied to the model of a feeder employed in manufacturing plants and the increase of performances is assessed through numerical simulations. The effectiveness of the method is also confirmed by the usage of a multibody simulator developed in MSC Adams.

2 Theoretical Background

2.1 Model of the Underactuated Linear Vibratory Feeder

Let us consider the linear vibratory feeder sketched in Fig. 1, employed in manufacturing plants to convey products (see e.g. [3]). The feeder tray is modeled through 4 Euler-Bernoulli beam elements. The vertical displacements and the nodal rotations of the tray are denoted respectively by: y_i, φ_i, $i = 1, \ldots, 5$. The horizontal displacement of the tray is x_t. The actuators are modeled by considering the displacements of the actuators, defined through the generalized coordinates: $x_{a_i} = x_t + s_i \cos\theta_f$ and $y_{a_i} = y_i + s_i \sin\theta_f$ where s_i is the relative displacement of the i-th actuator along the direction defined by the angle $\theta_f = 2\pi - \alpha_f$. The model of the actuators, whose mass is m_{ai} and k_{ai} the leaf spring stiffness, is here omitted for brevity (for more details about the system model refer to [3, 5]).

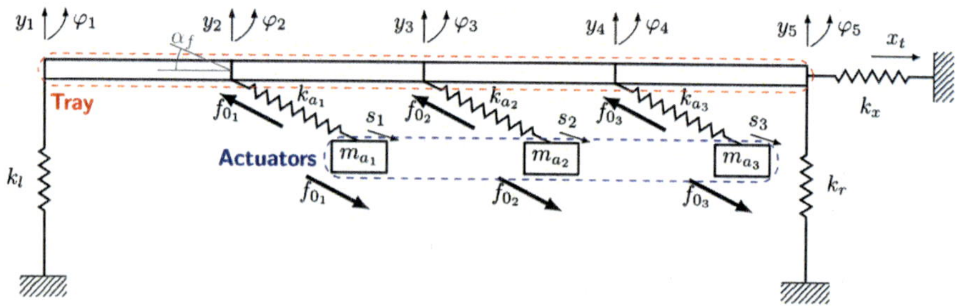

Fig. 1. Sketch of the linear vibratory feeder.

The equations of motion of the tray and of the actuators can be merged together and a second-order model of the *14*-DOFs undamped system is obtained:

$$\mathbf{M}\ddot{\boldsymbol{q}}(t) + \mathbf{K}\boldsymbol{q}(t) = \mathbf{B}\boldsymbol{f}(t) \tag{1}$$

where $\mathbf{M} \in \mathbb{R}^{N \times N}$ and $\mathbf{K} \in \mathbb{R}^{N \times N}$ are respectively the mass and stiffness matrices of the system with $N = 14$. Damping can be neglected [5]. The displacements and accelerations are collected in vectors $\boldsymbol{q}(t) \in \mathbb{R}^N$ and $\ddot{\boldsymbol{q}}(t) \in \mathbb{R}^N$, with $\boldsymbol{q}(t) = \left\{ y_1\ \varphi_1 \ldots y_5\ \varphi_5\ s_1\ s_2\ s_3\ x_t \right\}^T \in \mathbb{R}^{14}$. The device is excited through three independent ($N_A = 3$) in-phase harmonic forces: $\boldsymbol{f}(t) = \boldsymbol{f_0} \cos(\omega_f t)$ whose amplitudes are f_{0i}, $i = 1, 2, 3$, and whose frequency is ω_f. The forces exert by the motors are distributed along the system through the actuator influence matrix $\mathbf{B} \in \mathbb{R}^{N \times N_A}$:

$$\mathbf{B} = \begin{bmatrix} 0\ 0\ \sin\theta_f & 0 & 0 & 0 & 0 & 0\,0\,0\,1\,0\,0\,\cos\theta_f \\ 0\ 0 & 0 & 0\ \sin\theta_f & 0 & 0 & 0\,0\,0\,0\,1\,0\,\cos\theta_f \\ 0\ 0 & 0 & 0 & 0 & 0\ \sin\theta_f & 0\,0\,0\,0\,0\,1\ \cos\theta_f \end{bmatrix}^T \quad (2)$$

The system performances are optimized once the tray, over which the products flow, behaves as a rigid one in the throw angle direction α_f. However, the system is flexible due to the presence of the springs and due to the low stiffness of the long tray. Hence, $\boldsymbol{f_0}$ must be designed to obtain a rigid behavior even in presence of flexibilities.

2.2 Underactuation Issues: The Subspace of the Allowable Motion

Since the excitation is harmonic and the system is linear, its steady-state response is harmonic, and it is denoted through $\boldsymbol{q}(t) = \boldsymbol{q_0} \cos(\omega_f t)$ where $\boldsymbol{q_0}$ is the amplitudes vector. Under these hypotheses, Eq. (1) is recast in the frequency domain as:

$$\left(-\omega_f^2 \mathbf{M} + \mathbf{K}\right)\boldsymbol{q_0} = \mathbf{B}\boldsymbol{f_0} \quad (3)$$

where $\boldsymbol{f_0}$ is the force amplitude vector (all along the paper the subscript $\boldsymbol{0}$ denotes the amplitudes of the forces or displacements). The system is underactuated, i.e., $N_A < N$, and the three actuation forces are assumed to be independent, i.e., rank($\mathbf{B}$) $= N_A = 3$. Hence, not all the desired displacements $\boldsymbol{q_0^d}$ are attainable, but only those belonging to the subspace of the allowable motion [5]. Such subspace is characterized, in this paper through the following mathematical framework. First, let us partition the influence matrix $\mathbf{B}$ through QR-decomposition in the following form: $\mathbf{B} = \mathbf{QR} = \mathbf{Q}\begin{bmatrix} \mathbf{B_A} \\ \mathbf{0} \end{bmatrix}$, where $\mathbf{Q}$ is orthonormal, i.e., $\mathbf{Q}^T\mathbf{Q} = \mathbf{Q}\mathbf{Q}^T = \mathbf{I}$ and $\mathbf{B_A}$ is upper-triangular and invertible. A new coordinate vector is defined as follows: $\boldsymbol{y} = \mathbf{Q}^T\boldsymbol{q}$. The substitution of the QR-decomposed $\mathbf{B}$ together with $\boldsymbol{y}$ in Eq. (3) yields:

$$\left(-\omega_f^2 \mathbf{Q}^T\mathbf{M}\mathbf{Q} + \mathbf{Q}^T\mathbf{K}\mathbf{Q}\right)\boldsymbol{y_0} = \begin{bmatrix} \mathbf{B_A} \\ \mathbf{0} \end{bmatrix}\boldsymbol{f_0} \quad (4)$$

The new coordinates vector $\boldsymbol{y}$ might have no straightforward physical meaning. However, its introduction simplifies the solution of the force shaping problem. $\boldsymbol{y}$ can be partitioned into N_A actuated coordinates, $\boldsymbol{y_A}$, and N-N_A unactuated coordinates, $\boldsymbol{y_U}$. It is possible to partition $\mathbf{M}$ and $\mathbf{K}$ accordingly exploiting Eq. (4) obtaining:

$$\left(-\omega_f^2 \begin{bmatrix} \mathbf{M_{AA}} & \mathbf{M_{AU}} \\ \mathbf{M_{AU}^T} & \mathbf{M_{UU}} \end{bmatrix} + \begin{bmatrix} \mathbf{K_{AA}} & \mathbf{K_{AU}} \\ \mathbf{K_{AU}^T} & \mathbf{K_{UU}} \end{bmatrix}\right)\begin{Bmatrix} \boldsymbol{y_{A_0}} \\ \boldsymbol{y_{U_0}} \end{Bmatrix} = \begin{bmatrix} \mathbf{B_A} \\ \mathbf{0} \end{bmatrix}\boldsymbol{f_0} \quad (5)$$

The second matrix equation in Eq. (5) reveals that the motion of the unactuated coordinates is fully imposed by the actuated coordinates, i.e., $y_{U_0} = -\mathbf{G}_{\mathbf{UU}}^{-1}\mathbf{G}_{\mathbf{AU}}^{T}y_{A_0}$, where: $\mathbf{G}_{\mathbf{AU}}^{T} = -\omega_f^2\mathbf{M}_{\mathbf{AU}}^{T}+\mathbf{K}_{\mathbf{AU}}^{T}$ and $\mathbf{G}_{\mathbf{UU}} = -\omega_f^2\mathbf{M}_{\mathbf{UU}}+\mathbf{K}_{\mathbf{UU}}$. Hence, the transmission matrix $\mathbf{L}$ is defined such that:

$$\begin{Bmatrix} y_{A_0} \\ y_{U_0} \end{Bmatrix} = \begin{bmatrix} \mathbf{I}_{N_A \times N_A} \\ -\mathbf{G}_{\mathbf{UU}}^{-1}\mathbf{G}_{\mathbf{AU}}^{T} \end{bmatrix} y_{A_0}^{d} = \mathbf{L}y_{A_0}^{d} \tag{6}$$

Equation (6) reveals that the achievable displacements must belong to the N_A-dimensional subspace spanned by $\mathbf{L}$, i.e., $\begin{Bmatrix} y_{A_0} \\ y_{U_0} \end{Bmatrix} \in \text{span}(\mathbf{L})$. The subspace of the allowable motion depends on the inertial and elastic properties of the system (through the mass and stiffness matrices) and on the excitation frequency. In this light, structural modification is here adopted to improve the allowable motion subspace. The second matrix equation in Eq. (5) is recast in the following structural modification problem:

$$\min_{p} \left\| (\mathbf{G}_{\mathbf{AU}} + \Delta\mathbf{G}_{\mathbf{AU}}(p))y_{A_0}^{d} + (\mathbf{G}_{\mathbf{UU}} + \Delta\mathbf{G}_{\mathbf{UU}}(p))y_{U_0}^{d} \right\|_2^2 \quad \text{subj} \ \ \boldsymbol{p_L} \leq \boldsymbol{p} \leq \boldsymbol{p_U} \tag{7}$$

where $\boldsymbol{p}$ collects the n_p design parameters, i.e., the inertial and elastic properties to be modified with $\boldsymbol{p_L}$ and $\boldsymbol{p_U}$ the lower and upper bounds respectively. Once Eq. (7) has been solved (a solution method based on homotopy optimization has been proposed by the Authors in [5]), the structural modification parameters $\boldsymbol{p_{mod}}$ are obtained.

2.3 The Force Shaping Strategy

A strategy to compute the actuation forces $\boldsymbol{f_0}$ is here introduced and denoted Partial Force Shaping (PFS). Hereafter, $\mathbf{M}$ and $\mathbf{K}$ will be adopted to simplify the notation; however, the following relations holds also for $\mathbf{M_{mod}}$ and $\mathbf{K_{mod}}$. It is common, for large scale systems with several DOFs, that the desired displacements are imposed on just some entries of $\boldsymbol{q}$ [5, 6]. Hence, let us assume that only $n_d < N$ displacements, $\boldsymbol{q_s^d}$ are "set" to assume desired values. While, N-n_d coordinates collected into $\boldsymbol{q_f}$ are "free" to assume arbitrary values. Accordingly, to such coordinate partitioning, $\mathbf{Q}$ is partitioned into $\mathbf{Q_s} \in \mathbb{R}^{n_d \times N}$ and $\mathbf{Q_f} \in \mathbb{R}^{(N-n_d)\times N}$. By recalling that $\boldsymbol{y} = \begin{Bmatrix} y_A \\ y_U \end{Bmatrix} = \mathbf{Q}^T\boldsymbol{q}$ it is possible to recast Eq. (6) in the following form:

$$\mathbf{Q_s}^T\boldsymbol{q_{s_0}^d} = \begin{bmatrix} \mathbf{L} & -\mathbf{Q_f}^T \end{bmatrix} \begin{Bmatrix} \tilde{y}_{A_0}^{d} \\ q_{f_0} \end{Bmatrix} \tag{8}$$

Hence, the projection of $\boldsymbol{q_{s_0}^d}$ onto the subspace of the allowable motion spanned by the columns of $\mathbf{P} = \begin{bmatrix} \mathbf{L} & -\mathbf{Q_f}^T \end{bmatrix}$ yields:

$$\tilde{y}_{A_0}^{d} = \begin{bmatrix} \mathbf{I}_{N_f \times N_f} & \mathbf{0}_{N_f \times (N-n_d)} \end{bmatrix} \left(\mathbf{P}^T\mathbf{P}\right)^{-1}\mathbf{P}^T\mathbf{Q_s}^T\boldsymbol{q_{s_0}^d} \tag{9}$$

Finally, the optimal excitation vector $\boldsymbol{f_0}$ is obtained by introducing $\tilde{\boldsymbol{y}}^d_{\mathbf{A}_0}$ in lieu of $\boldsymbol{y}_{\mathbf{A}_0}$ in the first matrix equation in Eq. (5), leading to:

$$\boldsymbol{f}_0 = \mathbf{B}_\mathbf{A}^{-1}\left(\mathbf{G}_\mathbf{AA} - \mathbf{G}_\mathbf{AU}\mathbf{G}_\mathbf{UU}^{-1}\mathbf{G}_\mathbf{AU}^T\right)\tilde{\boldsymbol{y}}^d_{\mathbf{A}_0} \tag{10}$$

2.4 Selection of the Structural Modification Through Sensitivity Analysis

The choice of the design parameters to be employed and of the upper and lower bounds to be adopted in Eq. (7) is not trivial. In this paper the choice is performed to achieve low actuation effort by adopting as index the sensitivity $S_p = \sum\left(\frac{\partial \boldsymbol{f_0}}{\partial p}p_0\right)$, where p denotes either a mass or stiffness in $\boldsymbol{p}$. The original value of the parameter is p_0. The derivative $\frac{\partial \boldsymbol{f_0}}{\partial p}$ is obtained through the derivative of Eq. (10) with respect to p:

$$\begin{aligned}\frac{\partial \boldsymbol{f}_0}{\partial p} = \mathbf{B}_\mathbf{A}^{-1}\Bigg(&\left(\frac{\partial \mathbf{G}_\mathbf{AA}}{\partial p} - \left(\frac{\partial \mathbf{G}_\mathbf{AU}}{\partial p}\mathbf{G}_\mathbf{UU}^{-1} + \mathbf{G}_\mathbf{AU}\frac{\partial \mathbf{G}_\mathbf{UU}^{-1}}{\partial p}\right) - \mathbf{G}_\mathbf{AU}\mathbf{G}_\mathbf{UU}^{-1}\frac{\partial \mathbf{G}_\mathbf{AU}^T}{\partial p}\right)\tilde{\boldsymbol{y}}^d_{\mathbf{A}_0}\\ &+ \left(\mathbf{G}_\mathbf{AA} - \mathbf{G}_\mathbf{AU}\mathbf{G}_\mathbf{UU}^{-1}\mathbf{G}_\mathbf{AU}^T\right)\frac{\partial \tilde{\boldsymbol{y}}^d_{\mathbf{A}_0}}{\partial p}\Bigg)\end{aligned} \tag{11}$$

where $\frac{\partial \tilde{\boldsymbol{y}}^d_{\mathbf{A}_0}}{\partial p}$ can be computed from Eq. (9) as:

$$\begin{aligned}\frac{\partial \tilde{\boldsymbol{y}}^d_{\mathbf{A}_0}}{\partial p} = \left[\mathbf{I}_{N_A\times N_A}\ \ \mathbf{0}_{N_A\times(N-n_d)}\right]\Bigg(\left(\mathbf{P}^\mathbf{T}\mathbf{P}\right)^{-1}\Bigg(&\frac{\partial \mathbf{P}^T}{\partial p}\left(\mathbf{I}_{N\times N} - \mathbf{P}\left(\mathbf{P}^T\mathbf{P}\right)^{-1}\mathbf{P}^T\right)^T\\ &- \mathbf{P}\frac{\partial \mathbf{P}}{\partial p}\left(\mathbf{P}^T\mathbf{P}\right)^{-1}\mathbf{P}\Bigg)\Bigg)\mathbf{Q}_\mathbf{s}^T\boldsymbol{q}^d_{\mathbf{s}0}\end{aligned} \tag{12}$$

The derivatives $\frac{\partial \mathbf{L}}{\partial p}$ and $\frac{\partial \mathbf{P}}{\partial p}$ can be computed from Eqs. (6) and (8), leading to:

$$\frac{\partial \mathbf{P}}{\partial p} = \left[\tfrac{\partial \mathbf{L}}{\partial p}\ \mathbf{0}_{(N-n_d)\times N}\right],\qquad \frac{\partial \mathbf{L}}{\partial p} = \begin{bmatrix}\mathbf{0}_{N_f\times N_f}\\ -\left(\frac{\partial \mathbf{G}_\mathbf{UU}^{-1}}{\partial p}\mathbf{G}_\mathbf{AU}^\mathbf{T} + \mathbf{G}_\mathbf{UU}^{-1}\frac{\partial \mathbf{G}_\mathbf{AU}^\mathbf{T}}{\partial p}\right)\end{bmatrix} \tag{13}$$

3 Application of the Integrated Method

3.1 Numerical Simulations

The proposed method is applied to the model of a linear vibratory feeder employed in manufacturing plants whose parameters are tray flexural stiffness $EI = 1.93\text{e}5$ Nm^2, linear mass density $\rho A = 22.87$ kg/m, length L = 3.6 m, actuators masses and stiffnesses $m_{ai} = 23$ kg and $k_{ai} = 4.6\text{e}5$ N/m, springs stiffnesses $k_l = k_r = k_x = 1.8\text{e}5$ N/m.

A rigid behavior of the tray along its throw angle $\alpha_f = 20°$ when the system is excited at 35 Hz is desired. Hence, specifications are just set for y_i, φ_i, and x_t as: $q^d_{s_0} = \{y^d_0\ 0\ y^d_0\ 0\ y^d_0\ 0\ y^d_0\ 0\ y^d_0\ 0\ x^d_{t0}\}^T$ where the oscillation amplitude in the throw angle direction is $z^d_0 = 5\,\text{mm}$, hence $y^d_0 = z^d_0 \sin(\pi - \alpha_f)$ and $x^d_{t_0} = z^d_0 \cos(\pi - \alpha_f)$. Uniformity is enforced by imposing a rigid motion of the beam with null elastic nodal rotations, i.e., $\varphi^d_0 = 0$.

Table 1. Parametric sensitivities

p	m_{a1}	m_{a2}	m_{a3}	m_1	m_2	m_3	m_4	m_5
S_p	−1.5e4	2.2e3	−1.5e4	−9e2	−1e3	−1e3	−1e3	−9e2
p	k_{a1}	k_{a2}	k_{a3}	k_l	k_r	k_x		
S_p	2.1e4	3.2e3	2.1e4	1e2	1e2	1e3		

The performances of the proposed method, i.e. PFS with structural modification (SM) exploiting also sensitivity analysis (SA) hereafter denoted as PFS-SM-SA, will be compared with those achieved by computing $\boldsymbol{f}_0$ from Eq. (3) exploiting the pseudoinverse (†) of $\mathbf{B}$: $\boldsymbol{f}_0 = \mathbf{B}^\dagger\left(-\omega_f^2\mathbf{M} + \mathbf{K}\right)\boldsymbol{q}^d_0$. The force vector $\boldsymbol{f}_0$ computed through the pseudoinverse method will be denoted as PsI. Additionally, it will be compared with the sole application of the PFS and with the PFS method exploiting also structural modification (SM) without SA hereafter denoted as (PFS-SM).

The design variables initially chosen for the structural modification in the PFS-SM are 14 and consists of 5 nodal lumped masses to be placed to the tray, the actuators masses that can be modified by modifying the counterweight masses and all the lumped springs.

The parametric sensitivities of the forces with respect to the design variables, i.e., S_p, are listed in Table 1. These suggests that it can be beneficial to avoid reductions of m_{a1} and m_{a3} and to reduce the most the variations of k_{a1} and k_{a3}. Additionally, to simplify the mechanical design the lumped masses m_1, ..., m_5 are omitted in the PFS-SM-SA. This approach yields to structural modification parameters listed in Table 3 leading to the results summarized in Table 2.

The cosine between the obtained and the desired displacements $\cos\left(\boldsymbol{q}^d_{s_0}, \boldsymbol{q}_{s_0}\right)$ approaches one with the PFS-SM-SA method, hence the obtained displacements are a tight approximation of the desired ones. The selection of the design variables together with the wise choice of the modification bounds enables to lower the actuation effort as demonstrated by the Euclidean norm of $\boldsymbol{f}_0$, indeed the PFS-SM-SA is the method that requires the lower actuation effort. The maximum variation in the throw angle $\left|\max(\alpha_{fi}) - \min(\alpha_{fi})\right|$ approaches zero, hence the tray behaves as a rigid one. These results are confirmed by the tray displacements shown in Fig. 2.

Table 2. Comparison of the displacements and forces obtained through various methods.

	PsI	PFS	PFS-SM	PFS-SM-SA
$\cos\left(q_{s0}^{d}, q_{s0}\right)$ [-]	0.4535	0.9949	0.9998	0.9998
$\lvert\max(y_{0_i}) - \min(y_{0_i})\rvert$ [mm]	5.53	0.388	0.057	0.057
$\max(\lvert\varphi_{0_i}\rvert)$ [rad]	3.8e−3	2.6e−4	7.2e−5	7.2e−5
$\lvert\max(\alpha_{f_i}) - \min(\alpha_{f_i})\rvert$ [°]	56.07	4.20	0.61	0.61
f_{01} [N]	−3390.9	−6756.0	2723.2	−520.7
f_{02} [N]	−3390.9	3609.3	2761.3	2573.6
f_{03} [N]	−3390.9	−6756.0	2723.2	−520.7
$\lVert f_0\rVert$ [N]	5873.1	10213	4738.9	2676.9

Table 3. Structural modification parameters.

	PFS-SM		PFS-SM-SA	
	[pL; pU]	p_{mod}	*[pL; pU]*	p_{mod}
m_{a1} [kg]	[−5; + 5]	−4.88	[0; +5]	1.07
m_{a2} [kg]	[−5; +5]	4.78	[−5; +5]	4.59
m_{a3} [kg]	[−5; +5]	−4.88	[0; +5]	1.07
m_1 [kg]	[0; +3]	0	–	–
m_2 [kg]	[0; +3]	0.48	–	–
m_3 [kg]	[0; +3]	1.09	–	–
m_4 [kg]	[0; +3]	0.48	–	–
m_5 [kg]	[0; +3]	0	–	–
k_{a1} [Nm^{-1}]	[−2.3e5; 4.6e5]	2.34e5	[−2.3e5; 2.3e5]	1.12
k_{a2} [Nm^{-1}]	[−2.3e5; 4.6e5]	2.41e5	[−2.3e5; 4.6e5]	2.04
k_{a3} [Nm^{-1}]	[−2.3e5; 4.6e5]	2.34e5	[−2.3e5; 2.3e5]	1.12
k_l [Nm^{-1}]	[−9e4; 1.8e5]	1.8e5	[−9e4; 1.8e5]	1.8e5
k_r [Nm^{-1}]	[−9e4; 1.8e5]	1.8e5	[−9e4; 1.8e5]	1.8e5
k_x [Nm^{-1}]	[−9e4; 5.4e5]	5.4e5	[−9e4; 5.4e5]	5.4e5

3.2 The MSC Adams Multibody Simulator: Analysis of the Parts Flow

A preliminary evaluation of the effectiveness of the proposed force shaping method in feeding objects is performed through a multibody simulator developed on MSC Adams and shown in Fig. 3. The feeder model has been developed with the goal of emulating the system sketched in Fig. 1. The feeder tray has been created by meshing a beam through finite elements. The simulation time is set equal to 10 s and the step size used is 1 ms.

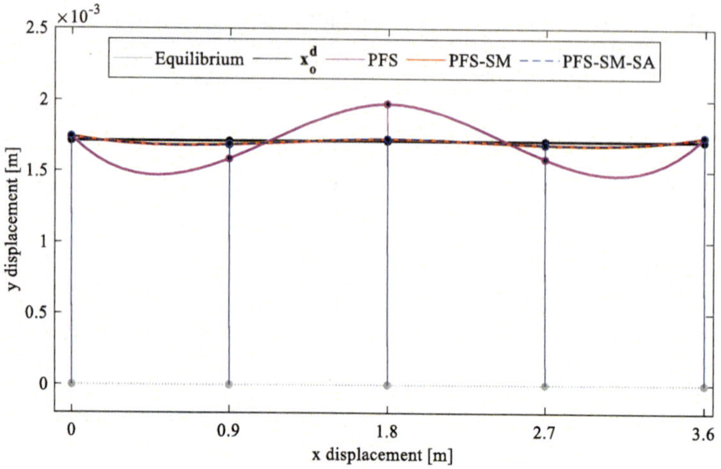

Fig. 2. Tray displacements obtained with different methods.

The actuation forces adopted in the simulations vary accordingly to those reported in Table 2 as well as the lumped spring and masses parameters that are modified through the structural modifications summarized in Table 3.

The simulations performed evidence that the PsI method is not capable of moving the parts along the feeder while the PFS and the PFS with SM are both effective. The most effective technique is the PFS-SM-SA method as shown in Fig. 4, where, by means of example, the time needed to convey along the feeder two parts denoted respectively Part A and B is shown. Figure 4 highlights that the PFS requires respectively 6.26 s and 6.32 s to convey parts A and B while the PFS-SM-SA requires 5.27 s and 6.10 s. Obviously, requiring different amplitude of displacements would change the feeding speed as well as the type of motion, i.e., sliding or hopping motion regime for the conveyed parts [7, 8], however such analysis is out of the scope of this work which focuses on providing a rigid motion of the feeder through the shaping of the excitation forces and structural modification of the allowable subspace of motion.

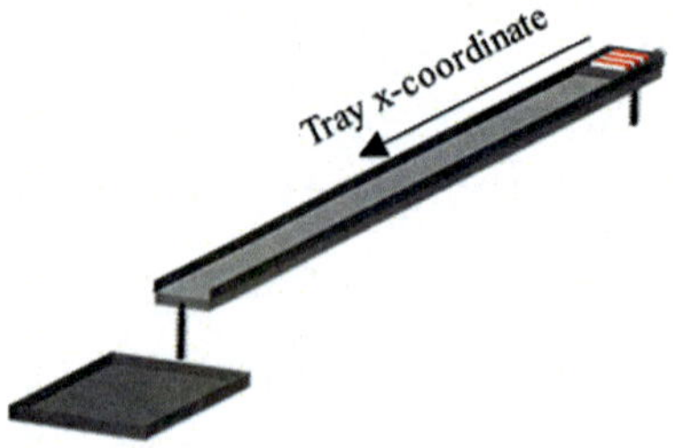

Fig. 3. The multibody simulator in MSC Adams.

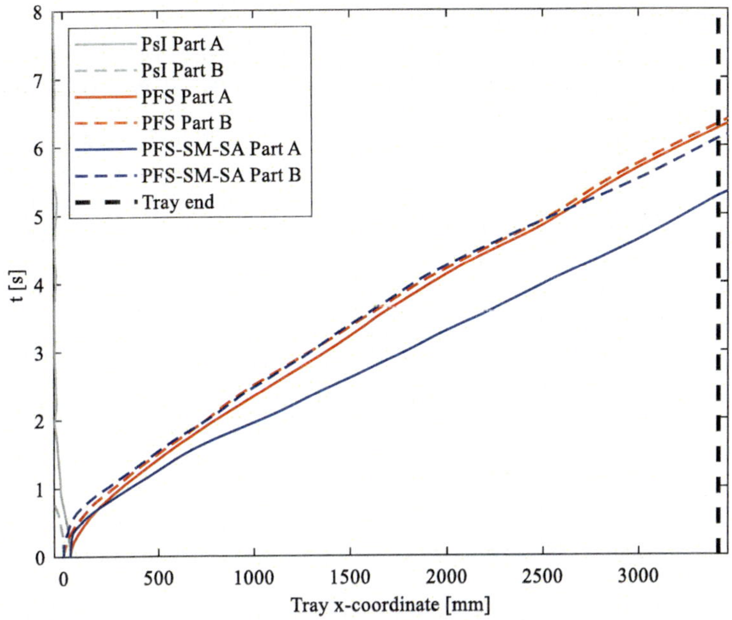

Fig. 4. Displacement of the center of mass of two cylindric parts with three different methods.

4 Conclusions

This paper proposes an extension of the force shaping method, proposed by the Authors in [5], to compute the harmonic excitation forces to be applied to underactuated linear vibratory feeders to convey products. The method exploits QR-decomposition and coordinate change to actuated and unactuated coordinates. This reveals that only the displacements that belongs to the subspace of the allowable motion are attainable. Such subspace is improved through structural modification to achieve higher performances. The design variables together with their feasible bounds are here chosen exploiting an ad-hoc sensitivity index developed in this work which aims at suggesting the structural modifications that both enables to improve the subspace of the allowable motion and simultaneously to reduce the actuation effort here computed through the Euclidean norm of the actuation force amplitudes. The method is numerically validated also by means of the multibody model of linear vibratory feeder implemented in MSC Adams. The effectiveness of the partial force shaping strategies in achieving only the displacements of interest of the system, i.e., those of the tray where the product flows is confirmed and enhanced once structural modification is employed, additionally the actuation effort is reduced by selecting only some structural modifications and through a wise selection of their lower and upper bounds.

References

1. Parameswaran, M.A., Ganapathy, S.: Vibratory conveying—analysis and design: a review. Mech. Mach. Theory **14**(2), 89–97 (1979)
2. Buzzoni, M., Battarra, M., Mucchi, E., Dalpiaz, G.: Motion analysis of a linear vibratory feeder: dynamic modeling and experimental verification. Mech. Mach. Theory **114**, 98–110 (2017)
3. Caracciolo, R., Richiedei, D., Trevisani, A., Zanardo, G.: Designing vibratory linear feeders through an inverse dynamic structural modification approach. Int. J. Adv. Manuf. Technol. **80**(9–12), 1587–1599 (2015). https://doi.org/10.1007/s00170-015-7096-0
4. Seifried, R.: Two approaches for feedforward control and optimal design of underactuated multibody systems. Multibody Syst. Dyn. **27**(1), 75–93 (2012)
5. Belotti, R., Richiedei, D., Tamellin, I., Trevisani, A.: Response optimization of underactuated vibration generators through dynamic structural modification and shaping of the excitation forces. Int. J. Adv. Manuf. Technol. **112**(1–2), 505–524 (2020). https://doi.org/10.1007/s00170-020-06083-2
6. Palomba, I., Richiedei, D., Trevisani, A.: Mode selection for reduced order modeling of mechanical systems excited at resonance. Int. J. Mech. Sci. **114**, 268–276 (2016)
7. Ramalingam, M., Samuel, G.L.: Investigation on the conveying velocity of a linear vibratory feeder while handling bulk-sized small parts. Int. J. Adv. Manuf. Technol. **44**(3), 372–382 (2009)
8. Lim, G.H.: On the conveying velocity of a vibratory feeder. Comput. Struct. **62**(1), 197–203 (1997)

Modal Properties and Modal-Coupling in the Wind Turbines Vibrational Characteristics

Ingrid Lopes Ferreira(✉) and Marcela R. Machado(✉)

Faculdade de Tecnologia Asa Norte, University of Brasília - Campus Universitário Darcy Ribeiro, Brasília - DF 70910-900, Brazil
ilf.ingrid@gmail.com, marcelam@unb.br

Abstract. Wind energy is one of the renewable sources in fast development and implementation around the world. The development of competitive renewable energies and the energy supply networks, e.g. wind turbine (WT), performance are essential to guarantee a sustainable power supply in cities and megacities. In these scenarios, a reliable energy supply is crucial. The WT selected is a National Renewable Energy Lab (NREL) monopile 5 MW baseline wind turbine. This WT is a conventional three-bladed variable-speed, pitch-to-pitch, upwind controlled turbine. In this paper, we explore the dynamic characteristic of the WT monopile 5 MW by mean of the spectral element method, which is a highly accurate method with low computational cost. A tapered spectral element couple to a lumped mass is formulated and used as the WT model. The results reveal a varied of modal properties and mode-coupling instability in the vibrations, which is essential information to the turbine's monitoring and control.

Keywords: Dynamic analysis · Spectral element method · Tapered beam

1 Introduction

Wind energy is one of the worldwide renewable sources in expanding development and implementation. Brazil has shown extraordinary growth in installing wind farms in the last decade. To date, it is a consolidated source of energy in the country. Wind energy production relies on constant wind incidences with the right intensity and out of unseen changes in speed or direction. Brazil's northeast region offers an ideal environment for wind farms. So far, more than 600 wind farms are in operation, with more than 7,532 wind turbines in 12 states [1]. A wind-power plant is considered a source of clean energy generation, which in Brazil is still considered a complement electrical source generation over the hydroelectric system.

The selected wind turbine (WT) is a 5 MW baseline designed by the National Renewable Energy Laboratory (NREL), used as a reference by research teams

Instability in the wind turbines vibrational characteristics.

Z. Dimitrovová et al. (Eds.): WMVC 2022, MMS 125, pp. 259–266, 2023.
https://doi.org/10.1007/978-3-031-15758-5_25

worldwide to standardize the specifications of baseline offshore wind turbines and quantify the benefits of advanced onshore and offshore wind energy technologies. The vibration of the WT is a complex multi-coupling phenomenon that is affected by several types of vibrations feature. Nevertheless, the important ones impact the blades, directly affecting the gearbox and nacelle. Multiple loads induce blade vibrations, such as wind turbulence, wind shear, gravity, mass and aerodynamic imbalances. Unsteady loads might lead to blade structural resonance, fatigue damage, and lifetime reduction, contributing to possible structural failure. The tower vibration originates from the coupled wind-rotor-tower system, mechanical transmission twist vibration, and rotor rotation, resulting from the loads, as mentioned above. Additionally, the effect of the waves on offshore wind turbines (OWT) should be considered. Wind turbine vibrations might represent a potential threat to the environment, community business interests, and land itself [2], which must suffer continuous monitoring.

Despite numerous techniques to model the WT, we chose the spectral element method (SEM) because of its high accuracy in the dynamic analysis combined with low computational cost. Thus, problems solved using the spectral element model can be used as benchmark problems to evaluate the accuracy and performance of a newly developed solution method [3]. The SEM was used in the structural dynamics analysis, including rod, beam, [7] plates, cables [8–10] and composite structures [11,12], wind turbine [4], and ongoing research proposes new and improved elements [13–15]. In this paper, we explore the dynamic characteristic of the 5 MW WT monopile modeled by SEM and analyse two different models configuration being a continuous beam coupled to a lumped mass, and a tapered beam coupled to a lumped mass. Both dynamic response are compared to other NREL commercial software models presented in [4]. The study the modal properties and the coupling mode instability in the WT vibration characteristics is essential to control designs and integrity monitoring.

2 Spectral Element Method

SEM is a mesh method similar to the finite element method with its shape function established from the exact solution of the wave equation formulated in the frequency domain [3]. The dynamic responses are assumed to be the superposition of a finite number of wave modes of different discrete frequencies based on discrete Fourier transform (DFT) theory. Consequently, the exact dynamic stiffness matrix calculation must be repeated at all discrete frequencies up to the frequency of greatest interest. The mesh can be applied when there are geometric or material discontinuities in the spatial domain of concern and also when there are any externally applied forces [3].

2.1 Beam Spectral Element

The two-node beam has two-degree-of-freedom per each node is illustrated in Fig. 1. The Bernoulli-Euler beam model assumes that the deflection of the centerline v (x, t) is small and only transverse. Although this theory assumes the

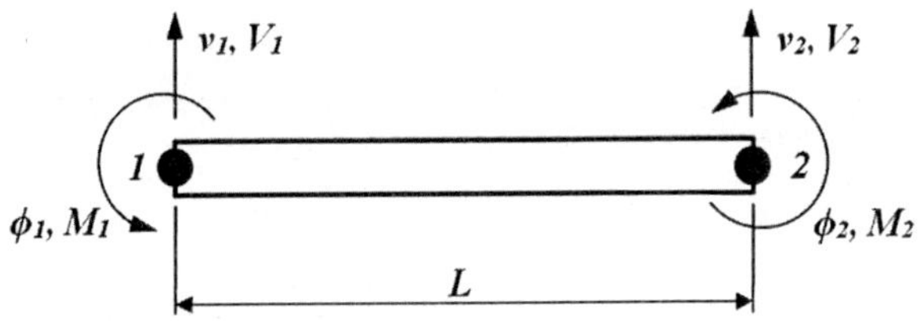

Fig. 1. Two-node spectral beam element.

presence of a transverse shear force, it neglects any shear deformation [6]. The fundamental equation for the flexural motion in a beam.

The undamped and unforced Euler-Bernoulli beam equation of motion under bending vibration is given by

$$EI\frac{\partial^4 v(x,t)}{\partial x^4} + \rho A\frac{\partial^2 v(x,t)}{\partial t^2} = 0 \tag{1}$$

and spectral representation gives

$$EI\frac{\partial^4 \hat{v}}{\partial x^4} + \omega^2 \rho A\hat{v} = 0 \tag{2}$$

where, E, I, ρ A are Young's modulus, inertia moment, mass density per unit length, and cross-section area, respectively; $v(x,t)$ is the transverse displacement, and $\hat{v}(x,w)$ is the displacement in frequency domain. The hysterestic damping is considered in the beam formulation by the damping factor (η) assumed into the Young's modulus (E), so that $E = E(1+i\eta)$, where $i = \sqrt{-1}$. By considering the homogeneous differential equation with constant properties along the beam length, the spectral form of equation (1) became

$$\frac{\partial^4}{\partial x^4} - k^4\hat{v} = 0, \tag{3}$$

where k is the wavenumber expressed by

$$k^4 = \omega^2\frac{\rho A}{EI} \tag{4}$$

By assuming the solutions of the form $e^{-i\beta x}$, the wavenumber related with propagating and evanescent waves are $k = \pm\beta$ and $k = \pm i\beta$, respectively. For the spectral Bernoulli-Euler beam element of length L, the general solution of Eq. (3) is assumed of the form

$$\hat{v}(x,\omega) = a_1 e^{-ikx} + a_2 e^{-kx} + a_3 e^{-ik(L-x)} + a_4 e^{-k(L-x)} = \mathbf{s}(x,\omega)\mathbf{a} \tag{5}$$

where

$$\mathbf{s}(x,\omega)a = \{e^{-ikx}, e^{-}kx, e^{-ik(L-x)}, e^{-k(L-x)}\}$$
$$\mathbf{a}(x,\omega) = \{a_1, a_2, a_3, a_4\}^T$$

Beam spectral nodal displacements and slopes are obtained at the edges, so that at node $1(x=0)$ and at node 2 it has $(x=L)$. Applying the boundary conditions and organize the displacement in a matrix for one has

$$d = \begin{bmatrix} v_1 \\ \phi_1 \\ v_2 \\ \phi_2 \end{bmatrix} = \begin{bmatrix} v(0) \\ v'(0) \\ v(L) \\ v'(L) \end{bmatrix} = \begin{bmatrix} s(0,\omega) \\ s'(0,\omega) \\ s(L,\omega) \\ s'(L,\omega) \end{bmatrix} \mathbf{a} = \mathbf{G}(\omega)\mathbf{a} \tag{6}$$

where

$$\mathbf{G} = \begin{bmatrix} 1 & 1 & e^{-ikL} & e^{-kL} \\ -ik & -k & ie^{ikL}k & e^{-kL}k \\ e^{-ikL} & e^{-kL} & 1 & 1 \\ -ie^{-ikL}k & -e^{-kL}k & ik & k \end{bmatrix}$$

The frequency-dependent displacement within an element is interpolated from the nodal displacement vector $\mathbf{d}$ by eliminating the constant vector $\mathbf{a}$ from Eq. (6) it is expressed as

$$v(x,\omega) = \mathbf{g}(x,\omega)\mathbf{d} \tag{7}$$

$\mathbf{g}(x,\omega)$ is shape function described by

$$\mathbf{g}(x,\omega) = \mathbf{s}(x,\omega)\mathbf{G}^{-1}(\omega) = \mathbf{s}(x,\omega)\Gamma(\omega) \tag{8}$$

Considering $\rho A\omega^2 = EIk^4$ and subtituting in Eq. (1), the weak form can be derived from weight-integral to obtain the dynamic stiffness matrix for the two-nodes beam element by,

$$S_b(\omega) = EI\left[\int_0^L g''(x)^T g''(x)\,dx - k^4\int_0^L g(x)^T g(x)dx\right] \tag{9}$$

where $''$ express the spatial partial derivative.

2.2 Formulation of the Spectral Tower Model

The spectral model of the WT proposed in [4] has been modeled within the NREL 5-MW baseline monopile OWT. It consist of a spectral beam element coupled to a lumped mass representing the slender tower and the rotor-nacelle, respectively. For the model considering a continuous tower only a spectral element is assumed, and for the tapered case, the tower is meshed into 11 beam spectral elements, where each beam element has a reduction in the cross-section area from element one up to the top.The equation of motion that represents the OWT in the spectral representation yields

$$EI\frac{d^4\hat{v}}{dz^4} + \omega^2\left[\rho A + \tilde{m}_n\delta(z-L)\right]\hat{v} = \hat{q}$$

where $\tilde{m}_n = m_n EI\beta^3$, $\delta(z-L)$ is Dirac's delta at $z = L$. The presence of the mass makes the reflection and transmission frequency dependent. Therefore, the mass will act as a filter. Low frequencies will not overpass the mass while very high frequency terms will be attenuated. By assembling the OWT it was used a beam and a lumped mass in its end by using the spectral form of each element, which gives the following global OWT spectral matrix

$$\mathbf{S}_{\mathrm{OWT}}(\omega) = \mathbf{S}_b(\omega) + S_{m_n}(z = L, \omega)$$
$$= \begin{bmatrix} S_b(1,1) & S_b(1,2) & S_b(1,3) & S_b(1,4) \\ S_b(2,1) & S_b(2,2) & S_b(2,3) & S_b(2,4) \\ S_b(3,1) & S_b(3,2) & S_b(3,3) + S_{m_n} & S_b(3,4) \\ S_b(4,1) & S_b(4,2) & S_b(4,3) & S_b(4,4) + S_{m_n} \end{bmatrix} \quad (10)$$

where $S_{m_n} = -\omega^2 \tilde{m}_n$.

3 Numerical Results

Based on the 5 MW baseline wind turbine monopile the two different models configuration being a continuous beam coupled to a lumped mass and the tapered model are illustrated in Fig. 2.

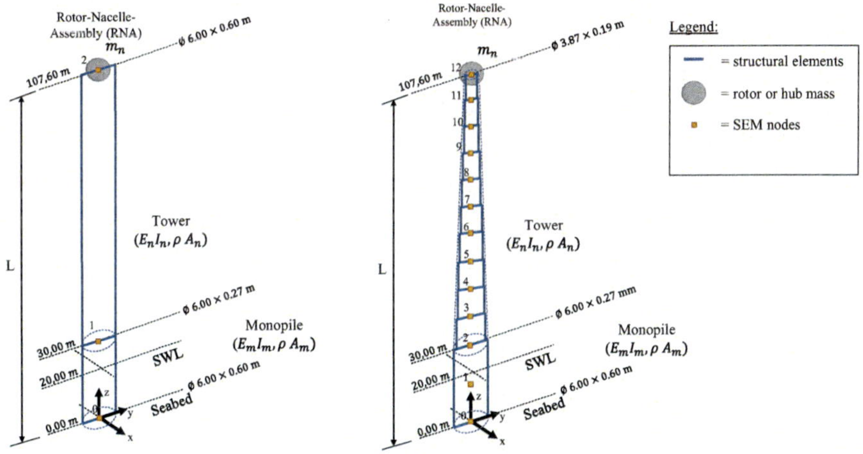

Fig. 2. Graphic representation of continuous WT (a) and Tapered WT (b)

Mechanical properties are assumed as Young's modulus of 210 GPa, and the effective density of the steel 8,500 kg/m^3 [4,5], and structural-damping factor 0.01. The mass attached on the top m_n represents the nacelle+rotor is assumed 350 tons. The continuous tower has a 6.60 m and thickness equals 0.60 m. The tapered beam tower has a land-based diameter of 6.00 m with a thickness of

0.027 m and a top diameter of 3.87 m with a thickness equal to 0.019 m. The eleven spectral elements' geometrical characteristics, length and thickness are listed in Table 1.

Table 1. Elevation coordinate z_n(m), diameter ϕ_n(m) and thickness t_n(m) of assemble Monopile-Tower related to node n.

n	0	1	2	3	4	5	6	7	8	9	10	11	12
z_n	0	20	30	37.8	45.5	53.3	61.0	68.8	76.6	84.3	92.1	99.8	107.6
ϕ_n	6.00	6.00	6.00	5.78	5.57	5.36	5.15	4.94	4.72	4.51	4.30	4.08	3.87
t_n	0.06	0.06	0.06	0.262	0.254	0.246	0.238	0.230	0.222	0.214	0.206	0.198	0.190

Dynamic analysis aiming to explore the optimised model for this study compares the continuous and tapered WTs. Both are excited by a unitary force and flexural and rotational response estimation on top of the tower and validated by comparison with the work of Colherinhas et al. [4]. Figure 3 shows the comparison of flexural receptance frequency response function (FRF) for the continuous and tapered tower, and Fig. 4 presents the rotational response estimation on top of the tower.

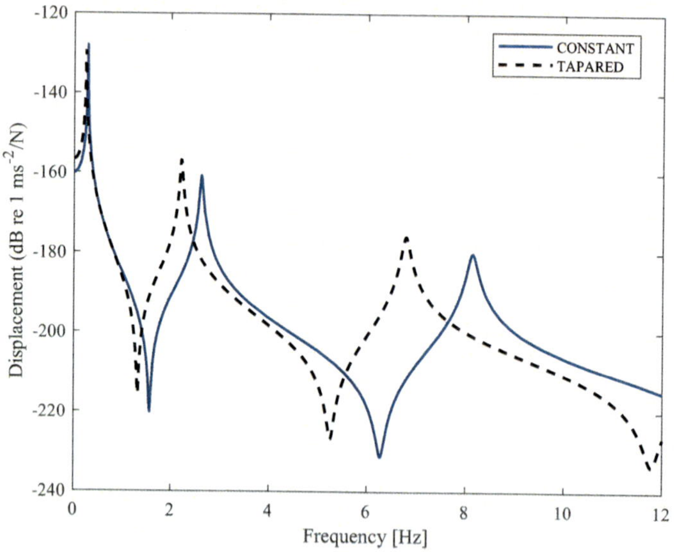

Fig. 3. Frequency response comparison between model WT Constant and Tapered with excitement at the top in ϕ_3 and $\phi_2 1$

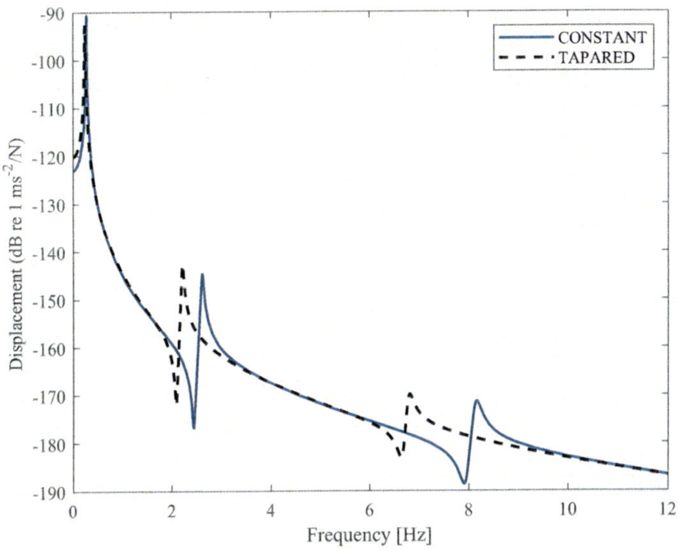

Fig. 4. Frequency response comparison between model WT Constant and Tapered with excitement at the top in v_4 and $v_2 2$

Comparing the continuous and tapered FRFs makes clear the shift of the frequencies to the left because the tapered WT has a reduced stiffness compared to the continuous. Therefore according to the referee paper's first and second resonance frequencies are 0.27 and 2.25 Hz, similar to the tapered WT model, which is shown to be a better representation of the OTW. Further studies in the implementation stage will model the OWT in 3-dimensional objecting the analysis of modal-coupling.

4 Final Remarks

This paper explored the in-plane vibrational characteristics of a selected National Renewable Energy Lab (NREL) monopile 5 MW baseline wind turbine. SEM was used to model a continuous and a tapered tower to verify the optimum choice for further studies. Both repentance responses were validated within the literature. The next step of this study is the implementation of the OWT in 3-dimensional objecting the analysis of modal-coupling. It is in progress and will be present at the conference.

References

1. The birth of wind power in Brazil. http://abeeolica.org.br/energia-eolica-o-setor/lncs. Accessed 15 Jan 2021

2. Awada, A.; Younes, R.; Ilinca, A. S.: Review of vibration control methods for wind turbines. Energies **14**(11), 1–35 (2021). https://doi.org/10.3390/en14113058
3. Lee, U.: Specral Element Method in Structural Dynamics. Wiley, Location (2009)
4. Colherinhas, G.B., Girão de Morais, M.V., Machado, M.R.: Spectral model of offshore wind turbines and vibration control by pendulum tuned mass dampers. Int. J. Struct. Stab. Dyn. **22**(5), 22500535 (2022). https://doi.org/10.1142/S0219455422500535
5. Jonkman, J., Butterfield, S., Musial, W., Scott, G.T.: Definition of a 5-MW Reference Wind Turbine for Offshore System Development. National Renewable Energy Laboratory NREL, Location (2009)
6. Doyle, J.F.: Wave Propagation in Structures. Springer, Cham (1997). https://doi.org/10.1007/978-3-030-59679-8
7. Machado, M.R., Adhikari, S., Dos Santos, J.M.C.: Spectral element-based method for a one-dimensional damaged structure with distributed random properties. J. Brazil. Soc. Mech. Sci. Eng. **40**, 415 (2018). https://doi.org/10.1007/s40430-018-1330-2
8. Machado, M.R., Dutkiewicz, M., Matt, C.F.T., Castello, D.A.: Spectral model and experimental validation of hysteretic and aerodynamic damping in dynamic analysis of overhead transmission conductor. Mech. Syst. Signal Process. **136**, 106483 (2020). https://doi.org/10.1016/j.ymssp.2019.106483
9. Dutkiewicz, M., Machado, M.R.: Spectral approach in vibrations of overhead transmission lines. IOP Conf. Ser. Mater. Sci. Eng. **471**(5), 052029 (2019). https://doi.org/10.1088/1757-899x/471/5/052029
10. Dutkiewicz, M., Machado, M.R.: Dynamic response of overhead transmission line in turbulent wind flow with application of the spectral element method. IOP Conf. Ser. Mater. Sci. Eng. **471**(5), 052031 (2019). https://doi.org/10.1088/1757-899x/471/5/052031
11. Song, Y., Kim, T., Lee, U.: Vibration of a beam subjected to a moving force: frequency- domain spectral element modeling and analysis. Int. J. Mech. Sci. **113**, 162–174 (2016). https://doi.org/10.1016/j.ijmecsci.2016.04.020
12. Gopalakrishnan, S.: Wave Propagation in Materials and Structures. CRC Press (2016)
13. Machado, M.R., Appert, A., Khalij, L.: Spectral formulated modelling of an electrodynamic shaker. Mech. Res. Commun. **97**, 70–78 (2019). https://doi.org/10.1016/j.mechrescom.2019.04.014
14. Kim, T., Lee, U.: Modified one-element method for exact dynamic responses of a beam by using the frequency domain spectral element method. Int. J. Mech. Sci. **119**, 333–342 (2016). https://doi.org/10.1016/j.ijmecsci.2016.10.029
15. Kim, T., Lee, B., Lee, U.: State-vector equation method for the frequency domain spectral element modeling of non-uniform one-dimensional structures. Int. J. Mech. Sci. **157–158**, 75–86 (2019). https://doi.org/10.1016/j.ijmecsci.2019.04.030

New Model to Characterize the Cyclostationarity of Walking and Running Biomechanical Signals

Mourad Lamraoui[1(✉)], Firas Zakaria[2,3], Mohamed El Badaoui[1], and Mohamad Khalil[4]

[1] University of Lyon, Jean Monnet Saint Etienne University, IUT de Roanne, LASPI, Roanne, France
mourad.lamraoui@univ-st-etienne.fr
[2] School of Engineering, Lebanese International University LIU, Bekaa, Lebanon
[3] School of Engineering, The International University of Beirut BIU, Beirut, Lebanon
[4] EDST, AZM Center for Research in Biotechnology, Lebanese University, Tripoli, Lebanon

Abstract. Sports and physical activities cause different effects within the body systems. They may affect the equilibrium of the internal environment. The relationship between physical exercise and muscle fatigue has particularly attracted attention, by many researchers for more than a century. Such a relationship is very complex and still needs further research.

Human locomotion analysis is used extensively by physicians for pre-treatment analysis. In general, human locomotion can be defined as sequences of cyclic and repeated gestures. Previous analyses have proved the cyclostationarity behavior of vertical ground reaction forces (VGRF) recorded during walking and running. In this article, the Random Slope Modulation (RSM) is introduced as a new model for characterizing cyclostationarity. This model presents the impact of random slope variation on the cyclic spectrum of the signal and, hence, it lends itself the ability to extract information about the cyclostationary properties and structure of the signals in question. The model is applied in the study of biomechanical signals where it considers the slope as a specific measure extracted from the VGRF. We show that the calculated slope is random and different for every passive peak which visualizes the random character of VGRF signals. This randomness introduces a cyclostationarity of order 2. Hence, the obtained signal presents a random phenomenon through a slope that varies randomly, but is repeated periodically. The results show that the slope and polynomial random coefficients of the passive component of the VGRF can provide interesting information concerning biomechanics running and concerning fatigue associated with long-distance running.

Keywords: Vertical ground reaction force (VGRF) · Passive force component · Cyclostationarity (CS) · Random slope modulation (RSM) · Polynomial with random coefficients

Z. Dimitrovová et al. (Eds.): WMVC 2022, MMS 125, pp. 267–276, 2023.
https://doi.org/10.1007/978-3-031-15758-5_26

1 Introduction

Spectral analysis is an efficient and adequate description of a signal. Its basic purpose is to decompose the function into spectral components, i.e., into sum of weighted sinusoidal functions. If the statistical parameters of a signal such as the mean and autocovariance functions fluctuate periodically with time, then the process is said to be cyclostationary (CS). There are numerous models used in telecommunication and in industrial and control systems that result in CS processes [1]. Module signals including the amplitude modulated signals (AM), FM, PM and PAM signals, etc., have particularly maintained attention because they have interesting properties consisting of a specific spectral correlation density function [1].

Every CS process can be represented as a coupling of a periodically modulated part together with a random stationary part. The periods of CS in modulated signals may correspond to carrier frequencies, repetition or pulse rates, time division multiplexing rates... Some interesting examples of the types of modulation that produce CS waveforms, i.e., waveforms with mixtures of periodicity and randomness are: the periodically modulated stationary noise (noise with periodically varying characteristics); pulses with random amplitudes on a periodic schedule; and pulses on a periodic schedule with randomly jittered timing [2]. These examples have received great attention in literature.

In this paper, a new modulation type producing cyclostationarity is presented. This model may be used to characterize signals with random slopes and is referred to as "the random slope modulation (RSM)". With this model, it is possible to extract information about the CS properties and structure of the signals. For many real CS signals, we hypothesize that the origin of cyclostationarity might come from the random variation of the slope.

The CS properties of the proposed model (RSM) can be suitably exploited to analyze biomechanical signals such as signals recorded during human walking and running, since these signals were proved to be CS. The CS nature of biomechanical signals was previously treated and characterized in references [4–8]. On the other hand, we prove that the passive component is the part that contains information; we propose new original indicators, i.e. the random variation of the slope and of the polynomials coefficients of passive peaks.

The slope has important implications in the signal processing domain. Here, we consider it as a specific measure extracted from the VGRF signals. We were interested to measure its value and to monitor its evolution with time. For such signals, we can explain the origin of CS by the random variation of the slope. The latter may play an important role and provide interesting information concerning fatigue and concerning running and walking performance. The random variation of polynomials coefficients of the passive peaks may also provide interesting and important information.

The paper is organized as follows: Sect. 2 presents a study of RSM vs. cyclostationarity. Section 3 presents an introduction to VGRF signal and its components; also proves the randomness of passive peaks. Section 4 details and describes "experimentally" the proposed methodology. Section 5 presents the results and discussions. Finally, Sect. 6 provides the conclusion and perspectives.

2 The Random Slope Modulation (RSM Model)

What we need to accomplish in this section is to provide a new model useful for characterizing cyclostationarity. Such a model may be quite valuable for studying numerous mechanical and biomechanical signals. We consider a new example of modulation type. We will study the impact of random slope variation on the cyclic spectrum of a CS signal.

Modulating a signal simply means to vary one parameter of the signal. In this section, we vary each time the first slope of a trapezoid. Moreover, the signal model is a trapezoid repeated periodically, but the slope of its first lateral side is generated randomly. So, we obtain a CS signal composed of a coupling of periodic and random phenomena. The slope is random and different for every peak. This randomness introduces a cyclostationarity of order 2. In this section, experimentations are performed using various values of slope variance.

Therefore, we consider a trapezoid signal having a slope that is modulated randomly (see Fig. 1). We change the first slope of the trapezoid randomly to simulate the walking and running signal. Such signal could be expressed by the following signal model:

$$S''(t) = \sum\nolimits_k p_k\{\delta(t - kT) - \delta(t - kT - \tau_k)\} = \frac{d^2X(t)}{dt^2} \tag{1}$$

Here, we used the second derivative of the proposed signal so the signal can be more readily modeled and analyzed. It also exhibits how much change occurs in the first slope. $\{\tau_i\}$ is set to be an independent and identically distributed (i.i.d.) random stationary sequence consisting of random variables produced every T s. we interpret these variables as the time samples of a random waveform. τ_i are normally distributed ~ $N(\mu_i, \sigma_i)$, and are confined within the interval $0 \le \tau_i \le \frac{T}{2}$. The slopes are given by:

$$p_1 = \frac{A}{\tau_1}; p_2 = \frac{A}{\tau_2}; p_3 = \frac{A}{\tau_3}, \ldots. \quad p_i = \frac{A}{\tau_i} \quad \infty \le p_i \le \frac{2A}{T} \tag{2}$$

$\delta(t)$ is the Kronecker delta. It's an indicator vector (for each t) having one element equal to unity and the rest equal to zero, and $k = 0, \pm 1, \pm 2, \ldots.$

It can be noticed that if p_k is stationary in the strict sense, then $S''(t)$ is generally CS. It is a random slope CS signal with the fundamental cycle 1/T.

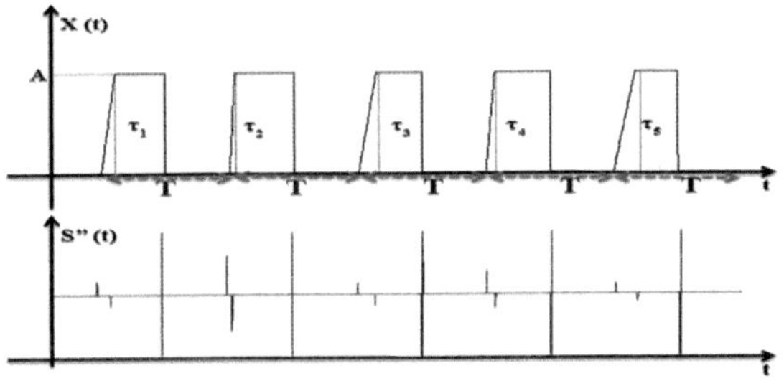

Fig. 1. X(t) is a trapezoid signal with random slope modulation. S″(t) is the second derivative (T: Constant cyclic period).

Figure 2 presents the slope variance of 2 simulated data such that one has low slope variation and another has big slope variation. For each case, we want to analyze the cyclic spectrum of the second derivative of the original trapezoids.

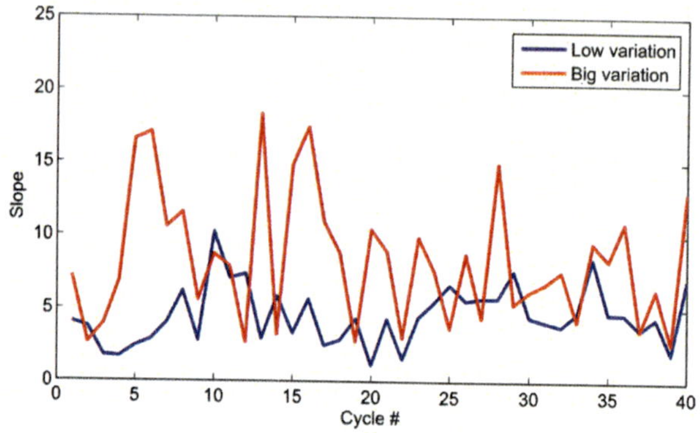

Fig. 2. Slope variance of 2 simulated data: one with low slope variation and another with high slope variation

Knowledge of the cyclic spectrum or spectral correlation density function for specific modulation types is important in performing evaluation and detection of different signal processing systems and in characterizing random processes that are CS. Figure 3 and 4 represent the magnitude plot of the cyclic autocorrelation functions, as a function of α and τ, calculated for different slope variances. The results show that the cyclic correlation increases for higher slope variance, also the fundamental frequency and its harmonics are higher. So, the slope could be used as a new indicator of CS.

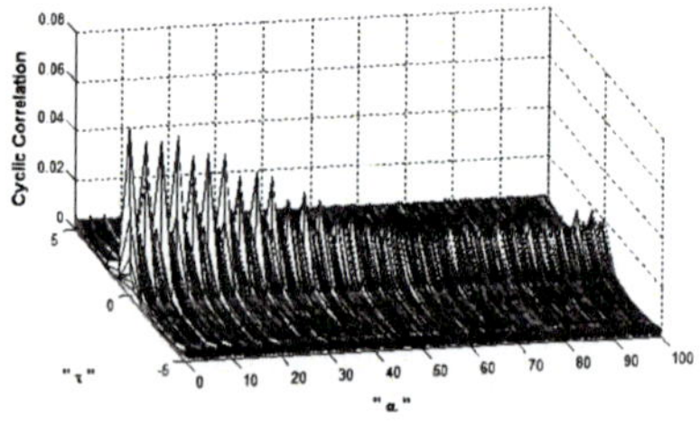

Fig. 3. Cyclic autocorrelation function for a small variance.

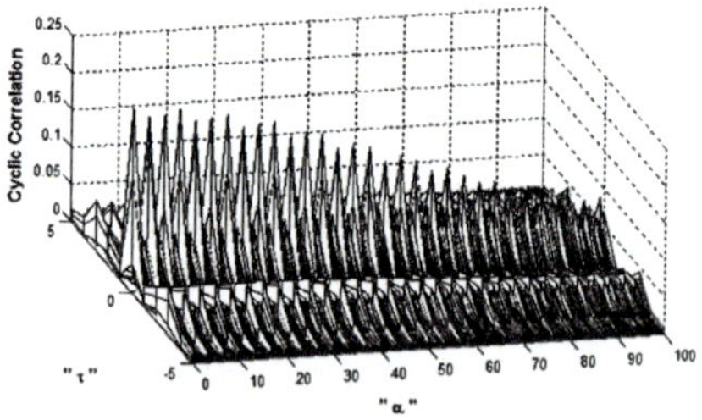

Fig. 4. Cyclic autocorrelation function for a big variance.

3 Introduction to VGRF Signal and Its Components

GRF signals are usually composed of two distinct peaks (see Fig. 5): an active force peak representing the propulsive force in addition to a passive force peak representing the impact force. The impact force is mainly passively generated and is due to the deceleration of the body mass at the instant of touch down. This implies that at least part of the energy was transferred passively into the bone-ligament system. Thus, the impact force is a major factor indicating the reaction of muscle that may reflects the fatigue state and performance of the muscle.

The simplistic model of biomechanical signals is in general characterized by the presence of a periodic component (CS1), a random component (CS2) in addition to i.i.d additive noise component. The CS aspect of a VGRF signal can be modeled mathematically as:

$$\mathrm{VGRF(t)} = \sum_n a_n \cos(2\pi n f_0 t + \varphi_n(t)) + b(t) \tag{3}$$

This equation presents the nature of cyclic variations of a VGRF signal [13]. Where $\varphi(t)$ represents the randomness, which varies cyclically. $\varphi(t)$ is very small ($\ll 1$) and is random, Gaussian and has a mean equal to zero. $b(t)$ is the i.i.d. random noise. f_0 is the step frequency. a_n is the amplitude representing principally the runner's weight. $\varphi(t)$ is very small, and thus Eq. (9) can be simplified to the following:

$$\mathrm{VGRF(t)} = \sum_n a_n \cos(2\pi n f_0 t) - \sum_n a_n \varphi_n(t) \sin(2\pi n f_0 t) + b(t) \tag{4}$$

$\{a_n \cos(2\pi n f_0 t)\}$ represents the CS of order 1, and $\{a_n \varphi_n(t) \sin(2\pi n f_0 t)\}$ represents the CS part of order 2. To characterize the variability from stride to another, we have to go through the CS 2, which is included as part of passive component.

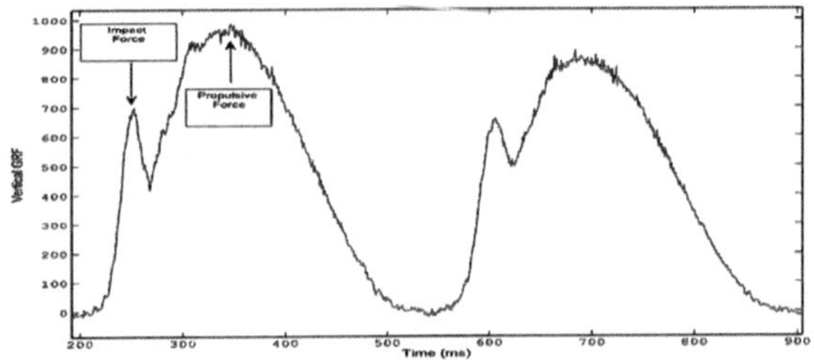

Fig. 5. VGRF active (propulsive force) and passive (impact force) peaks.

3.1 Synchronous Mean and Synchronous Variance

The synchronous mean consists in resynchronizing the VGRF signal using the correlation function estimated between the stride cycles, then, synchronously averaging the cycles according to the cyclic frequency where the cycle is being taken to be equal to the stride period.

The synchronous variance is calculated using the envelope analysis of the signal. However it requires a good estimation of the synchronous mean that is needs full information about the cyclic frequency. As a result, to estimate the most accurate cyclic period and to remove the low speed fluctuations in order to make the cyclic period constant, signal synchronization is used. We can make use of two different synchronization procedures: the synchronization method with maximization of the intercorrelation function or signal synchronization by means of angular resampling. By calculating the synchronous mean and synchronous variance of the VGRF signal (see Fig. 6), we notice that the first-order CS corresponds to the synchronous mean i.e. it can identify the deterministic periodic contribution of each leg. In addition, the second-order CS corresponds to the synchronous variance that visualizes the random character of the VGRF signals.

Figure 6 clearly shows that the active and passive peaks could be represented by the synchronous mean and synchronous variance respectively. Thus, the passive component is proved to be the part which contains the important information in a GRF signal. The active component is more periodic.

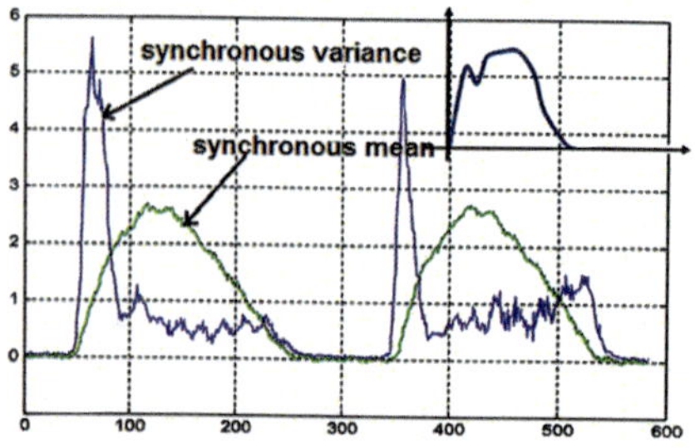

Fig. 6. Synchronous mean and synchronous variance of a VGRF signal.

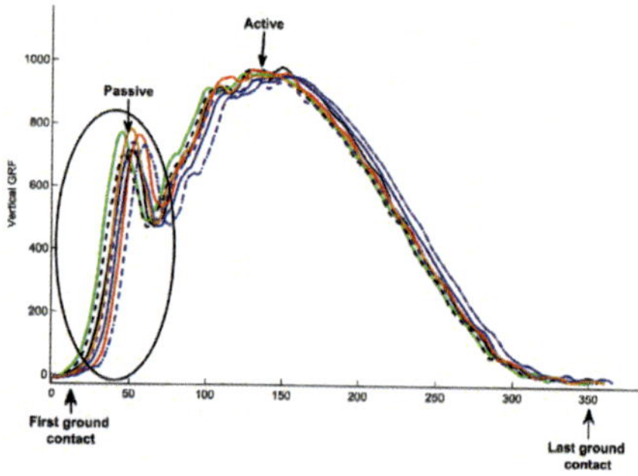

Fig. 7. Random variation of passive peaks for different gait cycles.

4 Methodology

The slope has important implications in the signal processing domain. A polynomial with random coefficients may also give important implications in signal processing. In this paper, we consider the slope and random coefficients of the passive polynomial (impact force) as specific measures extracted from the vertical ground reaction forces. As illustrated in Fig. 7, such coefficients are random and different for every gait cycle. This randomness introduces a cyclostationarity of order 2.

The passive peaks carry information on the random part in the signal while the active peaks are more impulsive. We now propose a method based on the analysis of the random variation of slope and of polynomial coefficients which characterize the passive peaks. Our main objective is to examine the biomechanical changes during a very long running exercise. Also, to study how such parameters have changed over time.

We analyzed the data from the original series of the study by the exercise physiology laboratory (LPE) of Jean Monnet St-Etienne University. This study conducted a recording experiment providing data suited to the analysis of running performance and fatigue [10]. During such an experiment, the VGRF signals were recorded at a sampling rate of 1000 Hz with electrodes (accelerometer sensors) placed at the four corners of a treadmill ergometer. This allowed accurate measurements of VGRF. The database was described in references [10] and [11] and a primary analysis was done in reference [9].

The database is composed of 120 GRF signals recorded from 10 experienced ultra-runners during an extreme ultra-long duration of running. The subjects were asked to run 24 h continuously with a short rest period every 2 h. A measurement period of 20 s was done every 2 h.

Figure 8 represents the proposed methodology used to calculate the coefficients of the polynomial representing the passive peaks of the signal. At first, the signal is normalized then denoised by means of the wavelet decomposition. Then, the global maximum and minimum are found using the derivative methods (first and second derivatives) while the false points are removed. The true minima are the accurate location of minimum edge points which can be detected using a comparison method. In the comparison method, simply, starting from the global minimum point, we compare the next series of points to the current point. The corresponding minimum position is precised when the series of points continue to increase without decreasing. The position with the minimum edge point is defined as the true minimum. The global minima and true minima are plotted as red and black points on Fig. 8. The minimum and maximum of the passive peak are plotted as green and blue points on Fig. 8.

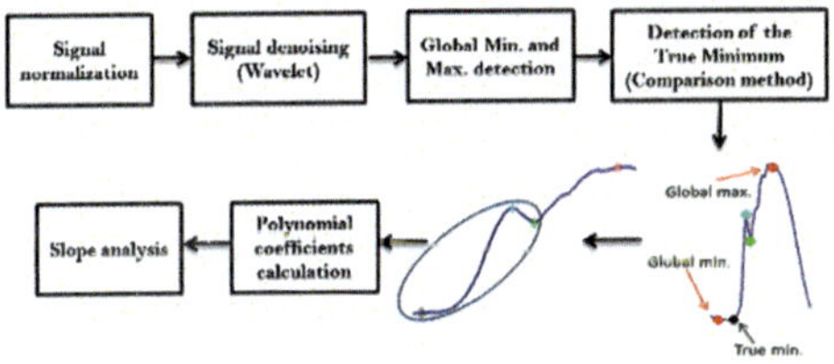

Fig. 8. System methodology.

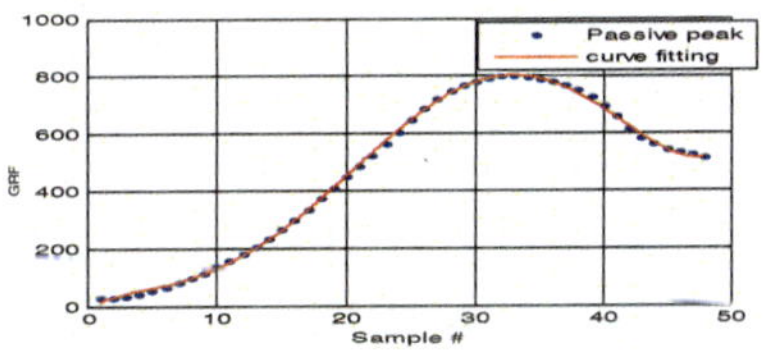

Fig. 9. Curve fitting (polynomial of degree 6).

Next, we try to find the curve that has the best fit to the passive peak and that passes through the maximum number of points (see Fig. 9). We found that the polynomial of degree 6 is the best. Then, for calculating the coefficients of polynomial function, we consider an algebraic polynomial given by:

$$P_n(t) = a_0 + a_1 t + a_1 t^2 + a_3 t^3 + \cdots + a_n t^n = \sum_{j=0}^{n} a_j t^j \qquad (5)$$

With random coefficients a_j, j = 0, 1, 2, ..., n

$\{a_j\}_{j=0}{}^{n}$ is a sequence of independent identically distributed random variables defined on R.

a_1 is the slope to be analyzed in the next section. This slope value measures the sensitivity or rate of change in VGRF as a result of a change in time. It is calculated as the rate of change of the impact force.

5 Results and Discussions

In this section, we hypothesized that the randomness in impact force peaks would increase with time running, i.e., with the degree of fatigue. We anticipated that the passive component changes with time running and is significantly different before and after fatigue. We analyzed the data described in the previous section. The slope was determined in each gait cycle, and then the variance of these slopes was calculated.

We aim to quantify the variation in GRF signals with time after 2 h and after 24 h of running. In Fig. 10, it appears that there exist some differences in slope variation after 2 h and after 24 h of running that is before and after fatigue.

In Fig. 11, we calculate the medians (of 10 subjects) of the slope variance after 2 h and after 24 h of running. Notice the significant changes observed between the two groups. Compared to level running, the slopes increase with time running and were dramatically larger after 24 h running (POST fatigue). These results were partly due to 10 subjects.

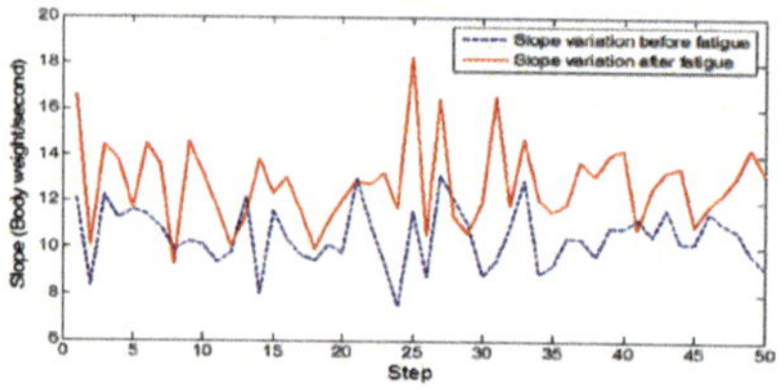

Fig. 10. Slope variation for one subject after 2 h (in blue) and after 24 h (in red) of running. It appears that there exist some differences in slope variation after 2 h and after 24 h of running that is before and after fatigue.

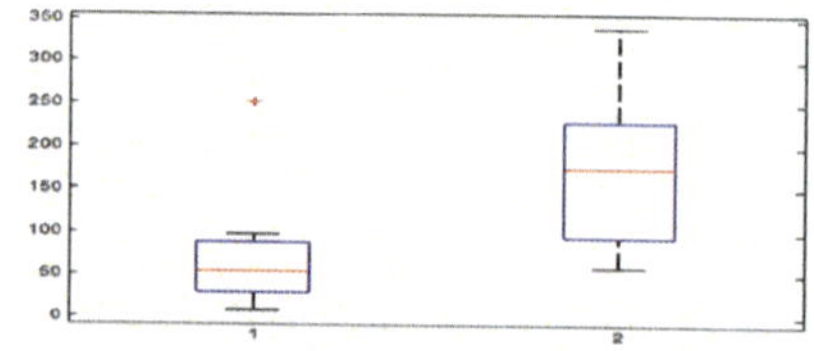

Fig. 11. The difference between the medians of the Slope variance after 2 h[1] and after 24 h[2] of running (The median of 10 subjects). Notice the significant changes observed between the two groups.

Polynomial with Random Coefficients

The polynomial with random coefficients may also shed light onto important implications in signal processing. The analysis of all random coefficients of the polynomial, is also an interesting field of study. Our polynomial of degree 6 can be written as follows:

$$P_n(t) = a + bt + ct^2 + dt^3 + et^4 + ft^5 \tag{6}$$

First of all, we calculated the polynomial coefficients (coefficients a, b, c, d, e, and f) between "true-min" and "Max. passive" and that for each cycle of the VGRF signal (Fig. 12). We then estimated the variance of each coefficient vector. We identified a slight increase in the variance of all coefficients after 24 h[2] of running i.e., in post fatigue.

Secondly, we calculate the polynomial coefficients of the polynomial presented in Fig. 13. This polynomial is composed of the points between "true-min" and "Min. passive".

The results provide significant differences between PRE and POST fatigue. We reported that the coefficients increased by 3 to 4 times after 24 h of continuous running ($P < 0.05$). In addition, we performed Welch's t-test to analyze the differences between PRE and POST fatigue. Significance was defined as $P < 0.05$. The tests revealed a significant result. Overall, our results, are better than the studies of the past [12, 13].

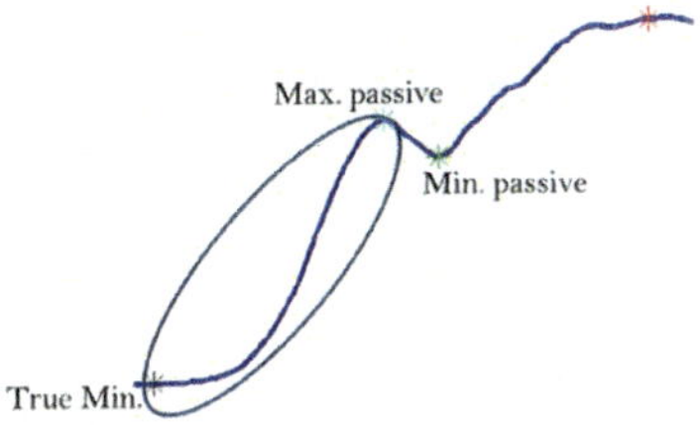

Fig. 12. Polynomial coefficients between "true-min" and maximum of the passive peak

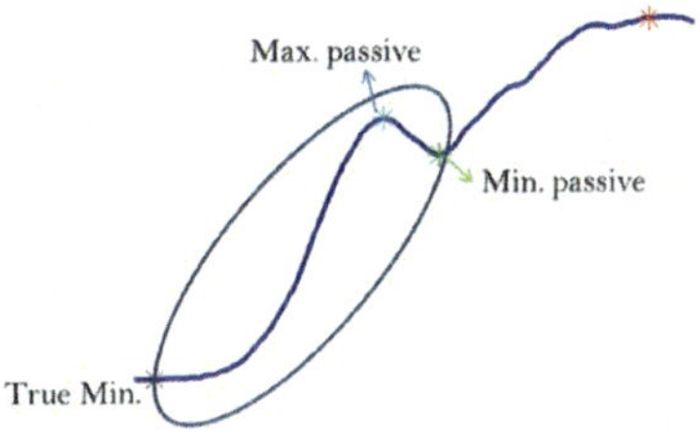

Fig. 13. Polynomial coefficients between "true-min" and the minimum of the passive peak

6 Conclusion

Human fatigue studies stretches back many decades and still a significant part of medical discourse. The purpose of this chapter was to examine the possible relationship between fatigue development in long-distance running, and the accompanying changes in the passive force peak (impact variables).

The article proved that the passive peaks of GRF signals contain the randomness, i.e. they contain useful information to be extracted using specific parameters. The random variation of the slope at the passive peak was proved to be the origin of CS of order 2 in the signal. This is considered a very important finding to the introduction of a new indicator of CS, which is very simple and easy to calculate i.e. the slope of passive peak (the slope of the polynomial between the true minimum and maximum of the passive peak). This puts forth evidence of a new CS model called "the random slope modulation".

In addition, a question arises whether from such an indicator, can the fatigue be estimated. Furthermore, signals recorded from 10 experienced runners over a 24 h period were treated in order to examine the changes during long distance running. The results showed that the proposed parameter seemed to vary with time during ultra-long running. These changes could be directly related to fatigue.

Results proved that such an indicator evolved significantly with time during running, so it seemed to be clearly affected by fatigue following fatiguing exercise. The random variation of the polynomial coefficients of passive peaks also significantly varied with time during running.

In future works, we are planning to reinforce our study by a mathematical model i.e., to find a robust mathematical model that proves the validity of a new CS model called: Random Slope Modulation. Moreover, the vertical component of the VGRF signal quickly rises and falls, forming the passive peak, then more slowly increases to a second peak at mid-stance, termed the active peak, before decreasing prior to toe-off. One can benefit from these changes and calculate the slope and polynomial coefficients for the different polynomials. The analysis of all random coefficients of each polynomial, is also an interesting field to study. These original model and indicators provide new insight, and open new perspectives and new possibilities for simple analysis of fatigue and falling of elderly.

Acknowledgment. The thesis was funded partially by the Agence Universitaire de la Francophonie (AUF). We appreciatively acknowledge their financial support. We also appreciate the financial aid given by the Rhône-Alpes International Cooperation and Mobilities (CMIRA). Also, we gratefully thank the CHU of Saint-Etienne for providing us with the database.

References

1. Gardner, W.A.: Cyclostationarity in communications and signal processing, Part I. IEEE Press (1993)
2. Gardner, W.A.: Spectral Correlation of modulated signals: Part I-analog modulation. IEEE Trans. Commun. **35**(6), 584–594 (1987)
3. Ginnakis, G.B.: Cyclostationary signal analysis. In: Digital Signal Processing Handbook. CRC Press LLC (1999)
4. Sabri, K., El Badaoui, M., Guillet, F., Adi, A.: Blind separation of ground reaction force signals. Appl. Math. Sci. **6**(53), 2605–2624 (2012)
5. Sabri, K., El Badaoui, M., Guillet, F., Belli, A., Millet, G., Morin, J.-B.: Cyclostationary modeling of ground reaction force signals. Signal Process. **90**(4), 1146–1152 (2010)
6. Zakaria, F.A., Toulouse, C.V., El Badaoui, M., Serviere, C., Khalil, M.: Contribution of the cyclic correlation in gait analysis: variation between fallers and non-fallers. In: IEEE 2014 16th International Conference on e-Health Networking, Applications and Services (HealthCom) (2014)
7. Maiz, S., El Badaoui, M., Leskow, J., Serviere, C.: Subsampling-based method for testing cyclostationarity: application to biomechanical signals. In: WOSSPA. IEEE (2013)
8. Zakaria, F.A., El-Badaoui, M., Khalil, M.: The random slope modulation: a new cyclostationarity model applied to biomechanical signals. In: 2015 International Conference on Advances in Biomedical Engineering (ICABME), pp. 297–300 (2015)
9. Lepers, P.N., Millet, G.Y.: Time course of neuromuscular alterations during a prolonged running exercise. Med. Sci. Sports Exerc. **36**, 1347–1356 (2004)
10. Millet, G.Y., Banfi, J.C., Kerherve, H., et al.: Physiological and biological factors associated with a 24 h treadmill ultra-marathon performance. Scand. J. Med. Sci. Sports **21**, 54–61 (2011)
11. Belli, A., Bui, P., Berger, A., Geyssant, A., Lacour, J.-R.: A treadmill ergometer for three-dimensional ground reaction forces measurement during walking. J. Biomech. **34**, 105–112 (2001)
12. Cavanagh, P.R., Lafortune, M.A.: Ground reaction forces in distance running. J. Biomech. **13**(5), 397–406 (1980)
13. Munro, C.F., Miller, D.I., Fuglevand, A.J.: Ground reaction forces in running: a reexamination. J. Biomech. **20**(2), 147–155 (1987)

Numerical and Experimental Study of Forced Undamped Vibrations of 2DOF Discrete Systems from Seismic Impact

Peter Pavlov(✉)

University of Architecture, Civil Engineering and Geodesy, 1 Hr. Smirnenski Blvd., 1046 Sofia, Bulgaria
matheus.quartaroli@unesp.br

Abstract. A combined - analytical, numerical and experimental study of forced undamped vibrations of 2DOF discrete system from seismic impact is presented in the paper. The study refers to a horizontal vibrating system, consisting of two translational moving bodies, connected by three springs. The seismic action is applied to the fixed point of the first of the three springs. With such an idealized scheme, the dynamic behavior of two-stories one-bay structures can be studied at an initial stage. The mathematical model of the vibrating system is presented in a matrix form. Numerical studies are conducted in two ways. Firstly, in the Simulink environment, a simulation model was composed, with a geometrically oriented approach in simulating the vibrating process. Then, in the MATLAB environment, an animation model was developed using the third animation method offered by the programming system. The experimental studies were conducted by Stand for study the vibrations of discrete planar systems. There were conducted numerical studies on vibrating systems with varied inertial and elastic characteristics and in varied seismic impact. For such systems an experimental verification of the results was also carried out. All models - the dynamic model and its corresponding mathematical, simulation, animation and experimental model are open to additional bodies to obtain discrete vibrating systems with a greater number of degrees of freedom. The models also allow a change of the point and nature of the impact.

Keywords: Discrete system · Force undamped vibrations · Mathematical model · Experimental model

1 Introduction

Forced vibrations of the material objects and systems are caused by various influences. The main ones are force, kinematic (geometric) and inertial. For example, a wind load on building structures is simulated by force influence, inertial impacts simulate the influence of the rotation of unbalanced masses. Kinematic effects on vibrating material objects simulate seismic loading on structures.

Force and inertial influences can be described by harmonic or non-harmonic mathematical expressions. This allows for some simpler tasks to conduct a precise analytical

Z. Dimitrovová et al. (Eds.): WMVC 2022, MMS 125, pp. 277–284, 2023.
https://doi.org/10.1007/978-3-031-15758-5_27

solution of the inverse problem of the vibrating processes. In the case of kinematic influences simulating seismic loading, an exact mathematical description of the influence is practically impossible, and therefore a precise solution of the inverse problem is also impossible even in simple models.

In the right part of the differential equations of the vibrations of material objects, caused by seismic impacts, real or synthetic vibrograms are applied, which have a random character. The solution of such a problems is performed mainly in a numerical or experimental setting.

The complete dynamic study of a certain class of vibrations is associated with the conduct of interrelated analytical, numerical and experimental studies. When considering forced vibrations, caused by seismic impacts, analytical research is reduced to compiling a dynamic and mathematical model of the vibrating system. Numerical studies are related to the development of appropriate models for solving composite systems of differential equations, describing the considered vibrations. Experimental studies verify the results, obtained by numerical models.

The described sequence, including partial analytical study, numerical solution and visualization of the results in animation mode and experimental verification of the solution, was performed for a discrete system with two degrees of freedom.

2 Dynamic Study of Forced Undamped Vibrations of 2DOF Discrete Systems from Seismic Impact

2.1 Dynamic Model

The dynamic model, describing the forced undamped vibrations of 2DOF discrete systems from seismic impact, is shown on Fig. 1. Two translationally moving bodies, connected by three springs, vibrate around the position of a stabile equilibrium, corresponding to unstressed springs. The kinematic action is applied by non-deterministic displacement of the fixed end of the first spring.

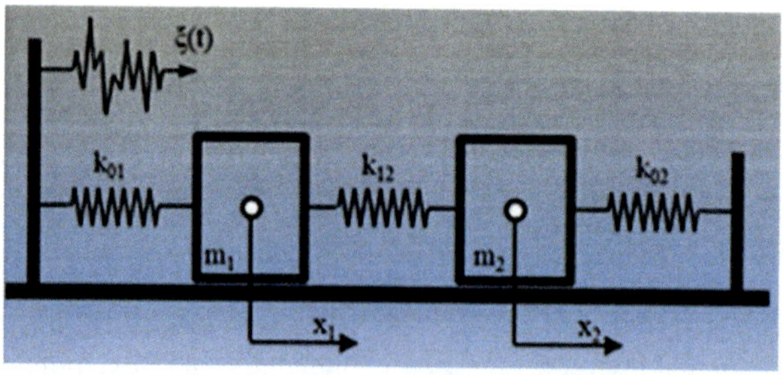

Fig. 1. Dynamic model of forced undamped vibrations of 2DOF discrete system from seismic impact.

2.2 Mathematical Model

The differential equations, describing the vibrations of the two masses, can be derived based on the Basic Equation of Dynamics or the second order Lagrange equations. The matrix form of the equations will be

$$[M]_{2\times 2} * \left\{x''\right\}_{2\times 1} + [C]_{2\times 2} * \{x\}_{2\times 1} = \left[Q\right]_{2\times 1} \quad (1)$$

The mass and elastic matrices of "Eq. (1)" will have the same form as in free vibrations

$$[M]_{2\times 2} = [m_1, 0; 0, m_2];\ [C]_{2\times 2} = [k_{01} + k_{12}, -k_{12}; -k_{12}, k_{12} + k_{02}]. \quad (2)$$

The kinematic action from seismic impact is included in the first row of the vector of generalized forces. The three vectors of "Eq. (1)" will have the form

$$\left\{x''\right\}_{2x1} = \left[x_1''; x_2''\right];\ \{x\}_{2x1} = [x_1; x_2];\ \{Q\}_{2\times 1} = [k_{01} * \xi; 0]. \quad (3)$$

2.3 Numerical Solution

The modern computational mathematics software packages offer in practice unlimited possibilities for solving systems of differential equations of the class, described in the previous subsection.

It is accepted in the article to compile a simulation model in a graphical environment for solving the system. The MATLAB software system and block diagram environment Simulink were used to perform the calculations.

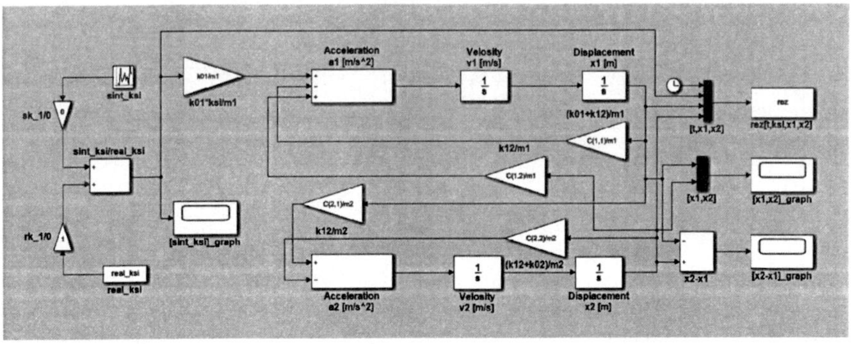

Fig. 2. Simulation model of forced undamped vibrations of 2DOF discrete system from seismic impact.

The simulation model presented in Fig. 2 is developed based on the modified "Eq. (1)", solved with respect to the accelerations of the two masses. In this way, the path of integration can be easily traced and an analysis of the influence of the various parameters on the nature of the motion can be made at each stage.

$$x_1'' = -(k_{01} + k_{12})/m_1 * x_1 + k_{12}/m_1 * x_2 + k_{01}/m_1 * \xi, \quad (4)$$

$$x_2'' = k_{12}/m_2 * x_1 - (k_{12} + k_{02})/m_2 * x_2. \quad (5)$$

The kinematic impact, modeling seismic loading, can be realized in two ways: by synthetic or by real vibrogram, corresponding to recorded seismic event. The vibrograms are scaled depending on the parameters of the studied model.

The real vibrogram, used in the numerical experiment is based on a real accelerogram - Charles F. Richter Seismological Laboratory, University of California acceleration data from October 17, 1989 Loma Prieta earthquake in the Santa Crz Mountains - Fig. 3.

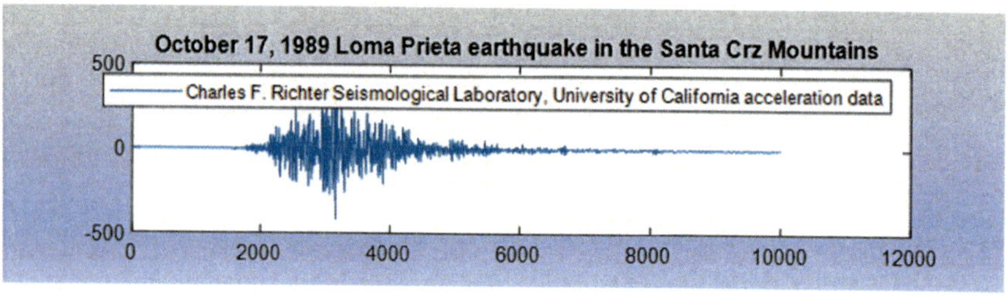

Fig. 3. Acceleration data from real earthquake event.

The scaled, in accordance with the parameters of the standard dynamic model, synthetic and real vibrogram are shown in Fig. 4a and Fig. 4b.

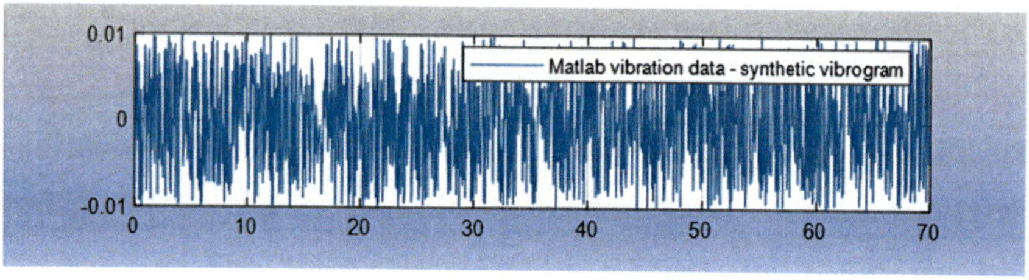

Fig. 4a. Scaled MATLAB vibration data – synthetic vibrogram.

With the help of the created simulation model, a numerical study of the forced undamped vibrations of the system was carried out under the influence of the synthetically generated vibrogram and under real vibrogram. The following values for the parameters of the vibration system are accepted: $m_1 = 0.5$ kg; $m_2 = 0.75$ kg; $k_{01} = 20$ N/m; $k_{12} = 30$ N/m; $k_{02} = 30$ N/m and zero initial conditions of motion of the both masses.

The results of the motion of the two masses are given in Fig. 5a and Fig. 5b.

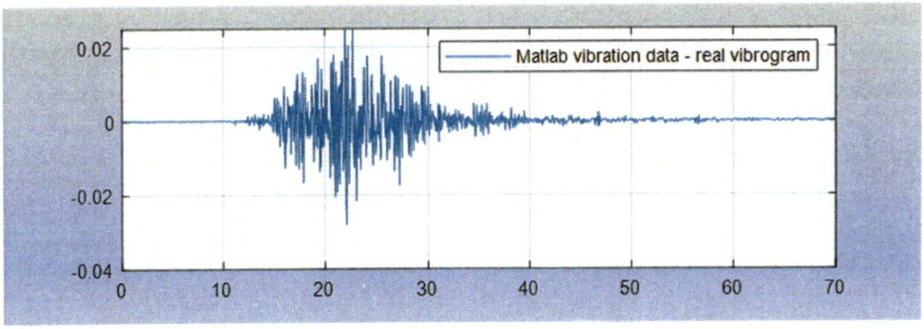

Fig. 4b. Scaled MATLAB vibration data – real vibrogram.

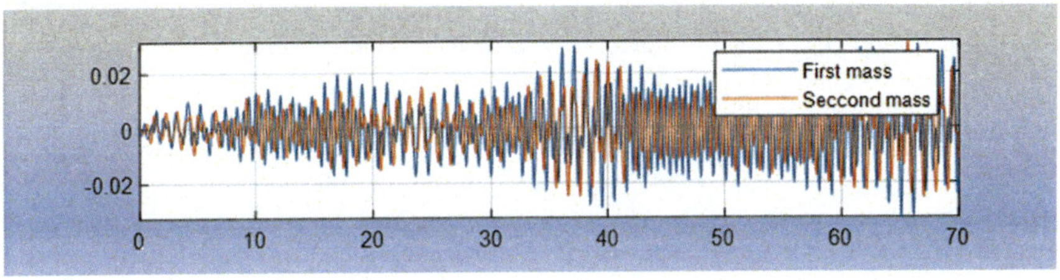

Fig. 5a. Results of the motion of the two masses – synthetic vibrogram.

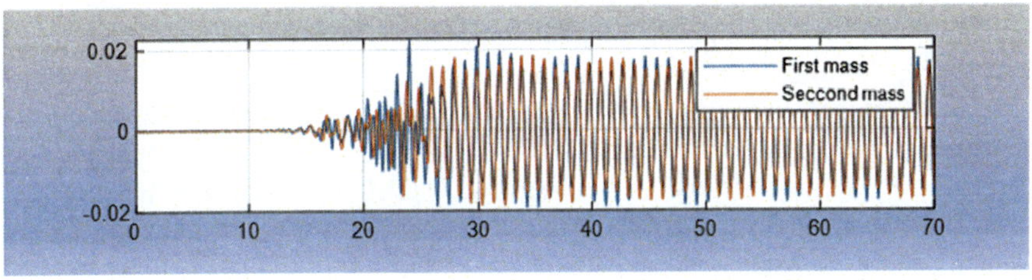

Fig. 5b. Results of the motion of the two masses – real vibrogram.

2.4 Animation of the Motion

More and more software packages offer options for animation of the motion of the material objects, which presents the latter in a very attractive way. In this paper was realized such a application, using the third method of animation [1], offered by the MATLAB program. The motion the left end of the first spring follow the generated synthetic or real vibrogram. The motion of both masses is based on the results obtained in Simulink.

The animation model shows the behavior of the vibrating system within 70 s, which is the duration of the solution in Simulink. In the process of animation changes the position of the movable end of the first spring, the two masses, as well as the deformation of the springs connecting the masses with fixed points.

If it is necessary to change the animation data, the simulation and related programs should be restarted to load new data into the system workspace.

The animation program also generates a video file. This allows the motion of the vibrating system to be shown by hardware that does not have installed MATLAB software.

The animation model of the vibrating system and a frame of the generated video file are shown in Fig. 6.

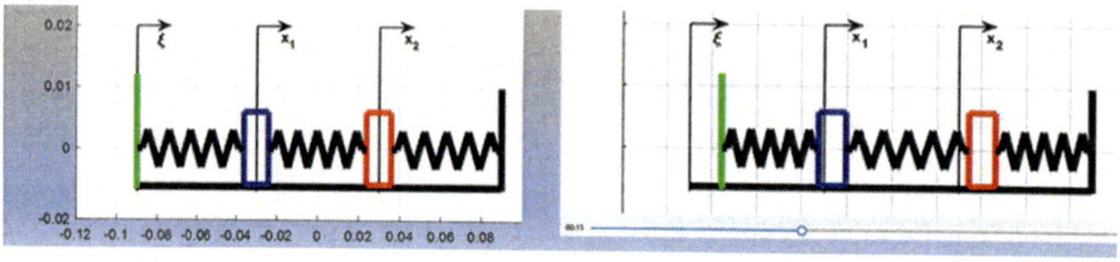

Fig. 6. Animation model of forced undamped vibrations of 2DOF discrete system from seismic impact and frame from the generated video file.

2.5 Experimental Study

To perform an experimental study, a simplified version of the designed stand for study of small vibrations of planar discrete systems is used. The rotating bodies and the dampers that dissipate vibration of the system have been removed from the standard stand configuration. The simplified configuration includes two translationally moving metal bodies and the springs between them and the fixed points.

The simulation of the motion of the fixed end of the first spring is realized by means of a stepper motor and a rotational-translational kinematic pair. The stepper motor is controlled by a program developed in the area of MATLAB, which converts the necessary translational movement of the fixed end of the first spring into rotational movement of the motor.

The displacements of the two bodies are measured with a developed light system of lamps and photoresistors, which through a transformed Arduino controller direct the numerical data to a computer configuration.

The simplified experimental configuration is shown in Fig. 7.

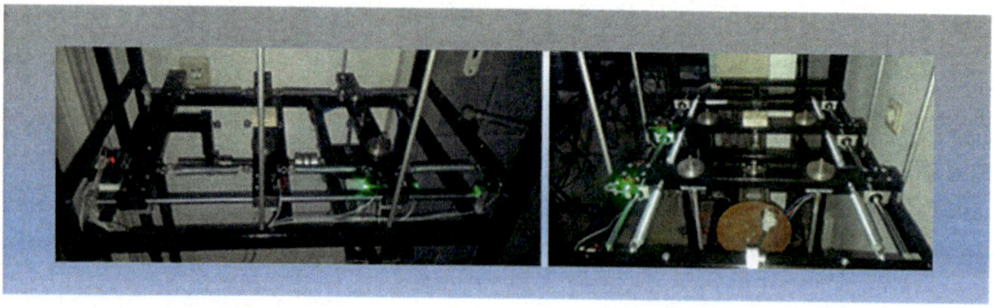

Fig. 7. Simplified experimental configuration of forced undamped vibrations of 2DOF discrete system from seismic impact.

The results of the experimental studies were obtained both under the influence of synthetically modeled vibrogram and under the influence of real vibrogram. The results are shown in Fig. 8a and 8b.

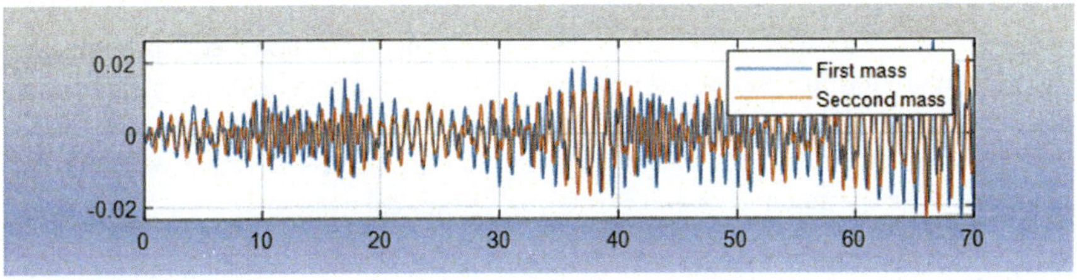

Fig. 8a. Results of the motion of the two masses in experimental mode – synthetic vibrogram.

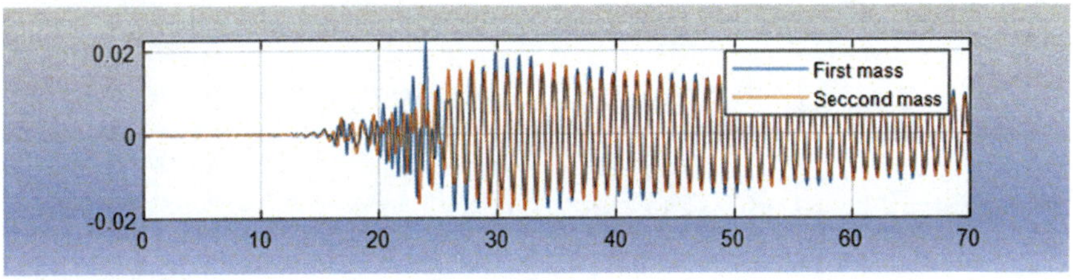

Fig. 8b. Results of the motion of the two masses in experimental mode – real vibrogram.

There is a certain decrease in the amplitudes of the vibration in the experimental studies. The reason is the presence of even small forces of friction between the bodies and the guide axles.

3 Transformation of Vibrating Two-Stories One-Bay Structures into a Discrete System with Two Degrees of Freedom

The bending moment diagrams from single displacements of vibrating two-stories one-bay structures with the necessary parameters for transformation into a discrete vibrating system with two dynamics degrees of freedom are shown in Fig. 9.

With some approximation, for masses of the equivalent vibrating discrete system with two degrees of freedom (see Fig. 1), the masses of the floor slabs and half of the masses of the columns below and above them can be assumed. The total horizontal reactions in floor 1 and floor 2 should be taken as the coefficients of the springs of the equivalent system.

The formulas for determining the stiffness of the springs of the equivalent model are

$$k_{01} = 12 * EI_1/h_1^3;\ k_{12} = 12 * EI_2/h_2^3;\ k_{02} = 0; \quad (6)$$

Formulas in "Eq. (6)" involve the stiffnesses of the two storey levels and the respective heights.

The equations of motion of the floor masses, with a given seismic impact can be obtained on the basis of the methods of structural dynamics. The motion of the equivalent masses of the dynamic model (see Fig. 1) with the same seismic impact can be obtained numerically through the developed simulation model and the related programs in the area of MATLAB.

In a numerical solution with parameters of the equivalent model equal to the parameters of the real vibrating system, results of close order are obtained. In an experimental study, where the parameters take into account the possibilities of the experimental setting, the factor of conformity can be found, which converts the data on the seismic behavior of the experimental model into those of the real system.

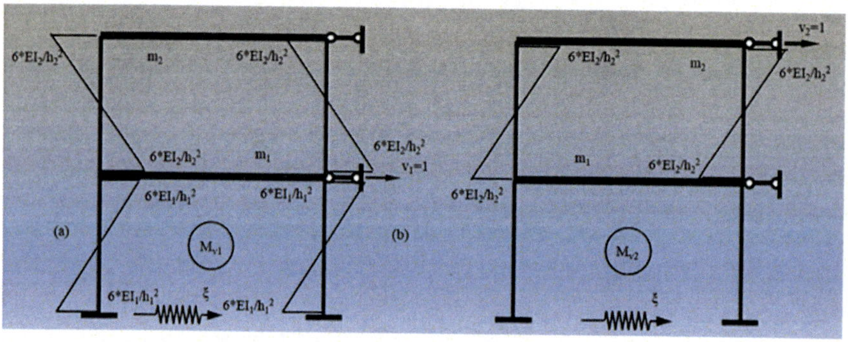

Fig. 9. Bending moment diagrams from single displacements of vibrating two-stories one-bay structures.

4 Conclusion

The models, that were created (dynamic, mathematical, simulation, animation, experimental), are part of the full dynamic study of the forced vibration of the discrete system. The study of the forced undamped vibrations is a necessary step which is carried out before the study of forced damped vibrations under the same effects. The latter type of vibrations is closest to the actual dynamic behavior of a vibrating structure that can be discredited to a system with two dynamic degrees of freedom.

References

1. Tyagi, A.: MATLAB and Simulink for Engineers. Oxford University Press, Oxford (2012)
2. Palm, W.: Introduction to MATLAB for Engineers. McGray Hill Education, New York (2011)
3. Pavlov, P.: Nonlinear damped vibrations of planar discrete systems – numerical and experimental modeling. MATEC Web Conf. **211**, 02006 (2018)
4. Pavlov, P., Evlogiev, D.: Factor of conformity in modular dynamic study of frame type constructions. In: IJTEE – ICEDM 2016 Fullpapers E-Book, pp. 64–69
5. Pavlov, P., Lilkova-Markova, S., Ivanova, G.: Modular approach in creating the matrix equations, describing the free vibrations of discrete plane systems. In: IJTEE – ICEDM 2016 Fullpapers E-Book, pp. 70–75

Optimal LQR Control for Longitudinal Vibrations of an Elastic Rod Actuated by Distributed and Boundary Inputs

Alexander Gavrikov[1,2] and Georgy Kostin[2(✉)]

[1] Pennsylvania State University, University Park, PA 16802, USA
Alexander.Gavrikov@psu.edu

[2] Ishlinsky Institute for Problems in Mechanics RAS, 119526 Moscow, Russia
kostin@ipmnet.ru

Abstract. We consider a vibrating system consisting of a thin rectilinear elastic rod actuated by external loads applied at the ends as well as by a normal force, which is distributed piecewise constantly in space. Such a force may be implemented by piezoelectric actuators. The intervals of constancy of this normal force are equal in length, and the force value on each of these sections is considered as an independent control input. We study the longitudinal motions of the rod and the means of control optimization. Based on the eigenmode decomposition, it is shown in the case of uniform rod that the original continuous system is split into several infinite vibrating subsystems each of which is controlled by a certain linearly independent combination of control inputs. It follows that if any of these combinations is taken equal to zero, then the corresponding subsystem becomes uncontrollable. Next, an optimal control problem on a finite time horizon is considered, where the terminal mechanical energy of the rod and energy losses in the control circuit are minimized with some weighting coefficients. We show that for a fixed number of actuators distributed along the rod, approximation of the problem is reduced to the design of linear-quadratic regulators. An example of a uniform rod is presented where finite expressions for the optimal control functions are obtained. Amplitudes of controlled and affected but not minimized modes are derived for approximated suboptimal control.

Keywords: Optimal control · Longitudinal vibrations · LQR · Distributed control

1 Introduction

The classical problem of active suppression of vibrations has many practical applications and has been studied for years, mainly as vibration control of discrete structures. Nowadays, continuous systems, e.g. elastic, are a substantial

The study has been done under financial support of the Russian Science Foundation (grant 21-11-00151)

Z. Dimitrovová et al. (Eds.): WMVC 2022, MMS 125, pp. 285–295, 2023.
https://doi.org/10.1007/978-3-031-15758-5_28

field of ongoing research. However, practical implementation of any control strategy implies reduction of either the vibrating system or its control inputs to finite-dimensional objects [1]. In this paper, we assume from the beginning that the input is finite-dimensional in space. Such an assumption allows for splitting the original continuous system into a finite number of subsystems each actuated by its own combination of the original inputs. We consider an elastic rod for which inputs, piecewise constant in space, may be implemented by means of piezo elements [2,3] attached to the rod's sides. The piezoelectric actuators are used in a wide range of engineering applications from precision positioning on nano-scale to motion control and vibration suppression on large scale in the aerospace field [4,5]. In our study, the subsystems, in which the original continuous system is split, are groups of vibrational modes. Since these groups are still infinite-dimensional, we most likely have to approximate them. Here, we use a simple approach by optimizing the motion of some lowest modes while higher modes are actuated by derived optimal inputs. We also estimate the influence of proposed strategy on all the modes by considering a suboptimal control law. This allows us to obtain exact expressions for amplitudes of both controlled and actuated modes.

In Sect. 2 we introduce the mechanical system and describe how it splits into subsystems (groups of modes) and state the optimal control problem. In Sect. 3, the original problem is reformulated in terms of these groups. We construct an LQ-optimal control strategy for the lowest mode in each group in Sect. 4 and propose an asymptotic approximation of the optimal feedback law. In Sect. 5 numerical results on the LQ-optimal and suboptimal strategies are presented.

2 Mechanical System

We consider a thin elastic rod that undergoes longitudinal vibrations. It is assumed that the rod is actuated by boundary forces $f^{\pm}(t)$ and a distributed force $f(t,x)$. The force f acts in normal direction to the cross-section and is distributed along the rod's length, so that it has N space intervals of constancy with equal length λ. Such a force may be implemented via a series of identical piezoelements placed symmetrically on the side surface of the rod. We suppose that the rod has a length of $2L$ and its center is at the point $x = 0$. Denoting the intervals of constancy of the distributed force f as $I_k := (x_{k-1}, x_{k+1})$, $k \in J_s$, we introduce auxiliary functions $f_k(t)$ (see also Fig. 1) such that

$$\begin{gathered} f_k(t) := f(t,x), \quad x \in I_k, \quad k \in J_s, \quad f_{\pm N \pm 1}(t) := f^{\pm}(t), \\ x_n = \tfrac{n\lambda}{2}, \quad \lambda = \tfrac{2L}{N}, \quad n \in J_x, \quad x_{\pm N} = \pm L, \\ J_s = \{1-N, 3-N, \dots, N-1\}, \quad J_x := \{-N, -N-2, \dots, N\}. \end{gathered} \tag{1}$$

In the dimensionless variables $x^* = x/L, t^* = t/\tau$ (the asterix is further omitted) the rod's motion is described by the following PDE system

$$\begin{gathered} \rho(x)\ddot{v}(t,x) = (\kappa(x)v'(t,x) + f(x,t))', \quad x \in (-1,1), \quad t \in (0,T), \\ \kappa(\pm 1)v'(t,\pm 1) = f^{\pm}(t), \quad v(0,x) = v^0(x), \quad \dot{v}(0,x) = \dot{v}^0(x). \end{gathered} \tag{2}$$

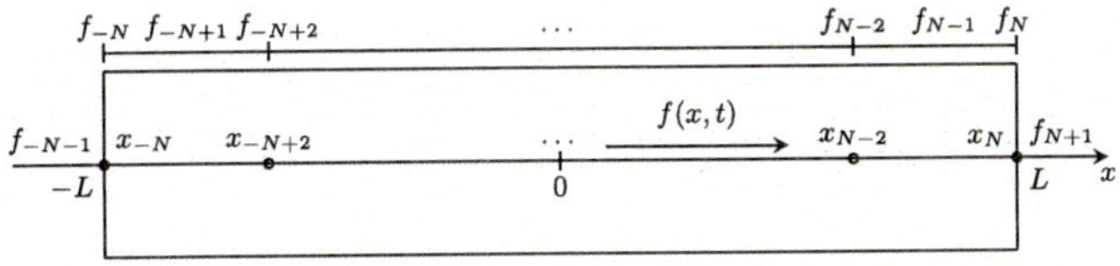

Fig. 1. Schematic representation of the elastic rod and applied forces.

Here, v is the longitudinal displacements of the cross-section, ρ and κ are dimensionless density and stiffness of the rod's material, τ is a characteristic time.

2.1 Control Problem

Let us formulate an optimal control problem of suppressing the rod's vibrations while minimizing the terminal internal mechanical energy E and energy losses F

$$E := \frac{1}{2}\int_{-1}^{1}(\rho\dot{v}^2 + \kappa(v')^2)|_{t=T}dx, \quad F := \frac{1}{T}\int_0^T\int_{-1}^{1} R(f_-, f_+, f)\,dxdt, \tag{3}$$

where R is a positive defined quadratic function w.r.t. forces $f^{\pm}, f$. That is, the aim is to minimize the cost function

$$\Phi := E + \gamma F \to \min_{f^{\pm}, f} \tag{4}$$

subject to constraints (2). In (4), $\gamma > 0$ is a weighting coefficient. We later assign different weights for forces $f_k(t)$ on each interval I_k in (3), but for simplicity of presentation we keep F in the form (3) in this section.

We decompose (2) by projecting this system in the weighted space $L_2(\rho(x), (-1, 1))$ onto eigenfunctions $\{w_n(x)\}$, where w_n, $n = 0, 1, \ldots,$ are found by solving the eigenproblems

$$(\kappa(x)w_n'(x))' = \eta_n\rho(x)w_n(x), \quad \kappa(x)w_n'(\pm 1) = 0, \quad n = 0, 1, \ldots. \tag{5}$$

As a result, the infinite-dimensional ODE system is obtained according to

$$\ddot{v}_n = -\eta_n v_n + \sum_{j\in J_x} w_n(x_j)f_j, \quad v_n(0) = v_n^0, \quad \dot{v}_n(0) = \dot{v}_n^0. \tag{6}$$

In (6), v_n, v_n^0 and $\dot{v}_n^0$ are components of projections of $v(t,x)$, $v^0(x)$ and $\dot{v}^0(x)$ onto $w_n(x)$, respectively. Additionally, we introduce in (6) the control functions

$$f_j(t) := f_{j+1}(t) - f_{j-1}(t), \quad j \in J_x. \tag{7}$$

That is, the rod is actuated only by the jumps of the forces f_k. Therefore, any forces $f(t,x), f^{\pm}(t)$ resulting in the same jumps yield the same motion. In what follows, we consider such forces equivalent. Moreover, the rod is uncontrollable if for at least one n we have

$$\sum_{j\in J_x} w_n(x_j)f_j(t) \equiv 0. \tag{8}$$

Therefore, the forces resulting in linear dependent jumps f_j s.t. (8) holds for some n do not actuate this specific mode. We exclude such forces from consideration.

2.2 Controlled Groups for a Uniform Rod

If the values of eigenfunctions are such that $w_i(x_k) = w_j(x_k)$ for some $i \neq j$ and all $k \in J_x$, the vibrational modes split in groups. It may happen if the rod consists of homogeneous pieces or, in the simplest case, both the density ρ and the rigidity κ are constant. Let $\rho = \kappa = 1$ in dimensionless units. Then

$$\begin{aligned} w_n(x) = \cos\left(\tfrac{\pi}{2}n(x+1)\right), \quad \eta_n = \tfrac{\pi^2 n^2}{4}, \quad n = 1, 2, \ldots, \\ w_0 = c_0 = \sqrt{2}/2, \quad \eta_0 = 0, \end{aligned} \tag{9}$$

and, for example, for $N = 2$ the system (6) takes the form

$$\ddot{v}_n = -\eta_n v_n + w_n(-1)f_{-2} + w_n(0)f_0 + w_n(1)f_2. \tag{10}$$

Since the functions $w_n(x_k) = \cos\left(\frac{\pi}{2}n\left(\frac{k}{N}+1\right)\right)$ take only a finite number of values for fixed k and N, the number of possible linear combinations $\sum_{j\in J_x} w_n(x_j)f_j$ is also finite. We introduce the effective control functions

$$u_n = \sum_{j\in J_x} w_n(x_j)f_j. \tag{11}$$

Then all vibrational modes split into $N+1$ independent groups such that each i-th group, $i = 0, \ldots, N$, with mode numbers $n = 2Nj \pm i \geq 0$, $j = 0, 1, \ldots$, is controlled by a specific u_i:

$$\ddot{v}_n = -\eta_n v_n + u_i. \tag{12}$$

For example, denoting for $N = 2$

$$\begin{aligned} u_0 = f_2 + f_0 + f_{-2} = f_3 - f_{-3}, \quad u_1 = -f_2 + f_{-2} = -f_3 + f_1 + f_{-1} - f_{-3}, \\ u_2 = f_2 - f_0 + f_{-2} = f_3 - 2f_1 + 2f_{-1} - f_{-3}, \end{aligned} \tag{13}$$

we obtain 3 groups, namely,

$$\begin{aligned} \ddot{v}_0 = c_0 u_0, \quad \ddot{v}_{4j+4} = -\eta_{4j+4}v_{4j+4} + u_0, \\ \ddot{v}_{2j+1} = -\eta_{2j+1}v_{2j+1} + u_1, \quad \ddot{v}_{4j+2} = -\eta_{4j+2}v_{4j+2} + u_2, \end{aligned} \tag{14}$$

where $j = 0, 1, \ldots$. Similarly, there are 4 groups for $N = 3$ with $n = 6j$, $6j \pm 1$, $6j \pm 2$, $6j \pm 3$ where $n > 0$. Likewise, the groups are formed for any N.

3 Transforming the Cost Function

Since the vibrational modes are split into independent groups, it is worth transforming the cost function Φ in (4) accordingly. By using the eigenfunction expansion of the displacements v, the internal mechanical energy E and the energy losses F in (3) take the form

$$E = \frac{1}{2}\sum_{n}\left(\dot{v}_n^2(T) + \eta_n v_n^2(T)\right), \quad F = \frac{1}{2T}\int_0^T \bar{f}^* R_f \bar{f} dt, \tag{15}$$

where the vector of control jumps $\bar{f} = (f_{1-N}, \ldots, f_{N-1}, f_{idle})^*$ is introduced, and $\bar{f}^*$ means the transpose of the vector $\bar{f}$; R_f is a positive weighting matrix. The component f_{idle} is equal to the sum of all original inputs f_k. Since the rod is effectively controlled by jumps of forces, any forces $f^{\pm}$, f having the same jumps yield the same control. Any "excessive" input does not change the state of the system. Minimizing $\Phi = E + \gamma F$, we imply that the function $f_{idle}(t)$ representing this "excessive" input is taken equal to zero: $f_{idle}(t) \equiv 0$.

Let us introduce the vector of effective control functions $\bar{u} = (u_0, \ldots, u_N)^*$. The energy losses F can be written as

$$F = \frac{1}{2T}\int_0^T \bar{f}^* R_f \bar{f} dt = \frac{1}{2T}\int_0^T \bar{u}^* R_u \bar{u} dt, \tag{16}$$

where the weighting matrix $R_u = C^* R_f C$ corresponds to linear transformation of $\bar{f}$ to $\bar{u} : \bar{f} = C\bar{u}$. For the following decomposition, we suppose further that R_u is equal to the identity matrix I since this can be achieved by assigning specific weights to each input f_j in (3).

As a result, the cost function Φ is split into $N+1$ independent terms

$$\begin{gathered}\Phi = \Phi_0 + \Phi_1 + \ldots + \Phi_N, \\ \Phi_0 = \frac{\dot{v}_0^2(T)}{2} + \frac{1}{2}\sum_{j=1}^{\infty}\left(\dot{v}_{2Nj}^2(T) + \eta_{2Nj} v_{2Nj}^2(T)\right) + \frac{\gamma}{2T}\int_0^T u_0^2(t)dt, \\ \Phi_k = \frac{1}{2}\sum_{j=0}^{\infty}\left(\dot{v}_{2Nj\pm k}^2(T) + \eta_{2Nj\pm k} v_{2Nj\pm k}^2(T)\right) + \frac{\gamma}{2T}\int_0^T u_k^2(t)dt,\end{gathered} \tag{17}$$

with $k \in J_1$, $J_1 := \{1, \ldots, N\}$. Therefore, the problem of vibration suppression (4) is transformed into $N+1$ independent control problems. Since each group is still infinite-dimensional, the direct solution of the corresponding problems is not straightforward, although an approximate solution is still possible [6]. In the next section we consider a finite-dimensional approximation of the optimal control problem.

4 Controlling a Finite-Dimensional Approximation

We consider the following approximation: let us minimize the cost function

$$\begin{gathered}\tilde{\Phi} := \tilde{\Phi}_0 + \tilde{\Phi}_1 + \ldots + \tilde{\Phi}_N \to \min_{u_k}, \quad \tilde{\Phi}_0 = \frac{\dot{v}_0^2(T)}{2} + \frac{\gamma}{2T}\int_0^T u_0^2(t)dt, \\ \tilde{\Phi}_k = \frac{1}{2}\left(\dot{v}_k^2(T) + \eta_k v_k^2(T)\right) + \frac{\gamma}{2T}\int_0^T u_k^2(t)dt, \quad k \in J_1.\end{gathered} \tag{18}$$

That is, we control only the lowest vibrational mode in each group, while the higher modes are actuated, but their motion is not optimized. For definiteness, we call the modes with $n \leq N$ the controlled modes, and the ones with $n > N$—the actuated modes. After denoting

$$\begin{gathered} y_1^{(0)}(t) = v_0(t), \quad y_2^{(0)}(t) = \dot{v}_0(t), \quad y_1^{(0)}(0) = v_0^0, \quad y_2^{(0)}(0) = \dot{v}_0^0, \\ y_1^{(k)}(t) = \mu_k v_k(t), \; y_2^{(k)}(t) = \dot{v}_k(t), \; y_1^{(k)}(0) = \mu_k v_k^0, \; y_2^{(k)}(0) = \dot{v}_k^0 \\ \mu_k = \sqrt{\eta_k}, \; y^{(0)} = (y_1^{(0)}, y_2^{(0)})^*, \; y^{(k)} = (y_1^{(k)}, y_2^{(k)})^*, \; k \in J_1, \end{gathered} \tag{19}$$

the cost functions $\tilde{\Phi}_k$ are rewritten as follows

$$\begin{gathered} \tilde{\Phi}_k = (y^{(k)}(T))^* Q^{(k)} y^{(k)}(T) + \int_0^T \bar{u}_k^*(t) R \bar{u}_k(t) dt, \quad \bar{u}_k = (0, u_k)^*, \\ R = \begin{pmatrix} 1 & 0 \\ 0 & \frac{\gamma}{2T} \end{pmatrix}, \quad Q^{(0)} = \frac{1}{2} \begin{pmatrix} 0 & 0 \\ 0 & 1 \end{pmatrix}, \quad Q^{(j)} = \frac{1}{2} I, \quad I = \begin{pmatrix} 1 & 0 \\ 0 & 1 \end{pmatrix}, \quad j \in J_1. \end{gathered} \tag{20}$$

Therefore, instead of problem (4) we consider $N+1$ minimization problems:

$$\tilde{\Phi}_k \to \min_{u_k(t), t \in [0,T]}, \quad k \in J_0, \quad J_0 := \{0, \ldots, N\}, \tag{21}$$

subject to the constraints

$$\begin{gathered} \dot{y}^{(k)} = A^{(k)} y^{(k)} + B \bar{u}_k, \quad k \in J_0, \quad A^{(0)} = \begin{pmatrix} 0 & 1 \\ 0 & 0 \end{pmatrix}, \quad B^{(0)} = \begin{pmatrix} 0 & 0 \\ 0 & c_0 \end{pmatrix}, \\ A^{(j)} = \begin{pmatrix} 0 & \mu_j \\ -\mu_j & 0 \end{pmatrix}, \quad B^{(j)} = \begin{pmatrix} 0 & 0 \\ 0 & 1 \end{pmatrix}, \quad j \in J_1, \end{gathered} \tag{22}$$

with initial conditions introduced in (19).

4.1 Feedback Optimal and Suboptimal Controls

Each subproblem (21), (22) can be solved in a standard way by means of the LQR theory [7]. The optimal feedback control functions are expressed as follows

$$\begin{gathered} u_0(t) = -\frac{T c_0 y_2^{(0)}(t)}{c_0^2 T (T-t)^2 + \gamma}, \\ u_k(t) = \frac{2\mu_k \alpha_T y_1^{(k)}(t) + (\mu_k T^2 \sin(2\mu_k (T-t)) - 2\mu_k^2 \beta_T T) y_2^{(k)}(t)}{\alpha_T + \mu_k^2 \beta_T^2}, \\ \alpha_T = T^2(\cos^2(\mu_k(T-t)) - 1), \quad \beta_T = (T(T-t) + 2\gamma). \end{gathered} \tag{23}$$

The behavior of the controlled modes is described by substituting (23) into (22) and integrating the result. For the zeroth mode, the solution is found explicitly

$$\begin{gathered} y_1^{(0)}(t) = y_1^{(0)}(0) + \left(t - \frac{T c_0^2 t^2}{2(T^2 c_0^2 + \gamma)}\right) y_2^{(0)}(0), \\ y_2^{(0)}(t) = \left(1 - \frac{T c_0^2 t}{T^2 c_0^2 + \gamma}\right) y_2^{(0)}(0). \end{gathered} \tag{24}$$

The other modes from the zeroth group with $k = 2Nj$, $j > 0$, are actuated by u_0 from (23) and expressed as

$$y_1^{(k)}(t) = \frac{T c_0 (\cos(\mu_k t) - 1)}{\mu_k (T^2 c_0^2 + \gamma)} y_2^{(0)}(0), \quad y_2^{(k)}(t) = -\frac{T c_0 \sin(\mu_k t)}{\mu_k (T^2 c_0^2 + \gamma)} y_2^{(0)}(0). \tag{25}$$

That is, their amplitudes decay as $\frac{1}{kT}$ for large k or T. It is worth mentioning that choosing T as a multiple of 4 in dimensionless units we obtain $y_1^{(k)}(T) = y_2^{(k)}(T) = 0$. Therefore, the proposed control completely suppresses excitation of higher modes with $n = 2Nj$ for such T at the terminal time instant.

Note that (23) holds for any $N > 0$. That is, this solution is applicable to any number N of piezoelements utilized. It is worth noting that it is possible to enhance approximation (18) by controlling several modes in each group. However, since we have to control several pendulums by means of one input in this case, we encounter a classical control problem when the number of inputs is much smaller than the number of controlled variables [8].

The behavior of actuated but not suppressed modes $y_{1,2}^n$, $n > N$, can be investigated numerically as done in the next section. However, under the assumption that N is large, and $k \gg 1$, we can estimate the modes amplitudes explicitly. Indeed, by introducing an approximate control

$$u_k^a(t) = -\frac{T}{\beta_T} y_2^k(t) \tag{26}$$

we obtain explicit expressions for amplitudes:

$$\begin{aligned} y_{1a}^{(k)}(t) &= \left(\tfrac{\beta_T y_2^k(0)}{T^2+2\gamma} + \tfrac{T(\mu_k(T^2+2\gamma)y_1^k(0)+Ty_2^k(0))}{\mu_k^2(T^2+2\gamma)^2} \right) \sin(\mu_k t) \\ &+ \left(\tfrac{\beta_T(\mu_k(T^2+2\gamma)y_1^k(0)+Ty_2^k(0)}{\mu_k(T^2+2\gamma)^2} - \tfrac{Ty_2^k(0)}{\mu_k(T^2+2\gamma)} \right) \cos(\mu_k t), \\ y_{2a}^{(k)}(t) &= -\mu_k \beta_T \left(\tfrac{(\mu_k(T^2+2\gamma)y_1^k(0)+Ty_2^k(0)) \sin(\mu_k t)}{\mu_k^2(T^2+2\gamma)^2} - \tfrac{y_2^k(0)\cos(\mu_k t)}{\mu_k(T^2+2\gamma)} \right). \end{aligned} \tag{27}$$

Note that $y_{2a}(T) = 0$ for T that are multiples of 4. The higher actuated modes, $n > N$, also can be derived analytically. We omit these lengthy expressions stating that their amplitudes behave as $\frac{1}{nT}$. In the next section, we show numerically that these approximations are close to the numerical (optimal) solution.

5 Example of Control

5.1 Numerical Results

In this section, we consider a numerical implementation of the strategy (23). Since the equations of motion (12) are linear, to estimate the influence of the control on the higher modes v_n with $n > N$, it is enough to consider zero initial conditions for such modes:

$$v_n^0 = \dot{v}_n^0 = 0 \text{ or } y_1^{(n)}(0) = y_2^{(n)}(0) = 0, \quad n > N. \tag{28}$$

For the lowest controlled modes, we choose initial conditions (ICs) such that energy of each mode is equal to 1 at $t = 0$:

$$\begin{aligned} E_0(0) = \tfrac{(\dot{v}_0^0)^2}{2} &= \tfrac{(y_1^{(0)}(0))^2}{2} = 1, \\ E_k(0) = \tfrac{1}{2}\left((\dot{v}_k^0)^2 + \eta_k(v_k^0)^2\right) &= \tfrac{1}{2}\left((y_2^{(k)}(0))^2 + (y_2^{(k)}(0))^2\right) = 1, \\ v_0^0 = \sqrt{2}, \quad \dot{v}_0^0 = 0, \quad v_k^0 &= \tfrac{1}{\sqrt{\eta_k}} = \tfrac{1}{\mu_k}, \quad \dot{v}_k^0 = 1, \\ y_1^{(0)}(0) = \sqrt{2}, \; y_2^{(n)}(0) = 0, \quad y_1^{(k)}(0) &= y_2^{(k)}(0) = 1, \quad k \in J_1. \end{aligned} \tag{29}$$

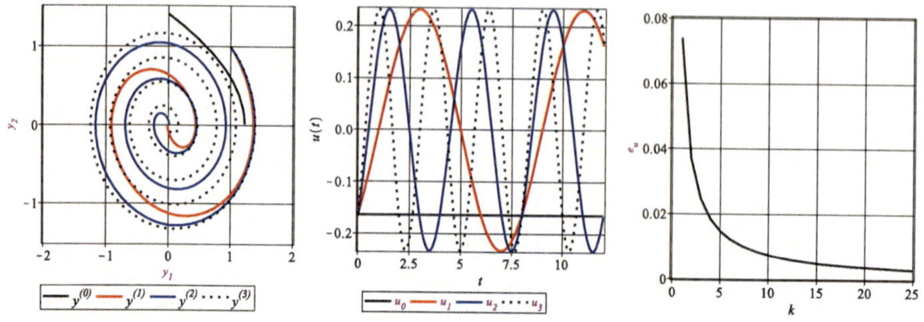

Fig. 2. The amplitudes $y_{1,2}^k$ of the controlled modes with $k = 0, 1, 2, 3$ (left). The optimal control inputs (middle). The relative error of the approximate control input (right).

Let us take $N = 3$. That is, we control the four lowest modes and the other modes are actuated but not optimized. The optimal control u_k acts on the modes with $n = 6j \pm k > 0$. We fix $\gamma = 1$, and $T = 12$, which corresponds to three periods of the first mode v_1. Fig. 2(left) shows the phase portrait of the controlled modes, and Fig. 2(middle) presents the controlled inputs. Here, the zeroth mode $y_{1,2}^{(0)}$ and the input u_0 are computed analytically according to (23), (24), while the amplitudes $y_{1,2}^{(k)}$, $k = 1, 2, 3$ are obtained numerically solving (22), with ICs (29) and u_k from (23). In Fig. 2(left), $y_1^{(0)}$ is scaled by 7 and at $t = T$ we have $y_1^{(0)} \approx 8.6$, $y_2^{(0)} \approx 0.02$, $y_1^{(1)} \approx y_2^{(1)} \approx -0.014$, $y_1^{(2)} \approx y_2^{(1)} \approx 0.014$, $y_1^{(3)} \approx y_2^{(3)} \approx -0.014$.

Figure 3(left) shows the mechanical energy during the controlled motion. This figure presents the energy of the controlled modes with $k = 0, 1, 2, 3$ with the terminal value $E(T) = 7 \cdot 10^{-4}$, and the energy of several actuated modes with $n = 4, \ldots, 9$. Here, 4th and 8th modes are actuated by u_2, 5th and 7th—by u_1, 6th—by u_0, and 9th—by u_3. The terminal energy of these five actuated modes is $E(T) \approx 10^{-11}$. During the motion, the amplitudes of the actuated modes do not exceed 0.15 for $y_{1,2}^{(4)}$ and 0.06 for $y_{1,2}^{(n)}$, $n = 5, \ldots, 9$. The terminal value of the approximate cost functional (18) is $\tilde{\Phi} = 5.5 \cdot 10^{-2}$.

Next, the time horizon T is varied. In Fig. 3(middle), we present the terminal mechanical energy of the first ten modes $E(T)$ (the lowest four are controlled while the rest are actuated) depending on $T \in [2, 102]$ and the corresponding value of the energy loss F.

Further, we perform analysis of the approximate analytical solution. Figure 2 (right) shows the relative errors in $L_2(0, T)$ of the approximation (26) depending of the number k of controlled mode: $e_u = \frac{\|u_k^a - u_k\|_{L_2}}{\|u_k\|_{L_2}}$ for $T = 12$. Here, the estimate for a specific k is valid for any number of control elements $N \geq k + 1$. The relative errors $e_{yi} = \frac{\|y_{ia}^{(k)} - y_i^{(k)}\|_{L_2}}{\|y_i^{(k)}\|_{L_2}}$, $i = 1, 2$ of approximated mode amplitudes have the similar value and dependence on k.

In Fig 3(right), the approximate mechanical energy E_a and the approximate energy losses F_a are given depending on T for $N = 3$ for the first ten modes computed analytically. The zeroth mode is controlled by u^0 (23) and for $k = 1, 2, 3$ we use u_a^k (26). The other modes with numbers $n > 3$ are actuated by one of these inputs.

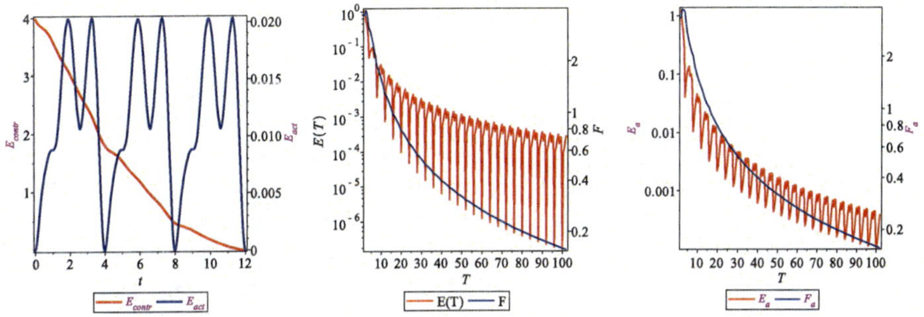

Fig. 3. The mechanical energy of the controlled and actuated modes during the motion (left). The terminal mechanical energy and the energy losses (middle). The approximate terminal mechanical energy and the energy losses (right)

5.2 Discussion

It follows from the presented results (see Figs. 2, 3) that the proposed control strategy allows for achieving a state relatively close to zero except for the zeroth mode. The velocity of this mode is rather small at $t = T$, but its amplitude, which corresponds to the motion of the rod as a rigid body, grows in time. This is due to the cost functional Φ (17) not including $v_0(T)$ since the change of the rod's position does not affect the mechanical energy of the rod. To minimize the displacement of the rod as a rigid body, the initial setup should be modified, e.g. by adding elastic springs at rod's ends. Then, all the modes will be described by similar pendulum equations.

Choosing the time horizon T larger, we allow for more precise control, see Fig. 3(middle). As can be expected for a vibrational system, for some T, specifically, multiples of 4—the period of the first mode—the proposed strategy is distinctively more effective. Although the actuated modes are not excited significantly in general, for the above-mentioned T their terminal amplitudes are several orders smaller than the terminal perturbation of the controlled modes.

To estimate analytically the terminal amplitudes of both controlled and actuated modes, we introduce the asymptotic control u_a^k. As seen from Fig. 2(right), this control law provides an approximation that becomes more accurate as k grows, that is if we utilize more control elements (make N larger). The relative error of the energy losses F (see Fig. 3, right) achieves about 10^{-4} as T grows for fixed $N = 3$. However, the approximate control (26) does not contain feedback in position. Therefore, the terminal values of displacement amplitudes are not zeros: $y_{1a}^{(k)}(T) \sim \frac{1}{kT}$ while $y_{2a}^{(k)}(T) = 0$, providing larger values of the terminal energy (of the order $\sum_{k=1}^{N} \frac{1}{k^2T^2} + \sum_{n>N} \frac{1}{n^2T^2}$) than numerical (optimal) solution. It is worth mentioning that this estimate is stable: the approximate terminal energy is about 10 times larger than the exact one for all T except for multiples of 4, when the exact energy is much smaller.

6 Conclusions

In this paper, we study a vibrating system controlled by distributed and boundary inputs, where the distributed force is piecewise constant in space. We have shown that the continuum system is split into a finite number of subsystems each actuated by certain linear combination of inputs. We consider an optimal control problem of suppressing the vibrations by minimizing the terminal mechanical energy and energy losses. Taking an approximation of the original control problem, the motion of the lowest mode in each group is optimized and the optimal feedback control law is found explicitly by means of LQR theory. For the subsystem containing the zeroth mode, the amplitude of both controlled and actuated modes are obtained analytically. Amplitudes of the other subsystems are described numerically. To show that the amplitudes of the higher modes are not excited significantly, we have derived an approximate feedback control and the corresponding mode behavior. We present a numerical example and analyze the optimal solution and its analytical approximation.

We plan to derive rigorous estimates of quality of suboptimal control and combine the proposed feedback strategy with recently developed feedforward approach providing analytical solution [9]. Also, since the piezoelements may serve both as actuators and sensors, it is promising to add observes in the model.

References

1. Glowinski, R., Lions, J.L.: Exact and approximate controllability for distributed parameter systems. Acta Numer. **3**(269378), 269–378 (1994)
2. Preumont, A.: Vibration Control of Active Structures, 3rd edn. Springer, Berlin (2011). https://doi.org/10.1007/978-94-007-2033-6
3. Tzou, H.S.: Piezoelectric Shells: Sensing, Energy Harvesting, and Distributed Control. Springer, Dordrecht (2019). https://doi.org/10.1007/978-94-024-1258-1
4. Mohith, S., et al.: Recent trends in piezoelectric actuators for precision motion and their applications: a review. Smart Mater. Struct. **30**, 013002 (2021)
5. Shivashankar, P., Gopalakrishnan, S.: Review on the use of piezoelectric materials for active vibration, noise, and flow control. Smart Mater. Struct. **29**, 053001 (2020)
6. Akulenko, L.D.: Reduction of an elastic system to a prescribed state by use of a boundary force. J. Appl. Math. Mech. **45**(6), 827–833 (1981)
7. Sontag, E.: Mathematical Control Theory: Deterministic Finite Dimensional Systems. Springer, New York (1998). https://doi.org/10.1007/978-1-4612-0577-7

8. Ovseevich, A., Ananievski, I.: Robust feedback control for a linear chain of oscillators. J. Optim. Theory Appl. **188**, 307–316 (2021). https://doi.org/10.1007/s10957-020-01765-z
9. Kostin, G., Gavrikov, A.: Optimal motions of an elastic structure under finite-dimensional distributed control (submitted, 2022)

Receptance-Based Robust Assignment of Natural Frequency in Vibration Systems

Zhang Lin[1], Zhang Tao[1](✉), Ouyang Huajiang[2], Li Tianyun[1], and Shang Baoyou[1]

[1] School of Naval Architecture and Ocean Engineering, Huazhong University of Science and Technology, Wuhan, China
zhangt7666@hust.edu.cn
[2] School of Engineering, University of Liverpool, Liverpool, UK

Abstract. This work proposes a novel receptance-based robust structural modification method of the natural frequency assignment. Potential perturbation of the target natural frequency, which arises from uncertainties in physical parameter modifications, is quantified by utilizing analytical sensitivity formulae derived in this research. The proposed sensitivity formula is introduced to the natural frequency assignment optimization calculation as an extra term penalizing poor-robust solutions. Such an improvement can boost obtaining the nominal values of structural modifications with high robustness in preserving the model-free superiority of receptance-based techniques. The numerical validation of the proposed method on a five-degree-of-freedom system demonstrates its capability to compute high-robust modifications in meeting the prescribed requirements and satisfying all the constraints.

Keywords: Natural frequency assignment · Receptance method · Robust design · Sensitivity analysis

1 Introduction

Nowadays, the energy density of a machine has significantly increased because such a machine tends to transfer a great amount of power or operate more efficiently than before. As a result, the dynamic strength failure caused by the resonance between the operating frequency and natural frequency is the most outstanding vibration control problem for a machine system [1]. From a structural dynamics point of view, employing inverse structural modification to shift its natural frequency away from the operating frequency is a useful way to avoid resonance and achieve better dynamic performance [2]. In particular, receptance method is the most fascinating and popular for researchers since it neither involves the theoretical model of structures nor requires complete modal information [3]. However, a limitation of the receptance method is that robustness is not addressed, and hence, it is neglected how the assigned natural frequencies are affected by perturbations of the modification parameters about the nominal ones [4].

In the frame of natural frequency assignment, robustness is often interpreted as finding the robust modifications, ensuring the effectiveness of the assigned natural frequency

Z. Dimitrovová et al. (Eds.): WMVC 2022, MMS 125, pp. 296–303, 2023.
https://doi.org/10.1007/978-3-031-15758-5_29

even under parameter modifications with systematic error [5]. As far as this work is concerned, a novel receptance-based method for the robust natural frequency assignment is proposed. The potential perturbation in assigned natural frequency, which arises from parameter modifications with uncertainty, is quantified by utilizing analytical sensitivity formulae derived in this research. The sensitivity formulae are computationally efficient, even for a large number of parameter modifications, and require only measured receptances, thereby preserving the model-free superiority of receptance-based techniques. A global optimization procedure is used to perform the robustness assignment of natural frequency. The proposed technique is tested numerically on a multi-degree-of-freedom system. The results show that the obtained modifications by the proposed method can improve the robustness of the desired natural frequency to the uncertainty in parameter modifications and reduce the potential perturbation of the assigned natural frequency.

2 Method Formulation

2.1 Receptance-Based Structural Modification

Considering an N-degrees of freedom (DOFs) un-damped vibrating system, its free vibration equation can be expressed as

$$(\mathbf{K} - \omega_i^2\mathbf{M})\boldsymbol{u}_i(\omega_i) = \mathbf{0} \; i = 1 \cdots N, \tag{1}$$

where $\mathbf{K}$, $\mathbf{M} \in \Re^{N\times N}$ are mass stiffness and matrix matrices of the original system, respectively. $(\omega_i, \mathbf{u}_i)$ is i^{th} eigenpair, made by i^{th} natural frequency and its related mode shape.

The problem of assigning the desired natural frequency is aimed at finding the additive structural modification matrices $\Delta\mathbf{K}$, $\Delta\mathbf{M} \in \Re^{N\times N}$ satisfying [2]

$$((\mathbf{K} + \Delta\mathbf{K}) - \tilde{\omega}_i^2(\mathbf{M} + \Delta\mathbf{M}))\tilde{\mathbf{u}}_i(\tilde{\omega}_i) = \mathbf{0}, \tag{2}$$

where $\tilde{\omega}_i$ is the ith natural frequency of the modified system, also a desired natural frequency; $\tilde{\mathbf{u}}_i$ is the corresponding i^{th} mode shape of the modified system. Since

$$\mathbf{H}(\tilde{\omega}_i) = (\mathbf{K} - \tilde{\omega}_i^2\mathbf{M})^{-1} = \sum_{j=1}^{N} \frac{\mathbf{u}_j\mathbf{u}_j^{\mathrm{T}}}{(\omega_j^2 - \tilde{\omega}_i^2)} \tag{3}$$

is the receptance matrix of the original system at the frequency $\tilde{\omega}_i$. After pre-multiplying the receptance matrix, Eq. (2) can be rewritten as [3]

$$(\mathbf{I} + \mathbf{H}(\tilde{\omega}_i)(\Delta\mathbf{K}(\theta) - \tilde{\omega}_i^2\Delta\mathbf{M}(\theta)))\tilde{\mathbf{u}}_i(\tilde{\omega}_i) = \mathbf{0} \; i = 1 \cdots N_a, \tag{4}$$

where the dependence of $\Delta\mathbf{K}$, $\Delta\mathbf{M}$ on the design variables is stressed and collected in vector $\boldsymbol{\theta} = [\theta_1, \ldots, \theta_x]^{\mathrm{T}} \in \Re^x$; N_a is the number of desired natural frequencies.

The key to assigning the desired natural frequency is to compute the design variable vector $\boldsymbol{\theta}$. However, the solvability of Eq. (4) is not ensured since no assumption has been made on the rank-matching conditions. If the specification on mode shape is introduced, the number of design variables is required to be increased. More especially, the constraint on the design variables due to physical or technical limitations extremely limits the attainable natural frequency and often imposes more design variables to assign just one eigenvalue. Hence, Eq. (4) can be conveniently recast as a constrained least-square minimization [3].

$$\min_{\boldsymbol{\theta}}\left\{\sum_{i=1}^{N_a}\alpha_i\|(\mathbf{I}+\mathbf{H}(\tilde{\omega}_i)\Delta\mathbf{B}(\tilde{\omega}_i,\theta))\tilde{\mathbf{u}}_i(\tilde{\omega}_i)\|_2,\ \theta\in\Gamma_\theta\right\},\tag{5}$$

where α_i is the positive weighing scalar, $\Delta\mathbf{B}(\tilde{\omega}_i,\boldsymbol{\theta})=\Delta\mathbf{K}(\boldsymbol{\theta})-\tilde{\omega}_i{}^2\Delta\mathbf{M}(\boldsymbol{\theta})$ is structural modification matrix, and $\|\cdot\|_2$ denotes the Euclidean norm of a vector. $\boldsymbol{\Gamma}_{\boldsymbol{\theta}}$ represents the feasible domain of those design variables, which includes lower and upper bounds (denoted by $\boldsymbol{\theta}^L$ and $\boldsymbol{\theta}^U$, respectively).

Suppose mode shapes (eigenvector) are not required to be assigned or just partially assigned. In that case, those unassigned mode shapes are treated as additional unknown entries, which are also constrained to the additional feasible domain $\boldsymbol{\Gamma}_{\tilde{\mathbf{u}}}$: $\tilde{\mathbf{u}}^L\leq\tilde{\mathbf{u}}\leq\tilde{\mathbf{u}}^U$. As a consequence, Eq. (5) can be further rewritten as.

$$\min_{\boldsymbol{\theta}}\left\{\sum_{i=1}^{N_a}\alpha_i\|(\mathbf{I}+\mathbf{H}(\tilde{\omega}_i)\Delta\mathbf{B}(\tilde{\omega}_i,\theta))\tilde{\mathbf{u}}_i(\tilde{\omega}_i)\|_2,\ \begin{Bmatrix}\theta\\ \tilde{\mathbf{u}}\end{Bmatrix}\in\Gamma\right\},\tag{6}$$

where the $\boldsymbol{\Gamma}=\boldsymbol{\Gamma}_{\boldsymbol{\theta}}\cup\boldsymbol{\Gamma}_{\tilde{\mathbf{u}}}$ represents the union of the feasible domain of $\boldsymbol{\theta}$ and $\tilde{\mathbf{u}}$.

It should be noted that the nonlinear programming problem in Eq. (6) can be converted as a non-convex optimization problem, whose least-square solutions can be computed by employing MATLAB built-in function *fmincon* [4].

2.2 Sensitivity-Based Robust Assignment of Natural Frequency

From the perspective of optimization, the least-square solutions admitted by Eq. (6) are not unique, which means that Eq. (6) can provide a series of alternative modification schemes for designers. Even if all of the modification schemes can theoretically realize the accurate assignment of the desired natural frequency, the real assignment results obtained by implementing different modification schemes may occur perturbation error and then have different assignment accuracy due to the presence of uncertainty of modification parameters.

In order to evaluate the potential perturbation of the target natural frequency assigned by different modification schemes, the sensitivity of the target natural frequency with respect to those variable modification parameters is exploited and redefined as follows.

$$\frac{\partial\tilde{\omega}_i}{\partial\theta_j}=-\frac{\tilde{\mathbf{u}}_i^{\mathrm{T}}\mathbf{H}(\tilde{\omega}_i)\left.\frac{\partial\Delta\mathbf{B}}{\partial\theta_j}\right|_{\theta_j}\tilde{\mathbf{u}}_i}{\tilde{\mathbf{u}}_i^{\mathrm{T}}(\Delta\mathbf{B}(\tilde{\omega}_i,\theta_j)\left.\frac{\partial\mathbf{H}}{\partial\omega}\right|_{\tilde{\omega}_i}+\mathbf{H}(\tilde{\omega}_i)\left.\frac{\partial\Delta\mathbf{B}}{\partial\omega}\right|_{\tilde{\omega}_i})\tilde{\mathbf{u}}_i},\tag{7}$$

By exploiting the derived expression about sensitivity, the potential perturbation of the target natural frequency can be evaluated by employing the total differential approach and expressed as

$$\mathrm{d}\tilde{\omega}_i=\sum_{j=1}^{N_\theta}\frac{\partial\tilde{\omega}_i}{\partial\theta_j}\mathrm{d}\theta_j.\tag{8}$$

Extending the differential of each parameter modification leads to

$$\delta\tilde{\omega}_i(\tilde{\omega}_i, \theta) = \sum_{j=1}^{N_\theta} \frac{\partial\tilde{\omega}_i}{\partial\theta_j}\delta\theta_j(\theta_j). \tag{9}$$

It should be noted that $\delta\theta_j$ is the maximum systematic error of the modification parameter θ_j caused by the uncertainty, which is known or can be obtained by estimating.

With the perturbation expression of the target natural frequency derived, it is possible to improve the problem in Eq. (6) by maximizing the robustness of the target natural frequency to the uncertainty in parameter modifications. In order to do so, Eq. (9) can be introduced to the optimization problem in Eq. (6) as an extra term penalizing poor-robust solutions and boosting the obtaining of the modifications with high robustness. As a consequence, the robust assignment problem of natural frequency becomes the following bi-criterion optimization problem.

$$\min_{\boldsymbol{\theta}}\left\{\sum_{i=1}^{N_a}\alpha_i(\|(\mathbf{I}+\mathbf{H}(\tilde{\omega}_i)\Delta\mathbf{B}(\tilde{\omega}_i,\theta))\tilde{\mathbf{u}}_i(\tilde{\omega}_i)\|_2+\lambda\|\delta\tilde{\omega}_i(\tilde{\omega}_i,\theta)\|_2),\ \begin{Bmatrix}\theta\\ \tilde{\mathbf{u}}\end{Bmatrix}\in\Gamma\right\}, \tag{10}$$

where $\lambda > 0$ denotes the penalizing scalar and is selected to trade between the cost of using parameter modifications with low sensitivity and the cost of missing the natural frequency assignment specification. As far as this work is concerned, λ is set as 10^{-4}.

Clearly, such a strategy can provide assistance in handling the inaccuracy problem of the assignment result due to the uncertainty of the modification parameters and realizing the robust assignment of natural frequency.

3 Numerical Assessment

In order to demonstrate the working of the proposed method, a numerical example is presented in this section. It should be noted that the needed receptance data in the example are all obtained from simulation, which is usually measured by experiment technology in real.

The employed numerical example is a five-DOFs system with five lumped masses, which has been adopted widely as a benchmark for eigenvalue assignment methods [1, 4, 6, 7]. The five-DOFs system is shown in Fig. 1, and its corresponding system parameters are collected in Table 1.

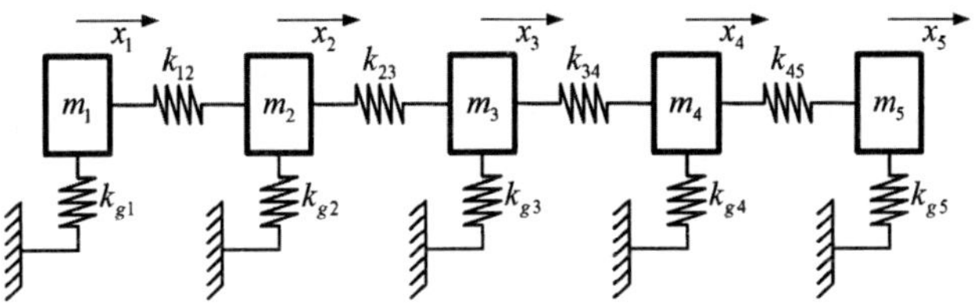

Fig. 1. A five-DOFs lump mass system in Ref [6].

Table 1. System parameter nominal values.

Parameters	m_1 (kg)	m_2 (kg)	m_3 (kg)	m_4 (kg)	m_4 (kg)
Nominal values	1.73	5.21	8.21	2.61	1.34
Bounds	[0, 2]	[0, 2]	[0, 2]	[0, 2]	[0, 2]
Parameters	k_{12} (kN/m)	k_{23} (kN/m)	k_{34} (kN/m)	k_{45} (kN/m)	k_g (kN/m)
Nominal values	73.6	68.2	73.5	82.1	98.9
Bounds	N/A	N/A	N/A	N/A	[0, 283]

In this work, structural modifications are only admitted for the five lumped masses as well as the five grounding springs, whose feasible domains have been shown in Table 1, leading to the design variable vector being a 10-dimensional vector ($\boldsymbol{\theta} = [\Delta k_{g1}, \Delta k_{g2}, \Delta k_{g2}, \Delta k_{g3}, \Delta k_{g4}, \Delta k_{g5}, \Delta m_1, \Delta m_2, \Delta m_3, \Delta m_4, \Delta m_5]^{\mathrm{T}}$). In order to evaluate the effectiveness of the proposed method, the uncertainty is introduced to all stiffness modifications of the ground spring used to implement, and the implemented stiffness modification with uncertainty is assumed to obey the normal distribution $\mathrm{N}(\Delta k_{gi}, (6.25\%\Delta k_{gi})^2)$.

The assignment task is to assign a natural frequency at 25 Hz. In this case, the mode shape of the lumped mass m_2 at the desired natural frequency is required as 1, and all other mode shape entries at this frequency are prescribed to belong to interval $[-1, 1]$.

In order to demonstrate the working of the proposed robust assignment method, a typical partial eigenstructure assignment method without robustness constraints [7] is employed for comparing with the proposed method. The nominal values of structural modifications obtained by the two methods are reported in Table 2. The five variable stiffness modifications and the sensitivities of the desired natural frequency to those stiffness modifications are instead highlighted in Fig. 2.

Table 2. Nominal values of structural modifications.

Parameters	Δm_1 (kg)	Δm_2 (kg)	Δm_3 (kg)	Δm_4 (kg)	Δm_4 (kg)
Nonrobust	2	1.40	1.39	0.11	2
Robust	0	0	0.23	0	0
Parameters	Δk_{g1} (kN/m)	Δk_{g2} (kN/m)	Δk_{g3} (kN/m)	Δk_{g4} (kN/m)	k_{g5} (kN/m)
Nonrobust	53.79	67.18	52.95	91.58	59.89
Robust	244.56	56.68	13.91	157.07	185.58

It is interesting to notice that different nominal values of the stiffness modifications are provided by the two natural frequency assignment methods in the case of similar sensitivities of the desired natural frequency to stiffness modifications. The stiffness modifications obtained by the classical (nonrobust) assignment method have no apparent correlation with the sensitivities; in contrast, those obtained by the proposed (robust) assignment method have a negative correlation with the sensitivities. It is evident that the

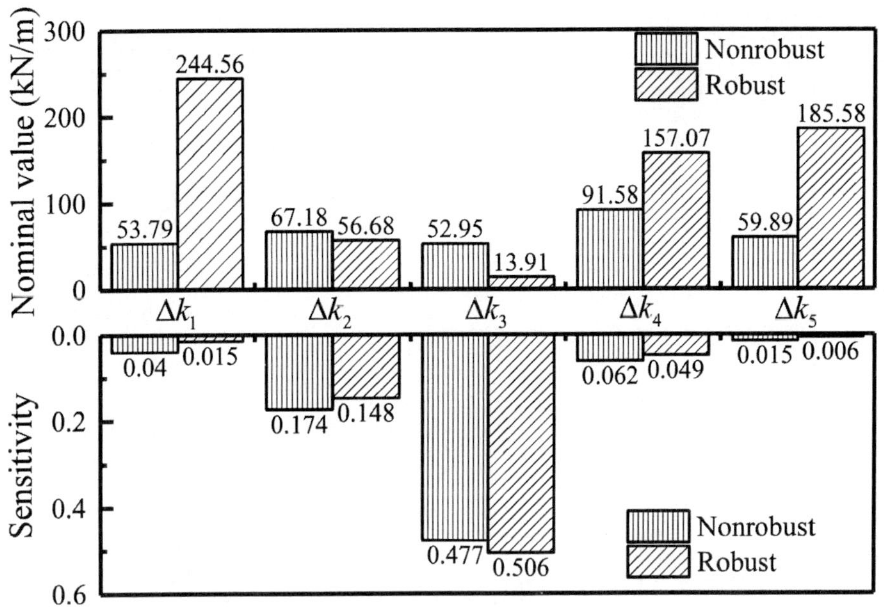

Fig. 2. Stiffness modifications and sensitivities of the two methods.

prescribed natural frequency by employing the proposed method has a lower sensitivity to the uncertainty of the stiffness modifications compared with the typical method.

In order to further demonstrate the superiority of the proposed method, the Monte-carlo statistical analysis is implemented: normally distributed random errors are imposed to the nominal values of Δk_{g1}–Δk_{g5}, and the stiffness modifications with uncertainties are simulated; the variability of the modified receptances is evaluated; the probability density functions of the frequency perturbations of the assigned natural frequency are obtained. The results are shown in Fig. 3.

What stands out in Fig. 3 is that the natural frequency assigned by the proposed method has a smaller perturbation interval than that assigned by the typical method. More especially, the probability of the assigned natural frequency being between [24.85, 25.15] Hz is 0.98 for the proposed method, much higher than 0.68 for the typical method. Such impressive comparisons corroborate the effectiveness of the proposed method in improving the robustness of the target natural frequency to the uncertainty in modification parameters and reducing the potential perturbation of the assigned natural frequency.

4 Conclusion

This paper proposes a novel receptance-based robust assignment method of natural frequencies. The sensitivity of the desired natural frequency with respect to those parameter modifications with uncertainty is derived and introduced to the optimization calculation of the natural frequency assignment as an extra term penalizing poor-robust solutions. Such a treatment boosts the obtaining of the nominal values of structural modifications with high robustness and reduces the potential perturbation of the assigned natural

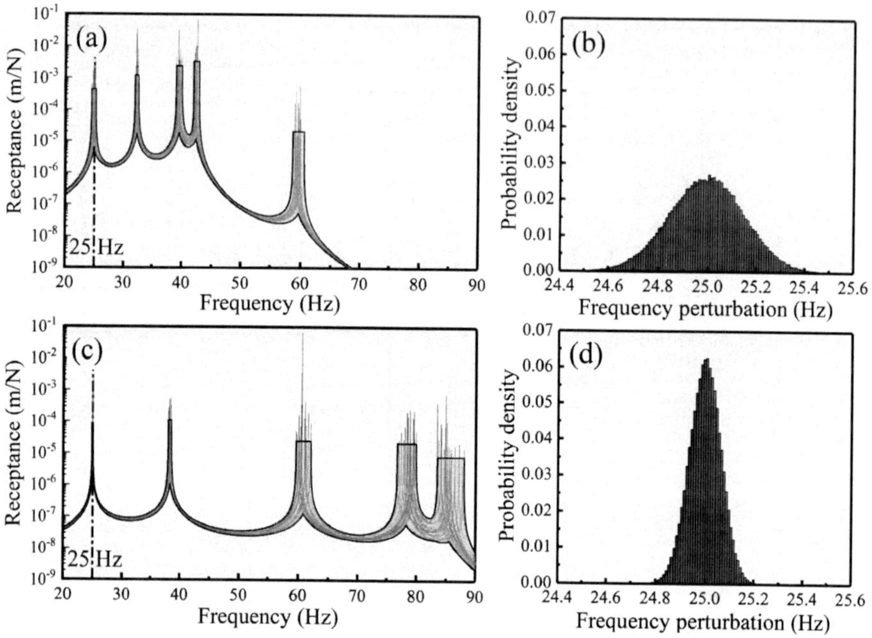

Fig. 3. (a) the modified receptance h_{15} obtained by the typical (nonrobust) method; (b) the probability density functions of the frequency perturbations of the nonrobust method; (c) the modified receptance h_{15} obtained by the proposed (robust) method; (d) the probability density functions of the frequency perturbations of the robust method.

frequency due to the uncertainties in the implemented stiffness modifications. The effectiveness of the proposed method has been demonstrated by comparing its performance with a typical eigenstructure assignment method proposed in literature. The natural frequency assignment of a five-DOF system has been performed, and it has been shown that the robust assignment of natural frequency is achieved by the proposed method, hence overcoming the shortcoming with poor robustness of the typical method.

References

1. Ouyang, H., Richiedei, D., Trevisani, A.: Eigenstructure assignment in undamped vibrating systems: a convex-constrained modification method based on receptances. Mech. Syst. Signal Process. **27**(1), 397–409 (2012)
2. Mottershead, J.E., Ram, Y.M.: Inverse eigenvalue problems in vibration absorption: passive modification and active control. Mech. Syst. Signal Process. **20**(1), 5–44 (2006)
3. Mottershead, J.E., Tehrani, M.G., Ram, Y.M.: Assignment of eigenvalue sensitivities from receptance measurements. Mech. Syst. Signal Process. **23**(6), 1931–1939 (2009)
4. Caracciolo, R., Richiedei, D., Tamellin, I.: Robust assignment of natural frequencies and antiresonances in vibrating systems through dynamic structural modification. Shock. Vib. **2021**, 1–20 (2021)
5. Adamson, L.J., Fichera, S., Mottershead, J.E.: Receptance-based robust eigenstructure assignment. Mech. Syst. Signal Process. **140**, 106697 (2020)

6. Zhang, L., Zhang, T., Ouyang, H., Li, T.: Receptance-based antiresonant frequency assignment of an uncertain dynamic system using interval multiobjective optimization method. J. Sound Vib., 116944 (2022)
7. Belotti, R., Richiedei, D., Trevisani, A.: Optimal design of vibrating systems through partial eigenstructure assignment. J. Mech. Des. **138**(7), 2–9 (2016)

GVB: Ground Vibration

Application of an Indirect Trefftz Method (Wave Based Method) for the Spectral Analysis of 2D Unbounded Saturated Porous Media

Mirjam Lainer(✉) and Gerhard Müller

Chair of Structural Mechanics, TUM School of Engineering and Design, Technical University of Munich, Arcisstraße 21, 80333 Munich, Germany
{mirjam.lainer,gerhard.mueller}@tum.de
http://www.cee.ed.tum.de/bm

Abstract. The Wave Based Method (WBM) uses weighted wave functions in order to model boundary value problems. These wave functions have to satisfy the underlying differential equations and usually violate the boundary conditions of a WBM element. A weighted residual formulation permits to minimize this error by determining the weighting factor for each wave function. By applying the WBM to a 2D saturated soil structure, the propagation of two longitudinal waves and one shear wave is modeled. In compliance with Biot's theory, the field variables consist of displacement components in the solid phase and seepage field components, which influence the decay of the wave amplitude within the structure. In order to fulfill the Sommerfeld radiation condition, this model is extended by an absorbing boundary condition, which transmits incident waves. The model is tested for different excitation frequencies and hydraulic conductivities in order to observe the radiated power within one period.

Keywords: Wave Based Method · Biot's theory · Absorbing boundary condition · Porous saturated medium

1 Introduction

The objective of this paper is to evaluate the application of the Wave Based Method (WBM) to an elastodynamic saturated soil domain with a harmonic load. Biot's theory is applied in order to extend the Lamé equations for an elastodynamic problem. For this, the displacement components are divided into solid phase deformations $\mathbf{u}^s$ and a seepage field $\mathbf{U} = n^w(\mathbf{u}^w - \mathbf{u}^s)$, with the fluid displacements $\mathbf{u}^w$ and the porosity n^w. The obtained wave pattern for a 2D problem is then described by a superposition of two P-waves and one S-wave [1,2].

According to the WBM concept, convex domains and subdomains are approximated by a finite number of weighted wave functions, which fulfill the underlying

Z. Dimitrovová et al. (Eds.): WMVC 2022, MMS 125, pp. 307–316, 2023.
https://doi.org/10.1007/978-3-031-15758-5_30

Helmholtz equations. The introduced weighting factors represent the unknown variables, which can be determined by a weighted residual approach. For this, residual errors along the boundaries of each subdomain are evaluated.

The Wave Based Method was introduced by Desmet [3] who applied it to 2D vibroacoustic problems. By comparing this method with a FEM approach, it could be shown that the WBM converges rather fast, the result quality depends less on the excitation frequency and the computational time can be decreased significantly, as i.a. less unknown quantities are necessary for the same problem description. Within the last two decades, the WBM has been extended to further boundary value problems, such as 3D bounded and unbounded vibroacoustic problems [4,5], plate bending and membrane problems [6], as well as the coupling of WBM and FEM domains [4]. Moreover, a WBM approach was developed in order to apply it to poroelastic materials that comply with Biot's theory [7]. In the latter approach, the focus was on the application of the WBM to low and mid-frequency problems for saturated acoustic materials. Additionally, geometrical aspects of given domains were investigated, for example to decrease stress singularities [8] and to adapt the method for axisymmetric structures [9].

These researches laid the ground for extending the WBM to a fully saturated soil structure, which is modeled by an elastodynamic halfspace and excited by harmonic loading. In order to fulfill the Sommerfeld condition, an absorbing boundary condition has been implemented. This absorbing boundary condition transmits normally incident waves and was introduced by Degrande who applied a spectral and finite element approach for the analysis of poroelastic media [10, 11]. Within another Trefftz approach - the so-called hybrid Trefftz finite element method - Moldovan successfully applied this absorbing boundary condition to saturated poroelastic soil [12]. In this paper, selected results by Degrande are used in order to assess the obtained simulation data with the WBM.

2 Problem Definition

For a saturated elastodynamic structure the displacement field consists of the solid phase displacements $\mathbf{u}^s = [u_x, u_y]^T$ and the fluid seepage field $\mathbf{U} = [U_x, U_y]^T$, which describes the amount of water flowing out of the pores. In the following compatibility condition, ζ describes the change of the fluid content.

$$\begin{pmatrix} \varepsilon_{xx} \\ \varepsilon_{yy} \\ \gamma_{xy} \\ \zeta \end{pmatrix} = \begin{bmatrix} \frac{\partial}{\partial x} & 0 & 0 & 0 \\ 0 & \frac{\partial}{\partial y} & 0 & 0 \\ \frac{\partial}{\partial y} & \frac{\partial}{\partial x} & 0 & 0 \\ 0 & 0 & \frac{\partial}{\partial x} & \frac{\partial}{\partial y} \end{bmatrix} \begin{pmatrix} u_x \\ u_y \\ U_x \\ U_y \end{pmatrix} \tag{1}$$

Considering the first and second Lamé coefficients λ, μ, the E-Modulus E and the Biot coefficients α and M, the plane strain constitutive relation is formulated as shown next. The field variable σ_f refers to the fluid pore pressure described as total stresses.

$$\begin{pmatrix} \sigma_{xx} \\ \sigma_{yy} \\ \tau_{xy} \\ \sigma_f \end{pmatrix} = \begin{bmatrix} \lambda + 2\mu + \alpha^2 M & \lambda + \alpha^2 M & 0 & \alpha M \\ \lambda + \alpha^2 M & \lambda + 2\mu + \alpha^2 M & 0 & \alpha M \\ 0 & 0 & \mu & 0 \\ \alpha M & \alpha M & 0 & M \end{bmatrix} \begin{pmatrix} \varepsilon_{xx} \\ \varepsilon_{yy} \\ \gamma_{xy} \\ \zeta \end{pmatrix} \tag{2}$$

By applying a Helmholtz approach, the deformation fields in the solid phase $\mathbf{u}^s$ and in the seepage field $\mathbf{U}$ are decomposed into two irrotational potentials $\Phi_{p_{1/2}}$ and one solenoidal potential Ψ_s:

$$\mathbf{u}^s = \boldsymbol{\nabla}\Phi_{p_1} + \boldsymbol{\nabla}\Phi_{p_2} + \widetilde{\boldsymbol{\nabla}}\Psi_s \tag{3}$$

$$\mathbf{U} = \gamma_{p_1}\boldsymbol{\nabla}\Phi_{p_1} + \gamma_{p_2}\boldsymbol{\nabla}\Phi_{p_2} + \gamma_s\widetilde{\boldsymbol{\nabla}}\Psi_s \tag{4}$$

Here $\boldsymbol{\nabla} = [\frac{\partial}{\partial x}, \frac{\partial}{\partial y}]^T$ is the gradient operator and $\widetilde{\boldsymbol{\nabla}} = [\frac{\partial}{\partial y}, -\frac{\partial}{\partial x}]^T$ indicates the curl operator. The parameters γ_{p_1}, γ_{p_2} and γ_s are derived within a decoupling procedure of the homogenous parts of the underlying differential equations, as shown by Biot [1].

For this paper, Molsand soil was used in order to validate the generated results with the WBM. This material was introduced by Degrande [10] and was also assumed by Moldovon in order to validate simulation results [12]. Table 1 shows the applied material parameters.

Table 1. Material parameters for Molsand soil

Symbol	Value	Definition
ρ_w	1000 kg/m^3	Fluid density
ρ	2009.8 kg/m^3	Mixture density
E	$2.98 \cdot 10^8$ N/m^2	Young's modulus of drained rock
K	$5.97 \cdot 10^9$ N/m^2	Bulk modulus of the mixture
α	1.0	1st Biot coefficient
M	$5.67 \cdot 10^9$ N/m^2	2nd Biot coefficient
n^w	0.388	Liquid volume fraction
ν	0.333	Poisson's coefficient of the drained rock
k_c*	0.01 m/s	Hydraulic conductivity
λ	$\nu E/[(1+\nu)(1-2\nu)]$	1st Lamé coefficient
μ	$E/[2(1+\nu)]$	2nd Lamé coefficient

*Originally, the hydraulic conductivity for Molsand soil was measured with $k_c = 0.0001$m/s [10].

3 Application of the Wave Based Method

According to the WBM, the two irrotational potentials $\Phi_{p_{1/2}}$ and the solenoidal potential Ψ_s are described by a finite number of wave functions, which fulfill the underlying differential equations. The displacement fields from the equations (3) and (4) are approximated by a predefined number of wave functions for the 1st and 2nd P-wave (n_{p_1}, n_{p_2}) as well as for the S-wave (n_s).

$$\mathbf{u}_h^s \cong \widehat{\mathbf{u}}_h^s = \sum_{i=1}^{n_{p1}} c_{p_1,i} \boldsymbol{\nabla} \Phi_{p_1,i} + \sum_{j=1}^{n_{p2}} c_{p_2,j} \boldsymbol{\nabla} \Phi_{p_2,j} + \sum_{k=1}^{n_s} c_{s,k} \widetilde{\boldsymbol{\nabla}} \Psi_{s,k} \tag{5}$$

$$\mathbf{U}_h \cong \widehat{\mathbf{U}}_h = \sum_{i=1}^{n_{p1}} \gamma_{p_1} c_{p_1,i} \boldsymbol{\nabla} \Phi_{p_1,i} + \sum_{j=1}^{n_{p2}} \gamma_{p_2} c_{p_2,j} \boldsymbol{\nabla} \Phi_{p_2,j} + \sum_{k=1}^{n_s} \gamma_s c_{s,k} \widetilde{\boldsymbol{\nabla}} \Psi_{s,k} \tag{6}$$

The wave functions for the 1st and 2nd P-wave as well as for the S-wave are described by the wave function sets according to Vanmaele and Deckers [6,7] and are weighted by the unknown values $c_{p_1,i}, c_{p_2,j}$ and $c_{s,k}$. All wave functions depend on the geometry of the single convex domains and on the wave numbers for the two P-waves and the S-wave (k_{p_1}, k_{p_2}, k_s). These wave numbers are influenced by the material parameters of the solid and fluid phase as well as by their interaction. The relation between the two phases is described by the coupling damping value ρ_{w2}. This consists of a real part, which is related to the coupling of both phases, and a frequency dependent, imaginary part that is characterized by a dissipation parameter ξ.

$$k_{p_{1/2}} = \sqrt{\frac{\rho_w + \gamma_{p_{1/2}} \rho_{w2}}{(\alpha + \gamma_{p_{1/2}}) M}} \cdot \omega, \tag{7}$$

$$\text{with } \rho_{w2} = \frac{\rho_w a}{n^w} - \frac{i\xi}{\omega (n^w)^2}, \quad \xi = \frac{\rho_w \cdot 9.81\,\mathrm{m/s^2}}{k_c} \cdot (n^w)^2,$$

$$k_s = \sqrt{\left(1 + \gamma_s \frac{\rho_w}{\rho}\right) \frac{\rho}{\mu}} \cdot \omega \tag{8}$$

In order to derive the unknown weighting values, residuals along the boundaries of each domain are defined. These residuals are necessary for a weighted residual formulation. In this procedure the predefined wave functions are weakly enforced to fulfill the boundary conditions along the adjoining edges. Additionally, a Galerkin approach is used in order to apply the wave functions as weighting functions. Figure 1 and Table 2 give an overview of possible boundary and coupling conditions, see also [7,10].

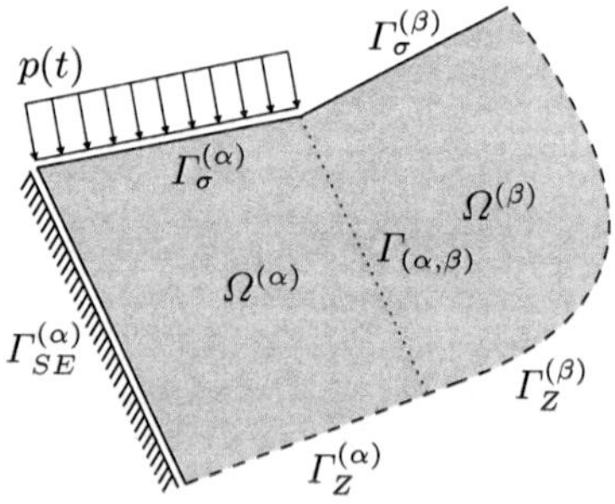

Fig. 1. Definition of boundaries and division into subdomains

4 Numerical Examples

4.1 Description of the 2D Boundary Value Problem

Table 2. Symbols for boundary conditions

Symbol	Definition
Γ_σ	Neumann boundary condition
Γ_u	Dirichlet boundary condition
Γ_{SE}	Mixed boundary condition (sliding edge)
Γ_Z	Absorbing boundary condition
$\Gamma_{(\alpha,\beta)}$	Coupling condition between the domains $\Omega^{(\alpha)}$ and $\Omega^{(\beta)}$

In order to validate the implemented absorbing boundary condition, a halfspace with Molsand soil described by a radius $L_{tot} = 0.5\,\mathrm{m} + L_0$ is chosen. Degrande and Moldovan used this structure as well in order to validate a spectral element and hybrid Trefftz finite element approach [10,12].

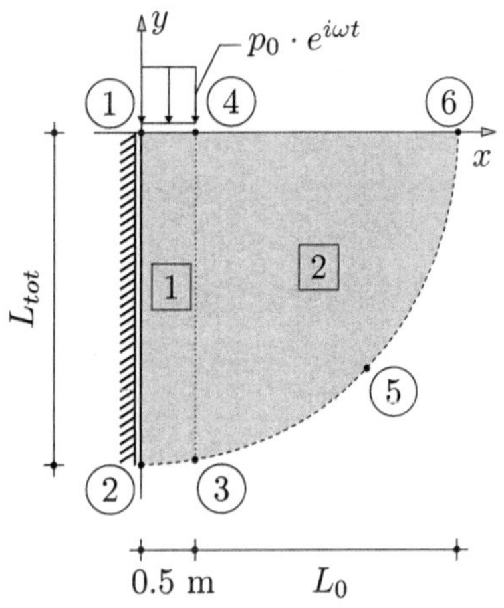

Fig. 2. Domain description

On the surface of this halfspace, the domain is vertically excited by a harmonically oscillating line load $p(x,t) = p_0 \cdot e^{i\omega t}$ for $x \in [-0.5\,\mathrm{m}, 0.5\,\mathrm{m}]$ with $p_0 = 1.0\,\mathrm{N/m^2}$. Due to the symmetry of this system with respect to the global y-axis, it can be reduced to a quadrant, as shown in Fig. 2. The edge along the y-axis ($x = 0\,\mathrm{m}$) is then simplified by a sliding boundary condition. The prescribed problem is divided into two WBM elements $\boxed{1}$ and $\boxed{2}$, whereby each of them is delimited by four edges and consequently by four nodes ⓘ, with $i \in \{1,2,3,4,5,6\}$. The two edges along the surface are described by Neumann boundary conditions. The shared edge between the two domains is defined by a coupling condition.

4.2 Average Displacement Amplitude

First of all, the average value for the vertical displacement amplitude of the solid skeleton under the loaded strip for $y = 0\,\mathrm{m}$ is evaluated. For this, the excitation frequency is varied with $f \in \{12.5, 25, 37.5, 50, 62.5, 75, 87.5, 100\}$ [Hz]. The mean displacement amplitude is computed as follows.

$$\overline{u}_y = \frac{1}{0.5\,\mathrm{m}} \int_0^{0.5\,\mathrm{m}} |u_y|\, dx \tag{9}$$

Figure 3 includes results for $L_{tot} = 10\,\text{m}$, $L_{tot} = 20\,\text{m}$ and reference solutions by Degrande with a spectral model. In general, the mean displacement under the loaded strip decreases for a growing excitation frequency while the amplitude p_0 remains constant. This indicates that the medium behaves stiffer under the loaded strip for higher frequencies. For $f = 12.5\,\text{Hz}$ and a halfspace radius $L_{tot} = 10\,\text{m}$, the value $\overline{u}_y$ for the WBM model ($5.68 \cdot 10^{-9}\,\text{m}$) is larger than for the spectral element model ($5.0 \cdot 10^{-9}\,\text{m}$). This probably results from reflections along the absorbing boundary, which transmits incident wave fronts that are perpendicular to it. Unlike the case of a point load, the loaded strip generates wavefronts, which are not normal to the absorbing edge, causing reflections. This undesired effect can be reduced by decreasing the ratio between the generated wavelengths and the domain dimensions. By increasing the halfspace radius for example, more deflections occur within the wave pattern of the domain, so that the wave front arrives with a lower intensity at the transmitting boundary. In Fig. 3, red points indicate simulation results for a halfspace radius of 20 m, which reveal an average displacement amplitude of $4.73 \cdot 10^{-9}\,\text{m}$ for $f = 12.5\,\text{Hz}$. This value is about 5.4% less than the reference solution.

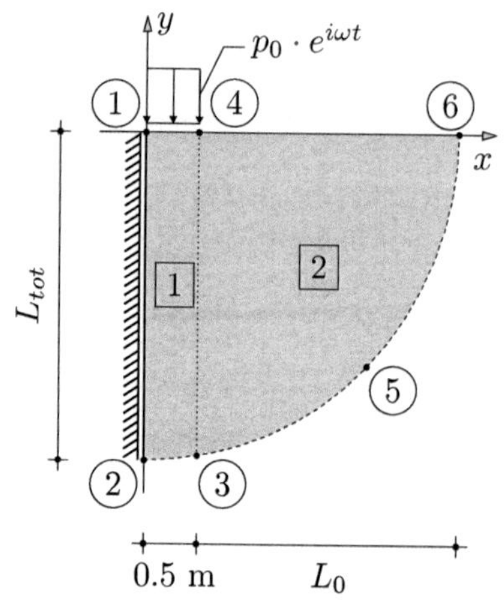

Fig. 3. Average displacement amplitude $\overline{u}_y$ under the loaded strip

4.3 Influence of the Hydraulic Conductivity k_c

In the following presented evaluation, the example from Fig. 2 has been investigated for $L_0 = 20\,\text{m}$, $k_c \in \left[1.0 \cdot 10^{-5}, 1.0\right]$ [m/s] and $f \in \{12.5, 25, 37.5, 50, 62.5, 75, 87.5, 100\}$ [Hz]. With these parameters, the ratio between the transmitted power along the absorbing boundaries (P_{out}) and the inserted power (P_{in}) was computed. The mean power, which flows through the absorbing boundary of the halfspace within one period T, is described by the intensity I, which is integrated along the considered edge. Here, $\widehat{\diamond}$ refers to a complex conjugate field variable.

$$P = \frac{1}{T} \int_T \int_\Gamma I \, d\Gamma \, dt \quad \text{with} \quad I = \frac{1}{2} \widehat{\boldsymbol{\sigma}}^T \cdot \frac{1}{2} \mathbf{v} \tag{10}$$

Figure 4 indicates with two diagrams for different values of the hydraulic conductivity, how the power ratio changes in dependency of the excitation frequency. In general, this value decreases for higher frequencies and consequently smaller wavelengths, as the influence of damping on the wave pattern becomes more dominant. Figure 4a shows that the power ratio tends to diminish for a hydraulic conductivity, which is raised from $k_c = 1 \cdot 10^{-5}$ m/s to $7.5 \cdot 10^{-3}$ m/s. After reaching a minimum value $P_{out}/P_{in} = 1.1\%$ for $f = 100$ Hz and $k_c = 7.5 \cdot 10^{-3}$ m/s, the hydraulic conductivity is increased.

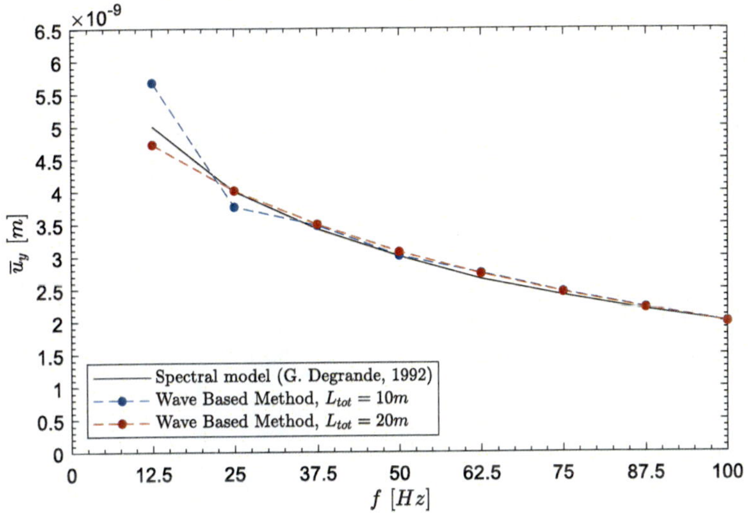

Fig. 4. Convergence of the power ratio to a minimum (—) for each frequency

Figure 4b depicts the development of the power ratio for $k_c \geq 7.5 \cdot 10^{-3}$ m/s and shows that this value predominantly rises for higher hydraulic conductivities. In this plot, a maximum value of 92.8% could be reached for $f = 12.5$ Hz and $k_c = 1$ m/s.

Independently of the chosen hydraulic conductivity, the radiated power of the system does not fall below a certain limit value. Degrande [10] and Garg et al. [13] show that a characteristic radial frequency ω_0 is introduced when solving the two Navier equations, which refer to the first and second P-wave. This value describes the transition between two types of system behavior. For $f < \omega_0/2\pi$ the structure acts nearly like one phase and heads for the state of a frozen mixture. The state of a frozen mixture represents in this context an ideal condition under which no seepage field exists. Whereas for excitation frequencies $f > \omega_0/2\pi$, relative displacements between the enclosed fluid and the solid skeleton become dominant. The simulations, which were performed for this paper, indicate that for each tested excitation frequency there is a minimum radiated power, described by a straight line (——) in Fig. 4. Each minimum power

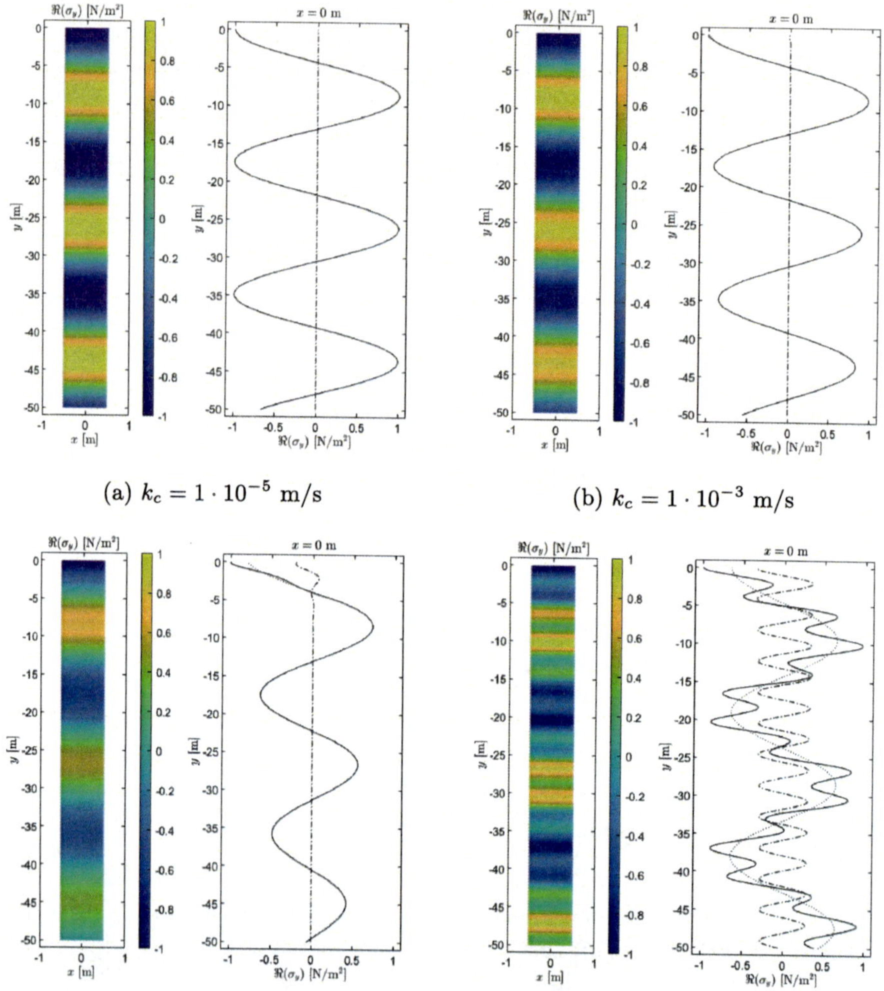

(a) $k_c = 1 \cdot 10^{-5}$ m/s

(b) $k_c = 1 \cdot 10^{-3}$ m/s

(c) $k_c = 7.5 \cdot 10^{-3}$ m/s

(d) $k_c = 1$ m/s

Fig. 5. Illustration of the stresses $\Re(\sigma_y)$ in a soil column for different k_c: total stresses (colored, ——), stresses due to the first P-wave (········), stresses due to the second P-wave (-·-·)

value is characterized by a determined hydraulic conductivity, which may be evaluated according the following equation, derived by Degrande [10]:

$$k_c = g \cdot \frac{\rho n^w}{\omega_0(\rho a - \rho_w n^w)} \tag{11}$$

In order to illustrate how the first and second P-wave propagate for different hydraulic conductivities, a soil column out of Molsand soil with a length $L = 50\,\text{m}$ and a width $B = 1.0\,\text{m}$ is tested. The surface of the column is excited by the constant line load $p(x,t) = 1.0\,\text{N/m}^2 \cdot e^{i\omega t}$, while the bottom is described by an absorbing boundary condition. The system is excited with a frequency $f = 100\,\text{Hz}$ and tested for $k_c \in \{1 \cdot 10^{-5}, 1 \cdot 10^{-3}, 7.5 \cdot 10^{-3}, 1\}$ [m/s].

Figure 5 presents for each hydraulic conductivity the distribution of the stresses $\Re(\sigma_y)$ as one colored surface plot and a line plot at $x = 0\,\text{m}$. The latter show stress contributions which are related to the first and second P-wave. It can be seen that for $k_c < 7.5 \cdot 10^{-3}\,\text{m/s}$, the influence of the second P-wave on the total stress distribution is marginal. For an increasing hydraulic conductivity from $1 \cdot 10^{-5}\,\text{m/s}$ to $7.5 \cdot 10^{-3}\,\text{m/s}$, the stress amplitude declines faster from $y = 0\,\text{m}$ to $y = -50\,\text{m}$. This indicates stronger damping within the structure for a growing k_c. Furthermore, the second P-wave is more distinctive for a hydraulic conductivity $k_c > 7.5 \cdot 10^{-3}\,\text{m/s}$. The solid and fluid phases are less coupled and cause two P-waves which seem to oscillate nearly independently from each other, as it is depicted in Fig. 5c for $k_c = 1\,\text{m/s}$.

5 Conclusion

In order to apply the Wave Based Method to boundary value problems in the field of porous saturated soil structures, an absorbing boundary condition has been introduced. This allows to fulfill the Sommerfeld radiation condition and to model halfspace problems. For this, the method was validated for a quadrant domain and was compared with reference solutions by Degrande [10]. Afterwards the model was tested for different hydraulic conductivities and excitation frequencies in order to illustrate how the coupling behavior between the solid and fluid phases changes. It could be shown that the radiated power within one period T does not go below a minimum value independently of the prevailing hydraulic conductivity.

References

1. Biot, M. A.: Theory of propagation of elastic waves in a fluid-saturated porous solid. I. Low-frequency range. J. Acoust. Soc. Am. **28**(2), 168–178 (1956)
2. Biot, M.A.: Mechanics of deformation and acoustic propagation in porous media. J. Appl. Phys. **33**(4), 1482–98 (1962)
3. Desmet, W.: A wave based prediction technique for coupled vibro-acoustic analysis. Ph.D. thesis 98D12, KU Leuven, Division PMA (1998)

4. van Hal, B.: Automation and performance optimization of the wave based method for interior structural-acoustic problems. Ph.D. thesis, Faculty of Engineering, Katholieke Universiteit Leuven (2004)
5. Pluymers, B.: Wave based modelling methods for steady-state vibro-acoustics. Ph.D. thesis 2006D04, KU Leuven, Division PMA, Leuven (2006)
6. Vanmaele, C.: Development of a wave based prediction technique for the efficient analysis of low- and mid-frequency structural vibrations. Ph.D. thesis 2007D11, KU Leuven, Division PMA (2007)
7. Deckers, E.: A wave based approach for steady-state Biot models of poroelastic materials. Ph.D. thesis 2012D12, Faculty of Engineering, Katholieke Universiteit Leuven (2012)
8. Deckers, E., Van Genechten, B., Vandepitte, D., Desmet, W.: Efficient treatment of stress singularities in poroelastic Wave Based models using special purpose enrichment functions. Comput. Struct. **89**, 1117–1130 (2012)
9. Deckers, E., Vandepitte, D., Desmet, W.: A wave based method for the axisymmetric dynamic analysis of acoustic and poroelastic problems. Comput. Methods Appl. Mech. Eng. **257**, 1–16 (2013)
10. Degrande, G.: A spectral and finite element method for wave propagation in dry and saturated poroelastic media. Ph.D Thesis. Katholieke Universiteit Leuven (1992)
11. Degrande, G., De Roeck, G.: A transmitting boundary condition for wave propagation in saturated poroelastic media. Soil Dyn. Earthq. Eng. **12**(7), 423–32 (1993)
12. Moldovan, D. I.: Hybrid-Trefftz elements for elastodynamic analysis of saturated porous media. Ph.D thesis, Universidade Técnica de Lisboa (2007)
13. Garg, S.K., Adnan, H.N., Good, A.J.: Compressional waves in fluid-saturated elastic porous media. J. Appl. Phys. **45**(5), 1968–1974 (1974)

Influence of Foundations Type on Traffic-induced Vibration Assessment Using an Experimental/Numerical Hybrid Methodology

Paulo J. Soares[1(✉)], Pedro Alves Costa[1], Robert Arcos[2,3], and Luís Godinho[4]

[1] Construct-FEUP, University of Porto, Porto, Portugal
up20305102@edu.fe.up.pt, pmbcosta@reit.up.pt

[2] Acoustical and Mechanical Engineering Laboratory (LEAM), Universitat Politécnica de Catalunya (UPC). c/ Colom, 11, 08222 Terrassa, Barcelona, Spain
robert.arcos@upc.edu

[3] Serra Húnter Fellow, Universitat Politécnica de Catalunya (UPC), Barcelona, Spain

[4] Universidade de Coimbra, Faculdade de Ciências e Tecnologia, Coimbra, Portugal
lgodinho@dec.uc.pt

Abstract. Prediction of vibrations in buildings due to railway traffic is challenging due to several aspects, namely the huge dimensions of the domain, the interaction between different systems, and the uncertainties inherent to the complexity of the system. Therefore, experimental/numerical hybrid prediction techniques are of great interest since they take advantage of the combination between experimental data and numerical resources. In this context, the type of foundations of a building are preponderant in the dynamic soil-structure interaction effect and may have an influence on the accuracy of the hybrid prediction. The presence of the building foundations could significantly modify the response of the local sub-soil of the building. This fact makes that the prediction of vibrations in buildings becomes a complex wave propagation problem. As different types of foundations behave differently, they lead to different dynamic responses in the structure for a prescribed incident wave field. Therefore, dynamic soil-structure interaction coupling is also affected by the building's foundation type. In order to investigate this phenomenon, this paper studies the influence that the consideration of three different foundation typologies, shallow, slab and deep foundations, have on the prediction of railway-induced vibrations on a building using an experimental/numerical hybrid methodology. A comparison is made between the results obtained through the hybrid method and the numerical results obtained through a numerical model based on the full system.

Keywords: Hybrid methodology · Railway traffic · Vibrations

Z. Dimitrovová et al. (Eds.): WMVC 2022, MMS 125, pp. 317–326, 2023.
https://doi.org/10.1007/978-3-031-15758-5_31

1 Introduction

A very important aspect to take into account when assessing the vibrations in buildings to be constructed close to an operational railway line is the foundations type. They have a preponderant role in the soil-structure interaction. Train traffic on railway lines near the site where a new building will be implemented induces vibrations in the soil that are transmitted to the building. When only soil-structure kinematic interaction is accounted for, the motion of the foundations is assumed to be the ground motion induced by the incident wave field without the building. However, there are some situations where this assumption is no longer valid due to the influence of the dynamic behaviour of the structure on the foundation's response. In these situations, dynamic soil-structure interaction should be considered.

Railway traffic in urban environments has become an important source of noise and vibration pollution these days. Railway-induced vibration is perceived by inhabitants of buildings in the range of frequency 1 Hz 80 Hz. In this range, dynamic soil-structure interaction should be considered if accurate predictions are required. This phenomenon not only depends on the geometrical configuration of the foundations but also on the local sub-soil mechanical properties and the specific characteristics of the train traffic in terms of axle loads, speed, etc. Methodologies for the prediction of railway-induced vibrations in the building has been proposed by various researchers. Papadopoulos et al. [1] studied the influence of uncertain local subsoil conditions on the response of buildings to ground vibration and most recently also Kuo et al. [2] presented a study with a hybrid model where the authors studied the effect of dynamic soil characteristics, surface foundation type, and building geometry on the building's response to railway induced vibrations [3]. The need to find simplified and equally accurate methodologies have led to developments in this field of research, Mendoza et al. [4] presented a new approach to predict building vibration based on a transfer function method. Along this work a 2.5D coupled FE-BE model was used to predict railway induced vibration in a building, this numerical model was calibrated with the modal characterisation determined experimentally.

The experimental/numerical hybrid methodology used in this work, previously described by the authors in [5–7], addresses soil-structure interaction problems accurately since it considers both kinematic and dynamic interactions through the numerical modelling of the foundations.

1.1 Hybrid Experimental/Numerical Methodology

The hybrid method for the prediction of railway-induced vibration in buildings proposed in [5] was thought and developed to address situations in which new structures are planned to be constructed next to operational railway lines. This method allows for the prediction of vibration levels at any point of the domain under study once the structure is already implemented. It is considered to be a hybrid method since it combines in situ experimental measurements at the ground surface and numerical models of the local sub-soil, foundations and the structure itself. The 2D FEM-PML [8] approach was the one used in this work.

This hybrid methodology is based on a similar approach to the method of fundamental solutions (MFS) [9] applied to radiation and scattering problems in elastodynamics. It uses a set of virtual forces calculated in order to represent the incident wave field that induces the vibration ground surface pattern experimentally determined in situ. The ground response is measured in a set of points at the surface called collocation points. The methodology presented consists of three steps. The first one is explained with the help of Fig. 1, where Ω represents the soil domain and $d\Omega$ represents the ground surface where the structure understudy will be built, also mentioned as the collocation points domain. The aim of this step is to determine the vibration field at the section of the ground surface where the building will be constructed. This vibration field is represented by vibration measurements at several collocation points located along $d\Omega$.

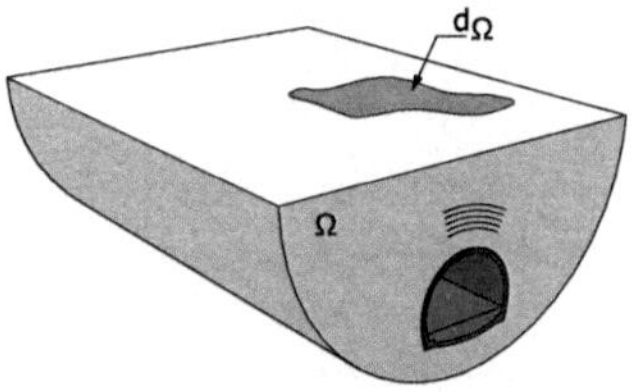

Fig. 1. Schematic representation of the first step of the hybrid methodology: evaluation of the boundary conditions on the ground surface where the building will be built.

Once the experimental component of the method is completed, the numerical part should be performed. The different domains and surfaces considered in the second step are represented in Fig. 2 (i). In this step, a virtual traction field that represents the incident wave field described by the measurements on the collocation points is determined. This virtual traction field is located at the auxiliary surface S, which encloses the local sub-soil domain Ω_s. The boundary integral equation along the virtual surface S for the elastodynamic interior problem in the frequency domain can be written according to Eq. (1) as

$$\mathbf{U}_r(\mathbf{r}, \omega) = \int_S \mathbf{H}(\mathbf{r}, \mathbf{r}_s, \omega)\, \mathbf{T}_v(\mathbf{r}_s, \omega)\, \mathrm{d}S(\mathbf{r}_s). \tag{1}$$

where $\mathbf{U}_r = \{U_{rx}\ U_{ry}\ U_{rz}\}^{\mathrm{T}}$ represents the ground displacements at a position $\mathbf{r} = \{x\ y\ z\}^{\mathrm{T}}$ restricted to the domain Ω_s in the three Cartesian directions and $\mathbf{H}(r, r_s, \omega)$ represents the soil Green's functions that relate the response at $\mathbf{r}$ to a force applied at $\mathbf{r}_s$, where $\mathbf{r}_s = \{x_s\ y_s\ z_s\}^T$ corresponds to the location of $\mathrm{d}S$ within the domain S. Finally, $\mathbf{T}_v(\mathbf{r}_s, \omega)$ represents the tractions at the virtual surface S considering a normal vector outward to the domain Ω_s.

Equation (1) is discretised on the basis of the MFS, reaching the simplified expression

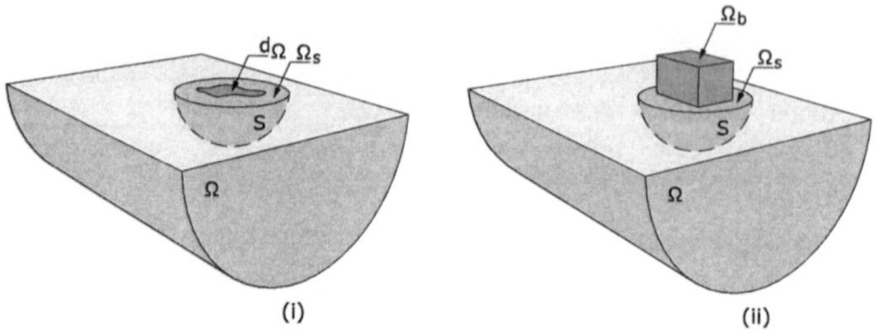

Fig. 2. Schematic representation of the second (i) and third (ii) steps of the hybrid methodology.

$$\mathbf{U}_c = \mathbf{H}_{cf}\mathbf{F}_v. \tag{2}$$

where $\mathbf{U}_c$ is the vector of all degrees of freedom in the collocation points, $\mathbf{F}_v$ represents the vector of all virtual force components (for all virtual sources) and $\mathbf{H}_{cf}$ is the matrix of receptances that relate $\mathbf{U}_c$ and $\mathbf{F}_v$ in the local sub-soil. Thus, virtual forces can be simply determined using the expression

$$\mathbf{F}_v = \mathbf{H}_{cf}^{-1}\mathbf{U}_c. \tag{3}$$

where $\mathbf{U}_c$ is collecting the experimentally measured response at the collocation points and the receptance matrix $\mathbf{H}_{cf}$ may be determined numerically through a model of the local sub-soil. Since the number of collocation points and virtual forces is adopted to be equal in this work, this leads to a square receptance matrix. It is considered that virtual forces are uniformly distributed along the surface S.

Once the virtual forces are obtained, they could be applied to a building-soil model to obtain the response on the building due to the targeted incident wave field. This building-soil model, as it is schematically represented in Fig. 2 (ii), should include the reduced local sub-soil domain Ω_s' (equal to Ω_s minus the foundations domain) and the building domain, Ω_b. Thus, it is possible to determine the response in the building according to the Eq. (4),

$$\mathbf{U}_b = \mathbf{H}_{bf}\mathbf{F}_v. \tag{4}$$

where $\mathbf{H}_{bf}$ is the receptance matrix of the building-soil system that relates the displacements at any evaluation point on the building with the virtual forces. The vector $\mathbf{U}_b$ represents the predicted response at the evaluation points. This methodology is based on the hypothesis that the coupling between the railway structure and the building is weak, which means that it is assumed that the transfer functions between the soil and the building are not affected by the presence of the railway structure, track, or tunnel. However, this assumption is invalid when the building and the railway structure are in close proximity. Coulier, et al. [10] presented a study on the influence of the source-receptor coupling on the conformity of the track response and the dynamic transfer functions between the source (railway infrastructure) and receiver (building). In recent work, the authors [5] have shown how the present method is affected by this phenomenon.

2 Numerical Study

2.1 Influence of Foundation Types on the Building Response

The influence that the building foundation type has on the building response and on the performance of the proposed hybrid method, for the simulation of the displacement response at a given point of evaluation, has been studied. In Fig. 3 it is possible to observe the building, the tunnel and the evaluation point, represented with a red dot. The vertical load F was applied on the tunnel invert, located 8 m depth and 19 m from the building. This study was carried out considering three types of foundations, shallow, slab, and deep, through the case studies presented in Fig. 4.

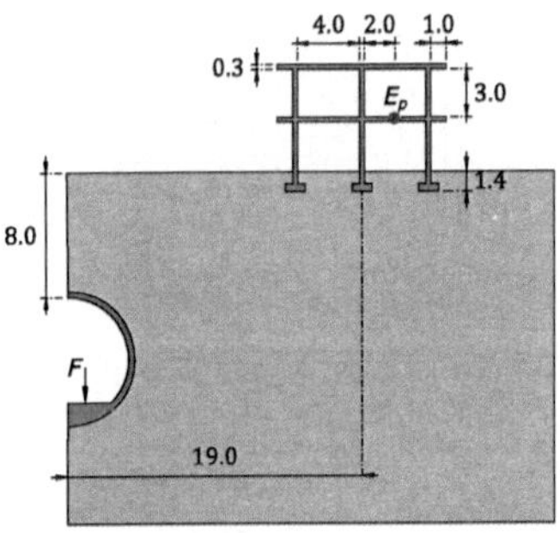

Fig. 3. Case study, tunnel location in relation to the building. The red dot denotes the position of the evaluation point. Distances in meters.

The response at the evaluation point was calculated with a 2D FEM-PML for the three different types of building foundations types, shallow, slab, and deep, corresponding to the schematization (i), (ii), (iii) in Fig. 4, respectively.

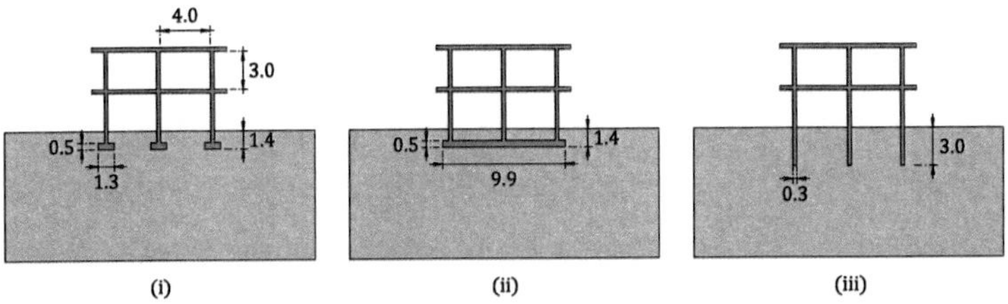

Fig. 4. Geometric scheme of the three types of foundations, (i) shallow, (ii) slab and (iii) deep. Distances in meters.

For this case study, a homogeneous half-space model of the ground has been considered, whose mechanical properties are presented in Table 1. The variables ρ, E, ν, and ξ represent the density, Young's modulus, Poisson's coefficient, and hysteric damping of both systems, respectively. The building structure is considered to be concrete with mechanical parameters also shown in Table 1.

Table 1. Mechanical properties of the soil and building.

Material	ρ $[kg/m^3]$	E [MPa]	ν $[-]$	ξ $[-]$
Soil	2000	195	0.3	0.04
Building	2500	30000	0.2	0.01

In Fig. 5, the responses of the reference model for each of the foundation types under study. It is shown that foundation type is not affecting significantly the response of the building at frequencies 20 Hz, while the effect is quite significant 20 Hz.

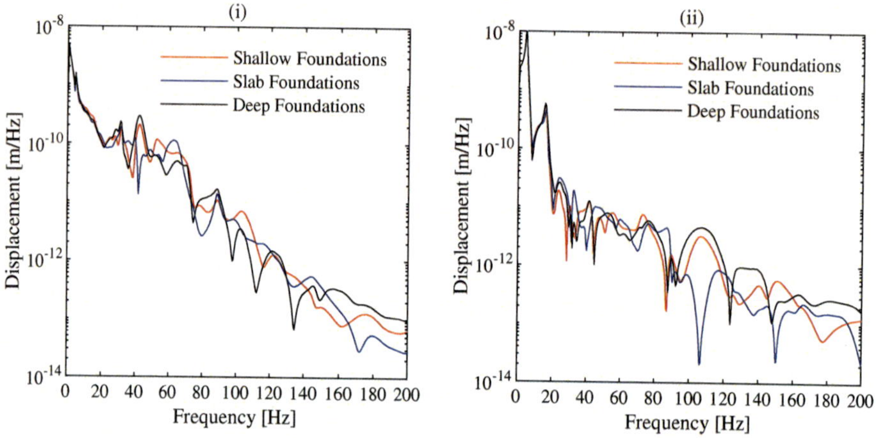

Fig. 5. Vertical (i) and horizontal (ii) displacements at the evaluation point for the three different foundation type cases considered in this study.

2.2 Influence of Foundation Types on the Accuracy of the Hybrid Methodology

In this section, the performance of the hybrid method depending on the foundation type is studied and evaluated. The response at the evaluation point was calculated in two ways: with the hybrid method and with a 2D FEM-PML of the full tunnel-soil-building system. To reduce the computational effort, the symmetry properties of the system were considered when 2D FEM-PML models of the systems were created. To have control over the results and ensure the accuracy of the hybrid approach, a synthetic field of vibration obtained with a 2D FEM-PML model of the tunnel-soil system was used instead of real measurements as input for the hybrid method. The reference response, at the evaluation point, was obtained with a 2D FEM-PML global model of the tunnel-soil-building system. The simulated displacement components at the evaluation point for both models are represented in Fig. 6. In this figure, the results for the hybrid methodology are shown for different arrangements of virtual forces. These virtual forces are uniformly distributed along a semicircle, the r is the radius of this semicircle and N is the number of virtual sources.

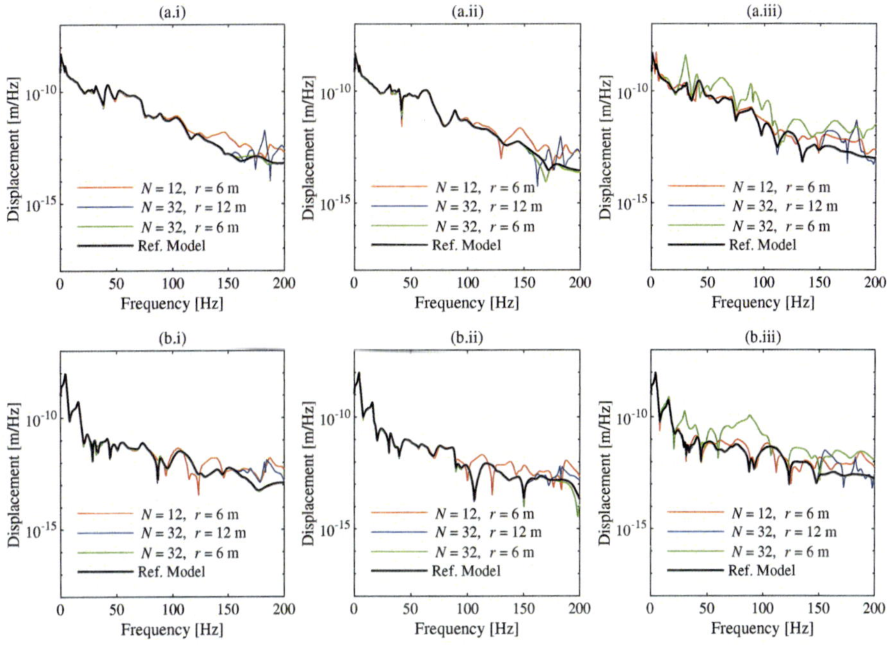

Fig. 6. Vertical (a) and horizontal (b) displacements at the evaluation point for the different foundation types: (i) direct foundations, (ii) foundation slab and (iii) deep foundations.

The agreement between the responses associated with each set of virtual forces and the type of foundation is significant up to the same frequency (20 Hz) in the three cases. The vertical component corresponding to the case of slab foundation presents some differences at the frequency of 42 Hz, coinciding with the peak response associated to the specific foundation type. A very important aspect is the fact that the set of virtual sources defined by $N = 32$ and $r = 6$, which for the cases of shallow and slab foundations coincide with the response of the reference model, is not coincident with the reference model in the case of deep foundations. This disagreement appears because, for this virtual forces set, their proximity between virtual source and the foundations is higher than in other cases. The graphical representation of the spatial distribution of virtual forces can be compared in Fig. 7.

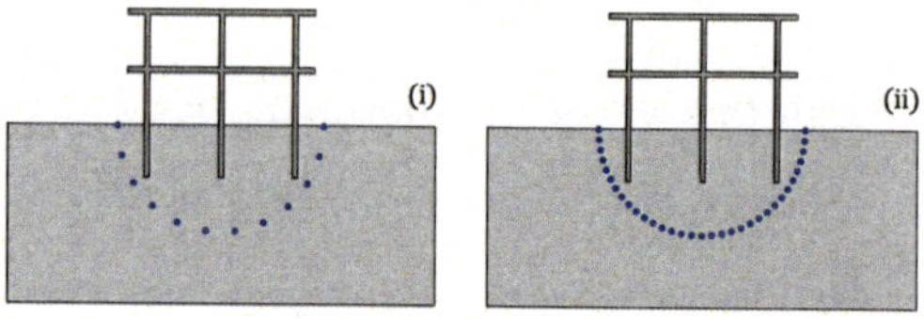

Fig. 7. Configuration of virtual forces distribution for cases $N = 12$ and $r = 6$, (i) and $N = 32$ and $r = 6$, (ii).

To have the global perception of the accuracy of the proposed hybrid method, the parameter ε has been used to evaluate the error, which expresses the accuracy of the response obtained by the hybrid method with respect to the numerical reference model. This parameter was calculated according to Eq. 5.

$$\varepsilon = \frac{1}{N_f} \sum_{i=1}^{N_f} \frac{\left|U_{bi}^r - U_{bi}^h\right|}{U_{bi}^r}. \tag{5}$$

Where N_f is the number of sampling frequency points corresponding to the calculation interval and U_{bi}^r is the displacement value, at the ith frequency sample, obtained with the reference model whilst U_{bi}^h is the displacement value calculated with the hybrid method. The frequency range considered for this study is 1 Hz to 80 Hz: these values were chosen because of the relevant frequency interval for the problem of vibrations induced in buildings [11]. The number of virtual sources considered varied between 4 and 32 and the radius of the semicircle varied between 6 and 12 meters. In Fig. 8 the results of the parametric study

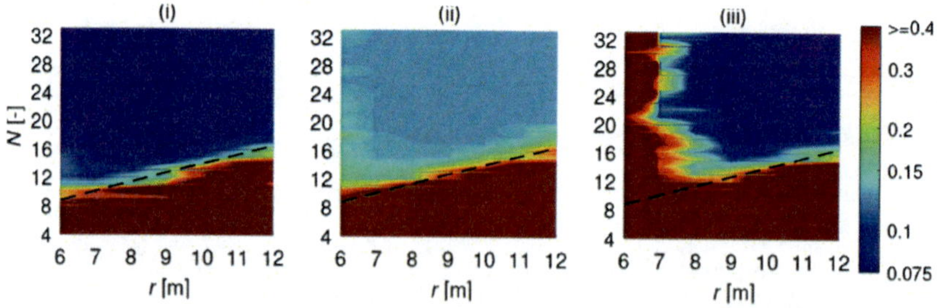

Fig. 8. Global error, ε, referring to the parametric study. The figure (i), (ii) and (iii) correspond to the case of shallow, slab and deep foundations, respectively.

are presented in terms of the parameter ε. The empirical criteria, $d < \lambda_s$, that established the rule to guarantee the methodology precision [5] has been also illustrated.

The geometry of this case study corresponds to the values $L_z = 8$ m and $L_y = 19$ m and it was considered a homogeneous geotechnical scenario (Soil A). The analysis of Fig. 8 shows a good agreement between the results of both methods and also demonstrates the adequacy of the empirical rule proposed. It is important to note that the discrepancies appearing for the deep foundations case, the results of the parametric study are presented in terms of the parameter ε is related with the proximity between the virtual forces and the building foundation.

For the deep foundations scenario, it is possible to note that the accuracy of the hybrid methodology is clearly unstable for the cases $r < 8$. A comprehensive analysis was performed to better understand this phenomenon. In this analysis, the distance from the building center to the foundations edge is 5.37 m, and the radius for the cases that lead to good predictions is 8 m. The difference between these values, 2.63 m, corresponds to the minimum distance between the virtual forces and the foundations. When this value is compared with the wavelength of the S-waves for maximum frequency of interest, 2.42 m, it is possible to conclude that the virtual forces should be at least one wavelength away from the foundations edge. Hus, to ensure a good performance of the hybrid methodology, in addition to the compliance with the empirical rule already studied, it is necessary to ensure a minimum separation between the edge of the building foundations and the virtual sources of at least one S-wavelength.

3 Conclusions

The main conclusions of this work are:

- The results obtained in all the case studies presented are demonstrating that the hybrid method performs accurately when the empirical rule is assured, meaning that the distance between sources is smaller than the wavelength of the S-waves for the highest frequency of interest. This condition is found to be crucial for the proper operation of the proposed hybrid methodology.
- For the studied cases, the building foundation type variation had no influence on the displacement at the evaluation up to a frequency of 20 Hz. Because up to this frequency value the response at the assessment point is the same for all three foundation types.
- To avoid numerical errors, the virtual forces should be placed at least one S-wavelength away from the building's foundations.
- Foundations with more rigidity have more impact on the error, the slab foundations have slightly more impact in the global method error than shallow foundations.

References

1. Papadopoulos, M., François, S., Degrande, G., Lombaert, G.: The influence of uncertain local subsoil conditions on the response of buildings to ground vibration. J. Sound Vib. **418**, 200–220 (2018)
2. Kuo, K.A., Papadopoulos, M., Lombaert, G., Degrande, G.: The coupling loss of a building subject to railway induced vibrations: Numerical modelling and experimental measurements. J. Sound Vib. **442**, 459–481 (2019)
3. Papadopoulos, M., et al.: Numerical prediction and experimental validation of railway induced vibration in a multi-storey office building. In: Degrande, G., et al. (eds.) Noise and Vibration Mitigation for Rail Transportation Systems. NNFMMD, vol. 150, pp. 529–537. Springer, Cham (2021). https://doi.org/10.1007/978-3-030-70289-2_57
4. López-Mendoza, D., Connolly, D.P., Romero, A., Kouroussis, G., Galvín, P.: A transfer function method to predict building vibration and its application to railway defects. Constr. Build. Mater. **232**, 117217 (2020)
5. Arcos, R., Soares, P.J., Costa, P.A., Godinho, L.: An experimental/numerical hybrid methodology for the prediction of railway-induced ground-borne vibration on buildings to be constructed close to existing railway infrastructures: Numerical validation and parametric study. Soil Dyn. Earthq. Eng. **150**, 106888 (2021)
6. Arcos, R., et al.: A hybrid methodology for the assessment of railway-induced ground-borne noise and vibration in buildings based on experimental measurement in the ground surface. In: INTER-NOISE and NOISE-CON Congress and Conference Proceedings, vol. 259. No. 4. Institute of Noise Control Engineering (2019)
7. Costa, P.A., et al.: Hybrid approach for the assessment of vibrations and re-radiated noise in buildings due to railway traffic: Concept and preliminary validation. In: Advances in Engineering Materials, Structures and Systems: Innovations, Mechanics and Applications: Proceedings of the 7th International Conference on Structural Engineering, Mechanics and Computation (SEMC 2019), September 2-4, 2019, Cape Town, South Africa. CRC Press (2019)
8. Lopes, P., Costa, P.A., Ferraz, M., Calçada, R., Cardoso, A.S.: Numerical modeling of vibrations induced by railway traffic in tunnels: From the source to the nearby buildings. Soil Dyn. Earthq. Eng. **61**, 269–285 (2014)
9. Fairweather, G., Karageorghis, A., Martin, P.A.: The method of fundamental solutions for scattering and radiation problems. Eng. Anal. Boundary Elem. **27**(7), 759–769 (2003)
10. Coulier, P., Lombaert, G., Degrande, G.: The influence of source-receiver interaction on the numerical prediction of railway induced vibrations. J. Sound Vib. **333**(12), 2520–2538 (2014)
11. Standard, I. S. O. (1989). ISO2631/2: Mechanical vibration and shock-evaluation of human exposure to whole body vibration-Part2: Continuous and shock induced vibration in buildings (1–80 Hz)

In-situ Measurements Frequency Analysis at a Site Scale. Application to Vibrations Induced by Tunnel Boring Machines

Antoine Rallu[1(✉)] and Nicolas Berthoz[2]

[1] University of Lyon, ENTPE, LTDS - UMR CNRS 5513 - CeLyA, Vaulx-en-Velin, France
antoine.rallu@entpe.fr
[2] French Centre for Tunnel Studies (CETU), Bron, France

Abstract. In this paper, a methodology is proposed in order to extract a relevant representation of a response spectrum considering numerous acquisitions from dynamic sensors located in different places and sollicitated under different sollicitations. This methodology, which is an extension of the Frequency Domain Decomposition method, is based on the intercorrelation of signals and the extraction of the singular values of the spectral density matrix. We illustrate it on the study of the impact of vibrations generated by the excavation of a tunnel by a tunnel boring machine (TBM), with measurements on the ground and inside a TBM under ambient noise then during the excavation.

Keywords: Response spectrum · Signal processing · In-situ measurements · Ground-borne vibrations

1 Introduction

Time-frequency duality is a concept that has been well known since Joseph Fourier in the XIXth century, the first who have had the ingenious idea of decomposing a phenomenon according to its eigenfrequencies. The scope of this theory is multidisciplinary, and the methods based on it have become powerful with the arising of computing tools and the fundamental paper of Cooley and Tukey [3] establishing the Fast Fourier Transform.

The interpretation of dynamic measurements in geophysics or infrastructure takes full advantage of these concepts, analysing on the one hand the temporal content (e.g. the amplitude of seismic waves propagating in a ground or in a building), and on the other hand the frequency content (e.g. response spectrum of a structure to a dynamic input) allowing to identify the frequencies for which the movements are preferential (e.g. structural eigenmodes). Operational modal analysis allows the extraction of the modal features of a structure on which several sensors acquire its response to ambient mechanical noise, the flagship method of which being the Frequency Domain Decomposition [1].

Z. Dimitrovová et al. (Eds.): WMVC 2022, MMS 125, pp. 327–336, 2023.
https://doi.org/10.1007/978-3-031-15758-5_32

The experimental study of wave propagation in a ground in the presence of an underground vibratory source may require the analysis of numerous acquisitions of dynamic phenomena: in the example proposed in Sect. 3, the experimental set-up was made of sensors on the ground and inside a tunnel boring machine; Moreover the acquisitions on the ground were carried out before and during the excavation: in total, three sets of measurements have to be analysed. In the time domain, it is fairly straightforward to evaluate the vibration levels at different points in space under various loads. However, it is much more difficult to extract a relevant information on the frequency contents at the site scale, that is considering all the acquisitions carried out during the in-situ campaign: indeed, it is possible to analyse the frequency content of each individual acquisition, but the response of each one depends on the close field on which it is located.

The goal of this paper is consequently to propose a concise representation of the frequency contents of all the sensors, the key idea being the intercorrelation between each sensor. In a first section, the general methodology is described, then its illustration on the Lyon-tunnel excavation is proposed.

2 Individual and Site Frequency Analysis

2.1 Assumptions on the Signals

In this paper we consider discrete (numerical) signals measured with digital sensors at a sampling frequency $f_s = 1/dt$, with dt the time step. That means (i) they take real values and (ii) they have a finite length $n \in \mathcal{N} = [\![0; L]\!]$ such as $T = Ldt$ with $I = [0, T]$ the time range of acquisition: numerical signals are described as $\{t_n = ndt\,;\, v_n = v(t_n) \in \mathbb{R}\}_{n\in\mathcal{N}}$. Moreover, in order to be processed by the proposed method they must check the following assumptions:

Stochasticity. Consequently a probabilistic approach must be followed: for example consider a set $\mathcal{K} = [\![1; K]\!]$ of realisations of a physical variable $\{\overset{k}{v}_n\}_{(k,n)\in\mathcal{K}\times\mathcal{N}}$. Denoting $E[.]$ the mathematical expected value, the statistical average at the given time $t_\ell \in I$, denoted $\mu_v(t_\ell)$, is defined as (see Fig. 1):

$$\mu_v(t_\ell) = E[v_\ell] = \lim_{K\to\infty} \frac{1}{K} \sum_{k=1}^{K} \overset{k}{v}_\ell$$

Moreover all the variables are considered centered, that is :

$$E[v_\ell - \mu_v(t_\ell)] = 0$$

In the same spirit, the intercorrelation of two variables (v, w) is defined as (we talk about autocorrelation if $v = w$) :

$$\forall p \in \mathcal{N},\, \mathscr{C}_{vw}(t_\ell, t_\ell - t_p) = E[v_\ell w_{\ell-p}] = \lim_{K\to\infty} \frac{1}{K} \sum_{k=1}^{K} \overset{k}{v}_\ell \overset{k}{w}_{\ell-p}$$

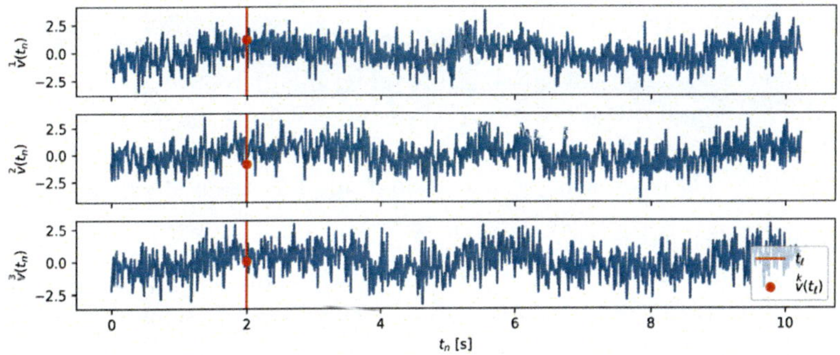

Fig. 1. Illustration of the statistical average.

Stationarity. Their statistical characteristics of the first and second order are constant:

$$\mu_v = \mu_v(t_\ell) \quad ; \quad \forall p \in \mathcal{N}, \, \mathscr{C}_{vw}(t_p) = \mathscr{C}_{vw}(t_\ell, t_\ell - t_p)$$

Ergodicity. All their statistical features can be determined from the temporal analysis of one unique realisation. This fundamental assumption allows to identify the temporal average (for a given representation) $\bar{v}$ (resp. intercorrelation) with the statistical average (for a given time) μ_v (resp. intercorrelation):

$$\mu_v = \bar{v} = \frac{1}{L+1} \sum_{n \in \mathcal{N}} v_n \, ; \, \mathscr{C}_{vw}(t_p) = \frac{1}{L+1} \sum_{n \in \mathcal{N}} v_n w_{n-p}$$

The last but not least consequence is the validity of the Wiener-Khintchine theorem, linking the Power Spectral Density (PSD) denoted $\mathscr{S}_{vw}$ and the intercorrelation of variables:

$$\forall m \in \mathcal{N}, \, (\mathscr{S}_{vw})_m = \mathcal{F}_m \left(\mathscr{C}_{vw} \right)$$

where $\mathcal{F}$ takes for the discrete Fourier operator, as $(\mathbf{i}^2 = -1)$:

$$\forall m \in \mathcal{N}, \, \mathcal{F}_m(v) = dt \sum_{n \in \mathcal{N}} v_n e^{-\mathbf{i} 2\pi \frac{mn}{L+1}}$$

Without Time Offset. Their time average must be null: $\bar{v} = 0$

Thus, the response of the system is considered characterized by the first and second stochastic moments because of the gaussian nature of (i) the multiple noise inputs (road and pedestrian traffic, ...) and of the TBM excitation, and of (ii) the response of a linear system to gaussian inputs. This latter can be checked by computing the distribution of output velocities.

The original framework of this method has been proposed in the previous paper [6] for continuous (analogical) signals. Here we derive it for numerical signals in order to propose a direct transposition to any computing language.

2.2 Power Spectrum

The frequency spectrum of a time-infinite integrable signal is well computed by a direct FFT [3] computation. However, the FFT of a finite length signal generates a very noisy spectrum. To deal with this issue, the Welch's method [7] allows to compute the PSD via an averaged periodogram. The main steps are as follow:

Tapering. The signal $\{v_n\}_{n\in\mathcal{N}}$ is split into N_S segments by a tapering window $\{w_n\}_{n\in\mathcal{N}_S}$ of size L_S, with or without overlap (an overlap of $L_S/2$ with a Hann window is recommanded in [1,7]), each segment signal being:

$$\{^s v_n, (n,s) \in \mathcal{N}_S \times \mathcal{S}\} \qquad \text{with } \mathcal{N}_S = [\![1; L_S]\!]\,;\, \mathcal{S} = [\![1; N_S]\!]$$

The Fourier Transform of a tapering window presents sidelobes that generate unwanted noise, which can be measured by the normalized equivalent noise bandwidth [1] $\text{NB}_s = 1 + \left(\frac{\sigma_w}{\bar{w}}\right)^2$ where $\bar{w}$ and σ_w^2 represent respectively the mean and the variance of the tapering window of length L_S. Finally the scaling is performed according to the effective noise bandwidth: $\text{EB}_s = \frac{f_s}{L_S}\text{NB}_s$. In the following, all the calculated PSD are scaled and the unit of the power spectrum is $((mm/s)^2)$.

Segment PS. On each segment, the power sepectrum is calculated for each discrete frequency $f_n = n\frac{f_s}{L_S}$ ($.^\star$ stands for the complex conjugate):

$$\forall n \in \mathcal{N}_S,\, (G_s)_n = \text{EB}_s \mathcal{F}_n(^s v)^\star . \mathcal{F}_n(^s v) = \text{EB}_s |\mathcal{F}_n(^s v)|^2$$

Averaging. Finally the averaged periodogram is calculated as the arithmetic mean of the PSD of each segment.

$$\forall n \in \mathcal{N}_S,\, G_n = \frac{1}{N_S} \sum_{s\in N_S} (G_s)_n$$

2.3 Frequency Domain Decomposition

In operational modal analysis (OMA), the Frequency Domain Decomposition (FDD) is a well-known method initially developped by Pr. Rune Brincker (see for example [1,2]) in order to extract the modal features of a structure (building, bridge, ...) sollicitated by ambiant mechanical noise, considered as a white noise.

Let consider a set of sensors measuring the same physical growth (velocity or acceleration), each with one or several acquisition channels (classically one or three orthogonal spatial directions (X, Y, Z)). All of these signals must be time synchronized and each set of data (same length $L+1$) must be represented at the same sampling frequency f_s (with a possible downsampling process if needed). In function of the goal of the study, we can take into account the whole acquisition channels, or select only one part (for example only channels in the vertical direction). Finally the considered set of $M+1$ acquisition channels is represented by: $\{v_n^{(m)}\}_{(n,m)\in\mathcal{N}\times\mathcal{M}}$ with $\mathcal{M} = [\![0; M]\!]$ the set of selected channels.

Welch's Method (see Sect. 2.2**).** Each signal $\{\underline{v}^{(m)}\}_{m\in\mathcal{M}}$ is tapered into N_S segments of length L_S, such as each data segment is:

$$\left\{{}^s v_n^{(m)}, (n,m,s) \in \mathcal{N}_S \times \mathcal{M} \times \mathcal{S}\right\}$$

Then, the averaged PS is a third-order tensor $\underline{\underline{\underline{G}}}$ of a size $M \times M \times L_S$ calculated by means on each segment $s \in \mathcal{S}$, such as

$$\forall(\alpha,\beta,n) \in M \times M \times \mathcal{N}_S,\ G_{\alpha\beta n} = \frac{1}{N_S} \sum_{s\in N_S} \mathrm{EB}_s \mathcal{F}_n({}^s v^{(\alpha)})^\star . \mathcal{F}_n({}^s v^{(\beta)})$$

Singular Value Decomposition (SVD). At each frequency $f_n = n\frac{f_s}{L_S}$, the SVD (for details see [1]) is performed on the second order square matrix $\underline{\underline{G}}_n$ of size $M \times M$, in order to factorize it as ($.^H$ stands for the Hermitian transposition):

$$\forall n \in \mathcal{N}_S,\ \underline{\underline{G}}_n = \underline{\underline{V}}_n . \underline{\underline{\Sigma}}_n . \underline{\underline{V}}_n^H \tag{1}$$

with $\underline{\underline{V}}_n$ an unitary matrix and $\underline{\underline{\Sigma}}_n$ a diagonal matrix with real positive values sorted in descending order. $(\underline{\underline{G}}_n)_{n\in\mathcal{N}_S}$ is called the **Spectral Matrix**.

In OMA, the singular values of $\{\underline{\underline{G}}_n\}_{n\in\mathcal{N}_S}$ represent the auto spectral powers of the modal coordinates and $\underline{\underline{V}}_n$ the associated mode shapes [1]. Transposing this idea into soil dynamics, the representation of the largest singular values

$$\left\{\sigma_n = (\underline{\underline{\Sigma}}_n)_{11} = \max_{(\alpha,\beta)\in M\times M} \left\{(\underline{\underline{\Sigma}}_n)_{\alpha\beta}\right)\right\}_{n\in\mathcal{N}_S} \tag{2}$$

in function of the frequency highlights the frequency where the sensors located on the whole site are the most correlated. This method is illustrated in the following section in the case of vibrations generated by the excavation of a tunnel by a tunnel boring machine.

3 Application to Tunnel Excavation in Lyon, France

3.1 Presentation of the Two Sites

The two measurement campaigns have been performed on the same project of extension of the line B of the Lyon subway, but in two different geotechnical contexts: the first one (september 2020) in granites, the second one (january 2021) in pre-glacial alluvium, see Figs. 2 and 3. In both cases, the tunnel boring machine was a variable density pressure TBM (Herrenknecht), whose the main features are presented on Table 1.

Table 1. Features of the TBM; Machines parameters during the two excavation campaigns.

Parameter	Granite	Alluvium
Chamber	Full of bentonite	
Excavation diameter	9.8m	
Cutting wheel speed	3 rpm	2 rpm
Penetration pitch	7 mm/revolution	15 mm/revolution
Torque on the cutting wheel	3 000 kN.m	1000 kN.m
Bentonite face pressure in the tunnel axis	180 kPa	200 kPa
Wheel/ground contact force	15 000 kN	3000 kN

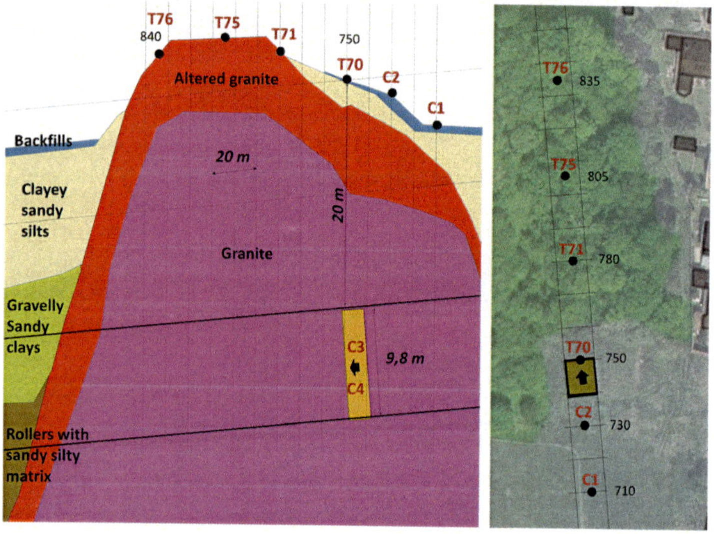

Fig. 2. Plan and longitudinal section of the Lyon-granites measurements (source [6]).

3.2 Instrumentation and Acquisition

The experimental set-up is made of ©TROMINO sensors (speed acquisition) of the Moho brand, denoted C1, C2, C3, C4, T70, T71, T75 and T76. These sensors get three orthogonal acquisition channels (X, Y, Z) of high sensitivity (sensors designed to measure ambient noise). Note that the X-direction is oriented in the tunnel axis in the excavation sens, the Y-direction in the transverse direction, and Z is oriented in the vertical direction.

Before each acquisition, all the sensors are synchronized by GPS at the same location, then moved according to the experimental set-up (see Figs. 2 and 3) on the surface and inside the two manlocks of the TBM. This allow to sensors inside he TBM to be synchronized, even if the connexion to GPS is impossible. Moreover, because of the low vibration levels and the heavy weight of the sen-

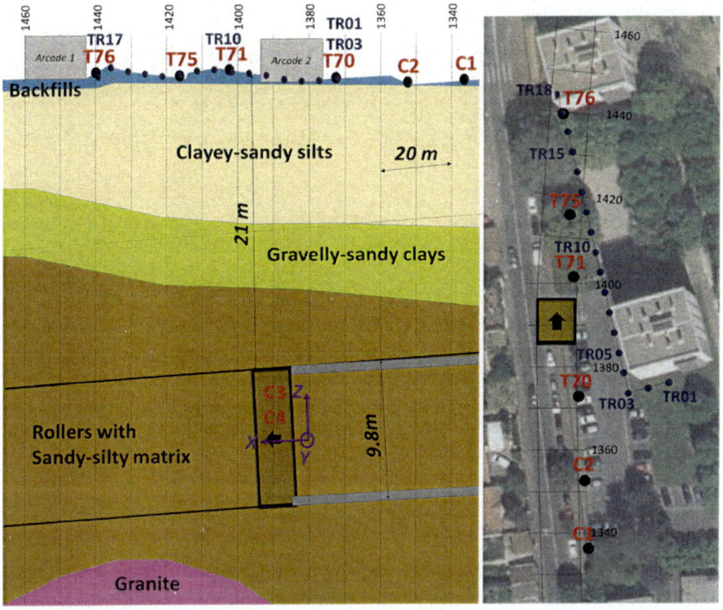

Fig. 3. Plan and longitudinal section of the Lyon-alluvium measurements (source [6]).

sors no sealement is apply to connect them to their support. All measurements were acquired at a sampling rate 512 Hz, allowing to investigate ground borne vibrations ($f \leq 40$Hz) and vibro-acoustic vibrations ($40 \leq f \leq 200$ Hz).

A measurement campaign is carried out in two steps:

Ambient Noise. Only for sensors on the surface, a measurement phase (about 30 min) under ambient mechanical noise (mainly road and pedestrian traffic), allowing to estimate (i) the characteristic ambient noise level and (ii) frequency content of the ground, that is the frequency ranges where the response of the ground is the highest.

Excavation. For sensors on the surface and inside the TBM, a synchronized acquisition phase during the excavation phase.

Each acquisition is pre-processed in order to clean the signals and to satisfy assumptions needed for the proposed methodology: (i) restriction of the time histories to a relevant (quasi)-stationary signal, (ii) suppression of the time-history offset, and (iii) band-pass filtering in the range [0.05–160] Hz corresponding to the usual frequencies of building engineering. All the raw data are available in [5] and in the dataset https://mycore.core-cloud.net/index.php/s/ybajakmrNwcRSqw.

During the excavation phase, the assumption of stationarity of the processed signal is based on (i) the continuous nature of the excavation and (ii) the small variations in the nature of the excavated ground during the time of the processed signal. In reality, although these assumptions are not strictly verified, assuming quasi-stationarity of the signal is relevant. To go further, non-stationary signal processing techniques [4] taking into account periodic phenomena (rotation of the cutting wheel, rotation of engines...) could also be used.

3.3 Representative Site Spectra

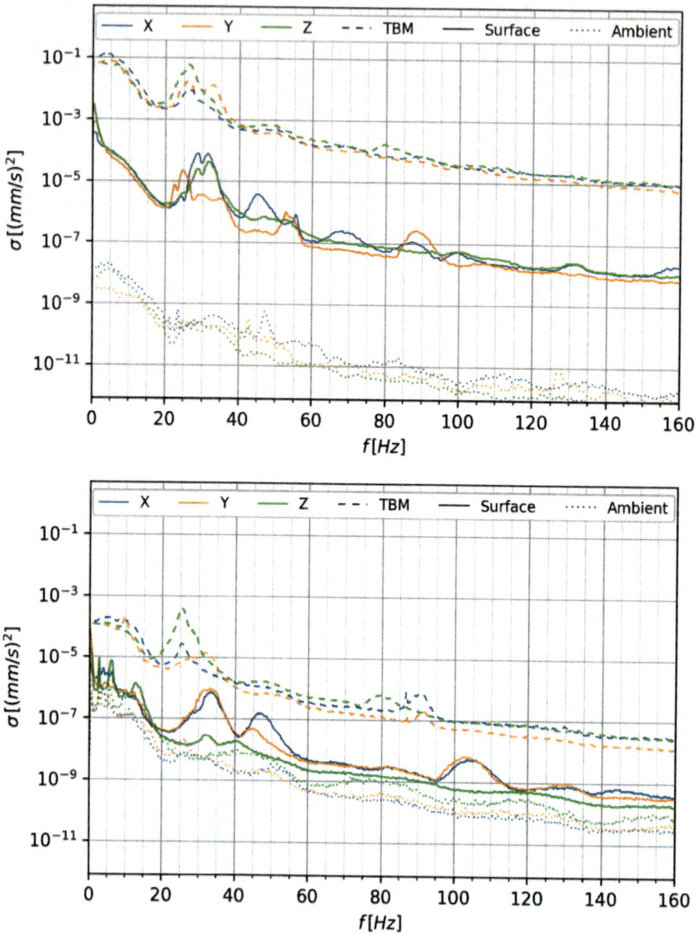

Fig. 4. Lyon-granite (top) and Lyon-alluvium (bottom) site-spectra in the three directions of acquisition (X - blue, Y - orange, Z - green) for sensors on the surface under ambient noise (dotted lines), for sensors on the surface during excavation (solid lines) and for sensors inside the TBM (dashed lines) (source [6]).

The methodology proposed in Sect. 2.3 is applied in order to get the site-spectrum of each campaign. In Fig. 4 are represented the largest singular values σ_n (2) in function on the frequencies $f_n = n\frac{f_s}{L_S}$ for three sets of 3D-sensors: for a given acquisition direction (X, Y, Z), (i) sensors on the surface under mechanical ambient noise before excavation, (ii) sensors on the surface during excavation, and (iii) sensors inside the TBM.

The goal of this paper is not the physical interpretation of these site-spectra, which had been done in [6], but a methodological framework allowing to interpret it. Firstly, at a given frequency f_n, a high value of σ_n corresponds simultaneously (i) to a common high intercorrelation between all the signals of the given dataset, and (ii) to a large amplitude in each local PSD at this frequency. Consequently a weak value can correspond to (i) a weak intercorrelation between signals or (ii) a weak amplitude in the local PSD.

Table 2. Representative magnitudes of characteristic velocities and PSD for three sets of sensors [6]

Sensors	Measurement	Characteristic velocity [mm/s]	σ_k [(mm/s)^2] (f = 0 Hz)
Surface sensors	Ambient noise	$<v_{\text{ambient}}> \approx 10^{-3}$	$\sigma_{0,\text{ambient}} \approx 10^{-8}$
	Excavation	$<v_{\text{surface}}> \approx 10^{0}$	$\sigma_{0,\text{surface}} \approx 10^{-3}$
TBM	Excavation	$<v_{\text{TBM}}> \approx 10^{1}$	$\sigma_{0,\text{TBM}} \approx 10^{-1}$

Moreover, following the characteristic values of velocities and PSD represented on Table 2, the order of magnitude of time histories (for example here characteristic velocities denoted with $<v>$) are linked to the maximum singular values at null frequency σ_0, as:

$$\frac{<v_{\text{TBM}}>}{<v_{\text{surface}}>} \approx \sqrt{\frac{\sigma_{0,\text{TBM}}}{\sigma_{0,\text{surface}}}} \quad ; \quad \frac{<v_{\text{surface}}>}{<v_{\text{ambient}}>} \approx \sqrt{\frac{\sigma_{0,\text{surface}}}{\sigma_{0,\text{ambient}}}}$$

This feature explains why, in terms of magnitude in the site spectra representations, the three sets of measurements in a rock site (Fig. 4 top) are more distant than in softer grounds (Fig. 4 bottom). This is coherent with the amplitudes of the phenomena involved in the three cases: sensors on the ground under ambient noise are less sollicitated than sensors on the ground during excavation which are less sollicitated than sensors in the TBM during the excavation phase.

4 Conclusion

In a measurement campaign involving many sensors, it is impossible to consider the spectrum of each sensor to extract information on the whole site. The proposed methodology provides a concise representation of response spectra at the site scale. It can be summarised in a few key steps:

- Depending on the goal of the in-situ campaign, one or several measurement phase with synchronous acquisitions.
- For each measurement phase and for all wanted signals:

1. Preprocessing of each signal and keeping only (quasi-)stationary parts of them;
2. Building the Spectral Density Matrix (Welch's method) with intercorrelation of signals;
3. At each frequency, computing the Singular Value Decomposition of the Spectral Density Matrix (for this frequency) then keep only the first (i.e. the maximum) value.

- Plot all the sets in the same figure, see Figs. 4.

In addition to the conciseness of the representation in frequency terms, it is interesting to note that the maximum singular values of the spectral density matrix are directly related to the characteristics amplitudes of the time histories. This signal processing can be seen as an average of the power spectra of a set of sensors. Therefore, local scale information is lost with this representation. Furthermore, the main limitation of the method is the stationarity of the signals; For example transient signals from shocks or explosions can't be processed.

This simple method can be applied in lots of fields involving stationary dynamic phenomena measured synchronously by different sensors: acoustics, electromagnetics,...

References

1. Brincker, R., Ventura, C.: Introduction to Operational Modal Analysis. Wiley, Hoboken (2015)
2. Brincker, R., Zhang, L., Andersen, P.: Modal identification from ambient responses using frequency domain decomposition. In: Proceedings of the International Modal Analysis Conference - IMAC, vol. 1, pp. 625–630 (2000)
3. Cooley, J.W., Tukey, J.W.: An algorithm for the machine calculation of complex Fourier series. Math. Comput. **19**(90), 297 (1965). https://doi.org/10.2307/2003354
4. Giannakis, G.B., Madisetti, V.: Cyclostationary signal analysis. In: Digital Signal Processing Handbook, vol. 31, pp. 1–17. Citeseer (1998)
5. Rallu, A., Berthoz, N.: Vibrations induced by tunnel boring machines in urban areas: Dataset of synchronized in-situ measurements inside the shield and on the surface. Data Brief **41**, 107826 (2022). https://doi.org/10.1016/j.dib.2022.107826
6. Rallu, A., Berthoz, N., Branque, D., Charlemagne, S.: Vibrations induced by tunnel boring machines in urban areas: in-situ measurements and methodology of analysis. J. Rock Mech. Geotech. Eng. (2022)
7. Welch, P.: The use of fast fourier transform for the estimation of power spectra: a method based on time averaging over short, modified periodograms. IEEE Trans. Electroacoust. **AU-15**(2), 70–74 (1967)

SSI Effect in Two Mining Regions for Low-Rise Traditional Buildings

Krystyna Kuzniar[1] and Tadeusz Tatara[2(✉)]

[1] Institute of Technology, Pedagogical University of Krakow, 30-084 Krakow, Poland
krystyna.kuzniar@up.krakow.pl

[2] Faculty of Civil Engineering, Cracow University of Technology, 31-155 Krakow, Poland
tadeusz.tatara@pk.edu.pl

Abstract. The article compared the transmission of underground vibrations caused by mines to foundations of representative low-rise masonry buildings situated in two different seismically active mining areas in Poland: the Legnica-Glogow Copperfield (LGC) and the Upper Silesian Coalfield (USC). The soil-structure interaction (SSI) effect has been investigated on experimental measurements of free-field (near to the building) and building foundation vibration accelerations occurring at the same time. Long-term (several years, hundreds of strong shocks), full-scale monitoring with the application of so-called 'an armed partition' measuring equipment, was carried out. The focus has been on the horizontal vibrations in the directions of the transverse and longitudinal axis of the buildings. It has been stated that records of the ground and building foundation vibrations registered at the same time can vary considerably in the case of the LGC, as well as the USC mining regions. The differences between the ground and building foundation vibrations have been analysed using the comparisons of dimensional and dimensionless response spectra from the building foundation and free-field vibrations, respectively. Furthermore, the relationship between the foundation and ground response spectra, the so-called ratio of response spectra (*RRS*), has been taken into account. Influences of epicentre distance, the magnitude of mining tremor energy, peak ground value of vibrations related to mining tremors on the SSI in the case of the same type of building but situated in mining regions with some differences concerning, e.g., site conditions, have also been analysed and compared.

Keywords: Soil-structure interaction · Mine-induced rockburst · Long-term experimental tests · Response spectra · Low-rise building

1 Introduction

The phenomenon of soil-structure interaction is important not only from a scientific point of view but also from a practical point of view. The results of the study of this phenomenon can be used in design procedures in seismic areas [1] and areas affected by so-called paraseismic vibrations, among others from the surface and underground mining, surface and underground car and train traffic, driving sealed walls, and pile

Z. Dimitrovová et al. (Eds.): WMVC 2022, MMS 125, pp. 337–344, 2023.
https://doi.org/10.1007/978-3-031-15758-5_33

driving. The study of the dynamic interaction of the building substrate system is carried out through theoretical analyses, separating the effect of kinematic interaction from inertial interaction [2, 3]. These two types of interactions significantly affect the average energy and damage spectra [2].

Some studies have undertaken the analysis of the SSI phenomenon through in situ studies, during which the registered vibration data from the free ground and building foundations performed in full scale long-term experimental monitoring are simultaneously used. Vibrations of the free-field close to the analysed buildings and on the building foundations are measured and registered concurrently with the application of the so-called 'an armed partition' accelerometers. The results of such studies are used to build and verify dynamic SSI models. This approach is the subject of the analyses shown in the articles, e.g. [4, 5].

The article presents the results of the transmission of the free-field vibrations induced by rockbursts to the foundations of low-rise buildings located in two mining districts in Poland. These regions – the Legnica-Glogow Copperfield (LGC) and the Upper Silesian Coalfield (USC) with underground coal mines – are the most seismically active zones in Poland. In particular, the study presents the results analysis of the ratio of response spectra (*RRS*) [6] as a function of dimensional and dimensionless response spectra, depending on mining shock parameters such as a maximum value of ground vibrations, an epicentre distance, a magnitude of energy. The *RRS* is used to evaluate and compare the differences between the simultaneously occurring free-field and building foundation vibrations regarding the same building type but situated in mining regions with some differences concerning, e.g. site conditions.

2 Buildings Investigated and the Scope of the Long-Term Monitoring

The long-term vibration monitoring concerned two selected representative low-rise buildings located in two different mining basins in Poland (LGC and USC). The magnitude of energy of the most intense tremors caused by underground mining of mineral deposits is comparable to that of weak earthquakes. The magnitude of these tremors is in the range of 4.1–4.6 on the Richter scale.

The first building, denoted B1, located in the LGC region, refers to a typical residential masonry structure with a basement. The dimension of the plan is 10 m × 10 m with a height equal to 7.6 m. The bearing system and the foundation consist of transverse-longitudinal masonry walls and reinforced strips at a depth equal to 1.5 m below ground level.

The second building, denoted as B2, is located at USC. This building is a two-story masonry building without a basement; it has a transverse-longitudinal load-bearing masonry wall system and concrete strip foundations at a depth equal to 1.4 m. The plan dimension of building B2 is 12.5 m × 23.8 m with a height of 7.3 m. Table 1 presents brief characteristics of sub-soils for the analysed buildings B1 and B2.

Full-scale measurements (in situ) referring to 24 and 484 pairs of horizontal component accelerations (free-field – foundation in the transverse and the longitudinal axis of each of the buildings), respectively for buildings B1 and B2 were analysed. Accelerations

Table 1. Sub-soil layers parameters in the vicinity of buildings B1 and B2

Building	Sub-soil layers
B1	The following layers are present in succession: the surface layer of soil, sand and gravel to a depth of 1.5 m, compact clay 1.5 m–4.5 m, fine and medium quartz sand 4.5 m–6.8 m, compact clay 6.8 m–23 m, below 350 m–bedrock
B2	The following layers are present in succession: the surface layer of soil, layers of fine and medium sand with yellow dust in some parts

in the time domain of free-field (at a distance of a few meters from the considered buildings) and building foundation vibrations were recorded concurrently with the application of special measuring equipment, the so-called'an armed partition' accelerometers. The number of pairs relating to the B1 building from LGC is small due to the sparse number of mining tremors during which it was possible to simultaneously record the vibrations of the free-field and the B1 building foundation without disturbance.

Registered pairs of vibrations were grouped according to parameters such as epicentre distance (*re*), energy magnitude (*En*), and peak ground acceleration (*PGA*). Table 2 presents data concerning the numerical balance of the analysed pairs of free-field and building foundation records referring to the B1 and B2 buildings, taking into account the adopted parameter range of the tremors.

Table 2. Numerical balance of the analysed pairs of free-field and building foundation records referring to the B1 and B2 buildings.

Rockburst parameter	Parameter range	Building	
		B1	B2
re [m]	to 700	18	88
	701–1500	3	374
	over 1500	3	22
En [J] – the magnitude of energy	E5, E6	11	460
	E7	9	22
	E8, E9	4	2
PGA [m/s^2]	to 0.300	12	367
	0.301–0.600	6	76
	0.601–0.900	2	22
	over 0.900	4	19
Sum of pairs – the whole range of *re, En, PGA*		24	484

3 Ratio of Response Spectra Depending on Rockburst Parameters

The main goal of the analysis is to investigate distinctions between free-field waves recorded on the ground surface near buildings B1 and B2 and accelerations simultaneously measured for their foundations.

The analysis focuses on the *RRS* ratio as a function of the dimensional and dimensionless response spectra, taking into account the values of the shock parameters in the assumed characteristic intervals (Table 2).

In the analysis, we used Eq. (1) to calculate standardized (dimensionless) spectral curves (β).

$$\beta = S_a/PGA \quad (1)$$

Appropriate values of individual dimensional spectra S_a and corresponding *PGA* values from free-field records were taken into consideration in this equation (Eq. (1)). The second variant of nondimensional response spectra β refers to foundation vibrations. These spectra were prepared in a way analogous to the way for ground spectra using proper components concerning foundations.

The ratios $RRS(S_a)$ and $RRS(\beta)$ were calculated by dividing the respective spectra (S_a or β) from the simultaneously measured building foundation vibrations by the spectra corresponding to nearby free-field vibrations.

Figure 1 shows the $RRS(S_a)$ curves and curves $RRS(\beta)$ corresponding to buildings B1 and B2 calculated considering the whole range of mining tremor parameters. The diagrams of the $RRS(S_a)$ ratios differ significantly for buildings B1 and B2. The differences may result from the influence of soil conditions on the waveforms of the applied components of the vibration acceleration, based on which individual dimensional response spectra are determined. The diagrams of $RRS(\beta)$ slightly differ for buildings B1 and B2 (Fig. 1). The differences result primarily from the dominant frequencies of the vibration waveforms in both analysed areas [5, 7] and the method for determining individual dimensionless spectra (β).

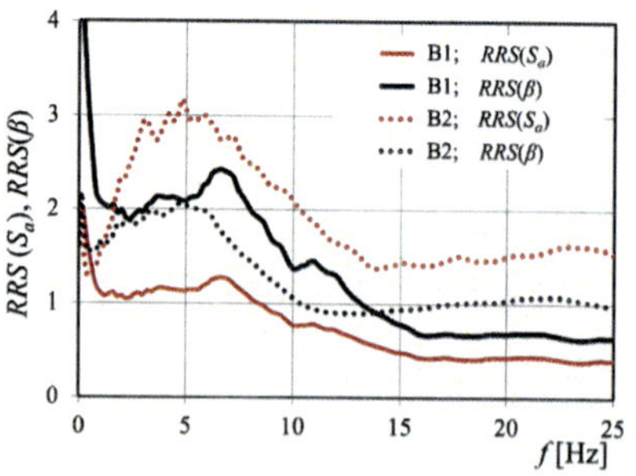

Fig. 1. Comparison of the $RRS(S_a)$ and $RRS(\beta)$ curves determined for B1 and B2 buildings with the whole range of mining tremor parameters considered.

The more developed comparison of the $RRS(S_a)$ and $RRS(\beta)$ curves determined for the buildings B1 and B2, with the ranges of shock parameters taken into account, is shown respectively in Fig. 2 in the case of *re*, in Fig. 3 in the case of *En*, and Fig. 4 in the case of *PGA*. For comparison, all the figures also show the curves prepared for the whole range of each of the parameters.

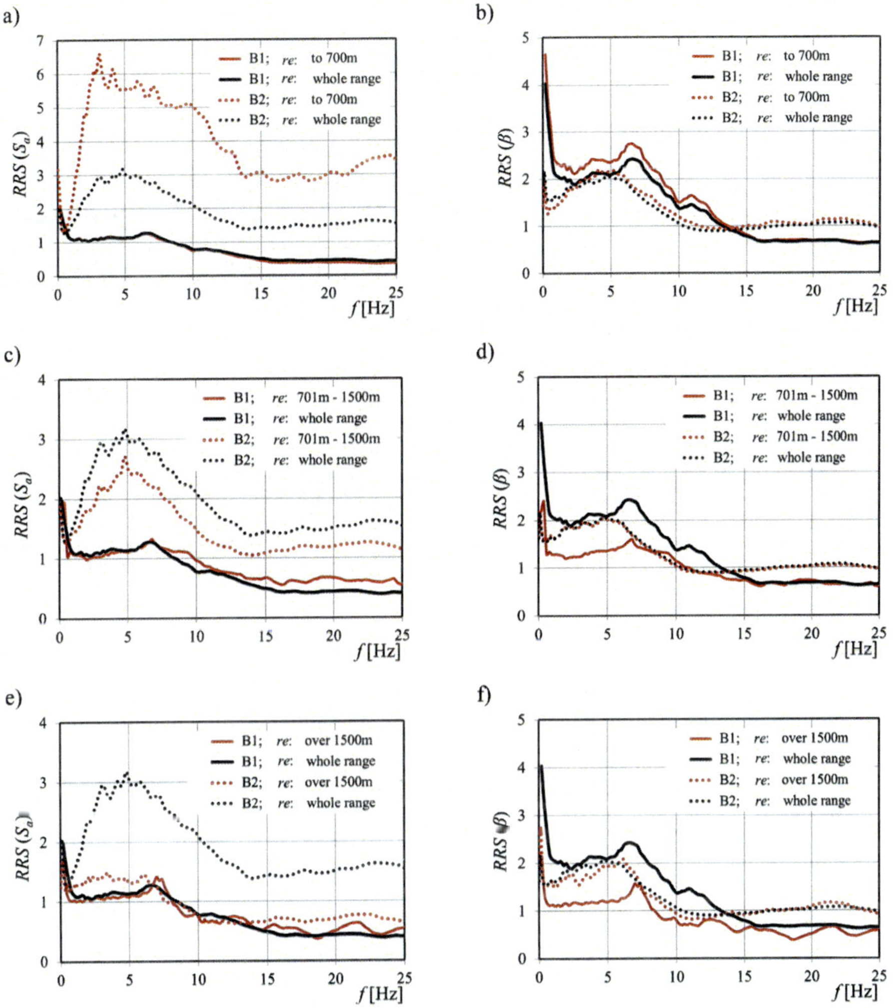

Fig. 2. $RRS(S_a)$ and $RRS(\beta)$ curve determined for B1 and B2 buildings in the ranges of *re*: a) and b) to 700 m; c) and d) 701–1500 m; e) and f) over 1500 m.

In Fig. 2, the differences in $RRS(S_a)$ can be seen in all of the ranges of *re*, especially for frequencies close to the natural frequencies of the analysed low-rise buildings B1 and B2. Only for shocks with very long epicentral distances, the $RRS(S_a)$ for both buildings (B1 and B2) are very close (Fig. 2e). The differences in the corresponding $RRS(\beta)$ are slightly smaller (Fig. 2b, 2d and 2f).

A similar conclusion applies to the dependence of $RRS(S_a)$ and $RRS(\beta)$ on the magnitude of *En* (Fig. 3).

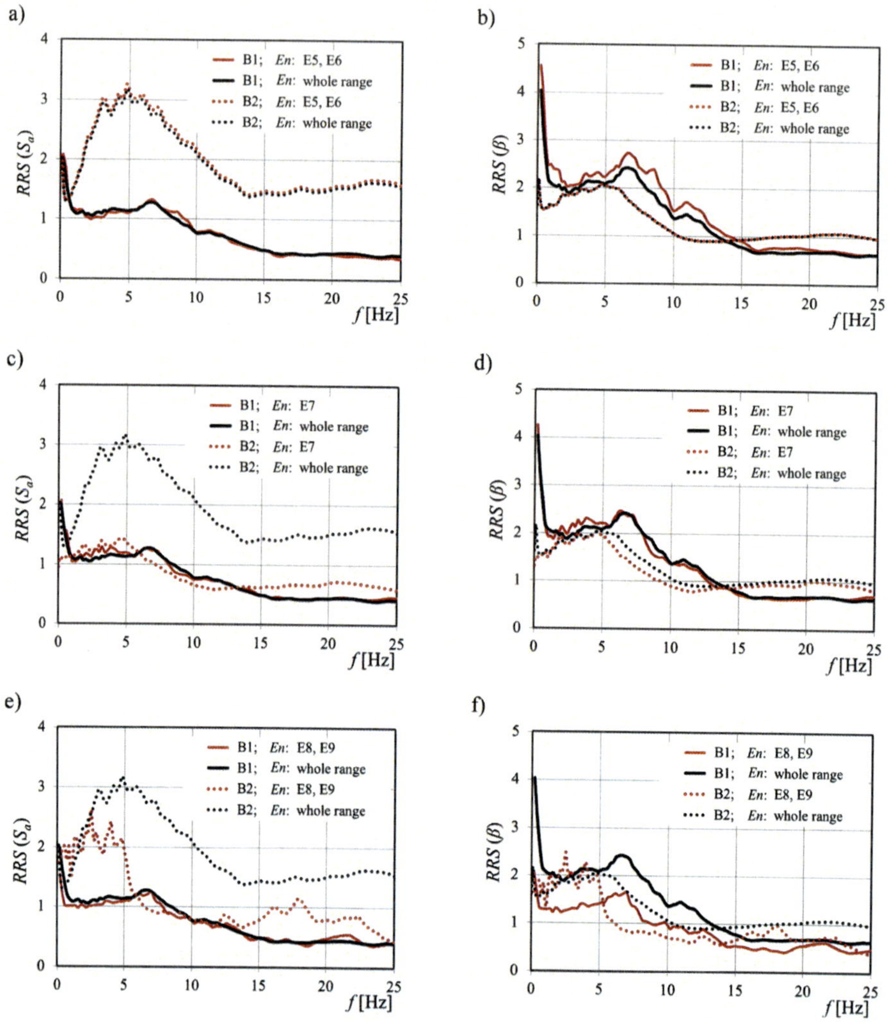

Fig. 3. $RRS(S_a)$ and $RRS(\beta)$ curve determined for B1 and B2 buildings in the magnitude of *En*: a) and b) E5, E6; c) and d) E7; e) and f) E8, E9.

On the other hand, Fig. 4 proves that in both mining regions (for buildings B1 and B2) the $RRS(S_a)$ curves are relatively close to each other in the case of vibrations with large *PGA*. Nevertheless, in the case of very small values of *PGA*, these differences are enormous (Fig. 4a).

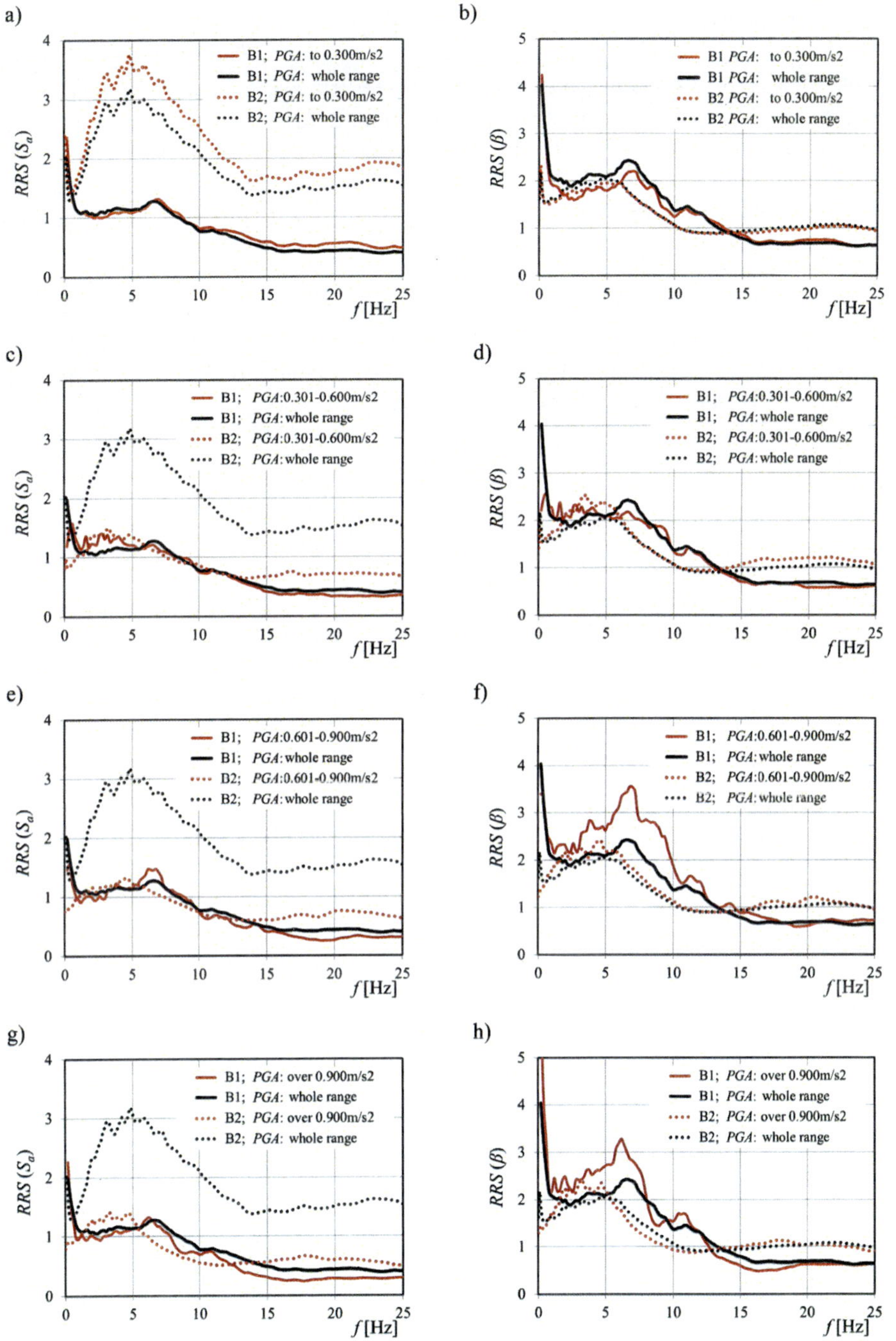

Fig. 4. $RRS(S_a)$ and $RRS(\beta)$ curve determined for B1 and B2 buildings in the ranges of *PGA*: a) and b) to 0.300 m/s^2; c) and d) 0.301–0.600 m/s^2; e) and f) 0.601–0.900 m/s^2; g) and h) over 0.900 m/s^2.

4 Conclusions

From many years of research on vibrations of free-field and foundations of two-storey masonry buildings and the results of *RRS* function, follows the values (magnitude) of parameters describing shocks have a significant impact on the phenomenon of dynamic interaction for this buildings' class. The relatively small differences between the *RRS* determined based on the dimensional and nondimensional spectra in the subsequent ranges of shock parameters, are obtained when the *PGA* value is considered. The comparison of the $RRS(S_a)$ curves obtained for buildings in the LGC and USC areas shows that the different ground conditions in both regions have an impact on the curves. The *RRS* function calculated from the nondimensional spectra gives the relatively minor differences when comparing the respective curves for buildings B1 and B2. The latter conclusion may be of great practical importance.

References

1. Aviles, J., Suarez, M.: Effective periods and dampings of building-foundation systems including seismic wave effects. Eng. Struct. **24**, 553–562 (2002)
2. Ahmadi, E.: On the structural energy distribution and cumulative damage in soil-embedded foundation-structure interaction systems. Eng. Struct. **182**, 487–500 (2019)
3. Ahmadi, E.: Concurrent effects of inertial and kinematic soil-structure interactions on strength-ductility-period relationship. Soil Dyn. Earthq. Eng. **117**, 174–189 (2019)
4. Kuzniar, K., Tatara, T.: The ratio of response spectra from seismic-type free-field and building foundation vibrations: the influence of rockburst parameters and simple models of kinematic soil-structure interaction. Bull. Earthq. Eng. **18**(3), 907–924 (2019). https://doi.org/10.1007/s10518-019-00734-w
5. Maciag, E., Kuzniar, K., Tatara, T.: Response spectra of ground motions and building foundation vibrations excited by rockbursts in the LGC region. Earthq. Spectra **32**(3), 1769–1791 (2016)
6. Stewart, J.P.: Variations between foundation-level and free-field earthquake ground motions. Earthq. Spectra **16**(2), 511–532 (2000)
7. Kuzniar, K., Tatara, T.: Full-scale long-term monitoring of mine-induced vibrations for soil-structure interaction research using dimensionless response spectra. Case Stud. Constr. Mater. **16**, e00801 (2022)

Validation of Periodic 3D Numerical Method for Analysis of Ground-Borne Vibrations

Alexandre Castanheira-Pinto[1(✉)], Pedro Alves Costa[1], and Luís Godinho[2]

[1] CONSTRUCT-FEUP, University of Porto, Rua Dr. Roberto Frias, 4200-465 Porto, Portugal
{amgcpinto,pacosta}@fe.up.pt

[2] ISISE, Department of Civil Engineering, University of Coimbra, Pólo II, Rua Luís Reis Santos, 3030-788 Coimbra, Portugal
lgodinho@dec.uc.pt

Abstract. Modern societies demand for efficient mass transportation systems, becoming the rail transport competitive for short to medium distance travels. The increasing rail network gives rise to new engineering challenges since the surrounding infrastructures will experience discomfort problems. To developed realistic studies in the pursue of mitigation measures design precisely to solve the previous problem, robust numerical tools needs to be developed. Several numerical methods were proposed in the last years, such as 2.5D models, capable of dealing, assuming some simplifications, with ground-borne vibrations problems. However, the application of such efficient methods is restricted to longitudinal invariant structures, which can represent a rough approximation. Since a rail track is clearly a tri dimensional structure, in this paper is presented a tri dimensional FEM-PML capable of modelling periodic structures by adopting special boundaries conditions.

Keywords: Numerical modelling · 3D periodic FEM · Numerical validation

1 Introduction

Rail transport is one of today's most efficient mass transportation systems and widely used for short to medium distance travels. Its energy efficiency as well as its intrinsic comfort are at the root of its competitive character when compared to road and air traffic [1]. With the increase of the rail and metropolitan traffic network, new challenges arise as a result of the discomfort caused to the inhabitants of buildings located in the vicinity of these networks. In order to solve these discomforts several analyses are required to find out which of the predicted solutions gives the best behaviour. For this, robust numerical models capable of accurately predicting the vibration field induced by a train passage are indispensable.

There are now several numerical methods capable of performing elastodynamic analyses which predicts the vibrations field induced by human activities. The multiplicity of numerical methods is explained by their intrinsic particularity, i.e. the suitability for dealing with the problem in question efficiently and accurately. A numerical method

Z. Dimitrovová et al. (Eds.): WMVC 2022, MMS 125, pp. 345–351, 2023.
https://doi.org/10.1007/978-3-031-15758-5_34

implemented on a finite element formulation is capable of modelling complex geometries compared to a boundary integration based formulations. On the other hand, this same formulation causes a new problem in elastodynamic analysis, which is not found in a formulation in boundary elements, the treatment of artificial boundaries and the need for discretization with an unbearable number of degrees of freedom. This dichotomy explains the need to adapt the numerical method to the case study in question.

As mentioned above, the increasing demand for solutions that solve the problems caused by the cyclic passage of trains creates the need to implement numerical tools capable of analysing the problem efficiently. Adding to the previous fact the increasing research in the field of periodic structures as mitigation solutions [2], also known as meta-materials, arises the need for numerical models that allow the simulation of the geometrical characteristics of these solutions. Thus, models based on the 2.5D approach, widely used in the ground-borne vibration field, aren't able to deal with such problems since it is impossible to considered longitudinal heterogeneity [7].

The aim of this paper is to present a three-dimensional model based on a finite element formulation by matching special boundary conditions, providing a numerical periodicity model, and thereby significantly reducing the computational effort required without losing the 3D character of the solution.

2 Methods

As can be perceived from the previous section, the numerical method here presented was developed to make possible elastodynamic analysis, such a train passage. Following a classic finite element formulation [3, 4] on the frequency-wavenumber domain, the system equilibrium can be described by the following equation:

$$\left[\boldsymbol{K}(k_1, \omega) - \omega^2 \boldsymbol{M}(k_1, \omega)\right] \cdot u(k_1, \omega) = F(k_1, \omega) \tag{1}$$

where K and M represent the stiffness and mass matrix respectively, k_1 the wavenumber, ω the angular frequency, F the external load vector and u the nodal displacement. Since the finite element method is not suitable for dealing with infinite domains, a mathematical technique must be used to treat spurious reflexions on the domain borders. "Perfect matched layers" was used, coupled with the finite element method, for treating the artificial limits in order to avoid undesirable reflexions [5, 6]. Moving loads can be achieved considering the angular frequency ω given by:

$$\omega = \Omega - k_1 \cdot v \tag{2}$$

where Ω is the load oscillating frequency and v the load travel velocity.

To obtained the vibration field produced by a point load, on the original domain, Eq. (1) must be solved for a range of sinusoidal waves defined by the wavenumber k_1, as it is expressed in Eq. (3).

$$F(k_1, \omega) = A * e^{ik_1 x} \tag{3}$$

where A is the loading amplitude and x the nodal distance to a reference point.

Finally, the vibration field on the original domain, spatial domain, is achieved by the combination of results for the range of wavenumbers considered. Such combination is accomplished using the inverse Fourier transformation. As it was mentioned previously, three dimensional finite element methods are time consuming, as well as computational demanding. To overcome such disadvantage, the present method takes into consideration the periodic character associated to a rail track in the form of special boundary conditions, expressed in Eq. (4).

$$u_{front\ border} = u_{back\ border} * e^{-ik_1 d} \tag{4}$$

where $u_{front\ border}$ and $u_{back\ border}$ represent the frontal and rear borders respectively and d is the cell thickness. To better understand the model reduction achieved by the present methodology, an illustrative scheme is present in Fig. 1 a), where complete 3D periodic model is simulated by considering only a unit slice (which is repeated infinitely). Figure 1 b) indicates the special boundary conditions which allow such reduction, writing the displacement of the front border as a function of the displacement of the back border multiplied by a coefficient related to the periodicity of the section.

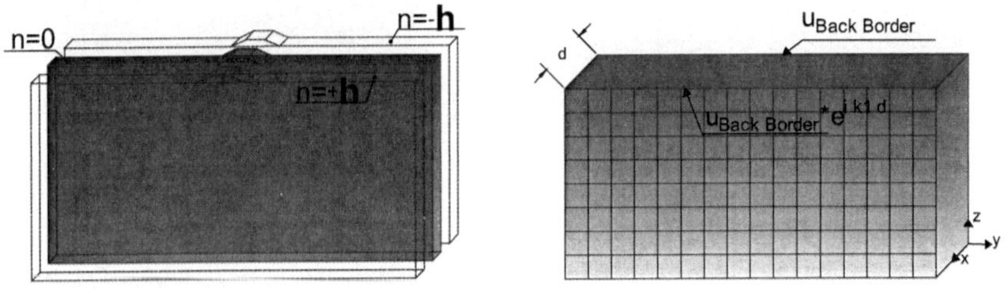

a) b)

Fig. 1. Illustrative scheme of the periodic 3D FEM-PML model: a) subdivision of global modal according to its periodicity; b) boundaries conditions imposed to the considered cell.

Since all the cells have the same properties and morphology, the global model can be obtained by solving just one section numerically and the remaining domain analytically, without additional computational effort, using Eq. (5)

$$u_n = u_{ref} * e^{ik_1 d * n} \tag{5}$$

where u_n is the nodal displacement of the desired cell and n the cell number

3 Results

To validate the numerical methodology two stages were considered, firstly the vibration field for a stationary load was evaluated, followed by the assessment for an oscillating moving load. Both stages, results are compared with results obtained from a validated 2.5D FEM-PML [5]. For that purpose, a realistic scenario was idealized, being the cross-section and the material properties illustrated in Fig. 2. The 3D periodic FEM-PML allows the reduction of larger models into small sections decreasing considerably the computational demand, making a 3D run almost as efficient as a 2.5D.

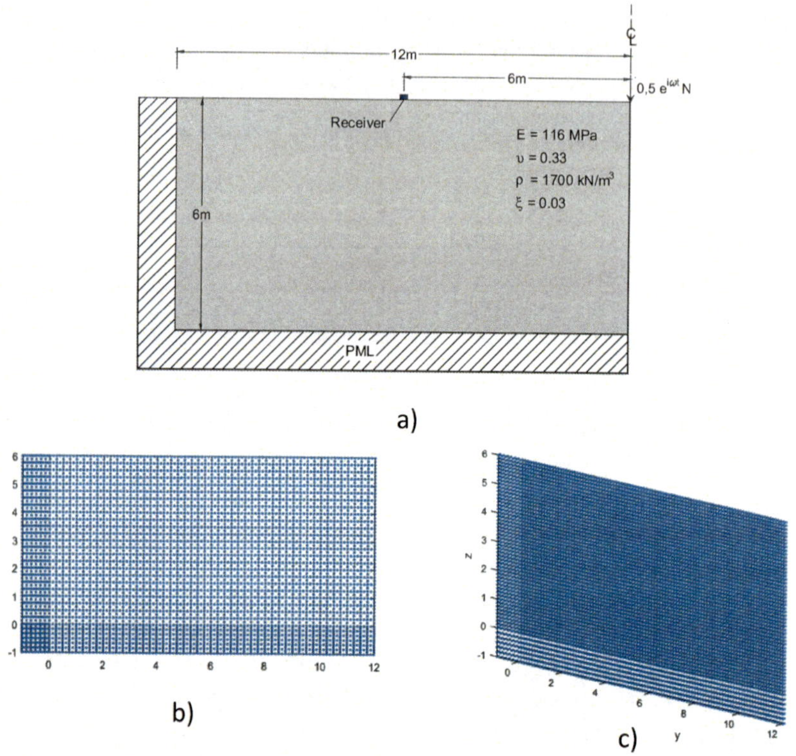

Fig. 2. Example description: a) morphology of the cross-section; b) 2.5D mesh; c) 3D mesh with thickness of 0.1 m.

3.1 Stationary Load Analysis

Both models were submitted to identical load conditions, namely a stationary load oscillating with three distinct frequencies, 25 Hz, 50 Hz, 75 Hz. Figure 3 presents the vertical displacement transfer function for the receiver point, on the transform domain, for both frequencies applied, being the result in red dots from the periodic 3D method and the solid blue line from the 2.5D FEM-PML model. Comparing the response obtained by both numerical methods a very good agreement can be seen, making the periodic 3D model validated for stationary oscillating loads.

3.2 Moving Load Analysis

For the purpose of developing train passage analysis, the moving character of the loading must be attended. Since it was assumed a load travel velocity of 25 m/s the oscillating frequency (ω) is no longer the load oscillating frequency itself (Ω), as can be seen by Eq. (2). On Fig. 4 it can be found similar results to the ones presented in Fig. 3, being expressed the real and imaginary vertical displacement transfer function for 50 Hz frequency at the receiver point. Once again a perfect fit occurs between results, verifying the validity of the periodic 3D methodology.

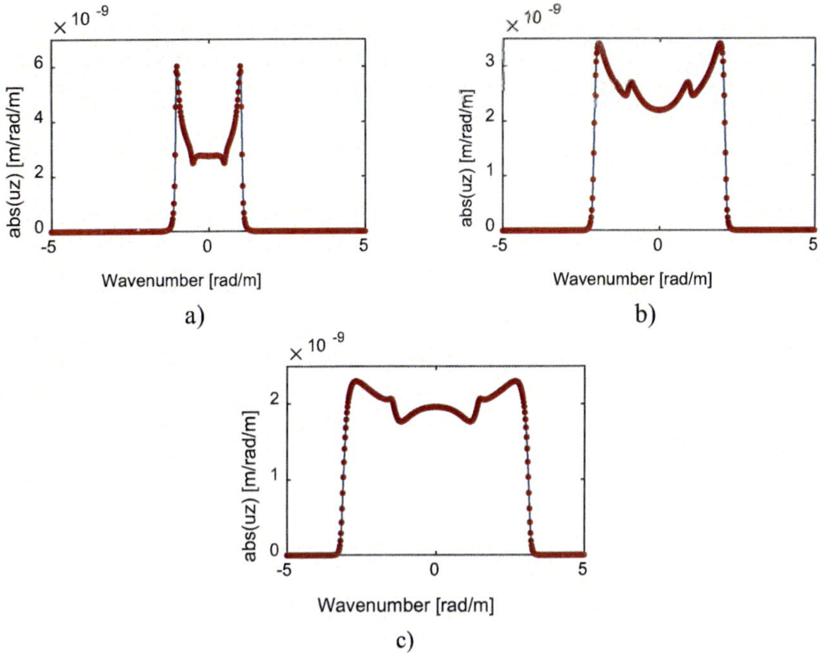

Fig. 3. Dynamic properties of the soil with depth: a) S-wave velocity profile; b) P-wave velocity profile.

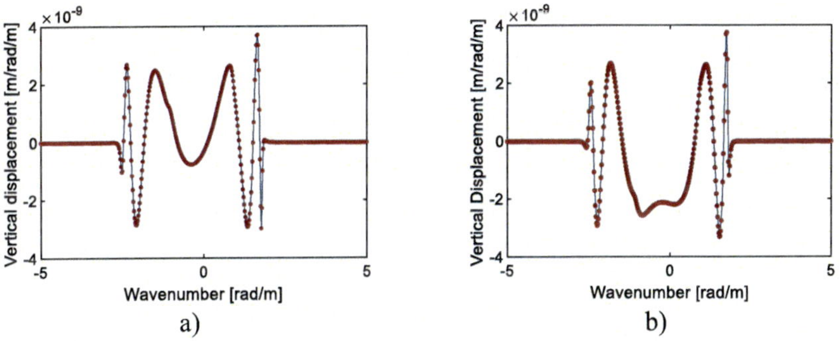

Fig. 4. Vertical displacement on wavenumber domain: a) 10 Hz; b) 50 Hz.

To obtain the results in the original domain an inverse Fourier transformation was applied to the previously results. Figure 5 presents the response for a longitudinal alignment that passes on top of the receiver for the 50 Hz frequency at the time instant zero. Black circles represent the points obtained numerically, being the rest of the domain calculated by taking the periodic condition into consideration, using Eq. (5), and represented using red dots. As can be seen the periodic 3D model is capable of predicting the vibration field for a moving oscillating load very accurately.

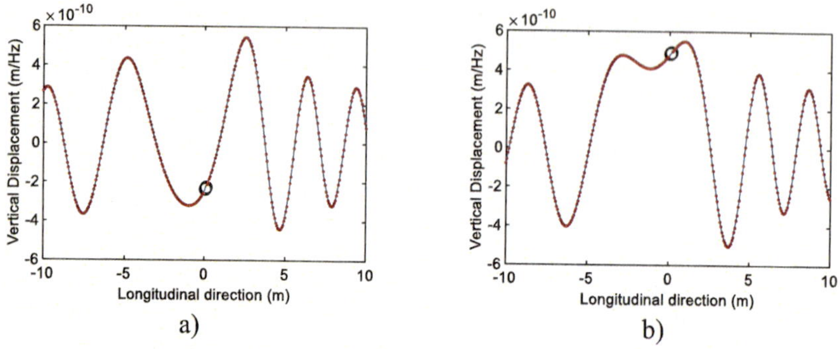

Fig. 5. Vertical displacement on spacial domain: a) Real component; b) Imaginary component.

4 Conclusions and Contributions

Consideration of variability in the longitudinal direction is one of the major advantage of the model presented here in comparison to the 2.5D models, which are restricted to an invariant longitudinal cross-section [8]. Such advantage has a drawback in terms of the computational strength required to perform an analysis since the size of the matrices are considerably higher when the third dimension is considered. However, the negative impact that comes from the use of 3D models is overcome by the adoption of boundary conditions capable of considering the periodicity of the analysed model. Thus, the resolution of the global model consists on a hybrid process, with a small portion of the model being determined numerically and the rest analytically. Transforming numerical operations in analytical procedure, to obtain the response for a 3D section, is one of the greatest advantages of the numerical model presented here, since part of the section previously calculated using a 3D numerical model is now obtained by means of a multiplication operation, and therefore more efficiently. Such transformation leads to a reduction in the computational effort required for the analysis in question since the matrices involved become smaller. To validate the periodic 3D FEM-PML numerical method, two examples were developed. Firstly, the models were submitted to a stationary oscillating load, being the results from a validated 2.5D FEM-PML method compared to the ones obtained by the periodic 3D model by the means of transfer functions. Lastly, a moving oscillating load was adopted, being the results compared following the same strategy as before. Both examples were characterized by good match between both results making the periodic 3D FEM-PML model suitable for predicting the vibration field induced by a dynamic loading.

Acknowledgments. This work was financially supported by: Programmatic funding - UIDP/04708/2020 of the CONSTRUCT - Instituto de I&D em Estruturas e Construções - funded by national funds through the FCT/MCTES (PIDDAC); Project PTDC/ECM-COM/1364/2014 – POCI-01-0145-FEDER-016783 – funded by FEDER funds through COMPETE2020 – Programa Operacional Competitividade e Internacionalização (POCI) and by national funds through FCT – Fundação para a Ciência e a Tecnologia, I.P; Project POCI-01-0145-FEDER-029557 – funded by FEDER funds through COMPETE2020 – Programa Operacional Competitividade e Internacionalização (POCI) and by national funds (PIDDAC) through FCT/MCTES; Individual Grant PD/BD/143004/2018.

References

1. Barron de Angoiti, I.: High speed rail: development around the world. In: Calçada, R. et al. (eds.) Noise and Vibration on High-Speed Railways, Porto, pp. 1–14 (2008)
2. Hussein, M., Leamy, M., Ruzzene, M.: Dynamics of phononic materials and structures: historical origins, recent progress, and future outlook. Appl. Mech. Rev. **66**(4), 040802 (2014)
3. Zienkiewicz, O.C., Taylor, R.L.: The Finite Element Method, vol. 1. Butterworth-Heinemann, Oxford (2000)
4. Onate, E.: Structural Analysis with Finite Element Method. Linear Statics. Springer, Dordrecht (2009). https://doi.org/10.1007/978-1-4020-8733-2
5. Lopes, P., et al.: Vibrations inside buildings due to subway railway traffic. Experimental validation of a comprehensive prediction model. Sci. Total Environ. **568**, 1333–1343 (2016)
6. Lopes, P., et al.: Numerical modeling of vibrations induced in tunnels: a 2.5D FEM-PML approach. In: Xia, H., Calçada, R. (eds.) Traffic Induced Environmental Vibrations and Controls: Theory and Application, Nova, pp. 133–166 (2013)
7. Alves Costa, P., Calçada, R., Silva Cardoso, A.: Track–ground vibrations induced by railway traffic: in-situ measurements and validation of a 2.5D FEM-BEM model. Soil Dyn. Earthq. Eng. **32**(1), pp. 111–128 (2012)
8. Yang, Y.B., Hung, H.H.: A 2.5D finite/infinite element approach for modelling viscoelastic body subjected to moving loads. Int. J. Numer. Methods Eng. **51**, pp. 1317–1336 (2001)

Wave Propagation from Hammer, Vibrator and Railway Excitation – Theoretical and Measured Attenuation in Space and Frequency Domain

Lutz Auersch(✉)

Federal Institute of Materials Research and Testing, Berlin, Germany
lutz.auersch-saworski@bam.de

Abstract. The attenuation of wave amplitudes is ruled by the planar, cylindrical or spherical geometry of the wave front (the geometric or power-law attenuation) but also by the damping of the soil (an exponential attenuation). Several low- and high-frequency filter effects are derived for the layering and the damping of the soil, for the moving static and the distributed train loads and for a homogeneous or randomly heterogeneous soil. Measurements of hammer- and train-induced vibrations at five sites have been analysed for these attenuation and filter effects. The measured attenuation with distance can be discribed by generalised power laws and some reasons will be discussed. The theoretical filter effects can well be found in the measurements.

Keywords: Hammer impact · Train passage · Layered soil · Attenuation · Filter effects · Randomly heterogeneous soil · Scattering

1 Theoretical Background of Wave Propagation and Attenuation

1.1 The Geometrical Wave Attenuation and the Hysteretical Damping of the Soil

The conservation of energy E for the elastic cylindrical surface wave yields.

$$E \sim A^2\, 2\pi r = \text{const so that}\, A \sim r^{-0.5} \tag{1}$$

which is a power law between Amplitude A and distance r. It is called a geometric attenuation as it follows from the geometry of the wave front. In the same way, the geometric (power-law) attenuation $A \sim r^{-1}$ of a body wave is derived.

A hysteretically damped soil is described by a complex shear modulus $G^* = G(1 + \mathrm{i}2D)$. The far-field asymptote is modified due to the complex Rayleigh and shear wave velocities $v_R^* \approx v_S^* = \sqrt{G^*/\rho} \approx v_R(1 + \mathrm{i}D)$, ρ the mass density, so that

$$A = A_1 r^{-0.5} \exp(-\mathrm{i}\omega r/v_R) \text{yields}\, A* = A_1 r^{-0.5} \exp(-\mathrm{i}\omega r/v_R) \exp(-D\omega r/v_R) \tag{2}$$

with $\omega = 2\pi f$ the circular frequency. The damping of the soil yields an additional attenuation factor $\exp(-D\omega r/v_R)$ which is an exponential function of distance r.

Z. Dimitrovová et al. (Eds.): WMVC 2022, MMS 125, pp. 352–359, 2023.
https://doi.org/10.1007/978-3-031-15758-5_35

1.2 Layering of the Soil

The compliances of a horizontally layered, hysteretically damped soil are calculated by a matrix methods in frequency-wavenumber domain. The particle velocity response at a distance r from a point force F follows as the wavenumber integral

$$\frac{v}{F}(r,f) = f\int_0^{\infty} N_{zz}(f,k)J_0(kr)k\,dk \tag{3}$$

where J_0 is the Bessel function and $N_{zz}(f, k)$ is the vertical compliance of the soil in frequency-wavenumber domain [1]. The amplitudes of the point-load admittance functions of a homogeneous and a layered soil are presented in Fig. 1 for some frequencies and distances. Generally, the admittance functions increase with the frequency except for the bending down of the curves at high frequencies which is due to the material damping. The layered soil (Fig. 1c) shows the low amplitudes of the underlying stiff soil at low frequencies and the high amplitudes of the soft top layer at high frequencies. Around 16 Hz a strong increase occurs from the low-frequency low amplitudes to the high-frequency high amplitudes. Thus, a soil has a low-frequency filter from the layering and a high-frequency filter effect from the damping.

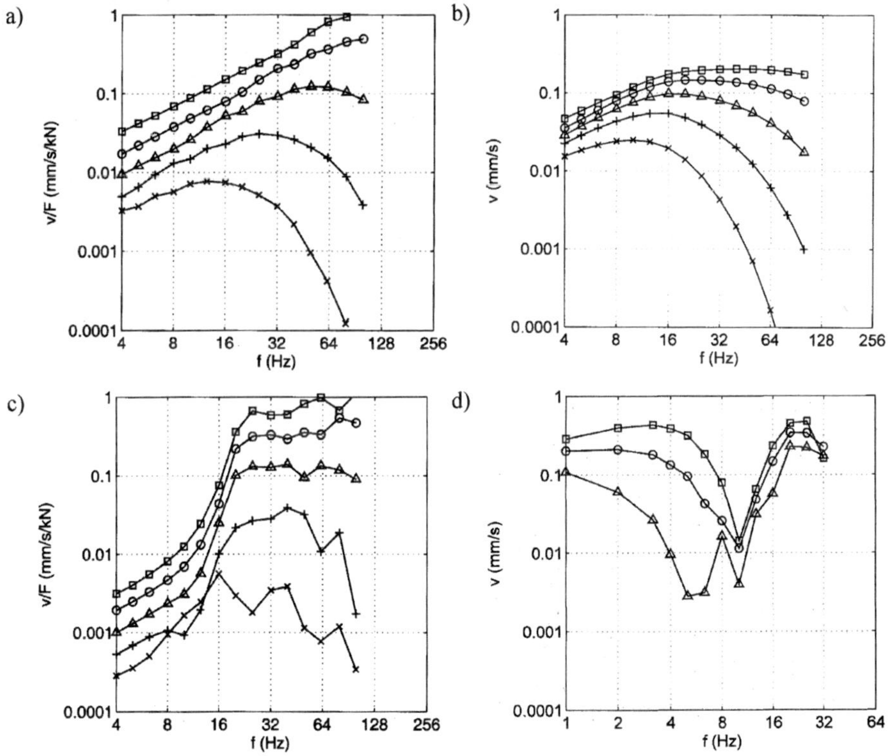

Fig. 1. Calculated wavefield of a homogeneous soil of $v_S = 100$ m/s, distances $r = 4 - 64$ m, a) from a point load, b) from a train load, c) point load on a layer $v_S = 100$ m/s over a half-space $v_S = 300$ m/s d) moving static train load on a randomly heterogeneous soil $v_S = 200$ m/s.

1.3 Railway Specific Attenuation and Filter Effects

The railway induced forces are distributed along and across the track. The train loads are idealized as a number of uncorrelated dynamic axle loads and all responses are summed up along the train length as the root of mean squares.

Figure 2 shows the amplitude-distance laws of a homogeneous soil with shear wave velocity $v_S = 200$ m/s and damping $D = 3\%$ for the frequencies $f = 10, 20, 40$, and 80 Hz. The attenuation becomes stronger with increasing frequency. The higher frequencies clearly show the curved exponential function which is bending down for longer distances in this double-logarithmic plot. The attenuation for the train load in Fig. 2c is somewhat weaker than the attenuation of the point load in Fig. 2a. Nevertheless, the exponential function is still dominating for the higher frequencies.

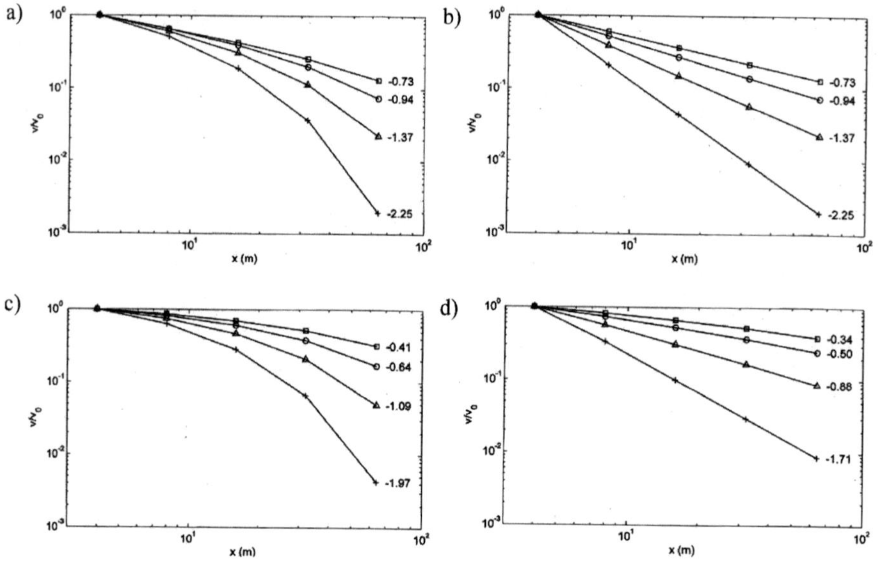

Fig. 2. Calculated attenuation laws for a point load (a, b) and for a train load (c, d), theoretical law $A\ r^{-0.5}\exp(-D\omega r/v_R)$ (a, c) and empirical law $A \sim r^{-q}$ (b, d).

The effect of the (uniform) load distribution across the track width $2a$ can be expressed by a frequency dependent factor which is derived by

$$\frac{v_a}{v_0} = \frac{\left|\int_{-a}^{+a} e^{j\omega x/v_S} dx\right|}{2a} = \frac{|\sin \omega a/v_S|}{\omega a/v_S} \leq \frac{1}{\omega a/v_S} \quad \text{for } \frac{\omega a}{v_S} > \frac{\pi}{2} \tag{4}$$

The response v_a of the soil to these load components has contributions with different phases and their superposition leads to a reduction compared to the response v_0 of a concentrated load. At higher frequencies, the smooth envelope of the strongly varying sin-function is used here. This high-frequency filter function is present in the transfer functions for a train load in Fig. 1b and acts in addition to the material damping.

1.4 Effects of a Randomly Heterogeneous Soil

Waves are scattered at the heterogeneities of a real soil and a part of the wave energy disappears in the depth of the soil. The scattering results in a similar exponential amplitude-distance relation A $\sim$ $\exp(-D_S\omega r/v_R)$ as the material damping. The attenuation can be linearly increasing with frequency similar to the hysteretic material damping, but also stronger dependencies on the frequency are possible [2, 3].

The moving static train loads result in the low-frequency quasi-static response of the soil which has a strong attenuation (f < 10 Hz in Fig. 1d) [4]. This quasi-static response could be strongly amplified for a train running with Rayleigh wave speed, but this effect (like a sonic boom) cannot occur in a heterogeneous soil [5]. On the other hand, the variation of the soil stiffness generates a "scattered" component of the quasi-static component at mid frequencies [6] which has a weaker attenuation and dominates the far field of the soil (f = 16 – 32 Hz in Fig. 1d).

2 Measurements and Evaluation of the Attenuation Laws

In the literature an empirical attenuation law can be found which is a power law $A \sim r^{-q}$ with a site and source specific power q. These attenuation laws are presented in Fig. 2b, d with the same overall attenuation as the theoretical exponential law.

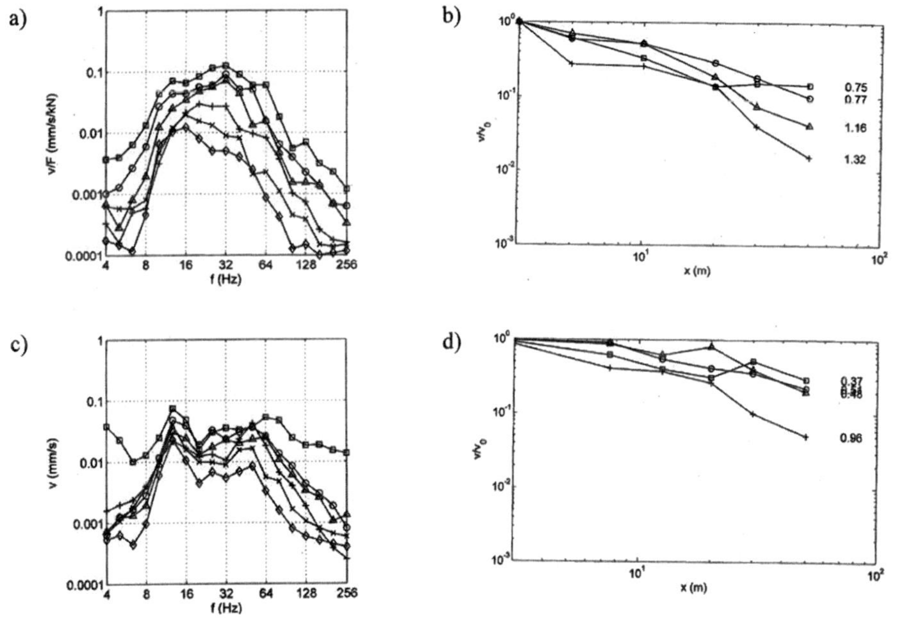

Fig. 3. Measured ground vibrations from hammer (a, b) and train (c, d) excitation at r = 3 – 50 m at Site W (v_S = 270 m/s, v_S = 1000 m/s), train speed v_T = 125 km/h, spectra (a, c), attenuation laws (b, d).

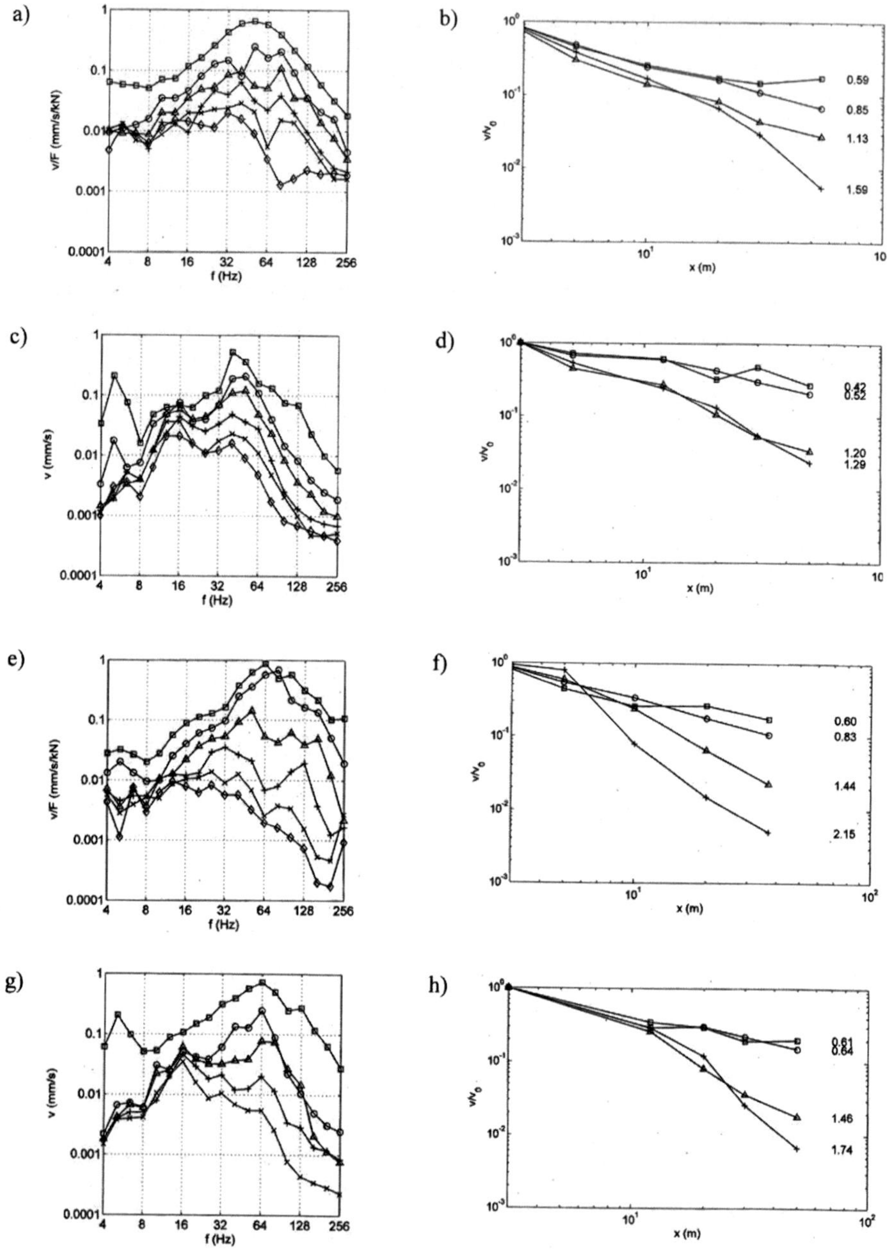

Fig. 4. As Fig. 3 but Site H (a – d) and H* (e–h), $v_S = 225$ m/s, $v_T = 160$ km/h.

The power laws are straight lines in the double logarithmic presentation which have higher powers and have a steeper descent for higher frequencies. The weaker attenuation

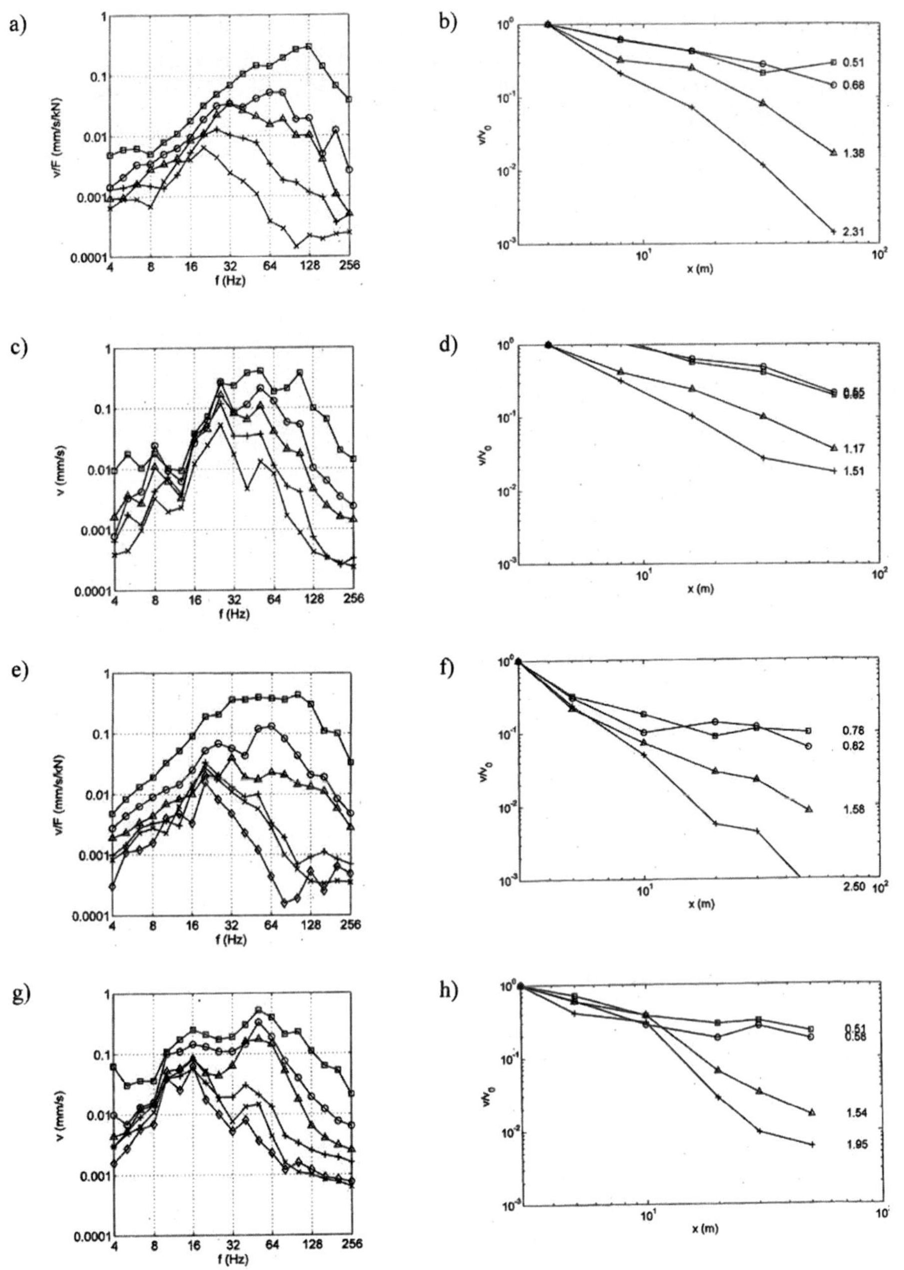

Fig. 5. As Fig. 3 but Site G* ($v_S = 200$ m/s, $v_T = 250$ km/h, a–d) and Site N ($v_S = 130$–200 m/s, $v_T = 150$ km/h, e–h).

for superposition of the train loads in Fig. 2c, d can be directly recognised by the lower exponents which are written on the right of the curves.

Measurements at five sites have been evaluated for spectra and attenuation laws (Fig. 3, 4 and 5). The one-third of octave spectra of the hammer impacts (Fig. 3, 4 and 5a, e) show smooth curves with maxima at 10 to 100 Hz where the lowest frequencies are for the longest distances. The high frequency filtering is due to the material and scattering damping. The spectra of train passages (Fig. 3, 4 and 5c, g) show some specific peaks. The high-frequency peaks at 50 Hz (for 125 km/h) to 100 Hz (250 km/h) are generated by the sleeper distance. At low frequencies, maxima can be found at 12 to 16 for normal speed and at 25 Hz for 250 km/h, particularly strong in Fig. 3c and 5g which are an indication of the scattered moving static loads. The low-frequency filter of a strictly layered soil can be observed in Fig. 3 whereas the other sites are more homogeneous with a continuously increasing stiffness of the soil and continuously decreasing amplitudes for low frequencies. The mid-frequency components are always dominating at the far field.

The relative amplitudes of the octaves around 10 (8–12), 20 (16–25), 40 (32–50), and 80 (64–100) Hz are plotted in the double logarithmic attenuation Figs. 3, 4 and 5b, d, f, h and evaluated for the overall attenuation power q. The measured attenuation at high frequencies is usually stronger with attenuation powers $q_I = 1.13$–2.5 for hammer impacts and $q_T = 0.48$–1.95 for trains. The attenuation at the lower frequencies is considerably smaller with $q_I = 0.51$–0.85 for impulses and $q_T = 0.37$–0.62 for trains. The difference of the attenuation powers for train and hammer excitation follows with $\Delta q = 0.2$–0.6 nearly the theoretical values of $\Delta q = 0.3$ and 0.5.

3 Discussion of the Attenuation Laws and Conclusion

Experiments with hammer impacts confirm the exponential attenuation law when all frequencies are considered in the evaluation scheme for the damping ratio, see the continuous line in Fig. 6a. The (normalised) amplitudes for all measurement points (normalised distances) and of a single frequency are plotted with a specific marker in a specific colour. These results of a single frequency resemble rather a pure power-law attenuation than an exponential law. The power law could have the following reasons.

The layering of the soil weakens the exponential law of the homogeneous soil at high frequencies when the less damped deeper soils become dominant over the strongly damped top layer, see Fig. 1c compared to Fig. 1a.

A broad-band constant spectrum A_0 at a near-field point yields an additional attenuation power of $\Delta q = 0.5$ instead of the different exponential laws for each frequency (Fig. 6b)

$$\sqrt{M\left(v^2\right)} \sim \sqrt{\int_0^\infty \frac{{A_0}^2}{r}\left(exp\left(-\frac{\omega D}{v_R}r\right)\right)^2 d\omega} = \frac{A_0}{\sqrt{r}}\sqrt{\frac{v_R}{2Dr}} \sim r^{-1.0} \tag{5}$$

This effect could also be expected for a single frequency if there is an interference of different wave modes of the layered soil.

The material damping of the soil is non-linear damping in general. The smaller the amplitudes in the far field the weaker is the damping and the corresponding attenuation. Therefore, the exponential attenuation of the constant damping would turn into a more straight-line (power-law) attenuation.

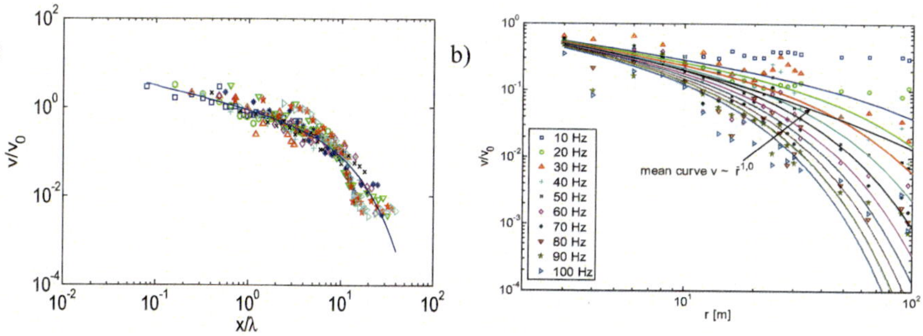

Fig. 6. Exemplary attenuation laws, a) overall exponential law fitted to single frequencies from measured hammer excitation, b) sum of exponential laws yield a power law from calculated broad-band excitation.

To conclude, the empirical power-law attenuation with frequency-dependent $q(f)$ seems to be appropriate for the measured hammer- and train-induced ground vibration. The power q increases with frequency and with hammer excitation (theoretically by $\Delta q = 0.3$ or 0.5), and has been found between 0.6 and 2.5 for hammer and between 0.37 and 1.95 for train excitation at five representative sites. The spatial attenuation and low- and high-frequency filter effects have been explained by the damping, layering and heterogeneity of the soil and by some distributed and moving-load effects of the train.

References

1. Auersch, L.: Wave propagation in layered soil: theoretical solution in wavenumber domain and experimental results of hammer and railway traffic excitation. J. Sound Vib. **173**, 233–264 (1994)
2. Sato, H., Fehler, M.: Seismic Wave Propagation and Scattering in the Heterogeneous Earth. Springer, Heidelberg (1998)
3. Auersch, L.: Technically induced surface wave fields, Part I: measured attenuation and theoretical amplitude-distance laws, Part II: measured and calculated admittance spectra. Bull. Seismol. Soc. Am. **100**(4), 1528–1550 (2010)
4. Auersch, L.: Ground vibration due to railway traffic—the calculation of the effects of moving static loads and their experimental verification. J. Sound Vib. **293**, 599–610 (2006)
5. Auersch, L.: Fast trains and isolating tracks on inhomogeneous soils. In: Krylov, V. (ed.) Ground Vibrations from High-Speed Railways: Prediction and Mitigation, pp. 27–76. ICE Publishing, London (2019)
6. Auersch, L.: Train-induced ground vibration due to the irregularities of the soil. Soil Dyn. Earthq. Eng. **140**, 106438 (2021)

IDF: Dynamic Response of Structures Interacting with Dense Fluids for Industrial Applications

Prediction of the Resonance Frequency of the Pipe Carrying Fluid Relative to the Fluid Velocity

H. Y. Ahmad[1](✉), M. J. Jweeg[2], and D. C. Howard[3]

[1] Safran Electrical and Power, Buckinghamshire, UK
Hayder.ahmad@safrangroup.com
[2] Al-Farahidi University, Baghdad, Iraq
[3] CDH Engineering Consultancy, Fareham, UK

Abstract. Pipes carrying fluid may resonate at certain fluid velocities which is called the critical fluid speed/velocity. This resonance may lead to pipe buckling or fatigue failure of welded pipes. Usually, such velocity is high in value and difficult to measure experimentally as this type of test will cause a severe damage to the pipes system. However, in this paper a new laboratory experimental approach was developed to measure fundamental natural frequencies of the pipe at different fluid velocity for laminar and/or turbulent fluid systems. The test method was carried out in the laboratory on sample pipes and the test results were validated with the theoretical/analytical results to establish the test accuracy.

Keywords: Natural frequency · Critical buckling velocity · Instability · Galerkin's method

Nomenclatures

$A_f.A_p$:	Fluid and pipe cross sectional area, respectively in (m^2)
CFD:	*Computational Fluid Dynamics*
D_o, $D_{i:}$:	Outer and inner pipe diameter, respectively in (m)
E:	Modulus of elasticity in (N/m^2)
FEA:	*Finite element analysis*
I:	Moment of inertia in (m^4)
L:	Pipe length in (m)
m_f, m_p:	Fluid and pipe mass per unit length, respectively in (kg/m)
Q_c:	Critical flow rate in (l/min)
U:	Dimensionless fluid velocity
U_c:	Dimensionless critical buckling velocity
V:	Fluid velocity in (m/s)
V_c:	Critical velocity of buckling in (m/s)
Ω:	Dimensionless frequency = $\omega L^2[(m_f + m_p) /EI]1/2$
ω:	Circular frequency in (rad/sec)
Ω_b:	Dimensionless natural frequency of beams.
ρ:	Fluid density in (kg/m^3)

Z. Dimitrovová et al. (Eds.): WMVC 2022, MMS 125, pp. 363–372, 2023.
https://doi.org/10.1007/978-3-031-15758-5_36

1 Introduction

The determination of instability of critical velocities of pipes of different boundary conditions and transmitting fluid is essential to avoid failure of pipes especially in the oil and gas industry. Number of researchers studied the dynamic response of the pipes to the fluid critical velocities for example; Jin [1] investigated the stability and dynamics of a cantilever pipe with an elastic support conveying fluid. The effect of the spring constant on the dynamic behaviour of the pipe was included in this study. On the other hand, Paidoussis [2] had shown that the natural frequencies of pipes conveying fluid with end constraints Pinned-Pinned (P-P), Clamped-Clamped (C-C) and Clamped-Pinned (C-P) decrease with increasing the fluid velocities. The absolute and convective bending instabilities in pipes conveying fluid was studied by Langre and Ouvrard [3]. The interaction of fluid-structure was taking into consideration the flow velocity and the stiffness of the elastic foundation. This research suggested that the results can be applied for studying the instabilities in infinitely extended one-dimensional system. Öz [4] investigated the vibration of tensioned pipes conveying fluid. This research was a combination of experimental work and numerical analysis to obtain the natural frequencies for different fluid velocities with the variation of the ratios of fluid mass to the pipe mass. Haung et al. [5] presented a new matrix method for solving the vibration and stability of curved pipes conveying fluid with minimum computing time. Aldraihem [6] analysed the dynamic stability of collar-stiffened pipes conveying fluid. In this study, the Euler-Bernoulli beam theory was applied and the finite element was utilized to investigate the dynamic stability of clamped-free collar-stiffened pipes. It was concluded that the collar-stiffened pipes have a unique stability compared to a uniform and the mode shape of the uniform beam is not necessarily the same as that of the collar-stiffened pipes.

In this work, a numerical approach will be introduced to calculate the critical velocity and the results will be validated experimentally by measuring the natural frequencies for low fluid velocities.

2 Theoretical Approach

The equation of the vibration motion of the pipe transmitting fluid according to the beam theory is expressed as follows [1]:

$$EI\frac{\partial^4 y}{\partial x^4} + m_f V^2\frac{\partial^2 y}{\partial x^2} + 2m_f V\frac{\partial^2 y}{\partial x \partial t} + m_f\frac{\partial V}{\partial t}\frac{\partial y}{\partial x} + (m_f + m_p)\frac{\partial^2 y}{\partial t^2} = 0 \quad (1)$$

Equation (1) is rewritten in a dimensionless form:

$$\eta^{IV} + \left(U^2 + \gamma\right)\eta'' + 2\beta U\dot{\eta}' + \dot{U}\eta' + \ddot{\eta} = 0 \quad (2)$$

where;

$$\eta = \frac{y}{L},\ \zeta = \frac{x}{L},\ U = VL\sqrt{\frac{m_f}{EI}},\ \beta = \sqrt{\frac{m_f}{m_f + m_p}} \text{ and } T = \frac{t}{L^2}\sqrt{\frac{EI}{m_f + m_p}} \quad (3)$$

The notations $(')$ and $(\cdot)$ are referred to the differentiation with respect to space and time, respectively. For steady flow, Eq. (2) is rewritten as follows:

$$\eta^{\mathrm{IV}} + \mathrm{U}^2\eta'' + 2\beta\mathrm{U}\dot{\eta}' + \ddot{\eta} = 0 \tag{4}$$

Equation (4) can be solved using Galerkin's method, where the continuous solution of the PDE can be discretized by assuming the following series solution:

$$\eta(\zeta,\tau) = \sum_{n=1}^{\infty} \Phi_n(\zeta)q_n(\tau) \tag{5}$$

where, $\Phi_n(\zeta)$ and $q_n(\tau)$ are the space (shape) functions and the generalised coordinates, respectively.

In case of beam like pipe, the normal modes of beams can be used as shape functions, which are good approximation for the pipe vibrations, since they can well describe the dynamical behavior as well as it satisfies all boundary conditions.

The harmonic nature of vibration motion implies that:

$$\mathrm{q_n(T) = u_n e^{i\Omega T}} \tag{6}$$

where u_n is arbitrary constants.

Substituting Eqs. (5) and (6) in Eq. (2):

$$\sum_{n=0}^{\infty}\left(\Phi_n^{IV} + U^2\Phi_n'' + 2i\beta U\Omega\Phi_n' - \Omega^2\Phi_n\right)u_n = 0 \tag{7}$$

$\Phi_n(\zeta)$ is replaced by Φ_{n} for simplicity, then multiplying Eq. (7) by the boundary residual value function Φ_{k} and integrating along the whole span of the pipe:

$$\int_0^1 \left\{\sum_{n=1}^{\infty}\left(\Phi_n^{IV} + U^2\Phi_n'' + 2i\beta U\Omega\Phi_n' - \Omega^2\Phi_n\right)\sum_{k=1}^{\infty}\Phi_k\right\}u_n = 0 \tag{8}$$

where;

$$\Phi_{\mathrm{n}}(\zeta) = \sin \eta_{\mathrm{n}}\zeta,\ \eta_{\mathrm{n}} = \mathrm{n}\pi, \quad \mathrm{for P - P}$$

$$\Phi_n(\zeta) = \cos \eta_{\mathrm{n}}\zeta - \cosh \eta_{\mathrm{n}}\zeta - \frac{\cos\eta_n - \cosh\eta_n}{\sin\eta_n - \sinh\eta_n}(\sin\eta_n\zeta - \sinh\eta_{\mathrm{n}}\zeta),$$

$$\eta_{\mathrm{n}} = 3.927, \quad 7.069 \text{ for } \mathrm{C - P}$$

$$\Phi_{\mathrm{n}}(\zeta) = \cos \eta_{\mathrm{n}}\zeta - \cosh \eta_{\mathrm{n}}\zeta - \frac{\cos\eta_n - \cosh\eta_n}{\sin\eta_n - \sinh\eta_n}(\sin\eta_n\zeta - \sinh\eta_n\zeta),$$

$$\eta_{\mathrm{n}} = 4.73, 7.853, \text{ for } \mathrm{C - C}$$

Equation (8) can be rearranged in matrix form as follows:

$$\left([\mathrm{C}] + \mathrm{U}^2[\mathrm{A}] + 2\mathrm{i}\beta\mathrm{U}\Omega[\mathrm{B}] - \Omega^2[\mathrm{I}]\right)\{\mathrm{u}\} = 0 \tag{9}$$

The elements of the matrices [A], [B] and [C] depend on the shape function chosen to satisfy the pipe end conditions while their size depend on the number of terms used.

The buckling instability occurs when the fundamental natural frequency is dropped to zero. In this regard, when only the fundamental mode ($n = 1, k = 1$) is considered in Eq. (9), the following relations between the fundamental natural frequencies and the fluid velocities for P-P, C-P and C-C pipes can be obtained as follows:

$$\Omega = \pi^2\sqrt{1 - \frac{U^2}{\pi^2}} \tag{10}$$

$$\Omega = 4.73^2\sqrt{1 - \frac{U^2}{6.378^2}} \tag{11}$$

$$\Omega = 3.927^2\sqrt{1 - \frac{U^2}{4.543^2}} \tag{12}$$

Squaring and rearranging Eqs. (10, 11 and 12):

When $\Omega = 0$, the solution of Eqs. (110, 11 and 12 gives the critical velocity of buckling U_c for the selected types of supporting conditions as follows:

$$U_c = \pi, 4.543, 6.378$$

In addition to that, when $U = 0$ the solution of, Eqs. (10, 11 and 12) yield to the fundamental natural frequencies of the beam:

$$\Omega_b = \pi^2, (3.927)^2$$

In this way, Eqs. (10, 11 and 12) can be arranged as follows:

$$\frac{\Omega^2}{\Omega_b^2} = 1 - \frac{U^2}{U_c^2} \tag{13}$$

Or:-

$$\Omega^2 = \Omega_b^2 - (\Omega_b/U_c)^2 U2 \tag{14}$$

Equation (14) represents a straight line in Ω^2–U^2.

Plotting Eq. (14) for a given fluid velocities Ω^2 versus U^2, a straight line is obtained from which the critical velocity is obtained. It can be concluded that the critical velocity is calculated from the measured the fundamental natural frequencies and plotting the relation Ω^2 versus U^2, which gives a straight line from which the mathematical form, may be written. By comparing, the fitted form with those obtained from Eq. (14), the critical velocity U^2 can be obtained.

When the fluid flow rate is relatively low the natural frequency variations will be so small which become insensitive.

Table 1. Specifications of the test pipe models [7]

Model no	Material	D_o (m)	t (m)	L (m)	ρ kg/m^3	E GN/m^2
1	Aluminum	0.01	0.001	1.25	2650	62.5
2	PVC	0.011	0.001	1.25	1000	7.12

3 Experimental Work

Table 1 shows the specifications of the two tested models used in this work.

To fulfil the requirements of the support pinned and clamped. The pinned support is ensured by a ball bearing (zero displacement and moment) and for clamped condition (zero displacement and slope), steel clamps are used to ensure this condition.

Figure 1 shows the testing rig with the water circuit. The main components of the rig are collecting tank 150 L centrifugal pump of velocity 60 l/min and height 20 m together with the control valve.

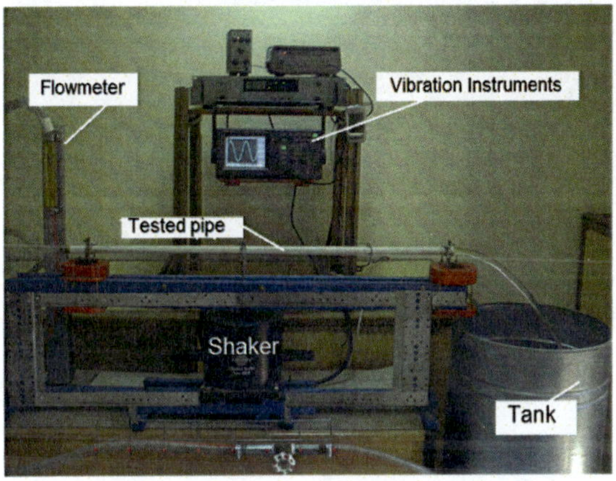

Fig. 1. Test rig setup extracted from the reference [8]

Prior to testing, all instruments were calibrated; the modulus of elasticity and density were measured accurately. The steps of the test are as follows:

I. The natural frequencies were measured for the selected models for zero flow and different flow rate velocities between 10 to 60 l/min by using different boundary conditions. The flow rate was adjusted using the controlling valve.
II. In case of turbulent flow, the water was allowed to settle for 10 to 15 min before measuring the fundamental natural frequencies.
III. The tested models were under forced vibration by using a harmonic force generated by a shaker.
IV. The response was picked up by using the accelerometer sensor with the output displayed by an oscilloscope.

4 Real Case Study

To validate the theoretical approach a real case study was performed on a pipe used for transporting oil in Al-Dura Refinery, Baghdad –Iraq. The pipe has the following specifications; Carbon steel CSA105 material, (4 m) span length, (30 cm) outer diameter and (9.25 mm) Schedule thickness. It transports a fluid with a density of 0.96 kg/m^3 and a flow rate ranging from 50 to 300 m^3/min. In this test, a hammer is used to initiate the transient vibration response. Five-flow rate values are used which are 0, 75, 150, 225 and 300 m^3/min. The data of the vibration response was measured and collected by using the vibration handheld instrument for every flow rate case. Finally, the collected data was imported to the PC and analysed by using LABVIEW software. In this program, the time response of the data is transformed to the frequency domain via the FFT spectrum transformation. From the FFT the fundamental frequencies are extracted for every flow rate value. The critical velocity of buckling is then estimated by using the fitting method. Figure 2 shows the testing method for the studied pipe.

Fig. 2. The testing method for the pipe

5 Results and Discussions

For the comparison purpose, the theoretical critical flow rates of the two model under P-P, C-P and C-C boundary conditions are calculated and summarized in Tables 2 and 3. For all cases and models, the pipe span length is 0.74 m.

Tables 2 and 3 show the measured natural frequencies for the each model.

The measured dimensionless fundamental natural frequencies Ω and the flow velocities U are plotted and compared against the calculated values from Eq. (3) as shown in Figs. 3 and 4 for the three types of boundary conditions. In these two Figs. (3 and 4) a fitted line is plotted to extract an equation from it which definite the relation between the natural frequencies Ω and the oil flow velocities U.

The equations of the fitted lines are detailed in Table 4.

Comparing the predicted dimensionless forms of the critical velocities listed in Table 4 with those calculated by using Eq. (14), the estimated critical velocities in the

Table 2. Experimental values of fundamental natural frequencies (Hz) for aluminum pipe

Q (l/min)	0	15	30	45	60	75	90	105
P-P frequency Hz	13.2	12.95	12.6	12.5	11.8	11.04	10.86	10.34
C-P frequency Hz	17.7	17.72	17.6	17.55	16.7	16.70	15.63	15.04
C-C frequency Hz	25.7	15.7	25.6	25.58	25.0	24.47	23.89	22.78

Table 3. Experimental values of fundamental natural frequencies (Hz) for PVC pipe

Q (l/min)	0	15	30	45	60	75	90	105
P-P Frequency Hz	7.2	6.56	5.9	4.22	2.5	2.42	1.29	0.57
C-P Frequency HZ	8.1	7.931	7.75	6.88	6.0	5.13	4.22	3.05
C-C Frequency HZ	12.00	11.60	11.20	10.55	9.9	9.10	8.25	7.49

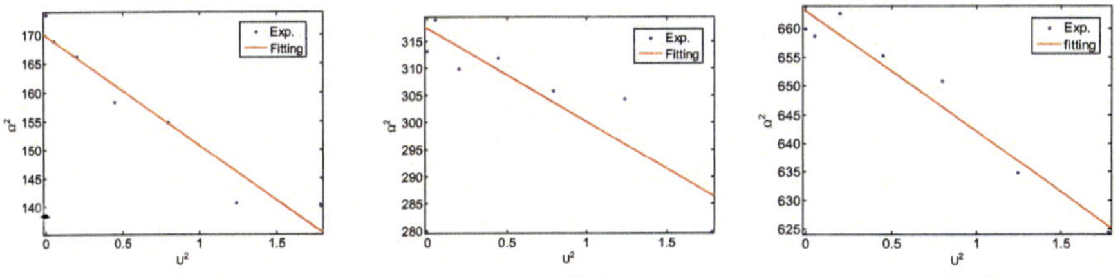

Fig. 3. Experimental plot of Ω^2 against U^2 for aluminum pipe at different boundary conditions

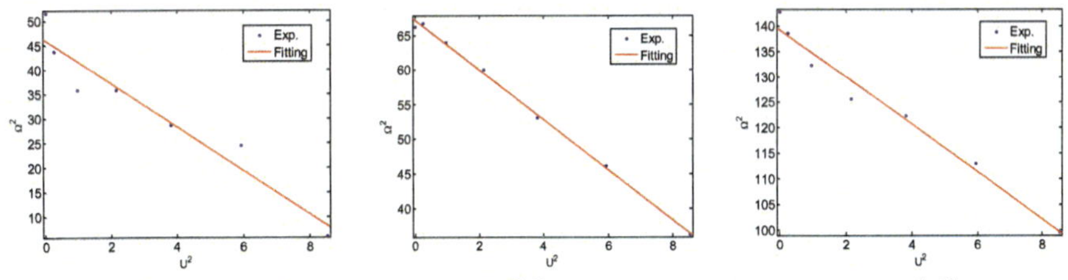

Fig. 4. Experimental plot of Ω^2 against U^2 for PVC pipe at different boundary conditions

dimensionless forms can be obtained. For example, the critical velocity of P-P Aluminum pipe is found from Table 4:

$$\Omega^2 = 169.91 - 19.033U^2$$

According to Eq. (14), $\Omega_b{}^2 = 169.91$ and $(\Omega_b/U_c)^2 = 19.029$. The solution of the above equations yields, Uc = 2.9881.

Table 4. Fitted lines Eqs. 7, 8, 9, 10, 11 and 12

Model	Fitting equation	Boundary conditions
Al	$\Omega^2 = 169.91–19.033U^2$	P-P
	$\Omega^2 = 317.5\ 1–17.272U^2$	C-P
	$\Omega^2 = 663.21–21.051U^2$	C-C
PVC	$\Omega^2 = 46.052–4.407U^2$	P-P
	$\Omega^2 = 67.291–3.605U^2$	C-P
	$\Omega^2 = 139.31–4.632U^2$	C-C

By inserting the Aluminum specifications (Table 1), in Eq. (3) the actual critical velocity becomes $V_c = 42.5031$ m/s.

The theoretical critical velocities of the two selected models together with the percentage of discrepancies are given in Tables 5 and 6.

Table 5. Theoretical and Estimated critical velocities for Aluminum pipe

Boundary conditions	Critical velocity (m/s)		Error%
	Estimated	Theoretical	
P-P	42.5	44.69	−4.91
C-P	61	64.01	−4.72
C-C	79.85	89.38	−10.11

Table 6. Theoretical and estimated critical velocities for PVC pipe

Boundary conditions	Critical velocity		Error %
	Estimated	Theoretical	
P-P	18.22	17.71	2.92
C-P	24.36	25.36	−3.96
C-P	31.92	35.41	−9.72

The results in Tables 5 and 6 show that the estimated and theoretical determination of the fluid flow velocities are close with a maximum discrepancy of 9.72% in the C-C pipe case for both models. This is because the clamped boundary cannot be ensured in the real tests. In order to achieve a good experimental prediction, a wider range of flow rate must be deployed for accurate measurements of natural frequencies.

Table 7 lists the measured fundamental frequencies of the case study pipe at the selected flow rates values. The other necessary dimensionless parameters associated to the velocity and natural frequencies are also calculated and displayed in this table. In order to estimate the critical velocity of buckling the values of U^2 and Ω^2 are plotted as shown in Fig. 5. The line of best fit of the measured data is also shown on the same figure. The Least Square Method is used for performing the fitting within the confidence limit where Sd = 1.86. The equation of the resulting line is obtained as follows:

$$\Omega^2 = 107.5 - 10.366U^2 \tag{15}$$

For $\Omega = 0$, Eq. (15) gives the estimated dimensionless critical velocity $U_c = 3.222$.

Now by using Eq. (3) the value of the estimated critical velocity is 3472.21 m/s and the corresponding critical flow rate is 216.11 m^3/s.

Table 7. Measured frequencies and dimensionless parameters of the case study pipe

Run no	Q m^3/min	Freq. Hz	U dimensionless	Ω dimensionless	U2	Ω2
1	0	564.7	0	10.44	0	114.49
2	75	503.6	1.02	9.87	1.04	99.80
3	150	441.6	1.43	9.24	2.05	85.43
4	225	386.2	1.74	8.65	3.02	71.36
5	300	329.4	2.01	7.98	4.07	57.43

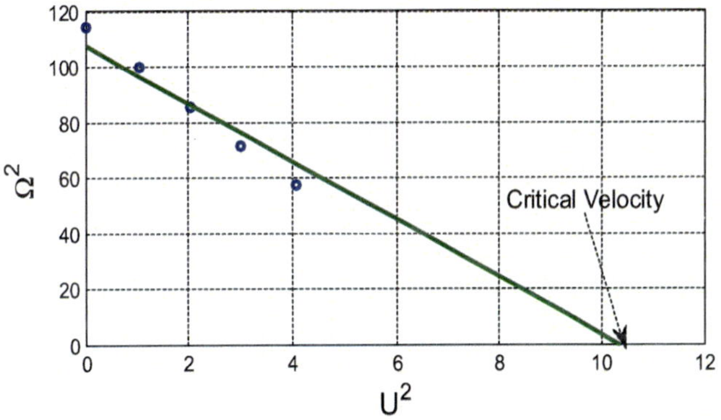

Fig. 5. Measured and fitted data of the case study pipe

6 Conclusions

The conclusions from this work are as follows:

1. The newly presented approach gives an effective, simple and low cost solution for an alternative to the experimental program to estimate the critical velocities of buckling.

2. From the experimental results of two pipe models under P-P, C-P and C-C boundary conditions, it can be deduced that; the suggested approach is more accurate for P-P and C-P pipes with a maximum discrepancy of 5% while for the C-C the maximum percentage discrepancy between the calculated and the measured natural frequencies is not more than 10%.
3. The presented method represents a nondestructive testing method for measuring the critical velocities of pipes operating in site without causing damage to the pipe system.

References

1. Jin, J.D.: Stability and chaotic motion of a restrained pipe conveying fluid. J. Sound Vib. **208**, 427–439 (1997)
2. Paidoussis, M.P.: Fluid–Structure Interactions: Slender Structures an Axial Flow, vol 1. Academic Press, New York (1998)
3. De Langre, E.: Absolute and convective bending instabilities in fluid conveying pipes. J. Fluids Struct. **13**, 663–680 (1999)
4. Öz, H.R.: Natural frequencies of fluid conveying tensioned pipes and carrying a stationary mass under different end condtions. J. Sound Vib. **253**, 507–517 (2002)
5. Haung, Y., Zeng, G., Wei, F.: A new matrix method for solving vibration and stability of curved pipes conveying fluid. J. Sound Vib. **251**, 215–225 (2002)
6. Aldraihem, O.J.: Analysis of the dynamic stability of collar – stiffened pipes conveying fluid. J. Sound Vib. **300**, 453–465 (2007)
7. ASHRAE Systems and Equipment Handbook (SI), pp. 41.3–41.4 (2000)
8. Jweeg, M.J., Yousif, A.E., Ismail, M.R.: Experimental estimation of critical buckling velocities for conservative pipes conveying fluid. Al-Khwarizmi Eng. J. **7**, 17–26 (2001)

The Pseudo-static Axisymmetric Problem for a Poroelastic Cylinder

Natalya Vaysfeld and Zinaida Zhuravlova(✉)

Odessa I.I. Mechnikov National University, Odessa 65082, Ukraine
{vaysfeld,z.zhuravlova}@onu.edu.ua

Abstract. The pseudo-static case for a poroelastic axisymmetric cylinder is considered in the terms of Biot's model. The cylinder is loaded by its boundary $r = a$, where the perfect drainage conditions are fulfilled, and the boundaries $z = 0$ and $z = h$ are in slide contact and undrained conditions. The initial conditions are zero. The initial problem is reduced to the one-dimensional problem with the help of Laplace and Fourier integral transforms applied by t and z variables respectively. The one-dimensional problem in transform space is formulated in a vector form, which is solved with the help of matrix differential calculation apparatus. According to it the corresponding matrix equation is considered, and the system of fundamental matrix solutions is constructed by the contour integral calculation. So, the exact solution of the vector boundary problem in Laplace-Fourier transform space is derived. The Heaviside expansion theorem is used to invert Laplace transforms, and the stress and pore pressure values are investigated for different ratios between a and h, and for different poroelastic materials.

Keywords: Pseudo-static poroelasticity · Axisymmetric cylinder · Integral transforms · Exact solution

1 Introduction

The poroelastic problems are widely used in many engineering and biological applications. Firstly, such problems were considered in the papers of Terzaghi [1] for a one-dimensional case and Biot [2] for a three-dimensional case.

The cylindric bodies are important models, especially in poroelastic statement. So, there are many papers dedicated to such problems. However, most of solutions are numerical. A pseudo-transient numerical method was used in solving the problem for an axisymmetric cylindrical model of fully coupled fluid flow [3]. The finite element method was used to solve the problem for mathematical model for bone reconstruction in the volume of a porous implant (scaffold) in [4]. The algorithm based on the spectral method was applied to solve the dispersion equation for cylindrical poroelastic structures [5]. The stochastic meshless local Petrov–Galerkin method for dynamic analysis of cylinders made of fully saturated porous materials was employed in [6].

The analytical and analytical-numeric solutions were derived in [7–12].

In this paper the analytical approach for the solving of the pseudo-static axisymmetric problem for a poroelastic cylinder is proposed.

Z. Dimitrovová et al. (Eds.): WMVC 2022, MMS 125, pp. 373–378, 2023.
https://doi.org/10.1007/978-3-031-15758-5_37

2 The Statement of the Problem

The poroelastic cylinder ($0 < r < a$, $-\pi < \varphi < \pi$, $0 < z < h$, $t > 0$) in the terms of pseudo-static Biot's model [2] is considered. The slide contact conditions are fulfilled at the cylinder's edges

$$
\begin{aligned}
&w|_{z=0} = 0,\ \tau_{rz}|_{z=0} = 0,\ \left.\frac{\partial p}{\partial z}\right|_{z=0} = 0,\\
&w|_{z=h} = 0,\ \tau_{rz}|_{z=h} = 0,\ \left.\frac{\partial p}{\partial z}\right|_{z=h} = 0
\end{aligned}
\tag{1}
$$

and the boundary $r = a$ the cylinder is loaded

$$
\sigma_r|_{r=a} = -l(z,t) - \alpha P(z,t),\quad \tau_{rz}|_{r=a} = 0,\quad p|_{r=a} = P(z,t) \tag{2}
$$

Here $u(r, z) = u_r(r, z)$, $w(r, z) = u_z(r, z)$ are displacements of the solid skeleton, $p(r, z)$ is pore pressure, $\sigma_r(r, z)$, $\tau_{rz}(r, z)$ are normal and shear effective stress.

The initial condition is the following

$$
\left.\left[\alpha\left(\frac{1}{r}\frac{\partial}{\partial r}(ru) + \frac{\partial w}{\partial z}\right) + S_p p\right]\right|_{t=0} = 0 \tag{3}
$$

The system of governing equations has the following form [13]

$$
\begin{cases}
\dfrac{1}{r}\dfrac{\partial}{\partial r}\left(r\dfrac{\partial u}{\partial r}\right) - \dfrac{1}{r^2}u + \dfrac{\kappa-1}{\kappa+1}\dfrac{\partial^2 u}{\partial z^2} + \dfrac{2}{\kappa+1}\dfrac{\partial^2 w}{\partial r\partial z} - \dfrac{\alpha}{G}\dfrac{\kappa-1}{\kappa+1}\dfrac{\partial p}{\partial r} = 0,\\
\dfrac{1}{r}\dfrac{\partial}{\partial r}\left(r\dfrac{\partial w}{\partial r}\right) + \dfrac{\kappa+1}{\kappa-1}\dfrac{\partial^2 w}{\partial z^2} + \dfrac{2}{\kappa-1}\dfrac{1}{r}\dfrac{\partial}{\partial r}\left(r\dfrac{\partial u}{\partial z}\right) - \dfrac{\alpha}{G}\dfrac{\partial p}{\partial z} = 0,\\
\dfrac{1}{r}\dfrac{\partial}{\partial r}\left(r\dfrac{\partial p}{\partial r}\right) + \dfrac{\partial^2 p}{\partial z^2} - \dfrac{\alpha}{k}\left[\dfrac{1}{r}\dfrac{\partial}{\partial r}\left(r\dfrac{\partial u}{\partial t}\right) + \dfrac{\partial^2 w}{\partial z\partial t}\right] - \dfrac{S_p}{k}\dfrac{\partial p}{\partial t} = 0
\end{cases}
\tag{4}
$$

where $\kappa = 3 - 4\mu$ is Muskhelishvili's constant, μ is Poisson ratio, G is shear modulus, α is Biot's coefficient, S_p is storativity of the pore space, k is permeability.

The displacements, stress and pore pressure that satisfy Eq. (4) and conditions (1)–(3) should be found.

3 The Reduction to the One-Dimensional Problem

The Laplace transform with parameter s is applied regarding the variable t to the initial problem (1)–(4). The problem in Laplace transform space can be written as

$$
\begin{cases}
\dfrac{1}{r}\dfrac{\partial}{\partial r}\left(r\dfrac{\partial u_s}{\partial r}\right) - \dfrac{1}{r^2}u_s + \dfrac{\kappa-1}{\kappa+1}\dfrac{\partial^2 u_s}{\partial z^2} + \dfrac{2}{\kappa+1}\dfrac{\partial^2 w_s}{\partial r\partial z} - \dfrac{\alpha}{G}\dfrac{\kappa-1}{\kappa+1}\dfrac{\partial p_s}{\partial r} = 0,\\
\dfrac{1}{r}\dfrac{\partial}{\partial r}\left(r\dfrac{\partial w_s}{\partial r}\right) + \dfrac{\kappa+1}{\kappa-1}\dfrac{\partial^2 w_s}{\partial z^2} + \dfrac{2}{\kappa-1}\dfrac{1}{r}\dfrac{\partial}{\partial r}\left(r\dfrac{\partial u_s}{\partial z}\right) - \dfrac{\alpha}{G}\dfrac{\partial p_s}{\partial z} = 0,\\
\dfrac{1}{r}\dfrac{\partial}{\partial r}\left(r\dfrac{\partial p_s}{\partial r}\right) + \dfrac{\partial^2 p_s}{\partial z^2} - \dfrac{\alpha s}{k}\left[\dfrac{1}{r}\dfrac{\partial}{\partial r}(ru_s) + \dfrac{\partial w_s}{\partial z}\right] - \dfrac{S_p s}{k}p_s = 0
\end{cases}
\tag{5}
$$

Here $u_s(r, z),\ w_s(r, z),\ p_s(r, z)$ are Laplace transforms of the displacements and pore pressure.

The sin-, cos- Fourier transform is applied to (5) by the scheme

$$\begin{bmatrix} u_{s\beta}(r) \\ w_{s\beta}(r) \\ p_{s\beta}(r) \end{bmatrix} = \int_0^h \begin{bmatrix} u_s(r, z) \\ w_s(r, z) \\ p_s(r, z) \end{bmatrix} \begin{bmatrix} \cos\beta z \\ \sin\beta z \\ \cos\beta z \end{bmatrix} dz,\ \beta_n = \frac{\pi n}{h},\ n = 0, 1, 2, \ldots$$

So, the one-dimensional problem in Laplace-Fourier transform domain is derived. It is formulated as vector boundary problem [14]

$$\begin{cases} L_2\vec{y}_{s\beta}(r) = 0,\ 0 < r < a, \\ A_{s\beta}\vec{y}\,'_{s\beta}(a) + B_{s\beta}\vec{y}_{s\beta}(a) = \vec{g}_{s\beta} \end{cases} \qquad (6)$$

where L_2 is differential operator of the second order $L_2 = \begin{pmatrix} \frac{1}{r}\frac{d}{dr}\left(r\frac{d}{dr}\right) - \frac{1}{r^2} - \frac{\kappa-1}{\kappa+1}\beta^2 & \frac{2\beta}{\kappa+1}\frac{d}{dr} & -\frac{\alpha}{G}\frac{\kappa-1}{\kappa+1}\frac{d}{dr} \\ -\frac{2\beta}{\kappa-1}\frac{1}{r}\frac{d}{dr}(r) & \frac{1}{r}\frac{d}{dr}\left(r\frac{d}{dr}\right) - \frac{\kappa+1}{\kappa-1}\beta^2 & \frac{\alpha\beta}{G} \\ -\frac{\alpha s}{k}\frac{1}{r}\frac{d}{dr}(r) & -\frac{\alpha\beta s}{k} & \frac{1}{r}\frac{d}{dr}\left(r\frac{d}{dr}\right) - \beta^2 - \frac{S_p s}{k} \end{pmatrix}$, $\vec{y}_{s\beta}(r) = \begin{pmatrix} u_{s\beta}(r) \\ w_{s\beta}(r) \\ p_{s\beta}(r) \end{pmatrix}$, $\vec{g}_{s\beta} = \begin{pmatrix} \frac{l_{s\beta}}{2G} \\ 0 \\ P_{s\beta} \end{pmatrix}$, $A_{s\beta} = \begin{pmatrix} \frac{\kappa+1}{2(\kappa-1)} & 0 & 0 \\ 0 & 1 & 0 \\ 0 & 0 & 0 \end{pmatrix}$, $B_{s\beta} = \begin{pmatrix} \frac{3-\kappa}{2a(\kappa-1)} & \frac{(3-\kappa)\beta}{2(\kappa-1)} & 0 \\ -\beta & 0 & 0 \\ 0 & 0 & 1 \end{pmatrix}$.

4 The Solving of the One-Dimensional Vector Boundary Problem

The solution of the vector boundary problem (6) is constructed with the help of the apparatus of matrix differential calculation [15]. According to it, the solution of the corresponding matrix equation $L_2 Y_{s\beta}(r) = 0,\ 0 < r < a$ should be found. Here $Y_{s\beta}(r)$ is the matrix 3×3 order. The correspondence $L_2 H(r, \xi) = -H(r, \xi) M_{s\beta}(\xi)$ is used, where [16].

$H(r, \xi) = \begin{pmatrix} J_1(\xi r) & 0 & 0 \\ 0 & J_0(\xi r) & 0 \\ 0 & 0 & J_0(\xi r) \end{pmatrix}$, $J_0(\xi r),\ J_1(\xi r)$ are Bessel functions,

$$M_{s\beta}(\xi) = \begin{pmatrix} \xi^2 + \frac{\kappa-1}{\kappa+1}\beta^2 & \frac{2\beta}{\kappa+1}\xi & -\frac{\alpha}{G}\frac{\kappa-1}{\kappa+1}\xi \\ \frac{2\beta}{\kappa-1}\xi & \xi^2 + \frac{\kappa+1}{\kappa-1}\beta^2 & -\frac{\alpha\beta}{G} \\ \frac{\alpha s}{k}\xi & \frac{\alpha\beta s}{k} & \xi^2 + \beta^2 + \frac{S_p s}{k} \end{pmatrix}.$$

The solution of the matrix homogenous equation is constructed by a formula [17] $Y_{s\beta}(r) = \frac{1}{2\pi i}\oint_C H(r, \xi) M_{s\beta}^{-1}(\xi) d\xi$, where $M_{s\beta}^{-1}(\xi)$ is the inverse matrix to the matrix $M_{s\beta}(\xi)$. The closed contour C covers all singularity points of the matrix $M_{s\beta}^{-1}(\xi)$.

The determinant of the matrix $M_{s\beta}(\xi)$ has 2 multiple poles of the second order $\xi = i\beta$, $\xi = -i\beta$ and 2 simple poles $\xi = i\sqrt{\frac{\frac{\alpha^2 s(\kappa-1)}{G(\kappa+1)} + S_p s}{k} + \beta^2}$, $\xi = -i\sqrt{\frac{\frac{\alpha^2 s(\kappa-1)}{G(\kappa+1)} + S_p s}{k} + \beta^2}$. The system of four fundamental matrix solutions is derived $Y_i(y),\ i = \overline{1, 4}$ with the help of the residual theorem.

The solution of the boundary problem (6) for the case when $\beta \neq 0$ has the following form

$$\vec{y}_\beta(r) = (Y_1(r) + Y_3(r))\begin{pmatrix} c_1 \\ c_2 \\ c_3 \end{pmatrix} \tag{7}$$

where constants $c_i,\ i = \overline{1, 3}$ are found from the boundary conditions in (6).

The case for $\beta = 0$ is considered separately, because the matrices' dimensions for this case are 2 by 2. The vector boundary problem is this case has the form

$$\begin{cases} \tilde{L}_2\vec{y}_0(r) = 0,\ 0 < r < a, \\ A_0\vec{y}_0'(a) + B_0\vec{y}_0(a) = \vec{g}_0 \end{cases} \tag{8}$$

where $\tilde{L}_2 = \begin{pmatrix} \frac{1}{r}\frac{d}{dr}\left(r\frac{d}{dr}\right) - \frac{1}{r^2} & -\frac{\alpha}{G}\frac{\kappa-1}{\kappa+1}\frac{d}{dr} \\ -\frac{\alpha s}{k}\frac{1}{r}\frac{d}{dr}(r) & \frac{1}{r}\frac{d}{dr}\left(r\frac{d}{dr}\right) - \frac{S_p s}{k} \end{pmatrix},\ \vec{y}_0(r) = \begin{pmatrix} u_0(r) \\ p_0(r) \end{pmatrix},\ \vec{g}_\beta = \begin{pmatrix} \frac{l_\beta}{2G} \\ P_\beta \end{pmatrix},$
$A_0 = \begin{pmatrix} \frac{\kappa+1}{2(\kappa-1)} & 0 \\ 0 & 0 \end{pmatrix},\ B_0 = \begin{pmatrix} \frac{3-\kappa}{2a(\kappa-1)} & 0 \\ 0 & 1 \end{pmatrix}.$

The solutions $Y_{0,i}(y),\ i = \overline{1, 3}$ are found analogically using the correspondence $\tilde{L}_2H_0(r, \xi) = -H_0(r, \xi)M_0(\xi)$ and the auxiliary matrices $H_0(r, \xi) = \begin{pmatrix} J_1(\xi r) & 0 \\ 0 & J_0(\xi r) \end{pmatrix},\ M_0(\xi) = \begin{pmatrix} \xi^2 & -\frac{\alpha}{G}\frac{\kappa-1}{\kappa+1}\xi \\ \frac{\alpha}{k}\xi & \xi^2 + \frac{S_p}{k} \end{pmatrix}.$

The solution of the vector boundary problem (8) has the following form

$$\vec{y}_0(r) = \left(Y_{0,1}(r) + Y_{0,3}(r)\right)\begin{pmatrix} c_{0,1} \\ c_{0,2} \end{pmatrix} \tag{9}$$

where constants $c_{0,i},\ i = \overline{1, 2}$ are found from the boundary conditions in (8).

5 The Inversion of Derived Transforms

The solutions (7) and (9) are uniting and the following formula is used to invert the Fourier transforms

$$\begin{bmatrix} u_s(r, z) \\ w_s(r, z) \\ p_s(r, z) \end{bmatrix} = \frac{1}{h}\begin{bmatrix} u_{s0}(r) \\ w_{s0}(r) \\ p_{s0}(r) \end{bmatrix} + \frac{2}{h}\sum_{n=1}^{\infty}\begin{bmatrix} u_{s\beta_n}(r) \\ w_{s\beta_n}(r) \\ p_{s\beta_n}(r) \end{bmatrix}\begin{bmatrix} \cos\beta_n z \\ \sin\beta_n z \\ \cos\beta_n z \end{bmatrix},\ \beta_n = \frac{\pi n}{h}. \tag{10}$$

The expressions for $u_s(r, z),\ w_s(r, z),\ p_s(r, z)$ have the same denominator

$$\left[As^2 + Bs + C + \left(Ds^2 + Es + F\right)\sqrt{as + b}\right](Gs + H) \tag{11}$$

multiplied by s^2 for $w_s(r, z)$ and by s for $u_s(r, z)$, $p_s(r, z)$. Here $A, B, C, D, E, F, G, H, a, b$ are known constants.

The denominator is multiplied by $As^2 + Bs + C - \left(Ds^2 + Es + F\right)\sqrt{as + b}$. As a result, the denominator transformed to the polynomial of the sixth order $a_6 s^6 + a_5 s^5 + a_4 s^4 + a_3 s^3 + a_2 s^2 + a_1 s + a_0$ multiplied by s^2 for $w_s(r, z)$ and by s for $u_s(r, z)$, $p_s(r, z)$. Besides the root $s = 0$ it has six roots s_i, $i = \overline{1, 6}$, which are found numerically. So, the expressions for the displacements and pore pressure are found analytically using the Heaviside expansion theorem [18].

6 Conclusions

The exact solution for the problem for pseudo-static poroelastic cylinder is derived with the help of matrix differential calculation. The Laplace transforms were inverted analytically using the Heaviside expansion theorem. The stress and pore pressure were investigated.

The proposed method allows, within the framework of the Biot's model, to derive an explicit form for the displacements, stress and pore pressure. In this paper, the problem is formulated in the terms of a model that does not consider the wave propagation, however, the proposed approach allows to solve the problem, taking into account the inertial term for the hyperbolic type of equations. This will make it possible to study the propagation of waves in a poroelastic medium by analytical methods, which allow to reveal the features of the wave process in poroelastic media.

References

1. Terzaghi, K.: Erdbaumechanik auf bodenphysikalischer Grundlage. Deuticke, Wien (1925)
2. Biot, M.A.: General theory of three-dimensional consolidation. J. Appl. Phys. **12**, 155–164 (1941)
3. Ishbulatov, S., Yarushina, V., Podladchikov, Yu.: A numerical simulation of poroelastic cylinder decompression problem on CUDA in an axisymmetric domain. In: EGU21-1955 (2021). https://doi.org/10.5194/egusphere-egu21-11955
4. Maslov, L.B.: Biomechanical model and numerical analysis of tissue regeneration in the volume of a porous implant. Appl. Math. Mech. **83**(5–6), 834–860 (2019). (in Russian)
5. Karpfinger, F., Gurevich, B., Bakulin, A.: Modeling of axisymmetric wave modes in a poroelastic cylinder using spectral method. J. Acoust. Soc. Am. **124**, EL230 (2008). https://doi.org/10.1121/1.2968303
6. Kazemi, H., Shahabian, F., Hosseini, S.M.: Shock-induced stochastic dynamic analysis of cylinders made of saturated porous materials using MLPG method: considering uncertainty in mechanical properties. Acta Mech. **228**, 3961–3975 (2017). https://doi.org/10.1007/s00707-017-1898-0
7. de Leeuw, E.H.: The theory of three-dimensional consolidation applied to cylindrical bodies. In: Proceedings of 6th International Conference on Soil Mechanics and Foundation Engineering, Montreal, vol. 1, pp. 287–290 (1965)
8. Hosseini, N., Namazi, N.: Acoustic scattering of spherical waves incident on a long fluid-saturated poroelastic cylinder. Acta Mech. **223**, 2075–2089 (2012). https://doi.org/10.1007/s00707-012-0697-x

9. Abousleiman, Y., Cheng, A.H.-D., Jiang, C., Roegiers, J.-C.: Poroviscoelastic analysis of borehole and cylinder problems. Acta Mech. **119**, 199–219 (1996). https://doi.org/10.1007/BF01274248
10. Detournay, E., Cheng, A.H.-D.: Fundamentals of poroelasticity. In: Fairhurst, C. (ed.) Comprehensive Rock Engineering: Principles, Practice and Projects, Analysis and Design Method, vol. 2, pp. 113–171. Pergamon Press (1993)
11. Fellah, Z., Groby, J.-P., Ogam, E., Scotti, Th., Wirgin, A.: Acoustic identification of poroelastic cylinder. OR 22, HAL Id: hal-00014654 (2005)
12. Auton, L.C., MacMinn, Ch.W.: From arteries to boreholes: transient response of a poroelastic cylinder to fluid injection. Proc. R. Soc. A **474**, 20180284 (2018). https://doi.org/10.1098/rspa.2018.0284
13. Verruijt, A.: An Introduction to Soil Dynamics. Theory and Applications of Transport in Porous Media, vol. 24. Springer, Dordrecht (2010). https://doi.org/10.1007/978-90-481-3441-0
14. Reut, V., Vaysfeld, N., Zhuravlova, Z.: Non-stationary mixed problem of elasticity for a semi-strip. Coupled Syst. Mech. **9**(1), 77–89 (2020). https://doi.org/10.12989/csm.2020.9.1.077
15. Popov, G.Ya.: Exact Solutions of Some Boundary Problems of Deformable Solid Mechanic. Astroprint, Odessa (2013). (in Russian)
16. Protserov, Yu.S.: Axisymmetric problems of elasticity theory for a cylinder of finite length with free cylindrical surface and with tacking into account its own weight. Math. Mech. **18**, **3**(19), 69–81 (2013). The Herald of Odessa I.I. Mechnikov National University
17. Grantmakher, F.R.: Theory of Matrices. Nauka, Moscow (1967).(in Russian)
18. Abramowitz, M., Stegun, I.: Handbook of Mathematical Functions with Formulas Graphs and Mathematical Tables. U.S. Government Printing Office, Washington (1972)

MCA: Modeling, Simulation and Control of the Dynamical Behavior of Aerospace Structures

Development of a Modular Metal Pallet for Transportation and Stationary Conditions: Numerical Analyses and Experimental Characterizations

Enrico Zacchei[1,2(✉)], Antonio Tadeu[1,3], João Almeida[1,2], Miguel Esteves[1,2], Maria Inês Santos[1], and Samuel Silva[4]

[1] Itecons, Coimbra, Portugal
enricozacchei@gmail.com
[2] CERIS, University of Coimbra, Coimbra, Portugal
[3] CERIS, Department of Civil Engineering, University of Coimbra, Coimbra, Portugal
[4] Portimpact, Rua Zona Industrial 1080, 4580-565 Lordelo PRD, Portugal

Abstract. Due to the massive number of products transported every day, the use of lightweight pallets made of recyclable materials, easy to clean, resistant and cheap should be ideal. In this paper, new modular metal pallets that combine blocks and deck boards to produce different configurations for use in transportation/stationary conditions are proposed. Due to the lack of specific codes to design this structure, a more complete structural analysis (analytical and numerical) to evaluate stresses and deformations has been carried out. Then, laboratory experimental tests were performed to validate the model. A comparative life cycle analysis (LCA) was also carried out to identify the main environmental impacts with respect the wood/plastic/aluminium pallets. Numerical and experimental results provided new ultimate loadings between about 10.0–100.0 kN with a maximum deformation of about 35.0 mm. This indicates that steel pallets perform satisfactorily in terms of resistance and stiffness. Results from LCA suggested that steel pallets perform better than their wood, plastic, and aluminium pallets, mainly due to high recyclability and less need of repairing. Therefore, for a specific analysis of metal pallets, the values and procedures provided in the current codes should be updated. Because the different demands of dimensions and weight, the development of modular pallets could be a good choice to fulfil the functional requirements for several industries.

Keywords: Steel pallet · Pallet design · Numerical analyses · LCA

1 Introduction

World trade in goods has increased over the past 30 years and it is estimated to be worth more than half of the total world economy [1, 2]. Pallets are the most used unit-load portable platform in the global market since goods can be transported in a very efficient way by using standard devices.

Z. Dimitrovová et al. (Eds.): WMVC 2022, MMS 125, pp. 381–389, 2023.
https://doi.org/10.1007/978-3-031-15758-5_38

Without pallets, goods should be lifted manually or moved by using more complex equipment, thus increasing handling costs. As mentioned by the National Wooden Pallet and Container Association [3] "pallets move the world". Also, pallets are particularly relevant, for instance, in the air and aerospace transport of different products.

There are different types of pallets in the market to find various loadings, costs, and other specific requirements. Pallets can be made by solid wood (most popular), plastic, paper, recycled paper, corrugated paperboard, metal, wood-based composites, and wood-plastic [4, 5].

In accordance with previous studies [4–7], metal pallets perform better in terms of durability, mechanical resistance, stiffness, and lifespan. However, they tend to be more expensive and heavier. The authors have not found significant studies on modular metal pallets, therefore similar structures, e.g., stringer pallets, panel deck pallets, pallet beams, steel pallet racks [8, 9] could be used as reference. Also, no specific codes exist to design metal pallets; the standard references to be used are Eurocodes [10, 11] and ISO 8611 [12–14].

In this paper, analytical, numerical, and experimental analyses are proposed for steel pallets to find a structure with good performance in terms of resistance. Life cycle analysis (LCA) is also carried out to find a sustainable pallet. For this, the main environmental impacts associated with the life cycle of the new proposed steel pallet have been identified and compared with the impacts produced by reference pallets made of wood, plastic, and aluminium [15].

2 Case Study

The new modular steel pallet under study is formed of 8 transverse deck boards (800 mm × 120 mm), 3 longitudinal bottom deck boards (1200 mm × 120 mm), 3 longitudinal top deck boards (1200 mm × 120 mm), and 9 regulars cubic blocks (96.20 mm for each side) (see grey elements in Fig. 1).

This pallet is considered modular since its components are interchangeable with each other and are removable. This is possible since the decks profiles have a common geometry that can be connected to the blocks elements.

The pallet design is based on three testing setups used for laboratory evaluations to simulate real loading conditions given by ISO 8611 [12–14]: (i) stacking test, which simulates compressive loads by placing pallets on top of one another without intermediate shelves; (ii) racking test, which simulates the storage of unit compressive loads in beam racks with free unsupported spans; (iii) forklifting test, which simulates the impact force through the load carried by the pallet during the moving use.

Figure 1 shows, only in transversal side, the configuration of these tests where the metal pallets, restraints, loadings, and linear variable differential transducers (LVDT) are indicated.

The metal pallet is made by cold-formed steel DX51D with yield strength f_y = 140.0 MPa, elastic modulus E = 210.0 GPa, Poisson's ratio $\upsilon = 0.30$, and mass density ρ = 78.50 kN/m^3 [10]. The pallet mass is 13.86 kg, whereas the total weight of the orange elements, which transmit the vertical loadings on the metal pallets, ranges between 0.15–0.67 kN.

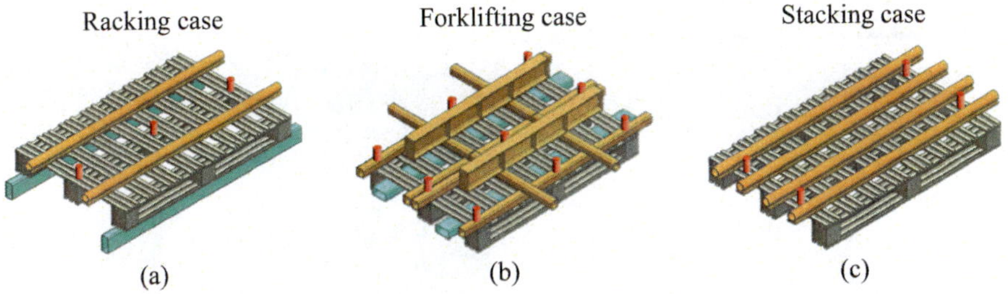

Fig. 1. Configuration of the experimental tests in transversal side for the (a) racking, (b) forklifting, (c) stacking case. Elements: grey = metal pallet; turquoise = restraints; orange = loadings; red = LVDT.

ISO 8611-2 [13] provides some examples of ultimate loads, U_t: 28.40 kN, 35.0 kN, 44.20 kN for racking, forklifting, and stacking, respectively. These values have been used as reference loadings to analyse the proposed steel pallet.

3 Analyses and Results

3.1 Design and FEM Models

The purpose of the structural analysis is to evaluate the flexural strength of the decks and the compressive strength of the blocks to estimate the pallet's maximum load capacity in accordance with Eurocode [10, 11].

For this, Eurocode defines the reduction of the gross cross-sectional area (gross section: $A_c = 384.90\ mm^2 \rightarrow$ effective section: $A_{c,eff} = 244.80\ mm^2$) of cold-form steel block under compressive loads accounting for the possible plate buckling phenomena. Also, the effective section modulus (i.e., $W_{eff} = 399.05\ mm^3$) should be estimated for deck boards.

Figure 2 shows the $A_{c,eff}$ of the block (Fig. 2(a)) and deck (Fig. 2(c)) indicating the effective gravity centre, $CG_{,c,eff}$, and the dimensions.

From these values, it is possible to estimate the axial compression resistance of the block, $N_{c,Rd}$ ($= A_{c,eff} \times f_y = 34.27$ kN), and the design moment resistance of a cross-section for bending, M_{Rd} ($= W_{eff} \times f_y = 55.87$ kN m $\rightarrow 55.87/0.12 = 0.46$ kN). Both resistance values have been applied in the model by finite element method (FEM) by SolidWorks software (Figs. 2(b)–(d)) where it is possible to see that the f_y is not reached.

The 3D analyses of the global metal pallet have been carried out. Solid elements models were used for all components (i.e., deck boards, blocks, supports, load boards), assuming fixed bases. For the racking and forklifting case, the fixed restraints coincide with the supports, whereas for the stacking case they coincide with the contact points between the deck boards and the ground.

Figure 3 shows the von Mises stresses, only in transversal side, when subjecting the pallets to the U_t loadings listed in Table 1. In this full analysis, it is possible to see that the von Mises stresses reach the f_y value (red regions) for the racking case, whereas for other cases the stresses are low (up to 93.33 MPa).

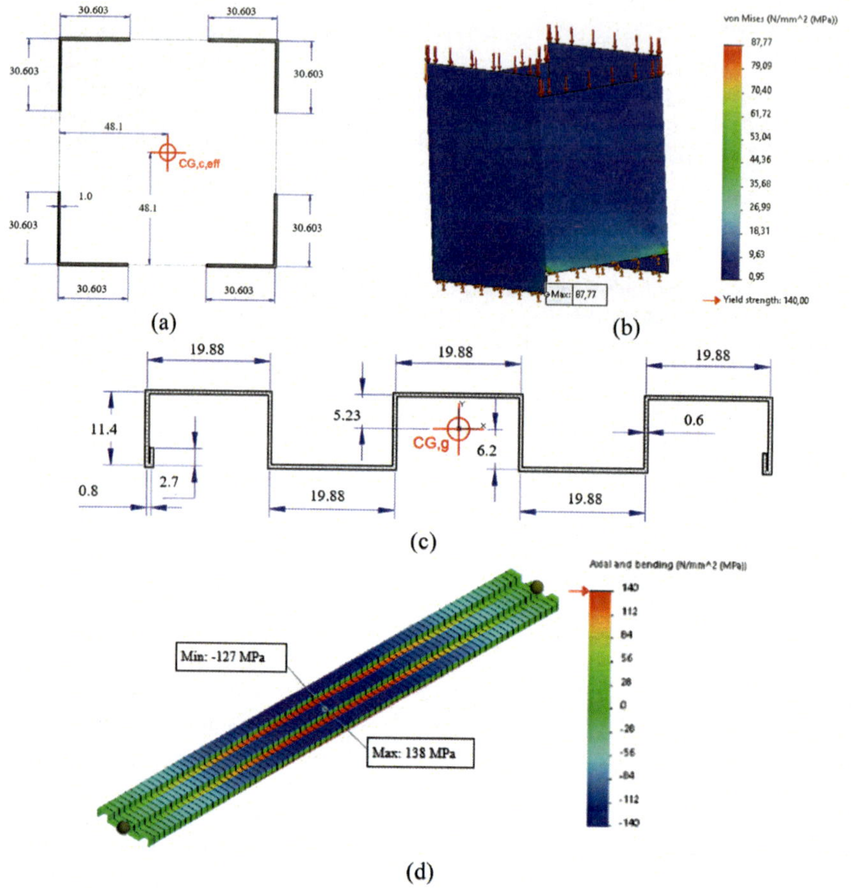

Fig. 2. Effective cross-section and FEM model of the (a, b) block and (c, d) transversal deck (dimension in mm).

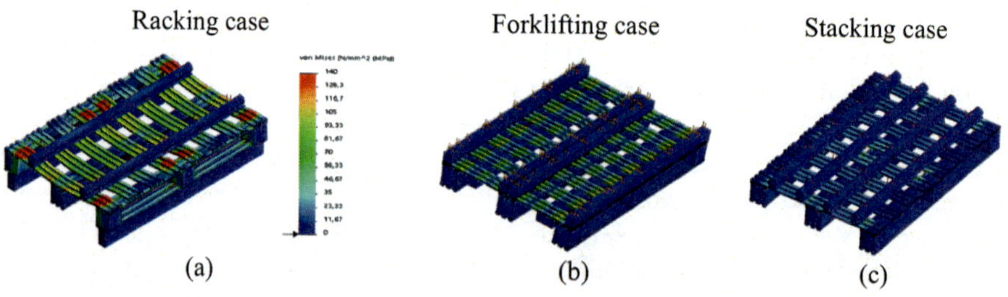

Fig. 3. FEM analyses for (a) racking, (b) forklifting, (c) stacking case in the transversal side.

3.2 Experimental Validations

To validate the design and FEM models and to understand the real mechanical behaviour, experimental laboratory tests of the whole steel pallet and its components have been carried out. Also, relationships between the pallet and structural components could be found.

Figures 4 and 5 shows the results and the used equipment for compression, i.e., Instron 600 RD with a 3.0 MN load cell, and hydraulic Instron Satec KN600K3965 with a 600 kN load cell. LVDT transducers, model HBM type WAT-100, were used to measure the vertical displacements. Two HBM MX840 A/B data loggers set up with HBM Catman AP 4.03 software were used to collect data.

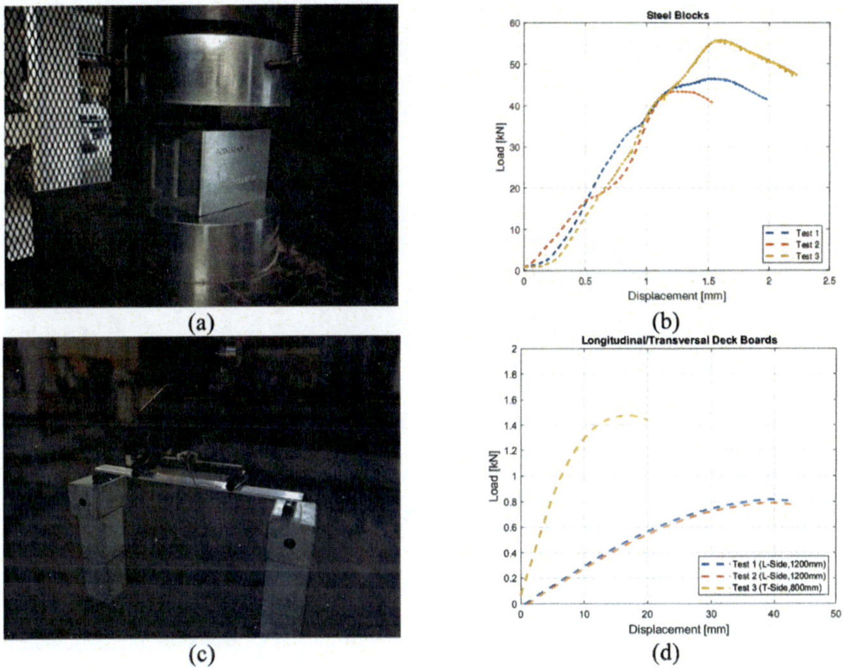

Fig. 4. Laboratory tests (3 + 3 tests) and results of the (a, b) block and (c, d) deck boards (L-Side = Longitudinal side. T-Side = Transversal side).

In Fig. 4(b), it is possible to see the ductile behaviour of the blocks since the curves do not strongly decrease after the peaks. However, the variability between the curves indicates the difficulty in evaluating this type of element. The experimental peak values of the block range between 43.61 and 55.95 kN, which are greater than the analytical values of 34.27 kN (or 53.88 kN by considering A_c) as estimated in Sect. 3.1.

For the deck boards (Fig. 4(d)), the initial inclination of the curves (i.e., structural stiffness) and the peak values between the longitudinal and transverse deck boards are different. For the longitudinal deck board, the mean peak value is 0.83 kN, which is 1.88 times smaller than for the transverse deck board (i.e., 1.56 kN). As expected, these peaks reached satisfactory values since they are greater than the analytical value 0.46 kN (Sect. 3.1).

The measurements for the whole pallet have been carried out in accordance with ISO 8611 [12–14] where the geometrical data regarding the application points of the restraints and loadings are provided. For brevity, here only some examples are shown. Figure 5 shows the used equipment (the same already described for Fig. 4) and static and dynamic results for the forklifting case in transversal side (see also Fig. 1(b)).

The forklifting test can be seen as a "dynamic" test due to the impact force of the moving supports therefore, a dynamic fatigue test has been also carried out. For this, the pallet was subjected to 2.0×10^5 load sinusoidal cycles with an amplitude of ~6.0 kN, at a frequency of 5.0 Hz. Given that first natural frequency of the pallet is ~20.0 Hz, evaluated experimentally, a steady-state amplitude response like a static response is expected.

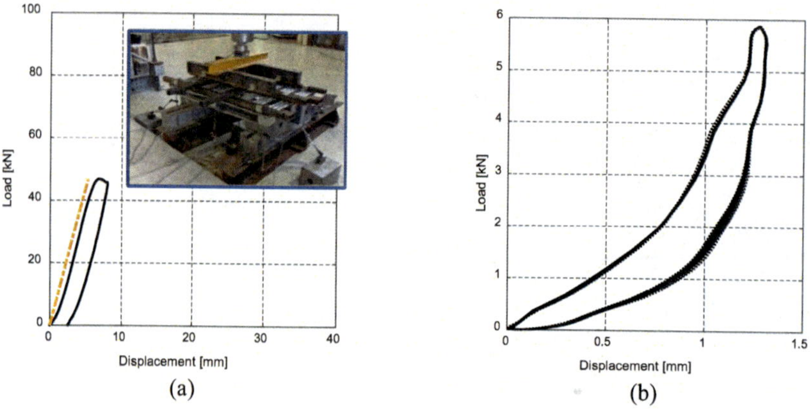

Fig. 5. Laboratory test measurements for forklifting case in transversal side: (a) static and (b) dynamic results. The used equipment is also shown.

In Fig. 5 it is possible to see the experimental elasto-plastic curves (black lines) and the numerical linear curve (yellow line). The differences between the experimental and numerical curves quantify the real contribution of the material and the connection between pallet elements. The slopes of the numerical lines are like the slopes of the first part of the experimental curves, where the pallets maintain an elastic-linear behaviour.

Table 1 shows the experimental and numerical results in terms of maximum vertical displacements, y_{max}, and U_t for all cases.

It is possible to see that the U_t values range between ~10.0–100.0 kN, which are very high with respect the reference values provide by ISO 8611-2 [13] (i.e., >44.20 kN). This because probably the code refers only to wood pallets.

The y_{max} values range between ~5.0–35.0 mm. The differences between numerical and experimental displacements are correlated to the linear behaviour assumed by the FEM model, while the experimental tests consider the actual capacity of the material. Also, in the experimental tests the support and ground stiffness effect can alter these results.

Table 1. Numerical and experimental results for all tests.

Case	y_{max} (mm)		U_t (kN)
	FEM results	Experimental results	
Racking (longitudinal side)	10.60	18.54	21.0
Racking (transversal side)	11.60	23.42	9.86
Stacking (long. side)	5.14	8.44	46.82
Stacking (trans. side)	7.61	8.50	19.03
Forklifting (long. side)	12.99	34.62	97.80
Forklifting (trans. side)	5.32	7.99 (1.39)[a]	44.13 (5.95)[a]

[a]In brackets the dynamic results (Fig. 5(b)).

In Fig. 5(b) it is possible to see that the load-displacement loops are similar for each cycle (i.e., the cycles overlap between them), indicating a stable behaviour during the fatigue test. Also, the pallet did not show any loss of strength after the fatigue test. Therefore, from all these results, Table 1 should provide new limits for metal pallets.

4 Life Cycle Analysis (LCA)

LCA analyses were performed by using the midpoint approach in accordance with the SimaPro software, ISO 14040 [16] and ISO 14044 [17]. The goal is to compare the steel pallet (proposed in this paper) with wood, plastic, and aluminium pallets.

For brevity only some results have been shown. More results and details (e.g., inventory, data, methodology) are available in [1]. Figure 6 shows the impact assessment and environmental performance for 4 pallet types by considering the eutrophication potential (EP).

Considering the product stage and other stages of the life cycle, the steel pallet under study is the solution offering best performance for all impact categories mainly due to high recyclability and less need of repairing/replacement during the defined service lifetime, while the plastic pallet performs worst for most of the impact categories.

5 Conclusions

In this paper a new modular metal pallet has been proposed. The main conclusions of this study are the following. A 3D FEM models are useful to estimate the stresses and deformations for different load conditions in global way. The analyses of each element (block and deck) provide more complete results. Laboratory tests, used to validate the models, provided new ultimate and test loads, for total pallet failure and performance limits (see Table 1). Also, the dynamic fatigue test shown a good performance. Therefore, for a specific analysis of metal pallets, the presented values and design procedures could be used to update existing codes. A comparative LCA study suggests that steel pallets perform better than wood, plastic, and aluminium pallets, mainly due to high recyclability and less need of repairing or replacement during the defined service lifetime.

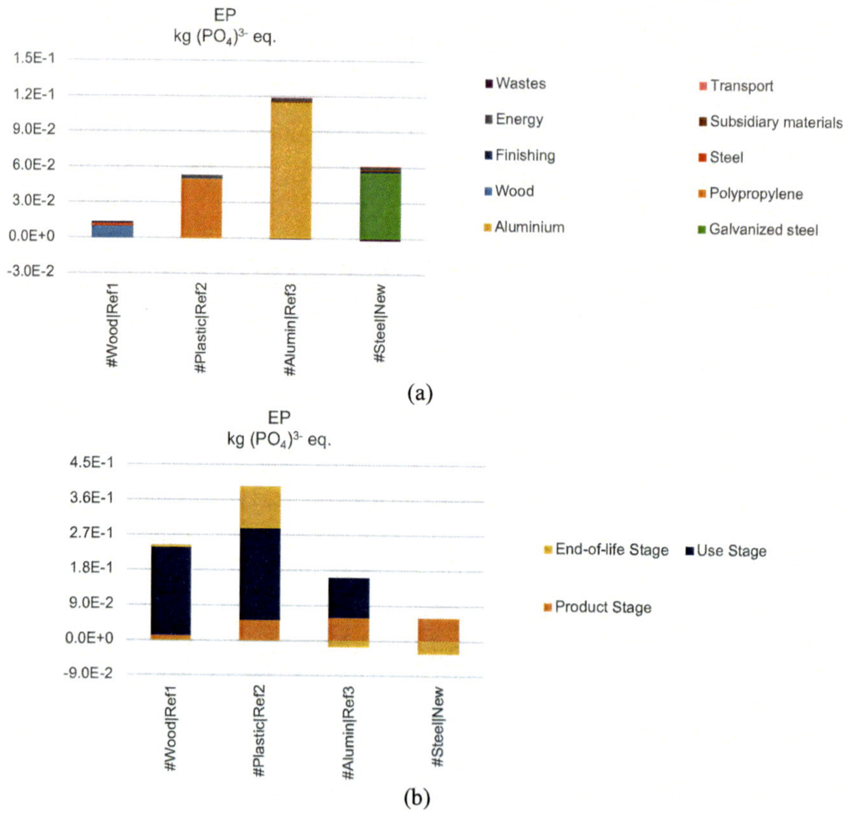

Fig. 6. LCA (some) results: (a) impact assessments and (b) environmental performances. Steel pallet = #Steel|New. Wood pallet = #Wood|Ref1. Plastic pallet = #Plastic|Ref2. Aluminium pallet = #Alumin|Ref3.

References

1. Zacchei, E., Tadeu, A., Almeida, J., Esteves, M., Santos, M.I., Silva, S.: Design of new modular metal pallets: experimental validation and life cycle analysis. Mater. Des. **241**, 1–18 (2022)
2. Alvarez, S., Rubio, A.: Compound method based on financial accounts versus process-based analysis in product carbon footprint: a comparison using wood pallets. Ecol. Ind. **49**, 88–94 (2015)
3. National Wooden Pallet and Container Association (NWPCA). https://www.palletcentral.com/. Accessed Oct 2021
4. Trevisan, A., Iaccheri, E., Fabbri, A., Guarnieri, A.: Pallet standards in agri-food sector: a brief survey. J. Agric. Eng. **220**, 90–95 (2014)
5. Chavan, S.J., Patil, R.N., Dhanrale, N.A.: Design and static stress analysis of returnable pallet for steel wheel. Int. J. Sci. Eng. Technol. Res. **5**, 2101–2103 (2016)
6. Amin, S.H., Wu, H., Karaphillis, G.: A perspective on the reverse logistics of plastic pallets in Canada. J. Remanuf. **8**, 153–174 (2018)
7. Soury, E., Behravesh, A.H., Rouhani, E.E., Zolfaghari, A.: Design, optimization and manufacturing of wood-plastic composite pallet. Mater. Des. **30**, 4183–4191 (2009)
8. Shah, S.N.R., Ramli Sulong, N.H., Jumaat, M.Z., Shariati, M.: State-of-the-art review on the design and performance of steel pallet rack connections. Eng. Fail. Anal. **66**, 240–258 (2016)
9. Adamakos, K., Sesana, S., Vayas, I.: Interaction between pallets and pallet beams of steel storage racks in seismic areas. Int. J. Steel Struct. **18**, 1018–1034 (2018)
10. European Committee for Standardization (CEN): Eurocode 3 – Design of steel structures – Part 1-1: General rules and rules for buildings. BS EN 1993-1-1:2005, Brussels, Belgium (2005)
11. European Committee for Standardization (CEN): Eurocode 3 – Design of steel structures – Part 1-3: General rules – Supplementary rules for cold-formed members and sheeting. BS EN 1993-1-3:2006, Brussels, Belgium (2006)
12. International Organization for standardization (ISO): ISO 8611-1:2011 – Pallets for materials handling – Flat pallets – Part 1: Test methods (2011)
13. International Organization for standardization (ISO): ISO 8611-2:2011 – Pallets for materials handling – Flat pallets – Part 2: Performance requirements and selection of tests (2011)
14. International Organization for standardization (ISO): ISO 8611-3:2011 – Pallets for materials handling – Flat pallets – Part 3: Maximum working loads (2011)
15. Yazdani, M., Kabirifar, K., Frimpong, B.E., Shariati, M., Mirmozaffari, M., Boskabadi, A.: Improving construction and demolition waste collection service in an urban area using a simheuristic approach: a case study in Sydney, Australia. J. Clear Prod. **280**, 1–17 (2021)
16. International Organization for Standardization (ISO): ISO 14040 - Environmental management: Life Cycle Assessment - Principles and framework (2006)
17. International Organization for Standardization (ISO): ISO 14044/Amd1 - Environmental management - Life cycle assessment - Requirements and guidelines (2006)

An FRF-Based Interval Multi-objective Model Updating Method for Uncertain Vibration Systems

Haotian Chen[1], Tianfeng Xu[2], Tao Zhang[1](✉), and Lin Zhang[1]

[1] School of Naval Architecture and Ocean Engineering, Huazhong University of Science and Technology, Wuhan, China
zhangt7666@hust.edu.cn

[2] China Ship Development and Design Center, Wuhan 430064, China

Abstract. Model updating plays an important role in dynamics modeling with high accuracy, which is widely used in mechanical engineering. However, the uncertainty of the system will greatly increase the difficulty of updating the model. In this paper, an interval model updating method for uncertain vibration systems using the frequency response function is proposed. This method can realize simultaneous updating of multiple frequency bands. Firstly, the uncertain parameters of the model are replaced through the intervals where they should belong, and the frequency response function of the model is replaced through the interval accordingly. Then multiple frequency bands are selected to update the model simultaneously, and the optimal solution of the updated parameters is calculated by the NSGA-II algorithm. To verify the effectiveness of the proposed method, a simulated 3-DoFs system is utilized in this paper. The results prove that the experimental results are totally covered in the frequency response function interval of the updated model, which proves that the method proposed has great effectiveness.

Keywords: Model updating · Frequency response function · Uncertain parameter · Multi-objective optimization

1 Introduction

In many engineering design problems, the accuracy of the simulation model is very important for the design and analysis of structures, especially for the prediction of dynamic performance, vibration control, and structural optimization. There may be some difference between the dynamic performance of the simulation model and the actual structure. In order to improve the reliability of the simulation model, it is necessary to correct the simulation model (model updating) by employing the measured data. Since the model updating method was proposed, many methods have been developed for various problems. Ha [1] proposed a finite element model updating method based on the closed-loop strain mode shapes. Zhang [2] proposed a Bayesian structural model updating method based on natural frequency and mode of vibration. Jones [3] proposed a finite element model updating method based on antiresonance frequency. All the above

Z. Dimitrovová et al. (Eds.): WMVC 2022, MMS 125, pp. 390–397, 2023.
https://doi.org/10.1007/978-3-031-15758-5_39

model updating methods exploit the dynamic characteristic parameters to evaluate the merits of the model updating results. Such methods cannot avoid the errors caused by modal fitting, the calculation efficiency is not high and the reliability of the updated model is not enough. Therefore, many scholars proposed the model updating method based on the frequency response function to solve these problems. Wu [4] proposed a model updating method using the frequency response function (FRF) based on the Sherman-Morrison formula. Asish [5] proposed a random finite element model updating perturbation method based on FRF, which can be identified with the parameters of mass matrix, stiffness matrix, and damping matrix. Tong [6] proposed a submarine finite element model updating method based on FRF.

Considering the manufacturing error, damage, and degeneration of the real structure, the system parameters may change, leading to the variability of the dynamic behaviors of the system. It is desirable to develop a new model updating technology for the uncertain system. However, due to due to data deficiencies, it is challenging to describe the probability distribution of the experimental data, and only the perturbation bounds of the experimental FRFs are obtained directly. The deterministic model could not meet the requirement, and the updating technology of the uncertain model needs to be developed to solve this kind of problem. Generally speaking, the uncertain parameters of the system fluctuate in a certain range, and the uncertainty of its FRF can be described as an interval form [7]. Therefore, in this paper, the experimental, uncertain FRF is defined as an interval form, and a model updating method for uncertain systems based on the interval FRF is proposed. The NSGA-II algorithm is employed to solve the norm minimization problem.

2 Method Description

2.1 Theory of the Model Updating

An undamped linear vibrating system with n-degree of freedom (DOFs) is considered. The Laplace transform of its motion equation can be written as

$$(-\omega^2\mathbf{M} + \mathbf{K})\mathbf{q}(\omega) = \mathbf{f}(\omega) \tag{1}$$

where $\mathbf{M}$ and $\mathbf{K}$ denote the $n \times n$ mass and stiffness matrix, respectively; $\mathbf{q}(\omega)$ is the displacement vector; $\mathbf{f}(\omega)$ denotes the input force vector. By modifying some structural parameters of the original system (denoted by the design variable vector $\mathbf{x} = [x_1, x_2, x_3, \cdots, x_p]^{\mathrm{T}}$), the updated matrices can be expressed as $\Delta\mathbf{M}_{n\times n} = \Delta\mathbf{M}(\mathbf{x})$, $\Delta\mathbf{K}_{n\times n} = \Delta\mathbf{K}(\mathbf{x})$. The motion equation of the updated system can be then written as [8]

$$(-\omega^2(\mathbf{M} + \Delta\mathbf{M}(\mathbf{x})) + (\mathbf{K} + \Delta\mathbf{K}(\mathbf{x})))\mathbf{q}(\omega) = \mathbf{f}(\omega) \tag{2}$$

$$((-\omega^2\mathbf{M} + \mathbf{K}) + \Delta\mathbf{Z}(\mathbf{x}, \omega))\mathbf{q}(\omega) = \mathbf{f}(\omega) \tag{3}$$

where $\Delta\mathbf{Z}(\mathbf{x}, \omega) = (-\omega^2\Delta\mathbf{M}(\mathbf{x}) + \Delta\mathbf{K}(\mathbf{x}))$ is called the addictive dynamic stiffness matrix.

Premultiplying both sides of Eq. (3) by the receptance matrix $\mathbf{H}(\omega) = (-\omega^2\mathbf{M} + \mathbf{K})^{-1}$ leads to

$$\mathbf{q}(\omega) = (\mathbf{I} + \mathbf{H}(\omega)\Delta\mathbf{Z}(\mathbf{x}, \omega))^{-1}\mathbf{H}(\omega)\mathbf{f}(\omega) \tag{4}$$

The receptance matrix $\mathbf{H}^*(\mathbf{x}, \omega)$ of the updated system can be expressed as

$$\mathbf{H}^*(\mathbf{x}, \omega) = (\mathbf{I} + \mathbf{H}(\omega)\Delta\mathbf{Z}(\mathbf{x}, \omega))^{-1}\mathbf{H}(\omega) \tag{5}$$

The FRF $h_{ij}^*(\mathbf{x}, \omega)$ of the updated system can be written as

$$h_{ij}^*(\mathbf{x}, \omega) = \mathbf{e}_i^T\mathbf{H}^*(\mathbf{x}, \omega)\mathbf{e}_j \tag{6}$$

where e_i and e_j are the ith and jth unit vector of n dimensions.

Suppose the measured FRF $\hat{h}_{ij}(\omega)$ is known, the model updating problem can be cast as an optimization problem. The objective function can then be written as

$$\min_{\mathbf{x}}\left\{\sum_{n=1}^{n_\omega}\alpha_n\left\|h_{ij}^*(\mathbf{x}, \omega_n) - \hat{h}_{ij}(\omega_n)\right\|_2^2\right\}, \mathbf{x} \in \mathbf{\Gamma}_\mathbf{x} \tag{7}$$

where n_ω denotes the number of the selected frequency bands, α_n denotes the level of importance of the nth frequency band, $\mathbf{\Gamma}_\mathbf{x}$ represents the feasible domain of the design variables.

2.2 Uncertain Dynamic System and Interval Algorithm

Here, the model updating problem of the uncertain system is considered. Assume that there are q uncertain structural parameters in the system. Generally, the uncertain factors are constrained by feasible domains. Hence, the uncertain structural parameter x_s $(1 \le s \le q)$ can be written as an interval $\left[x_s^L, x_s^U\right]$. The FRF of the uncertain system can then be written in the interval form

$$H_{ij}(\omega) = \left[h_{ij}^L(\omega), h_{ij}^U(\omega)\right] \tag{8}$$

where $H_{ij}(\omega)$ is the interval FRF. The L and U represent the lower and upper bounds, respectively.

Assume that all the uncertain parameters are updated after the model updating and some determined parameters are also updated after the model updating, the design variable vector $\mathbf{x}$ can then be written as

$$\mathbf{x} = \left[x_1, x_2, \cdots, x_q, x_{q+1}, x_{q+2}, \cdots, x_p\right]^{\mathrm{T}} \tag{9}$$

In order to separate the determined parameters and the uncertain parameters, the design variable vector $\mathbf{x}$ is divided into the uncertain design variable vector $\mathbf{x}_u \in \Re^q$ and the determined design variable vector $\mathbf{x}_d \in \Re^{p-q}$, which can be written as

$$\mathbf{x}_u = \left[x_1, x_2, \cdots, x_q\right]^{\mathrm{T}} \qquad \mathbf{x}_d = \left[x_{q+1}, x_{q+2}, \cdots, x_p\right]^{\mathrm{T}} \tag{10}$$

In order to express the uncertain design variable vector $\mathbf{x}_u$ of the uncertain system, some key information needs to be introduced

$$\begin{cases} x_s^C = \frac{x_s^L + x_s^U}{2} \\ \Delta x_s^R = \frac{\left|x_s^U - x_s^L\right|}{2} \\ \theta_s = \left|\frac{\Delta x_s^R}{x_s^C}\right| \times 100\% \end{cases} \tag{11}$$

where x_s^C is the mid-value of the interval $\left[x_s^L, x_s^U\right]$, Δx_s^R is the radius of the interval $\left[x_s^L, x_s^U\right]$, and θs is the uncertainty level of the interval $\left[x_s^L, x_s^U\right]$.

The uncertain design variable vector $\mathbf{x}_u$ can then be divided into the mid-value vector $\mathbf{x}_u^C \in \Re^q$ and the uncertainty vector $\boldsymbol{\theta} \in \Re^q$

$$\mathbf{x}_u^C = \left[x_1^C, x_2^C, \cdots, x_q^C\right]^{\mathrm{T}} \qquad \boldsymbol{\theta} = \left[\theta_1, \theta_2, \cdots, \theta_q\right]^{\mathrm{T}} \tag{12}$$

By combining the mid-value vector $\mathbf{x}_u^C$ and the determined design variable vector $\mathbf{x}_d$, the generalized mid-value vector $\tilde{\mathbf{x}} \in \Re^p$ can be written as

$$\tilde{\mathbf{x}} = \left[\mathbf{x}_u^{C\mathrm{T}}, \mathbf{x}_d^{\mathrm{T}}\right]^{\mathrm{T}} = \left[x_1^C, x_2^C, \cdots, x_q^C, x_{q+1}, x_{q+2}, \cdots, x_p\right]^{\mathrm{T}} \tag{13}$$

The generalized design variable vector $\mathbf{y} \in \Re^{p+q}$ of the uncertain system can then be written as

$$\mathbf{y} = \left[\tilde{\mathbf{x}}^{\mathrm{T}}, \boldsymbol{\theta}^{\mathrm{T}}\right]^{\mathrm{T}} = \left[x_1^C, x_2^C, \cdots, x_q^C, x_{q+1}, x_{q+2}, \cdots, x_p, \theta_1, \theta_2, \cdots, \theta_q\right]^{\mathrm{T}} \tag{14}$$

The design variable vector $\mathbf{x}$ of the uncertain system is uncertain but bounded, the bounds of $\mathbf{x}$ depend on the generalized design variable vector $\mathbf{y}$. The lower and upper bounds of the vector $\mathbf{x}$ can be expressed as

$$\mathbf{x}^L(\mathbf{y}) = \left[\left((\mathbf{I} - \boldsymbol{\theta}(\sum_{i=1}^{q} \mathbf{e}_i)^{\mathrm{T}})\mathbf{x}_u^C\right)^{\mathrm{T}}, \mathbf{x}_d^{\mathrm{T}}\right]^{\mathrm{T}} \quad \mathbf{x}^U(\mathbf{y}) = \left[\left((\mathbf{I} + \boldsymbol{\theta}(\sum_{i=1}^{q} \mathbf{e}_i)^{\mathrm{T}})\mathbf{x}_u^C\right)^{\mathrm{T}}, \mathbf{x}_d^{\mathrm{T}}\right]^{\mathrm{T}} \tag{15}$$

Like the interval FRF $H_{ij}(\omega)$, the FRF after updating is also an interval. The lower and upper bounds of the interval FRF after updating can be written as $h_{ij}^{*U}(\omega, \mathbf{y})$

$$\begin{cases} h_{ij}^{*L}(\omega, \mathbf{y}) = \min\left\{h_{ij}^*(\omega, \mathbf{x}^L(\mathbf{y})), h_{ij}^*(\omega, \mathbf{x}^U(\mathbf{y}))\right\} \\ h_{ij}^{*U}(\omega, \mathbf{y}) = \max\left\{h_{ij}^*(\omega, \mathbf{x}^L(\mathbf{y})), h_{ij}^*(\omega, \mathbf{x}^U(\mathbf{y}))\right\} \end{cases} \tag{16}$$

The interval FRF after updating can then be written as

$$H_{ij}^*(\omega, \mathbf{y}) = \left[h_{ij}^{*L}(\omega, \mathbf{y}), h_{ij}^{*U}(\omega, \mathbf{y})\right] \tag{17}$$

Due to the uncertainty of the system, it is necessary to measure the FRF of the system plenty of times. The measured interval FRF of the system can be written as

$$\hat{H}_{ij}(\omega) = \left[\hat{h}_{ij}^{L}(\omega), \hat{h}_{ij}^{U}(\omega)\right] \tag{18}$$

where $\hat{h}_{ij}^{L}(\omega)$ and $\hat{h}_{ij}^{U}(\omega)$ are the lower and the upper bound of the measured interval FRF, which means the minimum and the maximum value of all the measured FRF at the radian frequency ω.

The objective function can then be written as

$$\min_{\mathbf{y}}\left\{\sum_{n=1}^{n_\omega}\alpha_n\left(\left\|h_{ij}^{*L}(\omega_n,\mathbf{y})-\hat{h}_{ij}^{L}(\omega_n)\right\|_2^2+\left\|h_{ij}^{*U}(\omega_n,\mathbf{y})-\hat{h}_{ij}^{U}(\omega_n)\right\|_2^2\right)\right\}, \mathbf{y}\in\boldsymbol{\Gamma}_{\mathbf{y}} \tag{19}$$

where $\boldsymbol{\Gamma}_{\mathbf{y}}$ represents the union of the feasible domain of the $\mathbf{y}$.

2.3 Model Updating Method Based on the NSGA-II Algorithm

The NSGA-II algorithm is a non-dominated multi-objective optimization algorithm with an elite retention strategy. In this paper, the NSGA-II algorithm is utilized to calculate the optimal solution (the updated parameters) of Eq. (19).

The procedure of the model updating of the uncertain system is shown in Fig. 1.

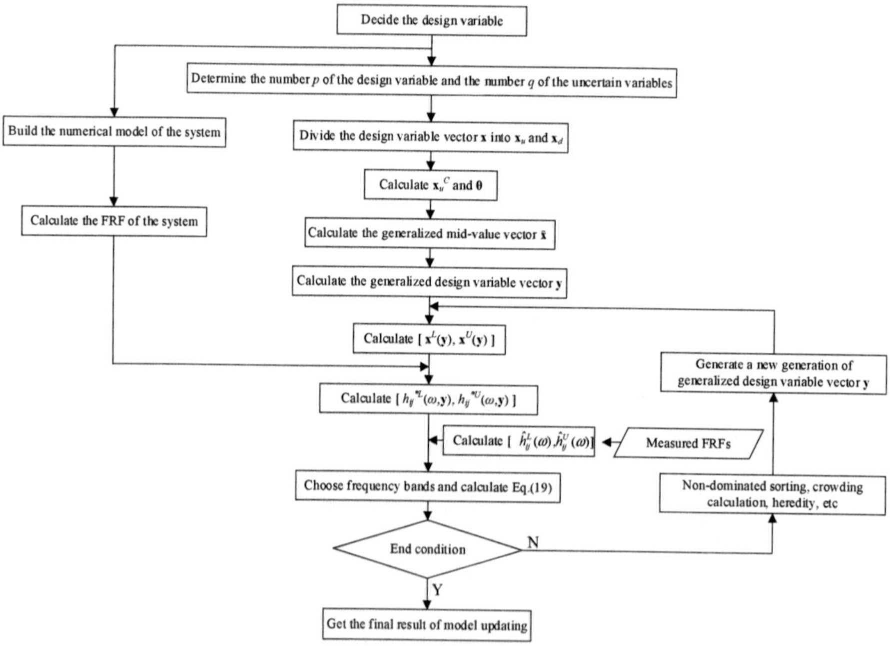

Fig. 1. The procedure of the model updating of the uncertain system

3 Method Assessment

To verify the effectiveness of the proposed method, it has been applied to the update of the vibrating system sketched in Fig. 2 which is a three-degree-of-freedom system with three lumped masses m_i (i = 1, 2, 3). The desired values of the system parameters are listed in Table 1. In this vibrating system, the value of mass m_1 is uncertain. The mass m_1, the springs k_2 and k_5 are assumed to be updated. The three updated parameters and the uncertainty of the mass m_1 lead to the four unknowns. The design variable vector $\mathbf{x}$ of the system is $\mathbf{x} = [m_1, k_2, k_5]$ and the generalized design variable vector $\mathbf{y}$ of the system is $\mathbf{y} = [m_1, k_2, k_5, \theta_1]$. The original values of the system parameters are listed in the second row of Table 2. The lower and upper bounds of the updated parameters are listed in the third and fourth rows of Table 2.

The interval FRF of the system is shown in Fig. 3. The parameters α_n ($n = 1, 2, 3$) in Eq. (13) have been set equal to 1 in order to impose the three frequency bands with the same level of concern. The number N of the iteration of the NSGA-II algorithm has been set to 500. The values of the parameters after updating and the relative errors are listed in Table 3. The interval FRF of the updated system and the experimental results of the real system are sketched in Fig. 4. As is shown in the figure, the experimental results are totally covered in the interval FRF of the updated model, which means that the dynamic performance of the updated model is almost the same as those of the real system. The comparison between the updated model and the real system proves that the proposed method can accurately calculate the updated values of the structural parameters.

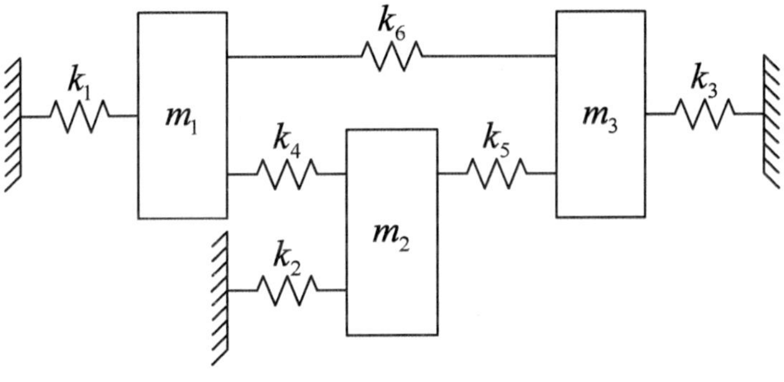

Fig. 2. Simulated three-DOF system

Table 1. The desired value of the system parameters

System parameter	k_1 [N/m]	k_2 [N/m]	k_3 [N/m]	k_4 [N/m]	k_5 [N/m]	k_6 [N/m]	m_1 [kg]	m_2 [kg]	m_3 [kg]
Desired value	2.0	2.0	1.0	1.0	2.0	3.0	1.0 ± 0.1	1.0	1.0

Table 2. System parameters original values and updated bounds

	k_1 [N/m]	k_2 [N/m]	k_3 [N/m]	k_4 [N/m]	k_5 [N/m]	k_6 [N/m]	m_1 [kg]	m_2 [kg]	m_3 [kg]
Original	2	1	1	1	1	3	0.6	1	1
Update lower bound	–	0	–	–	0	–	0	–	–
Update upper bound	–	3.8	–	–	3.8	–	1.6	–	–

Table 3. Value of the parameters after updating

Updated parameter	k_2 [N/m]	k_5 [N/m]	m_1 [kg]	θ_1 [kg]
Value after updating	1.0 + 1.0130	1.0 + 1.0323	0.6 + 0.4076	0.1072
Relative error	0.65%	1.615%	0.76%	7.2%

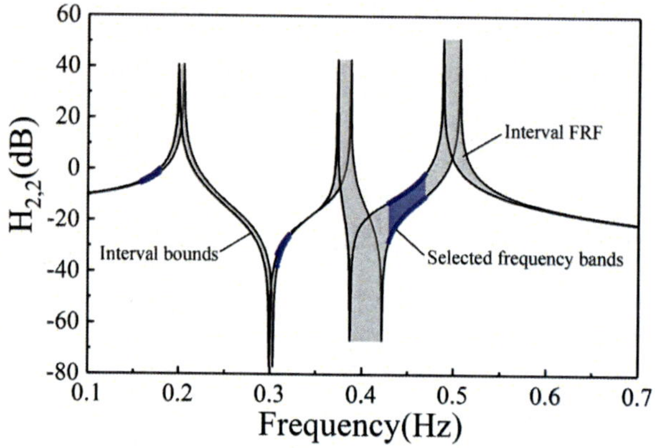

Fig. 3. Frequency bands selection on the interval FRF

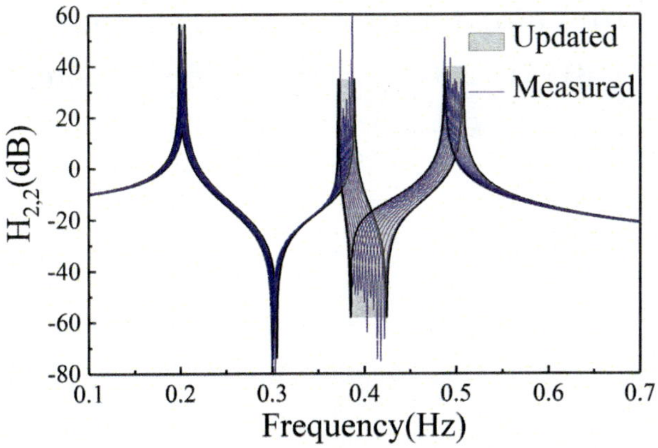

Fig. 4. Comparison between the updated interval FRF and measured FRFs

4 Conclusion

In this paper, an interval method for updating the uncertain undamped vibrating system is proposed. This method describes the uncertainty of the system in an interval form, thus it is unnecessary to collect a large number of data to construct a data sample, which can solve the problem of computational difficulties caused by insufficient data samples. Furthermore, the method proposed can avoid the error by model fitting and match of the predicted modal shapes with the measured data are not necessary, especially for structure updating with relatively dense modal distribution. The numerical analysis in this paper verifies the effectiveness of the proposed method. The results prove that the method proposed can effectively update the uncertain model.

References

1. Ha, J., Park, Y., Park, Y.: Model updating with closed-loop strain mode shapes. J. Guidance Control Dyn. **30**(4), 1206–1209 (2007)
2. Zhang, F., Ni, Y., Lam, H.: Bayesian structural model updating using ambient vibration datacollected by multiple setups. Struct. Control Health Monit. **24**(12), e2023 (2017)
3. Jones, K., Turcotte, J.: Finite element model updating using antiresonant frequencies. J. Sound Vibr. **252**(4), 717–727 (2002)
4. Wu, Y., Zhu, R., Cao, Z., Liu, Y., Jiang, D.: Model updating using frequency response functions based on Sherman-Morrison formula. Appl. Sci. **10**, 4985 (2020)
5. Panda, A., Modak, S.: An FRF-based perturbation approach for stochastic updating of mass, stiffness and damping matrices. Mech. Syst. Signal Process. **166**, 108416 (2022)
6. Tong, Z., Zhang, Y., Shen, R., Hua, H.: Submarine finite element model updating method based on frequency response functions of vibration. J. Shanghai Jiaotong Univ. **39**(11), 1847–1850 (2005)
7. Richiedei, D., Tamellin, I., Trevisani, A.: A homotopy transformation method for interval-based model updating of uncertain vibrating systems. Mech. Mach. Theory **160**, 104288 (2021)
8. Zhang, L., Zhang, T., Ouyang, H., Li, T.: Receptance-based antiresonant frequency assignment of an uncertain dynamic system using interval multiobjective optimization method. J. Sound Vibr. **529**, 116944 (2022)

Numerical Evaluation of Parametric Updating by Genetic Algorithm Implementation

Lucas Costa Arslanian[1], Lucas Fernandes Camargos[1](✉), Ariosto Bretanha Jorge[2], Gino Bertolluci Colherinhas[3], and Marcus Vinicius Girão de Morais[1,2]

[1] Mechanical Engineering Department, University of Brasilia, Brasília, Brazil
lucas.a108@gmail.com
[2] PPG Integridade de Materiais, Universidade de Brasília, Gama, Brazil
[3] Instituto Federal de Educação Ciência e Tecnologia de Goiás, Goiania, Brazil

Abstract. The present work compares the numerical performance of parametric updating between a proprietary tool and the GA/Matlab toolbox, using a genetic algorithm. The precision and processing time of parameter identifications are evaluated for two structural systems: (a) a numerical cantilever beam; and (b) the experimental results of a free-free beam. Parametric updating uses classic metrics for cost functions, for example: natural frequency, modal form, and the linear weights of the above. This work is part of the Demonstration Platform on Structural Integrity, a project developed collaboratively between the PPG Integrity (FGA-FT/UnB) and Embraer.

Keywords: Updating · Genetic algorithm · Modal identification · Discrete

1 Introduction

Screws and rivets are mechanical joining elements used routinely in the aeronautical industry. A simple way of modeling fastener elements is to describe them by dynamic discrete elements [3].

Using experimental modal techniques, the dynamic parameters of discrete joint models are determined by updating techniques, for example, genetic algorithm updating. Colherinhas *et al.* [5] present a GA toolbox that maximises a fitness function to minimise frequency response function peaks of a tower coupled to a pendulum reduced as 2DOF model. Friswell Penny and Garvey [7] show an application for damage detection using experimental vibration data. They minimise a weighted metric using frequency and modal shape (least-square) relative

Supported by UnB, FAP-DF & CNPq.

Z. Dimitrovová et al. (Eds.): WMVC 2022, MMS 125, pp. 398–407, 2023.
https://doi.org/10.1007/978-3-031-15758-5_40

error with reasonable results. This technique has the potential to be applied to identify dynamic parameters of discrete joint elements.

The present study compares the numerical performance of parametric updating using a genetic algorithm between a proprietary tool and the GA/Matlab optimization toolbox. The precision and processing time of parameter identifications are evaluated for two structural systems: (a) a cantilever insulated beam; and (b) a free-free beam, using numerical models and experimental results. Parametric updating uses classic metrics for cost functions, for example: natural frequency, modal form and the linear weights of the modal parameters above. This study is part of the Demonstration Platform on Structural Integrity [8] a project developed collaboratively between the PPG Integrity (FGA-FT/UnB) and Embraer.

2 Genetic Algorithm (GA) Optimization

Genetic algorithm [7] is meta-heuristic algorithm, inspired by the mechanism of Darwinian natural evolution and genetic recombination, maximising problems given an objective function. This GA toolbox was used to minimise the fitness function, by maximising its inverse J:

$$f_{\text{fitness}} = \frac{1}{J_i},\ i = 1, 2, \ldots, N_{\text{ind}} \tag{1}$$

where i is the chromosome of the population with N_{ind} individuals.

Using a population composed by N_{ind} individuals, several generations are performed to define the best parameters using the GA toolbox. For the present study:

- $N_{\text{gen}} = 100$, number of generations;
- $N_{\text{ind}} = 100$, number of individuals in the population;
- $p_{\text{c}} = 60\%$, crossover probability;
- $p_{\text{m}} = 2\%$, mutation probability;
- $p_{\text{elit}} = 2\%$, elitism probability;
- $p_{\text{dec}} = 20\%$, decimation probability;
- $N_{\text{dec}} = 20$, step of generation for the occurrence of decimation.

These parameters are more detailed in Colherinhas *et al.* [4,5].

The GA toolbox performs fifty optimum results and superimposes these on the response map of fitness J_i, function of mass density ρ and Young modulus E. This feature defines the GAs as a versatile tool that can be used to optimize several engineering problems. The precision and performance of the proprietary and Matlab GA toolboxes are compared.

3 Fitness Function for Parameters Identification

In this study, the fitness function implemented is based on the reference data (modelised or measured) of the modal parameters: natural frequency and mode shapes. The fitness function involves two expressions, one relating to the error in natural frequencies, J_ω, and another relating to the error in mode shapes, J_φ:

$$J = W_\omega J_\omega + W_\varphi J_\varphi \tag{2}$$

inspired by the objective function proposed by Friswell Penny and Garvey [7]. The fitness functions J_ω and J_φ are described below.

Modal Frequency Error - J_ω: The error function used to determine the modal frequency error was the squared magnitude of the resultant vector $\boldsymbol{f_n} - \boldsymbol{f_a}$ normalized with respect to $\boldsymbol{f_n}$, with an element-wise right division:

$$J_\omega = [(\boldsymbol{f_n} - \boldsymbol{f_a})\ ./\ \boldsymbol{f_n}] \cdot [(\boldsymbol{f_n} - \boldsymbol{f_a})\ ./\ \boldsymbol{f_n}] \tag{3}$$

where $\boldsymbol{f_a}$ is the reference natural frequencies array, and $\boldsymbol{f_n}$ is the individual natural frequencies array obtained by the genetic algorithms in a numerical model.

Mode Shape Error -J_φ: The double dot product of the resultant matrix $\varphi - \varphi_a$ with itself was used to determine the error function for the mode shape error:

$$J_\varphi = [\varphi - \varphi_a] : [\varphi - \varphi_a] = \mathrm{tr}\left([\varphi - \varphi_a]^T [\varphi - \varphi_a]\right) \tag{4}$$

where φ_a is the reference matrix of natural mode shapes column vectors, and φ is an individual matrix of mode shapes obtained with the genetic algorithms in a numerical model.

Relative Weighted Modal Parameters - J: In order to evaluate the combined error function, with respect to frequency and mode shapes, equations (3) and (4) are applied to the weighted error function (2), as described below:

$$J = 0.99\, J_\omega + 0.01\, J_\varphi \tag{5}$$

Through the testing of different combinations, the weighting factors W_ω e W_φ are determined by the combination that yields the best results. After previous analysis, a 99% weight on frequency and a 1% weight on the mode shape was established.

4 Results

In this section, two examples of parameters identification are presented. The parameters of mass density ρ and Young modulus E are identified for (a) a cantilever beam using a numerical model, and (b) a free-free beam using experimental results obtained by modal analysis. These two examples were used to compare the preciseness and computational cost of the parameters estimation by the GA Matlab Toolbox and a proprietary GA.

4.1 Numerical Cantilever Beam

A numerical model for a cantilever Euler-Bernoulli beam was discretised by ten equal Hermitian plane frame elements. Considering free vibration, natural frequencies and mode shapes were obtained through Matlab's eigensolver. The reference beam has density $\rho = 3000\,\mathrm{kg/m}^3$ and Young's modulus $E = 210$ GPa, length $L = 1.00\,\mathrm{m}$, beam cross section $A = 0.001\,\mathrm{m}^2$ and its area moment of inertia $I = 8.333\ 10^{-11}\,\mathrm{m}^4$. Three eigencouples, $(f_n, \varphi)_i$, $i = 1, 2, 3$, are used to compare the proprietary and Matlab GA results, using the reference cantilever beam model.

Modal Frequency Error. The fitness function of modal frequency (3) was implemented on Matlab's and the proprietary GA toolbox. Figure 1 shows the response map of modal frequency error J_ω as function of mass density ρ and Young's modulus E. After running the GA codes fifty times, the optimized results were superimposed to response map. The optimized results for both GA codes are aligned. Therefore, the modal frequency error J_ω on its own cannot establish an exact prediction to couple density ρ and Young's modulus E.

This linear dependency between ρ and E can be explained by the eigenfrequency formula for a cantilever beam, Blevins [2]:

$$f_i = \frac{\lambda_i^2}{2\pi L^2}\sqrt{\frac{EI}{m}} = \mathcal{C}(\lambda, A, I)\sqrt{\frac{E}{\rho}} \tag{6}$$

where, a constant relationship between these parameters for a given frequency is observed.

Mode Shape Error. Subsequently, the same procedure was repeated for the fitness function of modal shape error (4). Figure 2 shows the response map of modal shape error J_ω. Superimposed on the response map, the optimized results are a horizontal line for the expected density $\rho = 3000\,\mathrm{kg/m}^3$. Due to the mass matrix orthogonality relationship $\varphi^T \mathbf{M} \varphi = 1$, *i.e.* $\mathbf{M} = \varphi^{-T}\varphi^{-1}$, and the FE

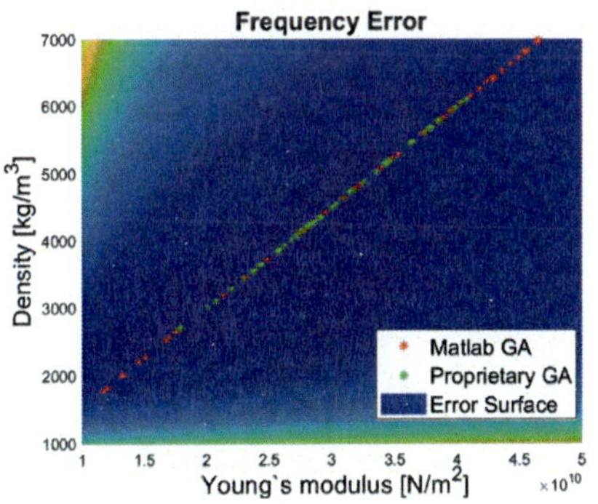

Fig. 1. Response map of numerical cantilever frequency error.

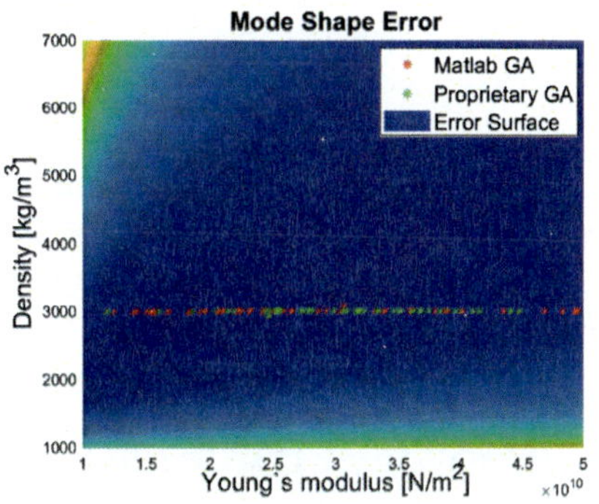

Fig. 2. Response map of numerical cantilever mode shape error.

consistent mass matrix $\mathbf{M} = \int_0^L \rho A \mathbf{N}^T \mathbf{N}\, dx$, there is a relationship between mass matrix $\mathbf{M}$ and mass density ρ.

Relative Weighted Modal Parameters Error. The fitness function of relative weighted modal parameters error (5) was implemented to combine the error functions with respect to the modal frequencies and mode shapes. Figure 3 shows the response map of relative weighted modal parameters error J as function of mass density ρ and Young modulus E. As expected, optimal values are located at the intersection of the lines generated in each of the previous individual results. These optimal values are located around the expected values for mass density and Young's modulus. In Fig. 3, they are located near $E = 200\,Pa$ and $\rho = 3000\,\mathrm{kg/m^3}$.

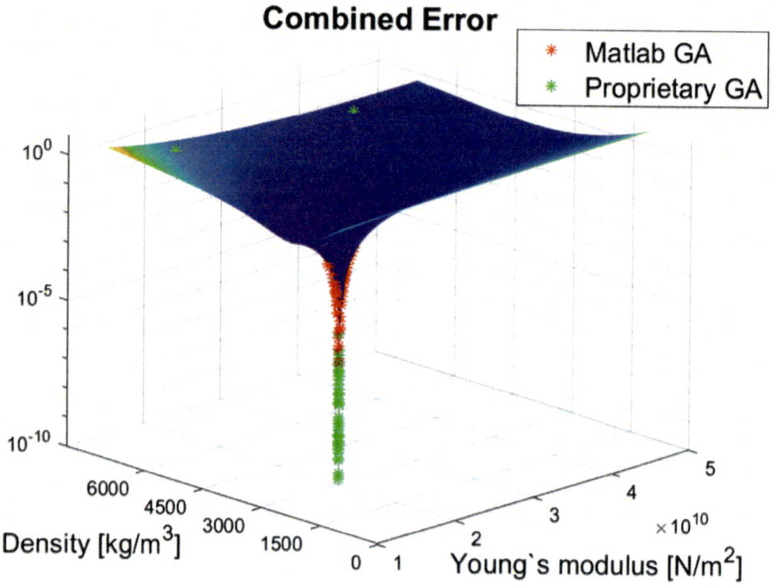

Fig. 3. Optimal values found through the implementation of the relative weighted fitness function on both genetic algorithms.

Time Performance Analysis. The mean time, standard deviation, and total time were stored for each example below. These computational performance results are shown in Table 1. In Fig. 4, the Matlab and Proprietary GA simulation times are shown in a box-plot representation. The processing time for fifty simulations using the Matlab GA is shorter than the same number of simulations using the proprietary GA. However, the standard deviation was much smaller for the proprietary GA due to several outliers presented in the Matlab GA.

Table 1. Computational cost for each fitness function

Fitness Function	Mean Time [s]	Total Time [s]
Frequency Error J_ω - Matlab	6.20 ± 1.27	328.24
Frequency Error J_ω - Proprietary	10.24 ± 0.42	521.05
Mode Shape Error J_φ - Matlab	9.28 ± 4.71	532.55
Mode Shape Error J_φ - Proprietary	10.50 ± 1.74	593.29
Combined Error J - Matlab	10.43 ± 5.67	603.47
Combined Error J - Proprietary	11.10 ± 0.43	540.44

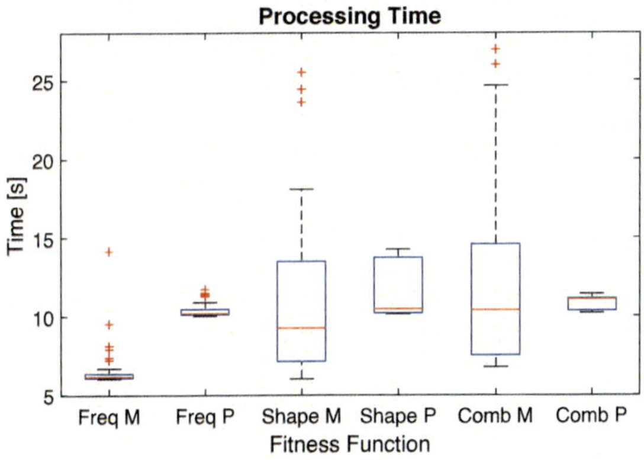

Fig. 4. Processing time box-plot of numerical cantilever comparing Matlab and proprietary GA simulation.

4.2 Experimental Free-Free Beam

A SAE 1045 steel free free beam was tested by experimental modal impact technique to determine the modal parameters of natural frequencies and mode shapes, as seen in Fig. 5. The dimensions of free-free beam are length $L = 500$ mm, rectangular cross section 5×25.5 mm ($3/16 \times 1$"). The total mass of the free-free beam is $m_{tot} = 473.06 \pm 0.01$ g. The estimated mass density is therefore $\rho \simeq 7569\,\text{kg/m}^3$. Finally, Young's modulus $E \simeq 181.84 \pm 12.75$ GPa is estimated by *Sonelastic*© (https://www.atcp-ndt.com/en/products/sonelastic.html), a impulse excitation technique based on ASTM-E1876 [1].

With the frequency response function obtained by the accelerometer and modal hammer, it is possible to determine the inertance for ten nodes in the free-free beam. The modal parameters of natural frequencies and mode shapes were determined by *EasyMod*'s [6] circle fit identification technique. Three eigencouples, $(f_n, \varphi)_i$, $i = 1, 2, 3$, are used to compare the proprietary and Matlab GA results, using the reference free-free beam model.

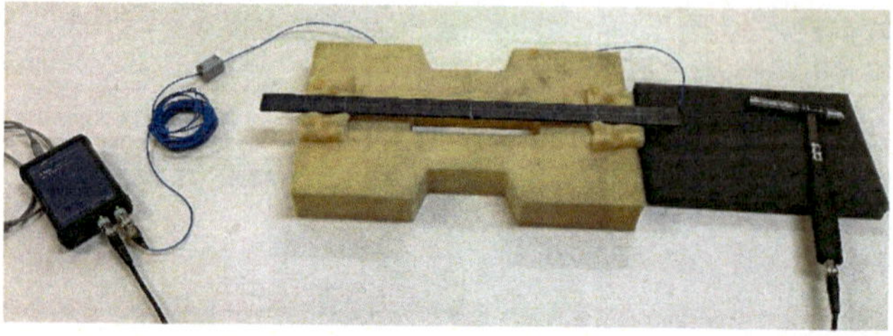

Fig. 5. Experimental modal testing setup: SAE 1045 steel free-free beam used for impact test configuration using modal hammer and accelerometer.

Figure 6 shows the response map of modal frequency error J_ω for the free-free experimental beam. The linear dependency between mass density ρ and Young's modulus E is also observed. Figure 7 shows the response map of modal shape error J_φ for the free-free experimental beam. In spite of a similar result to that of the numerical model, the mass density identified was not equal to the estimated value $\rho \simeq 7569$ kg/m^3.

Figure 8 presents a box-plot analysis of the processing times of the Matlab and proprietary GAs in order to determine the parameters (ρ, E) for a free-free beam. The proprietary GA is better optimized for this task. Its run time is less than half of the Matlab GA and has no outliers.

Young's Modulus Estimation. A simplified objective function is implemented using Young's modulus as an optimizable one-dimensional element. Assigning a fixed mass density $\rho = 7569$ kg/m^3, the frequency error fitness function (3) is used. Figure 9 shows the modal frequency error curve as function of Young's modulus.

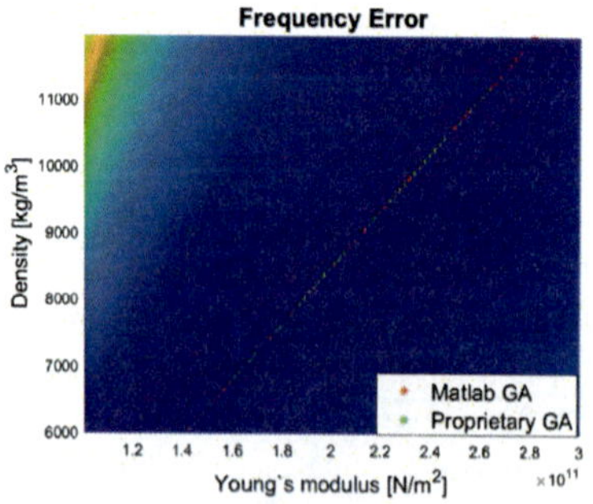

Fig. 6. Response map of experimental free-free beam frequency error.

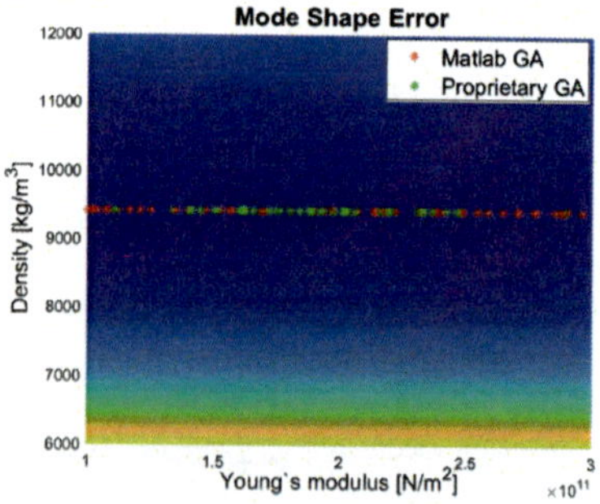

Fig. 7. Response map of experimental free-free beam mode shape error.

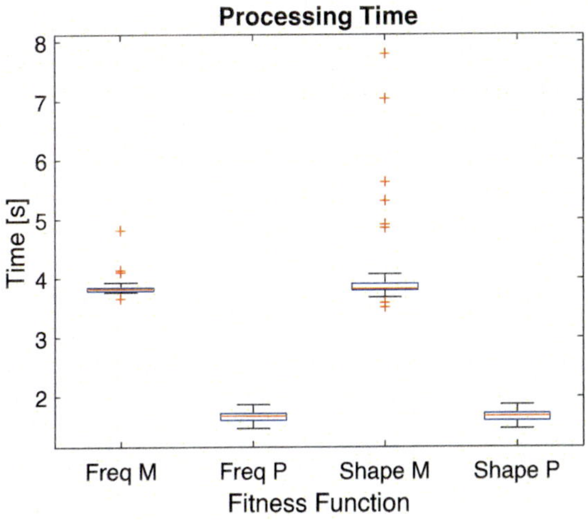

Fig. 8. Processing time box-plot analysis of the free-free experimental beam for parameter identification by Matlab and proprietary GA simulations.

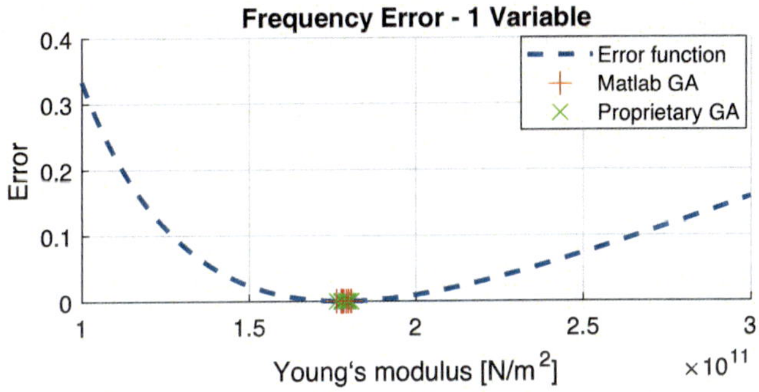

Fig. 9. Modal frequency error curve as function of Young's modulus E.

The GA codes obtain ten optimised values. Both the Matlab and the proprietary GA results find the minimum modal frequency error with reasonable precision. The Matlab GA detects a minimal error at $E = 178.7 \pm 1.4$ GPa, and the proprietary GA at $E = 177.9 \pm 0.003$ GPa. Young's modulus was estimated as $E = 181.84 \pm 12.75$ GPa by the *Sonelastic*© test, based on impulse excitation technique ASTM-E1876 [1].

Comparing to the Matlab's and the proprietary's GAs, errors of 1.73% and 2.17%, respectively, are determined to the mean value. The computational cost required to run each code was 3.56 ± 0.42 s for Matlab GA and 1.61 ± 0.19 s for proprietary GA.

5 Conclusion

In this paper, numerical performance of parametric updating using a proprietary GA was compared to the GA/Matlab optimization toolbox. A numerical cantilever Euler-Bernoulli beam model and experimental free-free beam results are used to compare the precision and processing time of parameter identifications. Parametric optimization was performed with fitness functions based on natural frequencies and modal shape forms.

Matlab and proprietary GA codes identified the mass density and Young's modulus parameters for the numerical model of a cantilever beam using a relative weighted fitness function of modal frequency and modal shape errors. The proprietary GA shows an inferior computational cost to that of the Matlab GA toolbox.

However, the response map of relative weighted error seems to be sensitive to noise pollution. Both Matlab and proprietary GA codes are not able to identify the same parameters of the experimental free-free beam model. Modal shape error pollution is responsible for this difficulty. Therefore, using a modal frequency error metric on a simpler one variable fitness function can identify Young's modulus for the experimental beam. This simpler fitness function is able to obtain an elasticity modulus similar to that of the standard test ASTM-E1876 (*Sonelastic*© system).

Acknowledgements. The authors acknowledge the Thermo-mechanical and Microstructure Characterization of Smart Materials (UnB - FT/ENM/LabMatI), led by prof. Edson Paulo da Silva, for letting them use the *Sonelastic*© system for experimental test. The authors would also like to acknowledge the funding received from the Research Support Foundation of the Federal District (FAP-DF) and CNPq that made this research possible.

References

1. Standard Test Method for Dynamic Young's Modulus, Shear Modulus, and Poisson's Ratio by Impulse Excitation of Vibration. ASTM Standart E1876, 27 December 2016
2. Blevins, R.: Formulas for Dynamics, Acoustics and Vibration. 1st edn. Wiley (2015)
3. Brake, M. R. W.(ed.): The Mechanical of Jointed Structures: Recent Research and Open Challenges for Developing Predictive Models for Structural Dynamics. Springer (2018)
4. Colherinhas, G.B.: Genetic algorithm toolbox and its applications in engineering, Master's thesis, University of Brasilia (2016). (in Portuguese)
5. Colherinhas, G.B., De Morais, M.V.G., Shzu, M.A.M., Avila, S.M.: Optimal pendulum tuned mass damper design applied to high towers using genetic algorithms: two-DOF modeling. Int. J. Struct. Stab. Dyn. **19**(10), 1950125 (2019)
6. EasyMod Homepage. https://hosting.umons.ac.be/html/mecara/EasyMod/index.html. Accessed 4 Feb 2022

7. Friswell, M.I., Penny, J.E.T., Garvey, S.D.: A combined genetic and eigensensitivity algorithm for the location of damage in structures. Comput. Struct. **69**, 547–556 (1998)
8. Jorge, A.B.: Technological demonstrator platform for inverse methods and modeling of uncertainties in the integrity of structures and components (port. Plataforma demonstradora tecnológica para métodos inversos e modelagem de incertezas em integridade de estruturas e componentes). Graduate Program in Engineering Materials Integrity (PPG Integridade), Universidade de Brasília, Brasília, p. 70 (2020)

Sensitivity Analysis Regarding the Impact of Intentional Mistuning on Blisk Vibrations

Oleg Repetckii[1], N. V. Vinh[1], and Bernd Beirow[2](✉)

[1] Irkutsk State Agrarian University, Irkutsk, Russia
[2] Brandenburg University of Technology, Cottbus, Germany

Abstract. The effect of different intentional mistuning (IM) patterns is investigated with respect to the forced response of an academic axial blisk. It could be shown in numerical analyses that a preliminary use of sensitivity algorithms helps to understand the feasibility and efficiency of introducing geometric changes of the blades. The implementation of IM patterns requires conducting intensive sensitivity studies based on FE simulations in order to identify the consequences of slight geometrical blade modifications on natural frequencies. Typical changes might be a modification of fillet radii or partial modifications of blade thickness, which are most suitable to adjust a target natural frequency without a severe loss of aerodynamic performance. A software tool developed at Irkutsk SAU is employed to evaluate the impact of mass and stiffness contributions, and with that, geometric deviations on blade natural frequencies.

Intensive blade vibration due to aerodynamic excitation of blisks is known as major source of high cycle fatigue, which may cause severe failures of turbine and compressor wheels during operation. The problem is relevant for several sectors of industry such as power generation, aviation or vehicle manufacturing. In consequence, there is a broad request of preventing any inadmissible vibration at any time. The application of IM can be regarded as powerful tool to avoid both, large forced responses and self-excited vibration. However, there is a lack of knowledge about how to implement mistuning without strong distortions of the flow passage. The main objective of this work is to close this gap based on comprehensive numerical analyses with regard to the effects of intended geometric modifications of blades on modal quantities.

Using FE models, the effectiveness of the proposed block models of mistuning is analyzed with and without taking into account the operational speed of the axial impeller. In conclusion, the consequences of different IM implementations on the forced response of an academic blisk are discussed. In particular, the most promising IM patterns are identified yielding the least forced response.

Keywords: Intentional mistuning · Sensitivity · Blisk · Forced response

The increasing demand for more efficient, more economic and environmentally friendly operation of turbomachines has led to an increasing application of integrally bladed wheels, meaning that blades and disks of turbo-machine wheels are manufactured as one piece. This technology satisfies the request for introducing light weight solutions in various fields of industries and allows for higher rational speeds, higher aerodynamic

Z. Dimitrovová et al. (Eds.): WMVC 2022, MMS 125, pp. 408–415, 2023.
https://doi.org/10.1007/978-3-031-15758-5_41

efficiencies and stage pressure ratios compared to the conventional design with separated blades and disk. However, integral designs are faced to higher stress levels and extremely low mechanical damping due to the lack of friction damping. Additionally, the negative consequences of unpreventable, manufacturing random mistuning are becoming worse, which further facilitates the susceptibility of blisk towards vibration.

Vibrations analyses of mistuned rotors are most frequently addressing the effect of mistuning on dynamic characteristics such as the maximum forced response and consequently on fatigue strength and durability. Blade mistuning manifests itself in small differences between the blades in terms of mass, geometry, material, etc., which violate the cyclic symmetry of rotor. According to [6] the magnitude of mistuning may be quantified by means of

$$\Delta f_i = \frac{f_{j,i} - \overline{f}_j}{\overline{f}_j} \tag{1}$$

where $\overline{f}_j$ represents the arithmetic mean of the blade dominated frequencies assigned to the jth blade mode shape, and $f_{j,i}$ denotes the blade alone frequency of the ith blade ($i = 1, \ldots, N$; N - number of blades).

A significant effect regarding the vibration of mistuned systems is an increase of the maximum forced response in terms of displacements and stresses compared to the ideal system. The factor γ is introduced for a quantitative assessment the maximum increase of amplitudes, which connects the maximum response amplitude of mistuned system with that of tuned system

$$\gamma = \frac{u_{mistuned\,(max)}}{u_{tuned\,(max)}} \tag{2}$$

where amplitude is understood as the maximum displacement or maximum dynamic stress during forced vibrations. The amplitude amplification factor substantially depends on both mistuning magnitude and shape of mistuning distribution. In theoretical calculations Ewins simulated the influence of different mistuning distributions on maximum vibration amplitude, which can vary from 130% to 210% [10] compared to the 100% tuned counterpart. Whitehead introduced a formula to estimate the maximum magnification of the forced response amplitude [5].

Contrary, the application of intentional mistuning has proved to mitigate the negative impact of random mistuning on the maximum forced response [4, 9, 11, 12]. Commonly, sensitivity analyses are carried out to identify suitable intentional mistuning patterns. The basic idea in this regards is to evaluate the consequences of changing an original parameter X_0 by means of ΔX. Hence, the modified system is characterized by $X = X_0 + \Delta X$ which has to be substituted in the static or dynamic equations. The main theory and algorithms for optimum design of turbine blades based on sensitivity analyses have been worked out e.g. by Kaneko, Mase and Fujita [1], Repetski and Zainchkovski [2], or Repetckii, Ryzhikov and Nguyen [3].

1 Numerical and Experimental Analyses of the Blisk

Numerous studies of different authors have shown that the phenomenon of mistuning may significantly affect the operation of power and transport turbine engines since the maximum displacement amplitudes and dynamic stresses can sharply increase during forced vibrations. An academic blisk with 10 blades, manufactured at Brandenburg University of Technology, was chosen as the object of study. The blisk is made of steel featuring a Young's modulus of $2.1 \cdot 10^5$ MPa, a Poisson's ratio of 0.3, and a density of 7850 kg/m^3. The general view of the rotor and one sector is shown in Figs. 1a and b. Supported by experimental data of authors, numerical studies in this work were carried out using both ANSYS and ABAQUS software packages. Figure 1c shows a finite element model of one sector, which uses ANSYS TET10 triangular finite elements, each having 3 degrees of freedom per node with a total of 2515 finite elements (FE) and 14616 degrees of freedom (DOF). The designated model was rigidly fixed at the rim of disk.

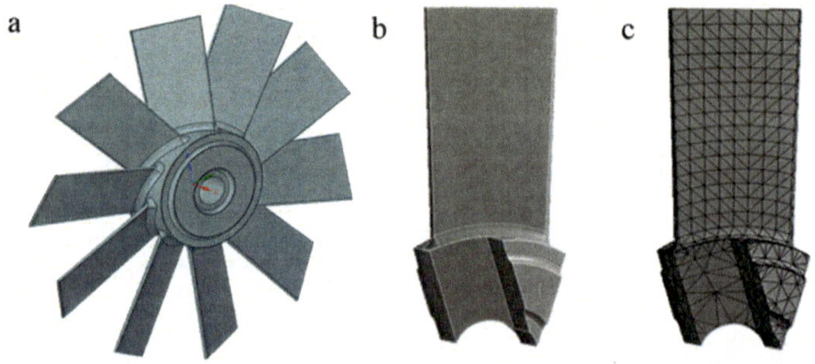

Fig. 1. Academic blisk with 10 blades (a) full disk; (b) one sector; (c) sector with FEM.

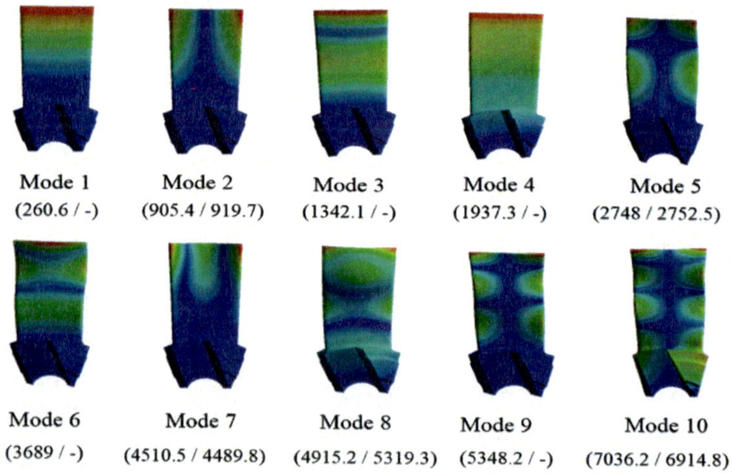

Fig. 2. Vibrational modes of one blisk sector without rotation (FEM/Test) in Hz.

Both natural frequencies and vibrational modes represent dynamic properties of blades. Thus, numerical analyses of natural frequencies and vibrational mode of the blades is a key task in designing turbomachines. Figure 2 shows the first ten sector mode shapes.

In order to prepare the design of suitable intentional mistuning patterns yielding a mitigated forced response, the so-called sensitivity function has been investigated to control and improve the efficiency of intentional mistuning. Some authors [2, 3, 6, 8, 13] show that blade sensitivity analysis can help to determine the location of mistuning zone being useful for the design of promising intentional mistuning patterns. Consequently, stress levels in the blade can be reduced by means of the results of sensitivity analyses, and therefore, the vibration susceptibility is decreased. For this reason, additional masses for detuning the blisk are employed in order to determine locations of suitable mistuning zones affecting the maximum effect on the forced response of the structure (Fig. 3). Here it is shown how the location of the mass affects the change of natural frequencies (bold blue points represent the zone of maximum frequency decrease, whereas bold red points represent the maximum frequency increase due to employing the mass in this zone. The sensitivity analysis of free and forced vibration blades for changing of thickness is performed in [6].

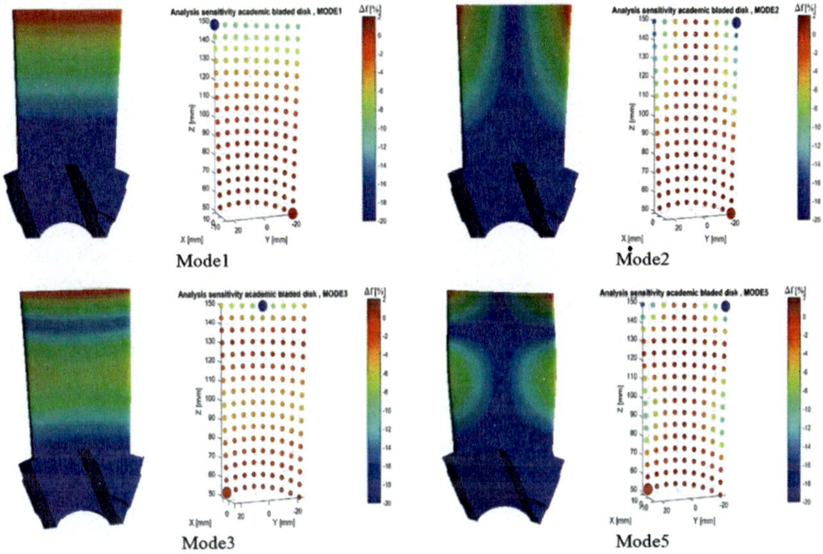

Fig. 3. Mode shapes (left) and sensitivity analysis (right) of academic blisk.

The analyses of the sensitivity of the mode shapes in Fig. 3 show that the zones of maximum sensitivity regarding the location of the masses on the blade coincide with the zones of maximum displacement during vibrations. The arrangement of the masses on nodal lines of the blade does not actually change the frequency of its natural frequencies. The location of the masses close to the blade tip reduces the natural frequencies (blue), and to maximize the frequency of natural vibrations, it is necessary to place the mass in the root part (red) of the effect of mass, in this case, is equivalent to a change in stiffness. The mistuning values are random values. Using experimental method to assess the effect of mistuning on the dynamics of rotor is a difficult task because it is necessary to analyze a large number of blisk hardware for experimentally determining mistuning patterns during the experiment. Numerical methods such as the Monte Carlo method can be used to study random mistuning [9]. The measurement system for experimentally determining mistuning is shown in Fig. 4a. The experimental setup consists of blisk, control device, laser vibrometer, modal hammer and foam pad [6]. The part of experimental data is shown in Fig. 2.

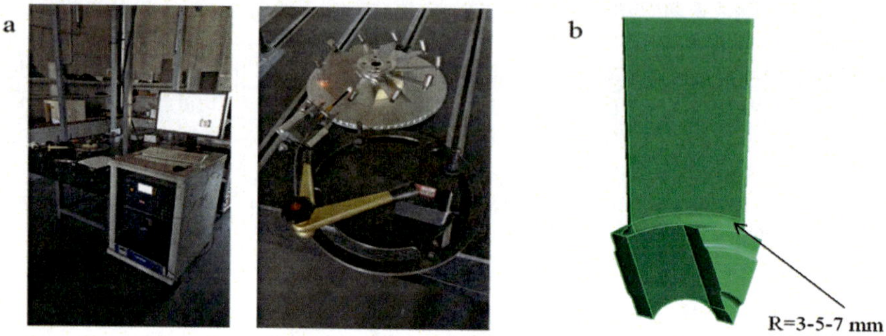

Fig. 4. (a) Experimental setup; (b) The changing of the radius.

2 Investigation of Intentional Mistuning

Following analyzes aim at analyzing the impact of intentional mistuning, which is accomplished such that the flow around the blades is hardly. As such types of wheel mistuning, changes in the radius of rounding of the transition of the blade to the disk and changes in the thickness of the blade are considered. Figure 4b shows a sector model of the academic blisk and the changing of the fillet radius from 3 mm to 7 mm. Such a measure basically enables the fine-tuning of real parts of turbomachines with respect to an attenuation of the forced response [7].

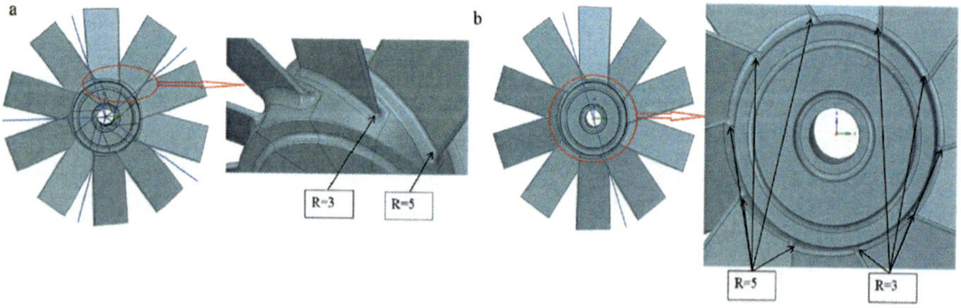

Fig. 5. (a) Block model No. 1 of the blisk mistuning; (b) Block model No. 2 of the blisk mistuning.

Following, we consider the first block mistuning model (BM) of alternate blade mistuning featuring blade root fillet radii of R = 3 mm or R = 5 mm, respectively (Fig. 5a, the variant without mistuning features R = 5 mm). Table 1 indicates the values of the maximum magnification factors due to changing fillet radii between blade root and disk, taking into account the intentional mistuning. The results show that the second block model (5 blades in one mistuning group, Fig. 5b) with intentional mistuning works most effectively. The second BM is effective for both at rest and at maximum speed (n = 100 1/s). The rows of the table show the value of the maximum magnification coefficient for the first forms in torsion mode (1T) and in bending mode (1B) when changing the initial radius of R = 5 mm to radii of 3, 1 and 7 mm.

Table 1. Calculation of the value of the maximum magnification factor due to changing fillet radius at the blade root

Block model (Number)	Mode shapes	R5 - R3 (n = 0 1/s)		R5 - R1				R5 - R7 (n = 0 1/s)	
				(n = 0 1/s)		(n = 100 1/s)			
		γmax	%	γmax	%	γmax	%	γmax	%
1	1T	1,63	−21,63	1,72	−17,31	1,88	−9,61	2,15	+3,34
	1B	1,10	−47,12	1,08	−48,10	1,10	−47,11	1,95	−6,25
2	1T	1,53	−26,44	1,58	−24,04	1,87	−10,10	2,19	+5,30
	1B	1,07	−48,55	1,03	−50,51	1,09	−47,60	1,90	−8,65

Analysis results of Table 1 shows that with the change radius from 5 to 1 mm, it will be possible to reduce the maximum magnification factor of the blades by −50,5% for the bending mode (1B). With increasing radius up to R = 7 mm, an increase of the maximum magnification factor on +5,30% is obtained for the torsion mode (1T). The influence of rotation on the γ_{max} is essential for the torsional vibrations.

3 Fatigue Life Prediction

Mistuning parameters affect the value of static and dynamic stresses, and the fatigue life. Figure 6a shows the process of fatigue life calculation and the place of mistuning in this process. The calculations on the basis of this diagram are presented in the program package BLADIS+ [7]. Figure 6b displays the schematic diagram of fatigue life calculation. A dynamic stress is determined following the scheme load classification, stress amplitude, number of cycles and fatigue life. An analysis of forced vibrations of one sector of an academic bladed disk (Fig. 1) was carried out. This blade is harmonically excited by 10 nozzles for time period 0–5 s [6, 13]. Figure 6b represents the results of dynamical stresses and estimation of fatigue life in time domain that were obtained using the method of schematization of random loading processes, the Rain flow method and a fatigue life computation according to the Palmgren-Miner hypothesis of one of the previously studied blisks. Figure 7 shows the calculation of the dynamic stress and fatigue life of the model with and without mistuning in the change radius from 5 to 3 mm. Table 2 indicates the values of the blisk fatigue life due to changing fillet radius at the blade root.

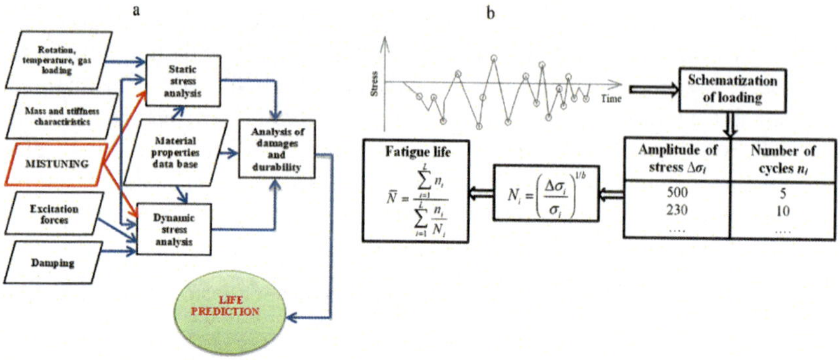

Fig. 6. (a) Mistuning in the life prediction; (b) Diagram of the fatigue life calculation.

Table 2. Calculation of the blisk fatigue life due to changing fillet radius at the blade root with rotational speed (n = 100 1/s)

Block model (Number)	R5 - R3		R5 - R1		R5 - R7	
	$\overline{N}$, cycles	%	$\overline{N}$, cycles	%	$\overline{N}$, cycles	%
1	$1,414 \cdot 10^6$	+2,11	$1,394 \cdot 10^6$	+0,64	$1,361 \cdot 10^6$	−1,72
2	$1,457 \cdot 10^6$	+5,25	$1,442 \cdot 10^6$	+4,12	$1,350 \cdot 10^6$	−2,53

Analysis results of Table 2 shows that the BM 2 of the change radius from 5 to 3 mm gives maximum fatigue life of the blisk, increasing by +5,25%. And the BM2 of the change radius from 5 to 7 mm gives minimum fatigue life, decreasing by −2,53%.

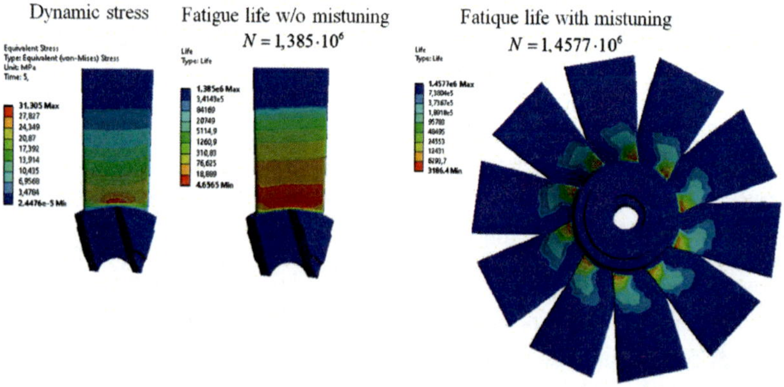

Fig. 7. Dynamic stresses and fatigue of one sector (w/o mistuning) and blisk (BM 2).

4 Conclusion

This paper presents the results of a numerical analysis of the effects of intentional mistuning in order to reduce the maximum amplitude magnification of an axial blisk. Intentional mistuning was implemented by means of small geometric changes namely by changing fillet radii at the blade roots (initial radius R = 5 mm). The results of the study show the reliability and effectiveness of the use of intentional mistuning used for blisks. The blisk fatigue life may be increased by +5,25% due to a reduction of the fillet radius down to 3 mm, and increasing the radius to 7 mm gives a reduction in the blisk fatigue life of −2,53% (BM 2). In addition, variants of introducing intentional mistuning in the form of changing the blade thickness, cropping of the trailing edge and drilling into the blade were investigated. The results of these studies for axial and radial bladed disks will be presented in the following scientific papers.

References

1. Kaneko, Y., Mase, M., Fujita, K.: Optimal design of turbine blade using sensitivity analysis. JSME Int. J. Ser. C: Dyn. Control Robot. Des. Manuf. **36**(2), 258–263 (1993)
2. Repetski, O., Zainchkovski, K.: The sensitivity analysis for life estimation of turbine blades. In: Proceedings of the 1997 ASME ASIA Congress & Exhibition, p. 136 (1997)
3. Repetckii, O., Ryzhikov, I., Nguyen, T.Q.: Dynamics analysis in the design of turbomachinery using sensitivity coefficients. J. Phys.: Conf. Ser. **944**, 012096 (2018)
4. Nakos, A., Beirow, B., Zobel, A.: Mistuning and damping of a radial turbine wheel. Part 1: fundamental analyses and design of intentional mistuning pattern. In: Proceedings of ASME Turbo Expo GT2021-59283 (2021)
5. Whitehead, D.S.: Effect of mistuning on the vibration of turbomachine blades induced by wakes. J. Mech. Eng. Sci. **8**, 15–21 (1966)
6. Repetckii, O.V., Nguyen, V.V.: Dynamics of turbomachine impellers using sensitivity functions. In: Radionov, A.A., Gasiyarov, V.R. (eds.) ICIE 2021. LNME, pp. 581–588. Springer, Cham (2021). https://doi.org/10.1007/978-3-030-54814-8_67
7. Repetckii, O., Ryzhikov, I., Springer, H.: Numerical analysis of rotating flexible blade-disk-shaft systems. In: Proceedings of the ASME Turbo Expo (1999)
8. Repetckii, O.V., Nguyen, V.V., Ryzhikov, I.N.: Numerical research of vibrational characteristics for compressor bladed disks of an energy turbomachine with intentional mistuning of blades. In: Proceedings of the International Multi-Conference on Industrial Engineering and Modern Technologies, FarEastCon 2020, p. 9271444 (2020)
9. Beirow, B., Figaschewsky, F., Kühhorn, A., Bornhorn, A.: Modal analyses of an axial turbine blisk with intentional mistuning. J. Eng. Gas Turb. Power **140**(1), 012503–012503-11 (2018)
10. Ewins, D.J.: The effects of detuning upon the forced vibrations of bladed disks. J. Sound Vibr. **9**, 65–79 (1969)
11. Griffin, J.H., Hoosac, T.M.: Model development and statistical investigation of turbine blade mistuning. ASME J. Vibr. Acoust. Stress Reliab. Des. **106**(2), 204–210 (1984)
12. Castanier, M.P., Pierre, C.: Using intentional mistuning in the design of turbomachinery rotors. AIAA J. **40**(10), 2077–2086 (2002)
13. Repetckii, O.: Development of mathematical models for assessment of damage accumulation and predictions of life for turbomachine blade. Mach. Build. **1–3**, 45–53 (1995). Proceedings of Higher Educational Institutions

NDC: Nonlinear Dynamics and Control of Engineering Systems

A Comparative Quantification of Existing Creep Models for Piezoactuators

Shabnam Tashakori[1,2,3(✉)], Vahid Vaziri[1], and Sumeet S. Aphale[1]

[1] Centre for Applied Dynamics Research, School of Engineering, University of Aberdeen, Aberdeen AB24 3UE, UK
[2] Department of Mechanical and Aerospace Engineering, Shiraz University of Technology, Shiraz, Iran
[3] Rahesh Innovation Center, Shiraz, Iran
shabnam.tashakori@alum.sharif.edu

Abstract. Piezoactuators are popularly employed in precise positioning applications at the micro- and nanometer scales. Their positioning performance, especially for low-frequency responses, is significantly impacted by creep - a phenomenon where the actuator deformation gradually changes in the presence of a persistently applied constant voltage. This change in deformation manifests itself in the gradual drifting of the end-effector position that the piezoactuator is driving. A significant research effort has therefore focussed on the accurate modelling of creep. This paper compares three popularly employed creep models against experimentally measured creep data obtained from a piezo-drive nanopositioner axis and quantifies their modelling accuracy. The quantification demonstrates that the fractional-order model (double logarithmic model) outperforms the other two integer-order models (Logarithmic and LTI models) along multiple, key performance indices.

Keywords: Creep dynamics · Nanopositioning · Piezoelectric actuator · Fractional-order modelling

1 Introduction

Nanopositioning encompasses a number of technologies that deliver nanometer-scale mechanical displacement with high precision. It underpins many conventional and emerging technical advances, e.g., scanning probe microscopy, disk-drive data storage, and nano-surgery. Piezoelectric actuation is most popularly employed to deliver nanopositioning due to several desirable characteristics, e.g., repeatability, lack of friction and stiction due to the absence of moving parts, easy control, and system integration. However, the positioning performance of piezoactuated nanopositioners is severely limited by the inherent linear resonance and nonlinear dynamics, i.e., hysteresis and creep.

Creep is a nonlinear phenomenon inherent to piezoactuators that results in drift over time in the output displacement despite fixed applied voltage; severely

Z. Dimitrovová et al. (Eds.): WMVC 2022, MMS 125, pp. 419–426, 2023.
https://doi.org/10.1007/978-3-031-15758-5_42

impacting slow-pace/start-from-previous-stop operations. To reduce this drift, the piezoelectric actuator should operate relatively fast and in a relatively short time interval [11]. However, even more accuracy is needed in many applications, e.g., while the measuring operation in scanning probe microscopes, the measuring sample should completely stay still without small movement, otherwise, the measured image will be distorted [6]. Another remedy can be using feedback methods. Nevertheless, these are not implementable in many situations as several displacement sensors should be mounted on the system, which is not always possible [11]. Therefore, precise modelling of the creep phenomenon is of great importance, whereas feedback methods can still be employed for further improvement in positioning. While Hysteresis is typically affecting all trajectories (operation speeds) and therefore is well studied [5,16], the creep is typically left unmodelled/unquantified in most nanopositioning literature [7,15]. With the advent of soft actuators, creep modelling has also seen renewed interest [8,14].

Extensive research efforts have focused on the mathematical modelling and open-/closed-loop compensation of the performance-limiting creep dynamics. Most modelling and control techniques proposed thus far lie within the span of integer-order calculus, categorized as: (i) logarithmic model [12], and (ii) linear time-invariant (LTI) model [17]. Recently, fractional-order (FO) calculus has emerged as a viable candidate capable of furnishing further improvements in the state-of-the-art in nanopositioning by allowing the formulation of improved models and controller designs [13]. Considering the PEA as a resistocaptance, a fractional-order model for creep phenomenon is proposed in [11]. In [10], another fractional-order model is presented by employing a physics-based fractional-order Maxwell resistive capacitor approach, and in [9], a simplified phenomenon-based fractional-order creep model is proposed.

In this paper, the integer-order and fractional-order modelling approaches describing the creep phenomenon are presented and numerically compared. Furthermore, experimental results are employed to further validate the fractional-order models, and demonstrate the superiority of this kind of modelling. Summarizing, the main contributions of this paper are (i) making a comparison between integer- and fractional-order models for creep, and (ii) validating a creep fractional-order model with experimental data.

The paper is organized as follows. In Sect. 2, two integer-order creep models as well as a fractional-order creep model are presented. Section 3 describes the experimental platform employed to record creep data that is subsequently used to quantify the accuracy of the three creep models being quantified in this work. Furthermore, the parameter choices of each model are explained and illustrative comparison is presented. Section 4 concludes this paper.

2 Creep Models

2.1 Integer-Order Logarithmic Model

In the logarithmic model, the output displacement $y(t)$ at time t is described with the following linear equation as a function of the logarithmic scaled time [11]:

$$y(t) = y_1(1 + \gamma log\frac{t}{t_1}), \tag{1}$$

with y_1 the displacement at time t_1 that is the time after which the creep occurs, and the constant γ specifies the creep rate.

2.2 Integer-Order Linear Time Invariant (LTI) Model

The LTI model considers a series of $n + 1$ springs and n dampers to model the creep response. Consequently, the transfer function between the output response $Y(s)$ and the input voltage $U(s)$ is given by [11]:

$$G_c(s) = \frac{Y(s)}{U(s)} = \frac{1}{k_0} + \sum_{i=1}^{n} \frac{1}{c_i s + k_i}, \tag{2}$$

where c_i, $i = 1, ..., n$ are the damping coefficients and k_i, $i = 0, ..., n$ are the stiffness constants of the springs.

2.3 Fractional-Order Double Logarithmic Model

A fractional-order integrator, with an order between 0 and 1, causes drift, and hence, can be exploited to model the creep phenomenon. This is concluded based on two facts: (i) the PEA is a distributed-parameter component with memory effect, and (ii) it is proved that a fractional-order system is able to successfully model distributed-parameter systems which have memory effect [4]. Motivated by these facts, a piezoelectric actuator can be described as a resistocaptance (RC) as follows [2,4]:

$$\frac{Q(s)}{U(s)} = \frac{K}{s^\alpha}, \tag{3}$$

where $Q(s)$ and $U(s)$ are the input charge and driving voltage in frequency domain, respectively, and K and $0 < \alpha < 1$ are constants. The output displacement $y(t)$ is given by the following dynamics:

$$m\ddot{y}(t) + c\dot{y}(t) + ky(t) = F_p + F_{ext}, \tag{4}$$

with m, c, and k the mass, damping, and stiffness constants, respectively. Herein, $F_p = Tq(t)$, where T is the electro-mechanical transformer ratio and $q(t)$ is the input charge. When the external force F_{ext} is zero, the transfer function $G(s)$ is given by:

$$G(s) := \frac{Y(s)}{U(s)} = \frac{b}{s^\alpha(1 + a_1 s + a_2 s^2)}, \tag{5}$$

where $a_1 = \frac{c}{k}$, $a_2 = \frac{m}{k}$, and $b = \frac{TK}{k}$. For sufficiently large time, the above PEA model in (5) melts down to a fractional-order integrator, as follows:

$$lim_{s\to 0}G(s) = \frac{b}{s^{\alpha}}, \tag{6}$$

which actually shows that the mechanical response can be disregarded after a certain time t_c, so (5) can be written as:

$$G(s) = \frac{b}{s^{\alpha}}, \quad t \geq t_c. \tag{7}$$

Let $U(s) = \frac{1}{s}$, the an approximate displacement would be [3]:

$$y(t) = \mathscr{L}^{-1}\{G(s)U(s)\} = \frac{bt^{\alpha}}{\alpha\Gamma(\alpha)}, \quad t \geq t_c, \tag{8}$$

with $\Gamma(\alpha)$ the gamma function, which gives the following "double-logarithmic" creep model:

$$log(y(t)) = \alpha log(t) + log(\frac{b}{\alpha\Gamma(\alpha)}) \quad t \geq t_c. \tag{9}$$

3 Experimental Validation and Comparative Quantification

3.1 Experimental Setup

The experimental results in this paper are obtained by using a two-axis piezo-actuated serial kinematic nanopositioning stage, shown in Fig. 1. The actuation voltage, which is in the range of 0 V to 200 V, is supplied by two voltage amplifiers, PDL200 and Piezodrive, with an amplification factor of 20. The voltage off-set is 100 V, and its consecutive full-range displacement is $\pm 20\mu$m along each axis. To gauge the real-time displacement in the matter of voltage signal, ranging from -10 V to 10 V, Microsense 4810, probe 2805, range of $\pm 50\mu$m, is used, which is a high resolution capacitive sensor. The data acquisition system is performed via the National Instruments card (PCI-6621) on a PC that has a real-time module, also equipped with OPTIPLEX 780 with an Intel Core (TM)2 Duo Processor running at 3.167 GHz and 2 GB of DDR3 RAM memory. The cross-coupling between the two scanning axes is down to -40 dB. Therefore, two axes can be assumed decoupled and, consequently, treated as independent single-input-single-output systems. Moreover, the axis with its resonance at 716 Hz (i.e., the fast axis) is chosen as the test platform with 20 kHz sampling frequency [1].

3.2 Simulation Results

The logarithmic model (1) includes one parameter γ that needs to be identified. As shown in Fig. 2b, the steady-state error is zero for $\gamma = 0.56$. Moreover, the

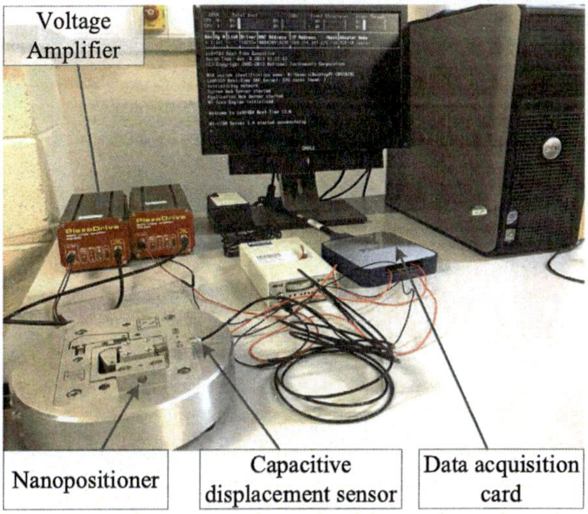

Fig. 1. Two-axis piezo-actuated serial kinematic nanopositioning stage, designed by the EasyLab, University of Nevada, Reno, USA [1].

minimum of the Root Mean Square (RMS) error and zero mean error occurs at $\gamma = 0.65$, as illustrated. The displacement for each of these γ are demonstrated in Fig. 2a in comparison to the experimental data.

Figure 2c illustrates the following reduced-order LTI model (n is set to one in (2)):

$$G_c(s) = \frac{1}{k_0} + \frac{1}{c_1 s + k_1}, \tag{10}$$

in which three constants k_0, k_1, and c_1 are to be identified. Since the displacement starts at 0µm, the constant k_0 should be selected large enough, e.g., 1000. The values for the stiffness constant k_1 can also be computed such that the final value at $t = 600$ s reaches 19.4191µm, which leads to $k_1 = 0.0515$. Therefore, only one parameter c_1 has to be selected. This constant specifies the sharpness of the unit response. Figure 2d shows the errors with respect to different values of c_1, from which $c_1 = 0.1$ is selected since it makes the mean error zero. Note that the steady-state error is zero for all values of c_1 since this error, as discussed above, is only influenced by k_1.

The fractional-order double logarithmic model (8) is illustrated in Fig. 2e. Two constants α and b are involved in this model, which should be identified. The constant α specifies the creep rate and the constant b can be selected afterwards such that the final model error at $t = 600$ s is zero. Consequently, as shown in Fig. 2f, the steady-state error is zero for all values of α. Therefore, this model also includes one independent constant α. The mean error is zero at $\alpha = 0.003$ where the RMS error is also minimum (with $b = 4.5684$). Using this α, the result of the double logarithmic model perfectly fits the experimental result, as can be seen in Fig. 2e.

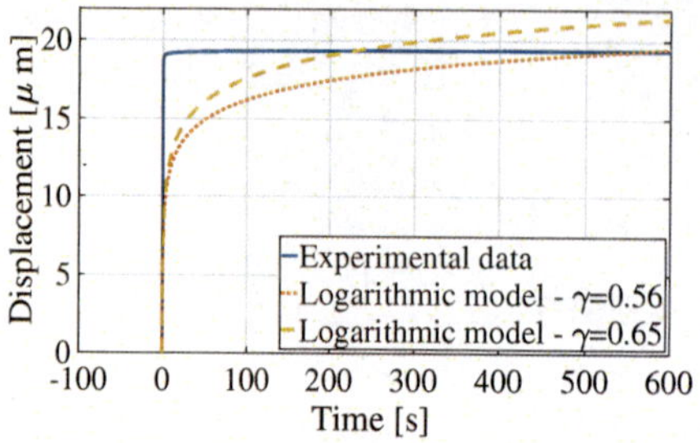

(a) The integer-order logarithmic model.

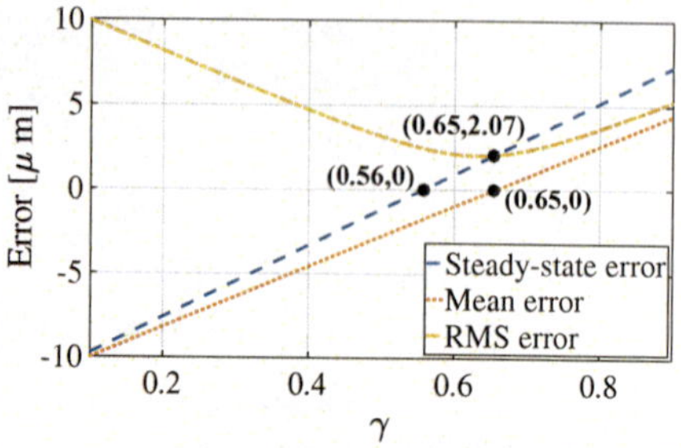

(b) Errors of the logarithmic model for different parameters γ.

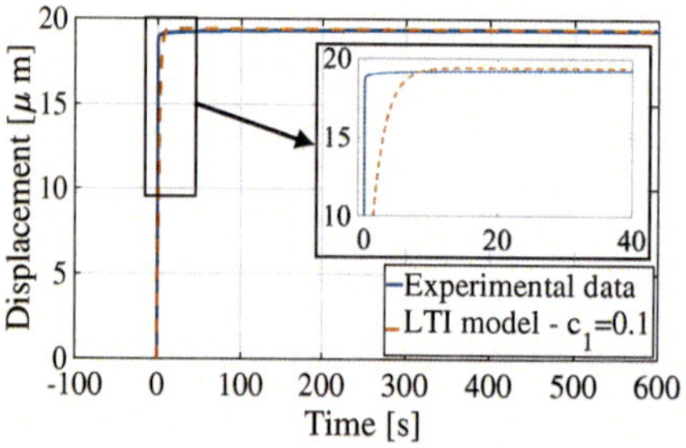

(c) The integer-order LTI model.

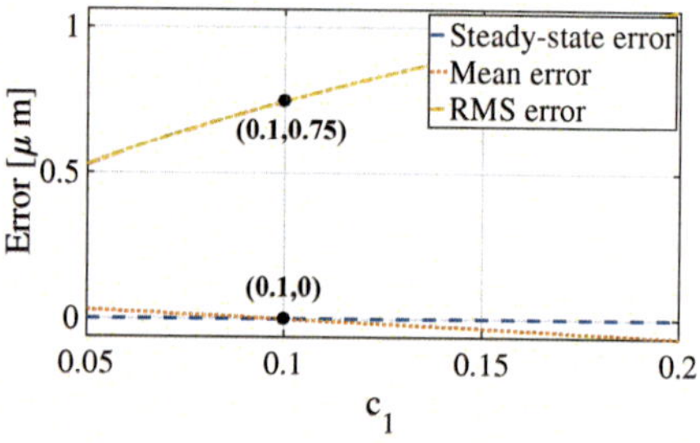

(d) Errors of the LTI model for different parameters c_1.

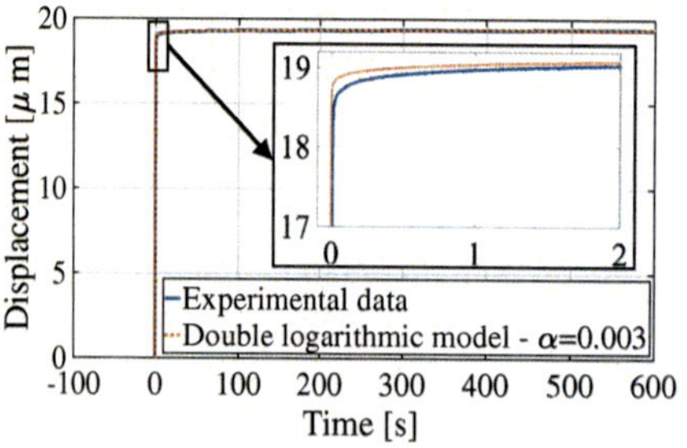

(e) The fractional-order double logarithmic model.

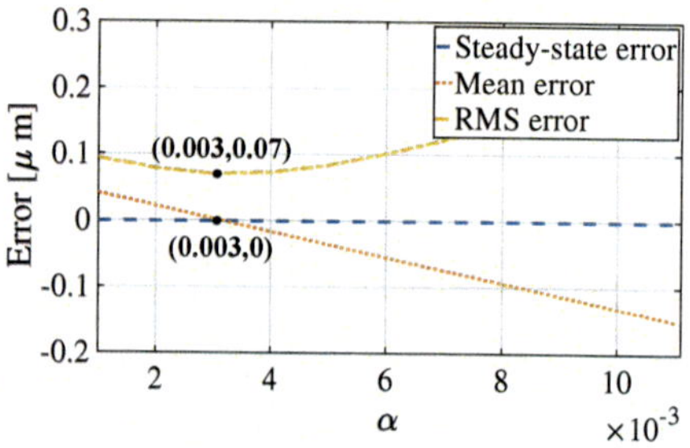

(f) Errors of the double logarithmic model for different parameters α.

Fig. 2. Choosing the model parameter that minimizes the modelling error and comparing the results with the experimental data.

Remark 1. Although the selected values for c_1 and α makes the mean error of both the LTI and double logarithmic model zero, the corresponding RMS error is smaller in the latter (RMS error is 0.75 in LTI model and 0.07 for double logarithmic model, see Figs. 2d and 2f). Thus, the fractional-order double logarithmic model fits more precisely to the experimental data in comparison to the integer-order LTI model. To support this, an illustrative comparison is shown in Fig. 3.

Remark 2. Assuming a higher order LTI model ($n > 1$) may lead to a more accurate modelling but it also increases the number of parameters ($2n + 1$) that need to be identified. Whereas, the fractional-order double logarithmic model results in an excellent accuracy, as shown in Fig. 3, with only one independent parameter.

Summarizing, in this paper, three different modelling schemes were studied and their modelling errors were compared based on measured system dynamics data for a specific input amplitude. These models can capture the creep dynamics for different input amplitudes.

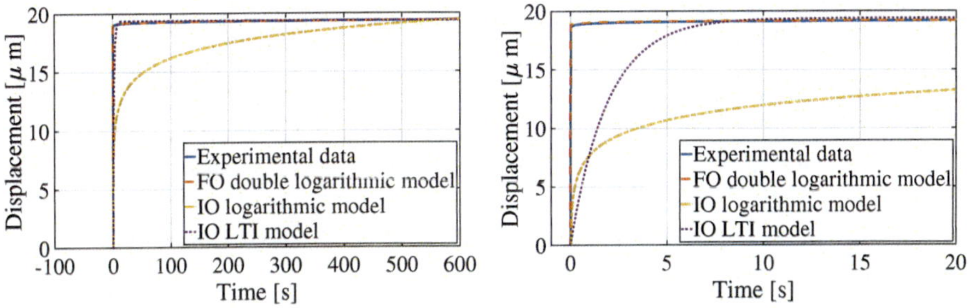

(a) Time frame $t = (0 : 600)s$.　(b) Zoomed at $t = (0 : 20)s$.

Fig. 3. Comparison between two integer-order models and a fractional-order model to capture creep phenomenon.

4 Conclusions

Creep is a key limiting factor in the controlled performance of artificial muscles/soft actuators/dielectric actuators/electroactive polymers that have significant additional nonlinearities such as hysteresis. Establishing accurate creep models will allow for improved control schemes for these actuators which enable a spectrum of exciting research avenues such as bioinspired robotics, flexible prosthetics and life-like haptic interfaces.

References

1. Babarinde, A.K., Li, L., Zhu, L., Aphale, S.S.: Experimental validation of the simultaneous damping and tracking controller design strategy for high-bandwidth nanopositioning-a pavpf approach. IET. Control. Theor. Appl. **14**(20), 3506–3514 (2020)
2. Bohannan, G., Hurst, S., Spangler, L.: Electrical component with fractional order impedance (Nov 30), uS Patent App. 11/372,232(2006)
3. Chen, Y., Petras, I., Vinagre, B.: A list of laplace and inverse laplace transforms related to fractional order calculus (2001). World Wide Web site at the address: http://people.tuke.sk/ivo.petras/foclaplace.pdf
4. Das, S.: Functional fractional calculus, vol. 1. Springer (2011)
5. Ding, C., Cao, J., Chen, Y.Q.: Fractional-order model and experimental verification for broadband hysteresis in piezoelectric actuators. Nonlinear Dyn. **98**(4), 3143–3153 (2019). https://doi.org/10.1007/s11071-019-05128-w
6. El-Rifai, O.M., Youcef-Toumi, K.: Creep in piezoelectric scanners of atomic force microscopes. In: Proceedings of the 2002 American Control Conference (IEEE Cat. No. CH37301), vol. 5, pp. 3777–3782. IEEE (2002)
7. Ge, R., Wang, X., Long, J., Chen, Z., Zhang, X.: Creep modeling and control methods of piezoelectric actuators based on fractional order theory. In: Sixth International Conference on Electromechanical Control Technology and Transportation (ICECTT 2021), vol. 12081, pp. 71–80. SPIE (2022)
8. Gu, G.Y., Gupta, U., Zhu, J., Zhu, L.M., Zhu, X.: Modeling of viscoelastic electromechanical behavior in a soft dielectric elastomer actuator. IEEE Trans. Robotics **33**(5), 1263–1271 (2017)
9. Liu, L., Yun, H., Li, Q., Ma, X., Chen, S.L., Shen, J.: Fractional order based modeling and identification of coupled creep and hysteresis effects in piezoelectric actuators. IEEE/ASME Trans. Mechatron **25**(2), 1036–1044 (2020)
10. Liu, Y., Shan, J., Gabbert, U., Qi, N.: Hysteresis and creep modeling and compensation for a piezoelectric actuator using a fractional-order maxwell resistive capacitor approach. Smart Mater. Struct. **22**(11), 115020 (2013)
11. Liu, Y., Shan, J., Qi, N.: Creep modeling and identification for piezoelectric actuators based on fractional-order system. Mechatronics **23**(7), 840–847 (2013)
12. Pesotski, D., Janocha, H., Kuhnen, K.: Adaptive compensation of hysteretic and creep non-linearities in solid-state actuators. J. Intell. Mater. Syst. Struct. **21**(14), 1437–1446 (2010)
13. San-Millan, A., Feliu-Batlle, V., Aphale, S.S.: Fractional order implementation of integral resonant control-a nanopositioning application. ISA Trans. **82**, 223–231 (2018)
14. Tominaga, K., et al.: Suppression of electrochemical creep by cross-link in polypyrrole soft actuators. Phys. Procedia **14**, 143–146 (2011)
15. Voda, A., Charef, A., Idiou, D., Machado, M.M.P.: Creep modeling for piezoelectric actuators using fractional order system of commensurate order. In: 2017 21st International Conference on System Theory, Control and Computing (ICSTCC), pp. 120–125. IEEE (2017)
16. Yang, C., Verbeek, N., Xia, F., Wang, Y., Youcef-Toumi, K.: Modeling and control of piezoelectric hysteresis: a polynomial-based fractional order disturbance compensation approach. IEEE Trans. Industr. Electron. **68**(4), 3348–3358 (2020)
17. Yang, Q., Jagannathan, S.: Creep and hysteresis compensation for nanomanipulation using atomic force microscope. Asian J. Control **11**(2), 182–187 (2009)

Adaptive Time-Delayed Feedback Control Applied to a Vibro-Impact System

Dimitri Costa(✉), Vahid Vaziri, Ekaterina Pavlovskaia, and Marian Wiercigroch

Centre for Applied Dynamics Research, School of Engineering, University of Aberdeen, King's College, Scotland, Aberdeen AB24 3FX, UK
dimitri.costa@abdn.ac.uk

Abstract. Multi-stable systems can be found in many fields of engineering from vibro-impact energy harvesters to origami structures. They are mainly characterised by the presence of co-existing attractors in their dynamic response. This feature can be beneficial when the system is required to adapt to the environment or unfavourable as the system may diverge from the desired behaviour. Hence, these systems usually require actuation or control that can be activated to maintain the desired response or switch between configurations. Recently, the idea to use the time-delayed feedback method to switch between co-existent attractors was proposed focusing on the controlled system's ability to exchange between attractors or maintain a specific type of behaviour. However, the control method is still required to have its gains set on a trial and error basis when dealing with stable solutions. Furthermore, it requires a deeper analysis of the system dynamics when dealing with unstable orbits. To address the beforementioned challenges and eliminate requirements of a priori knowledge of the system evolution function, this work uses the adaptive gain time-delayed feedback method to switch between co-existing attractors. Numerical results show that the proposed controller can be a viable low energy control option in multi-stable systems.

Keywords: Impact oscillator · Control · Nonlinear dynamics · Multi-stability

1 Introduction

Multi-stability is a double-edged sword phenomenon that can bring adaptability to an engineering system and expand its range of applications or be a safety risk to the usual operation of a device if not taken into consideration. In any case, its analysis is essential for most of nonlinear systems and is the topic of several studies in the literature.

Multi-stable system can be found in fields like energy harvesting where there is a preferred behaviour over the others due to differences in energy output [1,2], seismic mitigation where the preferred behaviour dissipates most of the energy of

Z. Dimitrovová et al. (Eds.): WMVC 2022, MMS 125, pp. 427–436, 2023.
https://doi.org/10.1007/978-3-031-15758-5_43

the system [3], in machining where the preferred behaviour reduces the vibrations of the system [4,5], and origami structures where configurations and behaviours need to be interchangeable [6]. In all cases, the ability to influence the system dynamics and choose between a desired response is crucial for operation.

In multi-stability investigations, impact oscillators are extensively studied as simple systems that present such phenomena. Thus, they can be very useful to test nonlinear control. A significant contribution to the experimental dynamics of impact oscillators was made in [7,8] where grazing phenomena were investigated. Recently, a new impact oscillator rig suitable for control was developed [9,10] which presents multi-stability between impacting and nonimpacting orbits, as well as between chaos and periodic orbits. Other works numerically investigate control methods such as the Time-Delayed Feedback method (TDF) near grazing [11] and methods to switch between co-existent attractors [12,13] in the impact oscillator systems.

Some nonlinear control methods aim to take advantage of a system dynamics to describe a control law or save energy following the original works of discrete control by Ott *et al.* [15] and continuum control by Pyragas [16]. The latter was found to be applicable in several systems to achieve different types of objectives. The original proposal of the TDF [16] aimed for the stabilization of Unstable Periodic Orbits (UPOs) embedded in a chaotic attractor. Several works studied this application and modified the controller to extend its capabilities. In [1] the TDF method was expanded to consider previous states. In [17], the authors created an adaptive delay method modification to the TDF, while in [18] a similar idea to define an adaptive gain control based on the TDF is used. Some works also studied the application of TDF in bifurcation control finding that the controller could be used to expand the stability of some of the system responses [19–21], and as a way to control grazing bifurcation [11].

Recently, the idea to switch between co-existing attractors was explored in [12,13] using new types of control. Following this objective, in [14] TDF was applied to switch between co-existing attractors. In that work, the authors analysed the performance of the TDF method in several scenarios and conclude that it can perform the exchange in most of the cases studied with a much lower knowledge requirement than previously proposed methods. However, in the chaotic case, the authors set the control gains by analysing the stability of UPOs embedded in the attractor, requiring knowledge of the system dynamics and parameters. The current study aims to improve upon previously proposed method [14] by applying the Adaptive gain Time-Delayed Feedback control (ATDF) [18]. Our results show that by using ATDF one can successfully control the system and reduce the knowledge requirements to set control gains by eliminating the need to perform a stability analysis when switching between UPOs. However, to avoid high energy consumption, control signal and high gains, the controller should have its gains reset at each target switch.

2 Adaptive Gain Time-Delayed Feedback Control Method

This work uses the ATDF control method initially proposed in [18]. It is a modification of the original TDF method proposed by Pyragas [16] that takes advantage of the original controller properties to define a dynamic equation for the gain. The original TDF method can be described as:

$$\dot{\mathbf{x}} = \mathbf{f}(\mathbf{x}) + \mathbf{K}(\mathbf{y}(t-\tau) - \mathbf{y}(t)) \tag{1}$$

$$\mathbf{y} = \mathbf{C}(\mathbf{x}) \tag{2}$$

where $\mathbf{x}$ is the state of the system, $\mathbf{f}$ is the system evolution function, $\mathbf{y}$ is the observation of the system, $\mathbf{C}$ is the observation function, $\mathbf{K}$ is the gain matrix, t is the time, dot indicates derivatives in relation to time t, and τ is the delay of the controller that is set to be the period of the targeted unstable periodic orbit. When dealing with UPOs, the TDF method requires its gain $\mathbf{K}$ to be defined by an analysis of the controlled orbit stability. This can be performed through the analysis of Lyapunov or Floquet exponents and requires knowledge of the system dynamics. However, Lehenert *et al.* [18] developed a way to set the gain $\mathbf{K}$ in an adaptive manner which eliminates the stability analysis and reduce knowledge requirements. Their idea was to find the value of $\mathbf{K}$ that would optimize a cost function linked to the control signal given by:

$$Q(t) = \frac{1}{2}(\mathbf{y}(t) - \mathbf{y}(t-\tau)) \cdot (\mathbf{y}(t) - \mathbf{y}(t-\tau)) \tag{3}$$

where $\cdot$ represents the scalar product. Note that $Q(t)$ is at its minimum point if the system response is the targeted periodic orbit of period τ, as $\mathbf{y}(t) = \mathbf{y}(t-\tau)$ implies $Q(t) = 0$. Thus, if $\mathbf{K}$ is defined such as to minimize the cost function Q, one can find a value of $\mathbf{K}$ that will stabilize the target orbit. The dynamics of $\mathbf{K}$ that would optimize the cost function Q can be defined by the speed gradient method [22]:

$$\dot{\mathbf{K}} = -\gamma \nabla_{\mathbf{K}} \dot{Q} \tag{4}$$

where γ is a parameter that regulates how fast or slow the dynamics of $\mathbf{K}$ is and $\nabla_{\mathbf{K}}$ is the gradient in relation to the gain matrix $\mathbf{K}$.

By substituting Eq. 3 into Eq. 4, the evolution for the gain matrix can be calculated as:

$$\dot{\mathbf{K}} = -\gamma \nabla_{\mathbf{K}} \left((\mathbf{y}(t) - \mathbf{y}(t-\tau)) \cdot \nabla_{\mathbf{x}} \mathbf{C}(\dot{\mathbf{x}}(t) - \dot{\mathbf{x}}(t-\tau)) \right) \tag{5}$$

where $\nabla_{\mathbf{x}}$ indicates the gradient in relation to the state $\mathbf{x}$. For the sake of simplicity, we consider that the observation of the system is the state itself

$\mathbf{y} = \mathbf{x}$ and that the gain matrix $\mathbf{K}$ is diagonal. In this case the, $\dot{\mathbf{x}}$ is substituted in Eq. 5, resulting in:

$$\dot{k_i} = \gamma(x_i(t) - x_i(t-\tau))(x_i(t) - 2x_i(t-\tau) - x_i(t-2\tau)) \tag{6}$$

where k_i is the ith gain in the diagonal of $\mathbf{K}$, and x_i is the ith component of the state vector. It is important to highlight that the control is adaptive as the gains k_i have defined dynamics. If the system state diverges from the targeted orbit by a slow change in the system parameters or some other effect the controller will try to change its gains accordingly to stabilize the desired orbit.

Finally, the evolution of the controlled system can be calculated by solving jointly Eqs. (1), (2) and (6).

3 System Modelling Using Dynamics

The controller is tested using an archetypal piece-wise linear impact oscillator model studied in [10] and shown in Fig. 1. It consists of a mass, m, connected to a leaf spring that has a spring coefficient k_1 and linear damping coefficient c. The mass resting position is at a distance g from an impact beam with elastic coefficient k_2. External excitation is provided to the system by a coil that generates a varying magnetic field that interacts with a magnet attached to the mass. This setup provide a direct sinusoidal excitation to the system $F_{ext} = aI_0 \sin(\omega t)$. In these conditions the equations of motion for the displacement of the mass X can be expressed by Newton's law:

$$m\ddot{X} = -k_1 X - c\dot{X} - k_2 H(X-g)(X-g) + F_{ext} \tag{7}$$

where H is the Heaviside's step function. By defining the system state $\mathbf{x} = [X, \dot{X}]$, the evolution equation can be obtained as:

$$\dot{\mathbf{x}} = \begin{bmatrix} \dot{X} \\ -\frac{k_1}{m}X - \frac{c}{m}\dot{X} - \frac{k_2}{m}H(X-g)(X-g) + \frac{F_{ext}}{m} \end{bmatrix} \tag{8}$$

Initially, the dynamics of the system is explored to identify cases where multistability is present. This is performed by tracing numerical bifurcation diagrams with frequency, ω as branching parameter for several initial conditions and the parameters shown in Table 1. Two forward (black and blue) and one backward (red) diagram are shown in Fig. 2. The diagrams display several windows of co-existence. Between $f = 6.80$ Hz and $f = 7.05$ Hz there are co-existing non-impacting period-1 and impacting period-2 orbits, after that, a window from $f = 7.05$ Hz to $f = 7.20$ Hz displays co-existence between period-2 and chaos. Two windows of co-existence between period-5 and period-2 orbits can be found between $f = 7.20$ Hz and $f = 7.35$ Hz and from $f = 7.35$ Hz to $f = 7.50$ Hz. Also, there is a narrow band of co-existence between period-7 and period-2 orbits from f = 7.66 Hz to f = 7.71 Hz, and two regions of co-existence between chaotic behaviour and period-3 orbit from f = 7.74 Hz to f = 7.95 Hz and from f = 8.10 Hz to f = 8.19 Hz. As originally the ATDF was designed to stabilize UPOs in a chaotic attractor, the frequency of 8.18 Hz is chosen as it is in the chaotic window of the diagram.

As a chaotic attractor is selected to perform the control, the identification of its UPOs is necessary to find the possible control targets. Hence, the system is simulated for 10000 period of excitation and its chaotic attractor is shown in Fig. 3(a). Afterwards, the close return method [23] is used to identify the periodic orbits embedded in the attractor. A period-1 and period-2 identified UPOs, shown in Fig. 3(b), are selected to be the targets of the controller as they had the smallest periods in the group of identified UPOs and are usually the most favourable to be stabilized.

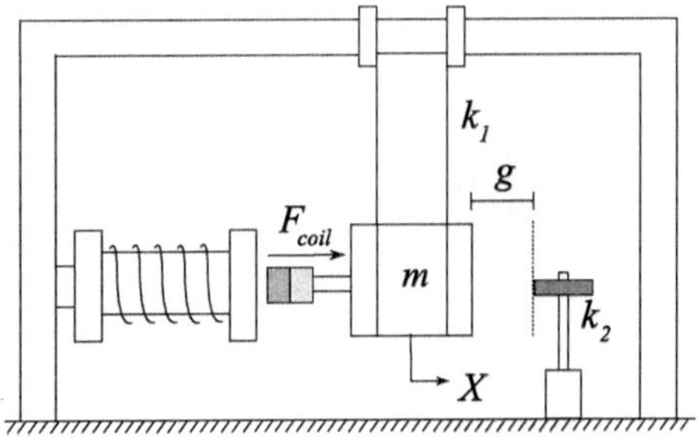

Fig. 1. Schematics of the impact oscillator model showing: gap g, main mass m, leaf spring linear stiffness k_1, impact beam linear stiffness k_2, main mass displacement X and magnetic force applied F_{coil} by the coil.

Table 1. Piece-wise linear model parameters values.

Symbol	Value	Unit	Symbol	Value	Unit
m	1.325	kg	c	0.27	kg/s^2
k_1	4331	N/m	g	0.7410^{-3}	m
k_2	87125	N/m	a	0.799	N/A
I_0	1.45	A	-	-	-

4 Control

We assume that the total force applied by the coil F_{coil} can be given by the sum of excitation, F_{ext}, and actuation F_{act} forces, resulting in:

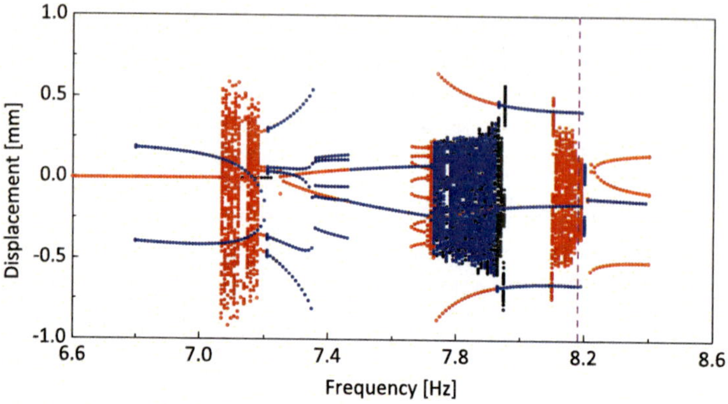

Fig. 2. Bifurcation diagram exploring the system dynamics to select the relevant case for control. The dashed magenta line highlights the selected frequency at f = 8.18 Hz.

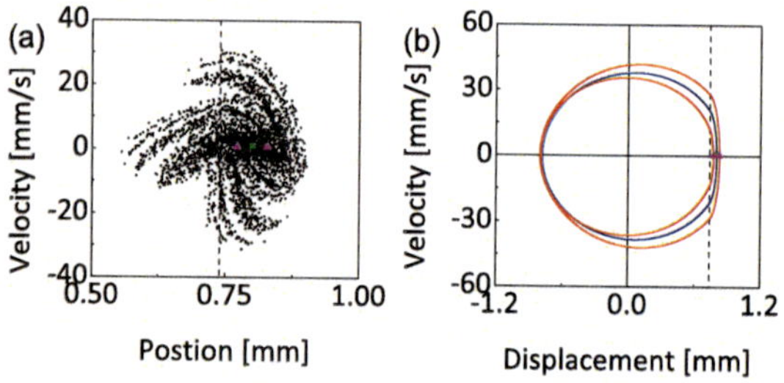

Fig. 3. Chaotic behaviour of the impact oscillator at excitation frequency f = 8.18 Hz. (a) Chaotic attractor and identified recurring point for period-1 and period-2 orbits. The full attractor is portrayed in ●, the period-1 orbit in ■, and the period-2 orbit in ▲ (b) Identified period-1 and period-2 orbits. The period-1 orbit is depicted in blue and the period-2 orbit in red.

$$F_{coil} = F_{ext} + F_{act} = aI_0 \sin(\omega t) + mk_v \left(\dot{X}(t-\tau) - \dot{X}(t)\right) \tag{9}$$

where k_v is the gain related to velocity.

In this context, it is assumed that velocity is the observable of the system, and actuation can only be applied as a force, the gain matrix $\mathbf{K}$ becomes:

$$\mathbf{K} = \begin{bmatrix} 0 & 0 \\ 0 & k_v \end{bmatrix} \tag{10}$$

By rearranging Eq. 7 to include the gain k_v dynamics we obtain:

$$\begin{bmatrix} \dot{X} \\ \ddot{X} \\ \dot{k}_v \end{bmatrix} = \begin{bmatrix} \dot{X} \\ -\frac{k_1}{m}X - \frac{c}{m}\dot{X} - H(X-g)\frac{k_2}{m}(X-g) + k_v\left(\dot{X}(t-\tau) - \dot{X}(t)\right) + \frac{aI_0}{m}\sin(\omega t) \\ \gamma\left(\dot{X}(t) - \dot{X}(t-\tau)\right)\left(\dot{X}(t) - 2\dot{X}(t-\tau) - \dot{X}(t-2\tau)\right) \end{bmatrix} \tag{11}$$

Originally the ATDF method was designed to stabilize UPOs in a chaotic attractor. Hence, we choose the frequency of f = 8.18 Hz, where the system presents a chaotic response, as the controller have the best conditions to perform the stabilization of the UPOs and switch between them.

The test is constructed to analyse if the control can stabilize the UPOs regardless of the previous behaviour. Hence, the system is initialized in a chaotic response. Afterwards, the control is turned on targeting the period-2 UPO by setting its delay to the target orbit period. After the stabilization of the system response is reached, the control changes its target to the period-1 UPO by changing its delay τ. In sequence, the control moves its target again to the period-2 UPO and finally is turned off after stabilization is reached. Before each switch the control gain k_v is always set to zero. If the reset of k_v is not performed, the gain will build-up over time for each orbit switch as long as it stays in the range of values that stabilize an orbit. This build-up is undesired as higher gains usually lead to higher control signals and higher energy consumption.

Figure 4 shows the numerical simulation of the performed test. The system is initialized in a chaotic response, seen in the Poincaré time history shown in Fig. 4(a). Afterwards, the controller is turned on and stabilizes the system in a period-2 behaviour after about 11 seconds. The stabilized behaviour is identical to the identified UPO, which can be verified by the state space diagram in Fig. 4(b). The gain time history seen in Fig. 4(e) shows that after the orbit is stabilized, the adaptive gain raises its value up to k_v = 0.460 s^{-1} where it stabilizes as the system reaches the desired orbit. In sequence, the controller changes its target to the period-1 UPO and quickly bring the system to the desired periodic orbit. The gain k_v also quickly raises to the value of 0.567 s^{-1} and stays constant after stabilization. It is important to notice here that the value of the gain reached is not the one that optimizes the orbit stability [14]. Afterwards, the controller changes its target to the period-1 UPO and quickly brings the system to the desired behaviour. This can also be verified in Fig. 4(c) where both the targeted and stabilized orbit are shown. The gain k_v has another increase up to the value of 3.20 s^{-1} as the system migrates to the desired behaviour. Finally, the control changes again its target to the period-2 UPO and successfully brings the system to the desired periodic orbit. In this transition, the control signal, despite being very small, is not zero, but this does not affect the constant value of the gain k_v after stabilization at 0.418 s^{-1}. It is also important to notice that, as discussed in [18], the stabilized value of k_v is dependent on initial conditions, thus, the stabilized values of k_v for the first transition between chaos and period-2 UPO and the third transition between period-1 and period-2 UPO are different from each other as well as the time that the controller requires to stabilize the orbit.

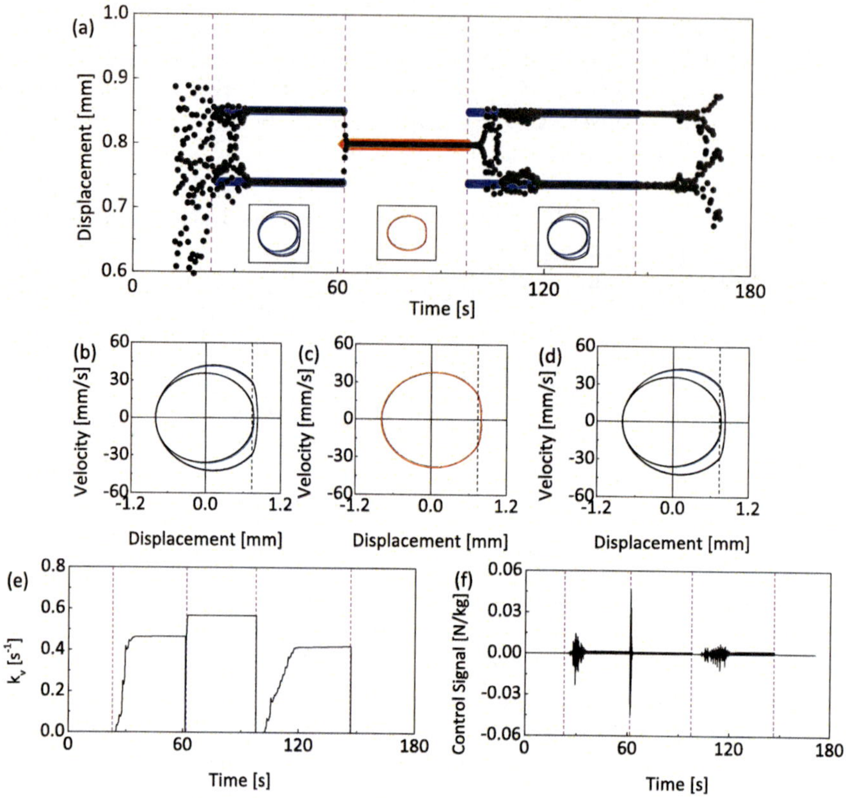

Fig. 4. Adaptive control of the impact system. (a) Time history of position taken on the Poincaré section mod $(\omega t, 2\pi) = 0$, where the system response is represented by ●, target period-2 UPO by ● and target period-1 UPO by ●. (b) Period-2 orbit targeted in the first phase (blue) and stabilized orbit (black). (c) Period-1 orbit targeted in the second phase (red) and stabilized orbit (black). (d) Period-2 orbit targeted in the third phase and stabilized orbit. (e) Time history of the gain k_v. (f) Control signal time history. Dashed magenta lines represent the control phase boundaries, dashed black lines represent the impact boundary in phase diagrams.

It is worth mentioning that, the TDF method has difficulties switching from the period-1 to the period-2 UPO as pointed out in [14]. If the system response is a period-1 UPO of period τ_s the control signal of the TDF method will still be negligible if the delay of the controller is changed to stabilize a period-2 behaviour, $\tau = 2\tau_s$. As $y(t - 2\tau_s) = y(t - \tau_s) = y(t)$, when a period-1 response is given by the system, the control signal becomes zero while forgetting a period-2 orbit. However, the ATDF control does not suffer from the same problems as TDF due to the reset of its gains, which allows the system to move away from the period-1 behaviour, helping the controller to move the system to the desired orbit and increase its control signal faster.

5 Conclusions

The ability to switch between stabilized UPOs embedded in a chaotic attractor can bring adaptability to a mechanical system. However, only a few studies deal with such controllers. This work improves previously proposed control methods by eliminating the stability analysis required to set control gains. This is demonstrated by the application of ATDF to an impact oscillator that presents multi-stability and chaos.

The idea of ATDF, to optimize a cost function Q based on the control signal, is presented and an evolution equation for the controller's gains, initially derived in [18], was developed to the applied case. Then, one of the chaotic windows in the bifurcation diagram of the system was chosen to test the controller. Results show that the ATDF method was able to stabilize and switch between the targeted UPOs without the need of a stability analysis to set control gains. This reduces the knowledge necessary to apply the control, greatly facilitating its use in real applications where the knowledge about the system is limited and its analysis can be costly and time consuming. It was also observed that ATDF presents a better performance and lower gains if the adaptive gain is set to zero when a target switch happens. This avoided the build-up of the gain and maintained its value on the lower end of the stability region required to stabilize the targeted orbits. ATDF also performed better than TDF when dealing with the transfer from a period-1 UPO to a period-2 UPO due to its fast divergence from the former.

The study of the ATDF method to switch between co-existing behaviours still requires a better understanding of its convergence mechanisms, experimental verification and analysis of robustness so it can finally be implemented in real applications. These will be investigated in future works.

References

1. Arrieta, A.F., Hagedorn, P., Erturk, A., Inman, D.J.: A piezoelectric bistable plate for nonlinear broadband energy harvesting. Appl. Phys. Lett. **97**(10), 104102 (2010)
2. Ai, R., Monteiro, L., Monteiro, P., Pacheco, P., Savi, M.A.: Piezoelectric vibration-based energy harvesting enhancement exploiting nonsmoothness. Actuators **8**(1), 25 (2019)
3. Nucera, F., Vakakis, A.F., McFarland, D.M., Bergman, L.A., Kerschen, G.: Targeted energy transfers in vibro-impact oscillators for seismic mitigation. Nonlinear Dyn. **50**(3), 651–677 (2007)
4. Wiercigroch, M., Krivtsov A.M.: Frictional chatter in orthogonal metal cutting. Philosophical Trans. Roy. Soc. London. Series A: Math. Phys. Eng. Sci. **359**(1781), 713-738 (2001)
5. Budd, C., Dux, F.: Chattering and related behaviour in impact oscillators. Philosophical Trans. Roy. Soc. London. Series A: Phys. Eng. Sci. **347**(1683), 365-389 (1994)
6. Salerno, M., Zhang, C., Menciassi, A., Dai, J.S.: A Novel 4-DOF origami grasper with an SMA-actuation system for minimally invasive surgery. IEEE Trans. Rob. **32**(3), 484–498 (2016)

7. Ing, J., Pavlovskaia, E., Wiercigroch, M., Banerjee, S.: Experimental study of impact oscillator with one-sided elastic constraint. Philosophical Trans. R. Soc. A: Math. Phys. Eng. Sci. **366**(1866), 679–705 (2008)
8. Ing, J., Pavlovskaia, E., Wiercigroch, M., Banerjee, S.: Bifurcation analysis of an impact oscillator with a one-sided elastic constraint near grazing. Physica D **239**(6), 312–321 (2010)
9. Wiercigroch, M., et al.: Versatile mass excited impact oscillator. Nonlinear Dyn. **99**(1), 323–339 (2019). https://doi.org/10.1007/s11071-019-05368-w
10. Costa, D., et al.: Chaos in impact oscillators not in vain: dynamics of new mass excited oscillator. Nonlinear Dyn. **102**(2), 835–861 (2020). https://doi.org/10.1007/s11071-020-05644-0
11. Zhang, Z., Páez Chávez, J., Sieber, J., Liu, Y.: Controlling grazing-induced multistability in a piecewise-smooth impacting system via the time-delayed feedback control. Nonlinear Dyn. **107**, 1595–1610 (2012)
12. Liu, Y., Wiercigroch, M., Ing, J., Pavlovskaia, E.: Intermittent control of coexisting attractors. Philosophical Trans. R. Soc. A: Math. Phys. Eng. Sci. **371**(1993), 20120428 (2013)
13. Zhang, Z., Chávez, J.P., Sieber, J., Liu, Y.: Controlling coexisting attractors of a class of non-autonomous dynamical systems. Physica D **431**, 133134 (2012)
14. Costa, D., Vaziri, V., Pavlovskaia, E., Savi, M.A., Wiercigroch, M.: Switching between periodic orbits in impact oscillator by time-delayed feedback methods. (Submitted article in March 2022)
15. Ott, E., Grebogi, C., York, J.A.: Controlling chaos. Phys. Rev. Lett. **64**(11), 1196–1199 (1990)
16. Pyragas, K.: Continuous control of chaos by self-controlling feedback. Phys. Lett. A **170**, 421–428 (1992)
17. Pyragas, V., Pyragas, K.: Adaptive modification of the delayed feedback control algorithm with a continuously varying time delay. Phys. Lett. A **375**(44), 3866–3871 (2011)
18. Lehnert, J.; Hovel, P.; Flunkert, V.; Guzenko, P. Y.;Fradkov, A. L. and Schöll, E.: Adaptive tuning of feedback gain in time-delayed feedback control. Chaos: Interdisciplinary J. Nonlinear Sci. **21**(4), 043111 (2011)
19. de Paula, A.S., Savi, M.A., Pavlovskaia, E., Wiercigroch, M.: Bifurcation control of a parametric pendulum. Int. J. Bifurcat. and Chaos **22**(05), 1250111 (2012)
20. de Paula, A.S., dos Santos, M.V.S., Savi, M.A., Bessa, W.M.: Controlling a shape memory alloy two-bar truss using delayed feedback method. Int. J. Struct. Stab. Dyn. **14**(08), 1440032 (2014)
21. Costa, D.D.A., Savi, M.A., de Paula, A.S., Bernardini, D.: Chaos control of a shape memory alloy structure using thermal constrained actuation. Int. J. Non-Linear Mech. **111**, 106–118 (2019)
22. Fradkov, A.L., Pogromsky, A.Y.: Introduction to Control of Oscillations and Chaos. World Scientific, Singapore (1998)
23. Pawelzik, K.: Schuster: Unstable periodic orbits and prediction. Phys. Rev. A **43**(4), 1808 (1991)

Constrained Control of Impact Oscillator with Delay

Mohsen Lalehparvar, Sumeet S. Aphale, and Vahid Vaziri(✉)

University of Aberdeen, Aberdeen, UK
vahid.vaziri@abdn.ac.uk

Abstract. Often, the system nonlinearities adversely impact system behaviour/performance. A deeper understanding of these nonlinearities and the parametric dependencies thereof has the potential to enable the formulation of robust, energy-efficient control strategies. Moreover, input constraints and/or delays significantly impact the controlled system performance and overall system stability. In this work, we present a systematic investigation to test the performance of control methodology in the presence of delay and control input constraints. Impact oscillators have been proved to be suitable testbeds to implement nonlinear control methods considering the system's dynamic rather than using high, control-effort-consuming gains. Therefore, a mass-excited impact oscillator is chosen as a candidate system to demonstrate the proposed control scheme's effectiveness. As a case study, the coexistence of impacting and non-impacting behaviour of the system is considered. A time-delayed feedback (TDF) scheme is employed in closed loop with the impact oscillator to achieve a non-impacting solution. Subsequently, the closed-loop performance of the overall system with and without constraints and delay are simulated and contrasted. It has been observed that the controller successfully exchanges the chosen attractors. However, the controller fails to stabilise the system with higher delays and constraints, even with higher gains. Therefore a new approach would be required to deal with delays and constraints in such scenarios.

Keywords: Impact oscillator · Nonlinear dynamics · Another keyword · Time delay · Saturation · Constraint · TDF control · Bifurcation · Phase plane

1 Introduction and Motivation

Exchange between coexisting attractors can be used to get the desired response of the system by switching to the desired attractor [18]. For instance, in drill string, when having two coexisting attractors of forward and backward whirls, based on the desired behaviour, the exchange between attractors could be accomplished [10].

Time delay, which is an integral part of real systems, can be induced by sensors or/and the inherent delay in actuators. Time delays are mainly caused

Z. Dimitrovová et al. (Eds.): WMVC 2022, MMS 125, pp. 437–446, 2023.
https://doi.org/10.1007/978-3-031-15758-5_44

by a limitation in bandwidth or the inherent lags in the systems [19]. Although, sometimes, the time delay can be used to design controllers [26], mostly it has a negative effect on the stability of the system, leading to limit cycles [13] or unstability of the systems [1]. Hence, the inherent time delay should be considered.

To study the effects of time delay and constraint and assess the controller performance, the impacting phenomenon has been taken into account, representing a class of frequent nonlinearities in the world. An impact oscillator can study this nonlinearity caused by the system's discontinuity. Mass excited impact oscillator, [22], which has been developed in the Centre for Applied Dynamics Research at the University of Aberdeen, provides researchers with a testbed to address impacting phenomenon as well as to examine controllers to exchange between the attractors or to stabilise the unstable periodic orbits of a chaotic system.

1.1 State-of-the-Art in Exchange Impacting and Non-impacting Attractors

Impact is a common phenomenon in many mechanical systems [27]. Depending on the application, it might be avoided or desired. For instance, in machining, impact-induced nonlinearities cause imperfections and should be avoided [3,23]. On the other hand, impacts are desired in some other applications like energy harvesting [21], Resonance Enhanced Drilling [17,24,25], microelectromechanical devices [2] and seismic mitigation [16].

An impact oscillator which exhibits both impacting and non-impacting behaviour, can provide means to exchange coexisting attractors in the multi stable systems. In fact, multistability, which is an inherent characteristic of such systems, can be used to achieve the desired response [5]. For instance, in self-propelled capsules in which stationary and forward-motion attractors coexist, the exchange between period one, forward-motion and period one, stationary-motion [9] requires a control scheme for the exchange. Another example could be the exchange between forward and backward motion [12]. In both applications, regardless of the control aim, impact provides the actuation for exchange between coexisting attractors. An exchange between coexisting attractors can be used in rotor dynamics [4,15] where impacting and non-impacting attractors coexist. Depending on the application, the control goal can be to perform an exchange from impacting to non-impacting attractors to avoid damage caused by impacting in rotary motion, or it can be the exchange between non-impacting to impacting, for instance, to free a stuck drill string [10].

In many mechanical tools, the impact could result in damage, degeneration of the system's performance or imperfections. Hence, avoiding impact is of interest in such applications. The point of transition between impacting and non-impacting orbits in which impact surfaces touch with zero velocity is called grazing bifurcation. At this critical point, if controlled, the exchange between impacting and non-impacting orbits become viable. Rotary systems are examples of such systems in which impact can lead to the degeneration of the system's performance and even failure. For instance, in Jeffcott rotors, the impact between the rotor and the snubber ring, which is an unwanted phenomenon [4], could be avoided at the grazing point by exchanging the non-impacting and impacting coexisting attractors.

Double grazing bifurcations can happen in railway wheelset systems with the flange contact [14]. The coexistence of attractors in near grazing dynamics could result in hard or soft impacts. In the grazing, it might be of interest to exchange to non-impacting behaviour in order to avoid damages to the wheelset system. The coexistence of impacting and non-impacting attractors in grazing can happen in elastic beams [11,20] which could be desired to exchange to non-impacting behaviour to avoid damage in the system. Grazing bifurcation, has been characterized by the interaction between ships through ship roll motion and icebergs [6–8]. The interest in exchanging coexisting attractors in grazing is not always to avoid the damage or degeneration of the system. It could be the exchange to non-impacting as a means to inverse the actuation in impact microactuators [28,29].

In this paper, firstly, the coexistence of impacting and non-impacting behaviour in system dynamics will be investigated. Then the use of closed-loop system for exchanging to the non-impacting solution will be tested. Finally, the effect of delay and constraint on the controller will be illustrated and discussed.

2 System Dynamics and Coexisting Attractors

Due to the discontinuity in the system, impact oscillators represent a type of nonlinearity which is called boundary or contact nonlinearity. This type of nonlinearity can be categorised among the most frequent nonlinearities in real-world systems. In mass-excited impact oscillator [22], shown in Fig. 1, a mass m is held using two leaf springs. A coil generates the excitation by applying electromagnetic forces to the mass. In the initial position, there is a gap between the impacting point and the mass. However, when the mass displacement X is equal or larger than the gap g, the mass collides with the impacting point. Applying model parameters as shown in Table 1 and using the Heaviside function, $H(X-g)$, to represent discontinuity on impacting point, the dynamics of the system can be represented as follows:

$$\ddot{X} = -\frac{k_1}{m}X - \frac{k_2}{m}(X-g)H(X-g) - \frac{c}{m}\dot{X} + \frac{F_{coil}}{m}. \tag{1}$$

Table 1. Model parameters

Symbol	Value	Unit
m	1.325	kg
k_1	4331	N/m
k_2	87125	N/m
c	0.27	kg/s^2
g	0.74	mm
a	0.799	N/A
I_0	1.45	A

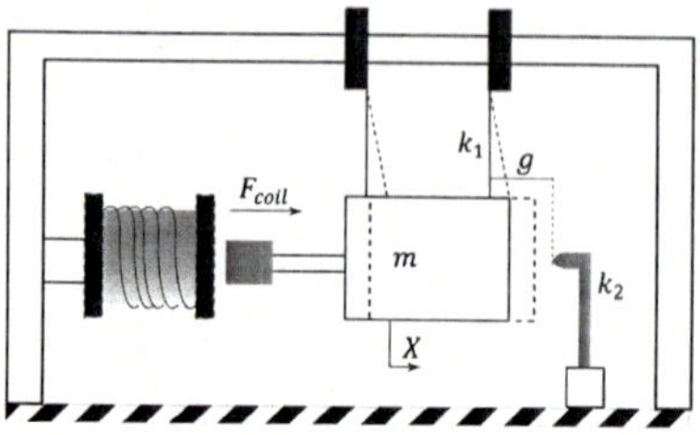

Fig. 1. Schematic of mass-excited impact oscillator.

The impact oscillator has complex dynamics which have been investigated previously [5,22]. Here, we focus on two types of system behaviours; Period-1 non-impacting (P1) and Period-2 (P2) impacting. These two solutions coexist at excitation frequency of 6.8 Hz and presented in Fig. 2.

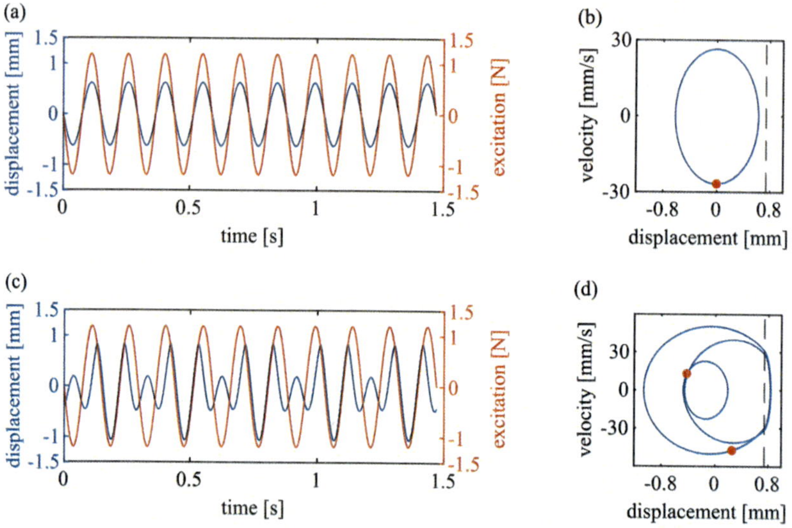

Fig. 2. Example of coexistence of attractors in the system dynamics. 10 period of time history of (a) Period-1 and (c) Period-2 with their phase planes in (b) and (d), respectively. Dashed line represents the initial gap between the mass and the impacting beams which is 0.74 mm. When the displacement is equal or bigger than the gap, impact happens.

2.1 Closed-Loop Control

Coexistence of P1 and P2 attractors provides the possibility of exchanging between the impacting attractor to the non-impacting one. To this purpose, the TDF controller is applied to the system to exchange to P1, non-impacting. The TDF control signal is as follows:

$$\begin{bmatrix} 0 \\ aI_{act} \end{bmatrix} = \begin{bmatrix} 0 & 0 \\ K_p & Kv \end{bmatrix} \begin{bmatrix} x(t-\tau_s) - x(t) \\ \dot{x}(t-\tau_s) - \dot{x}(t) \end{bmatrix} \tag{2}$$

Figure 3 presents the simulation results of the control exchange between impacting and non-impacting attractors. The system is in P2 and after 50 periods (0.1471 s) the controller with aim at P1 is switched on. At $t = 10.2956$ s the system fully settled at P1. In order to detect the settling time, two error levels have been defined for the displacement and velocity, 1×10^{-6} and 1×10^{-1}, consequently. The last 200 poincaré sections have been used for the comparison of displacement and velocity of each point. Then, 20 last periods have been eliminated, and the mean displacement and velocity of the last 20 of the remainder have been obtained. Finally, the poincaré sections have been compared to the means to find the first point, which has a difference to the mean more than our criteria.

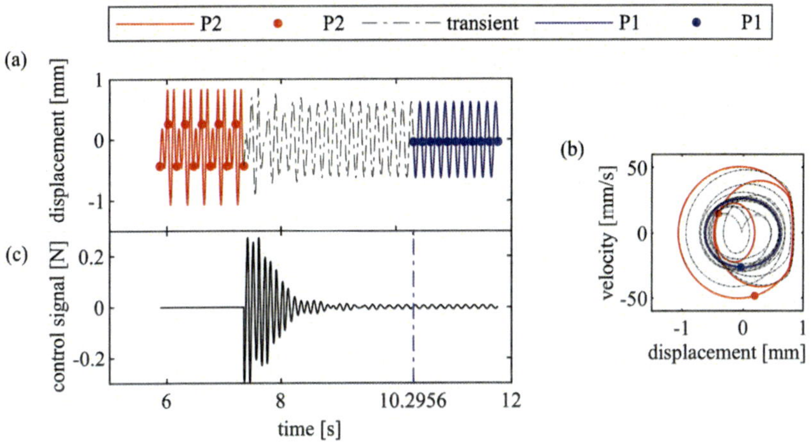

Fig. 3. (a) Time history of the system beginning from 10 periods of oscillation before the start of the controller, up to 10 period of the settled response. (b) Phase plane of P2 to P1 closed-loop control. (c) Control signal.

So far, the controller is applied at the beginning of the fifty period of the response. However, as the velocity and resultantly the kinetic energy of the system is varied in different phases of each period, the application moment of the control signal has been delayed gradually with time steps of $0.01T$ s, where T is the period of excitation equal to 0.1471 s, up to fifty first period. This analysis demonstrated the effect of the phase in which the control signal is applied to the performance of the TDF controller. Three selected cases of $t = 0$ s, $t = T$ s and $t = 2T$ s are shown in Fig. 4.

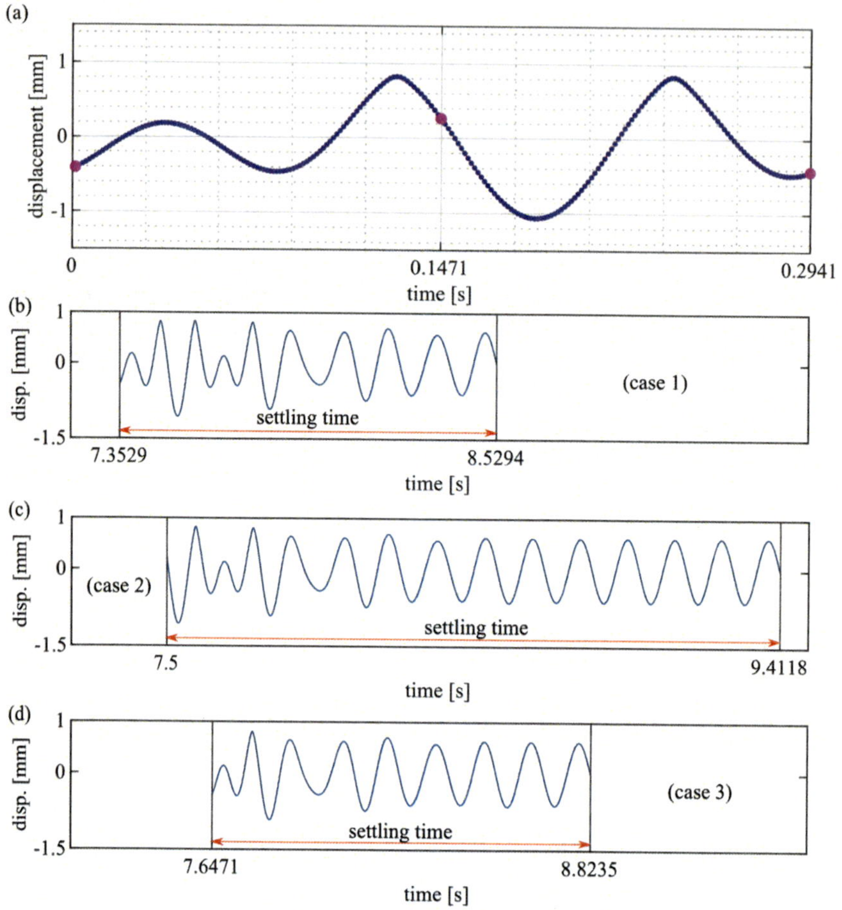

Fig. 4. (a) Two periods of time history of Period-2 response. The controller can start at any time, including 3 cases indicated by red dots. (b), (c) and (d) transient response of the system from the moment the control signal is applied to the moment the system settles in P1 for three sample cases of $t = 0$ s, $t = T$ s and $t = 2T$ s, respectively. The X axis of (b), (c) and (d) are the same to explicitly illustrate the difference of the application moment of control signal and how it manipulated the settling time.

3 Time Delay and Constraint

Actuator delay and sensor delay are applied to the system as shown in Eqs. 3 and 4, consequently:

$$\begin{bmatrix} 0 \\ aI_{act_{ad}}(t+\tau_a) \end{bmatrix} = \begin{bmatrix} 0 & 0 \\ K_p & Kv \end{bmatrix} \begin{bmatrix} x(t-\tau_s) - x(t) \\ \dot{x}(t-\tau_s) - \dot{x}(t) \end{bmatrix} \tag{3}$$

$$\begin{bmatrix} 0 \\ aI_{act_{id}}(t) \end{bmatrix} = \begin{bmatrix} 0 & 0 \\ K_p & Kv \end{bmatrix} \begin{bmatrix} x(t-\tau) - x(t) \\ \dot{x}(t-\tau) - \dot{x}(t) \end{bmatrix}, \text{ where } \tau = \tau_s + \tau_a. \tag{4}$$

where $I_{act_{ad}}$ and $I_{act_{id}}$ are actuation current in presence of actuator delay and sensor delay, consequently.

To investigate the performance of the TDF controller in presence of the delay and constraint, its settling time has been assessed in four cases: (a) When the system has actuator delay from zero up to $0.2T$ s with steps of $5 \times 10^{-3}T$ s. (b) With sensor delay from zero up to $0.2T$ s with steps of $5 \times 10^{-3}T$ s. (c) The start time of the TDF controller with steps of $0.01T$ s. (d) when the actuator is constrained from 0.2 N to 0.06 N with steps of $5 \times 10^{-3}T$ N, Fig. 5. As shown in the figure, by increasing the delay, the settling time decreased up to about $0.1T$

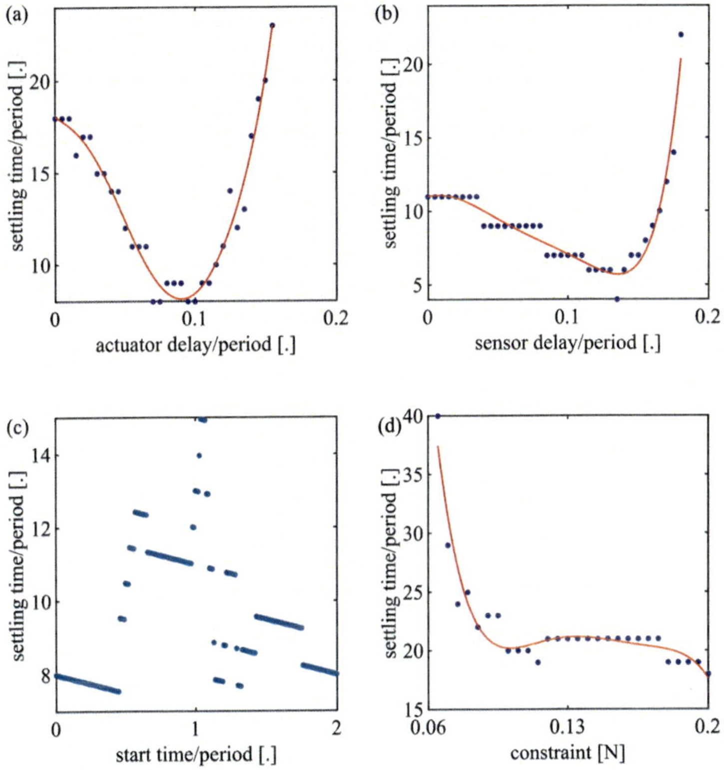

Fig. 5. (a) Settling time vs actuator delay. The actuator delay is increased from 0 s to $0.2T$ s with time steps of $5 \times 10^{-3}T$ s. The settling time is expressed as the portion of the period (from $10T$, $20T$ Et cetra). (b) Settling time vs sensor delay. The sensor delay is increased from 0 s to $0.2T$ s with time steps of $5 \times 10^{-3}T$ s. The settling time is expressed as the portion of the period. (c) Settling time vs start time, both axes are described as the portion of the system's period. The start time has been varied from 0 to $2T$ s with time steps of $0.01T$ s. (d) Settling time vs constraint. The constraint has been decreased (the effect of constraint on the system has been increased) from 0.2 N to 0.06 N with steps of $5 \times 10^{-3}T$ N while the settling time is illustrated as the fraction of the system's period.

and $0.15T$ for the cases of actuator and sensor delay, respectively. After these values of delay, settling time increases by increasing the delay up to some point in which the TDF fails to settle the system in the desired period (P1).

4 Concluding Remarks

Based on the simulation results, by increasing the actuator delay and sensor delay, the settling time of the system illustrated almost the same pattern; it starts to decrease up to some point and then it increases and finally, at certain points, the TDF fails to settle the system. By increasing the start time, however, the performance of the system becomes complicated; it has a decreasing tone. However, there are dramatic discontinuities in the system. Finally, decreasing constraint leads to increasing settling time up to some points, after which the controller fails. Due to the fact that the start of the controller illustrates a strange pattern, an analysis could be performed to figure out the reasons behind this pattern which would be helpful for the future control design. Moreover, by increasing delay, in both cases of actuator delay and sensor delay, the system's settling time decreases to some points. Those points and the system behaviour can be used for further optimisation investigations.

References

1. Asai, Y., Tasaka, Y., Nomura, K., Nomura, T., Casadio, M., Morasso, P.: A model of postural control in quiet standing: robust compensation of delay-induced instability using intermittent activation of feedback control. PLoS ONE **4**(7), e6169 (2009)
2. Bassinello, D.G., Tusset, A.M., Rocha, R.T., Balthazar, J.M.: Dynamical analysis and control of a chaotic microelectromechanical resonator model. Shock Vibr. **2018** (2018)
3. Budd, C., Dux, F.: Chattering and related behaviour in impact oscillators. Philos. Trans. Roy. Soc. Lond. Ser. A Phys. Eng. Sci. **347**(1683), 365–389 (1994)
4. Chávez, J.P., Hamaneh, V.V., Wiercigroch, M.: Modelling and experimental verification of an asymmetric Jeffcott rotor with radial clearance. J. Sound Vib. **334**, 86–97 (2015)
5. Costa, D., Vaziri, V., Pavlovskaia, E., Savi, M., Wiercigroch, M.: Switching between periodic orbits in impact oscillator by time-delayed feedback methods. Physica D (2022, accepted)
6. Grace, I., Ibrahim, R.: Modelling and analysis of ship roll oscillations interacting with stationary icebergs. Proc. Inst. Mech. Eng. C J. Mech. Eng. Sci. **222**(10), 1873–1884 (2008)
7. Grace, I., Ibrahim, R.: Vibro-impact interaction of ships with ice. In: ASME Pressure Vessels and Piping Conference, vol. 48272, pp. 433–440 (2008)
8. Grace, I., Ibrahim, R.: Elastic and inelastic impact interaction of ship roll dynamics with floating ice. In: Ibrahim, R.A., Babitsky, V.I., Okuma, M. (eds.) Vibro-Impact Dynamics of Ocean Systems and Related Problems. LNACM, vol. 44. Springer, Heidelberg (2009)

9. Guo, B., Liu, Y., Birler, R., Prasad, S.: Self-propelled capsule endoscopy for small-bowel examination: proof-of-concept and model verification. Int. J. Mech. Sci. **174**, 105506 (2020)
10. Kapitaniak, M., Vaziri, V., Chávez, J.P., Wiercigroch, M.: Experimental studies of forward and backward whirls of drill-string. Mech. Syst. Sig. Process. **100**, 454–465 (2018)
11. Liu, R., Yue, Y., Xie, J.: Existence and stability of the periodic orbits induced by grazing bifurcation in a cantilever beam system with single rigid impacting constraint (2021)
12. Liu, Y., Páez Chávez, J.: Controlling multistability in a vibro-impact capsule system. Nonlinear Dyn. **88**(2), 1289–1304 (2017)
13. MacDonald, N.: Time delay in prey-predator models. Math. Biosci. **28**(3–4), 321–330 (1976)
14. Miao, P., Li, D., Yin, S., Xie, J., Grebogi, C., Yue, Y.: Double grazing bifurcations of the non-smooth railway wheelset systems (2021)
15. Mora, K., Champneys, A.R., Shaw, A.D., Friswell, M.I.: Explanation of the onset of bouncing cycles in isotropic rotor dynamics; A grazing bifurcation analysis. Proc. Roy. Soc. A **476**(2237), 20190549 (2020)
16. Nucera, F., Vakakis, A.F., McFarland, D., Bergman, L., Kerschen, G.: Targeted energy transfers in vibro-impact oscillators for seismic mitigation. Nonlinear Dyn. **50**(3), 651–677 (2007)
17. Pavlovskaia, E., Hendry, D.C., Wiercigroch, M.: Modelling of high frequency vibro-impact drilling. Int. J. Mech. Sci. **91**, 110–119 (2015)
18. Pisarchik, A.N.: Controlling the multistability of nonlinear systems with coexisting attractors. Phys. Rev. E **64**(4), 046203 (2001)
19. Töpfer, J.D., Sigurdsson, H., Pickup, L., Lagoudakis, P.G.: Time-delay polaritonics. Commun. Phys. **3**(1), 1–8 (2020)
20. Wagg, D., Bishop, S.: Application of non-smooth modelling techniques to the dynamics of a flexible impacting beam. J. Sound Vib. **256**(5), 803–820 (2002)
21. Wei, S., Hu, H., He, S.: Modeling and experimental investigation of an impact-driven piezoelectric energy harvester from human motion. Smart Mater. Struct. **22**(10), 105020 (2013)
22. Wiercigroch, M., Kovacs, S., Zhong, S., Costa, D., Vaziri, V., Kapitaniak, M., Pavlovskaia, E.: Versatile mass excited impact oscillator. Nonlinear Dyn. **99**(1), 323–339 (2019)
23. Wiercigroch, M., Krivtsov, A.M.: Frictional chatter in orthogonal metal cutting. Philos. Trans. Roy. Soc. Lond. Ser. A Math. Phys. Eng. Sci. **359**(1781), 713–738 (2001)
24. Wiercigroch, M., Vaziri, V., Kapitaniak, M.: Red: revolutionary drilling technology for hard rock formations. In: SPE/IADC Drilling Conference and Exhibition. OnePetro (2017)
25. Wiercigrokh, M.: Resonance enhanced drilling: method and apparatus. world organization patent no. Technical report, WO/2007/141550, filed June 06, 2007, and published 13 December 2007
26. Youcef-Toumi, K., Wu, S.T.: Input/output linearization using time delay control (1992)
27. Zhang, Z., Chávez, J.P., Sieber, J., Liu, Y.: Controlling grazing-induced multistability in a piecewise-smooth impacting system via the time-delayed feedback control. Nonlinear Dyn. 1–16 (2021)

28. Zhao, X., Dankowicz, H., Reddy, C., Nayfeh, A.: Dynamic simulation of an electrostatically actuated impact microactuator. In: 2004 NSTI Nanotechnology Conference and Trade Show-NSTI Nanotech 2004, pp. 247–250. Citeseer (2004)
29. Zhao, X., Dankowicz, H., Reddy, C.K., Nayfeh, A.H.: Modeling and simulation methodology for impact microactuators. J. Micromech. Microeng. **14**(6), 775 (2004)

Deployment Feasibility Studies of Variable Buoyancy Anchors for Floating Wind Applications

Rodrigo Martinez[1], Sergi Arnau[1], Callum Scullion[2], Paddy Collins[2], Richard D. Neilson[1], and Marcin Kapitaniak[1](✉)

[1] The National Decommissioning Centre, School of Engineering, University of Aberdeen, Aberdeen, UK
marcin.kapitaniak@abdn.ac.uk

[2] Aubin Group, Castle Street, Castlepark Industrial Estate, AB41 9RF Ellon, UK

Abstract. To study the feasibility of deploying a novel type of anchor with variable buoyancy for mooring floating offshore wind turbines, a set of detailed modelling studies was performed in the state-of-the-art, Marine Simulator at the National Decommissioning Centre (NDC). The aim of the multi-physics simulations is to fully assess the proposed deployment method using a small tugboat fitted with a simple winch, thereby simplifying the process and reducing installation costs. The anchor has a 10 m square base, 4.5 m height and weight of 163 tonnes. The anchor is subjected to irregular waves with a JONSWAP spectrum with a significant wave height up to 5 m and peak period of 10 s. The analysis is divided in three sections: characterisation of the anchor buoyancy, positioning the anchor under the stern of the vessel and the controlled descent of the anchor to the seabed. An ideal winch speed of 0.35 m/s is identified, at which working load range on the winch cable decreases from 80 kN at the lowest winch speeds to about 30 kN. The sinking trajectory is similar at all winch speeds, however, the slower the descent, the further the anchor drifts. At this winch velocity, the descent from the resting position under the stern to the seabed takes roughly 5 min. In addition, the anchor's yaw range during the descent is below 10° at the optimal conditions.

Keywords: Offshore wind · Anchor dynamics · Virtual prototyping

1 Introduction

The economic drivers for lowering CapEx and OpEx of floating offshore wind technologies calls for innovation. In this study, the feasibility of deploying a novel type of anchor with variable buoyancy for mooring floating offshore wind turbines

Supported by EPSRC Supergen ORE Hub, ORE Catapult, FOW CoE, Aubin Group, Net Zero Technology Centre, the University of Aberdeen, Oceanetics Inc

Z. Dimitrovová et al. (Eds.): WMVC 2022, MMS 125, pp. 447–456, 2023.
https://doi.org/10.1007/978-3-031-15758-5_45

is presented. A set of detailed modelling studies are performed it the state-of-the-art, Marine Simulator at the National Decommissioning Centre (NDC). Using the multi-physics simulation allows for a more economical proof-of-concept approach, that will allow to fully assess the proposed deployment method and de-risk future offshore deployment. By using the proposed floating anchor, the use of heavy-lifting cranes and vessels could potentially be avoided, thereby reducing complexity and associated expenses. Instead, the anchor can be deployed from a smaller vessel, equipped with a simple winch. Once the anchor is towed to the deployment site, the anchor is pumped with liquid ballast and lowered with the winch. The proposed anchor (Fig. 1) has the shape of a truncated pyramid with a 10 m square base and is 4.46 m high (eyebolt inclusive). The empty anchor has a weight of 163 tonnes.

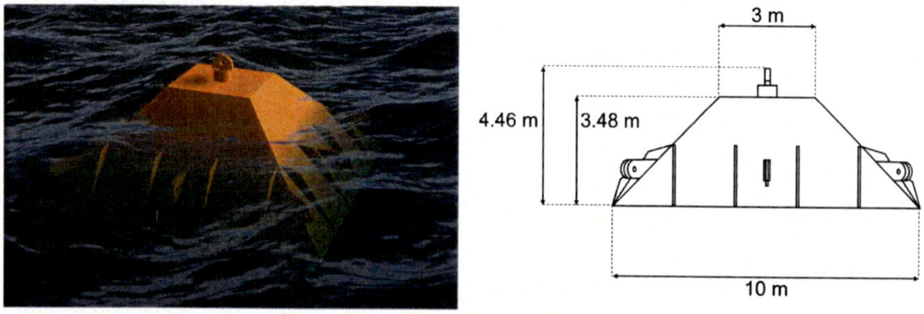

Fig. 1. Left: 3D representation of the anchor in the simulator's environment. Right: 2D diagram of the anchor.

2 Literature Review

This section provides an insight to the current state of the deployment of offshore wind turbines, their moorings and anchors as well as the typical weather conditions in which offshore deployments are performed.

2.1 Floating Offshore Wind Turbines

Offshore wind turbines can be divided in two categories: floating and fixed to the seabed. Much of the offshore wind energy resource worldwide is located over deep water and current fixed-bottom turbine technology may not be an economical solution for developing this deep water resource. Floating offshore wind turbines allow this resource to be harnessed [9]. Floating turbines are classified in four predominant types: semi-submersible, tension-leg-platform (TLP), spar and barge platforms. All of these require anchor(s) to be moored to the seabed. Regardless of the anchor type, deployments are usually proposed using established anchoring technologies/methodologies borrowed from the O&G industry [1]. At the time of writing, no literature was found on the development of novel anchoring technologies despite its potential for improvement [5,8,10].

2.2 Anchor Deployment

Little information is found related to the type of anchor employed for floating wind. However, knowledge from the Oil & Gas industry is adapted to meet the offshore wind requirements [3,4]. Figure 2 shows the most common anchors used for floating wind turbines [4]. In comparison with other types of anchors, gravity anchors (d) require medium to hard soils, their main loading direction is vertical but can perform at different angles. A drawback of gravity anchors is the weight for which they rely to work efficiently. This heavy weight increases installation costs and decreases the potential to recover the anchors upon decommissioning. Each type of anchor has its own deployment procedure. Drag-embedded (a) and gravity anchors (d) are simply lowered to the seabed. Whereas driven (b) and suction (c) piles need further interventions for their installation. The selection criteria for anchors is highly dependant on the seabed conditions of the deployment location. Hence, bathymetry surveys should be conducted as part of the planning process.

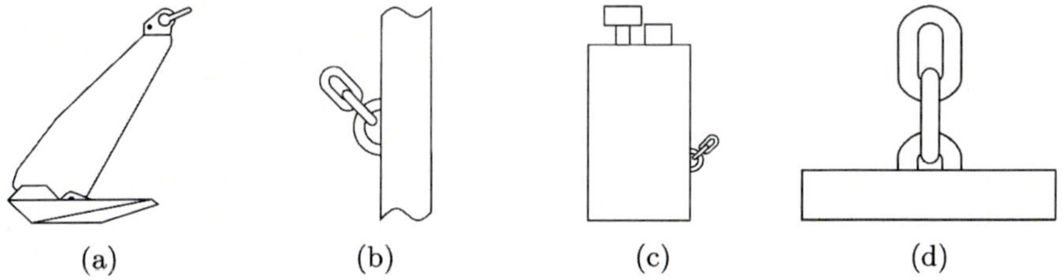

Fig. 2. Anchor types generally used in floating offshore turbines: a) drag-embedded, b) driven pile, c) suction pile, d) gravity anchor. Adapted from [4].

2.3 Flow Characteristics and Weather Window

Offshore deployment is limited by so-called weather windows. These weather windows are characterised by a series of environmental conditions that allow for the safe deployment of equipment [6,13]. The main characteristics associated to weather windows are the significant wave height (H_S) and flow velocity (U). Average flow velocity around the North Sea is usually below 1 m/s (∼2 kn) [2,12]; however, in certain areas characterised by channels, straits or some other land features, can reach up to 4 m/s (∼8 kn) [7,11].

3 Methodology

The proposed 3D anchor CAD model is imported into the OSC simulator software with the adequate collision model. The schematic of the anchor is depicted on Fig. 3 (left). The anchor has a base width of 10.00 m, body height of 3.48 m, overall height of 4.46 m and air weight of 163.20 t. The inertia properties of the steel anchor body are calculated based on the CAD drawing and are imported into the OSC simulator as part of the collision model generation process. Collision model generation is based on creating a mesh structure to represent the anchor and is generated using the software 3DS Max before importing it into the OSC simulator. Due to the simulator only taking into account the volume of the steel plates and internal bulkheads used to create the anchor shape, it is necessary to create a solid part that will represent the inner volume of the anchor (Fig. 3 (right)). This inner volume is filled/emptied to modify the anchor's buoyancy. The inner volume is fixed to the anchor within the simulator environment. The model of the anchor is assembled in the OSC simulator with the two components depicted in Fig. 3, where the inner volume representing the air/liquid is fixed rigidly inside the anchor. In order to ensure that the anchor floats, the inner volume is assigned a mass of 170 kg, which corresponds to the inner volume of 142.44 m^3 filled with air. The anchor is connected to the hose reel, using a standard 4′ hose and a 50 mm OD steel winch cable.

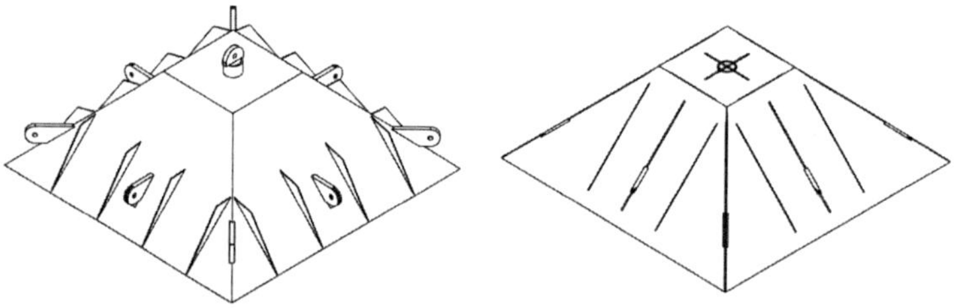

Fig. 3. 3D model of the anchor (left) and the internal representation (right).

The anchor deployment process is shown in Fig. 4. The developed scenario assumes that the floating anchor will be towed to the site and deployed from a support vessel equipped with a winch and a hose reel. As shown in Fig. 4, the floating anchor is connected to the winch using a cable and through a hose to the reel. Through the hose, the ballast fluid is pumped into the anchor. Once the anchor has negative buoyancy, the anchor starts sinking and positions itself under the stern of the vessel, eventually hanging from the winch cable (Fig. 4). At this point, the anchor starts its controlled descent to the seabed guided by the winch at the desired velocity (Fig. 4).

The simulation scenario assumes that the anchor is deployed in 100 m water depth with ocean conditions represented by irregular waves with a JONSWAP spectrum with significant wave height of 1 m, current of 0.1 kn and peak period of 10 s. Although the addition of wind is possible in the simulator, it was opted not to include this in the first round of tests.

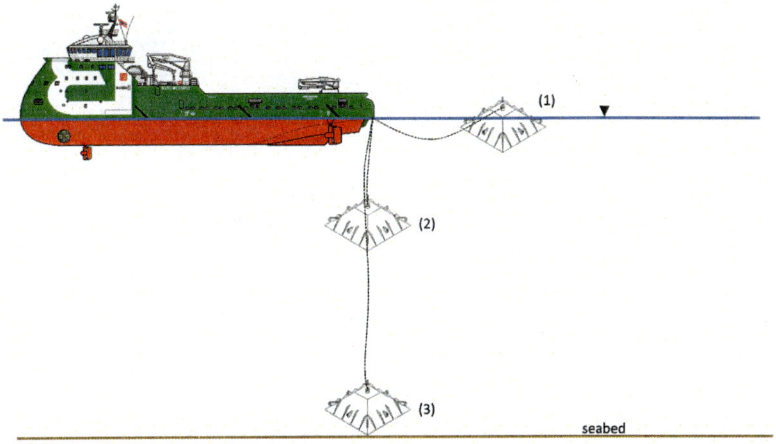

Fig. 4. Schematic of the deployment process of the anchor: (1) initial position of the anchor after towing. Pumping of ballast commences to generate negative buoyancy. (2) positioning of the anchor underneath the vessel's stern. (3) controlled descent of the anchor by means of a simple winch.

4 Results

The analysis of the deployment process is divided in three sections: 1.- characterisation of the anchor's buoyancy, 2.- positioning the anchor under the stern of the vessel by pumping ballast into the anchor to create negative buoyancy and, 3.- the controlled/guided descent of the anchor to the seabed by a winch.

4.1 Buoyancy

To determine the buoyancy limit of the anchor when filled with air, the mass of the anchor is increased until the anchor is fully submerged. The anchor buoyancy characterisation is shown in Fig. 5, where each curve corresponds to a specific pump rate. The vertical dashed line corresponds to the actual mass of the anchor (163.2 t), while the two horizontal dashed lines denote the anchor body height (3.48 m) and overall anchor height (4.46 m), respectively. As shown, the buoyancy characteristics of the anchor change when the draft reaches the lower dashed line (167.3 t), which means that the anchor body is fully submerged and only the lifting hook remains above the water level. After this point there is a sharp change in the buoyancy characteristics and the precise buoyancy limit can be determined from the crossing point between the curve and the top horizontal line (lifting hook at the water level). When the anchor mass reaches between 167 and 168 t, depending on the pump rate, the anchor becomes neutrally buoyant. This means that anchor possesses a gross buoyancy of about 4 t. The difference between the different pump rates is attributed to the inertia created by the speed at which the ballast is being pumped.

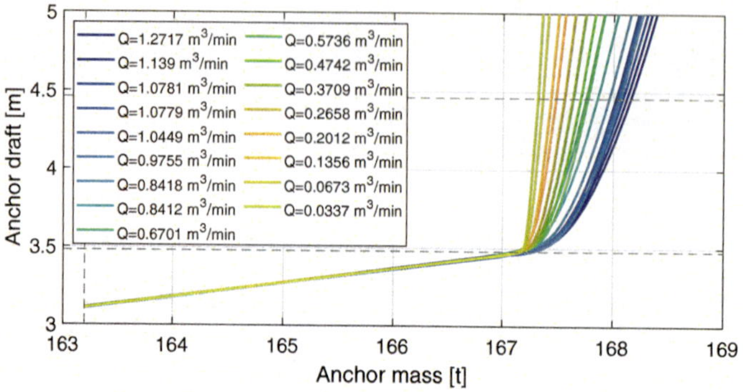

Fig. 5. Anchor buoyancy tests as a function of pump flow rate Q. Vertical line represents the weight of the anchor in air. The bottom horizontal line represent the height of the anchor without eye-bolt, the top line takes into consideration the height of the eye-bolt.

4.2 Positioning

The tests positioning the vessel underneath the vessel stern are analysed as a function of the pump rate Q (m^3/min). The variation of the winch force (F_W), anchor vertical position (Z_A) with the pump rate (Q) are shown in Fig. 6 (left). Vertical dashed lines indicate the time at which F_W and Z_A stabilise. At faster pump rates ($Q \approx 1\,m^3/min$), the anchor can be positioned underneath the stern of the vessel in under 5 min. In contrast, at slower pump rates ($Q \approx 0.05\,m^3/min$), the anchor takes up to 35 min to position under the vessel. Due to the length of the cable, the forces acting on the winch cable (F_W) remain constant until the anchor is roughly 18 m under the water surface. The anchor 3D trajectory from the surface to underneath the vessel is shown in Fig. 6 (right). At the lowest pump rate, the trajectory presents oscillations compared to the other pump rates. This is thought to be associated to the time it takes for the anchor to reach the bottom of the vessel, making it more susceptible to wave-induced effect.

Comparing the settling times of the winch force and anchor sinking velocity is shown in Fig. 7 for all pump rates Q. At $Q = 0.5\,m^3/min$, the anchor reaches its position under the stern around 10.5 min before the winch force stabilises. This means that once the anchor is in position, the ballast gets pumped for a further 10 min. In contrast, at higher Q, the pump stops a couple of minutes before the anchor has reached its position. At $Q = 0.4\,m^3/min$, both F_W and Z_A settle in 4.8 min. From this point, as pump rate increases, settling times decrease at a lower rate. At $Q = 1\,m^3/min$ F_W settles in 1.9 min and Z_A settles in 2.9 min.

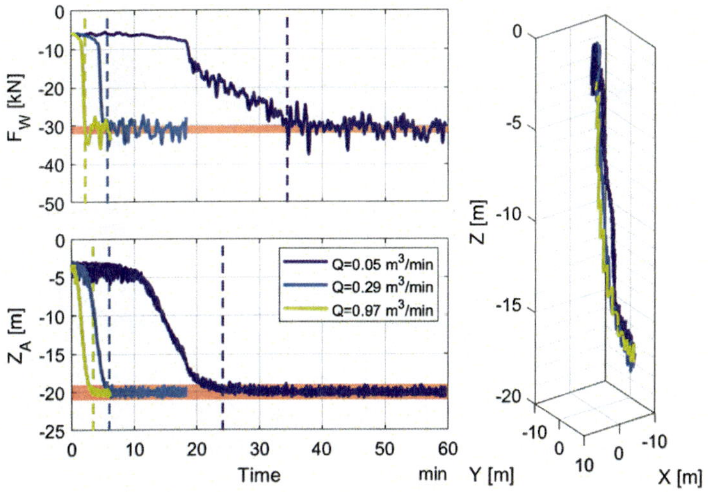

Fig. 6. Winch force (top) and anchor vertical position (bottom) for Q = 0.05, 0.29 and $0.97\,\mathrm{m}^3/\mathrm{min}$. Vertical dashed lines represent the time at which the signals stabilise.

4.3 Descent

In contrast to the positioning tests, the anchor descent tests are analysed as a function of the winch velocity V_W (m/s). In the preliminary results, a range of winch velocities, associated forces acting on the winch cable, along with the three-dimensional anchor descent trajectories and orientation are carefully analysed. Results from Fig. 8 (left) indicate that, in the presence of passive heave compensation, the working load amplitude on the winch cable decreases from 80 kN at the lowest winch velocities to about 30 kN for winch velocities above 0.35 m/s. The spread visible in the force data is associated to the wave-induced heave oscillations of the anchor and the vessel. Similar behaviour can be seen in Fig. 8 (right), where the deployment time stabilises above 0.35 m/s.

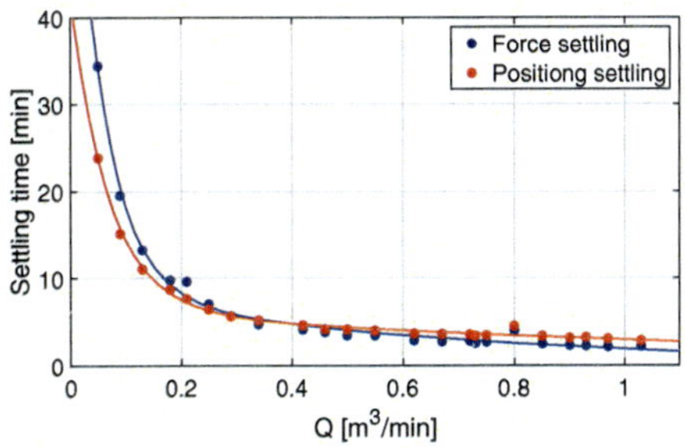

Fig. 7. Variation of winch force (top) and anchor positioning (bottom) settling time with pump rate Q. Exponential curves fitted.

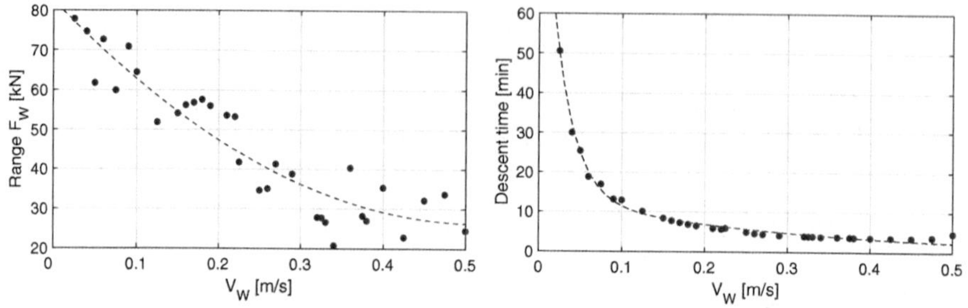

Fig. 8. Winch force range (left) and descent time (right) as a function of winch velocity.

The sinking trajectory is shown in Fig. 9 (centre). The trajectory is similar at all winch velocities. However, the slower the descent, the further the anchor drifts from its initial position (up to 10 m in the Y direction). As seen in Fig. 9 (bottom-left), at winch velocities above 3.5 m/s, the descent from the resting position under the vessel stern to the seabed takes under 5 min and the oscillations on the loading signal associated to the wave movement of the vessel is no longer an issue. The orientation range of the anchor is shown in Fig. 9 (right). The yaw range is the difference between the yaw angle at the time the anchor reaches the seabed and the angle the winch is released. For pitch and roll, which are only influenced by the waves, the range is the difference between the maximum and minimum values in the time series. During the anchor's descent, roll is kept almost constant at all winch velocities, with a standard deviation of 0.6°. Pitch has more variation as winch velocity increases, with a standard deviation of 1.7°. However, in the considered range of winch velocities, the anchor's rotation about its vertical axis (yaw) presents a standard deviation of 10°.

In Fig. 10 (left), the anchor descent velocity is plotted against winch velocities considered. It can be seen that for V_W values below 0.35 m/s the anchor descent is governed by the winch. However, at higher V_W values, the anchor reaches an equilibrium and it's descent velocity does not increase (free fall), regardless of the winch velocity.

Preliminary results using more ballast to sink the anchor are shown in Fig. 10 (centre/right). It appears that all weights take the same amount of time (4 min). However, this is explained looking at the winch force time series, where the tension in the cable increases with weight and the wave-induced oscillations are more visible. These oscillations have the same amplitude for all ballast weights. This means that for each anchor weight, there will be a case-specific winch speed at which descent speed and winch speed are the same (Fig. 10 (left)).

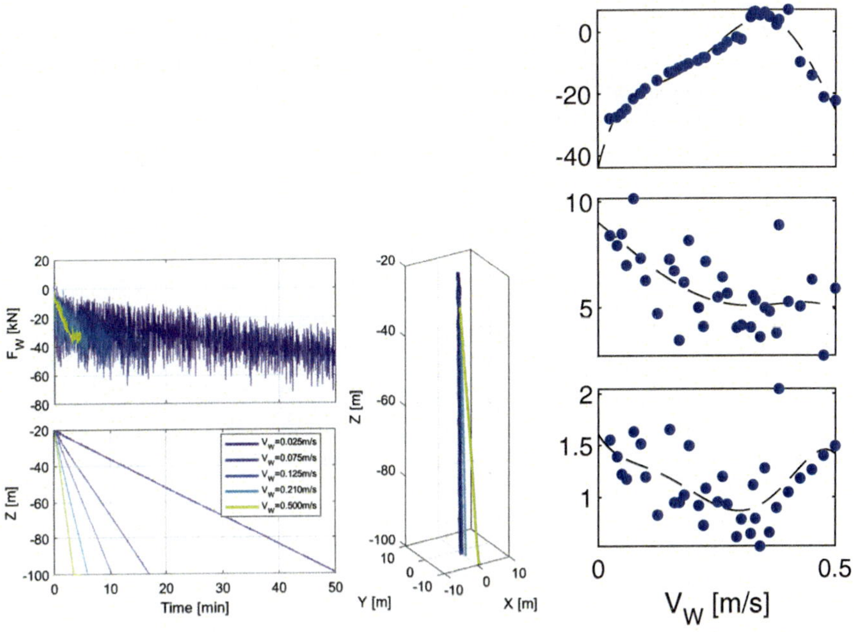

Fig. 9. Winch force range (top-left) and descent time (bottom-left) as a function of winch velocity and 3D descent trajectory (centre). Ranges of angular displacements of the anchor in yaw, pitch and roll directions (right)

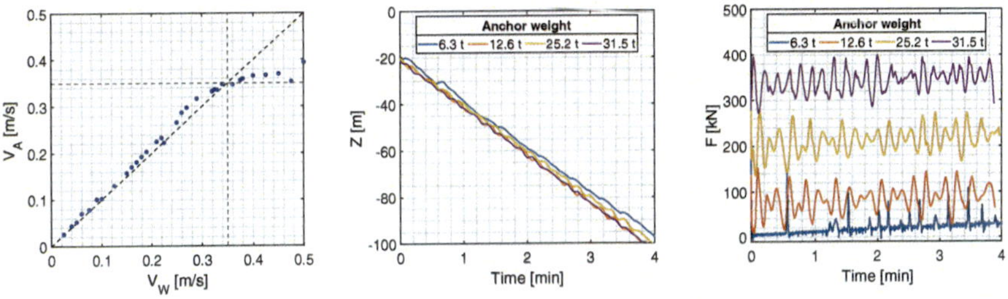

Fig. 10. Left: Variation of the anchor deployment speed (V_A) with the winch speed (V_W). Centre: Vertical position of anchors with different ballast weights. Right: Winch force (F_W) for different anchor ballast weights.

5 Conclusions

The deployment of a novel anchor design using a real-physics simulator is described in this work. The anchor's variable buoyancy means that it can float and be towed to the deployment site. This also allows for the deployment to be done by means of a simple winch.

Having established the anchor buoyancy limit (approximately 4 tonnes), we were able to study the effect of ballast pumping rate on the first deployment step (positioning under the stern). The rate at which the ballast is pumped into the anchor has a direct impact on the time it takes for the anchor to position itself under the stern of the vessel, at about 20 m depth. At $Q \approx 1\ m^3/min$ the anchor reaches the position in under 5 min, however, at the lowest Q values, it takes up to 25 min.

The analysis performed on the anchor dynamics during its descent, indicates that a winch velocity of 0.35 m/s is the best in this scenario as the anchor is allowed to descend almost at free-fall whilst still being controlled by the winch. At this winch velocity, the forces acting on the winch have a working range of 30 kN.

References

1. Bjerkseter, C., Agotnes, A.: Levelised costs of energy for offshore floating wind turbine concepts. Master's thesis, Norwegian University of Life Sciences, Department of Mathematical Sciences and Technology (2013)
2. Davies, A.M., Furnes, G.K.: Observed and computed m2 tidal currents in the north sea. J. Phys. Oceanogr. **10**, 237–257 (1980)
3. Ikhennicheu, M., et al.: D2.1 - review of the state of the art of mooring and anchoring designs, technical challenges and identification of relevant dlcs. Technical report, WindEurope & IREC (2020)
4. James, R., Ros, M.C.: Floating offshore wind: Market and technology review. Technical report, The Carbon Trust (2015)
5. James, R., et al.: Floating Wind Joint Industry Project - Phase I Summary Report. Technical report, The Carbon Trust (2018)
6. O'Connor, M., Lewis, T., Dalton, G.: Weather window analysis of Irish west coast wave data with relevance to operations & maintenance of marine renewables. Renewable Energy **52**, 57–66 (2013)
7. Sellar, B., Wakelam, G.: Characterisation of tidal flows at the European Marine Energy Centre in the absence of ocean waves. Energies **11**(1), 176 (2018)
8. Spearman, D.K., et al.: Floating Wind Joint Industry Project - Phase II Summary Report. Technical report, The Carbon Trust (2020)
9. Stewart, G., Muskulus, M.: A review and comparison of floating offshore wind turbine model experiments. Energy Procedia **94**, 227–231 (2016). 13th Deep Sea Offshore Wind R&D Conference, EERA DeepWind 2016
10. Strivens, S., Northridge, E., Evans, H., Harvey, M., Camp, T., Terry, N.: Floating Wind Joint Industry Project - Phase III Summary Report. Technical report, The Carbon Trust (2021)
11. Sutherland, D.R.J., Sellar, B.G., Harding, S., Bryden, I.: Initial flow characterisation utilising Turbine and seabed installed acoustic sensor arrays. In: Proceedings of the 10th European Wave and Tidal Energy Conference, pp. 1–8. Aalborg (2013)
12. Vindenes, H., Orvik, K.A., Søiland, H., Wehde, H.: Analysis of tidal currents in the north sea from shipboard acoustic doppler current profiler data. Cont. Shelf Res. **162**, 1–12 (2018)
13. Walker, R.T., Nieuwkoop-Mccall, J.V., Johanning, L., Parkinson, R.J.: Calculating weather windows: application to transit, installation and the implications on deployment success. Ocean Eng. **68**, 88–101 (2013)

Dynamic Response Analysis of Combined Vibrations of Top Tensioned Marine Risers

Dan Wang[1](✉), Zhifeng Hao[1], Ekaterina Pavlovskaia[2], and Marian Wiercigroch[2]

[1] School of Mathematical Sciences, University of Jinan, Jinan, China
danwang2014518@hotmail.com

[2] Centre for Applied Dynamics Research, School of Engineering, University of Aberdeen, King's College, Aberdeen, Scotland, UK

Abstract. A model including the cross-flow and in-line vibration of a top tensioned riser (TTR) under excitation from vortices and time-varying tension is proposed, where the Van der Pol wake oscillator is used to simulate the load caused by the vortex shedding. The governing partial differential equations describing the fluid-structure interactions are formulated and multi-mode approximations are obtained using the Galerkin projection method. The first-order approximation is analysed in this work. Dynamic responses of the in-line and cross-flow displacements are obtained by the numerical simulation for different parameter values. Results show that the large-amplitude vibration of the structure can be induced by the combined resonance. The results can be helpful to enhance design process of top tension risers.

Keywords: Dynamic response · Combined resonance · Top tensioned riser · Parametric excitation · Vortex-induced vibration

1 Introduction

The long slender structures such as risers are very important in oil and gas transportation, of which the working conditions (vortex, current, etc.) have significant risk on the lift time and safety of the structures. Vortex-induced vibrations (VIVs) and parametric resonances are two main phenomena occurred in practice, which can induce large-amplitude vibration of risers.

A number of comprehensive reviews [1–3] discussing VIV were focused on the single or multi-mode response, response at different Reynolds numbers, responses between multiple marine risers, impacts of the floating platforms and internal flows, etc. Previous work also highlighted that the parametric excitation of marine structures is encountered frequently, which can induce large-amplitude motions of structures as well.

With the development of technology, the impact of combined excitation from the VIV and time-varying tension has attracted significant attention in recent years. The numerical and experimental techniques are the main methods to study the response dynamics. While the semi-empirical wake oscillator models [4] are commonly used to describe the fluid-structure interactions with less cost.

Z. Dimitrovová et al. (Eds.): WMVC 2022, MMS 125, pp. 457–465, 2023.
https://doi.org/10.1007/978-3-031-15758-5_46

In order to understand the effects of combined excitation from the vortex and varying-tension, the coupled model of the riser as well as the fluid is constructed based on the two dimensional flows in this paper. Some numerical results are given and discussion.

2 Modelling and Methodology

2.1 Model of the Structure Motion

Working conditions for a TTR are complex which include excitation from currents and waves. As shown in Fig. 1, the TTR is connected to a floating platform and a wellhead, and it can move in the in-line (IL) and cross-flow (CF) directions. Herein, the TTR is modelled as a uniform Euler–Bernoulli beam simply supported at both ends as presented in Fig. 1.

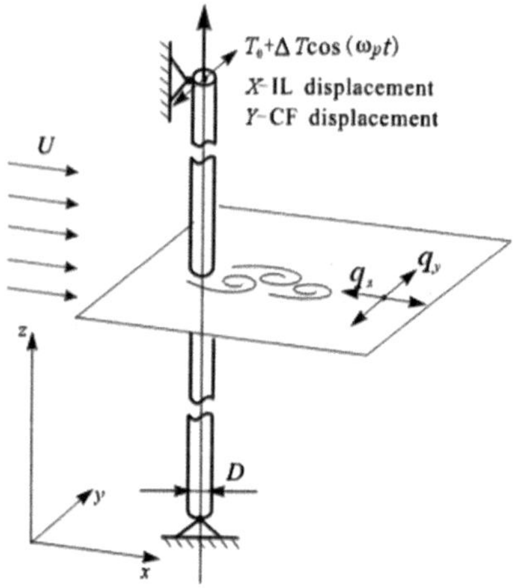

Fig. 1. Physical model of the TTR under combined excitation

Assuming $\vec{r}(z,t) = X(z,t)\vec{i} + Y(z,t)\vec{j}$ denote the displacement vector of the structure, the equation of motion can be written according to the Refs. [5, 6] as

$$m_* \frac{\partial^2 \vec{r}(z,t)}{\partial t^2} + \frac{\partial^2}{\partial z^2}\left(EI\frac{\partial^2 \vec{r}(z,t)}{\partial z^2}\right) - \frac{\partial}{\partial z}\left(T(z,t)\frac{\partial \vec{r}(z,t)}{\partial z}\right) = \vec{F}_F(z,t) \quad (1)$$

where $m_* = (\mu + C_a)\pi\rho_f D^2/4$ is the total mass per unit length including structural mass and added mass of fluid, μ is the mass ratio, C_a is the coefficient of fluid added mass, ρ_f is the density of the displaced fluid, D is the outer diameter of the riser, EI is the bending stiffness. $\vec{F}_F$ is the total fluid force acting on the structure (including the lift and drag force). $T(z,t)$ is the varying axial tension which is calculated as $T(z,t) = T_0 - (L-z)W_w + \Delta T\cos\omega_p t = T_b + W_w z + \Delta T\cos\omega_p t$, and T_0, T_b are the top and bottom tension of the riser.

2.2 The van der Pol Oscillator Model for the Fluid Force

In order to simulate the coupling between the structure and time-varying fluid force, the Van der Pol oscillators are introduced to represent the time-varying characteristics for the drag and lift force. The acceleration coupling is applied for the fluid oscillator.

Let $q_x = \frac{2C_D}{C_{D0}}$, $q_y = \frac{2C_L}{C_{L0}}$ (herein C_D and C_L denote the time-varying variables for the fluid damping and lift coefficients, C_{D0} and C_{L0} are the reference fluid damping and lift coefficients), and the wake oscillators can be written as

$$\frac{\partial^2 q_x}{\partial t^2} + 2\varepsilon_x \Omega_f (q_x^2 - 1)\frac{\partial q_x}{\partial t} + 4\Omega_f^2 q_x = A_x \frac{\partial^2 X}{\partial t^2}$$
$$\frac{\partial^2 q_y}{\partial t^2} + \varepsilon_y \Omega_f (q_y^2 - 1)\frac{\partial q_y}{\partial t} + \Omega_f^2 q_y = A_y \frac{\partial^2 Y}{\partial t^2}$$

2.3 Coupled Model of the TTR

According to Refs. [5, 6], the total fluid force $\vec{F}_F$ induced by the lift force and fluid damping force projects on the axes x and y are derived as

$$(\vec{F}_F)_X = \frac{\rho_f D}{4}\sqrt{\left(U - \frac{\partial X}{\partial t}\right)^2 + \left(\frac{\partial Y}{\partial t}\right)^2}\left(2C_{D0}U - 2C_{D0}\frac{\partial X}{\partial t} + q_x C_{D0}^{fl} U - q_x C_{D0}^{fl}\frac{\partial X}{\partial t} + q_y C_{L0}\frac{\partial Y}{\partial t}\right)$$
$$(\vec{F}_F)_Y = \frac{\rho_f D}{4}\sqrt{\left(U - \frac{\partial X}{\partial t}\right)^2 + \left(\frac{\partial Y}{\partial t}\right)^2}\left(-2C_{D0}\frac{\partial Y}{\partial t} - q_x C_{D0}^{fl}\frac{\partial Y}{\partial t} + q_y C_{L0}U - q_y C_{L0}\frac{\partial X}{\partial t}\right)$$

Introducing the expressions of the fluid forces into Eq. (1), the coupled equations of motion the TTR can be rewritten as

$$m_*\frac{\partial^2 X}{\partial t^2} + EI\frac{\partial^2 X}{\partial z^4} - W_w\frac{\partial X}{\partial z} - \left(T_b + W_w z + \Delta T\cos(\omega_p t)\right)\frac{\partial^2 X}{\partial z^2} =$$
$$\frac{\rho_f D}{4}\left(\begin{array}{l} 2C_{D0}U^2 - 4C_{D0}U\dfrac{\partial X}{\partial t} + C_{D0}^{fl}U^2 q_x - 2C_{D0}^{fl}Uq_x\dfrac{\partial X}{\partial t} + C_{L0}Uq_y\dfrac{\partial Y}{\partial t} + 2C_{D0}(\dfrac{\partial X}{\partial t})^2 \\ +C_{D0}^{fl}q_x(\dfrac{\partial X}{\partial t})^2 - C_{L0}q_y\left(\dfrac{\partial X}{\partial t}\right)\left(\dfrac{\partial Y}{\partial t}\right) + C_{D0}\left(\dfrac{\partial Y}{\partial t}\right)^2 \end{array}\right) \tag{2}$$

$$m_*\frac{\partial^2 Y}{\partial t^2} + EI\frac{\partial^2 Y}{\partial z^4} - W_w\frac{\partial Y}{\partial z} - \left(T_b + W_w z + \Delta T\cos(\omega_p t)\right)\frac{\partial^2 Y}{\partial z^2} =$$
$$\frac{\rho_f D}{4}\left(-2C_{D0}U\frac{\partial Y}{\partial t} - C_{D0}^{fl}Uq_x\frac{\partial Y}{\partial t} + C_{L0}U^2 q_y - 2C_{L0}Uq_y\left(\frac{\partial X}{\partial t}\right) + 2C_{D0}\left(\frac{\partial X}{\partial t}\right)\left(\frac{\partial Y}{\partial t}\right)\right) \tag{3}$$

$$\frac{\partial^2 q_x}{\partial t^2} + 2\varepsilon_x \Omega_f (q_x^2 - 1)\frac{\partial q_x}{\partial t} + 4\Omega_f^2 q_x = A_x \frac{\partial^2 X}{\partial t^2} \tag{4}$$

$$\frac{\partial^2 q_y}{\partial t^2} + \varepsilon_y \Omega_f (q_y^2 - 1)\frac{\partial q_y}{\partial t} + \Omega_f^2 q_y = A_y \frac{\partial^2 Y}{\partial t^2} \tag{5}$$

The nondimensional system can be obtained after introducing the nondimensional variables and parameters as

$$\frac{\partial^2 \bar{x}}{\partial \tau^2} + \frac{EI}{m_*\omega_0^2 L^4}\frac{\partial^4 \bar{x}}{\partial \zeta^4} - \frac{T_b(1+\beta\cos\bar{\omega}\tau)}{m_*\omega_0^2 L^2}\frac{\partial^2 \bar{x}}{\partial \zeta^2} - \bar{r}\frac{\partial \bar{x}}{\partial \tau} - \bar{r}\zeta\frac{\partial^2 \bar{x}}{\partial \zeta^2} =$$
$$\frac{a\Omega_R^2}{2\pi St} - 2a\Omega_R\frac{\partial \bar{x}}{\partial \tau} + \frac{b\Omega_R^2}{4\pi St}q_x - b\Omega_R q_x\frac{\partial \bar{x}}{\partial \tau} + \frac{c\Omega_R}{2}q_y\frac{\partial \bar{y}}{\partial \tau} + 2\pi a St\left(\frac{\partial \bar{x}}{\partial \tau}\right)^2 + \pi a St\left(\frac{\partial \bar{y}}{\partial \tau}\right)^2 \tag{6}$$

$$\frac{\partial^2\bar{y}}{\partial\tau^2}+\frac{EI}{m_*\omega_0^2L^4}\frac{\partial^4\bar{y}}{\partial\zeta^4}-\frac{T_b(1+\beta\cos\varpi\tau)}{m_*\omega_0^2L^2}\frac{\partial^2\bar{y}}{\partial\zeta^2}-\bar{r}\frac{\partial\bar{y}}{\partial\tau}-\bar{r}\zeta\frac{\partial^2\bar{y}}{\partial\zeta^2}=$$
$$-a\Omega_R\frac{\partial\bar{y}}{\partial\tau}-\frac{b\Omega_R}{2}q_x\frac{\partial\bar{y}}{\partial\tau}+\frac{c\Omega_R^2}{4\pi St}q_y-c\Omega_Rq_y\frac{\partial\bar{x}}{\partial\tau}+2\pi aSt\left(\frac{\partial\bar{x}}{\partial\tau}\right)\left(\frac{\partial\bar{y}}{\partial\tau}\right) \tag{7}$$

$$\frac{\partial^2 q_x}{\partial\tau^2}+2\varepsilon_x\Omega_Rq_x^2\frac{\partial q_x}{\partial\tau}-2\varepsilon_x\Omega_R\frac{\partial q_x}{\partial\tau}+4\Omega_R^2q_x=A_x\frac{\partial^2\bar{x}}{\partial\tau^2} \tag{8}$$

$$\frac{\partial^2 q_y}{\partial\tau^2}+\varepsilon_y\Omega_Rq_y^2\frac{\partial q_y}{\partial\tau}-\varepsilon_y\Omega_R\frac{\partial q_y}{\partial\tau}+\Omega_R^2q_y=A_y\frac{\partial^2\bar{y}}{\partial\tau^2} \tag{9}$$

where

$$\bar{x}=X/D,\ \bar{y}=Y/D,\ \tau=\omega_0t,\ \zeta=z/L,\ \Omega_f=2\pi StU/D,\ \Omega_R=\Omega_f/\omega_0,$$
$$\bar{\omega}=\omega_p/\omega_0,\ \beta=\Delta T/T_b,$$
$$\omega_1=\sqrt{\frac{EI\pi^4}{m_*L^4}+\frac{T_b\pi^2}{m_*L^2}},\ \omega_R=\omega_1/\omega_0$$
$$\bar{r}=W_w/(m_*L\omega_0^2),\ a=C_{D0}\rho_fD^2/(4\pi m_*St),\ b=(C_{D0}^{fl}\rho_fD^2)/(4\pi m_*St),$$
$$c=(C_{L0}\rho_fD^2)/(4\pi m_*St).$$

$\bar{x}$, $\bar{y}$ are the dimensionless displacements of in-line and cross-flow motion, τ is the dimensionless time, Ω_R is the dimensionless shedding frequency of vortices, ω_0 is the reference frequency, ϖ is the dimensionless frequency of the parametric excitation, β is the amplitude ratio of the time varying tension, $\bar{r}$, a, b, c are the dimensionless parameters. ω_1, ω_R denote the dimensional and nondimensional frequency of the first-order mode.

2.4 The Galerkin Approximation

Following earlier studies [5–7], the approximate solutions of Eqs. (6)–(9) are obtained by employing the Galerkin approach. Here, the displacements $\bar{x}$, $\bar{y}$ and drag and lift coefficients q_x, q_y are assumed as

$$\bar{x}=\sum_{n=1}^{\infty}\bar{x}_n(\tau)\bar{x}_n(\zeta),\qquad \bar{y}=\sum_{n=1}^{\infty}\bar{y}_n(\tau)\bar{y}_n(\zeta) \tag{10}$$

$$q_x=\sum_{n=1}^{\infty}q_{xn}(\tau)\bar{q}_{xn}(\zeta),\qquad q_y=\sum_{n=1}^{\infty}q_{yn}(\tau)\bar{q}_{yn}(\zeta) \tag{11}$$

And the mode functions are assumed as

$$\bar{x}_n(\zeta)=\bar{y}_n(\zeta)=\bar{q}_{xn}(\zeta)=\bar{q}_{yn}(\zeta)=\sin(n\pi\zeta)$$

By substituting Eqs. (10)–(11) into Eqs. (6)–(9), multiplying the equations by $\sin(n\pi\zeta)$, integrating it along the length of the riser, and applying the orthogonality of the mode functions, we can obtain the first-order approximation of the coupled system as

$$\begin{aligned}&\ddot{\bar{x}} + 2a\Omega_R\dot{\bar{x}} + \omega_R^2\bar{x} + \eta\beta\cos(\varpi\tau)\bar{x} = -\frac{\bar{r}\pi^2}{2}\bar{x} + \frac{2a\Omega_R^2}{\pi^2 St} + \frac{b\Omega_R^2}{4\pi St}q_x\\&-\frac{8b\Omega_R}{3\pi}\dot{\bar{x}}q_x + \frac{4c\Omega_R}{3\pi}\dot{\bar{y}}q_y + \frac{16aSt}{3}\left(\dot{\bar{x}}\right)^2 + \frac{8aSt}{3}\left(\dot{\bar{y}}\right)^2,\end{aligned} \tag{12}$$

$$\begin{aligned}&\ddot{\bar{y}} + a\Omega_R\dot{\bar{y}} + \omega_R^2\bar{y} + \eta\beta\cos(\varpi\tau)\bar{y} = -\frac{\bar{r}\pi^2}{2}\bar{y}\\&-\frac{4b\Omega_R}{3\pi}\dot{\bar{y}}q_x + \frac{c\Omega_R^2}{4\pi St}q_y - \frac{8c\Omega_R}{3\pi}\dot{\bar{x}}q_y + \frac{16aSt}{3}\dot{\bar{x}}\dot{\bar{y}}\end{aligned} \tag{13}$$

$$\ddot{q}_x + 2\varepsilon_x\Omega_R(\frac{3q_x^2}{4} - 1)\dot{q}_x + 4\Omega_R^2 q_x = A_x\ddot{\bar{x}} \tag{14}$$

$$\ddot{q}_y + \varepsilon_y\Omega_R(\frac{3q_y^2}{4} - 1)\dot{q}_y + \Omega_R^2 q_y = A_y\ddot{\bar{y}} \tag{15}$$

Herein the subscript of 1 which denotes the first-order approximation is omitted in Eqs. (12)–(15) for the simplification and $\eta = T_b\pi^2/(m_* L^2\omega_0^2)$.

3 Numerical Analysis

In this part, four sets of the dynamic responses of the first-order approximation (12)–(15) are calculated under the combined resonance condition for different amplitude ratio β. Moreover, due to excitation of the constant force, the response for the motion of in-line direction is asymmetrical. Nondimensional frequencies of Figs. 2, 3 and 4 are set as $\varpi = 2$, $\omega_R = 1$, which indicated the 1:2 parametric resonance. The other parameters are set as $St = 0.2$, $\bar{r} = 0.2$, $A_1 = A_2 = 12$, $\lambda_1 = 0.3$, $\lambda_2 = 0.2$, $\eta = 0.3$, $a = 0.4$, $b = 0.04$, $c = 0.04$, respectively. One set of initial values are selected as [0.1, 0, 3.8, 0, 1, 0, 5, 0]. The variables for the x-axis and y-axis are the in-line and cross-flow displacements respectively.

Figure 2 shows the varying trends of the displacements of the first-order approximation of the in-line and cross-flow vibration with respective different shedding frequency Ω_R for the amplitude ratio $\beta = 0$. As can be seen from Fig. 2 that the topological types of the responses exhibit different characteristics as the shedding frequency Ω_R increasing from 0.1 to 2.0. Specifically, for $\Omega_R = 0.1$ (which indicates the oncoming uniform flow is small), the displacements of the vibrations in the two directions are very small. While when Ω_R increases to 0.4, the values of amplitudes of the responses are increased by an order of magnitude. The similar characteristic occur for $\Omega_R = 0.4$ and 0.6. The more complicated responses for the motions in the two directions occur during the ‘lock-in’ frequency range, e.g. $\Omega_R = 0.9$ and 1.4. Continuing to increase Ω_R to 2.0, motions of the two directions also display the large-amplitude vibration, especially for the in-line vibration.

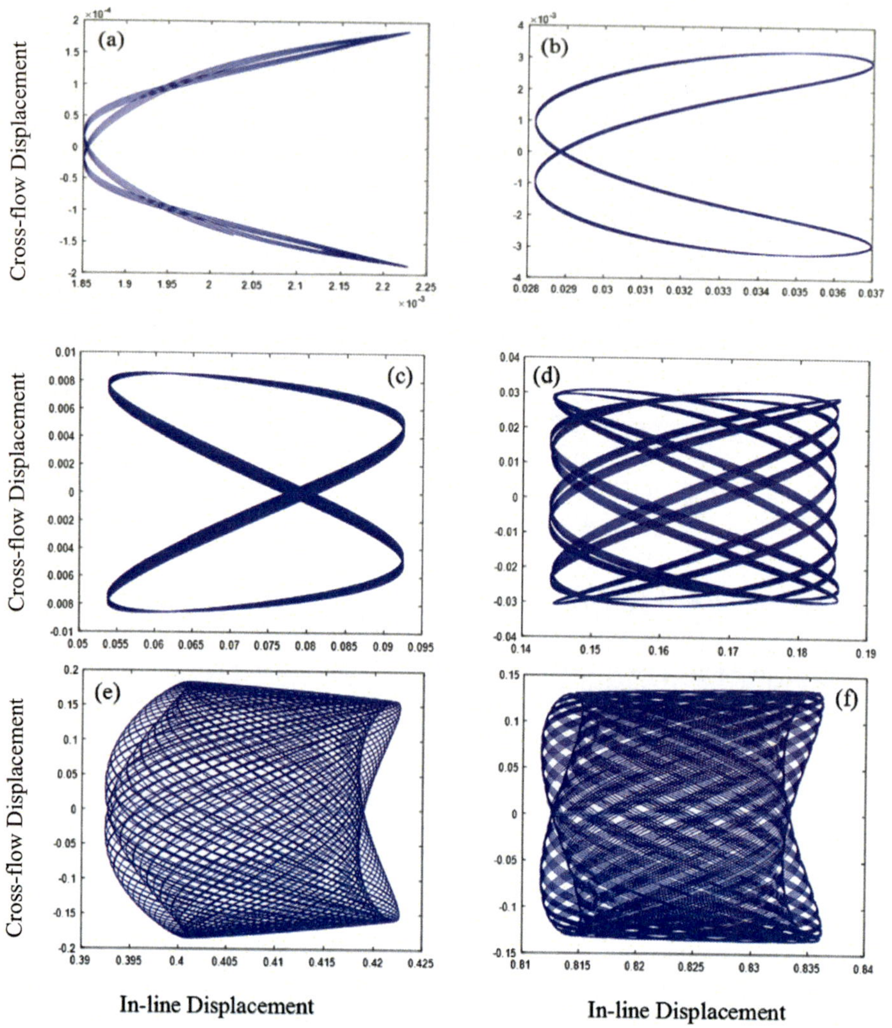

Fig. 2. The displacement orbits for the first-order approximation for $\beta = 0$: (a) $\Omega_R = 0.1$; (b) $\Omega_R = 0.4$; (c) $\Omega_R = 0.6$; (d) $\Omega_R = 0.9$; (e) $\Omega_R = 1.4$; (f) $\Omega_R = 2.0$

In contrast, the dynamic responses show different characteristics when considering the impact of the time-varying tension. As shown in Figs. 3 and 4. Two sets of displacements of the motions for the structure are calculated for $\beta = 0.2$ and 0.6, respectively as shown in Figs. 3 and 4. As shown in Fig. 3 that though the amplitude range for the two responses are similar to the case of $\beta = 0$, the responses exhibit more complicated features as the nondimensional shedding frequency increases for $\beta = 0.2$. Especially for $\Omega_R = 0.9$ and 1.4 that during the 'lock-in' frequency, the relations between the displacements of motions for the in-line and cross-flow directions show more complicated state.

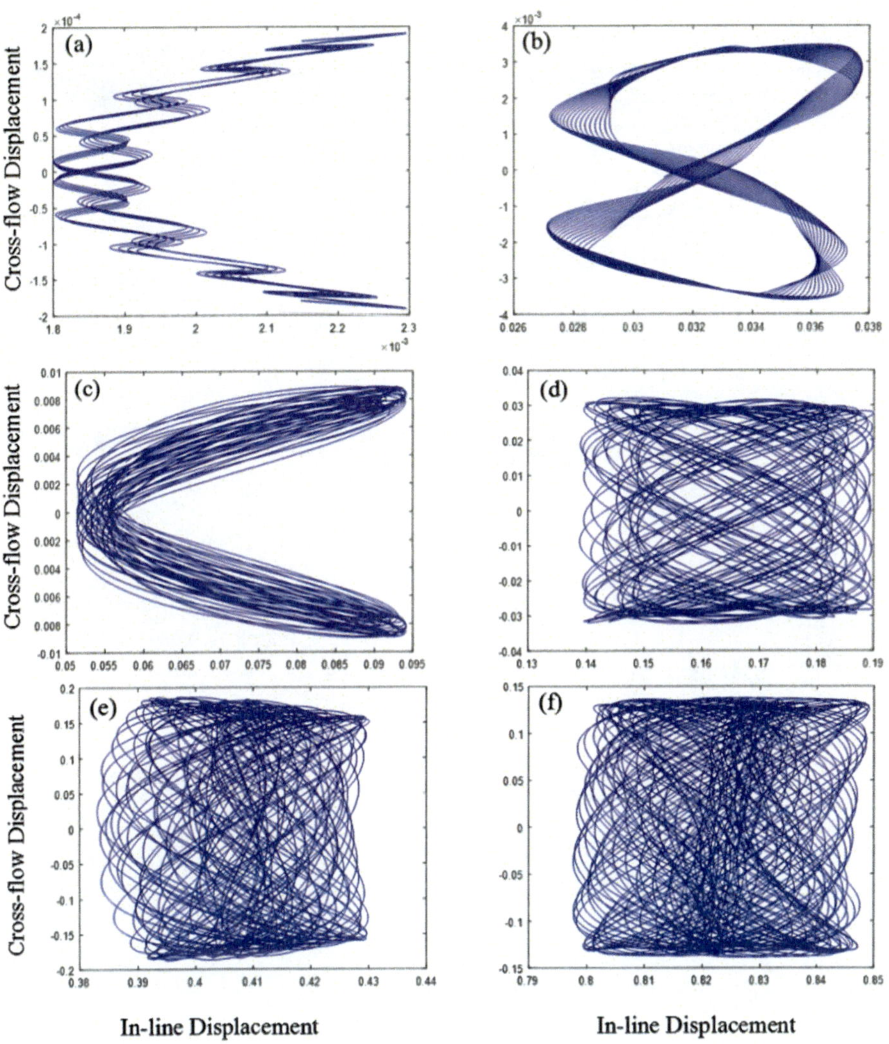

Fig. 3. The displacement orbits for the first-order approximation for $\beta = 0.2$: (a) $\Omega_R = 0.1$; (b) $\Omega_R = 0.4$; (c) $\Omega_R = 0.6$; (d) $\Omega_R = 0.9$; (e) $\Omega_R = 1.4$; (f) $\Omega_R = 2.0$

Figure 4 indicates the responses are more complicated for $\beta = 0.6$ and the coupling between the in-line and cross-flow vibration is more complicated. Moreover, the results also show that the varying tension has effect on the in-line and cross-flow vibrations. Comparing to the case for $\beta = 0.2$, the amplitude range for the responses of the in-line and cross-flow vibration are enlarged for $\beta = 0.6$. The combined excitation from the vortex and varying tension can induced large-amplitude vibrations of structure.

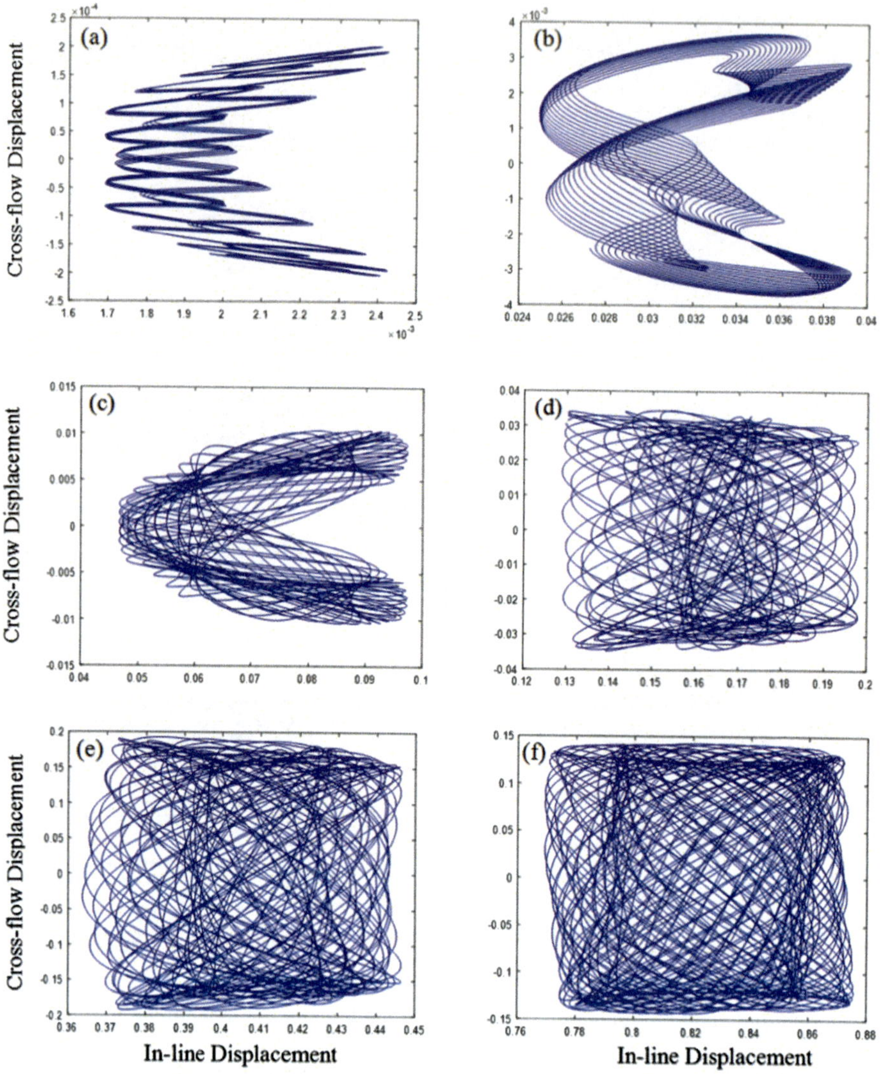

Fig. 4. The displacement orbits for the first-order approximation for $\beta = 0.6$: (a) $\Omega_R = 0.1$; (b) $\Omega_R = 0.4$; (c) $\Omega_R = 0.6$; (d) $\Omega_R = 0.9$; (e) $\Omega_R = 1.4$; (f) $\Omega_R = 2.0$

4 Conclusions

A coupled model of a TTR excited by the vortex and varying tension is constructed, where fluid forces are modelled using two Van der Pol oscillators. The first-order approximation motion equations are derived with the Galerkin projection method. Dynamic responses of motions for the in-line and cross-flow directions are studied by the numerical method. Three sets of parameters for the amplitude ratio β are chosen to investigated the varying trends of the displacements in two flow directions with respect to the shedding frequency.

As a comparison, $\beta = 0$ which indicates only the effect of vortex is studied. Results showed that for $\beta \neq 0$ (which indicates the combined excitation of the vortex and time-varying tension), the combined resonance can induced large-amplitude vibrations of the structure and the responses are more complicated. Further study about bifurcation and chaos of the system the will be followed.

References

1. Sarpkaya, T.: A critical review of the intrinsic nature of vortex-induced vibration. J. Fluids Struct. **19**, 389–447 (2004)
2. Williamson, C.H.K., Govardhan, R.: Abrief review of recent results in vortex-induced vibrations. J. Wind Eng. Ind. Aerodynam. **96**, 713–735 (2008)
3. Liu, G., Li, H., Qiu, Z., Leng, D., Li, Z., Li, W.: Review: a mini review of recent progress on vortex-induced vibrations of marine risers. Ocean Eng. **195**, 1–17 (2020)
4. Facchinetti, M.L., de Langrea, E., Biolley, F.: Coupling of structure and wake oscillators in vortex-induced vibration. J. Fluids Struct. **19**, 123–140 (2004)
5. Pavlovskaia, E., Keber, M., Postnikov, A., Reddington, K., Wiercigroch, M.: Multi-modes approach to modelling of vortex-induced vibration. Int. J. Non-Linear Mech. **80**, 40–51 (2016)
6. Kurushina, V., Pavlovskaia, E., Wiercigroch, M.: VIV of flexible structures in 2D uniform flow. Int. J. Eng. Sci. **150**, 1–24 (2020)
7. Wang, D., Hao, Z., Pavlovskaia, E., Wiercigroch, M.: Bifurcation analysis of vortex induced vibration of low-dimensional models of marine risers. Nonlinear Dyn. **106**, 147–167 (2021)

Dynamical Analysis of Pure Sliding and Stick-Slip Effect with a Random Field Friction Model

Han Hu[1], Anas Batou[2](✉), and Huajiang Ouyang[1]

[1] University of Liverpool, Liverpool L69 7ZF, UK
{han.hu,mehou}@liverpool.ac.uk
[2] Univ Gustave Eiffel, CNRS UMR 8208, Univ Paris Est Creteil, 77474 Marne-la-Vallée, France
anas.batou@u-pem.fr

Abstract. Friction-induced vibration problems possess inherent randomness due to factors such as the roughness of the contact interface, nondeterministic external load, and stochastic driven velocities. The spatial randomness in friction caused by rough contact is only addressed in a few papers in the existing literature and can significantly impact the dynamics in practical engineering. In this paper, the spatial randomness in planar friction is considered by modelling the coefficient of friction (COF) as a random field. The statistical properties of the frictional forces and torque resulting from uneven distributed COF are analysed for a disc in the pure sliding planar motion. It is found that the coefficient of variation of the frictional forces and torque are strongly dependent on the ratio of correlation length of the COF random field to the geometrical characterised length of the disc. A numerical case is presented to illustrate the impact of COF random field modelling on the stick-slip effects. New three-variable stick-slip transition criteria are proposed to accomplish the analysis.

Keywords: Dynamical analysis · Coefficient of friction · Random field · Stick-slip

1 Introduction

Friction-induced vibration of planar sliding motion is vital to be tackled properly for an accurate characterisation of the frictional behaviour in practical engineering problems. As a result, it has been extensively studied for centuries and various friction models have been proposed, among which the Coulomb's friction law is arguably the most widely used friction model. In most friction models, the coefficient of friction (COF) is modelled as a material-related constant. However, it has been found in recent decades that the COF possesses a significant level of randomness concerning the material heterogeneities and roughness of the contact interface [1,2]. Considerable studies have been conducted to introduce

Z. Dimitrovová et al. (Eds.): WMVC 2022, MMS 125, pp. 466–473, 2023.
https://doi.org/10.1007/978-3-031-15758-5_47

randomness into the friction model, most of which modelled friction as a random variable [6,7]. However, the COF can be spatially random when taken into account material heterogeneities and contact surface roughness [3,4]. In such case, it would be more appropriate to model the COF as a random field. The frictional response and dynamical behaviours of the contacting body are thus related to the properties of the COF random field. To the author's knowledge, theoretical analysis concerning random field friction modelling is rarely seen in the literature. In this work, the frictional response, e.g. the friction forces and frictional torque, is investigated for a pure sliding disc model (instead of a square plate model used in [3]) with a general planar motion, and the dynamical behaviours such as stick-slip effect are investigated through a spring-slider model (instead of a disc-on-belt model used in [4]).

2 Pure Sliding Model

In this section, a pure sliding model, depicted in Fig. 1 where a disc of a radius R with a flat bottom surface moves on an infinitely large rough plane, is investigated. The general motion of the disc can be decomposed as a translating velocity $\mathbf{v} = v\mathbf{e_x}$ aligned to the $\mathbf{e_x}$ direction and an angular velocity $\omega = \omega\mathbf{e_z}$ pointing out normal to the plane. The COF of the contact interface is modelled as a random field $\mathbb{H}(\mathbf{x};\theta), \mathbf{x} \in \Omega_D$ and θ is a latent random variable, attached to the plane Ω_D since only the plane has roughness. The COF random field is characterised by two factors, namely the probability distribution and the correlation structure. For the former, a lognormal distribution is used since it is inherently positive and has a bell shape as the normal distribution, and for the latter a correlation function of the Gaussian type is used, which can be expressed as

$$\mathrm{R}_{\mathbb{H}}(\mathbf{x},\mathbf{x}') = \mathrm{e}^{-\frac{\|\mathbf{x}-\mathbf{x}'\|^2}{l^2}}, \tag{1}$$

where l is the correlation length characterising the spectral property of the field.

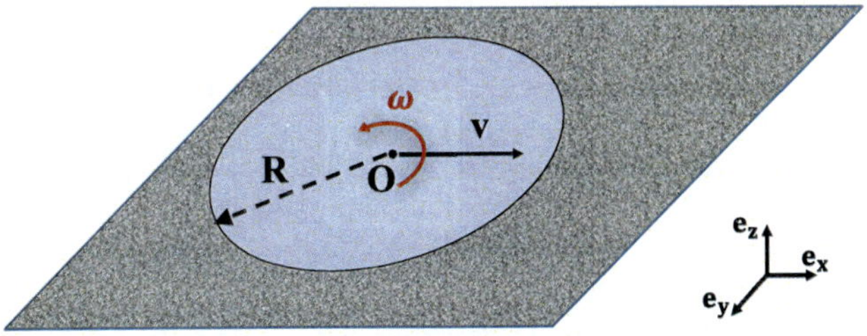

Fig. 1. The pure sliding model

The global friction force $\mathbf{F_O}$ and the frictional torque $\mathbf{T_O}$ at the geometry centre $\mathbf{O}$, which is set as the origin of the body-fixed local coordinate, can thus be formulated as

$$\mathbf{F_O} = -\iint_S P\mathbb{H}(\mathbf{x})\frac{\boldsymbol{\omega}\times\mathbf{x}+\mathbf{v}}{|\boldsymbol{\omega}\times\mathbf{x}+\mathbf{v}|}\mathrm{d}S$$
$$= \iint_S P\mathbb{H}(\mathbf{x})\frac{-(\rho^\star - y)}{\sqrt{(\rho^\star - y)^2 + x^2}}\mathrm{d}S\mathbf{e_x} + \iint_S P\mathbb{H}(\mathbf{x})\frac{-x}{\sqrt{(\rho^\star - y)^2 + x^2}}\mathrm{d}S\mathbf{e_y} \quad (2)$$

$$\mathbf{T_O} = -\iint_S P\mathbb{H}(\mathbf{x})\frac{\mathbf{x}\times(\boldsymbol{\omega}\times\mathbf{x}+\mathbf{v})}{|\boldsymbol{\omega}\times\mathbf{x}+\mathbf{v}|}\mathrm{d}S = \iint_S P\mathbb{H}(\mathbf{x})\frac{-(x^2 - \rho^\star y + y^2)}{\sqrt{(\rho^\star - y)^2 + x^2}}\mathrm{d}S\mathbf{e_z}, \quad (3)$$

where $P = mg/\pi R^2$ is a constant normal pressure applied to the disc, $\rho^* = v/\omega$, $\{\mathbf{e_x}, \mathbf{e_y}, \mathbf{e_z}\}$ are the directional basis vectors. Note that the normal pressure is generally not evenly distributed over the disc due to the roughness of the contact interface, but here P and $\mathbb{H}(\mathbf{x})$ are combined to represent effectively the fluctuations in the normal pressure and the fiction coefficient. In this case, $\mathbf{F_O}$ and $\mathbf{T_O}$ become random variables due to the incorporation of the COF random field. The coefficients of variation (CV) of these random variables are investigated with respect to varying correlation length under different normalised velocity ratio $\rho^*/R = v/(\omega R)$, as shown in Fig. 2. These results are obtained through 10,000 realisations for each case and using the COF random field with mean value $\mu_{\mathbb{H}} = 0.6$ and standard deviation $\sigma_{\mathbb{H}} = 0.1$.

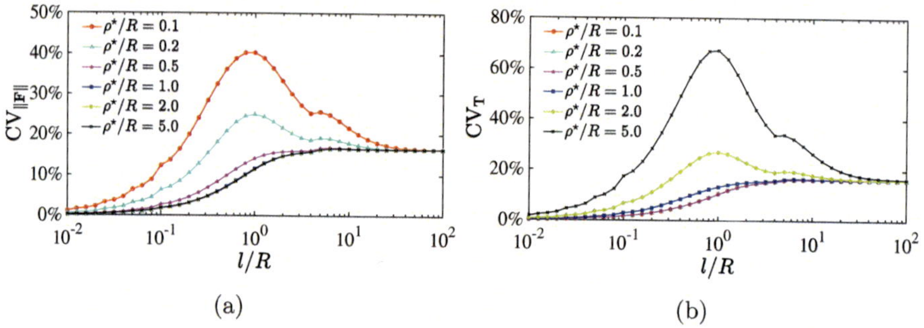

Fig. 2. Coefficient of variation of (a) friction force; (b) frictional torque.

A similar pattern can be observed from the results of $\mathrm{CV}_{\|\mathbf{F}\|}$ and CV_{T}: when the correlation length (l) is significantly smaller than the radius of the disc (R), both CVs tend to 0, which means in such situation the random friction model is effectively equivalent to a deterministic one due to the homogenisation effect. On the other side, both CVs converge to a value at $\sigma_{\mathbb{H}}/\mu_{\mathbb{H}} = 16.67\%$ as the ratio l/R is large (e.g. $l/R > 10$), for which case variables in the field are highly correlated and consequently the COF applies more like a random variable rather than a random field. The $\mathrm{CV}_{\|\mathbf{F}\|}$ monotonously increases with l/R when the ρ^*/R is larger than 0.5, while it has a peak value at l/R near 1 when translating dominates the motion, e.g. $\rho^*/R < 0.5$. On the opposite, the CV_T has its peak value at l/R near 1 when rotating dominates the motion, e.g. $\rho^*/R > 2.0$, otherwise the value increases with l/R from 0 to the converged value 16.67%.

3 Stick-Slip Effect

In this section, the stick-slip effect is investigated through a spring-slider model, which modifies the pure sliding model by connecting the geometry centre of the disc with a spring of stiffness k and imposing a constant velocity v at the end point of the spring aligned to the $\mathbf{e_x}$ direction, as shown in Fig. 3. The disc is initially stuck to the plane. As the endpoint of the spring moves, the friction force increases until it exceeds the threshold, after which the stick status no longer holds. The disc starts to slip until its velocity decreases to zero then the disc is stuck to the plane again. The local static COF μ_s and dynamical COF μ_d have the relationship as $\mu_d = 0.8\,\mu_s$. It is therefore vital to set effective stick-slip criteria to characterise the dynamical frictional responses. In what follows, the governing equation is discussed and the stick-slip transition criteria proposed in [4], is adopted here and briefly stated. Finally, stochastic analyses are performed concerning the friction force and the stick time duration.

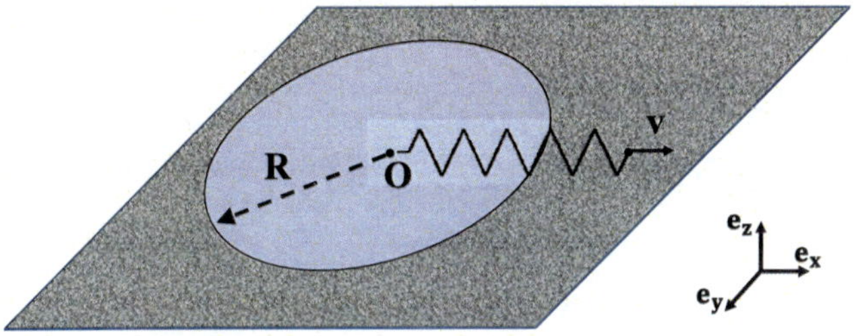

Fig. 3. The spring-slider model

3.1 Governing Equation

The governing differential equation of the system can be expressed as

$$\mathbf{M}\ddot{\mathbf{q}} + \mathbf{K}\mathbf{q} + \mathbf{Q_s} = \mathbf{0}, \tag{4}$$

where $\mathbf{M} = diag([m, m, \frac{1}{2}mR^2])$ is the mass matrix, $\mathbf{q} = [x, y, \phi]^{\mathrm{T}}$ is the generalised coordinate vector representing displacements in x- and y-direction and rotational angle, $\mathbf{K} = diag([k, 0, 0])$ is the stiffness matrix, and $\mathbf{Q_s} = [-(kvt + F_{sx}), -F_{sy}, -T_s]^{\mathrm{T}}$ is the vector containing the external forces and torque, where t is the elapsed time, $\{F_{sx}, F_{sy}\}$ are frictional forces and T_s is the frictional torque. Note that the friction forces and torque are random in time due to the spatial randomness of the COF and for which the values can be distinct concerning different states of the disc, such that for the state of stick $F_{sx} = -k(vt - x)$, $F_{sy} = 0$ and $T_s = 0$, and for the state of slip

$$\begin{aligned} F_{sx} &= \iint_S P\mathbb{H}_k(x,y)\frac{\dot{x} - \dot{\phi}y}{\sqrt{(\dot{x} - \dot{\phi}y)^2 + (\dot{y} + \dot{\phi}x)^2}}\mathrm{d}S, \\ F_{sy} &= \iint_S P\mathbb{H}_k(x,y)\frac{\dot{y} + \dot{\phi}x}{\sqrt{(\dot{x} - \dot{\phi}y)^2 + (\dot{y} + \dot{\phi}x)^2}}\mathrm{d}S, \\ T_s &= \iint_S P\mathbb{H}_k(x,y)\frac{\dot{\phi}(x^2 + y^2) + x\dot{y} - \dot{x}y}{\sqrt{(\dot{x} - \dot{\phi}y)^2 + (\dot{y} + \dot{\phi}x)^2}}\mathrm{d}S, \end{aligned} \tag{5}$$

where $P = mg/(\pi R^2)$ and $\mathbb{H}_k(x,y) = 0.8\mathbb{H}_s(x,y)$ is the dynamical COF random field.

3.2 Stick-slip Transition Criteria

To construct the stick-slip transition criteria, firstly the following nondimensionlised variables are defined

$$x^\star = x/R,\ y^\star = y/R,\ \dot{x}^\star = \dot{x}/(R\alpha),\ \dot{y}^\star = \dot{y}/(R\alpha),\ \dot{\phi}^\star = \dot{\phi}/\alpha, \alpha = \sqrt{k/m}$$

$$P^\star = PR^2/(mg) = 1/\pi,\ \mathrm{d}S^\star = \mathrm{d}S/R^2.$$

Secondly, a non-dimensional relative velocity magnitude $\lambda = \sqrt{\dot{x}^{\star 2} + \dot{y}^{\star 2} + \dot{\phi}^{\star 2}}$ is defined such that

$$\dot{x}^\star = \lambda\cos\theta\cos\psi,\ \dot{y}^\star = \lambda\cos\theta\sin\psi,\ \dot{\phi}^\star = \lambda\sin\theta. \tag{6}$$

Thirdly, the static limit of friction forces and torque can be nondimensionalised as

$$\begin{aligned} &F^\star_{sx}(\theta,\psi) = \\ &\quad\iint_{S^\star} P^\star\mathbb{H}_s(x,y)\frac{\cos\theta\cos\psi - y^\star\sin\theta}{\sqrt{(\cos\theta\cos\psi - y^\star\sin\theta)^2 + (\cos\theta\sin\psi + x^\star\sin\theta)^2}}\mathrm{d}S^\star, \\ &F^\star_{sy}(\theta,\psi) = \\ &\quad\iint_{S^\star} P^\star\mathbb{H}_s(x,y)\frac{\cos\theta\sin\psi + x^\star\sin\theta}{\sqrt{(\cos\theta\cos\psi - y^\star\sin\theta)^2 + (\cos\theta\sin\psi + x^\star\sin\theta)^2}}\mathrm{d}S^\star, \\ &T^\star_s(\theta,\psi) = \\ &\quad\iint_{S^\star} P^\star\mathbb{H}_s(x,y)\frac{(x^{\star 2} + y^{\star 2})\sin\theta + x^\star\cos\theta\sin\psi - y^\star\cos\theta\cos\psi}{\sqrt{(\cos\theta\cos\psi - y^\star\sin\theta)^2 + (\cos\theta\sin\psi + x^\star\sin\theta)^2}}\mathrm{d}S^\star. \end{aligned} \tag{7}$$

The non-dimesionalised triplet $(F^\star_{sx}, F^\star_{sy}, T^\star_s)$ thus forms an ellipsoidal-like domain Ω_l with boundary Γ_l with respect to the azimuth pair (θ, ψ). In contrast, the real friction triplet $\mathbf{Q}^{r\star}_s = (F^{r\star}_{sx}, F^{r\star}_{sy}, T^{r\star}_s)$ that indicates the current state of the friction is defined using $F_N = mg$ and $T_N = mgR$ as

$$F^{r\star}_{sx} = F_{sx}/F_N, \; F^{r\star}_{sy} = F_{sy}/F_N, \; T^{r\star}_s = T_s/T_N. \tag{8}$$

Slip-to-Stick Criteria. A criterion function $C(\dot{x}^\star, \dot{y}^\star, \dot{\phi}^\star) = \lambda - \varepsilon_\lambda$ is defined and a threshold value ε_λ is set such that the slip state remains when $C(\dot{x}^\star, \dot{y}^\star, \dot{\phi}^\star) \geq 0$ and breaks when $C(\dot{x}^\star, \dot{y}^\star, \dot{\phi}^\star) < 0$ and $\mathbf{Q}^{r\star}_s \in \Omega_l \backslash \Gamma_l$. As done in [5], after the detection of the earliest time t_r for which $C(\dot{x}^\star, \dot{y}^\star, \dot{\phi}^\star) < 0$ and the validation of the condition $\mathbf{Q}^{r\star}_s \in \Omega_l \backslash \Gamma_l$, a small velocity jump needs to be applied for the next time step t^+_r to improve the robustness of the algorithm: $\dot{x}\,(t^+_r) = 0$, $\dot{y}\,(t^+_r) = 0$, $\dot{\phi}\,(t^+_r) = 0$, where $\dot{x}\,(t^+_r)$, $\dot{y}\,(t^+_r)$ and $\dot{\phi}\,(t^+_r)$ are the initial velocity terms in the next stick state.

Stick-to-Slip Criteria. The real non-dimensionalised friction triplet $\mathbf{Q}^{r\star}_s = (F^{r\star}_{sx}, F^{r\star}_{sy}, T^{r\star}_s)$ is compared against the domain Ω_l. If $\mathbf{Q}^{r\star}_s$ is inside the domain Ω_l, the stick state maintains; if $\mathbf{Q}^{r\star}_s$ is on the boundary Γ_l or outside the domain Ω_l, the stick state then breaks and transits to the next slip state. After the break of the current stick state, a small velocity jump is applied at the initial time step t^+_l of the next slip state to the disc as $\dot{x}\,(t^+_l) = \varepsilon_\lambda \alpha R \cos\psi$, $\dot{y}\,(t^+_l) = \varepsilon_\lambda \alpha R \sin\psi$ and $\dot{\phi}\,(t^+_l) = 0$. Such velocity jump is performed to ensure the criterion function $C(\dot{x}^\star, \dot{y}^\star, \dot{\phi}^\star) = 0$ at time t^+_l and avoids possible oscillation problems when the velocity is extremely small.

3.3 Stochastic Analysis of the Frictional Responses

The dynamical responses of the system (4) can be solved numerically using e.g. backward-Euler method. The friction forces and the stick time duration are quantities of interest. In contrast, the frictional torque only fluctuates at the beginning of the simulation and then remains a non-significant level throughout the simulation and thus is not the focus. Investigations are conducted concerning varying the correlation length l and the driving velocity v. By varying one of these parameters, the other is set as $l/R = 0.2$ or $v = 0.1$ [m/s]. The total elapsed time is set to be 100 [s]. For each case, 100 time-related simulations are performed by Monte-Carlo method to obtain the statistical properties of the quantities of interest. As a comparison, all results are normalised by the corresponding results using a deterministic COF model, e.g. a constant equal to $\mu_{\mathrm{H}} = 0.6$.

The results are collected in Fig. 4. It can be observed that for a correlation length to radius ratio $l/R < 10^{-1}$, the frictional responses of the stochastic model are very close to the deterministic model. For a larger l/R, the mean value of the norm of the friction force starts to fluctuate, as shown in Fig. 4(a). The standard deviation of the friction force, shown in Fig. 4(c), indicating the variation of the friction force during the stick-slip transition has a peak value when l/R is near 2 and then fluctuates but is consistent with the deterministic model in the mean sense when $l/R > 10$. The stick time duration concerning varying correlation length remains stable with small fluctuations, as shown in Fig. 4(e). On the other side, the mean value of the friction force concerning varying driving velocity contains barely no fluctuation, as seen in Fig. 4(b). The variation of the friction force starts to fluctuate when the driving velocity exceeds 0.3 [m/s] and at the same time the stick time duration decreases significantly, as observed from Fig. 4(d) and (f).

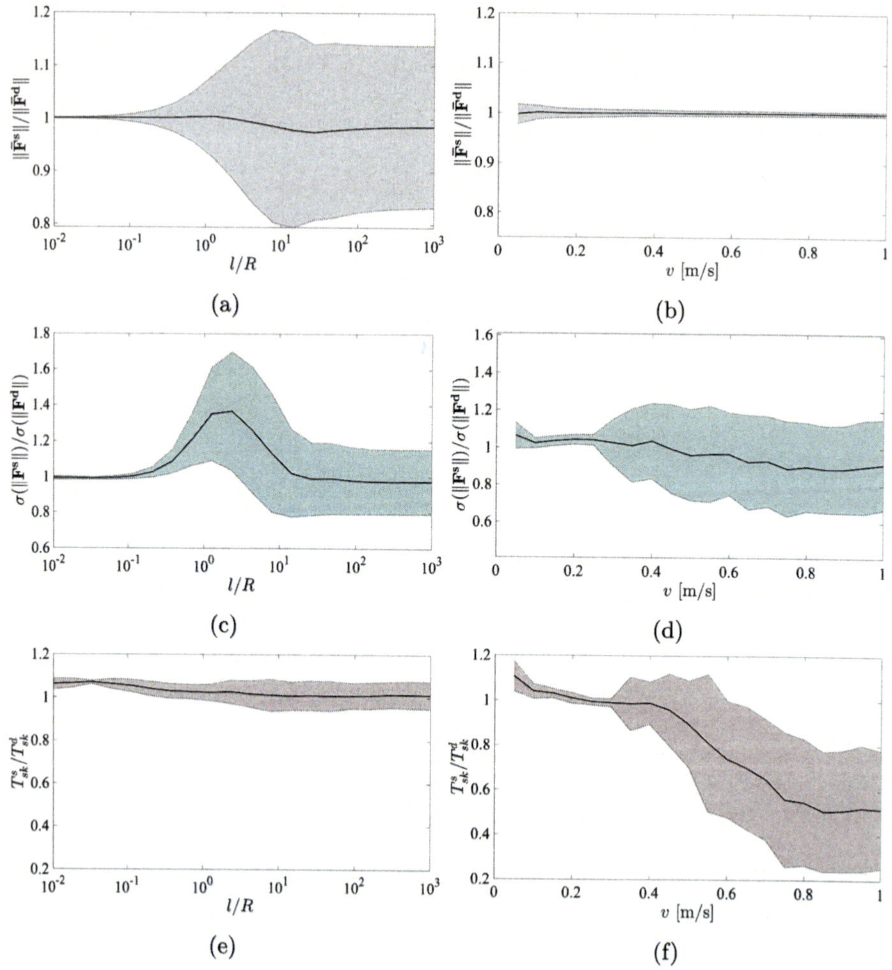

Fig. 4. Statistical properties of the (a), (b) mean value of the friction force; (c), (d) standard deviation of the friction force; (e), (f) stick time duration, in which (a), (c), (e) concerns varying l/R and (b), (d), (f) concerns varying v. The solid lines indicate the mean value taken from MC realisations and the shaded area the standard deviation.

4 Conclusion

In this work, dynamical analyses are performed by modelling COF as a lognormal random field through a pure sliding model and a spring-slider model. The results of the pure sliding model show that the COF model can be considered as a deterministic one when the correlation length is very small, and as a random variable when it is large compared with the characterised geometry length. Stochastic analyses are conducted concerning the stick-slip effect through the spring-slider model with three-dimensional stick-slip transition criteria. The results show that a correlation length comparable with the characterised geometry length can lead to significant fluctuations in both the mean value of the friction force and its variation. Moreover, increasing the driving velocity will affect the friction force variation and cause a shrink of the stick time duration.

References

1. Ben-David, O., Fineberg, J.: Static friction coefficient is not a material constant. Phys. Rev. Lett. **106**(25), 254301 (2011)
2. Feng, Q.: A discrete model of a stochastic friction system. Comput. meth. Appl. Mech. Eng. **192**(20–21), 2339–2354 (2003)
3. Hu, H., Batou, A., Ouyang, H.: Coefficient of friction random field modelling and analysis in planar sliding. J. Sound Vib. 116197 (2021)
4. Hu, H., Batou, A., Ouyang, H.: Friction-induced vibration of a stick-slip oscillator with random field friction modelling. Manuscript submitted for publication (2022)
5. Kudra, G., Awrejcewicz, J.: Bifurcational dynamics of a two-dimensional stick-slip system. Differ. Equ. Dyn. Sys. **20**(3), 301–322 (2012)
6. Ritto, T., Escalante, M., Sampaio, R., Rosales, M.: Drill-string horizontal dynamics with uncertainty on the frictional force. J. Sound Vib. **332**(1), 145–153 (2013). https://doi.org/10.1016/j.jsv.2012.08.007
7. Sarrouy, E., Dessombz, O., Sinou, J.J.: Piecewise polynomial chaos expansion with an application to brake squeal of a linear brake system. J. Sound Vib. **332**(3), 577–594 (2013). https://doi.org/10.1016/j.jsv.2012.09.009

Check for updates

Feedforward Control of a Nonlinear Underactuated Multibody System

Jason Bettega(✉), Dario Richiedei, and Alberto Trevisani

Department of Management and Engineering (DTG), University of Padova, Stradella San Nicola 3, 36100 Vicenza, Italy
jason.bettega@phd.unipd.it, {dario.richiedei, alberto.trevisani}@unipd.it

Abstract. An enhanced inverse dynamics approach is presented for feedforward control of a nonlinear underactuated multibody system. Its theoretical formulation is carefully explained and subsequently assessed through a numerical test case. In particular, the internal dynamics model is achieved and a stability analysis is performed in order to find the condition that makes the system a non-minimum phase one, obtaining an approximated internal dynamics equation characterized by eigenvalues that all reside in the left half of the complex plane. Knowing the desired time evolution of the system output coordinates, a stable computation of the required feedforward input term is performed.

Keywords: Inverse dynamics · Underactuated systems · Non-minimum phase · Internal dynamics · Feedforward control

1 Introduction

1.1 Motivations and State of the Art

Trajectory tracking solutions are a common topic since every multibody system is designed with the goal to fulfill a desired task. In order to achieve it, two degrees of freedom (DOF) controllers are normally adopted in mechatronic systems, where one degree of freedom is related to the feedforward control design while the second one concerns the feedback control implementation. Both of them cover an important role because the former normally ensures faster responses with lower tracking errors, while the latter is included in order to obtain a stable system capable to face model mismatches as well as external disturbances, leading to a more robust overall system. Focusing the attention on the feedforward term, its implementation requires the computation of the inverse dynamic model of the system under investigation. When rigid components are considered as well as the fully-actuation property, this computation is usually straightforward and it does not require any particular effort. However, when flexible components are taken into account, the evaluation of the feedforward input term become more challenging due to the increase of number of degrees of freedom and, consequently, the achievement of an underactuated system. More precisely, underactuation consists of

Z. Dimitrovová et al. (Eds.): WMVC 2022, MMS 125, pp. 474–482, 2023.
https://doi.org/10.1007/978-3-031-15758-5_48

having a number of degrees of freedom which is greater than the number of actuators. Furthermore, depending on the selection of the desired output, this kind of systems can also be characterized by the presence of right-half-plane zeros, which lead to two main issues: the presence of non-minimum phase systems and, consequently, the impossibility to perform inverse dynamics exactly. Since lightweight manipulators are becoming more and more common nowadays due to energy saving purposes, this issue represents an open problem in the research field. To overcome this limitation, several papers have been proposed in the recent years. In [1] an inversion-based approach to exact nonlinear output tracking control is proposed, together with the presence of a feedback term in order to stabilize the system along the desired trajectory; in particular, the solution is achieved by applying the Byrnes-Isidori regulator to a specific trajectory, introducing boundedness and integrability requirements. However, the outcome is noncausal and, as a consequence, an online procedure cannot be performed with this technique. In [2] stable inversion is extended to nonlinear time-varying systems, considering both minimum phase and non-minimum phase systems. Through this approach, the system dynamics is firstly linearized and, subsequently, partitioned into time-varying stable and unstable subsystems, which are used to design time-varying filters that ensure a bounded inverse input-state trajectory. However, this technique is local to the time-varying path and, additionally, it requires that internal dynamics varies slowly. An interesting analysis is presented in [3], where three stable approximate model inversion techniques, applied to non-minimum phase systems, are examined, which are: the non-minimum phase zeros ignore (NPZ-Ignore), the zero-phase-error tracking controller (ZPETC) and the zero-magnitude-error tracking controller (ZMETC). The proper choice of one of these techniques highly depends on the system under investigation and, therefore, this work also shows how the location of the zeros can indicate which could be the more effective technique to be used to guarantee good performances. With the goal to understand pros and cons coming from these approaches, experimental tests are performed on an AFM piezoscanner. Two further feedforward control techniques are proposed in [4]: in the first one, the model inversion is performed through the concept of nonlinear input-output normal form, which is achieved thanks to a coordinate transformation, while the second one exploits servo-constraints, leading to solve a set of differential algebraic equations (DAEs). Both techniques are exploited in an optimization problem and numerically assessed through an underactuated multibody system. An interesting comparison is presented in [5], considering two different strategies that are present in the literature: the standard stable inversion method, which is solved through the formulation of a two-point boundary value problem, and an alternative optimization problem formulation, which is defined without any boundary conditions. This comparison shows that both approaches can be formulated through DAEs and, additionally, numerical results are presented considering manipulators with passive and compliant joints. In [6] resonant vibration generators, such as vibratory feeders or ultrasonic sonotrodes, are considered and novel solutions are proposed in order to achieve the closest approximation of the desired vibrations, since these underactuated systems are usually excited in an open-loop manner. In particular, one of the proposed strategies exploits coordinate transformation and projection to solve an inverse dynamic problem that allows to achieve the optimal shaping of the harmonic forces exerted by the actuators.

1.2 Contributions of the Paper

In this paper a feedforward computation algorithm for a non-minimum phase underactuated multibody system is proposed by exploiting and extending the idea presented in [7]. In particular, since underactuated systems are characterized by less actuators than degrees of freedom, firstly the dynamic model is partitioned into two parts, an actuated subsystem and an unactuated one, thanks to the exploitation of the coordinate partitioning method. Secondly, the desired output trajectory of the system is described as a linear combination of the actuated and unactuated coordinates by introducing some tuning parameters due to the linearly combined output technique adopted. The feedforward input term is therefore computed through two consecutive steps: a differential part, followed by an algebraic one. With the goal to compute the former, the tuning parameters are properly selected in order to make the internal dynamics stable; once the differential part is achieved, the algebraic one is obtained exploiting the dynamic sub-model related to the actuated coordinates.

This paper is organized as follows: in Sect. 2 the proposed method is presented; in Sect. 3, the proposed technique is numerically assessed in the presence of an overhead cartesian crane with the goal to track a desired trajectory; in Sect. 4 conclusion remarks are highlighted.

2 System Model and Method Description

2.1 System Model

Let us consider the dynamic model of a 4-DOF overhead cartesian crane, that is here assumed as a meaningful example of an underactuated nonlinear multibody system, as sketched in Fig. 1. The following four independent coordinates are assumed to model the system: two translations describing the planar motion of the platform, that are the two actuated coordinates $\mathbf{q_A} = \begin{bmatrix} x_P \; y_P \end{bmatrix}^T \in \mathbb{R}^2$, and two unactuated components of the swing angle $\mathbf{q_U} = \begin{bmatrix} \theta_X \; \theta_Y \end{bmatrix}^T \in \mathbb{R}^2$. $\theta_X \in \mathbb{R}$ is the oscillation angle projected on the XZ plane, while $\theta_Y \in \mathbb{R}$ is the angle measured from the XZ plane. By assuming that the cable is taut with tensile stress and that hoisting is not allowed (i.e., the cable length, denoted with $h \in \mathbb{R}$, is constant), the following four nonlinear equations describe the system dynamic behaviour:

$$\begin{bmatrix} (M_X+m) & 0 & m\,h\cos(\theta_X)\cos(\theta_Y) & -m\,h\sin(\theta_X)\sin(\theta_Y) \\ 0 & (M_Y+m) & 0 & m\,h\cos(\theta_Y) \\ m\,h\cos(\theta_X)\cos(\theta_Y) & 0 & m\,h^2\cos^2(\theta_Y) & 0 \\ -m\,h\sin(\theta_X)\sin(\theta_Y) & m\,h\cos(\theta_Y) & 0 & m\,h^2 \end{bmatrix} \begin{bmatrix} \ddot{x}_P \\ \ddot{y}_P \\ \ddot{\theta}_X \\ \ddot{\theta}_Y \end{bmatrix} +$$

$$+ \begin{bmatrix} -m\,h\dot{\theta}_X^2\sin(\theta_X)\cos(\theta_Y) - m\,h\dot{\theta}_Y^2\sin(\theta_X)\cos(\theta_Y) - 2\,m\,h\,\dot{\theta}_X\dot{\theta}_Y\cos(\theta_X)\sin(\theta_Y) + c_X\dot{x} \\ -m\,h\dot{\theta}_Y^2\sin(\theta_Y) + c_Y\dot{y} \\ -2\,m\,h^2\,\dot{\theta}_X\dot{\theta}_Y\sin(\theta_Y)\cos(\theta_Y) + c_\theta\dot{\theta}_X \\ m\,h^2\,\dot{\theta}_X^2\sin(\theta_Y)\cos(\theta_Y) + c_\theta\dot{\theta}_Y \end{bmatrix} = \quad (1)$$

$$= \begin{bmatrix} 0 \\ 0 \\ -m\,h\,g\sin(\theta_X)\cos(\theta_Y) \\ -m\,h\,g\cos(\theta_X)\sin(\theta_Y) \end{bmatrix} + \begin{bmatrix} 1 & 0 \\ 0 & 1 \\ 0 & 0 \\ 0 & 0 \end{bmatrix} \begin{bmatrix} u_x \\ u_y \end{bmatrix}$$

The following definitions have been introduced in Eq. (1): u_X, $u_Y \in \mathbb{R}$ are the forces that are applied to the platform along the X and Y directions; $m \in \mathbb{R}$ is the lumped mass of the suspended load; c_X, c_Y, $c_\theta \in \mathbb{R}$ are the viscous friction coefficients characterizing the system; M_X, $M_Y \in \mathbb{R}$ are the translational masses of the platform along X and Y directions, respectively; $g \in \mathbb{R}$ is the gravity acceleration.

The goal of feedforward control is to make the load track a prescribed path in the cartesian plane, defined through the load cartesian absolute coordinates $\begin{bmatrix} x_L & y_L \end{bmatrix}^T \in \mathbb{R}^2$. Therefore, the vector of the controlled output $\boldsymbol{y} \in \mathbb{R}^2$ is defined as:

$$\boldsymbol{y} = \begin{bmatrix} x_P + h \sin \theta_X \cos \theta_Y \\ y_P + h \sin \theta_Y \end{bmatrix} \tag{2}$$

In this example, both the dynamic model and the output map are nonlinear.

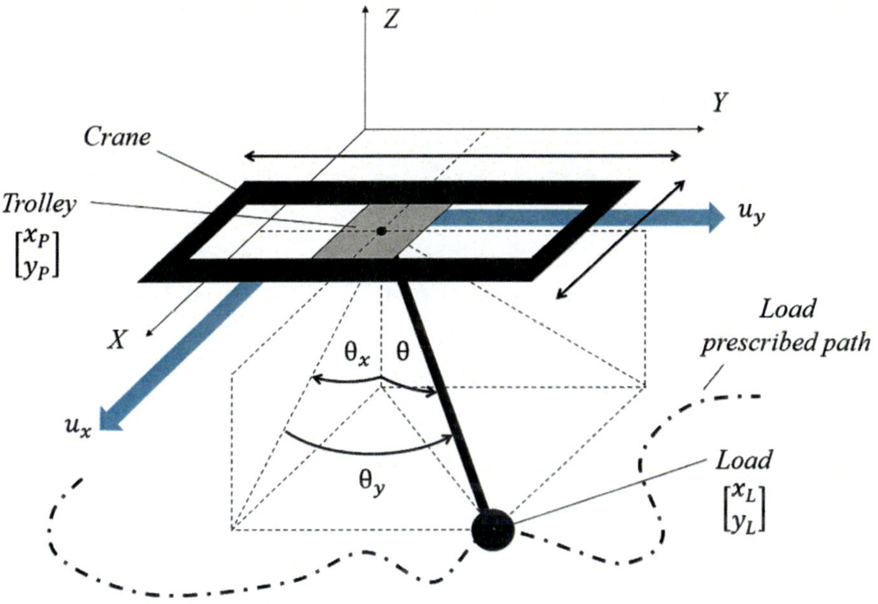

Fig. 1. Simplified scheme of the underactuated overhead cartesian crane.

2.2 Method Description

The force distribution matrix in Eq. (1) reveals that the coordinates of the trolley are the actuated ones, while the coordinates of the load are the unactuated ones. Therefore, the dynamic model can be directly partitioned as follows:

$$\begin{bmatrix} \mathbf{M_{AA}} & \mathbf{M_{AU}} \\ \mathbf{M_{AU}^T} & \mathbf{M_{UU}} \end{bmatrix} \begin{bmatrix} \ddot{\mathbf{q}}_\mathbf{A} \\ \ddot{\mathbf{q}}_\mathbf{U} \end{bmatrix} + \begin{bmatrix} \mathbf{C_A} \\ \mathbf{C_U} \end{bmatrix} = \begin{bmatrix} \mathbf{F}_A \\ \mathbf{F}_U \end{bmatrix} + \begin{bmatrix} \mathbf{B_A} \\ \mathbf{B_U} \end{bmatrix} \mathbf{u} \tag{3}$$

where $\ddot{\mathbf{q}}_\mathbf{A} = \left[\ddot{x}_P\ \ddot{y}_P\right]^T \in \mathbf{R}^2$, $\ddot{\mathbf{q}}_\mathbf{U} = \left[\ddot{\theta}_X\ \ddot{\theta}_Y\right]^T \in \mathbf{R}^2$, $\mathbf{B_A} = \mathbf{I}_2 \in \mathbb{R}^{2\times2}$ (i.e., a two-dimensional identity matrix), $\mathbf{B_U} = \mathbf{0}_2 \in \mathbb{R}^{2\times2}$ (i.e., a two-dimensional null matrix), $\mathbf{u} = \left[u_X\ u_Y\right]^T \in \mathbb{R}^2$. Considering only the unactuated subsystem, the internal dynamics is defined as:

$$\mathbf{M}_\mathbf{AU}^T \ddot{\mathbf{q}}_\mathbf{A} + \mathbf{M_{UU}}\ddot{\mathbf{q}}_\mathbf{U} + \mathbf{C_U} = \mathbf{F_U} \tag{4}$$

In order to study the stability of the internal dynamics, firstly Eq. (4) is linearized around the stable downward equilibrium point defined by $\theta_X = \theta_Y = 0\ [\deg]$ and, secondly, $\ddot{\mathbf{q}}_A$ is replaced thanks to the linearly combined output technique, with:

$$\ddot{\mathbf{y}}^\mathbf{des} = \ddot{\mathbf{q}}_\mathbf{A} + \Gamma_\mathbf{U}\, \ddot{\mathbf{q}}_\mathbf{U} \tag{5}$$

The approximation of $\mathbf{y}^\mathbf{des} \in \mathbb{R}^2$, as a linear combination of $\mathbf{q} = \left[\mathbf{q_A}\ \mathbf{q_U}\right]^T \in \mathbb{R}^4$, can be obtained by linearizing Eq. (2) about the vertical equilibrium configuration (i.e., $\theta_X = \theta_Y = 0\ [\deg]$), where $\Gamma_\mathrm{U} = h\mathbf{I}_2 \in \mathbb{R}^{2\times2}$.

However, if the exact Γ_U, arising from linearization, is used, the internal dynamics is unstable. Stabilization is achieved by replacing Γ_U with $\tilde{\boldsymbol{\Gamma}}_\mathrm{U} = \alpha\, \Gamma_\mathrm{U} \in \mathbb{R}^{2\times2}$, leading to the following model:

$$\mathbf{M_{ID}}\, \ddot{\mathbf{q}}_\mathbf{U} + \mathbf{C_{ID}}\, \dot{q}_\mathrm{U} + \mathbf{K_{ID}}\, \mathbf{q_U} = \mathbf{B_{ID}}\, \ddot{\mathbf{y}}^\mathbf{des} \tag{6}$$

where:

$$\begin{aligned} \mathbf{M_{ID}} &= \begin{bmatrix} m\,h^2(1-\alpha) & 0 \\ 0 & m\,h^2(1-\alpha) \end{bmatrix} \mathbf{C_{ID}} = \begin{bmatrix} c_\theta & 0 \\ 0 & c_\theta \end{bmatrix} \\ \mathbf{K_{ID}} &= \begin{bmatrix} m\,h\,g & 0 \\ 0 & m\,h\,g \end{bmatrix} \mathbf{B_{ID}} = \begin{bmatrix} -m\,h & 0 \\ 0 & -m\,h \end{bmatrix} \end{aligned} \tag{7}$$

The parameter $\alpha \in \mathbb{R}$ should be chosen as close as possible to the unitary value to ensure a good reconstruction of the desired output trajectory.

Once the internal dynamics is stable, $\mathbf{q_U}(t)$, $\dot{q}_\mathrm{U}(t)$ and $\ddot{\mathbf{q}}_\mathbf{U}(t)$ are computed by integrating the differential equations in Eq. (6) over the whole cycle time (and will be henceforth denoted as $\mathbf{q}_\mathbf{U}^\mathbf{des}(t)$, $\dot{\mathbf{q}}_\mathbf{U}^\mathbf{des}(t)$, $\ddot{\mathbf{q}}_\mathbf{U}^\mathbf{des}(t)$), as suggested in [7]. Subsequently, the control force vector $\mathbf{u}(t)$ is computed by solving the algebraic equations representing the inverse dynamics problem of the actuated subsystem. Rather than using again the approximations of the desired output provided by Eq. (5), this second step can be performed by inverting the actual map $\mathbf{y}^\mathbf{des} = \mathbf{h}(\mathbf{q_A}, \mathbf{q_U})$, together with its first and second time derivatives, to compute the reference trajectory for the actuated coordinates $\mathbf{q}_\mathbf{A}^\mathbf{des}(t)$, $\dot{\mathbf{q}}_\mathbf{A}^\mathbf{des}(t)$, $\ddot{\mathbf{q}}_\mathbf{A}^\mathbf{des}(t)$. By imposing such references, the required control forces can be inferred from the following algebraic computation:

$$\mathbf{u_{FFW}}(t) = \mathbf{B}_\mathbf{A}^{-1}\left(\mathbf{M_{AA}}\ddot{\mathbf{q}}_\mathbf{A}^\mathbf{des}(t) + \mathbf{M_{AU}}\ddot{\mathbf{q}}_\mathbf{U}^\mathbf{des}(t) + \mathbf{C_A} - \mathbf{F_A}\right) \tag{8}$$

where $\mathbf{M_{AA}} = \mathbf{M_{AA}}\left(\mathbf{q}^\mathbf{des}\right) \in \mathbb{R}^{2\times2}$, $\mathbf{M_{AU}} = \mathbf{M_{AU}}\left(\mathbf{q}^\mathbf{des}\right) \in \mathbb{R}^{2\times2}$, $\mathbf{C_A} = \mathbf{C_A}\left(\dot{\mathbf{q}}^\mathbf{des}, \mathbf{q}^\mathbf{des}\right) \in \mathbb{R}^2$, $\mathbf{F_A} = \mathbf{F_A}\left(\dot{\mathbf{q}}^\mathbf{des}, \mathbf{q}^\mathbf{des}, t\right) \in \mathbb{R}^2$.

3 Numerical Validation

A planar desired path is considered to verify the effectiveness of the feedforward computation algorithm. It is desired that the suspended load follows a rhombus in the XY plane, characterized by a side length of 0.5 $[m]$ and located at a vertical distance of 1 $[m]$ from the trolley; 1 $[m]$ also represents the length h of the cable. The entire path is set to be completed in 6 $[s]$ and, in order to assess also the behavior of the load at steady-state conditions, two idle intervals of 2 $[s]$ are introduced, at the beginning and at the end of the trajectory, leading to an overall time specification of 10 $[s]$. The displacement of the load and of the trolley are shown in Fig. 2. The reference trajectory is tracked with high precision, and this aspect is further corroborated by looking at the small tracking errors that have been achieved.

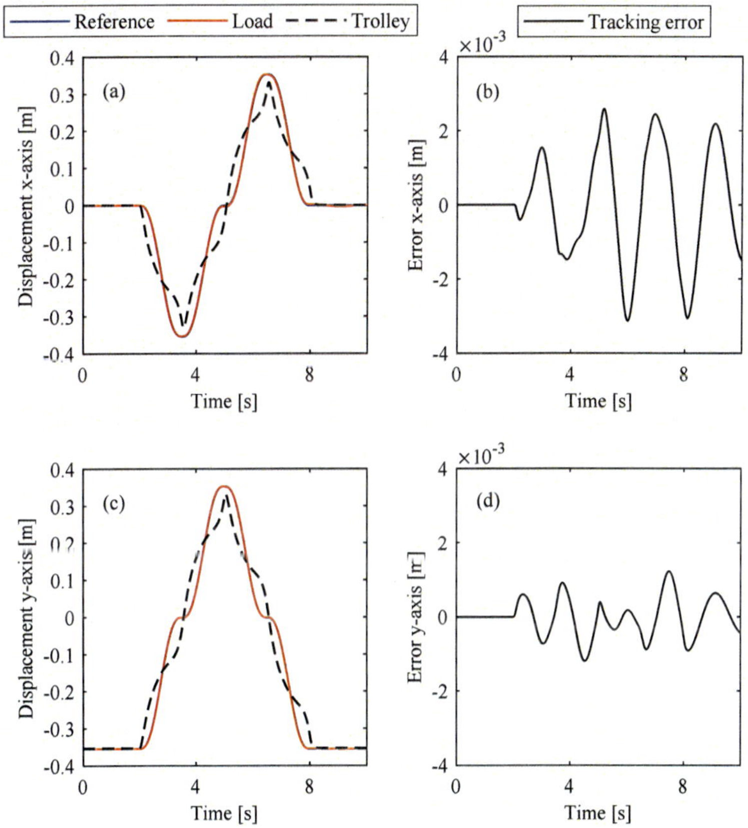

Fig. 2. Trajectory tracking responses (a, c) and trajectory tracking errors (b, d) in the presence of a rhombus reference.

The effectiveness of the feedforward computation algorithm is confirmed by the path tracking response shown in Fig. 3, where it can be noticed that very low contour error was achieved. Therefore, it can be stated that the presented inverse dynamic technique results to be very effective both in terms of trajectory and path tracking. This aspect covers huge importance since it leads to lower the gains of possible feedback controllers and hence increasing the robustness of the overall system.

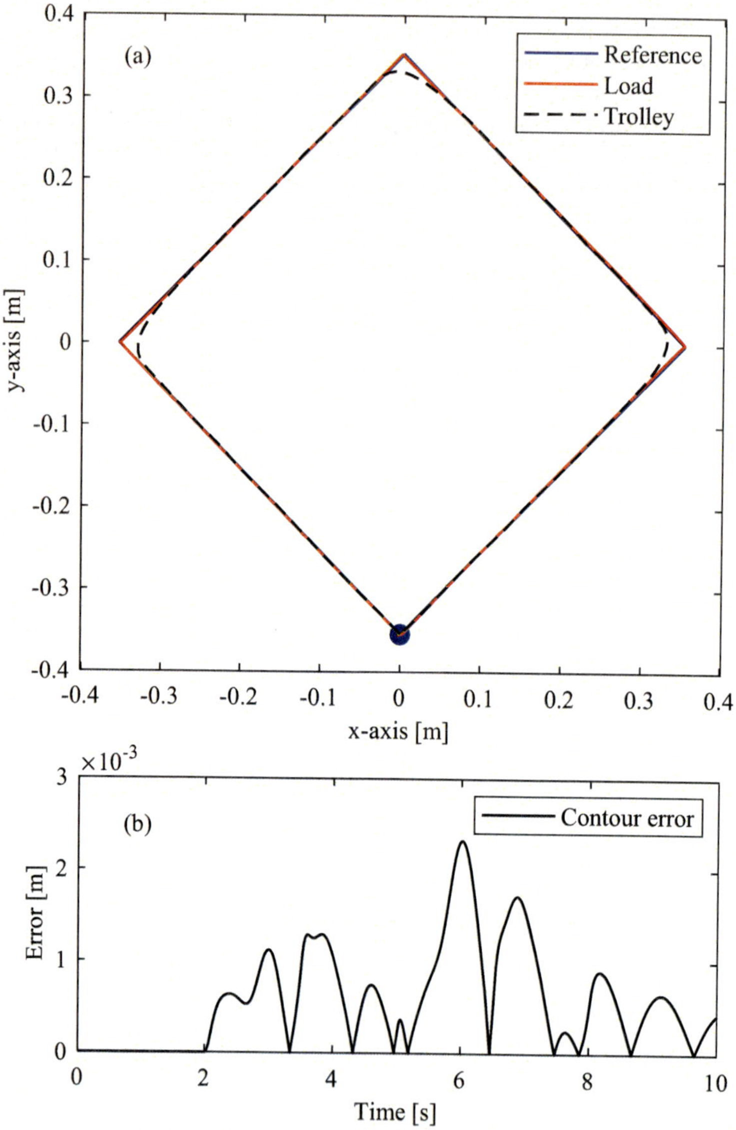

Fig. 3. Path tracking response (a) and contour error (b) in the presence of a rhombus reference.

Finally, Fig. 4 displays the forces along the two cartesian axes, computed to perform the desired task.

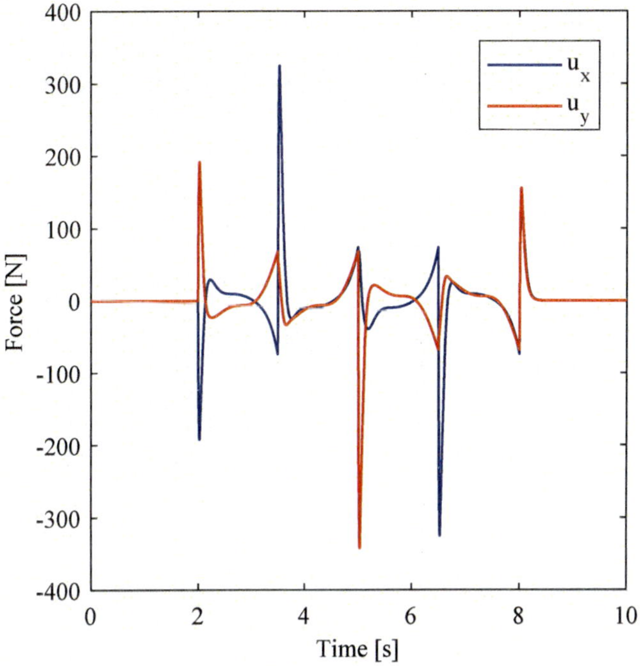

Fig. 4. Input forces in the presence of a rhombus reference.

4 Conclusions

This paper proposes an enhanced inverse dynamics algorithm for computing the feedforward forces in a non-minimum phase, underactuated multibody system. This kind of systems is usually characterized by the presence of right-half-plane zeros which prevents the exact inversion of the system model, since the internal dynamics is unstable. With the goal to overcome this limitation, firstly, the proposed formulation shows how it is possible to achieve a stable inversion by exploiting a state-of-the-art technique, called linearly combined output, and subsequently performing a stability analysis of the approximated internal dynamics. Secondly, this work improves the evaluation of the feedforward term proposed by the same state-of-the-art techniques by redefining the computation of the actuated acceleration terms thanks to the usage of the exact nonlinear model. The effectiveness of the proposed method has been assessed numerically considering an underactuated overhead cartesian crane, showing high performances both in trajectory tracking tasks and in path tracking ones, corroborated by the low tracking errors.

References

1. Devasia, S., Chen, D., Paden, B.: Nonlinear inversion-based output tracking. IEEE Trans. Automat. Contr. **41**, 930–942 (1996). https://doi.org/10.1109/9.508898

2. Devasia, S., Paden, B.: Stable inversion for nonlinear nonminimum-phase time-varying systems. IEEE Trans. Automat. Contr. **43**, 283–288 (1998). https://doi.org/10.1109/9.661082
3. Butterworth, J.A., Pao, L.Y., Abramovitch, D.Y.: Analysis and comparison of three discrete-time feedforward model-inverse control techniques for nonminimum-phase systems. Mechatronics **22**, 577–587 (2012). https://doi.org/10.1016/j.mechatronics.2011.12.006
4. Seifried, R.: Two approaches for feedforward control and optimal design of underactuated multibody systems. Multibody Syst. Dyn. **27**, 75–93 (2012). https://doi.org/10.1007/s11044-011-9261-z
5. Bastos, G., Seifried, R., Brüls, O.: Analysis of stable model inversion methods for constrained underactuated mechanical systems. Mech. Mach. Theory. **111**, 99–117 (2017). https://doi.org/10.1016/j.mechmachtheory.2017.01.011
6. Belotti, R., Richiedei, D., Tamellin, I., Trevisani, A.: Response optimization of underactuated vibration generators through dynamic structural modification and shaping of the excitation forces. Int. J. Adv. Manufact. Technol. **112**(1–2), 505–524 (2020). https://doi.org/10.1007/s00170-020-06083-2
7. Seifried, R.: Integrated mechanical and control design of underactuated multibody systems. Nonlinear Dyn. **67**, 1539–1557 (2012). https://doi.org/10.1007/s11071-011-0087-2

Modelling of Electromechanical Coupling Effects in Electromagnetic Energy Harvester

Krzysztof Kecik(✉)

Lublin University of Technology, Lublin, 20-618 Lisbon, Poland
k.kecik@pollub.pl

Abstract. Electromechanical coupling is a measure of the efficiency of conversion between electric and vibration energy. The crucial problem in the design of energy collection is the design and modelling of an electromechanical coupling. This paper shows that electromechanical coupling in electromagnetic energy harvesters is inherently nonlinear and can be easily modified. The design of the electromechanical coupling function through the special construction of coils and magnet-spacers is proposed. In order to demonstrate the effectiveness of the harvester with the proposed electromechanical models, numerical results are presented.

Keywords: Energy harvesting · Electromechanical coupling · Magnetic levitation

1 Introduction

1.1 Energy Harvesting

Current trends in electronics and MEMS technologies require a reduction in size and energy consumption. An alternative to implementing a battery as a power source is to make use of the reclaimed energy available in the environment. The process is called energy harvesting. Energy harvesting technologies are currently being researched as a means to recover energy from the environment. Of course, energy harvesting methods are not aimed at completely replacing batteries, but at supporting them or being an alternative. An energy harvester can be categorized based on the converted mechanism used. Mechanical vibration energy can be converted to electric energy using piezoelectric, electromagnetic and electrostatic transducer mechanisms.

This Research Was Financed Within the Framework of the Project: "Theoretical-experimental Analysis Possibility of Electromechanical Coupling Shaping in Energy Harvesting Systems" No. DEC-2019/35/B/ST8/01068, Funded by the National Science Centre, Poland

Z. Dimitrovová et al. (Eds.): WMVC 2022, MMS 125, pp. 483–490, 2023.
https://doi.org/10.1007/978-3-031-15758-5_49

The electromagnetic harvester converts kinetic energy in the form of vibration to electrical energy and has been designed for environmental frequency range. There are a large number of different sources from which vibrations can be acquired: oscillation of a car engine, washing machine, buildings, bridges, refrigerator, human motion and etc. Vibrations from the environment are typically in the range up to 200 (Hz) [1]. It has a simple construction, small internal resistance, high current, long time of use and easy processing compared to other types of energy harvesters. So these types of harvesters are very useful for practical engineering applications. The basic design of the electromagnetic harvester consists of a magnetic core and a coil and as a result of the Faraday law, the AC voltage on the coil is induced. Various electromagnetical harvester structures have been used with both single magnets [2] and multiple magnets to increase [3] their efficiency.

1.2 Electromechanical Coupling

In general, the energy recovered is maximized when the resonance frequency corresponds to the vibratory frequency. However, energy harvesting can be improved by properly configuring the harvester. An essential feature is the coupling between mechanical and electrical systems [4]. This coupling is referred to as *electromechanical coupling* or *transduction factor*. An electromechanical coupling is a parameter (function) describing the conversion of mechanical energy to electrical energy. Generally, the electromechanical coupling is estimated as a constant value because the magnetic flux density is treated as uniform over the coil volume [5]. Consequently, coupling is defined as the product between the number of turns in the coil, the mean magnetic induction field, and the length of a single coil turn. The issue is that this assumption can be used for small oscillation of a magnet. If the magnet executes a large amplitude of motion, a variation in the magnetic flux occurs. Moreover, its value depends on the average magnetic induction and the manner in which it is chosen. Hence, some researchers have experimentally determined an electromechanical coupling as function of the position of the magnet [6], by finite element analysis [7], the optimum load resistance [8] and from fitting measured data [9]. Therefore, a more accurate derivation of coupling factor may be imperative in certain situations.

This paper focused on improving the design to obtain a better coupling between the electrical and the mechanical domains and on the updating of the mathematical models of the magnetic levitation harvester. The motivation for this study is to enhance the energy harvesting effect, and this device is intended to be used as a pendulum damper [10].

2 Electromagnetic Harvester Design

2.1 Electromechanical Coupling Shaping

First method for shaping of the electromechanical coupling, the modular coil is proposed (Fig. 1a). It consists of the four separate terminals (modulus) connected in series. Each of coil's segment can be activated separately or together. The second method is the appropriately designed oscillating magnet in the coil. The magnet can have a different shape, consist of magnets stacked in a repulsive or attractive way or can be a set of magnets and spacers (Fig. 1b). The magnet-separator stacks were made in two variants: symmetric and asymmetrical.

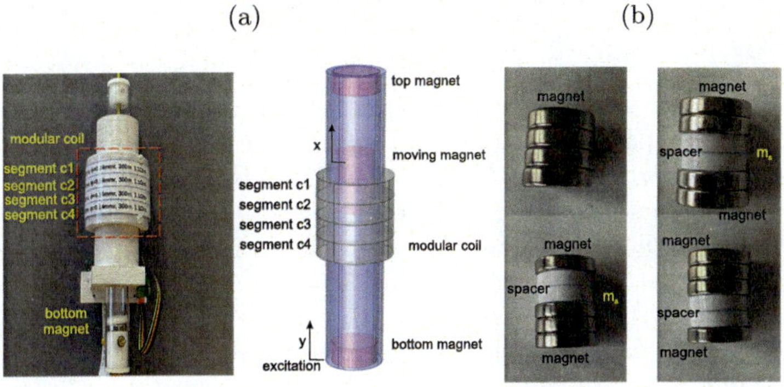

Fig. 1. Electromagnetic harvester with the modular coil consist of four separate terminals (a) and the different configuration of the moving magnet with spacers (b).

As regards determination of electromechanical coupling, a falling magnet through a coil is applied. The magnet was dropped through the modular coil and

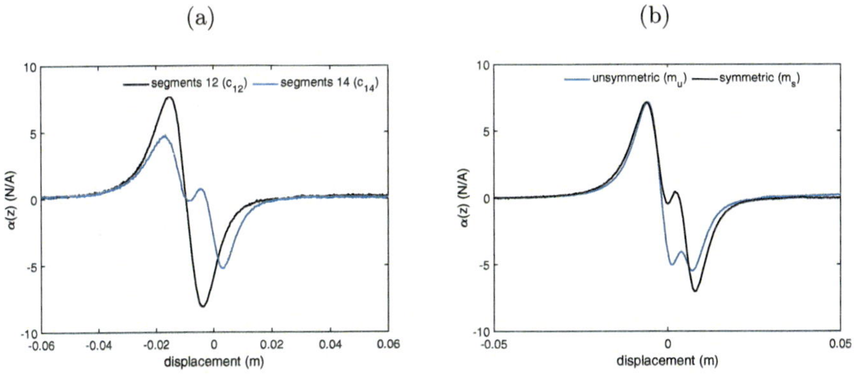

Fig. 2. Electromechanical coupling function obtained for different configurations of modular coil (a) and different configuration of the magnet (b).

the induced current was measured. Next, using the Kirchhoff's voltage law, the electromechanical coupling functions versus the magnet position were obtained. A detailed description of the methodology is available at [11].

In this study four exemplary electromechanical functions were obtained from the modular coil and from various configurations of the magnets-spacers (Figs. 2a and 2b). In Fig. 2a, the black line indicates the electromechanical coupling function for the active segments 1 and 2 (segments 3 and 4 are inactive). The configuration conforms to the classical coil, the highest values of the electromechanical coupling function occur close to the coil ends. In the middle of coil, the coupling function equals zero. The blue line marks result for active segments 1 and 4, which means that the gap distance in the middle of coil caused by inactive segments 2 and 3 occurs. Note that modifying the modular coil through the introduction of space has an extra effect (peak) near the center of the coil (Fig. 2a).

The second method to modify the electromechanical coupling is the magnets-spacers stacks. The exemplary stacks consist of four ring magnets and two very light nonmagnetic spacers interconnected together with opposite polarities facing each other are shown in (Fig. 1b). Two coupling functions for symmetric and asymmetric combinations of the magnet and spacers are illustrated in Fig. 2b.

Based on experimental results, the mathematical model of the electromechanical coupling $\alpha(\dot{z}, z, n)$ represented by trigonometric polynomials of order j formula is proposed. It is represented using sine and cosine functions of the form $sin(wz)$ and $cos(wz)$, where w is a nonnegative integer. A trigonometric polynomial representation appears to be quite useful, since it facilitates implementation. The electromechanical coupling model has the higher-order form

$$\alpha(\dot{z}, z, n) = a_0 + \sum_{j=1}^{20}(a_j cos(jwz) + b_j sin(jwz)), \tag{1}$$

where a_0 and a_j are constant terms associated with measured data. Electromechanical coupling denoted depends on the parameters: amounts of active coil segments n and/or configuration of the magnets and spacers. All fitting coefficients for various configurations are shown in Table 1.

Table 1. Coefficients of the Fourier electromechanical coupling models.

n	a_0	a_1	b_1	a_2	b_2	a_3	b_3	a_4	b_4	a_5	b_5
c12	0.1907	-1.2790	-1.6520	-2.5920	-0.6408	-2.0940	0.8981	-0.8861	1.3560	-0.0208	1.0420
c14	0.2426	-0.7538	-1.5990	-1.4900	-1.0280	-0.8760	-0.1482	-0.0785	-0.1259	0.2122	-0.3735
m_s	-0.0571	-0.1224	0.4423	-0.1511	0.6631	-0.1953	0.6701	-0.2178	0.5417	-0.1971	0.3756
m_a	0.0258	-0.0645	-0.3558	-0.0724	-0.6143	-0.0853	-0.6290	-0.0972	-0.5904	-0.1014	-0.4663
n	a_6	b_6	a_7	b_7	a_8	b_8	a_9	b_9	a_{10}	b_{10}	a_{11}
c12	0.3705	0.5264	0.2844	0.1271	0	0	0	0	0	0	0
c14	0.1808	-0.5032	0.0094	0.5009	-0.0939	-0.3074	0	0	0	0	0
m_s	-0.1450	0.1992	-0.0700	0.0654	0.0129	-0.0205	0.0746	-0.0626	0.1256	-0.0720	0.1506
m_a	-0.1115	0.3548	-0.1128	-0.2307	-0.1216	-0.139	-0.1284	-0.0609	-0.1338	-0.0202	-0.1305
n	b_{11}	a_{12}	b_{12}	a_{13}	b_{13}	a_{14}	b_{14}	a_{15}	b_{15}	a_{16}	b_{16}
c12	0	0	0	0	0	0	0	0	0	0	0
c14	0	0	0	0	0	0	0	0	0	0	0
m_s	-0.0586	0.1538	-0.0336	0.1447	-0.0108	0.1232	0.0098	0.0975	0.0240	0.0706	0.0292
m_a	0.0086	-0.1224	0.0164	-0.1126	0.0230	-0.1001	0.0162	-0.0843	0.0113	-0.0679	0.0017
n	a_{17}	b_{17}	a_{18}	b_{18}	a_{19}	b_{19}	a_{20}	b_{20}	w		
c12	0	0	0	0	0	0	0	0	66.66		
c14	0	0	0	0	0	0	0	0	67.63		
m_s	0.0484	0.0320	0.0293	0.0284	0.0160	0.0237	0.0077	0.0173	50.00		
m_a	-0.0522	0.0002	-0.0399	-0.0056	-0.0283	-0.0066	-0.0190	-0.0096	50.00		

The c_{12} and c_{14} denote coefficients for models obtained from the modular coil, while m_s and m_a from the symmetric and asymmetric magnet-spacer configurations. Note that the electromechanical coupling function from modular coil c_{14} is similar to model from magnet-spacer m_s. However, the mathematical models are different due to different peak height and width.

2.2 Mathematical Model of Harvester

The dynamics of the oscillating moving magnet was mathematically modeled by nonlinear spring-mass-damper system with an externally applied harmonic base excitation and the electrical circuit were derived from the forces acting on a system as [11]

$$m\ddot{z}(t) + c\dot{z}(t) + kz(t) + k_1 z(t)^3 + \alpha(\dot{z}, z, n)i(t) = mA\omega^2 sin(\omega t) + mg, \quad (2)$$

$$\alpha(\dot{z}, z, n)\dot{z}(t) = Ri(t) + L\dot{i}(t), \quad (3)$$

where m is the moving mass, c is the damping constant, k and k_1 are the linear and nonlinear stiffness constants. Coordinate $z(t)$ is the relative displacement of the magnet ($z(t) = y(t) - x(t)$, where $y(t)$ and $x(t)$ are the base and moving mass displacements respectively. The frequency of excitation denoted as ω, while amplitude displacement of excitation A, so that the amplitude of excitation force is $mA\omega^2$.

The electrical circuit is described by the coil inductance L and the total resistances (sum of load and internal) R. The induced current in the harvester's circuit is $i(t)$. Detailed description of the harvester model, identification of parameters and the derivation of the equations of motion have been presented in detail in [6,11].

3 Results and Discussion

The comparison of the proposed models is the main goal of this section. The values of parameters are taken from the experimental rig (Fig. 1): $m = 0.047$ (kg), $k = 20.4359$ (N/m), $k_1 = 150$ (kN/m^3), $c = 0.047$ (Ns/m), $L = 0.01$ (H), $R = 1$ (kΩ). All calculations are performed with Auto07p, for the continuation method.

The resonance curves of the maximum magnet responses z as well as the maximum recovered currents i versus the frequency of excitation ω are shown in Figs. 3a and 3b. The response of the system for active coil segment 1 and 2 (c_{12}) is denoted by the black line and the blue line shows the results for the active segments 1 and 4 (c_{14}). The dashed line means unstable, while solid denotes a stable numerical solution. The SN notation denotes the saddle-node bifurcation point, while period doubling bifurcation is marked as PD.

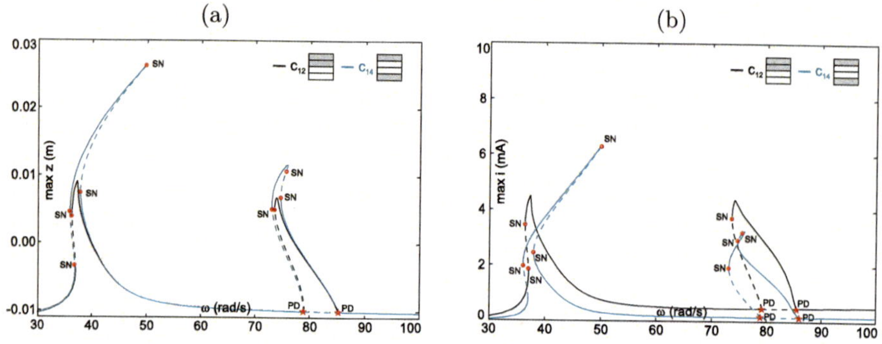

Fig. 3. Frequency response curves of the harvester: magnet's oscillation (a) and induced current (b), for different configuration of the coil segments. The numerical calculation have performed for varying ω and for fixed displacement amplitude A = 0.001 (m).

An analysis result reveals two resonance regions. The first main resonance occurs close to $\omega \approx 40$ (rad/s), the second around $\omega \approx 80$ (rad/s). The second resonance is caused by PD bifurcation which means double the period. As you can see, modification in the active coil segments influences the shape of resonance curve. The active segments c_{14} significantly expands the first resonance operation bandwidth. The resonance curve in the main resonance is strong bending and hardening effect is observed. In this region the two stable solutions occur (top and bottom branches). However, the recovered current at principal resonance close to the resonance peak is reduced. This means, that the activation of segments 1 and 4 causes expand the harvester operating range, but reduces the recovered energy close the resonance peak. In the second resonance region, the configuration c_{14}, does not extend the operation bandwidth, but the magnet vibration increased with a simultaneous reduction of induced current.

Figures 4a and 4b show the simulated maximum magnet response and induced current from the different configuration of the magnets-spacers. The black line represents a symmetric configuration denoted m_s, while the blue line is asymmetric m_a (see Fig. 1b).

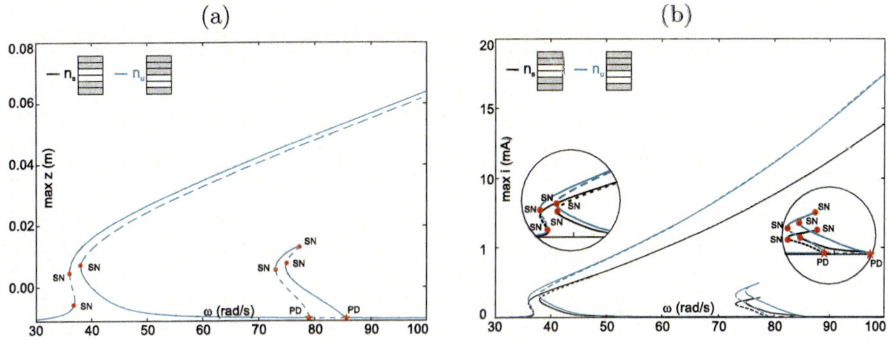

Fig. 4. Frequency response curves of the harvester: magnet's oscillation (a) and induced current (b), for different configuration of the magnets-spacers. The numerical calculation have performed for varying ω and for fixed displacement amplitude A = 0.001 (m).

In this case the first resonances is very wide, the resonance curves exhibits a strong hardening effect. Both configuration of the magnets-spacers do not affect to any resonance curves of the magnet displacements. However, the magnet-spacer influences the recovered current. The asymmetric configuration improve the induced current. The second resonance curve especially induced current is significantly reduced.

4 Conclusions

This paper presents the design and fabrication of a magnetic levitation harvester. In the harvester an electromechanical coupling function has been effectively modified by the special designed coil terminal and by the proper design of the levitating magnet. The various coupling models have been implemented in the numerical model to demonstrate its potential in induced current.

The obtained results have concluded that the electromechanical coupling can be effectively shaping by the configuration of the modular coil and/or the stacked magnets. The electromechanical coupling wave form is affected by the spacers each magnet and gap between the coils. The shape of electromechanical coupling can change profile and additional peaks can be introduced. The current induced from the interaction between the coil and the stacked magnets-separators differ significantly from the case of a single magnet. However, the response of the magnet is practically the same. The proper configuration of the modular coil cause expand the resonance bandwidth due to the higher nonlinearity of the electromechanical coupling. So, in summary, is possible to design the electromechanical coupling function with additional peaks in anywhere.

References

1. Whelan, M.J., Gangone, M.V., Janoyan, K.D., Jha, R.: Real-time wireless vibration monitoring for operational modal analysis of an integral abutment highway bridge. Eng. Struct. **31**(10), 2224–2235 (2009). https://doi.org/10.1016/j.engstruct.2009.03.022
2. Keck, K., Mitura, A.: Energy recovery from a pendulum tuned mass damper with two independent harvesting sources. Int. J. Mech. Sci. **174**, 105568 (2020). https://doi.org/10.1016/j.ijmecsci.2020.105568
3. Li, X., Meng, J., Yang, C., Zhang, H., Zhang, L., Song, R.A: Magnetically coupled electromagnetic energy harvester with low operating frequency for human body kinetic energy. Micromachines **12**(11), 1300 (2021). https://doi.org/10.3390/mi1211130
4. Mosch, M., Fischerauer, G.: A comparison of methods to measure the coupling coefficient of electromagnetic vibration energy harvesters. Micromachines **1012**, 826 (2019). https://doi.org/10.3390/mi10120826
5. Bedekar, V., Oliver, J., Priya, P.: Pen harvester for powering a pulse rate sensor. J. Phys. D Appl. Phys. **4210**, 105105 (2009). https://doi.org/10.1088/0022-3727/42/10/105105
6. Kecik, K., Kowalczuk, M.: Effect of nonlinear electromechanical coupling in magnetic levitation energy harvester. Energies **149**, 2715 (2021). https://doi.org/10.3390/en14092715
7. O'Donnell, T., Saha, C., Beeby, S., Tudor, J.: Scaling effects for electromagnetic vibrational power generators. Microsyst. Technol. **1311**, 1637–1645 (2007). https://doi.org/10.1007/s00542-006-0363-0
8. Stephen, N.G.: On energy harvesting from ambient vibration. J. Sound Vib. **293**, 409–425 (2006). https://doi.org/10.1016/j.jsv.2005.10.003
9. Spreemann, D., Hoffmann, D., Folkmer, B., Manoli, Y.: Numerical optimization approach for resonant electromagnetic vibration transducer designed for random vibration. J. Micromech. Microeng. **1810**, 104001 (2008). https://doi.org/10.1088/0960-1317/18/10/104001
10. Kecik, K.: Dynamics and control of an active pendulum system. Int. J. Non-Linear Mech. **70**, 63–72 (2015). https://doi.org/10.1016/j.ijnonlinmec.2014.11.028
11. Kecik, K., Mitura, A., Lenci, S., Warminski, J.: Energy harvesting from a magnetic levitation system. Int. J. Non-Linear Mech. **94**, 200–206 (2017). https://doi.org/10.1016/j.ijnonlinmec.2017.03.021

Multiple Regenerative Effects of the Bit-Rock Interaction in a Distributed Drill-String System

Mohammad Amin Faghihi[1(✉)], Shabnam Tashakori[2,3,4], Ehsan Azadi Yazdi[1], and Mohammad Eghtesad[1]

[1] Department of Mechanical Engineering, Shiraz University, Shiraz, Iran
aminfaghihi93@gmail.com
[2] Department of Mechanical and Aerospace Engineering, Shiraz University of Technology, Shiraz, Iran
[3] Center of Applied Dynamics Research, School of Engineering, University of Aberdeen, Aberdeen AB24 3UE, UK
[4] Rahesh Innovation Center, Shiraz, Iran

Abstract. In this paper, a distributed model in terms of neutral-type time-delay equations is presented to investigate the global non-linear axial-torsional dynamical behavior of a drilling string. A rate-independent bit-rock interaction law is employed for both cutting and frictional forces at the bit. A model is proposed for the estimation of the depth of cut which is valid in the case of bit bouncing and the bit reverse rotation. Illustrative simulation results are presented for a representative case study, which demonstrate the existence of the bit-bounce and reverse-rotation in some practical operating conditions, and indicates the need for taking the resulting multiple regenerative effects into account.

Keywords: Drill-string · Bit-rock interaction · Multiple regenerative effect · Bit-bounce · Bit reverse rotation

1 Introduction

Drill strings, which are used to dig oil and gas wells, are extremely susceptible to undesirable vibrations because of two main features, namely, (i) the very high length to diameter ratio, and (ii) irregular cutting forces. A schematic view of a drill string is illustrated in Fig. 1. Two well-known detrimental phenomena in oil-well drilling are the stick-slip and the bit-bounce. Stick-slip is a cyclic rotational oscillation with two phases, (i) the stick phase, during which the angular velocity becomes zero, and (ii) the slip phase, during which the angular velocity can be several times larger than the nominal angular velocity. In extreme torsional vibrations, the frictional component of the Torque On Bit (TOB) is not sufficient to prevent the bit from reverse rotation. Hence, the bit rotates in the

Z. Dimitrovová et al. (Eds.): WMVC 2022, MMS 125, pp. 491–500, 2023.
https://doi.org/10.1007/978-3-031-15758-5_50

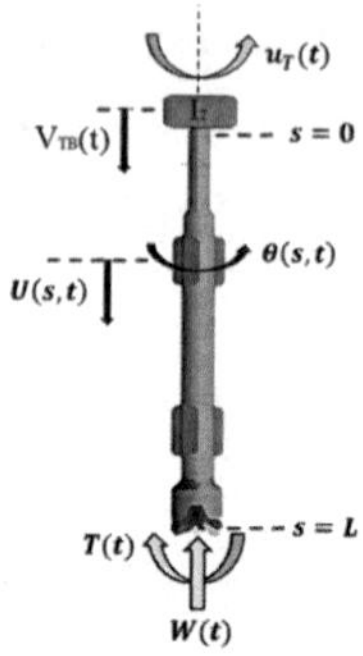

Fig. 1. A schematic of the drill-string [11].

reverse direction of the nominal angular velocity. The other adverse phenomenon is the bit-bounce. In this case, as the amplitude of axial vibrations increases, the bit losses contact with the formation. Both of these undesired phenomena have significant adverse effects on the drilling operation, such as bit wear, reduced rate of penetration, and down-hole tool failure, which ultimately increases the time and cost of the drilling operations. Accordingly, it is necessary to understand the root causes of these destructive vibrations to prevent and suppress them.

The main source of undesired axial-torsional vibrations is the interaction force between the bit and the formation. The interaction forces between the formation and Poly-crystalline Diamond Compact (PDC) bits could be divided into two independent processes: (i) the pure cutting process that occurs on the cutting face, and (ii) the frictional contact that acts between the underside of the bit and the formation [3]. Cutting forces increase with increasing the depth of cut. The instantaneous depth of cut is determined by subtracting the preceding position from the current position of the drill bit. This introduces state-dependent delays to the equations of motion. State-dependent delay model, proposed in [9], has motivated many researchers to investigate the roots of these unwanted vibrations [1,2,5,7,8,10]. However, this model assumes that the cutting operation is in progress whenever the bit rotates, and hence, cannot capture the multiple regenerative effects.

Multiple regenerative effects occur while the bit loses contact with the formation, i.e., when the bit rotates without cutting the formation. There are two scenarios for this situation, namely, bit-bounce and bit reverse rotation. In these cases, the instantaneous depth of cut is determined by the current and multiple preceding positions of the blade. There exist two different approaches to deal with the multiple regenerative effects in the bit-rock interaction [4,6]. In [4], the rock surface pattern is generated with the bit trajectory function, governed by a first-order Partial Differential Equation (PDE) with one corresponding boundary condition. By employing complex time delays in [6], multiple preceding positions of the bit are compared, and the one with the deepest penetration is considered the previous position. Then, the depth of cut is determined by subtracting the previous position from the current position of the drill bit. However, both of the above mentioned models are only valid for the bit-bounce scenario and not for the reverse-rotation scenario since the cut surface profile may be discontinuous and not solvable in the PDE framework.

In this paper, a formulation for the regenerative effects caused by the bit-bounce or the reverse-rotation is developed. The regeneration of the rock surface between two successive blades is employed to determine the instantaneous depth of cut. To determine the well-surface trajectory, an implicit function is employed rather than solving a PDE [4]. This enables the model to capture the discontinues well-surface trajectories. Thus, the main contribution of this paper is that it studies the entire global dynamics of the coupled axial-torsional motion of the bit considering (i) bit-bounce, (ii) stick-slip, and (iii) reverse-rotation. Note that to illustrate the existence of multiple regenerative effects, an NTD model, inspired by [10,12], is employed to simulate the string, which is a distributed model in terms of neutral-type time delay equations.

The paper is organized as follows: Sect. 2 presents the NTD model for the coupled axial-torsional motion of a drill-string and the boundary conditions. Simulation results for a representative case study is presented in Sect. 3. Conclusions are presented in Sect. 4.

2 Mathematical Modeling

By neglecting lateral vibrations, the string is considered as a slender structure acting as a rod (under axial forces) and a shaft (under torsional torques) simultaneously. The material properties and cross-sections are assumed constant along the string, and the internal and external damping are neglected. Based on these assumptions, the governing equations of motion are the torsional and axial wave equations given by (for the detailed derivation of the following equations, readers may refer to [12]):

$$\frac{\partial^2 \Phi}{\partial x^2}(x,t) = c_t^2 \frac{\partial^2 \Phi}{\partial t^2}(x,t), \tag{1}$$

$$\frac{\partial^2 U}{\partial x^2}(x,t) = c_a^2 \frac{\partial^2 U}{\partial t^2}(x,t), \tag{2}$$

where Φ and U are the angular and axial displacements of the pipe element located at a distance x from the top of the well at time t. The wave constants c_t and c_a are the reciprocal of the torsional and axial wave propagation speeds, given by:

$$c_t = \sqrt{\frac{\rho}{E}}, \quad c_a = \sqrt{\frac{\rho}{E}}, \tag{3}$$

with ρ, E, and G the density, Young modulus, and the shear modulus of the string, respectively. The boundary conditions are defined as follows:

$$\frac{\partial \Phi}{\partial t}(0,t) = \Omega_0(t), \tag{4}$$

$$GJ\frac{\partial \Phi}{\partial x}(L,t) = -I_b\frac{\partial^2 \Phi}{\partial t^2}(L,t) - T(t), \tag{5}$$

in the torsional direction and

$$\frac{\partial U}{\partial t}(0,t) = V_0(t), \tag{6}$$

$$EA\frac{\partial U}{\partial x}(L,t) = -M_b\frac{\partial^2 U}{\partial t^2}(L,t) - W(t), \tag{7}$$

in the axial direction, where J and A are the drill string polar moment of area, and cross-sectional area, and M_b and I_b are the mass and the polar moment of inertia of the bottom hole assembly (BHA), respectively. Note that $x = 0$ and $x = L$ denote the top and the bottom of the string. On the top boundary, $\Omega_0(t)$ and $V_0(t)$ are the imposed angular and axial velocities at the rotary table. On the other extremity, T(t) and W(t) are the TOB and the Weight On Bit (WOB). Employing the D'Alembert's solution, the axial and torsional displacements can be represented in terms of Riemann variables as follows:

$$\Phi(x,t) = \eta_t(t + c_t x) + \xi_t(t - c_t x), \tag{8}$$

$$U(x,t) = \eta_a(t + c_t x) + \xi_a(t - c_t x), \tag{9}$$

with η and ξ representing the up-traveling and down-traveling waves, and subscripts t and a denoting torsional and axial dynamics, respectively. According to (8), the boundary conditions (4) and (5) can be rewritten as follows:

$$\dot{\eta}_t(t) + \dot{\xi}_t(t) = \Omega_0(t), \tag{10}$$

$$GJc_t(\dot{\eta}_t(t + c_t L) - \dot{\xi}_t(t - c_t L)) = -J_b\ddot{\Phi}_b(t) - T(t). \tag{11}$$

The bit angular velocity is given by:

$$\dot{\Phi}_b(t) := \dot{\Phi}(t,L) = \dot{\eta}_t(t + c_t L) + \dot{\xi}_t(t - c_t L). \tag{12}$$

Solving (11) and (12) and shifting the time leads to:

$$\dot{\eta}_t(t) = -J_b/(2GJc_t)\ddot{\Phi}_b(t-\tau_t) - 1/(2GJc_t)T(t-\tau_t) + 1/2\dot{\Phi}_b(t-\tau_t), \tag{13}$$

$$\dot{\xi}_t(t) = J_b/(2GJc_t)\ddot{\Phi}_b(t+\tau_t) + 1/(2GJc_t)T(t+\tau_t) + 1/2\dot{\Phi}_b(t+\tau_t), \tag{14}$$

where $\tau_t = c_t L$ and $\tau_a = c_a L$ are, respectively, the time required for the torsional and axial waves to travel from one extremity of the string to the other end. Substituting (13) and (14) in (10) and then shifting the time, the following NTD model for the torsional dynamics is obtained [11]:

$$\begin{aligned}\ddot{\Phi}_b(t) - \ddot{\Phi}_b(t-2\tau_t) = &-\frac{GJc_T}{J_b}(\dot{\Phi}_b(t) + \dot{\Phi}_b(t-2\tau_t)) + \frac{1}{J_b}(-T(t)\\ &+T(t-2\tau_t)) + \frac{2GJc_t}{J_b}\Omega_0(t).\end{aligned} \tag{15}$$

Similarly, the axial NTD model is obtained as follows:

$$\ddot{U}_b(t) - \ddot{U}_b(t-2\tau_a) = -\frac{EAc_a}{M_b}(\dot{U}_b(t) + \dot{U}_b(t-2\tau_a)) + \frac{1}{M_b}(-W(t) \\ +W(t-2\tau_a)) + \frac{2EAc_a}{M_b}V_0(t), \tag{16}$$

where $\dot{U}_b(t)$ and $\ddot{U}_b(t)$ are the axial velocity and acceleration of the bit, respectively. The procedure to acquire axial NTD model (16) is skipped for the sake of brevity.

2.1 Bit-Rock Interaction

$T(t)$ and $W(t)$ in (15) and (16) can be decomposed to the cutting and frictional components, denoted with c and f subscripts as follows [9]:

$$T(t) = T_c(t) + T_f(t), \tag{17a}$$

$$W(t) = W_c(t) + W_f(t), \tag{17b}$$

where

$$T_c(t) = \frac{1}{2}\epsilon a^2 \mathbf{R}\left(d(t)\right)\mathbf{H}\left(\dot{\Phi}_b(t)\right), \tag{18a}$$

$$T_f(t) = \frac{1}{2}\mu\gamma a^2 \sigma l \mathbf{Sign}\left(\dot{\Phi}_b(t)\right)\mathbf{H}\left(d(t)\right)\mathbf{H}\left(\dot{U}_b(t)\right), \tag{18b}$$

$$W_c(t) = \epsilon a \zeta \mathbf{R}\left(d(t)\right)\mathbf{H}\left(\dot{\Phi}_b(t)\right), \tag{18c}$$

$$W_f(t) = \sigma a l \mathbf{H}\left(d(t)\right)\mathbf{H}\left(\dot{U}(t)\right), \tag{18d}$$

where ϵ is the rock intrinsic specific energy required to disintegrate a unit volume of the rock, μ is the friction coefficient between the blades and the formation, ζ is the cutter face inclination number, σ is the maximum contact stress at the wear-flat rock interface, d is the depth of cut, and l is the length of the drill bit wear-flat. Moreover, $\mathbf{R}(y)$, $\mathbf{H}(y)$, and $\mathbf{Sign}(y)$ are, respectively, the Ramp, Heaviside, and Sign functions of an arbitrary variable y [12]. Accordingly, determining the depth of cut is a crucial point in determining the TOB and the WOB.

2.2 Depth of Cut Estimation

Depth of cut is the thickness of the rock layer that is being cut by the cutting blades during the cutting process. In other words, it is the difference between the axial position of the well-surface in the front and the back of the blade, see Fig. 2. If the cutting process is continuously in progress, the depth of cut can be obtained from the following delayed equation [9]:

$$d(t) = U_b(t) - U_b\left(t - \tau_n(t)\right), \tag{19}$$

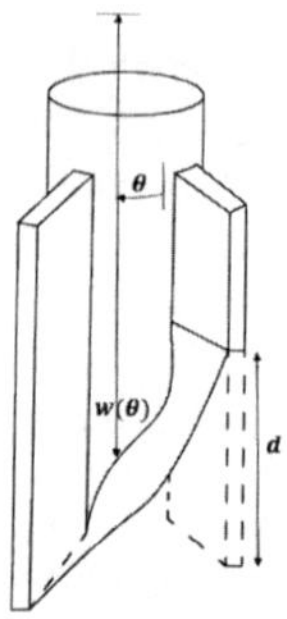

Fig. 2. Schematic of the cutting blades on the bit and the cutting surface.

where the state-dependent time delay τ_n is governed by:

$$\Phi_b(t) - \Phi_b(t - \tau_n) = \frac{2\pi}{n}. \tag{20}$$

According to (19) and (20), $d(t)$ introduces a state-dependent time delay to the equations of motion. When the drill bit rotates without cutting the formation, the time delay calculation from (20) will be no longer valid since multiple regenerative effects take place. Then, the right-hand side of (20) may be an integer multiple of $\frac{2\pi}{n}$ rather than $\frac{2\pi}{n}$. In this following, a modified model is proposed to obtain the depth of cut which enables capturing the multiple regenerative effects in both bit-bounce and reverse-rotation cases.

According to the same axial position of the n cutting blades, the well pattern is a periodic function of Φ with the period $\frac{2\pi}{n}$. Therefore, $\Phi_f(t)$ which is the angular position of the cutting blade that is located in first sector, $\Phi_{c_i} \in [0, \frac{2\pi}{n}]$ at time t is given by ($\Phi_{c_i}, 1 \leq i \leq n$ represents the angular position of the ith cutter):

$$\Phi_f(t) = \Phi_{abs}(t) - \frac{2\pi}{n}[\frac{\Phi_{abs}(t)}{2\pi/n}], \tag{21}$$

where [.] denote the floor function and

$$\Phi_{abs}(t) = \Phi_b(0) + \int_0^t \dot{\Phi}_b(t)dt. \tag{22}$$

Intuitively, $\Phi_{abs}(t)$ is the absolute angular displacement of the bit during the operation since the start of the process. When $\dot{\Phi}_b(t) > 0$, the following implicit function w governs the well-trajectory time evolution:

$$w\big(\Phi_f(t)\big) = max\Big(U_b(t), w\big(\Phi_f(t)\big)\Big). \tag{23}$$

which indicates that the cutting is in process and the well-surface trajectory is evolving by the bit that is cutting the well-surface at the point $\theta = \Phi_f(t)$ if (i) the bit is penetrated to the well-surface, i.e., $U_b(t)$ is greater than $w\big(\Phi_f(t)\big)$, and

(ii) the bit is rotating in the cutting direction ($\dot{\Phi}_b(t) > 0$). Otherwise, the well-trajectory remains unchanged. Regarding the well-surface trajectory determined by (23), the depth of cut is given by:

$$d(t) = \lim_{\varepsilon \to 0+} \Big(w\big(\Phi_f(t) - \varepsilon\big) - w\big(\Phi_f(t) + \varepsilon\big)\Big), \tag{24}$$

when $0 < \Phi_f(t) < \frac{2\pi}{n}$ and

$$d(t) = \lim_{\varepsilon \to 0+} \Big(w\big(\frac{2\pi}{n} - \varepsilon\big) - w\big(\varepsilon\big)\Big), \tag{25}$$

when $\Phi_f(t) = 0$ or $\Phi_f(t) = \frac{2\pi}{n}$.

The general idea behind the proposed model is as follows. If the cutting is in progress and the bit is penetrated to the formation, the well-surface depth behind the cutting blade is equal to the axial position of the blade. Otherwise, the depth of cut is zero, and the well-surface pattern remains unchanged. Accordingly, depth of cut can be obtained from the well-surface function, like the strategy in [4]. Note that, here, unlike [4], the formulation is not in the differential framework. This enables the model to capture the cases with discontinues and non-differentiable well-surface trajectories, e.g., when the reverse-rotation occurs. For more insight, see Fig. 3 which illustrates the discontinuity of the well-surface trajectory in the case of reverse-rotation.

3 Results and Discussion

The parameter values used for the simulations are given in Table 1. Figure 4 compares the behavior of the system modeled by the state-dependent delay model and the proposed model in this paper. In this figure, the nominal velocities and the rock specific energy are considered $\Omega_0 = 60\,\text{rpm}$, $V_0 = 5\,\text{m/h}$, and $\epsilon = 8\,\text{Mp}$, respectively. As can be seen, the velocities are positive in most of the time, i.e., the bit-bounce and reverse-rotation occur in very short time intervals. Consequently, both models predict almost the same behavior for the string, as shown.

Table 1. Parameter values [11].

Parameter	Value	Parameter	Value	Parameter	Value
J_b	$89\,\text{kgm}^2$	E	$200 \times 10^9\,\text{N/m}^2$	G	$79 \times 10^9\,\text{N/m}^2$
ρ	$8000\,\text{kg/m}^3$	A	$35^{-4}\,\text{m}^2$	J	$1.9 \times 10^{-5}\,\text{m}^4$
L	$1172\,\text{m}$	M_b	$40000\,\text{kg}$	a	10.8×10^{-2}
ζ	.6	σ	60×10^6	l	1.2×10^{-3}
μ	.6	γ	1	n	4

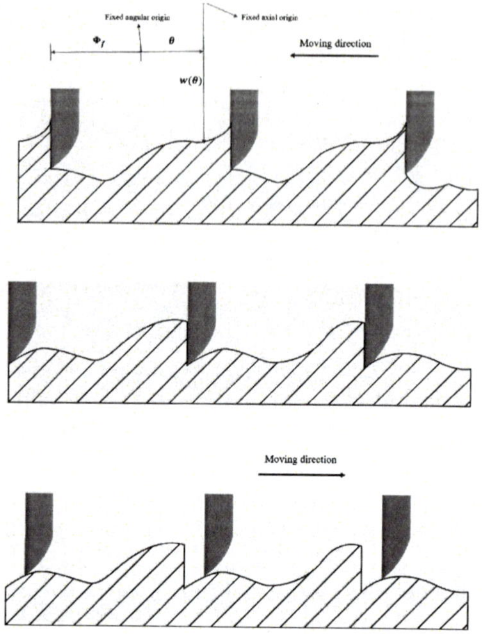

Fig. 3. A planar representation of a portion of the full cut surface by unfolding the cylinder when the bit (a) is rotating in the nominal direction, (b) stops rotating, (c) is rotating in the reverse direction.

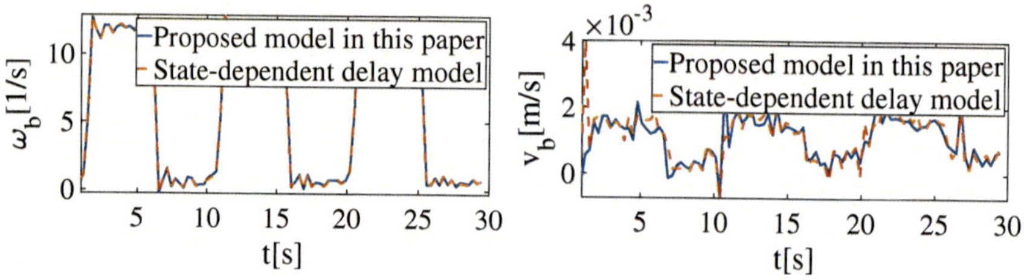

Fig. 4. Bit velocities for $V_0 = 5\,\text{m/h}$, $\Omega_0 = 60\,\text{rpm}$, and $\epsilon = 8\,\text{Mpa}$ simulated by our model and the state-dependent delay model presented in [10].

By increasing the rock-specific energy and the nominal penetration rate, the oscillations become more severe. The simulations for this case with $\Omega_0 = 60\,\text{rpm}$, $V_0 = 10\,\text{m/h}$, and $\epsilon = 60\,\text{Mp}$ are depicted in Fig. 5. Since the bit-bounce and the reverse-rotation take place in longer time intervals, the difference between the results of models is significant. Note that the simulation results of the model presented in [4] diverge since the well-trajectory becomes discontinuous when the reverse-rotation occurs. In this case, differentiating the well-trajectory for the governing PDE give infinite values.

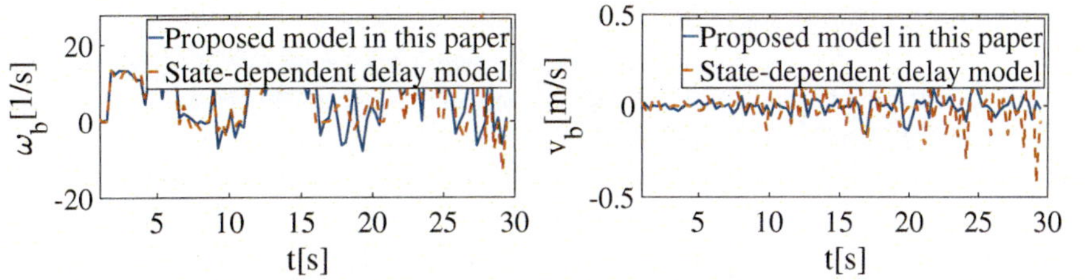

Fig. 5. Bit velocities for $V_0 = 10\,\text{m/h}$, $\Omega_0 = 60\,\text{rpm}$, and $\epsilon = 60\,\text{Mpa}$ simulated by our model and the state-dependent delay model presented in [10].

These simulation results, which are presented for practical field condition, illustrates the necessity of using the proposed model in this paper that captures the reverse-rotation phenomenon as well as the bit bouncing. In Fig. 5, bit bouncing happens due to the severity of the fluctuations, which itself excites the torsional vibrations considerably and causes the bit reverse rotation.

In addition, a comparison between simulation results of the lumped model in [4] and the NTD model presented in this paper is depicted in Fig. 6. These results emphasize the importance of considering the higher axial and torsional modes in string vibrations.

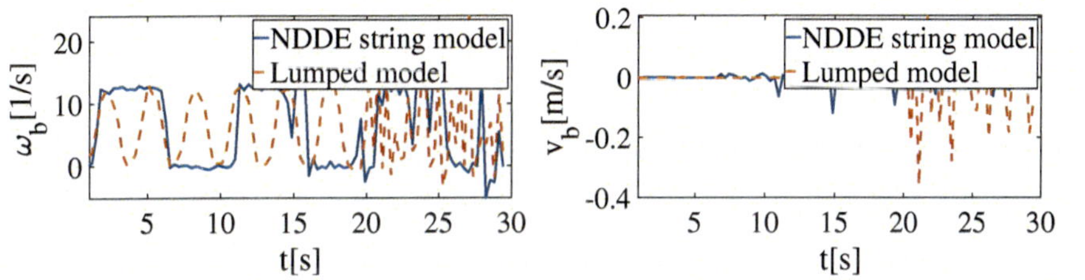

Fig. 6. Bit velocities for $V_0 = 5\,\text{m/h}$, $\Omega_0 = 60\,\text{rpm}$, and $\epsilon = 60\,\text{Mpa}$ simulated by the NDDE model in this paper and the lumped model in [4].

4 Conclusion

In this paper, coupled axial-torsional dynamics of a distributed drill string has been studied. A modified model has been proposed for the estimation of the depth of cut in the bit-rock interaction. The cut-surface profile has been determined by providing an implicit function of the bit torsional and axial trajectories. This modification enables the model to capture the multiple regenerative effects caused by the bit reverse rotation in addition to the bit bouncing. The proposed model is illustratively compared with other models in the literature by employing a distributed drill-string model in terms of neutral-type delay differential equations. The simulation results illustrate the existence of the reverse-rotation and bit-bounce in a wide range of operating conditions, which indicates the necessity of considering multiple regenerative effects.

References

1. Aarsnes, U.J.F., van de Wouw, N.: Dynamics of a distributed drill string system: Characteristic parameters and stability maps. J. Sound Vib. **417**, 376–412 (2018)
2. Besselink, B., Vromen, T., Kremers, N., Van De Wouw, N.: Analysis and control of stick-slip oscillations in drilling systems. IEEE Trans. Control Syst. Technol. **24**(5), 1582–1593 (2015)
3. Detournay, E., Defourny, P.: A phenomenological model for the drilling action of drag bits. Int. J. Rock Mech. Min. Sci. Geomech. Abs. **29**(1), 13–23. Elsevier (1992)
4. Gupta, S.K., Wahi, P.: Global axial-torsional dynamics during rotary drilling. J. Sound Vib. **375**, 332–352 (2016)
5. Gupta, S.K., Wahi, P.: Criticality of bifurcation in the tuned axial-torsional rotary drilling model. Nonlinear Dyn. **91**(1), 113–130 (2018)
6. Liu, X., Long, X., Zheng, X., Meng, G., Balachandran, B.: Spatial-temporal dynamics of a drill string with complex time-delay effects: bit bounce and stick-slip oscillations. Int. J. Mech. Sci. **170**, 105338 (2020)
7. Liu, X., Vlajic, N., Long, X., Meng, G., Balachandran, B.: Nonlinear motions of a flexible rotor with a drill bit: stick-slip and delay effects. Nonlinear Dyn. **72**(1), 61–77 (2013)
8. Liu, X., Vlajic, N., Long, X., Meng, G., Balachandran, B.: State-dependent delay influenced drill-string oscillations and stability analysis. J. Vib. Acoust. **136**(5), 9 (2014)
9. Richard, T., Germay, C., Detournay, E.: A simplified model to explore the root cause of stick-slip vibrations in drilling systems with drag bits. J. Sound Vib. **305**(3), 432–456 (2007)
10. Tashakori, S., Vossoughi, G., Zohoor, H., van de Wouw, N.: Suppression of axial-torsional vibrations in drilling system described by neutral-type delay differential equations. In: NODYCON 2019: First International Nonlinear Dynamics Conference (2019). https://nodycon.org/2019/
11. Tashakori, S., Vossoughi, G., Zohoor, H., van de Wouw, N.: Prediction-based control for mitigation of axial-torsional vibrations in a distributed drill-string system. IEEE Trans. Control Syst. Technol. **30**(1), 277–293 (2021)
12. Tashakori, S., Vossoughi, G., Zohoor, H., Yazdi, E.A.: Modification of the infinite-dimensional neutral-type time-delay dynamic model for the coupled axial–torsional vibrations in drill strings with a drag bit. J. Comput. Nonlinear Dyn. **15**(8) (2020)

Nonlinear Dynamics Analysis for a Model-Reduced Rotor System with Nonlinear Galerkin Method

Yue Xu[1,2], Jin Huang[3], Yuefang Wang[3](✉), Cong Li[1,2], and Xuemin Wei[1]

[1] Aviation Industry Corporation of China Aerodynamic Research Institute, Harbin 150001, China

[2] Aviation Key Laboratory of Aerodynamics for Low Speed and High Renolds Number, Harbin 150001, China

[3] Dalian University of Technology, Dalian 116024, China

yfwang@dlut.edu.cn

Abstract. Rotor-bearing systems are important components of aircraft engines of which vibration response must be analyzed under normal and disturbed conditions. Due to the existence of strong nonlinearity from squeeze film dampers and ball bearings, it is often very time consuming to evaluate the vibration of the rotor excited by unbalanced mass, which makes efficient and reliable modeling reduction is pivotal to rotordynamic analyses. In this paper, the nonlinear vibration and dynamical behavior of a twin-spool rotor assembly supported on bearings and squeeze film dampers with centralizing springs is investigated through the post-processed nonlinear Galerkin method. The rotor is spatially discretized with the finite element method, and the nonlinear governing equation of motion of the system is derived. To reduce the scale of the full-sized model, the post-processed nonlinear Galerkin method is employed to project the original system onto a low-dimensional sub-manifold with a relatively smaller number of degrees of freedom. The convergence of the nonlinear Galerkin method is verified through comparisons energy between the original and the reduced systems. The optimal choice of the number of master modes is determined to form the reduced model for nonlinear vibration analyses. Through the NGM the vibration of the system is solved, and the amplitude-frequency response, bifurcation and the Poincaré map are obtained to demonstrate the complexity in the rotor dynamics.

Keywords: Rotordynamics · Nonlinear Galerkin method · Model reduction · SFD · Ball bearing

1 Introduction

So far many efforts have been devoted to nonlinear dynamics of flexible rotor systems consisting of squeeze film dampers (SFDs) and ball bearings [1]. It is worth pointing out that the damping force of a SFD is nonlinear with respect to the perturbed position and velocity of the rotor, and the force from the ball bearing is nonlinear due to the Hertzian ball-race contact. Therefore, the dynamical behavior the rotor assembly is substantially

Z. Dimitrovová et al. (Eds.): WMVC 2022, MMS 125, pp. 501–511, 2023.
https://doi.org/10.1007/978-3-031-15758-5_51

different from linear dynamic systems. To make the situation even more complicated, the force of the ball bearing usually pulsates with a varying compliance (VC) frequency due to the periodic change of stiffness when the shaft rotates [2]. Due to the combined loading of mass imbalance, bearing clearance, nonlinear Hertzian contact and VC-frequency related vibration [3, 4], it is generally difficult to assess the vibration response, steady or transient, of the rotor system in terms of consumption of computer resource and runtime of simulation. Hence, it is important to reduce the scale, or order of the system. The traditional fixed-interface component mode synthesis method is dependent on not only the complicacy of nonlinear terms in the system equation but also the scale of the nonlinear subsystem. An effective approach for model reduction is the Galerkin method with which the nonlinear system is projected into a subspace spanned by a set of assumed modes. In this way, a large dynamical system can be separated into a master subsystem and a slave subsystem [5]. In a later publication the approximate inertial manifold (AIM) was introduced that could be used to attract the dynamical flow of the original system to the neighborhood of the AIM [6]. Based on the AIM, the nonlinear Galerkin method (NGM) was developed with a lower order sub-system constructed through approximating the slave subsystem as a function of the master subsystem. In general, the order of the slave subsystem as the complimentary set which is smaller in scale than the master subsystem to save computational time [7].

In this paper, the nonlinear vibration of a twin spool rotor is investigated through model reduction technique based on the nonlinear Galerkin method. The rotor assembly is discretized through the finite element method considering nonlinear SFDs and ball bearing. The governing equation of motion is developed and is reduced using the nonlinear Galerkin method with approximate inertial manifold. The accuracy and efficiency of the model reduction is verified with various number of master modes. Using the reduced model, the nonlinear vibration response of the system is numerically solved. It is observed that the dynamical behavior is complicated and diverse for different ranges of rotation speed. Quasi-periodic motion, bifurcation and route to chaos can be identified from frequency spectra, orbital plots and Poincaré maps.

2 Problem Setup and Equations of Motion

The sketch of a typical twin-spool rotor assembly is shown in Fig. 1 which consists of a low pressure (LP) shaft, a high pressure (HP) shaft, and several disks. The shafts are connected with an inter-shaft bearing. Mass unbalances exist in the bodies of one or all disks which create centrifugal force when the assembly spins. The whole rotor assembly is supported on ground bearings fixed to the engine's housing. Squeeze film dampers are frequently installed at the ends of the LP shaft for additive viscous damping to absorb vibration energy of the shafts during normal and disturbed operations.

The equations of motion of the rotor assembly can be derived through the Lagrange's equations in terms of the nodal displacements $\mathbf{q}$ of the rotor

$$\mathbf{M}\ddot{\mathbf{q}} + \mathbf{C}\dot{\mathbf{q}} + \mathbf{K}\mathbf{q} = \mathbf{f}(\mathbf{q}, t) \tag{1}$$

where $\mathbf{M}$ and $\mathbf{K}$ are assembled mass and stiffness matrices of the system, $\mathbf{C}$ is the assembled matrix taking into account the damping of the bearings and the gyroscopic matrix of the rotor **f** stands for the assembled forcing vector of mass imbalance, and the nonlinear forces generated by the SFDs and the ball bearing respectively.

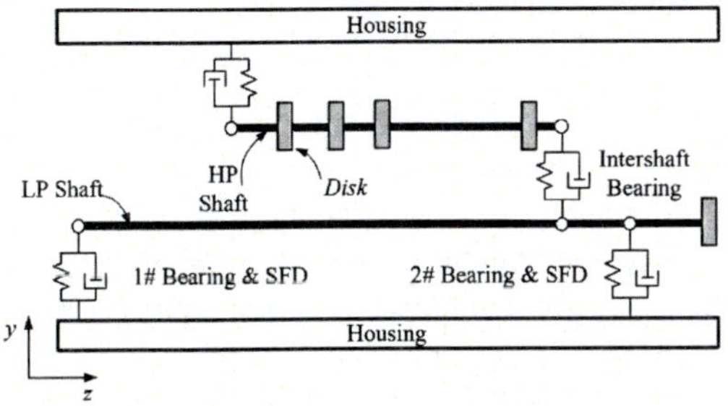

Fig. 1. Sketch of a twin-spool rotor assembly.

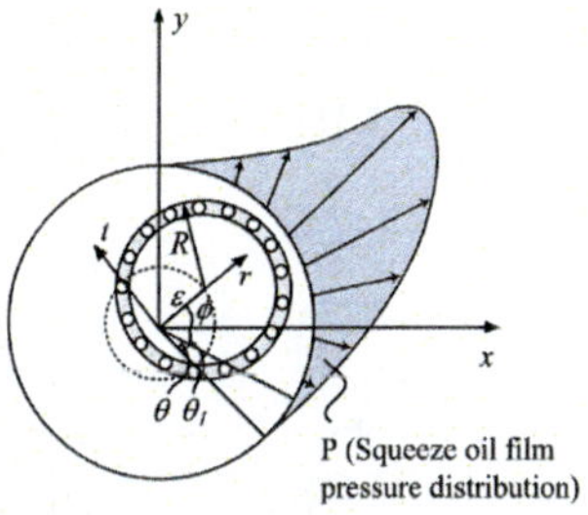

Fig. 2. Schematic oil pressure LP: low pressure, HP: high pressure distribution of a SFD.

The nonlinearity in the present rotor dynamics lies in the constitutive relation of the SFD and ball bearings with respect to the rotor motion. Among them the SFD force can be expressed following Ocvick's short bearing solution [8]

$$f_{Sx} = F_r\cos\phi - F_t\sin\phi, f_{Sy} = F_r\sin\phi + F_t\cos\phi \tag{2}$$

where $F_r = -\mu RL^3c^{-2}\left(I_1\dot{\varepsilon} + I_2\varepsilon\dot{\phi}\right)$, $F_t = -\mu RL^3c^{-2}\left(I_2\dot{\varepsilon} + I_3\varepsilon\dot{\phi}\right)$. μ, R, L and c are dynamic viscosity of lubricant, radius of out race of the attached bearing, length of the oil-film and nominal clearance of the damper, respectively. ε and ϕ are normalized radial offset and the attitude of the shaft journal (Fig. 2), respectively, and

$$I_1 = \int_{\theta_1}^{\theta_1+\pi}\frac{\cos^2\theta}{(1+\varepsilon\cos\theta)^3}d\theta,\ \ I_2 = \int_{\theta_1}^{\theta_1+\pi}\frac{\sin\theta\cos\theta}{(1+\varepsilon\cos\theta)^3}d\theta\ ,\ \ I_3 = \int_{\theta_1}^{\theta_1+\pi}\frac{\sin^2\theta}{(1+\varepsilon\cos\theta)^3}d\theta \tag{3}$$

where θ is the angle measured form the maximum film thickness. θ_1 denotes the angular position of the start of positive pressure region measured from θ.

For modeling the ball-bearing that has N_b rolling balls, the so-called passing frequency (or VC frequency) is defined in term of the rotation speed of the bearing cage

$$\omega_{VC} = \omega_{cage}N_b \tag{4}$$

The nonlinear contact force generated by the contact between the j-th rolling ball and the raceway can be obtained by the Hertzian theory in the same Cartesian frame, as

$$\begin{cases} F_{jx} = C_b(x\cos\theta_j + y\sin\theta_j - r_o)^{3/2}\cdot H(x\cos\theta_j + y\sin\theta_j - r)\cos\theta_j \\ F_{jy} = C_b(x\cos\theta_j + y\sin\theta_j - r_o)^{3/2}\cdot H(x\cos\theta_j + y\sin\theta_j - r)\sin\theta_j \end{cases} \tag{5}$$

where C_b is the Hertz contact stiffness related to the shape and materials of contact objects, $\theta_j = \omega_{cage}t + 2\pi N_b^{-1}(j-1)$ represents the angular position of the j-th rolling ball at time t. H is the Heaviside function. r_0 is the half of the bearing clearance. r is the radius of the inner race of the bearing. The supporting force of the entire bearing is obtained by summing up the contact force in Eq. (5) for each individual ball. The details will not be extended further due to the page limit of the paper.

3 Model Reduction Through the Nonlinear Galerkin Method

It can be observed that the present system dynamics is nonlinear due to the presence of both SFD and the ball bearing, and the numerical simulation of vibration response is expected to be quite time consuming. Herein, the nonlinear Galerkin method (NGM) is adopted to obtain a reduced model of the original rotor system.

First, the original system of Eq. (1) is transformed into a n-dimensional first-order system with the linear and nonlinear terms separated, expressed as

$$\dot{\mathbf{u}} + \mathbf{g}(\mathbf{q}, t) = \dot{\mathbf{u}} + \mathbf{A}\mathbf{u} + \mathbf{h}(\mathbf{u}, t) = 0, \quad \mathbf{u} = (\mathbf{x}, \dot{\mathbf{x}}), \tag{6}$$

where $\mathbf{u}$ is the state vector composed of displacements and velocities. Let $\mathbf{u} = \mathbf{Y}_m\xi + \mathbf{Z}_p\eta$ where the first m eigenvectors $\mathbf{Y}_m$ spans the subspace $\mathbf{Y}$(subsystem with m master modes) of $\mathbf{u}$, and the rest eigenvectors $\mathbf{Z}_p$ span the complement space $\mathbf{Z} = \mathbf{I} - \mathbf{Y}$($p$-dimensional slave subsystem), where $\mathbf{I}$ is identity. Usually $p \leq m, \quad p + m = n$. The original system can then be separated such that

$$\begin{cases} \dot{\xi} + \tilde{\mathbf{Y}}_m^T\mathbf{A}\mathbf{Y}_m\xi + \tilde{\mathbf{Y}}_m^T\mathbf{g}(\mathbf{Y}_m\xi + \mathbf{Z}_p\eta, t) = 0 \\ \dot{\eta} + \tilde{\mathbf{Z}}_p^T\mathbf{A}\mathbf{Z}_p\eta + \tilde{\mathbf{Z}}_p^T\mathbf{g}(\mathbf{Y}_m\xi + \mathbf{Z}_p\eta, t) = 0 \end{cases}. \tag{7}$$

Other than the standard Galerkin method where $\eta = \mathbf{0}$ is assumed, the nonlinear Galerkin method constructs an inertial manifold,$\eta = \Phi(\boldsymbol{\xi})$, to express the global attractors of the system. Eventually, this leads to

$$\dot{\boldsymbol{\xi}} + \tilde{\mathbf{Y}}_m^T\mathbf{A}\mathbf{Y}_m\xi + \tilde{\mathbf{Y}}_m^T\mathbf{g}(\mathbf{Y}_m\xi + \mathbf{Z}_p\Phi_{app}(\xi), t) = 0. \tag{8}$$

which can be solved numerically through the Runge-Kutta method. The $\Phi_{app}(\xi)$ is an approximate inertial manifold (AIM) since it is usually difficult to find the exact $\Phi(\xi)$. Once ξ is determined, the original solution can be reconstructed through

$$\mathbf{u} = \mathbf{Y}_m\xi + \mathbf{Z}_p\Phi_{app}(\xi) \tag{9}$$

More than one options can be used to construct approximate inertial manifolds. In this study, the quasi-stationary AIM (c.f. [7]) is adopted for better executive efficiency:

$$\eta = -\mathbf{J}^{-1}\tilde{\mathbf{Z}}_p^T\mathbf{g}(\mathbf{Y}_m\xi + \mathbf{Z}_p\eta, t) \tag{10}$$

where $\mathbf{J} = \tilde{\mathbf{Z}}_p^T\mathbf{A}\mathbf{Z}_p$. In practice, a fixed-point iteration method with a few iterations is proposed to solve for the mapping of η.

For the convergence of the NGM, it was proved by [9] that the NGM with AIM is convergent for large m. The test for the convergence performance of NGM will be presented in the next section.

4 Application of the NGM

The finite element (FE) model of the rotor is shown in Fig. 3. The length of the LP shaft is 1.6 m and divided into 6 segments, while the HP shaft is 1 m and divided into 8 segments. The total number of degrees of freedom of the full FE model is 60. The stiffnesses of the two SFDs are 70 MN/m and 30 MN/m, respectively. The lengths of shaft journals are both 20 mm. The radius of the journals are both 40 mm, and the radial clearances are both 0.1 mm. The dynamic viscosity of the SFD oil film is 0.02Pa·s. For the three linear ground bearings the supporting stiffnesses are 70 MN/m, 30 MN/m and 60 MN/m, respectively. For the intcr-shaft bearing (bearing #4 in Fig. 3), the stiffness is 20 MN/m. A mass imbalance of 4 kg·mm is mounted on the second impeller (node 11). The first node (bearing #1) is chosen as the observed point for studying the dynamic response of the rotor.

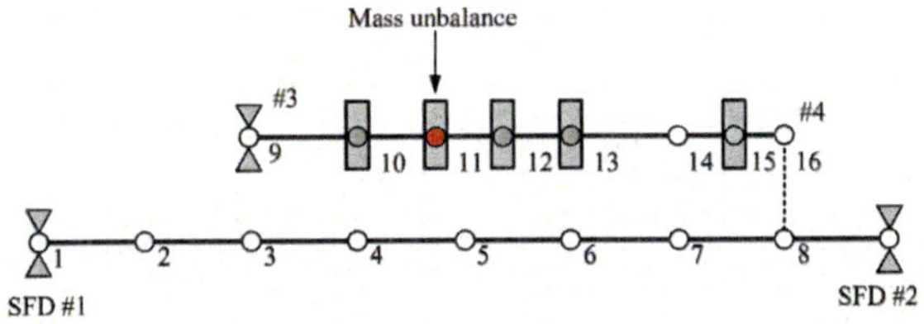

Fig. 3. Finite element model of the twin spool rotor.

4.1 Verification of the Convergence of NGM

The aforementioned model reduction based on the NGM with AIM are performed with various choices of master modes $\mathbf{Y}_m$. Each of the reduced-model solutions is compared with the full-model (of 60 DOFs) solution in term of the relative error of vibrational energy, displacement and velocity. For brevity, only the relative error of the kinetic energies is presented with respect to the number of master modes, Fig. 4, alongside the computation time consumed by each reduced model. The shadowed area indicates the possible choices of master modes corresponding to acceptable error and simultaneously, low computation cost. Hereinafter, 25 master modes of lower order eigenvectors will be chosen.

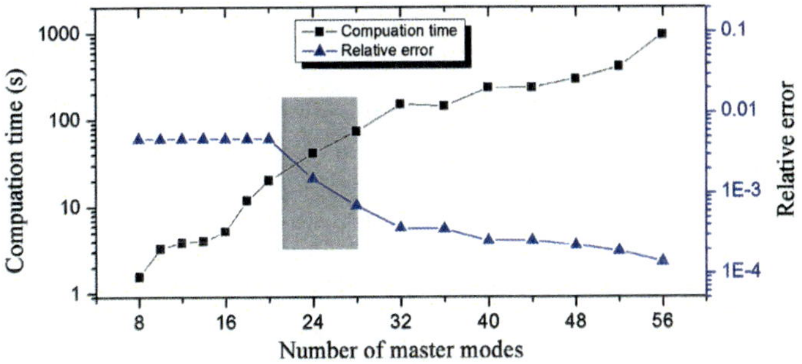

Fig. 4. Relative error of kinetic energy versus computation time

4.2 Results and Discussion

First, the vibration response is obtained to recognize the effect of the ball bearing. The frequency of the VC modulated vibration at Node 8 is depicted in Fig. 5 at rotation speed $\Omega = 10098$ rpm using time waveforms of displacement, frequency spectrum of the Y displacement and shaft's center orbit. Based on the frequency spectrum, the response is the combination of a series of harmonics with distinctive frequencies f_i and the harmonics carrying multiple folds of Ω. Owing to the subharmonic f_1, the superposed orbit of the rotor is close to a period-5 motion. That is to say, the period of the motion is approximately five times the period of the rotor spinning. Further, it is found that these f_i ($i = 1,\cdots,8$) are related to the VC frequency, i.e. the ball bearing's characteristic frequency, as well as the rotation speed in such a manner

$$\begin{aligned} & f_1 = VC - 3\Omega, f_2 = 2VC - 4\Omega, f_3 = 2VC - 2\Omega, f_4 = 4VC - 8\Omega, \\ & f_5 = 5\Omega, f_6 = VC + 2\Omega, f_7 = 4VC - 7\Omega, f_8 = 2VC + 2\Omega. \end{aligned} \tag{11}$$

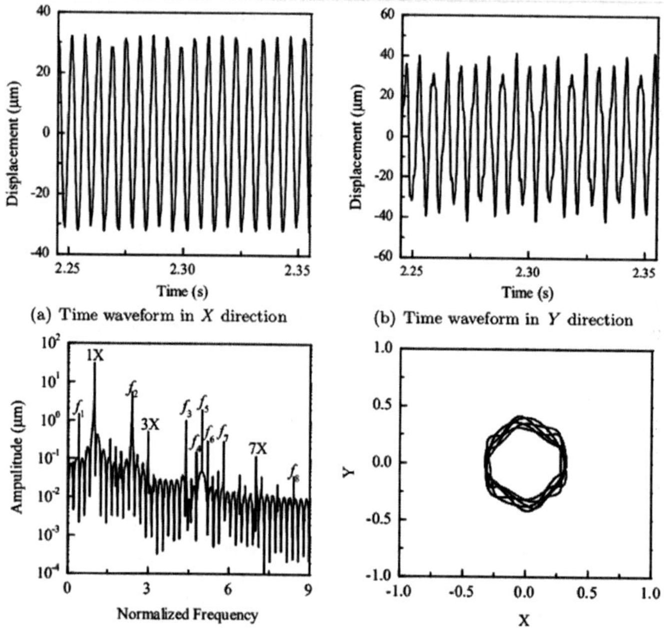

Fig. 5. Vibration in the X and Y directions for $\Omega = 10098$ rpm.

Figure 6 presents bifurcations of the normalized horizontal and vertical displacements at Node 8, respectively. The intermittency of motion can be seen in both directions which acts as the route to chaos. The influence of the VC frequency is prominent at close to the critical speeds. For better demonstration of the rotor-dynamic behavior, the bifurcation plots presented in Fig. 6 are divided into five regions with respect to the rotation speed Ω.

The characteristics of the rotor dynamics are summarized as follows:

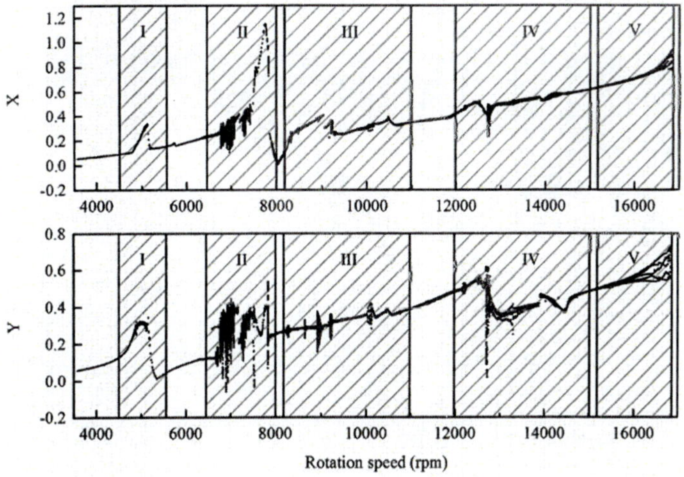

Fig. 6. Bifurcation of *X*- and *Y* displacements at Node 8

(1) Region I (Fig. 7, when Ω is between 4500 rpm and 5000 rpm). The vibrations are mostly periodic, and the resonance takes place when Ω approaches its first critical value. The responses become complicated when Ω = 4898 rpm. The Fast Fourier Transform performed on the *Y* displacement shows a 0.2 Ω component of frequency. The shaft's center moves in a tangled and twisted orbit of whirling. The Poincaré map of the (*x*, *y*) motions shows that vibration is dominated by a period-5 motion.
(2) Region II (Ω between 6500 rpm and 8000 rpm). Intermittent aperiodic motion can be found in both directions. The *Y*-displacement is quasi-periodic motion for speed Ω∈[6689, 6718] rpm. When Ω = 6704 rpm the power spectrum is composed of odd number folds of Ω and a band of other frequencies, Fig. 8. The flow on the Poincaré map is a patterned scattered points corresponding to the quasi-periodic motion. The motion remains period-1 until Ω = 6775 rpm, Fig. 9. For Ω larger than 6775 rpm, the system leaps to a larger amplitude state. From 6818 to 7076 rpm, the responses are a series of chaotic waveforms, Fig. 10. With the speed increased to a higher value of 7835 rpm, the frequency spectrum shows a white-band structure, which indicates the existence of chaotic motion, Fig. 11.
(3) Region III (Ω between 8500 rpm and 11000 rpm). For Ω∈[8294, 9024] rpm and [9210, 9239] rpm, the responses are quasiperiodic motion, Fig. 12.
(4) Region IV (Ω between 12000 rpm and 15000 rpm). For Ω∈[12705, 12748] rpm, the response becomes quasi-periodic again and the Poincaré map shows a closed loop, Fig. 13. The response is periodic for other rotation speeds.
(5) Region V (Ω between 15500 rpm and 17000 rpm). In this region, the motions in both X and Y directions are single- and multiple fold periodic regarding Ω. Nevertheless, details of the motion will not be presented due to the page limit of the paper.

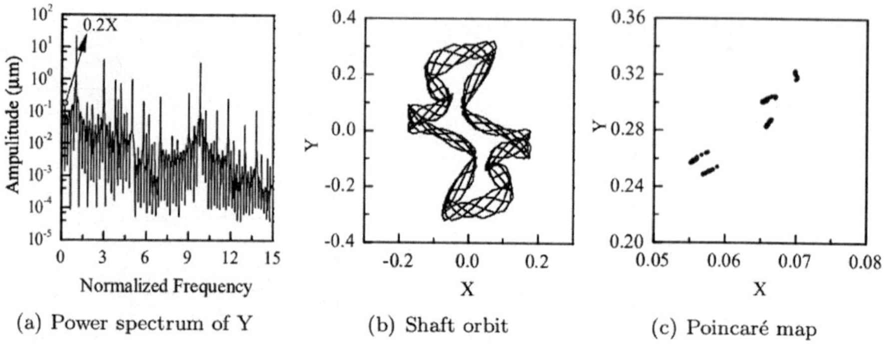

Fig. 7. Normalized Y displacement, shaft orbit and Poincaré map for $\Omega = 4898$ rpm.

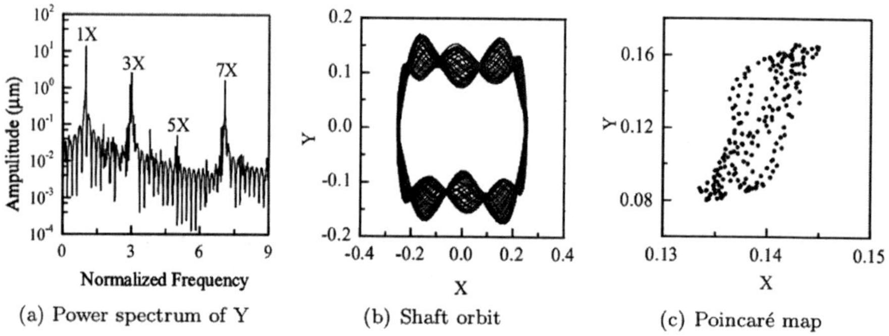

Fig. 8. Normalized Y displacement, shaft orbit and Poincaré map for $\Omega = 6704$ rpm.

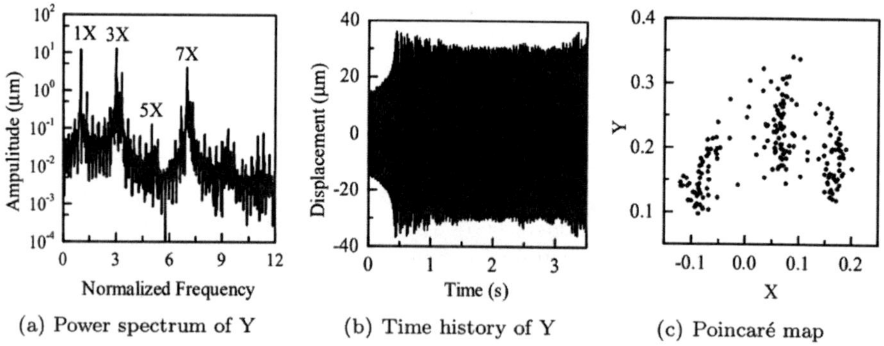

Fig. 9. Normalized Y displacement, its time history and Poincaré map for $\Omega = 6775$ rpm.

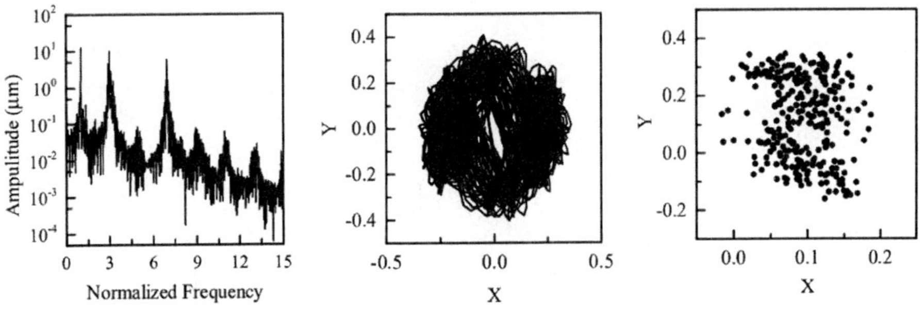

(a) Power spectrum of Y (b) Shaft orbit (c) Poincaré map

Fig. 10. Normalized Y displacement, shaft orbit and Poincaré map for $\Omega = 6904$ rpm.

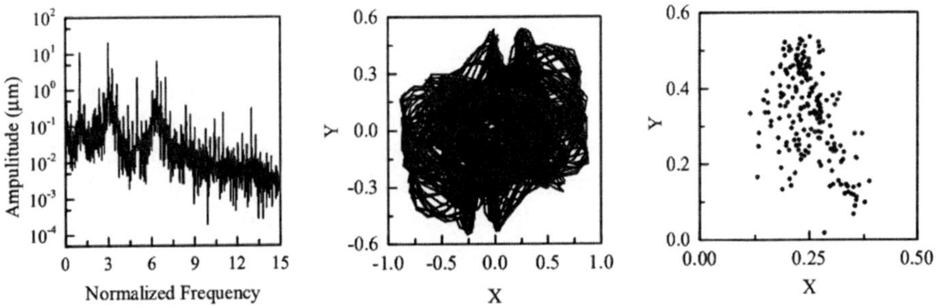

(a) Power spectrum of Y (b) Shaft orbit (c) Poincaré map

Fig. 11. Normalized Y displacement, shaft orbit and Poincaré map for $\Omega = 7835$ rpm.

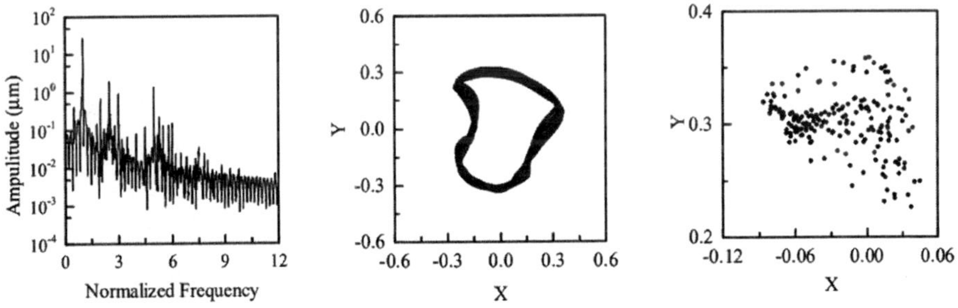

(a) Power spectrum of Y (b) Shaft orbit (c) Poincaré map

Fig. 12. Normalized Y displacement, shaft orbit and Poincaré map for $\Omega = 9225$ rpm.

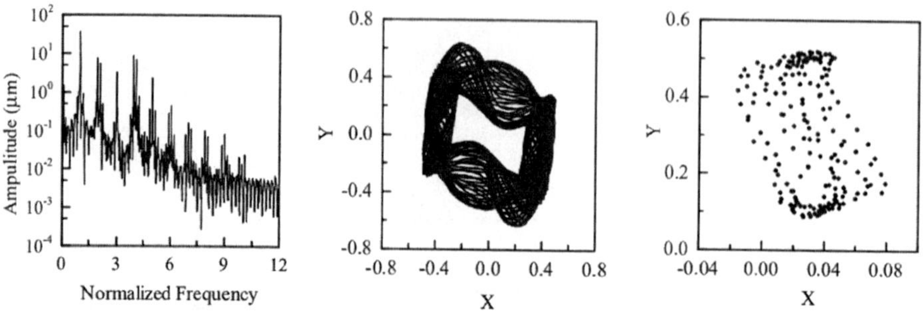

(a) Power spectrum of Y (b) Shaft orbit (c) Poincaré map

Fig. 13. Normalized Y displacement, shaft orbit and Poincaré map for $\Omega = 12705$ rpm.

5 Conclusion

In this paper, the nonlinear vibration characteristics of a twin spool rotor assembly is presented. The rotor is installed with squeeze film dampers and ball bearings, and is loaded by unbalanced mass excitation. The scale of the full-size model is reduced through the nonlinear Galerkin method with the approximate inertial manifold based on lower-order master modes. The number of the master modes is carefully chosen based on a convergence test to achieve a satisfactory balance between accuracy and computational time, which is vital to large scale nonlinear vibration analyses.

Using the NGM with the AIM, the vibration of the reduced system is studied to demonstrate the complexity in the rotor dynamics due to the nonlinear constitutive relations of the SFDs and ball bearings. Within certain range of rotation speed, periodic, quasiperiodic and chaotic motion can be observed through plots of bifurcation, orbital diagrams and the Poincaré plots. To sum up, the developed model reduction technique can provide an efficient tool for dynamic analysis of nonlinear rotor bearing systems.

Acknowledgements. The authors gratefully acknowledge the National Natural Science Foundation of China (Grant U1808214), the National Science and Technology Major Project (Grant 2019-IV-0019–0087), and the Industry-Education Fusion Project by the Bureau of Education of Liaoning Province for their sponsorships.

References

1. Zeidan, F.Y., San Andres, L., Vance, J.M.: Design and application of squeeze film dampers in rotating machinery. In: Proceedings 25th Turbomachinery Symposium, pp.169–188, Texas A&M University, College Station (1996)
2. Fukata, S., Gad, E.H., et al.: On the radial vibration of ball bearings computer simulation. Bull. JSME **28**(239), 899–904 (1985)
3. Tiwari, M., Gupta, K., Prakash, O.: Effect of radial internal clearance of a ball bearing on the dynamics of a balanced horizontal rotor. J. Sound Vib. **238**(5), 723–756 (2000)
4. Chen, G.: Study on nonlinear dynamic response of an unbalanced rotor supported on ball bearing. J. Vib. Acoust. **131**(6), 1980–1998 (2009)

5. Foias, C., Manley, O., Temam, R.: Modelling of the interaction of small and large eddies in two dimensional turbulent flows. RAIRO-Modélisation Mathématique et Analyse Numérique **22**(1), 93–118 (1988)
6. Foias, C., Sell, G.R., Temam, R.: Inertial manifolds for nonlinear evolutionary equations. J. Differ. Eqn. **73**(2), 309–353 (1988)
7. Titi, E.S.: On approximate inertial manifolds to the Navier-Stokes equations. J. Math. Anal. Appl. **149**(2), 540–557 (1990)
8. Childs, D.W.: Turbomachinery Rotordynamics: Phenomena, Modeling, and Analysis. John Wiley & Sons, New York (1993)
9. Marion, M., Temam, R.: Nonlinear galerkin methods. SIAM J. Numer. Anal. **26**(5), 1139–1157 (1989)

Numerical Simulation of Flow-Induced Forces on Subsea Structures in a Group Under Uniform and Sheared Flow

Henry Francis Annapeh[1] and Victoria Kurushina[1,2](✉)

[1] Industrial University of Tyumen, Tyumen 625000, Russia
v.kurushina@outlook.com
[2] Newcastle University, Newcastle upon Tyne NE1 7RU, UK

Abstract. Subsea production systems design requires estimates of hydrodynamic loads related to characteristics of structures and the external flow. The current work investigates flow-induced forces for a group of stationary rigid structures modelled in 2D including one structure with a squared cross-section and three smaller circular cylinders located in proximity of each other. Uniform flow and planar sheared flow conditions are considered in this work, with three different arrangements of smaller structures. Flow characteristics are obtained using CFD method and k-ω SST turbulence model. Simulation results include time histories of hydrodynamic coefficients, FFT data and velocity fields. Results for the planar sheared flow in the cases considered show a reduction of mean drag coefficients, increase of frequencies and amplitudes of the fluctuating drag and lift coefficients compared to values observed for the uniform flow.

Keywords: Flow-induced forces · Uniform flow · Sheared flow · CFD · Drag coefficient · Lift coefficient

1 Introduction

Offshore oil and gas industry during construction, drilling and production operations may encounter disruptions and delays as a result of subsea structures subjected to active vortex shedding and increased fluid loads. Exploration of natural resources discovered in deep waters leads to an increased demand for reliable, off-resonant safe designs in order to anticipate and prevent possible complications related to flow-induced forces. Subsea system layouts often involve arrangements of structures with different hydrodynamic properties in proximity to each other: pumping equipment, pipework, control units, supporting frames, jumpers, flow lines, umbilical lines, and risers. Design of a subsea system should account for various geometric configurations and statistically averaged velocity profiles of sea currents depending on the depth.

Z. Dimitrovová et al. (Eds.): WMVC 2022, MMS 125, pp. 512–522, 2023.
https://doi.org/10.1007/978-3-031-15758-5_52

Fundamental studies of a flow over a circular cylinder or a group of cylinders, in a fixed position or experiencing flow-induced vibration have mostly been focused on effects observed in uniform external flow conditions [1–7]. In parallel, a growing number of investigations are performed for a circular structure subjected to a sheared flow [8–15]. Investigations on the flow interference, hydrodynamic loads and vibration as a consequence of a wake superposition for a group of three structures of the same circular cross-section, placed in a tandem, were conducted by [16] and for four structures in a squared arrangement – by [17–20]. Systematic experimental and numerical studies of the dynamics of three and four flexible cylinders in tandem and side-by-side positions were performed by [21–25], and three structures in a triangular position were considered in [26]. Advancing this question further, a group of five risers of the same cross-section type and size was considered in [27], while the use of multiple smaller cylinders for the load mitigation purposes was extensively studied in [28].

Analysis of available studies indicates the existing gap related to arranging several structures of different cross-section types and dimensions in proximity, that would lead to a superposition of generated vortex street patterns. Also, there seem to be a lack of studies on the hydrodynamic forces acting on a group of pipelines submerged in sheared currents. Based on the literature survey, the present study aims to investigate hydrodynamic forces acting on three stationary structures of a circular cross-section placed near a larger piece of subsea equipment, represented by a cylinder of a squared cross-section. Structures are immersed in the planar uniform and two types of sheared flows corresponding to the developed turbulent flow regime. Computational fluid dynamics method is selected to achieve the goals of this investigation.

In this paper, Sect. 1 provides a brief theoretical background on the topic. Section 2 gives an overview of the numerical method and considered arrangements. Section 3 shows results of this study, and Sect. 4 provides conclusions for this work.

2 Numerical Model

A system of three identical circular structures of diameter $d = 0.3$ m is considered in this study in proximity of a squared cylinder with a side equal to D = 5d in a rectangular domain. CFD simulations are performed for the computational domain with a size of 30D x 16D, and three principal arrangements of smaller cylinders are illustrated in Fig. 1, with a different position of the downstream cylinder. Distance in between structures is $L/D = 0.6$, distance from the squared cylinder to the domain border is G = 20d, the incoming flow is entering the domain from the inlet, periodic and shadow conditions are used as the top and bottom boundary.

Uniform flow of the Reynolds number 3900 is considered in Cases 1, 2 and 3, as illustrated in Figs. 1(a)-1(c). Cases 4, 5, 6 correspond to the same structural arrangements as in Cases 1, 2, 3, but subjected to the planar sheared flow of type 1, as displayed in Fig. 1(d), where the maximum flow velocity is near the top boundary. Cases 7, 8, 9 are designed in the same manner, practically representing the structural arrangements in Cases 1, 2, 3, while subjected to the planar sheared flow of type 2, shown in Fig. 1(e), with the maximum flow velocity at the bottom boundary. Both sheared flows have a linear velocity profile U with the averaged velocity U_c consistent with the same Reynolds number of 3900 at the centreline of the computational domain:

$$U(y) = U_c + By, \tag{1}$$

where y is the vertical coordinate, and the gradient $B = 0.022\ \mathrm{c}^{-1}$.

Simulations are performed using the incompressible Navier-Stokes equation, k-ω SST turbulence model, PISO algorithm and the time step of 0.1 s. The triangular grid is used for simulation purposes. Mesh independence test results are reported in Table 1 for the uniform flow of the Reynolds number of 3900, and the mesh shown in Fig. 1(f), is selected for all calculations in the next section.

3 Results and Discussion

Numerical simulations are performed for three arrangements and three different flow types with an averaged flow speed corresponding to Re = 3900, which gives a matrix of nine considered cases in total. Maximum value of the lift and fluctuating drag coefficients, and the mean drag coefficient obtained for these cases are presented in Appendix for each structure. The drag coefficient C_D acting on the cylinder is defined as a sum of the mean drag coefficient C_{D0} and the fluctuating drag coefficient C_D^{fl}:

$$C_D = C_{D0} + C_D^{fl}. \tag{2}$$

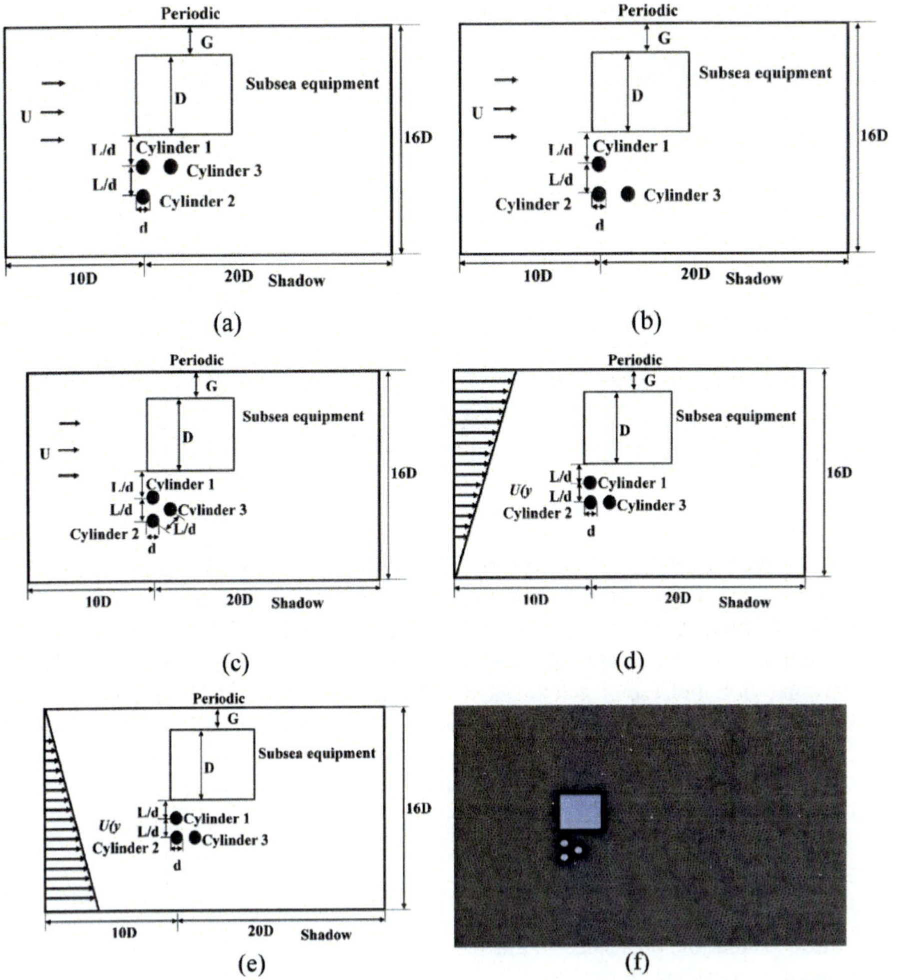

Fig. 1. Computational domain for the considered cases: (a) Case 1 with tandem and paired cylinders in uniform flow; (b) Case 2 with tandem and paired cylinders in uniform flow; (c) Case 3 with tandem and staggered cylinders in uniform flow; (d) Case 4 with tandem and paired cylinders in planar sheared flow of the type 1; (e) Case 7 with tandem and paired cylinders in planar sheared flow of the type 2; (f) Mesh of the computational domain for Cases 3, 6, 9.

Table 1. Mesh independency test results

Re = 3900			
Cases	C_{D0}	Number of cells	y_+
Current study			
Mesh 1	0.87	31 297	0.0133
Mesh 2	0.91	53 951	0.0132
Mesh 3	0.93	86 637	0.0131
Mesh 4	0.93	153 227	0.0131
Published data			
Experiment (Lourenco and Shih 1993)	0.985	–	–
RANS (Nguyen 2015)	0.920	78 000	–
VMS-LES (Stephen et al. 2011)	0.990	–	–

Among the Cases 1–3 with uniform flow, shown in Fig. 2 and summarized in Appendix, the highest mean drag coefficient of 0.51 is recorded in Case 3 for cylinder 1. The fluctuating drag coefficient amplitude for this structure is maximum in Case 2 and minimum in Case 1, and the maximum amplitude of the lift coefficient signal is demonstrated in both Cases 1 and 2. The mean drag coefficient of 0.47 for cylinder 2 is maximum in Case 1, while the maximum fluctuating drag coefficient of 0.26 is recorded for Case 2, and the maximum amplitude of the lift coefficient corresponds to Case 1. For cylinder 3, the maximum mean drag coefficient of 0.25 is recorded in Case 2. The maximum amplitude of the fluctuating drag and of the lift coefficient is observed in Case 2.

All signals obtained for the lift and drag force fluctuations, shown in Appendix, demonstrate very low and comparable frequency values. Superposition of wake patterns leads to a considerable presence of multiple frequencies in the signals of hydrodynamic forces, as shown in Figs. 2–4(c,d).

Figure 3 illustrates the signals obtained for the sheared flow of type 1. According to Appendix and Fig. 3, the maximum mean drag coefficient of 0.35 and the maximum fluctuating drag for cylinder 1 are observed in Case 5. In Case 6, the maximum lift coefficient is observed. Cylinder 2 exhibits the maximum mean and fluctuating drag coefficients, and also the maximum lift coefficient amplitude in Case 4.

Lift and drag coefficient signals for the sheared flow of type 2 are shown in Fig. 4. Here, Cylinder 2 demonstrates the maximum mean drag coefficient of 0.40 in Case 7. The maximum amplitudes of the fluctuating drag coefficient and the lift coefficient signal are observed for cylinder 3 in Cases 7 and 8 respectively.

Vortex formation process differs for cylinder 3, depending on its position and the flow type. Figure 5 demonstrates the differences experienced in Cases 3, 6 and 9 in comparison. Evenly paired vortices are generated at the far downstream side of cylinder 3 in the uniform flow, as in Fig. 5(a). In Fig. 5(c), uneven pair of vortices is generated at the immediate downstream side of cylinder 3. A single large vortex is formed just at the downstream side of cylinder 3 in Case 6, shown in Fig. 5(b) for the sheared flow.

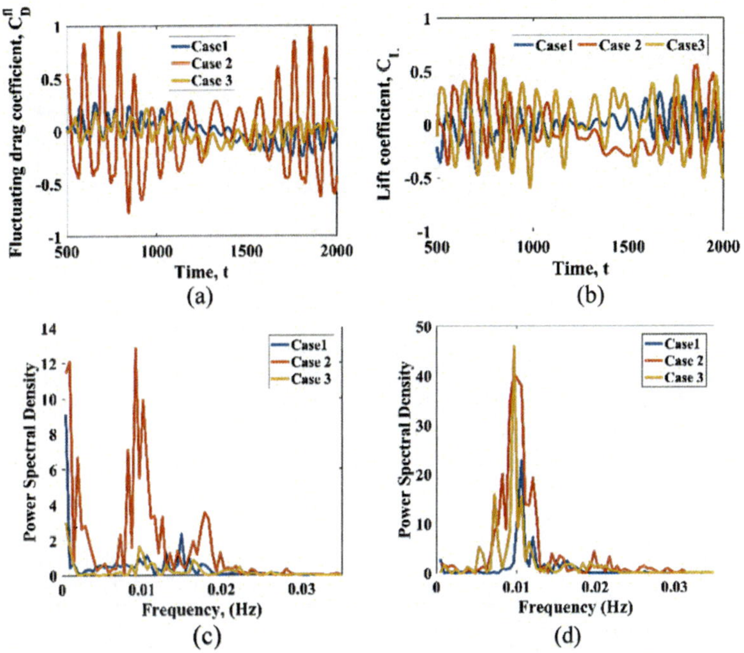

Fig. 2. Fluid force coefficients for cylinder 3 immersed in the uniform flow: (a) time history of the fluctuating drag coefficient; (b) time history of the lift coefficient; (c) the drag coefficient FFT; (d) the lift coefficient FFT.

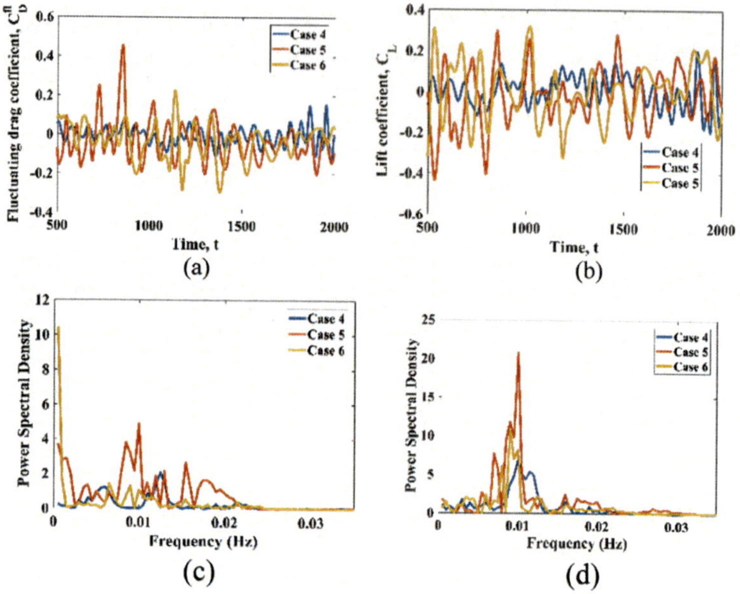

Fig. 3. Hydrodynamic coefficients for cylinder 3 in the planar sheared flow of type 1: (a) time history of the fluctuating drag coefficient; (b) time history of the lift coefficient; (c) the drag coefficient FFT; (d) the lift coefficient FFT.

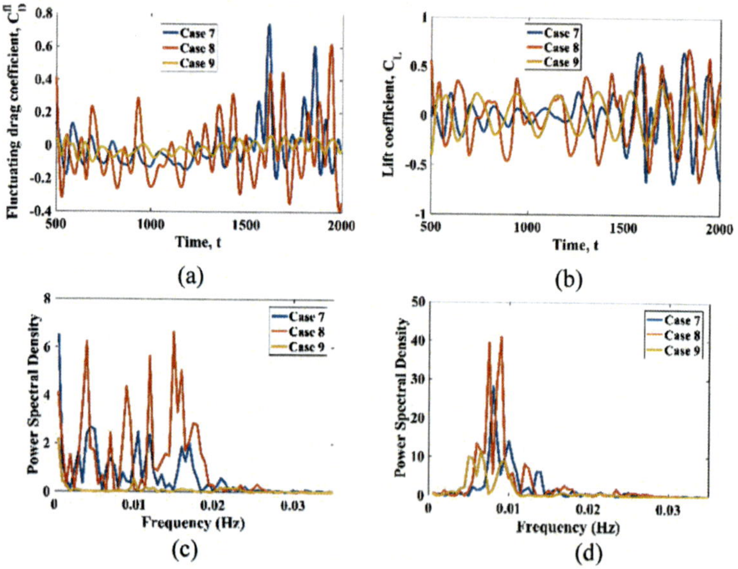

Fig. 4. Hydrodynamic coefficients on cylinder 3 in the planar sheared flow of type 2: (a) time history of the fluctuating drag coefficient; (b) time history of the lift coefficient; (c) the drag coefficient FFT; (d) the lift coefficient FFT.

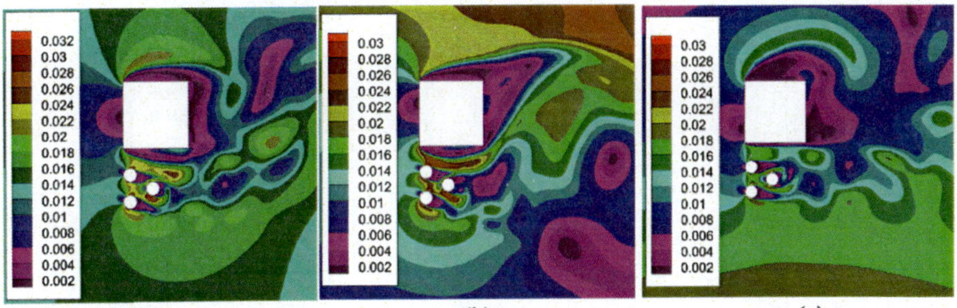

Fig. 5. Velocity contours (velocity magnitude, m/s) observed at 2000 s: (a) uniform flow – Case 3; (b) planar sheared flow of type 1 – Case 6; (c) planar sheared flow of type 2 – Case 9.

4 Conclusions

Numerical simulations are performed in this work for three circular cylinders placed in a close proximity to a piece of equipment modelled as a squared cylinder. Considered structures, especially, the downstream cylinder, experience the effects associated with the overlay of vortex shedding patterns. Generally, decreased mean drag coefficients of smaller cylinders are observed for the planar sheared flow compared to coefficients in the uniform flow cases. Higher amplitudes of the fluctuating drag coefficient and the lift coefficient are observed for the sheared flows.

Acknowledgements. The authors would like to acknowledge the support of the National Project "Science and Universities" of the Ministry of Science and Higher Education of the Russian Federation, grant number FEWN-2021–0012.

Appendix

Cases	Hydrodynamic coefficients				
	C_{D0}	$C_D{}^{fl}$	C_L	Dominant frequency, Hz	
				C_D	C_L
Uniform flow					
Cylinder 1					
Case 1	0.44	0.12	0.13	0.017	0.0025
Case 2	0.44	0.23	0.13	0.011	0.0175
Case 3	0.51	0.16	0.10	0.010	0.022
Cylinder 2					
Case 1	0.47	0.22	0.44	0.002	0.015
Case 2	0.44	0.26	0.32	0.002	0.009

(*continued*)

(*continued*)

Cases	Hydrodynamic coefficients				
	C_{D0}	C_D^{fl}	C_L	Dominant frequency, Hz	
				C_D	C_L
Case 3	0.40	0.18	0.24	0.003	0.013
Cylinder 3					
Case 1	0.14	0.3	0.34	0.015	0.011
Case 2	0.25	1.0	0.74	0.012	0.010
Case 3	0.16	0.24	0.6	0.01	0.010
Shear Flow 1					
Cylinder 1					
Case 4	0.30	0.18	0.11	0.0005	0.011
Case 5	0.35	0.25	0.29	0.0005	0.011
Case 6	0.31	0.19	0.18	0.0005	0.0165
Cylinder 2					
Case 4	0.37	0.39	0.45	0.011	0.011
Case 5	0.28	0.31	0.32	0.0005	0.01
Case 6	0.28	0.16	0.25	0.0005	0.0135
Cylinder 3					
Case 4	0.07	0.16	0.19	0.0125	0.01
Case 5	0.10	0.46	0.43	0.01	0.01
Case 6	0.11	0.28	0.32	0.0005	0.009
Shear Flow 2					
Cylinder 1					
Case 7	0.31	0.17	0.43	0.0015	0.008
Case 8	0.32	0.16	0.18	0.0015	0.012
Case 9	0.30	0.13	0.04	0.0015	0.020
Cylinder 2					
Case 7	0.40	0.21	0.48	0.012	0.013
Case 8	0.35	0.15	0.40	0.009	0.0085
Case 9	0.29	0.10	0.04	0.005	0.016
Cylinder 3					
Case 7	0.10	0.73	0.66	0.0045	0.008
Case 8	0.20	0.62	0.70	0.015	0.009
Case 9	0.21	0.06	0.32	0.010	0.007

References

1. Jordan, S.K., Fromm, J.E.: Laminar flow past a circle in shear flow. Phys. Fluids **15**, 972–976 (1972)
2. Williamson, C.H.K.: Vortex dynamics in the cylinder wake. Annu. Rev. Fluid Mech. **28**, 477–539 (1996)
3. Zdravkovich, M.M.: Review of flow interference between two circular cylinders in various arrangements. ASME J. Fluids Eng. **99**, 618–633 (1977)
4. Breuer, M.: A challenging test case for large eddy simulation: high reynolds number circular cylinder flow. Int. J. Heat Fluid Flow **21**, 648–654 (2000)
5. Lei, C., Cheng, L., Kavanagh, K.: Spanwise length effects on three-dimensional modelling of flow over a circular cylinder. Comput. Method Appl. Mech. **190**, 2909–2923 (2001)
6. Labbé, D.F.L., Wilson, P.A.: A numerical investigation of the effects of the spanwise length on the 3-d wake of a circular cylinder. J. Fluids Structure **23**, 1168–1188 (2007)
7. Xu, W., Haokai, W., Kun, J., Enhao, W.: Numerical investigation into the effect of spacing on the flow-induced vibrations of two tandem circular cylinders at subcritical reynolds numbers. Ocean Eng. **236**, 1–19 (2021)
8. Yoshino, F., Hayashi, T.: Numerical solution of flow around a rotating circular cylinder in uniform shear flow. Bull. JSME **27**, 1850–1857 (1984)
9. Tamura, H., Kiya, M., Arie, M.: Numerical study on viscous shear flow past a circular cylinder. Bull. JSME **23**, 1952–1958 (1989)
10. Ayukawa, K., Ochi, J., Kawahara, G., Hirao, T.: Effects of shear rate on the flow around a square cylinder in a uniform shear flow. J. Wind Eng. Ind. Aerodyn. **50**, 97–106 (1993)
11. Lei, C., Cheng, L., Kavanagh, K.: A finite difference solution of the shear flow over a circular cylinder. Ocean Eng. **27**, 271–290 (2000)
12. Wu, T., Chen, C.-F.: Laminar boundary-layer separation over a circular cylinder in uniform shear flow. Acta Mech. **144**, 71–82 (2000)
13. Sumner, D., Akosile, O.O.: On uniform planar shear flow around a circular cylinder at subcritical Reynolds number. J. Fluids Struct. **18**, 441–454 (2003)
14. Kang, S.: Uniform-shear flow over a circular cylinder at low Reynolds numbers. J. Fluids Struct. **22**, 541–555 (2006)
15. Shuyang, C., Shigehira, O., Yukio, T., Yaojun, G., Hironori, K.: Numerical simulation of Reynolds number effects on velocity shear flow around a circular cylinder. J. Fluids Struct. **26**, 685–702 (2010)
16. Gao, Y., Zhang, Y., Zhao, M., Wang, L.: Numerical investigation on two degree-of-freedom flow-induced vibration of three tandem cylinders. Ocean Eng. **201**, 107059 (2020)
17. Zhao, M., Cheng, L.: Numerical simulation of vortex-induced vibration of four circular cylinders in a square configuration. J. Fluids Struct. **31**, 125–140 (2012)
18. Han, Z., et al.: Flow-induced vibrations of four circular cylinders with square arrangement at low Reynolds numbers. Ocean Eng. **96**, 21–33 (2015)
19. Gao, Y., Yang, K., Zhang, B., Cheng, K., Chen, X.: Numerical investigation on vortex-induced vibrations of four circular cylinders in a square configuration. Ocean Eng. **175**, 223–240 (2019)
20. Gómez, H.A., Narváez, G.F., Schettini, E.B.: Vortex induced vibration of four cylinders configurations at critical spacing in 0° and 45° flow incidence angle. Ocean Eng. **252**, 111134 (2022)
21. Xu, W., Zhang, S., Liu, B., Wang, E., Bai, Y.: An experimental study on flow-induced vibration of three and four side-by-side long flexible cylinders. Ocean Eng. **169**, 492–510 (2018)
22. Wang, E., Xu, W., Yu, Y., Zhou, L., Incecik, A.: Flow-induced vibrations of three and four long flexible cylinders in tandem arrangement: an experimental study. Ocean Eng. **178**, 170–184 (2019)

23. Han, P., Pan, G., Zhang, B., Wang, W., Tian, W.: Three-cylinder oscillator under flow: Flow induced vibration and energy harvesting. Ocean Eng. **211**, 107619 (2020)
24. Xu, W., Zhang, S., Ma, Y., Liu, B.: Fluid forces acting on three and four long side-by-side flexible cylinders undergoing flow-induced vibration (FIV). Mar. Struct. **75**, 102877 (2021)
25. Fan, X., Wang, Z., Wang, Y., Tan, W.: The effect of vortices structures on the flow-induced vibration of three flexible tandem cylinders. Int. J. Mech. Sci. **192**, 106132 (2021)
26. Ma, Y., Xu, W., Liu, B.: Dynamic response of three long flexible cylinders subjected to flow-induced vibration (FIV) in an equilateral-triangular configuration. Ocean Eng. **183**, 187–207 (2019)
27. Liu, Y., Li, P., Wang, Y., Chen, X., Ren, X., Lou, M.: Dynamic response of five-riser group subjected to vortex-induced vibration in a cylindrical arrangement configuration. Ocean Eng. **254**, 111271 (2022)
28. Silva-Ortega, M., Assi, G.R.S.: Flow-induced vibration of a circular cylinder surrounded by two, four and eight wake-control cylinders. Exp. Therm. Fluid Sci. **85**, 354–362 (2017)

On the Longitudinal Vibration of a Driven Wheelset Running on Adhesion Limit

Marius-Alin Gheți(✉) and Traian Mazilu

University Politehnica of Bucharest, Bucharest, Romania
tr_mazilu@yahoo.com

Abstract. The driving wheelsets are used on the railway traction vehicles to support the vehicle's weight, for traction and braking purpose. Most commonly, a driving wheelset consists of the axle on which the two wheels and the driving gear are rigidly attached. The driving wheelset longitudinal vibrations are important for the locomotive performance from the traction point of view. This vibration may be caused by the friction forces at wheel level, the torque from the motor or from the drive system, and the steering system of the wheelset. In this paper, a basic mechanical model of a driving rigid wheelset, including the rotor dynamics and elasticity of the driven shaft and the steering system is considered to point out some aspect of the longitudinal vibration when the wheelset is running on adhesion limit. Using a numerical method for solving the nonlinear equations of motion, the limit cycle which appears at the adherence limit and the influence of the steering damping are presented.

Keywords: Driven wheelset · Locomotive · Three-phase squirrel-cage induction motor · Traction characteristic · Adhesion limit · Longitudinal vibration · Limit cycle

1 Introduction

The development of the locomotives with high specific power have led to increases of the tractive force, which is limited by the adhesion force between the wheel and the rail [1]. Variation of the friction force due to the excessive vibration can lead to complex dynamics problems with negative effects in terms of tread wear and even inability to develop all the possible necessary traction force that is available.

Through a rigorous theoretical study on the running dynamics and on how traction force is developed, traction performance can be improved. The studies can be extended by considering the influence of the constructive parameters of the bogie [2]. Longitudinal vibrations of the driven wheelset also contribute to the priming and development of corrugation wear, especially at the rail level [3].

This paper analyses the longitudinal vibrations of a driven wheelset of a locomotive when is running on adherence limit. For this purpose, a mechanical model describing the torsion and rebound vibration of the drive system and wheelset is developed. Based on the numerical simulations, configuration of the stationary position of the drive system and

Z. Dimitrovová et al. (Eds.): WMVC 2022, MMS 125, pp. 523–531, 2023.
https://doi.org/10.1007/978-3-031-15758-5_53

wheelset corresponding to the traction nominal behavior of the locomotive is determined and the stability of the stationary position and the existence and the stability of the periodic solutions are highlighted.

2 Mechanical Model

Figure 1 shows an overview of the mechanical model of the driven wheelset and the drive system from a locomotive, running at constant speed V on a straight rigid track.

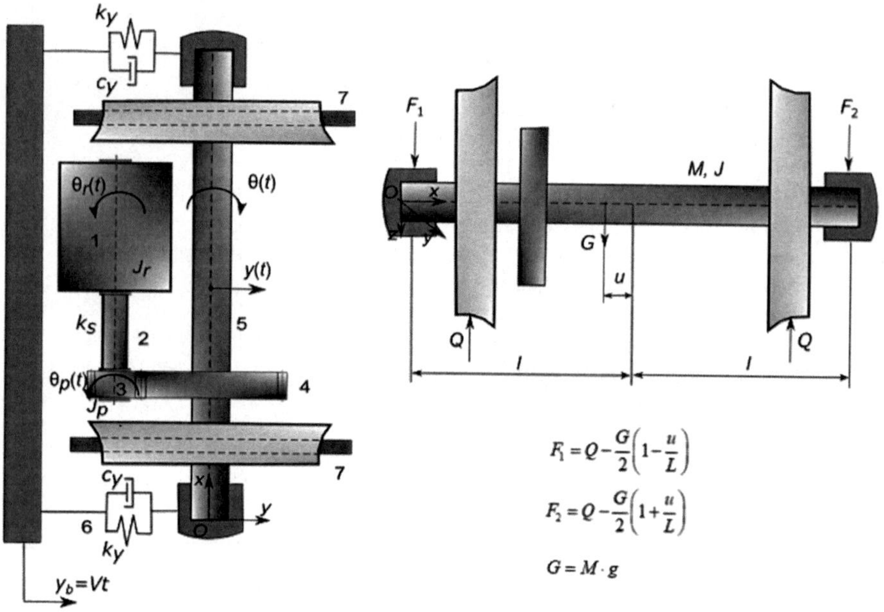

Fig. 1. Mechanical model of the driven wheelset: (1) rotor; (2) driven shaft; (3) pinion; (4) big cogwheel; (5) driven wheelset; (6) steering system; (7) rails.

The drive system of the wheelset is of unilateral type and consists of the rotor (1) of the electric traction motor which, by means of the driven shaft (2) and the single stage gear reducer formed by the pinion (3) and the big cogwheel (4), transmits the motor torque C_m to the driven wheelset (5) which is running on the rails 6. The driven wheelset, including the axle, wheels, axle boxes and toothed crown, is elastically longitudinal connected to a moving rigid base representing the bogie chassis via the steering system (7) consisting of two Kelvin-Voigt systems. The wheelset supports the static load $2Q$ due to the self-weight and the weight of its part of the suspended mass of the locomotive.

The rotor, pinion and the wheelset are modeled as rigid bodies in rotating motion – $\theta_r(t)$, $\theta_p(t)$ and $\theta(t)$ are the rotation angles, meanwhile the wheelset experiences also the longitudinal translation (rebound) $y(t)$, where t stands for the time.

The energy source that generates the movement of the system is a three-phase squirrel-cage induction motor.

Parameters associated to the presented model are as follows: M, J – the mass and the rotational inertia about the rotational axis of the wheelset, J_r, J_p – the rotational inertia about the rotational axis of the rotor and pinion, respectively, k_s the stiffness of the driven shaft and k_y, c_y the stiffness and constant damping of the steering system of the wheelset, wheel radius R, transmission ratio i_t,.

Motor torque has two components, the stationary component, C_{mo}, and the time dependent component, $\Delta C_m(t)$,

$$C_m(t) = C_{mo} + \Delta C_m(t). \tag{1}$$

The stationary component of the motor torque determines the rotation of the drive system of the wheelset at constant angular speed and the tensioning of the wheelset–drive system. The variable motor torque component induces torsional vibrations and the wheelset rebound; in other words, it is about of a perturbation about the stationary position.

The rotation angles of the wheelset, pinion and rotor can be written as

$$\theta(t) = (\Omega + \Omega_o)t + \Delta\theta(t),\ \theta_p(t) = i_t\theta(t),\ \theta_r(t) = i_t[(\Omega + \Omega_o)t] + \Delta\theta_{rp} + \Delta\theta_r(t), \tag{2}$$

where $\Omega = V/R$, Ω_o is the rotation speed corresponding to the wheels sleep on rail in the stationary position, $\Delta\theta_{rp}$ is the rotor-pinion angle in the stationary position and $\Delta\theta(t)$ and $\Delta\theta_r(t)$ are the perturbation of the wheelset and rotor rotation.

Longitudinal position of the wheelset is.

$$y(t) = Vt + \Delta y_o + \Delta y(t), \tag{3}$$

where Δy_o is the longitudinal bogie-wheelset displacement in the stationary position and $\Delta y(t)$ describes the rebound motion of the wheelset.

Applying Euler's rotation equations and Newton's second law, the nonlinear equations of motion result

$$\begin{aligned}
&J_r\Delta\ddot{\theta}_r(t) + k_s\left(\Delta\theta_{rp} + \Delta\theta_r(t)\right) - k_s i_t\Delta\theta(t) = C_m(t),\\
&\left(J + i_t^2 J_p\right)\Delta\ddot{\theta}(t) + k_s i_t^2\Delta\theta(t) - k_s i_t\left(\Delta\theta_{rp} + \Delta\theta_r(t)\right) = -2RF(t),\\
&M\,\Delta\ddot{y}(t) + 2c_y\Delta\dot{y}(t) + 2k_y(\Delta y_o + \Delta y(t)) = 2F(t),
\end{aligned} \tag{4}$$

where $F(t)$ is the wheel/rail friction force

$$F(t) = F_o + \Delta F(t), \tag{5}$$

including the stationary component F_o and the perturbation $\Delta F(t)$.

Wheel/rail friction force can be calculated using Polach's model [4].

$$F(t) = -\frac{2}{\pi}Q\mu(t)\left[\frac{k_A\varepsilon(t)}{1 + k_A^2\varepsilon^2(t)} + \mathrm{arctg}k_S\varepsilon(t)\right]. \tag{6}$$

where μ is the coefficient of friction, ε - the gradient of tangential stress in the area of adhesion, $k_{A,S}$ – the reduction factors in the area of adhesion/slip which depend on the wheel/rail surfaces condition (dry/wet).

Coefficient of friction depends on the wheel/rail slip velocity.

$$\mu(t) = \mu_s\Big((1 - A)e^{-B|w(t)|} + A\Big), \tag{7}$$

where μ_s is the coefficient of static friction, w – wheel/rail slip velocity, including the stationary value w_o and the perturbation $\Delta w(t)$.

$$w(t) = \dot{y}(t) - R\dot{\theta}(t) = -R\Omega_o + \Delta\dot{y}(t) - R\Delta\dot{\theta}(t) = w_o + \Delta w(t)\,, \tag{8}$$

$A = \mu_\infty / \mu_s$ with μ_∞ – friction coefficient when $w_o \to \infty$, and B – constant value [s/m].

Gradient of tangential stress in the area of adhesion is given by Polach.

$$\varepsilon(t) = \frac{\pi}{4}\frac{GabC_{11}}{Q\mu(t)}\nu(t), \tag{9}$$

where G is the transversal modulus of elasticity of the wheel/rail material, a, b semiaxis of the contact ellipse, C_{11} – Kalker's coefficient for longitudinal direction and υ – the wheel/rail creep.

$$\nu(t) = \frac{w(t)}{V} = \frac{-\Omega_o}{\Omega} + \frac{\Delta\dot{y}(t) - R\Delta\dot{\theta}(t)}{V} = \nu_o + \Delta\nu(t), \tag{10}$$

where $\upsilon_o = -(\Omega_o/\Omega)$ is the stationary component and $\Delta\upsilon\,(t)$ is the creep perturbation.

Motor torque of a three-phase squirrel-cage induction motor is given as [5].

$$C_m(t) = \frac{n_f p_1}{2\pi f}\frac{R_{21}U_1^2 s(t)}{(R_{21})^2 + (s(t)X)^2}\eta_m, \tag{11}$$

where n_f is the number of phases ($n_f = 3$), p_1 is the number of pairs of poles, R_{21} – the rotor resistance relative to the stator, X – the stator reactance, U_1 – the supply voltage, f – the supply voltage frequency, s – slip and η_m – the engine efficiency. It must underline that only the slip is time-dependent quantity due to the perturbation of the rotor rotation.

Slip is the ratio.

$$s(t) = \frac{\Omega_s - \dot{\theta}_r(t)}{\Omega_s} = \frac{\Omega_s - i_t\Omega(1 + \nu_o)}{\Omega_s} - \frac{\Delta\dot{\theta}_r(t)}{\Omega_s} = s_o + \Delta s(t), \tag{12}$$

where $\Omega_s = 2\pi f_1/p_1$ is the angular velocity of synchronism, s_o – the stationary slip, $\Delta s(t)$ – the perturbation of the slip.

Setting all time-dependent variables to zero in Eqs. (4), it results

$$i_t C_{mo} - 2RF_o = 0;\ \Delta\theta_{rp} = C_{mo}/k_s;\ \Delta y_o = F_o/k_y, \tag{13}$$

where the first equation is a nonlinear equation depending on the stationary component of the creep υo. This equation can be solved using the Newton-Ralphson method.

Inserting the results obtained from Eqs. (13) in Eqs. (4), the nonlinear equations of the perturbed motion around the stationary position hold

$$\begin{aligned} &J_r\Delta\ddot{\theta}_r(t) + k_s(\Delta\theta_r(t) - i_t\Delta\theta(t)) = \Delta C_m(t), \\ &\left(J + i_t^2 J_p\right)\Delta\ddot{\theta}(t) + k_s i_t(i_t\Delta\theta(t) - \Delta\theta_r(t)) = -2R\Delta F(t), \\ &M\,\Delta\ddot{y}(t) + 2c_y\Delta\dot{y}(t) + 2k_y\Delta y(t) = 2\Delta F(t). \end{aligned} \quad (14)$$

In the following section, Eqs. (13) are used to obtain the configuration of the stationary position of the drive system and wheelset at the adherence limit and Eqs. (14) allow to analyze the stability of the stationary position and the existence and the stability of the periodic solutions [6].

3 Numerical Application

In this section, based on the model above, the numerical results on the longitudinal vibrations of the drive axle drive system are analyzed.

The values of the numerical application correspond to the six-axle LEMA locomotive produced by the company SOFTRONIC from Craiova, which is equipped with asynchronous motors type MTA-ES-108 of 1000 kW. Traction limit characteristic of the locomotive is presented in Fig. 2, where the straight line corresponds to the constant tractive effort and the curve line means the constant power behavior. Point N corresponds to the nominal behavior: the nominal velocity of 71.4 km/h and the tractive effort of 292 kN or 48.667 kN/wheelset and the nominal motor torque at wheelset of 29.443 kNm for the wheel diameter of 1210 mm (semi wear).

Mechanical parameters: M = 2565 kg, J = 270 kgm^2, J_r = 55 kgm^2, J_p = 0.25 kgm^2, k_s = 340 kNm/rad, k_y = 20 MN/m, c_y = 16 kNs/m, μ_s = 0.55 – dry surfaces, k_A = 1.0, k_B = 0.4, A = 0.4, B = 0.6, G = 81 GPa, a = 8.169 mm, b = 5.140 mm, C_{11} = 4.883, i_t = 80/23, Q = 100 kN. Motor parameters: U_1 = 1200 V, $p_1 = 3$, f = 55 Hz, R_{21} = 0.0405 Ω, X = 0.681 Ω, s_o = 0.01 and η_m = 0.94.

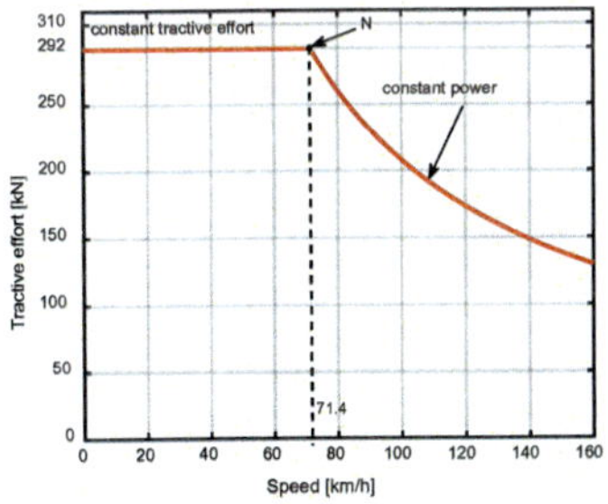

Fig. 2. Traction limit characteristic of the six-axle LEMA locomotive [7].

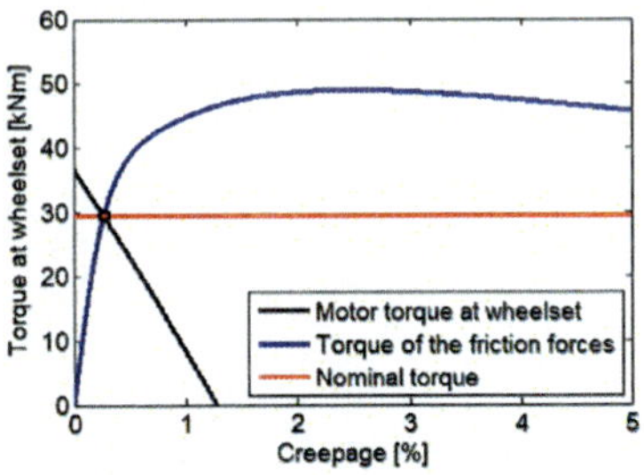

Fig. 3. Nominal torque at V = 71.21873 km/h and dry surfaces (μ_s = 0.55).

Traction limit characteristic is drawn considering ideal rolling condition without wheels creepage. However, when the creepage is considered, the stationary slip decreases and the torque is lower, and to compensate the creepage effect, the velocity must be lower than the nominal one. Figure 3 shows the motor torque at wheelset, the torque of the friction forces and the nominal torque of 29.443 kNm at velocity $V = 71.224$ km/h on dry rolling surface. The stationary creepage is 0.267%, but the motor slip is 0.01 corresponding to the nominal torque.

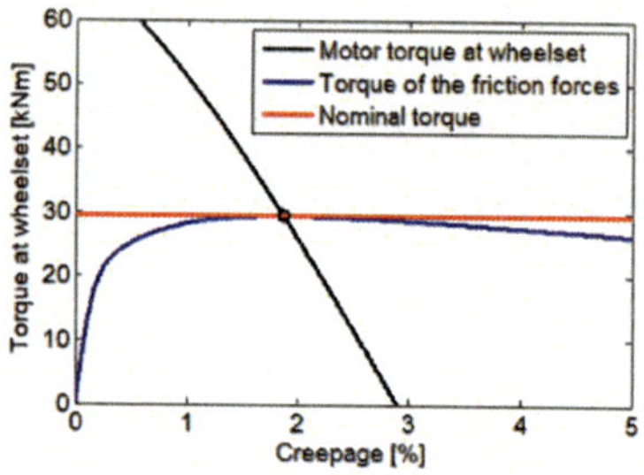

Fig. 4. Nominal torque at $V = 70.096$ km/h and dry surfaces ($\mu_s = 0.307372$).

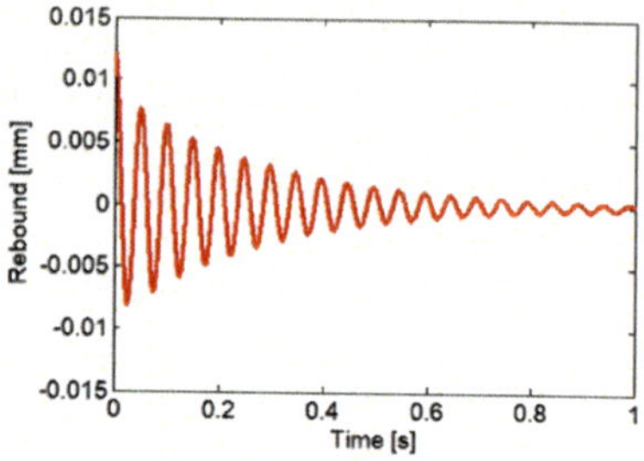

Fig. 5. Free vibration at $V = 71.21873$ km/h and $\mu_s = 0.55$.

When the friction coefficient is lower, the nominal motor torque is reached at higher creepage value and lower velocity. For instance, Fig. 4 presents what is happen when friction coefficient $\mu_s = 0.307372$ and the velocity $V = 70.096$ km/h. Driven wheelset develops the nominal torque at the stationary creepage of 1.873% and it is running on adherence limit. More precisely, it is about of the proximity of the adherence limit because this point is difficult to find.

Once the values of the stationary behavior parameters have been established, it is possible to proceed to the analysis of the stability of this behavior.

First, the natural frequencies of the rotor-wheelset torsion vibration and of the wheelset-bogie rebound should be known. To do this, the effects of the motor torque and friction forces are neglected. Natural frequencies of torsion and rebound vibrations are

$$f_{torsion} = \frac{1}{2\pi}\sqrt{\frac{J + i_t^2(J_r + J_p)}{J_r(J + i_t^2 J_p)}k_s}; \; f_{rebound} = \frac{1}{2\pi}\sqrt{\frac{2k_y}{M}},$$

respectively $f_{torsion} = 23.20$ Hz and $f_{rebound} = 19.87$ Hz.

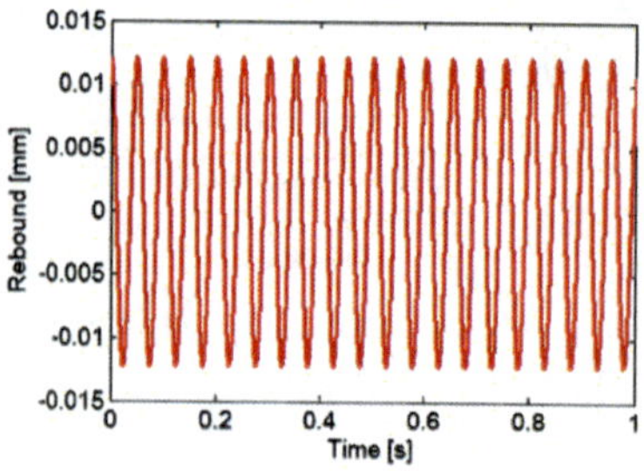

Fig. 6. Free vibration at $V = 70.096$ km/h and $\mu_s = 0.307372$.

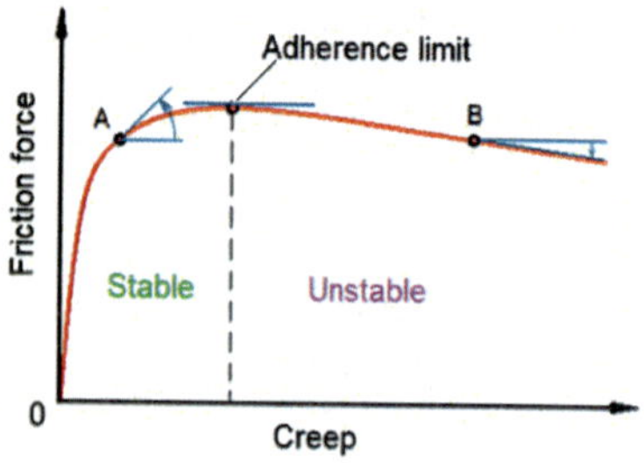

Fig. 7. Friction force versus creep.

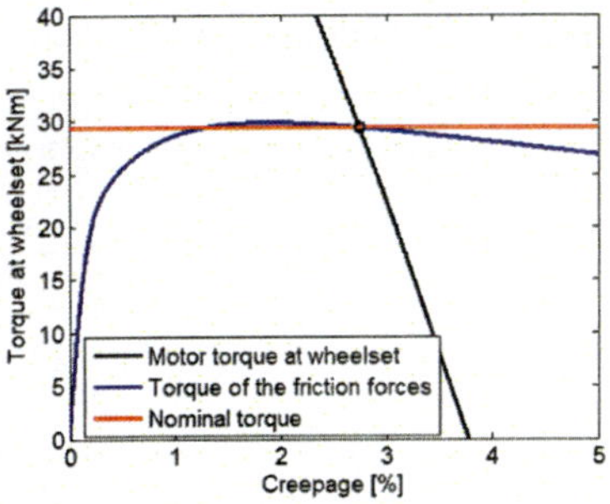

Fig. 8. Nominal torque at $V = 69.5$ km/h and dry surfaces ($\mu_s = 0.312331$).

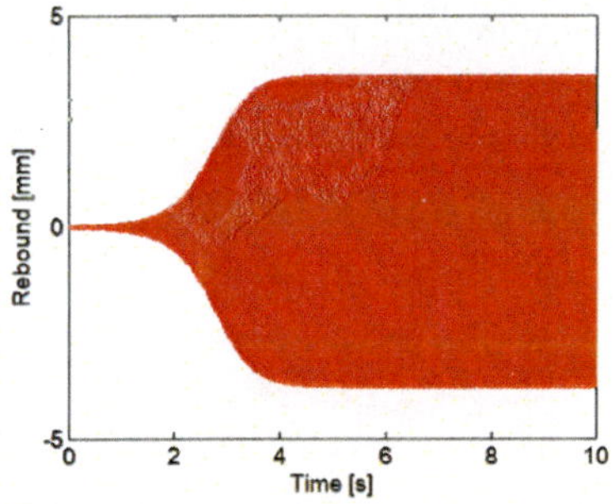

Fig. 9. Free vibration at $V = 69.5$ km/h and $\mu_s = 0.312331$.

Figure 5 and Fig. 6 show the free vibration of the wheelset around the stationary positions described in Figs. 3 and 4 respectively. In both cases, the damping has been neglected and the initial conditions are set to zero, excepting the rebound which is 1% of the stationary value. Despite that the damping of the system is disregarded, the free vibration decreases in amplitude (Fig. 5) because the variation of the friction force around the stationary position is positive (Fig. 7, point A). In this case, the stationary position is asymptotically stable.

However, when the adherence limit is reached, the contribution of the variation of the friction force to the system damping is zero because the tangent slope to the friction force curve is null (Fig. 7 – the point at the adherence limit). The stationary position is simple stable.

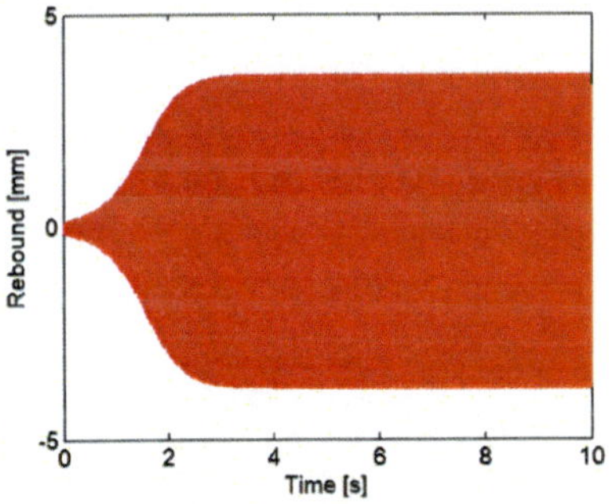

Fig. 10. Motor torque at wheelset at $V =$ 69.5 km/h and dry surfaces ($\mu_s = 0.312331$).

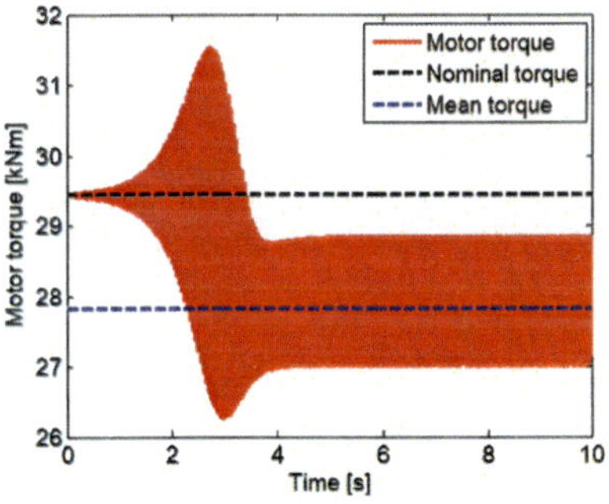

Fig. 11. Motor torque at wheelset at $V =$ 69.5 km/h and dry surfaces ($\mu_s = 0.312331$).

When the stationary position is beyond the adherence limit (Fig. 7, point B), the damping induced by the variation of the friction force is negative and the wheelset vibration becomes unstable. Figures 8 and 9 illustrate this situation, where the wheelset velocity of 69.5 km/h and the friction coefficient of 0.312331 are considered. The vibration amplitude increases gradually until it remains constant, and the vibration is turning out into a limit cycle. The frequency of the limit cycle is 19.83 Hz close to the natural rebound frequency – *the rebound limit cycle*. This cycle is stable as shows Fig. 10 where the initial longitudinal displacement of the wheelset is $0.1y_o$.

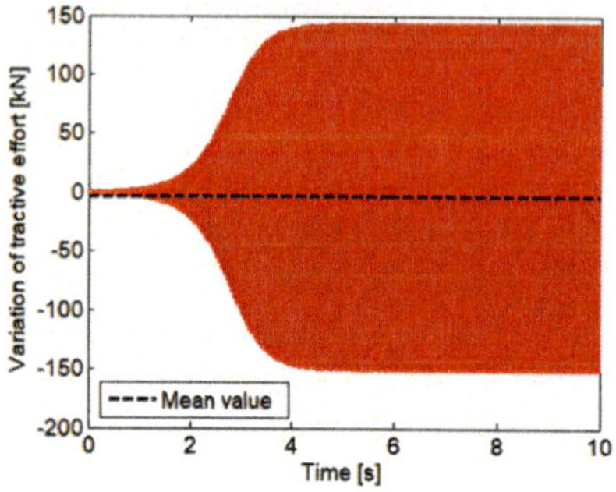

Fig. 12. Variation of the tractive effort at $V =$ 69.5 km/h and dry surfaces ($\mu_s = 0.312331$).

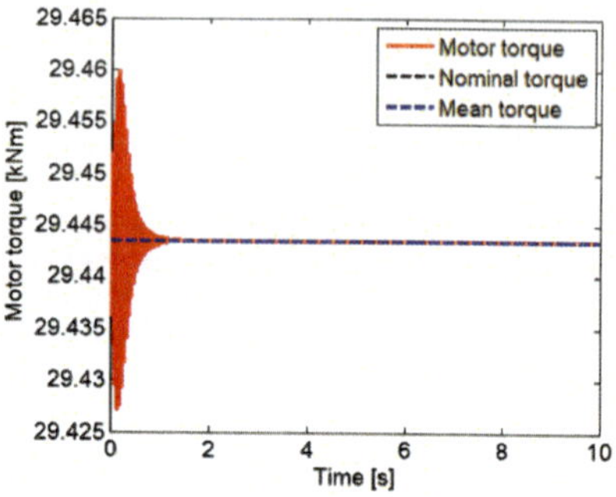

Fig. 13. Motor torque at wheelset at $V =$ 69.5 km/h and dry surfaces ($\mu_s = 0.312331$).

Figures 11 and 12 show the time-dependent motor torque at the wheelset and the variation of tractive effort and the corresponding mean values. Mean value of the motor torque is lower than the nominal value which correspond to the stationary position (Fig. 8), and the mean value of the variation of the tractive effort is –2.88 kN, which means weaker tractive performance. On the other hand, the variation of the tractive effort has very high amplitude even higher than the static load of the wheelset.

However, small damping in the steering system of the wheelset changes the character of the motion as it can see in Fig. 13, where the damping ratio of 0.05 is considered (c_y = 16 kNs/m).

4 Conclusions

In this paper, the longitudinal vibrations of a driven wheelset are studied based on a mechanical model composed from an asymmetrical rigid wheelset, the rotor of the electric traction motor together with the drive gear, and the wheelset steering system.

The main findings in the paper are:

1. When the creepage is taken into consideration, the stationary slip decreases and the torque is lower, and for the motor to achieve the nominal torque, the velocity must be lowered.
2. If the adhesion coefficient is lower, then the nominal motor torque is going to be reached for a higher creepage value and a lower velocity.
3. If the value of the adhesion coefficient is situated on the stable part of the friction force characteristic, even if the system has no damping, the free vibration of the wheelset decreases in amplitude. In this case, the stationary position is asymptotically stable. If the adhesion coefficient reaches the limit, the contribution of the variation of the friction force to the system damping is null, and for this case the stationary position is simple stable.

4. If the adhesion coefficient is beyond the adhesion limit, the wheelset vibrations become unstable due to the negative damping induced by the variation of the friction force. In this case, the amplitude of the vibration increases gradually until it remains constant, the vibration turns into a stable limit cycle. Also, the mean value of the tractive effort is less than the nominal value which correspond to the stationary position which translates into weaker tractive performance.
5. Introducing a small damping into the steering system of the wheelset makes the system stable.

References

1. Yamashita, M., Soeda, T.: Anti-slip re-adhesion control method for increasing the tractive force of locomotives through the early detection of wheel slip convergence. 17th European conference on power electronics and applications (2015)
2. Yao, Y., Zhao, S., Xiao, F., Liu, J.: The effects of wheelset driving system suspension parameters on the re-adhesion performance of locomotives. Veh. Syst. Dyn. **53**(12), 1935–1951 (2015)
3. Ciavarella, M., Barber, J.: Influence of longitudinal creepage and wheel inertia on short-pitch corrugation: a resonance-free mechanism to explain the roaring rail phenomenon. Proceedings of the Institution of Mechanical Engineers, Part J: J. Eng. Tribology **222**(3), 171–181 (2008)
4. Polach, O.: Creep forces in simulations of traction vehicles running on adhesion limit. Wear **258**, 992–1000 (2005)
5. Steimel, A.: Electric Traction – Motive Power and Energy Supply – Basics and Practical Experience. Oldenbourg Industrieverlag GmbH, Munich (2008)
6. True, H.: On the theory of nonlinear dynamics and its applications in vehicle systems dynamics. Vehicle Systems Dynamics **31**, 393–421 (1999)
7. https://fdocuments.in/document/catalog-engleza.html?page=7

Piecewise Analytical Solution for Rub Interactions Between a Rotor and an Asymmetrically Supported Stator

Heba El-Mongy[1,2], Tamer El-Sayed[1,2,3](✉), Vahid Vaziri[2], and Marian Wiercigroch[2]

[1] Department of Mechanical Design, Faculty of Engineering, Mataria, Helwan University, Helmeiat-Elzaton, P.O. Box 11718, Cairo, Egypt
tamer.el-sayed@abdn.ac.uk

[2] Centre for Applied Dynamics Research, School of Engineering, University of Aberdeen, Aberdeen AB24 3UE, UK

[3] School of Engineering, University of Hertfordshire, R5 New Garden city, Egypt

Abstract. Rotor-stator rubbing phenomenon is a very common fault in various engineering applications. Usually, the stator supports are assumed to be symmetric, however, in some conditions, the stator support stiffness is anisotropic in nature. In this paper, an approximate analytical solution for the rub interactions between a Jeffcott rotor and an asymmetrically supported stator is introduced. The present problem is a good example of piecewise nonlinear dynamics in which contact causes the system to switch between linear and nonlinear dynamic behavior. To evaluate an analytical solution to the contact equations, Taylor's expansion is used to linearize the contact forces. Then, an analytical solution is introduced based on this approximation. To investigate the validity of the present model, the proposed approximate analytical solution is compared with the numerical solution obtained using direct numerical integration for the same problem. The results show that the proposed piecewise analytical solution is efficient in evaluating the system response for both periodic and chaotic responses.

Keywords: Rotordynamics · Asymmetric stator · Jeffcott rotor · Analytical solution · Periodic · Chaotic · Piecewise nonlinear system

1 Introduction

Rub interactions between rotating and stationary parts of machinery have drastic effects on the reliability and safety of rotordynamic systems. Rotor-stator rubbing has been extensively investigated over the past decades as documented in several reviews [1–4]. However, the related phenomena need more investigation.

The dynamics of rotor-stator rubbing phenomenon is mostly explored using lumped parameter models. Owing to the inherent nonlinearity that characterizes rotor rubbing due to the discontinuous contact, the models are usually low-dimensional to keep the computational time reasonable. Most of the studies have solved the nonlinear differential

Z. Dimitrovová et al. (Eds.): WMVC 2022, MMS 125, pp. 532–539, 2023.
https://doi.org/10.1007/978-3-031-15758-5_54

equations of the system numerically using various techniques such as Runge-Kutta algorithm [5–7], Newmark-beta method as in [8] and finite element analysis as in [9]. The piecewise-smooth model is commonly used in most of the previous studies to describe the contact dynamics as a linear elastic stiffness that causes a normal restoring force [10–14].

The strong nonlinearity in contact systems typically generates different possible responses that include periodic, quasi-periodic and chaotic responses. Usually, a rich profile of subharmonics and super harmonics characterize the frequency response of rotor-stator rubbing systems [15–18]. The features of the rub response are greatly affected by changing the system parameters such as rotor speed, stator stiffness, friction coefficient, and system damping [15, 19, 20]. Varney and Green [12] studied the influence of support asymmetry on the response of a rotor-stator contact system. Direct stiffness asymmetry was shown to greatly influence the system response even for small values of stiffness asymmetry. However, rotor flexibility was not considered in the model. In [21], orbit asymmetry was observed experimentally during rubbing and this was attributed to lower bottom stator stiffness in comparison to that of the other sides. Several researchers have conducted experimental studies to validate the mathematical models. Different forms of rub response were observed experimentally [7, 16–18, 21–23] but more sophisticated behavior was exhibited compared to the mathematically obtained responses. Hence, it is inferred that an improvement of the mathematical models of rubbing is crucial to better represent the real phenomenon.

Very few analytical studies have been conducted owing to the strong non-linearity associated with the rubbing phenomenon. Childs [24] provided an explanation of the second (1/2 ×) and third (1/3 ×) order subharmonics based on parametric excitation analysis. Kim and Noah [25] employed the harmonic balance method (HBM) to obtain periodic solutions and conducted bifurcation analysis to predict chaotic motions. They also showed that increasing the cross-coupling stiffness term causes Hopf bifurcation that leads to quasi-periodic response. Lu et al. [26] used a semi-analytical approach to find periodic solutions of a rub-impact rotor system. Karpenko et al. [13] developed a piecewise approximate analytical solution to study the rotor-stator interactions using Taylor's series expansion. Most recently, El-Sayed et al.[27] extended the approximate analytical solution given in [13] to include Coulomb friction between the rotor and stator. The solution was applied to analyze whirling vibrations of drill-strings.

In this paper, an approximate piecewise analytical solution is developed to study the rotor-stator dynamic interactions where the stator support is considered anisotropic. Direct and cross-coupled stiffness terms are included in the model. A two degree of freedom rotor model is employed taking into consideration the static offset between the rotor and stator centers. A linear elastic contact model is used to represent the contact dynamics and Taylor's expansion is used to linearize the nonlinear contact forces. After this introduction, the piecewise analytical solutions for the non-contact and contact regimes are presented in Sect. 2. Then, the results of the present model are given in Sect. 3. Finally, the main findings are summarized in the conclusions section.

2 Analytical Model

A two degree of freedom Jeffcott rotor is considered here to rotate within a stator which is asymmetrically supported. The Jeffcott rotor mass is negligible in comparison to the disk mass M. The stiffness and damping of the rotor are denoted by K and C respectively. The stator is mounted elastically on an anisotropic support with the following stiffness parameters K_{xx}, K_{yy}, K_{xy}, K_{yx} where x, y are the cartesian coordinates as shown in Fig. 1.b. The center of the stator is denoted by O_{s0} and the geometrical center of the rotor O_r with static equilibrium position at O_{r0}. At static equilibrium, the radial difference between O_{s0} and O_{r0} in x and y directions are denoted as ε_x, ε_y respectively. The rotor spins by an angular frequency Ω and is subjected to unbalance where the unbalance mass is m and the eccentricity is ρ. The rotor center position is described by x and y components. The diametral clearance between the rotor and the stator is 2 γ. When $R = \sqrt{(x-\varepsilon_x)^2 + (y-\varepsilon_y)^2} > \gamma$ the rotating disk comes in contact with the stator.

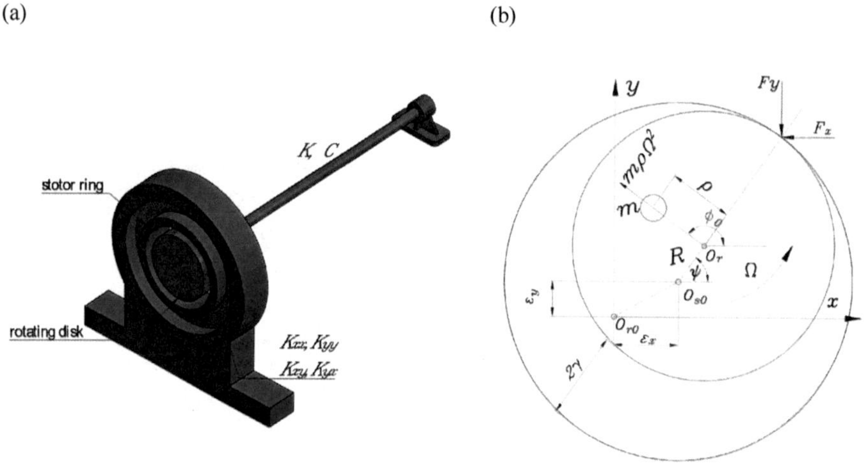

Figure 1 (a) Three-dimensional representation to the rotating disk and the stator ring mounted on anisotropic stiffness. (b) schematic drawing shows the coordinate systems and the contact forces. The rotor mass M is excited by the out of balance mass m. The out of balance force is $m\rho\Omega^2$. F_x, F_y are nonlinear contact force components which are in action when $R \geq \gamma$.

The equations of motion for the present model can be presented in the dimensionless form as follows:

$$\text{Nocontact}(\hat{z} < 1): \frac{d^2}{d\tau^2}\begin{bmatrix}\hat{x}\\ \hat{y}\end{bmatrix} + 2\nu\frac{d}{d\tau}\begin{bmatrix}\hat{x}\\ \hat{y}\end{bmatrix} + \begin{bmatrix}\hat{x}\\ \hat{y}\end{bmatrix} = \eta_m\hat{\rho}\eta^2\begin{bmatrix}\cos(\eta\tau+\varphi_0)\\ \sin(\eta\tau+\varphi_0)\end{bmatrix}, \quad (1)$$

$$\text{Contact}(\hat{z} > 1): \frac{d^2}{d\tau^2}\begin{bmatrix}\hat{x}\\ \hat{y}\end{bmatrix} + 2\nu\frac{d}{d\tau}\begin{bmatrix}\hat{x}\\ \hat{y}\end{bmatrix} + \begin{bmatrix}\hat{x}\\ \hat{y}\end{bmatrix} - \begin{bmatrix}\widehat{F}_x\\ \widehat{F}_y\end{bmatrix} = \eta_m\hat{\rho}\eta^2\begin{bmatrix}\cos(\eta\tau+\varphi_0)\\ \sin(\eta\tau+\varphi_0)\end{bmatrix}. \quad (2)$$

The contact forces in dimensionless form can be given as:

$$\widehat{F}_x = \frac{F_x}{\gamma} = \widehat{K}_{xx}\left(1 - 1/\hat{z}\right)\left(\hat{x} - \hat{\varepsilon}_x\right) + \widehat{K}_{xy}\left(1 - 1/\hat{z}\right)\left(\hat{y} - \hat{\varepsilon}_y\right) \tag{3}$$

$$\widehat{F}_y = \frac{F_y}{\gamma} = \widehat{K}_{yx}\left(1 - 1/\hat{z}\right)\left(\hat{x} - \hat{\varepsilon}_x\right) + \widehat{K}_{yy}\left(1 - 1/\hat{z}\right)\left(\hat{y} - \hat{\varepsilon}_y\right) \tag{4}$$

The analytical solution for the linear noncontact case can be written as:

$$\hat{x} = e^{-\nu\tau}(c_1\cos(\beta\tau) + c_2\sin(\beta\tau)) + c_{11}\cos(\eta\tau + \varphi_0) + \tag{5}$$

$$c_{22}\sin(\eta\tau + \varphi_0), \tag{6}$$

$$\hat{y} = \mathrm{e}^{-\nu\tau}(c_3\cos(\beta\tau) + c_4\sin(\beta\tau)) + c_{33}\cos(\eta\tau + \varphi_0) + c_{44}\sin(\eta\tau + \varphi_0).$$

The analytical solution for the nonlinear contact equations can be obtained after linearizing the nonlinear contact forces which results in contact equations as follows:

$$\text{Contact}(\hat{z} > 1): \frac{\mathrm{d}^2}{\mathrm{d}\tau^2}\begin{bmatrix}\hat{x}\\ \hat{y}\end{bmatrix} + 2\nu\frac{\mathrm{d}}{\mathrm{d}t}\begin{bmatrix}\hat{x}\\ \hat{y}\end{bmatrix} + \begin{bmatrix}F_{a1} & F_{b1}\\ F_{a2} & F_{b2}\end{bmatrix}\begin{bmatrix}\hat{x}\\ \hat{y}\end{bmatrix} = -\begin{bmatrix}F_{d1}\\ F_{d2}\end{bmatrix} + \eta_m\hat{\rho}\eta^2\begin{bmatrix}\cos(\eta\tau + \varphi_0)\\ \sin(\eta\tau + \varphi_0)\end{bmatrix} \tag{7}$$

where

$$F_{a1} = \left(1 + \widehat{K}_{xx}A_1 + \widehat{K}_{xy}B_2\right), F_{b1} = \left(\widehat{K}_{xx}B_1 + \widehat{K}_{xy}A_2\right), F_{D1} = \left(\widehat{K}_{xx}D_1 + \widehat{K}_{xy}D_2\right),$$

$$F_{a2} = \left(1 + \widehat{K}_{yy}A_2 + \widehat{K}_{yx}B_1\right), F_{b2} = \left(\widehat{K}_{yy}B_2 + \widehat{K}_{yx}A_1\right), F_{D2} = \left(\widehat{K}_{yy}D_2 + \widehat{K}_{yx}D_1\right),$$

where

$$A_1 = \left(\alpha^{1.5} - \alpha + \left(\tilde{x}_0 - \hat{\varepsilon}_x\right)^2\right)/\alpha^{1.5},$$

$$B_1 = B_2 = \frac{\left(\tilde{x}_0 - \hat{\varepsilon}_x\right)\left(\tilde{y}_0 - \hat{\varepsilon}_y\right)}{\alpha^{1.5}},$$

$$D_1 = \left(-\alpha\hat{\varepsilon}_x\left(\sqrt{\alpha} - 1\right) - \left(\tilde{x}_0 - \hat{\varepsilon}_x\right)\left(\tilde{x}_0\left(\tilde{x}_0 - \hat{\varepsilon}_x\right) + \tilde{y}_0\right) + \tilde{y}_0\left(\tilde{y}_0 - \hat{\varepsilon}_y\right)\right)/\alpha^{1.5}$$

$$A_2 = \left(\alpha^{1.5} - \alpha + \left(\tilde{y}_0 - \hat{\varepsilon}_y\right)^2\right)/\alpha^{1.5}$$

$$D_2 = \left(-\alpha\hat{\varepsilon}_y\left(\sqrt{\alpha} - 1\right) - \left(\tilde{y}_0 - \hat{\varepsilon}_y\right)\left(\tilde{x}_0\left(\tilde{x}_0 - \hat{\varepsilon}_x\right) + \tilde{y}_0\right) + \tilde{y}_0\left(\tilde{y}_0 - \hat{\varepsilon}_y\right)\right)/\alpha^{1.5}.$$

Then, the analytical solution for the contact case can be written as:

$$\hat{x} = e^{-\nu\tau}(c_1\cos(\gamma_1\tau) + c_2\sin(\gamma_1\tau) + c_3\cos(\gamma_2\tau) + c_4\sin(\gamma_2\tau)) + c_{11}\cos(\varphi_0 + \eta\tau) + c_{22}\sin(\eta\tau + \varphi_0) + c_{55}, \tag{8}$$

$$\hat{y} = e^{-\nu\tau}(s_1c_1\cos(\gamma_1\tau) + s_1c_2\sin(\gamma_1\tau) + s_2c_3\cos(\gamma_1\tau) + s_2c_4\sin(\gamma_1\tau)) + c_{33}\cos(\eta\tau + \varphi_0) + c_{44}\sin(\eta\tau + \varphi_0) + c_{66}, \tag{9}$$

where

$$s_1 = \frac{\gamma_1^2 + \nu^2 - F_{a1}}{F_{b1}},\ s_2 = \frac{\gamma_2^2 + \nu^2 - F_{a1}}{F_{b1}}$$

To derive the linear and nonlinear constants the reader is referred to [13, 27]. After obtaining the solution constants, the analytical solutions for the individual piecewise equations are obtained. To connect the two solutions, the switching condition $\hat{f}_z$ should be monitored precisely during the solution as:

$$\hat{f}_z = \sqrt{\left(\hat{x}\left(\tau;\tau_0,\tilde{x}_0,\tilde{y}_0,\tilde{x}_0',\tilde{y}_0'\right) - \hat{\varepsilon}_x\right)^2 + \left(\hat{y}\left(\tau;\tau_0,\tilde{x}_0,\tilde{y}_0,\tilde{x}_0',\tilde{y}_0'\right) - \hat{\varepsilon}_y\right)^2} - 1 \tag{10}$$

When $\hat{f}_z$ is greater than zero, the contact occurs and when $\hat{f}_z$ is smaller than zero the noncontact condition applies. Bisection algorithm is used to precisely evaluate the switching times.

3 Results and Discussion

To prove the validity of the presented analytical solution a selected example parameters based on the rotor-snubber ring test rig in the center for Applied Dynamics Research (CADR)of Aberdeen university is selected [23, 28]. Then, the model is solved using both analytical and numerical methods. The input data for the model in dimensionless form is listed in Table 1. In this study, the stator was assumed to be asymmetrically supported by the following stiffnesses $\widehat{K}_{yy} = 40 and\ \widehat{K}_{xx} = \widehat{K}_{xy} = \widehat{K}_{yx} = 2$.

Table 1. Dimensionless parameters of the system considered in the present analysis

Parameter	Value
Damping ratio, ζ	0.125
Mass ratio,η_m	0.005
Eccentricity ratio,$\hat{\rho}$	70
Normalized static displacement in x-direction,$\hat{\varepsilon}_x$	1
Normalized static displacement in y-direction,$\hat{\varepsilon}_y$	0

For the present model, the analysis in this paper is based on evaluating the system orbits at several frequency ratios ranging from $\eta = 1$ to $\eta = 3.5$ as shown in Fig. 2 (a) to (f). The results are evaluated using the present analytical solution. In addition, direct integration numerical solution is evaluated using Runge-Kutta MATLAB function ODE45. In both numerical and analytical solutions, a special code is prepared to detect the switching points. The red circle in Fig. 2 indicates the commencing of contact state and the green circles indicates the start of the non-contact state.

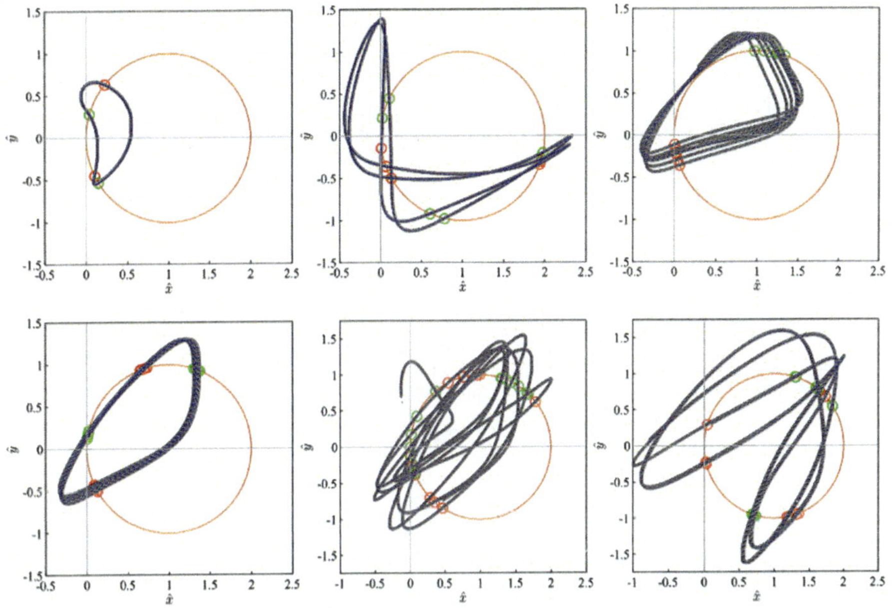

Fig. 2. Orbits for the nonlinear contact equations using numerical (ODE45) and analytical methods. $\widehat{K}_{yy} = 40$, $\widehat{K}_{xx} = \widehat{K}_{xy} = \widehat{K}_{yx} = 20$(a) $\eta = 1$, (b) $\eta = 1.5$, (c) $\eta = 2$, (d) $\eta = 2.5$, (e) $\eta = 3$, (f) $\eta = 3.5$, $[\tilde{x}_0 = 0.1, \tilde{x}_0' = 0, \tilde{y}_0 = 0.1, \tilde{y}_0' = 0]$ Analytical solution, Numerical solution indicates contact start indicates non-contact start.

Figure 2 shows that the rotor orbits are very sensitive to the frequency ratio. It can be realized that changing the rotational speed results in response changes as periodic, quasi periodic and chaotic. In the investigated cases, the results of the present analytical model are in good agreement with the numerical results as shown in Fig. 2.

4 Conclusions

In the present paper, an approximate piecewise analytical solution is introduced for the problem of rotor stator rubbing when the stator is mounted on an anisotropic support. The main novelty in this work is in including direct and cross-coupled stator stiffness asymmetry in the analytical solution of rotor-stator rubbing. The results show that the analytical solution is capable for evaluating the system response in the cases of periodic, quasi periodic and chaotic responses. The switching instances are calculated and compared with those captured by direct numerical integration method. The results of the analytical model are very promising and contribute towards more refined mathematical models of rotor-stator rubbing.

References

1. Muszynska, A.: Rotor-to-stationary element sub-related vibration phenomena in rotating machinery: literature survey. The Shock and Vibration Digest **21**, 3–11 (1989)
2. Ahmad, S.: Rotor casing contact phenomenon in rotor dynamics—literature survey. J. Vib. Control **16**, 1369–1377 (2010)
3. O. Alber, R.M.: Rotor-stator contact–overview of current research. In: MATEC Web of Conferences, EDP Sciences, p. 03001 (2014)
4. Prabith, K., Krishna, I.R.P.: The numerical modeling of rotor–stator rubbing in rotating machinery: a comprehensive review. Nonlinear Dyn. **101**(2), 1317–1363 (2020)
5. Karpenko, E.V., Wiercigroch, M., Cartmell, M.P.: Regular and chaotic dynamics of a discontinuously nonlinear rotor system. Chaos, Solitons Fractals **13**, 1231–1242 (2002)
6. Popprath, S., Ecker, H.: Nonlinear dynamics of a rotor contacting an elastically suspended stator. J. Sound Vib. **308**, 767–784 (2007)
7. Chávez, J.P., Hamaneh, V.V., Wiercigroch, M.: Modelling and experimental verification of an asymmetric Jeffcott rotor with radial clearance. J. Sound Vib. **334**, 86–97 (2015)
8. Patel, T.H., Darpe, A.K.: Coupled bending-torsional vibration analysis of rotor with rub and crack. J. Sound Vib. **326**, 740–752 (2009)
9. Behzad, M., Alvandi, M., Mba, D., Jamali, J.: A finite element-based algorithm for rubbing induced vibration prediction in rotors. J. Sound Vib. **332**, 5523–5542 (2013)
10. Choi, Y.-S.: On the contact of partial rotor rub with experimental observations. KSME international journal **15**, 1630–1638 (2001)
11. Patel, T., Darpe, A.: Use of full spectrum cascade for rotor rub identification. Adv. Vibration Eng. **8**, 139–151 (2009)
12. Varney, P., Green, I.: Nonlinear phenomena, bifurcations, and routes to chaos in an asymmetrically supported rotor–stator contact system. J. Sound Vib. **336**, 207–226 (2015)
13. Karpenko, E.V., Wiercigroch, M., Pavlovskaia, E.E., Cartmell, M.P.: Piecewise approximate analytical solutions for a Jeffcott rotor with a snubber ring. Int. J. Mech. Sci. **44**, 475–488 (2002)
14. Wang, J., Zhou, J., Dong, D., Yan, B., Huang, C.: Nonlinear dynamic analysis of a rub-impact rotor supported by oil film bearings. Arch. Appl. Mech. **83**, 413–430 (2013)
15. Sawicki, J.T., Padovan, J., Al-Khatib, R.: The dynamics of rotor with rubbing. Int. J. Rotating Machinery, **5** (1999)
16. Von Groll, G., Ewins, D.J.: A mechanism of low subharmonic response in rotor/stator contact—measurements and simulations. Journal of Vibration and Acoustics **124**, 350–358 (2002)
17. Chu, F., Lu, W.: Experimental observation of nonlinear vibrations in a rub-impact rotor system. J. Sound Vib. **283**, 621–643 (2005)
18. Beatty, R.F.: Differentiating rotor response due to radial rubbing. J. Vib. Acoust. Stress. Reliab. Des. **107**, 151–160 (1985)
19. Lin, F., Schoen, M., Korde, U.: Numerical investigation with rub-related vibration in rotating machinery. J. Vib. Control **7**, 833–848 (2001)

20. Zheng, Z., Xie, Y., Zhang, D., Ye, X.: Effects of Stator Stiffness, Gap Size, Unbalance, and Shaft's Asymmetry on the Steady-State Response and Stability Range of an Asymmetric Rotor with Rub-Impact, Shock and Vibration (2019)
21. Praveen Krishna, I., Padmanabhan, C.: Experimental and numerical investigations on rotor–stator rub. Proceedings of the Institution of Mechanical Engineers, Part C: Journal of Mechanical Engineering Science **232**, 3200–3212 (2018)
22. Gonsalves, D.H., Neilson, R.D., Barr, A.D.S.: A study of the response of a discontinuously nonlinear rotor system. Nonlinear Dyn. **7**, 451–470 (1995)
23. Karpenko, E., Wiercigroch, M., Pavlovskaia, E.E., Neilson, R.D.: Experimental verification of Jeffcott rotor model with preloaded snubber ring. J. Sound Vib. **298**, 907–917 (2006)
24. Childs, D.W.: Fractional-frequency rotor motion due to nonsymmetric clearance effects. J. Eng. For Power **104**, 533–541 (1982)
25. Kim, Y., Noah, S.: Bifurcation analysis for a modified Jeffcott rotor with bearing clearances. Nonlinear Dyn. **1**, 221–241 (1990)
26. Lu, Q.S., Li, Q.H., Twizell, E.H.: The existence of periodic motions in rub-impact rotor systems. J. Sound Vib. **264**, 1127–1137 (2003)
27. Elsayed, T., El-Mongy, H., Vaziri, V., Wiercigroch, M.: Analytical study of rotor-stator rubbing phenomenon. In: COBEM 2021: 26th International Congress of Mechanical Engineering - Virtual Event, Virtual Event (2021)
28. Chávez, J.P., Wiercigroch, M.: Bifurcation analysis of periodic orbits of a non-smooth Jeffcott rotor model. Commun. Nonlinear Sci. Numer. Simul. **18**, 2571–2580 (2013)

Stable Rotational Orbits of Base-Excited Pendula System

Alicia Terrero-Gonzalez[1,2](✉), Antonio S.E. Chong[1,3], Ko-Choong Woo[4], and Marian Wiercigroch[1]

[1] Centre for Applied Dynamics Research, School of Engineering, University of Aberdeen, Aberdeen AB24 3UE, UK
{a.terrerogonzalez.19,M.Wiercigroch}@abdn.ac.uk
[2] Curtin University, Perth, Australia
[3] Faculty of Natural Sciences and Mathematics, Escuela Superior Politécnica del Litoral, P.O. Box 09-01-5863 Guayaquil, Ecuador
[4] University of Nottingham Malaysia, Jalan Broga, 43500 Semenyih, Selangor Darul Ehsan, Malaysia

Abstract. The dynamics of a harmonically base-excited two pendula system, is investigated for the practical application of energy harvesting from rotatory motions [1,2]. The central aim of this study is to identify system parameter ranges for which pendulum rotations exist. The external harmonic excitation amplitude and frequency, and the difference in pendulums lengths are the system parameters which have been varied thresholds. Bifurcation analysis has been performed for the identification of values beyond which rotations exist, and the study of corresponding bifurcation points has been conducted with computational tool ABESPOL, developed at the Centre for Applied Dynamics Research (CADR) of the University of Aberdeen [3]. Direct simulations and one-parameter continuation analysis were performed with ABESPOL and some results were corroborated with direct numerical integration in Matlab, based on a Runge-Kutta algorithm. One parameter continuation results showed complex bifurcation scenarios for antiphase rotatory motions, presenting evidence of existence and form of representation. Further results showed that pendulum rotations, in phase and antiphase, co-exist with oscillatory motions. Therefore, the basins of attraction have been computed, enabling attractors to be targeted so as to enable antiphase rotatory motion.

Keywords: Pendula system · Rotations · Bifurcation analysis · Path following

1 Introduction

The application of pendula systems for energy harvesting from ambient vibration has been extensively investigated in the past years by the Centre for Applied

University of Aberdeen and Curtin University Alliance

Z. Dimitrovová et al. (Eds.): WMVC 2022, MMS 125, pp. 540–547, 2023.
https://doi.org/10.1007/978-3-031-15758-5_55

Dynamics and Research (CADR), University of Aberdeen. The motivation to harvest energy from sea waves led to novel concept of rotational coupled parametric pendulum for their application to Wave Energy Converters (WEC). Previous studies [4] proved the superiority of rotatory over oscillatory pendulum responses with regards to energy harvesting. Therefore, further understanding of system parameters and physics governing pendula rotatory motions are needed to enable rotations for a two-pendula parametrically excited system. The dynamics of a vertically excited parametric pendulum was studied by Clifford and Bishop [5] and the experimental research on this matter was conducted by Alevras *et al.* [6] to achieve rotations from a single parametric pendulum. Moreover, stochastic excitations were also considered by Andreeva *et al.* [7] for the specific application to sea waves energy harvesting. Further studies by Garira and Bishop [8] as well as Horton [9] limited the stability of pendula rotations, and associated period-1 rotations with saddle node bifurcations. Further studies from Lenci *et al.* [10] also presented saddle node bifurcations, with the birth of pendulum rotatory responses, identified by a perturbation method. The robustness of rotational solutions in terms of dynamics integrity was consolidated in later work [11]. Moreover, a classification of the double pendulum states of equilibrium, oscillations and rotations considering Lyapunov exponent was made by Dudkoski *et al.* [12], and the existence of self-sustained rotations was identified by Klimina, Lokshin and Samsonor [13].

This study is based on previous experimental studies of a two-pendula system coupled to an elastic base presented by Najdecka *et al.* under periodic [1] and stochastic excitation [14]. The analysis to identify rotational orbits in both pendulums excited parametrically and system parameters boundaries are studied by Marzal *et al.* [2]. The system parameters that will be considered in subsequent sections are length of pendulums, amplitude and excitation frequency.

The rest of the paper is organised as follows. Section 2 describes the physical model of the two-pendula system coupled to a common elastic base and harmonically excited in the vertical direction. System dynamic responses were obtained from direct numerical integration. Bifurcation analysis based on one-parameter continuation in the subsequent section is compared to system qualitative responses predicted by Runge-Kutta algorithm, Sect. 3. Concluding remarks are shown in Sect. 4, presenting the outcomes from the study.

2 Physical and Mathematical Modelling

The physical system, which can be deployed for energy harvesting, was analysed in the form of time histories and bifurcation diagrams by Najdecka, Kapitaniak and Wiercigroch [1], based on two parametric pendulums mounted on a flexible, common supporting structure excited in the vertical direction by a common harmonic force R_y, where ($R_y = \mathrm{A}\sin(\Omega \mathrm{t})$). Figure 1(a) shows the experimental rig used for model validation and experimental data acquisition. The two pendulums are able to rotate independently, and the aluminium elastic base sustains both pendulums, allowing their synchronization. Figure 1(b) shows typical phase plane responses when both pendulums are synchronized in antiphase rotatory motion, and also when the responses are close to heteroclinic cycles. Data presenting Pendulum 1 is drawn in black while that of Pendulum 2, in red. Therefore, the system is comprised of three masses: the supporting structure, M, and the two bob masses, m_1 and m_2 of pendulum 1 and pendulum 2 respectively; l_1 and l_2 represent the pendulum lengths and c_θ denotes the damping coefficient. k_x, k_y, c_x and c_y represent the total stiffness and damping properties of the supporting structure in the x and y directions. The system is modelled with four-degrees-of-freedom where X and Y denote the horizontal and vertical displacements respectively, and θ_1 and θ_2 depict the angular displacement of Pendulum 1 and 2 respectively measured from the vertical.

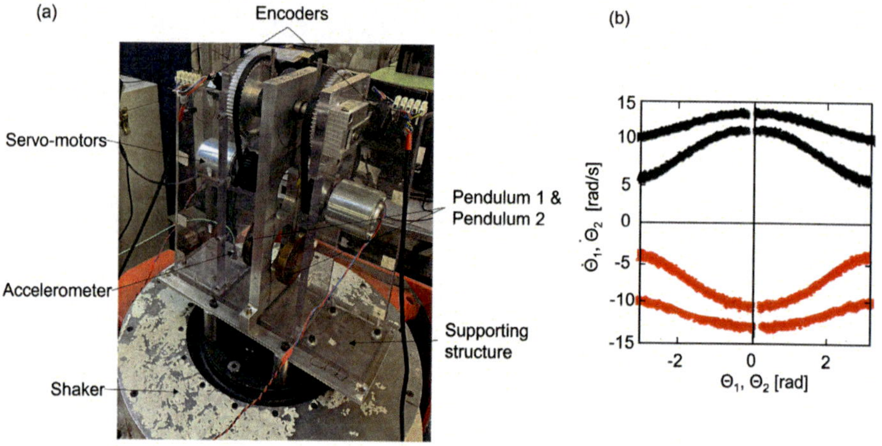

Fig. 1. (a) Two-pendula experimental rig. (b) Phase plane of typical rotatory and nearly heteroclinic orbit for Pendulum 1 (black) and Pendulum 2 (red).

The system governing equations of motion have been derived by the Lagrangian method for equal masses ($m_1 = m_2 = m$), and $l_1 = l$ and $l_2 = l + \delta$; the equations of motion are expressed in the form $\boldsymbol{Mu} + \boldsymbol{Cu} + \boldsymbol{Ku} = \boldsymbol{f(\theta_1, \theta_2)}$ where $\boldsymbol{u} = (\boldsymbol{X}, \boldsymbol{Y}, \boldsymbol{\theta_1}, \boldsymbol{\theta_2})^{\boldsymbol{T}}$ and f contains the nonlinear terms. The equations of motion in the dimensionless form can be written as follows:

$$
\begin{aligned}
x'' + &\gamma_x x' + \alpha_x x \\
&= -a[\theta_1'' cos(\theta_1) - (\theta_1')^2 sin(\theta_1)] - a\left(1 + \frac{\delta}{l}\right)[\theta_2'' cos(\theta_2) - (\theta_2')^2 sin(\theta_2)],
\end{aligned}
\tag{1}
$$

$$
\begin{aligned}
y'' + &\gamma_y (y' - \omega p cos(\omega\tau)) + \alpha_y [y - p sin(\omega\tau)] \\
&= -a[\theta_1'' sin(\theta_1) + (\theta_1')^2 cos(\theta_1)] - a\left(1 + \frac{\delta}{l}\right)[\theta_2'' sin(\theta_2) + (\theta_2')^2 cos(\theta_2)],
\end{aligned}
\tag{2}
$$

$$\theta_1'' + \frac{\gamma_\theta}{l}\theta_1' = -x''cos(\theta_1) - [1 - \omega^2 psin(\omega\tau)]sin(\theta_1), \tag{3}$$

$$\theta_2'' + \frac{\gamma_\theta}{(l+\delta)}\theta_2' = -x''lcos(\theta_2)\left(\frac{1}{l+\delta}\right) - [1 - \omega^2 plsin(\omega\tau)]sin(\theta_2). \tag{4}$$

where ' denotes differenciation with respect to the non-dimensionalized time τ and the non-dimensionalised parameters and variables are,

$$x = \frac{X}{l},\ \gamma_x = \frac{c_x}{(M+2m)\omega_n},\ \alpha_x = \frac{k_x}{(M+2m)\omega_n^2},\ a = \frac{m}{M+2m},\ \omega = \frac{\Omega}{\omega_n},$$
$$p = \frac{A}{l},\ \gamma_\theta = \frac{c_\theta}{m\omega_n},\ y = \frac{Y}{l},\ \gamma_y = \frac{c_y}{(M+2m)\omega_n},\ \alpha_y = \frac{k_y}{(M+2m)\omega_n^2}.$$

Equations (1) to (4) were non-dimensionalised with respect to a linear natural frequency, and solved by direct numerical integration to obtain a form of exact solution based on initial conditions initial conditions of $x(0) = x'(0) = 0$, $\theta_1(0) = 1.5708$, $\theta_1'(0) = 2.5$, $\theta_2(0) = 0.5708$ and $\theta_2'(0) = -2.5$. To give an overview of system qualitative responses, a three-dimensional graph was constructed for the three parameters in this study, which are the non-dimensionalised amplitude and frequency of the wave excitation, represented as p and ω, respectively and the parameter mismatch δ, describing the difference between pendulum lengths. System qualitative responses in the three-dimensional graph, were differentiated by colour. Pendulum 1 is oscillating, and Pendulum 2 is rotating, the plot colour is yellow. If Pendulum 1 is rotating and Pendulum 2 is oscillating, the plot is green. For both pendulums oscillating, black is used. Red is deployed for both pendulums rotating. Figure 2(b) present typical phase planes and trajectories, corresponding to different system qualitative responses.

Direct numerical integration predict the existence of rotations for amplitude values of p tending towards zero or larger values close to 1.4, when the frequency, ω, is 2.1. Values close to zero amplitude were studied further and identified to be values close to 0.05 when frequencies, ω, target between 2.0 and 2.2. The parameter mismatch δ ranged between 0.00 and 0.06. To identify the parameter values beyond which rotations cease to exist, a bifurcation analysis was performed with the computational tool ABESPOL [3] to identify attractors to be targeted in practice. This was based on the approach of continuation.

3 Bifurcation Analysis

To identify the ranges of parameter values for which rotations occur, numerical continuation was performed. One parameter numerical continuation has been performed with ABESPOL, which connects with the continuation core, COCO [17].

In addition, direct simulation has been carried out for values of p, ω and δ within the ranges computed by direct numerical integration. Results were shown in Fig. 2. The results obtained have been summarized for the case of $\omega = 2.00$ and $\delta = 0.00$ and one parameter continuation was implemented for $p \in [0.01274\ 1.40000]$. Figure 3 present the results obtained from one parameter continuation as well as the direct simulations required to initialize the continuations for

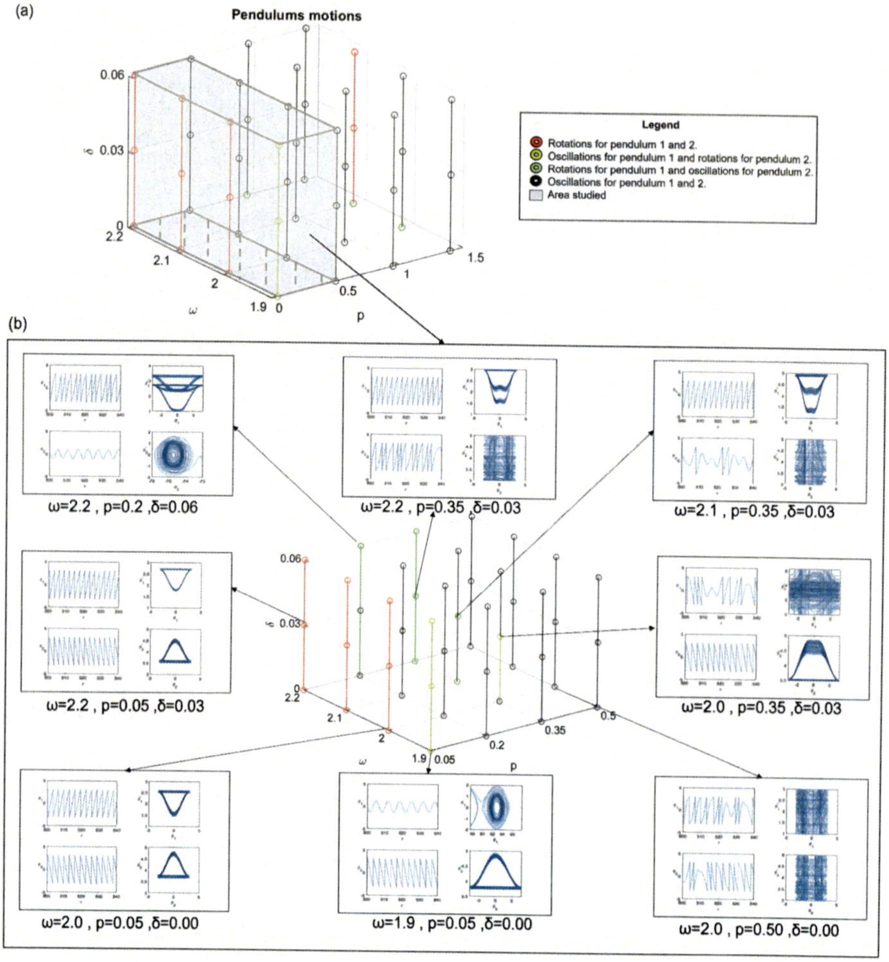

Fig. 2. Dynamical system responses, where yellow indicated Pendulum 1 oscillations and Pendulum 2 rotations, green denotes Pendulum 1 rotations and Pendulum 2 oscillations, black implies oscillations and red corresponds to rotations for ranges of system parameters of (a) $\delta \in$ [00.06], $\omega \in$ [1.9 2.2] and p∈[0.05 1.4]. (b) Area of future bifurcation analysis constrained between $\delta \in$[00.06], $\omega\in$ [1.9 2.2] and p∈ [0.05 1.5] and a selection of time histories and phase portraits for both pendulums.

each pendulum. Stable solutions are presented in green, and red lines correspond to unstable solutions. For both pendulums, the coexistence of phase rotations, antiphase rotations and oscillations has been identified. Basins of attraction are required to identify rotational motions for both pendulums so as to apply to energy harvesting. Some example phase planes are presented for antiphase and oscillatory motions, to prove the existence of this type of motion. Pendulum 1 experienced a saddle node bifurcation for the antiphase rotational motion,

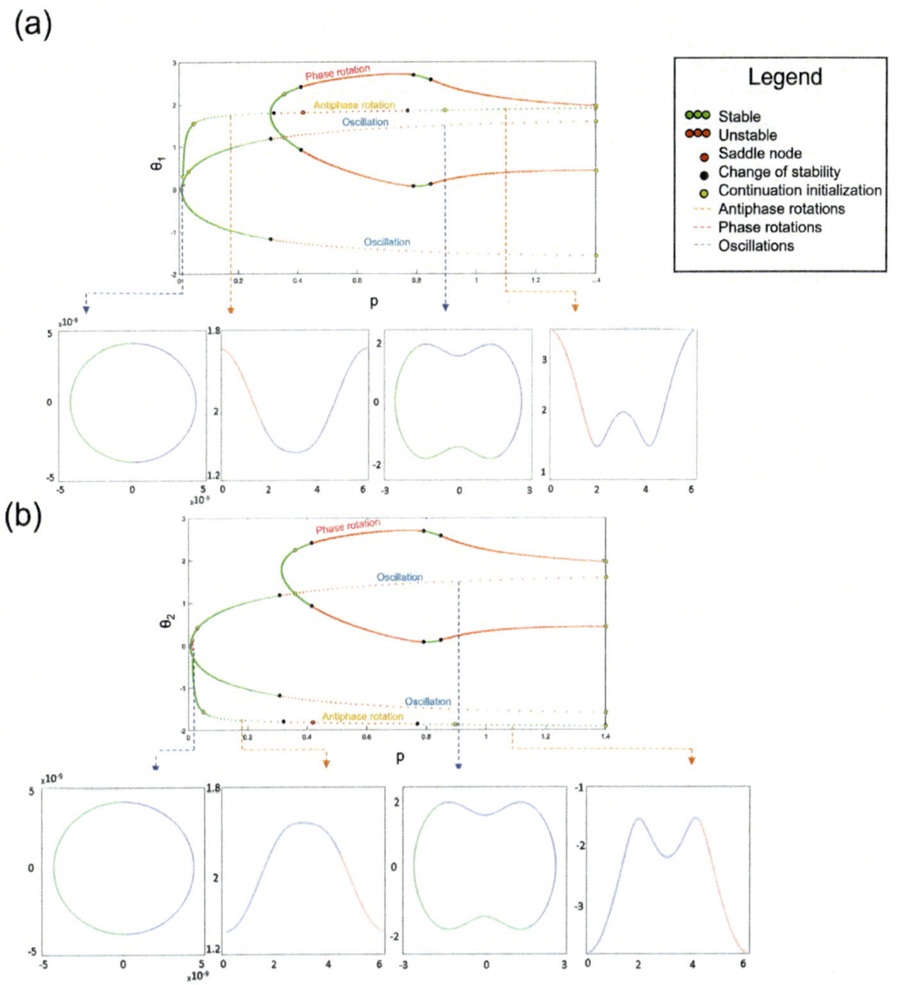

Fig. 3. Bifurcation analysis using path following with system initial conditions are $x(0) = x'(0) = 0$, $\theta_1(0) = 1.5708$, $\theta_1'(0) = 2.5$, $\theta_2(0) = 0.5708$ and $\theta_2'(0) = -2.5$. Case for $\omega = 2.00$ and $\delta = 0.00$ from $p = [0\ 1.40000]$. Stable and unstable solutions are presented in green and red respectively, saddle node bifurcations are represented as red dots. (a) and (b) Pendulum 1 and 2 bifurcation diagrams.

$p = 0.012739$, and a period doubling bifurcation at $p = 0.42600$ for $\omega = 2.00$ and $\delta = 0.00$. For Pendulum 2, only one saddle node was detected for antiphase rotations at $p = 0.012739$. Only the first saddle node bifurcation is located at the limiting values of system parameter values. These values distinguish between stable and unstable orbits. Further work would extend the identification of boundaries separating rotational and oscillatory motions.

4 Concluding Remarks

This paper studied a coupled pendula system excited vertically to identify ranges of system parameters for which pendulums exhibit stable rotations. This can be applied to wave energy harvesting in a novel WEC concept. Direct numerical integration showing that rotations for both pendulums exist when p $\rightarrow$ 0 and p $\rightarrow$ 1.4 for $\omega = 2.10$ and $\delta \in [0.00\ 0.06]$, and more specifically for ranges of p$\in$ [0.0500 0.500], $\omega \in [1.90\ 2.20]$ and $\delta \in [0.00\ 0.06]$, where p and ω denote non-dimesionalized amplitude and frequency of excitation respectively; whereas δ describes the difference in pendulum lengths.

Numerical continuation in ABESPOL has identified bifurcation scenarios and attractors to be targeted in the practice, and helped to identify ranges of system parameters for both Pendulum 1 and Pendulum 2. The analysis conducted for p $\in$ [0.012739 1.4], $\omega = 2.00$ and $\delta = 0.00$ has located a saddle node bifurcation at p = 0.012739 and period doubling bifurcations at p = 0.42600 for antiphase rotation. Oscilla-
tions coexist with anti-phase rotations for values of smaller than $p = 0.32000$. Chaotic oscillations coexist with period-2 oscillations for both Pendulum 1 and Pendulum 2.

References

1. Najdecka, A., Kapitaniak, T., Wiercigroch, M.: Synchronous rotational motion of parametric pendulums. Int. J. Non-Linear Mech. **70**, 84–94 (2015)
2. Marszal, M., Witkowski, B., Jankowski, K., Perlikowsk, P., Kapitaniak, T.: Energy harvesting from pendulum oscillations. Int. J. Non-Linear Mech. **94**, 251–256 (2017)
3. Chong, A.: Numerical Modelling and Stability Analysis of Non-smooth Dynamical Systems via ABESPOL. University of Aberdeen, Thesis (2016)
4. Terrero Gonzalez, A., Dinning, P., Howard, I., McKee, K., Wiercigroch, M.: Is wave energy untapped potential? Int. J. Mech. Sci. **205**, 106544 (2021)
5. Clifford, M., Bishop, S.: Rotating periodic orbits of the parametrically excited pendulum. Phys. Lett. A **201**(2–3), 191–196 (1995)
6. Alevras, P., Brown, I., Yurchenko, D.: Experimental investigation of a rotating parametric pendulum. Nonlinear Dyn. 201–213 (2015). https://doi.org/10.1007/s11071-015-1982-8
7. Andreeva, T., Alevras, P., Naess, A., Yurchenko, D.: Dynamics of a parametric rotating pendulum under a realistic wave profile. Int. J. Dyn. Control **4**(2), 233-238 (2016)
8. Garira, W., Bishop, S.: Rotating solutions of the parametrically excited pendulum. J. Sound Vib. **263**(1), 233–239 (2003)
9. Horton, B., Lenci, S., Pavlovskaia, E., Romeo, R., Rega, G., Wiercigroch, M.: Stability boundaries of period-1 rotation for a pendulum under combined vertical and horizontal excitation. J. Appl. Nonlinear Dyn. **2**(2), 103-126 (2013)
10. Lenci, S., Pavlovskaia, E., Rega, G., Wiencigroch, M.: Rotating solutions and stability of parametric pendulum by perturbation method. J. Sound Vib. **310**(1–2), 243–259 (2008)

11. Lenci, S., Rega, G.: Experimental versus theoretical robustness of rotating solutions in a parametrically excited pendulum: a dynamical integrity perspective. Phys. D Nonlinear Phenom. **240**(9–10), 814–824 (2011)
12. Dudkowski, D., Wojewoda, J., Czolczynski, K., Kapitaniak, T.: Is it really chaos? The complexity of transient dynamics of double pendula. Nonlinear Dyn. **102**, 759–770 (2020)
13. Klimina, L., Lokshin, B., Samsonov, V.: Bifurcation diagram of the self-sustained oscillation modes for a system with dynamic symmetry. J. Appl Math. Mech. **81**(6), 442–449 (2017)
14. Najdecka, A., Narayanan, S., Wiercigroch, M.: Rotary motion of the parametric and planar pendulum under stochastic wave excitation. Int. J. Non-Linear Mech. **71**, 30–38 (2015)
15. Chong, A., Yue, Y., Pavlovskaia, E., Wiercigroch, M.: Global dynamics for harmonically excited oscillator with a play: numerical studies. Int. J. Non-Linear Mech. **94**, 98–108 (2017)
16. Chong, A., Brzeski, P., Wiercigroch, M., Perlikowski, P.: Path-following bifurcation analysis of church bell dynamics. J. Comput. Nonlinear Dyn. **12**(061017), 1–8 (2017)
17. Dankowicz, H., Schilder, F.: Recipes for Continuation. SIAM (2013)

Surrogate Expressions for Dynamic Load Factor

Majid Aleyaasin(✉)

Engineering School, Aberdeen University, Aberdeen AB24 3UE, UK
eng780@abdn.ac.uk

Abstract. In this paper the elastic dynamic load factor in structural dynamic is revisited. The existing literature in which the response exists only for isosceles triangular pulse and shock load pulse is criticized.

A new pulse shape parameter is introduced by which both isosceles triangle impulse and shock load impulse and other unsymmetrical pulses can be expressed. This enables the elastic Dynamic Load Factor (DLF) to be computed versus the pulse shape parameter.

Thereafter a surrogate model is found by which the load factor can be computed via pulse duration, natural period and pulse parameter. The conservative values of the load factor extracted from the surrogate model and can be used for structural dynamic aspect of the design.

In numerical examples the Single Degree Of Freedom (SDOF)model subjected to blast loading is investigated. It is shown that numerical scheme for elastic dynamic load factor in this paper is very accurate. The accuracy is demonstrated in case when isosceles triangular pulse blast load is applied. Moreover, by introducing the pulse shape index parameter, any unsymmetrical pulse can be expressed and their response can be determined.

Two types of surrogate functions are introduced to substitute the elastic DLF data. It is concluded that nonlinear low order surrogate functions are not accurate enough to predict elastic DLF. However, higher order surrogate polynomials are very accurate and can be used in computational design of protective structures.

Keywords: Dynamic load factor · Nonlinear ODE · Protective structures

1 Introduction

Dynamic load factor (*DLF*) is a key parameter in damage evaluation in the structures subjected to dynamic loads like impact earthquake, etc. and is still a field of research [1]. The recent research in blast resistance structures also highly relies on determination of *DLF* via numerical methods [2] that sometimes are associated with experiments [3] .

The key step on the design of blast resistance structures, is determining the DLF via blast overpressure and the ratio of blast duration to the natural period (T_d/T) [4]. It originates from Single Degree of Freedom (SDOF) method. This SDOF is still used in preliminary design calculations [5]. It relies on design charts and graphs known as Bigg's charts [5].

To computerize the design procedure, an alternative surrogate formula, for Bigg's chart is required. A low fidelity expression exists in [5] that is erroneous because it is

Z. Dimitrovová et al. (Eds.): WMVC 2022, MMS 125, pp. 548–554, 2023.
https://doi.org/10.1007/978-3-031-15758-5_56

used for all types of pulses and this cannot be true and relied upon. This article is aimed at accurate determination of elastic DLF to be sued for design purpose.

This article initially highlights the two types of DLF, elastic (related to structure) and plastic (related to loading) . Such explicit statement is missed in [4] and [5]. The plastic DLF is straightforward and is known to the designer via the ratio of elastic resistance to maximum blast load. Comparison of the plastic DLF with elastic DLF determines if the structure is in elastic or elastic region.

The Bigg's chart [4, 5] includes both elastic and plastic region and is available only for symmetrical triangular explosion pulse. Moreover, there is not an expression for elastic DLF in it to be used by designers. This article analyses the elastic and plastic response resulted by unsymmetrical blast pulses, aiming at accurate determination of elastic DLF.

Using optimisation techniques, a four parameters nonlinear surrogate function is found for elastic DLF expression. Since there is noticeable error in this function, an alternative linear polynomial (order 17) is developed that results very accurate formula for elastic DLF. It is concluded that nonlinear surrogacy is successful only if initial function suggestion is appropriate. Otherwise higher order polynomials are preferred. Therefore a very accurate polynomial type surrogate expression is developed, by which the elastic DLF can be computed and relied upon.

2 Elastic and Plastic Dynamic Load Factor

The elastic dynamic load factor depends on $x_{\max}$ maximum deflection, is defined by:

$$DLF_E = \frac{x_{\max}}{x_{st}} \tag{1}$$

In (1) the x_{st} is the static deflection of the system which is given by:

$$x_{st} = \frac{F_{\max}}{k} \tag{2}$$

By substituting (2) into (1) we have:

$$DLF_E = \frac{k\, x_{\max}}{F_{\max}} \tag{3}$$

In (3) $F_{\max}$ and k are the maximum force and the stiffness per unit length of the protective structure, respectively. Therefore, $F_{\max}$ can be expressed by $F_{\max} = p_{\max} L_E$, where $p_{\max}$ is the maximum pressure and L_E is the equivalent length. The plastic dynamic load factor depends on maximum resistance R_m and is defined by:

$$DLF_R = \frac{R_m}{F_{\max}} \tag{4}$$

In (4) R_m depends on maximum elastic deflection x_{el} and is given by this equation:

$$R_m = k\, x_{el} \tag{5}$$

The structure remains in elastic status if the following inequality is true:

$$x_{el} > x_{\max} \tag{6}$$

The inequality (6) can be expanded as follows:

$$k\,x_{el} > k\,x_{\max} \Rightarrow \frac{k\,x_{el}}{F_{\max}} > \frac{k\,x_{\max}}{F_{\max}} \tag{7}$$

Substituting (5) into (4) and the result into (7), also using (3) and (1) in right side of (7) yields to:

$$DLF_R > DLF_E \tag{8}$$

By numerical simulation we can find the maximum deflection and we can check if (6) or (8) holds, then we can find if the structure is in elastic or plastic status.

3 SDOF Response to Unsymmetrical Pulse Force

When an unsymmetrical triangular pulse (Fig. 1 left) is applied to a mechanical system with mass M and the stiffness k (Fig. 1 right) the equations of motion in SDOF approach:

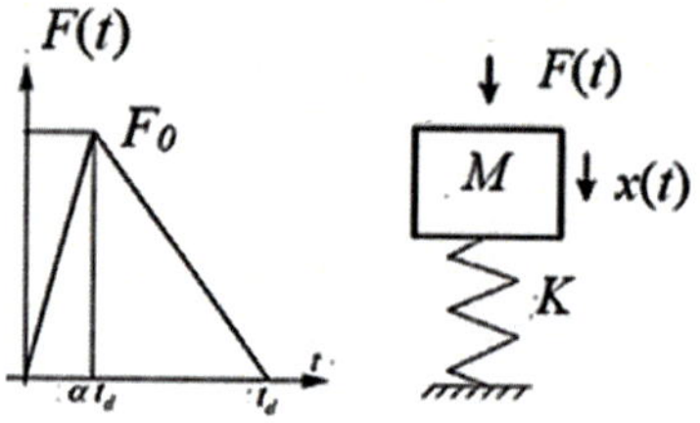

Fig. 1. Unsymmetrical triangular pulse shape (left) system (right)

$$M\,\ddot{x} + k\,x = \frac{F_{\max} t}{\alpha\,t_d} \qquad t < \alpha\,t_d \tag{9}$$

$$M\,\ddot{x} + k\,x = \frac{(t - t_d)F_{\max}}{(\alpha - 1)\,t_d} \qquad \alpha\,t_d < t < t_d$$

The equations in (9) can be changed to:

$$\frac{M}{k}\ddot{x} + x = \frac{F_{\max}}{k} \cdot \frac{t}{\alpha\,t_d} = x_{st} \cdot \frac{t}{\alpha\,t_d} \qquad t < \alpha\,t_d \tag{10}$$

$$\frac{M}{k}\ddot{x} + x = \frac{F_{\max}}{k} \cdot \frac{(t - t_d)}{(\alpha - 1)\,t_d} = x_{st} \cdot \frac{(t - t_d)}{(\alpha - 1)\,t_d} \qquad \alpha\,t_d < t < t_d$$

However, $\frac{M}{k}$ can be expressed in terms of T natural period of structure as follows:

$$\frac{M}{k} = \frac{T^2}{4\pi^2} \tag{11}$$

Substituting (11) into (10) changes it to:

$$\frac{T^2}{4\pi^2}\ddot{x} + x = x_{st} \cdot \frac{t}{\alpha\, t_d} \qquad t < \alpha\, t_d \tag{12}$$

$$\frac{T^2}{4\pi^2}\ddot{x} + x = x_{st} \cdot \frac{(t - t_d)}{(\alpha - 1)\, t_d} \qquad \alpha\, t_d < t < t_d$$

Considering dimensionless parameter $\overline{x} = \frac{x}{x_{st}}$, the Eqs. (12) can be changed to:

$$\frac{T^2}{4\pi^2} \cdot \frac{d^2\overline{x}}{dt^2} + \overline{x} = \frac{t}{\alpha\, t_d} \qquad t < \alpha\, t_d \tag{13}$$

$$\frac{T^2}{4\pi^2} \cdot \frac{d^2\overline{x}}{dt^2} + \overline{x} = \frac{(t - t_d)}{(\alpha - 1)\, t_d} \qquad \alpha\, t_d < t < t_d$$

Further dimensionless parameters $\tau = \frac{t}{T}$ and $\tau_d = \frac{t_d}{T}$ introduced which yields to:

$$\tau = \frac{t}{T} \Rightarrow dt = T\, d\tau \Rightarrow dt^2 = T^2 d\tau^2 \tag{14}$$

Considering (14) the Eqs. (13) will change to:

$$\frac{1}{4\pi^2} \cdot \frac{d^2\overline{x}}{d\tau^2} + \overline{x} = \frac{\tau}{\alpha\, \tau_d} \qquad \tau < \alpha\, \tau_d \tag{15}$$

$$\frac{1}{4\pi^2} \cdot \frac{d^2\overline{x}}{d\tau^2} + \overline{x} = \frac{(\tau - \tau_d)}{(\alpha - 1)\, \tau_d} \qquad \alpha\, \tau_d < \tau < \tau_d$$

Via numerical simulation of Eqs. (15) the history of $\overline{x}$ versus τ can be found and $\overline{x}_{\max}$ can be picked up easily. If we look at (3) it is obvious that we have:

$$DLF_E = \overline{x}_{\max} \tag{16}$$

For the symmetrical pulse shape (isosceles triangular). There is analytical solution for history of $\overline{x}$, in [5, 6] that can be expressed in notation used in this paper as follows:

$$\overline{x} = 2\left(\frac{2\pi\tau - \sin 2\pi\tau}{2\pi\, \tau_d}\right) \qquad 0 < \tau < 0.5\tau_d$$

$$\overline{x} = 2\left(\frac{2\pi\, \tau_d - 2\pi\tau - \sin 2\pi\tau}{2\pi\, \tau_d} + \frac{\sin\pi(2\tau - \tau_d)}{\pi\, \tau_d}\right) \qquad 0.5\tau_d < \tau < \tau_d$$

$$\overline{x} = 2\left(2\sin^2\frac{\pi}{2}\tau_d\right)\frac{\sin\pi(2\tau - \tau_d)}{\pi\, \tau_d} \qquad \tau > \tau_d \tag{17}$$

In Fig. 2, the elastic DLF is determined by using $\overline{x}$ from both (16) and (17). It shows that the numerical method is verifiable and there is not any error.

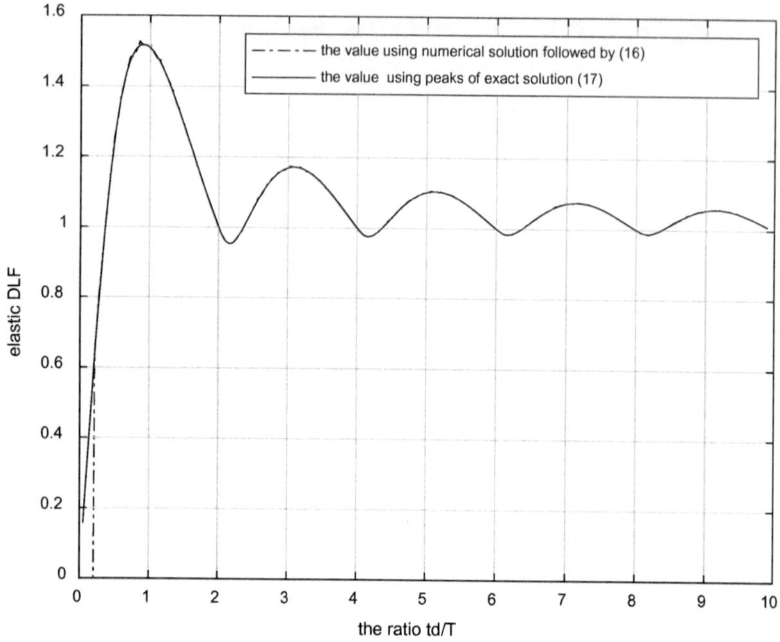

Fig. 2. Elastic DLF using Numerical and analytical method

4 Linear and Nonlinear Surrogate Models

Through the shape of the shock spectra in Fig. 2, we can suggest that the following form seems suitable for the elastic DLF.

$$DLF_E \cong \left(1 - a^{-b\frac{t_d}{T}} \cos\left(c\frac{t_d}{T} + d\right)\right)$$

Using direct search nonlinear optimisation method, the parameters a, b, c and d is found and the surrogate formula for Elastic DLF can be found via this:

$$DLF_E \cong \left(1 - 5.7286^{-0.5301\frac{t_d}{T}} \cos\left(2.4901\frac{t_d}{T} + 0.4854\right)\right) \quad (18)$$

In Fig. 3, it is shown that the suggested surrogate formula in (18) is not accurate, since substantial error (up to 17%) can be observed through. Therefore, an alternative polynomial form is suggested to represent the elastic DLF. Using the curve fitting tools in MATLAB, it is found that a 17th degree polynomial in terms of the ratio t_d/T via (19) and (20) summarized in Table 1, is an accurate surrogate expression:

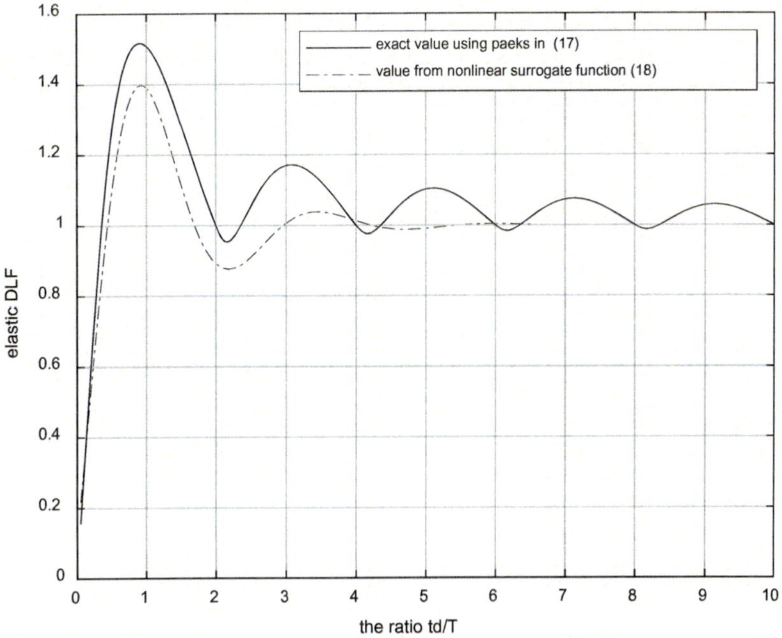

Fig. 3. Elastic DLF via numerical and nonlinear surrogate function

$$DLF_E \cong \sum_{i=1}^{18} \left(\frac{t_d}{T}\right)^{18-i} a_i \tag{19}$$

Table 1. Numerical values of the coefficients in (19)

Coefficient	Value	Coefficient	Value	Coefficient	Value
$a1$	-5.3453×10^{-10}	a_7	−0.1284	a_{13}	−134.6187
a_2	5.2161×10^{-8}	a_8	0.8806	a_{14}	127.3124
a_3	-2.3092×10^{-6}	a_9	−4.5012	a_{15}	−75.3144
a_4	6.1438×10^{-5}	a_{10}	17.0873	a_{16}	22.739
a_5	−0.0011	a_{11}	−47.6367	a_{17}	−0.1595
a_6	0.0139	a_{12}	95.6827	a_{18}	0.1548

In Fig. 4, it is shown that the suggested surrogate polynomial in (19) is accurate enough. Since when we examine Fig. 4 the error level is below 2%.

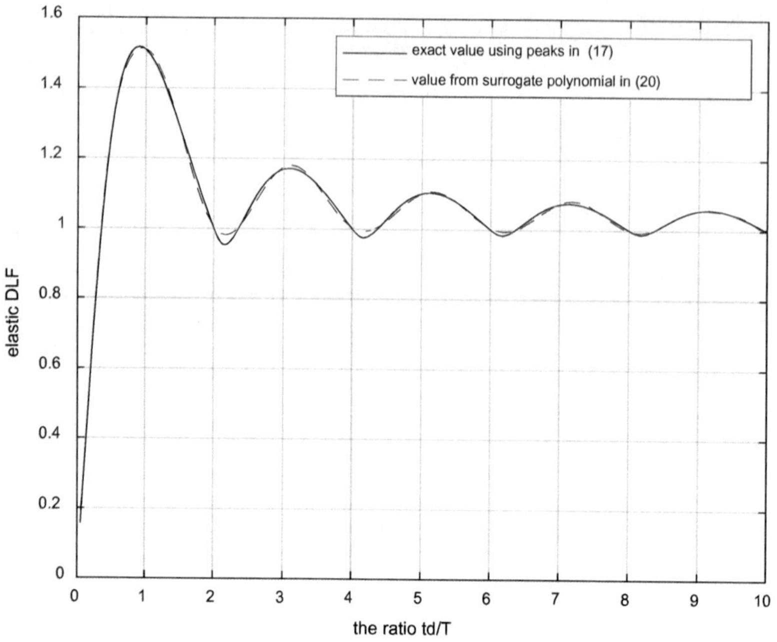

Fig. 4. Elastic DLF via numerical and polynomial surrogate function

5 Conclusions

The elastic DLF is a key factor in computational design of protective structures. Surrogate models are required to fulfil this objective. The nonlinear surrogate functions that seem suitable are not accurate enough. However, the higher order surrogate polynomials are very accurate in determination of the elastic DLF.

References

1. Chao, Z., Hong, H., Kaiming, B., Xueyuan, Y.: Dynamic amplification factors for a system with multiple-degrees-of-freedom. Earthq. Eng. Eng. Vib. **19**, 363–375 (2020)
2. Geng, S., Wei, Y., Wang, W.: Dynamic increase factor of an equivalent SDOF structural system for beams with different support conditions under conventional blast loading. J. Eng. Mech. **147**(4), 06021002 (2021)
3. Li, G., Ji, T., Chen, J.: Determination of the dynamic load factors for crowd jumping using motion capture technique. Eng. Struct. **174**, 1–9 (2018)
4. Task Committee on Blast-Resistant Design of the Petrochemical Committee of the Energy Division of ASCE: Design of Blast-Resistant Buildings in Petrochemical Facilities, 2nd edn. (2010)
5. Krauthammer, T.: Modern Protective Structures. CRC Press, New York (2008)
6. Jacobson, L.S., Ayre, R.S.: Engineering Vibrations: With Applications To Structures And Machinery. McGraw-Hill, Boca Raton (1958)

Virtual Prototyping of a Floating Wind Farm Anchor During Underwater Towing Operations

Rodrigo Martinez[1], Sergi Arnau[1], Callum Scullion[2], Paddy Collins[2], Richard D. Neilson[1], and Marcin Kapitaniak[1](✉)

[1] The National Decommissioning Centre, School of Engineering, University of Aberdeen, Aberdeen, UK
marcin.kapitaniak@abdn.ac.uk

[2] Aubin Group, Castle Street, Castlepark Industrial Estate, Ellon AB41 9RF, UK

Abstract. The aim of this paper is to present the initial results of feasibility studies aimed at optimising the towing configuration of a novel, complex shape (pyramid based) and thereby untested design of floating wind farm anchor during underwater towing. The study was carried out in the real physics Marine Simulator, at the National Decommissioning Centre. This enables us to study in detail, the drag/lift forces acting on the towed anchor/s, determine the optimal anchor installation arrangement (orientation, depth, position of towing cables, number of anchors towed together in an array) and establish the effects of the operational (towing velocity, drag) and environmental conditions (sea states, significant wave height, peak wave period) on the anchor's trajectory. The model developed is validated with computational fluid dynamics analysis to obtain representative drag and lift coefficients for the anchor during towing. Thus, this paper focuses on the calibration process to ensure robustness and relevance of the developed model in the simulator. Consistent results for the drag and lift coefficient were obtained for a range of towing speeds (0,25–3 m/s). The towing dynamics, forces acting on the anchor and the final configuration (e.g. water depth, offset angle) were obtained which in turn will allow the optimal conditions and requirements (e.g. equipment, vessel type etc.) to be recommend-ed in future studies.

Keywords: Marine simulator · Virtual prototyping · Underwater towing operation · Anchor · Floating wind farm

1 Introduction

Novel installation methods of anchors for offshore wind are currently being developed worldwide. Four of the most common floater types of offshore wind are

Supported by Energy Technology Partnership, EPSRC Supergen ORE Hub, ORE Catapult, FOW CoE, Aubin Group, Net Zero Technology Centre, Oceanetics Inc.

Z. Dimitrovová et al. (Eds.): WMVC 2022, MMS 125, pp. 555–564, 2023.
https://doi.org/10.1007/978-3-031-15758-5_57

shown in Fig. 1: semi-submersible, spar-buoy, tension-leg platform (TLP) and barge [1]. All four types of floating wind structures use mooring and anchoring systems to keep the structure in place. A state-of-the art review of the installation phase of offshore wind turbines is presented in [4,6,7]. Various installation methods for floating wind are critically discussed and the main challenges identified. It is concluded that the complex anchoring systems used presently are expected to evolve with new foundations technologies. A technology review conducted by Jiang [7] also concludes that considerable CapEx and OpEx savings are possible from developing and optimising new mooring and anchoring systems.

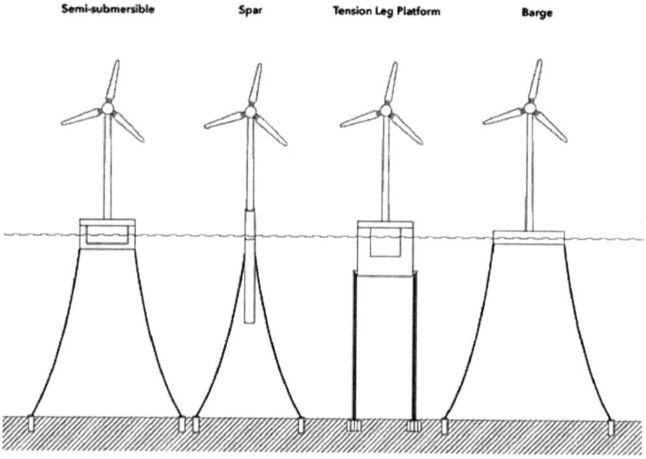

Fig. 1. Four common floater types for wind turbines [6]

In addition to optimising the installation process of anchors, Jiang [7] also identifies the need to develop new advanced modelling software to accurately simulate coupled behaviours of floating wind systems. In this context, the simulator at the National Decommissioning Centre (NDC) has been used to conduct virtual field trials, with the aim to prototype new installation method of anchors for offshore wind, with a primary focus on the installation challenges of floating wind farm anchors and mooring systems [10]. This will lead to the development of novel techniques for the deployment of wind farm anchors and mooring systems, which will be tested comprehensively through simulation of different scenarios to establish the applicability of the proposed methods in various weather conditions, sea states and ultimately study their benefits and limitations. This research will in turn enable a wider range of vessels (lower cost/less specialist) during anchorage/mooring installation and increased precision of the installation of the sub-sea equipment which could lead to the reduction of costs and CO_2 emissions.

In this study, the feasibility of an underwater towing operation that involves a novel type of anchor with variable buoyancy for mooring floating offshore wind turbines is presented (Fig. 2a). The proposed system reduces the need to use a Heavy Lift Vessels (HLV), thereby cutting overall installation costs, and reduces the installation lead times as the design of the operation is less dependent on sea and weather conditions. In this context, the industry is continuing to minimise installation costs by developing self-installing and port assembled systems.

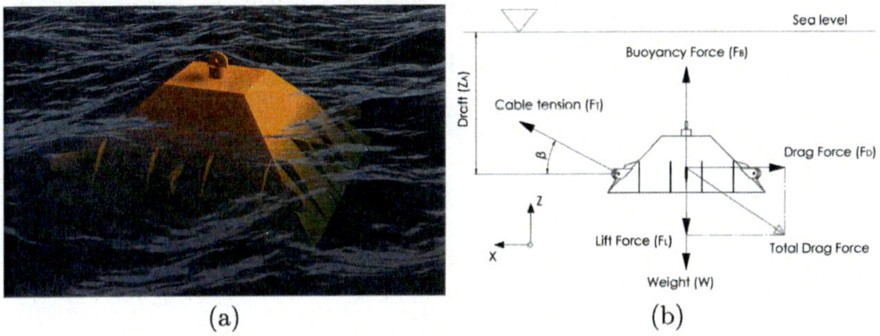

Fig. 2. (a) 3D render of the novel floating anchor concept. (b) free body diagram of the anchor.

A set of detailed modelling studies are performed in the state-of-the-art, Marine Simulator at the NDC. Using the multi-physics simulation allows for a more economical proof-of-concept approach, that will allow to fully assess the feasibility of underwater towing of anchors and de-risk future offshore deployment operations.

2 Virtual Prototyping

A review of virtual prototyping (VP) of offshore operations can be found in [9]. The importance of working with a tool that can integrate the different project phases from concept/tender, engineering, mobilisation, through operations is vital for future offshore operations to reduce cost and minimise risks. An example of virtual prototyping using the marine simulator supplied by the Offshore Simulator Centre (OSC) is presented in [11]. In this example, the real-time VP model is used to simulate the process of a riser pulled in from an installation vessel to a jacket platform. With special attention being made to the maximum bending curvature along the flexible riser, as this is one of the critical aspects during the operation. Finally, the results are validated against finite element (FE) analysis resulting in good agreement. Another example study stating the importance of simulations based on multi-body dynamics is presented in [8], in which a comprehensive analysis of potential collisions between objects during offshore lifting operations is made.

The study for a new concept of transportation of sub-sea templates is presented in [5], in which an underwater towing method is proposed. In this study, the dynamical effects involved during the towing operations are discussed and analysed by means of: (1) a mathematical mass-spring model with forced excitation, (2) a multi-body time domain simulation using the software SIMO [2] and (3) a physical model test in the laboratory. A good agreement was found between the three methodologies, specially between the SIMO analysis and the experimental results. Another example to study the underwater towing of a subsea module is presented in [12], in which the results of a computational fluid dynamics (CFD) model are compared with experimental results. Gu et al. [3] presents a study of a large caisson during wet-towing transportation. The caisson resistance coefficient was simulated via Ansys Fluent software and resultant CO_2 emissions under various conditions were calculated.

3 Methodology

3.1 Test Set-up

To investigate the variation of the towing force (drag and lift force) with the towing velocity, a set of simulations using both marine simulations and CFD analysis using Ansys Fluent are conducted. The parameters used in both simulations are shown in Table 1.

Table 1. CFD parameters

	Speed case (m/s)	Angle of attack (deg)	Fluid
1	0.25	0	Sea water
2	0.50	0	Sea water
3	0.75	0	Sea water
4	1.00	0	Sea water
5	1.25	0	Sea water
6	2.00	0	Sea water
7	3.00	0	Sea water

Towing direction: +X (see Fig. 3)

The model remains underwater at all times (wet towing). The general dimensions of the anchor are 10 m wide, 10 m long and 3,48 m high and it has a truncated pyramid shape as indicated in Fig. 2b. The length of the towing line is 22.1 m and the wet weight of the anchor for this set of simulations is 10.5 kN. Further information on the methodology used to carry out the marine simulations can be found in [10].

3.2 Physical Model

The free body diagram of the anchor with submerged weight W, buoyancy force F_B, towing force F_T, hydrodynamic drag F_D and lift F_L forces and offset angle

β is shown in Fig. 2b. In the steady state, the system can be simplified to Eq. 1 and 2, where the weight and drag force of the wire have been ignored. At any given vertical position Z_A, the system satisfies the following equilibrium in the Y and Z directions (see Fig. 3)

$$F_T \cos\beta - F_D = 0, \tag{1}$$

$$F_T \sin\beta + W - F_L = 0. \tag{2}$$

3.3 CFD

To verify the results obtained in the marine simulator, a series of CFD simulations in Ansys Fluent are carried out to determine the drag and lift coefficient for a range towing speeds (0,25–3 m/s) and at an angle of attack of 0°. Furthermore, a detailed study of the flow field is made to visualise the vortex pattern around the anchor during towing.

Figure 3 shows the computational domain used in the simulations. The leading edge of the anchor is 1L away from the velocity inlet and the aft is 3L away from the pressure outlet. In addition, the lateral, top and bottom distance between the anchor and the walls is 1L, 1.5L and 1L, respectively. In this case, the characteristic length L is 10 m. Each case was simulated using the Reynolds-Average Navier-Stokes equation (RANS) method. In the current work, the k-ω STT turbulence model was applied. The mesh size is selected so that the simulation produces consistent results regardless of the size and the number of cells yet producing results efficiently. The inflation layer meshing technique is used around the anchor to capture the boundary layer in the near-wall region. The resistance created by the towing rope is not included in the CFD simulation, as it is considered to be small in comparison to the drag force.

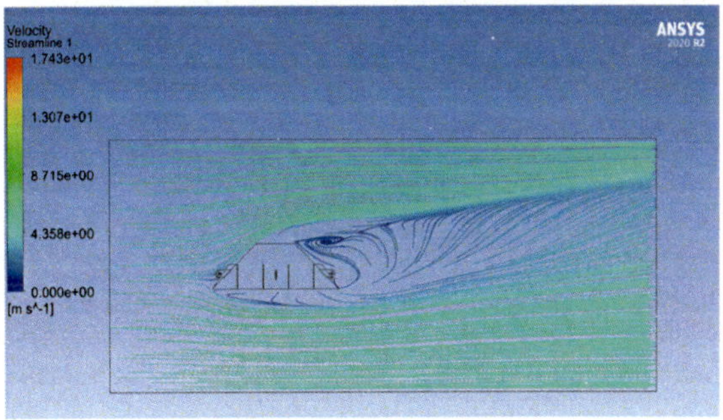

Fig. 3. CFD simulations flow velocity streamlines around the submerged anchor.

The output of each simulation was the drag force in the direction of the flow (X) and the lift force, perpendicular to the direction of the flow (Z). Once the drag and lift force have been obtained for each scenario, they can then be used to calibrate the marine simulations. For each CFD analysis, the drag and lift coefficient are determined as:

$$C_D = \frac{2F_D}{\rho A V^2}, \tag{3}$$

$$C_L = \frac{2F_L}{\rho A V^2}, \tag{4}$$

where F refers to the drag or lift force, ρ is the density of sea water, A is the area of body incident to flow, and V is the velocity at which the vessel is moving.

3.4 Marine Simulator

The towing system consists of a tugboat, the new anchor model and a towing rope, as shown in Fig. 4. For this preliminary study, the calibration was done during the steady-state stage of towing to minimise the effects of acceleration and with a calm sea (e.g. without waves, wind and currents). In this stage, the vessel and the anchor have a constant speed and the anchor's position in the water column and its orientation (e.g. pitch, role and yaw) are also constant. This creates a scenario as close as possible to the one simulated in Ansys Fluent, at which point the results of the CFD analysis can be used to calibrate the marine simulator. Martinez et al. [10] describes the process and methodology to setup a simulation in the state-of-the-art marine simulator.

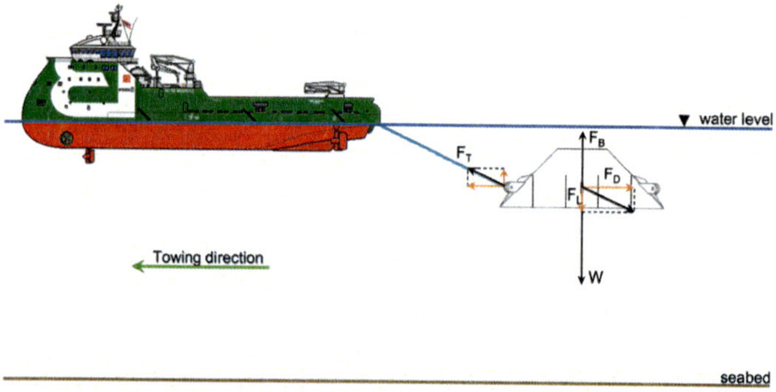

Fig. 4. Representation of the towing operation in the marine simulator.

4 Results

In this section, the results presented are divided in two sections: CFD simulations for which the drag coefficient C_D was calculated and, marine simulations in which the previously calculated C_D serves as an input to analyse the towing operations under different wave conditions and length of tow lines.

4.1 CFD Simulations

Figure 3 shows the flow velocity streamlines around the anchor when being towed at $v = 3.0$ m/s speed. Flow separation and vortex generation at the back of the anchor is observed.

Figure 5 (a) shows the calculated drag (blue) and lift (red) forces for different towing speeds while keeping the anchor horizontal. Drag and lift coefficients, C_D and C_L respectively, are then calculated using Eqs. 3 and 4. These values are then used as calibration input on the marine simulator. This process is performed for each towing scenario to get realistic drag and lift force data from the simulator.

4.2 Marine Simulations

Once the drag and lift coefficients are calibrated, the towing operations are tested in the marine simulator.

The variation of the calibrated drag and lift forces with towing speed are shown in Fig. 5 (b). The blue line is the drag force and the red line is the lift force. Green markers represent tests done with tow line length $L_{tow} = 22.1$ m, blue markers $L_{tow} = 41.6$ m and black markers $L_{tow} = 60.2$ m. From the figure it can be seen that, regardless of the tow line length, their respective drag and lift forces behave similarly at all towing speeds.

Figure 5 (c) shows the depth at which the anchor settles at all towing speeds for the three tow line lengths with same colour code as Fig. 5 (b). Although the tow line length does not have an impact in the drag and lift forces, it does have an impact on the depth (Z_A) at which the anchor is being towed. At lower towing speeds, the anchor sinks the furthest and at higher speeds the anchor gets closer to the surface. The length of the tow line is directly proportional to the equilibrium depth of the anchor: the longer the tow line, the deeper the anchor is being towed and the shorter the tow line, the closer to the water surface the anchor is being towed.

Next, for a fixed significant wave height of $H_S = 2.03$ m, parametric studies varying the peak wave period T_P are performed to establish the effect of wave loading on the towing dynamics, as shown in Fig. 6. In here, we consider only single tow line length of $L_{tow} = 22.1$ m, under towing velocity of $v = 1$ m/s and JONSWAP wave spectrum ($\gamma = 3.3$). Waves with $T_P < 4$ s have no noticeable effect on the anchor's position and orientation and towing force. For $T_P > 4$ s, wave-induced oscillations are more visible, maximum values for force and anchor depth amplitude are reached at $T_P \approx 15$ s. Yaw presents a local minima at $T_P \approx 5$ s, after which, its values start moving back to 0°. Pitch and roll, how ever, have a local minima at $10 < T_P > 11$ s.

Figure 7 shows example time histories for the anchor's dynamic response for towing with $T_P = 5.5$ s. In each of the time histories there is steady state response of regular periodic type (27 periods of excitation are plotted). Right hand side panels show the corresponding phase portraits that show the orbit of the response for each of the anchor parameters. Once the anchor reaches an equilibrium, its pitch, towing force and anchor depth reach a stable orbit

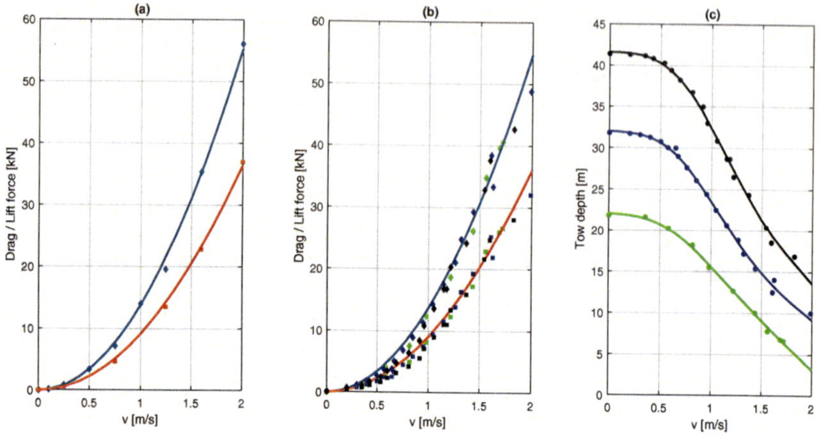

Fig. 5. (a) Drag and lift forces as a function of flow velocity from CFD analysis, (b) Calibrated response for towing drag and lift forces obtained from developed simulation for tow line lengths of L_{tow} = 22.1 m (green markers), L_{tow} = 41.6 m (blue markers), L_{tow} = 60.2 m (black markers), (c) The equilibrium tow depth of the anchor as a function of towing velocity for tow line lengths of L_{tow} = 22.1 m (green), L_{tow} = 41.6 m (blue), L_{tow} = 60.2 m (black). (Color figure online)

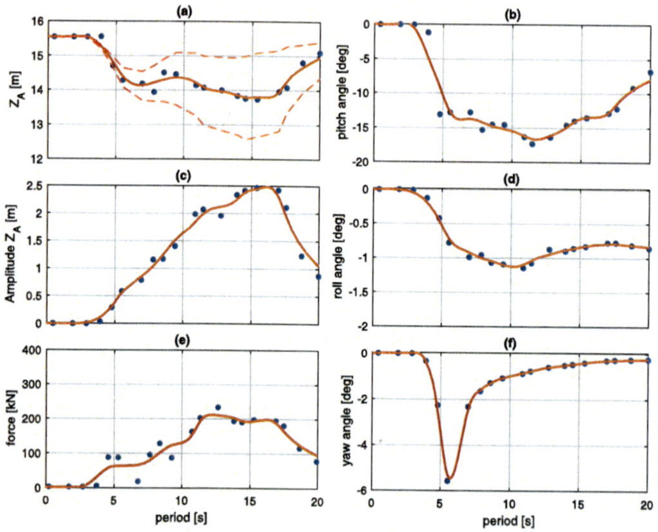

Fig. 6. Parametric plots depicting the effect of wave period on the anchor towing dynamics (using towing line of length L_{tow} = 22.1 m) for (a) anchor vertical position Z_A (dashed lines mark the min/max values), (b) amplitude of anchor's vertical position, (c) towing force, amplitude of anchor's pitch (d), roll (e) and yaw (f) angles.

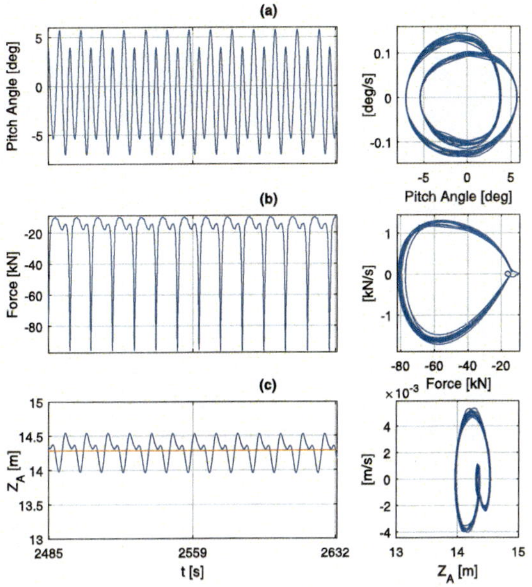

Fig. 7. An example of family of time histories for (a) anchor's pitch angle, (b) towing force, (c) anchor's vertical position Z_A and corresponding phase portraits (right panels) depicting anchors dynamics for peak wave period of $T_P = 5.5$ s

(period-2), that show the maximum/minimum variation of these quantities and their corresponding velocities.

5 Conclusions

The towing feasibility of a novel floating anchor design is described in this work. CFD simulations are initially performed to calculate drag and lift coefficients that would serve as input for realistic simulations in the marine simulator.

The analysis performed in the marine simulator shows that during towing operations, the drag and lift forces are not impacted by the length of the tow line. The towing depth of the anchor decreases as the towing speed increases but the towing depth of the anchor increases as tow line length increases. Simulations also show that wave period has a large impact in the anchor's depth and towing force at $T_P \approx 15$ s and at $T_P \approx 5$ s on the anchor's orientation (pitch, roll and yaw).

References

1. DNV-GL: DNVGL-ST-0119 - Floating wind turbine structures (2018). https://rules.dnv.com/docs/pdf/DNV/ST/2018-07/DNVGL-ST-0119.pdf

2. DNV-GL: Feature Description: SeSam software suite for hydrodynamic and structural analysis of renewable, offshore and maritime structures (2022). https://www.dnv.com/Images/Sesam-Feature-Description_tcm8-58834.pdf
3. Gu, H., Wang, H., Zhai, Q., Feng, W., Cao, J.: Study on the dynamic responses of a large caisson during wet-towing transportation. Water **13**(2), 126 (2021). https://doi.org/10.3390/w13020126
4. Ikhennicheu, M., et al.: Review of the state of the art of mooring and anchoring designs, technical challenges and identification of DLCs. Technical Report, COREWIND (2020)
5. Jacobsen, T., Leira, B.J.: Numerical and experimental studies of submerged towing of a subsea template. Ocean Eng. **42**, 147–154 (2012). https://doi.org/10.1016/j.oceaneng.2012.01.003
6. James, R., Ros, M.C.: Floating offshore wind: market and technology review. Technical Report, The Carbon Trust (2015). https://www.carbontrust.com/resources/floating-offshore-wind-market-technology-review
7. Jiang, Z.: Installation of offshore wind turbines: a technical review. Renew. Sustain. Energy Rev. **139**, 110576 (2021). https://doi.org/10.1016/j.rser.2020.110576
8. Lee, H.W., Roh, M.I.: Review of the multibody dynamics in the applications of ships and offshore structures. Ocean Eng. **167**, 65–76 (2018). https://doi.org/10.1016/j.oceaneng.2018.08.022
9. Major, P., Zhang, H., Hildre, H.P., Edet, M.: Virtual prototyping of offshore operations: a review. Ship Technol. Res. **68**(2), 84–101 (2021). https://doi.org/10.1080/09377255.2020.1831840
10. Martinez, R., Arnau, S., Scullion, C., Collins, P., Neilson, R.D., Kapitaniak, M.: Deployment feasibility studies of variable buoyancy anchors for floating wind applications. Mech. Mach. Sci. (2022).
11. Yuan, S., Major, P., Zhang, H.: Flexible riser replacement operation based on advanced virtual prototyping. Ocean Eng. **210**, 107502 (2020). https://doi.org/10.1016/j.oceaneng.2020.107502
12. Zan, Y., Guo, R., Yuan, L., Wu, Z.: Experimental and numerical model investigations of the underwater towing of a subsea module. J. Mar. Sci. Eng. **7** (2019). https://doi.org/10.3390/jmse7110384

RML: Recent Advances in Railway Mechanics and Moving Load Problems

An Iterative Approach for Analyzing Wheel-Rail Interaction

Aditi Kumawat(✉), Francesca Taddei, and Gerhard Müller

Chair of Structural Mechanics, TUM School of Engineering and Design, Technical University of Munich, Arcisstraße 21, 80333 Munich, Germany
{aditi.kumawat,francesca.taddei,gerhard.mueller}@tum.de

Abstract. The steady-state vehicle-track interaction or the interaction of the moving train with rail defects may result in unstable vibrations. The interaction of the moving train with such track defects induces additional dynamic stresses in the track system that may prove harmful for the structural health of the track. In this paper, a new iterative approach is proposed for analyzing the coupled equations of the vehicle-track system. The proposed approach can account for the wheel/rail contact loss. The results show that the proposed approach is computationally efficient and can be employed to study the effect of a wide range of track defects on the vehicle-track response. As an example, the vehicle-track response is obtained for the case where the wheel is traversing a rail-head corrugation. A loss of contact is observed when the wheel encounters the rail-head corrugation. The wheel-rail contact loss results in high impact loads over the rail beam leading to a sudden increase in rail beam deflections (by up to 85% and 57% for the undamped and damped cases, respectively).

Keywords: Railway track · Oscillator · Track defects

1 Introduction

In operation, the railway track structure is subjected to motion-induced steady-state interactions as well as the arbitrarily time-varying loads caused by the interaction of a moving vehicle with various track defects. Such track defects include, e.g., vertical rail imperfections, rail discontinuities, and local track irregularities. To study the effects of time-dependent loads caused by track defects, the railway track structure is commonly analyzed with the help of various numerical and analytical models [6,7].

Further aspects of vehicle-track interaction modelling are the idealization of the vehicle system and the evaluation of the time-dependent loads caused by the interaction of a moving vehicle with the track. Usually, two frameworks are followed for the idealization of the vehicle system. In the first framework, the vehicle is idealized as a point force with a magnitude equal to the axle load. This idealization, where the inertial and internal degrees of freedom of

Z. Dimitrovová et al. (Eds.): WMVC 2022, MMS 125, pp. 567–580, 2023.
https://doi.org/10.1007/978-3-031-15758-5_58

the vehicle are neglected, is more suited for studying the dynamic behaviour of the track itself or wave propagation analysis in the soil medium underlying the track [11]. In the second framework, the degrees of freedom of the vehicle are duly considered by modelling the vehicle as a moving mass [16,20], a single or multi-degree-of-freedom (SDOF or MDOF) moving oscillator system [3,17], or as a multi-degree-of-freedom lumped mass model [11,19].

A commonly used vehicle idealization that can account for the inertia and the degrees of freedom of the vehicle is the 'moving oscillator models'. It may be mentioned that the more accurate representation of the vehicle system is by MDOF systems (in comparison to SDOF systems), where the masses can represent both the unsprung (wheel) and sprung masses of the vehicle (bogie, car body). Further, the primary and secondary suspension systems of the vehicle can be represented via spring-damper systems. For evaluating the time-dependent loads due to the moving oscillator, a coupling is established between the governing equations of the oscillator and track model via a pre-defined wheel-rail contact model (e.g., permanent, linear, nonlinear contact models). In some studies, the coupled equations of motion associated with the vehicle-track system are solved by using analytical methods [3], modified numerical integration techniques [13–15], and finite element method [4,17]. In addition to that, this problem has also been investigated by employing the Green's function [10,12,21] derived using the conventional frequency-domain approach.

A recent study by Dimitrovová [3] has shown that it is also possible to analyze the moving oscillator problem using a semi-analytical approach. In their study, the beam deflections are evaluated for a beam on the viscoelastic-Pasternak model under one and two mass uniformly moving oscillators. The solution is presented as a sum of the steady-state part (derived analytically), induced harmonic part (derived semi-analytically), and transient part (derived numerically). However, it may be noted that the proposed solution is only applicable to the permanent wheel-rail contact model. For more complex vehicle-track models that take into account the possibility of contact loss, most studies tend to use numerical approaches [1,18]. Those numerical approaches are in general cumbersome and, in some cases, computationally intensive [1].

In this study, a new analytical iterative approach is presented to analyze the coupled equation of motion of the vehicle-track system. The vehicle is modeled as a multi-degree-of-freedom system moving on a one-dimensional track model. The proposed approach is validated with the above-described analytical study [3]. The results show the vehicle-track response for three cases. In the first case, the wheel (unsprung mass) remains in permanent contact with the rail beam. In the second case, the wheel interacts with the rail via a nonlinear Hertzian spring. In the third case, in addition to modeling the wheel-rail interaction using the Hertzian spring, the wheel is considered to be moving over a rail-head corrugation.

2 The Iterative Approach

2.1 Track Model

The model used to idealize and study the behaviour of the railway track system is shown in Fig. 1. The rail beam is modeled as an infinite Euler-Bernoulli beam (with x denoting the space coordinate measured along the length of the rail beam) overlying a two-parameter foundation model consisting of (1) Pasternak shear layer of thickness H_P and shear modulus G_P per unit beam length and (2) viscoelastic Winkler spring layer with K_0 denoting the spring stiffness per unit beam length. The model is subjected to a time-varying vertical force $P(t)$ resulting due to interaction between the rail beam and a multi-degree-of-freedom (MDOF) oscillator system moving with uniform velocity v.

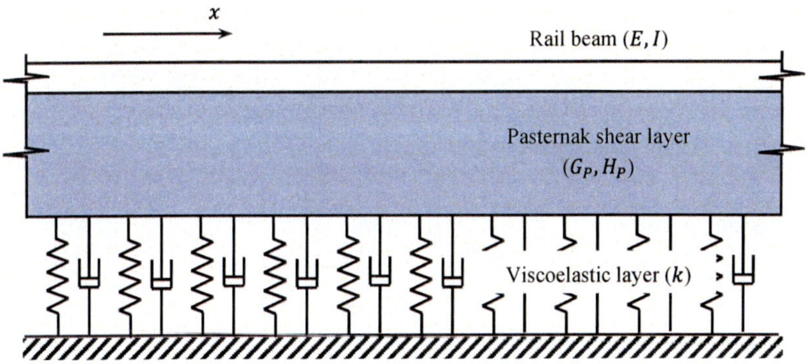

Fig. 1. Definition sketch of the model idealizing railway track section.

Under the above-described idealizations, the equation of motion of a rail beam is described by [8]

$$EI\frac{\partial^4 w}{\partial x^4} - K_1\frac{\partial^2 w}{\partial x^2} + K_0 w + c\frac{\partial w}{\partial t} + \rho\frac{\partial^2 w}{\partial t^2} = P(t)\delta(x - vt) \tag{1}$$

where $w(x,t)$ is the transverse deflection of the rail beam (considered positive downwards), E is the Young's modulus of rail beam material, I is the moment of inertia of the rail beam cross-section about the axis of bending, $K_1 = (G_P H_P)$ is the shear parameter associated with the Pasternak shear layer, ρ is the mass per unit length of the beam, c is the coefficient of viscous damping per unit beam length, and δ is the Dirac's delta function.

Denoting $\iota = \sqrt{-1}$ and $\hat{f}(\omega)$ as the Fourier transform of an arbitrary time-varying function $f(t)$ can be expressed as,

$$\hat{f}(\omega) = \int_{-\infty}^{\infty} f(t)e^{-\iota\omega t}dt \tag{2}$$

$$f(t) = \frac{1}{2\pi} \int_{-\infty}^{\infty} \hat{f}(\omega) e^{\iota \omega t} d\omega \tag{3}$$

On taking the Fourier transform of Eq. 1 (assuming that $w(x, t)$ and its time derivatives vanish at $t = \pm\infty$) using Eq. 2 and solving for the rail beam deflection $\widehat{w}(x, \omega)$ in space-frequency-domain, we obtain,

$$\widehat{w}(x, \omega) = \left(\frac{P(x/v) v^3}{EI\omega^4 + K_1 \omega^2 v^2 + K_0 v^4 - \rho \omega^2 v^4 + \iota c \omega v^4} \right) e^{-i\omega\left(\frac{x}{v}\right)} \tag{4}$$

Further, using Eqs. 4 and 3 the expression for the rail-beam deflection $w_0(t) = w(vt, t)$ at the point of contact with the load $P(t)$ can be written in a concise form as

$$w_0(t) = \frac{P(t)}{2\pi} I(v, t) \tag{5}$$

where

$$I(v, t) = \int_{-\infty}^{\infty} \frac{v^3 d\omega}{EI\omega^4 + K_1 \omega^2 v^2 + K_0 v^4 - \rho \omega^2 v^4 + \iota c \omega v^4} \tag{6}$$

here ω stands for the frequency. The mathematical technique exponential window method [5] is employed while evaluating the above integral.

2.2 Vehicle Model

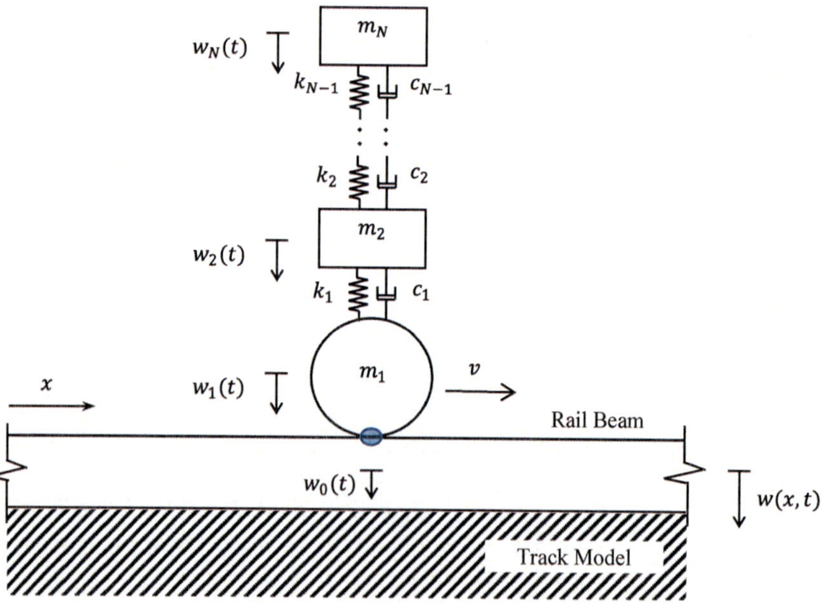

Fig. 2. MDOF oscillator moving over the track model with no loss of contact.

Figure 2 shows the considered oscillator system comprising N masses (m_1, m_2,...,m_N), connected via $N-1$ springs (of stiffness, $k_1, k_2, ..., k_{N-1}$), and $N-1$ dash-pot systems (with viscous damping coefficients, $c_1, c_2, ..., c_{N-1}$). Further, $(w_i)_{i=1\ to\ N}$ respectively represent the absolute displacements of the masses $(m_i)_{i=1\ to\ N}$. The contact between the oscillator and the rail beam is such that the unsprung mass m_1 always remains in contact with the rail beam (see Fig. 2),

$$w_1(t) = w_0(t) \tag{7}$$

On considering the vertical equilibrium of the mass m_1 and using Eqs. 5 and 7 the displacements w_1 (of the unsprung mass m_1 or rail beam) and w_2 (of the sprung mass m_2) can be related as

$$m_1\ddot{w}_1 + \left(k_1 + \frac{2\pi}{I(T_P, v)}\right) w_1 + c_1\dot{w}_1 = k_1 w_2 + c_1\dot{w}_2 + P_0 \tag{8}$$

here $P_0 = (m_1 + m_2 + ... + m_N)g$ (where, g is the acceleration due to gravity) is the static load which is considered equal to the total weight of the oscillator. Furthermore, on considering the vertical equilibrium of the forces acting on the sprung masses (m_2, $m_3, ..., m_N$) of the oscillator system following expression can be written

$$\mathbf{m}\ddot{\mathbf{w}}(t) + \mathbf{k}\mathbf{w}(t) + \mathbf{c}\dot{\mathbf{w}}(t) = \mathbf{f}(t) \tag{9}$$

where $\mathbf{w}(t) = \{w_2(t),\ w_3(t), \ldots, w_N(t)\}^T$ is the displacement vector and, $\dot{\mathbf{w}}(t)$ and $\ddot{\mathbf{w}}(t)$ respectively denote the corresponding velocity and acceleration vectors of the sprung masses (m_2, $m_3, ..., m_N$). Further, **m**, **k**, and **c** are respectively the mass, stiffness, and damping matrices, and the column vector $\mathbf{f}(t)$ representing the forces acting on the sprung masses is given by,

$$\mathbf{f}(t) = \begin{Bmatrix} k_1 w_1 + c_1\dot{w}_1 \\ 0 \\ \vdots \\ 0 \end{Bmatrix} \tag{10}$$

It may be noted that Eq. 9 is a set of $N-1$ coupled differential equations governing the displacements $\mathbf{w}(t)$ resulting due to forces $\mathbf{f}(t)$. This system possess $N-1$ classical natural modes ϕ_n corresponding to $N-1$ natural vibration frequencies ω_n, where n describes the mode number ($n = 2,\ 3, ...,\ N$). Furthermore, those $N-1$ natural modes can be written in the form of a modal matrix $\mathbf{\Phi}$ as,

$$\mathbf{\Phi} = \left[\phi_{jn}\right] \tag{11}$$

where j indicates the degrees of freedom ($j = 2,\ 3, ...,\ N$). Further, using the modal matrix, the displacement vector can be represented in the following form

$$\mathbf{w}(t) = \mathbf{\Phi}\mathbf{q}(t) \tag{12}$$

where $\mathbf{q}(t) = \{q_2(t),\ q_3(t), \ldots,\ q_N(t)\}^T$ are the modal coordinates.

Further, the classical modal analysis procedure [2] is used to transform the equation of motion for the sprung masses (see Eq. 9) to N uncoupled differential equations in modal coordinates $\mathbf{q}(t)$, described by

$$\mathbf{M}\ddot{\mathbf{q}}(t) + \mathbf{K}\mathbf{q}(t) + \mathbf{C}\dot{\mathbf{q}}(t) = \mathbf{F}(t) \tag{13}$$

where, $\mathbf{M}$, $\mathbf{K}$, and $\mathbf{C}$ are diagonal matrices given by

$$\mathbf{M} = \mathbf{\Phi}^T\mathbf{m}\mathbf{\Phi} \qquad \mathbf{K} = \mathbf{\Phi}^T\mathbf{k}\mathbf{\Phi} \qquad \mathbf{C} = \mathbf{\Phi}^T\mathbf{c}\mathbf{\Phi} \tag{14}$$

and the vector $\mathbf{F}(t)$ as,

$$\mathbf{F}(t) = \mathbf{\Phi}^T\mathbf{f}(t) \tag{15}$$

The uncoupled, modal differential equations (see Eq. 13) can be solved to find the modal responses $\mathbf{q}(t)$, which are then combined to determine the displacement response $\mathbf{w}(t)$ using Eq. 12.

It may be observed from Eqs. 9 and 10 that $\mathbf{w}(t)$ is dependent on $w_1(t)$. Furthermore, as seen from Eq. 8, $w_1(t)$ itself depends on $w_2(t)$ (and hence on $\mathbf{w}(t)$). To solve this coupled system of equations governing the displacement responses $w_1(t)$ (or $w_0(t)$) and $\mathbf{w}(t)$ a two-step iterative scheme is employed. In the first step, $\mathbf{w}(t)$ (or $w_2(t)$) is determined using Eqs. 9–15 for a given $w_1(t)$ (or $w_0(t)$). In the second step, the above-evaluated $w_2(t)$, is used to find $w_1(t)$ using Eq. 8. These two steps are repeated until the respective values of $w_1(t)$ and $\mathbf{w}(t)$ converges.

Figure 3 illustrates the steps of the proposed iterative scheme via a flowchart. In this figure, the variables k and $\epsilon(t)$ represent the iteration number and required precision, respectively. It may be noted that for the first iteration ($k = 1$), $w_0(t) = w_1(t) = 0$ is used.

3 Validation of the Proposed Approach

The results show the response analysis of a MDOF oscillator comprising three masses (m_1, m_2, and m_3) connected via two springs (of stiffness k_1 and k_2) and two dash-pot systems (with viscous damping coefficients, c_1 and c_2) moving uniformly over the rail beam overlying the Pasternak-viscoelastic model. The track and oscillator parameters used for this analysis are given in Table 1.

Before analysing the vibration responses of the MDOF oscillator, a validation exercise is carried out. The proposed approach is implemented to find the deflection response of a mass-spring system with mass m_1 and stiffness k_1 (see Table 1) traversing the rail beam overlying a viscoelastic layer ($G_P = 0$) with velocity v = 100 m/s (see Fig. 4). It is assumed that there is no loss of contact between the mass-spring system and rail beam. Figure 5 compares the normalized deflection values of the rail beam at the position of the moving mass-spring system ($\overline{w}_0(t)$), and that of mass m_1 ($\overline{w}_1(t)$) obtained using the proposed approach with those evaluated using a recently proposed analytical solution by Dimitrovová [3]. The deflection values (w_1 and w_0) are normalized with respect to the rail beam deflection value, $w_0(t) = (m_1 g/2\pi)I(T_P, v)$ (see Eq. (5)), where, g is the acceleration due to gravity. The agreement between the vibration responses evaluated using the two approaches lends confidence in the proposed approach.

4 Wheel-Rail Contact Loss

In previous sections it is assumed that the moving oscillator system (see Fig. 2) never loses contact with the rail beam. However, for a more realistic modelling of vehicle-track interaction it is essential to consider the wheel-rail contact loss. The complex vehicle-track models that take into account the possibility of contact loss tend to use numerical approaches. Those approaches are generally based on time integration (via Newmark-β, Runge-Kutta, or precise integration methods) of the equation of motion derived from the finite element vehicle-track models [1,18]. In this section, it is assumed that the wheel interacts with the rail via a nonlinear Hertzian spring (see Fig. 6). The track model is same as shown in Fig. 1.

For analyzing this system, the equation relating the rail beam deflection $w_0(t)$ to the load $P(t)$ acting on the rail beam (see Eq. 5) is modified as,

$$\left(\frac{P(t)}{C_H}\right)^{2/3} = \xi(t)H\left[\xi(t)\right] \tag{16}$$

where, C_H is the Hertz's constant, $H(.)$ is the Heaveside function, and

$$\xi(t) = w_1(t) - w_0(t) - z(t) \tag{17}$$

here $z(t)$ denotes the rail or wheel roughness.

Using the proposed iterative approach, the response analysis of the moving MDOF oscillator comprising three masses (see Fig. 6) is performed for three different cases. In the first cases, the wheel (unsprung mass) always remains in contact with the rail beam. In the second case, the wheel interacts with the rail via a nonlinear Hertzian spring. In the third case, in addition to modelling the wheel-rail interaction using the Hertzian spring, the MDOF oscillator is considered to be moving over a rail-head corrugation (see Eqs. 16 and 17) of the form,

$$z(t) = -\frac{e}{2}\left(1 - \cos(2\pi vt)\right), \qquad \frac{x_{z_1}}{v} \leq t \leq \frac{x_{z_2}}{v} \tag{18}$$

where e represents the depth of the rail indentation, and x_{z_1} and x_{z_2} respectively denote the start and end locations of the rail corrugations along the length of the rail beam. Further, each of the above case is analyzed considering both the undamped and damped oscillator systems. The track parameters and oscillator parameters are shown in Table 1. The vertical deflections of the rail beam ($w_0(t)$) and those of the masses m_1 ($w_1(t)$), m_2 ($w_2(t)$), and m_3 ($w_3(t)$) are evaluated at the location $x = vt$. It is assumed that the oscillator is at rest and all deflections values are zero at the location $x = 0$ and time $t = 0$. Further, the computed deflection values are normalized by the rail beam deflection value, $w_0(t) = (P_0/2\pi)I(T_P, v)$ (see Eq. 5), where $P_0 = (m_1 + m_2 + m_3)g$.

Figures 7 and 8 show the normalized deflection $\overline{w}_0(t)$), $\overline{w}_1(t)$, $\overline{w}_2(t)$), and $\overline{w}_3(t)$ with time t for the above-mentioned cases of wheel-rail contact undamped and damped oscillator systems, respectively. Here, the oscillator velocity is considered as $v = 100\,\text{m/s}$ and, for illustration, the rail corrugation depth, start and

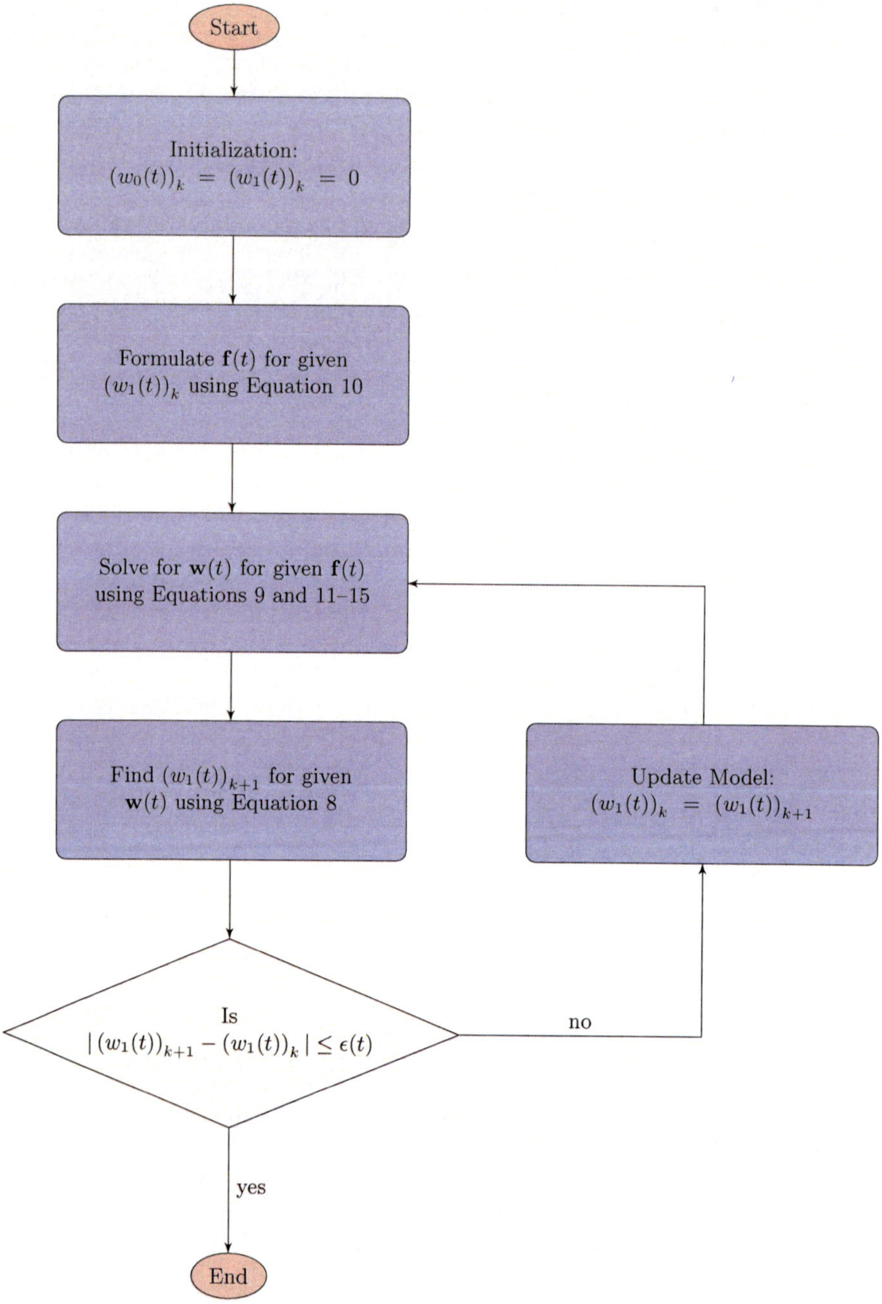

Fig. 3. Flow chart showing the steps of the iterative approach used for analyzing the track model traversed by a MDOF system (k and $\epsilon(t)$ represent the iteration number and required precision, respectively).

Table 1. Track and oscillator parameters [9,12]

Parameter	Value
Rail Beam	
Mass per unit beam length, ρ	60 kg/m
Modulus of Elasticity of rail, E	210 GPa
Central moment of Inertia of rail, I	3055 cm^4
Pasternak Layer	
Shear Modulus, G_P	43.3 MPa
Height, H_P	0.3 m
Viscoelastic Layer	
Stiffness, K_0	4.08 MPa
Coefficient of viscous damping, c	1.56 kNs/m
Oscillator	
Lower Mass, m_1	1125 kg kg
Middle Mass, m_1	2000 kg kg
Upper Mass, m_1	6875 kg kg
Stiffness of lower connecting spring, k_1	6.3 MN/m
Stiffness of upper connecting spring, k_2	390 kN/m
Coefficient of viscous damping of lower connecting dashpot, c_1	23 kNs/m
Coefficient of viscous damping of upper connecting dashpot, c_2	20 kNs/m
Hertz's constant, C_H	98.92 $GN/m^{3/2}$

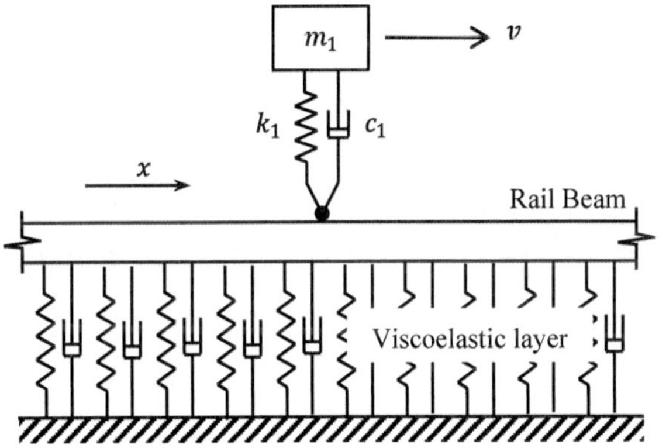

Fig. 4. MDOF oscillator moving over the track model with no loss of contact.

end locations are respectively chosen as, $e = 0.35$ mm, $x_{z_1} = 20$ m, and $x_{z_2} = 40$ m (see Eq. 18). For both figures, parts, (a), (b), and (c) respectively correspond to the cases of (1) no loss of wheel-rail contact, (2) Hertzian wheel-rail contact, and (3) wheel moving over a rail-head corrugation (see Eq. 18).

It may be observed for both the considered cases of wheel-rail contact (no loss of contact and Hertzian) that the introduction of damping into the oscillator

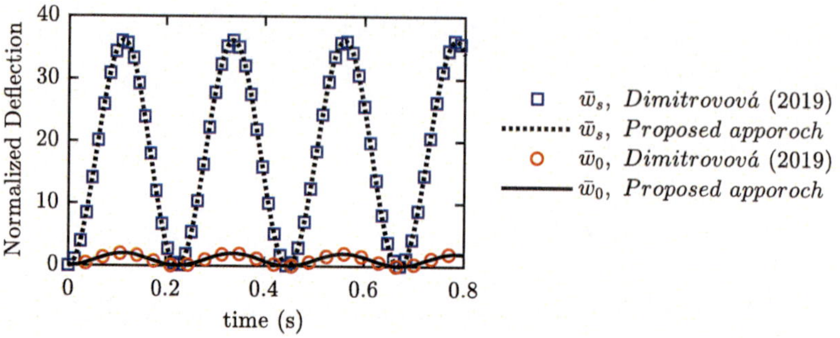

Fig. 5. Comparison of the normalized rail beam deflection ($\bar{w}_0$) and normalized oscillator deflections ($\bar{w}_1$) computed using the proposed approach and analytical solution by Dimitrovová [3].

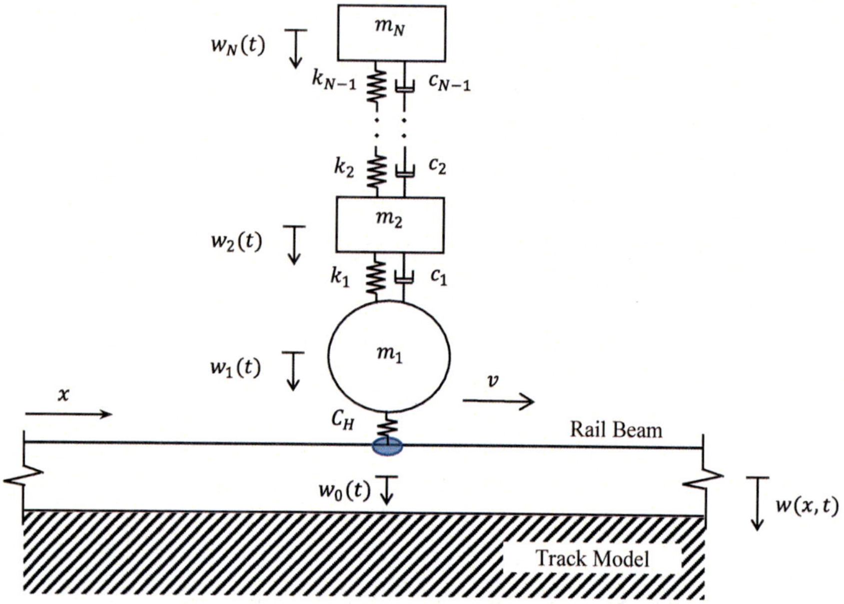

Fig. 6. MDOF oscillator moving over the track model with nonlinear Hertzian contact.

results in the decay of deflection amplitudes of the rail beam as well as that of the oscillator masses. Moreover, it is interesting to note that the damping brings the normalized rail beam deflection closer to $\overline{w}_0$ =1, i.e., the deflection value observed if a point load of magnitude P_0 moves over the rail beam with velocity $v = 100\,\text{m/s}$.

Further, it is found that the idealisation of wheel-rail contact via a Hertzian spring (see Figs. 7b and 8b) leads to slightly higher amplitudes (by up to 8%) of the rail beam deflection as compared to those observed for the case where

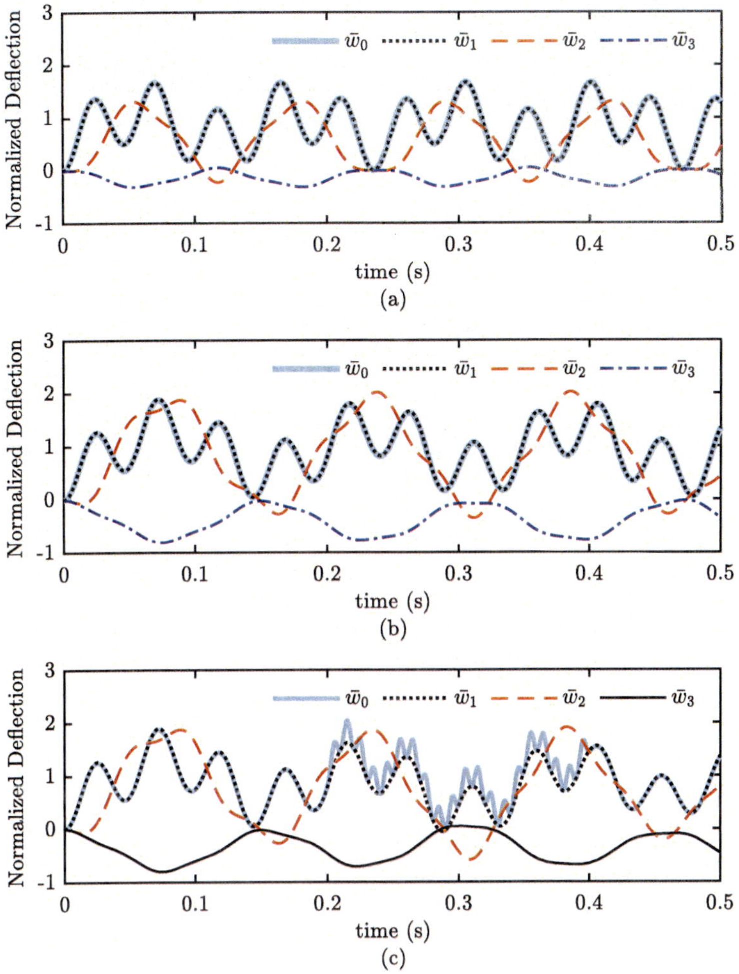

Fig. 7. Normalized deflections of the rail beam ($\overline{w}_0(t)$), and masses m_1 ($\overline{w}_1(t)$), m_2 ($\overline{w}_2(t)$), and m_3 ($\overline{w}_3(t)$) with time t at location $x = vt$ ($v = 100\,\mathrm{m/s}$) of an undamped oscillator system for the cases of (a) no loss of wheel-rail contact, (b) Hertzian wheel-rail contact, (c) rail-head corrugation.

there is no loss of wheel-rail contact (see Figs. 7a and 8a). A similar behaviour is observed for the deflection responses ($\overline{w}_1(t)$), $\overline{w}_2(t)$), and $\overline{w}_3(t)$) of the oscillator masses m_1, m_2, and m_3.

It may also be seen that similar to the case of no wheel-rail contact loss (see Figs. 7a and 8a), the unsprung mass m_1 remains in contact with the rail beam even for the case of Hertzian contact (see Figs. 7b and 8b). However, a loss of contact is observed when the oscillator encounters the rail-head corrugation at $t = 0.2\,\mathrm{s}$ (see Figs. 7c and 8c). The wheel-rail contact loss results in high impact

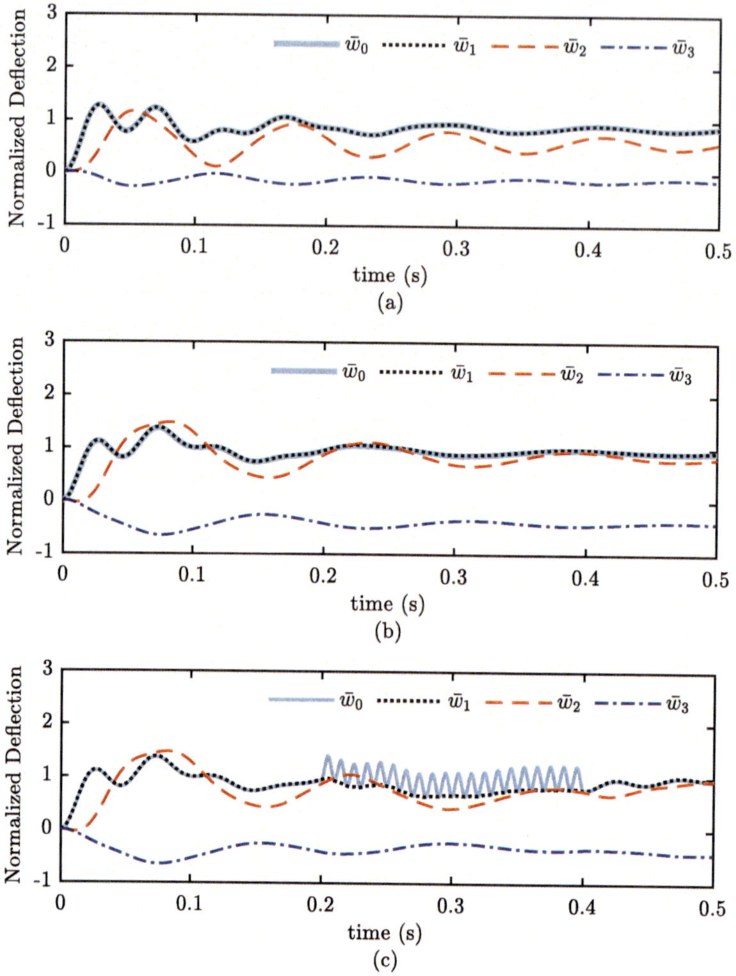

Fig. 8. Normalized deflections of the rail beam ($\overline{w}_0(t)$), and masses m_1 ($\overline{w}_1(t)$), m_2 ($\overline{w}_2(t)$), and m_3 ($\overline{w}_3(t)$) with time t at location $x = vt$ ($v = 100\,\text{m/s}$) of an damped oscillator system for the cases of (a) no loss of wheel-rail contact, (b) Hertzian wheel-rail contact, (c) rail-head corrugation.

loads over the rail beam leading to a sudden increase in rail beam deflections (by up to 85% and 57% for the undamped and damped cases, respectively). It may be noted that the time required for the evaluation of the deflection responses shown in Fig. 7c on a personal computer (with 8 GB RAM and 3.40 GHz quad-core Intel i5 processor) is ~40 s.

References

1. Baeza, L., Roda, A., Nielsen, J.C.: Railway vehicle/track interaction analysis using a modal substructuring approach. J. Sound Vib. **293**(1–2), 112–124 (2006)
2. Chopra, A.K., et al.: Dynamics of Structures. Pearson Education Upper Saddle River, NJ (2012)
3. Dimitrovová, Z.: Semi-analytical solution for a problem of a uniformly moving oscillator on an infinite beam on a two-parameter visco-elastic foundation. J. Sound Vib. **438**, 257–290 (2019)
4. Hino, J., Yoshimura, T., Konishi, K., Ananthanarayana, N.: A finite element method prediction of the vibration of a bridge subjected to a moving vehicle load. J. Sound Vib. **96**(1), 45–53 (1984)
5. Kausel, E., Roësset, J.M.: Frequency domain analysis of undamped systems. J. Eng. Mech. **118**(4), 721–734 (1992)
6. Knothe, K., Grassie, S.: Modelling of railway track and vehicle/track interaction at high frequencies. Veh. Syst. Dyn. **22**(3–4), 209–262 (1993)
7. Kouroussis, G., Connolly, D.P., Verlinden, O.: Railway-induced ground vibrations-a review of vehicle effects. Int. J. Rail Transp. **2**(2), 69–110 (2014)
8. Kumawat, A., Raychowdhury, P., Chandra, S.: Frequency-dependent analytical model for ballasted rail-track systems subjected to moving load. Int. J. Geomech. **19**(4), 04019016 (2019)
9. Kumawat, A., Raychowdhury, P., Chandra, S.: A wave number based approach for the evaluation of the green's function of a one-dimensional railway track model. Eur. J. Mech.-A/Solids **78**, 103854 (2019)
10. Mazilu, T.: Instability of a train of oscillators moving along a beam on a viscoelastic foundation. J. Sound Vib. **332**(19), 4597–4619 (2013)
11. Mazilu, T.: Interaction between moving tandem wheels and an infinite rail with periodic supports-green's matrices of the track method in stationary reference frame. J. Sound Vib. **401**, 233–254 (2017)
12. Mazilu, T., Dumitriu, M., Tudorache, C.: Instability of an oscillator moving along a timoshenko beam on viscoelastic foundation. Nonlinear Dyn. **67**(2), 1273–1293 (2012)
13. Muscolino, G., Benfratello, S., Sidoti, A.: Dynamics analysis of distributed parameter system subjected to a moving oscillator with random mass, velocity and acceleration. Probab. Eng. Mech. **17**(1), 63–72 (2002)
14. Muscolino, G., Palmeri, A.: Response of beams resting on viscoelastically damped foundation to moving oscillators. Int. J. Solids Struct. **44**(5), 1317–1336 (2007)
15. Muscolino, G., Palmeri, A., Sofi, A.: Absolute versus relative formulations of the moving oscillator problem. Int. J. Solids Struct. **46**(5), 1085–1094 (2009)
16. Patil, S.P.: Response of infinite railroad track to vibrating mass. J. Eng. Mech. **114**(4), 688–703 (1988)
17. Rodrigues, C., Simões, F.M., da Costa, A.P., Froio, D., Rizzi, E.: Finite element dynamic analysis of beams on nonlinear elastic foundations under a moving oscillator. Eur. J. Mech.-A/Solids **68**, 9–24 (2018)
18. Sadeghi, J., Khajehdezfuly, A., Esmaeili, M., Poorveis, D.: Investigation of rail irregularity effects on wheel/rail dynamic force in slab track: comparison of two and three dimensional models. J. Sound Vib. **374**, 228–244 (2016)
19. Sun, Y.Q., Dhanasekar, M.: A dynamic model for the vertical interaction of the rail track and wagon system. Int. J. Solids Struct. **39**(5), 1337–1359 (2002)

20. Visweswara Rao, G.: Linear dynamics of an elastic beam under moving loads. J. Vib. Acoust. **122**(3), 281–289 (2000)
21. Yang, B., Tan, C., Bergman, L.: Direct numerical procedure for solution of moving oscillator problems. J. Eng. Mech. **126**(5), 462–469 (2000)

Dynamic Laboratory Testing of Mechanically Stabilized Layers for Railway Applications

Leoš Horníček[1](✉), Zikmund Rakowski[2], Jacek Kawalec[2,3], and Slawomir Kwiecien[3]

[1] Faculty of Civil Engineering, Czech Technical University in Prague, Prague, Czech Republic
hornicek@fsv.cvut.cz

[2] Tensar International, s.r.o., Český Těšín, Czech Republic

[3] Silesian University of Technology, Gliwice, Poland

Abstract. The application of mechanically stabilized layers with stiff geogrids is an effective way for sustainable and durable construction of under sleeper structures in railways. The paper explains the basics of the theory of mechanical stabilization of granular material by geogrids. Determining the mechanical properties of mechanically stabilized layers is still a practical problem. The paper describes possible laboratory approaches combined with inverse analysis of the 3D finite element method. Smaller laboratory models are used to observe differences in the performance of different geogrids even with different grain fills. Parameters describing these differences were specified and demonstrated in specific laboratory models. The inverse finite element method was used to calibrate the measured parameters with those modelled. Some approximation of the mechanical properties of the mechanically stabilized material, such as the modulus of deformation and the Poisson's ratio, has been achieved. Determining the mechanical parameters of mechanically stabilized layers with geogrids is very useful for evaluating their possible contribution to the structure under the sleeper in comparison with unstabilized granular material. A new innovative generation of stiff geogrids was included in the research.

Keywords: Mechanically stabilized layer · Stiff geogrid · FEM · Dynamic loading · Mechanistic-experimental approach

1 Introduction

Geogrids have been used in infrastructure for decades, which has made it possible to gather remarkable knowledge about their application [2, 3, 6, 7, 10, 11]. Recently, a new innovative generation of multiaxial geogrids has been put into practical use. They are expected to be used in transport infrastructure, including railways [4] and potentially in the construction of high-speed railways (HSR). To obtain the necessary knowledge for their practical use, it was proposed to perform laboratory tests with different fractions of aggregates and focusing on the study of the properties of mechanically stabilized layers of different thicknesses.

Z. Dimitrovová et al. (Eds.): WMVC 2022, MMS 125, pp. 581–588, 2023.
https://doi.org/10.1007/978-3-031-15758-5_59

2 Mechanically Stabilized Soil/Layer

Mechanically stabilized soil is created as a result of the interaction of granular soil and geogrid. The grains penetrate into the geogrid aperture and are locked in a stable position due to loading in it, Fig. 1 [5, 13, 14]. The grains are strongly immobilized. This phenomenon is called interlocking [9, 12]. Due to the fact that the adjacent apertures are also filled with grains and interlocked, a composite solid layer called as geogrid-grain-composite (GGC) is formed near the geogrid [13]. It is approx. 5–10 cm thick and has specific mechanical properties: high modulus of elasticity and a very low Poisson's ratio, which means increased shear modulus compared to standard crushed stone. GGC is also manifested by tensile strength, which is impossible for granular materials [1]. Above and possibly also below the geogrid, if the geogrid is placed in a granular layer, a transition zone is created, in which the properties change up to the level where the granular material has "normal" properties (Fig. 1).

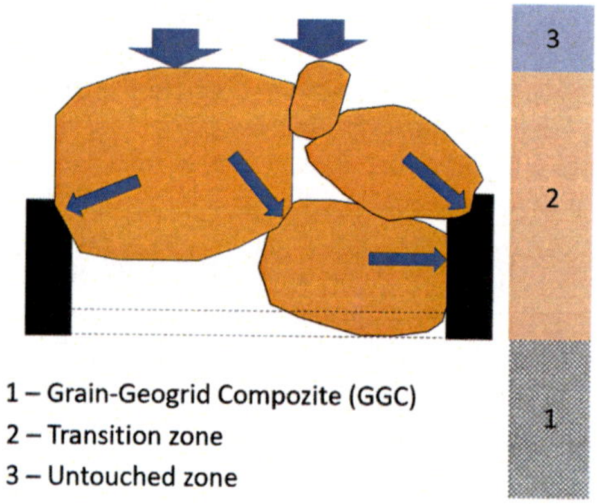

Fig. 1. Principle of mechanically stabilized layer.

3 Geogrids

A new innovative generation of multiaxial geogrids was used for testing (Fig. 2). A notable innovation is that the apertures do not have a uniform shape, as is standard on older generations of geogrids [8], but have three different geometric shapes: hexagonal, trapezoidal and triangular. The basic dimensions in mm are shown in Fig. 2. The triangles make up 23.3%, the hexagons 12.2% and the trapezoid 64.5% of the total geogrid area. The logical question is, how does such a profiled geogrid behave in combination with an aggregate layer?

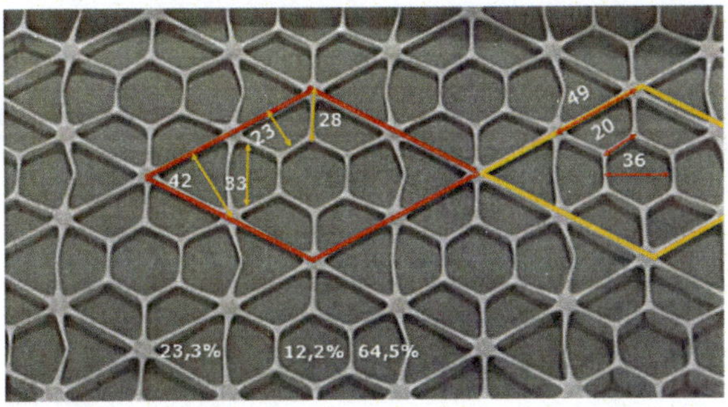

Fig. 2. New generation of geogrid - multiaxial geogrid.

4 Laboratory Equipment and Testing Regime

Special box of dimensions 800 × 400 × 300 mm was used for laboratory experiments. The scheme of the box and structure of its fill can be seen on Fig. 3. The model is based on a 10 cm block of extruded polystyrene (EPS), which simulates a rather weak subsoil and ensures uniform material properties under the geogrid for all tests. The geogrid is placed directly on the EPS and then covered with crushed stone. In the first stage, new multi-axis geogrid was tested in combination with crushed stone fraction 8–32, a model without geogrid was used as a reference. In the second stage, new geogrid was tested in combination with aggregates of fractions 8–16, 16–32 and 8–32. Those fractions were applied in 100 mm layer over geogrid. The rest 100 mm was filled with 8–32 fraction.

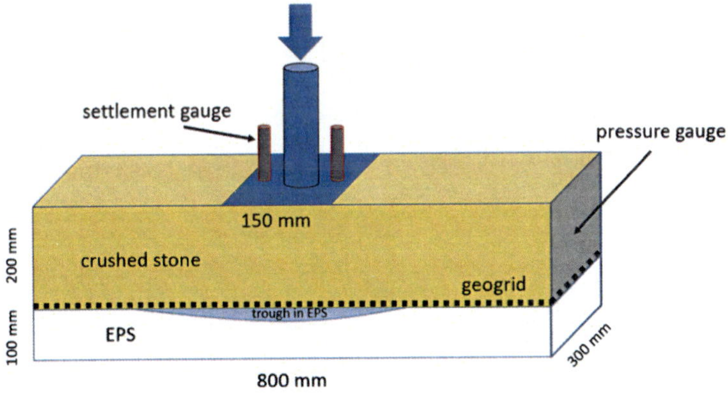

Fig. 3. The scheme of experimental box test.

The load was generated by a hydraulic actuator through a steel plate with an area of 150 × 300 mm situated in middle of the box. Special pressure gauges were placed on the side wall of the box to record the lateral pressure during loading of the model. Furthermore, the settlement of the steel plate and the dimensions of the trough created in the EPS during loading were recorded. The dynamic loading mode was as follows: frequency 3 Hz, force 1–12 kN, total 5,000 cycles.

5 Results of Testing

The settlement, side pressure and width of the trough in the EPS were the measured parameters. Side (lateral) pressure is the indicator of grain confinement in the geogrid. The lower lateral pressure the more efficient confinement. The width of the trough in the EPS is the indicator of load distribution. The wider trough the better load distribution.

In the first stage, three models with three different geogrids of new generation were compared with the model without geogrid at all with following results:

- Testing successfully recognized the difference between models with and without geogrid.
- The differences in settlement were very small, in the order of millimetres.
- The average lateral pressure of the geogrid models was 59% lower than in the case of reference mode.
- The width of the trough in EPS in reference model was 402.5 mm. The models with geogrids showed wider trough by 3–21%.

In the second stage, selected type of geogrid of new generation was tested with three different granular fills in the 100 mm thick layer positioned in the contact with the geogrid. The following results were achieved:

- The smallest settlement was measured in the model with 8–32 mm mixture above the geogrid, the other were 10–70% greater.
- The 8–32 mm mixture provided the smallest lateral pressure increase of 5.6 kPa. Other models showed 6.5 kPa, resp. 8.0 kPa.
- The widest trough was measured on a 16–32 mm mixture model. The 8–32 mm mixture had a smaller depression by 14% and the third by 30%.

The evaluation was done with so called positioning matrix (Table 1). The best model was given position 1, the worst position 3.

Table 1. Models matrix.

Model	Position (settlement)	Position (lateral pressure)	Position (trough width)	Average position
8–16 mm	3	2	3	2.7
8–32 mm	1	1	2	1.3
16–32 mm	2	3	1	2.0

The model with 8–32 mm mixture reached the lowest average position value, it means the best placement overall.

6 Inverse FEM Modelling

Direct measurement of mechanical properties of mechanically stabilized soil is not possible. Therefore, inverse 3D FEM modeling was used within the mechanistic-experimental approach defined in [13]. The goal of the modelling is to get values of E modulus and Poisson's ratio for the MSL. Expected high stiffness and high E modulus of the MSL allowed to use linear elasticity constitutive law in FEM. Results of laboratory experiments (static regime in this case) are the reference values (settlement, side pressure on box wall, trough width in EPS). Iterative FEM modelling looks for the closest possible laboratory values. Because the FEM is static, parameters from static tests were used for comparison. The settlement of the steel plate was 6–7 mm at a load of 10 kN. The lateral pressure was measured in the range of 16–21 kPa. These values were the reference for 10 iterative steps of FEM modelling.

Prior to all special laboratory model loading just EPS layer with the same load plate was done to get some approximation of the EPS properties. Three more or less linear phases were recognized by loading EPS, Fig. 4. "Deformation modulus" in the first stage up to 4 kN load is quasi elastic, after that long section of more or less plastic deformation follows. The last part represents hardening stage. First two phases were applicable for the range of loading in the FEM model.

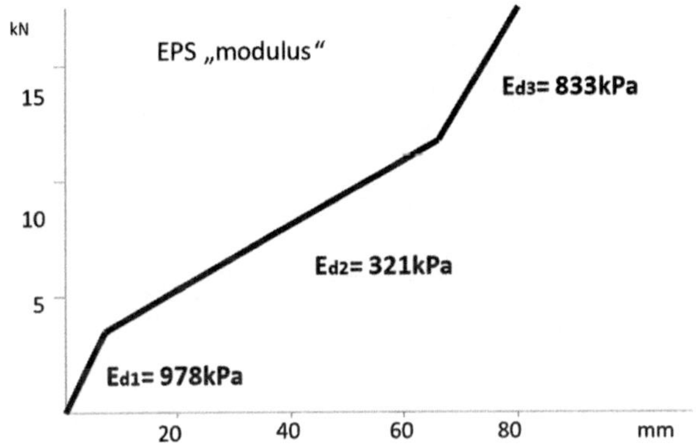

Fig. 4. Deformability of the EPS under plate load.

The 3D FEM model was structured similarly to the laboratory model, Fig. 5. At the bottom it has a 100 mm layer representing EPS. A layer of MSL was then placed, which was divided into 2 parts: the first part 60 mm and the second part 40 mm. The top layer was again made of stone in two parts: one 40 mm and the other 60 mm. The plate load was up to 10 kN.

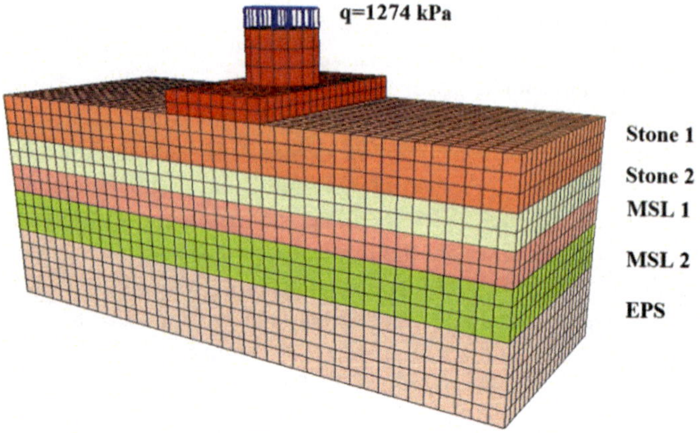

Fig. 5. The structure of 3D FEM model.

A total of 10 iteration steps were performed until the considered outputs were as close as possible to the measured values. An example of the output in the case of horizontal pressure in 3D is shown in Fig. 6.

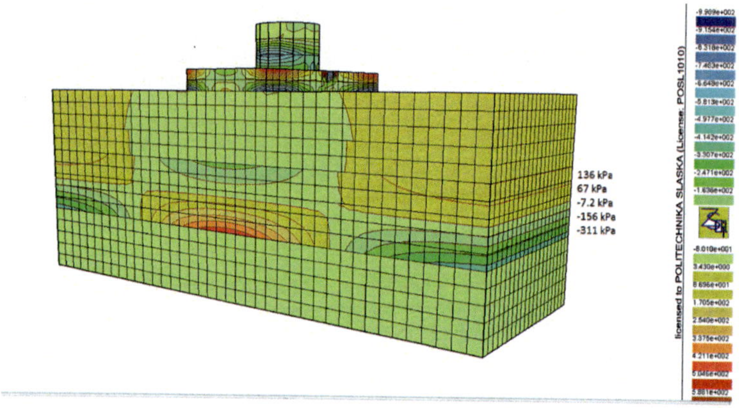

Fig. 6. 3D map of horizontal pressure.

In the model No. 9 the settlement of the steel plate was 6.8 mm (measured 6–7 mm) and average side pressure in MSL layer 20.4 kPa (measured 16–21 kPa), which is well within the range of measured values. The width of the tensile zone at the MSL base is in the same range as measured in the laboratory.

The parameters of the FEM model can be seen in Table 2. MSL has a significantly higher modulus of elasticity compared to a standard stone. On the other hand, the Poisson's ratio is very low, which is an expression of strong confinement and immobilization of grains in the geogrid apertures. In this respect, the shear modulus (G), which depends on E and Poisson's ratio, appears to be the appropriate parameter describing the deformation properties of the mechanically stabilized layer. From the point of view of railway applications of these geogrids, the most important is significantly higher stiffness and significantly reduced horizontal deformability of MSL.

Table 2. FEM model parameters.

Material	h [mm]	E [MPa]	ν [–]	γ [kN/m^3]
Stone 1	60	100	0.1	20
Stone 2	40	250	0.05	20
MSL 1	40	1000	0.01	20
MSL 2	60	2000	0.01	20
EPS	100	0.55	0.2	0.2

7 Conclusions

The following conclusions can be drawn from the performed experiments:

- Dynamic plate testing in small box with EPS layer appeared to be very useful test method.
- The new generation of multi-axial geogrids demonstrated their performance compared to the model without geogrid.
- Stone mixture 8–32 mm provided the best performance with new geogrids.
- Mechanically stabilized soil with stiff geogrids in combination with an appropriate stone mixture is characterized by significantly reduced horizontal pressure. This is a very important feature for the structural layers located under the superstructure on the railway, as it leads to an increase in the service life of the structure and a prolongation of the maintenance cycles.
- The high potential of the MSL application is also identified for high-speed railways, where an asphalt layer is used under the ballast.
- The mechanistic-experimental approach made it possible to derive the basic mechanical parameters of MSL.
- MSL is characterized by a remarkably higher modulus of elasticity and a reduced Poisson's ratio. The shear modulus describes increased stiffness and reduced deformability.

Acknowledgment. This conference paper was created with the state support of the Technology Agency of the Czech Republic within the Transport 2020+ Program. The authors also thank Tensar Corporation for supporting this research.

References

1. Giroud, J.P., Han, J., Tutumluer, E., Dobie, M.J.D.: The use of geosynthetics in roads. Geosynth. Int., 1–34 (2022).https://doi.org/10.1680/jgein.21.00046
2. Grygierek, M., Kawalec, J.: Selected laboratory and field research on geogrid impact on stabilization of unbound aggregate layer. In: Proceedings of the 11th International Conference on Geosynthetics, pp. 1–8. International Geosynthetics Society (2018)
3. Grygierek, M., Kawalec, J.: Selected laboratory research on geogrid impact on stabilization of unbound aggregate layer. Procedia Eng. **189**, 484–491 (2017)
4. Horníček, L., Kawalec, J.: Stabilisation of aggregates by geogrids in railway applications. In: Proceedings of XXVIII Convegno Nazionale Geositetici Reinforzo Filtrazione E Contenimento Nelle Opere Di Ingegneria Geotecnica, a cura di Danielle Cazzuffi Claudio Soccodato, Bologna (2016)
5. Badr, M., Lotfy, A. (eds.): GeoMEast 2018. SCI, Springer, Cham (2019). https://doi.org/10.1007/978-3-030-01884-9
6. Kawalec, J.: Stabilisation of unbound aggregte by geogrids for transport infrastructure applications. In: Proceedings of 1st Seminar on Transportation Geotechnics "Improvement, Reinforcement and Rehabilitation of Transport Infrastructures", pp. 36–37. Sociedade Portuguesa de Geotecnia, Lisboa (2017)
7. Kawalec, J.: Stabilisation with geogrids for transport applications - selected issues. In: Proceedings of International Geotechnical Symposium "Geotechnical Construction of Civil Engineering & Transport Structures of the Asian-Pacific Region", pp. 1–8. EDP Sciences (2019)
8. Koerner, R.: Designing with Geosynthetics, 6th edn. Xlibris US, Bloomington (2012)
9. Kwon, J., Tutumluer, E.: Geogrid base reinforcement with aggregate interlock and modeling of associated stiffness enhancement in mechanistic pavement analysis. Transp. Res. Rec. J. Transp. Res. Board **2116**, 85–95 (2009)
10. Marcotte, B., et al.: Full scale testing of geogrid-stabilised aggregate working platforms under controlled loading conditions. In: Proceedings of 4th Pan-American Conference on Geosynthetics, pp. 1–9. International Geosynthetics Society (2020)
11. Mazurowski, T., Zamara, K., Gewanlal, C., Kawalec, J.: Experiences with the use of stabilisation geogrids in demonstrating an improvement in bearing capacity of recycled materials. In: Proceedings of the Eleventh International Conference on the Bearing Capacity of Roads, Railways and Airfields. CRC Press-Taylor & Francis Group (2022)
12. Qian, Y., Tutumluer, E., Mishra, D., Kazmee, H.: Triaxial testing and discrete-element modelling of geogrid-stabilised rail ballast. Proc. Inst. Civ. Eng. Ground Improv. **171**(4), 223–231 (2018). https://doi.org/10.1680/jgrim.17.00068
13. Rakowski, Z., Kawalec, J., Horníček, L., Kwiecień, S.: Mechanistic-experimental approach for determination of basic properties of mechanically stabilized layers. In: Petriaev, A., Konon, A. (eds.) Transportation Soil Engineering in Cold Regions, Volume 2. LNCE, vol. 50, pp. 37–44. Springer, Singapore (2020). https://doi.org/10.1007/978-981-15-0454-9_5
14. Rakowski, Z.: An attempt of the synthesis of recent knowledge about mechanisms involved in stabilization function of geogrids in infrastructure constructions. Procedia Eng. **189**, 166–173 (2017)

Investigating the Effect of Pre-load on the Behavior of Rail Pads for Railway Tracks Under Quasi-static and Dynamic Loads

Hana Y. A. Shamayleh(✉) and Mohammed F. M. Hussien

Department of Civil and Architectural Engineering, Qatar University, Doha, Qatar
hs2009640@student.qu.edu.qa, mhussein@qu.edu.qa

Abstract. Numerical modelling of railway tracks is important to understand the true behaviour of tracks and define strategies to reduce the impact of the generated forces at the wheel-rail interface on ride quality and other effects such as ground-borne noise and vibration. Most modern railway tracks use rail pads to reduce the train's impact on the track infrastructure and to improve the dynamic behaviour of railway tracks. The rail pads provide other functions for the tracks as they modify and smooth the perceived roughness by the wheels due to tracks unevenness. Capturing the true behaviour of rail pads is significantly important to improve the accuracy of models of railway tracks. This paper utilizes a model of a track based on discretely supported beam to anlayse the behaviour of rail pads under quasi-static and dynamic loads by modelling the dynamic behaviour of the rail beam under the action of a wheel-mass and with harmonic excitation induced through relative displacement between the wheel-mass and the beam. Information and properties of rail pads as well as other track's components are taken from the literature. The quasi-static loads are calculated based on typical information of trains. The developed model is used to examine the extent of deformation of railpads and whether this deformation is associated with linear or non-linear behaviour of the railpads.

Keywords: Pre-load · Rail pad stiffness · Nonlinearity

1 Introduction

As known train carriages come in all shapes and sizes, and they generally weigh about 30 to 200 tons or more depending on their purpose. The loads transmitted to the rail depend on the number of axles in the train, while the load transmitted to one pad depends on the bending stiffness of rail and the normal stiffness of pads.

Commonly researchers assume the behaviour of rail pads to be linear. However, a number of researchers such as Thompson et al. 1998 [1], Maes et al. 2006 [2] and Koroma et al. 2015 [3] discuss nonlinearity of rail pads. According to Thompson. et al. 1998 [1], the rail pads were mostly assumed to be viscoelastic materials that have a complex stiffness depending on parameters like frequency, pre-load, temperature, etc.

Z. Dimitrovová et al. (Eds.): WMVC 2022, MMS 125, pp. 589–598, 2023.
https://doi.org/10.1007/978-3-031-15758-5_60

Wu. et al. 1999 [4] studied the effect of pre-load on the behavior of a railway track with nonlinear rail pads. A frequency based discretely supported track model was used after determining the stiffness distribution to obtain the track responses and decay rates. Koroma et al. 2013 [5] conducted similar research but in the time domain using a time integration scheme to study the effects of nonlinearity on track dynamics.

In this paper, a discretely supported beam has been modeled statically and dynamically for three different pre-loads to investigate the pre-load effect on pad's behaviour. Section 2 describes the used rail pad model and analyzes the behaviour of rail pads under quasi-static load i.e. pre-load. The stiffness of pads is defined from the load-displacement relationship from the literature Thompson et al. 1998 [1]. In Sect. 3, a static model based on the stiffness method is used to account for a beam discretely supported by springs having the non-linear load-displacement relationship, under a single pre-load applied on the top of the rails' beam. The model is used to calculate the resulting displacement and stiffness of pads under three different values of pre-loads. In Sect. 4, based on the static model, three cases have been investigated using the dynamic-stiffness method, i.e. following harmonic analysis, by modelling the dynamic behaviour of the rail beam under the action of a wheel-mass and with harmonic excitation induced through relative displacement between the wheel-mass and the beam. In the first case, the rail pads are modelled as homogeneous stiffness with value corresponding to an unloaded track. For the second case, pads with homogeneous stiffness are used with value corresponding to a loaded-track. For the third case, the stiffness is non-homogeneous with values calculated according to the amount of pre-load in each pad. The result are shown and discussed in Sect. 5, while the conclusion is given in Sect. 6.

2 Railpad Model

A Kelvin-Voigt visco-elastic model shown in Fig. 1 has been chosen to represent the railpad in which a spring in parallel with a dashpot is representing the stiffness k and damping c.

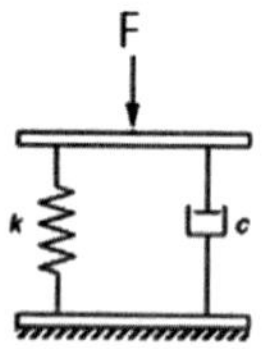

Fig. 1. Kelvin-Voigt visco-elastic model for rail pad.

The pre-load subjected to the rail accounts for the total weight transmitted through the wheel on one rail, this pre-load will be transmitted through the pads by specific load distribution F in which the pad under the pre-load directly will transmit the biggest portion of pre-load P_r and the portion of pre-load transmitted through the pad F will keep decreasing for the next pad along the beam. The relation between load transmitted to the rail pad F and the static displacement u is nonlinear according to Fig. 2 this relation has been characterized by curve fitting the experimental data extracted from Thompson et al. 1998 [1], in a similar process that Koroma et al. 2013 [5] followed. The relationship was found to be well described by a seventh degree polynomial accounting the load as a function of displacement as shown in Eq. (1).

$$F(u) = a_1 u + a_3 u^3 + a_5 u^5 + a_7 u^7 \tag{1}$$

Values of a_1, a_3, a_5, and a_7 are 20.818 MN/m, –2.3696*10^6 MN/m^3, 4.4938*10^{11} MN/m^5 and 5.954*10^{16} MN/m^7 respectively, where a_1 is the unloaded tangent stiffness of the rail pad, the even terms of the polynomial has been avoided to maintain anti-symmetrical relationship.

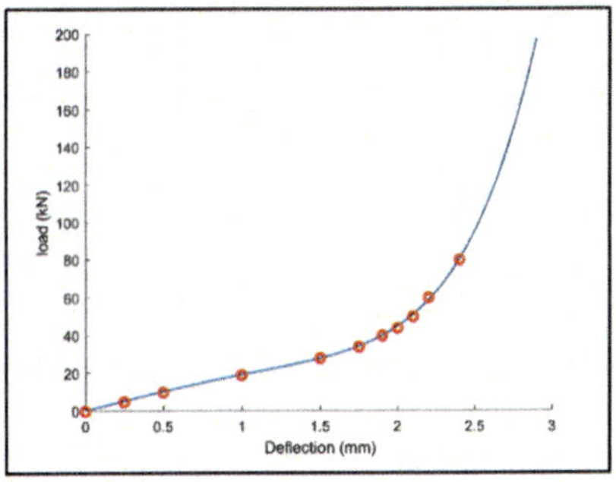

Fig. 2. Static deflection-load relation for a pad according to Thompson et al. 1998 [1] measurements.

The static stiffness of rail pad under load equals F_o is the tangent at the displacement u_o and can be obtained using the derivative of Eq. (1).

$$k_s(u) = a_1 + 3a_3 u^2 + 5a_5 u^4 + 7a_7 u^6 \tag{2}$$

3 The Static Rail Model

The rail was modelled as an Euler-Bernoulli beam, with mass per unit length m and bending stiffness EI. The rail is discretely supported on rail pads with spacing of d, and modeled as Kelvin-Voigt visco-elastic model, the layers under the track were assumed to be rigid since their stiffness is larger than the stiffness of rail pads, and the quasi-static load P_r is accounted for the pre-load, (see Fig. 3). Note that u(x) is the static beam displacement at distance x from the load, k_1, k_2, k_3...$k_{n/2+1}$ are the rail pads' stiffness and c is the damping coefficient.

In order to find the static displacement under each pad along the beam length, the stiffness-method has been used to solve the static model. The beam has been discretized for number of elements n = beam length/d.

First the system matrices were created such that the beam global stiffness matrix K_B with size of 2(n + 1) × 2(n + 1) generated using the local stiffness matrices K_{B_n} for all

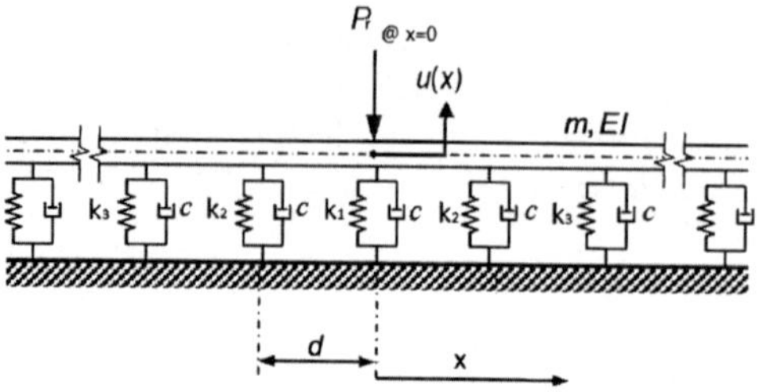

Fig. 3. A statically discretely supported track model on elastic foundation subjected to pre-load P_r.

n[th] elements in the beam, Eq. (3) shows K_{B_n}. The derivation of these matrices can be found in Bathe 2006 [6].

$$K_{B_n} = \frac{EI}{d}\begin{bmatrix} \frac{12}{d^2} & \frac{6}{d} & -\frac{12}{d^2} & \frac{6}{d} \\ \frac{6}{d} & 4 & -\frac{6}{d} & 2 \\ -\frac{12}{d^2} & -\frac{6}{d} & \frac{12}{d^2} & -\frac{6}{d} \\ \frac{6}{d} & 2 & -\frac{6}{d} & 4 \end{bmatrix} \tag{3}$$

Then Eq. (2) has been modified to accounts for all pads at the same time results in Eq. (4).

$$F_G = (K_B + A_1)U + A_3U^3 + A_5U^5 + A_7U^7 \tag{4}$$

Note that F_G is the force vector which has 2(n + 1) elements equals zero everywhere except for the vertical force for the node number n + 1 that equals the pre-load, and U is the static displacement vector with 2(n + 1) length. A_1, A_3, A_5, and A_7 are 2(n + 1) × 2(n + 1) null matrices in which the values of a_1, a_3, a_5 and a_7 have been added in the corresponding places for the vertical degree of freedom.

Using a nonlinear solver, Eq. (4) has been solved and gives the static displacement under each pad. The next step is to calculate the static stiffness distribution K_s, by using Eq. (2) for each pad displacement as Eq. (5) shows.

$$K_s = a_1 + 3a_3U^2 + 5a_5U^4 + 7a_7U^6 \tag{5}$$

4 The Dynamic Rail Model

In order to investigate the influence of the rail pad's nonlinearity on track's dynamics, a linear frequency based model has been used where the rail was modelled as an Euler-Bernoulli beam, with mass per unit length m and bending stiffness for one rail EI. The rail is discretely supported on rail pads with spacing of d, and modeled as Kelvin-Voigt visco-elastic model, the quasi-static load P_r is accounted for the pre-load. The rail beam is under the action of a wheel-mass with harmonic excitation induced through relative displacement between the un-sprung wheel mass M_u and the beam $\Delta e^{i\omega_0 t}$ using a pull through model as shown in Fig. 4, u(x) is the beam response at distance x from the load, k_1, k_2, $k_3 \ldots k_{n/2+1}$ are the stiffness of rail pads and c is the damping coefficient.

Note that for this model, the pre-load does not contribute directly in the calculations. Instead, its effect comes from the static model since it affects the values of stiffness of the rail pads.

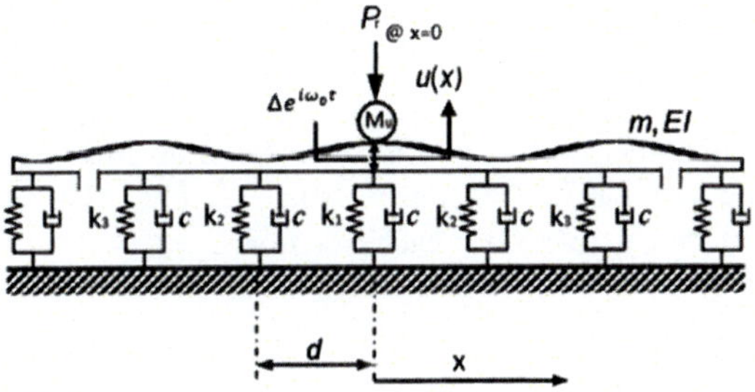

Fig. 4. Discretely supported track model on elastic foundation subjected to pre-load P_r with a relative displacement $\Delta e^{i\omega_0 t}$ between the beam and the un-sprung wheel mass M_u.

Accordingly the dynamic model has been solved for three cases, in the first case the rail pads having homogeneous un-loaded stiffness k, and the rail pads are assumed to have homogeneous loaded stiffness k_{loaded} in the second case, while the rail pads in the third case are assumed to have a non-homogeneous loaded stiffness's k_1, k_2, $k_3 \ldots k_{n/2+1}$ which have been extracted from solving the static model.

The dynamic stiffness of rail pads will be calculated as a function of the dynamic displacement and using a correction factor γ, where γ is appositive factor equals 3.6 according to Wu et al. 1999 [4]. By substitute the static stiffness extracted from the static model in to Eq. (6) the dynamic stiffness can be calculated for each pad.

$$k_{dyn} = \gamma k_s + i\omega c \tag{6}$$

where k_{dyn} and k_s is the dynamic stiffness and static stiffness for each pad respectively and ω is the frequency. Using the dynamic-stiffness method the dynamic responses can be calculated as in Talbot 2001 [7], a 1 unit amplitude dynamic load represented here as a relative displacement between the mass M_u and the beam is subjecting to the model described earlier. First for each frequency the global stiffness $2(n+1) \times 2(n+1)$ matrix K_G has been created using the nth local stiffness matrices K_L which can be assembled using Eq. (7) such that N and M are 4×4 matrices, see Eqs. (8), (9) and (10).

$$K_L = N \times M^{-1} \tag{7}$$

$$N = \begin{bmatrix} 0 & EIB^3e^{-(\frac{Bd}{2})} & -EIB^3e^{(\frac{Bd}{2})} & iEIB^3e^{(\frac{iBd}{2})} \\ 0 & EIB^2e^{-(\frac{Bd}{2})} & -EIB^2e^{(\frac{Bd}{2})} & EIB^2e^{(\frac{iBd}{2})} \\ 0 & -EIB^3e^{(\frac{Bd}{2})} & EIB^3e^{-(\frac{Bd}{2})} & -iEIB^3e^{-(\frac{iBd}{2})} \\ 0 & EIB^2e^{-(\frac{Bd}{2})} & EIB^2e^{-(\frac{Bd}{2})} & -EIB^2e^{-(\frac{Bd}{2})} \end{bmatrix} \tag{8}$$

$$M = \begin{bmatrix} 0 & e^{-(\frac{Bd}{2})} & e^{(\frac{Bd}{2})} & e^{(\frac{iBd}{2})} \\ 0 & Be^{-(\frac{Bd}{2})} & -Be^{(\frac{Bd}{2})} & -Be^{(\frac{iBd}{2})} \\ 0 & e^{(\frac{Bd}{2})} & e^{-(\frac{Bd}{2})} & e^{-(\frac{iBd}{2})} \\ 0 & Be^{(\frac{Bd}{2})} & -Be^{-(\frac{Bd}{2})} & -iBe^{-(\frac{Bd}{2})} \end{bmatrix} \quad (9)$$

$$B = \left(\frac{m\omega^2}{EI}\right)^{1/4} \quad (10)$$

where m is the rail mass per unit length, EI is the bending stiffness for one rail EI, d is the rail pads spacing and ω is the frequency. Using Eqs. (11), (12) and (13) the dynamic displacement U_d can be calculated for each frequency.

$$U_{d1} = K_G^{-1} * F_G \quad (11)$$

$$R = \frac{M_u\omega^2}{(1 - U_{d1}M_u\omega^2)} \quad (12)$$

$$U_d = R * U_{d1} \quad (13)$$

where the global force vector has been denoted as F_G with 2(n + 1) zero elements except the element representing the node number (n + 1) that should equals 1 unit, and the dynamic displacement due to 1 unit load vector is U_{d1} with a length of 2(n + 1). In order to calculate the dynamic response U_d due to the oscillating wheel mass M_u, U_{d1} should be multiplied by the Reaction R under the mass directly as in Eq. (13).

5 Results and Discussion

The parameters used for the numerical analyses are for 60E1 rail with mass of 60.2862 kg/m and bending stiffness of 8.8285 MN.m^2, rail pad spacing d = 0.5 m, un-loaded rail pad stiffness k = 20.818 MN/m, rail pad damping coefficient c = 5.0101 kN.s/m, un-sprung wheel mass = 1563.5 kg, Three pre-loads has been used 30 kN, 245 kN and 360 kN.

5.1 The Static Model Results

The static model shown in Fig. 3 has been solved for three values of pre-loads 30 kN, 245 kN and 360 kN such that the load transmitted to the pad is 7.813 kN, 94.318 kN and 168.263 kN respectively. As shown in Fig. 5 the pre-loads put the pads behaviour in different zones on the deflection-load curve.

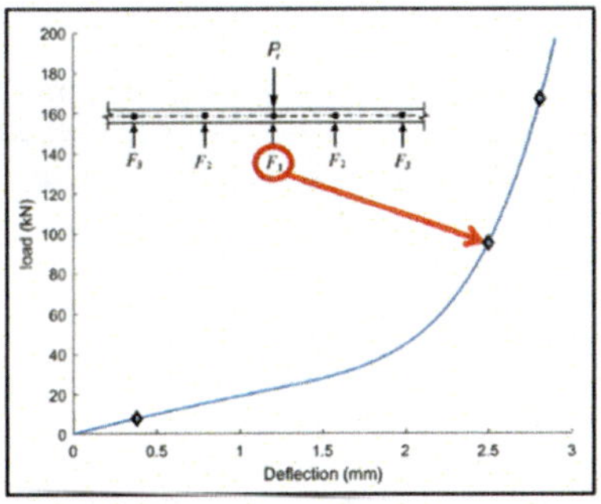

Fig. 5. The location of the three different pre-loads on the deflection-load curve for pad.

The static displacement has been plotted twice, linearly assuming the pad behaviour is linear with stiffness equals the un-loaded track stiffness, and nonlinearly according to the nonlinear relation extracted in Sect. 2 with the actual stiffness distribution in Fig. 6. The difference between linear and nonlinear based model displacement is increased with the increasing in pre-load since the pads under higher load becomes stiffer and experience lower deflection subsequently compared to the linear case.

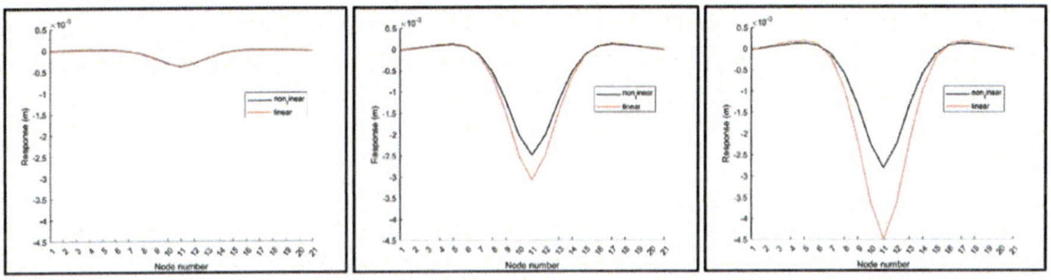

Fig. 6. Static displacement along the track for both cases linear and nonlinear static models, (a) 30 kN, (b) 245 kN and (c) 360 kN.

5.2 The Actual Stiffness Distribution

Figure 7 shows the actual stiffness distribution for the three pre-loads 30 kN, 245 kN and 360 kN. The loaded stiffness / un-loaded stiffness ratio for 30 kN, 245 kN and 360 kN pre-load cases are 0.956, 7.9, and 15 respectively. Hence, based on the used pad's model the higher pre-load affects the pad's stiffness more. Also it has been noticed that the loading effect on rail pad's stiffness extended around the applied load no more than 2 m i.e. 4 pads on each side.

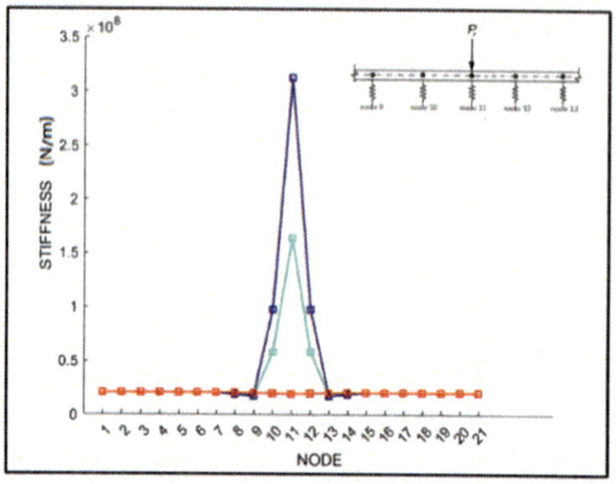

Fig. 7. Actual static stiffness distribution for three pre-loads red: 30 kN, cyan: 245 kN and blue: 360 kN.

5.3 Effect of Stiffness Distribution on the Track Dynamics

Looking at the results for the cases associated here with higher pre-loads i.e. 245 kN and 360 kN which is accounted for some very heavy freight trains plotted in Fig. 8, it is clearly seen that the dynamic response due to the fully loaded stiffness are over estimated as well as the un-loaded stiffness which are under estimated, so in such high pre-loads cases the nonlinearity of the rail pads stiffness should be taking into account. However Fig. 8 indicates that the nonlinearity effect tends to increase with the rise of pre-load as the three curves for the three cases of stiffness mismatched.

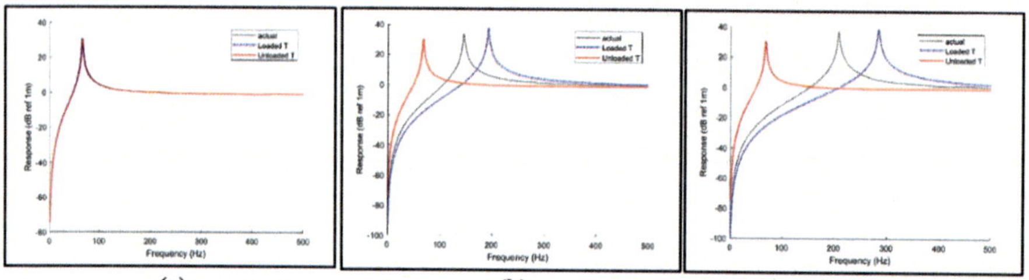

(a) **(b)** **(c)**

Fig. 8. Displacement amplitude plotted against the frequency for three different stiffness cases __ Red: un-loaded homogeneous stiffness, …black: actual stiffness distribution, -.-.blue: homogeneous fully loaded stiffness. (a) 30 kN pre-load, (b) 245 kN pre-load and (c) 360 kN pre-load.

5.4 Effect of Pre-load on Track Dynamics

Figure 9 illustrate the results of the linear dynamic model in case of the three pre-loads, The frequency and the response amplitude didn't change at all for the pre-load 30 kN because the load transmitted to the pad (7.813 kN) is within the linear zone in the deflection-load curve (see Fig. 5) which explains the linear behaviour of the 30 kN pre-load case. However, the response and resonance frequency increased for the second (245 kN) and third (360 kN) pre-loads which can be explained by their locations on the deflection-load curve within the nonlinear zone.

The most obvious trends here that the resonance frequency is shifted to the right as the pre-load increased, as it changed from 68 Hz for the first case 30 kN pre-load to reach 208 Hz for the third case of 360 kN pre-load.

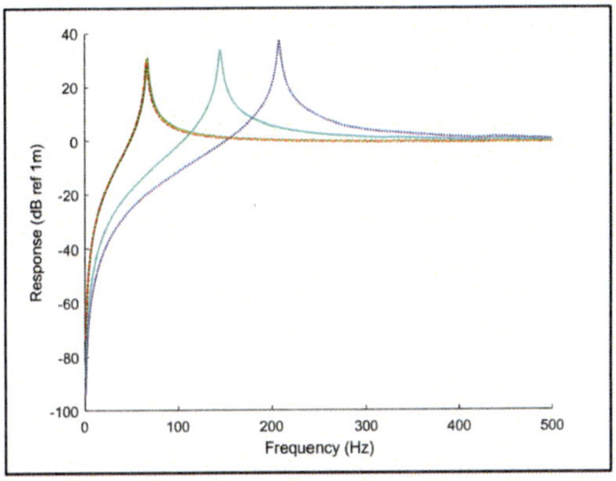

Fig. 9. Displacement amplitude plotted against the frequency for three different pre-loads with the un-loaded case. Green: un-loaded, ---red: 30 kN, __cyan: 245 kN, …blue: 360 kN

6 Conclusion

A discretely supported model subjected to static load on top of a mass with harmonic excitation induced through relative displacement between the wheel-mass and the beam has been modeled to investigate the effects of pads nonlinearity on track dynamics under three deferent pre-loads values 30 kN, 245 kN and 360 kN. This model is solved statically first using the stiffness-method to find the actual stiffness distribution, then dynamically in the frequency domain using the dynamic-stiffness method.

Based on the pad model used, the analysis of the pad behavior demonstrated that the actual pad stiffness affected by the value of load transmitted to the pad, and the effect of the pre-load does not extend more than 2 m from the load position i.e. about 4 pads on each side. Hence, the pre-load must be taken into account when modeling the rails.

The results presented for the track's dynamic response under the applied load showed that the pre-load influences the linearity of the track especially for high values of pre-loads, which is typically associated with the heavy freight trains. Furthermore, it has been observed that the resonance frequency has increased from 68 Hz for 30 kN pre-load to reach 208 Hz for the highest case of pre-load of 360 kN which corresponds to a pad out of the linear range.

Acknowledgement. The authors would like to acknowledge funding from Qatar Rail through grant number QUEX-CENG-QR-21/22–1.

References

1. Thompson, D.J., Van Vliet, W.J., Verheij, J.W.: Developments of the indirect method for measuring the high frequency dynamic stiffness of resilient elements. J. Sound Vib. **213**, 169–188 (1998)
2. Maes, J., Sol, H., Guillaume, P.: Measurements of Dynamic Railpad Properties. J. Sound Vib. **293**, 557–565 (2006)
3. Koroma, S.G., Hussein, M.F.M., Owen, J.S.: Influence of preload and nonlinearity of railpads on vibration of railway tracks under stationary and moving harmonic loads. J. Low Freq. Noise Vib. Active Control **34**, 289–306 (2015)
4. Wu, T.X., Thompson, D.J.: The effects of local preload on the foundation stiffness and vertical vibration of railway track. J. Sound Vib. **219**, 881–904 (1999)
5. Koroma, S.G., Hussein, M.F.M., Owen, J.S.: The effects of railpad nonlinearity on the dynamic behaviour of railway tracks. Proc. Inst. Acoust. **35**, 176–183 (2013)
6. Bathe, K.J.: Finite Element Procedures. Prentice Hall, New Jersey (2006)
7. Talbot, J.P.: On the Performance of Base-Isolated Buildings: A Generic Model. PhD thesis. University of Cambridge (2001)

Machine Learning Analysis in the Diagnostics of the Dynamics of Ball Bearing with Different Radial Internal Clearance

Bartłomiej Ambrożkiewicz[1,2](✉), Arkadiusz Syta[1], Alexander Gassner[2], Anthimos Georgiadis[2], Grzegorz Litak[1], and Nicolas Meier[2]

[1] Lublin University of Technology, Nadbystrzycka 36, 20-618 Lublin, Poland
b.ambrozkiewicz@pollub.pl

[2] Leuphana University of Lüneburg, Universitätsallee 1, 21335 Lüneburg, Germany

Abstract. Interpretation of acceleration time-series from rolling-element bearings is sometimes challenging if no prior knowledge of the system is available. The evaluation must adapt operational conditions or the actual value of operational parameters. In our analysis, we apply the machine learning methods and statistical indicators in the diagnosis of dynamical response in the self-aligning ball bearing with different radial internal clearance. Machine learning methods are applied for the quantification of the acceleration time-series with statistical indicators and their assignation to the specific state or clearance value. The results of the analysis allow recognizing the bearing's condition and the clearance value based on experimental acceleration time-series. Additionally, confusion matrices are presented for showing the accuracy of proposed methods. The results of applied Machine Learning methods are on the level of around 80% in classifying the dynamical response to the specific radial clearance. The motivation of the research is to introduce it to on-site practice in the test rig.

Keywords: Rolling-element bearing · Radial internal clearance · Machine learning · Statistical indicators

1 Introduction

On-line condition monitoring of rolling-element bearings gives the real-time information on actual condition of rotating elements helping in the early detection of its worrying behavior. Accurate monitoring is based on the measurement of dynamical response of bearings in form of velocity or acceleration signal, which is later on processed to cross-check its condition. The vibrations in rolling-element bearings can have different source, such as Hertzian contact stresses, geometrical imperfections, wear, tear & fatigue or operational parameters [1, 2]. While linear methods such as FFT are in common use in estimation of characteristic frequencies [3, 4], sometimes nonlinear methods need to be used for the detection of transient states caused by other factors occurring in bearing [5, 6].

Z. Dimitrovová et al. (Eds.): WMVC 2022, MMS 125, pp. 599–606, 2023.
https://doi.org/10.1007/978-3-031-15758-5_61

One of the most important parameters is the radial internal clearance, which value in the rotating bearing defines not only its dynamical behavior, but refers to the temperature, friction torque and primarily bearing's lifetime. Radial internal clearance in ball bearings is defined as the total distance through which one bearing ring can be moved to the other in the radial direction [7]. Its value is pre-defined by the bearing's manufacturer and on this base, the bearing with specific clearance is matched to the specific operating conditions like subjected load, velocity or predicted temperatures. Mentioned factors can cause the change of its value during bearing's operation, then the clearance is so called the operating one. The key to the proper operation of bearing is the clearance fitting to the real conditions, it can't be too small due to strong friction torque, while it can't be too big as well due to undesirable sliding of rolling elements. One of challenges is to control the value of changing radial clearance in time.

Following the previous approach [8, 9] using recurrence-based methods for the quantification of bearing's dynamics by specific value of radial internal clearance, we would like to extend this research with Machine Learning (ML) methods by the analysis of statistical indicators applied in the diagnostics of rotating machinery. The aim of the paper is to cross-check different ML methods in the identification of clearance classes and propose the most accurate. Studied Artificial Intelligence methods have been already tested with the success for the identification of different kind of faults in the rotating machines [10–13]. Referring to other research, we carry on our experiment with 4 different ML methods in PyChart Python's notebook.

The remainder of this paper is following. In Sect. 2, the proposed ML methods are described with its mathematical formulation. In Sect. 3, the experimental procedure and data processing are presented. In Sect. 4, the results of Machine Learning classification are discussed and Sect. 5 summarizes the paper.

2 Methods

For the analysis of the experimental acceleration signals for bearing's dynamical response, 4 following Machine Learning methods had been applied. Its definition is following:

1. Light Gradient Boosting Machine (LightGBM) – is a fast, distributed as well as high-performance gradient boosting framework that makes the use of a learning algorithm that is tree-based, and is used for ranking, classification as well as many other machine learning tasks. Its main advantages are high efficiency and faster training speed, usage of lower memory, better accuracy, handling of large-scale data [14].
2. Extreme Gradient Boosting (XGBoost) – in this algorithm, decision trees are created in sequential form, weights are assigned to all the independent variables, which are then fed into the decision tree predicting results. Weight of variables predicted wrong by the tree is increased and these variables are next fed into the second decision tree. Individual classifiers or predictors are collected to give a strong and more precise model. The algorithm works on regression, classification, ranking and user-defined prediction problems [15].

3. Extra Trees – is a type of ensemble learning technique, which aggregates the results of multiple de-correlated decision trees collected in a forest to output it's classification result. Each decision tree in the extra trees forest is constructed from the original training sample. Then, at each test node, each tree is provided with a random sample of k features from the feature-set from which each decision tree must select the best feature to split the data based on some mathematical criteria. This random sample of features leads to the creation of multiple de-correlated decision trees. To perform feature selection, each feature is ordered in descending order according to the Gini importance of each feature and the user selects the top k features according to the selected way [16].
4. Random Forest – is an algorithm used in supervised machine learning to solve regression and classification problems. Each random forest is comprised of multiple decision trees that work together as an ensemble to produce one prediction. A decision tree is a logical construct that resembles a flowchart and illustrates a series of if-else statements. Random forest algorithms can produce acceptable predictions even if individual trees in the forest have incomplete data. Statistically, increasing the number of trees in the ensemble will correspondingly increase the precision of the outcome [17].

3 Experiment

For the experiment, 10 brand-new self-aligning ball bearings NTN 2309SK were used. In the experimental part, two test rigs were applied i.e. automated system with the automatic clearance measurement for radial internal clearance estimation after mounting (Fig. 1) and test rig for the dynamical test (Fig. 2). At the first test rig, the value of radial internal clearance was set and measured after bearing's mounting on the shaft. Next, the bearing with pre-defined clearance is mounted in the plummer block connected with the propelling motor. The inverter allow to conduct the tests in the range of operational velocities up to 3000 rpm, however for the initial tests, the only one rotational velocity 600 rpm corresponding to 10 Hz is chosen. For each bearing and the specific value of radial clearance, the test was conducted for 10 min, while the sampling frequency is equal to 1562.5 Hz (0.00064 s – sampling time) for acceleration time-series acquisition. After collection of the experimental data, the acceleration signal is firstly normalized and next filtered with bandpass filter 0–100 Hz. As the last part of data processing, each feature is calculated in short-time windows consisting of 500 data points, then the ML analysis is conducted in PyCaret Python programming language notebook. Referring the analysis 9 features were selected, i.e. mean, median, variance, kurtosis, skewness, peak to peak, crest factor, FM4 and NA4, which mathematical formulation is described in following paper [18].

Fig. 1. Automated system with the automatic clearance measurement after mounting with seated ball bearing NTN 2309SK [8].

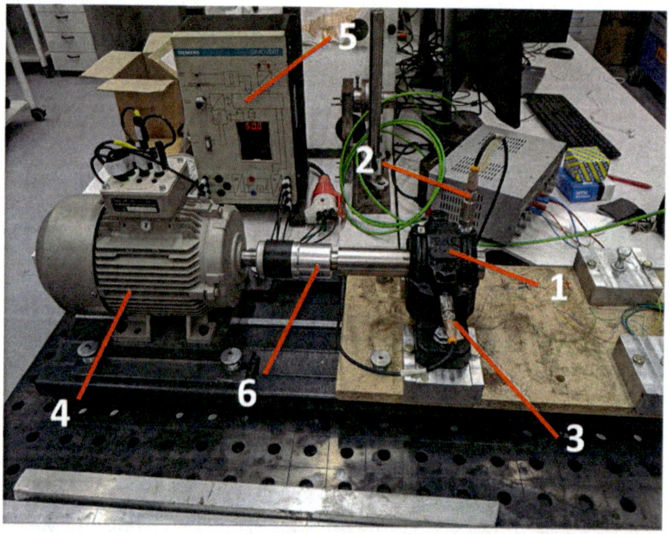

Fig. 2. Experimental setup: 1 – plummer block with mounted ball bearing, 2 – vertical accelerometer, 3 – horizontal accelerometer, 4 – three-phase motor, 5 – inverter, 6 - coupling [8].

4 Machine Learning Classification

Further analysis was based on classification of the radial internal clearance with same rotational velocity of 10 Hz. Below one can find brief description of data preparation and features selection (Fig. 3):

1. Signals of vertical acceleration for 10 new bearings where collected and filtered with bandpass filter from 0–100 Hz frequency.
2. Two data sets were prepared: training (90% of the data – approx. 20000 samples) and validating (10% of the data – approx. 2000 samples).
3. Moving windows statistics in both time and frequency domain was applied
4. resulting in 500 samples for each RIC with 9 following features: mean, median, variance, kurtosis, skewness, peak to peak, crest factor, FM4, NA4. The analysis was conducted with PyCaret Python notebook.
5. Feature selection included measures described in table Table 1.
6. Labeling of each observation has been applied with 5 labels based on RIC values: C2L – (7 μm – 11 μm), C2H – (14 μm – 16 μm), CNL – (20 μm – 24 μm), CNH – (30 μm – 36 μm), C3 – (38 μm – 45 μm).
7. The data set was unbalanced in classes due to less experimental data.
8. Different models based of artificial intelligence classification based on random forest were selected and compared in terms of accuracy and time.

The results of classification for optimal classifiers (in terms of accuracy) during training process are presented in table Table 1.

Table 1. The optimal classifiers and results of training.

Classifier	Accuracy	AUC	Recall	Precision	F1	Time [s]
Light Gradient Boosting	0.77	0.95	0.77	0.77	0.77	0.37
Extreme Gradient Boosting	0.77	0.95	0.77	0.77	0.76	2.46
Extra Trees	0.76	0.94	0.76	0.76	0.76	0.22
Random Forest	0.73	0.94	0.73	0.74	0.73	0.44

The trained model have been used for classification on testing data set. The results are presented in table Table 2.

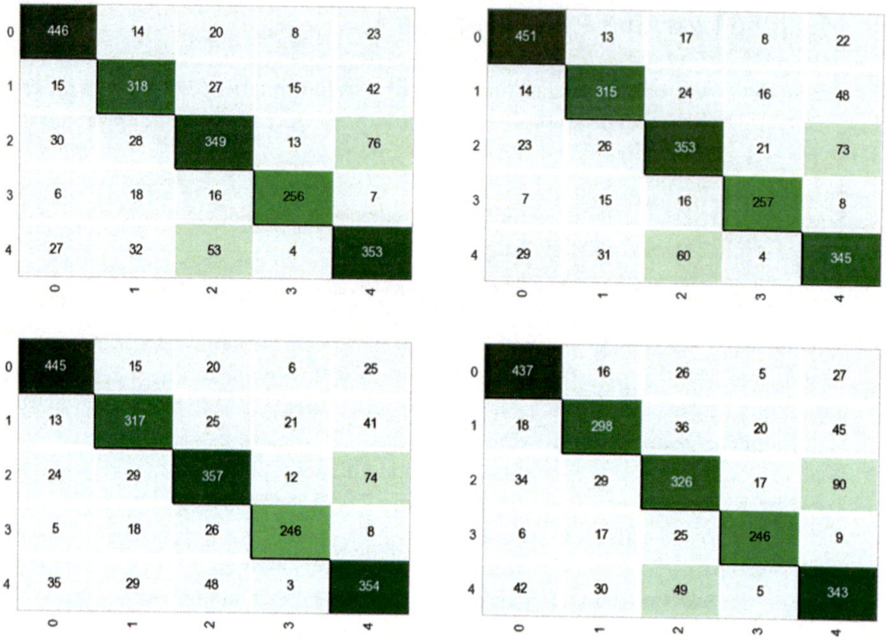

Fig. 3. The confusion matrices for chosen classificators (from left): Light Gradient Boosting, Extreme Gradient Boosting, Extra Trees and Random Forest. Class named 0 corresponds to C2L, 1 - C2H, 2 – CNL, 3 – CNH and 4 – C3, respectively.

Table 2. The results of classification for test data set.

Classifier	Accuracy	AUC	Recall	Precision	F1
Light Gradient Boosting	0.78	0.95	0.78	0.78	0.78
Extreme Gradient Boosting	0.78	0.95	0.78	0.78	0.78
Extra Trees	0.78	0.95	0.78	0.78	0.78
Random Forest	0.75	0.94	0.75	0.75	0.75

5 Discussion

In this paper, the linear analysis of the radial internal clearance of bearing were performed for fixed velocity 10 Hz. The goal of this work was to show the sensitivity of initial condition (RIC value) to dynamics of the system. Using windowing technique of time series analysis, the indicators in time and frequency domain were calculated with different RIC and split in two data sets: training and testing. Using machine learning methods, it was shown that with maximal accuracy of 0.78% one can distinguish between different class according to RIC values. That indicated changes in the statistics of underlying time series (system response) with respect to internal clearance. As predicted, the best results were obtained with tree-based methods. In particular, the algorithms using boosted decision trees (Light Gradient Boosting and Extreme Gradient Boosting) allowed for the highest classification accuracy. On the other hand, one can pick standard Extra Trees model with the same accuracy but the shortest time of learning. Obtained promising results allow for continuing the analysis with other operational cases, finally introducing methods in on-site in the practice.

References

1. Zhang, F., Huang, J., Chu, F., Cui, L.: Mechanism and method for the full-scale quantitative diagnosis of ball bearings with and inner race fault. J. Sound Vib. **488**, 115641 (2020)
2. Verstraete, D.B., Droguett, E.L., Meruane, V., Modarres, M., Ferrada, A.: Deep semi-supervised generative adversarial fault diagnostics of rolling element bearings. Struct. Health Monit. **19**(2), 390–411 (2020)
3. Zhao, Z., Yin, X., Wang, W.: Effect of the raceway defects on the nonlinear dynamic behavior of rolling bearing. J. Mech. Sci. Technol. **33**(6), 2511–2525 (2019). https://doi.org/10.1007/s12206-019-0501-0
4. Altaf, M., et al.: Automatic and efficient fault detection in rotating machinery using sound signals. Acoustics Australia **47**(2), 125–139 (2019)
5. Abhilash, S., Pradeep, R., Rejith, R., Bijudas, C.R.: Health monitoring of rolling element bearings using improved wavelet cross spectrum technique and support vector machines. Tribol. Int. **154**, 106650 (2021)
6. Zeng, G., et al.: Study on simplified model and numerical solution of high-speed angular contact ball bearing. Shock. Vib. **2020**, 1–17 (2020). https://doi.org/10.1155/2020/8843524
7. Zhang, Z., Chen, Y., Cao, Q.: Bifurcations and hysteresis of varying compliance vibrations in the primary parametric resonance for a ball bearing. J. Sound Vib. **350**, 171–184 (2015)
8. Ambrożkiewicz, B., Syta, A., Gassner, A., Georgiadis, A., Litak, G., Meier, N.: The influence of the radial internal clearance on the dynamic response of self-aligning ball bearings. Mech. Syst. Signal Process. **171**, 108954 (2022)
9. Ambrożkiewicz, B., Syta, A., Meier, N., Litak, G., Georgiadis, A.: Radial internal clearance in ball bearings. Eksploatacja i Niezawodnosc – Maintenance and Reliability **23**(1), 42–54 (2021)
10. Wang, F.K., Mamo, T.: Hybrid approach for remaining useful life prediction of ball bearings. Qual. Reliab. Eng. Int. **35**(7), 2494–2505 (2019)
11. Chen, X., Qi, X., Wang, Z., Cui, C., Wu, B., Yang, Y.: Fault diagnosis of rolling bearing using marine predators algorithm-based support vector machine and topology learning and out-of-sample embedding. Meas. J. Int. Meas. Confederation **176**, 109116 (2021)
12. Cerrada, M., et al.: A review on data-driven fault severity assessment in rolling bearings. Mech. Syst. Signal Process. **99**, 169–196 (2018)
13. Wang, X., Shen, C., Xia, M., Wang, D., Zhu, J., Zhu, Z.: Multi-scale deep intra-class transfer learning for bearing fault diagnosis. Reliab. Eng. Saf. Syst. **202**, 107050 (2020)
14. Fan, J., Ma, X., Wu, L., Zhang, F., Yu, X., Zeng, W.: Light gradient boosting machine: An efficient soft computing model for estimating daily reference evapotransportation with local and external meteorological data. Agric. Water Manag. **225**, 105758 (2019)
15. Minhas, A.S., Singh, S.: A new bearing fault diagnosis approach combining sensitive statistical features with improved multiscale permutation entropy method. Knowl.-Based Syst. **218**, 106883 (2021)

16. Saeed, U., Jan, S.U., Lee, Y.D., Koo, I.: Fault diagnosis based on extremely randomized trees in wireless sensor networks. Reliab. Eng. Syst. Safe **205**, 107284 (2021)
17. Hu, Q., Si, X.S., Zhang, Q.H., Qin, A.S.: A rotating machinery fault diagnosis method based on multi-scale dimensionless indicators and random forests. Mech. Syst. Signal Process. **139**, 106609 (2020)
18. Sharma, V., Parey, A.: A review of gear fault diagnosis using various condition indicators. Procedia Eng. **144**, 253–263 (2016)

Moving Element Analysis of Maglev Train Over Multi-span Elevated Bridge

Jian Dai[1(✉)], Joshua Guan Yi Lim[2], and Kok Keng Ang[2]

[1] Department of Civil Engineering and Energy Technology, Oslo Metropolitan University, 0166 Oslo, Norway
jiandai@oslomet.no
[2] Department of Civil & Environmental Engineering, National University of Singapore, Singapore 117576, Singapore
joshualimgy@u.nus.edu, kk_ang@nus.edu.sg

Abstract. Maglev trains are modern railway transport systems that utilize non-contact magnetic forces for levitation and propulsion. Owing to the advantage of being wear-free, they can travel at a speed that is much higher than conventional wheeled trains with lower noise emissions. For city and inter-city maglev lines, maglev trains often operate on multi-span elevated bridges. The dynamic response of the coupled maglev train-bridge system is of great interest to researchers and engineers to ensure safe travel with the ever-increasing demand for higher operating speeds from the society. This study employs the moving element analysis to efficiently compute the dynamic response of a maglev vehicle traversing a multi-span bridge with each span modeled as a simply supported beam with rotational springs connecting to the adjacent spans. The vehicle is modeled by employing the multi-body system. The vehicle and the bridge are coupled via electromagnetic forces. The accuracy of the proposed computational model is examined by comparison with results obtained using the equivalent finite element model. Parametric studies are carried out to investigate the effect of various factors on the dynamic response of the coupled system. These include the speed of the train, the bridge span length and the stiffness of the suspension system.

Keywords: Maglev train · Multi-span bridge · Moving load · Moving element method

1 Introduction

Maglev trains are modern modes of transportation that utilize non-contact magnetic forces for levitation and propulsion. Such a sophisticated transport system enjoys the advantages of being free of wear and rolling resistance. Thus, it can operate at a speed that is much higher than conventional trains which greatly shorten the travel time. This has led to a strong interest in both academia and industry related to R&D and construction activities of maglev train lines in several countries that have such a demand [1].

In a typical maglev train line, the trains are travelling along a guideway which is often in the form of a long, multi-span elevated bridge. The structural rigidity of the

Z. Dimitrovová et al. (Eds.): WMVC 2022, MMS 125, pp. 607–614, 2023.
https://doi.org/10.1007/978-3-031-15758-5_62

guideway span and connection stiffness between adjacent spans have a strong effect on the traveling dynamics and thus the ride comfort and safety of maglev trains. Several studies employing different methods and models have been conducted to examine the dynamic responses of the maglev train-guideway coupled system [1–4]. Such studies use the land-fixed coordinate systems which encounter two main complications. The first complication is associated with the artificial boundary effects when adopting a truncated domain to represent a nearly infinitely long guideway for the sake of computational efficiency. Furthermore, the train will soon move out of the truncated domain, especially at high speeds. When the domain size is enlarged to mitigate the aforementioned complication, the second complication of increased computational cost arises. An accurate and computationally efficient method for the dynamic response analysis of the coupled maglev train-guideway system is thus highly desirable.

In this paper, a numerical scheme in conjunction with the moving element method is developed to efficiently compute the dynamic response of a maglev vehicle traversing an infinitely long multi-span guideway bridge. Each guideway span is modeled as simply a supported beam with rotational springs connecting to the adjacent spans. The maglev vehicle is modeled by employing the multi-body system. The vehicle and the guideway are coupled via the electromagnetic force. The accuracy of the proposed computational model is examined by comparison with results obtained using the equivalent finite element model. Parametric studies are carried out to investigate the effect of various factors on the dynamic response of the coupled system. These include the speed of the train, the bridge span length and the stiffness of the suspension system.

2 Methodology and Models

2.1 Train Model

A maglev passenger car is modeled as a multi-body system comprising a rigid car body and eight magnets, interconnected via suspension systems represented by using spring-dashpot units, as shown in Fig. 1. The car body of mass m_c has two degrees of freedom (DOFs), namely vertical DOF y_c and pitching DOF θ_c. Each magnet of mass m_m has only the vertical DOF. In total, there are 10 DOFs. For sake of conciseness, the governing equations of motion for the train model can be written as

$$\mathbf{M}_t\ddot{y}_t + \mathbf{C}_t\dot{y}_t + \mathbf{K}_t y_t = \mathbf{F}_t \tag{1}$$

where $\mathbf{M}_t$, $\mathbf{C}_t$, $\mathbf{K}_t$ and $\mathbf{F}_t$ denote the mass, damping and stiffness matrices and external force vector of the train model, respectively. $\ddot{y}_t$, $\dot{y}_t$, y_t denote the acceleration, velocity and displacement vector of the train, respectively.

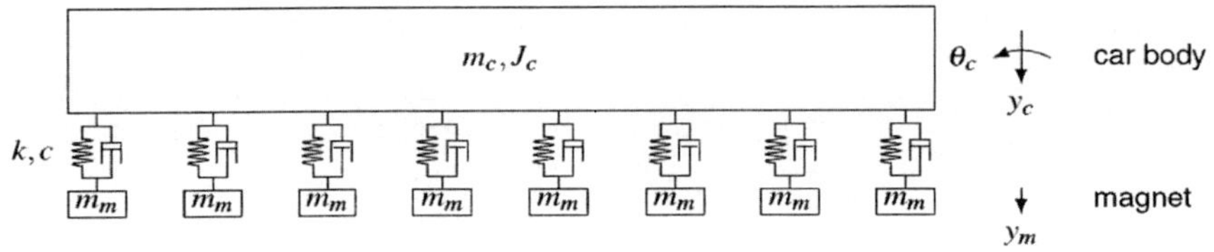

Fig. 1. Train model.

2.2 Guideway Model

Figure 2 schematically illustrates a two-span guideway bridge model for sake of simplicity. Note that in this study a multi-span bridge representing an infinitely long guideway is considered. Each span is modeled as a prismatic Euler beam is individually supported by bridge columns idealized as vertical springs of stiffness k_b at its two ends. Between two adjacent beams, a rotational spring of stiffness k_c connects the two beams. By adopting the land-fixed coordinate system (X, T) and Rayleigh damping, the equation of motion of a single guideway beam is written as

$$EI\frac{\partial^4 u}{\partial X^4} + \alpha_1 EI\frac{\partial^5 u}{\partial X^4 \partial T} + \alpha_0 \rho A\frac{\partial u}{\partial T} + \rho A\frac{\partial^2 u}{\partial T^2} = q \tag{2}$$

where EI and ρA denote the flexibility and mass per unit length of the guideway, respectively, q denotes the load density on the guideway, u is the vertical displacement of the guideway, T refers to the time, and α_0 and α_1 are the Rayleigh damping coefficients.

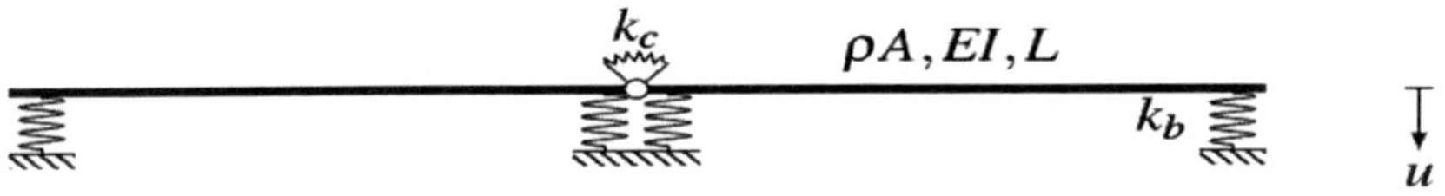

Fig. 2. Guideway model.

2.3 Electromagnetic Force Model

The maglev train interacts with the guideway bridge through the electromagnetic force. The nonlinear electromagnetic force F_{en} is a function of the electric current i and the air gap h_n between the n-th magnet and the guideway, which is given as

$$F_{en} = \frac{\mu_0 N^2 A}{4}\left(\frac{i}{h_n}\right)^2 = K_0\left(\frac{i}{h_n}\right)^2 \tag{3}$$

where $\mu 0 = 4\pi \times 10^{-7}$ H/m denotes the magnetic permeability of vacuum, N is the number of turns of the coil, A is the magnetic pole area, and $K0 = (\mu 0N^2A)/4$ is the coupling factor.

Considering i_0 and h_0 as the nominal current and air gap, respectively, the electromagnetic force at static equilibrium state F_{e0} is

$$F_{e0} = K_0\left(\frac{i_0}{h_0}\right)^2 = \frac{1}{8}(m_c + 4m_b + 8m_m)g \tag{4}$$

A reasonable simplification can be made by linearizing the electromagnetic force about the nominal static equilibrium state. Then, the electromagnetic force F_{en} can be rewritten as

$$F_{en} = F_{e0} + C_i\Delta i - C_h\Delta h_n \tag{5}$$

where $C_i = 2K_0 i_0/(h_0)^2$, $C_h = 2K_0 i_0 2/(h_0)^3$, $\Delta i = i_n - i_0$, and $\Delta h_n = h_n - h_0$. According to the current control law [4], $\Delta i = k_p\Delta h_n + k_v\dot{y}_{mn} + k_a\ddot{y}_{mn}$, where k_p, k_v and k_a are the feedback gains corresponding to the air gap change, velocity and acceleration of the magnet, respectively.

2.4 Moving Element Method

The moving element method (MEM) is employed for the analysis of the coupled train-guideway system. In the moving element model, the governing equations of motion for the guideway bridge are formulated using a moving coordinate (r, t) whose origin is moving at the same speed as the train. A segment of the multi-span guideway in the moving coordinate system is truncated and discretized into a series of moving elements. Such a numerical scheme has been successfully adopted to deal with various train-track problems and it has shown its superiority over the conventional finite element method (FEM) especially when the train is traveling at a high speed [5–10]. Note that when the moving coordinate is employed, the equations of motion for the guideway beam need to be reformulated according to the chain rule. Then, Galerkin's approach is adopted to treat the weighted residual. In this study, the Hermitian cubic polynomials are employed as the shape and weighting functions. Owing to the page constraint, the detailed formulation of the coupled train-guideway system in the moving coordinate is omitted. However, the readers are referred to [5–10] for more details.

In the moving coordinate system, the train is stationary while the guideway is moving in the opposite direction of the train. In view of the fact that the guideway beams are discretely supported, the motion of the supports needs to be tracked and updated at each time step. Figure 3 schematically illustrates a truncated multi-span guideway model under a train load. The domain size is assumed to be sufficient such that the artificial boundary effects are negligible. Thus, the fixed boundary conditions are employed. In this study, the guideway has uniform spans. Therefore, the motion of the discrete supports is completely periodic. Under such a circumstance, the dynamic guideway element matrices for the duration when the train travels across one span can be stored and retrieved conveniently for the computation for subsequent periods [7, 8].

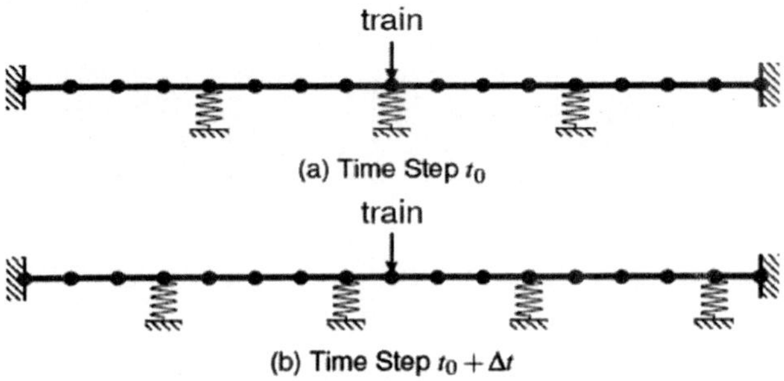

Fig. 3. Motion of discrete supports in moving coordinate.

3 Numerical Studies

3.1 Model Partial Verification

To verify the numerical model described in Sect. 2, the case of a single DOF train model traversing a multi-span flexible guideway is considered, as shown in Fig. 4. In this model, the train body and the magnet are rigidly connected. The train body has a mass m_v = 39,000 kg. The mass of the magnet is neglected. The guideway beams have a uniform length $L = 25$ m, mass per unit length ρA = 6,000 kg/m and flexural rigidity $EI = 8.58 \times 10^{10}$ Nm^2. The bear support has a stiffness $k_b = 3.2 \times 10^{10}$ N/m. The rotational coupling connection between adjacent guideways is $k_c = 1 \times 10^7$ Nm/rad. In the electromagnetic model, the nominal current is taken as 25 A and the nominal air gap is 0.01 m. The feedback gains k_h, k_v and k_a is set to 6,500, 40 and 0.2, respectively.

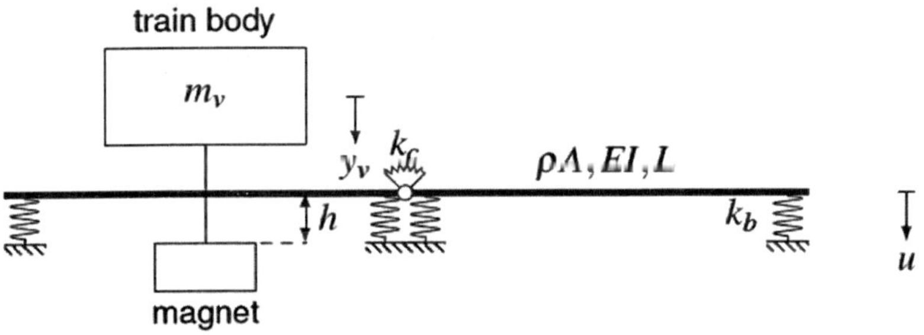

Fig. 4. Single DOF train model on multi-span guideway.

Figure 5 shows the time history of train displacement, velocity and acceleration over 10 spans at a constant speed of 300 km/h. Note that in the moving element model a total of 6 spans are considered. A uniform element size of 0.5 m and a time step size of 0.006 s are employed. As can be seen, the results generated by using the MEM agree very well with those by using the FEM.

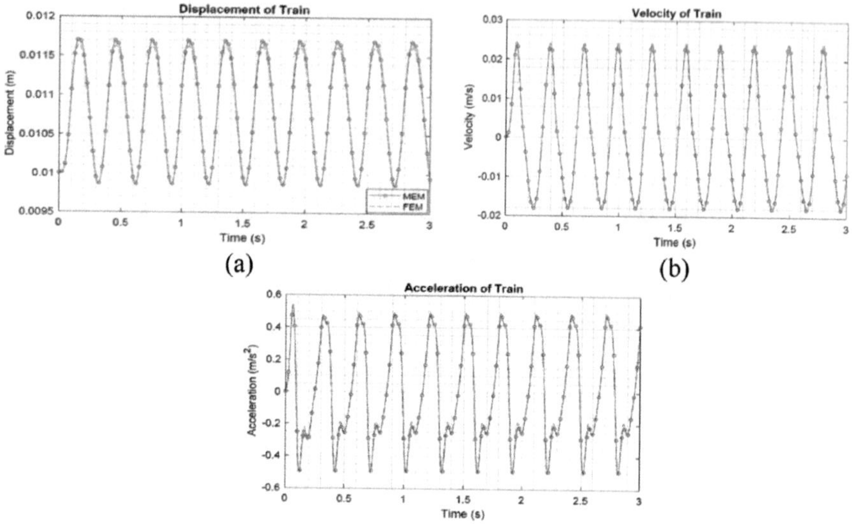

(a) (b) (c)

Fig. 5. Comparison between MEM and FEM: (a) displacement, (b) velocity, and (c) acceleration of train.

3.2 Parametric Studies

The 10-DOF train model shown in Fig. 1 is next employed to examine the effect of various factors on the dynamic responses of the maglev-guideway system. The same mass of the car body employed in Sect. 3.1 is adopted. The pitching moment of inertia of the car body is $J_c = 2 \times 10^6$ kgm^2. Each magnet has a mass m_m = 1,000 kg. Table 1 lists the guideway parameters adopted in this study. Note that these values correspond to the Shanghai Transrapid test line [1]. In the moving element model, a total of 6 guideway spans are modeled. The element size is taken to be 0.124 m. The time step size is set to 0.018 s.

Table 1. Shanghai Transrapid test line guideway parameters.

Parameter	Bottom	Value
Length	L	12.368 m
Mass per unit length	ρA	4,500 kg/m
Flexural rigidity	EI	1.62×10^{10} Nm^2
Stiffness of bearing support	k_b	1×10^{20} N/m
Stiffness of rotational coupling	k_c	0 Nm/rad

Figure 6 shows the effect of stiffness of suspension and train speed on the maximum vertical acceleration of the train body. Note that in Fig. 6(a), the train is traveling at a constant speed of 300 km/h. As can be seen, the stiffness of suspension is found to be most effective in reducing the magnitude of the vertical acceleration of the train when it is below 10^5 N/m, while resonance is found to occur when the stiffness is increased to 10^8 N/m. When the stiffness is set to 4×10^5 N/m, the first peak in the vertical train acceleration is found to occur when the train speed is around 150 km/h and a monolithic increase in the acceleration is observed for train speed between 400 km/h and 600 km/h. Nevertheless, the maximum vertical acceleration observed in this study is below around 0.06 m/s2, which is much lower than the limit to ensure passenger ride comfort by the German Maglev Design Guide [11].

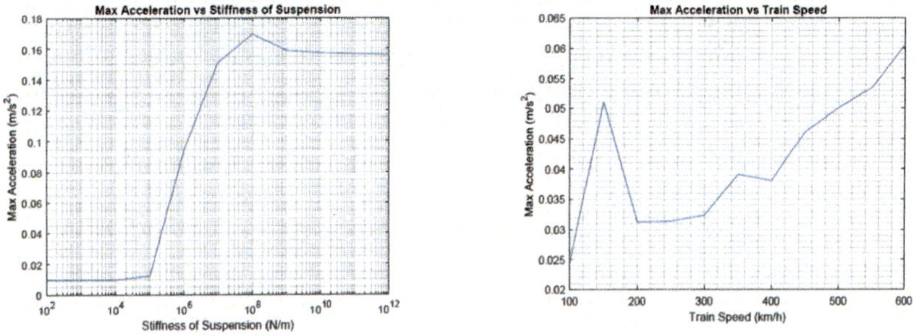

(a) (b)

Fig. 6. Effect of (a) stiffness of suspension and (b) train speed on maximum vertical acceleration of train body.

Figure 7 shows the effect of train speed on maximum vertical train acceleration when the train is traveling on guideways of different span lengths. Also shown in this figure are the theoretical resonant speeds V_{res} which are the product of the fundamental train frequency times the span length of the guideway. As can be seen, the train speed-induced resonant responses increase with the increase in the guideway span length. Moreover, a monolithic increase in the resonant acceleration is found with the increase in the span length. This is expected in view of the amplified guideway displacements when its span

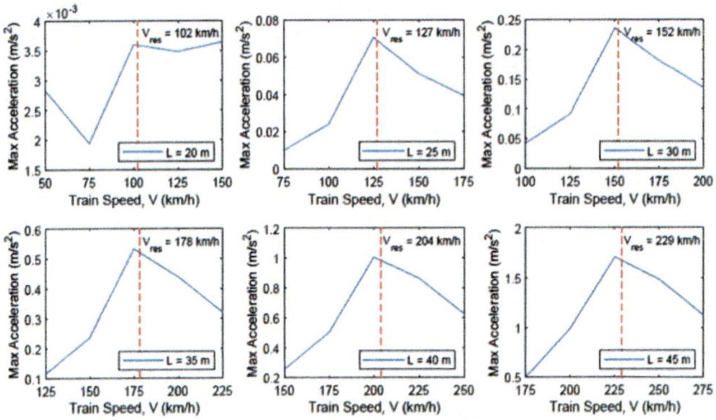

Fig. 7. Resonant speeds corresponding to different guideway span lengths.

length becomes too long. In this case, the span length should not exceed 35 m to ensure the ride comfort of the passengers. Otherwise, a stiffer guideway should be considered.

4 Conclusions

In this paper, a numerical scheme in conjunction with the moving element method is developed for the response analysis of a maglev train traveling on a multi-span guideway bridge. The accuracy of the moving element model is examined by comparison with the equivalent finite element model. Parametric studies are carried out to investigate the effect of various factors on the dynamic responses of the train. Results show that the dynamic response of the train body is sensitive to the traveling speed, suspension stiffness and the guideway span length. Further detailed studies are needed to ensure a comfortable, safe and cost-effective design and operation of the system.

Acknowledgement. This work is partially supported by Konnekt, the national competence center for transportation in Norway. Any opinions, findings, and conclusions or recommendations expressed in this paper are those of the authors and do not reflect the views of Konnekt.

References

1. Huang, J., Zhang, L.: Dynamic simulation and analysis of a high-speed maglev train/guideway interaction system. In: Tutumluer, E., Chen, X., Xiao, Y. (eds.) Advances in Environmental Vibration and Transportation Geodynamics. LNCE, vol. 66, pp. 505–526. Springer, Singapore (2020). https://doi.org/10.1007/978-981-15-2349-6_32
2. Cai, Y., Chen, S.S.: Vehicle/guideway dynamic interaction in Maglev systems. J. Dyn. Syst. Meas. Contr. **118**(3), 526–530 (1996)
3. Talukdar, R.P., Talukdar, S.: Dynamic analysis of high-speed MAGLEV vehicle–guideway system: An approach in block diagram environment. Urban Rail Transit **2**, 71–84 (2016). https://doi.org/10.1007/s40864-016-0039-8
4. Shi, J., Wang, Y.-J.: Dynamic response analysis of single-span guideway caused by high speed maglev train. Lat. Am. J. Solids Struct. **8**(3), 1–14 (2011)
5. Ang, K.K., Dai, J.: Response analysis of high-speed rail system accounting for abrupt change of foundation stiffness. J. Sound Vib. **332**(12), 2954–2970 (2013)
6. Luong, V.H., Cao, T.N.T., Reddy, J.N., Ang, K.K., Tran, M.T., Dai, J.: Static and dynamic analyses of Mindlin plates resting on viscoelastic foundation by using moving element method. Int. J. Struct. Stab. Dyn. **18**(11), 1850131 (2018)
7. Dai, J., Ang, K.K., Tran, M.T., Luong, V.H., Jiang, D.: Moving element analysis of discretely supported high-speed rail systems. J. Rail Rapid Transit **232**(3), 783–797 (2018)
8. Dai, J., Ang, K.K., Jiang, D., Luong, V.H., Tran, M.T.: Dynamic response of high-speed train-track system due to unsupported sleepers. Int. J. Struct. Stab. Dyn. **18**(10), 1850122 (2018)
9. Dai, J., Ang, K.K., Luong, V.H., Tran, M.T., Jiang, D.: Out-of-plane responses of overspeeding high-speed train on curved track. Int. J. Struct. Stab. Dyn. **18**(11), 1850132 (2018)
10. Dai, J., Han, M., Ang, K.K.: Moving element analysis of partially filled freight trains subject to abrupt braking. Int. J. Mech. Sci. **151**, 85–94 (2019)
11. Magnetschnellbahn. Ausführungsgrundlage Fahrweg: Teil II. Bonn. Eisenbahn-Bundesamt, Germany (2007)

Nonlinear "Beam Inside Beam" Model Analysis by Using a Hybrid Semi-analytical Wavelet Based Method

Piotr Koziol(✉)

Cracow University of Technology, ul. Warszawska 24, 31-155 Kraków, Poland
piotr.koziol@pk.edu.pl

Abstract. The previously developed model "beam inside beam" used for rail head vibrations analysis is extended by assumptions regarding viscoelastic nonlinear properties of the system. The coupled system of dynamic differential equations describes the beam-foundation structure subjected to distributed forces travelling along a beam. The upper beam, although arbitrarily distinguished, is not separated from the whole beam. This layer is considered as a vibration generator for the whole beam longitudinal axis. An assumption of nonlinear foundation stiffness is introduced as an important extension compared to existing solutions. This nonlinear system is solved by using a hybrid method based on the Adomian's decomposition combined with a wavelet based approximation. Although description of the problem was already presented before, a new solution along with computational examples are important contribution to the field and the main novelty of the paper. The theoretical parameters used in the analysis are taken from the literature and although being close to real structures they should be validated by experimental measurements. The aim of this parametrical study is to compare the new "beam inside beam" model with classical system of double-beam resting on nonlinear foundation. This work is left however for future study.

Keywords: Nonlinear "Beam Inside Beam" Model · Rail head vibrations · Semi-analytical solution

1 Introduction

There exist various approaches to modelling railway track dynamic behaviour. The most valued are analytical solutions due to their ability of effective parametrical analysis, although these need to be combined with heuristic or Delphi techniques in order to better explain real phenomena. Other group of models mainly consists of problems solutions based on semi-numerical FEM and related approaches. These, although giving solution to very complex problems, are time and power consuming. Hybrid methods appearing relatively rarely in the literature are kind of combination containing different computational methods. When taking into account analytical solutions, rail is usually modelled as a beam which leads to serious simplifications when it comes to the investigations of

Z. Dimitrovová et al. (Eds.): WMVC 2022, MMS 125, pp. 615–621, 2023.
https://doi.org/10.1007/978-3-031-15758-5_63

real structures. So, e.g. calculation of rail dynamic stresses fails when using such simplified model [1]. Hence other representation of rail is needed for better description of its dynamic features. One of recently proposed models is so called "beam inside beam" concept which treats the rail head as a beam separated from the whole rail but, at the same time, describes it as a part of rail submerged in its whole profile. The idea of nonlinear "beam inside beam" model was mentioned for the first time in [2], although it was not solved yet. The linear problem solution is presented in [3] showing that the head on web effect [4] can be confirmed by this model.

The solution is obtained by using the Fourier transform and several simplifying assumptions leading to semi-analytical results with regard to the dynamic response to loads generated by particular rail vehicles. In current paper, the solution is obtained by using a semi-analytical wavelet based approximation combined with the Adomian's decomposition applied to nonlinear term describing the rail support stiffness. This nonlinearity is represented by classical cubic monomial and shows mechanical properties of all layers under rail, including fastening system which also possesses nonlinear properties of stiffness [5]. Due to the complex variability of rail track components, difficult to experimentally determined, one can say that the assumed values of nonlinear coefficient are theoretical and should be estimated in future taking into account measurements of real structures and experimental studies, including numerical experiments. However this is left for further investigations, while in the present paper the convergent solution derivation allowing parametrical analysis is shown and considered as an important contribution to the field.

2 Model

The model considered in this paper is based on so called "double beam" which has been used in several earlier studies. The difference is that the change of parameters is strictly related to rail mechanical properties and the layer connecting two beams becomes very stiff as related to the rail head connection with the rest of rail. The model is described as a coupled system of two fourth order differential Eqs. (1a-b):

$$
\begin{aligned}
&EI_{rh}\frac{\partial^4 w_{rh}}{\partial x^4} + N_{rh}\frac{\partial^2 w_{rh}}{\partial x^2} + m_{rh}\frac{\partial^2 w_{rh}}{\partial t^2} \\
&+ c_{rh}\left(\frac{\partial w_{rh}}{\partial t} - \frac{\partial w_r}{\partial t}\right) + k_{rh}(w_{rh} - w_r) - k_{Nr}{w_r}^3 = P(x,t) \\
&EI_r\frac{\partial^4 w_r}{\partial x^4} + N_r\frac{\partial^2 w_r}{\partial x^2} + m_r\frac{\partial^2 w_r}{\partial t^2} + c_r\frac{\partial w_r}{\partial t} + k_r w_r \\
&+ k_{Nr}{w_r}^3 = c_{rh}\left(\frac{\partial w_{rh}}{\partial t} - \frac{\partial w_r}{\partial t}\right) + k_{rh}(w_{rh} - w_r)
\end{aligned} \tag{1}
$$

$w_{rh}(x,t)$ [m] and $w_r(x,t)$ [m] are the vertical vibrations of rail head and the whole rail, respectively; EI_{rh} [Nm^2], m_{rh} [kg/m] – bending stiffness and unit mass of rail head; EI_r [Nm^2], m_r [kg/m] – bending stiffness and unit mass of rail; k_{Nr} [N/m^4], k_r [N/m^2], c_r [Ns/m^2] – nonlinear part of stiffness, linear stiffness and viscous damping of the rail foundation (i.e. rail support including fastening system, sleepers, track bed and so on);

k_{rh} [N/m^2], c_{rh} [Ns/m^2] – linear stiffness and viscous damping of the rail head; $P(x, t)$ [N/m] – a set of loads $Q_k(x, t)$ generated by axles of train moving uniformly along rail with constant speed V [m/s]:

$$Q_k(x, t) = (P_C + \Delta P \cdot \exp(i(\Omega t + \varphi_k)))\frac{1}{2a}cos^2\left(\frac{\pi(x - Vt - s_k)}{2a}\right) \cdot H\left(a^2 - (x - Vt - s_k)^2\right) \quad (2)$$

where $H(.)$, $2a$, s_k and V are the Heaviside function, the span of single load, the distance of consecutive forces produced by wheels from the first axle and the speed of train, respectively. The frequency of load $\omega = \Omega + \varphi_k/t$ depends on position of wheel on the rail head irregularity and the term φ_k represents the phase shift associated with the particular axle position on the wavy imperfection that could be e.g. a consequence of rail deflection between sleepers. The term $P_C(x, t)$, constant in time, is generated by weight of train, while $\Delta P(x, t)$ represents forces arising from regular imperfections of rail track, like e.g. changing of foundation stiffness due to the rail deflection between sleepers. The single load density is described by the cos^2 function.

Generally, the single force can be described as a sum of three different terms:

$$P(x, t) = P_C(x, t) + P_D(x, t) + P_R(x, t, \gamma) \quad (3)$$

The terms $P_C(x, t)$ and $P_D(x, t)$ represent quasi-static load and a part of load changing in time in terms of irregularities possible to describe in deterministic analytical way, respectively, while the third term $P_R(x, t, \gamma)$ is a stochastic load generated by unexpected and unable to predict mechanical changes leading possibly to additional dynamic interactions dependent on random variable γ. This third factor is the subject of other studies and has been omitted in this paper.

3 Solution

The system of Eqs. (1a-b) has been solved by using a semi-analytical method based on wavelet approximation using coiflet filters [6–8]. The nonlinear term is represented by the Adomian series [9, 10]:

$$w_r{}^3(x, t) = \sum_{k=0}^{\infty} P_k(x, t) \quad (4)$$

where

$$P_k(x) = \frac{1}{k!}\left[\frac{d^k}{d\lambda^k}\left(\sum_{j=0}^{\infty}\lambda^j w_r(x, t)\right)^3\right]_{\lambda=0} \quad (5)$$

are the Adomian polynomials [8] for $k = 0, 1, 2, \ldots$.

After applying the Adomian approximation and the Fourier transform to each term of the Adomian series, the solution in the physical domain can be found as an appropriate term of approximating sequence defining a semi-analytical inversion of the Fourier transform $\tilde{w}$ [5–7]:

$$w_r(x) = \lim_{n\to\infty} w_{rn}(x), w_{rh}(x) = \lim_{n\to\infty} w_{rhn}(x) \tag{6}$$

$$w_{rn}(x) = \frac{1}{2^{n+1}\pi}\prod_{k=1}^{K(N)}(\sum_{j=0}^{3N-1} p_j e^{ijx/2^{n+k}})\sum_{k=-\infty}^{+\infty} \tilde{w}_r((k+M(N))2^{-n})e^{ixk2^{-n}} \tag{7}$$

$$w_{hrn}(x) = \frac{1}{2^{n+1}\pi}\prod_{k=1}^{K(N)}(\sum_{j=0}^{3N-1} p_j e^{ijx/2^{n+k}})\sum_{k=-\infty}^{+\infty} \tilde{w}_{hr}((k+M(N))2^{-n})e^{ixk2^{-n}} \tag{8}$$

where $M = \sum_{j=0}^{3N-1} jp_j$ and N is a degree of accuracy for applied set of coiflet coefficients (p_k).

Although the convergence of the sequences (6)–(8) is secured based on the Hilbert space theory, each of these solutions contains a truncated series produced by application of the Adomians decomposition. Therefore the solution error should be controlled as it depends on several mechanical and computational parameters. This can be done with regard to the Adomian's part by the error index [5].

$$w_{hr\,n}^{\ er} = \frac{\left|\|S_n^{w_{hr}}\| - \|S_{n-1}^{w_{hr}}\|\right|}{\|S_n^{w_{hr}}\|}, w_{r\,n}^{\ er} = \frac{\left|\|S_n^{w_r}\| - \|S_{n-1}^{w_r}\|\right|}{\|S_n^{w_r}\|} \tag{9}$$

$$0 \le \frac{\|w_{hrj+1}\|}{\|w_{hrj}\|} < 1, 0 \le \frac{\|w_{rj+1}\|}{\|w_{rj}\|} < 1 \tag{10}$$

For the terms of the Adomian series $j = 0, 1, 2, \ldots$. $S_n^{w_{hr}}$ and $S_n^{w_r}$ denote the nth partial sum of the approximate solution in physical domain. Conditions (9)–(10) allow controlling the solution with regard to the assumed level of accuracy.

The proper number of function in sequences (6)–(8) possible to consider as an approximate solution representing the system response closely enough must be chosen on the basis of the stabilization criterion in relation to all parameters involved. One must remember that the applied wavelet based approach gives a shape approximation and matching checking at single point might not secure the whole function usefulness. Therefore several computational examples must be done before selecting the appropriate approximation order. One can show that for the considered mechanical system, the 4th terms of sequences (6)-(8) are sufficiently accurate, i.e. the 5th order of the wavelet approximation can be found suitable for the system of parameters used in in this paper.

4 Computational Examples

The mechanical parameters, especially those describing rail head and its foundation properties, used in computational analysis are estimated theoretically and should be validated in order to reflect real structures behaviour [3, 11]:

1. Rail 60E1: Young modulus $E = 2.1*10^8$ kN/m^2; moment of inertia in the vertical plane $I_r = 3055*10^{-8}$ m^4; unit mass of rail $m_r = 60$ kg/m; longitudinal force in rail axis $N_r \equiv 0$.
2. Rail head: moment of inertia in the vertical plane $I_{rh} = 43.84*10^{-8}$ m^4; unit mass of the rail head $m_{rh} = 23.3$ kg/m, longitudinal force in rail head axis $N_{rh} \equiv 0$;
3. Rail head foundation: stiffness $k_{rh} = 0.19*10^{10}$ N/m^2, damping coefficient $c_{rh} = 20$ Ns/m^2;
4. Rail foundation: stiffness $k_r = 51*10^6$ N/m^2, damping coefficient $c_r = 3950$ Ns/m^2, nonlinear stiffness $k_{Nr} = 1.5*10^{18}$ N/m^4;
5. The load constant in time: generated by the coal wagon Falns 441 VA with quasi-static load $P_C = 112\,500$ N and axles configuration 0 m, 1.8 m, 5.04 m, 6.84 m;
6. The load changing in time: $P_D(x, t) = \Delta P(x, t) = 0.18 \cdot P_C(x, t)$ which corresponds to irregularities similar to a rail deflection between sleepers [12].
7. The speed of the vehicle $V = 50$ km/h.

Figure 1 is an example of the system vertical vibration for the considered nonlinearity, showing that the nonlinear solution is possible. One can notice that the response of the rail head and the whole rail differ significantly in terms of maximal deflections at the forces application points and the dynamic behaviour, especially in the area between axles.

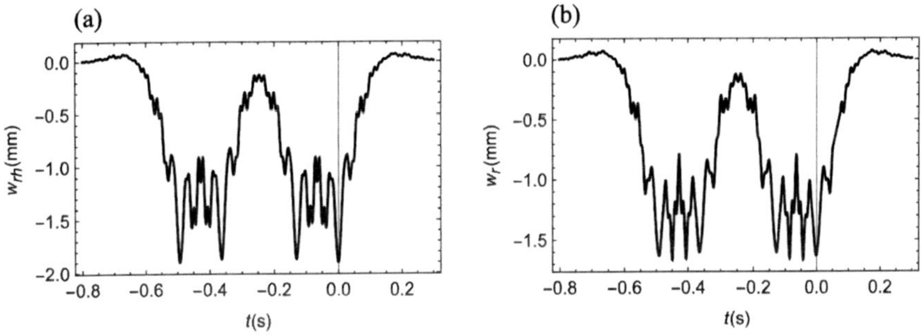

Fig. 1. Vertical vibrations of the system: (a) rail head; (b) whole rail.

Figure 2 presents a direct comparison of linear and nonlinear solutions for the first two axles of the considered vehicle. The differences between linear and nonlinear responses can be easily seen. The "head on web" effect is also confirmed with an assumption of the nonlinear stiffness of rail foundation. It is worth to answer if this effect is increased or diminished along with nonlinearity introduction. For this purpose, the differences of the layers (beams) maximal responses are examined in both cases (Fig. 3). One can see that the "head on web" effect becomes stronger in the case of nonlinear system, again especially in the time when two closest axles of the vehicle are passing by the rail surface. This is with accordance to recognized previously phenomenon of the nonlinearity accumulation in the case of decreasing distance between moving forces, when one considers multilayer rail track systems. In the case of linear system, the "head on web" effect is significantly stronger directly in place of the force application, i.e. at the contact point between wheel and rail (see Figs. 1, 2 and 3).

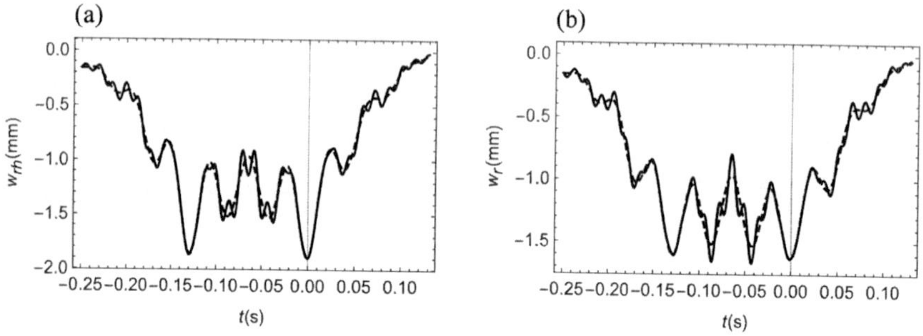

Fig. 2. "Zoomed" vertical vibrations of the system for the first two axles (dashed – linear; solid – nonlinear): (a) rail head; (b) whole rail.

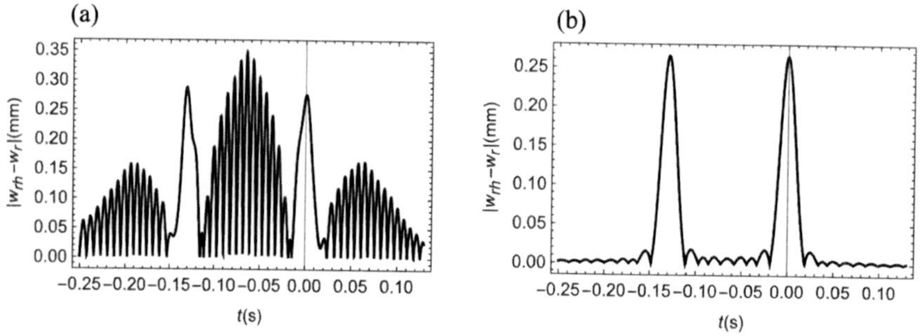

Fig. 3. "Head on web" effect reflected in the maximal response differences between the rail head and the whole rail: (a) nonlinear; (b) linear.

5 Conclusions

The two layer system representing the rail with distinguished rail head is solved under assumption of nonlinear rail foundation stiffness. The "head on web" effect has been confirmed in both cases, linear and nonlinear. Computational examples show that the considered nonlinearity makes the investigated phenomenon stronger and both solutions, linear and nonlinear, differ significantly one from another. These differences become stronger when the distance between moving forces decreases. Although the deflection differences are relatively small for the considered system of parameters, further studies are needed for better recognition of the nonlinearity influence on the system response. The obtained semi-analytical solution makes possible the parametrical analysis of the proposed "beam inside beam" model.

References

1. Koziol, P.: Analytical approximation of rail bending stress. MATEC Web of Conferences **148**, 05002 (2018)
2. Czyczula, W., Koziol, P., Urbanek, M.: Effect of Rail Head Vibrations: Nonlinear "Beam Inside Beam" Model. Proceedings of ICEDyn2017. International Conference on Structural Engineering Dynamics. Ericeira. Portugal, July 3–5 (2017)
3. Czyczuła, W., Błaszkiewicz, D., Urbanek, M.: Rail with separated head, New Approach to Analysis of Railway Track Dynamics – Rail Head Vibrations, Open Engineering (2021)
4. Orringer, O., Morris, J.M., Joeng, D.Y.: Detail fracture growth in rails: test results. Theoret. Appl. Fract. Mech. **5**(2), 63–95 (1986)
5. Koziol, P., Pilecki, R.: Nonlinear double-beam system dynamics. Arch. Civ. Eng. **67**(2), 337–353 (2021)
6. Koziol, P.: Wavelet Approach for the Vibratory Analysis of Beam-Soil Structures: Vibrations of Dynamically Loaded Systems. VDM Verlag Dr. Müller, Saarbrucken (2010)
7. Koziol, P.: Experimental validation of wavelet based solution for dynamic response of railway track subjected to a moving train. Mech. Syst. Signal Process. **79**, 174–181 (2016)
8. Monzon, L., Beylkin, G., Hereman, W.: Compactly supported wavelets based on almost interpolating and nearly linear phase filters (coiflets). Appl. Comput. Harmon. Anal. **7**, 184–210 (1999)
9. Adomian, G.: Nonlinear Stochastic Systems Theory and Application to Physics. Kluwer Academic Publishers, Dordrecht (1989)
10. Wazwaz, A.M., El-Sayed, S.M.: A new modification of the Adomian decomposition method for linear and nonlinear operators. Appl. Math. Comput. **122**, 393–405 (2001)
11. Pietrus M.: Comparative analysis of rail vertical deflection during train run, obtained using single-layer and „beam to beam" models, under assumption of wavy wear of the rail running Surface (in Polish) BSc thesis, Cracow University of Technology, (supervisor: Piotr Koziol) (2021)
12. Czyczula, W., Koziol, P., Kudla, D., Lisowski, S.: Analytical evaluation of track response in the vertical direction due to a moving load. Journal of Vibration and Control, Volume: 23 issue: 2989–3006 (2017)

Reconstruction of Road Defects from Dynamic Vehicle Accelerations by Using the Artificial Neural Networks

Kais Douier[1](✉), Mohammed F. M. Hussein[1], and Jamil Renno[2]

[1] Department of Civil and Architectural Engineering, Qatar University, Doha, Qatar
kd1204127@student.qu.edu.qa, mhussein@qu.edu.qa
[2] Department of Mechanical and Industrial Engineering, Qatar University, Doha, Qatar
jamil.renno@qu.edu.qa

Abstract. Monitoring of roads is considered the first step in establishing a successful road maintenance program, which includes scheduling adequate maintenance to a certain road section at the right time. Road monitoring assesses the road's condition for later analysis, which helps the prioritization of maintenance activities. Optimizing the process of scheduling and prioritizing maintenance activities is driven by cost-efficiency and is critical for preserving the overall road networks' health and smooth operation. However, the implementation of road monitoring, using traditional equipment such as profilometers, total stations, and automated road meters, could be exceedingly time-consuming. This research proposes the usage of Artificial Neural Networks (ANNs) and dynamic vehicle accelerations to reconstruct road defects in a time/cost-effective manner. A multi-degree of freedom numerical model is used to simulate the dynamics of a vehicle passing on various road defects such as potholes and speed bumps. These road defects are reconstructed by using two different methods. The first method employs a frequency domain approach to inversely reconstruct the road defects numerically. The second method uses ANNs to reconstruct the road defects with time histories of the vehicle's acceleration as the input to the network. The ANN model was trained by dynamic vehicle accelerations as inputs and various road defect profiles as outputs. Both methods were later compared on the basis of their accuracy and efficiency, and the ANN method was found to be more promising for implementation on experimental data.

Keywords: Road defects · Vibrations · Artificial neural network · Machine learning

1 Introduction

Road networks are the arteries of any country, that provide the residents the support they need to transport goods and move freely within its boundaries. Maintaining the health and sustainability of such huge infrastructure is a critical task to ensure safe and smooth living conditions in the country. However, these road networks are destined to deteriorate over the years due to normal usage and other external environmental conditions [1]. Thus,

Z. Dimitrovová et al. (Eds.): WMVC 2022, MMS 125, pp. 622–629, 2023.
https://doi.org/10.1007/978-3-031-15758-5_64

effective constant monitoring is needed to ensure the road's safe functionality is properly maintained. Also, by constantly monitoring the health of these road networks, proper timely actions could be taken to provide the correct cost-effective maintenance required [2].

Manual monitoring of the surface pavement is very time-consuming and extremely dependent on the skill of the observer [3–5]. While automated monitoring methods, such as automated profilographs, might cut down the monitoring time, they remain an expensive substitute to manual monitoring [6–8]. Thus, highway agencies and researchers have attempted to figure out a method that can be less time consuming by using specially designed trucks that have high-resolution cameras and lasers to monitor the roads for road defects such as potholes, transverse cracking, fatigue, rutting, etc. [9–12]. Nevertheless, mounting high-resolution cameras and sensors on a specially designed truck requires capital and operational expenditure which makes it expensive when compared to manual monitoring. Researchers have also tried using unmanned air vehicles (UAVs) to train artificial neural networks (ANNs) and convolutional neural networks (CNNs) to monitor certain surface defects [13, 14]. Although these methods have the potential to detect surface distress in a time and cost-effective way, these methods still lack the degree of accuracy to be implemented in highway management systems.

A study was conducted by Zhang et al. [15] to classify road roughness classes from connected cars' dynamic accelerations, where the authors concluded that the use of neural can achieve a 90% accuracy in classifying road roughness's.

A numerical study was developed by Ngwangwa et. al [16] in an attempt to reconstruct the road roughness profile and road defects using dynamic vehicle accelerations and ANNs. The authors concluded that the use of ANNs requires less system characterization to be defined and fewer analytical skills to be implemented.

This paper aims towards predicting road defects, such as speed bumps and potholes, on a randomly generated road roughness profile using ANNs trained with dynamic vehicle accelerations. We will also reconstruct the road profile with its defects using the frequency domain method and compare the accuracy and computational time between both methods.

2 Methodology

This research proposes the use of ANNs by utilizing the dynamic vehicle accelerations to reconstruct road defects such as potholes and speed bumps. The ANN used in this research was trained by the accelerations exported from a simulated quarter car model (two degrees of freedom), that passes on randomly generated road roughness profiles with speed bumps and/or potholes. The random road roughness profiles were generated in accordance with the ISO 8608 standard [17], while the surface defects were adapted from reference [16]. The simulated sprung and unsprang masses' accelerations, exported from the quarter car model, were then used as inputs to the ANN and the road profiles with their surface defects were taken as the outputs/targets to the ANN.

2.1 Two Degrees of Freedom Numerical Model

The quarter car model was chosen in this study due to its simplicity, and since it has been used reliably to examine the dynamic behavior of a moving vehicle [18]. The quarter car model used in this study consists of sprung and unsprang masses, as shown in Fig. 1, these masses are used to simulate the behavior of the vehicle's body and wheel, respectively.

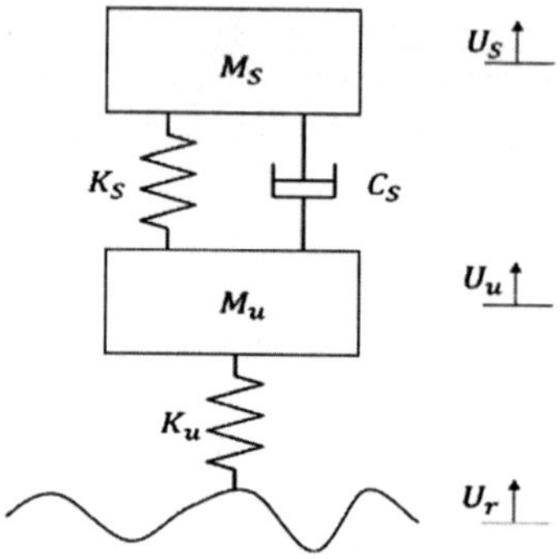

Fig. 1. Illustration of a quarter car model moving on random roughness profile.

The quarter car model's simplicity helps in preserving computational time and power, as shown by its governing equations presented in Eqs. (1) and (2) [18].

$$M_S\ddot{U}_S + C_S(\dot{U}_S - \dot{U}_u) + K_S(U_S - U_u) = 0 \tag{1}$$

$$M_u\ddot{U}_u - C_S(\dot{U}_S - \dot{U}_u) - K_S(U_S - U_u) + K_u U_u = K_u U_r \tag{2}$$

where C, K, and M correspond to the damping, stiffness, and mass of the system, respectively, while the subscripts s and u relate each system property to the sprung and unsprang division of the system. The terms $\ddot{U}$, $\dot{U}$, and U correspond to the dynamic acceleration, velocity, and displacement of the system, while the term U_r describes the road profile in the time domain. The above-mentioned equations, Eqs. (1) and (2), are then transformed to a set of ordinary differential equations in the form of state-space matrix representation as shown in Eq. (3), where $Z_1 = [U_S\ U_u]^T$, $Z_2 = [\dot{U}_S\ \dot{U}_u]^T$, $\dot{Z}_1 = Z_2$, $\dot{Z}_2 = [\ddot{U}_S\ \ddot{U}_u]^T$ and $F = K_u U_r$. Equation (3) was then solved using MATLAB ODE 45 function.

$$\begin{bmatrix} \dot{Z}1 \\ \dot{Z}2 \end{bmatrix} = \begin{bmatrix} 0_{2X2} & I_{2X2} \\ -M^{-1}K & -M^{-1}C \end{bmatrix} \begin{bmatrix} Z1 \\ Z2 \end{bmatrix} + \begin{bmatrix} 0_{2X1} \\ -M^{-1}F \end{bmatrix} \tag{3}$$

2.2 Artificial Neural Network (ANN)

A Levenberg-Marquardt trained neural network was used in this study to reconstruct a randomly generated road roughness profile with surface defects, such as speed bumps and potholes. The Levenberg-Marquardt training algorithm was chosen because it is generally faster in training ANNs when compared to other methods, due to approximating the Hessian matrix rather than solving for it [19].

The ANN utilized consists of 100 input neurons, 20 hidden tan-sigmoidal neurons, and 50 output tan-sigmoidal neurons, as shown in Fig. 2. The input neurons are fed by the sprung and unsprang acceleration ($[\ddot{U}_S\ \ddot{U}_u]^T$) of the simulated quarter car model traveling on a road profile (U_r) with random roughness, speed bumps, and/or potholes. While the output neurons try to predict the road profile with its roughness and surface defects. This ANN could be represented numerically by $O = f^a\ (W^h f^a + b^h)$, where W^h and b^h represent the weight and bias matrices of the ANN, O represents the output neurons, and f^a represents the tan-sigmoidal activation function [19].

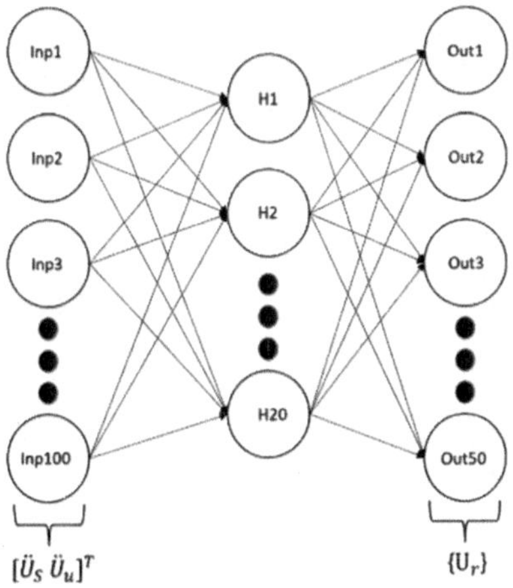

Fig. 2. Illustration of artificial neural network structure.

A library of inputs (dynamic vehicle accelerations) and output (road profile) was generated to train the ANNs on predicting the road profile with random roughness, speed bumps, and potholes. Each road profile consists of 50 data points, which were then plugged into the quarter car model to yield 50 data points for each dynamic vehicle accelerations (sprung and unsprang accelerations).

Two road roughness classes (class A and C) were chosen in accordance with ISO 8608 [17] to initially investigate the effect of the road roughness on the ANN ability to reconstruct a road profile with surface defects. A total of 2000 road profiles were generated for each road roughness class, and each profile contains either a speed bump or a pothole randomly placed throughout the profile. The speed bumps had a span of 1 m, and a height of 0.3 m, while the pothole had a span of 0.5 m and a depth of 0.04 m.

2.3 Frequency Domain Method

This method simply transforms the governing equations, Eq. (2), of the quarter car model from time to frequency domain using Fourier transformation. Thus, simplifying the equation to only having one unknown, which is the road profile ($\tilde{U}_r$) in the frequency domain, as shown in Eq. (4), where $\tilde{U}_u$, $\tilde{U}_s$, and $\tilde{U}_r$, are the sprung, unsprang, and road profile displacement in the frequency domain respectively, and ω is the angular frequency.

$$-\omega^2 M_u \tilde{U}_u - i\omega C_S\left(\tilde{U}_S - \tilde{U}_u\right) - K_S\left(\tilde{U}_S - \tilde{U}_u\right) + K_u \tilde{U}_u = K_u \tilde{U}_r \tag{4}$$

3 Results and Discussion

A single ANN was trained for each road roughness class to predict the road profile with its surface defects, such as speed bumps and potholes. Each network was then tested by a randomly generated profile containing either a speed bump or a pothole at a distance of 5 m. The randomly generated profile consisted of 1000 data points, that were divided into 20 sections, and each section was fed into the same ANN to produce the final full-length prediction of the road profile. Figure 3 presents the results of testing the ANN trained by class A road roughness profiles with speed bumps and potholes. The ANN was able to predict these road profiles having a pothole and a speed bump with correlation factors of 97.09% and 99.66%, respectively, and with root mean square error (RMSE) of 0.18% and 0.24%, respectively.

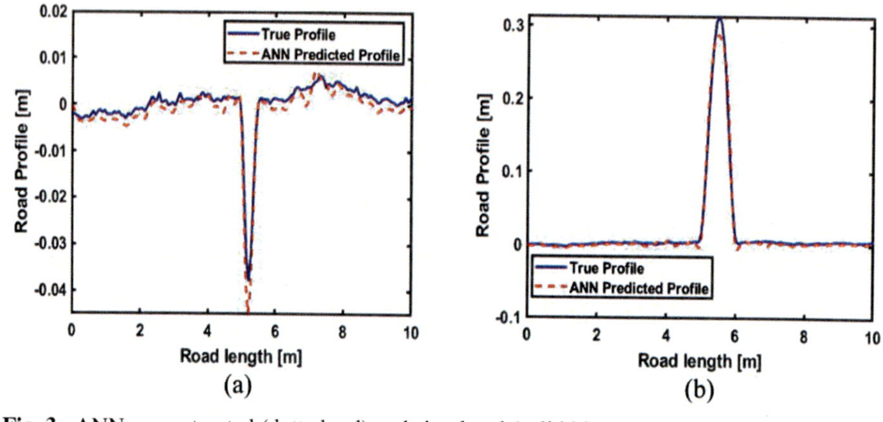

Fig. 3. ANN reconstructed (dotted red) and simulated (solid blue) class A road profile with: (a) pothole, and (b) speed bump.

Figure 4 presents the results of testing the ANN trained by class C road roughness profiles with speed bumps and potholes. The ANN was able to predict these road profiles containing a pothole and a speed bump with correlation factors of 98.23% and 99.48%, respectively, and with RMSE of 0.27% and 0.31%, respectively. The ANN was able to yield out the previous results with an average computational time of 0.9983 s while running on a CPU with 8 cores and a base processor speed of 2.9 GHz.

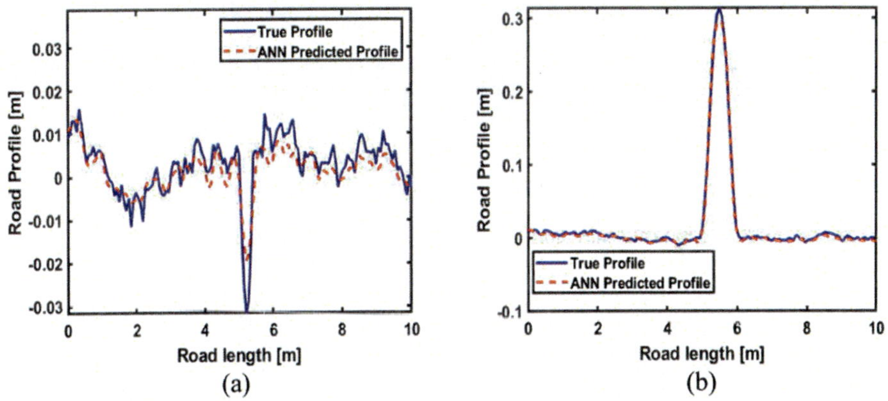

Fig. 4. ANN reconstructed (dotted red) and simulated (solid blue) class C road profile with: (a) pothole, and (b) speed bump.

Figure 5 demonstrates the frequency domain method's ability in reconstructing road profiles with class A road roughness, speed bumps and potholes. The frequency-domain method was able to reconstruct a pothole and a speed bump road profile with a correlation factor of 99.57%, and 99.98%, respectively, and with RMSE of 0.12%, and 0.06%, respectively.

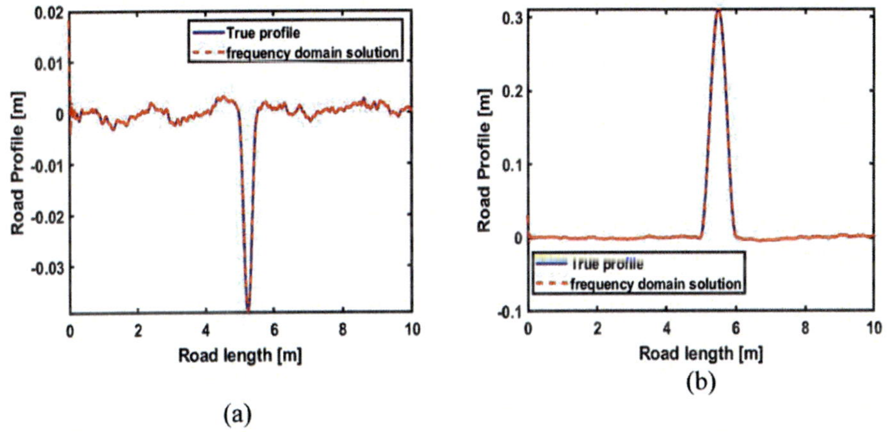

Fig. 5. Frequency domain method reconstructed (dotted red) and simulated (solid blue) class A road profile with: (a) pothole, and (b) speed bump.

Figure 6 shows the frequency domain method's ability in reconstructing road profiles with class C road roughness, speed bumps and potholes. The frequency-domain method was able to reconstruct a pothole and a speed bump road profile with a correlation factor of 96.30%, and 99.98%, respectively, and with RMSE of 0.56%, and 0.05%, respectively. The frequency-domain method was able to yield out the previous results with an average computational time of 2.3253 s while running on the same setup as the ANN.

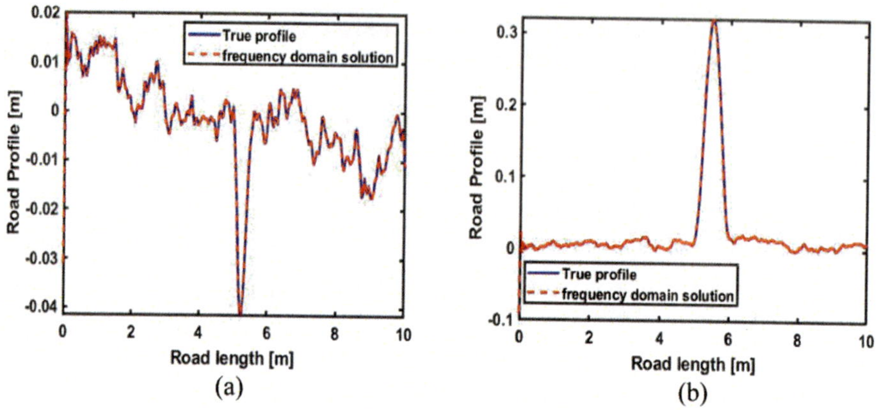

Fig. 6. Frequency domain method reconstructed (dotted red) and simulated (solid blue) class C road profile with: (a) pothole, and (b) speed bump.

4 Conclusion

The viability of using ANN to inversely reconstruct road profiles using dynamic vehicle accelerations was investigated and demonstrated in this paper. The frequency-domain method was also used to reconstruct the road profiles for later comparison. Mainly two surface defects, a pothole and a speed bump, were used in this paper. Although the frequency domain method showed dominance in reconstructing the road profiles for both surface defects accurately, the computational time utilized by the frequency domain method was 2.3 times larger than the computational time utilized by the ANN. Also, the effect of road roughness classes on the ANN ability to reconstruct road profiles with surface defects was barely noticeable in this research. Further investigation is needed to quantify the effect of different road roughness classes and various geometrically shaped surface defects on the ANN ability to reconstruct road profiles using dynamic vehicle accelerations.

Acknowledgement. The authors would like to acknowledge funding from Qatar University through a Graduate Assistantship. The authors would also like to acknowledge funding from Qatar Rail through grant number QUEX-CENG-QR-21/22–1.

References

1. Du, Y., Liu, C., Wu, D., Li, S.: Application of vehicle mounted accelerometers to measure pavement roughness. Int. J. Distrib. Sens. Networks (2016)
2. Chin, M.J., Babashamsi, P., Yusoff, N.I.M.: A comparative study of monitoring methods in sustainable pavement management system. IOP Conf. Ser. Mater. Sci. Eng. **512**(1), 012039 (2019)
3. Shtayat, A., Moridpour, S., Best, B., Shroff, A., Raol, D.: A review of monitoring systems of pavement condition in paved and unpaved roads. J. Traffic Transp. Eng. (English Ed. **7**(5), 629–638 (2020)

4. Markow, M.J.: Highway management systems: state of the art. J. Infrastruct. Syst. **1**(3), 186–191 (1995)
5. Benson, K.R., Elkins, G.E., Uddin, W., Hudson, W.R.: Comparison of Methods and Equipment to Conduct Pavement Distress Surveys (1999)
6. Haas, R., Hudson, W.R., Falls, L.: Pavement Management Development from 2010. In: Pavement Asset Management. John Wiley & Sons, Ltd, pp. 15–20 (2015)
7. Kulkarni, R.B., Miller, R.W.: Pavement management systems: past, present, and future. Transp. Res. Rec. **1853**(1), 65–71 (2003)
8. Peraka, N.S.P., Biligiri, K.P.: Pavement asset management systems and technologies: A review (2020)
9. Amador-Jimenez, L., Matout, N.: A low cost solution to assess road's roughness surface condition for Pavement Management (2014)
10. Becerik-Gerber, B., Masri, S.F., Jahanshahi, M.R.: An Inexpensive Vision-Based Approach for the Autonomous Detection, Localization, and Quantification of Pavement Defects. Innov. Deserving Explor. Anal. **169** (2015)
11. Sreedhar, S., Coleri, E.: The effect of long-term aging on fatigue cracking resistance of asphalt mixtures. Int. J. Pavement Eng. 1–13 (2020)
12. Miller, J.S., Bellinger, W.Y.: FHWA, Distress Identification manual for the Long-Term Pavement Performance Program. Report FHWA-HRT-13–092 (2014)
13. Zhang, C., Elaksher, A.: An unmanned aerial vehicle-based imaging system for 3D measurement of unpaved road surface distresses. Comput. Aided Civ. Infrastruct. Eng. **27**, 118–129 (2012)
14. Khoudeir, M., Brochard, J., Legeay, V., Do, M.-T.: Roughness characterization through 3D textured image analysis: contribution to the study of road wear level. Comput. Civ. Infrastruct. Eng. **19**(2), 93–104 (2004)
15. Zhang, Z., Sun, C., Bridgelall, R., Sun, M.: Application of a machine learning method to evaluate road roughness from connected vehicles. J. Transp. Eng. Part B Pavements **144**(4), 04018043 (2018)
16. Ngwangwa, H.M., Heyns, P.S., Labuschagne, F.J.J., Kululanga, G.K.: Reconstruction of road defects and road roughness classification using vehicle responses with artificial neural networks simulation. J. Terramechanics **47**(2), 97–111 (2010)
17. I. 8608:2016(E): International Standard ISO Mechanical Vibration — Road surface profiles — Reporting of measured data (2016)
18. Gillespie, T.: Fundamentals of Vehicle Dynamics. Engineers, Society of Automotive, INC, Warrendale (1992)
19. Mathworks INC, MATLAB Help Tutorial, 2021 https://ch.mathworks.com/help/deeplearning/ref/trainlm.html?s_tid=srchtitle_trainlm_1 Accsessed 12 Jan 2022

Solving Groan Noise Problems in a Metro Braking System

Gianluca Megna(✉) and Andrea Bracciali

Dipartimento di Ingegneria Industriale, Università di Firenze, Florence, Italy
gianluca.megna@unifi.it

Abstract. A relatively standard brake system used on a metro vehicle showed abnormal noise and vibrations in service. Beyond passenger's annoyance, this led to both brake calliper levers failures and premature wear of brake pads.

No similar problems appeared during test benches performed by the manufacturer before and a reinforced version of the levers prevented failures nevertheless leaving unchanged high noise and vibration effects during braking. Several pads were tested without success.

The paper shows how the problem was solved by means of a dynamic analysis of all the components involved and a minor redesign of a component. The impact on the existing fleet in terms of retrofit costs and times in described as well.

Keywords: Braking · Groan noise · Modal analysis · Structural modification · Stiffening · Softening

1 Introduction

Friction braking introduces specific features in any vehicle. Degraded dynamics, thermal effects, higher noise and vibrations and wear are amongst the factors that are a side effect of the desired energy dissipation through the sliding contact between a dedicated component (the brake pad) and a rotating counterpart (the wheel or a disc brake).

The intrinsically nonlinear behaviour of the brake pad/disc is well described in the available literature for both the road and rail vehicles. Specifically, the falling friction characteristics of the friction force vs. sliding speed curve above the adhesion limit (maximum force without macro slip) leads to typical stick-slip phenomena (sometimes also called sprag-slip phenomena). A simple example of such motions can be easily observed when writing with a long (i.e. low stiffness) chalk on a blackboard. In this case the screech produced is due to the sprag-slip phenomenon.

Braking noise can be grouped in two main classes [1]:

- squeal noise, with high frequency content (typically > 1 kHz);
- groan noise, with frequency content at lower frequencies (typically < 1 kHz).

Z. Dimitrovová et al. (Eds.): WMVC 2022, MMS 125, pp. 630–637, 2023.
https://doi.org/10.1007/978-3-031-15758-5_65

The two mechanisms are very different in terms of generation and consequences. Squeal noise involves an out-of-plane movement of the disc, which vibrates at its eigenfrequencies generating an extremely high pitch and high level noise. Typically, squeal noise is extremely annoying but has limited consequences on structural components as the amplitude of vibrations is tiny and not such to generate noticeable stress increase in the braking components.

Groan noise is instead a motion in which the disc shows an in-plane movement as the brake pad is the responsible for the motion. In many cases, the disc can be considered as perfectly rigid and all the flexibility can be attributed to brake pads, pad holders and brake calliper levers.

A low-frequency noise was observed in a trailer bogie of a metro vehicle. The characteristics of the highly tonal noise, centred around 250 Hz, led to the conclusion that the noise was of the groan type. High vibrations led after a short time to failures (cracks) in the calliper components, while brake pads material often crumbled, showing inadmissible life shortening. This paper deals with the activities performed to individuate the origin of such high abnormal vibrations and describes how the problem was solved.

2 Experimental Data Analysis

Available data from already performed tests on a brake calliper mounted on a dynamometric bench resulted in useless conclusions, as no abnormal vibrations were observed. This is not surprising as the system created on the bench does not reproducing the actual bogie.

A thorough vibration measurement campaign was then designed and performed on the vehicle with the sensor arrangement shown in Fig. 1. Four components potentially are involved in the generation of the phenomenon:

- the bogie frame, which is made of welded steel sheets and profiles, with the interfaces necessary to connect all the attached equipment, brake unit included;
- the so-called brake support, i.e. a steel casted components bolted to the bogie frame with an "adjuster" function, i.e. making the bogie frame connectable to different brake units;
- the brake unit, suspended with links bolted to the brake support;
- the wheelset, including axle, wheels and brake discs.

Regardless of operating conditions the same behaviour was observed, i.e. very loud noise and high vibrations even reaching the carbody. Selected time histories showed that abnormal vibrations in the braking elements are completely uncoupled from axlebox vibrations (Fig. 2). As the primary suspension filters out effectively wheel vibrations, abnormal brake unit vibrations cannot be ascribed to track irregularities.

Fig. 1. Top left: general view of the vehicle components. Top right: view of the braking unit with indication of accelerometers used during testing. Bottom: Accelerometers mounted on the brake unit.

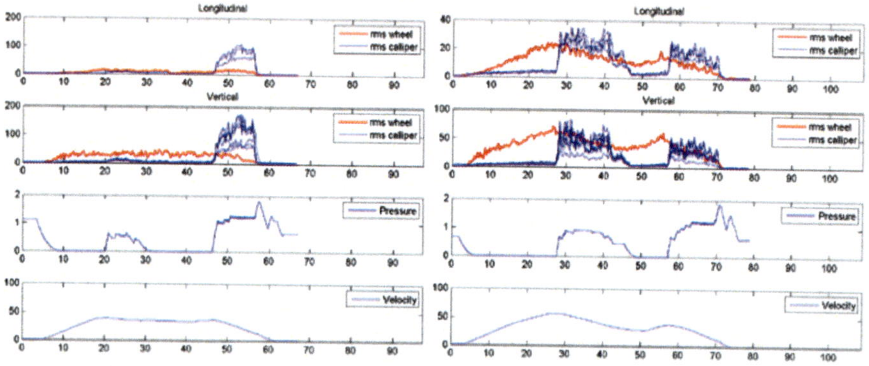

Fig. 2. Accelerations recorded during test runs. Left: good quality track. Right: bad quality track. From top to bottom: longitudinal RMS vibrations on wheel and calliper, vertical RMS vibrations on wheel and calliper, brake cylinder pressure, speed.

A first analysis conducted using the EN 61373 standard [2] showed that vibration levels recorded are below the limits stated in the standard except for a high peak at around 230 Hz. The limit of the frequency range imposed by the standard (250 Hz), made any higher frequency analysis impossible. This is a well-known limit of these curves, as the standard considers only the resonances of the fixtures during testing of railway components, and in this case it is not easy to define the fixture: all the components may exhibit (and in fact they do exhibit) an elastic behaviour and are possibly subjected to resonances.

Further analyses were then performed looking at narrow-band spectra rather than at power spectral densities. Acceleration data was integrated twice to estimate displacements. If in the original data the contribution of the harmonics following the fundamental frequency looks important, after the integration process it can be said that the only dominant frequency is the fundamental one, around 220 ÷ 230 Hz. Short clips obtained from calliper geometry and global displacements time histories showed that the two sides of the brake calliper move in counterphase (Fig. 3).

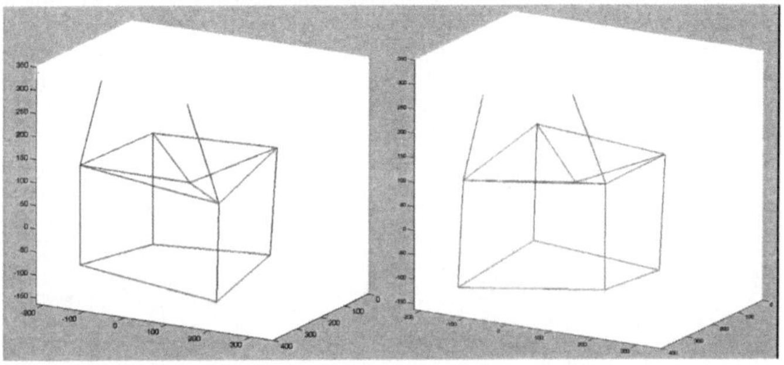

Fig. 3. Two frames of the animation of the points recorded on the calliper, seen from the brake cylinder side. The vertical segments are the links connecting the pad holders to the brake support.

3 Modal Analysis of the Wheelset

Simplified dynamic response measurements, similar to those shown in [3], were performed on a spare wheelset resting on soft rubber blocks placed under the axleboxes by using an impact hammer in vertical and longitudinal direction in the centre of the axle (Fig. 4), allowing to identify the first natural modes of the free wheelset.

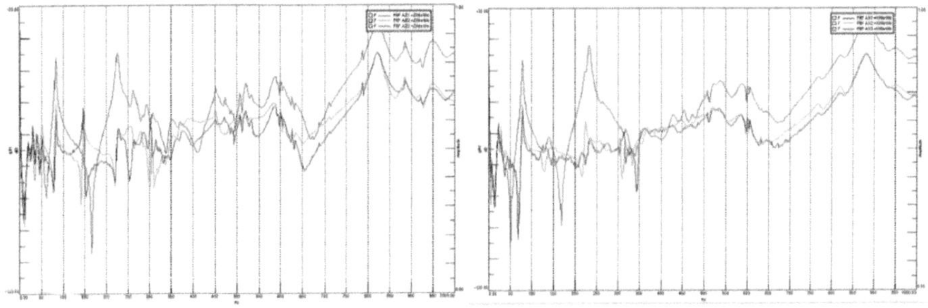

Fig. 4. Vertical (left) and longitudinal (right) point FRFs. The first two peaks are at around 80 Hz and 230 Hz in both FRFs.

A finite element model of the wheelset was then prepared, including the details of the five-sectors brake discs. The obtained numerical modal basis was used to estimate with two strategies the harmonic response up to 1 kHz (see the comparison with the measured point FRF in Fig. 5). The two estimates consider a damping of $\xi = 0.001$ (red) and $\xi = 0.01$ (blue), while the frequency resolution was kept *variable* (optimized, red) and *fixed* to $\Delta f = 1$ Hz (blue). The latter case found antiresonances that were "skipped" by the optimized method. No attempt was performed to tune the amplitude by changing the damping, as for the scope of the present research the knowledge of resonance frequencies was considered sufficient. The numerical model reproduces sufficiently well the real behaviour of the wheelset, at least for nominal conditions (i.e. new wheels and new brake discs). The forced response analysis allowed to identify all the modes that are potentially involved in the problem.

4 Identification of the Root Cause

The brake unit is designed to give the same force to either side of the brake disc during all its life. Inertial and elastic forces at even low frequencies break this equilibrium. Three motions of the disc and the resulting interaction with the pads, the pad holders and the links connecting the brake unit to the brake support were identified (Fig. 6):

- when the disc moves in the longitudinal direction, there is no change in the braking force;
- when the disc moves in the vertical direction, the change of the braking force on both sides of the calliper is identical (in phase) and of limited amplitude unable to stick-slip phenomena;
- when the disc moves in the lateral direction, the pads and the pad holders are shaken by the movement leading to large variations of the normal force. As a results, similar changes on the tangential forces are generated, with opposite phase ("sharing forces").

The vibrations observed during tests are compatible with the latter movement. It was therefore preliminary concluded that the generation of counterphase vibrations on the brake support is given by sharing forces which, in turn, are generated by out-of-plane (axial or lateral) motion of the brake disc. Modal shapes with high lateral movements of the brake disk surface are therefore the best candidates to generate sharing forces, depending on actual forces and the correct damping.

The root cause is therefore the coincidence of the 3rd natural bending frequency of the wheelset and the "twisting" mode of the braking systems that are "short-circuited" by the pads during braking.

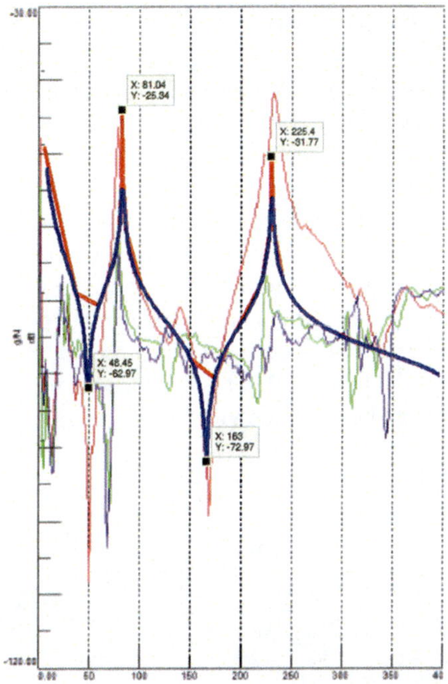

Fig. 5. Comparison of measured (thin red line) and estimated (thick red and blue line) $H_{11}x$ point FRFs.

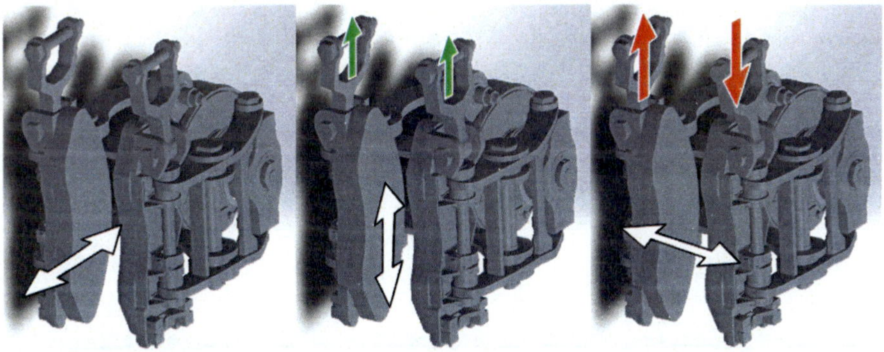

Fig. 6. Possible motions of the disc w.r.t. the brake pads (white arrows). Vertical arrows indicate the change in the instantaneous braking force.

5 Response of the Bogie Frame to Sharing Forces

As a counterphase vibration was observed on the bogie frame, of which the brake support is a structurally connected part, it was interesting to investigate the response of the combined bogie frame + brake support + brake unit to the set of alternate forces at the frequencies described in the previous chapter.

The contribution of the mass of brake unit was considered inserting heavy "plugs" in the "eyes" of the brake support. The model was tuned and applied to the model of the braking bar (Fig. 7). The analysis was limited to the braking bar, the welded bracket and the brake support. While the number of natural frequencies obtained by performing the analysis of the full bogie frame is large, most likely the involved modes are difficult to excite. The force exciting the brake calliper is produced at the pad-disc interface, and it is reasonable to suppose that bogie frame vibrations are limited.

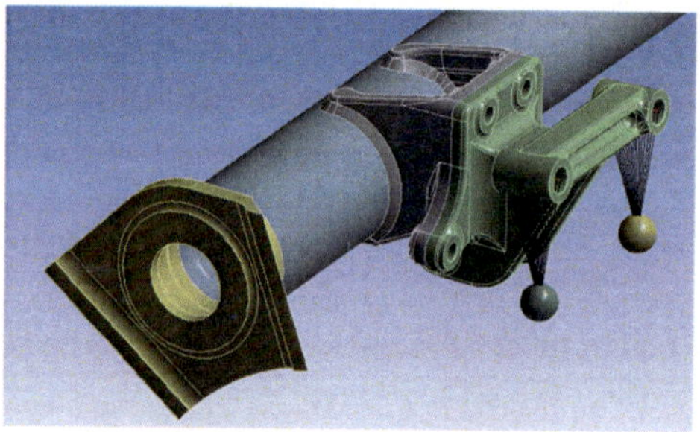

Fig. 7. Model of the brake support, the braking bar and the welded bracket.

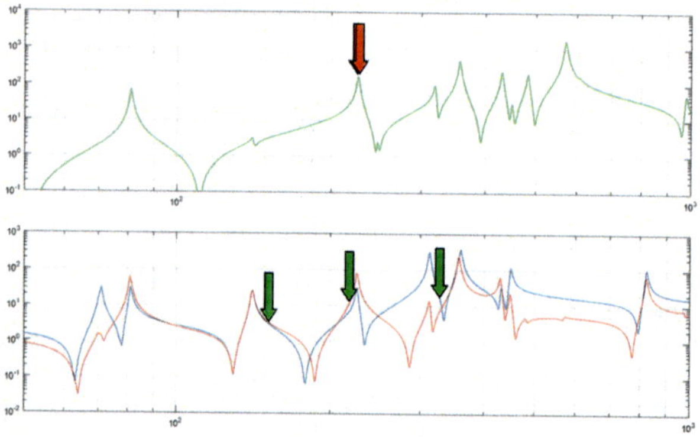

Fig. 8. Point FRFs of the wheelset compared to natural frequencies of the braking system. Response in the axial direction (above) may generate "twist" forces in the braking system, while response in the radial/longitudinal direction (below) may generate "pitch" forces in the braking system.

6 Structural Modification, Effects and Conclusions

Brake support mass increasing, stiffness decreasing and stiffness increasing were all considered (Fig. 9). Only stiffness increasing was adopted and tested. Robustness of this solution considering wheel, disc and brake pads wear will be shown in the full paper.

Modified supports were successfully tested in July 2017, showing no noise and vibrations. It can be concluded that the application of a very simple and low-cost modification solved the problem completely (Fig. 9 and Fig. 10).

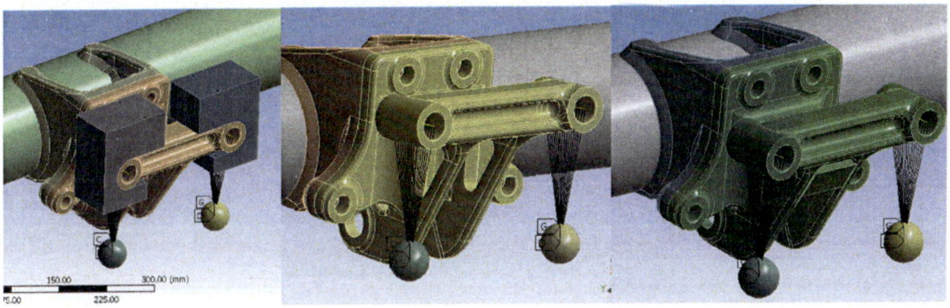

Fig. 9. Mass increase, flexible support and stiffened support (40 mm rib).

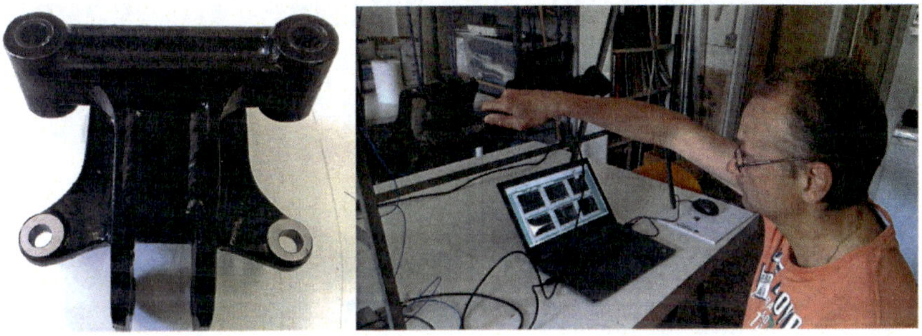

Fig. 10. Left: modified support (80 mm rib). Right: measurement of the point FRF on both original a modified supports.

References

1. Brennan, M.J., Shin, K.: Brake noise prediction and control. In: Crocker, M.J. (ed.) Handbook of Noise and Vibration Control, pp. 1133–1137. John Wiley and Sons (2007). ISBN 978–0–471–39599–7
2. EN 61373, Railway applications - Rolling stock equipment - Shock and vibration tests, CENELEC, September 2010
3. Bracciali, A., Rissone, P.: Analisi Modale Sperimentale di una Sala Portante del Convoglio ETR500. Ingegneria Ferroviaria **7**(8), 394–407 (1994)

RWP: Recent Advances in Wave Propagation in Periodic Media and Structures

Blocking Masses Applied to Surface Propagating Waves

Andrew Peplow(✉) and Mathias Barbagallo(✉)

Engineering Acoustics, Department of Construction Sciences, Lund University, 221 00 Lund, Sweden
{andrew.peplow,mathias.barbagallo}@construction.lth.se

Abstract. Ground vibration generated by rail and road traffic is a major source of environmental noise and vibration pollution in the low-frequency range. A promising and cost-effective mitigation method could be the use of heavy masses placed as an array on the ground surface near the road or track, these could be concrete or stone blocks, or specially designed brick walls, for example. This work concerns the effectiveness of such "blocking" masses. We consider propagating waves in a finite depth elastic layer assuming plane strain conditions. Given that the masses are considered solid objects we may place these on the surface, embedded or submerged in the elastic medium where a finite element method is considered as a computational solution technique. Next, we may assume the masses are aligned as a periodic infinite array and enforce periodic boundary conditions around a "unit-cell". By consideration of propagating waves via Floquet-Bloch theory we shall investigate the existence or lack thereof of propagating surface and body waves. This analysis supports a semi-analytical lumped-parameter method assuming the blocking masses are point masses situated on an elastic waveuide. The work is enhanced by an example highlighting advantages and disadvantages of multiple-mass scatterers in terms of possible stopband intervals related to propagating surface waves.

Keywords: Blocking masses · Surface elastic waves · Periodic array

1 Introduction

Ground vibration from traffic, construction work and rotating machinery cause annoyance to people living near railways, roads, construction sites, factories and workshops. Therefore, many researchers have proposed mitigation of vibration by various methods, including trenches, barriers and wave-impeding blocks (WIBs). Barriers can be made of materials that are ei-ther soft and light or stiff and heavy relatively to the soil. The governing principle is in any case to produce a mechanical-impedance mismatch between the barrier material and the soil, leading to reflection rather than transmission of energy. A wave-impeding block

Z. Dimitrovová et al. (Eds.): WMVC 2022, MMS 125, pp. 641–650, 2023.
https://doi.org/10.1007/978-3-031-15758-5_66

can be placed on the ground surface in which case it mainly contributes by its mass, or it can be embedded in the ground, adding stiffness to the soil. Possibly one of the earliest studies of the harmonic response of a single heavy mass placed on the surface of an elastic half-space was performed by Warburton [1]. The efficiency of such masses for mitigation of harmonic vibration from traffic was also studied by Krylov [2]. Recently, Dijckmans et al. [3] introduced the idea of placing an array of heavy masses along a railway track in order to mitigate vibration. The solution was found to have a significant effect.

Peplow et al. [4] introduced the idea of embedding a rigid block in the soil below a source in order to mitigate ground vibration by stiffening the soil. This so-called wave-impeding block (WIB) was found to have a significant impact on wave propagation in a layered ground, depending on the properties of the soil. More recently, a similar study was performed by Peplow and Finnveden [5], but this time based on a wave-based finite-element formulation. Takemiya [6] studied WIBs formed by piles placed in a honeycomb pattern around the foun-dation for a pylon suorting a railway bridge. The chosen WIB geometry was found advantageous in terms of mitigating waves in the considered frequency range. Along the same line, Masoumi et al. [7] demonstrated the efficiency of wave-impeding blocks regarding the mitiga-tion of ground vibration from railway traffic. They alied a numerical model validated by experimental work. Andersen and Nielsen [8] conducted a study of ground vibration from a moving force on a railway track, employing a coupled finite-element/boundary-element (FE/BE) model. They analysed various vibration mitigation methods, including soil stiffening under the rail and trenches or barriers along the track.

Andersen and Liingaard [9] performed a detailed study of vibration barriers formed by one or more sheet pile walls. Especially, the influence of barrier position and inclination was analysed using a two-dimensional FE/BE model. Their work was followed up by Andersen and Augustesen [10] who used a three-dimensional FE/BE model to study the efficiency of inclined, embedded barriers placed along a railway. They concluded that inclined barriers might perform better than vertical barriers when a concrete lid needs to be placed on top to prevent people or water from entering the trench. A further study of optimization of wave barriers was performed by van Hoorickx et al. [11].

Instead of relying on the screening effect of a single barrier or the scattering provided by a single WIB, the possibly destructive interference caused by repeated inclusions can be utilized. Mead [12] studied the effect of periodicity in a structure, finding that bandgaps occur in certain frequency ranges depending on the mechanical properties of the materials and the geometry of the cell which is repeated along the periodic array. Not necessarily a characteristic of periodicity but observations of the *locking* effect analysed by in [13] occurs here where branches of attenuating oscillating waves join together in a stopband region which later unlock at an adjacent propagating wave. Based on this idea, Andersen [14] proposed the inclusion of periodic structures in the ground to mitigate vibration. Using a two-dimensional FE model and Floquet theory, he demonstrated that stopbands are produced by arrays of WIBs or ground surface changes in the form of small hills and valleys repeated in a periodic manner. Persson et al. [15] also studied the concept of ground-surface landscaping to mitigate vibration. Finally, Andersen et al. [16] conducted a number of small-scale experiments and FE analyses with the aim to quantify the insertion loss provided by placing one or more blocks on or in the ground within the transmission path between a source and a receiver.

2 Methodology

The geometry of the periodic structure studied in this work is depicted in Fig. 1. The layered periodic structure consists of a series of typical cells Fig. 2, which are infinitely extended in the x direction. Each typical cell includes two different materials, which are upper and lower layered strate. A third concrete material defines the blocking mass. The widths of the components and the total width of the typical cell is defined as $a > 0$. The width of the typical cell is also referred to as the periodic constant. The cells are completely invariant along the z direction, so a 2D plane–strain problem was established to calculate the dispersion curves of the periodic structures. As indicated in Fig. 2, the free surface is perpendicular to the y direction and the surface wave propagates along the x direction.

2.1 Governing Equations

For plane strains the behaviour of the elastic material is described by Navier's elastodynamic equations. Without loss of generality, in the absence of body forces, these equations apply to any layer and are written in vector form as follows:

$$(\lambda + \mu)\nabla(\nabla.\mathbf{u})(\mathbf{r}) + \mu\nabla^2\mathbf{u}(\mathbf{r}) = \rho\ddot{\mathbf{u}}, \tag{1}$$

where $\mathbf{u} = (u, w)$ represent the components at position $(\mathbf{r}) = (x, y)$ of the displacement in the x and y directions and Δ is the dilatation, ρ is the density of the material and λ, μ are complex Lamé constants. To study the behavior of

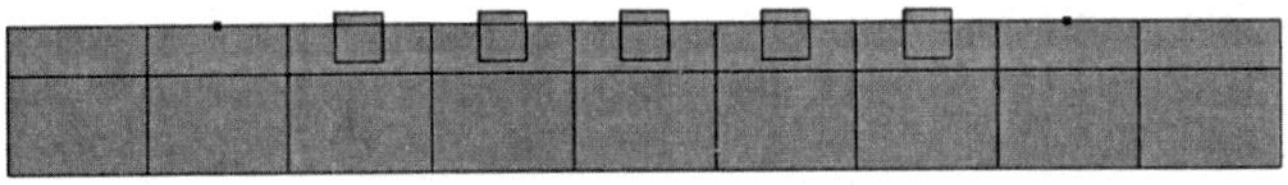

Fig. 1. Geometry of a plane–strain wave propagation model. Two end elements modelled by a PML element. Coordinates are (x, y) with top surface free and bottom a bedrock layer, fixed by zero displacement.

wave propagation in an infinite unit cell structure, the Floquet-Bloch theorem is applied. The theory was originally developed to solve differential equations of

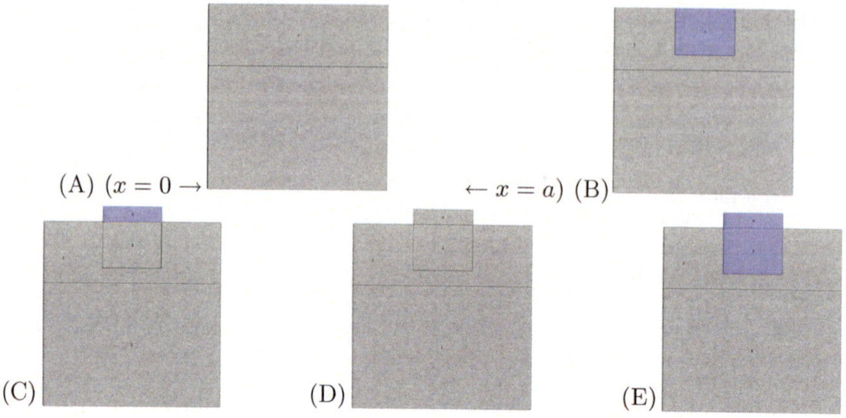

Fig. 2. Five unit cell–types of width a. Lilac shaded areas depict the third concrete material.

wavelike particles in physical sciences. According to this theory, the solution to Eq. (1) can be expressed as

$$\mathbf{u}(\mathbf{r}, t) \;=\; \mathbf{u}_\gamma(\mathbf{r})e^{\mathrm{i}(\gamma x - \omega t)}, \tag{2}$$

Eq. (2) is the periodic displacement boundary condition. It implies that the phase difference between the input and output wave is given by a scalar parameter γ. By factoring out the harmonic time dependence $exp(\mathrm{i}\omega t)$, the periodic boundary condition for a typical cell along $x = 0$ and $x = a$, see Fig. 2(A), can be expressed as follows:

$$\mathbf{u}(\mathbf{r} + a) \;=\; e^{\mathrm{i}\gamma x}\mathbf{u}(\mathbf{r}). \tag{3}$$

Due to this characteristic, the wave modes in an infinite periodic unit cell structure is actually the repetition of free vibration modes of a single unit cell of width a. In the present study we have applied the periodic boundary condition on two vertical sides of the built–up section, i.e. at $x = 0$ and $x = a$. Also, a fixed zero displacement at the lower boundary and a free boundary at the surface complete the boundary conditions necessary.

2.2 Finite Element Formulation for the Dispersion Equation

A dispersion relation is an implicit function between wavenumber γ and frequency ω. A finite element framweork can provide this relation and moreover the "modes" of wave propagation with passband and stopband can be observed from a dispersion plot. In this work the dispersion relation can be obtained by substituting the periodic boundary condition of Eq. (2) into Eq. (1) and solving the eigenvalue problem that can be expressed as

$$\mathbf{\Phi}(\gamma) - \omega^2\mathbf{M} \;=\; \mathbf{0}, \tag{4}$$

As mentioned above, the FE technique was employed to study the elastodynamic properties of the typical cell. The FE discretization of the typical cell and the interaction with its neighbors are modelled by cubic-sernedipity plane–strain elements and periodic boundary conditions must be applied on an identical grid along the extremal boundaries in order to use a typical cell to represent an infinite periodic structure Fig. 3 where $\mathbf{\Phi}$ is the stiffness matrix which is a function of the wavenumber γ, and $\mathbf{M}$ being the mass matrix. For a given scalar γ, there exists the eigenvalue, or eigenfrequency ω, that can be obtained by using a numerical eigenvalue solver.

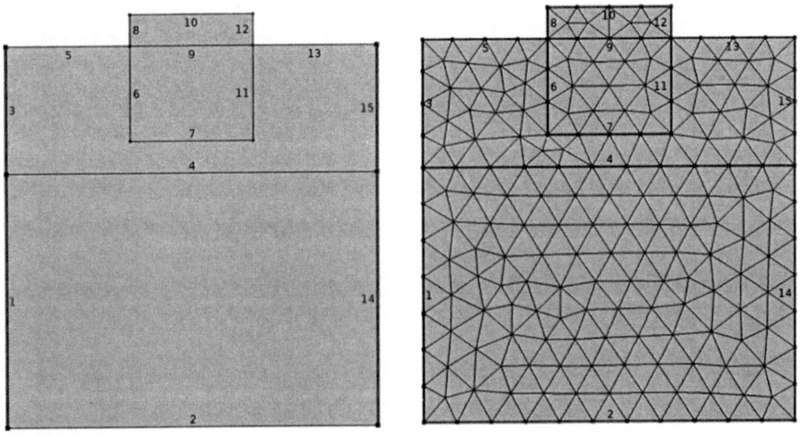

Fig. 3. Illustrating periodic boundary conditions highlighted and mesh used in finite element discretization.

3 Results

In our analysis we provide the value for the wavenumber γ (either real or imaginary) and determine the eigenvalue $\omega > 0$. We are particularly interested in values where $\pi/a \geq \gamma_R > 0$, ($\overline{\gamma_R} = \gamma_R/(\pi/a)$) representing **all** the propagating wavenumbers. As emphasized before, the most important characteristics for periodic structures in this application are "zones" where the surface-wave propagation is prohibited. To further explain this feature, complex wavenumbers will be defined as $\gamma = \pi/a + \mathrm{i}\gamma_I$ ($\overline{\gamma_I} = \gamma_I/(\pi/a)$) to calculate the surface modes. Subsitituting the complex wavenumber into Eq. (4) the displacement vector can be written as $\mathbf{u}(x+a) = e^{\mathrm{i}\pi}e^{-\gamma_I}e^{\mathrm{i}\gamma x}\mathbf{u}(x)$ but also imaginary wavenumbers which represent the point at which the real wavenumbers become complex, $\gamma = \pi/a + \mathrm{i}\gamma_I$. These wavenumbers can describe "leaky" modes and can give some insight into stopband phenomena. Qualitatively, we are seeking the value ω which is the cut-off frequency where the nature of the wave modes change,

Table 1. Material properties.

	Blocking mass	Upper soil	Lower layer
Young's modulus, E (Pa)	1.0×10^{10}	5.0×10^{7}	5.0×10^{8}
Density ρ (kgm^{-3})	2400	2000	2200
Poissons ratio, ν	0.2	0.48	0.45
c_R (ms^{-1})	1200	90	265
c_2 (ms^{-1})	1320	90	280
c_1 (ms^{-1})	2150	470	930

from propagating wave to an attenuating oscillatory one. Thus, all wave modes propagating in an infinite periodic structure with passbands and soptbands can be identified. For a wavevector where no frequency exists, it is termed the stop-band, where no wave propagation occurs. In contrast, a wavevector for which frequency has definite value is a passband and wave propagates in the medium. Since a dispersion relation is obtained based on the assumption of an infinite unit cell structure, but practically all structures are finite. Therefore, in a finite domain analysis, very weak wave propagation can be noticed corresponding to bandgap frequencies as compared to passband frequencies. In the present study, the commercial FEM software COMSOL Multiphysics 5.6 is used to solve the eigenvalue equation and dispersion relation. The Floquet-Bloch periodicity condition is aligned to the vertical sides of the unit cell which represent the geometry and finite depth layered soil material, (blue highlighted boundaries, Fig. 3).

Figure 4 shows the dispersion curves which are strongly dependent on the properties of the material and geometry of the periodic structures. For each real wavenumber a maximum number of 10 eigen–frequencies are found. We shall not perform a parametric study here, but focus on the stopband behaviour for the specific periodic array outlined here. The width of the cell is $a = 6$ m, the upper layer is 2 m thick while the lower layer is 4m deep. Note that the curves, represented by dots, on the left hand side of the figure represent propagating waves and the red curves (dots) represent specific attenuation factors. Possibilities of veering and locking, [13] are displayed in the dispersion diagrams shown here.

(A) **Uniform layers.** Due to the unit cell periodicity condition the dispersion relation for a uniform layer appears unconventional to dispersion relations for an infinite uniform transversal isotropic layered material. Neverthess, the periodic problem differs substantially from the uniform layer situation. It does show though that at each frequency, above cut–on (6 Hz) at least one propagating wave exists, Fig. 5. The non–dimensional complex wavenumbers which appear

(B) **Embedded masses.** For a blocking masses, the dispersion relation diagram Fig. 5 has developed into a significant qualitative change. This is not surprising since the mass is significantly large in geometry and impedance change is prominent. At around 16 Hz there now appears a slight stop–band and

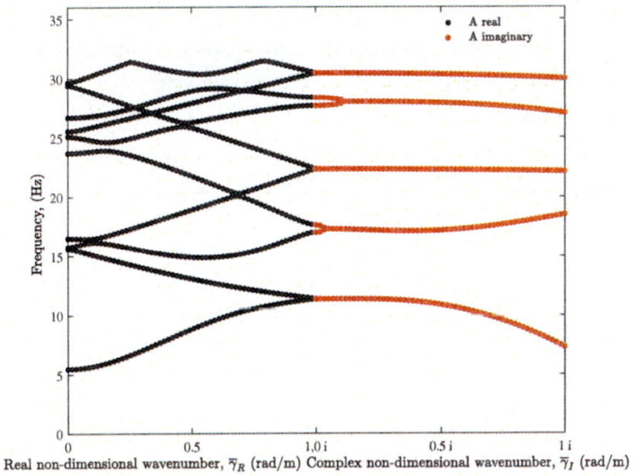

Fig. 4. Dispersion curves for the periodic uniform layer case, (A). Black dots represent real non–dimensional wavenumbers $\overline{\gamma_R}$, red-dots appropriate complex non–dimensional wavenumbers, $\overline{\gamma_I}$.

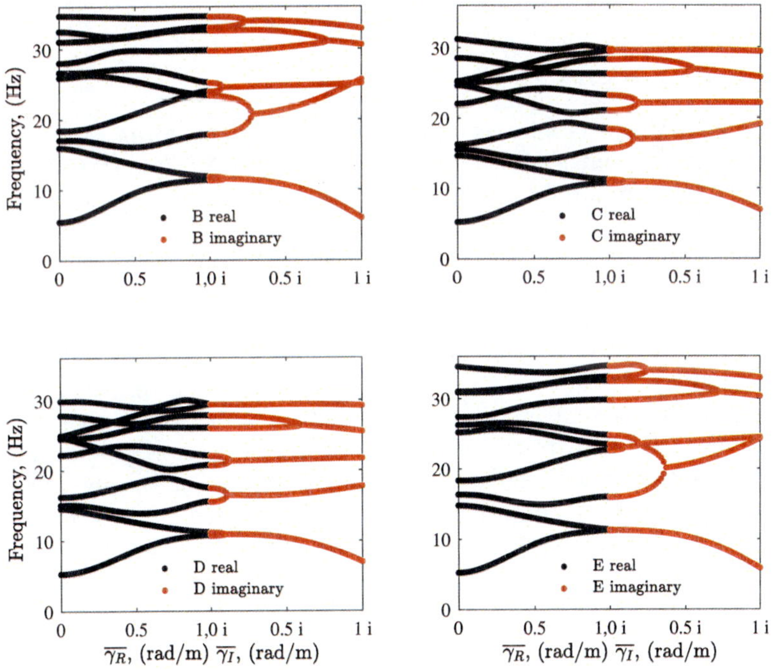

Fig. 5. Dispersion curves related to the four cases (B)–(E)

also noticeable is the pairing of the attenuating (red) curves which bifurcate into a single attenuating mode.

(C) **Extruded masses.** Whereas the embedded masses are geometrically not dissimilar to uniform layering, it is rather surprising to observe a lack of clear stop–band behaviour.

(D) **Periodic profile.** Similar to Extruded masses it appears the propagation wavenumbers up to 30 Hz are not affected greatly by the material properties of the extruded mass.

(E) **Submerged & extruded masses.** Evidence of stop–band beahviour is shown in this example where the blocking mass is submerged and also extruded from the supporting soil layers. At aroundf 16, 18, 26, &30 Hz wave propagation appears inhibited at these frequencies which could support possible engineering solution if a periodic array of blocks is a tangible option in construction.

While it is interesting to observe stop–band behaviour throughout the body of a cell, it is of special (civil engineering) interest to determine frequency bands for which surface wave propagation is limiting. And establishing a criterion, either qualititatively or numerically would be of a great advantage for engineers. This naturally means that the underlying physics to establish this, in the form of the modes related to the dispersion diagrams in Fig. 5, needs further investigation. This work is currently underway.

3.1 Insertion Losses

To provide some evidence, in the form of a forced response, for possible vibration reduction a unit point–source located was located at 3.0 m distance from the first block, and a receiver 3.0 m from the fifth block, giving a total distance 24 m between the two, Fig. 6. Up to 30 Hz the insertion loss, in decibels, due to the inclusion of the blocks, was calculated in unitary frequency steps. A PML element was included at each end of the geometry. Figure 6 shows the range of insertion loss for motion in the vertical direction for the cases (B)–(E). Each case responds approximately similarly, a large rise in insertion loss at the first *stop–band* 20 Hz followed by a dip, where there appears many propagating waves cut–on, which is then followed by a large rise again for case (E) where a solid mass, submerged and extruded above the surface, with a stop–band at 30 Hz yields the best insertion loss. Generally, insertion loss at certain frequencies may be explained by propagation dispersion diagrams, but the absolute level or relative insetion loss level between cases is perhaps too complex to establish physical explanations currently.

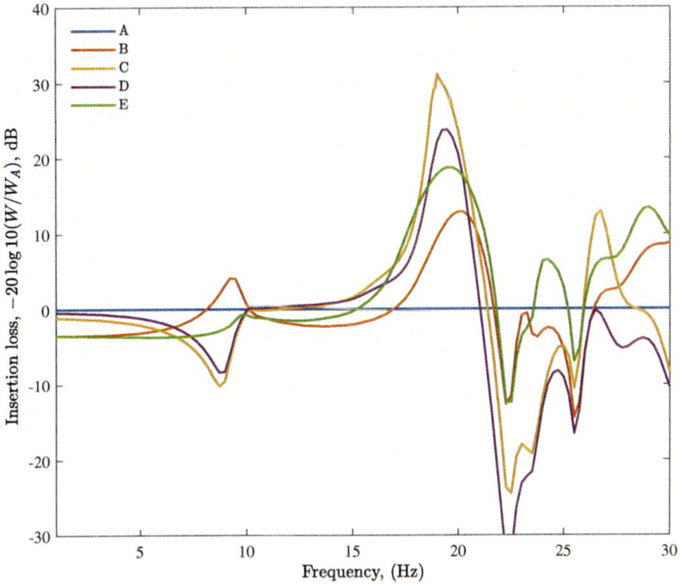

Fig. 6. Insertion loss of vertical motion for the geometry illustrated in Fig. 1.

4 Conclusion

We have assumed that large masses aligned as a periodic infinite array can be modelled by enforcing periodic boundary conditions around a "unit-cell". By consideration of propagating waves via Floquet-Bloch theory we have investigated the existence of propagating waves and observed the phenomena of mode conversion via an evanescent wave where bifurcation of the dispersion curve is evident. Future research will be directed towards investigating the following points:

- A parametric study of depth of layers, material properties, relative material properties for qualitative check on establishing the mechanism that determines the width of a stop–band.
- Investigate a quantification method which isolates or distinguishes surface wave modes from other *body modes*
- Consider topology optimization to determine shape orientation
- Consider the problem where the lower layer is an elastic halfspace

References

1. Warburton, G., Richardson, H., Webster, J.: Harmonic response of masses on an elastic half-space. J. Eng. Ind. Trans. ASME **94**(1), 193–200 (1972)
2. Krylov, V.: Scattering of Rayleigh waves by heavy masses as method of protection against traffic-induced ground vibrations. In: Environmental Vibrations: Prediction, Monitoring, Mitigation and Evaluation (ISEV 2005), pp. 393–398 (2005)

3. Dijckmans, D., et al.: Mitigation of railway induced ground vibration by heavy masses next to the track. Soil Dyn. Earthq. Eng. **75**, 158–170 (2015)
4. Peplow, A., Jones, C., Petyt, M.: Surface vibration propagation over a layered half-space with an inclusion. Appl. Acoust. **56**, 283–296 (1999)
5. Peplow, A., Finnveden, S.: Calculation of vibration transmission over bedrock using a waveguide finite element model. Int. J. Numer. Anal. Methods Geomech. **32**(6), 701–719 (2008)
6. Takemiya, H.: Field vibration mitigation by honeycomb WIB for pile foundations of a high-speed train viaduct. Soil Dyn. Earthq. Eng. **24**(1), 69–87 (2004)
7. Masoumi, H., Van Leuven, A., Urbaniak, S.: Mitigation of train induced vibrations by wave impeding blocks, numerical prediction and experimental validation. In: EURODYN 2014, pp. 863–870 (2014)
8. Andersen, L., Nielsen, S.: Reduction of ground vibration by means of barriers or soil improvement along a railway track. Soil Dyn. Earthq. Eng. **25**, 701–716 (2005)
9. Andersen, L., Liingaard, M.: Vibration screening with sheet pile walls. In: Environmental Vibrations: Prediction, Monitoring, Mitigation and Evaluation (ISEV 2005), pp. 429–437 (2005)
10. Andersen, L., Augustesen, A.: Mitigation of traffic-induced ground vibration by inclined wave barriers - a three-dimensional numerical analysis. In: 16th International Congress on Sound and Vibration 2009, (2009)
11. van Hoorickx, C., Sigmund, O., Schevenels, M., Lazarov, B., Lombaert, G.: Topology optimization of two-dimensional elastic wave barriers. J. Sound Vib. **376**, 95–111 (2016)
12. Mead, D.: Free wave propagation in periodically suorted, infinite beams. J. Sound Vib. **11**(2), 181–197 (1970)
13. Mace, B., Manconi, E.: Wave motion and dispersion phenomena: veering, locking and strong coupling effects. J. Acoust. Soc. Am. **131**(2), 1015–1028 (2012)
14. Andersen, L.: Using periodicity to mitigate ground vibration. In: COMPDYN 2015 - 5th ECCOMAS Thematic Conference on Computational Methods in Structural Dynamics and Earthquake Engineering (2015)
15. Persson, P., Persson, K., Sandberg, G.: Reduction in ground vibrations by using shaped landscapes. Soil Dyn. Earthq. Eng. **60**, 31–43 (2014)
16. Andersen, L., Bucinskas, P., Persson, P., Muresan, M., Muresan, L.-I., Paven, I.-O.: Mitigating ground vibration by periodic inclusions and surface structures. In: Proceedings of the INTER-NOISE 2016 - 45th International Congress and Exposition on Noise Control Engineering: Towards a Quieter Future (2016)

Dynamic Amplification in a Periodic Structure Subject to a Moving Load Passing a Transition Zone: Hyperloop Case Study

Andrei B. Fărăgău(✉), Andrei V. Metrikine, and Karel N. van Dalen

Delft University of Technology, Stevinweg 1, 2628, CN Delft, The Netherlands
A.B.Faragau@tudelft.nl

Abstract. Hyperloop is an emerging high-speed transportation system in which air resistance is minimised by having the vehicle travel inside a de-pressurised tube supported by columns. This design leads to a strong *periodic* variation of the stiffness (among other parameters) experienced by the vehicle. Also, along its route, the Hyperloop will encounter so-called *transition zones* (e.g., junctions, bridges, etc.), where the properties (e.g., support stiffness) are different than for the rest of the structure. In railway engineering, increased degradation is seen in the vicinity of these transition zones, leading to increased frequency of maintenance. This work investigates response amplification mechanisms in a Hyperloop system that arise due to the combination of a transition zone and the structure having a periodic nature. The amplification mechanisms investigated here can help prevent degradation of the Hyperloop tube close to transition zones as well as fatigue and wear of the vehicle.

Keywords: Periodic structure · Moving load · Hyperloop · Transition radiation · Wave interference

1 Introduction

Periodic systems under the action of moving loads have been extensively studied by researchers in the past century. These problems do not only pose academic challenges but are also of high practical relevance due to their application in railway, road, and bridge engineering, among others. Despite numerous studies on periodic systems (e.g., [1,2]), only few of them investigate the influence of a local inhomogeneous region, a so-called transition zone, on the dynamic response. In railway and road engineering, increased degradation is seen in the vicinity of these transition zones [3–6], leading to increased maintenance requirements.

Hyperloop is a new emerging transportation system that is in the development stage. Its design minimises the air resistance by having the vehicle travel inside a de-pressurised tube (near vacuum) supported by columns. This design

Z. Dimitrovová et al. (Eds.): WMVC 2022, MMS 125, pp. 651–661, 2023.
https://doi.org/10.1007/978-3-031-15758-5_67

may lead to a strong periodic variation of the stiffness experienced by the vehicle. Moreover, along its route, it will cross bodies of water or come across junctions/switches, all of which are transition zones.

The current study investigates the potential response amplification in a Hyperloop system that results from the combination of (i) a transition zone and (ii) the structure having a periodic nature. In a previous study [7], three such phenomena have been identified in a system representative of a catenary structure (overhead wires in railway tracks). In this work, the three phenomena identified in Ref. [7] are investigated using a model representative for a Hyperloop transportation system. Accounting for these phenomena in the design can help prevent degradation of the Hyperloop tube close to transition zones as well as fatigue and wear of the transportation pod.

2 Model Description and Solution Method

The system formulated here consists of an infinite Euler-Bernoulli beam with mass per unit length ρ, bending stiffness EI, and material damping ratio ζ. The beam is discretely supported by springs with stiffness $k_{\mathrm{s}}(x)$ and dashpots with damping coefficient $c_{\mathrm{s}}(x)$, and a point mass m_{s} is located at each support. $x \in [nd, (n+1)d]$ is the generic cell where n is the cell number and d is the cell width, and the spring-dashpot-mass element is located in the middle of the cell at $x = \bar{n}d$ with $\bar{n} = n+\frac{1}{2}$. The system is acted upon by a moving constant load of amplitude F_0 and velocity v. A zone (stiff zone) of length l has the stiffness and damping of the supports p times larger than for the rest of the infinite domain; the region covering the stiff zone and its close vicinity is called the transition zone. Figure 1 visually describes the system, while its equation of motion reads

$$EI(1+\zeta\tfrac{\partial}{\partial t})w''''+\rho\ddot{w}+\sum_{n=-\infty}^{\infty}\left(m_{\mathrm{s}}\frac{\partial^2}{\partial t^2}+c_{\mathrm{s}}(x)\frac{\partial}{\partial t}+k_{\mathrm{s}}(x)\right)w\delta(x-\bar{n}d) = -F_0\delta(x-vt), \tag{1}$$

where primes and overdots denote partial derivatives in space and time, respectively. The stiffness $k_{\mathrm{s}}(x)$ is a piecewise-defined function in space and it reads

$$k_{\mathrm{s}}(x) = \begin{cases} k_{\mathrm{s}}, & x < x_{\mathrm{a}}, \\ p\,k_{\mathrm{s}}, & x_{\mathrm{a}} \leq x \leq x_{\mathrm{b}}, \\ k_{\mathrm{s}}, & x > x_{\mathrm{b}}. \end{cases} \tag{2}$$

For simplicity, the same spatial distribution is assumed for the damping.

There are multiple designs of the Hyperloop transportation system; here, a typical Hyperloop design is considered. The steel tube has a thickness of 19 mm and an inner diameter of 2.5 m, leading to $\rho = 1331$ kg/m (a 10% increase was considered to account for additional equipment) and $EI = 2.5 \times 10^{10}$ Nm^2. The support stiffness is tuned using a FEM analysis of the 3D structure (excluding the soil) such that the displacement at the location of the supports match when a static load is applied in the middle of the span. Note that the displacement

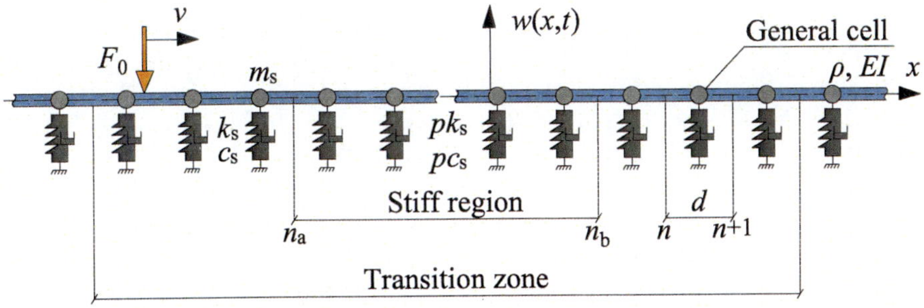

Fig. 1. Model schematics: infinite Euler-Bernoulli beam discretely supported by an inhomogeneous foundation, subjected to a moving constant load.

is at the rail level which is located at the top of the tube (the vehicle is suspended from the top in this design); therefore, the stiffness of the support (in our phenomenological model) accounts not only for the stiffness of the column, but also for the flexibility of the connection between the tube and the rail and, most importantly, for the flexibility introduced by the ovalization of the tube. The concrete columns supporting the tube have a spacing of $d = 16$ m and are assumed to be of 1.5 m diameter and 5 m height; the point mass in the model represents the mass of the columns that is activated by the vehicle and is chosen here as 10% of the overall mass of the column $m_s = 2332$ kg (such a small value is chosen because most flexibility at the supports comes from the ovalization of the tube and thus, the columns are not deformed much). When it comes to the damping, a very small amount is assumed to be conservative, namely $\zeta = 5 \times 10^{-6}$ and $c_s = 10$ kNs/m. Although the damping seems small, it originates mostly from the tube itself and not from the soil (since the columns are not deformed much) and the metal tube is not expected to have high damping.

2.1 Solution Method for the Homogeneous System

The approach to determine the steady-state solution for the system without a transition zone is based on the Floquet theory [8]. The procedure is explained in detail for a string in Ref. [7] and is summarized in the following. After applying the Fourier transform over time to Eq. (1), the analysis can be restricted to one cell without loss of generality. The relation between the states (displacement, slope, bending moment, and shear force) at the two boundaries of the generalized cell reads (see [7] for a detailed derivation)

$$\tilde{\mathbf{w}}_{n+1} = \mathbf{F}\tilde{\mathbf{w}}_n + \tilde{\mathbf{w}}_{n+1}^{\mathrm{ML}}, \tag{3}$$

where the 4×4 matrix $\mathbf{F}$ is the Floquet matrix and $\tilde{\mathbf{w}}_{n+1}^{\mathrm{ML}}$ includes the influence of the particular solution to the equation of motion. Performing the eigenvalue (α) and eigenvector ($\mathbf{u}$) decomposition of $\mathbf{F}$ leads to the following expression:

$$\tilde{\mathbf{w}}_n = a_1 e^{-\mathrm{i}k_1^{\mathrm{F}} nd}\mathbf{u}_1 + a_2 e^{-\mathrm{i}k_2^{\mathrm{F}} nd}\mathbf{u}_2 + a_3 e^{-\mathrm{i}k_3^{\mathrm{F}} nd}\mathbf{u}_3 + a_4 e^{-\mathrm{i}k_4^{\mathrm{F}} nd}\mathbf{u}_4 + \tilde{\mathbf{w}}_{n+1}^{\mathrm{ML}}, \tag{4}$$

where a_1, a_2, a_3, and a_4 are unknown amplitudes and $k^{\mathrm{F}} = \frac{\mathrm{i}\ln(\alpha)}{d}$ are the Floquet wavenumbers. To fully determine the solution, the so-called *periodicity condition* [7,9] is used, which imposes the response inside each cell to be exactly the same as in the previous one but shifted in time by $\frac{d}{v}$. The periodicity condition leads to four conditions (displacement, slope, moment, and shear force) with which all four unknown amplitudes are determined. The steady-state solution in the Fourier domain is now determined, and to obtain the time-domain solution, the inverse Fourier transform is performed numerically.

The dispersion characteristics of the system without the transition zone are presented in Fig. 2. There are infinitely many wavenumbers k^{F} corresponding to one frequency ω and the distance between subsequent wavenumbers is $\frac{2\pi}{d}$. These repeating zones are called Brillouin zones [10] (just three zones are presented in Fig. 2, but there are infinitely many). For discrete systems, all information about wave propagation is contained in the first Brillouin zone. Unlike discrete systems, continuous ones allow for wave propagation at all wavenumbers. Consequently, the response $\tilde{w}(x, \omega)$ will contain wavenumbers from all Brillouin zones and the continuous wavenumber reads $k = k^{\mathrm{F}} + m\frac{2\pi}{d}$ with $m = \pm 1, \pm 2, \ldots$. Also, Fig. 2 shows that the periodic system exhibits multiple (actually infinitely many [10]) frequency ranges where no propagation is possible; these frequency ranges are called *stop bands*, while the frequency ranges in which propagation is possible are called *pass bands*. Strictly speaking, stop/pass bands only exist if the system does not have dissipation; however, for small values of dissipation, the wave propagation is strongly attenuated in the stop bands.

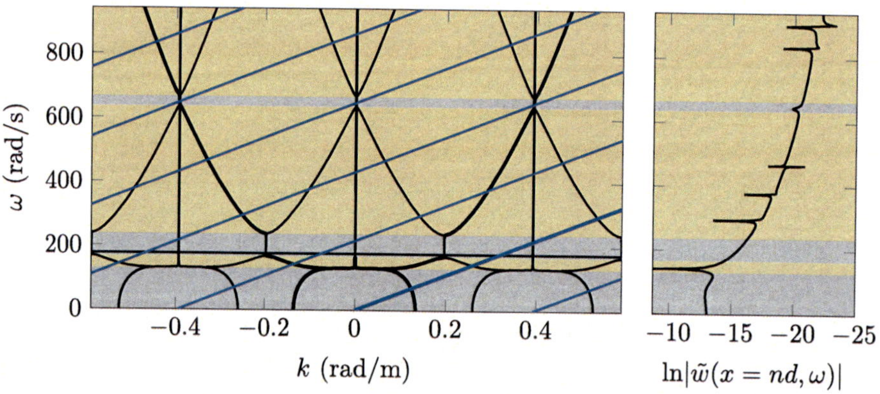

Fig. 2. The dispersion curves in three Brillouin zones (black lines) and the kinematic invariants (blue lines) (left panel), and the frequency spectrum of the steady-state displacement (right panel); the grey/yellow background represents a stop/pass band.

To determine the frequency and wavenumbers of the waves generated by the moving load, we need another equation next to the dispersion equation that expresses a relation between the frequency, wavenumber, and the load velocity, namely the kinematic invariant. There are infinitely many kinematic invariants (seven of them are presented in Fig. 2). The 0th-order kinematic invariant is given by $\omega = kv$ while the higher order kinematic invariants are given by $\omega = kv + m\frac{2\pi v}{d}$ with $m = \pm 1, \pm 2, \ldots$. The intersections between one of the kinematic invariants and the dispersion curve represent waves emitted by the moving load in the steady state, as can be seen in the right panel of Fig. 2. Moreover, it is important to observe that the emitted waves form a discrete frequency spectrum and that all generated waves have frequencies in the pass bands of the system.

2.2 Solution Method for the Inhomogeneous System

The transient solution is obtained by firstly determining the response of the system to the moving load acting inside each individual cell separately, and then superimposing the individual solutions. To determine the response of the system to the moving load acting inside only one cell, the forward Fourier transform is applied over time to Eq. (1). The obtained equation of motion is then divided in 5 domains; when, for example, the moving load is applied to the left of the stiff zone, the 5 domains are (1) left of the loaded cell, (2) the loaded cell, (3) right of the loaded cell and left of the stiff zone, (4) inside the stiff zone, and (5) to the right of the stiff zone. When the moving load is applied to the stiff zone or to the right of the stiff zone, an analogous division is made. The solutions of the 5 domains are analogous to Eq. (4) where $\tilde{\mathbf{w}}_{n+1}^{\mathrm{ML}}$ appears only in the loaded domain. The 5 solutions have 16 unknown amplitudes (after applying the boundary conditions at infinite distance from the load), which are determined from the interface conditions between the domains (i.e., continuity in displacements, slopes, bending moments, and shear forces). This procedure is repeated until the moving load is applied inside each individual cell of interest.

To obtain the response of the system to the moving load acting on all cells, the individual solutions are superimposed as follows:

$$\tilde{\mathbf{w}}_n = \sum_{n_\xi = N_{\mathrm{left}}}^{N_{\mathrm{right}}} \tilde{\mathbf{w}}_{n,n_\xi}, \tag{5}$$

where $\tilde{\mathbf{w}}_{n,n_\xi} = [\tilde{\mathbf{w}}_{1,n,n_\xi}, \tilde{\mathbf{w}}_{2,n,n_\xi}, \tilde{\mathbf{w}}_{3,n,n_\xi}, \tilde{\mathbf{w}}_{4,n,n_\xi}, \tilde{\mathbf{w}}_{5,n,n_\xi}]$ is the solution for all the cells when the load is applied at n_ξ, N_{left} is the first cell on which the load acts (at $t = 0$) and N_{right} represents the last cell. N_{left} needs to be chosen sufficiently to the left of the transition zone such that the response is in the steady state prior to reaching the transition zone. N_{right} can be chosen based on the maximum time of the simulation and it does not introduce any unwanted transients in the response. The solution is now determined at the interfaces between cells. To determine the solution inside the cells, one needs to use the equations that lead to the Floquet matrix (see Eq. (8) in [7]).

Next, three phenomena that lead to amplifications of the response in the transition zone are investigated.

3 Wave Interference Phenomenon

The increased support stiffness of the stiff zone causes an upward shift in frequency of its stop bands compared to the soft zones. Consequently, a wave generated by the moving load outside the transition zone can be reflected at the stiff region provided that the frequency of the wave is in a stop band of the stiff region. This leads to wave interference between the reflected wave and the wave field radiated by the approaching load, which in turn can cause an amplification of the response in the transition zone.

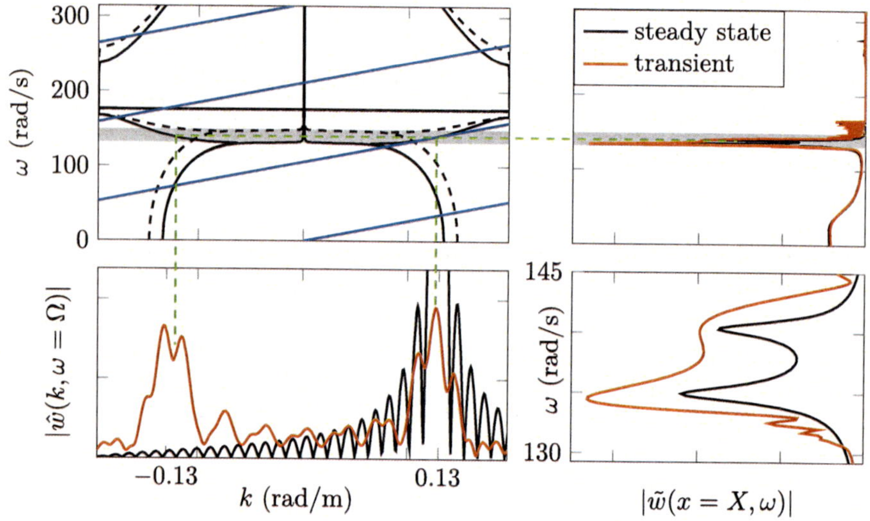

Fig. 3. The dispersion curves for the soft (black solid line) and stiff (black dashed line) regions and the kinematic invariants (blue lines) (top left panel), the displacements frequency spectra to the left of the stiff zone (top right panel; $X = x_a - 5$ m), and the wavenumber spectra of the displacements (bottom left panel) evaluated at $\Omega = 140$ rad/s (indicated by the horizontal green dashed line); the bottom right panel is a zoom in of the top right panel; the grey background indicates the overlapping region of the pass-band of the soft zone and the stop-band of the stiff one; $p = 1.3$ and $v = 269$ m/s.

The frequency-domain response in Fig. 3 shows that there are two harmonics of large amplitude generated in the steady state, which are in one of the stop-bands of the stiff zone. These waves are, as can be seen, amplified in the transient response (at the left of the stiff zone) due to the wave interference between the incoming and reflected waves. The reflection of one of the two harmonics (the one at $\omega = 140$ rad/s) can be seen in the wavenumber-domain response through the presence of an additional peak (compared to the steady state) at wavenumber equal in magnitude but opposite in sign (i.e., opposite direction of propagation) to that of the forward propagating wave.

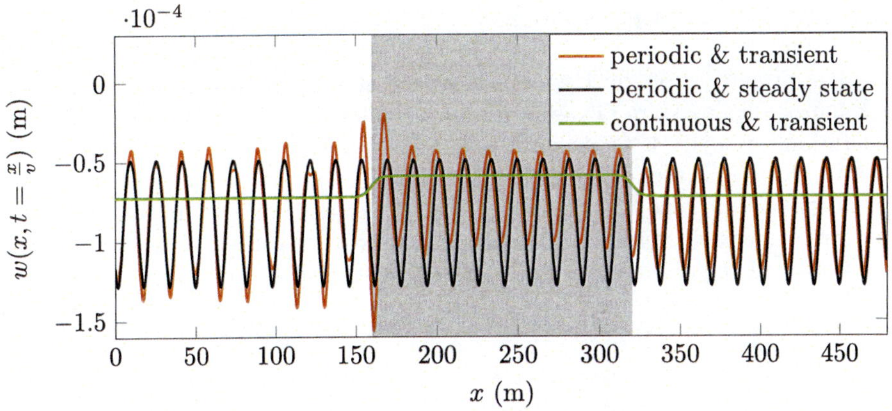

Fig. 4. The displacement evaluated under the moving load for the wave-interference phenomenon; the location of the stiff zone is indicated by the grey background.

To quantify the amplification, the time-domain response under the moving load is presented in Fig. 4. The response is evaluated under the moving load because it is governing. A considerable amplification can be observed at the left of the stiff zone that, at its maximum, is of about 20%. The response of the equivalent continuously supported system with a transition zone is also presented to show that, in that case, there is no visible amplification. Clearly, this significant amplification is caused by the periodicity of the system together with the transition zone; if any of these two characteristics are removed, the amplification vanishes.

It is important to note that a larger difference in stiffness p does not cause a significant increase in the amplification; the important factor for this phenomenon is that p is such that the generated harmonics are in one of the stop bands of the stiff zone. Also, even though the velocity is in the operational range for Hyperloop, it is chosen specifically for this mechanism to occur (see Sect. 4.1 in Ref. [7] for the choice of velocity); for other velocities, the generated waves are either of low amplitude or inside the pass-bands of the stiff zone, making this mechanism not to occur. Finally, the larger the damping (especially the tube's internal damping), the smaller the amplification observed because the generated waves cannot propagate sufficiently far before being attenuated.

4 Passing from Non-resonance Velocity to a Resonance Velocity

The velocity at which one of the kinematic invariants is tangential to one of the dispersion curves is called critical velocity. At such velocities, resonance of the system occurs. From a physical point of view, resonance occurs because the group velocity of the generated wave is equal to the load velocity, which makes that the wave cannot propagate away and leads to a build up of energy around the moving load.

The properties of the Hyperloop system should be chosen such that these resonance velocities are far away from operational speeds. However, even if the operational velocity is far from resonance velocities outside transition zones, it can be close to a resonance velocity inside it. In this section, we investigate the situation in which the load passes from non-resonance velocity in the soft region to a resonance velocity inside the stiff region. Note that the load velocity is kept constant and the resonance velocity just changes due to higher support stiffness.

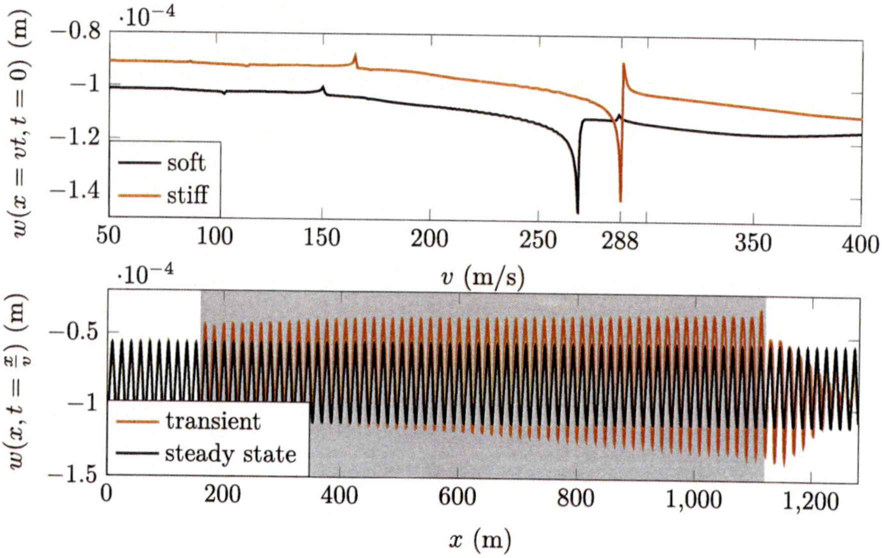

Fig. 5. Response under the load vs velocity in the soft and stiff regions (top panel) and the displacements evaluated under the moving load for resonance in the stiff zone (bottom panel); the stiff zone is marked by the grey background; $p = 1.3$.

The top panel in Fig. 5 presents the resonance velocities for the soft and stiff regions. For $v \approx 288$ m/s, the stiff zone resonates while the soft one does not. The bottom panel in Fig. 5 presents the displacement under the moving load for $v = 288$ m/s. The amplification in the stiff zone is clearly observed with a drastic increase compared to the steady state. At its maximum, the amplification of the displacement is of more than 20% while the amplification of the bending moment (not presented here for brevity) is more than 25%. The increase in response requires many cell lengths to develop, characteristic to resonance; for short stiff zones, resonance might not have time to develop, but for longer ones strong response amplification can develop. It must be mentioned that increasing the damping diminishes the amplification, as expected for resonance.

5 Wave Trapping Inside the Stiff Zone

The stiff zone has a finite length l, and consequently the incoming waves generated by the moving load in the soft region could get trapped inside. Wave trapping could lead to response amplification inside the stiff zone even when the moving load is relatively far away. The conditions for a wave to get trapped in the stiff zone are described in detail in Ref. [7] and are summarized in the following. An approximate condition for wave trapping is that q half-wavelengths of the wave inside the stiff zone is equal to or an integer fraction of l. From this conditions, the wavenumber k_{tr} of the wave trapped in the stiff zone can be determined, and from the dispersion curves (Fig. 2 with the properties of the stiff zone), the corresponding frequency ω_{tr} can be obtained. The incoming wave from the soft zone needs to have the same frequency ω_{tr} and the corresponding wavenumber $k_{\text{tr},2}$ can be determined from the dispersion curves with the properties of the soft zone. So, the incoming wave from the soft zone with wavenumber $k_{\text{tr},2}$ and frequency ω_{tr} will lead to a wave of wavenumber k_{tr} and frequency ω_{tr} in the stiff zone that will get trapped. The load velocity that excites a wave of wavenumber $k_{\text{tr},2}$ and frequency ω_{tr} can be obtained from the kinematic invariant expression.

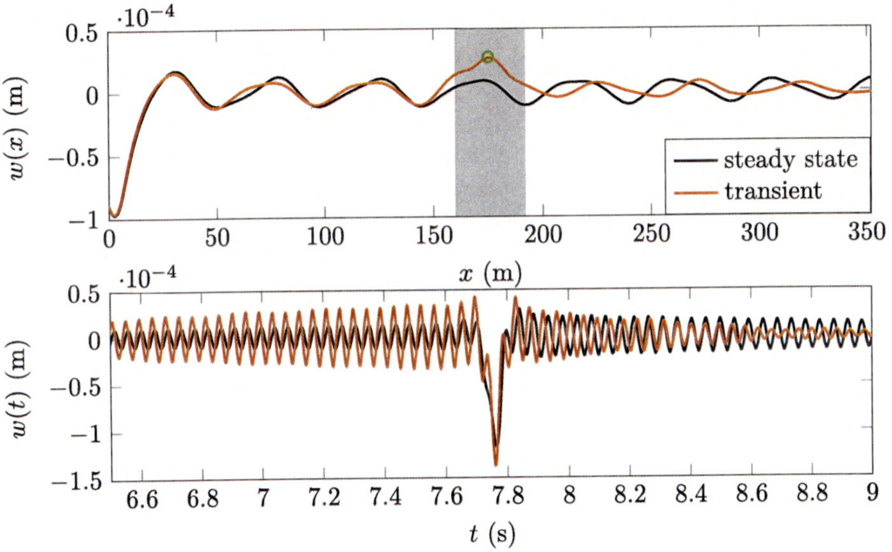

Fig. 6. Snapshot of the time-domain displacements (top panel) and the displacements time-history at the point marked by the green circle (bottom panel) for the situation when the wave is trapped in the stiff zone; the stiff zone is indicated by the grey background; $p = 1.2$ and $v = 270$ m/s.

The top panel in Fig. 6 presents a snapshot of the displacement field where the trapped wave can be clearly observed even though the load is relatively far away. The bottom panel in Fig. 6 presents the displacement time-history at a point inside the stiff zone. The amplitude inside the stiff zone is more than double the one of the steady state when the load is relatively far away from the transition zone; under the moving load, at its maximum, the amplification is around 17%. For slightly different velocities or different lengths of the stiff zone (except for integer fractions of half the wavelength), the amplification vanishes.

6 Conclusions

This paper investigated three phenomena that can lead to response amplification in a continuous and periodic system with a transition zone (described by an increase in support stiffness); the system is representative of a Hyperloop transportation system. The phenomena are the product of a periodic system and a local inhomogeneity, and if one of these characteristics is omitted, the phenomena will not occur.

The wave-interference phenomenon leads to response amplification at the edges of the stiff zone while the passing-to-critical-velocity and wave-trapping phenomena cause amplification inside the stiff zone. While the wave-interference and wave-trapping phenomena lead to response amplification also for short transition zones, the passing-to-critical-velocity requires a long stiff zone for the amplification to develop. Results show that all three phenomena can lead to a response amplification which, at its maximum, is about 20%. Also, the phenomena have been observed at envisioned operational velocities of the Hyperloop vehicles.

Although this amplification would not cause failure of the structure, over time it can lead to increased degradation of the structure as well as discomfort for passengers. Moreover, all three phenomena are diminished with increased damping; therefore, if the designed system does not have sufficient inherent damping, additional passive or active damping measures may be needed. Finally, the three investigated phenomena can be considered as additional constraints for the design parameters at transition zones such that amplifications are avoided.

Acknowledgements. This research is supported by the Dutch Technology Foundation TTW [Project 15968], part of the Netherlands Organisation for Scientific Research (NWO), and which is partly funded by the Ministry of Economic Affairs.

References

1. Mead, D.J.: Wave propagation in continuous periodic structures: research contributions from southampton. J. Sound Vib. **190**, 495–524 (1996)
2. Jezequel, L.: Response of periodic systems to a moving load. J. Appl. Mech. Trans. ASME **48**(3), 613–618 (1981)
3. Steenbergen, M.J.M.M.: Physics of railroad degradation: The role of a varying dynamic stiffness and transition radiation processes. Comput. Struct. **124**, 102–111 (2013)

4. Fărăgău, A.B., Metrikine, A.V., van Dalen, K.N.: Transition radiation in a piecewise-linear and infinite one-dimensional structure-a Laplace transform method. Nonlinear Dyn. **98**, 2435–2461 (2019)
5. Fărăgău, A.B., Mazilu, T., Metrikine, A.V., Lu, T., van Dalen, K.N.: Transition radiation in an infinite one-dimensional structure interacting with a moving oscillator-the Green's function method. J. Sound Vibration **492** (2021)
6. Fărăgău, A.B., Keijdener, C., de Oliveira Barbosa, J.M., Metrikine, A.V., van Dalen, K.N.: Transition radiation in a nonlinear and infinite one-dimensional structure: a comparison of solution methods. Nonlinear Dyn. **103**(2), 1365–1391 (2021). https://doi.org/10.1007/s11071-020-06117-0
7. Fărăgău, A.B., de Oliveira Barbosa, J.M., Metrikine, A.V., van Dalen, K.N.: Dynamic amplification in a periodic structure with a transition zone subject to a moving load: three different phenomena. Math. Mech. Solids **27**(9), 1740–1760 (2022)
8. Floquet, G.: Sur les Equations Differentielles Lineaires a Coefficients Periodiques. Annales Scientifics de l'Ecole Normale Superieur **12**, 47–88 (1883)
9. Vesnitskii, A.I., Metrikin, A.V.: Transition radiation in mechanics. Phys. Usp. **39**(10), 983–1007 (1996)
10. Brillouin, L.: Wave Propagation in Periodic Structures. Dover Publications, Inc., ii ed. (1953)

Gyroscopic Periodic Structures for Vibration Attenuation in Rotors

André Brandão[1,2](✉), Aline Souza de Paula[1], and Adriano Fabro[1]

[1] University of Brasília, Department of Mechanical Engineering, 70.910-900 Brasília, DF, Brazil
[2] Petrobras, Petróleo Brasileiro S.A., Rio de Janeiro, RJ, Brazil
andrebrandao@petrobras.com.br
http://www.unb.br, http://www.petrobras.com.br

Abstract. Rotordynamics and periodic mechanical structures - or mechanical metastructres - are two widely explored fields of knowledge in dynamical systems, each being its own universe with very specific practices and interests. The latter has been of particularly growing interest to the scientific community over the past years, and its expansion has found several intersections with the most diverse fields of study, such as acoustics, optics and seismology. However, the combination of mechanical metastructures and rotating axially symmetric systems - rotordynamics - is still a largely unexplored subject. In this work we propose the study of a simple rotor system with periodically attached resonators. Its dynamical behavior is investigated using the system's transfer matrix. The gyroscopic effects acting both on the rotor and the resonators create interesting phenomena such as polarization of the whirl modes and 'invisibility' effects. A rainbow-type arrangement of resonators is also explored for translational and flexural resonators, showing good enhancement of attenuation performance. The interaction between resonators modes and rotor modes is explored and the impacts on vibration transmissibility is evaluated. The translational resonators showed significantly superior attenuation performance, both in amplitude and bandwidth. However, unique effects such as polarized invisibility and speed dependent bandgap formation where only achievable for the flexural resonators.

Keywords: Periodic structures · Rotordynamics · Vibration attenuation · Rainbow matematerials · Flexural resonators

1 Introduction

Beam models traditionally used for the study of periodic structures for vibration attenuation, often referred to as metamaterial beams, are considered to develop planar motion, i.e. a single transverse direction of wave propagation is evaluated. In either analytic descriptions of infinite structures, usually using the Plane Wave Expansion method (PWE) [9] or the Bloch-Floquet Theorem approach, or in numerical analyses of finite components applying FEM with 2 degrees of freedom

Z. Dimitrovová et al. (Eds.): WMVC 2022, MMS 125, pp. 662–671, 2023.
https://doi.org/10.1007/978-3-031-15758-5_68

(DoF) elements, only one of the orthogonal transverse directions is included in the equations of motion. This approach is useful for most applications, since the displacements in orthogonal directions - horizontal and vertical - can be considered to be independent.

If, however, we consider the beam to be rotating around its axial direction, gyroscopic effects will couple the vertical and horizontal motion equations. This gives rise to interesting effects that were not yet fully explored in the literature. The work of [2] has shed some light on the wave propagation phenomena arising from spinning beams due to the gyroscopic effects. It has been shown that the wave propagation in such structures generates polarized helical wave structures whose dispersion properties diverge with increasing rotation speed. The effects of polarization are studied in deeper detail in the present work by evaluating the effect of backward and forward excitation in a finite rotating structure with periodically attached resonators.

Although the dynamics of rotating shafts has been extensively studied in the field of rotordynamics, not much effort has yet been put on the description of wave propagation phenomena and bandgap formation mechanisms in these structures. The system's speed dependent dynamic properties open new paths for the application of vibration attenuation mechanisms. Using the concept of locally resonant periodic structures and exploring the flexural and translational degrees of freedom, we are able to create interesting dynamical behaviors with promising vibration attenuation characteristics.

These unique dynamic behaviors are especially interesting when local resonance phenomena involve the system's flexural degrees of freedom. It has been shown by [7] that translational resonators have superior overall attenuation performance for conventional metamaterial beams. However the inclusion of flexural resonators coupled to the main structure enables interesting flexibility in bandgap tuning and targeted vibration attenuation in a single whirl direction.

This work describes a rotor system with attached translational and flexural resonators. The model was built using the software Ross, an open source, Python based rotordynamics tool [8]. The system's vibration response to synchronous forward and backward excitation is calculated and analyzed. The application of graded resonators, using the rainbow metamaterial concept, for enhanced attenuation performance is also explored. The mode structures are evaluated, providing a preliminary understanding of the interaction between resonators and main structure across different frequencies. Finally, a visual evaluation of the modes distribution for different resonator arrangements is presented providing further insight on the nature of the observed behaviors.

2 System Description

The picture representation in Fig. 1 depicts the rotor model studied in this article, with local resonators represented by the blue attachments. The spring-damper structures shown in the figure represent the elastic supports (bearings) and the red markers are lumped mass and inertia elements that represents impellers or other mechanical components commonly attached to rotors.

The FEM model was build using ROSS - Rotordynamic OpenSource Software, an open-source Python based library specific for rotordynamic analyses [8]. The original model was then modified to account for the introduction of local resonators in different configurations.

Table 1 shows the parameters used for the rotor model. The values of k_{xx}, k_{yy}, k_{xy} and k_{yx} represent the direct and cross-coupled stiffness coefficients of the rotor supports; c_{xx} and c_{yy} are the support's direct damping coefficients and I_t and I_p are the transverse and polar inertia of the disk elements.

Table 1. Rotor parameters.

Diameter	50 mm	N^o of disks	3
N^o of Elements	30	Disk mass	32,59 kg
Length	1.5 m	I_t	1.78 $kg.m^2$
Mass	120 kg	I_p	0.33 $kg.m^2$
Material	Steel	$k_{xx} = k_{yy}$	$9 * 10^6 N/m$
$k_{xy} = k_{yx}$	0	$c_{xx} = c_{yy}$	$500 Ns/m$

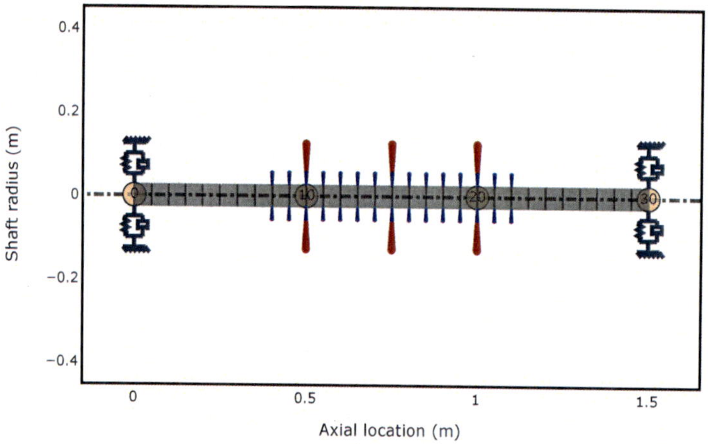

Fig. 1. Rotor model including local resonators in blue.

The original rotor's Campbell Diagram, considering the resonators to be rigid lumped mass attachments, is shown in Fig. 2 in which the 8 first natural frequencies can be identified. The synchronous component is shown as a dashed blue line. The first two modes, both occurring close to $60 rad/s$, are the cylindrical rigid body modes, which is why the separation of its forward and backward components do not become clearly visible. The second set of modes occurring around $200 rad/s$ corresponds to the conical rigid body modes, which, having significant amplitudes of the angular degrees of freedom, generate clearly visible separation of forward and backward components. The remaining 4 modes are actual flexural modes, or bending modes, of the rotor shaft and also clearly show the separation between forward and backward modes.

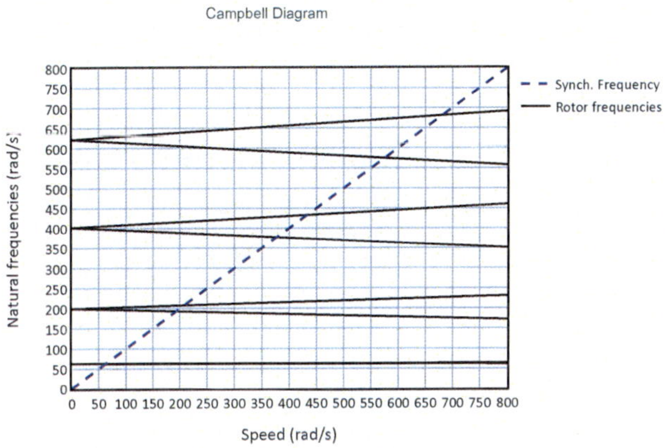

Fig. 2. Campbell Diagram for the original rotors.

The points in which the dashed line intersects each of the modes are known as the system's critical speeds. These speeds usually represent the occurrence of the maximum vibration amplitudes in a rotor system, for either forward or backward excitation.

In this work we explore the neutralization of the first flexural critical speed which is located around $375 rad/s$. This mode was chosen because, when compared to the rigid body modes, it tends to have greater energy concentration in the shaft's bending motion, which makes it a better target for flexural resonators.

As it was briefly explored by [7], the effectiveness of bandgap formation when using flexural resonators is limited if compared to translational elements. However, there are specific characteristics of the flexural resonators that can be explored when considering rotor-like structures. In such structures, the system's motion equations are a function of the rotation speed Ω and it contains components that couple the orthogonal angular directions. The system's equation becomes as described in Eq. 1.

$$[\mathbf{M}]\ddot{\mathbf{x}} + ([\mathbf{C}] + \Omega[\mathbf{G}])\dot{\mathbf{x}} + [\mathbf{K}]\mathbf{x} = 0 \tag{1}$$

where $[\mathbf{M}]$, $[\mathbf{C}]$ and $[\mathbf{K}]$ are the mass, damping and stiffness matrices respectively, $\mathbf{x}$ is the vector of displacements with time derivatives represented by the dotted letters. Each rotor n^{th} node, containing 4 DoFs, has its displacement $\mathbf{x}_n$ described by in the form of 2.

$$\mathbf{x}_n = \begin{Bmatrix} x \\ y \\ \theta_x \\ \theta_y \end{Bmatrix} \tag{2}$$

The matrix $[\mathbf{G}]$ is the gyroscopic matrix which appears multiplied by the rotation speed Ω. The $[\mathbf{G}]_n$, gyroscopic matrix of the n_{th} shaft node, is an antisymmetric matrix in the form shown in Eq. 3, considering counter-clockwise (X to Y) rotation.

$$[\mathbf{G}]_n = \begin{bmatrix} 0 & 0 & 0 & 0 \\ 0 & 0 & 0 & 0 \\ 0 & 0 & 0 & I_p^n \\ 0 & 0 & -I_p^n & 0 \end{bmatrix} \tag{3}$$

As it becomes clear from Eq. 3, when $\Omega > 0$ the $[\mathbf{G}]$ matrix has the effect of coupling the angular velocity $\dot{\theta}_x$ to a moment M_y in the $-\theta_y$ direction and the angular velocity $\dot{\theta}_y$ creates a moment M_x in the θ_x direction. This is the source of the natural frequency separation observed in the Campbell Diagram from Fig. 2.

Equations 4 and 5 show the resonator element matrices, which have 8×8 shape representing the coupling between the 4 DoFs from the resonator and the 4 DoF from the respective shaft node.

$$[\mathbf{M}]_r = \begin{bmatrix} 0&0&0&0&0&0&0&0 \\ 0&0&0&0&0&0&0&0 \\ 0&0&0&0&0&0&0&0 \\ 0&0&0&0&0&0&0&0 \\ 0&0&0&0&m_r&0&0&0 \\ 0&0&0&0&0&m_r&0&0 \\ 0&0&0&0&0&0&I_{t_{res}}&0 \\ 0&0&0&0&0&0&0&I_{t_{res}} \end{bmatrix} \quad ; \quad [\mathbf{G}]_r = \begin{bmatrix} 0&0&0&0&0&0&0&0 \\ 0&0&0&0&0&0&0&0 \\ 0&0&0&0&0&0&0&0 \\ 0&0&0&0&0&0&0&0 \\ 0&0&0&0&0&0&0&0 \\ 0&0&0&0&0&0&0&0 \\ 0&0&0&0&0&0&0&I_{p_{res}} \\ 0&0&0&0&0&0&-I_{p_{res}}&0 \end{bmatrix} \tag{4}$$

$$[\mathbf{K}]_r = \begin{bmatrix} k_0&0&0&0&-k_0&0&0&0 \\ 0&k_0&0&0&0&-k_0&0&0 \\ 0&0&k_1&0&0&0&-k_1&0 \\ 0&0&0&k_1&0&0&0&-k_1 \\ -k_0&0&0&0&k_0&0&0&0 \\ 0&-k_0&0&0&0&k_0&0&0 \\ 0&0&-k_1&0&0&0&k_1&0 \\ 0&0&0&-k_1&0&0&0&k_1 \end{bmatrix} \quad ; \quad [\mathbf{C}]_r = c_p[\mathbf{K}]_r \tag{5}$$

The proportional damping parameter was considered as $c_p = 10^{-4}$. To define the mass and inertia parameters m_r, $I_{t_{res}}$ and $I_{p_{res}}$, each resonator was considered to be a cylindrical element with outer diameter $d_o = 140\,\text{mm}$, inner diameter $d_i = 60$ mm and length $L = 40$ mm. The stiffness parameters k_0 and k_1 were calculated using Eq. 6 as a function of the desired frequency to which the resonators would be tuned. The translational natural frequency was defined as ω_0 and the flexural natural frequency ω_1.

$$k_0 = m_r \omega_0^2 \quad ; \quad k_1 = I_{t_{res}} \omega_1^2 \tag{6}$$

The effects of variability on resonators' mass properties on the final systems' attenuation performance was also evaluated. This has been investigated in numerous other related works that usually refer to the strategy as "Rainbow Metastructures" [1,3–6,9]. For this purpose, the $[\mathbf{M}]_r^i$ matrix of the ith resonator element was multiplied by a factor ε_i given by Eq. 7.

$$\varepsilon_i = 1 + 0.3 \left(\frac{2i}{N} - 1 \right)^n \tag{7}$$

where N is the total number of resonators and $n = 1, 3, 5...$ is a variable that enables manipulation of the mass gradient distribution.

3 Results for Translational Resonators

In this section we shall consider $\omega_0 = 375 rad/s$ in such a way that the third critical speed is attenuated by the attached resonators. The flexural natural frequency of the resonators was set in such a way that $\omega_1 \gg \omega_0$ and their motion can be considered as purely translational.

Using the system's transfer matrix, the system's response was calculated for two different load cases. Each case consists of forward and backward unitary rotating forces applied to the shaft's first node. The amplitude and phase of the orthogonal components of these forces can be written, in polar notation, as

$$F_{x_f} = 1\underline{/0^\circ} \quad \text{and} \quad F_{y_f} = 1\underline{/-90^\circ} \tag{8}$$

for the forward excitation case and as

$$F_{x_b} = 1\underline{/0^\circ} \quad \text{and} \quad F_{y_b} = 1\underline{/90^\circ} \tag{9}$$

for the backward excitation case.

The two loads represent, respectively, forward and backward rotating forces applied to the shaft. The force was applied to the leftmost shaft node and the response is evaluated on the rotors opposite side.

Figure 3 shows the response heatmap with superimposed Campbell Diagram, illustrating the effect of the addition of translational resonators. The grey lines show the modes in which the relative translational displacement of the resonators are, in average, greater then the absolute displacement of the rotor nodes. It becomes clear that around the chosen frequency $\omega_0 = 375 rad/s$ the resonators' displacement dominate all existing modes and the response is strongly attenuated both for forward and backward excitation.

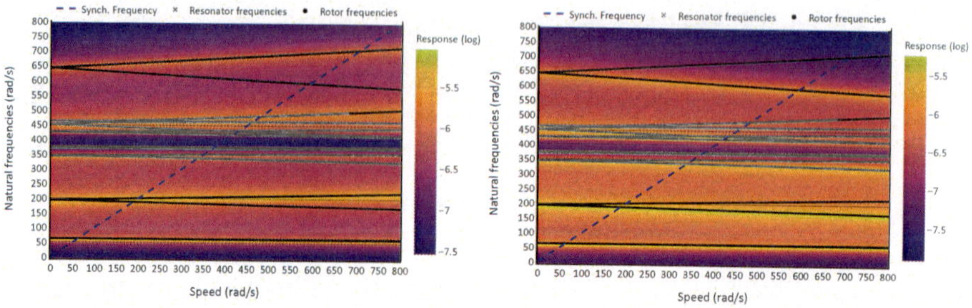

(a) Forward excitation. (b) Backward excitation.

Fig. 3. Campbell diagram and response heatmap for identical translational resonators.

Including the variability in the resonators' mass properties with $n = 1$ (see Eq. 7), the same response heatmap and Campbell diagram is shown in Fig. 4.

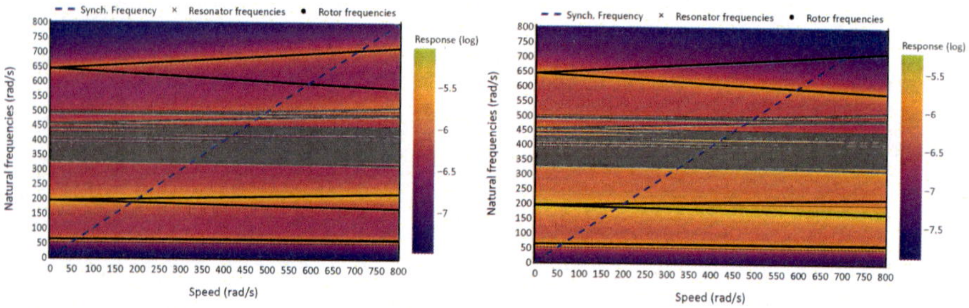

(a) Forward excitation. (b) Backward excitation.

Fig. 4. Campbell diagram and response heatmap for translational rainbow resonators.

It is evident that the translational resonator arrangement produces consistent, broadband vibration attenuation around the chosen tuned frequency. The added variability produced significant increase on the attenuated frequency range, almost eliminating the effect of edge frequencies occurring around the target frequency, although the maximum attenuation seems to be reduced.

4 Results for Flexural Resonators

A very similar approach was followed to tune the resonators flexural natural frequency to attenuate the desired frequency range, setting at this time $\omega_1 \ll \omega_0$ to isolate the translational DoFs. However, in this case the actual natural frequency of the resonators s a function of the rotation speed Ω, and the tuning parameter ω_1 represents the resonators natural frequency when $\Omega = 0$.

Considering that we intend to neutralize the rotor's synchronous vibration at target speed $\Omega = \Omega_t$ using the resonators' backward natural frequency, the resonator's base natural frequency must be set to $\omega_1 > \Omega_t$. For the target frequency of $\Omega_t = 375 rad/s$ we must then tune the flexural resonators to $\omega_1 = 634.4\,\mathrm{rad/s}$.

The Campbell Diagram superimposed to the response heatmap is shown on Fig. 5 provides a graphical explanation for the different tuning method for flexural resonators. The lower branch of resonators' natural frequencies decreases with rotation speed intersecting the synchronous line at the desired target frequency.

A few interesting behaviors arise in this configuration and can be clearly observed in these plots. First of all, there seems to be no interaction at all between the resonators and the forward components of modes and excitation. The forward modes, the ones pointing upwards in the Campbell diagram, shows virtually no interaction with the resonator modes, as they intersect each other with no visible change. Similar absence of interaction can be noticed for the response to forward excitation on Fig. 5a, in which the amplification areas are not affected when crossing the resonator modes.

The backward modes, on the other hand, exhibit strong interaction with the resonator dynamics. We can see a consistent lower amplitude zone around the resonators' backward modes on Fig. 5b, and they strongly interact with the rotor modes around the target frequency of $375 rad/s$. In this intersection area, we can see that the edge frequencies, identifiable by the brighter amplification zones on the response heatmap, are created around the attenuation zone that follows the resonator modes. Therefore, however the target frequency is significantly attenuated, the frequency bandwidth of attenuation is relatively narrow and the synchronous line still crosses regions with higher amplification.

Including the variability in the resonators mass properties with $n = 3$ (see Eq. 7), the Campbell Diagram becomes as shown in Fig. 6, in which it becomes evident the significantly broader range of frequencies affected by the resonator modes.

The mass gradient distribution with $n = 3$ gives significantly enhanced attenuation performance, being able to neutralize even the edge modes on both sides

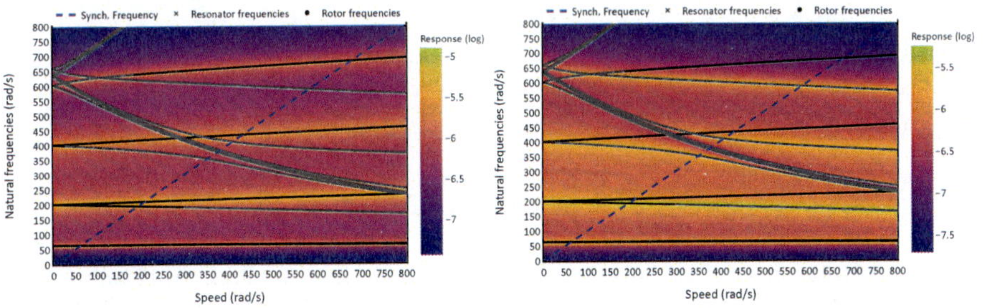

(a) Forward excitation. **(b)** Backward excitation.

Fig. 5. Campbell diagram and response heatmap for identical flexural resonators.

of the 375 rad/s target frequency, as it creates a broader pathway to the synchronous line with no sever response amplification.

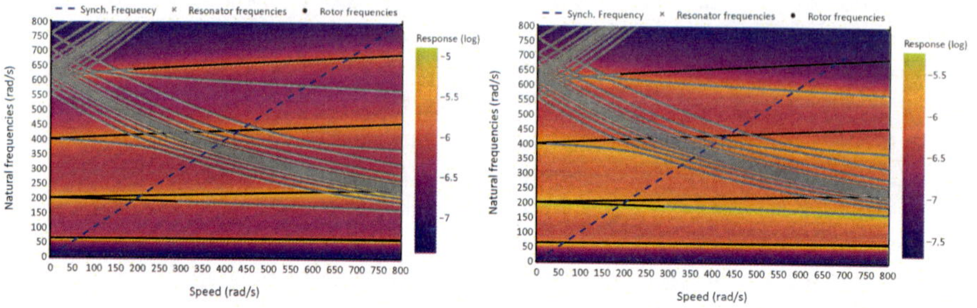

(a) Forward excitation.

(b) Backward excitation.

Fig. 6. Campbell diagram and response heatmap for flexural rainbow resonators.

As a general observation we may point out that the resonators are invisible to forward excitation, while strongly restricting the vibration transmissibility with backward excitations, especially in the rainbow arrangement. Of course, the opposite result should be achieved by tuning the forward mode of the resonators to a specific desired frequency. We could also imagine a flexural resonator arrangement tuned to arbitrary forward and backward excitation frequencies if we are free to set the rotating speed of each resonator to an independent value.

5 Conclusions

The work has explored the concept of periodic locally resonant metastructures applied to a simple rotor system. Application of flexural and translational resonators revealed interesting phenomena not yet reported in past research.

The translational resonators showed to provide strong attenuation capabilities for both backward and forward excitations. The application of a mass gradient in the array of resonators, a rainbow-type arrangement, generated significantly enhanced bandgap width, with a slight decrease in its depth.

The flexural resonators, in the specific settings shown in this work, interacted with the backward whirl oscillations, but seemed to be virtually invisible to the forward excitations. The application of a rainbow-type arrangement to this setting created almost perfect flattening of the targeted critical speed, without the creation of significant edge frequency amplification regions.

Acknowledgements. The authors are grateful to the Brazilian National Council of Research (CNPq - Brazil), processes number 310850/2019-3, 433730/2018-8, 420304/2018-5 and 314168/2020-6, the São Paulo Research Foundation (FAPESP - Brazil), project number 2018/15894-0, the Federal District Foundation for Research Support (FAPDF -Brazil), project number 00193-00000766/2021-71 and Petrobras - Petróleo Brasileiro S.A. for the support.

References

1. Cai, G., Lin, Y.: Localization of wave propagation in disordered periodic structures. AIAA J. **29**(3), 450–456 (1991). https://doi.org/10.2514/3.10599
2. Chan, K., Stephen, N., Reid, S.: Helical structure of the waves propagating in a spinning timoshenko beam. Proc. Roy. Soc. A: Math. Phys. Eng. Sci. **461**(2064), 3913–3934 (2005)
3. Hodges, C.: Confinement of vibration by structural irregularity. J. Sound Vib. **82**(3), 411–424 (1982). https://doi.org/10.1016/S0022-460X(82)80022-9
4. Hodges, C., Woodhouse, J.: Vibration isolation from irregularity in a nearly periodic structure: theory and measurements. J. Acoustical Soc. Am. **74**(3), 894–905 (1983). https://doi.org/10.1121/1.389847
5. Meng, H., Chronopoulos, D., Bailey, N., Wang, L.: Investigation of 2d rainbow metamaterials for broadband vibration attenuation. Materials **13**(22), 1–9 (2020). https://doi.org/10.3390/ma13225225
6. Meng, H., et al.: 3d rainbow phononic crystals for extended vibration attenuation bands. Sci. Rep. **10**(1), 1–9 (2020)
7. Sun, H., Du, X., Pai, P.F.: Theory of metamaterial beams for broadband vibration absorption. J. Intell. Mater. Syst. Struct. **21**(11), 1085–1101 (2010)
8. Timbó, R., Martins, R., Bachmann, G., Rangel, F., Mota, J., Valério, J., Ritto, T.G.: Ross - rotordynamic open source software. Journal of Open Source Software **5**(48), 2120 (2020). https://doi.org/10.21105/joss.02120, https://doi.org/10.21105/joss.02120
9. Xiao, Y., Wen, J., Wen, X.: Broadband locally resonant beams containing multiple periodic arrays of attached resonators. Phys. Lett. A **376**(16), 1384–1390 (2012). https://doi.org/10.1016/j.physleta.2012.02.059

Improving Locally Resonant Metamaterial Performance Predictions by Incorporating Injection Moulding Manufacturing Process Simulations

Kristof Steijvers[1,2](✉), Claus Claeys[2,3], Lucas Van Belle[2,3], and Elke Deckers[1,2]

[1] Department of Mechanical Engineering, KU Leuven, Campus Diepenbeek, Wetenschapspark 27, 3950 Diepenbeek, Belgium
[2] DMMS Core Lab, Flanders Make, Brussels, Belgium
[3] KU Leuven, Department of Mechanical Engineering, Celestijnenlaan 300B box 2420, 3001 Heverlee, Belgium

Abstract. Metamaterials are artificial structures which have been engineered to have properties that are not common in nature. So-called locally resonant metamaterials are comprised of a host structure in or onto which resonant structures are added on a sub-wavelength scale. Their interaction leads to stopbands, which are frequency ranges in which no free wave propagation is allowed, causing strong vibration reduction. These stopbands are typically predicted using the finite element method, assuming that geometry and material properties are known.

At present, locally resonant metamaterials are mainly produced as ad-hoc demonstrators to validate academic findings. However, the currently used manufacturing approaches are not suited for mass production. Additionally, manufacturing induced changes in LRM geometry and material properties are hard to account for in the early design process, resulting in an off-design metamaterial performance.

To enable widely applicable locally resonant metamaterial solutions with robust performance by design, injection moulding is a promising manufacturing process, as it comes with low cycle times and costs when used for mass production. Although good process repeatability in terms of the dynamic response of produced resonators recently has been shown, significant off-design performance can still be present due to imprecise knowledge on geometry and material properties of the realisations after manufacturing. To address this gap, this paper proposes the use of dedicated injection moulding process simulations to update structural dynamic finite element metamaterial models. Based on three resonator realisations it is shown that incorporating the outcomes of process simulations can clearly improve the resonance frequency predictions of manufactured resonators.

Keywords: Vibro-acoustics · Metamaterials · Injection moulding · Manufacturing · Finite element modelling · Injection moulding simulation

Z. Dimitrovová et al. (Eds.): WMVC 2022, MMS 125, pp. 672–680, 2023.
https://doi.org/10.1007/978-3-031-15758-5_69

1 Introduction

Locally resonant metamaterials (LRMs) have come fore as promising engineering solutions which can combine lightweight properties with pronounced noise and vibration attenuation capabilities [1–4]. Therefore, frequency zones of strongly reduced vibration response, stopbands, are created by attaching local resonant structures on a subwavelength scale. These zones can be predicted by means of dispersion diagrams, which are typically calculated by combining unit cell modelling with Bloch-Floquet boundary conditions in a Finite Element (FE) model [5].

Although the potential of LRMs is demonstrated through a large number of demonstrators [2–6] these are currently adhoc realisations since the addition or inclusion of the resonant elements is not straightforward. On the one hand, add-on approaches have been used wherein pre-manufactured resonators are added to the host structure. For example, separate laser cut resonators [7, 8] are produced whereas additive manufacturing (AM) is used to produce entire patches of resonators [4] which both need to be manually glued onto the host structure. On the other hand, integrated structures incorporating both host structure and resonators have been realised. Here, for instance thermoforming [3] and punching [6] have been used to fabricate LRM panels with integrated resonators while a post-production milling or bending step is required to create the resonators. Additionally, AM has also been used to create integrated LRMs [9]. As these manual gluing or post-production treatment steps are cumbersome and time consuming, these realization approaches do not reach the throughput needed for mass production [10].

Additionally, during the realisation of LRMs, manufacturing processes influence the geometry and material properties accounted for during the design phase, often leading to off-design LRM performance. In case of AM, the obtained geometric inaccuracies, material property variations and anisotropy lead to off-design LRM stopbands [2]. For thermoforming and bending processes, the hard to control local thinning of the sheets also highly influences the LRM behavior [3]. Currently, off-design metamaterial performance after manufacturing is typically dealt with by updating manufactured geometry and material parameters in iterative production steps. Therefore, a clear need arises to anticipate for manufacturing effects during the LRM design.

In view of robust mass-production of LRMs, the high throughput and affordable injection moulding (IM) process is recently explored. Although the repeatability of the IM process for the manufacturing of resonators is already shown to be promising [11], a significant difference between predicted and experimentally measured resonance frequency is still present. In order to achieve accurate performance predictions, accounting for the IM process, this work proposes the use of IM simulations (Moldex3D [12]) to predict the resulting resonator geometry and mass density distribution after the moulding process. These results will then be used in the resonator FE models to accurately predict the resonance frequency by incorporating the manufacturing process.

In what follows, three different resonator designs are introduced and their IM manufacturing is described. Next, the injection moulding simulations are explained and the resonance frequency of each design was numerically predicted using two different approaches. On the one hand, the calculations start form the original geometry and datasheet material properties. On the other hand, the geometry and mass density distribution resulting from IM simulations are used. Both approaches are compared to experimentally obtained resonance frequencies.

2 Resonator Designs and Injection Moulding

To demonstrate the advantage of linking IM manufacturing process simulations to structural dynamic performance predictions in view of IM LRM design, three types of resonators are introduced which are suitable for IM. Next, the IM equipment and process settings used for resonator production are discussed.

2.1 Resonator Designs

Resonators in LRMs are typically designed to have a desired eigenmode at a specific frequency, in view of creating corresponding stopband behavior. In this paper, three resonator designs are considered which have an out-of-plane eigenmode, often used to create bending wave stopbands (Fig. 1): a cantilever beam (Type A), a ring (Type B) [11] and a double beam (Type C) resonator. The resonator geometries are tailored to suit the IM process: a maximal wall thickness of 3 mm is maintained and small draft angles are foreseen to safeguard proper part ejection out of the mould.

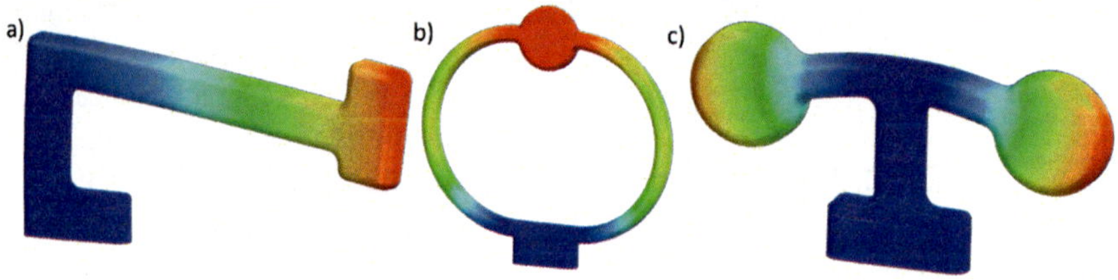

Fig. 1. 1st out-of-plane mode of the a) type A, b) type B and c) type C resonator.

2.2 The Injection Moulding Process

The IM process cycle consists of four typical steps. First, the polymer granulate is heated and mixed by reciprocating screw. Second, the polymer melt is injected in the mould by a speed controlled process. When the cavity is almost filled (switch over point), the remaining cavity volume is filled at a controlled packing pressure which is maintained for a specified time to compensate for the material shrinkage during cooling. After the cooling time, the mould opens and the product is ejected after which the moulding cycle starts again. In this work, a Demag Ergotech IntElect 50/330–100 electric IM machine is used, with a 440 mm long screw of 22 mm diameter and a maximum mould closing force of 500 kN.

For the resonator manufacturing, an in-house available mould is used [11], which makes use of inserts containing the product cavities. This way different products can be created with minimal mould changes, as only the inserts must be replaced. For each resonator, the corresponding insert is designed and machined in an in-house workshop (Fig. 2 a). Two different classes of polymers commonly used in IM are considered for the resonators: semi-crystalline Polypropylene (PP) and amorphous acrylonitrile-butadiene-styrene (ABS). The most important IM process settings are listed in Table 1. A decreasing injection speed profile (Fig. 2 b) is used to allow proper air venting during the end of the filling stage.

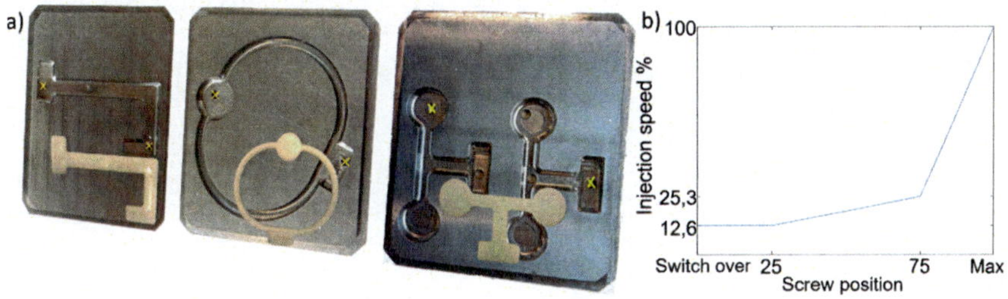

Fig. 2. a) Corresponding mould insert and resonator part for the type A, B and C resonator respectively, yellow crosses indicate the possible injection locations, b) relative injection speed profile.

Table 1. IM process settings.

	Type 1 & 2		Type 3	
	PP	ABS	PP	ABS
Injection temperature [°C]	240	240	240	240
Mould temperature [°C]	40	70	40	70
Packing pressure [MPa]	30	30	20	40
Packing time [s]	10	10	10	10
Cooling time [s]	10	15	10	15
Dosing volume [cm^3]	9,5	9,5	5,32	5,32
Switch over point [cm^3]	3,8	3,8	3,8	3,8
Max. Vol. Inj. Rate [cm^3/s]	19	19	15,2	15,2

3 Numerical Performance Predictions

To be able to more accurately predict the LRM performance accounting for IM manufacturing process influences during the design phase, a forward linked simulation modelling scheme is proposed (Fig. 2). Traditionally, a resonator CAD model is combined with uniform material properties obtained from suppliers' datasheets or material characterization to construct a structural dynamic finite element (FE) model which allows predicting e.g. eigenfrequencies and – when combined with a host structure model – also stopbands (Approach 1). However, this often leads to significant differences with experimental results after manufacturing and/or requires multiple simulation and characterization iterations. To overcome this, instead, this work proposes to first perform IM simulations to obtain the final manufactured resonator geometry and material properties distributions after manufacturing (Approach 2). In this paper specifically, a numerical IM simulation in Moldex3D 2021 is used to produce a manufacturing-induced deformed resonator mesh and corresponding material density property distribution. These are next used to construct a structural dynamic resonator FE model in Simcenter 3D, which is used to compute the clamped-foot eigenfrequencies of the resonator. In what follows both the IM and structural dynamic simulations are further explained (Fig. 3).

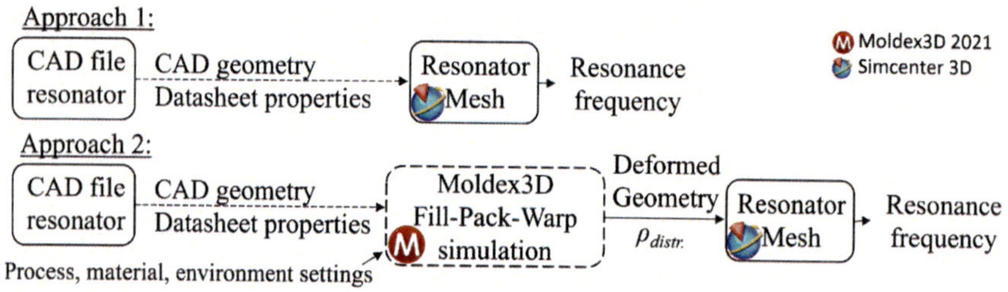

Fig. 3. Flowchart of the IM LRM performance prediction procedure.

3.1 Injection Moulding Simulations

To account for IM process induced manufacturing effects such as shrinkage, warping and non-uniform mass density distribution in the resonator, IM process simulations allow accounting for the underlying process settings such as packing pressure and melt temperature which affect the cavity filling behaviour.

In this work, the IM simulations are used to calculate the deformed part geometry and density distribution by a numerical fill-pack-warp analysis (Fig. 4a). For each resonator type, a boundary layer mesh (BLM) is created with a node seed of 0.5 mm and 5 boundary layers based on the resonator CAD file. The BLM mesh type is used to predict the filling behavior while accurately capturing the high velocity gradients at the interface between the part and the mould by the boundary layers. The simulations are performed using the PP [13] and ABS [14] datasheet properties in Moldex3D, together with the IM settings of Table 1 as input parameters. After the simulations, a dedicated FEA interface in Moldex3D is used to transfer the deformed resonator mesh resulting from the warp simulation and corresponding element density properties to Simcenter 3D. Although a separate material property can be exported for every element, to keep the computational cost limited, an automatically clustered density distribution with 30 material densities is considered in this work.

3.2 Structural Dynamic Simulations

In order to compare the numerical resonance frequency calculations with and without incorporated manufacturing effects to experimentally obtained frequencies, both calculation approaches are performed in Siemens NX. For approach 1, a quadratic solid mesh with an element size of 0.5 mm is created from the resonator CAD file and uniform material properties are assigned (PP: $\rho = 930\ kg/m^3$, $E = 1350$ MPa; ABS: $\rho = 1040\ kg/m^3$, $E = 2240$ MPa). In approach 2, the deformed BLM mesh and corresponding density distribution exported out of Moldex3D (Fig. 4b) are used. The number of elements for each case are summarized in Table 2. In accordance with the measurements of Sect. 4, the resonator foots are clamped using a fixed constraint where after the first out-of-plane eigenfrequency is calculated for each resonator and for both materials (Fig. 4c) (Table 3).

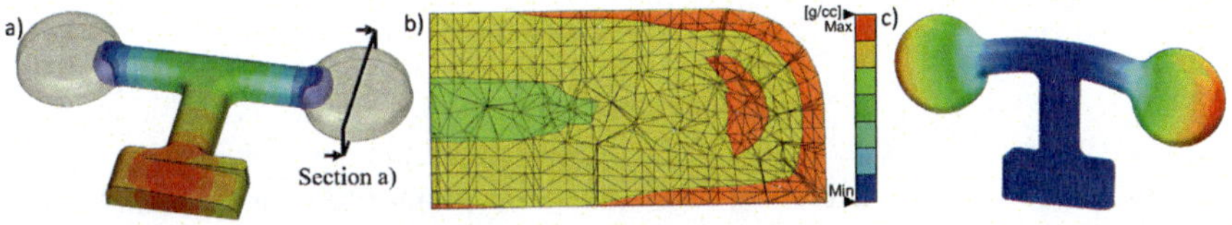

Fig. 4. a) Moldex3D filling simulation of resonator Type 3, b) density distribution result of the resonator in section a, c) resulting 1st out-of-plane mode of the resonator calculated in siemens NX.

Table 2. Number of elements for each approach and resonator type.

Resonator	Number of elements	
	Approach 1	Approach 2
Type 1	75621	124350
Type 2	100582	170406
Type 3	57015	106229

4 Experimental Verification Procedure

For each resonator type and material, 50 resonators are manufactured. Next, the resonance frequencies of their targeted out-of-plane modes are measured. To this end, a dedicated platform is used (Fig. 4a) to clamp the resonators onto an electromagnetic shaker, which excites the resonators at the base using a white noise force input signal. The end-tip velocity responses of each of the clamped resonators are measured using a Polytec PSV-500 scanning laser doppler vibrometer (Fig. 4b) with 625 MHz resolution. Next, the resonance frequencies are determined from the end-tip velocity frequency responses using Simcenter Testlab (Fig. 4c) (Fig. 5).

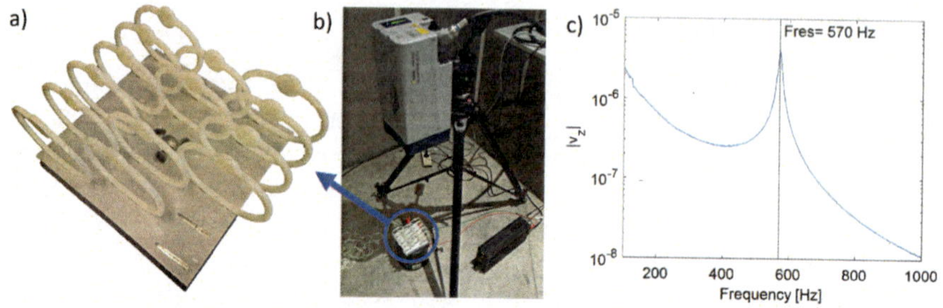

Fig. 5. a) Resonator clamping platform, b) measurement setup, c) example measured velocity frequency response of an ABS ring resonator, with its first resonance frequency occurring at 570 Hz.

5 Results and Discussion

The mean measured and numerically predicted resonance frequencies of the clamped resonators, as well as the relative error between them are summarized in Table 2.

The predicted eigenfrequencies using either approach are observed to systematically underestimate the measured resonance frequencies, for all resonator types and for both materials. The underestimation is attributed to the polymer shrinkage which occurs in the IM process. As the polymer shrinks, the resonator legs become shorter, which causes the resonance frequency of the produced samples to increase. For PP, this underestimation is always larger than for ABS as the semi-crystalline PP material has the highest shrinkage. By incorporating the IM simulation in Approach 2, the relative error is consistently lowered. Using Approach 2, the shrinkage is incorporated in the resonance frequency prediction, which significantly reduces the errors. However, additional geometrical measurements also revealed that the manufactured samples still were smaller than resulting from the IM simulation, which requires further investigation. Nonetheless, combining IM process simulations with structural dynamic simulations consistently reduces the errors on the predicted resonance frequencies of the manufactured resonators, paving the way to improved LRM performance predictions.

Table 3. Comparison of numerically predicted resonance frequencies [Hz] using both default and simulated geometry and mass density to the experimentally measured frequencies.

		Experimental		FEM approach 1		FEM approach 2	
		$F_{res-mean}$	STDEV	F_{res}	Error %	F_{res}	Error %
Type 1	PP	240	2,52	217,8	9,3	228,8	4,7
	ABS	281,8	1,96	265,3	5,9	272,9	3,2
Type 2	PP	475	3,67	432,9	8,9	453,4	4,5
	ABS	565,7	1,42	527,3	6,8	543,7	3,9
Type 3	PP	1240	9,5	1045	15,7	1080	12,9
	ABS	1401	1,9	1272	9,2	1310	6,5

6 Conclusions

Current LRM manufacturing methods are typically not suited for mass-production and also often lead to off-design LRM performance. In this work, the mass-production capable IM process is considered and a first step is made to incorporate manufacturing effects during the LRM design phase. By proposing an approach which links numerical IM process simulations and structural dynamic simulations, improved accuracy of the LRM performance predictions is targeted. Considering three resonator designs, which have been produced using IM and of which the resonance frequencies have been measured, it is shown that the error on the numerically predicted resonance frequencies can strongly be reduced by accounting for the IM process induced shrinkage and mass density distribution. These promising results are also first steps towards improved accuracy for stopband and vibration attenuation predictions of mass-manufacturable IM LRMs.

7 Responsibility Notice

The authors are the only responsible for the printed material included in this paper.

Acknowledgements. Internal Funds KU Leuven are gratefully acknowledged for their support. This research was also partially supported by Flanders Make, the strategic research centre for the manufacturing industry. VLAIO (Flanders Innovation & Entrepreneurship Agency) is also acknowledged for its support. The Research Foundation – Flanders (FWO) is gratefully acknowledged for its support through research grants no.1271621N and no. G0F9922N.

References

1. Liu, Z., et al.: Locally resonant sonic materials. Science (80), **289**(5485), 1734–1736 (2000)
2. Claeys, C.: Design and analysis of resonant metamaterials for acoustic insulation, KU Leuven (2014)
3. de Melo Filho, N.G.R., et al.: Realisation of a thermoformed vibro-acoustic metamaterial for increased STL in acoustic resonance driven environments. Appl. Acoust. **156**, 78–82 (2019)
4. Sangiuliano, L., et al.: Reducing vehicle interior NVH by means of locally resonant metamaterial patches on rear shock towers, SAE Tech. Pap, pp. 1–10 (2019)
5. Claeys, C., et al.: On the potential of tuned resonators to obtain low-frequency vibrational stop bands in periodic panels. J. Sound Vib. **332**(6), 1418–1436 (2013)
6. Droste, M., et al.: Design and validation of production-suited vibroacoustic metamaterials for application in a vehicle door, in Fortschritte der Akustik - DAGA, p. 1647 (2021)
7. Claeys, C., et al.: Design and validation of metamaterials for multiple structural stop bands in waveguides. Extrem. Mech. Lett. **12**, 7–22 (2017)
8. Rocha de Melo Filho, N.G.: Vibro-acoustic resonant metamaterials: from concept to engineering solution. KU Leuven (2020)
9. Van Belle, L., et al.: On the impact of damping on the dispersion curves of a locally resonant metamaterial: modelling and experimental validation. J. Sound Vib. **409**, 1–23 (2017)
10. Wu, X., et al.: Perspective of additive manufacturing for metamaterials development, Smart Mater. Struct. **28**(9), 093001 (2019)

11. Steijvers, K., et al.: On the potential of injection moulding for the production of locally resonant metamaterials. In: COBEM 2021, p. 9 (2021)
12. Moldex3D. https://www.moldex3d.com/en/products/software/moldex3d/
13. SABIC PHC31–81. www.sabic.com/en/products/polymers/polypropylene-pp/sabic-pp
14. NOVODUR P2H-AT- https://www.ineos-styrolution.com/Product/Novodur_Novodur-P2H-AT_SKU401200140442.html

Labyrinth Resonator Design for Low-Frequency Acoustic Meta-Structures

Giuseppe Catapane[1(✉)], Dario Magliacano[1,2], Giuseppe Petrone[1,2], Alessandro Casaburo[2], Francesco Franco[1,2], and Sergio De Rosa[1,2]

[1] PASTA-Lab (Laboratory for Promoting Experiences in Aeronautical Structures and Acoustics), Department of Industrial Engineering - Aerospace Section, Università degli Studi di Napoli "Federico II", Via Claudio 21, 80125 Naples, Italy
giuseppe.catapane@unina.it

[2] WaveSet S.R.L., Via A. Gramsci 15, 80122 Naples, Italy
info@wavesetconsulting.com

https://www.pastalab.unina.it/, https://wavesetconsulting.com

Abstract. Acoustic meta-structure represents a class of composite structure characterized by local resonators that improve the sound absorption. In recent years, the interest for these complex systems is tremendously risen, above all for their capacity to alter waves in low-frequency ranges. Local resonators are generally based on quarter-wavelength resonance, which leads to not negligible problems: the narrow bandwidth of influence and the fact that for low-frequency design, the quarter-wavelength means a wave path too big for conventional problems where limited thickness is a requirement.

Labyrinth resonators (LRs) can be the perfect solution in order to overcome these problems: the tube follows a labyrinth path that enables the resonance effect without increasing too much the sample thickness. The whole length can be stretched in other directions rather than along the thickness.

Even though the resonance behavior of quarter-wavelength tubes and labyrinth resonators is quite similar, the analytical formula of quarter-wavelength tubes is not overall able to predict the natural frequency of LRs. The scope of this project is to prove that labyrinth resonance frequency cannot be predicted through conventional quarter-wavelength formula, maynly because natural frequency depends just on tube length. Moreover, a new analytical formula is proposed by considering the main parameters of labyrinth resonators: number of labyrinth branches, dimension of the single-port air-inlet, and the total length of the labyrinth.

Keywords: Acoustics · Meta-material · Meta-structure · Sound absorption · Labyrinth · Resonance

Z. Dimitrovová et al. (Eds.): WMVC 2022, MMS 125, pp. 681–694, 2023.
https://doi.org/10.1007/978-3-031-15758-5_70

1 Introduction

The acoustic wave propagation and interaction with structures are interesting and challenging engineering field, since a significant part of the research community has been focusing its attention to absorb sound energy, avoid transmission or enhance the sound emissions produced by an object.

Sound absorption is important in many applications and absorption improvement is necessary for vehicles, railways and generally transports, but also for computers or household appliances.

Sound energy can be dissipated thanks to porous media [1,2] or acoustic resonators [3,4]. Porous media are materials made of channels, cracks or cavities, which let the sound waves to travel inside the foam, thus dissipating their energy by viscous and thermal losses; these energy consumption dynamics allow sound absorption over wide frequency ranges [5,6]. Nevertheless, they show low value of sound absorption at low-frequency applications [7].

On the contrary, acoustic resonators maximize the dissipation of the sound energy in correspondence of their resonance frequencies; thus, a source that is very localized in frequency can be easily suppressed through their introduction. Among acoustic resonators, perforated panels [8], Helmholtz resonators [9–11] and quarter-wavelength tubes [12] are the most used solutions. One degree-of-freedom systems like these are tunable, but they perform narrow frequency band-gaps of influence [13].

During the past decades, acoustic meta-materials and meta-surfaces [14–17] have shown several functionalities in the manipulation of sound such as negative refraction [18–20], sub-wavelength imaging [21–23], cloaking [24,25], one-way transmittance [26], and even highly efficient sound absorption within a compact volume.

Meta-materials are made of conventional materials, such as metal and plastic, and some periodic micro-structures, that are shaped, sized and arranged in a way that affect transmission of energy and create unnatural and unconventional material properties. In this way, they can attenuate, stop or guide an elastic wave propagating in a desired path. While meta-materials are generally the combination of several media with particular shape and geometry, a meta-structure only relies on the arrangement of the local micro-structure. An acoustic meta-structure needs to be designed taking into account each acoustics phenomenon, to maximize its reliability. Furthermore, coupling a great amount of structures, like Helmholtz resonators or perforate panels, to reach high value of sound absorption in various frequency regions may result in a complex process of design and production.

In contrast, quarter-wavelength tubes are easy to tune and manufacture: indeed, they are basically cylindrical or square-section tubes with their length as main parameter for the prediction of their resonance frequencies; unfortunately, their length increases with decreasing tuning frequency, then they would be too bulk for a low-frequency industrial application. To overcome this limit, several authors have studied labyrinth resonators (LRs), where the hollow structure is made with a labyrinth shape which enables the resonance effect without increasing too much the sample thickness [27,28]. These interesting solutions are demonstrated to be effective and smart, but the effects due to the change in shape cannot be neglected: the sound wave path is different, then the working frequency of the tubes is not the same as a classical quarter-wavelength one. On the base of this hypothesis, this work wants to demonstrate the shift in frequency of LRs respect to a single branch quarter-wavelength tube, and to formulate an improved and more reliable formula for the prediction of the resonance frequency in LRs. It is important to notice that labyrinth resonators are generally used for the absorption of a source; thus, herein, their study is limited to a single cell, made with one or more labyrinth resonators, assuming that the study of a single cell brings to the same results in the case of a more complex analysis carried out on a complete periodic structure. After all, an acoustic meta-material made with LRs with similar resonance frequencies is designed thanks to the new formulation, showing also a model with great performance in the whole audible range of frequency.

In Sect. 2, the problem is explained, discussing about the geometrical properties of the analyzed configurations; then, the Finite Elements implementation is examined. Afterwards, in Sect. 3, the results are plotted and reported.

2 Definition of the Problem

Quarter-wavelength tubes (Fig. 1) exhibit maximum absorption when the length of the tube L is equal to the quarter of the wavelength $\lambda = c/f$, where c is the wave speed and f is the frequency.

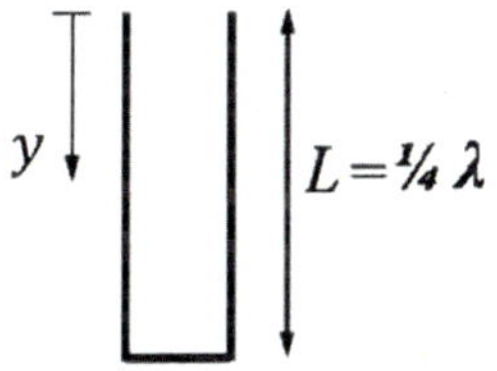

Fig. 1. Example of a quarter-wavelength resonator.

Physically explained, a hard surface can provide absorption through re-radiation of sound that is out of phase with the incident sound. When the tube length is an odd-integer multiple m of the quarter of the wavelength, the reflected wave from the bottom returns a half wavelength that is 180° out of phase respect to the incident wave, thus showing perfect absorption [12]. Therefore, a system like this reaches its maximum absorption when the exciting frequency is equal to one of its resonance frequencies, written as:

$$f_{res} = \frac{(2m-1)c}{4L} \tag{1}$$

where c is the sound wave speed, almost equal to 343 $\frac{m}{s}$ for air. This system can be very useful in order to absorb noise in low-frequency ranges, because foams are generally not so effective in such regions. The main problem is that λ is of the order of meters for low-frequency ranges, leading to bulky systems that are not suitable for most of acoustic applications. For instance, an exciting tone of 100 Hz can be absorbed with a tube 860 mm long; this cannot be applied in any transport configuration, but it would not be suitable for civil applications as well.

For this aim, labyrinth resonators (LRs) represent an efficient solution: a labyrinth resonator is a tube of constant width d, where the sound wave can enter, following a labyrinth path. A solution like this allows to increase the length of the tube without affecting the thickness t of the sample (Fig. 2).

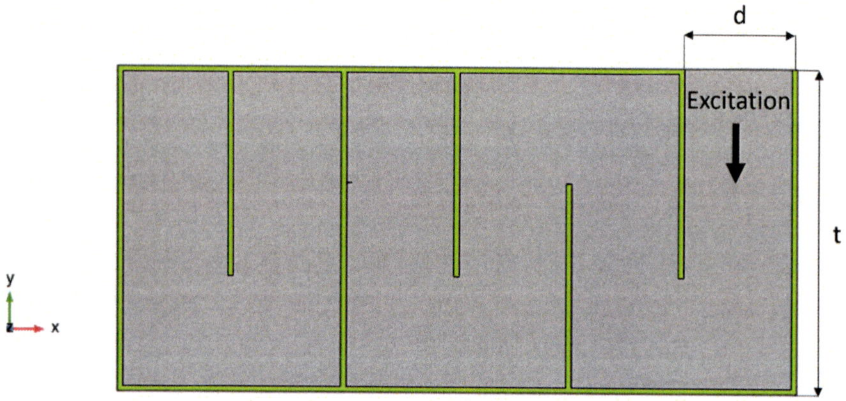

Fig. 2. Example of a labyrinth resonator.

A labyrinth resonator is a good compromise, and the physical phenomenon of quarter-wavelength resonance is still present; nevertheless, the wave path is changed: while the sound wave is linearly crossing the tube in the case of a single tube, the labyrinth shape forces the wave to change direction. It is reasonable to think that the resulting effective length of the wave path is not exactly equal to the whole length of the labyrinth: it may be easier to understand this concept thanks to Fig. 3, where the isobar curves of the total acoustic pressure inside a straight tube (Fig 3a)), and inside a 2D labyrinth resonator (Fig. 3b)) are represented.

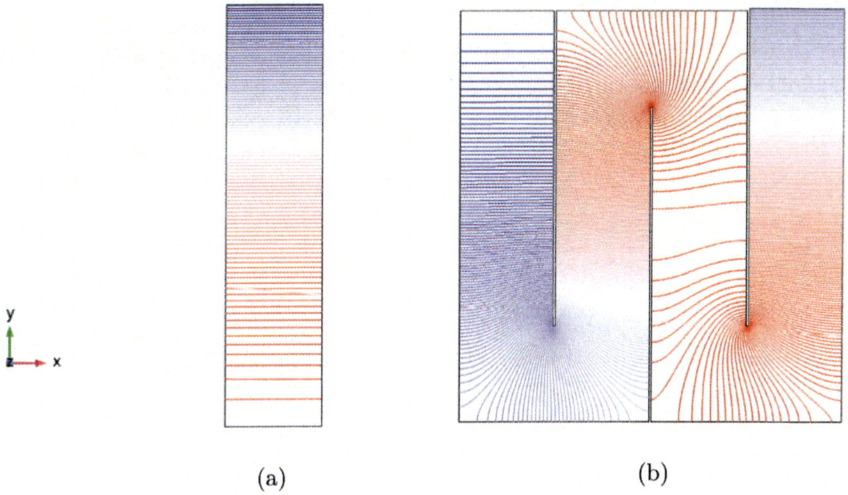

Fig. 3. a) Isobar curves inside a quarter-wavelength tube; b) Isobar curves inside a labyrinth resonator.

The path is clearly affected by the labyrinth; in detail, this effect mainly depends on the number of labyrinth branches n and on the dimension of the tube cross-section d. Thus, an effective length and the respective resonance frequency should be evaluated with a more accurate formula. Following the wave-path along the isobars centreline, it is possible to draw a semi-circumferential pattern in the contact zone of two adjacent branches.

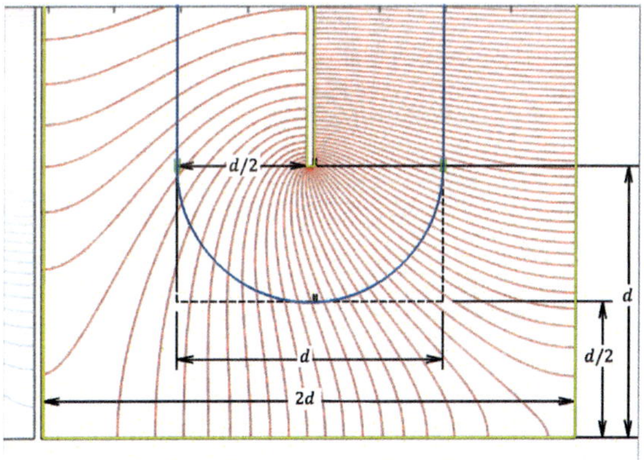

Fig. 4. Zoom in the corner zone of a labyrinth resonator.

This phenomenon is more clear by looking to Fig. 4: with the assumption of a 2D circular path in proximity of the connection between branches, the effective length of the labyrinth is different. Every time there is a corner, the hypothetical length was assumed as $2d$; physically, it is more correct to assume wave-path length equal to $\pi d/2$. This correction is necessary for each corner; for n labyrinth branches, there are $n-1$ corners. In the end, the effective length and the working frequency for $m=1$ have to be written respectively as:

$$L_{eff} = L - (4-\pi)\frac{d}{2}(n-1) \tag{2}$$

$$f_{res} = \frac{c}{4\left[L - (4-\pi)\frac{d}{2}(n-1)\right]} \tag{3}$$

Hereinafter, the resonance formula will be always considered for $m=1$. This is mainly motivated by the fact that this work is focused on low-frequency ranges, and a m-value greater than 1 leads to higher frequency values.

2.1 Geometrical Properties

The typical geometry of a labyrinth resonator is shown in Fig. 2: the plane wave radiation can easily pass through an inlet of width d, following the tube path. The passage between two branches is done with junction of height d, in order to not variate the section of the tube.

In order to prove that the new formula is more correct than the classical quarter-wave length one, four lengths (and then four resonance frequencies) are analyzed with several numbers of branches (from 2 to 10) and for three different width values d. These configurations are reported in Table 1 and compared with the 12 basic configurations with just one branch (standard quarter-wavelength tube). These 12 basic configurations are studied in order to verify if the numerical results are coherent with the analytic formula.

Table 1. Geometrical parameters of the analyzed configurations. The overall number of studied configurations is 97, in addition to the 12 configurations with standard tube.

Conf.	f_{res} [Hz]	L [mm]	d [mm]	N
C1	62.5	1372	5	2,3,4,5,6,7,8,9,10
C2	62.5	1372	10	2,3,4,5,6,7,8,9,10
C3	62.5	1372	20	2,3,4,5,6,7,8,9,10
C4	125	686	5	2,3,4,5,6,7,8,9,10
C5	125	686	10	2,3,4,5,6,7,8,9,10
C6	125	686	20	2,3,4,5,6,7,8,9,10
C7	250	343	5	2,3,4,5,6,7,8,9,10
C8	250	343	10	2,3,4,5,6,7,8,9,10
C9	250	343	20	2,3,4,5,6,7,8
C10	500	172	5	2,3,4,5,6,7,8,9
C11	500	172	10	2,3,4,5,6,7,8
C12	500	172	20	2,3,4

It is possible to notice that, from *C1* to *C8*, the analyses are carried out for n that goes from 2 to 10, while from *C9* the number of branches is limited by the length of the tube. Indeed, the height of the branches must be greater than two times the junction height $2d$. Two examples of labyrinth resonators studied in this paper are reported in Fig. 5, with a nomenclature that make them easier to recognize. Indeed, $CiNj$ stands for the i–th configuration with j–th number of labyrinth branches.

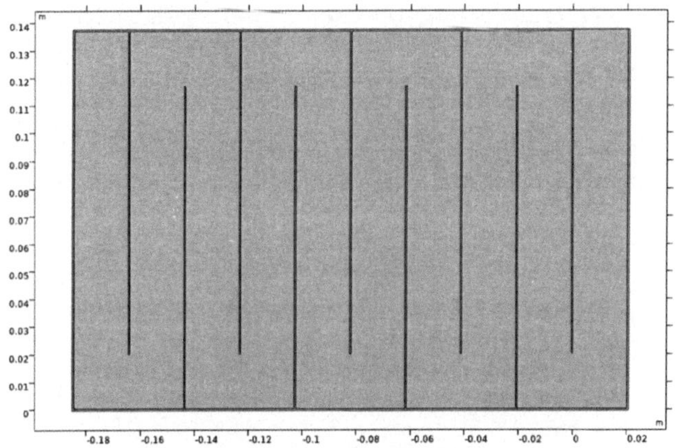

(a) Conf. *C3* with 10 branches (C3N10)

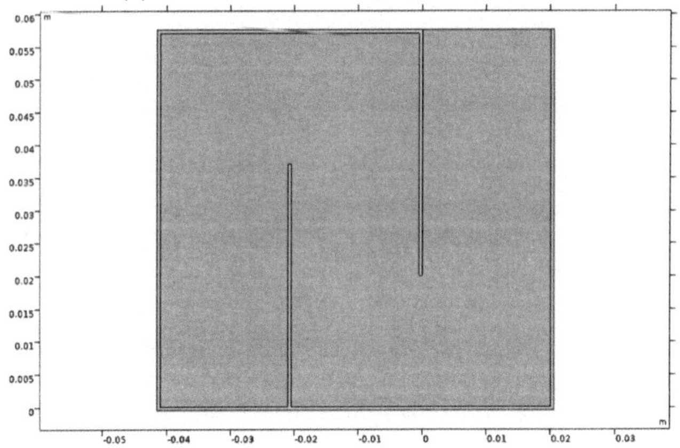

(b) Conf. *C12* with 3 branches (C12N3)

Fig. 5. 2D configuration examples.

2.2 Finite Element Implementation

For what concerns the FE implementation, the module "Pressure Acoustics and Frequency Domain" of COMSOL MultiPhysics is used both as modeling environment and numerical solver.

An example of a meshed LRs is represented in Fig. 6, where walls are defined as rigid (yellow colored). The 2D numerical model (xy-plane) is developed as follows: a plane wave radiation of intensity 1 Pa acts along the negative verse of y-axis in the opened branch of the labyrinth, then following the labyrinth path. Each wall has Sound Hard Boundary Wall (SHBW) as boundary condition, which means that the whole sound wave energy is reflected when it goes against the wall. The propagation of the sound wave is hence normal to the x-axis; an angle of excitation different from the one selected does not affect the labyrinthine resonance frequency; of course with a different angle, LRs harvest less energy respect to a normal excitation. Since the present study is aimed to demonstrate that the QWT resonance frequency cannot be used for LRs, a normal excitation is preferred to diffuse acoustic field or different angle excitation. A detailed description of classical FE formulation and equations can be easily found in the context of the relevant literature [29].

For all configurations reported in Table 1, the mesh consists of triangular elements generated through physics controlled algorithms that are pre-implemented in the software [30]. Nevertheless, the maximum element size of each LR meshed is always lower then 1/4 of the minimum wavelength λ.

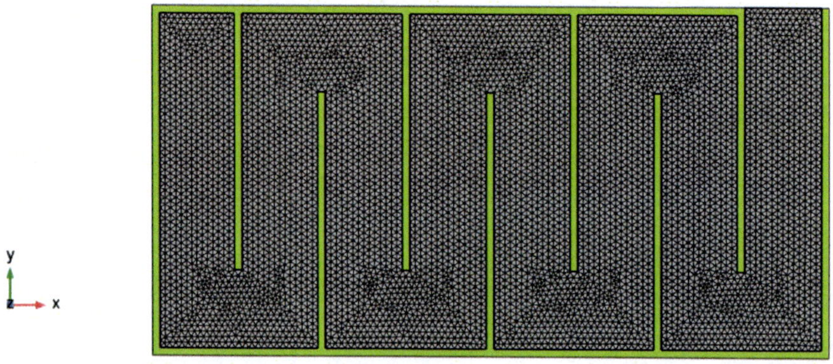

Fig. 6. Mesh example for the 2D numerical models. In detail, this is a LR with first resonance frequency at 500 Hz, with 8 branches and $d = 5$ mm (*Configuration C10N8*).

3 Discussion of Results

In this section, the results of the 12 configurations are compared respect to the analytical formulations taken by Eq. (1) and Eq. (3). Starting from the comparison between Eq. (1) and numerical results (Fig. 7), it is possible to notice that the higher the width d and the number of branches n, the bigger is its relative error respect to the Finite Element (FE) result. The numerical results are then compared with the new formula in Eq. (3), and they are reported in Fig. 8, where the maximum relative error is pair to 3.98%, much lower respect to the maximum relative error obtained with Eq. (1), which is equal to 26.26%.

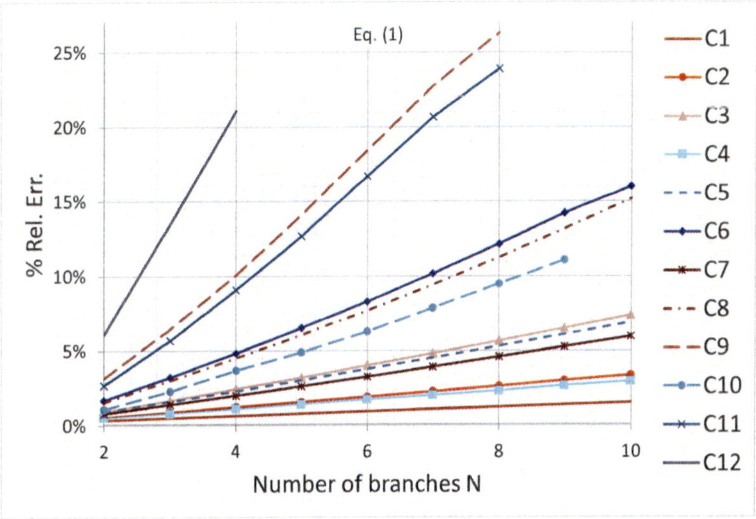

Fig. 7. Relative error of Eq. (1) respect to FEM results, plotted for each configuration.

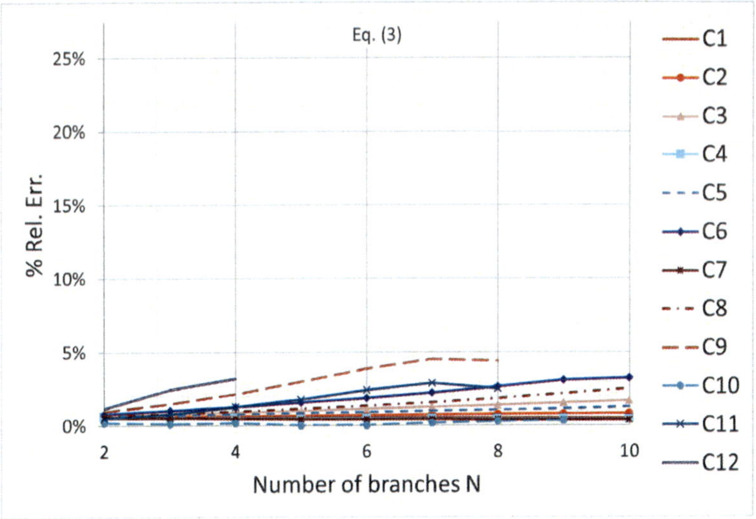

Fig. 8. Relative error of Eq. (3) respect to FEM results, plotted for each configuration.

Four contour plots are represented in Fig. 9; each one is a contour plot of the FE resonance frequency of Labyrinthine Resonators with the same overall length, and thus same resonance frequency according to Eq. (1), properly tuned respectively at $f_{res} = 62.5$ in Fig. 9a, $f_{res} = 125$ in Fig. 9b, $f_{res} = 250$ in Fig. 9c, $f_{res} = 500$ in Fig. 9d. It is noteworthy to highlight how each parameter markedly influence the resonance frequency of LRs, up to shift the resonance frequency of a system even above 100 Hz.

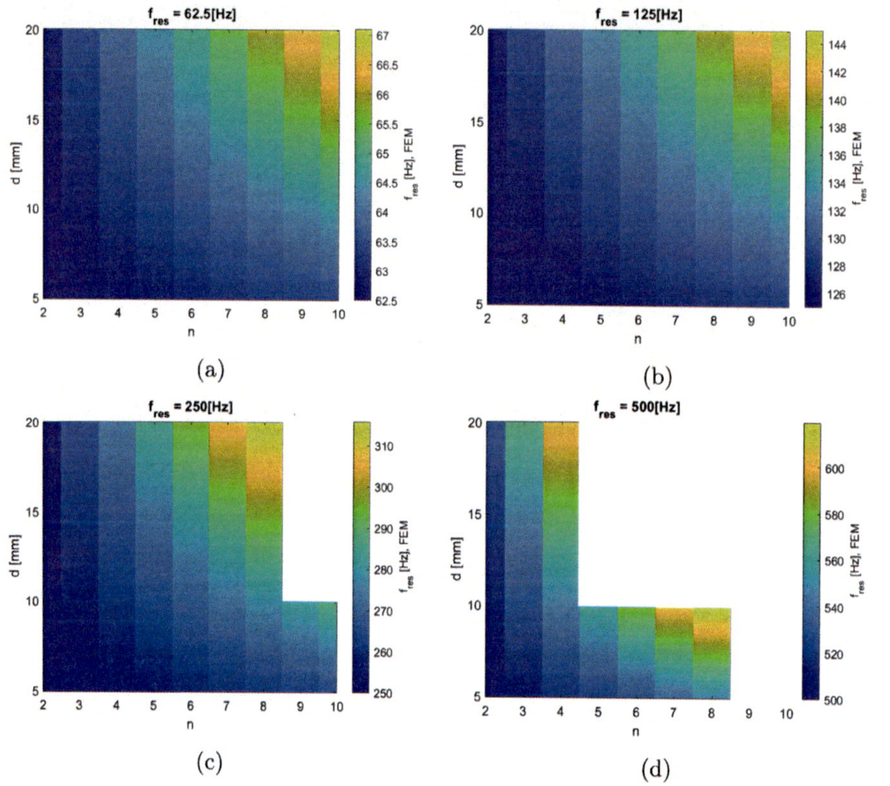

Fig. 9. Contour plot of resonance frequencies of LRs, varying the number of branches n ad the width of the labyrinth d.

Furthermore, plots in Fig. 10 represent relative errors while the number of branches and the width of channels are changing. Plots on the left can be used as carpet plots; in detail, it is possible to design a labyrinth resonator making use of the subplots on the left of Fig. 10 combined with Eq. (1), instead of directly using Eq. (3). The result is the same, but contour plots may be faster to use. In contrast, plots on the right are reported in order to underline the improvement in terms of accuracy with the new formula; indeed, they are considered with the same colormap range of the respective plot on the left.

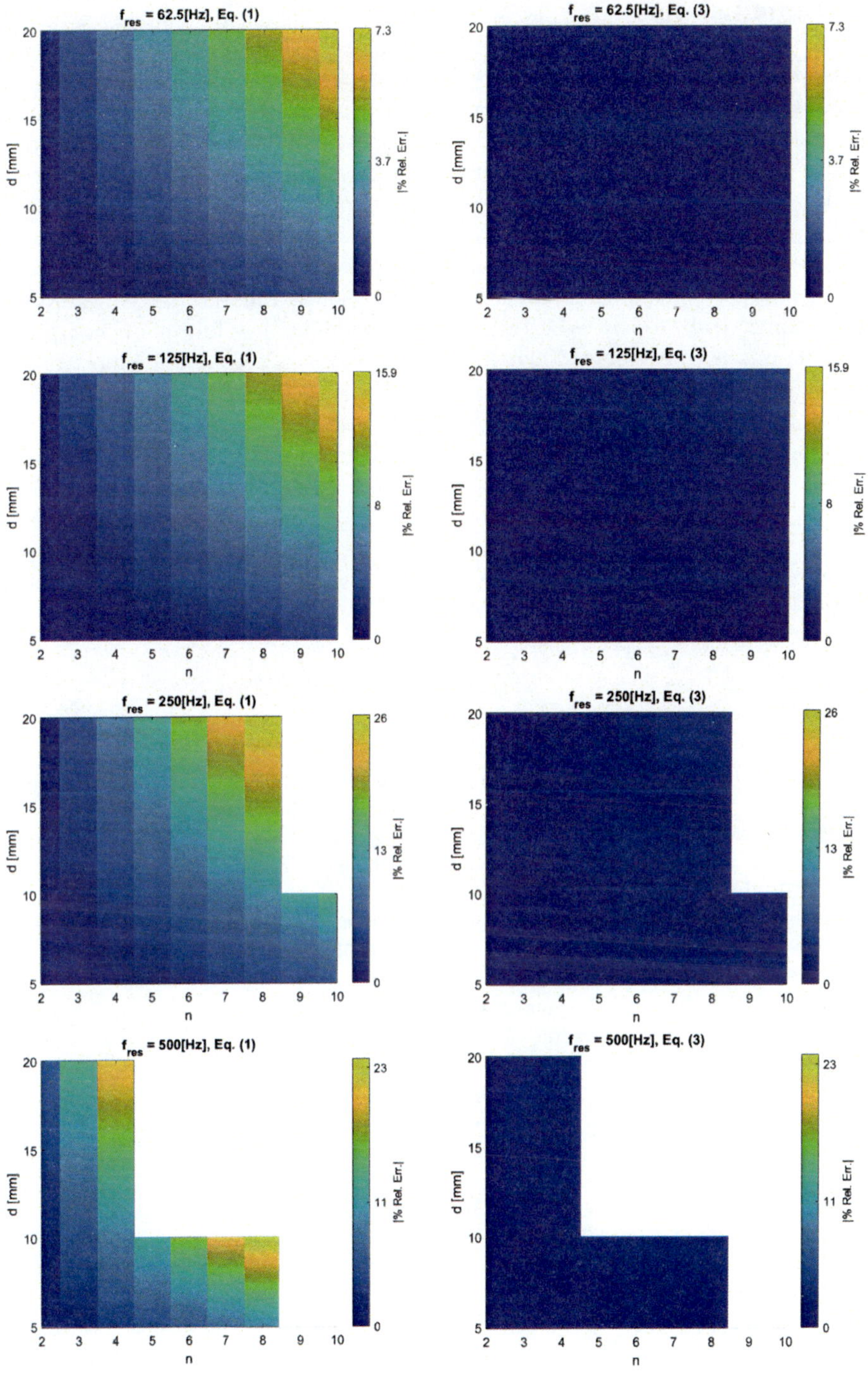

Fig. 10. Relative error patterns of Eq. (1) and Eq. (3) respect to FEM.

4 Conclusions

In this work, labyrinth resonators are deeply studied because they are considered as valid replacements of quarter-wavelength tubes, the latter being very interesting objects for acoustic purposes, but limited by their inconvenient length for low-frequency applications. Going from classical quarter-wavelength tubes to LRs, it is noticed that quarter-wavelength resonance formula is not reliable for an accurate evaluation of the labyrinth resonator working frequencies. An improved formula for the prediction of their resonance frequency is proposed and demonstrated as valid, thanks to the study of 97 different cases, where the main geometrical properties of LRs are changed. The new formula is compared with the quarter-wavelength formula, proved as not acceptable for most of the LR configurations. Thus, thanks to the formula derived herein, it would be possible to properly design periodic acoustic solutions based on labyrinth resonators. A further development of the work presented herein may consist in an experimental campaign, which would constitute the final verification of the proposed formula, to make it suitable for a real application for sound absorption purposes.

Acknowledgments. The authors acknowledge the support of the Italian Ministry of Education, University and Research (MIUR) through the project DEVISU, funded under the scheme PRIN-2107 - grant agreement No. 22017ZX9X4K006.

References

1. Allard, J.F., Atalla, N.: Propagation of sound in porous media: modelling sound absorbing materials (2009)
2. Magliacano, D., et al.: Computation of dispersion diagrams for periodic porous materials modeled as equivalent fluids. Mech. Syst. Signal Process. **142** (2020). https://doi.org/10.1016/j.ymssp.2020.106749
3. Ingard, U.: On the theory and design of acoustic resonators. J. Acoust. Soc. Am. **25**(6), 1037–1061 (1953). https://doi.org/10.1121/1.1907235
4. Rayleigh, J.W.S.: The theory of sound (1898)
5. Berardi, U., Iannace, G.: Acoustic characterization of natural fibers for sound absorption applications. Build. Environ. **94**, 840–852 (2015). https://doi.org/10.1016/j.buildenv.2015.05.029
6. Xinzhao, X., Guoming, L., Dongyan, L., Guoxin, S., Rui, Y.: Electrically conductive graphene-coated polyurethane foam and its epoxy composites. Compos. Commun. **7**, 1–6 (2018). https://doi.org/10.1016/j.coco.2017.11.003
7. Yang, M., Sheng, P.: Sound absorption structures: from porous media to acoustic metamaterials. Annu. Rev. Mater. Res. **47**(1), 83–114 (2017). https://doi.org/10.1146/annurev-matsci-070616-124032
8. Maa, D.Y.: Potential of microperforated panel absorber. J. Acoust. Soc. Am. **104**(5), 2861–2866 (1998). https://doi.org/10.1121/1.423870
9. Catapane, G., Magliacano, D., Petrone, G., Casaburo, A., Franco, F., De Rosa, S.: Evaluation of improved correction factors for the prediction of Helmholtz resonances. In: Proceedings of CEAS Aerospace Europe 2021 Conference, Warsaw, Poland, no. 1 (2021)

10. Tang, P., Sirignano, W.: Theory of a generalized Helmholtz resonator. J. Sound Vib. **26**(2), 247–262 (1973). https://doi.org/10.1016/S0022-460X(73)80234-2
11. Mekid, S., Farooqui, M.: Design of Helmholtz resonators in one and two degrees of freedom for noise attenuation in pipe lines. Acoust. Aust. **40**, 194–202 (2012)
12. Long, M.: Architectural acoustics (2006)
13. Lv, L., Bi, J., Wei, C., Wang, X., Cui, Y., Liu, H.: Effect of micro-slit plate structure on the sound absorption properties of discarded corn cob husk fiber. Fibers Polym. **16**(07) (2015)
14. Tang, Y., et al.: Hybrid acoustic metamaterial as super absorber for broadband low-frequency sound. Sci. Rep. **259** (2017). https://doi.org/10.1038/srep43340
15. Ma, G., Sheng, P.: Acoustic metamaterials: from local resonances to broad horizons. Sci. Adv. **2**(2), e1501595 (2016). https://doi.org/10.1126/sciadv.1501595
16. Huang, H., Sun, C.: Theoretical investigation of the behavior of an acoustic metamaterial with extreme young's modulus. J. Mech. Phys. Solids **59**(10), 2070–2081 (2011). https://doi.org/10.1016/j.jmps.2011.07.002
17. Peng, H., Frank Pai, P.: Acoustic metamaterial plates for elastic wave absorption and structural vibration suppression. Int. J. Mech. Sci. **89**, 350–361 (2014). https://doi.org/10.1016/j.ijmecsci.2014.09.018
18. Liang, Z., Li, J.: Extreme acoustic metamaterial by coiling up space. Phys. Rev. Lett. **108**, 114301 (2012). https://doi.org/10.1103/PhysRevLett.108.114301
19. Xie, Y., Popa, B.I., Zigoneanu, L., Cummer, S.A.: Measurement of a broadband negative index with space-coiling acoustic metamaterials. Phys. Rev. Lett. **110**, 175501 (2013). https://doi.org/10.1103/PhysRevLett.110.175501
20. Kaina, N., Lemoult, F., Fink, M., Lerosey, G.: Negative refractive index and acoustic superlens from multiple scattering in single negative metamaterials. Nature **525** (2015). https://doi.org/10.1038/nature14678
21. Li, J., Fok, L., Yin, X., Bartal, G., Zhang, X.: Experimental demonstration of an acoustic magnifying hyperlens. Nature **8** (2009). https://doi.org/10.1038/nmat2561
22. Zhu, J., Christensen, J., Jung, J., Martin-Moreno, L., Yin, X.E.A.: Experimental demonstration of an acoustic magnifying hyperlens. Nat. Phys. **7** (2011). https://doi.org/10.1038/nphys1804
23. Christensen, J., De Abajo, F.J.G.: Anisotropic metamaterials for full control of acoustic waves. Phys. Rev. Lett. **108**, 124301 (2012). https://doi.org/10.1103/PhysRevLett.108.124301
24. Zigoneanu, L., Popa, B., Cummer, S.: Three-dimensional broadband omnidirectional acoustic ground cloak. Nat. Mater. **13** (2014). https://doi.org/10.1038/nmat3901
25. Faure, C., Richoux, O., Félix, S., Pagneux, V.: Experiments on metasurface carpet cloaking for audible acoustics. Appl. Phys. Lett. **108**(6), 064103 (2016). https://doi.org/10.1063/1.4941810
26. Fleury, R., Sounas, D.L., Sieck, C.F., Haberman, M.R., Alù, A.: Sound isolation and giant linear nonreciprocity in a compact acoustic circulator. Science **343**(6170), 516–519 (2014). https://doi.org/10.1126/science.1246957
27. Zhang, C., Hu, X.: Three-dimensional single-port labyrinthine acoustic metamaterial: perfect absorption with large bandwidth and tunability. Phys. Rev. Appl. **6**, 064025 (2016). https://doi.org/10.1103/PhysRevApplied.6.064025
28. Li, Y., Assouar, B.M.: Acoustic metasurface-based perfect absorber with deep subwavelength thickness. Appl. Phys. Lett. **108**(6), 063502 (2016). https://doi.org/10.1063/1.4941338

29. Isaac, C., Wrona, S., Pawełczyk, M., Roozen, N.: Numerical investigation of the vibro-acoustic response of functionally graded lightweight square panel at low and mid-frequency regions. Compos. Struct. **259** (2021). https://doi.org/10.1016/j.compstruct.2020.113460
30. COMSOL Multiphysics Reference Manual (2021). https://doc.comsol.com/

On the Effect of Multiple Incident Waves on the Reflected Waves in a Semi-infinite Rod with a Nonlinear Boundary Stiffness

Moein Abdi(✉), Vladislav Sorokin, and Brian Mace

Acoustics Research Centre, Department of Mechanical Engineering,
The University of Auckland, Auckland, New Zealand
mabd314@aucklanduni.ac.nz

Abstract. This paper concerns a potential application of vibration control using nonlinearity. In this paper, a nonlinear boundary problem for a thin rod is considered. An incident wave propagates along the rod at frequency ω giving rise to an infinite number of reflected waves with frequencies $n\omega$. This paper concerns the reflected waves produced by multiple waves at frequencies $n\omega$ incident on the boundary of a semi-infinite rod with linear and cubic nonlinear stiffnesses. Equations are truncated including the third harmonic and solved using the harmonic balance method for the case with two reflected and two incident waves. The effects of different parameters on the magnitudes of the reflected waves are studied. It is shown that the phase difference between the incident waves affects the reflected waves' behaviour. Numerical examples are presented to find the conditions at which the magnitude of the reflected wave of the 1st harmonic is minimum and the maximum energy leaks from the 1st harmonic to the 3rd harmonic. The results show that the presence of the second incident wave can decrease the magnitude of the reflected wave of the 1st harmonic.

Keywords: Reflection coefficient · Multiple incident waves · Nonlinearity · Nonlinear boundary · Harmonic balance method

1 Introduction

Exploiting nonlinearity to control vibration is of interest from both purely theoretical and applied perspectives. Vibrations at one frequency can be scattered into higher frequencies by a nonlinear boundary support and the higher frequency vibrations generally dissipate more rapidly. In this paper, a wave propagation approach is used to study the interaction between the incident and reflected waves in a rod with a nonlinear boundary stiffness. Wave reflection in a rod from a linear boundary has been studied previously and results are well established [1, 2]. However, in the real world, systems behave nonlinearly. The nonlinearity might be in the material properties [3, 4] or at boundaries. In this paper, a nonlinear boundary problem is studied. In previous studies on axial vibration and wave reflection in a rod from a nonlinear boundary stiffness, it was shown that the behaviour of the reflected waves depends on the nonlinear boundary condition. Nayfeh

Z. Dimitrovová et al. (Eds.): WMVC 2022, MMS 125, pp. 695–702, 2023.
https://doi.org/10.1007/978-3-031-15758-5_71

et al. [5], showed that a single time harmonic incident wave produces an infinite number of reflected waves. However, the physical behaviour of the system mainly depends on the reflected lower harmonic waves, i.e. vibrational energy is mostly in the lower harmonics. A perturbation method was used to study the difference between the amplitudes of the incident wave and reflected waves. Brennan et al. [6], retained only the first harmonic and studied softening and hardening type nonlinear stiffnesses. Under this assumption, it was shown that the nonlinearity only changes the phases of the reflected waves, not their magnitudes. Retaining higher harmonics shows that, in contrast to the linear case, nonlinearity affects the magnitudes of the reflected waves. As the number of harmonics increases the magnitudes of the reflected waves decrease [7]. Following [6], Tang et al. [8], considered the calculation of the natural frequencies of the system using the phase closure principle. Chouvion [9] used a wave approach to study the forced response of beam-like systems by retaining waves with higher harmonics. The approach was based on the assembly of incident and reflected wave matrices. In a subsequent study, the author applied the method to study the forced response of a beam with a nonlinear energy sink [10]. These two studies focused on the receptance function of the continuous structures, e.g. rods and beams, not reflection coefficients. Hence, the effects of the phase and magnitude of the incident waves on the reflection coefficients were not explicitly studied. Furthermore, the minimum reflection coefficient of the 1st harmonic or the maximum leakage of energy from the 1st harmonic to the higher harmonics was not investigated. Santo et al. [11] studied forced vibration of a rod with multiple springs. The parameters were defined with respect to the stiffness of the rod. The backbone curves were obtained and compared to the case with one spring. In previous works [5–8], the nonlinearity was assumed to be weak, and parameters were defined in terms of the linear spring stiffness or the rod stiffness; hence, the case with zero linear boundary stiffness, i.e. essential nonlinearity, was not considered. It is evident that, as the linear boundary stiffness decreases, the nonlinearity becomes more pronounced. In this paper, the case with essential nonlinearity is also studied. For this purpose, a new definition of dimensionless parameters and the strength of nonlinearity in terms of a reference frequency is presented. Additionally, in previous studies, only one incident wave at the boundary was retained. Although Chouvion [9, 10] considered incident and reflected waves with higher harmonics and studied the case of essential nonlinearity, the effects of multiple incident waves on the reflection coefficients were not studied. In this paper, the effect of the second incident wave on the reflected waves is studied. It is shown that the second incident wave has a significant effect on the reflected waves. First, the general formulation retaining all incident and reflected waves is derived, then the system of equations is truncated and solved using the harmonic balance method. Finally, numerical examples are presented.

2 Problem Statement

As shown in Fig. 1, a nonlinear boundary stiffness is attached to the end of a thin, uniform rod. The governing equation of motion of the semi-infinite rod is $\partial^2 u(x,t)/\partial x^2 = \left(1/c_g^2\right)\left(\partial^2 u(x,t)/\partial t^2\right)$ with the wave speed $c_g = \sqrt{E/\rho}$ and, for time harmonic behaviour at frequency ω, wavenumber $\kappa = \omega/c_g$, where ρ, E are the density and elastic modulus of the rod, and $u(x, t)$ represents the axial displacement [1].

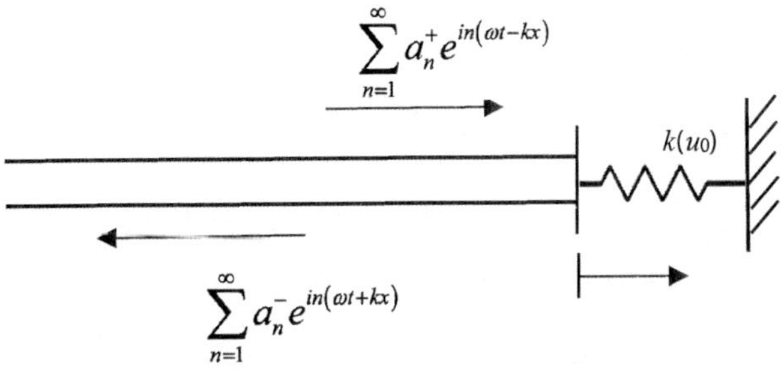

Fig. 1. Waves in a semi-infinite rod with incident and reflected waves at the boundary.

The boundary force is written as a polynomial function of u_0 with coefficients that do not depend on time. The boundary condition at $x = 0$ can thus be written as

$$P(0, t) = ES\frac{\partial u(0, t)}{\partial x} = -\sum_{m=1}^{M} k_m u_0^m(t); \quad u_0(t) = u(0, t) \tag{1}$$

where $P(0, t)$ is the axial force in the rod at $x = 0$ and S is the cross-sectional area of the rod. Waves of amplitudes a_n^+ with frequencies $n\omega$ are incident on the boundary and an infinite number of reflected waves with amplitudes a_n^- with the same frequencies $n\omega$ are produced. The displacement of the rod at x is the superposition of the incident and reflected waves, and can thus be written as

$$u(x, t) = \sum_{n=1}^{\infty} a_n^+ e^{in(\omega t - \kappa x)} + \sum_{n=1}^{\infty} a_n^- e^{in(\omega t + \kappa x)} + CC, \tag{2}$$

where CC denotes the complex conjugate of the preceding terms. In Eq. (2), $Z = S\sqrt{E\rho}$ is the characteristic impedance of the rod. The reflection coefficient of the nth reflected wave and the relative magnitudes of the incident waves of the nth harmonic to the 1st harmonic are defined as $r_n = a_n^-/a_1^+$ and $\alpha_n = a_n^+/a_1^+$. Note that the reflection coefficients depend on all the incident waves, in contrast to the linear case where r_n is independent of the incident wave amplitude. It is now assumed that the nonlinearity is cubic and of the form $P(0, t) = -k_1 u_0 - k_3 u_0^3$, where k_1 and k_3 are the linear and nonlinear stiffnesses of the spring. Under the assumption of cubic nonlinearity of this form, it is known that reflected waves of only odd harmonics exist in the system [7]. It is also assumed that only odd harmonics are incident on the boundary. Putting $x = 0$ in

Eq. (2), and substituting in the boundary condition results in

$$\left[i\omega Z\left(\sum_{n=1,3,5,\ldots}^{\infty} n\left(a_n^+ - a_n^-\right)e^{in\omega t}\right)\right] + CC$$
$$= k_1\left[\sum_{n=1,3,5,\ldots}^{\infty} \left(a_n^+ + a_n^-\right)e^{in\omega t} + CC\right] + k_3\left[\sum_{n=1,3,5,\ldots}^{\infty} \left(a_n^+ + a_n^-\right)e^{in\omega t} + CC\right]^3. \tag{3}$$

A system of equations with an infinite number of harmonics is obtained. This system has to be truncated and solved numerically. From Eq. (3) it can be noted that the amplitudes of the reflected waves depend on k_1, k_3, ωZ and the amplitudes of the incident waves. In Eq. (3), a_n^+ is complex and has a magnitude and phase, $a_n^+ = \left|a_n^+\right|e^{-i\Phi_n}$. For simplicity, we consider $\Phi_1 = 0$ so that a_1^+ is pure real. A reference frequency ω_0 is defined and dimensionless parameters $\Omega = \omega/\omega_0$, $K = k_1/\omega_0 Z$, $A = k_3 a_1^{+2}/\omega_0 Z$are introduced, where A denotes the nonlinear effects. These are strongest when $A \gg K$ and $A \gg \Omega$. It is expected that the effects of nonlinearity are strongest when $K = 0$. Hence, the case of essential nonlinearity is of interest. Equation (3) can be truncated at some finite value of $n = N$ and solved by using the harmonic balance method, yielding equations which have to be solved numerically. For example, for the case of two reflected and incident waves, taking $N = 3$ and collecting the coefficients of $e^{i\omega t}$ and $e^{3i\omega t}$ in Eq. (3) and substituting the dimensionless parameters results in

$$i\Omega[1 - r_1] = K[1 + r_1]$$
$$+3A\left[(r_1 + 1)^2(\bar{r}_1 + 1) + 2(r_1 + 1)(\alpha_3\bar{\alpha}_3 + r_3\bar{r}_3 + \alpha_3\bar{r}_3 + \bar{\alpha}_3 r_3) + (\alpha_3 + r_3)(\bar{r}_1 + 1)^2\right], \tag{4}$$

$$3i\Omega[\alpha_3 - r_3] = K[\alpha_3 + r_3]$$
$$+A\left[(r_1 + 1)^3 + 6(r_1 + 1)(\bar{r}_1 + 1)(\alpha_3 + r_3) + 3(\bar{\alpha}_3 + \bar{r}_3)(\alpha_3 + r_3)^2\right]. \tag{5}$$

3 Numerical Examples

This section presents numerical examples for the case of two incident and two reflected waves. Equations (4) and (5) are solved numerically for various values of the parameters and the effects of magnitude and phase of α_3 on the magnitudes of the reflection coefficients are studied in this section.

3.1 The Effect of Φ_3

In Fig. 2, the magnitudes of the reflection coefficients of the 1st and 3rd harmonics are shown as functions of Ω with $K = 0.1$, $A = 0.01$, $|\alpha_3| = 1$, and various Φ_3. As seen in Fig. 2a, for these specific values, the magnitude of r_1 is maximum and minimum when $\Phi_3 \simeq 0, \pi$, respectively, since when a_1^+ and a_3^+ have the same phases, a larger incident displacement is produced which leads to producing a larger reflected wave, while when

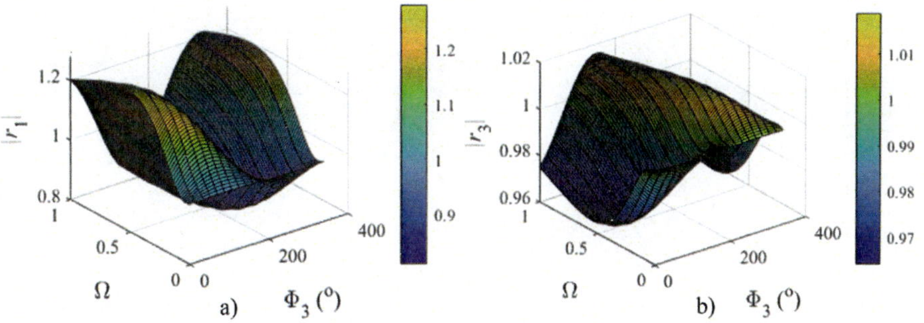

Fig. 2. The magnitudes of the reflection coefficients for the 1st and 3rd reflected harmonics, with K=0.1, A=0.01, $|\alpha_3| = 1$ and various Φ_3 as functions of Ω: : a) $|r_1|$; b) $|r_3|$.

$\Phi_3 \simeq \pi$, the incident waves are in anti-phase. In this case, the incident waves produce a smaller reflected wave; therefore, the maximum magnitude of the reflection coefficient of the 1st harmonic decreases.

When Φ_3 changes from 0 to π, I II $|r_1|$ decreases, while when it changes from π to 2π, $|r_1|$ increases. Additionally, it can be noted from Fig. 2b that the magnitude of the reflection coefficient of the 3rd harmonic is minimum when $|r_1|$ is maximum. This is due to conservation of energy i.e., $1 + |3\alpha_3|^2 = |r_1|^2 + |3r_3|^2$. It can be noted that as $|r_1|$ increases, $|r_3|$ decreases. Consequently, the physical behaviour of the magnitude of the reflected wave of the 3rd harmonic is opposite to that of the 1st harmonic.

3.2 The Effect of $|\alpha_3|$

In Fig. 3, $|r_1|$ and $|r_3|$ are shown as functions of Ω for $K = 0.1, A = 0.01$, and various $|\alpha_3|$. In Fig. 3a and Fig. 3b, $\Phi_3 = 0$ while in Fig. 3c and Fig. 3d, $\Phi_3 = \pi$. The results of Fig. 3 lead to conclusion that for these specific values, when $|a_1^+|$ and $|a_3^+|$ are almost equal and $\Phi_3 \simeq \pi$, then I then II then $|r_1|$ is minimum and $|r_3|$ is maximum.

It can be noted from Fig. 3 that when $\alpha_3 \neq 0$, the behaviour of the reflected waves depends on the phase of the incident wave of the 3rd harmonic. For example, for $\Phi_3 = \pi$, as $|\alpha_3|$ increases, $|r_1|$ first decreases and reaches a minimum value and then increases; while when $\Phi_3 = 0$, as $|\alpha_3|$ increases, $|r_1|$ first increases and reaches a maximum value and then decreases. The results of these figures show that for these specific values, the magnitude of the reflection coefficient of the 1st reflected harmonic is minimum or maximum when $|\alpha_3| \simeq 1$. Additionally, Fig. 3b and Fig. 3d show that when $|\alpha_3| > 1$, $|r_3| \simeq |\alpha_3|$, i.e. the magnitudes of the incident and reflected waves of the 3rd harmonic are almost identical since the reflected 3rd harmonic is dominated by the contribution of the direct reflection of the incident third harmonic. The magnitude of α_3 affects both $|r_1|$ and $|r_3|$. However, the variation of $|r_1|$ is more substantial than $|r_3|$, especially when α_3 is large.

With essential nonlinearity, $|r_n|$ depends on the parameter ratio $\lambda = A/\Omega$, where A is a measure of nonlinearity and Ω represents the dynamic stiffness of the rod. Hence, λ denotes the relative effects of the nonlinear boundary force compared to that of the linear internal force of the rod when $K = 0$.

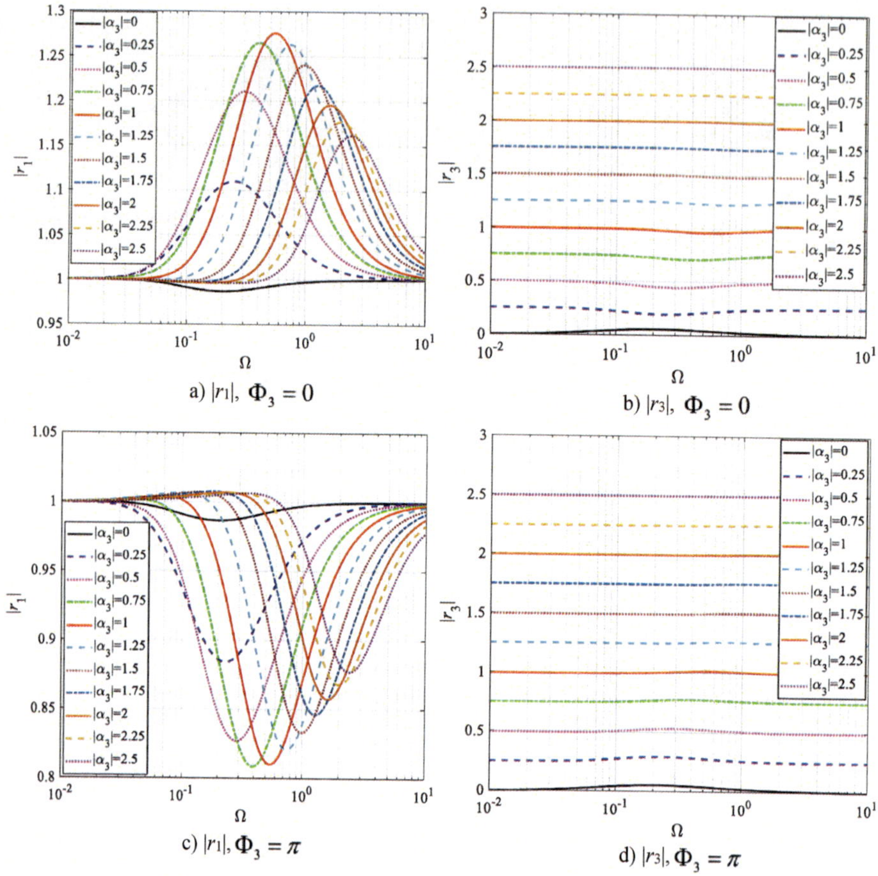

Fig. 3. The magnitudes of the reflection coefficients for the 1st and 3rd reflected harmonics, with $K = 0.1$, $A = 0.01$, and various $|\alpha_3|$ as functions of Ω: a) $|r_1|$, $\Phi_3 = 0$; b) $|r_3|$, $\Phi_3 = 0$; c) $|r_1|$, $\Phi_3 = \pi$; d) $|r_3|$, $\Phi_3 = \pi$.

Results are shown for the case with essential nonlinearity and $|\alpha_3| = 1$ in Fig. 4 as functions of λ. As can be seen in Fig. 4, when $K = 0$, the maximum and minimum magnitudes of the reflection coefficients depend only on Φ_3. When $\Phi_3 = 0, |r_1|_{max} \simeq 1.4$ and $|r_3|_{min} \simeq 0.945$, while with $\Phi_3 = \pi, |r_1|_{min} \simeq 0.75$ and $|r_3|_{max} \simeq 1.024$.

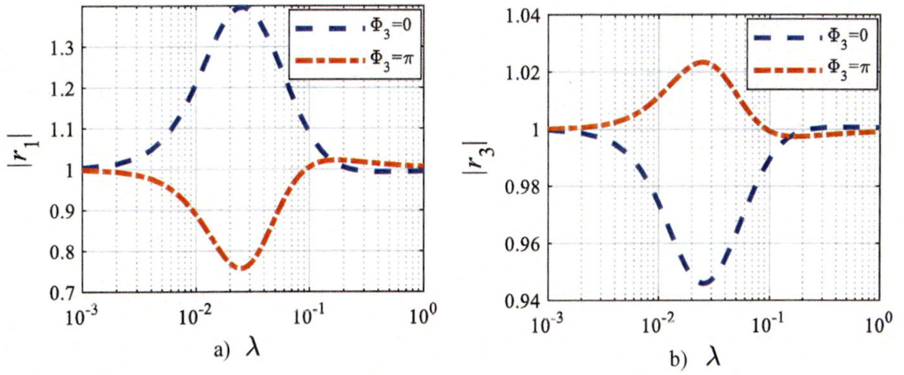

Fig. 4. The magnitudes of the reflection coefficients for the 1st and 3rd reflected harmonics, with $|\alpha_3| = 1$, as functions of λ a) $|r_1|$; b) $|r_3|$.

4 Conclusion

This paper presented a numerical study of the effect of multiple incident waves on the reflected waves for a thin rod with a nonlinear boundary stiffness. Equations were derived, retaining all incident and reflected waves. Then, for the case of cubic nonlinearity, the harmonic balance method was used to truncate and solve the equations. Equations were truncated at $N = 3$, including two incident and reflected waves. It was shown that the presence of the second incident wave significantly changes the reflected waves' behaviour. With essential nonlinearity, the magnitudes of the reflection coefficients are minimum or maximum and take values independent of the strength of nonlinearity. With two incident waves, in some specific conditions the leakage of energy to the third harmonic is more than in the case with one incident wave. The magnitude of the reflection coefficient of the 1st harmonic can be decreased by up to 25%. The effects of the parameters on the reflection coefficients were also studied. The phase of the incident wave of the 3rd harmonic affects the magnitude of r_n. From numerical examples, it was seen that with $N = 3$, the leakage of energy is maximum when $K = 0$, $\Phi_3 \simeq \pi$, and $|a_1^+| \simeq |a_3^+|$.

References

1. Graff, K. F.: Wave motion in elastic solids. Dover Publications (2012)
2. Cremer, L., & Heckl, M.: Structure-borne sound: structural vibrations and sound radiation at audio frequencies. Springer Science & Business Media (2013)
3. Dreiden, G.V., Porubov, A.V., Samsonov, A.M., Semenova, I.V.: Reflection of a longitudinal strain soliton from the end face of a nonlinearly elastic rod. Tech. Phys. **46**(5), 505–511 (2001)
4. Autrusson, T.B., Sabra, K.G., Leamy, M.J.: Reflection of compressional and Rayleigh waves on the edges of an elastic plate with quadratic nonlinearity. The Journal of the Acoustical Society of America **131**(3), 1928–1937 (2012)
5. Nayfeh, A.H., Vakakis, A.F., Nayfeh, T.A.: A method for analyzing the interaction of nondispersive structural waves and nonlinear joints. The Journal of the Acoustical Society of America **93**(2), 849–856 (1993)

6. Brennan, M. J., Manconi, E., Tang, B., & Lopes Jr, V.: Wave reflection at the end of a waveguide supported by a nonlinear spring. In EURODYN 2014, the 9th International Conference on Structural Dynamics, Porto, Portugal, 30 June-02 July. (2014)
7. Vakakis, A.: Scattering of structural waves by nonlinear elastic joints. J. Vib. Acoust. **115**(4), 403–410 (1993)
8. Tang, B., Brennan, M.J., Manconi, E.: On the use of the phase closure principle to calculate the natural frequencies of a rod or beam with nonlinear boundaries. J. Sound Vib. **433**, 461–475 (2018)
9. Chouvion, B.: Vibration analysis of beam structures with localized nonlinearities by a wave approach. J. Sound Vib. **439**, 344–361 (2019)
10. Chouvion, B.: A wave approach to show the existence of detached resonant curves in the frequency response of a beam with an attached nonlinear energy sink. Mech. Res. Commun. **95**, 16–22 (2019)
11. Santo, D.R., Mencik, J.-M., Gonçalves, P.J.P.: On the multi-mode behavior of vibrating rods attached to nonlinear springs. Nonlinear Dyn. **100**(3), 2187–2203 (2020). https://doi.org/10.1007/s11071-020-05647-x

On the Formation of a Super Attenuation Band in a Mono-coupled Finite Periodic Structure Comprising Asymmetric Cells

Vinicius Germanos Cleante[1](✉), Michael John Brennan[1], Paulo José Paupitz Gonçalves[2], and Jean Paulo Carneiro Jr[1]

[1] School of Engineering, São Paulo State University (UNESP), Ilha Solteira, Brazil
vinicius.cleante@unesp.br

[2] School of Engineering, São Paulo State University (UNESP), Bauru, Brazil

Abstract. Metamaterials are employed to reduce vibration levels by exploiting the effects of structural periodicity. When structural elements are arranged in a periodic pattern, they act as mechanical filters, creating stop-bands. The term stop-band is often used for infinite structures, but a more appropriate term for a finite structure is attenuation band. A way of obtaining this effect is by attaching vibration absorbers, which create a local resonance stop-band plus a Bragg stop-band. The local resonance stop-band is controlled only by the properties of the attached device. The Bragg stop-band depends on the interaction between the host cell and the device. The combination of these two effects can create an attenuation zone – the so-called super attenuation band. Recent works on finite mono-coupled metamaterials have shown that asymmetric periodic structures have better attenuation properties when compared to the symmetric ones, if they are correctly orientated. This paper investigates the formation of a super attenuation band in a finite mono-coupled structure using vibration absorbers. The system is defined by the formation of a cell, which repeats along with the whole structure. The cell can be divided into sub-cells with equal or different dynamic properties. The dynamic features to form the super attenuation band are determined from the displacement transmissibility of a single cell. This analysis is extended to several cells. The results show that a super attenuation band can only occur when each attached vibration absorber is optimally tuned to its corresponding host cell in a structure comprising cells with dynamical asymmetry.

Keywords: Super stop-band · Super attenuation band · Asymmetric · Mono-coupled · Finite structure

1 Introduction

Advances in metamaterial technologies for reducing vibration has made periodic structures a hot research topic. Several configurations, including changes in the material [1] and the geometry [2, 3], have been extensively investigated in the formation of stop-bands, which are frequency ranges where waves do not propagate. However, it remains

Z. Dimitrovová et al. (Eds.): WMVC 2022, MMS 125, pp. 703–712, 2023.
https://doi.org/10.1007/978-3-031-15758-5_72

a challenge to design periodic structures with a large stop-band without adding too much mass. The use of an array of vibration absorbers has been investigated to overcome this issue.

The main advantage in the use of vibration absorbers is that these devices add two physical mechanisms responsible for stop-bands: Bragg scattering and local resonances [4,5]. The Bragg scattering mechanism forms the Bragg stop-band [5] and depends on the dynamic coupling between the host cell and the vibration absorber. The local resonance stop-band is due to an antiresonance formed by vibration absorber dynamics, and the frequency range where it occurs is controlled by the natural frequency of the vibration absorber [1,5]. This makes the local resonance attractive in terms of metamaterial design, and different approaches have been investigated, such as by attaching multi-degree-of-freedom absorbers [6], using arrays of absorbers with the same [7, 8] or different [9, 10] natural frequencies.

Another alternative is to use an array of vibration absorbers to form a super stop-band. This phenomenon occurs when the local resonance stop-band is combined with the Bragg stop-band [11] when the device is tuned appropriately. This phenomenon was exploited to suppress the vibration in pipelines [12], in rods [13] and beams [1, 14]. Using a finite chain of masses, Gonçalves et al. [5] showed that the magnitude of attenuation inside the super stop-band is high. Hao et al. [15] combined the Bragg scattering mechanism due to boundary conditions with the mechanisms due to Bragg and local resonance added by the vibration absorber to form a super stop-band in a periodically pinned supported beam.

Recent work in finite mono-coupled periodic structures has demonstrated that structural periodicity creates frequency ranges where the vibration is attenuated – an attenuation band. In structures comprised of asymmetric cells, the attenuation band is greater than the frequency range comprising the stop-band (infinite structure), unless the structure is of infinite length [2]. Furthermore, when correctly orientated, the vibration attenuation within the band of an asymmetric cell is greater than that of symmetric cells [2].

The aim of this paper is to investigate the formation of a super attenuation band in a finite mono-coupled structure consisting of asymmetric cells, combining the already existing attenuation band due to the asymmetry in the host cell with the attenuation bands added by the vibration absorber. It is assumed that the host cell consists of two sub-cells that are dynamically asymmetric. To achieve this objective, the transfer matrix method is used and a numerical investigation into the displacement transmissibility is performed.

2 Problem Statement

Figure 1a presents a generalised one-dimensional structure comprised of N cells which repeat along with the whole structure. The response of the system is analysed in terms of the displacement transmissibility of a single cell, which is the ratio between the displacements at the right- and left-hand side of the cell, represented by the n-th element in Fig. 1b, to investigate the formation of a super attenuation band. Later, extending this analysis to a finite number of cells.

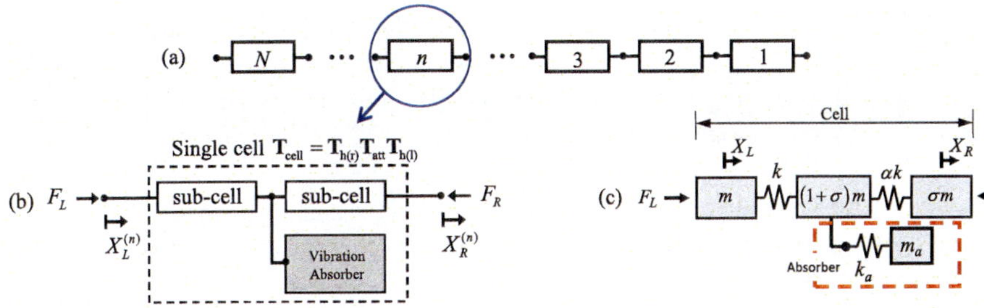

Fig. 1. Example of a mono-coupled periodic structure with N cells showing the details of the n-th cell.

The periodic cell considered is shown in Fig. 1c. It can be divided into two sub-cells. The left sub-cell consists of two end masses m connected by a spring with stiffness k and a material loss factor η. The right sub-cell consists of two end masses σm connected by a spring with stiffness αk and a material loss factor η. The host cell is then formed by connecting the respective right mass of the left sub-cell to the left mass of the right sub-cell. The vibration absorber is attached to the centre mass of the host cell. It consists of a single degree-of-freedom vibration absorber with mass m_a, stiffness k_a and a material loss factor η_a. The symmetry or asymmetry of the host cell is created by changing the stiffness parameter α and the mass parameter σ in the right sub-cell. When $\alpha = 1$ and $\sigma = 1$, the dynamic properties of both sub-cells are equal, and the host cell is symmetric. When $\alpha = \sigma$, but different to one, the host cell is asymmetric, but both sub-cells are dynamically symmetric as they have the same natural frequency. When $\alpha \neq \sigma$, the host cell is asymmetric and the dynamic properties of both sub-cells are also different, as the natural frequencies of the sub-cells are different. Of particular interest in this paper is an investigation into the vibration transmissibility for the case when $\sigma = 1$ and $\alpha > 1$. For this condition, the right sub-cell is stiffer than the left sub-cell, which gives a correct orientation regarding the vibration source [2].

The relationship between the state vectors, comprised of normalised forces, $\widehat{F} = F/\alpha k$, and displacements, X, on each side of the cell shown in Fig. 1c can be defined using the transfer matrix, as defined in [16], such that

$$\mathbf{q}_R^{[N]} = \mathbf{T}_g \mathbf{q}_L^{[1]}, \tag{1}$$

where $\mathbf{q}_R = \left\{\widehat{F}_R, X_R\right\}^T$, $\mathbf{q}_L = \left\{\widehat{F}_L, X_L\right\}^T$ and $\mathbf{T}_g$ is the global transfer matrix of the periodic structure. The subindex L and R denote left and right, respectively, and the superscript ^{T} denotes the matrix transpose.

There are alternative ways of determining the global transfer matrix. One way is to raise the N-th power the matrix of a single cell, i.e., by $\mathbf{T}_g = \mathbf{T}_{\text{cell}}^N$, where $\mathbf{T}_{\text{cell}} = \mathbf{T}_{\text{h(r)}} \mathbf{T}_{\text{att}} \mathbf{T}_{\text{h(l)}}$ is the cell transfer matrix, in which $\mathbf{T}_{\text{h(r)}}$, $\mathbf{T}_{\text{h(l)}}$ and $\mathbf{T}_{\text{att}}$ are the transfer matrices for the left and right host sub-cells and the vibration absorber, respectively. Note that all variables are frequency dependent. For the case shown in Fig. 1c, the corresponding sub-cell and absorber non-dimensional transfer matrices are given by

$$\mathbf{T}_{\mathrm{h(i)}} = \begin{bmatrix} -D_{1(\mathrm{i})}/D_{2(\mathrm{i})} & D_{2(\mathrm{i})} - D_{1(\mathrm{i})}^2/D_{2(\mathrm{i})} \\ -1/D_{2(\mathrm{i})} & -D_{1(\mathrm{i})}/D_{2(\mathrm{i})} \end{bmatrix} \text{ and } \mathbf{T}_{\mathrm{att}} = \begin{bmatrix} 1 & D_{\mathrm{att}} \\ 0 & 1 \end{bmatrix}, \tag{2}$$

where the subindex (i) indicates left or right sub-cell and the element D inside the matrices are the non-dimensional dynamic stiffness. For the left sub-cell $D_{1(\mathrm{l})} = (1+\mathrm{j}\eta)/\alpha - \Omega^2$ and $D_{2(\mathrm{l})} = -(1+\mathrm{j}\eta)/\alpha$. For the right sub-cell $D_{1(\mathrm{r})} = 1+\mathrm{j}\eta-\Omega^2$ and $D_{2(\mathrm{r})} = -(1+\mathrm{j}\eta)$. $\Omega = \omega/\omega_n$ is the non-dimensional frequency in which, $\omega_n = \sqrt{\alpha k/m}$ and $\mathrm{j} = \sqrt{-1}$. The non-dimensional dynamic stiffness of the attached vibration absorber is given by [17]

$$D_{\mathrm{att}} = \frac{-\mu\, \Omega^2(1 + \mathrm{j}\eta_a)}{1 - \Omega^2/\gamma^2 + \mathrm{j}\eta_{\mathrm{a}}}, \tag{3}$$

where $\mu = m_a/m$ and $\gamma = \omega_a/\omega_n$ are, respectively, the mass ratio and the tuning frequency parameter, in which $\omega_a = \sqrt{k_a/m_a}$. The displacement transmissibility from left to right is given by

$$\frac{X_R}{X_L} = \frac{1}{\mathbf{T}_{\mathrm{g}}(1, 1)}, \tag{4}$$

where $\mathbf{T}_{\mathrm{g}}(1, 1)$ is the element in the first line and the first column of the global transfer matrix.

3 Super Attenuation Band Formation in an Asymmetric Periodic Structure

In this section, the formation of the super attenuation band in a mono-coupled periodic structure comprised of asymmetric cells is discussed.

3.1 Single Unit Cell Analysis

Consider the unit asymmetric cell shown in Fig. 1c. With the objective of examining the conditions controlling the formation of a super attenuation band, the displacement transmissibility is calculated numerically for $\gamma = 0.8$; 1; and 1.2, assuming $\alpha = 2$ and $\mu = 1.5$. The modulus of Eq. (4) is plotted in Fig. 2. The corresponding displacement transmissibility in the absence of the absorber i.e., setting μ to 0, is also shown in this figure. Note that small damping ($\eta = \eta_{\mathrm{a}} = 0.01$) is considered to avoid sharp resonant peaks.

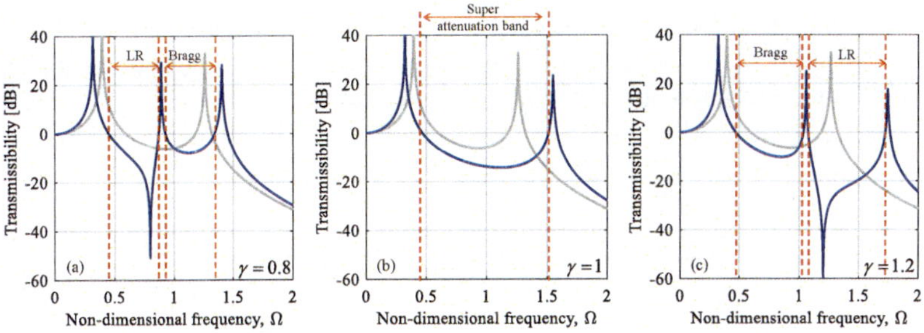

Fig. 2. Modulus of the displacement transmissibility of a single cell, Eq. (4), assuming $\alpha = 2$, $\mu = 1.5$ and $\eta = \eta_a = 0.01$ for: (a) $\gamma = 0.8$; (b) $\gamma = 1$; and (c) $\gamma = 1.2$ (the grey solid line is the displacement transmissibility of the host cell only, i.e., without the vibration absorber). LR – attenuation band due to local resonance mechanism; and Bragg – attenuation band due to Bragg mechanism.

It is clear in Fig. 2a and 2c that there are two frequency ranges where the transmissibility dips below 0 dB, which are the local resonance and the Bragg attenuation bands. Moreover, note that the level of attenuation in the Bragg attenuation band is greater than the case where no absorber is attached (in the absence of the vibration absorber, the attenuation band is created by the Bragg mechanism due to the host cell asymmetry [2]). This occurs because the Bragg mechanism added by the vibration absorber [5] is combined with the Bragg mechanism due to the host cell asymmetry and this forms a band with greater attenuation, however the bandwidth is smaller. Of particular interest is the case when $\gamma = 1$, shown in Fig. 2b. This is the condition where the super attenuation band is formed. In this case, the local resonance and the Bragg attenuation bands are combined to form a single but wide attenuation band, with vibration attenuation much greater than that for other tuning conditions. Moreover, note that the sharp dip and the sharp resonance peak, characteristic of the local resonance stop-band, vanishes from the displacement transmissibility, and a smooth attenuation band is formed.

3.2 Asymmetric Structure with *N* Periodic Cells

In the previous section, the formation of a super attenuation band was investigated in a single unit asymmetric cell. In this section, a structure with more than one periodic cell is considered. The system consists of two asymmetric cells, in which each cell is similar to the cell shown in Fig. 1c. The cells are concatenated from left to right, so that cell-1 is defined as the cell on the opposite side to the vibration source, similar to the schematic shown in Fig. 1a. Note that both vibration absorbers have the same tuned frequency to respect periodicity.

To investigate the formation of the super attenuation band in a structure with more than one cell, the displacement transmissibility is calculated numerically as a function of the non-dimensional frequency Ω, and the tuning frequency γ, when $\alpha = 2$ and $\mu = 1.5$. The non-dimensional frequency Ω and the parameters γ vary between 0.2 to 2 with an increment of 0.01, and the damping, η and η_a, are set 0. The top-view of the modulus of the displacement transmissibility is plotted in Fig. 3. The bounding frequencies as a function of the tuning frequency were determined numerically, and they are also plotted in Fig. 3 as a solid black line. Together it is also plotted, as a blue dashed line, the natural frequency of the vibration absorber.

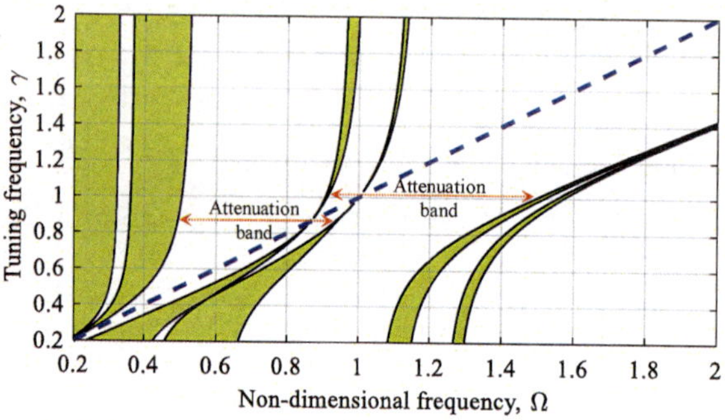

Fig. 3. Top view of the modulus of the displacement transmissibility (the colour is for $|T| > 0$ dB) as a function of Ω and γ. The black solid line lines indicate the bounding frequencies of the attenuation band of two periodic cells; the dashed line indicates the antiresonance introduced due to the vibration absorber.

Note in Fig. 3 that there are two tuning conditions where the bounding frequencies overlap and also coincide with the absorber natural frequency. These tuned frequencies correspond to $\gamma = 0.876$ and $\gamma = 1$. It can be seen that when tuned to one of these frequencies, the frequency range where there is attenuation is improved. To better examine the dynamics of the structure with these tuned conditions, the displacement transmissibilities are plotted in Fig. 4. In particular Fig. 4a is for $\gamma = 0.876$ and Fig. 4b for $\gamma = 1$. It is clear that the smooth super attenuation band is not formed in any of the two tuned conditions. The appearance of the resonance peak and the sharp dip, which is characteristic of the local resonance attenuation band, indicates that the local resonance attenuation band and the Bragg attenuation band are not coupled. In Fig. 4b, which is the same tuning condition in which the smooth super attenuation band is formed in a structure with a single cell (see Fig. 2b), the antiresonance occurs at $\Omega = 1$, which indicates

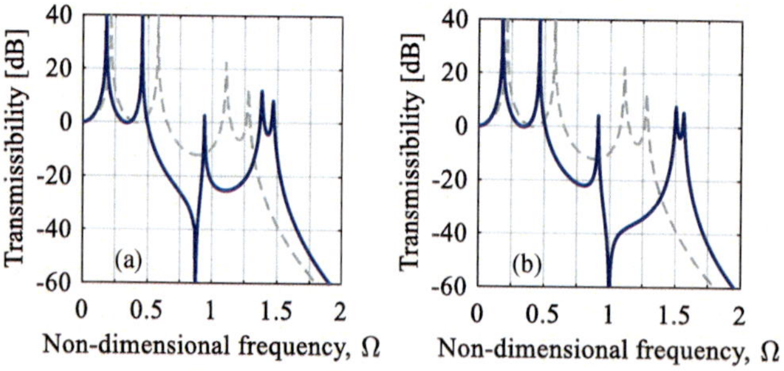

Fig. 4. The modulus of the displacement transmissibility for a periodic structure with two cells with the same tuning frequency (a) $\gamma = 0.876$ and (b) $\gamma = 1$. (The grey solid line is the displacement transmissibility of the host structure only, i.e., without the vibration absorbers).

that there is a mistuning in the vibration absorber attached to one of the cells. Despite adding more periodic cells improves the attenuation within the band, the smoothness of the super attenuation band observed for a single cell will not repeat in a finite periodic structure comprised of cells with dynamic asymmetry.

4 Discussion

It was demonstrated in the previous section that a smooth super attenuation band can be formed in a single cell, but it cannot be formed in a structure consisting of periodic cells. In this section, a strategy to form a smooth super attenuation band in a structure comprised of N asymmetric cells is discussed.

Consider again the structure with two asymmetric cells discussed in the previous section. However, now each host cell has a vibration absorber attached which is tuned to a particular frequency. This implies that the structure is aperiodic, as the cells do not repeat in the structure. The new global transfer matrix is calculated following the arrangement of the cells as in Fig. 1a, which is given by

$$\mathbf{T}_{\mathrm{g}} = \prod_{i=1}^{N} \mathbf{T}_{\mathrm{cell}(i)}, \tag{5}$$

where $\mathbf{T}_{\mathrm{cell}(i)}$ is the non-dimensional transfer matrix for each host cell with its corresponding vibration absorber tuned to a particular frequency. The dynamics of the cell on the opposite side of the vibration source, i.e., cell-1, behaves similar to the dynamics of a single cell alone. As discussed in Sect. 3.1, its tuning condition is equal to $\gamma_1 = 1$. A numerical investigation into the displacement transmissibility is performed to determine the optimum tuning condition for the vibration absorber associated with cell-2. Assuming the non-dimensional frequency, Ω, and the tuning frequency of the second absorber, γ_2, are set to vary between 0.2 to 2 with an increment of 0.01, the displacement transmissibility is calculated for $\alpha = 2$, $\mu = 1.5$ and $\eta = \eta_a = 0$. The top-view of the modulus of the displacement transmissibility as a function of Ω and γ_2 is plotted in Fig. 5. The bounding frequencies as a function of the tuning frequency are determined numerically and are also plotted in Fig. 5 as a solid black line. The natural frequency of the vibration absorber attached to the second cell is plotted as a dashed line.

It can be seen in Fig. 5 that there is only one tuned condition where the bounding frequencies overlap and also coincide with the absorber natural frequency, which occurs when $\gamma_2 = 0.876$. To better evaluate the dynamics of the aperiodic structure, the displacement transmissibility for the structure with each absorber tuned to $\gamma_1 = 1$ and $\gamma_2 = 0.876$, respectively, is plotted in Fig. 6 for $\alpha = 2$ and $\mu = 1.5$. Together is plotted the displacement transmissibility of a single cell tuned to $\gamma = 1$. Also plotted, as a grey shaded area, is the frequency range comprising the super attenuation band for an aperiodic structure with 10 cells.

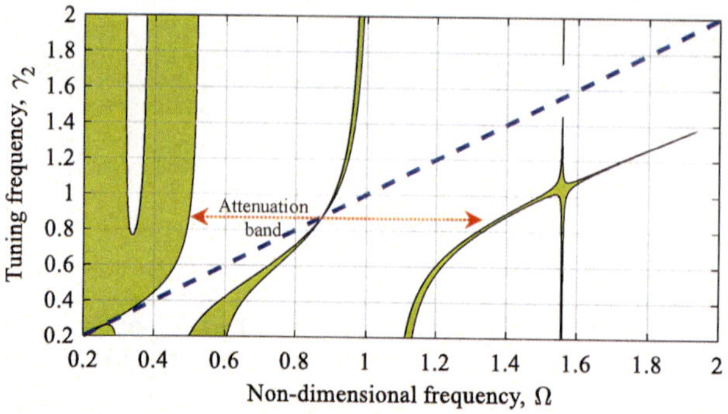

Fig. 5. Top view of the modulus of the displacement transmissibility (the colour is for $|T| >$ 0 dB) in the function of Ω and γ_2. The solid black lines indicate the bounding frequencies of the attenuation band; the dashed blue line indicates the vibration absorber natural frequency.

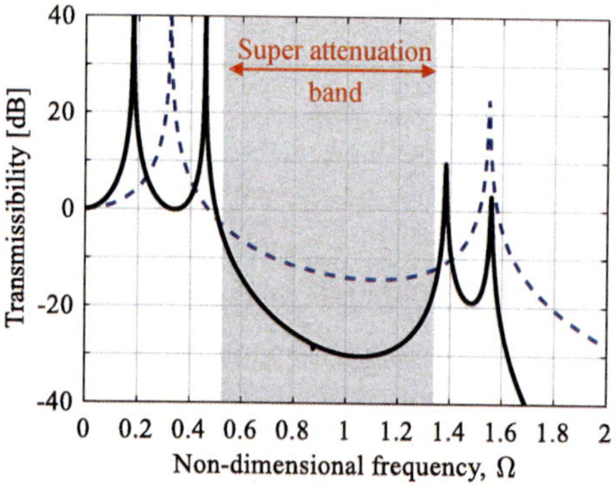

Fig. 6. The modulus of the displacement transmissibility of an *a*-periodic structure comprised of two asymmetric cells in which each cell has attached an optimally tuned vibration absorber to form the super attenuation band when $\mu = 1$ and $\eta = \eta_a = 0.01$. The black solid line is for the structure with two cells where $\gamma_2 = 0.876$ and $\gamma_1 = 1$; and the blue dashed line is for a single cell with $\gamma = 1$. The grey shaded area corresponds to the frequency range of the super attenuation band in a structure consisting of 10 cells.

It is clear in Fig. 6 that a smooth super attenuation band is formed. In the aperiodic structure comprising asymmetric cells, each of the optimally tuned vibration absorbers individually couples the local resonance attenuation band with Bragg attenuation band within its corresponding attached cell. This guarantees a complete coupling for the whole structure. Furthermore, it can also be seen that there is a considerable improvement in the attenuation within the band. However, there is a clear reduction in the bandwidth of the attenuation band, which is a characteristic effect in finite mono-coupled structures comprised of asymmetric cells [2]. If a large number of periodic cells is considered, the frequency range comprising the super attenuation band converge to the super stop-band in an infinite structure, as shown in Fig. 6 by the grey shaded area.

5 Conclusions

This paper has investigated the formation of a smooth super attenuation band in a mono-coupled structure comprised of asymmetric cells using an array of vibration absorbers. The asymmetry in the host cell is created by changing the stiffness of a sub-cell, resulting in dynamic asymmetry between the sub-cells. The analysis in a single unit cell showed that a frequency range comprising a smooth super attenuation band is formed when the absorber is correctly tuned. It was found that at this tuned condition, the Bragg mechanism inherent to the host cell asymmetry couples with the local resonance and Bragg mechanisms, introduced by the vibration absorber, forming a single but wide attenuation band. The analysis considering more than one periodic cell showed that the super attenuation band cannot exist. It was found that due to the dynamic asymmetry in the host cell, the mechanism to form the super attenuation band is uncoupled, giving rise to two distinct attenuation bands. It was also found that the smoothness within the super attenuation band only occurs when optimally tuned vibration absorbers are attached to each host cell.

Acknowledgements. The authors would like to acknowledge the financial support of the São Paulo Research Foundation (FAPESP) (grant numbers 2018/15894-0, 2019/19335-9 and 2020/00659-6), and National Council for Scientific and Technological Development (CNPq), grant number (406594/2021-0).

References

1. Liu, L., Hussein, M.I.: Wave motion in periodic flexural beams and characterization of the transition between bragg scattering and local resonance. J. Appl. Mech. **79**, 011003 (2012)
2. Carneiro Jr., J.P., Brennan, M.J., Gonçalves, P.J.P., Cleante, V.G., Bueno, D.D., Santos, R.B.: On the attenuation of vibration using a finite periodic array of rods comprised of either symmetric or asymmetric cells. J. Sound Vib. **511**, 116217 (2021)
3. Santos, R.B., Carneiro Junior, J.P., Gonsalez-Bueno, C.G., Lucca, B., Bueno, D.D.: On the number of cells for flexural vibration suppression in periodic beams. Meccanica **56**(11), 2813–2823 (2021)
4. Raghavan, L., Phani, A.S.: Local resonance bandgaps in periodic media: theory and experiment. J. Acoust. Soc. Am. **134**, 1950–1959 (2013)
5. Gonçalves, P.J.P., Brennan, M.J., Cleante, V.G.: Predicting the stop-band behaviour of finite mono-coupled periodic structures from the transmissibility of a single element. Mech. Syst. Signal Process. **154**, 107512 (2021)
6. Li, Z., Wang, X.: Wave propagation in a dual-periodic elastic metamaterial with multiple resonators. Appl. Acoust. **172**, 107582 (2021)
7. Xiao, Y., Wen, J., Wen, X.: Broadband locally resonant beams containing multiple periodic arrays of attached resonators. Phys. Lett. A **376**, 1384–1390 (2012)
8. Chen, J.S., Huang, Y.J.: Wave propagation in sandwich structures with multiresonators. J. Vib. Acoust. **138**, 041009 (2016)
9. Hu, G., Austin, A.C.M., Sorokin, V., Tang, L.: Metamaterial beam with graded local resonators for broadband vibration suppression. Mech. Syst. Signal Process. **146**, 106982 (2021)
10. Silva, T.M.P., Clementino, M.A., de Sousa, V.C., De Marqui, C.: An experimental study of a piezoelectric metastructure with adaptive resonant shunt circuits. IEEE/ASME Trans. Mechatron. **25**, 1076–1083 (2020)

11. Xiao, Y., Mace, B.R., Wen, J., Wen, X.: Formation and coupling of band gaps in a locally resonant elastic system comprising a string with attached resonators. Phys. Lett. A **375**, 1485–1491 (2011)
12. Yu, D., Wen, J., Zhao, H., Liu, Y., Wen, X.: Vibration reduction by using the idea of phononic crystals in a pipe-conveying fluid. J. Sound Vib. **318**, 193–205 (2008)
13. Xiao, Y., Wen, J., Wen, X.: Longitudinal wave band gaps in metamaterial-based elastic rods containing multi-degree-of-freedom resonators. New J. Phys. **14**, 033042 (2012)
14. Xiao, Y., Wen, J., Yu, D., Wen, X.: Flexural wave propagation in beams with periodically attached vibration absorbers: band-gap behavior and band formation mechanisms. J. Sound Vib. **332**, 867–893 (2013)
15. Hao, S., Wu, Z., Li, F., Zhang, C.: Numerical and experimental investigations on the band-gap characteristics of metamaterial multi-span beams. Phys. Lett. A **383**, 126029 (2019)
16. Gardonio, P., Brennan, M.J.: Mobility and impedance methods in structural dynamics. In: Advanced Applications in Acoustics. Noise and Vibration, pp. 389–447. Taylor and Francis, London (2004)
17. Brennan, M.J.: Vibration control using a tunable vibration neutralizer. Proc. Inst. Mech. Eng. C J. Mech. Eng. Sci. **211**, 91–108 (1997)

On Unified Formulation of Floquet Propagator in Cartesian and Polar Coordinates

A. Hvatov[1(✉)] and S. Sorokin[2]

[1] National Centre for Cognitive Research, ITMO University, 197101 49 Kronverksky pr., St Petersburg, Russia
alex_hvatov@itmo.ru

[2] Department of Materials and Production, Aalborg University, Fibigerstrade 16, DK9220 Aalborg, Denmark
svs@mp.aau.dk

Abstract. A modified formulation of Floquet propagator is proposed to analyze free wave motion in homogeneous and periodic waveguides both in the Cartesian and in the polar coordinates. For homogeneous waveguides, it substantiates the application of a Wave Finite Element Method in polar coordinates. For radially periodic waveguides, it facilitates the application of conventional (i.e., used in Cartesian coordinates) criterion to identify frequency-wise positions of pass- and stop-bands. The application of the proposed methodology is illustrated by a simple example of a wave propagation problem governed by the Helmholtz equation (a dilatation wave in a membrane).

Keywords: Floquet propagator · Polar coordinates · WFEM · Periodic structures

1 Introduction

The periodicity effects are well-known and understood for waveguides in Cartesian coordinates regardless of their dimensions (in terms of structural mechanics, a rod/beam, a plate, or a three-dimensional lattice). The underlying Floquet theory relies on the translational invariance of the problem formulation for an infinite periodic structure [2]. More specifically, all Floquet propagators in Cartesian coordinates are constants independent of the shape of a unit periodicity cell and its location in a waveguide.

A problem of wave propagation in a radially periodic waveguide was formulated a long time ago with reference to the guiding of light, and its approximate solution is well-known in optics as the theory of Bragg fiber [18]. Leaving aside numerous purely numerical studies (see, for instance, [8,9,13]), we notice that two approaches have been used so far to 'adjust' the Bloch theorem [7] for a cylindrically symmetric Bragg fiber. The first one is based on the use of far-field

Z. Dimitrovová et al. (Eds.): WMVC 2022, MMS 125, pp. 713–724, 2023.
https://doi.org/10.1007/978-3-031-15758-5_73

approximation of Bessel functions [6,17], and the second one implies special radial varying of material properties of cladding [1,14,15].

The idea to consider uniform, homogeneous waveguides as periodic is fundamental for the wave finite element method [10,12]. The propagators obtained using periodicity conditions allow one to retrieve wavenumbers and thus define the acoustical properties of the waveguide. For Cartesian coordinates, it is explained in [16]. However, to apply the wave finite element method to the structures in other coordinate systems, such as polar coordinates, tuning of the Floquet theory is required.

In our earlier paper on the application of Floquet theory in polar coordinates [5], the Floquet propagator has been introduced as a function of the radial coordinate, and several heuristically chosen approximations of this dependence have been compared with each other and verified via direct numerical integration. The novel contribution of the present paper is the formulation of a propagator, which is equally applicable for Cartesian and polar coordinates. Then its independence upon the axial Cartesian coordinate naturally emerges, and its dependence upon the radial polar coordinate becomes uniquely defined in agreement with findings presented in [5].

For clarity, we consider the simplest case possible: wave propagation governed by the Helmholtz equation. In Sect. 2, we consider a homogeneous waveguide with exact solutions readily available. First, we modify the formulation of periodicity conditions and prove its validity. Next, we propose the unified formulation of a propagator and assess its applicability for the identification of pass-bands. In Sect. 3, we consider a periodic waveguide, introduce the propagator in the same wave as in a homogeneous waveguide, and assess its applicability for the identification of stop and pass-bands. In Conclusion, we discuss reported results and future work.

2 An Uniform Membrane. Motivation and Problem Statement

The Helmholtz equation (Eq. 1) is the simplest mathematical model to formulate the Floquet theory for wave propagation in a periodic structure composed of continuous constituents. In Cartesian coordinates, it describes the time-harmonic propagation of a plane wave in a membrane, which in this case is reduced to a string. In polar coordinates, it describes the time-harmonic propagation of a cylindrical wave of dilatation in a membrane:

$$(k^2 + \Delta)u = 0 \tag{1}$$

In Eq. 1, $k = \frac{\omega}{c}$is the wavenumber, $c = \sqrt{\frac{T}{\rho}}$ is the speed of a dilatation wave, T is the membrane uniform pre-tension, ρ is the material density per unit length, Δ is the Laplacian operator in a given coordinate system.

In this section, we apply the Floquet theory for an authentically uniform, homogeneous waveguide. The existence of elementary analytical solutions for wave propagation problems both in the Cartesian and in the polar coordinates provides an incentive for tackling the case of a periodic waveguide.

2.1 Uniform Propagator in Cartesian Coordinates

For a uniform membrane in the Cartesian coordinate, general solution of Eq. 1 has a form Eq. 2.

$$u(s) = A\exp(iks) + B\exp(-iks) \tag{2}$$

In Eq. 2, s is the coordinate counted along the axis of a waveguide, i.e. in Cartesian coordinates $s = x$. For the segment of an arbitrary length L, the relation Eq. 3 holds.

$$\begin{aligned} u(s) - (\Lambda_1^{cart} A\exp(ik(s+L)) \\ + \Lambda_2^{cart} B\exp(-ik(s+L))) = 0 \end{aligned} \tag{3}$$

In Eq. 3, $\Lambda_1^{cart}, \Lambda_2^{cart}$ are propagators, which are extracted directly in form Eq. 4.

$$\begin{aligned} \Lambda_1^{cart} &= \tfrac{\exp(iks)}{\exp(iks+ikL)} = \exp\left(-ikL\right) \\ \Lambda_2^{cart} &= \tfrac{\exp(-iks)}{\exp(-iks-ikL)} = \exp\left(ikL\right) \end{aligned} \tag{4}$$

On the other hand, for the periodic structures, it is natural to use the Floquet formulation of the periodicity condition in form Eq. 5.

$$u(s) - \Lambda^{cart} u(s+L) = 0 \tag{5}$$

Obviously, the solution $u\left(x\right)$ still has the form Eq. 2, but the propagators $\Lambda_1^{cart}, \Lambda_2^{cart}$ are not the solutions of Eq. 5. To recover the correct result Eq. 4 for these parameters, we differentiate Eq. 5 and obtain the second equation Eq. 6.

$$u'(s) - \Lambda^{cart} u'(s+L) = 0 \tag{6}$$

In doing so, we take advantage of the translation invariance of the problem formulation and regard the parameter Λ^{cart} as constant (which, of course, follows from Eq. 4). Now we substitute general solution Eq. 2 in Eq. 5–Eq. 6 and set to zero the determinant of the system of linear algebraic equations with respect to the unknown amplitudes A, B.

$$D = (\Lambda^{cart})^2 - 2\cos(kL)\Lambda^{cart} + 1 = 0 \tag{7}$$

The roots of Eq. 7, $\Lambda_{1,2}^{cart} = \exp\left(\pm ikL\right)$ are given by Eq. 4, and, as follows from Floquet theorem, propagation condition is $|\Lambda_{1,2}^{cart}| = 1$. It is certainly held in a homogeneous membrane (string) at any frequency. For any given L, dimensional wavenumbers are easily recovered from $\Lambda_{1,2}^{cart}$.

The idea to apply periodicity conditions for a homogeneous waveguide, find the propagation constants, and convert them into wavenumbers is the essence of the wave finite element method. In its standard formulation, Eq. 5 is introduced as the periodicity of displacements, and Eq. 6 is referred to as the periodicity of forces. The ad hoc assumption that periodicity of forces and displacements is the same is proved by the derivation of Eq. 6 from Eq. 5. As already mentioned, the wave finite element method utilizes this approach for an arbitrarily shaped uniform (in Mead's terminology [11], multi-coupled) waveguide in Cartesian coordinates and yields excellent results.

2.2 Large Argument Approximation in Polar Coordinates

Next, we replicate this Wave Finite Element approach in the polar coordinates and solve the Helmholtz equation in the same way. With no loss of generality, we consider an axisymmetric case. First, we consider standard large argument approximation [3]. Then the solution has the form Eq. 8.

$$u\left(s\right)=\sqrt{\frac{2}{\pi ks}}\left[\hat{A}\exp\left(iks\right)+\hat{B}\exp\left(-iks\right)\right] \tag{8}$$

In Eq. 8, s stands for the radial coordinate, i.e., $s = r$. In polar coordinates case, propagators become functions of the spatial coordinate, and equations Eq. 5–Eq. 6 are written in form Eq. 9.

$$\begin{aligned} u(s)-\Lambda^{polar}(s)u(s+L)=0\\ \tfrac{d}{ds}\left[u(s)-\Lambda^{polar}(s)u(s+L)\right]=0 \end{aligned} \tag{9}$$

After substitution of the solution in form Eq. 8 to Eq. 9 and equating to zero the determinant of linear equations with respect to unknown amplitudes $\hat{A}$ and $\hat{B}$, we obtain:

$$\begin{aligned} D=&\Bigg[\left(\Lambda^{polar}(s)\right)^2\\ &-2\sqrt{\frac{s+L}{s}}\cos(kL)\Lambda^{polar}(s)+\frac{s+L}{s}\Bigg]\\ &+\frac{\sin(kL)}{kL}\Bigg[\frac{d}{ds}\left(\sqrt{\frac{s+L}{s}}\right)\Lambda^{polar}(s)\\ &\qquad-\sqrt{\frac{s+L}{s}}\frac{d\Lambda^{polar}(s)}{ds}\Bigg]=0 \end{aligned} \tag{10}$$

The first term in Eq. 10 has the same structure as Eq. 7, and the second term, if equated to zero yields the differential equation for $\Lambda^{polar}(s)$:

$$\frac{d}{ds}\left(\sqrt{\frac{s+L}{s}}\right)\Lambda^{polar}(s)-\sqrt{\frac{s+L}{s}}\frac{d\Lambda^{polar}(s)}{ds}=0 \quad (11)$$

Solution of the Eq. 11 gives the ansatz for the propagator $\Lambda^{polar}(s)$

$$\Lambda^{polar}(s)=\Lambda_0\sqrt{\frac{s+L}{s}} \quad (12)$$

with Λ_0 to be defined as the integration constant.
Substituting Eq. 12 to the first term in Eq. 10 gives the equation for Λ_0

$$D=\frac{s+L}{s}\left(\Lambda_0^2-2\Lambda_0\cos kL+1\right)=0 \quad (13)$$

Its solutions are the Cartesian propagators Eq. 4. Then the final form of Eq. 12 can be written in form Eq. 14.

$$\Lambda_{1,2}^{polar}=\sqrt{\frac{s+L}{s}}\exp\left(\pm ikL\right)=S\left(s\right)\Lambda_{1,2}^{cart} \quad (14)$$

The form of Eq. 14 suggests that the propagators in polar coordinates merge their Cartesian formulation Eq. 4 as soon as $s\rightarrow\infty$. This is certainly the expected result: cylindrical waves tend to plane waves in the far field so that propagators in polar coordinates match their counterparts in Cartesian coordinates. The propagation condition preserves its Cartesian form with the 'cylindrical wave-correction'. In the far-field, such formulation clearly agrees with the Bragg fiber theory, i.e. $\Lambda_{1,2}^{polar}\rightarrow\Lambda_{1,2}^{cart}, s\rightarrow\infty$. In the next section, we propose a refined formulation for the Floquet theory in polar coordinates.

2.3 Uniform Propagator in Polar Coordinates

As the next step, we use the exact representation of the displacement field. The solution is well-known and has the form Eq. 15.

$$u(s)=AH_0^{(1)}(ks)+BH_0^{(2)}(ks) \quad (15)$$

In this case, we again re-formulate the propagator with the correction factor pulled out: $\Lambda^{polar}\left(s\right)=S\left(s\right)\tilde{\Lambda}^{polar}\left(s\right)$. We note that the shape function, or 'envelope' $S\left(s\right)=\sqrt{\frac{s+L}{s}}$ could be induced ad hoc following the results presented in Sect. 2.2. To show, that the shape function is defined by the choice of a coordinate system rather, than by the particular problem, we introduce the periodicity condition with unknown function $S(s)$ in form Eq. 16.

$$u(s)-S\left(s\right)\tilde{\Lambda}^{polar}\left(s\right)u(s+L)=0 \quad (16)$$

Then we differentiate Eq. 16 with respect to s and assume that $\frac{d}{ds}\tilde{\Lambda}^{polar}(s) << \frac{d}{ds}S\left(s\right)$:

$$\begin{aligned}&\frac{d}{ds}\left[u(s) - S\left(s\right)\tilde{\Lambda}^{polar}\left(s\right)u(s+L)\right]\\&= u'(s) - (\tilde{\Lambda}^{polar}\left(s\right)\left(S'(s)u(s+L) + S(s)u'(s+L)\right))\\&\qquad\qquad\qquad\qquad\qquad\qquad\qquad\qquad\quad = 0\end{aligned} \tag{17}$$

Then, as in the previous case, we obtain the system of the homogeneous linear algebraic equations with respect to the two unknown amplitudes A, B. Equating its determinant to zero, the equation with respect to propagator becomes Eq. 18.

$$\begin{aligned}D = \left(\tilde{\Lambda}^{polar}(s)\right)^2 + a_1\left(s, S\left(s\right), S'\left(s\right)\right)\tilde{\Lambda}^{polar}(s)\\+ a_0\left(s, S\left(s\right), S'\left(s\right)\right) = 0\end{aligned} \tag{18}$$

To find shape function $S(s)$ we use following condition Eq. 19 that represents that wave should be "corrected" to satisfy condition abs$(\Lambda) = 1$ in pass-band and $\Lambda_1 \cdot \Lambda_2 = 1$ for stop-band in periodic membrane case.

$$a_0\left(s, S\left(s\right), S'\left(s\right)\right) \equiv 1 \tag{19}$$

For membrane in polar coordinates, Eq. 19 is a simple algebraic equation. The solution of Eq. 19 with respect to $S(s)$ has the form

$$S(s) = \pm\sqrt{\frac{s+L}{s}} \tag{20}$$

This correction function matches its counterpart obtained in Sect. 2.2. Substituting Eq. 20 into Eq. 21 gives the quadratic equation in $\tilde{\Lambda}^{polar}(s)$ in form Eq. 21.

$$D = \left(\tilde{\Lambda}^{polar}(s)\right)^2 + a_1\left(s\right)\tilde{\Lambda}^{polar}(s) + 1 = 0 \tag{21}$$

Unlike the large argument approximation Eq. 10, the coefficient a_1 in Eq. 21 remains dependent upon coordinate s. To simplify further computations, we introduce scaled dimensionless parameters Eq. 22.

$$\begin{aligned}p_1 &= kL + ks\\p_2 &= ks\end{aligned} \tag{22}$$

We can note that $a_1(s)$ is a symmetric function of two variables p_1, p_2 and it is more natural to write the function in terms Eq. 22 instead of parameter set k, L, s. Explicitly coefficient $a_1(s)$ for an uniform membrane in polar coordinates may be written in form Eq. 23.

$$a_1 = \frac{\pi}{4\sqrt{p_1 p_2}}\left[\begin{array}{c}-2p_1p_2J_1\left(p_2\right)Y_0\left(p_1\right) - 2p_1p_2J_1\left(p_1\right)Y_0\left(p_2\right) + J_0\left(p_2\right)\left(\left(p_1 - p_2\right)Y_0\left(p_1\right) + 2p_1p_2Y_1\left(p_1\right)\right)\\+J_0\left(p_1\right)\left(2p_1p_2Y_1\left(p_2\right) + \left(p_2 - p_1\right)Y_0\left(p_2\right)\right)\end{array}\right] \tag{23}$$

We can introduce the difference of Cartesian Eq. 7 and polar Eq. 23 coefficients a_1, and, using large-argument approximations [3], the majoring function may be found as:

$$\Delta_M = \frac{(p_1 - p_2)\,\mathrm{Sin}(p_1 - p_2)}{2p_1p_2} \tag{24}$$

Therefore, as soon as $p_1 \to +\infty$ or $p_2 \to +\infty$ the difference becomes infinitely small, which means that the both coefficients are equal for infinite large either x or k. This also gives an approximation for the coefficient $a_1(s)$ in the polar coordinates as:

$$a_1^{approx} = -2\cos(kL) + \frac{L\sin(Lk)}{2ks\,(L+s)} \tag{25}$$

It is clear, that roots of Eq. 21 do not match the Cartesian predictions Eq. 7 with the coefficient $a_1(s)$ defined neither by the exact formula Eq. 23 nor by the approximation Eq. 25.

The free term in Eq. 21 equals one, so that the condition $\tilde{\Lambda}_1^{polar} \cdot \tilde{\Lambda}_2^{polar} = 1$ always holds. However, it does not mean that $\tilde{\Lambda}_1^{polar}$ and $\tilde{\Lambda}_2^{polar}$ are always complex conjugates. As seen, for example, from Eq. 25, there are sets of parameters, when $|a_1^{appox}| > 2$, so that the propagators are purely real with $\left|\tilde{\Lambda}_1^{polar}\right| \neq \left|\tilde{\Lambda}_2^{polar}\right| \neq 1$. The Cartesian criterion of stop band generation has exactly this form, and, since there cannot be any stop bands in a homogeneous membrane, it becomes invalid - but only in the immediate vicinity of the origin of polar coordinates.

In Fig. 1, we present the zones, where $\left|\tilde{\Lambda}_1^{polar}\right| = \left|\tilde{\Lambda}_2^{polar}\right| = 1$ as shadowed, and zones, where $\left|\tilde{\Lambda}_1^{polar}\right| \neq \left|\tilde{\Lambda}_2^{polar}\right| \neq 1$ as white. Blue zones are found using the exact formulation Eq. 21 and green zones are found using the approximation Eq. 25. In this diagram, the radial coordinate is scaled as $r = \frac{s}{\lambda}$, where $\lambda = \frac{2\pi}{k} = \frac{2\pi\, c}{\omega}$ is the length of a wave, which propagates at a given frequency in membrane. As seen from Fig. 1, the zone, in which the conventional propagation condition for 'cylindrical wave-corrected' propagators is recovered, emerges at the distance equal to some fraction of the length of a propagating wave.

Not surprisingly, the exact formulation of the coefficient a_1 narrows down the artificial 'stop-band'. It should also be noted that the condition $\Omega = kL = \pi$ defines the position of a 'seed' of the first stop band both in Cartesian and in polar coordinates, where the quadratic equation for a propagator acquires a repeated root.

It is seen that using large-order approximation Eq. 25 for polynomial Eq. 21 is viable however, it introduces error with order that may be defined using majoring function Eq. 24.

These results are useful for adjustments of the Wave Finite Element Method for polar coordinates. To obtain reasonable predictions of wavenumbers, first, the correction factor $S\,(s) = \sqrt{\frac{s+L}{s}}$ should be taken into account, and, second,

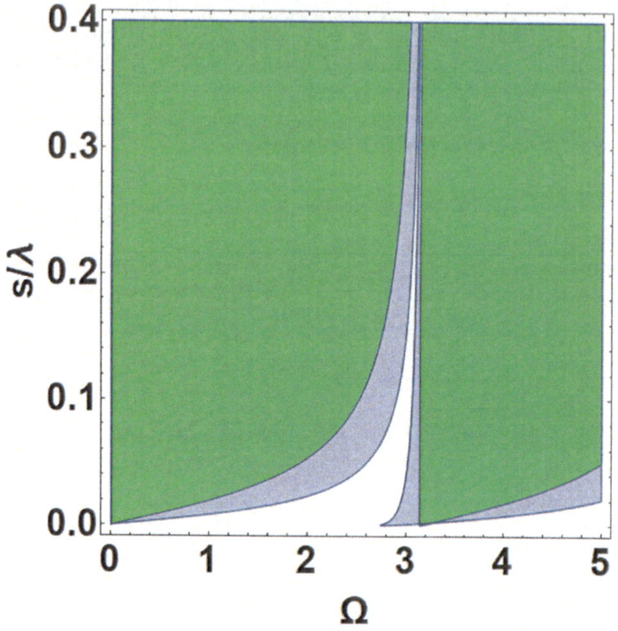

Fig. 1. Zones where $\left|\tilde{\Lambda}_1^{polar}\right| = \left|\tilde{\Lambda}_2^{polar}\right| = 1$: blue using Eq. 21 and green using Eq. 25.

a FE-modeled segment should be placed not closer to the origin of coordinates, than 0.2–0.4 of the longest propagating wave.

3 A Periodic Membrane

The results presented in the previous Section give an incentive for the formulation of propagators in either Cartesian or polar coordinates as authentically periodic waveguides and identification of stop-bands by these means. We consider an infinite periodic membrane shown in Fig. 2.

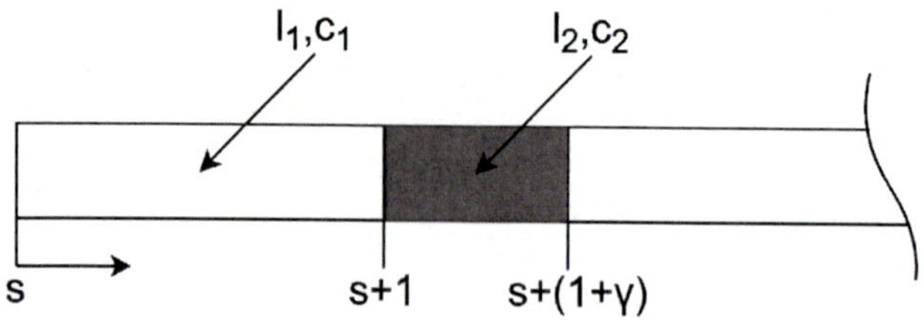

Fig. 2. The example of periodic structure - parts with different colour have different waveguide properties.

Dimensionless parameter $\sigma = \frac{c_2}{c_1}$ in this case is the ratio between dilatation wave propagation speeds in the segments, $\gamma = \frac{l_2}{l_1}$ is the length ratio, $\Omega = \frac{\omega l_1}{c_1}$ is the dimensionless frequency parameter. In Cartesian coordinates Helmholtz equation (Eq. 1) for every segment has solution in form Eq. 26.

$$u_i(x) = \begin{cases} A^{(i)} \exp(i\Omega s) + B^{(i)} \exp(-i\Omega s), \; x \in \text{“white” part} \\ A^{(i)} \exp(i\frac{\Omega}{\sigma} s) + B^{(i)} \exp(-i\frac{\Omega}{\sigma} s), \; x \in \text{“grey” part} \end{cases} \tag{26}$$

Solution in polar coordinates has analogous form shown in Eq. 27.

$$u_i(x) = \begin{cases} A^{(i)} H_0^{(1)}(\Omega s) + B^{(i)} H_0^{(2)}(\Omega s), \; x \in \text{“white” part} \\ A^{(i)} H_0^{(1)}(\frac{\Omega}{\sigma} s) + B^{(i)} H_0^{(2)}(\frac{\Omega}{\sigma} s), \; x \in \text{“grey” part} \end{cases} \tag{27}$$

In Eq. 26–Eq. 27 $A^{(i)}, B^{(i)}$ are the amplitudes of outgoing and incoming waves in the i-th segment. It is assumed that the first segment is chosen arbitrary and following segments have sequential numbers.

The interfacial conditions representing the continuity of a periodic structure have the form Eq. 28.

$$\begin{gathered} u_i(s+1) = u_{i+1}(s+1) \\ u'_i(s+1) = u'_{i+1}(s+1) \\ u_{i+1}(s+1+\gamma) = u_{i+2}(s+1+\gamma) \\ u'_{i+1}(s+1+\gamma) = u'_{i+2}(s+1+\gamma) \end{gathered} \tag{28}$$

Similar to the case of a homogeneous membrane, we formulate periodicity conditions ($\tau = 1 + \gamma$ is the scaled length of a unit periodicity cell of the membrane) in form Eq. 29.

$$\begin{gathered} u_i(s) - \Lambda(s) u_{i+2}(s+\tau) \\ \frac{\partial}{\partial s}[u_i(s) - \Lambda(s) u_{i+2}(s+\tau)] = 0 \end{gathered} \tag{29}$$

We note that the equations Eq. 28–Eq. 29 are not dependent on a type of the operator and can be used for an arbitrary second-order elliptic equation. Determinant of the system Eq. 28–Eq. 29 of linear algebraic equation with respect to the unknown amplitudes is the first-order differential equation in $\Lambda(s)$. As shown in [5], this equation cannot be solved analytically.

However, we can use same assumptions as in the uniform membrane case, meaning propagation parameter in form $\Lambda(s) = S(s)\Lambda$ to obtain second-order polynomial in Λ and condition Eq. 19 (free term $a_0 \equiv 1$) to recover the form of the shape function $S(s)$. In the Cartesian coordinates, like the uniform membrane the only possible form of $S(x)$ is:

$$S(x) = \pm 1 \tag{30}$$

Equation 30 gives the classical formulation of the Floquet theory for the membrane in the Cartesian coordinates equation [4] in form Eq. 31.

$$D = (\Lambda^{cart})^2 + \Lambda^{cart}\left(\frac{1}{\sigma} + \sigma\right)\sin(\Omega)\sin\left(\frac{\gamma\Omega}{\sigma}\right) - 2\Lambda^{cart}\cos(\Omega)\cos\left(\frac{\gamma\Omega}{\sigma}\right) + 1 = 0 \quad (31)$$

The conventional conditions $|\Lambda_1^{cart}| = |\Lambda_2^{cart}| = 1$ and $|\Lambda_1^{cart}| \neq |\Lambda_2^{cart}| \neq 1$ define location of pass-bands and stop-bands, respectively. Moreover, we see that the classical formulation is the only possible case for Floquet theory. We will not reproduce further classical analysis for brevity.

The propagator for a radially periodic membrane in polar coordinates is defined by the quadratic equation in $\tilde{\Lambda}_{1,2}^{polar}(s)$. The exact form of the propagator is obtained by substitution Eq. 27 in Eq. 28–Eq. 29 with assumption $\Lambda(s) = S(s)\Lambda$ and setting to zero the determinant of the system of linear algebraic equations with respect to amplitudes of outgoing and incoming waves in Eq. 27. The resulting polynomial has the same structure as Eq. 21, but the coefficient $a_1(s)$ is much more cumbersome and not presented here. The shape function may be obtained from equation Eq. 19 and has the form Eq. 32.

$$S(s) = \pm\sqrt{1 + \frac{\tau}{s}} \quad (32)$$

It matches the 'cylindrical wave-correction' function for a homogeneous membrane. Then the propagator for a periodic membrane in polar coordinates is $\Lambda^{polar}(s) = \tilde{\Lambda}^{polar}(s)\sqrt{\frac{s+\tau}{s}}$. We note that similar function obtained using different approach, which could be found in [12]. The condition $\left|\tilde{\Lambda}^{polar}(s)_1\right| = \left|\tilde{\Lambda}^{polar}(s)_2\right| = 1$ defines location of pass-bands. However, given the results presented in Sect. 2.3 for a homogeneous membrane, the opposite case, $\left|\tilde{\Lambda}^{polar}(s)_1\right| \neq \left|\tilde{\Lambda}^{polar}(s)_2\right| \neq 1$, does not necessarily specify a stop-band zone if these propagators are computed in the immediate vicinity of the origin of coordinate system.

Nevertheless, introduction of the shape function $S(s)$ permits to unify formulation of propagators in Cartesian and polar coordinates and recovers the conventional criterion of stop-band formation $\left|\tilde{\Lambda}^{polar}(s)_1\right| \neq \left|\tilde{\Lambda}^{polar}(s)_2\right| \neq 1$ 'almost everywhere' in the (Ω, s) plane.

An alternative unified formulation of Floquet propagator could be introduced in form Eq. 33.

$$\Lambda = \frac{\Lambda_{periodic}}{\Lambda_{uniform}} \quad (33)$$

In Eq. 33 $\Lambda_{uniform}$ is the root of the discriminant Eq. 21 for uniform membrane, $\Lambda_{periodic}$ is the corresponding root of the discriminant of the system Eq. 28–Eq. 29. In this case, using Eq. 33 we can preserve the stop-bands definition (abs$(\Lambda) \neq 1$). For Fig. 3 following set of the dimensionless parameters is taken $(\gamma = 2, \sigma = 0.1)$.

We note that like the homogeneous case, at the distances of order less than $\frac{\lambda}{5}$ the wave propagation picture may not be considered as correct.

The formulation of the Floquet-like theory Eq. 28–Eq. 29 and Eq. 14 shows the mechanism of Bragg fiber theory. Moreover, it may be used for the general case to obtain the stop-band picture for more advanced operators.

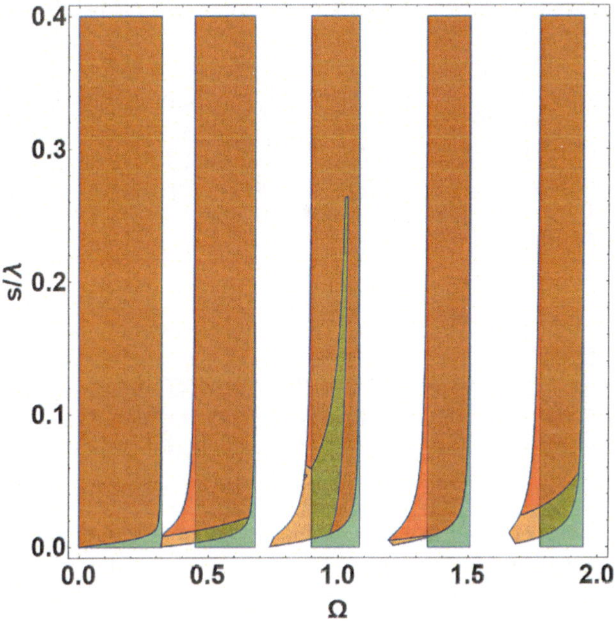

Fig. 3. Pass-bands(color) and stop-bands(white) with respect to the cartesian definitoion (is considered as pass-band) for the radial membrane obtained from determinant of the rod in the Cartesian coordinates problem Eq. 31 (green), determinamt of the system Eq. 28–Eq. 29 with Eq. 32 (orange), ratio Eq. 33 (red).

4 Conclusions

In this paper, we have demonstrated how the Cartesian formulation of the Floquet propagator may be tuned to be applicable in polar coordinates. The simple 'cylindrical wave correction function' facilitates the use of a Wave Finite Element Method to find wavenumbers of uniform waveguides in polar coordinates. The introduction of this function extends the applicability of conventional (i.e., used in Cartesian coordinates) criterion to identify frequency-wise positions of pass- and stop-bands for radially periodic waveguides. However, it is not entirely clear what stop-bands are mean in this case.

References

1. Arretche, I., Matlack, K.H.: Effective phononic crystals for non-cartesian elastic wave propagation. Phys. Rev. B **102**, 134308 (2020)
2. Brillouin, L.: Wave propagation in periodic structures: electric filters and crystal lattices, vol. 2. Dover publications (1953)
3. Heitman, Z., Bremer, J., Rokhlin, V., Vioreanu, B.: On the asymptotics of bessel functions in the fresnel regime. Appl. Comput. Harmon. Anal. **39**, 347–356 (2015)
4. Hvatov, A., Sorokin, S.: Free vibrations of finite periodic structures in pass-and stop-bands of the counterpart infinite waveguides. J. Sound Vib. **347**, 200–217 (2015)
5. Hvatov, A., Sorokin, S.: On application of the floquet theory for radially periodic membranes and plates. J. Sound Vib. **414**, 15–30 (2018)
6. Kitagawa, A., Sakai, J.: Bloch theorem in cylindrical coordinates and its application to a bragg fiber. Phys. Rev. A **80**, 033802 (2009)
7. Kittel, C., McEuen, P., McEuen, P.: Introduction to Solid State Phys., vol. 8. Wiley, New York (1996)
8. Li, Y., Chen, T., Wang, X., Yu, K., Chen, W.: Propagation of lamb waves in one-dimensional radial phononic crystal plates with periodic corrugations. J. Appl. Phys. **115**, 054907 (2014)
9. Ma, T., Chen, T., Wang, X., Li, Y., Wang, P.: Band structures of bilayer radial phononic crystal plate with crystal gliding. J. Appl. Phys. **116**, 104505 (2014)
10. Mace, B.R., Duhamel, D., Brennan, M.J., Hinke, L.: Finite element prediction of wave motion in structural waveguides. J. Acoustical Soc. Am. **117**, 2835–2843 (2005)
11. Mead, D.: Wave propagation and natural modes in periodic systems: Ii. multi-coupled systems, with and without damping. J. Sound Vib. **40**, 19–39 (1975)
12. Mencik, J.M., Ichchou, M.: Multi-mode propagation and diffusion in structures through finite elements. Eur. J. Mech.-A/Solids **24**, 877–898 (2005)
13. Shi, X., et al.: Research on wave bandgaps in a circular plate of radial phononic crystal. Int. J. Mod. Phys. B **30**, 1650162 (2016)
14. Torrent, D., Sánchez-Dehesa, J.: Radial wave crystals: radially periodic structures from anisotropic metamaterials for engineering acoustic or electromagnetic waves. Phys. Rev. Lett. **103**, 064301 (2009)
15. Torrent, D., Sánchez-Dehesa, J.: Acoustic resonances in two-dimensional radial sonic crystal shells. New J. Phys. **12**, 073034 (2010)
16. Waki, Y., Mace, B., Brennan, M.: Numerical issues concerning the wave and finite element method for free and forced vibrations of waveguides. J. Sound Vib. **327**, 92–108 (2009)
17. Xu, Y., Ouyang, G.X., Lee, R.K., Yariv, A.: Asymptotic matrix theory of bragg fibers. J. Lightwave Technol. **20**, 428 (2002)
18. Yeh, P., Yariv, A., Marom, E.: Theory of bragg fiber. JOSA **68**, 1196–1201 (1978)

Rainbow Smart Metamaterial to Improve Flexural Wave Isolation and Vibration Attenuation of a Beam

Braion B. Moura(✉) and Marcela R. Machado

Department of Mechanical Engineering, University of Brasilia, Brasília 70910-900, Brazil
braionbarbosa@gmail.com

Abstract. This article analyzes wave propagation isolation and vibration attenuation strategies of a beam coupled to piezoelectric sensors periodically arranged in a given frequency band. Each piezo sensor is connected to a resonant shunt circuit. The influence on the attenuation band is due to a tunable shunt impedance associated with the corresponding piezo. Hence, the piezo's resonate at different and neighbouring frequencies creates a tunable rainbow trap that can attenuate the energy within a bandgap characteristic. The smart metastructure is modelled by means of the spectral element method, which is a highly accurate method with a low computational cost. Flexural wave propagation is obtained using the Transfer Matrix Method with the scatter diagram plot. Results show the effect of broadband vibrations' attenuation and propagating waves isolation. Moreover, the spectral range over which attenuation is achieved with the rainbow arrangements is on average wider than the usual metamaterials configurations.

Keywords: Rainbow metamaterial · Smart metastructure · Spectral element method · Transfer matrix method

1 Introduction

The smart materials, metamaterials and metastructure are structures increasingly applied in advanced multi-physics systems. Piezoelectric materials (piezo) coupled to beams, bars and plates are a structural example of smart materials and metamaterials capable of exerting vibrational control and wave propagation. Such control depends on the operational configuration of the piezos that can convert energy from the mechanical to electrical domain and vice versa, that is, direct piezoelectric effect when the piezo converts the mechanical stress into an electric field, and reverse piezoelectric effect when the piezo converts the electric field into mechanical stress [13]. Each operational configuration allows the evolution of these smart materials and metamaterials, where the potential for on-demand ownership modulation can be achieved by passive, active

Supported by organization FAPDF.

Z. Dimitrovová et al. (Eds.): WMVC 2022, MMS 125, pp. 725–733, 2023.
https://doi.org/10.1007/978-3-031-15758-5_74

or hybrid control [3,10]. Active vibration control techniques are related to the reverse piezoelectric effect, where they use an electrical energy source to increase the mechanical energy needed by the system. Unlike active techniques, passive vibration control techniques use the direct piezoelectric effect, making changes in the electrical energy generated by the piezo to promote a specific dynamic property in the structure, without relying on an external source of electrical energy [8]. Hybrid vibration controls combine both active and passive control techniques.

In the literature, several studies address the use of passive vibrational control with piezos connected to external circuits composed of passive components such as resistors, inductors and capacitors [7]. These circuits, known by the term shunt, began to be explored by Foward (1979) with the aim of inducing vibrations. Later, Uchino and Ishii (1988) explored the direct piezoelectric effect, dissipating the resulting electrical energy through an external resistance. Thus, it was noticed that a significant variation in the structure's damping factor occurs when the value of the external resistance is changed. Hagood and Flotow (1988) continued their study of the resistive shunt circuit, but added the inductive element and realized the ability to adjust the attenuation effect in frequency, similar to a dynamic vibration absorber. From there, other works explore various combinations of circuits with resistive-inductive elements applied to vibrational control, and these circuits became known by the term resonant shunt circuits [1,9,16,17,19,21].

The application of passive control with the resonant shunt circuit is commonly performed identically and periodically along a structure. However, some works are using the resonant shunt circuit with the coupling configuration of 7 piezos tuned at different frequencies, but close to each other. This coupling configuration is called a rainbow trap and can provide attenuation in a given frequency band [2,11,20]. In this context, the present work aims to explore the Frequency Response Function (FRF) and the Scatter Diagram (DD) of a beam subjected to a vibrational control with resonant shunt in the rainbow trap configuration. With this, we intend to investigate the influence that each piezo causes on the attenuation width. For this, numerical models are developed based on SEM to perform accurate and computationally efficient analyses, without the need for large discretizations [4]. Furthermore, the Transfer Matrix Method (TMM) is used to estimate the DD of the structure.

2 Smart Material Theory Background

The SEM is considered a very efficient method for representing different types of geometries, boundary conditions and materials. Part of the efficiency of this method is due to the fact that the shape functions of the elements are obtained from the analytical solution of the governing differential equations and the solution of the dynamical system written in the domain frequency [5,15]. The SEM can represent the smart beam with the subdivision of beam (B) and beam-piezo-shunt (BPS) elements, as shown in Fig. 1. Each element is composed of two nodes and each node has three degrees of freedom based on the Euler-Bernoulli theory [12].

(a)

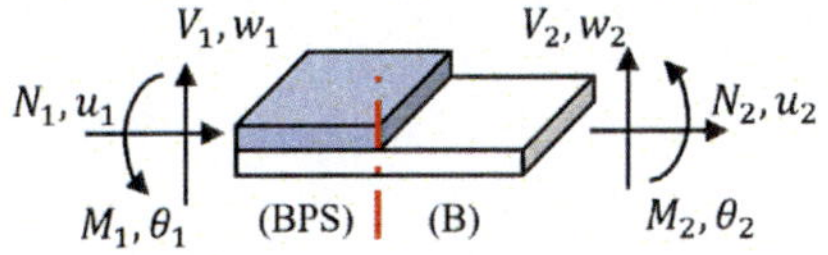

(b)

Fig. 1. Representation: a) Spectral model of the structure; b) Free-body diagram.

For the beam element (B), Doyle (1997) reports that through the nodal relations of force and displacement it is possible to express the spectral stiffness matrix $\mathbf{S}_B(\omega)$ for the Euler-Bernoulli beam element. Similarly, Lee (2000) reports how to obtain the spectral stiffness matrix $\mathbf{S}_{BP}(\omega)$ for the beam element with a uniformly coupled piezo layer. However, the mathematical representation of the beam-piezo element connected to a resonant shunt circuit (BPS) is given by the following equation of motion

$$\begin{aligned} &EIw'''' + \rho A\ddot{w} + cA\dot{w} + \Gamma V = -\alpha\ddot{u}_b' + \beta u''' + \gamma\ddot{w}'' + c_1\dot{w}'' - c_4\dot{u}_b' + Fw'' \\ &+ p(x,t) \\ &EAu_b'' - \rho A\ddot{u}_b - cA\dot{u}_b + \Gamma V = -\alpha\ddot{w}' + \beta w''' - c_4\dot{w}' - \Gamma(x,t) \\ &E_T\dot{x} + C_P^T\dot{V} = I_c(x,t) \end{aligned} \tag{1}$$

where

$$\begin{aligned} \alpha &= (1/2)\rho_p A_p h, & EI &= E_b I_b + E_p I_p + (1/4)\, E_p A_p h^2, & c_1 &= (1/4)c_p A_p h^2, \\ \beta &= (1/2)E_p A_p h, & EA &= E_b A_b + E_p A_p, & c_4 &= (1/2)c_p A_p h, \\ \gamma &= (1/4)\rho_p A_p h^2, & \rho A &= \rho_b A_b + \rho_p A_p, & cA &= c_b A_b + c_p A_p \end{aligned}$$

where ($\prime$) denotes space derivative, ($\dot{}$) denotes time derivative, and viscous damping coefficients is presented by c, and $p(x,t)$ and $\tau(x,t)$ are the external forces applied along the beam. The E, ρ, A and I are Young's modulus, mass density, transverse area and moment of inertia, respectively. Furthermore, I_c is the current, V is the voltage, C_P^T is the piezoelectric capacitance, Γ is the coupling term.

The global stochastic electromechanical equation of motion coupling the shunt circuit to the piezoelectric component is defined in terms of the spectral stiffness matrix as [14],

$$\begin{aligned} \mathbf{S}_{BP}(\omega)\mathbf{d} - \mathbf{S}_{SH}(\omega,\theta)V(\omega) &= \mathbf{f}(\omega) \\ i\omega\mathbf{S}_{SH}(\omega,\theta)\mathbf{d} + i\omega C_p^T V(\omega) &= I_c(\omega) \end{aligned} \tag{2}$$

where $\mathbf{S}_{SH}(\omega,\theta)$ is the shunt circuit spectral matrix, θ represents the variability, $\mathbf{d}$ is the generalized nodal displacement, $\mathbf{f}$ the generalized force. Similar to the nodal relationship of beam and beam-piezo elements, the behavior of the piezoelectric structure with the shunt circuit can be expressed with the equivalent nodal displacements and forces [6,12]. Therefore, $\mathbf{S}_{SH}(\omega,\theta)$ can be assembled as follows

$$\mathbf{S}_{SH}(\omega,\theta) = [N_{e1}W(x_0,\omega), 0, -M_{e1}W(x_0,\omega), -N_{e2}W(x_0,\omega), 0, M_{e2}W(x_0,\omega)]^T \tag{3}$$

where

$$N_{e1} = N_{e2} = \frac{k_{ij}^2 i\omega Z^{EL} b_p d_{31} E_p}{1 + i\omega C_p^T Z_{eq}}, \qquad M_{e1} = M_{e2} = \frac{k_{ij}^2 i\omega Z^{EL} h b_p d_{31} E_p}{2 + 2i\omega C_p^T Z_{eq}}$$

The nodal functions of the piezoelectric structure with shunt circuit are related to the piezoelectric coupling coefficient k_{ij}, the width b_p, and the piezoelectric constant d_{31}. Therefore, a general representation of the dynamic behaviour of the unimorph beam is

$$\left[\mathbf{S}_{BP}(\omega) + \omega^2 \mathbf{S}_{SH}^2(\theta) \frac{1}{i\omega + 1/Z_{eq}(\theta)}\right] \mathbf{d}(\omega) = \mathbf{f}(\omega) \tag{4}$$

where $Z_{eq} = -V/I_c$ is the general impedance given by the junction of the admittance of the shunt circuit with the internal admittance of the piezoelectric. The impedance for open circuit and short circuit cases is presented with $Z_{eq} = i\omega C_P^T$ and $Z_{eq} = 0$, respectively. For the case of series resistive-inductive (RL) shunt circuit, also known as resonant shunt circuit, we have the following general impedance

$$Z_{eq} = \frac{R + i\omega L_n}{1 - \omega^2 L_n C_P^T + i\omega R C_P^T} \tag{5}$$

where R is the resistor and L_n is the inductor. For experimental practical purposes, the inductor component can be replaced by an antoniou circuit-type synthetic inductor (shown in Fig. 2), and its tuning frequency ω_{SH} can be defined by

$$\omega_{SH} = \sqrt{\frac{1}{L_n C_P^T}} = \frac{1}{\sqrt{(C_1 R_1 R_3 R_4 / R_2) C_P^T}} \tag{6}$$

Once the matrices of the spectral elements $\mathbf{S}_B(\omega)$, $\mathbf{S}_{BP}(\omega)$ and $\mathbf{S}_{SH}(\omega)$ are defined, it is possible to obtain the global matrix by assembling the elements. This procedure is similar to the one used in the Finite Element Method. Therefore, the global equation can be written so that

$$\mathbf{S}_g(\omega) = \mathbf{d}_g(\omega) = \mathbf{f}_g(\omega) \tag{7}$$

where $\mathbf{S}_g(\omega)$ is the assembled global dynamic stiffness matrix, $\mathbf{d}_g$ is the global spectral nodal DOFs vector, and $\mathbf{f}_g$ is the global spectral nodal forces and moments vector.

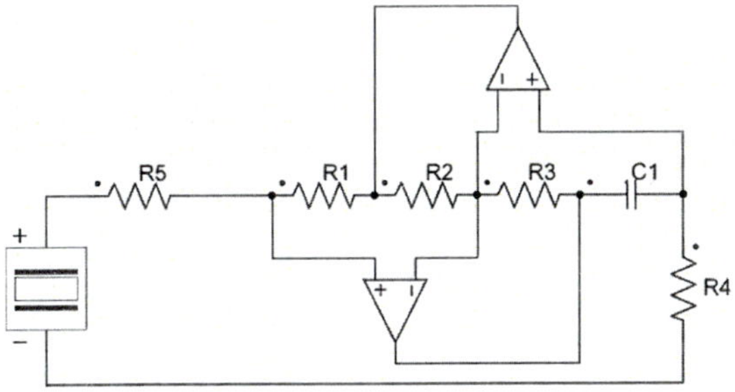

Fig. 2. Topography of a piezo connected in RL shunt circuit;

3 Numerical Results and Discurssion

The smart beam structure analyzed is an aluminum beam with seven piezoelectrics periodically coupled along the length L of the beam, as shown in Fig. 3. The boundary condition used is free-free. A unit impulse is performed at the penultimate degree of freedom (driving point).

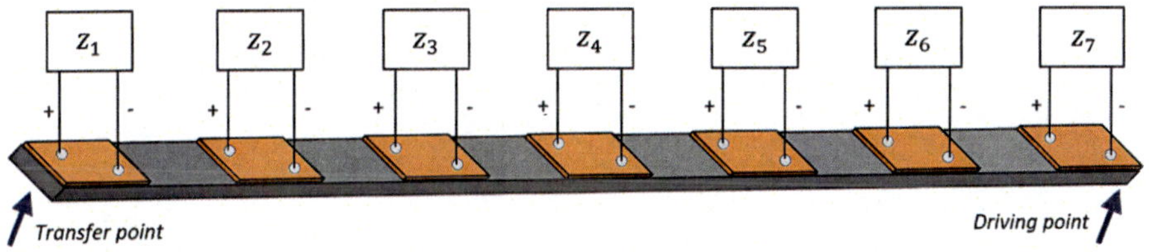

Fig. 3. Smart beam illustration.

The implementation of the structure model via SEM and TMM was performed using MatLab software. The properties and geometries considered for the beam are E_b = 71 GPa, ρ_b = 2700 kg/m^3, length L = 0.5 m, width 12.7 mm e thickness h_b = 2.286 mm. For piezo were considered E_p = 64.9 GPa, ρ_b = 7600 kg/m^3, length L = 38.46 mm, width 12.7 mm e thickness h_b = 0.762 mm. In addition, we considered the coupling coefficient $k_{31} = 0,31$, piezoelectric constant $d_{31} = -175$ m/Vx10^{-12}, dielectric constant $\beta_{33}^S = -350$ m/Vx10^{-12} and piezoelectric capacitance $C_P^T = 200$ nF were considered. Regarding the resonant shunt circuits, the components $C_1 = 100\ \mu$F, $R_5 = 50\Omega$ and $R_1 = R_3 = R_4 = 1\ K\Omega$, were used for all circuits. However, different resistors R_2 with values 90, 103, 117, 132, 148, 165 and 182 Ω were used in each circuit to tune the frequencies 420, 450, 480, 510, 540, 570 600 Hz, respectively.

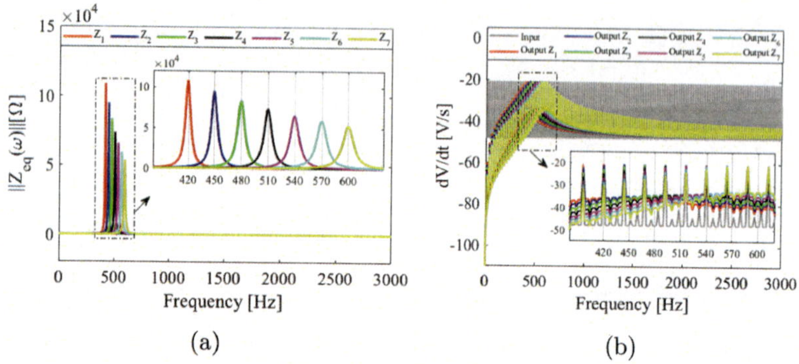

Fig. 4. Electrical behavior of shunt circuits a) Impedance; b) Voltage.

Figure 4a shows the real part of the impedances of each resonant shunt circuit and Fig. 4b shows the relationship between the voltage generated by the piezo and the voltage dissipated by each shunt circuit. These electrical relationships associated with each piezo coupled to the beam, result in the vibrational effect shown in Fig. 5.

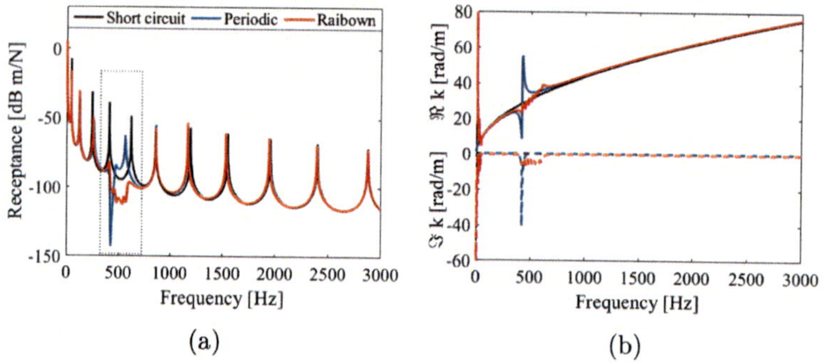

Fig. 5. Vibrational comparison of the beam with 7 piezos connected in the short circuit, periodic and rainbow trap configurations: a) FRF; b) Dispersion diagram;

Figure 5a shows a vibrational comparison of Frequency Response Function (FRF) between the short circuit (black line), periodic impedance RL (blue line) and RL rainbow trap (red line) configurations. The same settings are used for wavenumber comparison in the dispersion diagram of Fig. 5b.

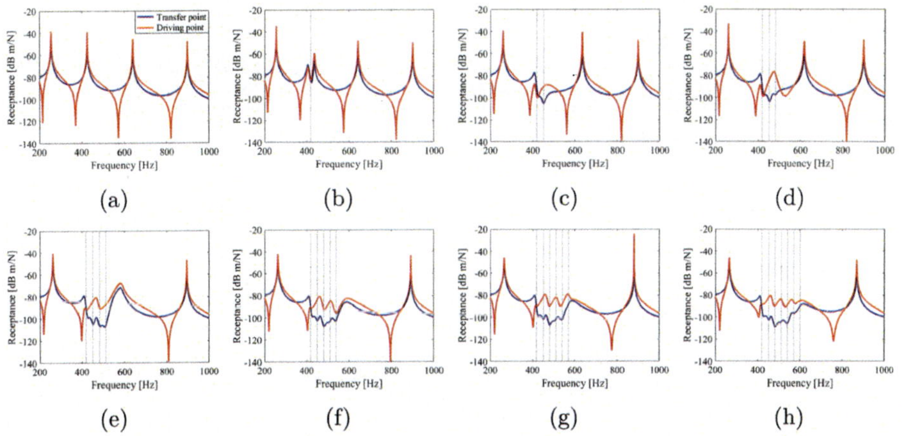

Fig. 6. Frequency response function of smart beam in rainbow trap setting; a) Short circuit; b) One connected shunt; c) Two connected shunt; d) Three connecteed shunt; e) Four connected shunt; f) Five connected shunt; g) Six coneected shunt; h) Seven connected shunt;

In Fig. 6, the FRF is presented in the GDL where the forcing takes place (driving point) and in the opposite GDL where the forcing takes place (transfer point). In this analysis, all seven piezos are coupled along the beam, but the connection of the resonant shunt circuit happens in a restricted way in piezo to piezo. In Fig. 6b, a resonant shunt circuit tuned 420 Hz is connected to the first piezo. In Fig. 6c a second resonant shunt circuit tuned 450 Hz is connected to the second piezo. Likewise, in Figs. 6d to 6h there is an addition of other resonant shunt circuits that are connected to each piezo, separately. However, each shunt circuit is tuned incrementally 30 Hz until 600 Hz.

Analyzing all the letters in Fig. 6, it is observed a vibrational effect caused by the addition of each shunt circuit. This effect, known by the term band gap, is characterized by the creation of a vibration isolation band, and as a new resonant shunt circuit is added there is an increase in the band gap width. This effect can also be observed with the dispersion diagram in Fig. 7.

In Fig. 7 the dispersion diagram corresponding to the addition of resonant shunt circuits are presented. The yellow and green lines represent the positive and negative propagation waves corresponding to the transverse displacement, respectively. The red and blue lines represent the positive and negative propagation waves corresponding to the shunt circuit impedance, in that order. In all the figures it is possible to notice changes in the waves exactly in the tuning frequency.

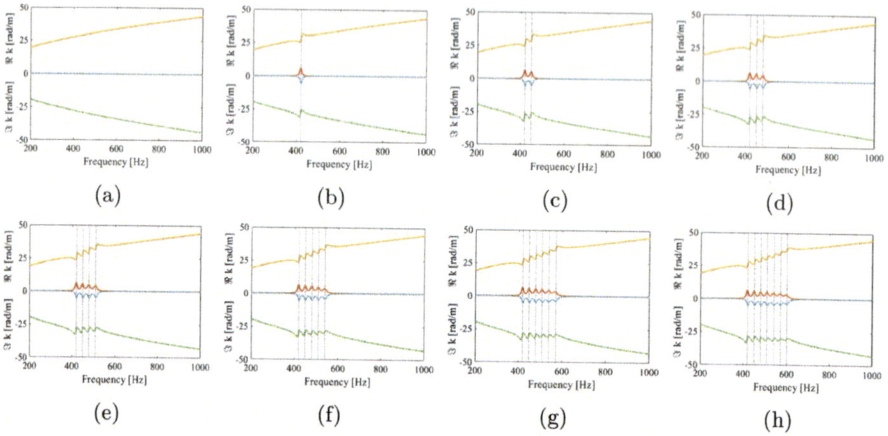

Fig. 7. Dispersion diagram of smart beam in rainbow trap setting; a) Short circuit; b) One connected shunt; c) Two connected shunt; d) Three connecteed shunt; e) Four connected shunt; f) Five connected shunt; g) Six coneected shunt; h) Seven connected shunt;

4 Conclusion

This work approached an investigation on the band gap generation in a beam with seven piezoelectrics coupled in periodicity. A SEM and TMM model was used to implement the structure in MatLab software. A connection comparison of RL shunt circuits in short circuit, periodic impedance and rainbow trap configuration was performed to investigate band gap generation. With this, it was identified that the rainbow configuration is more susceptible to band gap generation. Furthermore, it was noticed that as new shunt circuits are added to the rainbow configuration, there is an increase in the isolation bandwidth (band gap). In summary, although these vibrational effects have already been explored in the literature, the present work demonstrated that the band gap width in the rainbow configuration can also be efficiently observed by the scatter diagram.

References

1. Airoldi, L., Ruzzene, M.: Design of tunable acoustic metamaterials through periodic arrays of resonant shunted piezos. New J. Phys. **13**, 113010 (2011)
2. Cardella, D., Celli, P., Gonella, S.: Manipulating waves by distiling frequencies: a tunable shunt-enabled rainbow trap. Smart Mater. Struct. **25**, 085017 (2016)
3. Casadei, F., Ruzzene, M., Dozio, L., Cunefare, K.: Broadband vibration control through periodic arrays of resonant shunts: experimental investigation on plates. Smart Mater. Struct. **19**, 015002 (2009)
4. Dutkiewicz, M., Machado, M.R.: Spectral element method in the analysis of vibrations of overhead transmission line in damping environment. Struct. Eng. Mech. **71**(3), 291–303 (2019)

5. Dutkiewicz, M., Machado, M.R.: Measurements in situ and spectral analysis of wind flow effects on overhead transmission lines. Sound Vibr. **53**(4), 161–175 (2019)
6. Doyle, J.F.: Wave Propagation in Structures: A Spectral Analysis Approach, 2nd edn. Springer-Verlag, New York (1997). https://doi.org/10.1007/978-1-4684-0344-2_6
7. Gripp, J.A., Rade, D.A.: Vibration and noise control using shunted piezoelectric transducers: a review. Mech. Syst. Signal Process. **112**, 359–383 (2018)
8. Hagood, N.W., Flotow, A.V.: Damping of structural vibrations with piezoelectric materials and passive electrical networks. J. Sound Vib. **146**(2), 243–268 (1991)
9. Hollkamp, J.J.: Multimodal passive vibration suppression with piezoelectric materials and resonant shunts. J. Intel. Mat. Syst Str. **5**, 49–57 (1994)
10. Jaffe, B., Cook, R., Jaffe, H.: Piezoelecfric Ceramics. Academic Press, New York (1971)
11. Kaijun, Y., Matten, G., Ouisse, M., Sadoulet, E., Collet, M., Chevallier, G.: Programmable metamaterials with digital synthetic impedance circuits for vibration control. Smart Materm Struct. **29**, 035005 (2020)
12. Lee, U.: Spectral Element Method in Structural Dynamics. Wiley, Singapore (2009)
13. Leo, D.J.: Engineering Analysis of Smart Material Systems, pp. 1–7. John Wiley & Sons, New Jersey (2007)
14. Machado, M.R., Fabro, A.T., Moura, B.B.: Spectral element approach for flexural waves control in smart material beam with single and multiple resonant impedance shunt circuit. J. Comput. Nonlinear Dyn. **1** (2020). https://doi.org/10.1115/1.4047389
15. Machado, M., Dutkiewicz, M., Matt, C., Castello, D.: Spectral model and experimental validation of hysteretic and aerodynamic damping in dynamic analysis of overhead transmission conductor. Mech. Syst. Signal Process. **136**(1), 106483 (2020)
16. Moura, B.B., Machado, M.R., Mukhopadhyay, T., Dey, S.: Dynamic and wave propagation analysis of periodic smart beams coupled with resonant shunt circuits: passive property modulation. Eur. Phys. J.-Spec. Top. **1**, 1–18p (2022)
17. Sugino, C., Ruzzene, M., Erturk, A.: Design and analysis of piezoelectric metamaterial beams with synthetic impedance shunt circuits. IEEE/ASME Trans. Mechatron. **23**(5), 2144–2155 (2018)
18. Uchino, K., Ishii, T.: Mechanical damper using piezoelectric ceramics. Nippon Seramikkusu Kyokai Gakujutsu Ronbunshi/J. Ceramic Soc. Jpn. **96**(8), 863–867 (1988)
19. Viana, F.A., Steffen, J.V.: Multimodal vibration damping through piezoelectric patches and optimal resonant shunt circuits. J. Braz. Soc. of Mech. Sci. Eng. **28**, 293–310 (2006)
20. Zhang, W., Cardella, D., Gonella, S.: A disorder-based strategy for tunable, broadband wave attenuation. In: Proceedings of the SPIE 10170, Health Monitoring of Structural and Biological Systems, 101700F (2017)
21. Wang, G., Cheng, J., Chen, J., He, Y.: Multi-resonant piezoelectric shunting induced by digital controllers for subwavelength elastic wave attenuation in smart metamaterial. Smart Mater. Struct. **26**(2), 025031 (2017)

Ranking the Contributions of the Wave Modes to the Sound Transmission Loss of Infinite Inhomogeneous Periodic Structures

Vanessa Cool[1,2](✉), Régis Boukadia[1,2], Lucas Van Belle[1,2], Wim Desmet[1,2], and Elke Deckers[1,2]

[1] KU Leuven, Department of Mechanical Engineering, Leuven, Belgium
vanessa.cool@kuleuven.be

[2] DMMS Core Lab, Flanders Make, Lommel, Belgium

Abstract. In recent years, periodic structures such as metamaterials and phononic crystals have come to the fore in the search for innovative lightweight and compact noise and vibration solutions. The vibro-acoustic performance of these structures is often analyzed by means of dispersion curves and/or sound transmission loss calculations, using unit cell modeling. However, the link between these two performance indicators is not always straightforward. Recently, a first step to bridge this gap was taken by Yang et al. [12] who proposed a method which allows decomposing the sound transmission of infinite in-plane homogeneous media into a sum of wave mode contributions. Despite providing useful insights in the sound transmission of homogeneous structures, this method is limited to meshes with corner degrees-of-freedom only and hence not readily applicable to arbitrarily complex, inhomogeneous periodic structures. To expand the potential of this method towards identifying the most important transmission mechanisms in periodic media, this work extends the method towards periodic inhomogeneous structures represented by unit cells with arbitrarily complex meshes. The proposed methodology is applied to a periodic double panel partition thereby demonstrating its ability to provide new insights into the sound transmission mechanisms of periodic media/structures.

Keywords: Sound transmission loss · Wave and finite element method · Wave modes · Unit cell modeling · Periodic structures

1 Introduction

In the search for designs which satisfy the typically conflicting requirements of being lightweight and compact as well as having favorable noise and vibration characteristics, periodic structures such as phononic crystals and locally resonant metamaterials have come to the fore [4]. These periodic structures are often investigated based on the infinite Floquet-Bloch theory, using the unit cell (UC)

Z. Dimitrová et al. (Eds.): WMVC 2022, MMS 125, pp. 734–742, 2023.
https://doi.org/10.1007/978-3-031-15758-5_75

to characterize the structure. Both dispersion curves and sound transmission loss (STL) computations are generally used to analyze their vibro-acoustic performance. Dispersion curves contain information on the wave propagation in the periodic media, often without accounting for any interaction with the surrounding acoustic medium. STL calculations provide insight into the acoustic insulation performance of these infinite periodic structures and different approaches exist in the literature, ranging from analytical to fully numerical ones. One commonly used approach combines the wave and finite element (WFE) method, in which the UC is discretized with a FE model, with a semi-analytical modeling of the incident, reflected and transmitted acoustic fields [5,10].

Although the dispersion curves and STL computations are widely applied, the link between these two performance indicators is often not straightforward. To reach a better understanding of the periodic structures under investigation, a method to identify the dominant contributing wave modes to the sound transmission is required. Recently, a first attempt towards such approach was introduced by Yang et al. [12]. However, the methodology is limited to in-plane homogeneous media represented by meshes with corner degrees-of-freedom (DOFs) only.

To investigate arbitrarily complex periodic inhomogeneous media, which require meshes with both interior and boundary DOFs, this work extends the method of Yang et al. [12]. To achieve this goal, a new decomposition of the STL in terms of the wave modes is proposed [1]. The novelty of the decomposition is three-fold: i) the STL is decomposed while including all DOFs at the transmitted side of the structure in the fluid-structure interaction; ii) information of the higher order acoustic harmonics is included in the decomposition of the STL to correctly represent the periodic phenomena; iii) a new method is proposed to decompose the unit cell DOFs in terms of the wave modes, which is applicable to unit cell meshes with a large number of DOFs.

This paper is structured as follows. Section 2 describes the problem formulation. In Sect. 3, the developed methodology is discussed. In Sect. 4, the developed methodology is applied to a periodic double panel partition. Section 5 summarizes the main conclusions.

2 Problem Description

This work investigates 2D infinite periodic structures in the x-y plane which are coupled to two infinite acoustic halfspaces on either side (Fig. 1(a)). The periodic structure is modeled with a UC with dimensions L_x, L_y and L_z and is discretized using the FE method. The acoustic medium is air with a density $\rho_a = 1.225\ \text{kg/m}^3$ and speed of sound $c_a = 340$ m/s. At the bottom side, the structure is excited by an acoustic plane wave with incident angles (θ,ϕ), amplitude P_i and wavenumber $k_a = \omega/c_a$, with ω the angular frequency: $P_i \mathrm{e}^{-\mathrm{i}(k_x x + k_y y + k_z z)}$ with $\mathrm{i}^2 = -1$, $k_x = k_a \sin(\theta)\cos(\phi)$ and $k_y = k_a \sin(\theta)\sin(\phi)$. Time harmonic motion is assumed, whereby the $\mathrm{e}^{\mathrm{i}\omega t}$-dependency is suppressed for clarity.

3 Methodology

This section gives an overview of the proposed methodology to determine the contributions of the wave modes to the sound transmission of infinite periodic inhomogeneous structures.

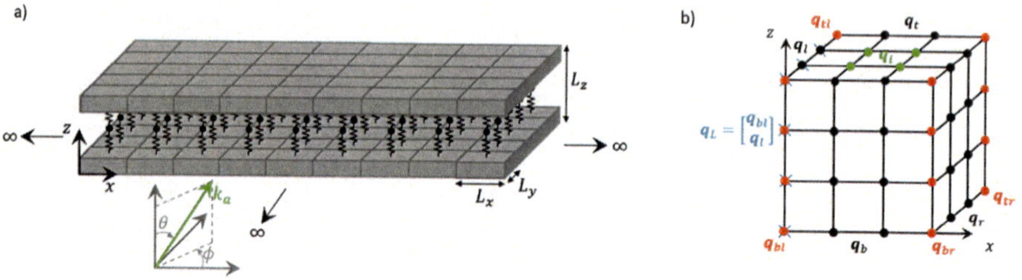

Fig. 1. a) Problem description. b) Visualization of node groups.

3.1 WFE Method

The WFE method combines the FE model of the UC with the Bloch-Floquet periodicity theorem [5,6]. The time harmonic equations of motion of an FE UC are given by:

$$(\mathbf{K}(\omega) - \omega^2\mathbf{M})\mathbf{q} = \mathbf{f} + \mathbf{e}, \tag{1}$$

with $\mathbf{M}, \mathbf{K}$ the mass and stiffness matrices, and $\mathbf{f}, \mathbf{e}$ the internal and external nodal forces, respectively. The UC DOFs $\mathbf{q}$ are partitioned as visualized in Fig. 1(b), a similar partitioning for the forces is applied. After applying the Bloch-Floquet theorem along the y-direction with propagation constant $\mu_y = \mathrm{e}^{-\mathrm{i}k_y L_y}$, the $\mathbf{q}_i$ and $\mathbf{q}_b$ are condensed out and finally the Bloch-Floquet theorem is applied along the x-direction with $\mu_x = \mathrm{e}^{-\mathrm{i}k_x L_x}$ (see [5,6] for details):

$$\Bigg(\underbrace{\begin{bmatrix} \mathbf{D}_{LL} & -\mathbf{I} \\ \mathbf{D}_{RL} & \mathbf{0} \end{bmatrix}}_{\mathbf{A}} - \mu_x \underbrace{\begin{bmatrix} -\mathbf{D}_{LR} & \mathbf{0} \\ -\mathbf{D}_{RR} & -\mathbf{I} \end{bmatrix}}_{\mathbf{B}}\Bigg) \begin{bmatrix} \mathbf{q}_L \\ \mathbf{f}_L \end{bmatrix} = \begin{bmatrix} \mathbf{e}_{L,\mathrm{tot}} \\ \mathbf{e}_{R,\mathrm{tot}} \end{bmatrix}, \tag{2}$$

in which $\mathbf{e}_{L,\mathrm{tot}}, \mathbf{e}_{R,\mathrm{tot}}$ are the external forces after applying the Bloch-Floquet theorem and condensation step. For in-vacuo free wave propagation, the system becomes an eigenvalue problem in $\lambda_j = \mathrm{e}^{-\mathrm{i}k_j L_x}$:

$$(\mathbf{A} - \lambda_j \mathbf{B}) \begin{bmatrix} \phi_q \\ \phi_f \end{bmatrix}_j = \mathbf{0}, \tag{3}$$

with $\phi_j = [\phi_q^T, \phi_f^T]^T$ the right eigenvectors. The eigensolutions of this problem come in pairs of left and right going wave modes, i.e. λ_j^+ and $\lambda_j^- = 1/\lambda_j^+$. To overcome any numerical issues due to the presence of large and small eigenvalues, the eigenvalue problem is solved in this work using Zhong's method [9].

3.2 Wave Mode Decomposition of the UC DOFs

The unknown $\mathbf{q}_L$ and $\mathbf{f}_L$ of Eq. (2) are written as a sum of the $2n_{q_L}$ wave modes:

$$\begin{bmatrix} \mathbf{q}_L \\ \mathbf{f}_L \end{bmatrix} = \Phi \mathbf{a} = \sum_{j}^{2n_{q_L}} a_j \phi_j, \tag{4}$$

in which ϕ_j is the jth column of Φ. The coefficients a_j are determined using the left eigenvectors ξ_j of Eq. (3), which can be computed using the right eigenvector $\phi_{q,\lambda_j^{-1}}$ of Eq. (3) corresponding to eigenvalue $1/\lambda_j$ [1]:

$$a_j = \frac{\xi_j^T \begin{bmatrix} \mathbf{e}_{L,\text{tot}} \\ \mathbf{e}_{R,\text{tot}} \end{bmatrix}}{(\lambda_j - \mu_x)(\xi_j^T \mathbf{B} \phi_j)}, \qquad \xi_j = \begin{bmatrix} \phi_{q,\lambda_j^{-1}} \\ \frac{1}{\lambda_j} \phi_{q,\lambda_j^{-1}} \end{bmatrix}. \tag{5}$$

Next, a decomposition for all DOFs $\mathbf{q}$ can be obtained by reversly applying the Bloch-Floquet theorem and condensation:

$$\mathbf{q} = \mathbf{E}_1 \mathbf{T}_1 \begin{bmatrix} \mathbf{q}_L \\ \mathbf{f}_L \end{bmatrix} + \mathbf{E}_0 \mathbf{e}_O, \tag{6}$$

in which $\mathbf{E}_1$ and $\mathbf{E}_0$ contain the information of the condensation and Bloch-Floquet theorem, while $\mathbf{T}_1$ is a matrix with ones and zeros to select the $\mathbf{q}_L$ DOFs. Finally, the transverse displacement of the wetted surface $w_s(x, y)$ is computed from the nodal DOFs $\mathbf{q}$ using the FE shape functions $\mathbf{N}(x, y, z = L_z)$:

$$w_s(x, y) = \mathbf{N}(x, y, z = L_z) \mathbf{T_2} \mathbf{q}, \tag{7}$$

in which $\mathbf{T_2}$ extracts the transverse displacement DOFs at the output side.

3.3 Fluid-structure Interaction

Since the UC is coupled to two infinite acoustic halfspaces, a fluid-structure interaction takes place. This interaction is modeled with the analytical approach of Yang et al. [11]. At the interfaces ($z = 0$, $z = L_z$), the structure is described with a discrete FE model in terms of $w_s(x, y)$. The fluid is described with a continuous model with the normal fluid displacement $w_f(x, y)$ and the incident $p_i(x, y)$, reflected $p_r(x, y)$ and transmitted $p_t(x, y)$ pressure field.

At the input interface, the acoustic pressure excitation is translated into equivalent nodal forces by lumping them at the input nodes [10]. In this work, only direct acoustic loading is modeled, neglecting the back-coupling such that the structure is excited with a blocked pressure wave with amplitude $2P_i$ [3]. The external normal force field reads:

$$e_j(x, y) = 2P_i \alpha_j \mathrm{e}^{-\mathrm{i}(k_x x + k_y y)}, \tag{8}$$

in which α_j relates the distributed pressure excitation to the nodal forces [10].

At the output interface, a relation between the normal structural displacement and pressure is obtained using the continuity conditions [10]:

$$P_{t,mn} = D_{f,mn} W_{mn}, \tag{9}$$

with W_{mn} and $P_{t,mn}$ the amplitude of the $(m,n)^{\text{th}}$ acoustic wave harmonic and harmonic pressure component respectively, $D_{f,mn}$ is the fluid dynamic stiffness:

$$D_{f,mn} = \frac{\mathrm{i}\rho_a \omega^2}{k_{z,mn}}, \quad W_{mn} = \frac{1}{L_x L_y} \int_0^{L_x} \int_0^{L_y} w_f(x,y) \mathrm{e}^{\mathrm{i}(k_{x,m}x + k_{y,n}y)} dxdy, \tag{10}$$

$$k_{x,m} = k_x + \frac{2m\pi}{L_x}, \qquad k_{y,n} = k_y + \frac{2n\pi}{L_y}, \qquad k_{z,mn} = \sqrt{k_a^2 - k_{x,m}^2 - k_{y,n}^2}, \tag{11}$$

in which the displacement continuity is enforced by projecting $w_s(x,y)$ onto $w_f(x,y)$ [11]. Note that the higher order wave harmonics with $(m,n) \neq (0,0)$ are induced due to the periodicity of the structure [5].

3.4 STL in Terms of Wave Mode Contributions

In this section, the STL is written in terms of the wave mode contributions [1]. For an incident wave at angles (θ, ϕ), the STL is defined using the ratio of transmitted (P_t) to incident (P_i) pressure amplitudes obtained in the methodology of Sect. 3.3:

$$\text{STL} = -10 \log_{10}(\tau(\theta,\phi)), \qquad \tau(\theta,\phi) = \sum_m^{\infty} \sum_n^{\infty} \frac{\text{Re}(k_{z,mn})}{k_{z,mn}} \left| \frac{P_{t,mn}}{P_i} \right|^2, \tag{12}$$

with τ the sound transmission coefficient. The infinite double sum changes to a finite double sum between $\pm m_{\max}$ and $\pm n_{\max}$ using a truncation rule [8]. The unknown $P_{t,mn}$ are determined using Eq. (9), while the double integral for the coefficients W_{mn} is computed using a numerical integration method:

$$P_{t,mn} = D_{f,mn} \mathbf{M}_{\text{int}}^T \mathbf{w}_s, \tag{13}$$

with

$$\mathbf{M}_{\text{int}}(j) = \frac{1}{L_x L_y} A_j^e \mathrm{e}^{\mathrm{i}(k_{x,m}x_j + k_{y,n}y_j)}, \qquad \mathbf{w}_s(j) = w_s(x_j, y_j), \tag{14}$$

with A_j^e the nodal area of element j and (x_j, y_j) the coordinates of the center of mass of element j. In this work, weakly periodic structures are considered. Therefore, $\mathbf{w}_s$ is only determined for $(m,n) = (0,0)$ and used for all W_{mn} terms. The $\mathbf{w}_s$ vector is written in terms of the wave modes, using consecutively Eqs. (7), (6), (4):

$$\mathbf{w}_s = \mathbf{N}\mathbf{T}_2\mathbf{q} = \mathbf{N}\mathbf{T}_2 \left(\mathbf{E}_1 \mathbf{T}_1 \begin{bmatrix} \mathbf{q}_L \\ \mathbf{f}_L \end{bmatrix} + \mathbf{E}_o \mathbf{e}_o \right) = \mathbf{N}\mathbf{T}_2 \left(\mathbf{E}_1 \mathbf{T}_1 \Phi \mathbf{a} + \mathbf{E}_o \mathbf{e}_o \right). \tag{15}$$

Both a contribution of the wave modes as well as a direct force contribution ($\mathbf{e}_O$-term) appear. For the computation of the STL, both terms are generally required. However, in this work, the main focus is on identifying the dominant wave mode contributions, making the $\mathbf{e}_O$-term irrelevant. Therefore, the total STL will be computed with and without the $\mathbf{e}_O$-term. If these coincide, the direct force contribution is not dominant and can be neglected. Substituting Eq. (15) into Eq. (13), dividing by P_i and extracting one wave mode, the ratio of the transmitted to incident pressure amplitude of the jth wave mode for the $(m,n)^{\text{th}}$ harmonic is:

$$\left(\frac{P_{t,mn}}{P_i}\right)_j = 2D_{f,mn}\mathbf{M}_{\text{int}}^T\mathbf{N}\mathbf{T}_2\mathbf{E}_1\mathbf{T}_1\phi_j a_j. \tag{16}$$

For periodic structures, several harmonic wave functions need to be included. Therefore, an equivalent τ is defined per wave mode j for all contributing (m,n)-pairs (τ_j) and the relative contribution of mode j to the STL is defined as Z_j:

$$\tau_j(\theta,\phi) = \sum_{-m_{\max}}^{m_{\max}} \sum_{-n_{\max}}^{n_{\max}} \frac{\text{Re}(k_{z,mn})}{k_{z,mn}} \left|\left(\frac{P_{t,mn}}{P_i}\right)_j\right|^2, \qquad Z_j = \left|\frac{\tau_j(\theta,\phi)}{\tau(\theta,\phi)}\right| \tag{17}$$

These contributions contain the information of which wave modes dominate the sound transmission.

4 Numerical Case

In this section, the proposed methodology is used to investigate the STL of a periodic double panel partition coupled to two infinite acoustic halfspaces (Fig. 2(a)). A double panel, with a UC of 20 mm in the x- and y-direction, is envisaged with each panel 1 mm steel (Young's modulus $E = 210$ GPa, density $\rho = 7800$ kg/m^3 and Poisson's ratio $\nu = 0.3$) resulting in the mass of each plate $m = 3.12$ g. The two plates are connected with a spring-mass-spring connection [13]. The point mass represents a 25% mass addition to the UC, i.e. $m_r = 1.56$ g and is tuned such that the grounded spring-mass-spring system resonance equals 778 Hz, i.e. $k_r = 18629.87$ N/m. The structure is exited by an oblique plane wave with $\theta = 45°$ and $\phi = 0°$. The frequency range of interest goes from 20 Hz till 2000 Hz. The UC is meshed with ten linear solid elements in the x- and y-direction, while three elements are used in the z-direction of each plate. The spring is modeled with a 1D CBUSH in Siemens NX [7].

A good correspondence of the STL is seen between the hybrid WFE method [2] and the proposed technique (Fig. 2(b)). Two dips occur in the considered frequency range. A first dip occurs around the mass-spring-mass frequency of the double panel and corresponds to the analytical prediction: $f_{\text{msm1}} = 389$ Hz [3]. The second dip is caused by the local resonance of the point mass in the middle of the spring-mass-spring connection. A second mass-spring-mass resonance of the double panel is introduced, whereby the mass m_r is distributed between the plates according to the spring stiffnesses. The observed dip corresponds with the analytical prediction: $f_{\mathrm{msm2}} = 870$ Hz [13]. Note that, since back-coupling is neglected, the proposed method goes sub 0 dB at these dips.

Figures 2(c,d) show the dispersion curves colored according to the wave mode contributions. The wave modes with a positive real part are right-going waves and the ones with a negative real part are left-going waves. Because a double panel partition is investigated, both a bending mode in which the plates move in-phase and out-of-phase is present. The in-phase bending mode (A) cuts-on at 0 Hz, while the out-of-phase bending mode first has an imaginary part (B) and cuts-on at f_{msm1} (C). At f_{msm2}, the local resonance of the mass-spring core is able to hybridize with the in-phase bending mode of the plates resulting in two spatially attenuating waves (D).

Looking at the wave mode contributions, following observations can be made per region: (i) at the lower frequencies starting from 0 Hz, the propagating right-going in-phase bending wave (A) dominates the STL. In this frequency range, the pure imaginary and left-going bending wave are of importance as well. (ii) Going up in frequency, towards f_{msm1}, the out-of-phase bending wave (B) starts dominating the sound transmission. This corresponds with the double panel theory, namely due to the decoupling of both plates, an increase in STL is achieved [3]. The out-of-phase bending wave (C) dominates the STL till f_{msm2}, while the importance of the evanescent waves drops. (iii) When the local resonance hybridizes with the in-phase bending mode, the corresponding spatially attenuating waves (D) start to dominate the STL. The in-phase bending

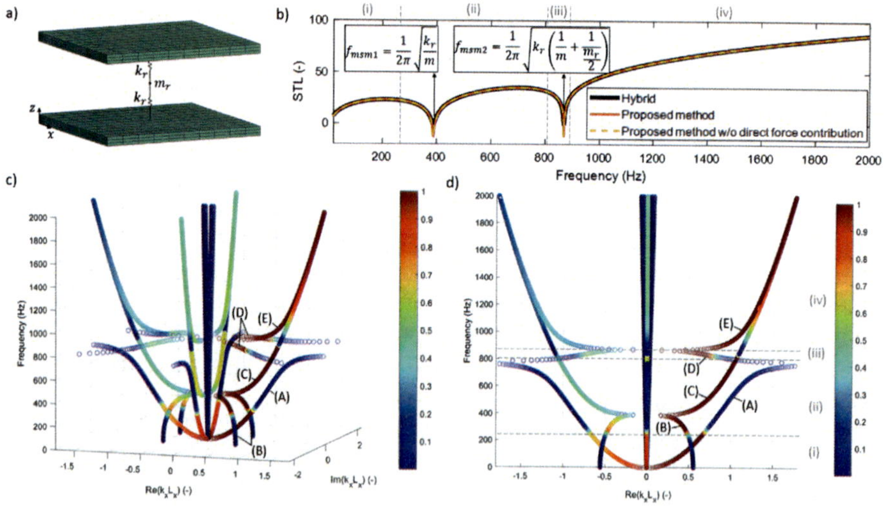

Fig. 2. Results of the numerical case. a) UC FE model. b) STL in dB calculated with the hybrid WFE method [2] and proposed methodology. c) Dispersion curves with as coloring scheme the contributions per mode branch normalized per frequency. d) 2D view of dispersion curves.

mode (E) is cut-on at f_{msm2} and dominates the STL. (iv) Above f_{msm2}, the in-phase bending wave remains dominating the STL, however, with increasing frequency, the importance of the out-of-phase bending wave quickly increases.

This numerical example shows that the proposed methodology is able to identify the dominating wave modes to the sound transmission and enables a thorough understanding of the periodic structure under investigation.

5 Conclusion

This work extends the WFE based method of Yang et al. [12] for the decomposition of the sound transmission in terms of the wave mode contributions. The methodology is extended from homogeneous media represented by meshes with corner DOFs only to inhomogeneous periodic structures represented by arbitrarily complex meshes. This is done by writing the STL as a sum of wave mode contributions while including all DOFs and higher order acoustic harmonics in the fluid-structure interaction. Additionally, a new decomposition of the UC DOFs in terms of the wave modes is proposed. The proposed methodology is demonstrated on a numerical case of a periodic double panel partition and is able to identify the dominant contributing wave modes to the sound transmission of periodic structures which enables a thorough understanding of their behavior and phenomenology.

Acknowledgments. The research of V. Cool (no. 11G4421N) and L. Van Belle (no. 1271621N) is funded by a grant from the Research Foundation - Flanders (FWO). The Research Fund KU Leuven is gratefully acknowledged for its support.

References

1. Cool, V., et al.: Contribution of the wave modes to the sound transmission loss of inhomogeneous periodic structures using a wave and finite element based approach. Submitted to J. Sound. Vib. **537**, 117183 (2022). https://doi.org/10.1016/j.jsv.2022.117183
2. Deckers, E., et al.: Prediction of transmission, reflection and absorption coefficients of periodic structures using a hybrid wave based-finite element unit cell method. J. Comp. Phys. **356**, 282–302 (2018)
3. Fahy, F., et al.: Sound and Structural Vibration: Radiation, Transmission and Response, 2nd edn. Elsevier, UK (2007)
4. Hussein, M., et al.: Dynamics of phononic materials and structures: Historical origins, recent progress, and future outlook. Appl. Mech. Rev. **66**(4), 040802 (2014)
5. Mace, B., et al.: Modelling wave propagation in two-dimensional structures using finite element analysis. J. Sound Vib. **318**(4–5), 884–902 (2008)
6. Renno, J., et al.: Calculating the forced response of two-dimensional homogeneous media using the wave and finite element method. J. Sound Vib. **330**(24), 5913–5927 (2011)
7. Software, S.I.: NX Nastran 10, Quick Reference Guide (2014). https://docs.plm.automation.siemens.com

8. Van Genechten, B., et al.: An efficient Wave Based Method for solving Helmholtz problems in three-dimensional bounded domains. Eng. Anal. Bound. Elem. **36**(1), 63–75 (2012)
9. Waki, Y., et al.: Numerical issues concerning the wave and finite element method for free and forced vibrations of waveguides. J. Sound Vib. **327**(1–2), 92–108 (2009)
10. Yang, Y., et al.: Prediction of sound transmission through, and radiation from, panels using a wave and finite element method. J. Acoust. Soc. Am. **141**(4), 2452–2460 (2017)
11. Yang, Y., et al.: Vibroacoustic analysis of periodic structures using a wave and finite element method. J. Sound Vib. **457**, 333–353 (2019)
12. Yang, Y., et al.: Ranking of sound transmission paths by wave and finite element analysis. J. Sound Vib. **492**, 115765 (2021)
13. Zuo, S., et al.: Low-frequency band gap of locally resonant phononic crystals with a dual-base plate. J. Acoust. Soc. Am. **143**(3), 1326–1332 (2018)

Strain Energy Approach for Nonlinear Stiffness Coeffcients in the Design of Periodic Structures

Rodrigo dos Santos Cruz(✉) and Marcos Silveira

School of Engineering, State University of São Paulo (UNESP), Bauru, São Paulo, Brazil
{rodrigo.s.cruz,marcos.silveira}@unesp.br

Abstract. In this work, we explore the dynamic behaviour of a discrete model of a periodic structure under harmonic input with nonlinear stiffness. The periodic structure has a unit cell with three degrees of freedom. We devise an approach that replaces the linear stiffness characteristic of the structure with a nonlinear one in which the nonlinear stiffness coefficients provide the same strain energy. The effect of this approach on the frequency response is analysed using numerical simulation, focusing on band gaps. The approach to determining nonlinear stiffness coefficients is based on the concept of equivalent elastic strain energy. This is different from the common approach found in the literature of adding a cubic term to the linear one, resulting in an increase in the elastic deformation energy of the system. Once the strain energy of the linear system is determined, a family of possible nonlinear stiffness coefficients is found, parameterised by the ratio between the linear and cubic coefficients. This approach can be used with hardening or softening stiffness characteristics. With the nonlinear stiffness coefficients defined, the dynamic response of the metastructure shows the usual shift to high and low frequencies. In addition, some frequency ranges are shown where vibration levels can be greatly reduced when the ratio of nonlinear stiffness coefficients is increased, compared to the case where there are only linear springs. Also, it is shown that the addition of the nonlinear component in the structure can increase or decrease the distance between the resonant frequencies.

Keywords: Metastructure · Vibration analysis · Nonlinear stiffness

1 Introduction

In recent decades, ecological and environmental requirements, combined with the search for energy efficiency, have given rise to the need for lightweight, resistant

Supported by Coordenação de Aperfeiçoamento de Pessoal de Nível Superior – Brazil (CAPES), grants # 88882.432818/2019-01

Z. Dimitrovová et al. (Eds.): WMVC 2022, MMS 125, pp. 743–750, 2023.
https://doi.org/10.1007/978-3-031-15758-5_76

and versatile materials. Metamaterials present interesting wave filtering characteristics, such as pass bands and stop bands or band gaps, which are not found in classical structures. They are applicable to many engineering structures such as trains, bridges, planes, satellites and automobiles [1]. These characteristics can be manipulated by the macro geometrical arrangement of its fundamental components, or unit cells, in a way that characteristics of mass, stiffness and or damping are spatially repeated and the resulting band gaps are in a desired frequency range [2,3]. The band gaps can be predicted by the natural frequencies of these unit cells with free and fixed boundary conditions [4], and by the transmissibility of the complete structure, which can be expressed in terms of the transmissibility of a unit cell using the transfer matrix approach [5].

A common approach when comparing linear and nonlinear stiffness is simply adding a nonlinear term to the restoring force [6–8]. This can be useful in situations in which one compares small deflections approximation (linearised stiffness) to large deflections (nonlinear stiffness). However this approach results in an overall stiffer spring, i.e., larger strain energy for the same deformation. Another approach is to consider same strain energy in the range of operation, in order to compare the linear and nonlinear stiffness characteristics in this range [9]. The specific procedure proposed here is to consider that linear and nonlinear strain energy are equal within the range of operation of the spring.

In this context, in this paper we analyse the influence of including a nonlinear cubic stiffness, with the same strain energy as a linear one, on the dynamical behaviour of a finite DOF model of a periodic structure under harmonic excitation. Specifically, we investigate the frequency response of a unit cell of the structure and quantify the response in three cases, corresponding to the placement of the nonlinear stiffness in the unit cell. The mathematical model of the structure and the description of the approach to define the nonlinear stiffness coefficients are given in Sect. 2. The dynamical behaviour of the structure is analysed numerically in Sect. 3, and conclusions are given in Sect. 4.

2 Mathematical Model

Figure 1 shows the schematic model of the lumped parameter unit cell and respective periodic structure. Only two cells are shown for convenience. The unit cell is comprised of two elements with mass m_1 and one element with mass m_2. Both dampers have damping coefficient c. The stiffness element can be linear with stiffness coefficient k, or nonlinear with coefficients k_l and k_n, which will be detailed in the following paragraph. Harmonic force is applied at the rightmost element.

As stated before, in this work we use an approach to define the coefficients of the nonlinear restoring force based on the stiffness coefficient of a linear restoring force which has the same strain energy in the range of operation of the spring, in order to compare the linear and nonlinear stiffness characteristics in this range. In this study, the range of operation ($[0, \Delta_1]$) is chosen as the maximum deformation at first resonance of the spring. The approach is described in the following steps. First, define the linear and nonlinear restoring forces, respectively, as:

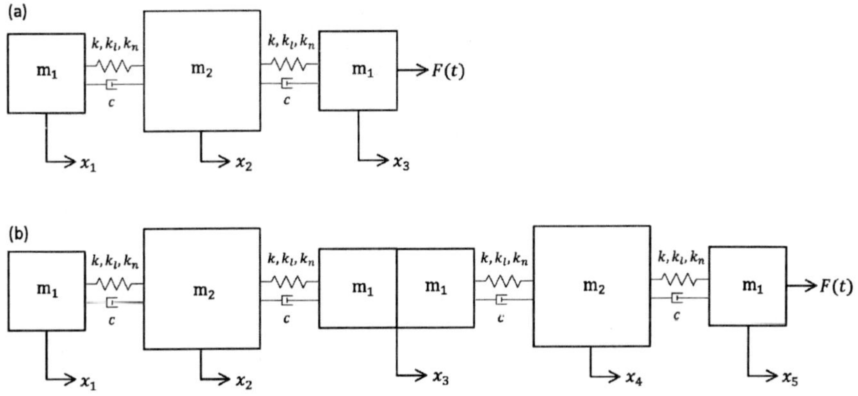

Fig. 1. Schematic model of (a) unit cell and (b) respective periodic structure.

$$F_L = k\Delta \qquad F_{NL} = k_l\Delta + k_n\Delta^3 \tag{1}$$

in which F_L and F_{NL} are the linear and nonlinear restoring forces, Δ is the spring deformation. Note that k_n can be negative (softening spring) or positive (hardening spring). The ratio between k_n and k_l is defined as:

$$b = \frac{k_l}{k_n} \tag{2}$$

Second, denote the strain energy related to the linear force as S_L, and the strain energy related to the nonlinear force as S_{NL}, and enforce the equality:

$$S_L = S_{NL} \tag{3}$$

Using the definitions of F_L and F_{NL} given by Eq. 1, relation 3 becomes:

$$\int_0^{\Delta_1} k\Delta d\Delta = \int_0^{\Delta_1} (k_l\Delta + k_n\Delta^3)\, d\Delta \tag{4}$$

in which $[0,\ \Delta_1]$ is the known range of operation.

Finally, solving this relation for k_l, it is possible to relate the coefficients of the linear and nonlinear restoring forces by:

$$k_l = \frac{2k}{b\Delta_1^2 + 2} \tag{5}$$

in which b is given by 2.

The dynamical equations can finally be obtained by applying Newton's second law, giving the general form:

$$\begin{aligned}
&\mu m_2\ddot{x}_1 + k_l(x_1 - x_2) + k_n(x_1 - x_2)^3 + c(\dot{x}_1 - \dot{x}_2) = 0 \\
&m_2\ddot{x}_2 + k_l(x_2 - x_1 - x_3) + k_n[(x_2 - x_1)^3 + (x_2 - x_3)^3] \\
&\quad + c(\dot{x}_2 - \dot{x}_1 - \dot{x}_3) = 0 \\
&\mu m_2\ddot{x}_3 + k_l(x_3 - x_2) + k_n(x_3 - x_2)^3 + c(\dot{x}_3 - \dot{x}_2) = F(t)
\end{aligned} \tag{6}$$

in which x_i is the displacement of each block, $\mu = m_1/m_2$, dot denotes differentiation over time t, and $F(t)$ is the external force applied to the structure, given by:

$$F(t) = F_0\cos(\omega t) \tag{7}$$

in which F_0 is the amplitude and ω is the frequency of excitation.

3 Dynamical Response of Metastructure

In this section, modal analysis is used to determine eigenfrequencies and frequency response of the base structure and the metastructure with linear and nonlinear stiffness, both with n = 2 elements. Note that it is a semi-definite system where one of its natural frequencies is equal to zero, due to a rigid body mode. Semi-definite systems are also known as unrestricted or degenerate systems [10].

Three spring arrangements are considered in the unit cell, denoted as cases I, II and III. In case I both springs are nonlinear; in case II the left spring is nonlinear while the right one is linear; and in case III the left spring is linear while the right one is nonlinear. Note that Eq. 6 is the general case, with both nonlinear springs. Carefully choosing values of k_l and k_n according to the procedure given in the previous section, one can represent all three cases.

The dynamical response is obtained by solving Eq. 6. In the case of linear system ($k_n = 0$), the analytical solution is well known [6,10]. In the case of nonlinear stiffness ($k_n \neq 0$), numerical solution is used with a fixed step fourth-order Runge-Kutta method, calculated for 100 periods of the external harmonic force ($T = 2\pi/\omega$) with 500 points per period. The amplitude in steady state response is considered in all the following analyses. The values of parameters used in this work are: $m_2 = 4kg$; $k = 1N/m$; $c = 0.1Ns/m$; $F_0 = 1N$.

Figure 2 shows influence of the mass ratio in the of the leftmost (X_1) and rightmost (X_n) blocks of the unit cell, and the displacement transmissibility $T_d = X_1/X_n$. The dotted vertical lines indicate the three undamped natural frequencies to $\mu = 0.25$, $\mu = 0.50$ and $\mu = 1.00$.

The metastructure with nonlinear spring is analysed, in which the nonlinear restoring force given by Eq. 1, with the maximum deformation Δ_1 as (x_1 - x_2) at first resonance of the leftmost absorber's spring. Figure 3 shows the nonlinear restoring force as function of for varying b for this approach, in which it is clear that the strain energy is maintained in the interval. The black line is for the

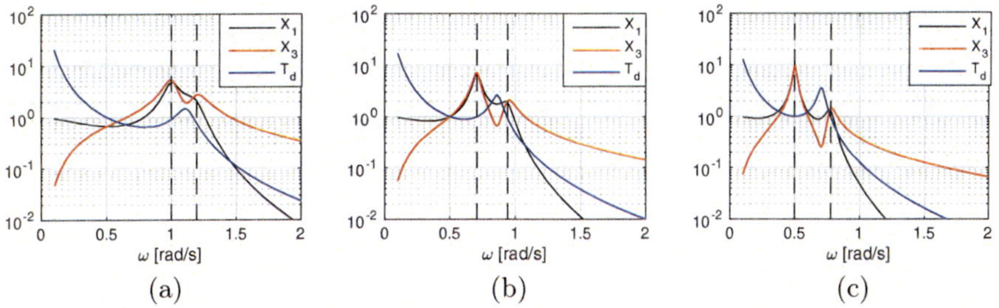

Fig. 2. Frequency response of the leftmost (X_1) and rightmost (X_3) blocks of the base structure, and the displacement transmissibility $T_d = X_1/X_n$, with dotted vertical lines indicating the undamped natural frequencies. (a) $\mu = 0.25$, (b) $\mu = 0.50$ and (c) $\mu = 1.00$.

system with linear stiffness only, it is possible to notice in the other curves, different slopes in relation to the linear system. Note that all curves intersect at the value of Δ^*, using the approach to defining the coefficients, they all have the same strain energy S, i.e., the same value of area under each curve described in the Eq. 3.

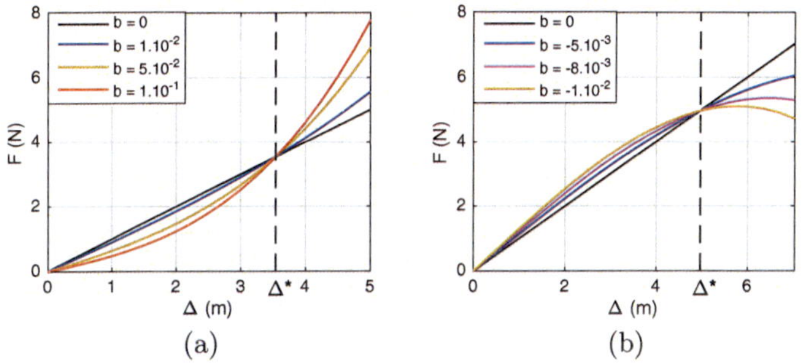

Fig. 3. (a) Nonlinear restoring force as function of strain Δ for varying nonlinear coefficient ratio b for constant strain energy (a) hardening stiffness for mass ratio $\mu = 0.25$ and (b) softening stiffness $\mu = 1.00$

Figure 4 some comparisons between the linear system and the nonlinear system, according to the Δ_1 the stiffness values of the hardening spring, $b > 0$ and $k_l > 0$ and softening spring, $b < 0$ and $k_n < 0$. The Table 1 shows the values of the norm $\mathcal{H}_2$ and the stiffness coefficients of the nonlinear force (k_l and k_n) calculated for each value of b using with $\mu = 0.25$ for hardening and $\mu = 1.00$ for softening.

Figure 4 shows the comparison of transmissibilities between the three cases. In case I there is nonlinear stiffness in both springs, in II the cubic nonlinearity is present only in the elastic force present in the first block, more to the left

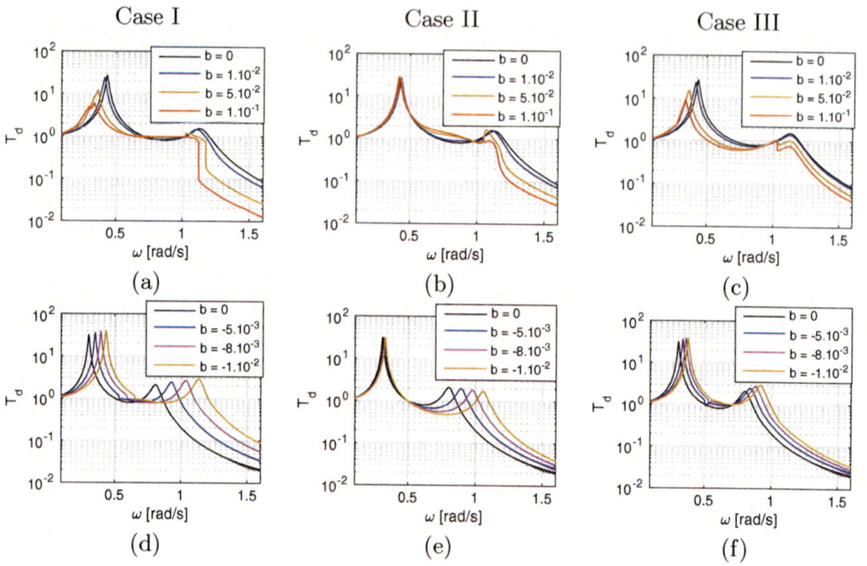

Fig. 4. Frequency response of the metastructure shown in the Fig. 1, for cases I, II and III, where (a), (b) and (c) is for mass ratio $\mu = 0.25$ with stiffness nonlinear hardening type, (d), (e) and (f) for $\mu = 1.00$ with nonlinear softening stiffness. Frequency responses are shown with different values of linear and nonlinear stiffness coefficient ratios (b).

Table 1. Stiffness coefficients and $\mathcal{H}_2$ norm

	b	k_l	k_n	$\mathcal{H}_2$		
	$[m^2]$	[N/m]	[N/m^3]	I	II	III
Hardening	0	1.0000	0	2.8264	2.8264	2.8264
($\mu = 0.25$)	1×10^{-2}	0.8953	0.0090	2.6617	2.8343	2.6516
	5×10^{-2}	0.6311	0.0316	2.0625	2.9099	2.1034
	1×10^{-1}	0.4611	0.0461	1.6705	2.9785	1.6018
Softening	0	1.0000	0	2.0591	2.0591	2.0591
($\mu = 1.00$)	-5×10^{-3}	1.1411	−0.0057	2.4349	2.0906	2.3603
	-8×10^{-3}	1.2466	−0.0100	2.7756	2.1132	2.5938
	-1×10^{-2}	1.3285	−0.0133	3.0817	2.7781	2.7563

and in case III there is cubic nonlinearity only in the elastic force present in the last block, according to the coefficients k_l and k_n, shown in the Table 1 and in the Fig. 3 with their respective values b and norm $\mathcal{H}_2$. The norms of a system are important, because with them it is possible to measure the intensity of the response of a structure under some excitation [2].

When the proposed approach is used with the three cases it is possible to notice that the transmissibility curves shift to the left, when the stiffness is of the hardening type, or to the right when it is of the softening type, the coefficient b increases. Note that this is valid for the same strain energy.

4 Conclusions

This work was developed with the objective of exploring the dynamic behaviour of a periodic structure, Fig. 1 under harmonic input with linear and nonlinear stiffness. Furthermore, an approach was presented that replaces the linear stiffness characteristic of the structure by a nonlinear one in such a way that the nonlinear stiffness coefficients provide the same strain energy and the dynamic response was numerical analysis.

Through the results it was possible to prove that the vibration could be attenuated as the ratio of the linear and nonlinear stiffness coefficients increased, and the resonance peaks were shifted to the left when the nonlinear stiffness was of the hardening type and was shifted right to the softening type. Furthermore, it was possible to verify that the higher the mass ratio, the lower the $\mathcal{H}_2$ norm and that the influence of the ratio ratio of linear and nonlinear stiffness coefficients depends on the case.

It was also possible to investigate that the addition of a cubic stiffness spring is capable of decreasing the $\mathcal{H}_2$ norm, but it depends on the case and the mass ratio. For case III, where the hardening nonlinear characteristic is present only in the rightmost spring, the $\mathcal{H}_2$ norm was lower. As for the softening nonlinear characteristic, the $\mathcal{H}_2$ norm proved to be lower for the case II in which the cubic nonlinearity is present only in the most to the left.

Acknowledgements. The first author wants to thanks CAPES (grant # 88882.432818/2019-01) for financial support for this project.

References

1. Hussein, M.I., Leamy, M.J., Ruzzene, M.: Dynamics of phononic materials and structures: historical origins, recent progress, and future outlook. Appl. Mech. Rev. **66**, 1–38 (2014)
2. Cveticanin, L., Zukovic, M., Cveticanin, D.: Influence of nonlinear subunits on the resonance frequency band gaps of acoustic metamaterial. Nonlinear Dyn. **93**(3), 1341–1351 (2018). https://doi.org/10.1007/s11071-018-4263-5
3. Lamarque, C. H., Savadkoohi, A.T., Charlemagne, S.: Experimental results on the vibratory energy exchanges between a linear system and a chain of nonlinear oscillators. J. Sound Vibr. **437**, 97–109 (2018)
4. Mead, D. J.: Wave propagation and natural modes in periodic systems: I. mono-coupled systems. J. Sound Vibr. **40**, 1–18 (1975)
5. Gonçalves, P.J.P., Brennan, M.J., Cleante, V.G.: Predicting the stop-band behaviour of finite mono-coupled periodic structures from the transmissibility of a single element. Mech. Syst. Signal Process. **154**, 107512 (2021)

6. Nayfeh, A.H., Balachandran, B.: Applied Nonlinear Dynamics: Analytical, Computational, and Experimental Methods. John Wiley and Sons, Hoboken (2008)
7. Chakraborty, G., Mallik, A.K.: Dynamics of a weakly non-linear periodic chain. Int. J. Non-Linear Mech. **4**, 375–389 (2001)
8. Marathe, A., Chatterjee, A.: Wave attenuation in nonlinear periodic structures using harmonic balance and multiple scales. J. Sound Vib. **289**, 871–888 (2006)
9. Vasconcellos, D.P., Silveira, M.: Optimization of axial vibration attenuation of periodic structure with nonlinear stiffness without addition of mass. ASME. J. Vib. Acoust **142**(6), 061009 (2020)
10. den Hartog, J. P.: Mechanical Vibrations. McGraw-Hill, 4 edition (1956)

Vibration Attenuation in Plates with Periodic Annuli of Different Thickness

Matheus M. Quartaroli[1], Elisabetta Manconi[2](✉), Fabrício C. L. De Almeida[1], and Rinaldo Garziera[2]

[1] Department of Mechanical Engineering, UNESP, Bauru, São Paulo 17033-360, Brazil
matheus.quartaroli@unesp.br
[2] Dipartimento di Ingegneria e Architettura, Università di Parma, 43124 Parma, Italy
elisabetta.manconi@unipr.it

Abstract. The ability of periodic structures to create stop-bands has shown great potential application in noise and vibration reduction. In mechanical and civil engineering, the studies on periodic structures typically deal with waveguides in Cartesian coordinates, while the effects of polar periodicity have been rarely analysed. However, the passive reduction of the vibrations generated from a localised excitation source on a large flexible structure can be obtained by exploiting radial periodicity effects. This paper investigates the vibration isolation properties of a plate made of a sequence of annuli with periodic alternating thickness. Two approaches are proposed and compared: the first implements a Wave Finite Element method to a small slice of the structure approximated by piecewise rectangular segments; the second deals with a standard FE model and the evaluation of the difference in the vibration before and after the insertion of *n* consequent periodic annuli in the finite uniform plate.

Keywords: Polar periodicity · Periodic structures · Stop-band · Vibration attenuation · Finite element

1 Introduction

Periodic structures attenuate vibration transmission in specific frequency ranges called stop-bands, which are frequency intervals where waves cannot propagate [1]. A comprehensive review of the research in periodic materials and structures has been presented in [2], where some of the main progresses in the numerical and experimental results in this field, including acoustic/elastic metamaterials, are presented up to 2014.

Due to their potential applications in Noise and Vibration Harshness mitigation in various engineering areas, modelling periodic structures has become an active and prolific research subject in structural dynamics. In particular, effective solution in creating vibration attenuation zones over a broadband frequency spectrum have been investigated, e.g., [3–5], and the interest has grown in the last years due to the new technological possibility of creating multi-scale periodic structures with local resonance phenomena, which has made the design of these structures very attractive for new metamaterials [6, 7].

Z. Dimitrovová et al. (Eds.): WMVC 2022, MMS 125, pp. 751–760, 2023.
https://doi.org/10.1007/978-3-031-15758-5_77

The Bloch-Floquet theorem, typically applied to predict the behaviour of a whole structure from the knowledge of the dynamic properties of a single periodic cell, relies on the translational invariance of the problem formulation. Consequently, most of the studies on periodicity effects in structural dynamics have concerned Cartesian periodicity, while polar periodicity has been rarely analysed, e.g. [8–10]. However, circular bidimensional structures are used in many engineering constructions, and there are many practical applications in which a plate, or a large flexible structure, is excited by a localised excitation source. Since periodic structures can act as a vibroacoustic filter when the dominant excitation frequency follows within a stop-bands, it naturally follows that radial periodicity may be used to attenuate the vibration transmission in these cases.

The work presented in this paper follows a previous work by the same authors [11], where the response of a radially periodic plate was evaluated for a low-frequency harmonic excitation. Compared to [11], the study herein presented extends the analysis, and verification of the results, to higher frequencies for predicting the periodicity effects. The paper is organised as follows. In Sect. 2, the technique to evaluate the plate response is summarised. The method is based on the Wave Finite Element (WFE) approach [12, 13] and relies on Floquet-theory and standard FE analysis. The adaptation of this approach to polar structures is accomplished considering only a small slice of the structure, which is approximated using piecewise Cartesian segments. Wave characteristics (wavemodes and dispersion curves) are obtained by the WFE for each segment, while waves amplitude changes are accommodated in the model, assuming that the energy flow through each periodic cell is the same as the wave propagates in the radial direction. In Sect. 3, numerical results are presented for a plate made of a sequence of periodic annuli with alternating thickness. A standard FE model of a free circular plate with the insertion of a sufficient number of periodic annuli is also analysed to support the results. Conclusions are then given in Sect. 4.

2 Wave and Finite Element Model Adaptation to Polar Periodic Structures

The WFE method is a technique to predict wave propagation in periodic structures. It has been presented in many papers and studies, and several benchmark cases have shown its practical applicability to periodic and continuous structures [14]. The method models a single period, or cell, using standard FE analysis, and it applies the Floquet/Bloch theory [1] to evaluate the behaviour of the whole structure in terms of wave propagation. The response of a radially periodic plate to a low-frequency harmonic excitation was studied in [11] using the WFE approach. This section will shortly resume the procedure already presented in this paper, extending the application of the approach to higher frequencies where stop-band occurs.

A lossless and linear elastic plate with radial periodicity is considered, Fig. 1(a). The structure is infinite. For this periodic structure, the unit cell (period) is identified by an integer number j, which represents the radial position of the periodic annulus. The periodic constant, which is the length of the unit cell in the corresponding Cartesian plate, is here denoted by $\overline{R}$. To avoid numerical errors and singularities, the left nodes of the first cell are shifted at an arbitrarily small distance R_0 from the origin of the coordinates.

As shown in Fig. 1(b), a slice of the plate is taken, and it is approximated as a piecewise rectangular waveguide in Cartesian coordinates, Fig. 1(c). Since the periodicity is in the radial direction, the angle θ of the slice in Fig. 1(b) can be arbitrarily small. Therefore, we can assume that the geometrical properties of the waveguide do not vary rapidly. The plate's periods are numbered increasingly in the radial direction as $j = 1 \ldots n$.

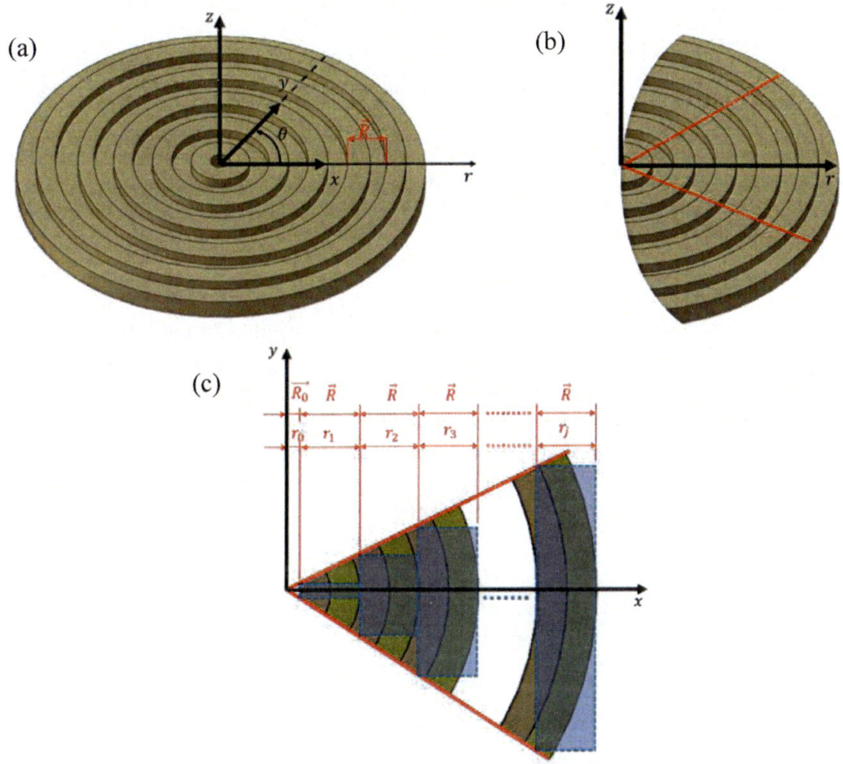

Fig. 1. (a) infinite radial periodic structure; (b) slice of the infinite radial periodic structure; (c) approximation in piecewise Cartesian periodic waveguides.

Wave characteristics of periodic segment j in Fig. 1(c) are evaluated using the WFE method as described in [12, 14]. The degrees of freedom $\mathbf{q}_j$, Eq. (1) in [1], are ordered and condensed as $\mathbf{q}_j = [\mathbf{q}_{jL}^T \ \mathbf{q}_{jR}^T]^T$, where the subscripts L and R are associated with the right and left nodal degrees of freedom of the cell. A similar expression is used for the nodal forces $\mathbf{f}_j = [\mathbf{f}_{jL}^T \ \mathbf{f}_{jR}^T]^T$. The internal Dofs of the unit cell must be reduced using a dynamic condensation as in [11] or more refined methods, e.g., [15].

The waves obtained by WFE are numerically represented by the dispersion curves, $(\mathbf{k}_j, \omega)$, which give the information on the wave vector $\mathbf{k}_j$ available for each frequency ω, and the corresponding FE nodal displacements and forces, $\mathbf{\Phi}_{j\mathbf{q}}$ and $\mathbf{\Phi}_{j\mathbf{f}}$, occurring under the passage of a wave. In predicting the characteristics of the waves propagating in the periodic segments in Fig. 1(c), attention must be paid to numerical issues related to FE discretisation and condensation of internal nodes, in particular as the frequency increases.

In this paper, the interest is in the out-of-plane response of the plate. Although solid elements are recommended at higher frequencies, the numerical example is studied using shell elements to evaluate the proposed approach's ability and robustness in predicting the plate's response. Therefore, 2D shell elements are considered, and only the wave modes with significant time average kinetic energy content in the out-of-plane direction are considered.

Using the wavemodes as basis functions, the total displacement and force of a segment j can be described by the sum of a finite number of the positive and negative wavemodes so that

$$\begin{bmatrix} \mathbf{q}_{jL} \\ \mathbf{f}_{jL} \end{bmatrix} = \begin{bmatrix} \mathbf{\Phi}_{j\mathbf{q}}^{+} & \mathbf{\Phi}_{j\mathbf{q}}^{-} \\ \mathbf{\Phi}_{j\mathbf{f}}^{+} & \mathbf{\Phi}_{j\mathbf{f}}^{-} \end{bmatrix} \begin{bmatrix} \mathbf{a}_{j}^{+} \\ \mathbf{a}_{j}^{-} \end{bmatrix}, \tag{1}$$

where $\mathbf{a}$ is the wave amplitude and superscripts refer to positive and negative propagating waves.

Supposing that a concentrated harmonic force $\mathbf{f}_e$ is applied to the left DOFs of the segment $j = 1$, only positive going waves of amplitude $\mathbf{a}^{+}$, propagating away from the excitation point, will be generated and $\mathbf{\Phi}_{1\mathbf{f}}^{+}\mathbf{a}_{1}^{+} = \mathbf{f}_e$, see Eq. (9) in [11]. To recover the wave amplitude, it is advantageous to exploit the left eigenvectors of the WFE eigenvalue problem

$$\mathbf{\Psi} = \begin{bmatrix} \mathbf{\Psi}_{j\mathbf{f}}^{+} & \mathbf{\Psi}_{j\mathbf{q}}^{+} \\ \mathbf{\Psi}_{j\mathbf{f}}^{-} & \mathbf{\Psi}_{j\mathbf{q}}^{-} \end{bmatrix}, \tag{2}$$

which are orthogonal to the wavemodes and can be normalised, $\mathbf{\Psi}\mathbf{\Phi} = \mathbf{I}$ [14]. Therefore the excited wave amplitude for the first segment can be easily obtained without pseudo-inversion of the matrix in Eq. (1), that is

$$\mathbf{a}_{1}^{+} = \mathbf{\Psi}_{1\mathbf{q}}^{+}\mathbf{f}_{e}. \tag{3}$$

Once the forced wave amplitudes in the first segment are found, the wave amplitudes in the next segment are obtained through the following steps.

In the corresponding Cartesian waveguide, the nodal displacements and forces between two adjacent cells are related by the periodicity conditions $\mathbf{q}_{(j+1)L} = \mathbf{T}_{j}^{+}(\overline{R})\mathbf{q}_{jL}$ and $\mathbf{f}_{(j+1)L} = \mathbf{T}_{j}^{+}(\overline{R})\mathbf{f}_{jL}$, where $\mathbf{T}_{j}^{+}(\overline{R}) = \mathrm{diag}\left[\exp\left(-\mathrm{i}\mathbf{k}_{j}^{+}\overline{R}\right)\right]$ is the transfer matrix.

Therefore, using Eq. (1) and Eq. (3), the left nodal displacements and forces of cell 2 are

$$\begin{aligned} \mathbf{q}_{2L} &= \mathbf{T}_{1}^{+}(\overline{R})\mathbf{\Phi}_{1\mathbf{q}}^{+}\mathbf{a}_{1}^{+}; \\ \mathbf{f}_{2L} &= \mathbf{T}_{1}^{+}(\overline{R})\mathbf{\Phi}_{1\mathbf{f}}^{+}\mathbf{a}_{1}^{+}. \end{aligned} \tag{4}$$

Using the left eigenvectors as in Eq. (3), the estimated wave amplitudes are

$$\tilde{a}_{2}^{+} = \mathbf{\Psi}_{2\mathbf{f}}^{+}\mathbf{q}_{2L} + \mathbf{\Psi}_{2\mathbf{q}}^{+}\mathbf{f}_{2L}. \tag{5}$$

The time average energy flow density in each segment can then be obtained as

$$\Pi_j = \frac{1}{2}\mathbf{f}_{jL}^H \dot{\mathbf{q}}_{jL} = \frac{1}{2}\mathbf{f}_{jL}^H i\omega \mathbf{q}_{jL}, \tag{6}$$

where the superscript H denotes the Hermitian transpose. The energy flow in Eq. (6) is complex-valued [16]: its real part, $\mathrm{Re}(\Pi_j)$, being the average power; its imaginary part, $\mathrm{Im}(\Pi_j)$, being the peak of the reactive power, which can be seen as the stored averaged energy density in analogy with circuit theory.

In a uniform and lossless waveguide, the amplitudes of waves propagating in the structure do not change; this is not true in non-uniform waveguides, where waves amplitude is a function of the position. In our case, we can assume that the power associated with each propagating wave is preserved along the waveguides. Therefore, the changes in the waves amplitude (as they travel in the radial direction) can be accommodated in the model by requiring that the time-average energy flow density through the interfaces of each Cartesian segment is the same [17].

Therefore, we deduce that the ratio between $\mathrm{Re}(\Pi_1)$ and $\mathrm{Re}(\Pi_2)$ gives the wave amplitude decay ξ_2

$$\xi_2^2 = \frac{\mathrm{Re}(\Pi_1)}{\mathrm{Re}(\Pi_2)} = \frac{\mathrm{Re}\left\{\mathbf{f}_{1L}^H i\omega \mathbf{q}_{1L}\right\}}{\mathrm{Re}\left\{\mathbf{f}_{2L}^H i\omega \mathbf{q}_{2L}\right\}}. \tag{7}$$

In the stop-bands, where only evanescent and highly decaying waves exist, there is no significant energy flow but mainly reactive power and $\xi_2 = 1$.

Accurate knowledge of the dispersion curves, including pass- and stop-bands related to each wavemode, is fundamental to apply the approach correctly. Consequently, for waveguides with complex characteristics, efficient methods for following the dispersion curve become essential as the frequency range of the analysis increases.

Once Eq. (7) calculates the wave amplitude decay, the waves amplitude in waveguide 2 are approximated to

$$\mathbf{a}_2^+ = \xi_2 \tilde{\mathbf{a}}_2^+. \tag{8}$$

The waves amplitude decay up to cell $j = n$ are then evaluated following the same passages. When $\mathbf{a}_n^+$ is predicted, the nodal displacement of the nth segment at the radial position $r = R_0 + n\overline{R}$ can be predicted as

$$\mathbf{q}_{nR} = \mathbf{T}_n^+(\overline{R})\mathbf{\Phi}_{n\mathbf{q}}^+ \mathbf{a}_n^+. \tag{9}$$

3 Numerical Example

The section of the plate is depicted in Fig. 2(a), and the geometrical characteristics of the unit cell are indicated in Fig. 2(b). The material properties are given in Table 1.

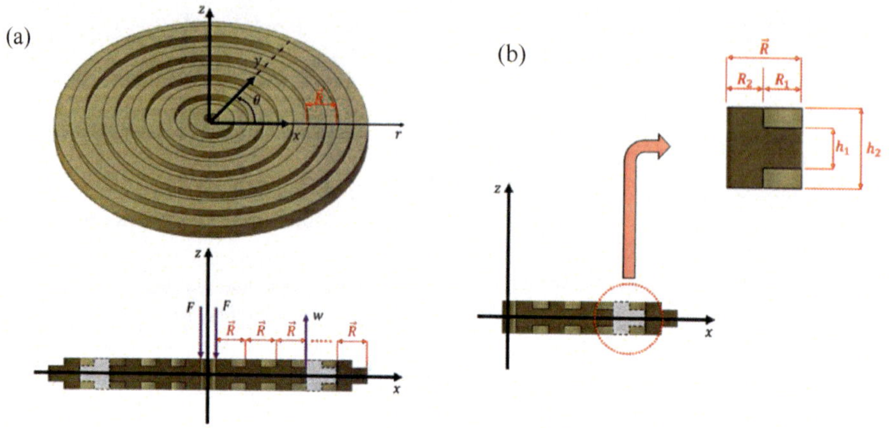

Fig. 2. (a) cross-section of the radial periodic plate with an internal hole of radius R_0; (b) period, or unit cell.

Table 1. Material properties

Young's modulus (GPa)	Density (kg/m^3)	Posisson's ratio
210	7850	0.3

In the following, the non-dimensional frequency $\Omega = \omega\overline{R}^2\sqrt{\rho h_1/D_1}$ is introduced, where $\Omega = 1$ corresponds to 34.17 Hz. Here ρ, h_1 and D_1 are the density, thickness and bending stiffness of the corresponding homogeneous plate of thickness h_1, ω is the frequency in radiant, and $\overline{R}$ is the periodic length of the structure. We assume that a localised excitation source acts at the centre of the plate. The excitation is modelled as a transverse harmonic force distributed around the internal circumference, subsequently discretised in FE nodal forces, Fig. 2(a). Waves are induced only in the positive r direction (radial direction). The response w_j, at all the external circumferential nodes in the j periodic annulus, is calculated following the procedure described in the previous section.

As already introduced in Sect. 2, a very small angle θ is used to study the polar problem. Because of this, since the length of the period is fixed, the first and last periodic segments could result in too small or too large aspect ratio. This limits the application of the approach in terms of the distance of the evaluation point from the centre (or in terms of the maximum number of periodic segments that can be included in the model).

In this example, the opening angle is $\theta = 9°$, and the number of segments considered in the analysis is $n = 5$. The five segments were modelled using 4-noded plane elements in bending with three degrees of freedom per node: translation in the z-direction and rotations around the x- and y-directions. Therefore, the stiffness matrix used to apply the WFE model was a 12×12 matrix after the dynamic reduction, resulting in a WFE model of 6 DOFs.

Compared to the previous work presented in [11], the dynamic response in the pass-bands and stop-bands is investigated here. Figure 3 shows the dispersion curves of flexural waves propagating in the corresponding Cartesian periodic plate. In this example, the frequency range of the analysis is limited to the second pass-band (FE plane elements cannot capture the behaviour of the structure at high frequency when higher order wavemodes cut-off).

The diagram in Fig. 3(a) shows the relation between the wavenumber k and the frequency ω for the corresponding periodic Cartesian plate. Vertical dashed lines indicate the stop-band. The stop-band can also be evaluated by the absolute value of the propagation constant $\lambda = e^{-ik\overline{R}}$, where $|\lambda| = 1$ indicates the presence of a pass-band, while $|\lambda| \neq 1$ indicates a stop-band, Fig. 3(b).

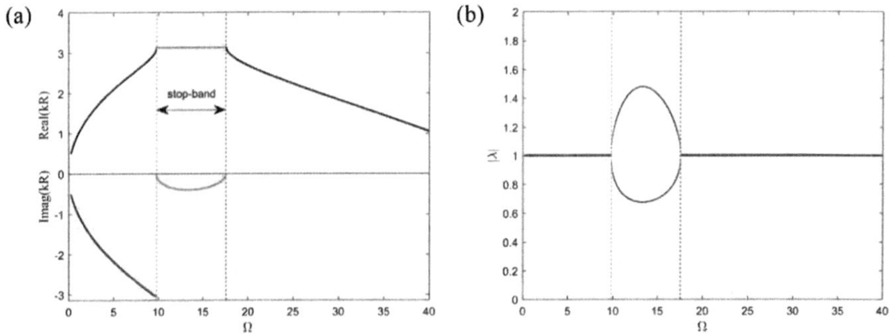

Fig. 3. Radially periodic plate. WFE prediction of the flexural dispersion curves: (a) ($\mathbf{k}$, ω), black line: propagating waves; blue line: evanescent waves; magenta lines: attenuated waves (complex conjugate wavenumbers); (b) ($\omega|\lambda|$). Vertical dashed lines show the pass-band.

The dispersion curves and wavemodes are then used as basis functions to predict the plate's wave amplitude and forced response, as presented in Sect. 2. The WFE results in Fig. 4 confirmed the vibration attenuation in the stop-band. The figure shows the absolute value of the transverse response (nodal displacement in the z-direction, Fig. 2), calculated for $j = 1$ and $j = 5$, at the end of the first and fifth periodic annuli - at a distance $r = R_0 + \overline{R}$ and $r = R_0 + 5\overline{R}$ from the centre of the plate.

The forced response of the radial periodic plate was also numerically studied using a standard FE approach to verify the wave filtering properties predicted by the WFE method. This standard FE model was realised in COMSOL Mutiphysics®. The model is a homogeneous circular plate of thickness h_1 and radial dimension $R = 10\overline{R}$ to which periodic annuli were inserted from R_0. It is worth mentioning that the reduction in the vibration level in a finite structure depends on the total number of the periodic cells, and a minimum number of cells must be inserted to obtain a proper vibration attenuation in the desired stop-bands. This number can be predicted by the *insertion loss*, which gives the difference in dB between the response before and after the insertion of n periodic annuli. In this numerical case, it was found that eight periodic annuli were sufficient to obtain a high vibration attenuation in the stop-band. The plate was assumed to be free, and the same forcing conditions of the WFE model were applied. The FE harmonic

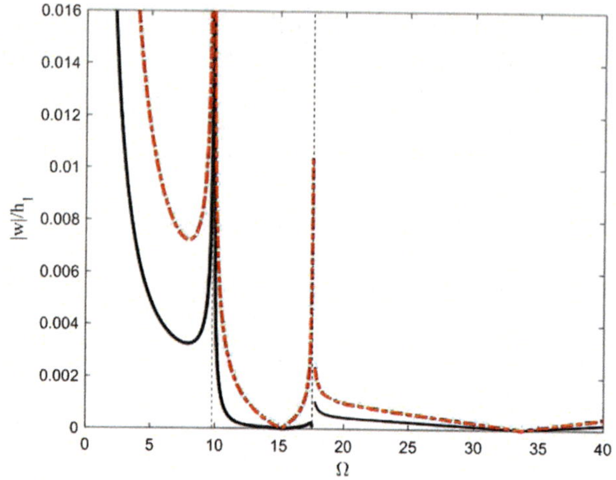

Fig. 4. WFE forced response. Black line: $|w_5(\Omega)|/h_1$; red dot-dashed lines: $|w_1(\Omega)|/h_1$. Thin vertical dashed lines show the stop-band in Fig. 3.

analysis in COMSOL was solved using triangular shell elements having six degrees of freedom per node (translations and rotations in the z, x, and y directions). Figure 5 shows the effectiveness of this insertion by comparing the FE results before and after the insertion of the eight periodic annuli in the circular plate.

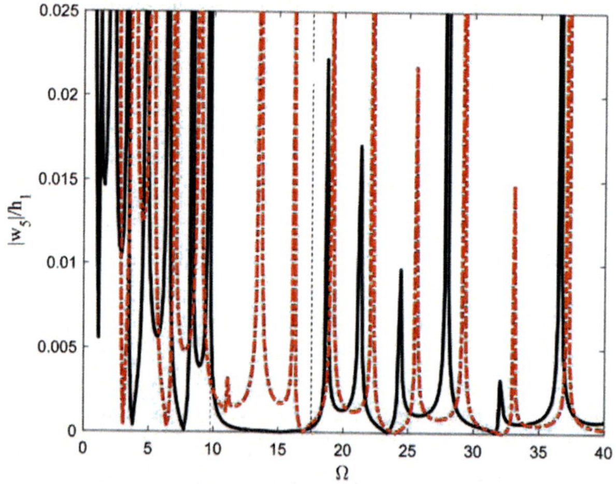

Fig. 5. FE forced response, $|w_5(\Omega)|/h_1$. Black line: circular plate with eight periodic annuli; red dot-dashed lines: circular homogeneous plate. Thin vertical dashed lines show the stop-band in Fig. 3.

The FE and the WFE results are compared in Fig. 6. It can be seen that both models show a significant vibration attenuation in the stop-band. The figure clearly shows the ability of the WFE model to give information on the dynamic behaviour of the structure at a very low computational cost. Although there could be different choices for the FE discretisation, one of the distinct advantages of the WFE approach is the incredible reduction in the design and computational time due to the simplified structure and the significantly reduced size of the model.

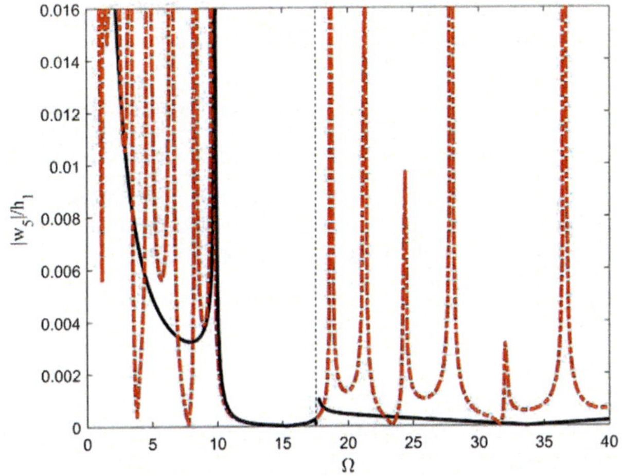

Fig. 6. Forced response, $|w_5(\Omega)|/h_1$. Comparison between the WFE - radially periodic plate - and the FE results - circular homogenous plate with eight periodic annuli. Black line: WFE results; red dot-dashed lines: FE results. Thin vertical dashed lines show the stop-band in Fig. 3.

4 Conclusion

This work was concerned with the forced response of a radially periodic plate excited by a harmonic transverse vibration source. In particular, the vibration attenuation characteristics of the plate were investigated using a numerical approach based on finite element analysis and the theory of wave propagation in periodic structures, the Wave Finite Element (WFE) method. The method was adapted to radial periodic waveguides. Only a small slice of the plate, discretised using the standard FE method, was studied, while wave characteristics were obtained using Floquet theory. Assuming that the power associated with each propagating wave is preserved, the waves amplitude changes, as they propagate in the radial direction, were evaluated from the ratio between the energy flowing through each subsequent unit cell. The method's ability to predict the vibration response and attenuation in the pass- and stop-bands was presented for a plate with a sequence of annuli with alternating thickness. The results were compared with those obtained from a standard FE model of a uniform finite circular plate with the insertion of a limited number of periodic annuli. The numerical results were in good agreement, showing the advantages of the approach in terms of computational time, model design and efficiency.

References

1. Brillouin, L.: Wave Propagation in Periodic Structures, 1st edn. McGraw-Hill, New York (1946)
2. Hussein, M.I., Leamy, M.J., Ruzzene, M.: Dynamics of phononic materials and structures: historical origins, recent progress, and future outlook. Appl. Mech. Rev. **66**(4), 040802 (2014)
3. D'Alessandro, L., Belloni, E., Ardito, R., Corigliano, A., Braghin, F.: Modeling and experimental verification of an ultra-wide bandgap in 3D phononic crystal. Appl. Phys. Lett. **109**(22), 221907 (2016)
4. Poggetto, V.F., Arruda, J.R.F.: Widening wave band gaps of periodic plates via shape optimisation using spatial Fourier coefficients. Mech. Syst. Signal Process. **147**, 107098 (2021)
5. Carneiro Jr., J.P., Brennan, M.J., Gonçalves, P.J.P., Cleante, V.G., Bueno, D.D., Santos, R.B.: On the attenuation of vibration using a finite periodic array of rods comprised of either symmetric or asymmetric cells. J. Sound Vib. **511,** 116217 (2021)
6. Bastawrous, M.V., Hussein, M.I.: Theoretical band-gap bounds and coupling sensitivity for a waveguide with periodically attached resonating branches. J. Sound Vib. **514**, 116428 (2021)
7. Claeys, C.C., Vergote, K., Sas, P., Desmet, W.: On the potential of tuned resonators to obtain low-frequency vibrational stop bands in periodic panels. J. Sound Vib. **332**(6), 1418–1436 (2013)
8. Torrent, D., Sanchez-Dehesa, J.: Acoustic resonances in two-dimensional radial sonic crystal shells. New J. Phys. **12**, 073034 (2010)
9. Arretche, I., Matlack, K.H.: Effective phononic crystals for non-Cartesian elastic wave propagation. Phys. Rev. B **102**, 134308 (2020)
10. Hvatov, A., Sorokin, S.: On application of the Floquet theory for radially periodic membranes and plates. J. Sound Vib. **414**, 15–30 (2018)
11. Manconi, E., Sorokin, S.V., Garziera, R., Quartaroli, M.M.: Free and forced wave motion in a two-dimensional plate with radial periodicity. Appl. Sci. **11**, 10948 (2021)
12. Mace, B.R., Duhamel, D., Brennan, M.J., Hinke, L.: Finite element prediction of wave motion in structural waveguides. J. Acoust. Soc. Am. **117**, 2835–2843 (2005)
13. Mace, B.R., Manconi, E.: Modelling wave propagation in two-dimensional structures using finite element analysis. J. Sound Vib. **318**, 884–902 (2008)
14. Renno, J.M., Manconi, E., Mace, B.R.: A finite element method for modelling waves in laminated structures. Adv. Struct. Eng. **16**, 61–75 (2013)
15. Boukadia, R.F., Droz, C., Ichchou, M.N., Desmet, W.A.: Bloch wave reduction scheme for ultrafast band diagram and dynamic response computation in periodic structures. Finite Elem. Anal. Des. **148**, 1–12 (2018)
16. Auld, B.A.: Acoustic Fields and Waves in Solids. Krieger Publishing Company (1990)
17. Fabro, A.T., Ferguson, N.S., Mace, B.R.: Wave propagation in slowly varying waveguides using a finite element approach. J. Sound Vib. **442**, 308–329 (2019)

Wave Transmission and Reflection Analysis Based on the Three-dimensional Second Strain Gradient Theory

Bo Yang[1], Mohamed Ichchou[1(✉)], Christophe Droz[2], and Abdelmalek Zine[3]

[1] LTDS - CNRS UMR 5513, Vibroacoustics and Complex Media Research Group, Ecole Centrale de Lyon, Lyon, France
mohamed.ichchou@ec-lyon.fr

[2] Inria, COSYS/SII, I4S Team, University Gustave Eiffel, Paris, France
christophe.droz@inria.fr

[3] Institute Camille Jordan – CNRS UMR 5208, Ecole Centrale de Lyon, Lyon, France
abdel-malek.zine@ec-lyon.fr

Abstract. The scattering of guided waves through a coupling region is a crucial information when studying waveguides. In this paper, the second strain gradient theory (SSG) is used to describe wave transmission and reflection in a three-dimensional micro-sized medium. First, the constitutive relation of 3D SSG model is derived while six quintic Hermite polynomial shape functions are used for the displacement field. Then Hamilton's principle is used for the weak formulation of the unit-cell's stiffness matrix finite element stiffness, mass matrices and force vector. Eventually the wave diffusion (i.e. including reflection and transmission coefficients) are computed and discussed for various coupling conditions.

Keywords: Second strain gradient theory · Wave finite element method · Wave transmission · Wave reflection

1 Introduction

The dynamical properties of guided waves such as wave transmission and reflection have been widely studied over the past decade especially in the field of acoustics, earthquake and electromagnetic. Initially, studies focused on the guided waves interaction at interfaces between different macro-medias. But for the micro-sized structure with size effects, the micro-particles such as atoms with associated energy on the free surface of the structure has a significant influence on the structure's behavior. This energy related to surface atoms is called surface free energy which produces surface tension. The surface tension can not be ignored due to the very large ratio between the surface and the volume of structure. On the other hand, long-range or non-local interaction between micro-particles has also an indispensable effect on the micro-sized structure's dynamical behaviors [1,2,20]. The wave propagation and diffusion in micro-medias can no longer be reasonably predicted by Classical Theory (CT) of continuum mechanics [3].

Z. Dimitrovová et al. (Eds.): WMVC 2022, MMS 125, pp. 761–772, 2023.
https://doi.org/10.1007/978-3-031-15758-5_78

Therefore, the non-classical continuum theories of elasticity that can interpret the properties of micro-sized structures have been proposed. Generally, these non-classical theories can be categorized into non-local elasticity theory [4], micro-continuum theory [5,6], surface elasticity theory [7] and strain gradient family [8]. The strain gradient family is composed of the couple stress theory, the first and second strain gradient theories and the modified couple stress theory. Mindlin established one of the strain gradient family called First Strain Gradient (SG) theory [9] in which the constitutive relations is composed of strain and the first gradient of strain. The atomic structure with the nearest and next nearest interactions between different particles is used to describe the SG theory in the framework of lattice spring model, but only in noncentro-symmetric materials [10]. In order to explore the properties of centro-symmetric materials, the Second Strain Gradient (SSG) theory [11] was put forward, which offers a reasonable description of the strain and surface tension properties on the micro-structure's surface by introducing the high-order parameters. The constitutive relations in SSG theory is a function of strain, first gradient of strain and second gradient of strain. The connections between the SSG theory and lattice spring model with the nearest, next nearest and next-next nearest neighbor interactions for 1D structures can be confirmed through the Fourier series transform [3,22].

On the other hand, in order to study the wave diffusion in complex structures, the numerical methods such as Spectral Finite Mlement (SFE) method [12] and Semi-Analytical Finite Element (SAFE) [13] can be used. In the past decade, the Wave Finte Element Method (WFEM) [14,15,23] has attracted many works. The advantage of WFEM is the convenient application in engineering field. Since it can be developed from the Finte Element Method (FEM) packages which allows the current element library and grid generation procedures to be applied for the modelling of different waveguide structures. In addition, WFEM can reduce a global periodic structure into a single substructure or unit cell based on the periodic structures theory [16,24]. The resulting stiffness and mass matrices are post processed to offer the dynamic stiffness matrix. The dynamical properties of the periodic structure can be reflected through the spectral analysis of the unit cell [17,21]. The main objective of this work is, firstly, to calculate the multi mode propagation in a 3D periodic waveguide by SSG theory, and, secondly, to confirm the wave diffusion under a complex coupling condition.

The article's structure is the following: in Sect. 2, the constitutive relations of 3D micro-sized model are introduced in the SSG theory framework and the weak formulations including element stiffness, mass matrices and force vector are calculated. Afterwards, in Sect. 3, free wave propagation characteristics are expressed by solving eigenvalue problems in the direct WFEM framework, diffusion matrix for a complex coupling condition are confirmed. In Sect. 4, wave diffusion is introduced. Ultimately, some useful conclusions are presented in Sect. 5.

2 A Brief of Second Strain Gradient Elasticity

2.1 3D Constitutive Relations Calculation

The strain energy density $\overline{\mathcal{U}}$ composed of strain $\boldsymbol{\varepsilon}=sym(\nabla\mathbf{W})$, first gradient of strain $\boldsymbol{\xi}=\nabla\nabla\mathbf{W}$ and second gradient of strain $\boldsymbol{\zeta}=\nabla\nabla\nabla\mathbf{W}$ in the SSG theory framework was put forward by Mindlin [11], as below:

$$\begin{aligned}\overline{\mathcal{U}} =&\frac{1}{2}\lambda\varepsilon_{ii}\varepsilon_{jj} + \mu\varepsilon_{ij}\varepsilon_{ij} + a_1\xi_{ijj}\xi_{ikk} + a_2\xi_{iik}\xi_{kjj} + a_3\xi_{iik}\xi_{jjk} + a_4\xi_{ijk}\xi_{ijk} + a_5\xi_{ijk}\xi_{jki} + b_1\zeta_{iijj}\zeta_{kkll}\\ &+ b_2\zeta_{ijkk}\zeta_{ijll} + b_3\zeta_{iijk}\zeta_{jkll} + b_4\zeta_{iijk}\zeta_{llkj} + b_5\zeta_{iijk}\zeta_{lljk} + b_6\zeta_{ijkl}\zeta_{ijkl} + b_7\zeta_{ijkl}\zeta_{jkli} + c_1\varepsilon_{ii}\zeta_{jjkk}\\ &+ c_2\varepsilon_{ij}\zeta_{ijkk} + c_3\varepsilon_{ij}\zeta_{kkij},\end{aligned} \tag{1}$$

where $\mathbf{W}$ is the displacement vector, symbol ∇ means the gradient operator, λ and μ represent the *Lamé* parameters which are related to the Young's modulus E, the Poisson's ratio ν and the shear modulus G [3]. a_i, b_i and c_i denote the higher order parameters [10] in SSG theory. The higher order parameters for Al and Cu are shown in Table 1. Based on the 3D elasticity theory, the vector of displacement field defined in the Cartesian coordinate system (x, y, z) is given as:

$$\mathbf{W}(x,y,z,t) = \big(u(x,y,z,t),\ v(x,y,z,t),\ w(x,y,z,t)\big)^{\mathrm{T}}, \tag{2}$$

where u, v and w are the the displacements along x, y and z direction.

Table 1. Higher order material parameters a_i $(eV/\text{Å})$, b_i $(eV{\cdot}\text{Å})$, c_i $(eV/\text{Å})$.

Material	a_1	a_2	a_3	a_4	a_5	b_1	b_2	b_3	b_4	b_5	b_6	b_7	c_1	c_2	c_3
Al	0.140	0.002	-0.008	0.096	0.258	0.792	0.064	-0.194	-0.001	0.001	16.156	48.529	0.504	0.357	0.178
Cu	0.183	0.010	0.001	0.072	0.189	0.661	0.066	0.206	0.002	0.002	12.625	37.940	0.845	0.573	0.347

The relations between strains and displacement components can be defined by introducing the vectors of first, second and third order derivatives of displacement components:

$$\boldsymbol{\varepsilon} = \mathbf{Q}_1\mathbf{W},\ \boldsymbol{\xi} = \mathbf{Q}_2\mathbf{W},\ \boldsymbol{\zeta} = \mathbf{Q}_3\mathbf{W}, \tag{3}$$

where

$$\mathbf{Q}_1 = \begin{bmatrix}\partial_1 & 0 & 0 & 0 & \partial_3 & 0\\ 0 & \partial_2 & 0 & \partial_3 & 0 & \partial_2\\ 0 & 0 & \partial_3 & \partial_2 & \partial_1 & \partial_1\end{bmatrix}^{\mathrm{T}},\ \mathbf{Q}_2 = \begin{bmatrix}\mathbf{t}_1 & \mathbf{0} & \mathbf{0}\\ \mathbf{0} & \mathbf{t}_1 & \mathbf{0}\\ \mathbf{0} & \mathbf{0} & \mathbf{t}_1\end{bmatrix} \otimes \big(\partial_{11}\ \partial_{22}\ \partial_{33}\ 2\partial_{12}\ 2\partial_{13}\ 2\partial_{23}\big)^{\mathrm{T}},$$

$$\mathbf{Q}_3 = \begin{bmatrix}\mathbf{t}_2 & \mathbf{0} & \mathbf{0}\\ \mathbf{0} & \mathbf{t}_2 & \mathbf{0}\\ \mathbf{0} & \mathbf{0} & \mathbf{t}_2\end{bmatrix} \otimes \big(\partial_{111}\ \partial_{222}\ \partial_{333}\ 3\partial_{112}\ 3\partial_{113}\ 3\partial_{221}\ 3\partial_{223}\ 3\partial_{331}\ 3\partial_{332}\ 6\partial_{123}\big)^{\mathrm{T}},$$

symbol $\otimes$ stands for the Kronecker product, $\mathbf{t}_1$ with size 6×1 and $\mathbf{t}_2$ with size 10×1 are the matrices whose element value is 1.

Then, the constitutive relations for 3D model by SSG theory can be defined as:

$$\boldsymbol{\tau}_1 = \mathcal{L}\boldsymbol{\varepsilon} + \mathcal{C}\boldsymbol{\zeta},\ \boldsymbol{\tau}_2 = \mathcal{A}\boldsymbol{\xi},\ \boldsymbol{\tau}_3 = \mathcal{B}\boldsymbol{\zeta} + \mathcal{C}^{\mathrm{T}}\boldsymbol{\varepsilon}, \tag{4}$$

in which $\mathcal{L}$ is matrix including classical parameters, $\mathcal{A}$, $\mathcal{B}$ and $\mathcal{C}$ are matrices including higher order parameters [18]. Finally, the strain energy density for SSG theory can be rewritten as the matrix form:

$$\overline{\mathcal{U}} = \frac{1}{2}\left(\varepsilon^{\mathrm{T}}\mathcal{L}\varepsilon + \xi^{\mathrm{T}}\mathcal{A}\xi + \zeta^{\mathrm{T}}\mathcal{B}\zeta + 2\zeta^{\mathrm{T}}\mathcal{C}^{\mathrm{T}}\varepsilon\right). \tag{5}$$

Here should be noted that, Eq. 5 is the basic form of building a 3D model using partial differential equations (PDE) weak form in some commercial numerical simulation software (e.g., COMSOL).

2.2 Finite Element Discretization

After obtaining the strain energy density, the next step is to calculate the weak form including stiffness and mass matrices and force vector. The definition of node degree of freedoms (DOFs) for 1D and 3D Hermite elements. Firstly, in order to ensure the continuity of higher derivatives up to the second order between 1D elements, the six-term polynomial function is considered to interpolate the scalar field $\mathrm{W}_1 = u(x,t)$ inside a 1D element, as follows:

$$\mathrm{W}_1 = \left[1\ x\ x^2\ x^3\ x^4\ x^5\right]\left[\mathrm{s}_0\ \mathrm{s}_1\ \mathrm{s}_2\ \mathrm{s}_3\ \mathrm{s}_4\ \mathrm{s}_5\right]^{\mathrm{T}} = \mathbf{xs}. \tag{6}$$

The evaluation of the nodal displacement vector $\overline{\mathbf{w}}_1^{(e)}$ of 1D element gives:

$$\overline{\mathbf{w}}_1^{(e)} = \left(\mathbf{d}_1\ \mathbf{d}_2\ \mathbf{d}_3\ \mathbf{d}_4\ \mathbf{d}_5\ \mathbf{d}_6\right)^{\mathrm{T}}\mathbf{s} = \mathbf{ds}, \tag{7}$$

where $\mathbf{d}_1 = (1\ -l_e\ l_e^2\ -l_e^3\ l_e^4\ -l_e^5)$, $\mathbf{d}_2 = (0\ 1\ -2l_e\ 3l_e^2\ -4l_e^3\ 5l_e^4)$, $\mathbf{d}_3 = (0\ 0\ 2\ -6l_e\ 12l_e^2\ -20l_e^3)$, $\mathbf{d}_4 = (1\ l_e\ l_e^2\ l_e^3\ l_e^4\ l_e^5)$, $\mathbf{d}_5 = (0\ 1\ 2l_e\ 3l_e^2\ 4l_e^3\ 5l_e^4)$, $\mathbf{d}_6 = (0\ 0\ 2\ 6l_e\ 12l_e^2\ 20l_e^3)$. Then, submitting Eq. 7 into Eq. 6, the displacement vector within the 1D element can be derived by employing the six quin-tic Hermite polynomial shape function and nodal displacement vector, as follows:

$$\mathrm{W}_1 = \mathbf{x}\mathbf{d}^{-1}\overline{\mathbf{w}}_1^{(e)} = \mathbf{S}(x)\overline{\mathbf{w}}_1^{(e)}, \tag{8}$$

in which the shape function $\mathbf{S}(x)$ is written as:

$$\mathbf{S}(x) = \left(\mathrm{S}_1^0(x), \mathrm{S}_1^1(x), \mathrm{S}_1^2(x), \mathrm{S}_2^0(x), \mathrm{S}_2^1(x), \mathrm{S}_2^2(x)\right), \tag{9}$$

where $\mathrm{S}_1^0(x) = \frac{5x^3}{8l_e^3} - \frac{15x}{16l_e} - \frac{3x^5}{16l_e^5} + \frac{1}{2}$, $\mathrm{S}_1^1(x) = \frac{5l_e}{16} - \frac{7x}{16} - \frac{3x^2}{8l_e} + \frac{5x^3}{8l_e^2} + \frac{x^4}{16l_e^3} - \frac{3x^5}{16l_e^4}$, $\mathrm{S}_1^2(x) = \frac{l_e^2}{16} - \frac{l_e x}{16} - \frac{x^2}{8} + \frac{x^3}{8l_e} + \frac{x^4}{16l_e^2} - \frac{x^5}{16l_e^3}$, $\mathrm{S}_2^0(x) = \frac{15x}{16l_e} - \frac{5x^3}{8l_e^3} + \frac{3x^5}{16l_e^5} + \frac{1}{2}$, $\mathrm{S}_2^1(x) = \frac{3x^2}{8l_e} - \frac{7x}{16} - \frac{5l_e}{16} + \frac{5x^3}{8l_e^2} - \frac{x^4}{16l_e^3} - \frac{3x^5}{16l_e^4}$, $\mathrm{S}_2^2(x) = \frac{l_e x}{16} + \frac{l_e^2}{16} - \frac{x^2}{8} - \frac{x^3}{8l_e} + \frac{x^4}{16l_e^2} + \frac{x^5}{16l_e^3}$. $\mathbf{S}(y) = \mathbf{S}(x)|_{x=y} = [\mathrm{N}_1^0(y), \mathrm{S}_1^1(y), \mathrm{S}_1^2(y), \mathrm{S}_2^0(y), \mathrm{S}_2^1(y), \mathrm{S}_2^2(y)]$, $\mathbf{S}(z) = \mathbf{S}(x)|_{x=z} = [\mathrm{S}_1^0(z), \mathrm{S}_1^1(z), \mathrm{S}_1^2(z), \mathrm{S}_2^0(z), \mathrm{S}_2^1(z), \mathrm{S}_2^2(z)]$. The shape function of hexahedral element:

$$\mathbb{S}(x,y,z) = \left(\mathbf{S}_1(x,y,z)\otimes\mathbf{e}_1, \mathbf{S}_2(x,y,z)\otimes\mathbf{e}_2, \mathbf{S}_3(x,y,z)\otimes\mathbf{e}_3\right)^{\mathrm{T}}. \tag{10}$$

The element in $\mathbf{S}_p(x, y, z)$ and $\mathbf{e}_p$ ($p = 1, 2, 3$) are defined as:

$$\mathrm{S}_p^{i(j,k,l)}(x, y, z) = \mathrm{S}_{i'}^{j}(x)\mathrm{S}_{i''}^{k}(y)\mathrm{S}_{i'''}^{l}(z), \ \mathbf{e}_p = \left(\epsilon_{p1}\ \epsilon_{p2}\ \epsilon_{p3}\right), \tag{11}$$

where $i = 1, ..., 8$. $j, k, l = 0, 1, 2$. $\mathrm{S}_p^{i(j,k,l)}(x, y, z)$ is associated with the DOFs $\partial^{j+k+l}u_1^i/(\partial x^j \partial y^k \partial z^l)$ of node i of the hexahedron element. $i', i'', i''' = 1, 2$ relate to the node number in the corresponding 1D element and they take values of 1 or 2 if the coordinate value of node i is $-l_e$ or l_e. The displacement vector $\mathbf{W}(x, y, z)$ within 3D element can be expressed as:

$$\mathbf{W}(x, y, z, t) = \mathbb{S}(x, y, z)\mathbf{w}^{(e)}(t), \tag{12}$$

where $\mathbf{w}^{(e)}=[(\mathbf{w}_1^{(e)})^{\mathrm{T}},\ (\mathbf{w}_2^{(e)})^{\mathrm{T}},\ (\mathbf{w}_3^{(e)})^{\mathrm{T}}]^{\mathrm{T}}$, $(\mathbf{w}_p^{(e)})^{\mathrm{T}}=[(\mathbf{w}_p^{1(e)})^{\mathrm{T}},\ (\mathbf{w}_p^{2(e)})^{\mathrm{T}},\ ...,\ (\mathbf{w}_p^{8(e)})^{\mathrm{T}}]^{\mathrm{T}}$ ($p = 1, 2, 3$). Then, integrating the strain energy density over its volume to obtain the strain potential energy $\mathcal{U}$ as:

$$\mathcal{U} = \int_V \overline{\mathcal{U}} \mathrm{d}V = \frac{1}{2}\mathbf{w}^{(e)\mathrm{T}} \int_V \left(\mathbb{S}^{\mathrm{T}}\mathbf{Q}_1^{\mathrm{T}}\mathcal{L}\mathbf{Q}_1\mathbb{S} + \mathbb{S}^{\mathrm{T}}\mathbf{Q}_2^{\mathrm{T}}\mathcal{A}\mathbf{Q}_2\mathbb{S} + \mathbb{S}^{\mathrm{T}}\mathbf{Q}_3^{\mathrm{T}}\mathcal{B}\mathbf{Q}_3\mathbb{S} + 2\mathbb{S}^{\mathrm{T}}\mathbf{Q}_3^{\mathrm{T}}\mathcal{C}\mathbf{Q}_1\mathbb{S}\right) \mathrm{d}V\mathbf{w}^{(e)}. \tag{13}$$

On the other hand, the beam kinetic energy is expressed as:

$$\mathcal{T} = \frac{1}{2}\int_V \left(\frac{\partial \mathbf{W}}{\partial t}\right)^{\mathrm{T}} \rho \left(\frac{\partial \mathbf{W}}{\partial t}\right) \mathrm{d}V = \frac{1}{2}\left(\frac{\partial \mathbf{w}^{(e)}}{\partial t}\right)^{\mathrm{T}} \int_V \left(\mathbb{S}^{\mathrm{T}}\rho\mathbb{S}\right) \mathrm{d}V \left(\frac{\partial \mathbf{w}^{(e)}}{\partial t}\right), \tag{14}$$

where ρ denotes the linear mass density, l_1 and l_2 are higher-order length-scale parameters. Meanwhile, the work done $\delta\mathcal{W}$ by external force can be expressed as:

$$\delta\mathcal{W} = \int_V \delta\mathbf{W}^{\mathrm{T}}\mathbf{f}_V \mathrm{d}V + \int_S \delta\mathbf{W}^{\mathrm{T}}\mathbf{f}_S \mathrm{d}S = \delta\mathbf{w}^{(e)\mathrm{T}} \left(\int_V \left(\mathbb{S}^{\mathrm{T}}\mathbf{f}_V\right) \mathrm{d}V + \int_S \left(\mathbb{S}^{\mathrm{T}}\mathbf{f}_S\right) \mathrm{d}S\right), \tag{15}$$

where $\mathbf{f}_V$ is the volume force, $\mathbf{f}_S$ means the face force. The 3D element stiffness, mass matrices and force vector can be confirmed as:

$$\begin{aligned}\mathbf{K}^{(e)} &= \int_V \left(\mathbb{S}^{\mathrm{T}}\mathbf{Q}_1^{\mathrm{T}}\mathcal{L}\mathbf{Q}_1\mathbb{S} + \mathbb{S}^{\mathrm{T}}\mathbf{Q}_2^{\mathrm{T}}\mathcal{A}\mathbf{Q}_2\mathbb{S} + \mathbb{S}^{\mathrm{T}}\mathbf{Q}_3^{\mathrm{T}}\mathcal{B}\mathbf{Q}_3\mathbb{S} + 2\mathbb{S}^{\mathrm{T}}\mathbf{Q}_3^{\mathrm{T}}\mathcal{C}\mathbf{Q}_1\mathbb{S}\right) \mathrm{d}V,\\ \mathbf{M}^{(e)} &= \int_V \left(\mathbb{S}^{\mathrm{T}}\rho\mathbb{S}\right) \mathrm{d}V,\ \mathbf{F}^{(e)} = \int_V \left(\mathbb{S}^{\mathrm{T}}\mathbf{f}_V\right) \mathrm{d}V + \int_S \left(\mathbb{S}^{\mathrm{T}}\mathbf{f}_S\right) \mathrm{d}S.\end{aligned} \tag{16}$$

3 Diffusion Analysis Through Wave Finite Element Method

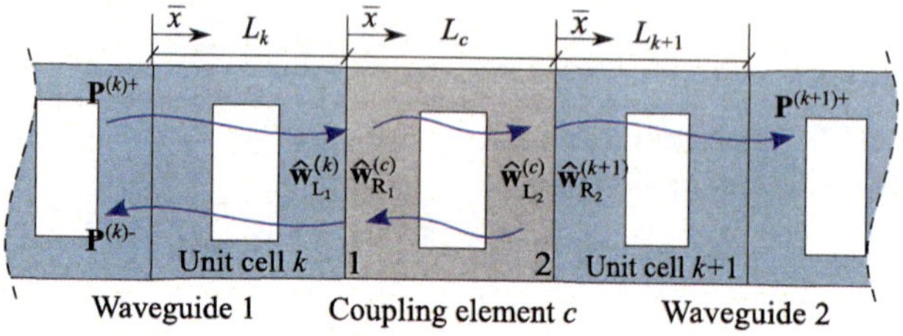

Fig. 1. Two coupled waveguides through a coupling element ($\overline{x}$ is the local coordinate).

This section is concerned with the characterization of coupling conditions between two semi-infinite periodic waveguides which are connected through an elastic coupling element at surfaces 1 and 2 as shown in Fig. 1. The reflection coefficients ($\mathbf{R}$) and transmission coefficients ($\mathbf{T}$) are confirmed through a complex coupling condition: the wave modes calculation in waveguides is based on the CT but SSG theory in coupling element (CT-SSG). The propagation constants $\mathbf{\Lambda}$ and eigenvectors $\mathbf{\Psi}_{\mathrm{u}}$ in a unit cell can be solved by direct Bloch formulation [3,15,17] as:

$$\left[\mathbf{D}_{\mathrm{RL}}(\omega)\mathbf{\Lambda}^{-1} + (\mathbf{D}_{\mathrm{RR}}(\omega) + (\mathbf{D}_{\mathrm{LL}}(\omega)) + \mathbf{D}_{\mathrm{LR}}(\omega)\mathbf{\Lambda}\right] \mathbf{\Psi}_{\mathrm{u}} = 0, \tag{17}$$

where $\mathbf{\Lambda}$=diag $\{\lambda_j\}_{j=1,\dots,2p}$, $\mathbf{\Psi}_{\mathrm{u}} = \{\phi_j\}_{j=1,\dots,2p}$ which can be divided into $\mathbf{\Psi}_{\mathrm{u}}^{+} = \{\phi_j^{+}\}_{j=1,\dots,p}$ and $\mathbf{\Psi}_{\mathrm{u}}^{-} = \{\phi_j^{-}\}_{j=p+1,\dots,2p}$, in which p=m for SSG, p=n for CT. The waves propagate to positive when $|1/\lambda_j|$ <1. The waves propagate to negative when $|\lambda_j|$ >1. Here, λ_j take the form λ_j=exp$(-\mathrm{i}\kappa_j L_{(q)})$, subscript q=c for coupling element, q=k, k+1 for unit cells. It should be noted that the spectral analysis method of the coupling element and unit cell k+1 is the same as that of unit cell k. The displacement field of coupling element (c) and unit cells (k, k+1) can be represented by the superposition of the eigenmodes:

$$\hat{\mathbf{w}}^{(q)\pm} = \mathbf{\Psi}_{\mathrm{u}}^{(q)\pm}\boldsymbol{\upsilon}^{(q)\pm}\mathbf{P}^{(q)\pm}, \tag{18}$$

where $\mathbf{P}^{(q)\pm}$ is the amplitudes of wave modes. $\boldsymbol{\upsilon}^{(q)\pm}$=diag $\left\{\exp\left(\mp\mathrm{i}\kappa_j^{(q)}\overline{x}\right)\right\}$ $(0 \le \overline{x} \le L_{(q)})$, symbol q=c for coupling element, q=k, k+1 for unit cells. On the other hand, the force components of coupling element (c) and unit cells (k, k+1) can be expressed as:

$$\hat{\mathbf{F}}^{(q)\pm} = \mathbf{\Psi}_{\mathrm{F}}^{(q)\pm}\boldsymbol{\upsilon}^{(q)\pm}\mathbf{P}^{(q)\pm}, \tag{19}$$

in which $\mathbf{\Psi}_{\mathrm{F}}^{(q)\pm}$=$\mathbf{D}_{\mathrm{LL}}^{(q)}\mathbf{\Psi}_{\mathrm{u}}^{(q)\pm}$+$\mathbf{D}_{\mathrm{LR}}^{(q)}\mathbf{\Psi}_{\mathrm{u}}^{(q)\pm}$diag$\left\{\exp\left(\mp\mathrm{i}\kappa_j^{(q)}L_{(q)}\right)\right\}$, symbol q=c for coupling element, q=k, k+1 for unit cells. The eigensolutions will be the same when waveguides 1 and 2 have the same cross-section and material. The state vector on the right side of surface 1, represented by R_1:

$$\begin{pmatrix} \hat{\mathbf{w}}_{R_1}^{(c)} \\ \hat{\mathbf{F}}_{R_1}^{(c)} \end{pmatrix} = \begin{bmatrix} \boldsymbol{\Psi}_{u}^{(c)+}\mathbf{P}^{(c)+} + \boldsymbol{\Psi}_{u}^{(c)-}\mathbf{P}^{(c)-} \\ \boldsymbol{\Psi}_{F}^{(c)+}\mathbf{P}^{(c)+} + \boldsymbol{\Psi}_{F}^{(c)-}\mathbf{P}^{(c)-} \end{bmatrix}. \tag{20}$$

In addition, the state vector on the left side of surface 2, represented by L_2, is written as:

$$\begin{pmatrix} \hat{\mathbf{w}}_{L_2}^{(c)} \\ \hat{\mathbf{F}}_{L_2}^{(c)} \end{pmatrix} = \begin{bmatrix} \boldsymbol{\Psi}_{u}^{(c)+}\boldsymbol{\upsilon}^{(c)+}|_{\overline{x}=L_c}\mathbf{P}^{(c)+} + \boldsymbol{\Psi}_{u}^{(c)-}\boldsymbol{\upsilon}^{(c)-}|_{\overline{x}=L_c}\mathbf{P}^{(c)-} \\ \boldsymbol{\Psi}_{F}^{(c)+}\boldsymbol{\upsilon}^{(c)+}|_{\overline{x}=L_c}\mathbf{P}^{(c)+} + \boldsymbol{\Psi}_{F}^{(c)-}\boldsymbol{\upsilon}^{(c)-}|_{\overline{x}=L_c}\mathbf{P}^{(c)-} \end{bmatrix}. \tag{21}$$

Combining Eq. 20 and Eq. 21, the relation between state vector on left and right side of coupling element will be conformed:

$$\left(\hat{\mathbf{w}}_{L_2}^{(c)},\ \hat{\mathbf{F}}_{L_2}^{(c)}\right)^{\mathrm{T}} = \mathbf{X}^{(c)}\left(\hat{\mathbf{w}}_{R_1}^{(c)},\ \hat{\mathbf{F}}_{R_1}^{(c)}\right)^{\mathrm{T}}, \tag{22}$$

with

$$\mathbf{X}^{(c)} = \begin{bmatrix} \boldsymbol{\Psi}_{u}^{(c)+}\boldsymbol{\upsilon}^{(c)+}|_{\overline{x}=L_c},\ \boldsymbol{\Psi}_{u}^{(c)-}\ \boldsymbol{\upsilon}^{(c)-}|_{\overline{x}=L_c} \\ \boldsymbol{\Psi}_{F}^{(c)+}\boldsymbol{\upsilon}^{(c)+}|_{\overline{x}=L_c},\ \boldsymbol{\Psi}_{F}^{(c)-}\ \boldsymbol{\upsilon}^{(c)-}|_{\overline{x}=L_c} \end{bmatrix} \begin{bmatrix} \boldsymbol{\Psi}_{u}^{(c)+} & \boldsymbol{\Psi}_{u}^{(c)-} \\ \boldsymbol{\Psi}_{F}^{(c)+} & \boldsymbol{\Psi}_{F}^{(c)-} \end{bmatrix}^{-1}. \tag{23}$$

Next, assuming that the incident waves come from the infinity of waveguide 1 and there is no reflection from the end of waveguide 2. The initial boundary of waveguide 1 is also non-reflecting. The state vector on the left side of surface 1, represented by L_1, is expressed as:

$$\begin{pmatrix} \hat{\mathbf{w}}_{L_1}^{(k)} \\ \hat{\mathbf{F}}_{L_1}^{(k)} \end{pmatrix} = \begin{bmatrix} \boldsymbol{\Psi}_{u}^{(k)+}\boldsymbol{\upsilon}^{(k)+}|_{\overline{x}=L_k}\mathbf{P}^{(k)+} + \boldsymbol{\Psi}_{u}^{(k)-}\boldsymbol{\upsilon}^{(k)-}|_{\overline{x}=L_k}\mathbf{P}^{(k)-} \\ \boldsymbol{\Psi}_{F}^{(k)+}\boldsymbol{\upsilon}^{(k)+}|_{\overline{x}=L_k}\mathbf{P}^{(k)+} + \boldsymbol{\Psi}_{F}^{(k)-}\boldsymbol{\upsilon}^{(k)-}|_{\overline{x}=L_k}\mathbf{P}^{(k)-} \end{bmatrix}. \tag{24}$$

The state vector on the right side of surface 2, represented by R_2, is written as:

$$\begin{pmatrix} \hat{\mathbf{w}}_{R_2}^{(k+1)} \\ \hat{\mathbf{F}}_{R_2}^{(k+1)} \end{pmatrix} = \begin{bmatrix} \boldsymbol{\Psi}_{u}^{(k+1)+}\boldsymbol{\upsilon}^{(k+1)+}|_{\overline{x}=0}\mathbf{P}^{(k+1)+} \\ \boldsymbol{\Psi}_{F}^{(k+1)+}\boldsymbol{\upsilon}^{(k+1)+}|_{\overline{x}=0}\mathbf{P}^{(k+1)+} \end{bmatrix}. \tag{25}$$

Here should be noted that the size of state vector for coupling element is $2m \times 1$, but $2n \times 1$ for unit cells k and k+1. The higher order parts in state vector for coupling element is $2(m - n) \times 1$. In order to ensure the continuity on surfaces 1 and 2, defining new state vectors including higher order parts $(2(m - n) \times 1)$ for unit cells k and k+1:

$$\begin{aligned} \left(\hat{\mathbf{w}}_{L_1}^{*(k)},\ \hat{\mathbf{F}}_{L_1}^{*(k)}\right)^{\mathrm{T}} &= \left(\hat{\mathbf{w}}_{L_1}^{(k)},\ \hat{\mathbf{w}}_{L_1}^{'(k)},\ \hat{\mathbf{F}}_{L_1}^{(k)},\ \hat{\mathbf{F}}_{L_1}^{'(k)}\right)^{\mathrm{T}}, \\ \left(\hat{\mathbf{w}}_{R_2}^{*(k+1)},\ \hat{\mathbf{F}}_{R_2}^{*(k+1)}\right)^{\mathrm{T}} &= \left(\hat{\mathbf{w}}_{R_2}^{(k+1)},\ \hat{\mathbf{w}}_{R_2}^{'(k+1)},\ \hat{\mathbf{F}}_{R_2}^{(k+1)},\ \hat{\mathbf{F}}_{R_2}^{'(k+1)}\right)^{\mathrm{T}}, \end{aligned} \tag{26}$$

where $\hat{\mathbf{w}}_{\mathrm{L}_1}^{'(k)}$, $\hat{\mathbf{F}}_{\mathrm{L}_1}^{'(k)}$ and $\hat{\mathbf{w}}_{\mathrm{L}_1}^{'(k+1)}$, $\hat{\mathbf{F}}_{\mathrm{L}_1}^{'(k+1)}$ are unknown higher order displacements and forces vectors for unit cell k and k+1 respectively. The continuity on surfaces 1 and 2 is:

$$\left(\hat{\mathbf{w}}_{\mathrm{R}_1}^{(c)},\ \hat{\mathbf{F}}_{\mathrm{R}_1}^{(c)}\right)^{\mathrm{T}} = \left(\hat{\mathbf{w}}_{\mathrm{L}_1}^{*(k)},\ \hat{\mathbf{F}}_{\mathrm{L}_1}^{*(k)}\right)^{\mathrm{T}},\ \left(\hat{\mathbf{w}}_{\mathrm{L}_2}^{(c)},\ \hat{\mathbf{F}}_{\mathrm{L}_2}^{(c)}\right)^{\mathrm{T}} = \left(\hat{\mathbf{w}}_{\mathrm{R}_2}^{*(k+1)},\ \hat{\mathbf{F}}_{\mathrm{R}_2}^{*(k+1)}\right)^{\mathrm{T}}. \tag{27}$$

Combining Eq. 22, Eq. 26 and Eq. 27, assume that higher order parts for k and k+1 are $\mathbf{0}$. Define $\mathbf{P}^{(k)+}$=$\mathbf{I}$, $\mathbf{P}^{(k)-}$=$\mathbf{R}$, $\mathbf{P}^{(k+1)+}$=$\mathbf{T}$, the $\mathbf{R}$ and $\mathbf{T}$ can be conformed as:

$$\mathbf{G} = \mathbf{X}^{(c)}\mathbf{H}, \tag{28}$$

where $\mathbf{G}$=$[\boldsymbol{\Psi}_{\mathrm{u}}^{(k+1)+}\upsilon^{(k+1)+}|_{\overline{x}=0}\mathbf{T}, \hat{\mathbf{w}}_{\mathrm{R}_2}^{(k+1)}, \boldsymbol{\Psi}_{\mathrm{F}}^{(k+1)+}\upsilon^{(k+1)+}|_{\overline{x}=0}\mathbf{T}, \mathbf{0}]^{\mathrm{T}}$, $\mathbf{H}$=$[\boldsymbol{\Psi}_{\mathrm{u}}^{(k)+}\upsilon^{(k)+}|_{\overline{x}=L_k}\mathbf{I} + \boldsymbol{\Psi}_{\mathrm{u}}^{(k)-}\upsilon^{(k)-}|_{\overline{x}=L_k}\mathbf{R}, \hat{\mathbf{w}}_{\mathrm{L}_1}^{(k)}, \boldsymbol{\Psi}_{\mathrm{F}}^{(k)+}\upsilon^{(k)+}|_{\overline{x}=L_k}\mathbf{I} + \boldsymbol{\Psi}_{\mathrm{F}}^{(k)-}\upsilon^{(k)-}|_{\overline{x}=L_k}\mathbf{R}, \mathbf{0}]^{\mathrm{T}}$.

4 Numerical Applications

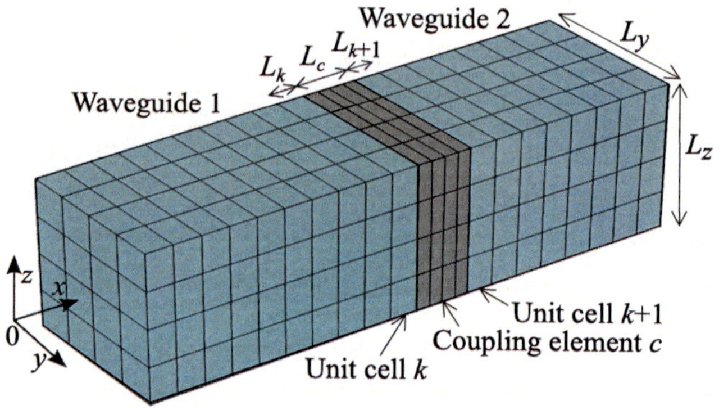

Fig. 2. Finite element model of two waveguides coupled by a coupling element.

In this part, the WFEM is applied to analyze the multi-mode diffusion. The unit cells k and k+1 with $L_{k/k+1} = 50a_0$, $L_y = 300a_0$ and $L_z = 300a_0$ (a_0 is the lattice parameter) as presented in Fig. 2. The coupling element with $L_c = 100a_0$ as shown in Fig. 2. Materials Aluminum (Al) and Copper (Cu) are used here. The Young's modulus E is 70 GPa for Al and 110 GPa for Cu, linear mass density ρ is 2.7 g/cm^3 for Al and 8.96 g/cm^3 for Cu. he damping lose factor η=1e^{-4}. Unit cells k and k+1 are meshed into 16 3D elements, coupling element is meshed 64 3D elements.

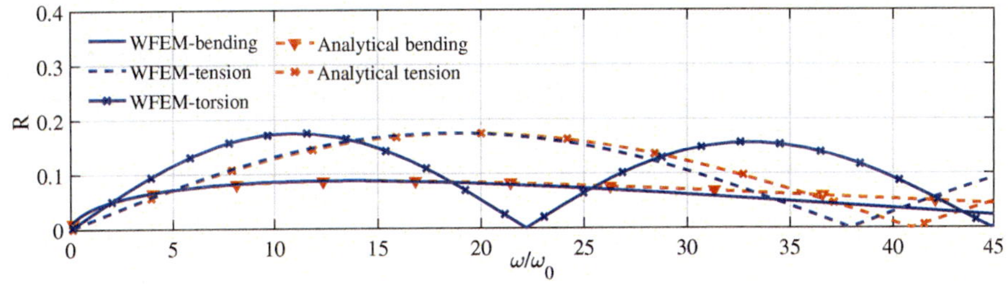

(a) Reflection coefficients (R).

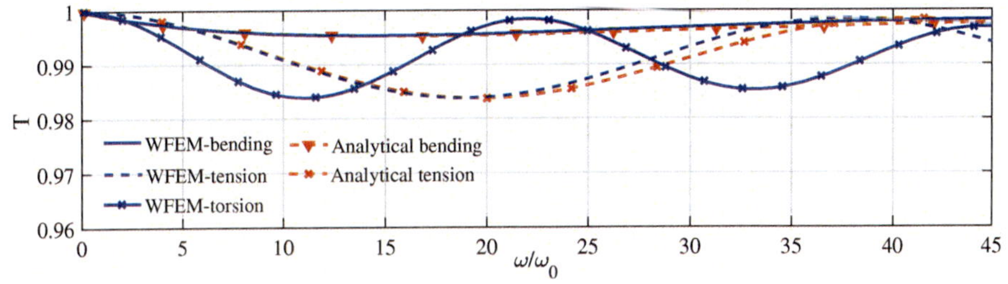

(b) Transmission coefficients (T).

Fig. 3. Absolute values of diffusion coefficients (the materials of coupling element, unit cell k and $k+1$ are Al).

In order to study the wave diffusion under a complex coupling condition, as shown in Fig. 2, first of all, defining the the materials of coupling element, unit cell k and $k+1$ are Al, the diffusion model of coupling element is built by SSG theory with higher-order parameters, the diffusion models of unit cell k and $k+1$ are built by CT. The $\mathbf{R}$ and $\mathbf{T}$ coefficients can be calculated from Eq. 28 including bending, tension and torsion modes. As shown in Fig. 3, the black lines denote the WFEM results. The value of $\mathbf{R}$ representing the non-classical part of reflection is no longer 0, and the value of $\mathbf{T}$ representing the non-classical part of transmission is no longer 1. The influence of non-local interactions caused by higher-order parameters can be reflected by this model. On the other hand, an analytical method [19], shown by the red lines, is used to valid the WFEM results for bending and tension. As we can see, for the reflection coefficient, the result obtained by WFEM is very close to the one by the analytical method at low frequency, but the results are different at high frequencies. For the transmission coefficient, the results by WFEM matches the results by analytical method well.

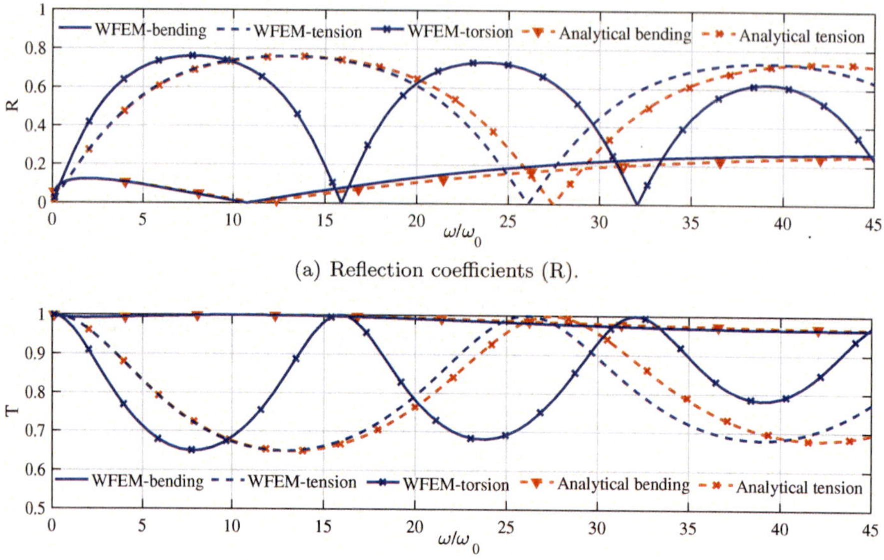

(a) Reflection coefficients (R).

(b) Transmission coefficients (T).

Fig. 4. Absolute values of diffusion coefficients (the material of coupling element is Cu, unit cell k and $k+1$ are Al).

In addition, the joint influence of classical parameters (i. g., Young's modulus, Poisson's ratio and mass density) and higher-order parameters of material on diffusion is also a very meaningful study. Defining the material of coupling element is Cu, unit cell k and $k+1$ are Al, the diffusion model of coupling element is built by SSG theory with higher-order parameters, the diffusion models of unit cell k and $k+1$ are built by CT. The **R** and **T** coefficients including bending, tension and torsion modes can be illustrated by Eq. 28 as well. As shown in Fig. 4, the black lines denote the WFEM results and the red lines represent the results from an analytical method [19]. The diffusion is different from the case presented in Fig. 3, this shows that the impedance mismatch is not only due to the non-local interactions caused by higher order parameters in the SSG theory model but also the local interactions caused by classical parameters.

5 Conclusions

In this paper, SSG theory is used for the multi-mode diffusion analysis within the WFEM framework. The diffusion is confirmed through two different cases. For the first case, the value of **R** representing the non-classical part of reflection is no longer 0, and the value of **T** representing the non-classical part of transmission is no longer 1. The influence of non-local interactions caused by higher-order parameters can be reflected by this model. The second case shows that the impedance mismatch is not only due to the non-local interactions caused by higher order parameters in the SSG theory model but also the local interactions caused by classical parameters such as Young's modulus, Poisson's ratio and mass density.

Acknowledgments. This work is supported by LabEx CeLyA (Centre Lyonnais d'Acoustique, ANR-10-LABX-0060) of Université de Lyon. The research of B. Yang is funded by the China Scholarship Council (CSC).

References

1. Miller, R.E., Shenoy, V.B.: Size-dependent elastic properties of nano-sized structural elements. Nanotechnology **11**(3), 139 (2000)
2. Lim, C.W., He, L.H.: Size-dependent nonlinear response of thin elastic films with nano-scale thickness. Int. J. Solids Struct. **46**(11), 15–26 (2004)
3. Yang, B., Droz, C., Zine, A.M., Ichchou, M.N.: Dynamic analysis of second strain gradient elasticity through a wave finite element approach. Compos. Struct. (2020). https://doi.org/10.1016/j.compstruct.2020.11342
4. Kroner, E.: Elasticity theory of materials with long range cohesive forces. Int. J. Solids Struct. **3**, 731–742 (1967)
5. Eringen, A.: Simple microfluids. Int. J. Eng. Sci. **2**, 205–217 (1964)
6. Eringen, A.: Linear theory of micropolar elasticity. J. Appl. Math. Mech. **15**, 909–923 (1966)
7. Gurtin, M.E., Murdoch, A.I.: A continuum theory of elastic material surfaces. Arch. Ration. Mech. Anal. **57**(4), 291–323 (1975)
8. Lazar, M., Maugin, G.A., Aifantis, E.C.: Dislocation in second strain gradient elasticity. Int. J. Solids Struct. **43**, 1787–1817 (2006)
9. Mindlin, R.D.: Micro-structure in linear elasticity. Arch. Ration. Mech. Anal. **16**, 51–78 (1964)
10. Shodja, H.M., Ahmadpoor, F., Tehranchi, A.: Calculation of the additional constants for fcc materials in second strain gradient elasticity: behavior of a nano-size bernoulli-euler beam with surface effects. Appl. Mech. **72**(2), 021008 (2010)
11. Mindlin, R.D.: Second gradient of strain and surface tension in linear elasticity. Int. J. Solids Struct. **1**, 147–438 (1965)
12. Mahapatra, D., Gopalakrishnan, S.: A spectral finite element for analysis of wave propagation in uniform composite tubes. J. Sound Vibr. **268**, 429–463 (2003)
13. Innvedan, S., Fraggstedt, M.: Waveguide finite elements for curved structures. J. Sound Vibr. **312**, 644–671 (2008)
14. Mencik, J.M., Ichchou, M.N.: Wave finite elements in guided elastodynamics with internal fluid. Int. J. Solids Struct. **44**, 2148–2167 (2007)
15. Droz, C., Lainé, J.P., Ichchou, M.N., Inquiété, G.: A reduced formulation for the free-wave propagation analysis in composite structures. Compos. Struct. **113**, 134–144 (2014)
16. Mead, D.: A general theory of harmonic wave propagation in linear periodic systems with multiple coupling. J. Sound Vibr. **27**, 235–260 (1973)
17. Duhamel, D., Mace, B.R., Brennan, M.: Finite element analysis of the vibrations of waveguides and periodic structures. J. Sound Vibr. **294**, 205–220 (2006)
18. Torabi, J., Ansari, R., Bazdid-Vahdati, M., Darvizeh, M.: Second strain gradient finite element analysis of vibratory nanostructures based on the three-dimensional elasticity theory. Iran. J. Sci. Technol. Trans. Mech. Eng. **44**(3), 631–645 (2020)

19. Zhu, G., Zine, A., Droz, C., Ichchou, M.: Wave transmission and reflection analysis through complex media based onthe second strain gradient theory. Eur. J. Mech. A Solids (2021). https://doi.org/10.1016/j.euromechsol.2021.104326
20. Rosi, G., Placidi, L., Auffray, N.: On the validity range of strain-gradient elasticity: a mixed static-dynamic identification procedure. Eur. J. Mech. A Solids **69**, 179–191 (2018)
21. Ahsani, S., Boukadia, R., Droz, C., Claeys, C., Deckers, E., Desmet, W.: Diffusion based homogenization method for 1D wave propagation. Mech. Syst. Signal Process. **136**, 106515 (2020). https://doi.org/10.1016/j.ymssp.2019.106515
22. Khakalo, S., Niiranen, J.: Form ii of mindlin's second strain gradient theory of elasticity with a simplification: For materials and structures from nano- to macro-scales. Eur. J. Mech. A Solids **71**, 292–319 (2018)
23. Zhao, J.Q., Zeng, P., Pan, B.: Improved hermite finite element smoothing method for full-field strain measurement over arbitrary region of interest in digital image correlation. Opt. Lasers Eng. **50**, 1662–1671 (2012)
24. Singh, R., Droz, C., Ichchou, M., Franco Bareille, F., De Rosa, O.S.: Stochastic wave finite element quadratic formulation for periodic media: 1D and 2D. Mech. Syst. Signal Process. **136**, 106431 (2020)

SDI: Structural Damage Identification

A Conceptual Design for Underground Hydrogen Pipeline Monitoring System

Jae-Woo Park and Dong-Jun Yeom(✉)

Korea Institute of Civil Engineering and Building Technology, Goyang-si, South Korea
djyeom@kict.re.kr

Abstract. Recently, the interest in using hydrogen energy in city planning has been increasing due to the demand for environmental preservation. The primary objective of this study is to develop a conceptual design of an underground hydrogen pipeline monitoring system that improves the underground pipeline in safety, stability, and responsiveness. For this, the following research works are conducted sequentially; 1) definition of sensing information, 2) literature review, 3) deduction of conceptual design and its work process. As a result, DAS (Distributed Acoustic Sensing), DTS (Distributed Temperature Sensing) are selected as core technologies. Furthermore, a conceptual design and work process of the underground hydrogen pipeline monitoring system is developed based on the core technologies. It is expected that the application range and impact on the construction industry will be enormous due to the increasing interest in using hydrogen energy.

Keywords: Hydrogen · Underground · Pipeline · Monitoring system · Conceptual design

1 Introduction

Recently, as the importance of eco-friendly energy has been increased, hydrogen energy is in the spotlight as future energy. Due to its special properties, hydrogen gas is more difficult to detect than similar combustible gases (CH4, C3H3, etc.), requiring more precise sensing technology. Hydrogen gas has the fastest flame rate, so if even a small amount leaks into the air, it can easily explode even in small ignition heat, so it is essential to secure stability against hydrogen (Han 2010).

To detect hydrogen gas, low concentration/high concentration detection should be possible, gases other than hydrogen should not be affected, and water vapor and temperature should be durable. In addition, it must have a lifespan of more than 5 years, high accuracy, rapid reactivity, high reliability, low operating temperature, mass production, small size, low power consumption, and low price (Lee 2007; Gerard and Meiger 2008; Grimes 2006).

The purpose of this study is to deduct the conceptual design of the sensor system for safety monitoring of the underground buried hydrogen pipeline. For this, the following research works are conducted sequentially. 1) Definition of sensing information, 2) literature review related to the detection of underground pipeline interference and

Z. Dimitrovová et al. (Eds.): WMVC 2022, MMS 125, pp. 775–782, 2023.
https://doi.org/10.1007/978-3-031-15758-5_79

leakage locations, 3) analysis of related regulations, 4) case study of measurement data (temperature, vibration, noise, etc.) sensing technology.

2 Literature Review

2.1 Definition of Sensing Information for Underground Pipeline Monitoring

In cities, an underground pipeline is an essential social resource linked to the lives of citizens. However, urban disasters can occur due to invisible factors such as leakage of underground pipelines, deformation of underground structures, changes in groundwater level, and ground deformation (road subsidence, under-ground buildings). In particular, as the number of accidents as above has increased rapidly in recent years, it is urgent to prepare measures to prevent human casualties and economic losses (NST 2014).

To prevent accidents due to hydrogen gas, it is important to first know the properties of hydrogen gas. Hydrogen gas is a single molecule formed by bonding two hydrogen atoms, the smallest gas molecule present on Earth, and may explode by a rapid reaction with oxygen. It is eco-friendly energy that generates high temperatures without pollution after combustion. However, the boiling point is low at −253 °C, so it exists as a gas at room temperature and atmospheric pressure. That is, even if it is transported in a liquid form in the pipeline, it is likely to leak in the gas, and the risk of explosion is high even in fine static electricity, handling difficult. Therefore, before handling hydrogen energy, securing a safety monitoring system for production, storage, transportation, and use must be preceded.

The following two pieces of information are very important for monitoring the underground hydrogen pipeline. The first is the various external factors that can cause damage to the pipeline. In other words, when an external factor approaches the pipeline, the location and impact strength should be identified through the monitoring system. The second is to identify changes caused by damage to the pipeline itself. The information to be identified is damage or deformation of the pipeline, above all, it is important to detect hydrogen gas leaking from the pipeline. In other words, it is necessary to find the location of the leak, stop hydrogen transport immediately, and quickly repair the pipeline.

2.2 Literature Review of Underground Pipeline Interference Sensing

DFOS (Distributed Fiber Optic Sensing) is typically used as a technology for detecting underground pipeline interference. It is cost-effective because sensing is possible only by the install of additional cables and has high durability. In particular, since it is specialized for sensing long-distance linear objects, the entire pipeline can be monitored in real-time.

In particular, DAS (Distributed Acoustic Sensing)-based vibration detection technology is specialized in pipeline impact and interference monitoring. However, DAS-based signals are difficult to process algorithmically due to various noises, weak signals, and signal fluctuations in the processing. Low frequencies enable rapid processing but require high specifications for algorithms due to limitations in data transmitted per unit time. Since Tanimola and Hill (2009) proposed DAS technology first to improve the above problems, continuous research has been conducted applying machine learning algorithms (Kabir et al. 2016; Gou et al. 2018; Wu et al. 2019; Yang et al. 2019).

In other words, DAS-based technology has been studied to detect underground pipeline interference, and it was analyzed that machine learning-based detection methods were applied to improve sensing algorithms.

2.3 Literature Review of Underground Leakage Locations Sensing

Gas sensors commonly used for monitoring, measuring, and controlling various parameters are devices that detect the concentration of gas in a specific environment. To this end, electrochemical, MOS (Metal Oxide Semiconductors), catalysts, infrared rays, lasers, and MEMS (Micro-Electronic Mechanical Systems) sensing are used (INNOPOLIS 2020).

The global gas sensor market is expected to increase from $1.042 billion in 2019 to an average annual growth rate of 6.4%, reaching $1.418 billion by 2024. In particular, the hydrogen gas sensing market grows 2.4% annually, forming a market size of more than $20 million per year (MarketsandMarkets 2019). In this study, the types of sensors applicable to identify the leakage location of underground hydrogen pipelines are analyzed as shown in Fig. 1 below.

Meanwhile, DTS (Distributed Temperature Sensing) can be applied to identify the gas leakage location of the underground pipeline. Therefore, DTS and gas sensors were used in combination to strengthen the identification rate of the gas leakage location.

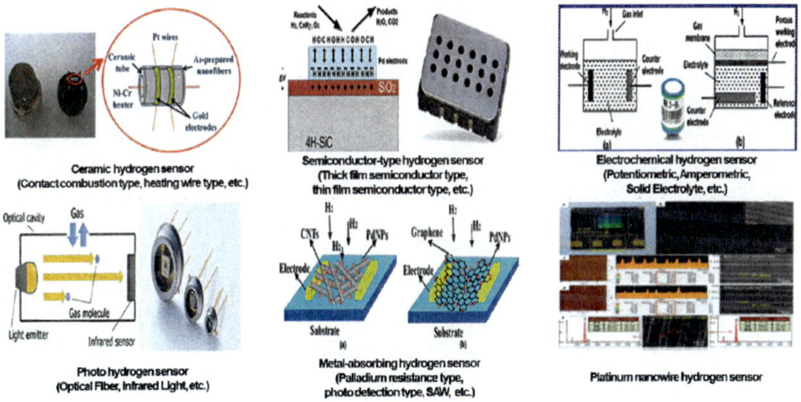

Fig. 1. Example of gas sensor type.

3 Analysis of the Underground Pipeline Monitoring Technology

3.1 Current Status of Underground Pipeline Construction in South Korea

According to the government audit data submitted from KGS (Korea Gas Safety Corporation), about 80% (3030) of the 3,825 gas pipeline buried works conducted in Korea from 2017 to 2019 were constructed differently from the design. The case where the depth of the pipeline was buried shallower than it was about 30 cm on average (KGS 2019).

In November 2019, the Ministry of Trade, Industry, and Energy established an enforcement rule that requires excavators to check whether or not pipelines are buried 24 h before construction to confirm the situation of urban gas pipes. However, 32 urban gas-related accidents occurred in 2019, of which 41% (13 cases) were excavation-related accidents. Accidents related to excavation work are classified as other construction and cannot manage their safety, so they are only at the level of informing the risk of preventive safety.

In Korea, the number of excavation accidents did not decrease to 7 in 2016, 7 in 2017, and 6 in 2018, and surged to 13 in 2019. In particular, the number of constructions received by the excavation work information support center was 334,561 in 2018 and 351,948 in 2019, an increase of 17,387 in a year, and the risk of related accidents is not small (Joo 2020).

In other words, there are currently many cases of construction error of underground pipelines, and more active safety management methods should be prepared than existing laws and regulations to improve such problems.

3.2 Case Study of Underground Pipeline Monitoring Technology

Recently, as DFOS technology has been actively used, pipeline safety management technology has been put into practical use. DFOS technology is not affected by the flow state in the pipe, so it is possible to obtain measurement data with minimized noise.

Some technology leaders provide solutions for multi-phase pipe monitoring and location recognition such as gas, water, liquid hydrocarbons, LNG, and LPG. Detailed technologies include third-party interference monitoring to prevent damage or theft of pipes, detection of pipeline rupture and fire, monitoring of pipeline overheat, detection of multi-phase pipeline leaks, and tracking of residues.

3.2.1 Fotech Solutions DAS Technology

DAS technology of Fotech Solutions monitors the entire pipeline by converting communication single-mode optical fibers into thousands of acoustic and vibration sensors (Fig. 1). A DAS connected to one end of the optical fiber uses a laser to send thousands of short pulses of light per second along with the optical fiber. Light moving from the optical fiber is reflected again by a Rayleigh backscatter.

The vibration of the surrounding environment is observed by DAS interlocutors because it interferes with the light of the optical fiber. At this time, since the data is processed in real-time, the algorithm recognizes the unique features of each event type and reports an alarm to the server based on sensing data that may be a problem. The system expresses the exact location of the threat to the operator and provides detailed information about the event.

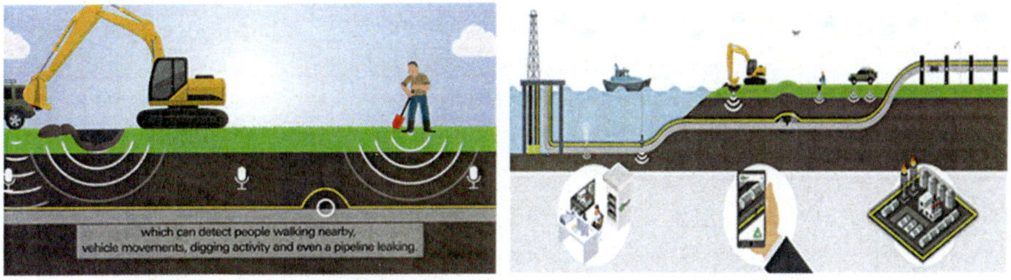

Fig. 2. Fotech solutions DAS technology (Fotech 2021).

3.2.2 SMARTEC : Pipeline Safety Monitoring Systems

The pipeline safety monitoring systems of SMARTEC utilizes DTS technology to detect and localize gas leakage location before pipeline damage occurs (Fig. 2). It can detect pipeline deformation of 30 km or more and erosion inside the soil. It provides an integrated solution of distributed optical fiber cables, measuring devices, and software. Also, it can identify leaks, intrusion, and deformation within a 1m accuracy range throughout the pipeline. This prevents accurate leakage area identification and major structural defects through temperature distribution analysis of underground pipelines. Since optical fibers are protected by stainless steel cables, the ease of installation and durability of cables are secured.

Fig. 3. SMARTEC: pipeline safety monitoring systems (SMARTEC 2021)

3.2.3 AP Sensing: Pipeline Monitoring (Leak Detection, Flow Assurance, Third Party Interference)

AP Sensing provides a solution that utilizes DAS and DTS systems in combination (Fig. 3). It is used to ensure safety in a wide range of fields, including oil refining and gas tank monitoring, fire detection, power cable monitoring, etc. It utilizes Smart Vision, a management software, to provide an integrated leak detection system with other sensors such as Mass Balance and Negative Pressure Wave LDS. High-accuracy data can be obtained in real-time from all sections of optical sensor cables, excellent durability without affecting EMI, long-distance measurements are possible, and there is no effect on flow conditions in the pipeline.

DTS accurately detects the location of hot spots/cold spots through temperature analysis and detects leakage of pipeline based on the data. DAS analyzes changes in noise/vibration, temperature, and low-pressure expansion waves to detect gas leaks, and detects third-party interference such as excavation, construction, and pipe drilling work (Fig. 4).

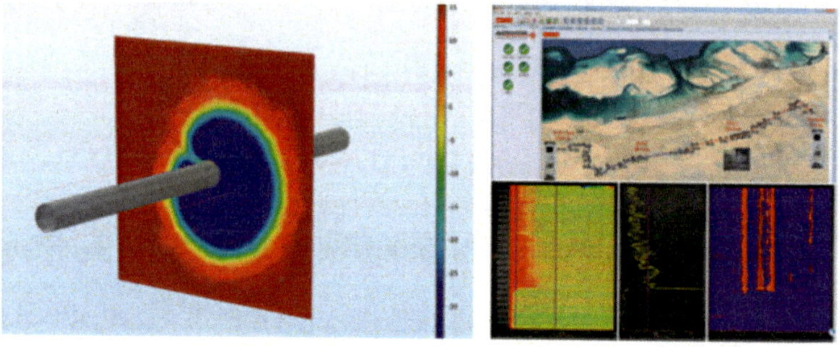

Fig. 4. AP sensing: pipeline monitoring (AP sensing 2021)

4 Conceptual Design for Underground Hydrogen Pipeline Monitoring System

In this study, two types of information: 1) underground hydrogen pipeline interference and 2) identification of the location of gas leakage in underground hydrogen pipelines were defined as main sensing information. Literature review and technology analysis result, the development direction was set by fusing DFOS (DAS, DTS) technology with additional sensing information.

4.1 Underground Hydrogen Pipeline Interference Monitoring

In this study, the technology to be applied to detect underground hydrogen pipeline interference is DAS. In order to detect pipeline interference using DAS, an accurate distinction between noise and interference is required. The underground pipeline is affected by various noises such as vehicle driving, pedestrian walking, and surrounding vibration. Therefore, to monitor the underground pipeline with DAS, a machine learning algorithm is applied to increase the recognition rate of interference.

4.2 Underground Hydrogen Pipeline Gas Leakage Location Monitoring

In this study, the technology to be applied to detect underground hydrogen pipeline leakage location is DTS. DTS has less impact on noise than DAS and has a higher recognition rate, so its usability is excellent. However, since hydrogen has a higher risk than other gases, the gas sensor is combined to strengthen the identification.

4.3 Conceptual Design for Underground Hydrogen Pipeline Monitoring System

The conceptual design for Underground Hydrogen Pipeline Monitoring System derived based on the above analysis results is shown in Fig. 5 below.

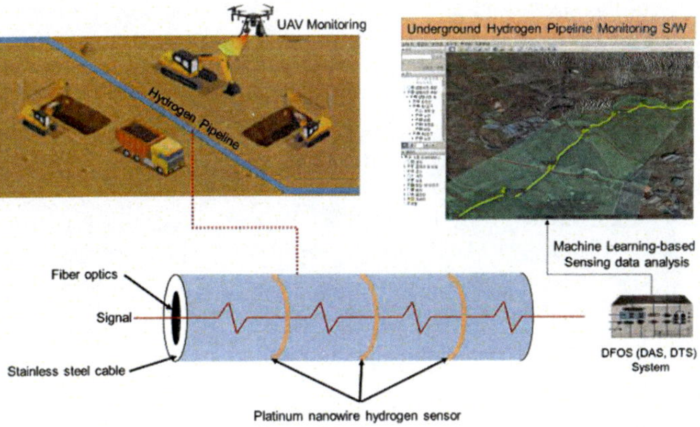

Fig. 5. Conceptual design for underground hydrogen pipeline monitoring system

5 Conclusion

The conclusions of this study are as follows.

1) Two types of information: 1) underground hydrogen pipeline interference and 2) identification of the location of gas leakage in underground hydrogen pipeline were defined as main sensing information.
2) Literature review: 1) underground pipeline interference sensing, 2) underground leakage locations sensing were conducted. It was analyzed that machine learning-based detection methods were applied to improve sensing algorithms. Also, DTS and gas sensors were used in combination to strengthen the identification rate of the gas leakage location.
3) The concept of the Underground Hydrogen Pipeline Monitoring System was designed by reflecting the results of the case study. In addition, monitoring technology using UAV was applied together.

The results of this study will contribute to activating the supply of hydrogen energy, which is attracting attention as eco-friendly energy. Further research is needed to improve DAS and DTS systems based on field sensing data.

References

Han, S.D.: Review and new trends of hydrogen gas sensor technologies. J. Korean Sens. Soc. **19**(2), 67–86 (2010)

Lee, D.D.: Air Environment Sensor Technology. Kyungpook National University Press (2007)

Meijer, G.C.M.: Smart Sensor Systems. Wiley, Hoboken (2008)

Grimes, C.A., Dickey, E.C., Pishko, M.V.: Encyclopedia of Sensors. American Scientific Publishers (2006)

National Science and Technology Research Association: Internet of Things (IoT)-based urban underground facility monitoring and management system technology development (2014)

Tanimola, F., Hill, D.: Distributed fibre optic sensors for pipeline protection. J. Nat. Gas Sci. Eng. **1**(4–5), 134–143 (2009)

Kabir, G., Sadiq, R., Tesfamariam, S.: A fuzzy Bayesian belief network for safety assessment of oil and gas pipelines. Struct. Infrastruct. Eng. **12**(8), 874–889 (2016)

Guo, X., Zhang, L., Liang, W., Haugen, S.: Risk identification of third-party damage on oil and gas pipelines through the Bayesian network. J. Loss Prev. Process Ind. **54**, 163–178 (2018)

Wu, H., Liu, X., Xiao, Y., Rao, Y.: A dynamic time sequence recognition and knowledge mining method based on the hidden Markov models (HMMS) for pipeline safety monitoring with OTDR. J. Lightwave Technol. **37**(19), 4991–5000 (2019)

Yang, Y., Li, J., Tian, M., Zhou, Y., Dong, L., He, J.X.: Signal analysis of distributed optic-fiber sensing used for oil and gas pipeline monitoring. In: Proceedings of the 2019 International Symposium on Signal Processing Systems, pp. 21–25 (2019)

INNOPOLIS: Market of Gas Sensor (annual report) (2020)

MarketsnasdMarkets: Gas Sensors Market (2019)

Korea Gas Safety Corporation: Trade, industry and energy small and medium venture business committee, the government audit report (2019)

Joo, B.: Urban gas piping construction, 80% of the last three years, unlike the design, have been constructed. The Environmental Transportation Times (2020)

Damage Index Implementation for Structural Health Monitoring

Alaa Diab(✉) and Tamara Nestorović

Mechanics of Adaptive Systems, Ruhr University Bochum, Bochum, Germany
{alaa.diab,tamara.nestorovic}@rub.de
https://www.ruhr-uni-bochum.de/mas/

Abstract. Since the requirements for constructing more stable and durable structures are increasing, structural health monitoring (SHM) is getting more importance nowadays. SHM relies on various methods to monitor a specific structure even live online or through direct observation of some structural parameters to decide whether a structure is damaged or not. Besides the need of getting reliable updates about the structure status, the need of making it more affordable for all types of projects is getting decisive. One of the simple implementations that are used for this purpose is to employ wave excitation at certain position of the structural part of interest and to observe the response at another predefined location of this part. After that, the observed wave propagation signal is used to calculate the damage index (DI) parameter which can be implemented in different algorithms for monitoring the current status and possibly predicting the remaining reliable operation of the structure. The signal processing procedure implemented in this work is based on the wavelet decomposition of the wave signal for the purpose of determining the energy of the propagating wave. This research is done as a part of a larger project that will study the possibility of detecting damages in different structures in general and in massive structures in specific. The paper investigates the possibilities and benefits of implementing the DI metric for monitoring the structural condition. The method will be demonstrated through a case study. Several models will be invoked to clarify its applicability. For the sake of clarity, the implementation will be done on 2D models. In addition, a 3D example based on finite element modeling should give the scope of using DI in a general case.

Keywords: Damage index (DI) · Wave propagation · Damage detection. · Wavelet decomposition

1 Introduction

Due to the increased demand of establishing sustainable structures, structural health monitoring (SHM) gets a great interest in the field of construction. SHM is

Funded by the Deutsche Forschungsgemeinschaft (DFG, German Research Foundation) – 448696650; 77309832

Z. Dimitrovová et al. (Eds.): WMVC 2022, MMS 125, pp. 783–791, 2023.
https://doi.org/10.1007/978-3-031-15758-5_80

an interdisciplinary field of engineering based on innovative methods of monitoring structural safety, integrity and performance without changing the structure performance or its operation [8]. In addition to the visually observable damages in structures, another art of hidden damages can occur [2]. Therefore, it is very important to get actual and definite information about the structure damage state. The most important category of the structural health monitoring relies on non-destructive methods (NDM) for damage detection within a specific structure either civil structures like hotels, bridges, or service projects such as factories, reservoirs, etc. [3].

Many non-destructive methods were implemented in the last decade such as acoustic emission testing (AE), electromagnetic testing (ET), laser testing methods (LM), microwave testing, ultrasonic testing (UT) [4]. These methods have a great contribution to the structural monitoring field with a remarkable performance. However, most of these methods are relatively expensive and demand special tools. Thus, the research is mainly focusing on developing a comparatively cheap method that provides a convenient detection of structural damage if exists. The method used in this work is based on using piezoelectric transducers (PZT) as actuators and sensors recording a wave propagating through the structure. After that, the wave record will be used to calculate the damage index which will give the notification of existing damage if its value is changed [7].

In this paper, the planned experiments are done as numerical simulations using finite element method (FEM) to get a simple implementation check of the Damage Index (DI) metric being used as an effective damage evaluation technique. The wave propagation simulation will be done based on the critical parametric conditions of choosing the element size and the time increment [1]. Through a case study, the calculation of the damage index will be demonstrated.

2 Methodology

In this paper, the wave propagation will be modeled as excitation force applied at a specific position which refers to an approximated piezoelectric transducer (PZT) due to the linear relationship between the electrical charge and displacement of the PZT [5]. This approximation provides reduction in the analysis cost.

Furthermore, the discrete wavelet decomposition will be implemented within the algorithm for calculating the one dimensional damage index along a signal path. A signal path is defined as the direct line connecting between a PZT point used as actuator and the receiving PZT point defined as a sensor. Finally, the algorithm for calculating the one-dimensional damage index will be presented.

2.1 Analysis and Modeling of the Wave Propagation

In this research, the wave propagation is generated at points where the PZT actuator is located in the real experiment. The excitation will be applied subsequently. Every propagated wave is mainly composed of bulk wave and surface wave. The bulk waves will propagate as a longitudinal wave (L) through the interior of the experimental models, while the surface waves (S) propagate on the surface of the model or between two surfaces of the model.

Since in the wave propagation based SHM, particularly in ultrasonic range, the lateral dimensions of the experimental models are greater than the propagated wavelength, the velocities of elastic waves and the mechanical properties of concrete are related through the following equation [5]:

$$c_L = \sqrt{\frac{\lambda + 2\mu}{\rho}}, \qquad c_S = \sqrt{\frac{\mu}{\rho}} \tag{1}$$

where, the Lame constants (μ, λ) in the previous equation can be calculated as follows:

$$\mu = \frac{E}{2(1+\nu)}, \qquad \lambda = \frac{E\mu}{(1--2\nu)(1+\nu)} \tag{2}$$

where the following notation is used:

c_L, c_S the velocities of the longitudinal and shear waves, respectively
E, ν, ρ Young's modulus, Poisson's ratio and material density respectively

For the modeling of the wave propagation, it is important to mention that the choice of the time increment and the element size play an important role in capturing the correct response, a detailed investigation regarding this is done previously [1].

2.2 Wavelet Decomposition

The wavelet decomposition can be performed based on two main methods: the continuous wavelet transform (CWT) and the discrete wavelet transform (DWT). The main difference between the two methods is that the DWT has a restricted choice of wavelets that apply 2-band reconstruction such as Daubechies, Coiflets, Spine, etc., while the CWT allows a huge range of admissible wavelets being chosen as a mother wavelet. However, one advantage of the DWT against the CWT is that its wavelets are often orthogonal or close to orthogonal, which simplifies the computations and inversions analysis.

In this research, the DWT is to be applied on the observed signals. In the DWT, the signal is to be decomposed into several levels. For each level, the signal $x(n)$ is passed through a series of filters including a low-pass filter g that gives the approximation coefficients A, and a high-pass filter h that results in the details coefficients D. Each decomposition level leads to a half of the time resolution and double of the frequency resolution. Furthermore, each decomposition level ends with down-sampling by 2 (see Fig. 1).

As the decomposition is repeated in multilevel, a filter bank is assembled from a binary tree of a sub-space with different time-frequency localization [9].

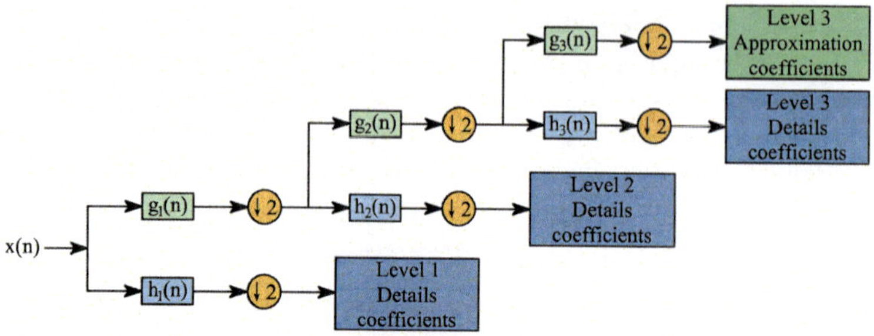

Fig. 1. A three-level filter bank (signal decomposition)

The decomposition process requires the input signal to be a factor of 2^n where n is the number of levels.

2.3 One-Dimensional Damage Index

The one-dimensional damage index calculation technique is based on the energy difference between the intact case of the specimen and its damaged case. This difference is calculated according to the root mean square deviation (RMSD) definition [6]. The energy is to be calculated from the decomposition of each input signal. Therefore, the starting point is by decomposing the input signal using n-level wavelet decomposition into 2^n signal sets where the mother wavelet is taken as Deubechies wavelet base db9:

$$\{X_1, X_2, \cdots, X_{2^{n-1}}, X_{2^n}\} \tag{3}$$

$$X_j = [x_{j,1}, x_{j,2}, \cdots, x_{j,m}] \tag{4}$$

where the vector of Xs includes the decomposed signals in total of 2^n level, and each of them is shown in Eq. (4) where m is the number of sampling data and x is the reading data.

Then the energy of each decomposed signal can be calculated using the following:

$$E_{d,k} = \|X_k\|_2^2 = x_{k,1}^2 + x_{k,2}^2 + \cdots + x_{k,m-1}^2 + x_{k,m}^2 \tag{5}$$

where d is the time index and k is the number of decomposed signals ($k = 1, \cdots, 2^n$). Then the resulting energy of each decomposition can be presented in a vector:

$$E_d = E_{d,1}, E_{d,2}, \cdots, E_{d,2^n} \tag{6}$$

$$E_h = E_{h,1}, E_{h,2}, \cdots, E_{h,2^n} \tag{7}$$

where the index d represent the damaged state at time index, and the index h refers to the intact state of the specimen. After that, applying the RMSD rule, we can get the damage index as follows:

$$DI = \sqrt{\frac{\sum_{p=1}^{2^n}(E_{d,p} - E_{h,p})^2}{\sum_{p=1}^{2^n} E_{h,p}^2}} \tag{8}$$

So we can see that the increase of the damage index value means that there is a damage increase in the specimen. Furthermore, the value of the damage index is ranged between 0 for the intact case and 1 for the completely damaged case.

3 Numerical Simulation

All the models in this research are designed using the finite element analysis (FEA) software ABAQUS. An explicit FEA is implemented for the wave propagation analysis. As a summary of the research process, the models are excited by the same signal many times from different positions on the boundary of the specimen. After that, for each excitation, the response in all other predefined points will be recorded. These will be processed to get the damage index for each direction using a MATLAB code. This code will be responsible for calculating the signals decompositions, the energy of all decompositions for both intact and damaged case, and subsequently, for calculating the corresponded damage index. Finally, the program will collect all the values of damage index of all directions in one matrix, which will be called the global damage index matrix.

A simplified determination of damage location will be detected as presented in the following paragraph. For the sake of clarity, the implementation on 2D models will be shown.

3.1 Building the Models

Two models were built due to the calculation requirements of the damage index The first model represents the intact state of the specimen, while the second model includes a damaged specimen with identical characteristics (geometry and material properties) as in the intact case except having a whole inside.

The concrete material defined in our 2D concrete plates models is considered as an isotropic elastic material. Both models have the same dimensions of (40 × 40 x 4) cm. For the signal generation and observation, 16 PZTs are attached to each of the specimens, see (Fig. 2).

3.2 Signal Generation

The excitation waves are considered as a 3.5-cycle Hanning-windowed tone burst applied for the duration of 3.5×10^{-5} s, which is derived from the following equation

$$P_{(t)} = \begin{cases} \left[1 - cos\left(\frac{2\pi}{N}t\right)\right] sin(2\pi f t), & \text{for} \quad 0 \leq t \leq \frac{N}{f} \\ 0, & \text{for} \quad t > \frac{N}{f} \end{cases} \tag{9}$$

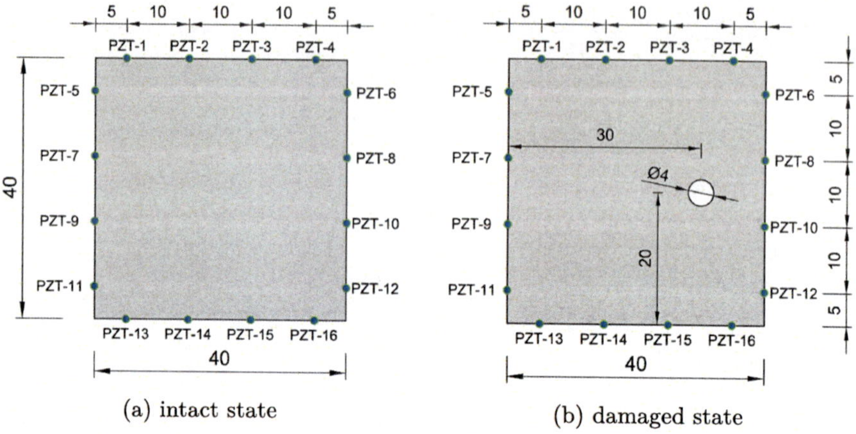

Fig. 2. Geometry of the concrete specimens + PZTs (dimensions in cm)

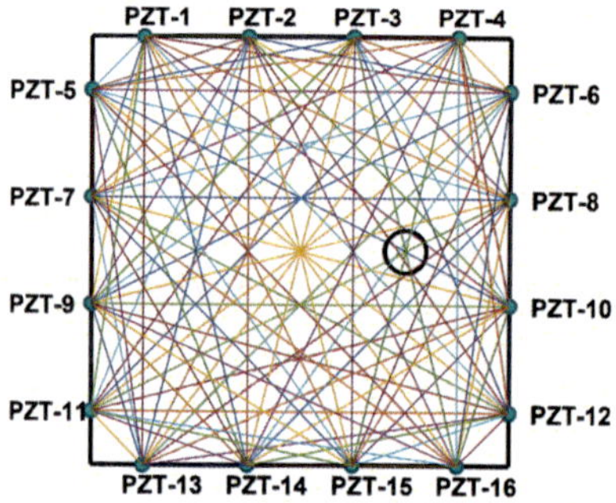

Fig. 3. Actuator-sensor directions.

where f is the signal frequency (100 kHz), N is the number of cycles (3.5 cycles), t the time in seconds, and $P_{(t)}$ is the excitation by the actuator.

The signal is applied first at PZT-1, and all other PZTs will record its propagated response. The same process will be applied for all other PZTs subsequently. In Fig. 3, all the actuator-sensor directions being observed through the experiment are shown.

3.3 Decomposition Implementation

As mentioned in the methodology, the DWT decomposition is applied for the signal decomposition. However, the Deubechies wavelet base db9 is used as a mother wavelet for the decomposition.

Each of the signals will be first subjected to a check, in which the first non zero value of both cases (intact and damaged) are not having the same amplitude. If this check is satisfied, it is assumed that a damage is present in the sample.

Thus the signal is passed through to be decomposed into 3 decomposition levels (see Fig. 1).

It is important to mention that the number of data samples of each signal direction of the intact and damaged model should be identical to guarantee a correct calculation of the signal decomposition.

4 Damage Index Calculation and Localization Technique

Along each actuator-sensor direction, the signal decompositions is used to calculate the energy of both models (intact model and damaged model) of the concrete plate.

According to Eq. (8), the damage index is then obtained and assembled in the global damage index matrix. A heatmap is presented to give a view on the values being calculated (see Fig. 4).

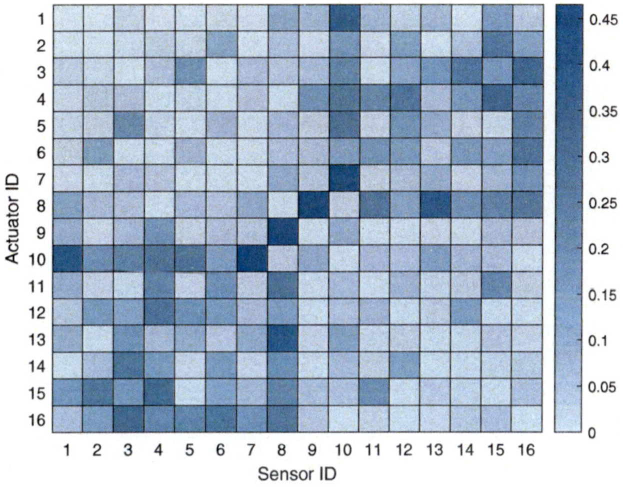

Fig. 4. Golbal damage index matrix

From the heatmap the largest value of the DI is 0.46 which refers to the DI being calculated from the signal sent from PZT-7 to PZT-10. Furthermore, an AS-pair from PZT-7 to PZT-10 performs the same as the AS-pair from PZT-10 to PZT-7 which proves that the global DI matrix is symmetric.

A simplified approach to make a first guess of the damage location is developed as followed. Since the values of the damage index can only be between 0 and 1, all the resulting values withing the global DI matrix are normalized so that the maximum calculated DI is equal to one. Furthermore, after the normalization is done, a threshold has been set to a specific value so that only the AS-pairs satisfying this threshold will be considered. Different threshold values are examined and the most reasonable values are presented (see Fig. 5).

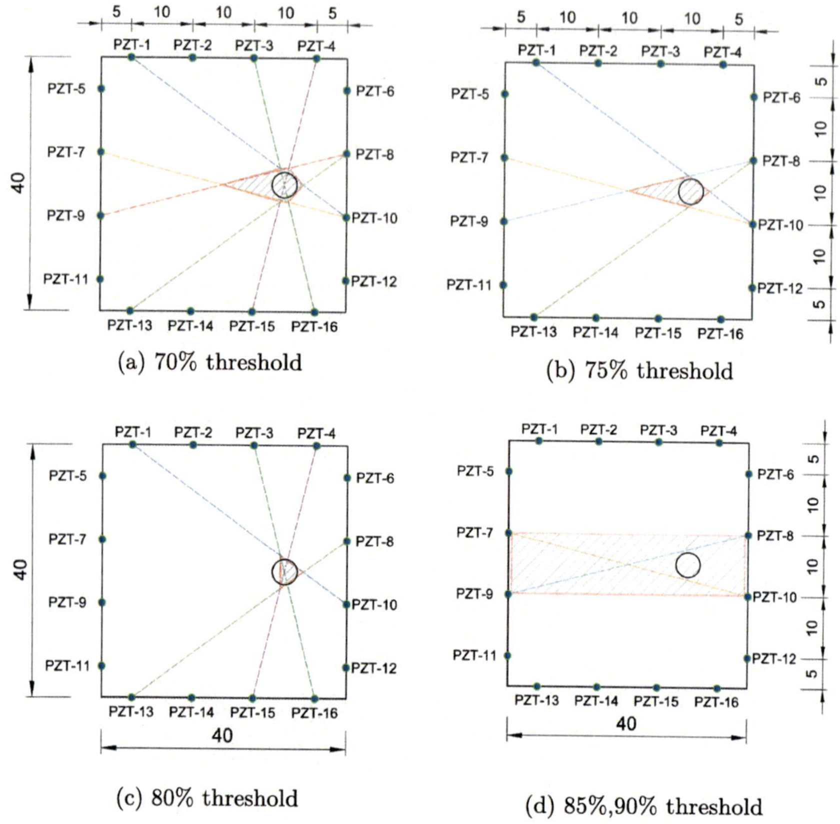

(a) 70% threshold (b) 75% threshold (c) 80% threshold (d) 85%,90% threshold

Fig. 5. A-S Directions satisfying the threshold for DI

After getting the filtered A-S directions, the intersection points between these directions are to be connected creating a polygon. This polygon is to be considered as the approximate location of the damage in the specimen.

From Fig. 5a and Fig. 5b it can be noticed that the resulting polygon (hashed) is including the location of the real damage. Furthermore, in Fig. 5c, the resulting polygon is even closer to the location and size of the real damage. However, using a high value of the threshold, in Fig. 5d, leads to uncertainty detecting the damage location.

5 Conclusion and Further Research Suggestions

Applying the 1D-damage index approach, we have got a detection of damage existence. For this method, it is always required to have the specimen observed before being damaged as a baseline.

Another useful result by the presented simplified technique is to get an initial damage localization. However, this method does not give an accurate information about the damage size and shape.

Further investigations is recommended regarding the threshold being considered for this technique. Examination of the method with different shapes and number of damages is also to be further experimented.

One drawback of this method is that it is time consuming when it comes to the signal recording of the numerical simulation. However, this method can be faster when the signals are directly collected from the laboratory experiment.

References

1. Diab, A., Nestorović, T.: Convergence study on wave propagation in a concrete beam. In: 11th International Workshop NDT in Progress (2021)
2. Vinores., C., F. Hezel II, R.: Hidden defects in concrete masonry construction: it's what's on the inside that counts (2021). https://www.envistaforensics.com/knowledge-center/insights/articles/hidden-defects-in-concrete-masonry-construction-it-s-what-s-on-the-inside-that-counts/, Accessed: 26 Feb 2022
3. Kot, P., Muradov, M., Gkantou, M., Kamaris, G.S., Hashim, K., Yeboah, D.: Recent advancements in non-destructive testing techniques for structural health monitoring. Appl. Sci. **11**(6), 2750 (2021)
4. The Welding Institute Ltd: What is non-destructive testing (ndt)? methods and definition (2019). https://www.twi-global.com/technical-knowledge/faqs/what-is-non-destructive-testing Accessed: 26 Feb 2022
5. Markovic, N., Nestorovic, T., Stojic, D.: Numerical modeling of damage detection in concrete beams using piezoelectric patches. Mech. Res. Commun. **64**, 15–22 (2015)
6. Marković, N., Nestorović, T., Stojić, D., Marjanović, M., Stojković, N.: Hybrid approach for two dimensional damage localization using piezoelectric smart aggregates. Mech. Res. Commun. **85**, 69–75 (2017)
7. Nestorović, T., Stojić, D., Marković, N.: Active structural health monitoring of reinforced concrete structures using piezoelectric smart aggregates. In: 8th European Workshop On Structural Health Monitoring (2016)
8. Stepinski, T., Uhl, T., Staszewski, W.: Advanced Structural Damage Detection: From Theory to Engineering Applications, Wiley (2013). https://books.google.de/books?id=VR-1MQEACAAJ
9. Wikipedia: Discrete wavelet transform (2004). https://en.wikipedia.org/wiki/Discret-e_wavelet_transform Accessed: 26 Feb 2022

Investigation of Tensile Behavior of SA 387 Steel Using Acoustic Emission Monitoring

Swadesh Dixit and Vikas Chaudhari(✉)

Department of Mechanical Engineering, BITS Pilani, Goa Campus, Goa, India
vikas@goa.bits-pilani.ac.in

Abstract. This work provides structural integrity of SA 387 pressure vessel steel subjected to tensile loading using acoustic emission (AE) monitoring. AE signals were detected and investigated using specialized system developed by Physical Acoustics. To detect the elastic waves released by the material under stress and owing to dislocation motion, two sensors were mounted on the surface of the specimen. AE parameters such as amplitude, energy, hits, counts, rise time and average frequency are considered in this study. Variations of these AE parameters are investigated across different regions on tensile curve such as micro-plastic deformation, yield, strain-hardening, necking and fracture by analyzing the overall acoustic signals. The results of this study revealed that intense AE signals are detected in micro-plastic as well as fracture regions. AE cumulative energy (AECE) and AE cumulative count (AECC) curves have shown sharp increment in micro-plastic deformation and fracture regions, whereas in rest of regions remained stable. The AE source is located using time difference of arrival (TDOA) method.

Keywords: Acoustic emission · Acoustic signal analysis · Tensile testing · Pressure vessel steel

1 Introduction

Low alloy steels are widely used as structural engineering material for pressure vessels [1]. The pressure vessels are generally operated under extreme conditions. The most of pressure vessels are designed to last several decades. During service life of pressure vessels, deformation damage is occurred due to material defects and overloaded operating conditions. Therefore, it's crucial to keep track and evaluate deformation damage of pressure vessel materials.

Acoustic emission (AE) monitoring is non-destructive method that can be effectively utilized for tracking and evaluating deformation damage of materials [2–4]. This technique is based on the physical response of structural member under external loading. AE is releasing of energy from discrete source such as

Supported by BITS Pilani

Z. Dimitrovová et al. (Eds.): WMVC 2022, MMS 125, pp. 792–798, 2023.
https://doi.org/10.1007/978-3-031-15758-5_81

a dislocation, on-set of crack and its propagation in materials which results in the formation of transient stress waves [5]. AE technique is widely utilized in construction, automotive and petroleum industries to monitor health of the structures. Great advantage of this technique over other Non destructive testing methods is, it is highly sensitive to damage of material and allows real-time monitoring of material degradation. Also, this technique is capable to investigate deformation and damage micro mechanism [6–8]. In middle of twentieth century, first attempt of AE investigations was performed to analyze plastic deformation and fracture of notched/un-notched specimens [7]. The different researcher have performed AE study for tensile tests to detect burst AE signals [9], identify AE activity generated due to dislocation movement [10] and micro-yielding [11]. AE features (energy, duration, frequency and amplitude) have also been linked to mechanical testing parameters (load, stress and growth of crack) [12–14].

The goal of this investigation is to assess AE monitoring for tensile test of SA 387 pressure vessel steel.The tensile test is carried out in conjunction with the associated AE monitoring system. Along with locating the AE source, AE parameters are linked with tensile data.

2 Experimental Details

In this study, SA 387 pressure vessel steel (quenched + tempered) was selected. Chemical composition of this steel was 0.16 C, 0.57 Mn, 0.021 P, 0.014 S, 0.54 Si, 1.37 Cr, 0.47 Mo and balance Fe. Flat tensile specimen was prepared as per ASTM E8 standard [15].

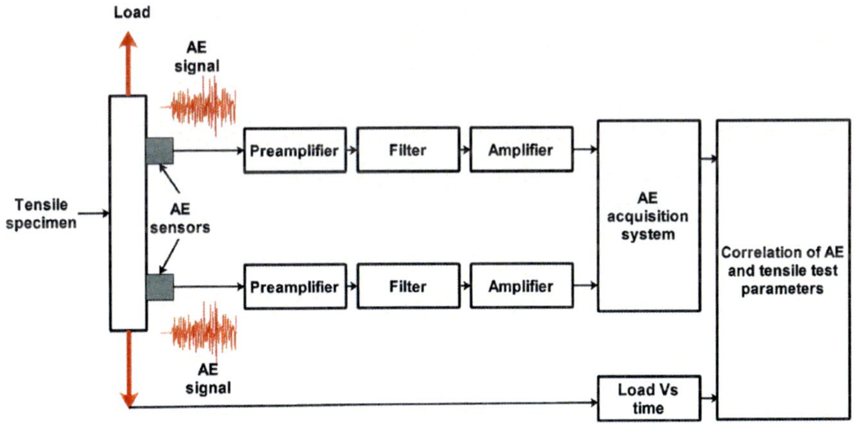

Fig. 1. Schematic representation of AE system for tensile test.

Schematic diagram of AE monitoring system is presented in Fig. 1. Sensor, pre-amplifier, filter, amplifier, and digital signal processing unit were incorporated in the AE monitoring system [5]. AE system developed by Physical Acoustic, USA was used for AE monitoring. AE monitoring setup used in current investigation is shown in Fig. 2. With the use of C-type clamps, two wide band sensors were mounted on the specimen at 80 mm distance (see inset image of Fig. 2). Sensors with a resonance frequency of 125 kHz and gain preamplifier of 40 dB were used to detect AE signals. Between sensors and specimen, vacuum grease was applied as coupling agent. Hit lock, hit definition and peak definition times were selected as 300 μs, 100 μs and 50 μs, respectively. Hsu and Nielsen's standard pencil-lead breakage test was used to acquire the time-driven parameters [16,17]. The dummy test was carried out by pre-loading tensile specimen. Based on findings of dummy test, threshold of 35 dB and suitable filter of 100–400 kHz were set to eliminate noise [18]. At room temperature, tensile test was performed at 1 mm/min strain rate.

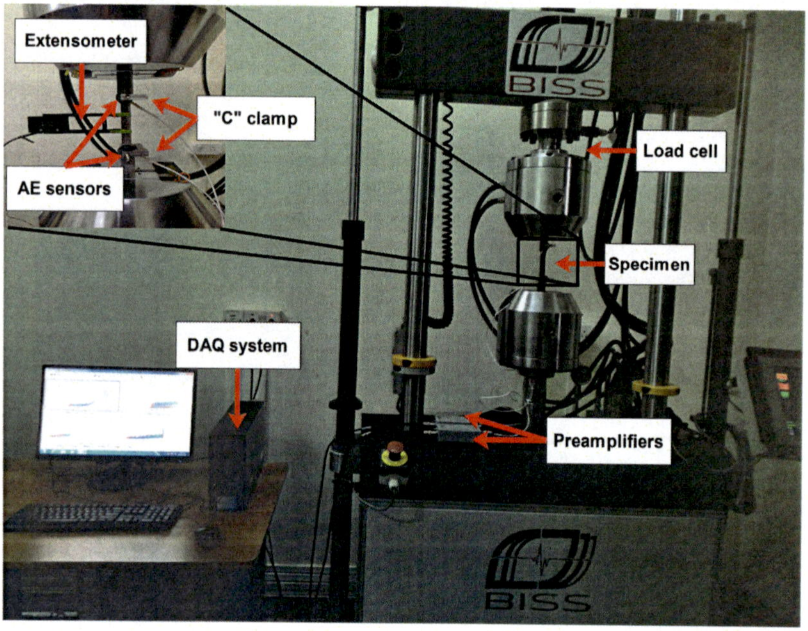

Fig. 2. AE experimental setup for tensile test.

3 Results and Discussion

Typical plot of stress, AECC and AECE versus time for the investigated steel is shown in Fig. 3. The tensile progress can be divided in to five regions: (1) micro-plastic deformation (2) yield (3) strain hardening (4) neck formation and (5) fracture. The rise of AECC and AECE are observed at micro-plastic deformation and fracture regions. However, AECC and AECE are remain stable in rest of the regions. Huang et al. [19] discussed variation of AECC and AE count rate across

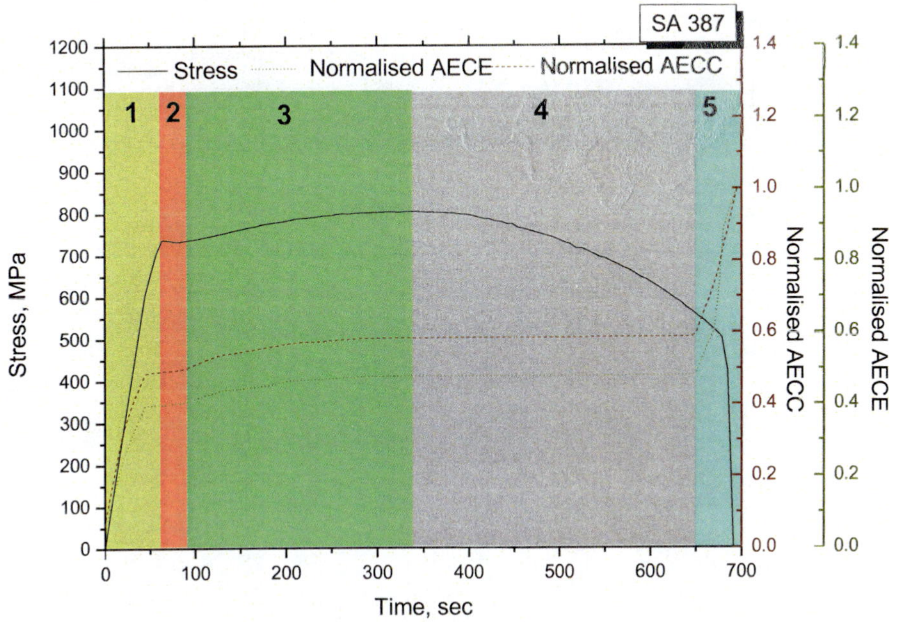

Fig. 3. Cumulative AE counts and energy Vs time curves in different tensile regions.

tensile regions. They observed peak values of AE parameters before yielding, however after that AE activities were decreased in rest of the stages till failure.

The change of AE parameters normalized values (i.e. amplitude, energy, rising time, average frequency and counts) across different tensile regions is presented in Fig. 4. In the micro-plastic deformation region, high values of rising time, and counts are detected, which could be attributed to heterogeneous dislocation movement [20,21]. Significantly high values of energy, counts, and rise time are obtained in the fracture region, which can be linked to void coalescence and damage [13]. Amplitude value has remained high in all regions.

AE source located based on time difference of arrival (TDOA) method [22]. The amplitude distribution of the recorded hits can be observed by analyzing the AE data. Histogram graph of AE hits versus amplitude versus position is shown in Fig. 5. Neck formation is observed at the center of specimen. Most of the hits occur at the place of neck formation which can be linked with movement of dislocation and damage. The occurrence of the most AE activity with the maximum amplitude values is demonstrated near to neck formation region.

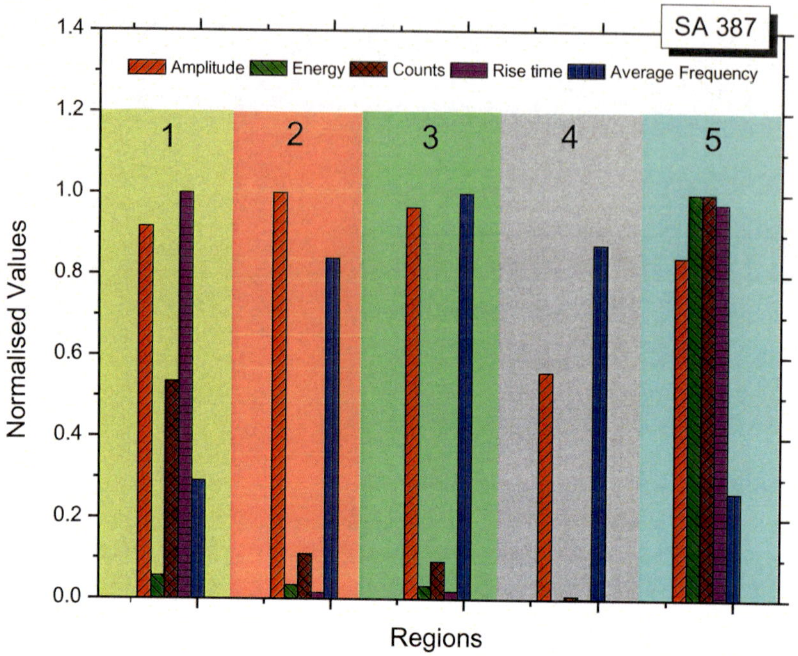

Fig. 4. AE parameters variation along different tensile regions.

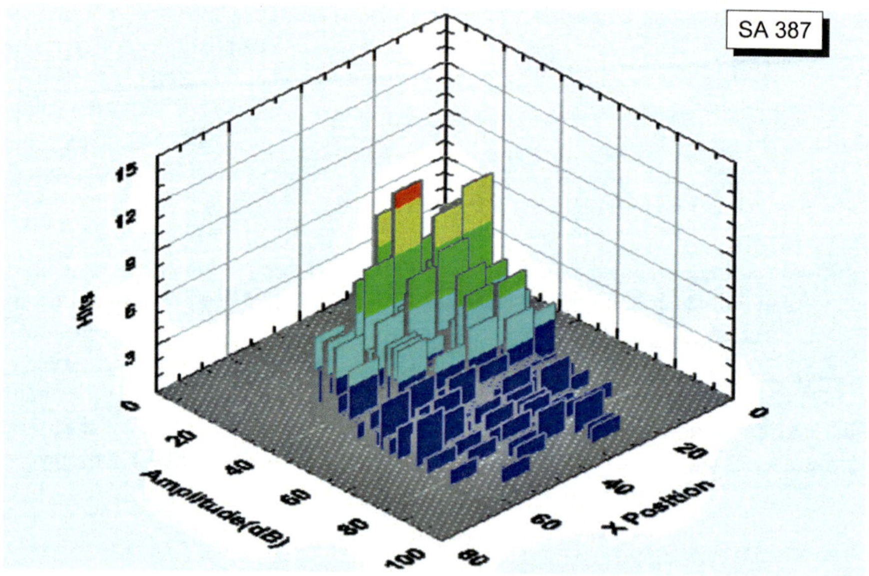

Fig. 5. Histogram plot representation of AE hits at a given position along given amplitude for tensile test.

4 Conclusion

In current study, acoustic emission monitoring is used to investigate deformation and damage process in SA 387 pressure vessel steel during tensile testing. The tensile plot divided in to five stages: micro-plastic deformation, yield, strain hardening, neck formation and fracture. The following conclusions are drawn:

- In the micro-plastic deformation region, there is a sharp rise in AE cumulative count and AE cumulative energy, as well as high values of rise time, and counts. This can be attributed to heterogeneous dislocation movement.
- Substantial increase of AECC and AECE along with significantly higher energy, count and rise time are detected in fracture region. This observation can be linked to void coalescence and damage.
- The AE activities are stable in rest of tensile regions.
- Most of AE activities were occurred near neck formation location with maximum amplitude.
- The AE monitoring has proven to be sensitive in detecting damage mechanisms at early stage.

References

1. Boiler, A.S.M.E., and Pressure Vessel Committee.: ASME Boiler and Pressure Vessel Code. American Society of Mechanical Engineers (2015)
2. Hwang, W., Bae, S., Kim, J., Kang, S., Kwag, N., Lee, B.: Acoustic emission characteristics of stress corrosion cracks in a type 304 stainless steel tube. Nucl. Eng. Technol. **47**(4), 454–460 (2015)
3. Babu, M.N., et al.: Study of fatigue crack growth in RAFM steel using acoustic emission technique. J. Constr. Steel Res. **126**, 107–116 (2016)
4. Krampikowska, A., Pała, R., Dzioba, I., Świt, G.: The use of the acoustic emission method to identify crack growth in 40CrMo steel. Materials **12**(13), 2140 (2019)
5. Nakamura, H., et al.: Practical Acoustic Emission Testing. Springer, Tokyo (2016). https://doi.org/10.1007/978-4-431-55072-3
6. Fallahi, A., Khamedi, R., Minak, G., Zucchelli, A.: Monitoring of the deformation and fracture process of dual phase steels employing acoustic emission techniques. Mater. Sci. Eng. A **548**, 183–188 (2012)
7. Lyasota, I., Kozub, B., Gawlik, J.: Identification of the tensile damage of degraded carbon steel and ferritic alloy-steel by acoustic emission with in situ microscopic investigations. Arch. Civil Mech. Eng. **19**(1), 274–285 (2018). https://doi.org/10.1016/j.acme.2018.09.011
8. Grosse, Christian U.., Ohtsu, Masayasu, Aggelis, Dimitrios G.., Shiotani, Tomoki (eds.): Acoustic Emission Testing. STCE, Springer, Cham (2022). https://doi.org/10.1007/978-3-030-67936-1
9. Ono, K.: Current understanding of mechanisms of acoustic emission. J. Strain Anal. Eng. Des. **40**(1), 1–15 (2005)
10. Mukhopadhyay, C.K., Rajkumar, K.V., Jayakumar, T., Raj, B.: Study of tensile deformation behaviour of M250 grade maraging steel using acoustic emission. J. Mater. Sci. **45**(5), 1371–1384 (2010)

11. Mukhopadhyay, C.K., Jayakumar, T., Raj, B., Ray, K.K.: The influence of notch on the acoustic emission generated during tensile testing of nuclear grade AISI type 304 stainless steel. Mater. Sci. Eng. A **276**(1–2), 83–90 (2000)
12. Kral, Z., Horn, W., Steck, J.: Crack propagation analysis using acoustic emission sensors for structural health monitoring systems. Sci. World J. (2013)
13. Han, Z., Luo, H., Wang, H.: Effects of strain rate and notch on acoustic emission during the tensile deformation of a discontinuous yielding material. Mater. Sci. Eng. A **528**(13–14), 4371–4380 (2011)
14. Roy, H., Parida, N., Sivaprasad, S., Tarafder, S., Ray, K.K.: Acoustic emissions during fracture toughness tests of steels exhibiting varying ductility. Mater. Sci. Eng. A **486**(1–2), 562–571 (2008)
15. ASTM E8, E8M,: Standard Test Methods for Tension Testing of Metallic Materials, p. 2016. PA, ASTM International, West Conshohocken (2016)
16. Chai, M., Zhang, J., Zhang, Z., Duan, Q., Cheng, G.: Acoustic emission studies for characterization of fatigue crack growth in 316LN stainless steel and welds. Appl. Acoust. **126**, 101–113 (2017)
17. Hsu, N.N.: Characterization and calibration of acoustic emission sensors. Mater. Eval. **39**, 60–68 (1981)
18. Chuluunbat, T., Lu, C., Kostryzhev, A., Tieu, K.: Investigation of X70 line pipe steel fracture during single edge-notched tensile testing using acoustic emission monitoring. Mater. Sci. Eng. A **640**, 471–479 (2015)
19. Huang, M., Jiang, L., Liaw, P.K., Brooks, C.R., Seeley, R., Klarstrom, D.L.: Using acoustic emission in fatigue and fracture materials research. J. Min. Met. Mater. Soc. **50**(11), 1–14 (1998)
20. Kotoul, M., Bilek, Z.: Acoustic emission during deformation and crack loading in structural steels. Int. J. Press. Vessels Pip. **44**(3), 291–307 (1990)
21. de Almeida, D.M., da Silva Maia, N., Bracarense, A.Q., Medeiros, E.B., Maciel, T.M., Santos, M.A.: Characterization of steel pipeline damage using acoustic emission technique. Soldagem and Inspeção **12**(1), 55–62 (2007)
22. Quy, T.B., Kim, J.M.: Crack detection and localization in a fluid pipeline based on acoustic emission signals. Mech. Syst. Sign. Process. **150**, 107254 (2021)

Modal Parameter Estimation in Transmissibility Functions from Digital Image Correlation Measurements

Ángel J. Molina-Viedma[1(✉)], Manuel Pastor-Cintas[1], Luis Felipe-Sesé[1], Elías López-Alba[2], José M. Vasco-Olmo[2], and Francisco Díaz[2]

[1] Departamento de Ingeniería Mecánica y Minera, Campus Científico Tecnológico de Linares, Universidad de Jaén, 23700 Linares, Spain
ajmolina@ujaen.es

[2] Departamento de Ingeniería Mecánica y Minera, Campus Las Lagunillas, Universidad de Jaén, 23071 Jaén, Spain

Abstract. High-speed digital image correlation is a camera-based displacement measurement technique that has experienced increasing popularity for modal analysis. It is due to a high spatial density of measurement that provides an outstanding interpretation of mode shapes or operational deflection shapes. However, its sensitivity is typically lower than other techniques and transducers such as accelerometers or laser vibrometry. With these measurements, the estimation of any transfer function in the frequency domain is noisier in valley regions of the response, including anti-resonances. Therefore, an accurate analytical model of the response should be used for modal identification. Modal analysis algorithms perform an identification of the modal parameters and a reconstruction of the response based on the transfer function between force excitation and displacement response. This is called the frequency response function. However, force measurements are not available in many experiments, where the motion excitation is recorded instead. This is the case when the vibration is applied to the specimen through a motion in its base. This transfer function, also known as the transmissibility function, relates the motion of excitation and response, whose shape differs from the frequency response function. This difference in the response function is relevant in a relatively noisy measurement like using digital image correlation. Hence, in this work, a transformation of the transmissibility function into equivalent frequency response function, based on theoretical models, is performed prior to modal identification in base motion tests. More suitable experimental response functions are obtained, increasing the accurateness of the modal parameter estimation.

Keywords: Modal analysis · Transmissibility functions · Full-field measurement

1 Introduction

In recent years, optical techniques have attracted the attention of the scientific community in vibration and modal analysis thanks to the development of high-speed cameras

Z. Dimitrovová et al. (Eds.): WMVC 2022, MMS 125, pp. 799–806, 2023.
https://doi.org/10.1007/978-3-031-15758-5_82

technology. The main advantage of these novel methodologies is that they perform contactless measurement without modifying the mechanical behavior of the system. With an appropriate image processing, full-field measurements are obtained. This represents an extremely high-density grid of virtual sensors, i.e., measurement points. Hence, useful data can be obtained with vision techniques which provide both alternative and complementary tools in modal analysis [1], quite useful for model updating [2].

One of the most used techniques is Digital Image Correlation (DIC). This technique has been widely employed for dynamic studies and also in the field of modal analysis. It has been employed for modal identification using both simplified single-degree-of-freedom [3] and multi-degree-of-freedom methods [4–6], and also for operational modal analysis [7, 8]. The modal identification algorithms are based in frequency response functions (FRF) [9], a transfer function between the displacement response and the force excitation. However, in many experimental situations is not possible to perform and measure pointwise forces to obtain FRFs at different coordinates. Applying a motion excitation in the base or support of the structure is one of the most common alternatives to force excitation. Hence, the vibration is transmitted from the base to the system and, now, the transfer function relates the motion of the measured degrees-of-freedom (DOF) of the structure and the motion of the base. This is known as the transmissibility function and its theoretical expression differs from the FRF's. Hence, using conventional identification without a convenient transformation of the data may entail erroneous modal parameter estimation [10].

This work evaluates modal identification on transmissibility functions where the full-field response is measured by 3D-DIC under base motion excitation. A transformation of the transmissibility functions is accomplished to suit classical identification procedure based on frequency response functions fitting [10, 11]. The comparison with the original transmissibility functions, as measured in the test, reveals how the obtained modal model fits better the experimental behaviour, especially concerning mode shapes and curve synthesis.

2 Base Excitation for Modal Analysis

When a mechanical system is subjected to motion excitation in its support, z, the governing equations for a system excited through the base are condensed in this matrix equation [10]:

$$[M]\{\ddot{x}\} + [C]\{\dot{x}\} + [K]\{x\} = -\ddot{z}\{g\}[M] \tag{1}$$

where [M], [C], [K] are the mass, damping, stiffness matrices [g] is a vector function of the geometry. The damping and stiffness forces, which depend on the relative motion between adjacent DOF, x, are defined by relative coordinates to the base frame as the base motion is a common term in the absolute motion of all the DOF. Conversely, inertial forces depend on the absolute motion, $\{\ddot{x}\} + \ddot{z}\{g\}$. However, it has been organized so that the left side of Eq. (1) is expressed in terms of the relative coordinates. Moreover, the right side is only function of the excitation.

On the other hand, when force excitation is applied at a certain coordinate, the transfer function between force magnitude, F, and displacement magnitude, X, depending on the harmonic term, ω, is known as frequency response function, $[H]$:

$$\{X\} = [H]\{F\} \tag{2}$$

This relation is the basis for experimental modal analysis, where FRFs are experimentally measured through the frequency-domain signals of displacements and loads.

In the base motion configuration, if the excitation is harmonic, the magnitude of the equivalent force can be expressed as:

$$\{F\} = -\ddot{Z}\{g\}[M] \tag{3}$$

and, therefore,

$$\{X\} = -[H]\{g\}[M]\ddot{Z} \tag{4}$$

Comparing Eqs. (2) and (4) and being $\{g\}$ and $[M]$ constant, it is found out that an FRF relating X and F is proportionally equivalent to the transmissibility function, $-[H]\{g\}[M]$, between the relative motion to the base, X, and the magnitude of the base motion acceleration, $\ddot{Z}$. Therefore, transmissibility functions in this form can be employed as regular FRFs in modal analysis.

3 Experimental Procedure

When base excitation is put into practice and DIC is employed to measure the response, the response coordinates are usually measured in terms of absolute motion. The aim of this work is to reveal the errors committed during the modal characterization in absolute response transmissibility functions using FRF-based identification algorithms. To carry out this, modal analyses for those transmissibility functions and those defined in terms of relative motion, equivalent to FRFs, were compared through modal parameters and synthesized response curves reconstruction.

In this study, a rectangular polycarbonate plate with dimensions 210 mm in length, 140 mm in width and 4 mm in thickness was designed through a modal analysis in a finite element model to contain several modes in a frequency range suitable for DIC. Experimentally, the plate was fixed at its central point to a shaker, through a rigid joint, as shown in Fig. 1 (a).

In this test, a random excitation signal was applied from 20–500 Hz. The base excitation was registered by an accelerometer placed on the shaker's armature. The response was recorded by a stereoscopic system consisting of two high-speed cameras (FastCam SA4 from Photron, 50 mm lenses, 1 megapixel) was employed. A schematic layout of the setup is shown in Fig. 1. The recording frame rate was 1000 fps. In total, two sequences of 5457 images each were recorded. A DAQ system (NI USB-6251 DAQ) was devoted to acquiring the accelerometer signal, synchronised with the high-speed cameras.

A commercial DIC algorithm was used (VIC-3D software by Correlated Solutions Inc) in order to process the images. This technique performs tracking of subsets of

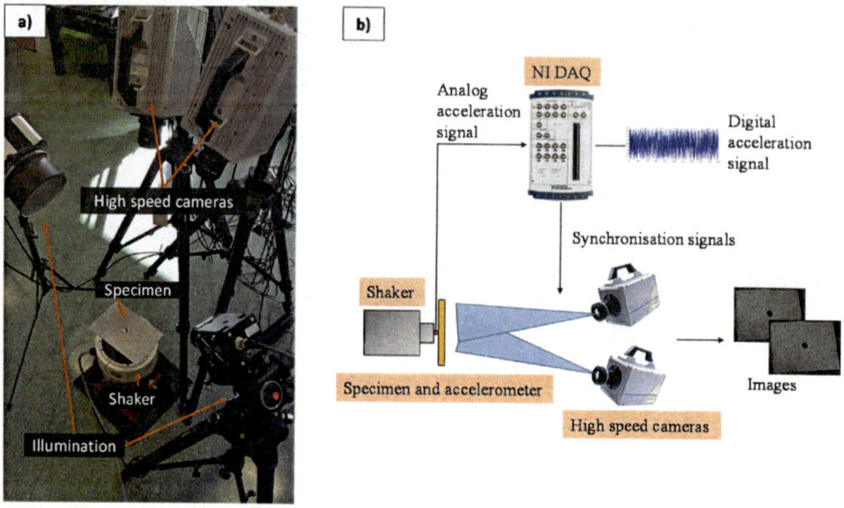

Fig. 1. (a) Picture and (b) scheme of the experimental set up.

the region of interest based on correlation criteria. For this purpose, a habitual procedure of specimen preparation was carried out, consisting in covering with a white paint background and, afterwards, adding black dots to make a random speckle pattern.

Afterwards, both the absolute and relative motion transmissibility functions were estimated and employed as input data for the characterisation of modal parameters using FRF-based method. Particularly, a standard identification method was considered in this study, the least-squares complex exponential (LSCE) [9].

At the first stage, the poles of the system, which contain the natural frequency and damping of the physical modes, were identified through the stability diagram.

Once the poles were extracted, a linear least-squares frequency-domain solver is used to obtain the residues in Eq. (1) by fitting the curve to the given experimental transmissibility functions. In this step, mode shapes are defined. In addition to the identification of the modal parameters, the synthetised transmissibility reconstruction was assessed.

4 Results

In view of the above, the absolute and the relative motion transmissibility functions were obtained. The first step of the analysis is the poles identification using a LSCE procedure. Maximum order of 15 was chosen to evaluate the stability of poles, avoiding the over-fitting of computational poles.

The resulting stability diagrams for each assumption are shown in Fig. 2. Three resonances can be visually identified in the average function from each data set. Preliminarily, certain differences are observed in the curves, more evident in the low-response regions between resonance peaks. It includes the low-frequency spectrum before the first resonance, which reveals the raising trend toward 0 Hz in the original transmissibility function, as a false pole. With both transmissibility functions sets, the three modes are quickly stabilized at a model order of about six using the identification algorithm. No great differences were found in the poles values between the relative motion data and the absolute one. It can be observed in the resulting natural frequency and damping ratio values, in Table 1.

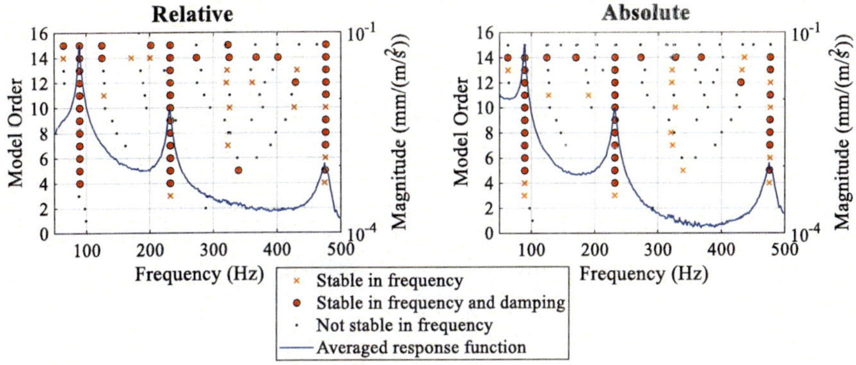

Fig. 2. Stabilisation diagrams for the relative and absolute motion transmissibility functions.

Table 1. Natural frequencies and damping ratios for the relative and absolute motion transmissibility functions.

	Natural freq. (Hz)		Damping ratio	
	Rel. motion	Abs. motion	Rel. motion	Abs. motion
Mode 1	89.616	89.666	1.30	1.31
Mode 2	231.579	231.576	1.01	1.01
Mode 3	475.331	475.387	1.02	0.99

Subsequently, using the identified poles, mode shapes could be obtained. The resulting maps are shown in Fig. 3, for each case, using amplitude normalization. As opposed to the natural frequencies and damping ratios, they show clearer differences between datasets. Despite the fact that there are differences in the first and second modes, the third mode, with its more complex and stiffer shape, exposes more clearly the differences between the original and adapted functions. This differences appear as an overestimation of the amplitude in the out-of-phase regions.

This observation can be confirmed by the curve synthesis analysis. In Fig. 4, the synthesized response curves of two points are depicted along with the experimental curves. One point is located at the corner, where a high response is expected in most of the modes. The second one is placed in the middle of the right half of the plate. It can be observed that the low-response areas are poorly fitted in the original data. For instance, the adapted data have allowed for a better fitting of the low-frequency spectrum, below the first mode. Moreover, it provided the appropriate representation of the anti-resonance before the second mode at the corner point. Furthermore, false

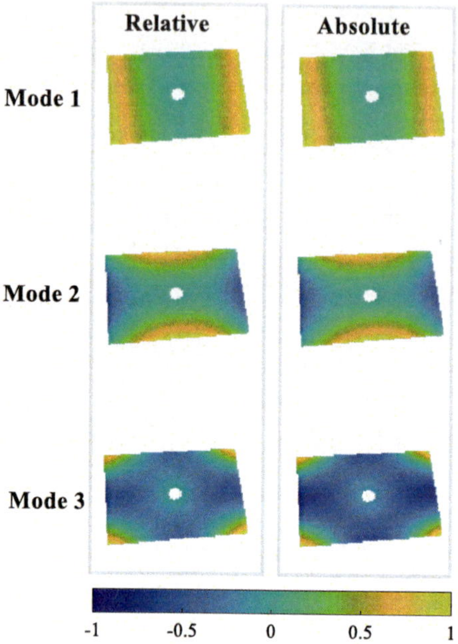

Fig. 3. Modal shapes obtained for the relative and absolute motion transmissibility functions.

anti-resonances are described by the synthesized curve for the original data around the third resonance peak. Besides these deficiencies, remarkable behavior is presented in the second plotted point regarding the third mode fitting. Whereas the peak amplitude is properly described for the adapted data, the estimation of the third mode amplitude in the original data is noticeable higher than experimental. This point is located in the region where the highest differences between the third mode estimation take place, according to Fig. 3. Therefore, it reveals the overestimation of amplitude in the third mode in these regions due to the non-suitable shape of the original curves for the FRF-based procedure.

Another indicator of the accuracy of the curve fitting is the error. In the corner point the minimum error is observed for both data set: 1.2% and 2.3% for the relative motion and absolute motion data. In the corners the higher displacement amplitudes are obtained and the committed error is low. Conversely, the error increases as approaching the fixation, where the lower displacements appear. The error in the second depicted point in Fig. 4 is 8,2% for the relative motion data whereas it reaches the 21,7% for the absolute motion data, up to 86% error in the fixation surroundings.

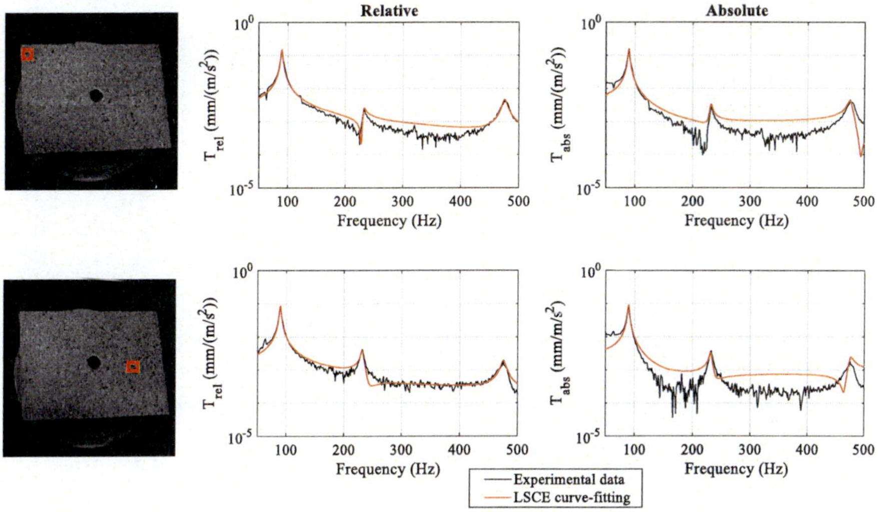

Fig. 4. Curve synthesis at two measurement points for the relative and absolute motion transmissibility functions.

5 Conclusions

This work addresses the analysis of the validity of the FRF-based algorithm for modal estimation for base excitation tests using DIC. It is motivated by the increasing popularity of vision techniques for their contactless full-field measurements. Moreover, base motion excitation is quite common in practice and requires specific treatment for the analysis.

3D-DIC was employed for the experimental test in order to determinate the full-field response of a plate. The convenient transformation of the transmissibility functions from absolute to relative response terms was performed to be employed for FRF-based identification methods. Namely, modal analysis was carried out applying one of the best-known modal identification procedure. The benefits of the transformation are not totally revealed in the natural frequencies and damping ratios estimation, as no major differences were found, but in the mode shapes estimation and the curve synthesis. The comparison of the mode shapes, enhanced by the full-field information of DIC, showed discrepancies in amplitude for the absolute response transmissibility functions. It was definitively confirmed by the analysis of the synthesized curves synthesis, where the error in the fitting procedure was found to be significantly higher for the absolute response transmissibility functions. This remarks the necessity of adapting the experimental data to obtain accurate modal models using FRF-based identification procedures.

References

1. Baqersad, J., Poozesh, P., Niezrecki, C., Avitabile, P.: Photogrammetry and optical methods in structural dynamics – a review. Mech. Syst. Signal Process, **86**, 17–34 (2017)
2. Wang, W., Mottershead, J.E., Ihle, A., Siebert, T., Reinhard Schubach, H.: Finite element model updating from full-field vibration measurement using digital image correlation. J. Sound Vib. **330**, 1599–620 (2011)

3. Trebuňa, F., Hagara, M.: Experimental modal analysis performed by high-speed digital image correlation system. Measurement **50**, 78–85 (2014)
4. Molina-Viedma, Á.J., Felipe-Sesé, L., López-Alba, E., Díaz, F.A.: Comparative of conventional and alternative digital image correlation techniques for 3D modal characterisation. Meas. J. Int. Meas. Confed. **151**, 107101 (2020)
5. Huňady, R., Hagara, M.: A new procedure of modal parameter estimation for high-speed digital image correlation. Mech. Syst. Signal Process **93**, 66–79 (2017)
6. Zanarini, A.: Full field optical measurements in experimental modal analysis and mod-el updating. J. Sound Vib. **442**, 817–42 (2019)
7. Chang, Y.-H., Wang, W., Chang, J.-Y., Mottershead, J.E.: Compressed sensing for OMA using full-field vibration images. Mech. Syst. Signal Process **129**, 394–406 (2019)
8. Uehara, D., Sirohi, J.: Full-field optical deformation measurement and operational modal analysis of a flexible rotor blade. Mech. Syst. Signal Process **133**, 106265 (2019)
9. Ewins, D.J.: Modal Testing: Theory, Practice, and Application, 2nd edn. Research Studies Press LTD., Baldock (2000)
10. Béliveau, J.G., Vigneron, F.R., Soucy, Y., Draisey, S.: Modal parameter estimation from base excitation. J. Sound Vib. **107**, 435–49 (1986)
11. Molina-Viedma, Á.J., Felipe-Sesé, L., Pastor-Cintas, M., López-Alba, E., Díaz, F.A.: Evaluation of modal identification under base motion excitation using vision techniques. Mech. Syst. Signal Process **179**, 109405 (2022)

Pounding Between High-Rise Buildings with Different Structural Arrangements

Mahmoud Miari(✉) and Robert Jankowski

Department of Construction Management and Earthquake Engineering, Faculty of Civil and Environmental Engineering, Gdańsk University of Technology, Gdańsk, Poland
mahmoud-miari@hotmail.com, mahmoud.miari@pg.edu.pl

Abstract. Earthquake-induced structural pounding has led to significant damages during previous earthquakes. This paper investigates the effect of pounding on the dynamic response of colliding high-rise buildings with different structural arrangements. Three 3-D buildings are considered in the study, including 5-storey building, 7-storey building and 9-storey building. Three pounding scenarios are also taken into account, i.e. pounding between 5-storey and 7-storey buildings, pounding between 5-storey and 9-storey buildings and pounding between 7-storey and 9-storey buildings. These three pounding scenarios are studied and compared with the no pounding case. The results show that the level of accelerations of colliding buildings significantly increases for all scenarios, as compared to the no pounding case. At the same time, displacements experience both increase and decrease, while the peak storey shears experience an increase due to pounding with few exceptions regarding the top storeys. Finally, pounding leads to the generation of dangerous impact forces with higher peak values experienced in taller buildings.

Keywords: Structural pounding · Buildings · Dynamic response · Earthquakes

1 Introduction

Pounding between adjacent structures is defined as the repeatable collisions that occur during earthquakes [1–5]. It leads to the generation of high impact pulses not considered in the engineering design. Pounding has been observed in many earthquakes, e.g. in the Mexico earthquake where 40% of the buildings experienced pounding and 15% of buildings with collapse or severe damage experienced pounding [6] where in 20–30% of them pounding was the major reason for damage [7]. Statistics of the Loma Prieta earthquake states that pounding was experienced in 200 buildings out of 500 surveyed structures [8]. Pounding was also observed in recent earthquakes, such as the Christchurch earthquake (New Zealand 2011) [9] and the Gorkha earthquake (Nepal 2015) [10].

Research on earthquake-induced pounding has been conducted for more than three decades (see, for example, two state-of-the-art papers [11, 12] for details). Pounding occurs as a result of the out of phase vibrations caused by the difference in the natural periods of colliding buildings due to the difference in their dynamic properties [13–15].

Z. Dimitrovová et al. (Eds.): WMVC 2022, MMS 125, pp. 807–816, 2023.
https://doi.org/10.1007/978-3-031-15758-5_83

It is enhanced when there is a significant difference in the dynamics properties (mass, height, periods) between the colliding buildings [16, 17]. Pounding was found to amplify the response of the flexible building and has insignificant effect on the response of the stiffer structure in some studies (see [18] for example) while it was found that it amplifies the response of the stiffer building and suppress the response of the flexible structure in other studies (see [19] for example). Indeed, the properties of the ground motion have a significant effect on buildings experiencing pounding [20]. The response of colliding buildings is significantly affected in the direction of pounding and unaffected in the other direction [21].

The aforementioned literature emphasizes the significance of pounding and illustrates that the effect of collisions on the response of buildings can be substantial. The aim of this paper is to study the effect of pounding on the dynamic response of the colliding buildings for different structural arrangements. In this paper, three 3-D buildings are considered, including 5-storey, 7-storey and 9-storey structures. Three pounding scenarios are also taken into account, i.e. pounding between 5-storey and 7-storey buildings, pounding between 5-storey and 9-storey buildings and pounding between 7-storey and 9-storey buildings. These three pounding scenarios have been studied and compared with the no pounding case.

2 Numerical Models of Colliding Buildings

All analysed buildings have a storey height of 3 m and the bay width is 4 × 4 m in each direction which leads to 16 m length and width. The models of buildings have been created in ETABS software [22] using the finite element (FE) method (see Fig. 1).

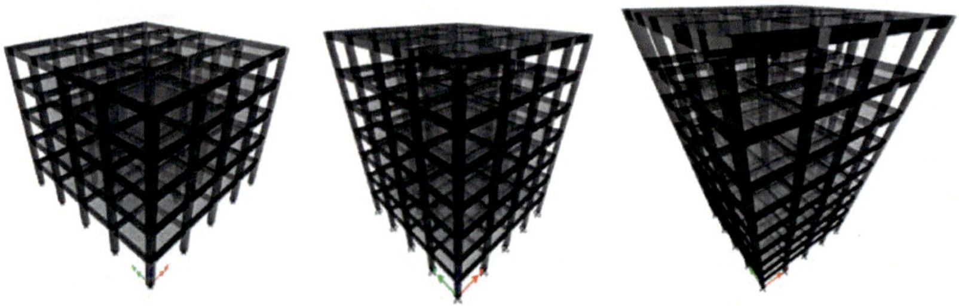

(a) 5-storey building (b) 7-storey building (c) 9-storey building

Fig. 1. FE models of the studied buildings

Three pounding scenarios have been considered in this study, i.e. pounding between 5-storey and 7-storey buildings (5–7 pounding), pounding between 5-storey and 9-storey buildings (5–9 pounding) and pounding between 7-storey and 9-storey buildings (7–9 pounding). The seismic gap has been considered as equal to 4 cm for all the cases. The collisions have been modelled in ETABS software using special gap elements. These gap elements are two-node compression-only link elements that are activated in the case of contact and de-activated elsewhere. These gap elements have been placed every 4 m along the collision length at all storeys (5 elements at each storey have been used).

The soil type B (rock), defined in the ASCE 7–10 code [23], has been used in all cases of pounding as well as in the no pounding cases. The soil type has been considered by defining the response spectrum for certain site class and matching the earthquake record with the target response spectrum. The soil parameters are selected as the parameters that lead to higher responses, which are 0.5 for S_1 (mapped risk-targeted MCE_R spectral response acceleration parameter at 1-s period), 1.25 for S_s (mapped risk-targeted MCE_R spectral response acceleration parameter at short period) and additionally 8 s for T_L (the long-period transition period) (see [24, 25]).

The nonlinear analysis has been carried out for the Parkfield earthquake of 1966 measured in San Luis Obispo station and with the peak ground acceleration of 0.01175 g (see PEER website database [26]). The structural response has been obtained applying the fast nonlinear analysis method developed by Ibrahimbegovic and Wilson [27]. In this method, the nonlinearity is considered for the gap and support elements (see [28] for details).

3 Pounding Between Buildings with Different Dynamic Properties Founded on the Same Soil Type

In this section, pounding is analysed for 5–7, 5–9, and 7–9 pounding scenarios and compared with the no pounding case. Figures 2, 3, and 4 show the acceleration time histories of the colliding buildings at the level of collision (top storey of the shorter building) for different scenarios. The results presented in these figures indicate that a significant increase in accelerations is experienced in all cases due to pounding. The amplification of the peak acceleration ranges between 5 and 21 times, as compared to that of the no pounding case which reveals the significance of pounding phenomenon. Figures 5, 6, and 7 show the displacement time histories of the colliding buildings at the level of collision for 5–7, 5–9, and 7–9 pounding scenarios, respectively, as compared with the no pounding case. It can be seen from the figures that displacements have experienced both amplification and de-amplification phases. The amplification and de-amplification of the peak displacement ranges between 1 and 2 times, as compared to that of the no pounding case (see Figs. 8, 9 and 10 which show the peak storey displacements of the 5–7, 5–9 and 7–9 pounding scenario, respectively). The increase and decrease trend is not only experienced at the level of collision but also at the level of all other storeys. The amplification of the displacement is due to the generated high accelerations. However, the decrease in the displacement in other cases is due to the fact that pounding blocks the motion of vibrating buildings.

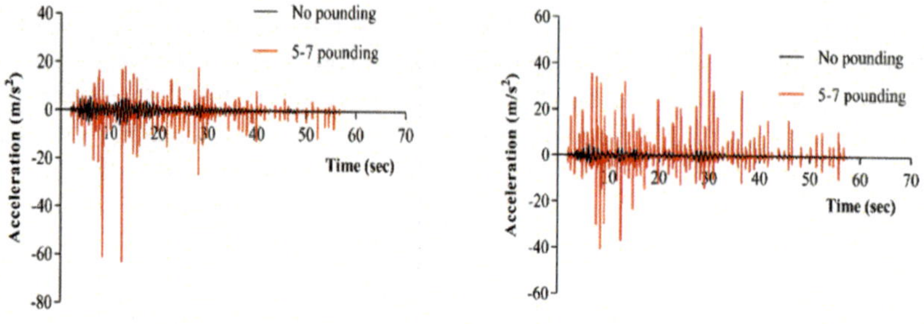

(a) 5-storey building (b) 7-storey building

Fig. 2. Acceleration time histories of the colliding buildings at the 5th storey for 5–7 pounding scenario, as compared to the no pounding case

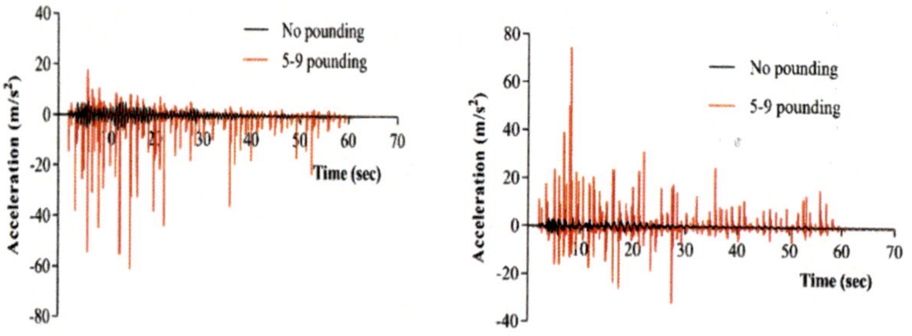

(a) 5-storey building (b) 9-storey building

Fig. 3. Acceleration time histories of the colliding buildings at the 5th storey for 5–9 pounding scenario, as compared to the no pounding case

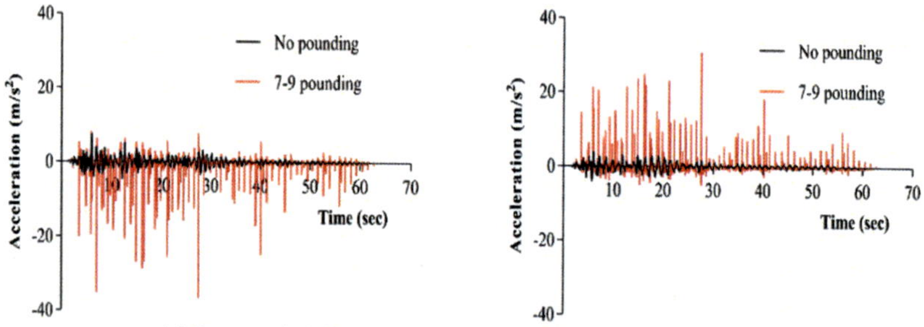

(a) 7-storey building (b) 9-storey building

Fig. 4. Acceleration time histories of the colliding buildings at the 7th storey for 7–9 pounding scenario, as compared to the no pounding case

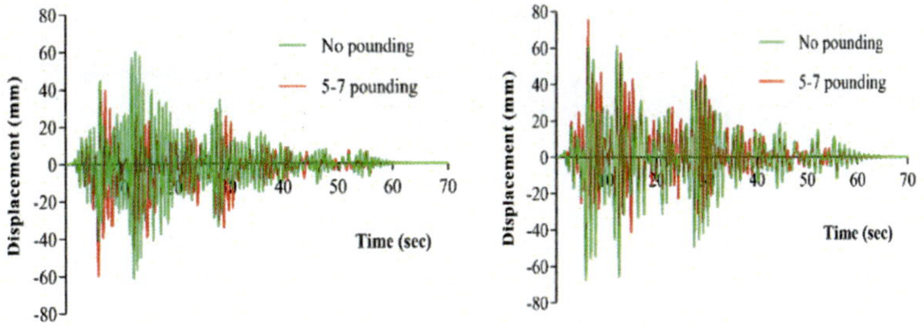

(a) 5-storey building (b) 7-storey building

Fig. 5. Displacement time histories of the colliding buildings at the 5th storey for 5–7 pounding scenario, as compared to the no pounding case

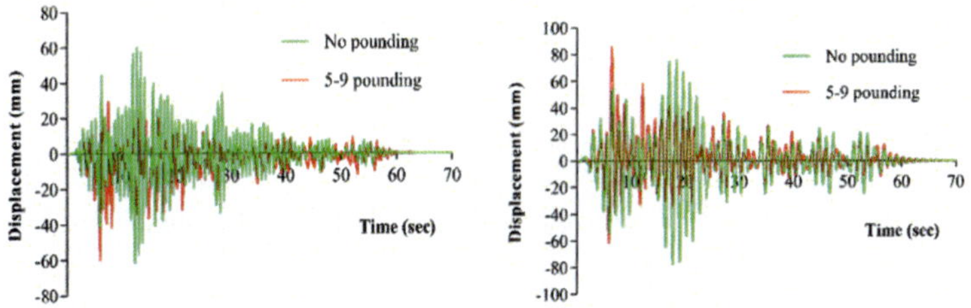

(a) 5-storey building (b) 9-storey building

Fig. 6. Displacement time histories of the colliding buildings at the 5th storey for 5–9 pounding scenario, as compared to the no pounding case

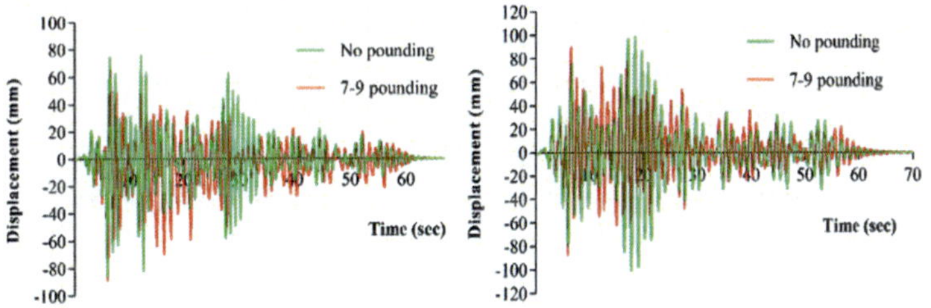

(a) 7-storey building (b) 9-storey building

Fig. 7. Displacement time histories of the colliding buildings at the 7th storey for 7–9 pounding scenario, as compared to the no pounding case

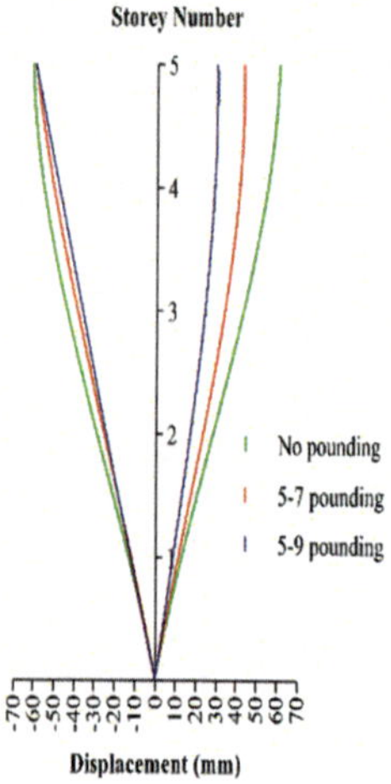

Fig. 8. Peak storey displacements of the 5-storey storey building for different pounding scenarios

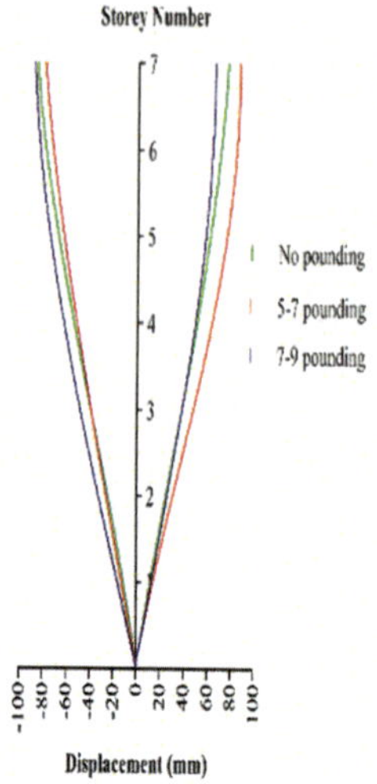

Fig. 9. Peak storey displacements of the 7-storey building for different pounding scenarios

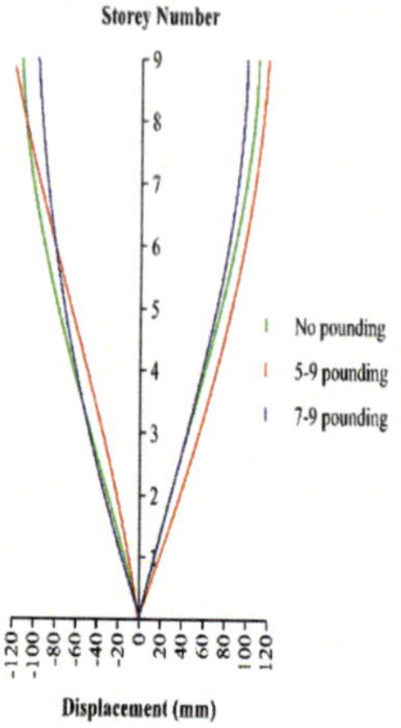

Fig. 10. Peak storey displacements of the 9-building for different pounding scenarios

Figures 11 and 12 show the peak storey shears of the 7-storey and 9-storey buildings under different pounding scenarios, as compared with that of the no pounding case. The results presented in these figures indicate that a significant increase in the peak storey shears is experienced due to pounding. It can be seen that amplification took place at all storeys with few exceptions at the top storeys. The amplification of the peak storey shears ranges between 1 and 3 times, as compared to that of the no pounding case. Similar level of changes concerns de-amplification of the peak storey shears at the top storeys. Since the storey shears at the bottom storeys are much higher than the storey shears at the top storeys, the de-amplification of the top storey shears is not considered as a significant issue while the amplification of the shears at the bottom storeys may cause problems.

Figures 13, 14 and 15 show the impact forces time history for 5–7, 5–9 and 7–9 pounding scenarios, respectively. The peak values of the pounding force for 5–7, 5–9 and 7–9 pounding scenarios are: 6215.93 kN, 11658.7 kN and 18858.3 kN, respectively. These high impact forces are not considered in the engineering design which may lead to the significant damage that has been experienced in previous earthquakes due to pounding. It can also be concluded that higher impact forces are experienced in taller buildings. These conclusions emphasize the significance of pounding phenomena and its effects on the dynamic response of colliding high-rise buildings.

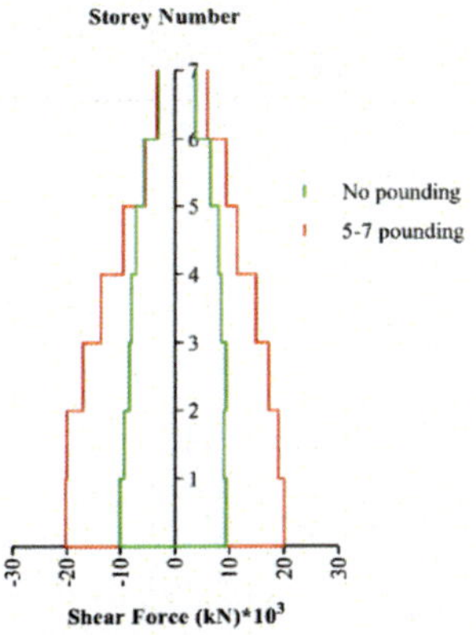

Fig. 11. Peak storey shears of the 7-storey storey building for different pounding scenarios

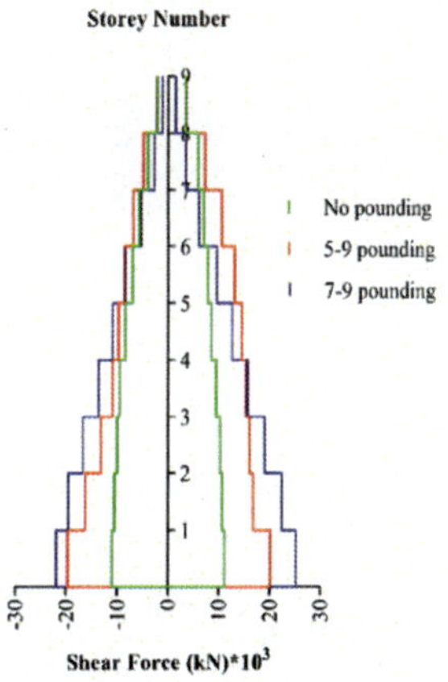

Fig. 12. Peak storey shears of the 9-building for different pounding scenarios

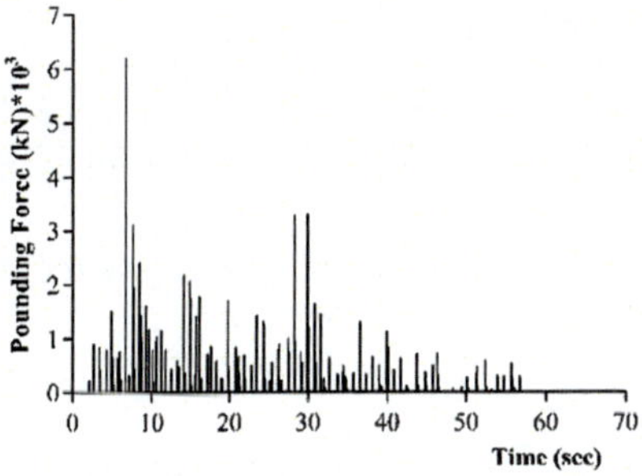

Fig. 13. Pounding force time history for 5–7 pounding scenario

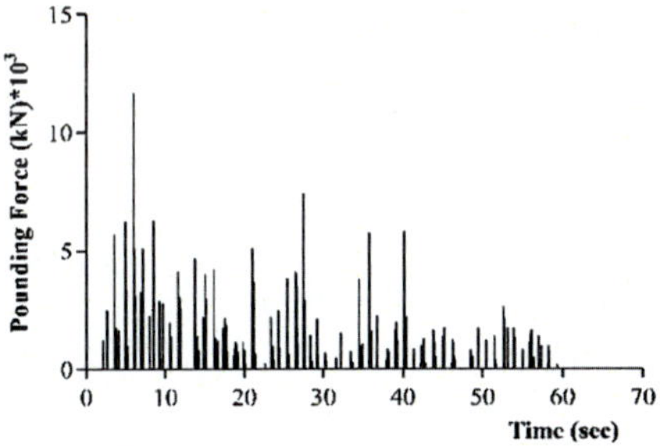

Fig. 14. Pounding force time history for 5–9 pounding scenario

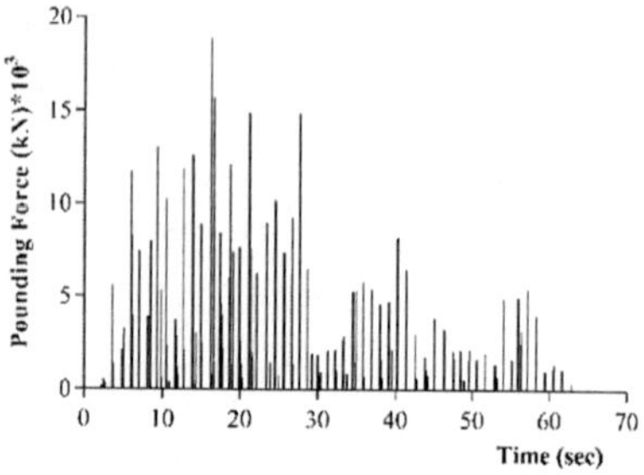

Fig. 15. Pounding force time history for 7–9 pounding scenario

4 Conclusions

This paper investigates the effect of pounding on the dynamic response of colliding high-rise buildings with different structural arrangements. Three 3-D buildings have been considered in the study (5-storey, 7-storey and 9-storey structures). Three pounding scenarios have been taken into account, including pounding between 5-storey and 7-storey buildings, pounding between 5-storey and 9-storey buildings and pounding between 7-storey and 9-storey buildings. These three pounding scenarios have been studied and compared with the no pounding case. The conclusions of this study are:

(1) The accelerations of the colliding buildings increase for all pounding scenarios, as compared to the no pounding case. The amplification of the peak acceleration of the colliding building ranges between 5 and 21, as compared to the no pounding case.
(2) The displacements experience both increase and decrease, as compared to the no pounding case. The amplification and de-amplification of the peak displacement ranges between 1 and 2 times that of the no pounding case.
(3) The peak storey shears experience increase due to pounding with few exceptions regarding the top storeys. The amplification of the peak shear ranges between 1 and 3 times that of the no pounding case. Similar level of changes concerns de-amplification of the peak storey shears at the top storeys. Considering the fact that higher shears are experienced at the bottom storeys and much smaller shears are experienced at the top storeys, the de-amplification of the top storey shears is not considered as a significant issue while the amplification of the shears at the bottom storeys may cause problems.
(4) Pounding leads to the generation of dangerous impact forces with higher peak values experienced in taller buildings.

Acknowledgements. The first author (Mahmoud Miari) gratefully acknowledges the financial support of this research from the "Doctoral Scholarship" awarded from Gdańsk University of Technology.

References

1. Mavronicola, E.A., Polycarpou, P.C., Komodromos, P.: Effect of ground motion directionality on the seismic response of base isolated buildings pounding against adjacent structures. Eng. Struct. **207**, 110202 (2020)
2. Favvata, M.J.: Minimum required separation gap for adjacent RC frames with potential inter-story seismic pounding. Eng. Struct. **152**, 643–659 (2017)
3. Rezaei, H., Moayyedi, S.A., Jankowski, R.: Probabilistic seismic assessment of RC box-girder highway bridges with unequal-height piers subjected to earthquake-induced pounding. Bull. Earthq. Eng. **18**(4), 1547–1578 (2019). https://doi.org/10.1007/s10518-019-00764-4
4. Miari, M., Jankowski, R.: Pounding between high-rise buildings founded on different soil types. In: 17th World Conference on Earthquake Engineering, Sendai, Japan (2021)
5. Miari, M., Jankowski, R.: Seismic gap between buildings founded on different soil types experiencing pounding during earthquakes. Earthq. Spectra (2022). https://doi.org/10.1177/87552930221082968. Accessed 07 Apr 2022
6. Rosenblueth, E., Meli, R.: The 1985 Mexico earthquake. Concr. Int. **8**(5), 23–34 (1986)
7. Anagnostopoulos, S.: Building pounding re-examined: how serious a problem is it. In: Eleventh world conference on Earthquake Engineering, p. 2108. Elsevier Science Oxford, Pergamon (1996)
8. Kasai, K., Maison, B.F.: Building pounding damage during the 1989 Loma Prieta earthquake. Eng. Struct. **19**(3), 195–207 (1997)
9. Cole, G.L., Dhakal, R.P., Turner, F.M.: Building pounding damage observed in the 2011 christchurch earthquake. Earthq. Eng. Struct. Dyn. **41**(5), 893–913 (2012)
10. Sharma, K., Deng, L., Noguez, C.C.: Field investigation on the performance of building structures during the April 25, 2015, Gorkha earthquake in Nepal. Eng. Struct. **121**, 61–74 (2016)
11. Miari, M., Choong, K.K., Jankowski, R.: Seismic pounding between adjacent buildings: identification of parameters, soil interaction issues and mitigation measures. Soil Dyn. Earthq. Eng. **121**, 135–150 (2019)
12. Miari, M., Choong, K.K., Jankowski, R.: Seismic pounding between bridge segments: a state-of-the-art review. Arch. Comput. Methods Eng., 1–10 (2020)
13. Degg, M.R.: Some implications of the 1985 Mexican earthquake for hazard assessment. In: McCall, G.J.H., Laming, D.J.C., Scott, S.C. (eds.) Geohazards. AGID Report Series, pp. 105–114. Springer, Dordrecht (1992). https://doi.org/10.1007/978-94-009-0381-4_11
14. Elwardany, H., Seleemah, A., Jankowski, R., El-khoriby, S.: Influence of soil–structure interaction on seismic pounding between steel frame buildings considering the effect of infill panels. Bull. Earthq. Eng. **17**(11), 6165–6202 (2019). https://doi.org/10.1007/s10518-019-00713-1
15. Kazemi, F., Miari, M., Jankowski, R.: Investigating the effects of structural pounding on the seismic performance of adjacent RC and steel MRFs. Bull. Earthq. Eng. **19**(1), 317–343 (2020). https://doi.org/10.1007/s10518-020-00985-y
16. Anagnostopoulos, S.A.: Pounding of buildings in series during earthquakes. Earthq. Eng. Struct. Dyn. **16**(3), 443–456 (1988)
17. Anagnostopoulos, S.A., Spiliopoulos, K.V.: An investigation of earthquake induced pounding between adjacent buildings. Earthq. Eng. Struct. Dyn. **21**(4), 289–302 (1992)
18. Jankowski, R.: Experimental study on earthquake-induced pounding between structural elements made of different building materials. Earthq. Eng. Struct. Dyn. **39**(3), 343–354 (2010)
19. Chau, K., Wei, X., Guo, X., Shen, C.: Experimental and theoretical simulations of seismic poundings between two adjacent structures. Earthq. Eng. Struct. Dyn. **32**(4), 537–554 (2003)

20. Abdel-Mooty, M., Al-Atrpy, H., Ghouneim, M.: Modeling and analysis of factors affecting seismic pounding of adjacent multi-story buildings. WIT Trans. Built Environ. **104**, 127–138 (2009)
21. Jameel, M., Islam, A., Hussain, R.R., Hasan, S.D., Khaleel, M.: Non-linear FEM analysis of seismic induced pounding between neighbouring multi-storey structures. Latin Am. J. Solids Struct. **10**(5), 921–939 (2013)
22. Computers and Structures, Inc.: Berkeley, California. ETABS
23. Minimum Design Loads for Buildings and Other Structures (ASCE/SEI 7-10), 078447785X (2013)
24. Miari, M., Jankowski, R.: Analysis of pounding between adjacent buildings founded on different soil types. Soil Dyn. Earthq. Eng. **154**, 107156 (2022)
25. Miari, M., Jankowski, R.: Incremental dynamic analysis and fragility assessment of buildings founded on different soil types experiencing structural pounding during earthquakes. Eng. Struct. **252**, 113118 (2021)
26. Pacific Earthquake Engineering Research Centre (PEER NGA DATABASE). http://peer.berkeley.edu/nga
27. Ibrahimbegovic, A., Wilson, E.L.: Simple numerical algorithms for the mode superposition analysis of linear structural systems with non-proportional damping. Comput. Struct. **33**(2), 523–531 (1989)
28. CSI Analysis Reference Manual For SAP2000, ETABS, SAFE and CSiBridge

Study of Machine Learning Techniques for Damage Identification in a Beam

Jefferson da Silva Coelho[1](✉), Amanda Aryda Silva Rodrigues de Sousa[2], Marcela Rodrigues Machado[1], and Maciej Dutkiewicz[3]

[1] Department of Mechanical Engineering, University of Brasília, Campus Universitário Darcy Ribeiro, Faculdade de Tecnologia - Asa Norte, Brasília, DF, Brazil
jeffersoncoelho@ufam.edu.br, marcelam@unb.br

[2] Post-Graduate Program - Integrity of Engineering Materials, University of Brasília, Campus Universitário Darcy Ribeiro, Faculdade de Tecnologia - Asa Norte, 70910-900 Brasília, DF, Brazil

[3] Faculty of Civil, Environmental Engineering and Architecture, University of Science and Technology, 85-796 Bydgoszcz, Poland
macdut@utp.edu.pl

Abstract. Damage is defined as any change to the material, geometry or boundary condition that can modify the structure's properties or response. In the past, damage identification was performed through periodical inspection, non-destructive testing/non-destructive evaluation, or visual observation. Structural health monitoring (SHM) has emerged to transition from offline damage identification to near real-time and online damage assessment. Hence, SHM is a damage detection strategy to monitor a structure for a period using a series of continuous measurement devices, which then collect the system's characteristics and subsequently perform statistical analysis to assess the current circumstances and health of the structure. One of the first steps for the study of SHM is damage detection followed by monitoring. Machine Learning (ML) techniques can be used to develop viable algorithms to make potential predictions. ML algorithms are therefore providing the tools needed to enhance the capabilities of SHM systems and provide intelligent solutions to past challenges. Preliminary studies have reviewed that the extension of ML into SHM has dramatically increased the system's capabilities, providing innovative solutions to different research challenges. This work conducts a study of the use of ML to identify damage in a beam and discuss the challenge of usage, performance and implementation of each technique.

Keywords: Decision tree · Random forest · Damage index · K-Nearest-Neighbor · Support vector machine

---

J. da Silva Coelho—Supported by organization x.

Z. Dimitrovová et al. (Eds.): WMVC 2022, MMS 125, pp. 817–826, 2023.
https://doi.org/10.1007/978-3-031-15758-5_84

1 Introduction

Mechanical systems such as rotating machinery, aircraft, bridges, power generation systems, offshore oil platforms, buildings, and defence systems work under different conditions that can induce damage that influence the system's performance and integrity over time [19]. Damage is defined as any change in the material, geometry, and boundary condition modifying the structure dynamic response. Early damage identification and periodic assessment of structural health are required so that the system operates properly and that damage is identified, monitored, and corrected. Damage identification is performed, including techniques such as condition monitoring (CM), nondestructive assessment (NDE), nondestructive testing (NDT), health and usage monitoring system (HUMS), statistical process control (SPC), damage prognosis (DP), and structural health monitoring (SHM).

Structural health monitoring (SHM) refers to the process of implementing a damage detection strategy for aerospace, civil, or mechanical engineering infrastructure. This area has grows interested in academic and industrial research, which has become important recently due to its ability to acquire, validate, and analyze technical data that facilitate the decision of the structure's life cycle [1,3,6,17]. The monitoring process involves observing a structure or mechanical system over time and acquiring measured data [20]. The extraction of damage-sensitive information and the statistical analysis from these measurements allows discriminating the actual structural condition for short or long-time periods [6,8] Therefore to classify data, machine learning has been trained to generate the most probable outcome and validate the model based on unseen set data [9].

Machine learning (ML) algorithms provide the necessary tools to expand the capabilities of SHM systems [10]. It offers efficient solutions to build models, or representations for mapping input patterns in measured sensor data to output targets for a damage assessment at different levels [9]. The concept of machine learning enters this paradigm of feature selection and statistical modelling for feature discrimination [11] described in [6]. This work conducts a study of K-Nearest-Neighbor, support vector machine, decision tree, and random forest ML techniques to identify damage in a beam discussing the challenges of usage, performance, and implementation of each method. Numerical results demonstrate that the four ML techniques performed a good classification and the beam's integrity indication.

2 Machine Learnings Techniques

For efficient damage classification, one must ensure data availability from damaged, undamaged, or even both conditions. Additionally, the success of assessing and predicting the system's damage, the selected damage-sensitive features relay on the algorithm used to learn from the system features. The most common learning algorithms in ML for the SHM framework are the supervised, unsupervised, and semi-supervised methods. In rare scenarios where both damaged and undamaged structure data are available for engineering structures, supervised learning is the most suitable method. In this case, group classification and regression analysis are the primary supervised learning methods. Many machine learning algorithms can be implemented to perform simple to complex tasks. The most supervised ML algorithms are the K-Nearest-Neighbor (KNN) classifier, Support Vector Machine (SVM), K-Means Clustering, Random Forest (RF), Neural Networks, Bayesian, and Decision Tree. This paper mainly focused on the KNN, SVM, RF, and decision tree classification supervised learning algorithms.

2.1 K-Nearest-Neighbor Classifier

K-Nearest neighbor (kNN) is one of the simplest supervised learner methods [4,13] and most widely used for pattern recognition. [2]. KNN can be used for both classification for data with discrete labels and regression for data with continuous labels. Neighbor-based classification is an instance-based type of learning, that is, there are no parameters to learn, it simply stores instances of the training data. The classification is calculated from a simple majority vote of the nearest neighbors of each point: a query point is assigned the data class that has more representatives within the nearest neighbors of the point, for this a metric between the points is used spaces [4]. The most regarded method is the classification based on estimating the Euclidian distance because of its ease in use, better productivity and efficiency.

The Euclidian distance between two vectors x_i and x_j can be calculated as shown in Eq. (1) [7].

$$d(x_i, x_j) = \sqrt{\sum_{l=1}^{d} (x_i^l - x_j^l)^2} \tag{1}$$

where x_i and x_j are objects represented by vectors in $\Re^d$ space, and x_i^l and x_j^l are elements of the vectors, which correspond to the values of the coordinate l(attributes).

The k-NN algorithm, in its simplest version, only considers exactly one nearest neighbor, which is the closest training data point to the point we want to predict. The prediction is then simply the known output for this training point. Depending on the value of 'k', each sample is compared to find similarity or closeness with 'k' surrounding samples. For example, when k = 5, the individual samples undergo comparison with the nearest five samples and hence the unknown sample is classified accordingly [4]. The optimal choice of the value of 'k' is highly data-dependent, in general, a larger suppresses the effects of noise but makes the classification boundaries less distinct.

2.2 Decision Tree and Random Forest

Decision tree supervised algorithm can target categorical variables such as classification of a damaged or undamaged statement and continuous variables as regression to compare the signal with the healthy state of the system [13]. Learning a decision tree means learning the sequence of if/else questions that gets us to the true answer most quickly. A tree contains a root node representing the input feature(s) and the internal nodes with significant data information. Each node (also called a leaf or terminal node) either represents a question that contains the answer. The interactive process is repeated until the last node (leaf node) is reached such that the node becomes impure [4]. The data get into the form of binary features in our application, and a classification procedure is performed.

Random Forest ML algorithm is an ensemble classifier that consists of many decision trees where the class outputs is the node composed of individual trees. The RF has high prediction accuracy, robust stability, good tolerance of noisy data, and the law of large numbers they do not overfit, and it has been used for structure damage detection and has shown a better performance [16].

2.3 Support Vector Machine

Support Vector Machines (SVM) are supervised machine learning techniques developed from the Statistical Learning Theory that can be used for the classification and regression of structured data [15]. In case of linear classification, with two classes, let $\{(x_i, y_i), ..., (x_n, y_n)\}$, a training dataset with n observations, where x_i represents the set of input vectors and $y_i(+1, -1)$ is the class label of x_i, the hyperplane is a straight line that separates the two classes with a marginal distance (as seen in Fig. 1). The purpose of an SVM is to construct a hyperplane using a margin, defined as the distance between the hyperplane and the nearest points that lie along the marginal line termed as support vectors [5].

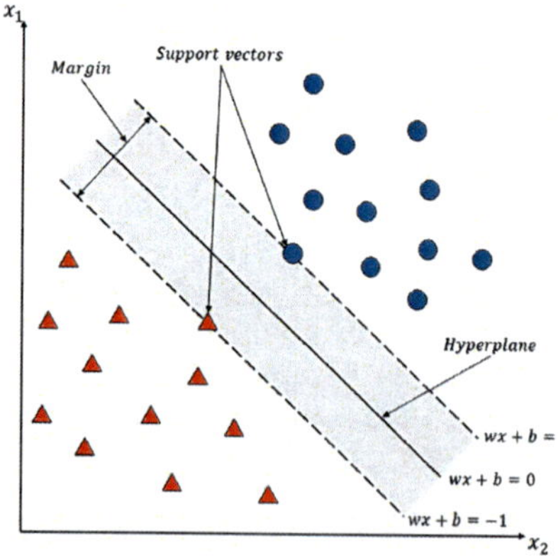

Fig. 1. Illustration of SVM.

SVM can be applied to nonlinear classification problems using kernel functions (e.g. linear, polynomial or Gaussian radial basis function), where it projects the sample space over a higher dimensional space where the data are linearly separable [2,14]. In this work, the linear classification strategy is used. SVM takes a set of data as input and predicts, for each input, which of two possible classes the input belongs to. By providing a training dataset, in 20% of the total data, the algorithm builds a model, which assigns the new test datasets to one of the two categories. The new test datasets are then mapped to the same space and predicted to belong to a category based on which side of the space they are placed. The advantage of SVM is its ability to form an accurate boundary from a smaller training dataset.

3 Numerical Results

Numerical studies performed on a beam investigate the feasibility and accuracy of three machine learning classifier techniques as the random forest, K-nearest, and support vector machine. Those techniques on its classifier application is applied to structural damage identification. A clamped undamaged beam representation followed by the modal shapes as shown in Fig. 2a has its three first natural frequencies at $\omega_1 = 25.15$ Hz, $\omega_2 = 158.81$ Hz, and $\omega_3 = 441.15$ Hz. Beam's length is 1m, width of 0.01 m, and height of 0.03 m. Young's modulus of 2.1 GPa and mass density of 7800 kg kg/m^3. Crack into the structure reduce its stiffness inducing a shift in the resonance frequencies, which can affect different modal shapes depending on the crack location. Figure 2b demonstrates the effect of a crack with different severity levels on the dynamic beam's response.

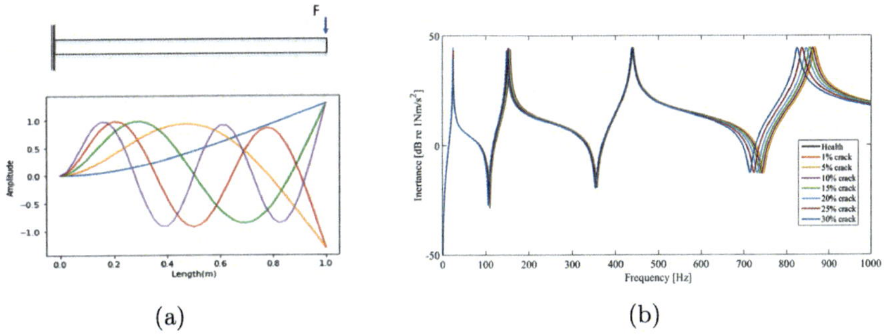

Fig. 2. Representation: a) Spectral model of the structure; b) Schematic representation of the smart beam SEM mesh.

Data generation consists of the natural frequencies estimation considering nominal parameters and set natural frequencies calculated assuming randomness of 5%. The change in natural frequency simulates the shift frequencies event caused by damage [12,18]. A damage index is calculated to create an indicator to help classify the beam's integrity. Figure 3 shows the firsts five lines of the generated data containing the random first natural frequency, deterministic natural frequency named "reference", damage index(DI), the DI's class, and the class name. For DI values between zero to a unity it accuses damage and a higher value up to unity indicates health system state.

	Medidas	Reference	DI	Class	Class name
0	19.92	25.15	0.79	0	Damaged
1	25.45	25.15	1.01	1	Health
2	30.16	25.15	1.20	1	Health
3	26.92	25.15	1.07	1	Health
4	27.97	25.15	1.11	1	Health

Fig. 3. Data generation

Out of the total data samples, as in most ML algorithm implementations, 80% of the data is generally used for training purposes, and 20% for testing the model, i.e., the training and testing split ratio is 0.20

The **K-Nearest-Neighbor Classifier** method used the KNeighborsClassifier algorithm available in the scikit-learn package to analyze the damage identification problem.

The first analysis was cross-validation, whose objective was to define the best value of k in each configurations. K values ranging from 1 to 30 were used and for each k value a cross-validation test was performed.

Figure 4 shows the results of cross-validation to determine the best value of 'k', having characteristic precision of 1, and the best value of k = 1.

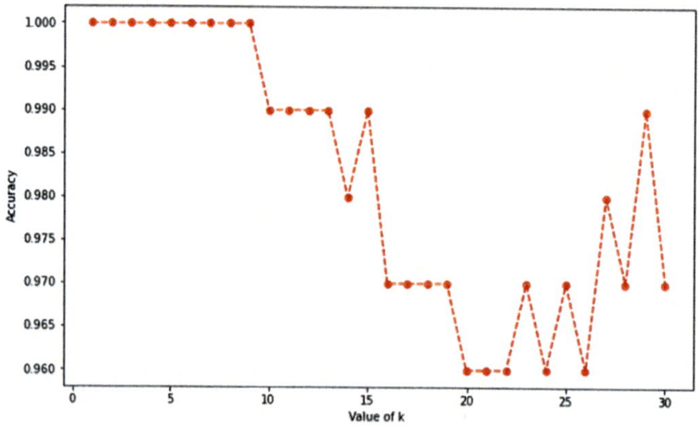

Fig. 4. Best 'k' value

For K = 1, the complex boundary regions are much more delimited, being classified correctly, as a value of k = 1 is used in Fig. 5a. In Fig. 5b, there are no defined limits, that is, all samples will be classified as being from class 1, which is the class with the highest number of samples. Whereas, class 1 is considered undamaged and class 0 is considered damaged. A large value can result in the overlapping of invalid samples from other classes, which in turn makes the classification technique much more difficult. So the performance of the KNN tuner is based on the selection of parameter 'k'.

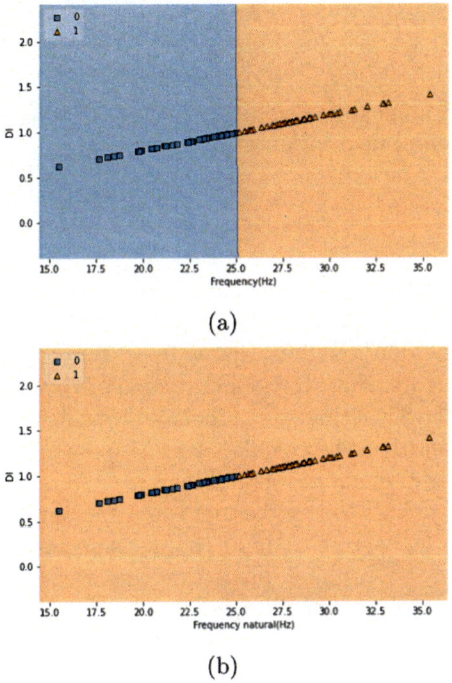

Fig. 5. Complexity boundaries: a) k = 1; b) k = 100

The **Random Forest** classify model used a combination of DI successfully discriminated in the data. In this paper, the analysis only the first natural frequency was considered to monitor the beam's integrity. Classification accuracy indicates 1.0, indicating a high detection rate. The random forest algorithm available on the scikity-learning package considered 80 samples out of the 100 metrics for the classification, where 39 samples were selected as damaged state and 41 undamaged as displayed in Fig. 6c. Because of the data simplicity the random forest algorithm used only a tree in the classification of damaged or health statement. Data classification according to DI and class is shown in Fig. 6a, and DI related to random natural frequency generation, named 'measurement data', is displayed in Fig. 6b, both were able to classify the data leading to a decision on the beam's integrity.

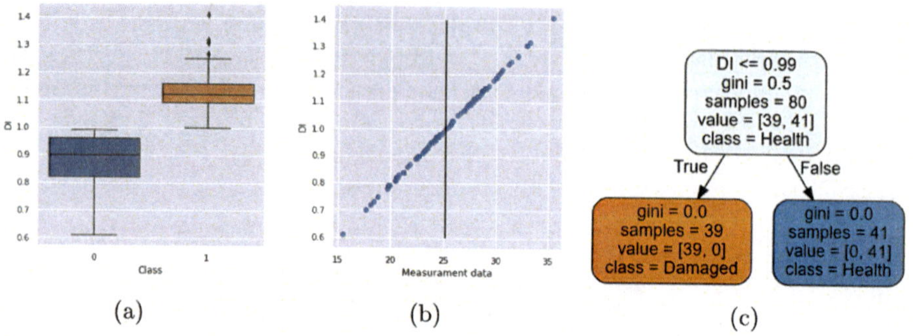

Fig. 6. a) Data classification according to DI and class; b) Data classification according to DI and random natural frequencies; c) A decision tree to distinguish the integrity of the beam.

For the damage detection problem with the **Support Vector Machine** algorithm, the linear kernel was used and the regularization parameter "C" was selected as 1. The parameter C is a penalty (cost) of errors. The higher the value of C, the lower the margin maximisation will be, thus, there is a classifier with fewer classification errors. However, a lower value of C favours smooth margins.

Figure 7a shows the SVM that was calculated considering the training dataset created. It can be seen that the calculated SVM algorithm can clearly separate the training datasets into undamaged (blue) and damaged (red) classes with 100% accuracy. It is possible to observe that there are margin violations, however, the instances are on the correct side of the decision limit, which did not lead to prediction errors. After training, predicting the integrity of the structure just involves figuring out which side of the decision boundary the test point is on.

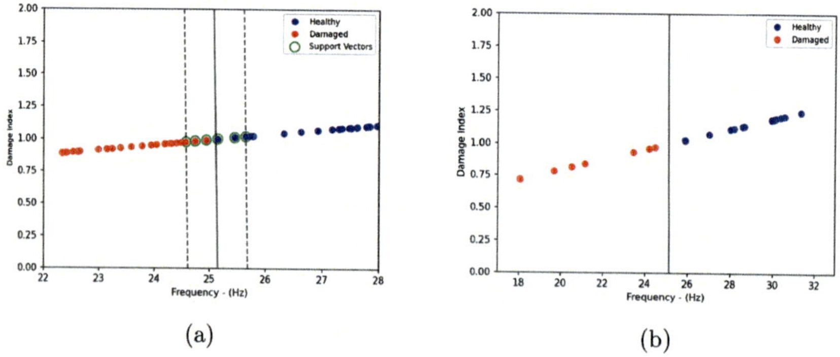

Fig. 7. Dataset: a) training; b) test.

The trained SVM was tested by classifying a test dataset in which the damaged class was created from a model of the framework with a 5% reduction in mode frequency. It can be seen in Fig. 7b that the trained SVM can classify the test dataset for damage with 100% accuracy, that is, the algorithm got all the elements right.

4 Conclusion

The development of this work resulted in the exposure of three machine learning objects of structural damage in a beam. For analysis, random damage was generated with changes in natural frequency to simulate the damage and the damage index as an indicator to help classify beam data. Through model evaluation tests, where tests were found for the test evaluation with one of the data for data evaluation, evaluation tests of new models were found, where they were found for data evaluation tests. The results presented show a good ability to accurately classify the damage data, being able to vary when there was presence or absence of structure.

References

1. Avci, O., Abdeljaber, O., Kiranyaz, S., Hussein, M., Gabbouj, M., Inman, D.J.: A review of vibration-based damage detection in civil structures: from traditional methods to machine learning and deep learning applications. Mech. Syst. Sign. Process. **147**, 107077 (2021). https://doi.org/10.1016/j.ymssp.2020.107077
2. Kurian, B., Liyanapathirana, R.: Machine learning techniques for structural health monitoring. In: Wahab, M.A. (ed.) Proceedings of the 13th International Conference on Damage Assessment of Structures. LNME, pp. 3–24. Springer, Singapore (2020). https://doi.org/10.1007/978-981-13-8331-1_1
3. Cheung, A., Cabrera, C., Sarabandi, P., Nair, K.K., Kiremidjian, A., Wenzel, H.: The application of statistical pattern recognition methods for damage detection to field data. Smart Mater. Struct. **17**(6), 065023 (2008). https://doi.org/10.1088/0964-1726/17/6/065023
4. Cutler, J., Dickenson, M.: Introduction to Machine Learning with Python. O'Reilly, Sebastopol (2020). https://doi.org/10.1007/978-3-030-36826-5-10
5. Otchere, D.A., Ganat, T.O.A., Gholami, R., Ridha, S.: Application of supervised machine learning paradigms in the prediction of petroleum reservoir properties: comparative analysis of ANN and SVM models. J. Petrol. Sci. Eng. **200**, 108182 (2021). https://doi.org/10.1016/j.petrol.2020.108182
6. Farrar, C.R., Worden, K.: Structural Health Monitoring: a Machine Learning Perspective. Chichester, West Sussex. Wiley, Hoboken (2013)
7. Kataria, A., Singh, M.D.: A review of data classification using K-nearest neighbour algorithm. Int. J. Emerg. Technol. Adv. Eng. **3**(6), 354–360 (2013)
8. Silva, M.F.M.D.: Machine learning algorithms for damage detection in structures under changing normal conditions. Dissertation (Masters) - Federal University of Pará, Institute of Technology, Post graduate Program in Electrical Engineering, Belém, Brazil (2017)
9. Rytter, A.: Vibrational Based Inspection of Civil Engineering Structures. Doctoral dissertation. Aalborg University, Department of Building Technology and Structural Engineering (1993)

10. Yuan, F.-G., Zargar, S.A., Chen, Q., Wang, S.: Machine learning for structural health monitoring: challenges and opportunities. 1137903 (2020). https://doi.org/10.1117/12.2561610
11. Machado, M.R., Adhikari, S., Dos Santos, J.M.C.: A spectral approach for damage quantification in stochastic dynamic systems. Mech. Syst. Sign. Process. **88**, 253–273 (2017). https://doi.org/10.1016/j.ymssp.2016.11.018
12. Machado, M.R., Adhikari, S., Dos Santos, J.M.C.: Spectral element-based method for a one-dimensional damaged structure with distributed random properties. J. Braz. Soc. Mech. Sci. Eng. **40**(9), 1–16 (2018). https://doi.org/10.1007/s40430-018-1330-2
13. Malekloo, A., Ozer, E., AlHamaydeh, M., Girolami, M.: Machine learning and structural health monitoring overview with emerging technology and high-dimensional data source highlights. Struct. Health Monit. **21**, 1906–1955 (2021). https://doi.org/10.1177/14759217211036880
14. Zouhri, W., Homri, L., Dantan, J.-Y.: Handling the impact of feature uncertainties on SVM: a robust approach based on Sobol sensitivity analysis. Expert Syst. Appl. **189**, 115691 (2022). https://doi.org/10.1016/j.eswa.2021.115691
15. Selvaraj, Y., Selvaraj, C.: Proactive maintenance of small wind turbines using IoT and machine learning models. Int. J. Green Energy **19**(5), 463–475 (2022). https://doi.org/10.1080/15435075.2021.1930004
16. Zhou, Q., Ning, Y., Zhou, Q., Luo, L., Lei, J.: Structural damage detection method based on random forests and data fusion. Struct. Health Monit. **12**(1), 48–58 (2013). https://doi.org/10.1177/1475921712464572
17. Moura, B.B., Machado, M.R., Mukhopadhyay, T., Dey, S.: Dynamic and wave propagation analysis of periodic smart beams coupled with resonant shunt circuits: passive property modulation. Eur. Phys. J.-Spec. Top. **1**, 1–18 (2022). https://doi.org/10.1140/epjs/s11734-022-00504-x
18. Dutkiewicz, M., Machado, M.R.: Spectral element method in the analysis of vibrations of overhead transmission line in damping environment. Struct. Eng. Mech. **71**(3), 291–303 (2019)
19. Dutkiewicz, M., Machado, M.R.: Measurements in situ and spectral analysis of wind flow effects on overhead transmission lines. Sound Vib. **53**(4), 161–175 (2019)
20. Machado, M., Dutkiewicz, M., Matt, C., Castello, D.: Spectral model and experimental validation of hysteretic and aerodynamic damping in dynamic analysis of overhead transmission conductor. Mech. Syst. Sign. Process. **136**(1), 106483 (2020)

Vibration Based Damage Detection of Beams by Supervised Learning Approach

Stanislav Stoykov[1(✉)] and Emil Manoach[2]

[1] Institute of Information and Communication Technologies, Bulgarian Academy of Sciences, Sofia, Bulgaria
stoykov@parallel.bas.bg
[2] Institute of Mechanics, Bulgarian Academy of Sciences, Sofia, Bulgaria
e.manoach@imbm.bas.bg

Abstract. A supervised learning algorithm that uses neural networks is developed for localizing damages of beam structures. The equation of motion of beam is derived by Timoshenko's theory and discretized by the finite element method. Damages are modelled by reducing the thickness of the beam at the damaged area. Training data for the supervised learning algorithm is generated by solving the equation of motion for variety of damages and external forces. Transverse displacements along the whole beam's length for several instants of time are stored into one-dimensional vector and used as training data. It is shown that neural networks can successfully localize damages even for excitation frequencies that are not used in the training set.

Keywords: Structural health monitoring · Neural network · Timoshenko theory · Finite element method

1 Introduction

Beam structures are widely used in many engineering applications like wind power generators, bridges, airplanes, etc. They are often exposed to large deformations and extreme weather conditions which lead to structural damages. The early detection of damages and their correct localization is essential for the operation and maintenance of these structures.

The current work proposes a damage localization method which is inspired from vibration based methods [1] and supervised machine learning approach [2]. The equation of motion of beam is derived by the principle of virtual work, assuming Timoshenko's beam theory. Geometrical type of nonlinearity is included in the model and the equation of motion is discretized by the finite element method. The damage of the beam is modelled by reducing the thickness of the beam on the damaged area.

Then, a set of dynamic responses is generated, considering variety of damages along the length of the beam and variety of excitation forces. This set serves as training data set for the supervised learning damage detection algorithm. Each simulation gathers transverse displacements at different points along beam's length, for variety of time

Z. Dimitrovová et al. (Eds.): WMVC 2022, MMS 125, pp. 827–833, 2023.
https://doi.org/10.1007/978-3-031-15758-5_85

steps within one period of vibration. This data is converted into a one-dimensional vector which represents the training input.

A neural network is used to "learn" from these training examples, i.e. the goal in training the neural network is to find weights that minimize an objective function. It is shown that the neural network can localize accurately damages of beams, by given input vector of transverse displacements. The proposed method localizes damages for excitation frequencies which were not used in the training data set.

2 Mathematical Model of Damaged Beams

Beams that can vibrate in transverse and longitudinal directions are considered in the model. The equation of motion is derived by assuming Timoshenko's hypothesis [3]. The displacements in transverse and longitudinal directions are expressed in the following way:

$$
\begin{aligned}
u(x, y, z, t) &= u_0(x, t) + z\phi_y(x, t), \\
w(x, y, z, t) &= w_0(x, t)
\end{aligned}
\tag{1}
$$

where $u_0(x, t)$ and $w_0(x, t)$ denote the longitudinal and transverse displacements on the middle line of the beam and $\phi_y(x, t)$ denotes the rotation of the cross section about y axis. Direct and shear strains, denoted by ε_x and γ_{xz} are expressed by considering geometrical type of nonlinearity:

$$
\begin{aligned}
\varepsilon_x &= \frac{\partial u}{\partial x} + \frac{1}{2}\left(\frac{\partial w}{\partial x}\right)^2, \\
\gamma_{xz} &= \frac{\partial w}{\partial x} + \frac{\partial u}{\partial z} + \frac{\partial w}{\partial x}\frac{\partial w}{\partial z}
\end{aligned}.
\tag{2}
$$

Homogeneous and isotropic materials are assumed and constitutive relations are expressed by Hooke's low:

$$
\begin{Bmatrix} \sigma_x \\ \tau_{xz} \end{Bmatrix} = \begin{bmatrix} E & 0 \\ 0 & \lambda G \end{bmatrix} \begin{Bmatrix} \varepsilon_x \\ \gamma_{xz} \end{Bmatrix},
\tag{3}
$$

where σ_x and τ_{xz} denote the direct and shear stresses, E is the Young's modulus, G – shear modulus and λ is shear correction factor.

The equation of motion is derived by the principle of virtual work.

$$
\delta W_V + \delta W_{in} + \delta W_E = 0,
\tag{4}
$$

δW_V represents the virtual work of internal forces, δW_{in} represents the virtual work of inertial forces and δW_E represents the virtual work of external forces. Space discretization is performed by the finite element method by using quadratic elements. The following system of ordinary differential equations is obtained:

$$
\mathbf{M}\ddot{\mathbf{q}} + \beta\mathbf{K}_\mathbf{l}\dot{\mathbf{q}} + \mathbf{K}_\mathbf{l}\mathbf{q} + \mathbf{K}_\mathbf{nl}(\mathbf{q})\mathbf{q} = \mathbf{f}
\tag{5}
$$

where $\mathbf{M}$ represents the mass matrix, $\mathbf{K_l}$ represents the stiffness matrix of constant terms, $\mathbf{K_{nl}}(\mathbf{q})$ represents the stiffness matrix that depends on the vector of generalized coordinates $\mathbf{q}$. $\mathbf{f}$ is the vector of generalized external forces. Stiffness proportional damping is introduced in the model with damping factor denoted by β.

Damage is introduced into the equation of motion by reducing the thickness of the beam on the damage area. In the current finite element space discretization, the reduced thickness is applied to whole finite elements. One can consider damage in more than one sequential elements, or one can reduce the element size, in order to model smaller damages.

3 Damage Detection by Supervised Learning Approach

Neural networks provide one of the best solutions to variety of problems in image recognition, speech recognition, and natural language processing. Similar techniques, but in damage detections, are applied here. The aim is to define a neural network which can learn from simulated data. The input data of the neural network is the time response vibrations of damaged beam. The aim of the neural network is to determine the weights and biases of the nodes so that the output correctly classifies the damage location.

A neural network is characterized by input layer, output layer and between them it can have several hidden layers. Each layer has several nodes, where at each of the nodes there are various mathematical manipulation of the data. The input layer of the network contains neurons encoding the time response of the beam. The output layer contains the output neurons which in the current case will be equal to the number of damages considered in the model.

The neural network needs training data, which is used to learn the network to classify damages. The training data is generated by solving the equation of motion (5), for different external forces and different locations of the damage. Each simulation has a label indicating the location of the damage.

3.1 Generation of Training Data

Beam with homogeneous, isotropic and elastic material is considered with the following material properties: Young modulus $E = 70\,\text{GPa}$, density $\rho = 2778\,\text{kg/m}^3$, Poisson ratio $\nu = 0.3$ and with dimensions: length $l = 0.58\,\text{m}$, width $b = 0.02\,\text{m}$ and thickness $h = 0.002\,\text{m}$. Damping factor $\beta = 0.0001$ is used.

In the current work, 32 finite elements are used for the space discretization. It results into 65 nodes. The equation of motion is solved by Newmark method in time domain. Newton's method is used to solve algebraic nonlinear system that result from the application of Newmark's method. Damage is assumed to be on two consecutive elements. In the current case 16 damages are considered, as shown in Fig. 1.

It was shown in [1] that damage can be successfully localized by using the curvature of the beam and appropriate selection criteria. Since the curvature is computed from the transverse displacements, the information regarding the damage location is already stored at the transverse displacement, even though it is difficult to see any differences, as shown in Fig. 2. The figure shows transverse displacements and curvature of beams with three different damages. One can easily see the indicators from the curvature related with the damage, while it is difficult to make any conclusions from the transverse displacements. The aim of the current work is to use transverse displacements for damage localization, thus only transverse displacements are taken into account for the training data, but longitudinal displacements and rotations of the cross section are included in the numerical simulations at the equation of motion.

Fig. 1. Beam with 16 possible damage locations along its length.

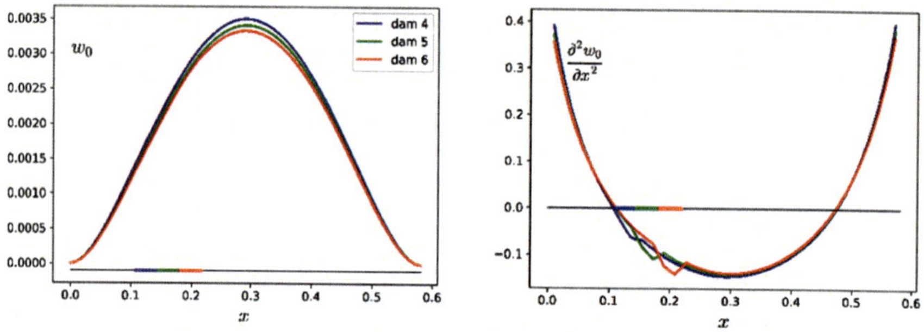

Fig. 2. Transverse displacement and its second derivative for damages 4, 5 and 6, $\omega = 150\,\text{rad/s}$, $t = 0.0419\,\text{s}$.

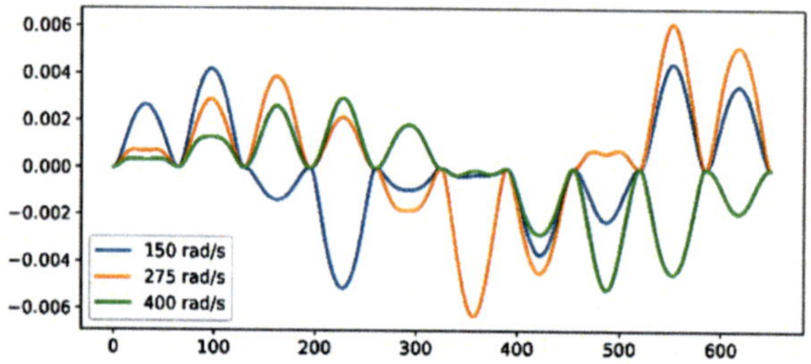

Fig. 3. Sequence of 10 transverse shapes of vibration for different excitation frequencies, damage is set at position 4.

The applied external force is assumed to be uniformly distributed with amplitude of 30 N/m. Several excitation frequencies are considered, from 150 rad/s till 400 rad/s by interval of 5 rad/s, i.e. 51 different excitation frequencies. The time integration is performed for one period of vibration, determined by the excitation frequency. Shapes of vibrations at 10 equally spaced time intervals within the transient time interval are taken for the training data. Since 65 nodes are used, this represents a vector with dimension of 650. Additionally, one can include the frequency of vibration. An example of training data for different excitation frequencies is shown in Fig. 3.

3.2 Definition of Neural Network

Keras library [4] is used for defining the neural network and training the data. The training data is first standardized and its dimension is reduced by using its principal components. This step introduces lower dimensional data, but at the same time it keeps most of variability and properties of the data. First three principal components are shown in Fig. 4. One can see that the data is in clusters which are characterized by excitation frequency. Figure 5 shows part of the data, decomposed by its first two principal components and colored by the location of the damage. In the current work 60 principal components are used for the dimensionality reduction.

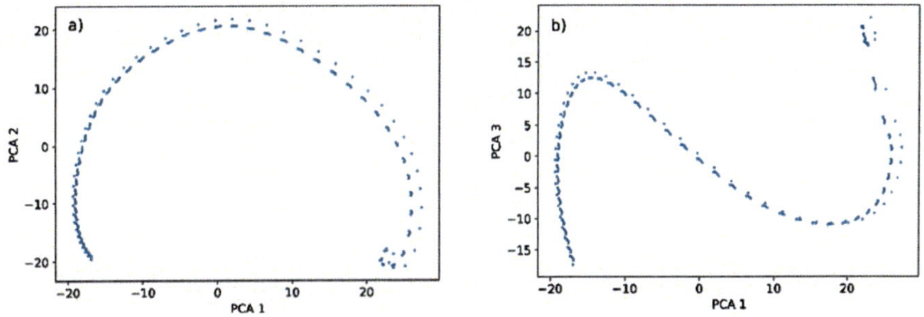

Fig. 4. First three principal components of generated data, a) PCA 1 and 2, b) PCA 1 and 3.

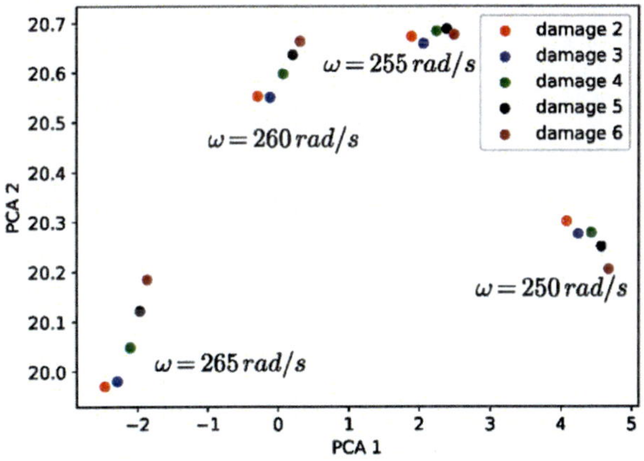

Fig. 5. First two principal components of selected data.

The input layer of the neural network contains 60 nodes. Four hidden layers, each with 100 nodes are introduces. The damages are considered as categorical labels, hence the output layer contains 16 nodes. A simplified version of the neural network is shown in Fig. 6. Rectified linear unit function is used as activation function and categorical cross entropy function is used as cost function for the optimization problem.

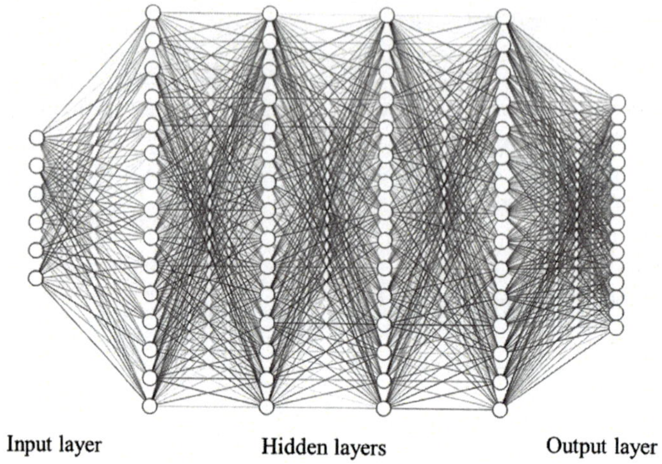

Fig. 6. Neural network with four hidden layers.

4 Results

The time responses of the last excitation frequency, i.e. $\omega = 450\,\text{rad/s}$, are excluded from the training data. They are used for evaluating of the network performance. The remaining data is separated into two groups. Randomly are selected 80% of the data which are used for training data and the remaining 20%, called test data, are used for evaluating the performance of the neural network.

The network predicts very well the test data. An accuracy of 97% is achieved. Furthermore, the neural network predicted all of the initially excluded time responses with excitation frequency, i.e. $\omega = 450\,\text{rad/s}$ which was not used in the training data.

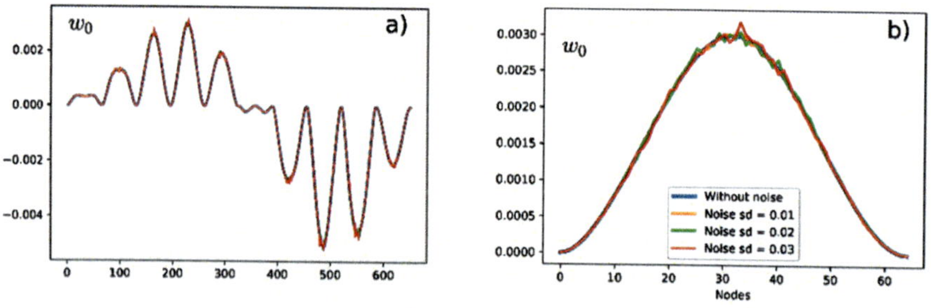

Fig. 7. a) Shocked test data from time response due to excitation frequency of 450 rad/s, b) zoom area for one shape of vibration.

In order to investigate the possibilities of neural networks for damage detection of real data, the test input data is shocked by multiplying with white noise with mean one and different standard deviations. An example of shocked time response is shown in Fig. 7. The algorithm for damage detection still manages to localize damages when the standard deviation of the white noise is small, as shown in Table 1.

Table 1. Number of correct damage localizations of shocked time responses.

Standard deviation of shocked response	Correct damage localizations (from 16)
Without shock	16/16
0.01	16/16
0.02	14/16
0.03	11/16
0.04	9/16
0.05	8/16

5 Conclusion

Damages of beam structures were successfully localized by supervised learning algorithm that uses neural networks. The training data represents essential component in the whole process. Transverse displacements along the beam's length for several instants of time were used as training data. The data was generated by numerically solving the equation of motion for different damage locations and excitation frequencies. An accuracy of 97% was achieved. The proposed algorithm managed to localize successfully damages from time responses due to excitation frequencies which were not used in the training data set. Furthermore, the algorithm was tested on shocked time responses and it behaved fairly well, in the cases of small shocks.

Acknowledgement. The support of this work through project No. KP-06KITAJ/3 is gratefully acknowledged.

References

1. Stoykov, S., Manoach, E.: Damage localization of beams based on measured forced responses. Mech. Syst. Signal Process. **151**, 107379 (2021)
2. Sutton, R.S., Barto, A.G.: Reinforcement Learning: An Introduction. MIT Press, Cambridge (1998)
3. Wang, C., Reddy, J., Lee, K.: Shear Deformable Beams and Plates. Elsevier, Oxford (2000)
4. Gulli, A., Pal, S.: Deep Learning with Keras, Packt Publishing Ltd. (2017)

SRM: Sustainable Railway Maintenance

Geosynthetics in the Renewal of East Railway Line

Julieta Ribeiro[1(✉)] and Madalena Barroso[2]

[1] Sener-Engivia, Consultores de Engenharia, S.A., Lisbon, Portugal
julieta.ribeiro@sener.pt
[2] Laboratório Nacional de Engenharia Civil (LNEC), Lisbon, Portugal
mbarroso@lnec.pt

Abstract. Geosynthetics have been widely used in various geotechnical engineering applications to enhance the strength of geosystems, as well as the fact that they are economical and relatively easy to install. Separation, filtration, drainage, reinforcement and stabilization can be referred as the main functions of the geosynthetics.

Regarding the reinforcement function, literature review shows that these materials can reduce the settlement and degradation of ballast. Placing geosynthetics under the ballast confines the aggregates via interlock or frictional resistance among ballast particles which then stiffen the surrounding aggregates and increase the shearing resistance of the composite system.

This paper aims to present a case study where a geotextile was used for improving the subgrade bearing capacity and, therefore, reduce the settlements. The case study concerns a railway section between Elvas and Caia (border with Spain), about 11 km long. It is located at East Railway Line, a 140 km long railway line that connects Abrantes to Caia, serving as an alternative to the Center - North corridor with regard to the transport of goods. This paper describes the design, construction and installation issues related with the geotextile used.

Keywords: Railroad · Rehabilitation · Case study · Geotextile · Reinforcement

1 Introduction

The construction of the railway network in Portugal began in the 19th century. The first section, between Lisbon and Carregado, finished in 1856, having had a very significant expansion in the following years. In 1861, the current East Railway Line was completed, reaching the Spanish border in Badajoz. Its main objective was the connections to the interior regions of the country, in order to facilitate the transportation of goods, as well as the transportation of passengers from Europe.

Z. Dimitrovová et al. (Eds.): WMVC 2022, MMS 125, pp. 837–846, 2023.
https://doi.org/10.1007/978-3-031-15758-5_86

In the second half of the 19th century, until the 30s of the 20th century, the expansion of the railway network was notorious and in the beginning of the 30s, the network reached 3424 km in length. From that moment onwards, the expansion of road transport took place and with the two world wars the decline of the railway was very significant. Currently, the national rail network is 2526 km long.

History shows that, as initially in the 19th century, the genesis of investment in railroads is still maintained, with the objective of transporting people and goods. We have also learned from experience and history that maintenance and modernization must be the subject of constant investment, otherwise it will require large occasional investments, as it happened in the last century. The railroad consumes, on average, seven times less energy and emits close to nine times less carbon dioxide than road transport, for the same volume of goods. In addition, rail transportation can be carried out with zero emissions when using electricity from renewable sources, in the case of electrified lines, or alternatively with battery-powered or hydrogen-powered railcars, as is already being developed in Spain. Thus, opening the real possibility of meeting the Paris agreement, the European Ecological Pact, and the Plan to achieve carbon neutrality by 2050.

In this sense, efforts are currently being made to improve the National Railway Network, which includes the Renewal Plan for the Southern International Corridor, which aims to improve the railway connections between Setúbal and Sines Ports, industrial and urban areas located in South of the country, with Spain and whole Europe.

One of the lines that underwent renewal in 2018/2019 was the East Line. This line has been the subject of works on several sections. The renewal included the improvement of the railway platform, the improvement of drainage, the stabilization of slopes and the construction of support and containment structures, the replacement of all track material, electrification, among others.

Regarding the platform improvement, the objective was to increase its load capacity, in order to avoid deformations of the platform and consequent interdictions for the regularization of the track.

This paper focuses on the works carried out between kilometers 269 + 200/270 + 000 and 274 + 850/end of the East Line, where a geotextile was used for improving the subgrade bearing capacity and, therefore, reduce the settlements.

2 The East Railway Line

The East Railway Line runs between Abrantes and Caia (Fig. 1), it is approximately 140.7 km long, on a single track, with ability to reach speeds of 90 km/h to 120 km/h, serving as an alternative to the Center – North corridor, in the concerning the transport of goods.

Fig. 1. East railway line location (IP 2021)

The mentioned renewal included several works, namely:

- Enlargement of the Elvas Railway Station, in order to allow the handling of freight trains 750 m long (adjustments of the station building to install new technical rooms; raising the platforms; suitability of pedestrian accesses in order to guarantee pedestrian mobility, mainly accesses improvement for people with reduced mobility);
- Full renovation of the track superstructure (rail, sleepers, and ballast);
- Track bed rehabilitation, drainage improvement, slope stabilization and construction of support and containment structures;
- Installation of infrastructure for future implementation of electronic signalling (ETCS and GSM-R) and catenary remote control;
- Construction of catenary foundation blocks for future electrification;
- Improvement of safety in road-rail traffic, through the construction of road levelling and reestablishments, to suppress the four existing level crossings along the section;
- Replacement of the decks and reinforcement of the abutments and pillars of the centuries-old railway bridges, Caiola Bridge and Caia Bridge.

Among the works indicated, those related to the track bed rehabilitation, namely where geosynthetics were used, are highlighted in this paper.

3 Studies Before the Rehabilitation Works

In order to defining the improvements to be made on track bed, a preliminary study was carried out to characterise the actual features of the track bed. Plate loading tests, trial pits and visual inspections were carried out.

The loading tests aimed to determine strain moduli (Ev1, Ev2 and ratio Ev2/Ev1). The modulus is an indicator for the bearing capacity of the subgrade soil. An Ev2 higher than 80 MPa was required by the construction owner specifications (CE 2016).

Trial pits, excavated at regular intervals along the track bed, intended to assess the soil conditions. The soils were classified using the international classification proposed by UIC719R (2008), which classifies soils according to their suitability for track beds:

- ***QS0***: *"Unsuitable" soils which do not form a suitable subgrade and therefore require improvement (replacement to a certain depth with better quality soil, stabilisation with binding agents, use of geotextiles, reinforcement with piles, etc.). For this reason, these soils are not considered here when dimensioning the track bed layers.*
- ***QS1***: *"Poor" soils which are acceptable in their natural condition subject to adequate drainage being provided and maintained in good order. These soils could be considered for upgrading by means of the appropriate treatment (e.g. stabilisation binding agents).*
- ***QS2***: *"Average" soils.*
- ***QS3***: *"Good" soils.*

Finally, the visual inspections, conducted by the construction owner, aimed to identify potential pathologies.

Results obtained for the two sections were geosynthetics were used (km 269 + 200 to km 270 + 000 and km 274 + 850 to km 275 + 611) are shown in Table 1.

Table 1. Plate loading tests before rehabilitation works.

Section	Strain moduli (MPa)			Depth (m)	Soil class
	Ev1	Ev2	Ev2/Ev1		
km 269 + 200 to km 270 + 000	26	40	1.5	0.37	QS0
km 274 + 850 to km 275 + 611	42	58	1.4	0.35	QS3

As can be seen, strain modulus obtained were lower than the minimum required by construction owner specifications. In section km 269 + 200/270 + 000, the subgrade is unsuitable for track bed. Also, the visual inspection undertaken revealed excessive gauge on several sections.

4 Design

In terms of the track bed, it was necessary to plan the interventions to guarantee the achievement of medium bearing capacity (5 < CBR < 20), *i.e.*, P3 type platform, according to the classification of the UIC719R (2008). In order to comply with this requirement, before the execution of the ballast and sub-ballast layers, different types of improvements were defined, according to the load capacity of the subgrade (Table 2).

Table 2. Design options.

Section	Design	
	Option 1	Option 2
km 269 + 200 to km 270 + 000	1 m soil replacement + replacement with QS3 soil class	Geogrid + 0.19 m of QS3 soil class
km 274 + 850 to km 275 + 611	Subgrade compaction (Ev2 ≥ 60 MPa) + 0.35 m of QS3 soil class	Geogrid + 0.15 m of QS3 class of soil

Option 2 was selected. It consisted of application of a geogrid over a layer of QS3 soil or ABGE (crushed aggregate with extensive grain size), with 0.19 m or 0.15 m thick, respectively, for the sections between km 269 + 200/270 + 000 and km 274 + 850/275 + 611.

For both sections over the geogrid, the design included a layer of sub-ballast, 0.15 m thick, and a layer of ballast, 0.30 m thick.

Geogrid function was subgrade reinforcement. It was designed based on the German EBGEO standard (DGGT 2011). The main required characteristics for geogrid were the following:

- Type: biaxial
- Raw material: PVA (Polyvinyl alcohol), with polymeric coating
- Mesh size: 30 × 30 mm^2, with a minimum open area of 75%
- Modulus: 1330 kN/m
- Deformation to nominal stress (EN ISO 10319): less than 6%

During the construction phase, the geogrid was replaced by a high-strength geotextile (woven material). The replacement of the geogrid by the geotextile was due to the difficulty in obtain this material in time for construction. It should be noted that the required mechanical properties for the geotextile were similar to those required for the geogrid.

Table 3 shows the main properties of the geotextile according to its data sheet.

Table 3. Main features of geotextile.

Properties	Test standard	Unit	Values
Tensile strength (machine direction and cross direction)	EN ISO 10319	kN/m	70
Elongation at nominal strength (machine direction and cross direction)	EN ISO 10319	%	10
Tensile strength at 5% (machine direction)	EN ISO 10319	kN/m	35

(continued)

Table 3. (*continued*)

Properties	Test standard	Unit	Values
Static puncture resitance CBR	EN ISO 12236	N	7000
Dynamic perforation (cone drop)	EN ISO 13433	mm	7
Characteristic opening size, O_{90}	EN ISO 12956	mm	0.100
Water permeability normal to the plane	EN ISO 11058	l/m^2s	5

Installation procedures were also defined in the project. As they were firstly defined for the geogrid, some adaptations were undertaken.

According the design specifications, the geotextile should be applied over a smooth surface. All sharp edges materials with should have to be removed *priori* the application of the geotextile. In addition, it was to be unrolled directly over the area where it was to be applied, either manually or using special equipment so that it does not damage the geotextile.

It was also required that the geotextile should be cut according to project drawings and applied taking into account the platform features such as wells, ditches, manholes or other openings in the area. No folds or wrinkles were allowed in its surface.

Seams between adjacent geotextile panels should be performed by overlapping, with a minimum overlap range of 0.50 m, according to the manufacturer's recommendations.

Regarding the placement of the overlying material, the design indicated that the material should be spread and compacted in order to minimize the risk of wrinkling. Under no circumstances, construction machinery and other equipment could operate in direct contact with the geotextile. Only after applying 0.30 m of soil (compacted), or covering material, vehicles were allowed to circulate on the geotextile.

In case of damage during installation, geotextile should be repaired immediately, according to manufacturer's recommendations.

5 Geotextile Installation

The installation of the geotextile, in general, occurred according to the design specifications described in previous section.

Prior to the installation, the following activities were performed (Fig. 2): disruption, removal of the track and ballast and subgrade preparation.

Fig. 2. Pre-installation activities

After performing the above activities, the geotextile was installed as shown in Fig. 3.

Fig. 3. Geotextile installation

Figure 4 shows the activities carried out after the installation of the geotextile, namely the placement of the layer of sub-ballast, with a thickness of 0.15 m, the placement of the superstructure (rails + sleepers) and the execution of the rail welds.

Fig. 4. Activities subsequent to the application of the geotextile

As example, Fig. 5 shows the final activities of the intervention carried out, namely the application of ballast and the tamping.

Fig. 5. Final intervention activities

6 Plate Loading Tests After the Track Bed Rehabilitation

After concluding the described activities, quality control tests were performed. They aim to determine the strain moduli (Ev2) of the infrastructure layers after the rehabilitation works. According to construction owner specifications, this modulus should be greater than 80 MPa.

Figure 6 shows ongoing tests on the sub-ballast layer (sections km 269 + 200/270 + 000 and km 274 + 850/275 + 611). Tests were carried out in October 2018 and April 2019, respectively.

Fig. 6. Plate loading test

Results obtained are shown in Table 4. As can be seen, strain modulus met the specification requirements, being much higher than those obtained in the same sections of the platform before the rehabilitation works.

These results indicate that the design solution, involving the placement of a high-strength geotextile, resulted in increase in strain modulus.

Table 4. Results of the tests performed.

Section	km	Ev2 (MPa)
km 269 + 200 to 270 + 000	269 + 393	153.2
km 274 + 850 to km 275 + 611	274 + 945	171

7 Conclusions

This paper presents a case study, carried out as part of the renewal of the East Railway Line, where a high-tensile strength geotextile was applied to reinforce the railway track bed of the existing Line, at the interface between sub-ballast and soil. This case study involved two sections of track, located at kilometres 269 + 200/270 + 000 and 274 + 850/275 + 611, respectively. The main issues regarding the design and the installation of the geotextile were addressed.

Results of plate loading tests, carried out before and after the installation of the geotextile, on both sections of the railway line, were presented. Results show that strain modulus was significantly higher after application of the geotextile and that they fulfilled the requirements set out in the specifications.

The application of the geotextile reduced the time spent on rehabilitation works and the number of disruptions. Also, greater area was achieved per night of work, allowing construction to be optimised. Finally, it reduced overall material costs.

References

1. UIC (Union Internationale des Chemins de Fer) 719R. Earthworks and track bed for railway lines, 3rd edn. UIC, Paris (2008)
2. DGGT (Deutsche Gesellschaft fur Geotechnik). Recommendations for Design and Analysis of Earth Structures using Geosynthetic Reinforcements – EBGEO, 1st edn. Ernst & Sohn, Berlin (2011)
3. IP (Infraestruturas de Portugal). Diretório de Rede 2023. Anexo 2.2.1. (2021). https://servicos.infraestruturasdeportugal.pt/pt-pt/parceiros/operacao-ferroviaria/os-nossos-servicos/diretorio-da-rede-ips
4. CE (Caderno de Encargos). Projeto de Execução da Nova Ligação Ferroviária entre Évora Norte e Elvas/Caia – Lote E – Linha do Leste e Ligações à nova Linha Évora-Caia (2016)
5. IP (Infraestruturas de Portugal). Projeto de Execução da Nova Ligação Ferroviária entre Évora Norte e Elvas/caia (Fronteira com Espanha) – E – Linha do Leste e Ligações à Nova Linha Évora-Caia. Edgar Cardoso/Intecsa, Portugal (2017)

Human Perception of Railway Vibration-Case Study

Alicja Kowalska-Koczwara(✉) and Filip Pachla

Cracow University of Technology, Krakow, Poland
akowalska@pk.edu.pl

Abstract. Environmental protection has several aspects from the most popular like air or water pollution to landscape protection. One of the aspects of pollution which is neglected is protection against vibration. Meanwhile, vibrations are a pollution not only by Polish legislation but also in EU directives. One of the most subjective parameters to judge this level of vibration pollution is the human perception of vibration. It depends on individual perception which could be cause of age, sex, high of a person. Vibration excitation in buildings comes mainly from external sources such as: industrial machinery (building machines) such as vibration road rollers, pile driving etc. or transport excitation from roads, railways, subways or trams. Vibrations that are transmitted from the ground to building may influence the building structure but more often can result in discomfort of the occupants. Especially unexpected vibrations coming from transport vibration could be annoying. This paper aims to investigate if there is an influence between the type, train speed passing next to a building in which people live with the use of linear or non-linear regression. The article summarizes a lot of information that may be useful for designers or rail transport managers.

Keywords: Vibration · Regression · Human perception of vibration · RMS method · Measurements in-situ

1 Introduction

Human perception of vibration could be the basic parameter in designing new buildings located close to transport sources. The human perception of vibrations in buildings, despite many investigations made in the past, despite standards and requirements in this area, is still a topic not fully understood mainly due to the subjective nature of the perception of vibrations by various people. This paper aims to analyze the human perception of railway vibrations based on an example building. There are some investigations available in the literature on railway induced vibration and its annoyance [ex. 1, 2, 3, 4], but they do not consider possibilities of protection against the harmful effects of vibration. It is worth noting that especially low-frequency vibrations in the range of 5-25 Hz could be even dangerous for human health because this frequency range is close to resonant frequencies of human internal organs [5]. Vibrations can cause sleep disorders, headaches and neurotic conditions [6, 7]. Moreover, it is not known exactly

Z. Dimitrovová et al. (Eds.): WMVC 2022, MMS 125, pp. 847–854, 2023.
https://doi.org/10.1007/978-3-031-15758-5_87

what influences human perception of vibration. Humans are more sensitive to unexpected low-frequency vibrations and providing vibrational comfort in buildings could be the basic element for buildings design criteria. Some of the authors show the relations between the high, sex or even age of the person on the perception of vibration [8, 9], but some of them do not [10]. This discrepancy alone causes that research on human perception of vibration is highly non-deterministic. It is also important how the vibrations are perceived in an active (whole-body vibration) or passive way. The active perception of vibrations occurs when vibrations are transmitted to our body directly from the source of vibrations. Such a situation takes place in the case of machine operators, drivers or even vehicle passengers [11, 12]. The passive perception of vibration occurs when vibrations are transmitted from the source to the human body through some medium, most often it is the ground. This kind of situation mostly occurs in buildings to which vibration is transmit-ted from transport sources [13, 14]. This article is focused on the passive perception of vibration in the building caused by train passing. In such a situation there are three methods of evaluation acc. [15]: root mean squared method (RMS), vibration dose value (VDV) and maximum transient vibration value (MTVV). The first RMS method is called in [15] basic method which is used for vibrations with a small value of so-called crest factor.

In-situ measurements require very good planning. They require reconnaissance (local vision), selection of the room in which the measurement will be carried out (mainly the residential located on the side of the dynamic excitation), selection of measurement points depending on the information to be obtained. Measurement point selection depends whether the information on human perception of vibration is needed or the information of vibration influence on building is required. In the first situation the measurement point should be located in the place where human perception of vibration is the highest – mainly it is the center of the floor. When the vibration influence on building is investigated the measurement point should be located on the ground level in the stiff node of the building (the corner) the closest one to the excitation source. It is worth emphasizing that the room in which the measurement will be performed is not accidental. For the ground excitation the best will be localization on the top floor of the building, for the underground excitation the best will be the lowest floor.

2 The Analyzed Building

The building chosen for analysis is masonry, a multi-family residential building located about 15 m from the high-speed railway. It is localized in Zawiercie, Wezilera 28St., close to the Central Railway Line leading from the south of Poland to the north. The building is cracked, in poor condition and what is important with wooden floors (see Fig. 1). This type of floor is very sensitive to vibrations coming through the ground and it is very popular in old masonry buildings in Poland.

Fig. 1. The front facade (from the side of the railway line) of the building.

3 Measurement Methodology

Human perception of vibrations in the building was analyzed using the most popular method of evaluation, known as a basic method acc. ISO standard [15], which is the Root Mean Squared method (RMS). It is worth noting that the duration of vibration in the Polish standard [16] is strictly defined. The duration of vibration determines the range in which the value of the amplitude of the vibration acceleration does not fall below 0.2 the value of the maximum amplitude in the recorded signal (Fig. 2).

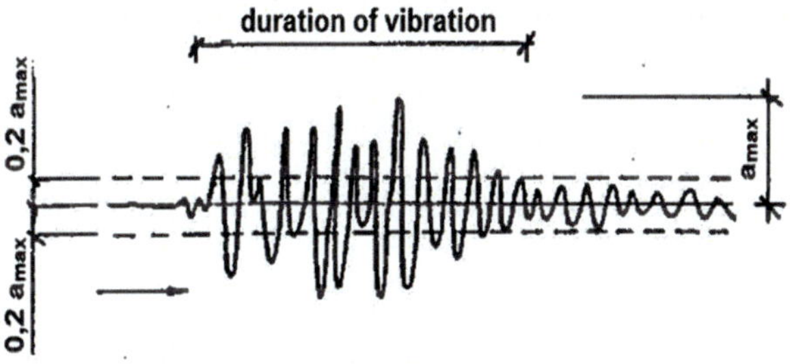

Fig. 2. Definition of vibration duration.

That is why the RMS analysis requires writing a procedure in Mathlab because the standard RMS analysis available in data analysis programs do not take into account the duration of vibration acc. [16]. RMS analysis was made in the 1/3 octave bands to make results more visible and to know in which frequency bands the comfort level is exceeded. Moreover acc. [16] the coefficient WODL was used to make results from the RMS method more visible:

$$\text{WODL} = \max\left(\frac{a_{RMS}}{a_z}\right) \tag{1}$$

where: a_{RMS} the acceleration RMS value, a_Z the acceleration RMS value equivalent to the threshold for the human perception of vibration in the Z direction.

The 87 dynamic events coming from the railway were recorded. At the same time, the train type was identified, its speed was measured, and the track on which it was passing was identified. Verification of recorded signals was carried out and all correctly recorded (without interference) were subjected to further analyzes. The selected waveforms of vibration acceleration from sensors in the vertical and horizontal directions located inside the building (in the structural center of the floor (Fig. 3) were subjected to analysis.

Fig. 3. Measurement point in the center of the floor.

The structure of the trains that passed during the 24 h vibration monitoring varied considerably (see Fig. 4).

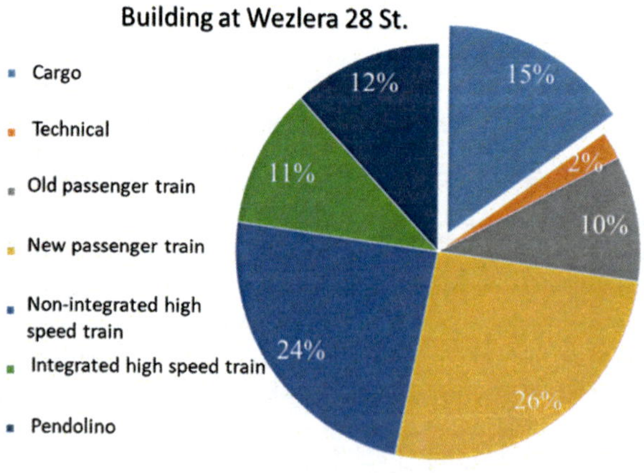

Fig. 4. Example of waveforms of vibration

4 Measurement Results

Example of recorded waveform of vibrations are shown in Fig. 5. It is a registration of an old passenger train ride, as a result of the analysis of which the highest value of exceeding the comfort threshold was obtained, equal to 16.65.

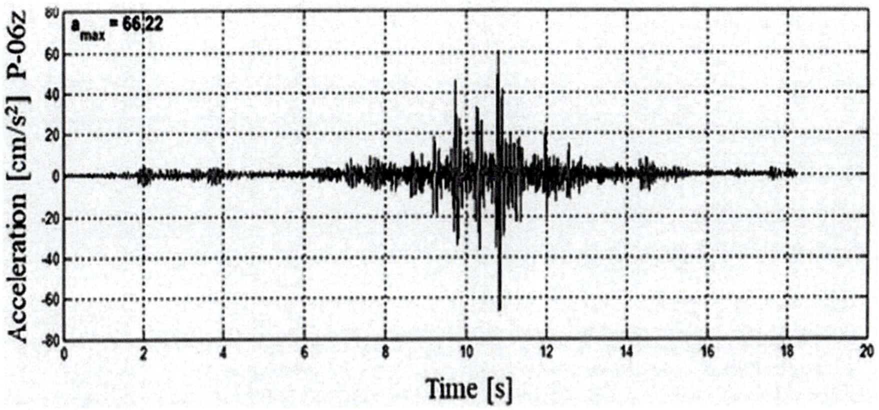

Fig. 5. Example of waveforms of vibration

After recording the vibration waveforms, the RMS procedure was carried out to assess the impact of vibrations on the building residents. In Fig. 6 example of RMS, made from waveforms are summarized.

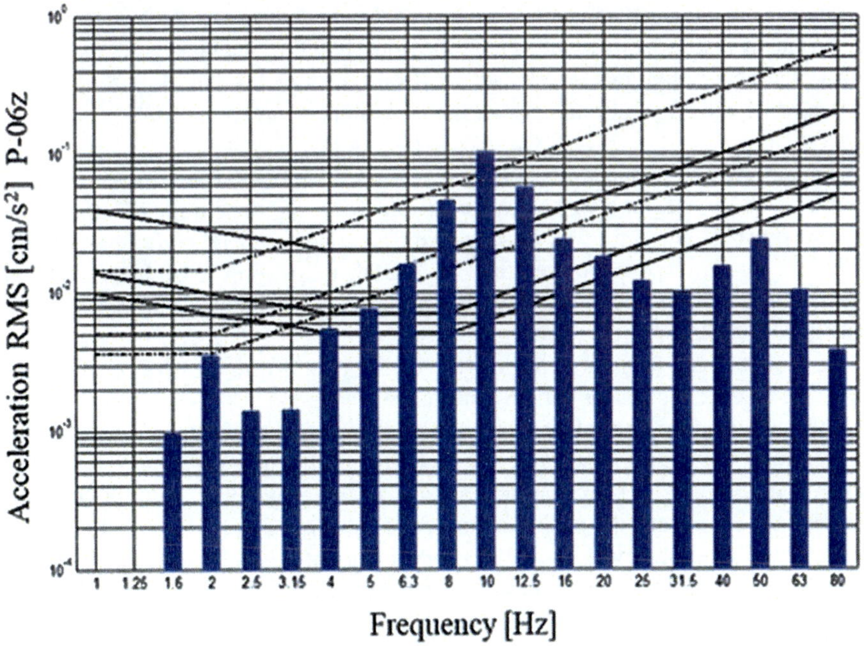

Fig. 6. Example of RMS analysis in 1/3 octave band

In the 50 of 87 dynamic events, the perception threshold of vibrations was exceeded. In seven of them also the vibrational comfort level was exceeded. The maximum value of the WODL coefficient is equal to 16.65 for a frequency band equal to 10 Hz. It is worth noting that exceedance of the comfort level of human perception of railway vibrations appeared for all types of trains (Table 1):

Table 1. Summary of the maximum value of WODL depending on the type of train.

Train type	Train speed [km/h]	Value of WODL [-]
Old passenger train	64	9.28
Old passenger train	82	16.65
Technical train	44	4.52
Non-integrated high-speed train	80	4.92
Cargo train	31	4.94
New integrated high-speed train	101	13.66
Old integrated high-speed train	92	6.21

As could be seen from Table 2 almost every type of train causes the exceedance of comfort level. To make possible predictions and because of a large amount of data, it was decided to perform a nonlinear regression analysis.

5 Discussion and Conclusion

There is many research on the subject of vibrational comfort, but mostly they concerned with whole-body vibrations in vehicles. For example, a very interesting research is in [18], in which the effect of road imperfections on human perception while driving a car is investigated.

Similar research was made in [19]. In this article, the relationship between the train speed and the WODL ratio was investigated. Only one, the same train, Pendolino type was investigated. The train was passing at different speed. It is worth noting that in Table 1 of this article can be seen that the WODL ratios mostly are higher with the higher train speed but there are some dynamic events that do not suit the trend (see Table 2).

As could be seen from these results (for the same train with controlled passing and speed) the tendency is visible but the correlation is not good.

The investigations show that there is a relationship between train speed and the WODL ratio, but the correlation is not good enough because of some events which outlier from the model. The better correlation is when the model of multiple regression is taken into account and not only the train speed but also its type is taken into account. The problem is still unsolved, although is very important especially in the prediction of annoyance in the buildings caused by trains passing nearby.

Table 2. Summary of the maximum value of WODL depending on the type of train [19].

No	Speed [km/h]	Central frequency [Hz]	WODL [-]
1	40	20	0.20
2	80	16	0.32
3	120	16	0.33
4	160	16	1.60
5	160	16	1.71
6	200	16	3.08
7	200	16	2.86
8	200	16	2.60
9	230	16	5.14
10	230	16	5.21
11	250	20	2.62
12	250	20	2.72
13	250	16	5.08
14	250	20	2.89
15	250	20	3.06

The measurements were carried out before the reconstruction of the railway line in order to select the appropriate vibration isolation. Well-selected vibration isolation improves the vibration comfort of people staying in buildings along the railway line under this study. It is worth noting here that poorly selected vibroinsulation may not only not affect vibrational comfort but even make it worse.

Acknowledgment. Scientific research results were financed by the European Union from the European Regional Development Fund within the Smart Growth Operational Programme 2014–2020. "Anti-vibration industrial floor system" project is implemented as a part of the Regional Science and Research Agendas (RANB) competition of the National Centre for Research and Development (NCRD).

References

1. Zouab, C., Wanga, Y., Moore, J.A., Sanayei, M.: Train-induced field vibration measurements of ground and over-track buildings. Sci. Total Environ. **575**, 1339–1351 (2017)
2. Howarth, H., Griffin, M.J.: Human response to simulated intermittent railway-induced building vibration. J. Sound Vib. **120**(2), 413–420 (1988)
3. Licitra, G., Fredianelli, L., Petri, D., Vigotti, M.A.: Annoyance evaluation due to overall railway noise and vibration in Pisa urban areas. Sci. Total Environ. **568**, 1315–1325 (2016)
4. Peris, E., Woodcock, J., Sica, G., Sharp, C., Moorhouse, A.T., Waddington, D.C.: Effect of situational, attitudinal and demographic factors on railway vibration annoyance in residential areas. J. Acoust. Soc. Am. **135**, 194 (2014)

5. Coermann, R.R.: The mechanical impedance of the human body in sitting and standing position at low frequencies. Hum. Factors: J. Hum. Factors Ergon. Soc. **4**, 227–253 (1962)
6. Arnberg, P.W., Bennerhult, O., Eberhardt, J.L.: Sleep disturbances caused by vibrations from heavy road traffic. J. Acoust. Soc. Am. **88**(3), 1486–1493 (1990)
7. Croy, I., Smith, M.G., Persson, W.K.: Effects of train noise and vibration on human heart rate during sleep: an experimental study. BMJ Open **3**, e002655 (2013)
8. Plumbj C. S., Meigs W., Human vibration perception. I. Vibration perception at different ages. Arch. Gen. Psychiatry **4**, 611–614 (1961)
9. Meh, D., Denislic, M.: Influence of age, temperature, sex, height and diazepam on vibration perception. J. Neurol. Sci. **134**(1–2), 136–142 (1996)
10. Tamura, Y., Kawana, S., Nakamura, O., Kanda, J., Nakata, S.: Evaluation perception of wind-induced vibrationin buildings. Struct. Build. **159**, 1–11 (2006)
11. Deo M.M., Mani A.K.: Whole body vibration on drivers seat and fender with fully loaded double axle tractor-trailers under different operating conditions: whole body vibration on drivers seat and fender. J. AgriSearch **8**(2), 149–154 (2021)
12. Sujatha C., Phaskara Rao P.V., Narayanan S.: Whole-body vibration exposure in Indian buses (1995)
13. Noviyanti A., Woodcock J.: A multidimensional valuation of the human perception of construction vibration. J. Phys. Conf. Ser. **1075**(1), 012044 (2018)
14. Schiavi, A., Rossi, L., Ruatta, A.: The perception of vibration in buildings: a historical literature review and some current progress. Build. Acoust. **23**(1), 59–70 (2016)
15. International Organization for Standardization, ISO 2631-1: mechanical vibration and shock: evaluation of human exposure to whole-body vibration - part 1: general requirements (1997)
16. PN-B-02171:2017-06: Evaluation of vibrations influence on people in buildings. Polish Standard (2017). (in Polish)
17. Pearson, K.: A Second Study of the Statistics of Pulmonary Tuberculosis: Marital Infection. Dulau & Co., London (1908)
18. Wenbo S., Ming L., Jingxuan G., Kaixuan Z., Evaluation of road service performance based on human perception of vibration while driving vehicle. J. Adv. Transp. **2020**(1), 1–8 (2020)
19. Pachla F.: The impact of the passenger train speed on the comfort of humans in a building close to the railway

VMI: Vibration and Acoustics of Musical Instruments

Correlation Between Dynamic Features of Unvarnished and Varnished New Violins and Their Acoustic Perceptual Evaluation

Mircea Mihalcica[1](✉), Alina Maria Nauncef[2], Vasile Ghiorghe Gliga[3], Mariana Domnica Stanciu[1], Silviu M. Nastac[4], and Mihaela Campean[5]

[1] Department of Mechanical Engineering, Transilvania University of Brasov, B-dul Eroilor 29, 500036 Brasov, Romania
mihalcica.mircea@unitbv.ro
[2] Faculty of Music, Transilvania University of Brasov, B-dul Eroilor 29, 500036 Brasov, Romania
[3] Gliga Instrumente Muzicale, Str. Pandurilor 120, Reghin, Romania
[4] Faculty of Engineering and Agronomy, "Dunarea de Jos" University of Galati, 810017 Braila, Romania
[5] Faculty of Furniture Design and Wood Engineering, Transilvania University of Brasov, B-dul Eroilor 29, 500036 Brasov, Romania

Abstract. This paper reports an investigation into correlations between dynamic features of unvarnished and varnished new violins and their acoustic perceptual evaluation. Seven violins with modified plate thicknesses were tested before varnishing (in white) and after varnishing. During the first stage, the violins were dynamically tested and then acoustically evaluated by specialists in the field using a blind test. In the second stage, the testing procedure was resumed, but on the varnished violins. Thus, in the dynamic analysis of the violins were identified their signature modes, the dominant frequency, the frequency spectrum, the quality factor and their damping. In using perceptual acoustic analysis, five acoustic criteria were investigated (sound clarity, sound warmth, brightness of tone, equal sound on strings, amplitude of sounds), based on the musical audition of each violin studied, without the subjects seeing the violin or knowing their physical characteristics.

Keywords: Frequency spectrum · Acoustic perception · Violin · Varnish

1 Introduction

Research on the acoustic quality of historical violins has shown that the type of varnish used by the old luthiers has a very important role [1]. Some researchers [2, 3] highlighted the fact that the lacquer influences the dynamic response of the violin boards, the magnitude of the influence being correlated with the direction of the wood grain. The greatest influence of the varnish occurs in the first 3 months after finishing, then the lacquer film undergoes changes in its amorphous structure, which leads to a decrease in negative

Z. Dimitrovová et al. (Eds.): WMVC 2022, MMS 125, pp. 857–864, 2023.
https://doi.org/10.1007/978-3-031-15758-5_88

effects on the acoustic quality of the wood. Investigations into the chemical composition of varnishes and the number of layers applied to old violins are being developed by non-destructive and non-invasive means, especially in the case of well-functioning old violins [4–6]. Pioneering work on the influences of primer on the acoustic quality of resonance wood [7] revealed that the samples covered with primer recorded improved features compared to those covered by lacquer. In the audible frequency range, the violin has numerous vibrational modes. The signature modes of these violins are referred to [8] as cavity modes (A0, A1), corpus modes (CBR or C bouts rhomboidal), and main body resonance (B1− and B1+). There are two low frequency modes associated with the pressure variation in the cavity of the violin box, A0 and A1. A0 and A1 are coupled modes. A0 occurs around 270 Hz and is called the Helmholtz resonance. A1 is a first standing wave along length of the box typically occurring in the range 470–490 Hz and is the response of the structure to the "inflating" and "deflating" of the upper and lower bouts. CBR - is the lowest corpus frequency mode at around 400 Hz [8].

The goal of this study was to examine the acoustic quality of white violins in comparison with varnished violins, using dynamic testing and a psychoacoustic survey.

2 Materials and Method

2.1 Materials

For the tests, 14 violins of the same anatomical quality class of wood were manufactured (class A - according to [8–10]). Of these, seven violins were analyzed in white (unfinished) and 7 violins were varnished with an oil-based varnish applied in approximately 10 to 12 layers (Fig. 1).

Fig. 1. Types of investigated violins.

The unvarnished violins have the codes: A00C1, AM2C1, AM4C1, AM6C1, AP2C1, AP4C1, AP6C1. The varnished violins have the codes: A00C1F, AM2C1F, AM4C1F, AM6C1F, AP2C1F, AP4C1F, AP6C1F.

The wooden material (both spruce and maple wood), used to manufacture the experimental violin plates, originated from the Gurghiu region situated in the Eastern Carpathians and it was selected so as to fit the requirements of Maestro (class A) violins [8]. Tests regarding the anatomical characteristics and density of the selected material were carried out. Thus, the spruce violin plates are characterized by a total ring width (TRWi) less than 2.5 mm, the latewood proportion within the annual ring was LWPi $\leq 20\%$; the width difference between two consecutive growth rings $\Delta i \leq 0.5$ mm. The maple violin plates for the back are very curly, having TRWi ≤ 1.3 mm and a wavelength of the curly grain of maple $\lambda \leq 3$ mm [9, 10].

2.2 Methods

Dynamic Test

Dynamic testing consists of exciting each violin supported on elastic elements with a B&K 8204 impact hammer, and the output signal was captured using a microphone. The generated signals were transmitted via a signal conditioning device to a NI USB-9233 dynamic data acquisition board produced by National Instruments (Austin, USA), connected to a laptop. The signal was viewed using a special application developed in NI-LabVIEW ©, and the graphical data was processed using the MATLAB © program. The procedure was presented in previous research [10]. After the analysis in the time and frequency of the signals, based on the exponential curve of the damping and the spectral composition of the signal for each violin, the values of the natural frequencies of the tested violins, the damping factor and the analysis in time were extracted, these being correlated with aspects of wood structure and violin geometry. For each violin, the signals from three successive tests were taken.

Psycho-Acoustic Survey

The survey allowed the violinists to rate all violins recordings on a continuous scale from 0 to 5 based on the most relevant acoustic features which were: sound clarity; sound warmness; brightness of tone; amplitude of sounds; equal sound on strings, criteria that were in turn established on the basis of a previous survey. In the first stage, for each parameter and each violin, the average of the marks given by the respondents was calculated, obtaining a ranking from the point of view of the audience experience classified by gender and age, for each violin and acoustic criterion assessed. Then, in order to achieve the ranking regarding the acoustic quality of the violins, the averages obtained by each violin in relation to each acoustic criterion were comparatively analyzed. Finally, the global ranking on the acoustic quality of the violins was calculated by summing the averages of the marks given to all the criteria for each violin [11, 12].

31 participants having experience in the musical domain took part in this survey (21 females; 10 males;), 15 participants having more than 21 years of experience and 16 participants with experience between 6–20 years. Their task was to listen to the musical recordings of the violins under study and to evaluate the acoustic parameters mentioned above, with marks from 1 to 5, where 1 represents the lowest score and 5 – the greatest score.

The music recordings were made under the same conditions for all violins and by the same violinist [11, 12]. The method of testing and analysis of the results was designed based on similar tests presented in references [13–15], but the study authors correlated the investigation method with the study objective.

3 Results and Discussion

3.1 Frequencies Spectrum

Following the dynamic test, it was found that the dynamic response of the white and lacquered violins has both similarities and differences. Figure 2 shows the frequency domain analyses for vibration of unvarnished and varnished violins. The frequency responses differ in the case of those two states: the white violins present more harmonics than varnished violins. Also, the dominant frequency depends on surface treatment of violins. It is generally observed that covering the violins with varnish leads to a decrease in the vibration amplitude of the first resonant frequency and an increase in the amplitude of the octave (mode 5).

Table 1 shows the values for the first, fifth and dominant frequency obtained for lacquered and unpainted violins. From a dynamic point of view, according to research [16], each violin mode is characterized by a shape, characteristic value(s) of frequency, damping parameter and impedance. In this study, the characteristic values of main frequencies are presented. The first frequency corresponds to first mode and the fifth frequency is related to fifth mode. The ratio between mode 5 and mode 1 is generally an octave. It can be noticed that there are some differences between the violins' frequencies. This means that a relatively small difference in frequency indicates a relatively large difference in stiffness which is due to the anatomical and physical properties of resonance spruce and maple wood used for the violin body [16].

Table 1. The values for the 1st, 5th and dominant frequency for lacquered and unpainted violins

Violins	1st frequency [Hz]		5th frequency [Hz]		Dominant frequency [Hz]	
	Unvarnished	Varnished	Unvarnished	Varnished	Unvarnished	Varnished
AM6	238.0	266.6	488.2	524.9	238.0	266.6
AM4	244.1	270.0	479.1	543.2	402.8	496.6
AM2	244.1	266.0	488.2	535.4	390.6	453.7
A00	250.2	273.5	496.6	543.2	396.7	496.6
AP2	256.3	273.8	517.2	543.2	256.3	497.4
AP4	256.3	273.8	494.3	543.9	256.3	434.1
AP6	250.2	273.5	488.2	548.7	250.2	273.5

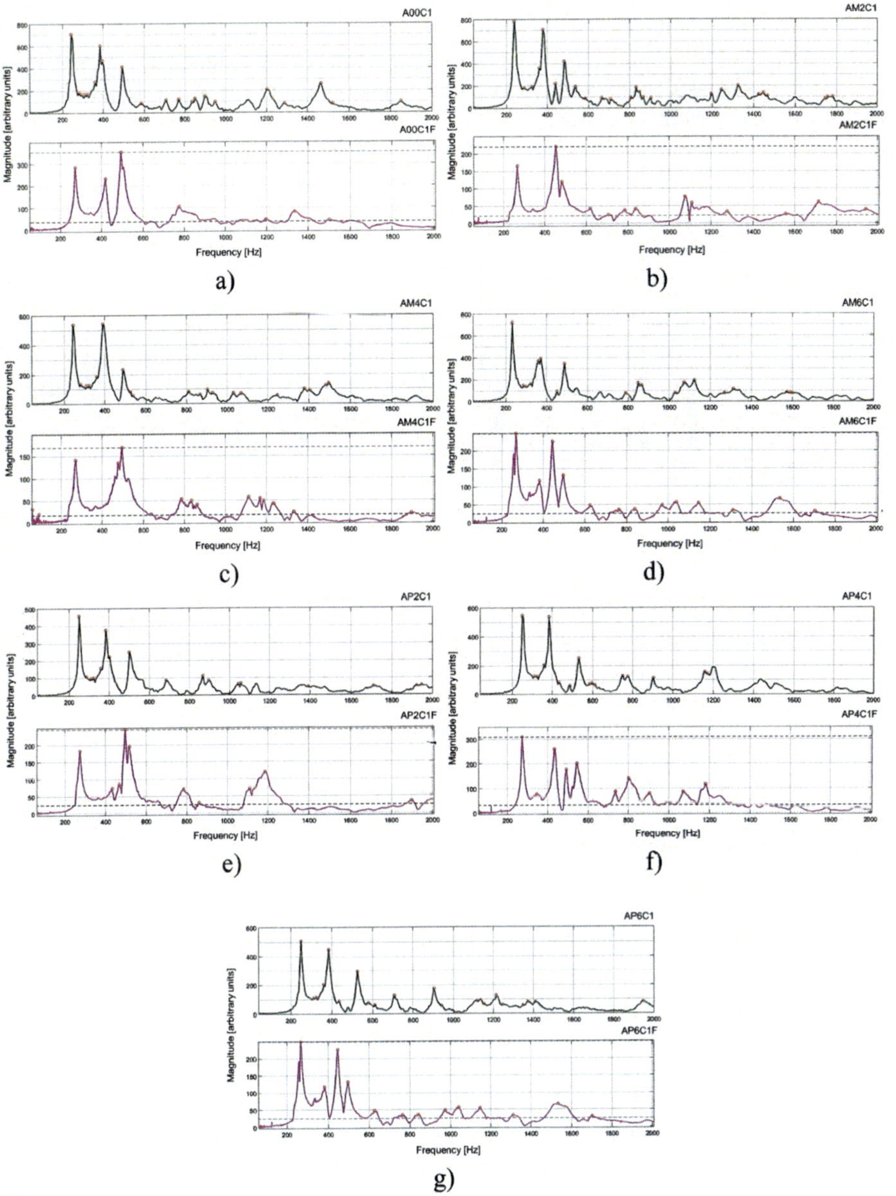

Fig. 2. Frequencies spectrum of tested violins: a) A00C; b) AM2C; c) AM4C; d) AM6C; e) AP2C; f) AP4C; g) AP6C.

The application of the lacquer led to an increase in resonant frequencies by about 5–13% (Fig. 3, a). According to [16], mode 2 reflects mainly the cross-grain stiffness of the wood, while mode 5 reflects the long-grain stiffness plus a significant amount of cross-grain stiffness. The unfinished violins AM2, A00 and AP2 have a ratio between mode 5 and mode 1 of one octave, compared to the others where the value varies by −0.07 ÷ +0.05. The application of the varnish also has the advantages of compensating for variations due to the structure and the elastic properties of the wood, in the sense of diminishing the differences between octaves. Thus, the ratio between mode 5 and mode 1 of finished violins varies between −0.03 and +0.016.

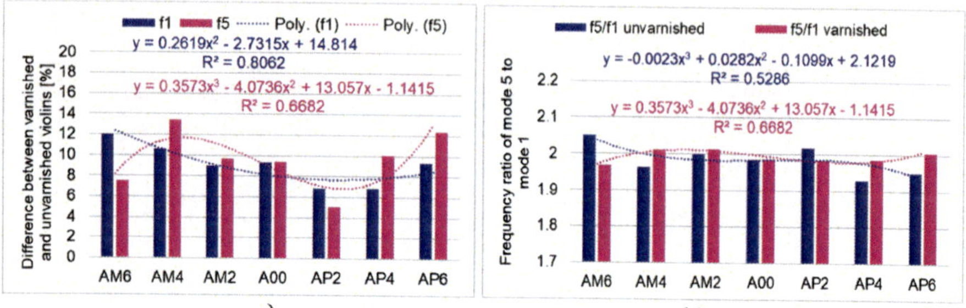

a) b)

Fig. 3. Comparison between analyzed violins: a) in terms of percentage increases of resonance frequency for mode 1 and mode 5, with surface coating modification; b) frequency ratio between mode 5 and mode 1 of unvarnished and varnished violins.

3.2 Overall Preference Ratings of the Unvarnished and Varnished Violins

In Fig. 4 are presented the overall preference ratings reported for the acoustic quality of violins. One can notice that varnished violins (tested after one month of varnish coating) recorded the lower values, regardless of the acoustic criteria. The brightness of tone scored more highly for violins AP6C1, AP2C1, AP6C1F, AM4C1, AP4C1, while the violins AM4C1F, A00C1F were evaluated with the lowest scores (2.84–2.9).

Violins AM4C1, AP2C1, AP6C1 are considered to have the highest clarity of sound compared to the others. In terms of sound amplitude, the highest values were obtained by violins AP6C1, AP6C1F, AP2C1, and the lowest, by violins A00C1F, AM4C1F, AM2C1F and AM6C1F. The evaluation for the warmness of sound for the violins led to the following result: the violins rated with warmest sound are: AP6C1, AP6C1F, A00C1, AP4C1, AP4C1F (over 3.5 points), and the weakest values were registered in the range 3.3–3.4. The equal sound on the strings was appreciated with highest scores for violins AP6C1 and AP6C1F, while the violins AM4C1F and AM6C1F obtained the lowest scores.

From the point of view of the total ranking, for the analyzed violins, the best ranked are violins AP6C1, AP6C1F, AP2C1, AP4C1, AM4C1 with scores over 70%, and the lowest rated are AM4C1F, A00C1F, AM6C1F, AM2C1F and AP2C1F.

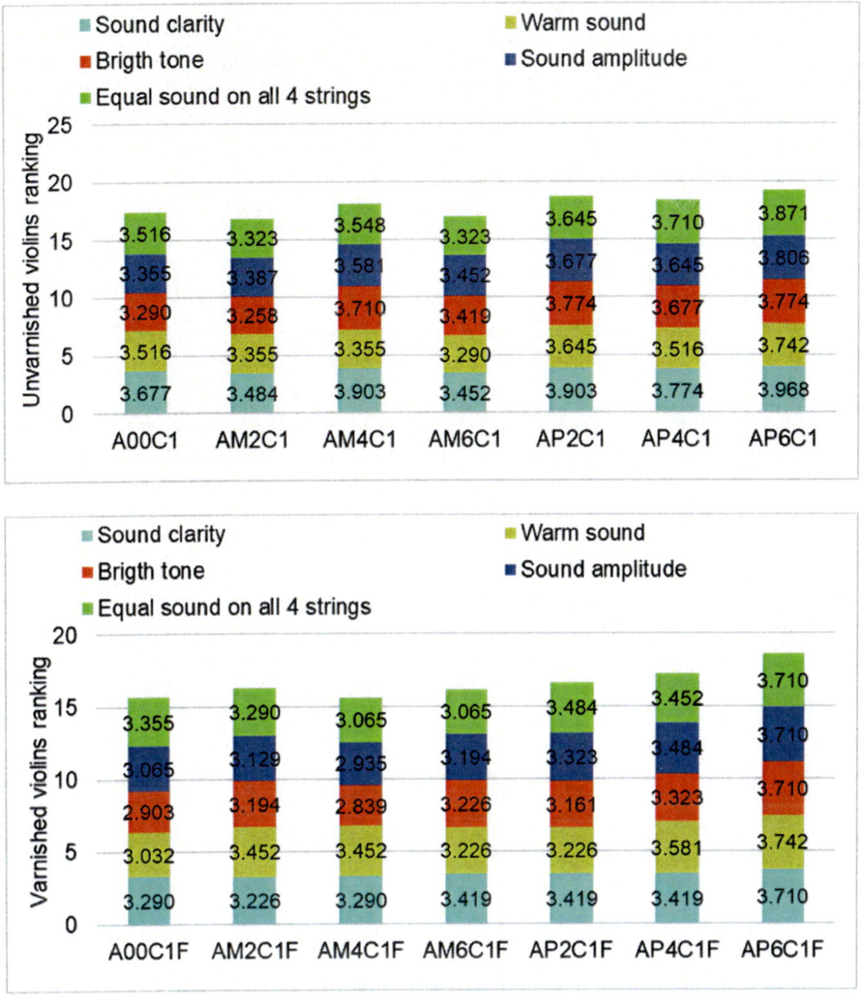

Fig. 4. The psycho-acoustic evaluation of unvarnished and varnished violins: brightness of tone, sound clarity, sound amplitude, warmness of sound, equal sound on all four strings

4 Conclusion

The paper focused on the dynamic and acoustic evaluation of violins before and after varnishing. From a dynamic point of view, the layers of lacquer led to a change in the frequency spectrum compared to white violins. Using only the objective method, it is difficult to decide whether the lacquer has improved or decreased the acoustic properties of the violins. For analysis of the psychoacoustic effect, a survey was prepared and different musicians answered the questions. Based on the acoustic impressions, it turned out that in most cases the violins finished with an age of a month have a lower quality than the unfinished ones. The studies will continue with recordings of the finished violins one year after the application of the varnish to determine if in time, the drying of the varnish improves the acoustic quality of the instrument.

Acknowledgments. This research was funded by a grant UEFISCDI, PN-III-P2-2.1-PED-2019-2148, project number. 568PED/2020 MINOVIS.

References

1. Schelleng, J.C: Acoustical effects of violin varnish. J. Acoust. Soc. Am. **44**(1175) (1968). https://doi.org/10.1121/1.1911243
2. Lämmlein, S.L., Van Damme, B., Mannes, D., Willis, F., Schwarze, M.R., Burgert, I.: Violin varnish induced changes in the vibromechanical properties of spruce and maple wood. Holzforschung **74**(8), 765–776 (2020). https://doi.org/10.1515/hf-2019-0182
3. Ono, T.: Effects of varnishing on acoustical characteristics of wood used for musical instrument soundboards. J. Acoust. Soc. Jpn. (E) **14**(6), 397–407 (1993)
4. Hudoba de Badyn, M.: The Effects of Violin Varnish on the Acoustic Properties of Picea Engelmannii. R. Science One Research Projects (2011). Accessed 26 Feb 2022. https://open.library.ubc.ca/collections/undergraduateresearch/51869/items/1.0107235
5. Fiocco, G., et al.: A micro-tomographic insight into the coating systems of historical bowed string instruments. Coatings **9**(81) (2019). https://doi.org/10.3390/coatings9020081
6. Sedighi Gilani, M., Pflaum, J., Hartmann, S., Kaufmann, R., Baumgartner, M., Schwarze, F.W.M.R.: Relationship of vibro-mechanical properties and microstructure of wood and varnish interface in string instruments. Appl. Phys. A **122**(4), 1–11 (2016). https://doi.org/10.1007/s00339-016-9670-1
7. Bissinger, G.: Structural acoustics of good and bad violins. J Acoust. Soc. Am. **124**, 1764–1773 (2008)
8. Toth, A.I.: The influence of the primer on the sound quality of the violins (in Romanian language). Industria Lemnului **1**, 33–45 (1979)
9. Dinulica, F., Albu, C.T., Borz, S.A., Vasilescu, M.M., Petritan, I.C.: Specific structural indexes for resonance Norway spruce wood used for violin manufacturing. BioResources **10**(4), 7525–7543 (2015). https://doi.org/10.15376/biores.10.4.7525-7543
10. Dinulică, F., Stanciu, M.D., Savin, A.: Correlation between anatomical grading and acoustic-elastic properties of resonant spruce wood used for musical instruments. Forests **12**, 1122 (2021). https://doi.org/10.3390/f12081122
11. Mihalcică, M., et al.: Signature modes of old and new violins with symmetric anatomical wood structure. Appl. Sci. **11**(23), 11297 (2021). https://doi.org/10.3390/app112311297
12. Nauncef, A.M., Mihălcică, M., Roșca, I.C., Gliga, G.V., Marc, R.: Interdisciplinary approach to assessing the acoustic quality of violins, Conferința internațională Contemporary challenges in artistic education, National University of Arts, George Enescu Iaşi, 11–13 November 2021 (2021)
13. Saitis, C., Fritz, C., Scavone, G.P.: Sounds like melted chocolates. How musicians conceptualize violin sound richness. In: Proceedings of ISMA 2019, International Symposium on Music Acoustics, Detmold, Germany, 13–17 September, pp. 50–57 (2019)
14. Saitis, C., Fritz, C., Scavone, G. P., Guastavino, C., Dubois, D.: Perceptual evaluation of violins: a psycholinguistic analysis of preference verbal descriptions by experienced musicians. J. Acoust. Soc. Am. **141**, 2746–2757 (2017). https://doi.org/10.1121/1.4980143
15. Zacharakis, A., Pastiadis, K., Reiss, J.D.: An interlanguage unification of musical timbre: bridging semantic, perceptual, and acoustic dimensions. Music Percept. Interdisc. J. **32**, 394–412 (2015). https://doi.org/10.1525/mp.2015.32.4.394
16. Curtin, J.: Tap tones and weights of old Italian violin tops. Violin Soc. Am. Pap. **1**(2), 1–13 (2006)

Dynamic Analysis of Musical Triangle

Mariana Domnica Stanciu[1](✉), Mihai Trandafir[1], Silviu M. Nastac[2], and Voichita Bucur[3]

[1] Department of Mechanical Engineering, Transilvania University of Brasov, B-dul Eroilor 29, 500036 Brasov, Romania
mariana.stanciu@unitbv.ro

[2] Faculty of Engineering and Agronomy, "Dunarea de Jos" University of Galati, 810017 Braila, Romania

[3] School of Science, RMIT University, GPO Box 2476, Melbourne, VIC 3001, Australia

Abstract. The paper deals with dynamic analysis of a two types of musical triangles, one made from stainless steel and other from alloy. The experimental method applied in this study consisted in exciting the structure with the impact hammer and recording the acoustic signals with the help of the microphone located near the sample, the acquired signals being subsequently processed in the program developed in Matlab. The time and frequency analysis showed for each drag, the frequency spectrum, the dominant frequency and the damping. Then, finite element analysis and damping simulation of the musical triangles was done with Simcenter 3D software, based on real properties of materials and experimental signals. The results between the experimental and the numerical analysis showed a good correlation between the two methods. Thus, it was observed that at the frequency of 878 Hz, the triangle vibrates in a torsional mode, followed by longitudinal vibrations (at the frequency of 1053 Hz), and at the dominant frequency, of 1877 Hz, the triangle vibrates in a transverse mode, in the plane or. The higher modes corresponding to the frequencies 3089 Hz and 3454 Hz are torsion modes.

Keywords: Triangle · Dynamic analysis · Numerical simulation · Vibration modes

1 Introduction

The modern triangle has several configurations, with a side usually of 20 cm or 30 cm and can reach up to 40 cm or 45 cm. The triangles are made of a bent rod with a circular cross section, produced by folding the metal around a template. Its shape can be an equilateral or isosceles triangle. The orchestral triangle is actually a combination of straight and curved bars, so it is an instrument with a relatively complex structure. The triangle is struck with a metal hammer and is by an elastic thread in the upper corner of the suspended instrument [1–3]. The triangle struck on different sides generates longitudinal and bending modes of vibration, which are excited perpendicularly or parallel to the plane of the triangle.

For transverse modes, the vibration modes of a triangle are almost identical to those of the straight bar from which the triangle is bent [1]. This statement supported the

Z. Dimitrovová et al. (Eds.): WMVC 2022, MMS 125, pp. 865–872, 2023.
https://doi.org/10.1007/978-3-031-15758-5_89

hypothesis that the shape of the triangle does not have a large effect on its modes of vibration. However, [2] noted that the bending of a bar influences the voltage distribution, which depends on the moments of inertia at the ends of a bar. For bending vibrations in the plane, there is an important discontinuity in impedance at each bending angle. The transverse vibration inside a side is coupled with the longitudinal vibration on the adjacent side. Therefore, each side of the triangle will have its own vibration modes and will be coupled to another side, to form sets of three own modes, for a side of the same length [3]. According to [4], the high quality triangle has the following dominant eigenfrequencies: $f0 = 1539.1$ Hz; $f16 = 4239.6$ Hz; $f17 = 5658.9$ Hz; $f21 = 7795.9$ Hz. In contrast, the low-quality triangle has a single dominant frequency [4]. Amplitude is the sound pressure relative to the excitation energy for 5 s. The spectrum of the high-quality triangle is characterized by its dominant frequencies. Also, the damping of the quality triangle has a lower value, so the sound persists longer. Previous studies confirm that the geometric aspects of the triangle have a decisive effect on the modes of vibration [5, 6]. The low frequency vibration modes are mainly bending and are strictly related to the size and mechanical properties of the material from which the instrument is made. At a constant length, the frequency increases with the vibration modes. As the side length increases, so does its frequency. This means that triangles with shorter sides vibrate at much higher frequencies than those with longer sides. Regarding the influence of the angles of the triangle, in general, the natural frequencies of the isosceles triangle with the side of about $L = 20$ cm, are higher than for other types of triangles [5–7].

The goal of this study was to determine the natural frequencies of two types of triangles by means of experimental method and numerical one.

2 Materials and Method

2.1 Materials

The study analyzed three triangles with a configuration similar to an open equilateral triangle (Fig. 1). The differences between the triangles were the material of the bar and the size of the sides. The physical characteristics of the triangles are shown in Table 1.

Table 1. The physical characteristics of the triangles.

Triangles	Dimensions [mm]			The angle [°]	Diameter [mm]	Mass [g]	Material
	A	B	C				
1	149	161	149	60	7.84	166.971	Stainless steel
2	162	170	162	60	9.60	88.895	Aluminum
3	214	222.5	214	60	9.60	118.644	Aluminum

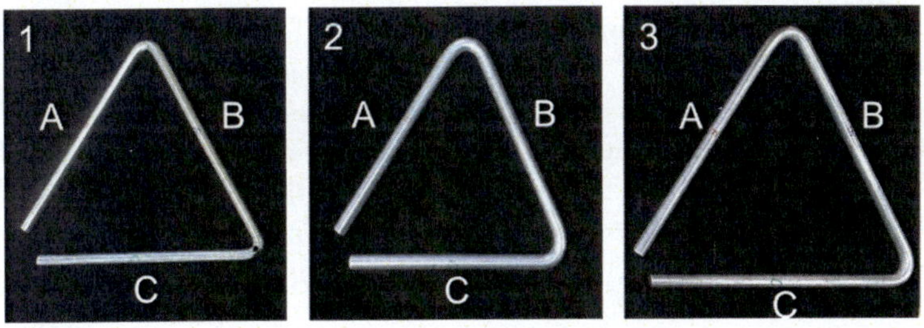

Fig. 1. Types of investigated triangles.

2.2 Methods

Experimental Modal Analysis

The test method consisted of recording the sound pressure measured by means of a microphone fixed near the triangle. The experimental stand shown in Fig. 2 was composed of: the triangle (3) attached to the support (1) by means of an elastic element (2) at a height of 250 mm from the surface on which the support is placed. The main microphone (4) is mounted on a stand (5) at a height of 230 mm and its role is to capture the sound of the triangle when a force is applied by means of the impact hammer (9). The recorded information is then transmitted via acquisition board (7) to a laptop (8) for processing. The structure was hit with a hammer on each side. Three tests were performed on each side of the triangle, resulting in a total of 27 records. The time interval in which the data was acquired was 0–3 s and the mechanical pulse was applied to the median area of each side using the impact hammer [7]. In Fig. 2 shows the components of the dynamic test system. After the acquisition of the signals, they were processed through a program developed in Matlab to determine the time and frequency response of the tested samples.

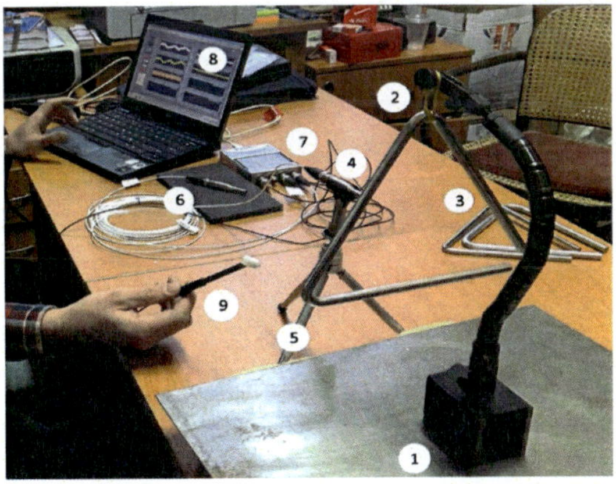

Fig. 2. The experimental stand.

Modal Analysis Using the Numerical Method

The Simcenter 12 software from the Mechanical Engineering Department, Transilvania University of Brasov, was used to create the geometric model. Considering that the geometric model of the triangle is made in three dimensions, for discretization, three-dimensional finite elements with 4 nodes, of tetrahedral shape were chosen: CTETRA (4). The structure was discretized in 1988972 finite elements and 38066 nodes [6, 7]. In the finite element analysis, the determination of the boundary conditions was done by blocking the translational movements on the three axes of a node in the real clamping area of the triangle, as can be seen in Fig. 3.

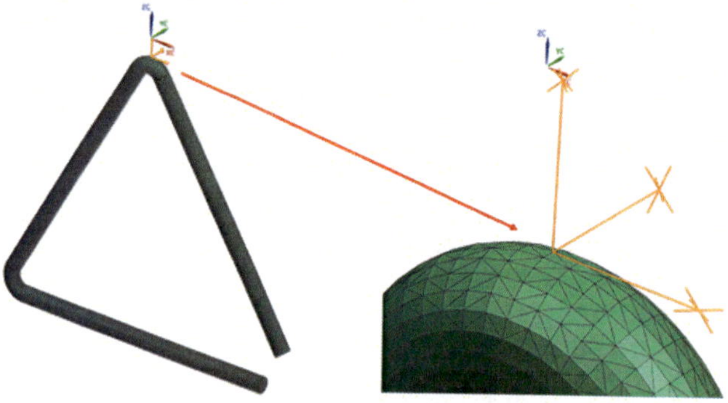

Fig. 3. The boundary conditions of triangle.

The values of the elastic characteristics of the materials were determined experimentally and are presented in Table 2. The experiment consists of tensile tests performed on the Lloyd Instruments LS100Plus universal mechanical testing machine with a maximum load capacity of 100 kN, from the Department of Mechanical Engineering, Transilvania University of Brașov. The elongation of the sample during the stress was measured with the EPSILON extensometer. The stress rate used during the test was 5 mm/min. The data acquisition was done with Nexygen Plus software, and the processing of the characteristic curves was done in excel.

Table 2. The experimental elastic characteristics of the materials.

Material type	Elasticity modulus E [MPa]	Density [kg/m^3]	Poisson coefficient
Stainless steel	184479	7813	0.30
Aluminum	68382	2706	0.33

3 Results and Discussion

After processing the experimentally acquired signals, the time and frequency response of the three types of tested rods was obtained, as can be seen in Fig. 4. It was found that the first proper frequency of triangles with similar dimensions, but made of different materials have relatively close values: 168.2 Hz (Triangle 2 - aluminum) and 159.5 Hz, (Triangle 1 - stainless steel). The aluminum triangle T2 has a mass about twice the mass of the stainless steel triangle, although the length of the sides is about 8% longer than that of the stainless steel triangle T1. It is interesting that the increase of the sides of the aluminum triangle by 32% (T3), led to the decrease of the value of the first natural frequency by about 42% (from 168 Hz - triangle T2, to 96.5 Hz - triangle T3) [6, 7].

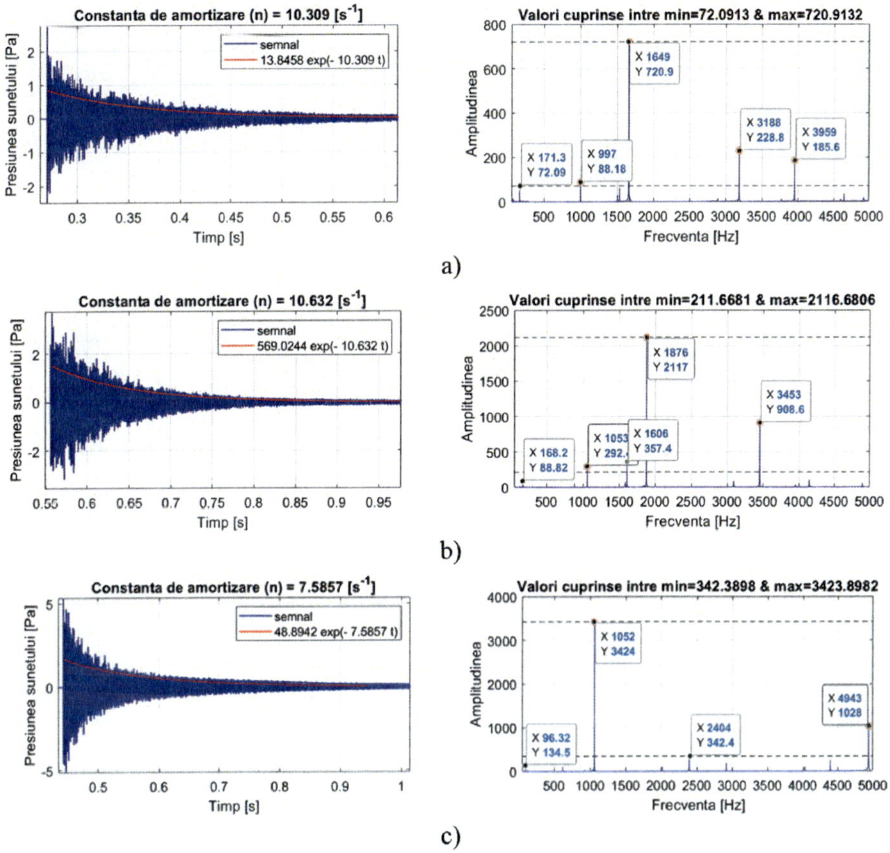

Fig. 4. Time and frequency analysis of triangles after excitation of the A side: a) triangle 1; b) triangle 2; c) triangle 3.

Analyzing the dominant frequency (the one with the highest amplitude in the frequency spectrum), it is found that the aluminum triangle type T2 has the highest dominant frequency (1876 Hz), compared to the stainless steel triangle (1657 Hz). Increasing the sides of the triangle (T3) in aluminum reduces the dominant frequency by about 43% compared to the triangle with the smaller sides (T2).

Bestle 2014 [4] considers that the sound of a triangle is pleasant if it contains a fundamental sound of several frequencies that are significantly louder than the remaining sound. This remaining sound is without a key, and the sound level is significantly lower than the dominant sound. The spectrum of the high-quality triangle contains four tones that are significantly louder than the others, namely: 1539.1 Hz; 4239 Hz; 5658 Hz; 7795 Hz (Table 3).

Table 3. The experimental and numerical results of modal analysis.

Modes/frequency [Hz]	Triangle 1		Triangle 2		Triangle 3	
	FEA	EXP	FEA	EXP	FEA	EXP
1	156.37	159.5	168.27	168.2	97.669	96.32
2	162.36	-	172.75	-	99.58	-
3	269.35	-	300	-	172.4	-
4	805.28	-	878.26	878.1	519.67	-
5	978.74	997.97	1056.66	1052.85	622.26	612.8
6	1465.83	-	1560.88	-	912.36	899.7
7	1499.43	1532	1601.97	1606	932.88	-
8	1627.43	-	1812.41	-	1050.32	1052.28
9	1645.37	1657.10	1857.84	1876	1072.29	-
10	2828.32	-	3060.17	3089	1866.02	-
11	3137.89	3188	3450.05	3454	2039.44	-
12	3696.39	-	3902.5	-	2362.94	-
13	3869.76	-	4126.86	-	2427.51	-
14	4465.99	-	4714.65	-	2875.17	-
15	4589.69	4652	4780.72	-	2966.86	2917
16	4829.16	-	5110.17	-	3349.27	-
17	5311.73	-	5411.21	-	3791.14	-
18	6295.22	-	6840.37	-	4074.8	-
19	6657.23	-	6915.81	-	4403.28	4402

Figure 5 selectively shows the vibration modes of the stainless steel triangle obtained from the numerical analysis. Dynamically, the modal shapes are similar for aluminum triangles. Thus, it can be seen that the first vibration mode occurs in the plane of the triangle (transverse mode), at a frequency of 156 Hz. The second mode, at 162 Hz, is a vibration mode that occurs through the vibrations of the free sides of the triangle outside the plane of the triangle, also called torsion mode. The third mode of vibration, produced at a frequency of 269 Hz, is manifested by longitudinal vibrations of the triangle [7]. These are the main vibration modes of the triangle, the following similar modes occurring at higher frequencies, as can be seen in Fig. 5.

Since the experimental dynamic analysis could not determine the vibration modes, in Fig. 6 represents the correlation between the numerically obtained eigenmodes and the experimentally obtained eigenvalues, especially since the error between the two methods is less than 2%. It can be seen that at 878 Hz, the triangle vibrates in a torsional mode, followed by longitudinal vibrations (at 1053 Hz), and at the dominant frequency of 1877 Hz, the triangle vibrates in a transverse mode, in its plane. The higher modes corresponding to 3089 Hz and 3454 Hz are torsion modes.

Comparing the results obtained experimentally with the numerical ones, it can be seen that the differences are less than 2%.

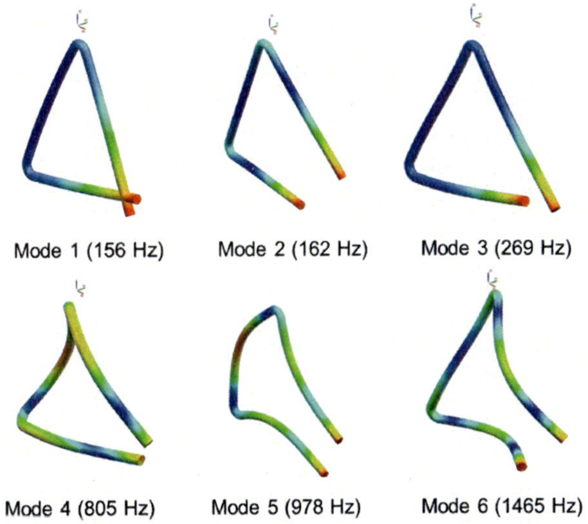

Fig. 5. The first modes of vibration of the stainless steel triangle

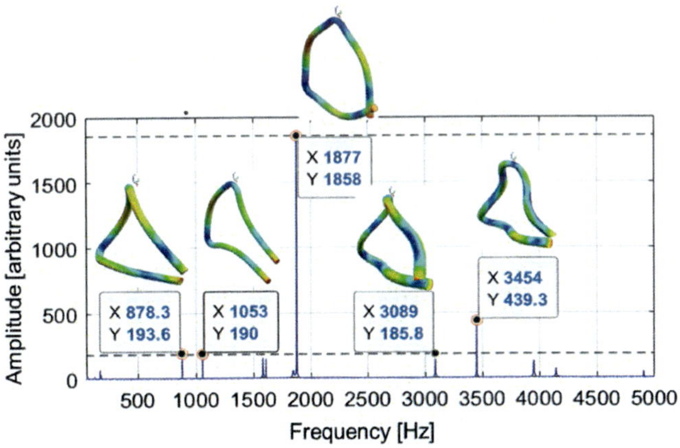

Fig. 6. The correlation between the values of the eigenfrequencies obtained experimentally and the eigenmodes determined numerically.

4 Conclusion

The aim of this paper was to determine the response in time and frequency of the three types of rods, resulting in the frequency spectrum and the damping factor. It was observed that the value of the first frequency of triangles with the same dimensions but from different materials (T1 - stainless steel, T2 - aluminum), differs by about 5.17%. On the other hand, the larger aluminum triangle (T3) has a natural frequency value 42.74% lower than T2. Also here it was observed that when the side B is excited, located between the open sides A and C, the first frequency has a different value than the case of hitting the side A or C, regardless of the analyzed structure. This is due to the fact that side B is fixed at both ends in terms of contour conditions, while sides A and C have a fixed end and a free end. However, the following determined natural frequencies have approximately identical values for all three excited sides: 996 Hz for T1, 878 Hz for T2 and 612 for T3.

Acknowledgments. This research was funded by a grant UEFISCDI, PN-III-P2-2.1-PED-2019-2148, project number. 568PED/2020 MINOVIS.

We are grateful to the Siemens - Brasov company that provided the SimCenter software within the master program in Simulation and Testing in Mechanical Engineering within the Faculty of Mechanical Engineering of the Transilvania University of Brasov, Romania.

References

1. Rossing, T.D.: Acoustics of percussion instruments. Am. Assoc. Phys. Teach. **14**, 546–556 (1976)
2. Dunlop, J.I.: Flexural vibration of triangle. Acoustica **55**(4), 250–253 (1984)
3. Gough, C.: Musical acoustics. In: Rossing, T. (ed.) Springer Handbook of Acoustics. Springer Handbooks, pp. 533–667. Springer, New York (2007). https://doi.org/10.1007/978-0-387-30425-0_15
4. Bestle, P., Hanss, M., Eberhard, P.: Experimental and numerical analysis of the musical behaviour of triangle instruments. In: Proceedings of the 5th European Conference of Computational Mechanics (ECCM V), Barcelona, pp. 3104–3114 (2014)
5. Bucur, V.: Handbook of Materials for Percussion Instruments. Springer International Publishing, Cham (2022). https://doi.org/10.1007/978-3-030-98650-6
6. Stanciu, M.D., Trandafir, M., Dron, G., Munteanu, M.V., Bucur, V.: Numerical modal analysis of kinked bars–triangle case of study. In: The 9th International Conference on Modern Manufacturing Technologies in Industrial Engineering, 23rd–26th June 2021. MODTECH (2021)
7. Stanciu, M.D., Nastac, S.M., Bucur, V., Trandafir, M., Dron, G., Nauncef, A.M.: Dynamic analysis of the musical triangles—experimental and numerical approaches. Appl. Sci. **12**, 6275 (2022). https://doi.org/10.3390/app12126275

The Effect of Resonance Wood Quality on Violins Vibration

Mircea Mihalcica[1](✉), Mariana Domnica Stanciu[1], Florin Dinulica[2], Adriana Savin[3], and Voichita Bucur[4]

[1] Department of Mechanical Engineering, Transilvania University of Brasov, B-dul Eroilor 29, 500036 Brasov, Romania
mihalcica.mircea@unitbv.ro
[2] Department of Forest Engineering, Forest Management Planning and Terrestrial Measurements, Transilvania University of Brasov, B-dul Eroilor 29, 500036 Brasov, Romania
[3] Institute of Research and Development for Technical Physics, B-dul Mangeron 47, 700050 Iasi, Romania
[4] School of Science, RMIT University, GPO Box 2476, Melbourne, VIC 3001, Australia

Abstract. The aim of this paper is to highlight the correlations between the anatomical quality of the resonant wood (spruce and maple) and the dynamic response of the violins. In this study were analyzed 28 violins, seven from each structural quality class of wood (A, B, C, D), according to the classification made by manufacturers. The violins were analyzed in terms of physical parameters (width of annual rings, proportion of late wood, proportion of early wood, wavelength of sycamore curly grains) after which they were dynamically tested, determining the dynamic parameters. The results of experimental investigations showed that the frequency response of violins is closely related to the anatomical quality of the wood in the structure of violin boards.

Keywords: Quality grade of resonance wood · Physical features of wood · Dynamic response · Frequencies spectrum

1 Introduction

From ancient times the violin makers select the resonant wood by qualitative verification of the anatomical structure of the wood, this identification being easy and without advanced technical means (it is based on the optical verification of the wood).

From ancient times the violin makers select the resonant wood by qualitative verification of the anatomical structure of the wood, this identification being easy and without advanced technical means (it is based on the optical verification of the wood). In the chord musical instrument industry, there is a classification of musical instruments according to the anatomical quality of the wood in their construction, this aspect being revealed in the final price of the instrument. The classification of quality raw materials for the manufacture of stringed instruments has been presented and analyzed in detail in various studies [1–4]. The anatomical quality classes of the resonant spruce from the Gurghiu

Z. Dimitrovová et al. (Eds.): WMVC 2022, MMS 125, pp. 873–881, 2023.
https://doi.org/10.1007/978-3-031-15758-5_90

Mountains and the main anatomical descriptors are presented in the publications [3, 5, 6]. In maple wood, the wavelength of the fiber was determined by the wavelength [7–10]. Studies in the literature highlight the values of acoustic and dynamic parameters of spruce and resonant maple wood, without a sensitive highlighting of values depending on the anatomical quality classes. For this reason, this paper aims to analyze the comparative dynamic response of violins of different anatomical quality classes.

2 Materials and Method

2.1 Materials

To carry out this study, spruce and maple boards belonging to the four structural quality classes of wood, A (master), B (professional), C (student), D (school), of thickness classes were analyzed different. In Fig. 1 can see the types of violin classes related to the anatomical quality of the mold and maple wood in the violin body component. 8 violins were analyzed, two violins with the same thickness of tiles from each class.

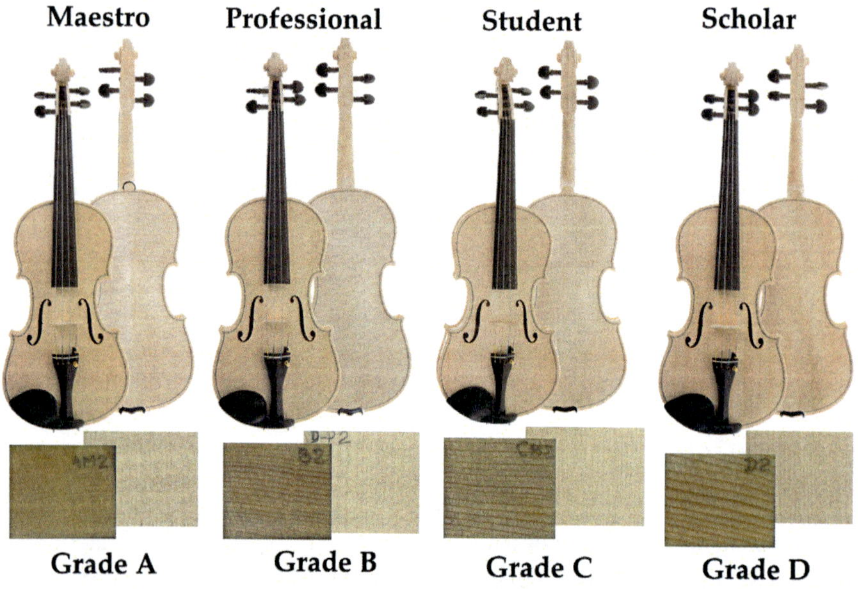

Fig. 1. The quality classes of violins analyzed.

2.2 Methods

Identification of the Anatomical Characteristics of the Quality Classes of Resonant Wood

In addition to the data known in the literature on the anatomical descriptors of resonant wood, the sensitive analysis of the descriptors according to the four classes of anatomical quality of spruce wood, respectively maple (class A, B, C, D) made on samples in the form of cubes with the size of 40 × 40 × 40 (mm^3). Anatomical characteristics were measured using the WinDENDRO Density system, determining the width of the annual rings marked TRW, the width of the early wood EWW and the width of the late wood LWW, the wavelength of the crested fiber to the maple wood (wavelength CWL), as well as the parameters wood color.

Dynamic Test

The generated signals were transmitted via a signal conditioning device to a NI USB-9233 dynamic data acquisition board produced by National Instruments (Austin, USA), connected to a laptop. The signal was viewed using a special application developed in NI-LabVIEW ©, and the graphical data was processed using the MATLAB © program. The procedure was presented in previous researches [10]. After the analysis in time and frequency of the signals, based on the exponential curve of the damping and the spectral composition of the signal for each violin, the values of the own frequencies of the tested violins, the damping factor and the analysis in time were extracted, these being correlated with aspects of wood structure and violin geometry. For each violin, the signals from three successive tests were taken.

3 Results and Discussion

3.1 Anatomic Features of Resonance Wood Classes

The measurement of the anatomical descriptors of the wood led to the clear highlighting of the physical differences between the four classes A, B, C, D, both for the resonant spruce wood and for the maple wood. Tables 1 and 2 show the values of the anatomical features.

Table 1. The anatomic features of resonance spruce wood

Variables	Grade			
Average values/STDV	A	B	C	D
Annual rings widths (mm)	0.71 0.005	1.38 0.018	1.69 0.045	2.28 0.005
Early wood width (mm)	0.54 0.011	1.07 0.029	1.33 0.039	1.74 0.029

(*continued*)

Table 1. (*continued*)

Variables	Grade			
Average values/STDV	A	B	C	D
Latewood width (mm)	0.18 0.013	0.30 0.013	0.36 0.022	0.54 0.026
Early wood proportion (%)	74.97 1.519	78.53 1.203	78.71 0.895	76.36 1.138
Latewood proportion (%)	25.03 1.519	21.47 1.203	21.29 0.895	23.64 1.136
Lightness L^* (%)	84.15 0.349	83.57 0.398	84.21 0.700	83.65 0.120
Green-red scale a^*	2.54 0.093	2.97 0.149	2.47 0.202	2.76 0.093
Blue-yellow b^*	19.78 0.573	19.62 0.163	20.02 0.727	20.85 0.223

Table 2. The anatomic features of resonance maple wood

Variables	Grade			
Average values/STDV	A	B	C	D
Annual rings widths (mm)	1.230 0.419	1.214 0.520	0.836 0.374	0.985 0.409
Wavelength of wavy fiber (mm)	5.448 0.513	6.295 0.698	7.418 1.565	∞
Lightness L^* (%)	83.48 0.654	86.626 0.384	86.46 0.309	82.142 1.342
Green-red scale a^*	3.637 0.245	2.215 0.0744	1.975 0.157	3.623 0.268
Blue-yellow b^*	18.105 0.542	16.675 0.377	17.893 0.682	18.123 0.246

Experimental data on the anatomical characteristics of resonant spruce wood show a trend of decreasing the proportion of late wood with the width of the annual ring, which, although statistically significant, is not consistent enough (R2 = 2.6%). The width of the annual rings decreases as the quality of the wood increases (Fig. 2, a). In the case of maple wood, there is a tendency to increase the width of the rings and decrease the wavelength with the improvement of the quality class. Quality class C stands out; in this class the amplitude of the values is also higher (Fig. 2, b).

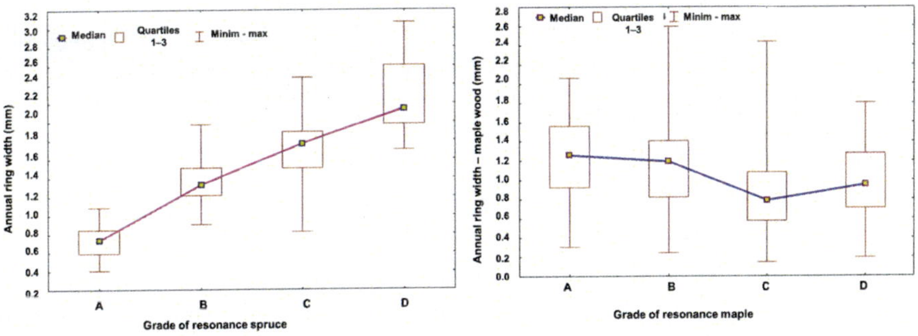

a) b)

Fig. 2. Variation of the width of the annual rings with quality class: a) spruce wood; b) maple wood.

Figure 3, a shows the correlation between brightness (color brightness), yellow/blue, and quality classes. For example, in the case of spruce wood, quality class A wood has the highest percentage of color brightness ($L* = 84\%$), a lower degree of red ($a* < 2.6$) and yellow ($b* < 20$). In quality class A, the color variables show the lowest amplitudes of variation. In grades B and D, spruce wood is darker ($L* < 83\%$) and has a higher red content ($a* > 3.2$). Class D also has the highest yellow content in color ($b* > 20.5$). Quality class C has the lowest red content ($a* < 2.6$).

In the case of resonant maple wood, the values of the color parameters are framed only in the positive range of the color space - no shades of green or blue have been identified. Based on the correlations between the anatomical descriptors of spruce wood, it can be stated that the variables of the annual rings have no influence on the brightness of the wood, but the size of the yellow shade is directly proportional to the width of the rings and their constituents (early wood and late wood). By correlating the color of the wood with the density, it turned out that the wood with a higher density is darker in color and has a weaker shade of red. The differences between the quality grades of maple wood in

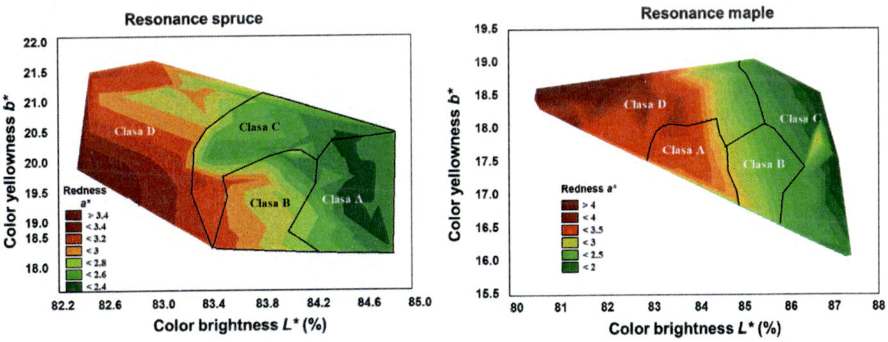

a) b)

Fig. 3. The color spectrum for resonance wood: a) spruce; b) maple.

terms of color are more pronounced than in spruce. Color stratification by quality class reveals the formation of two groups: classes A and D, respectively B and C (Fig. 3, b).

3.2 Frequencies Spectrum

Violin-specific vibration modes are referred to by [11–14] as A0, A1, CBR, B1– and B1+ modes. There are two low frequency modes associated with the variation of air pressure in the violin box cavity, A0 and A1. Modes A0 and A1 are coupling modes. The A0 mode is recorded around 270 Hz and is called the Helmholtz resonance mode, both plates vibrating in antiphase. Mode A1 is a first standing wave that vibrates along the violin body, in the range 470–490 Hz. CBR mode - known as center rotation or rhomboid vibration mode, is the lowest corpus frequency mode at about 400 Hz [14]. The first ways to bend the body, denoted B1- and B1+ are around 500 Hz both radiate energy. [14] noted that each violin has only five specific modes: the whole body, the top plate, the sides; of the back plate. The other structural elements of the violin (neck, cord, etc.) can be coupled to these modes, dividing them. The frequency ranges specific to the free violin strings 196–660 Hz (string: Sol, 196 Hz; Re, 293.6 Hz; La, 440a Hz; Mi, 659.3 Hz), is crucial for the sound of the violin and therefore these intervals of frequencies are studied very carefully in violins. In Fig. 4 are presented the frequencies responses of each main part of violin body (top, back and air). The spectral composition of the air contains both the frequencies of the top plate and of the back plate. In the case of the second eigenfrequency, there are differences between plates and those of the air: in most cases the plates vibrate with the same frequency.

Table 3 shows the values for the first specific vibration modes of violins and quality factor. With increasing frequency, the differences between the frequency spectra of the three components increase.

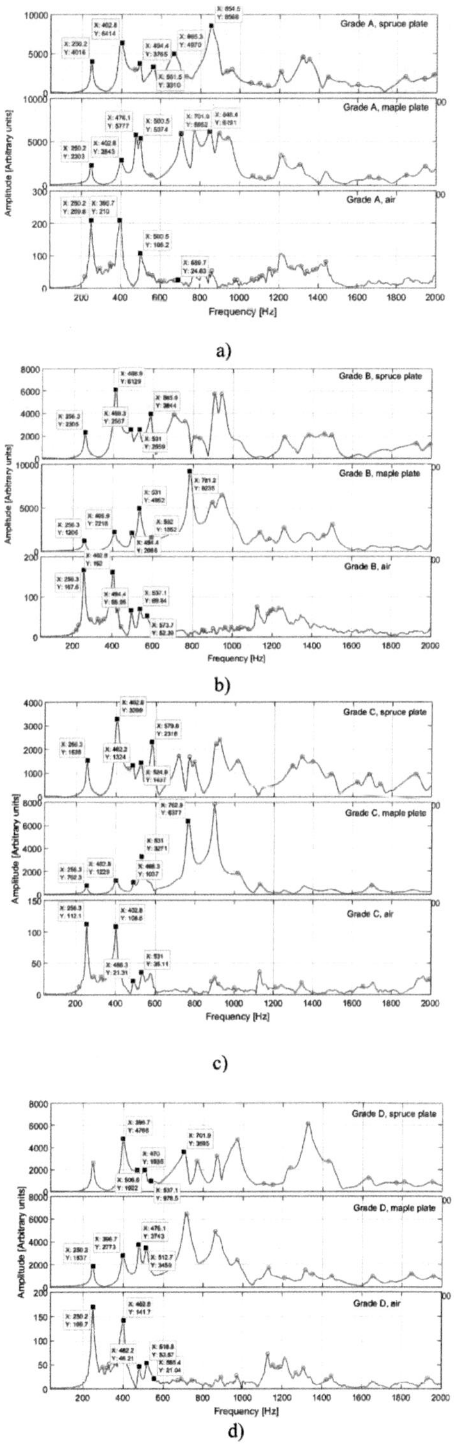

Fig. 4. Frequencies spectrum of tested violins: a) A00C; b) AM2C; c) AM4C; d) AM6C; e) AP2C; f) AP4C; g) AP6C.

Table 3. The frequencies spectrum and quality factors

Variables	Grade			
	A	B	C	D
1st resonance frequency	250	256	256	250
Quality factor Q(f1)	21	20	22	20
2nd resonance frequency	402	408	402	396
Quality factor Q(f2)	12	22	17	16
3rd resonance frequency	476	488	488	470
Quality factor Q(f3)	13	41	28	30
4th resonance frequency	500	531	531	537
Quality factor Q(f4)	13	17	33	-
5th resonance frequency	561	585	579	555
Quality factor Q(f5)	17	26	30	11

4 Conclusion

The paper focused on the dynamic responses of different violins graded from anatomical point of view in for classes. It was found that the dynamic response of the violins is to a small extent influenced by the anatomical descriptors of the wood in the construction of the front and back boards. An influence of these characteristics is observed on the quality factor. Future experimental investigations will deepen these issues.

Acknowledgments. This research was funded by a grant UEFISCDI, PN-III-P2-2.1-PED-2019-2148, project number. 568PED/2020 MINOVIS.

References

1. Gliga, V.G., Stanciu, M.D., Nastac, S.M., Campean, M.: Modal analysis of violin bodies with back plates made of different wood species. BioResources **15**, 7687–7713 (2020)
2. Rojas, J.A.M., Alpuente, J., Postigo, D., Rojas, I.M., Vignote, S.: Wood species identification using stress-wave analysis in the audible range. Appl. Acoust. **72**, 934–942 (2011)
3. Dinulică, F., Stanciu, M.D., Savin, A.: Correlation between anatomical grading and acoustic-elastic properties of resonant spruce wood used for musical instruments. Forests **12**, 1122 (2021). https://doi.org/10.3390/f12081122
4. Stanciu, M.D., Coşereanu, C., Dinulică, F., Bucur, V.: Effect of wood species on vibration modes of violins plates. Eur. J. Wood Wood Prod. **78**(4), 785–799 (2020). https://doi.org/10.1007/s00107-020-01538-5
5. Carlier, C., Brémaud, I., Gril, J.: Violin making "tonewood": comparing makers' empirical expertise with wood structural/visual and acoustical properties. In: International Symposium on Musical Acoustics ISMA2014, July 2014, Le Mans, France, pp. 325–330 (2014). ⟨hal-01233098⟩

6. Sonderegger, W., Martienssen, A., Nitsche, C., Ozyhar, T., Kaliske, M., Niemz, P.: Investigations on the physical and mechanical behavior of sycamore maple (Acer pseudoplatanus L.) Eur. J. Wood Prod. **71**, 91–99 (2013)
7. Spycher, M., Schwarze, F.W.M.R., Steiger, R.: Assessment of resonance wood quality by comparing its physical and histological properties. Wood Sci. Technol. **42**, 325–342 (2008). https://doi.org/10.1007/s00226-007-0170-5
8. Bucur, V.: Acoustics of Wood, 2nd edn., pp. 173–196. Springer, Heidelberg (2006). https://doi.org/10.1007/3-540-30594-7
9. Bremaud, I., Gril, J., Thibaut, B.: Anisotropy of wood vibrational properties: dependence on grain angle and review of literature data. Wood Sci. Technol. **45**, 735–754 (2011)
10. Carlier, C., Alkadri, A., Gril, J., Brémaud, I.: Revisiting the notion of "resonance wood" choice: a decompartementalised approach from violin makers' opinion and perception to characterization of material properties' variability. In: Perez, M., Marconi, E. (eds.) Wooden Musical Instruments. Different Forms of Knowledge: Book of End of WoodMusick COST ActionFP1302. Hal Documentation, Paris (2018)
11. Spycher, M., Schwarze, F.W.M.R., Steiger, R.: Assessment of resonance wood quality by comparing its physical and histological properties. Wood Sci. Technol. **42**, 325–342 (2008)
12. Wegst, U.G.K.: Wood for sound. Am. J. Bot. **93**(10), 1439–1448 (2006)
13. Viala, R., Placet, V., Cogan, S.: Simultaneous non-destructive identification of multiple elastic and damping properties of spruce tonewood to improve grading. J. Cult. Herit. (2019). https://doi.org/10.1016/j.culher.2019.09.004
14. Bissinger, G.: Structural acoustics of good and bad violins. J Acoust. Soc. Am. **124**, 1764–1773 (2008)

On the Vibrations of the Bowed String Instruments

Francesco Sorge(✉)

Department of Engineering, University of Palermo, Palermo, Italy
francesco.sorge@unipa.it

Abstract. The present note considers the bowed string instruments of the violin family and focuses on the string-soundbox dynamical coupling in the low-frequency range, carrying out a numerical and an analytical modal approach in parallel and aiming at a simple theoretical model of the sound production. The numerical results show just slight aperiodic fluctuations of the amplitude and very slow phase shifts in comparison with the simple analytical solutions, suggesting the latter as a fairly realistic description of the instrument performances.

Keywords: String instrument · Modal approach · Numerical solution · Analytical approximation

1 Foreword

The vibration analysis of the bowed string instruments started with the early studies of Helmholtz and Raman [1, 2] but the scientific interest began to grow greatly only from the middle of the last century up until today, as testified by many papers (see [3–9], just as a few examples). Here, we focus on the string-soundbox coupling in the low-frequency range, i.e. the range of the signature modes, more or less, carrying out a numerical and an analytical modal approach in parallel, to identify the coupling effects on the dynamical response of the instrument and to work out an acceptable approximate model of the sound production. The analysis takes into account proper functional relations between the force and displacement at the string-bridge contact and those at the bridge feet, for symmetric, antisymmetric and general modes. The characteristic equation of the coupled system is formulated and, assuming realistic values for the soundbox own frequencies, the coupling frequency spectrum is identified. The motion equations are then solved in the time domain numerically and compared with the analytical results obtainable assuming the pure Helmholtz motion as the string exciting motion.

2 Theoretical Model

Let us consider a single bowed string, for example, the string A4 of a violin as in Fig. 1 (440 Hz), and refer to a frame *Oxyz* with the origin *O* on the nut, the *x*-axis along the

F. Sorge—Retired.

Z. Dimitrovová et al. (Eds.): WMVC 2022, MMS 125, pp. 882–889, 2023.
https://doi.org/10.1007/978-3-031-15758-5_91

string and the y-axis along the bow motion direction. The string displacement may be expanded in a series of sinusoidal eigenfunctions $e_i(x) = a_i \sin(\omega_i x / v_w)$, multiplied by generalized coordinates $q_i(t)$, that is $y(x, t) = \sum_i q_i(t) e_i(x)$, where ω_i are the natural frequencies, $v_w = \sqrt{T/(\mu_s S_s)}$ is the propagation speed, T is the pre-tensioning, μ_s and S_s are the mass density and the cross-section area of the string. The eigenfunctions $e_i(x)$ are not orthogonal to each other in general, nor are the frequencies in arithmetic progression as for the string with fixed-fixed ends, because the extreme on the bridge vibrates together with the soundbox.

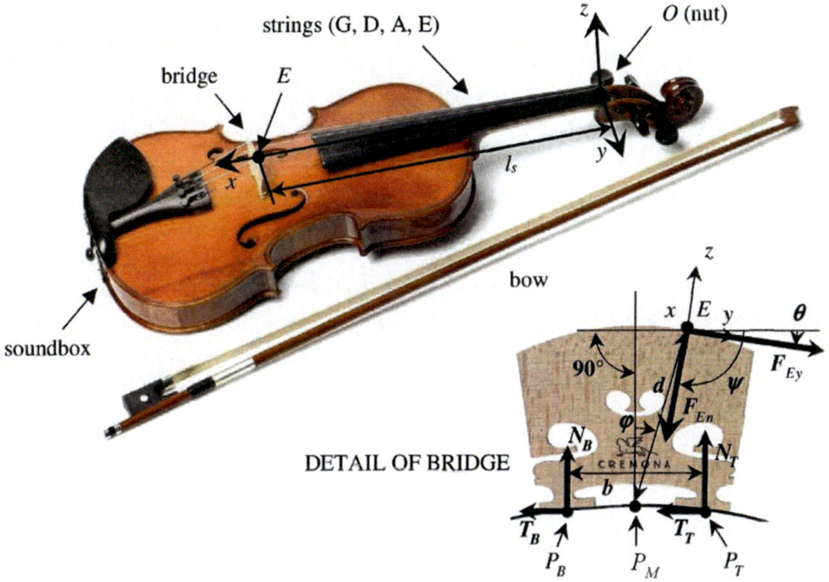

Fig. 1. Violin and reference frame. The detail shows the forces acting on the bridge: 1) $\boldsymbol{F}_{Ey}$, force applied by the string due to the bow thrust. 2) $\boldsymbol{F}_{En}$, force applied by the string due to the string tensioning. 3) $\boldsymbol{N}_B, \boldsymbol{N}_T, \boldsymbol{T}_B, \boldsymbol{T}_T$, normal and tangent forces applied by the soundbox top plate at the bass and treble feet of the bridge base, P_B and P_T, respectively.

The bridge translates along its axis in the symmetric modes of the plate and rotates around the base mid-point P_M in the antisymmetric modes.

Introduce the full symmetric matrix $[e_{ji}]$, where $e_{ji} = \mu_s S_s \int_0^{l_s} e_j(x)\, e_i(x) dx$, and impose that $\mu_s S_s \int_0^{l_s} e_i^2(x) dx = 1$, whence, indicating the string mass and length with m_s and l_s, one gets

$$a_i = \sqrt{\frac{2}{m_s\left[1 - \frac{v_w}{2\omega_i l_s} \sin\left(\frac{2\omega_i l_s}{v_w}\right)\right]}} \tag{1}$$

so, the physical dimensions of the eigenfunctions e_i are [kg$^{-1/2}$] and those of the generalized coordinates q_i are [kg$^{1/2}$ × m]. Moreover, introduce the one-dimensional

Dirac distribution $\delta(x - x_*)$ in order to manage generic concentrated forces F_*, so that $\int_0^{l_s} \delta(x - x_*) F_* e_j(x) dx = F_* e_j(x_*)$, and define the vectors $\{F_B e_j(x_B)\}$ and $\{F_{Ey} e_j(l_s)\}$, where x_B is the abscissa of the string-bow contact point *B*, F_B is the bow force and $F_{Ey} = -T \times \sum_i q_i (de_i/dx)_{x=l_s}$ is the bridge force (see Fig. 1). The motion equations of the string sub-system may be written in the matrix form

$$[e_{ji}]\left\{\frac{d^2 q_i}{dt^2} + 2\zeta_i \omega_i \frac{dq_i}{dt} + \omega_i^2 q_i\right\}^T = \{F_B e_j(x_B)\}^T - \{F_{Ey} e_j(l_s)\}^T \quad j = 1, 2, \ldots \tag{2}$$

where the damping effects are quite small and are here assumed uncoupled for the various modes. Observe that q_i must be considered in general as the sum of a variable part, $q_{i\sim}(t)$, which has the particular form $q_{i\sim}(t) = q_{ix} \sin(\omega_i t)$ when considering the natural modes, and a constant part q_{i0}, which is the static part due to the mean bow force.

The violin string could not emit a vigorous and harmonious sound by itself but needs the dynamic cooperation of the soundbox. Using capital letters for the quantities of the soundbox sub-system, we introduce the natural frequencies Ω_I, the two-dimensional eigenfunctions E_I and the modal coordinates Q_I, where the subscripts *I* refer to the single characterizing modes of the soundbox alone. It is presumed that all these parameters are obtainable by experimental tests, e. g. by holographic interferometry or impulse hammer and accelerometers or laser Doppler vibrometry. Besides, other experiments should also permit evaluating the bandwidths of the singular modes and then the damping factors Z_I. The present analysis is just limited to the low modes (<~1500 Hz), which may be clearly identified and characterized in the frequency response. As well known, the high-frequency range coincides with the so-called "bridge hill", where a large overlap of bandwidths occurs, the modal approach ceases to provide useful results and other methods should be applied (e. g. see [6]). The whole frequency response of the soundbox should be obtained by joining the two frequency ranges.

The detail of the bridge in Fig. 1 shows the forces applied to the bridge by the string and the soundbox top plate. As the bridge own frequencies are higher than the examined range, the bridge may be presumed rigid, while its mass may be neglected. Hence, disregarding the moment of $\boldsymbol{T}_T + \boldsymbol{T}_B$ with respect to P_M, the calculation of the normal reaction forces at the bridge feet is quite straightforward:

$$\begin{aligned} N_T &= F_{Ey}\left[\frac{\sin\theta}{2} + \frac{d}{b}\cos(\varphi - \theta)\right] + F_{En}\left[\frac{\sin\psi}{2} + \frac{d}{b}\cos(\psi - \varphi)\right] \\ N_B &= F_{Ey}\left[\frac{\sin\theta}{2} - \frac{d}{b}\cos(\varphi - \theta)\right] + F_{En}\left[\frac{\sin\psi}{2} - \frac{d}{b}\cos(\psi - \varphi)\right] \end{aligned} \tag{3a, b}$$

As the force F_{En} is induced by the constant string tensioning, it yields only invariant deflections and is irrelevant in the analysis of the vibratory motion, so the second terms on the right sides of Eqs. (3a, b) may be ignored. The concentrated forces on the soundbox surface, – $\boldsymbol{N}_T$ and – $\boldsymbol{N}_B$, may be dealt with using two-dimensional Dirac distributions, $\delta(X - X_*, Y - Y_*)$, where *X* and *Y* are coordinates on this surface. Applying the usual modal separation technique to the soundbox own motions, these forces turn out to be multiplied by $E_I(X_{PT}, Y_{PT})$ and $E_I(X_{PB}, Y_{PB})$ for each mode *I*, and moreover, $E_I(X_{PT},$

$Y_{PT}) = \pm E_I(X_{PB}, Y_{PB})$ for the symmetric and antisymmetric mode shapes, respectively. Therefore, using P_T as a reference point, the overall effect is $h_{fI} F_{Ey} E_I(X_{PT}, Y_{PT})$, where the force factor h_{fI} is sinθ or 2(*d*/*b*)cos($\varphi - \theta$) for the one or the other shape.

On the other hand, the soundbox vibration, even though very small compared with the string, implies small vibration components of the latter on the *xz* plane as well. Indicating the *y*-displacement of the string end on the bridge with y_E for each individual mode, the normal displacements of points P_T and P_B towards the box inside, identifiable by $Q_I E_I(X_{PT}, Y_{PT})$ and $Q_I E_I(X_{PB}, Y_{PB})$, are both equal to $y_E/\sin\theta$ for the symmetric modes, whereas they are opposite to each other and equal to $\pm y_E b/[2d\cos(\varphi - \theta)]$ for the antisymmetric ones. Hence, one may write $y_E = h_{dI} Q_I E_I(X_{PT}, Y_{PT})$, where the displacement factor h_{dI} is equal to sinθ or 2(*d*/*b*)cos($\varphi - \theta$) for the former and latter modes and is then equal to the force factor h_{fI}, in perfect accordance with the virtual work principle.

It must be clarified that the introduction of the sound post and the bass bar in the inside of the harmonic box modifies its symmetry characteristic with respect to the preliminary artefact with no additional elements, so each mode shape turns out to be a sort of combination of a symmetric and an antisymmetric shape. Therefore, the above factors, h_{fI} and h_{dI}, must be combined by proper weighting coefficients, somehow guided by the results from the experimentation, and a similar operation must be applied to the point values of the eigenfunctions $E_I(X_P, Y_P)$. In practice, indicating with p_s and p_a the values of any parameter p for the symmetric and antisymmetric deformation, respectively, we will set $p_I = p_{sI} w_{sI} + p_{aI}(1 - w_{sI})$ (with $0 < w_{sI} < 1$), where w_{sI} is the "weight" of the symmetric shape in the specific mode I. Also, it will be assumed for simplicity in the following that $d/b = 1$ and $\varphi = \theta$.

Assume normalized eigenfunctions, so that $\int_0^{S_P} \mu_p h_p E_R E_S dXdY = \delta_{RS}$ where μ_p, h_p and S_p are the mass density, the thickness and the surface area of the vibrating plates and δ_{RS} is the Kronecker delta. Then, the usual modal separation yields the motion equations of the soundbox sub-system when excited by the forces $-\boldsymbol{N}_T$ and $-\boldsymbol{N}_B$,

$$\frac{d^2 Q_J}{dt^2} + 2Z_J \Omega_J \frac{dQ_J}{dt} + \Omega_J^2 Q_J = h_{fJ} E_J\left(X_{P_T}, Y_{P_T}\right) F_{Ey} \quad J = 1, 2, \ldots \tag{4}$$

3 Natural Modes

Looking for the natural modes, one has to ignore the damping terms in Eqs. (4), replace $F_{Ey} = -T \times \sum_i q_i (de_i/dx)_{x=l_s}$, consider only the time-varying terms and solve for the Q_J. Then, taking into account the equality $T = \mu_s S_s v_w{}^2$ and using the correlation formulae reported in the previous section for the forces and displacements, one has $y(t, l_s) = \sum_J h_{dJ} Q_J(t) E_J\left(X_{P_T}, Y_{P_T}\right)$, that is

$$\sum_i a_i \sin\left(\frac{\omega_i l_s}{v_w}\right) q_{ix} \sin(\omega_i t)$$

$$= \mu_s S_s l_s \sum_i a_i \left(\frac{\omega_i l_s}{v_w}\right) \cos\left(\frac{\omega_i l_s}{v_w}\right) q_{ix} \sin(\omega_i t) \left[\sum_J \frac{h_{dJ} h_{fJ} E_J^2\left(X_{P_T}, Y_{P_T}\right)}{\left(\frac{\omega_i l_s}{v_w}\right)^2 - \left(\frac{\Omega_J l_s}{v_w}\right)^2}\right] \tag{5}$$

Since Eq. (5) must hold instant by instant, the summation concerning i and the time functions $q_{ix}\sin(\omega_i t)$ may be dropped, obtaining the characteristic equation:

$$\tan\frac{\omega_i l_s}{v_w} = \mu_s S_s l_s \sum_J h_{dJ} h_{fJ} E_J^2(X_{P_T}, Y_{P_T}) \frac{\frac{\omega_i l_s}{v_w}}{\left(\frac{\omega_i l_s}{v_w}\right)^2 - \left(\frac{\Omega_J l_s}{v_w}\right)^2} \tag{6}$$

Since $\mu_s S_s l_s$ is the string mass and the order of magnitude of $E_J{}^2$ is the reciprocal of the vibrating mass of the soundbox, which is much greater than the string mass, the coupled frequencies ω_i turn out to be very close to the uncoupled ones, either to $\omega_i = i\pi v_w/l_s$, when $\tan(\omega_i l_s/v_w) \cong 0$, or to Ω_J, when $\omega_i \cong \Omega_J$. The sequence Ω_J is chosen in the following calculation using verisimilar soundbox frequencies and realistic values of the weighting coefficients w_{sJ}. Once fixing the frequencies Ω_J and all other parameters, the exact values of the natural frequencies ω_i of the full system may be calculated numerically using Eq. (6). It is remarkable that some of the Ω_J are well separated from the angular frequencies of the string with fixed-fixed ends, so the soundbox is feebly excited, whereas some are close to the string frequencies, so the soundbox is resonant and a vigorous sound level is emitted to the surrounding environment.

4 Numerical and Analytical Solutions in the Time Domain

The time solutions may be obtained by use of a Euler-Cauchy solver, considering a finite but sufficiently large number n of modes.

Observe that Eqs. (2) and (4) refer in practice to the same modes, i. e. the common modes of the whole coupled system string + soundbox, but only the modes for which $e_j(l_s) \neq 0$ give their contribution to the vector $\{F_{Ey} e_j(l_s)\}^T$ at the right side of Eq. (2). These modes have nearly the same frequencies Ω of the soundbox and are characterized by $Q_I(t) = q_i(t) e_i(l_s)/[h_{dI} E_I(X_{PT}, Y_{PT})]$, according to Sect. 2. Therefore, it is possible to eliminate the force F_{Ey} between Eqs. (2) and Eqs. (4) obtaining

$$\left\{\frac{d^2 q_i}{dt^2} + 2\zeta_i \omega_i \frac{dq_i}{dt} + \omega_i^2 q_i\right\}^T = \left[e_{ji} + \frac{e_i^2(l_s)\delta_{ji}}{h_{dJ} h_{fJ} E_J^2(X_{P_T}, Y_{P_T})}\right]^{-1} \{F_B e_j(x_B)\}^T \tag{7}$$

where δ_{ji} is the Kronecker delta and the small damping factors of the two sub-systems were equalized for simplicity.

The sequential procedure consists in solving Eqs. (7) for the q_i first, replacing the q_i into F_{Ey} in Eqs. (4) and then solving Eqs. (4) for the Q's. This can be considered an "experimental" result and tends to become all the more exact the more correct the input data are.

For the damping factors ζ_i, the following laws were used, assuming $r_i = \omega_i l_s/(\pi v_w)$:

$$\begin{aligned} 2\zeta_i \omega_i &\cong 5\pi\left(2.9 + 0.3 r_i^2\right)\ \mathrm{s}^{-1} && \text{for } r_i \leq 3 \\ 2\zeta_i \omega_i &\cong 5\pi^{\frac{[5.6(10-r_i)+23(r_i-3)]}{7}}\ \mathrm{s}^{-1} && \text{for } r_i \geq 3 \end{aligned} \tag{8a, b}$$

As regards the bow force F_B, it depends on the state of slip or stick between the string and the bow. For the former state, it is possible to assume

$$F_B = F_{\mathrm{slip}} = F_s \times \left[0.3 + 0.7 \times \exp(-1.25 \times v_{\mathrm{rel.}})\right] \tag{9}$$

where $v_{rel.} = v_B - dy_B/dt$, v_B is the bow velocity and F_s is the maximum static friction force, a function of the normal force exerted by the violinist, who must control it in a very shrewd way. We here assume the formula $F_s = 0.036[l_s/(l_s - x_B)]^{1.43}$ N, which complies with Schelleng's diagram [3]. During the stick phase, on the other hand, one has $dy_B/dt = v_B =$ constant, whence $\sum_i e_i(x_B)d^2q_i/dt^2 = 0$ and Eq. (7) gives

$$F_B = F_{\text{stick}} = \frac{\sum_i e_i(x_B)\left[2\zeta_i\omega_i\frac{dq_i}{dt} + \omega_i^2(q_{i\sim} + q_{i0})\right]}{\sum_i e_i(x_B)\sum_j inv_{ji} \times e_j(x_B)} \tag{10}$$

where the coefficients inv_{ji} are those of the inverse matrix on the right side of Eq. (7). The mean temporal bow force F_m is found to be roughly equal to the constant slip force F_{slip}, so $q_{i0} \cong F_{\text{slip}}\sum_j \left[inv_{ji} \times e_j(x_B)\right]/\omega_i^2$ by Eq. (7),

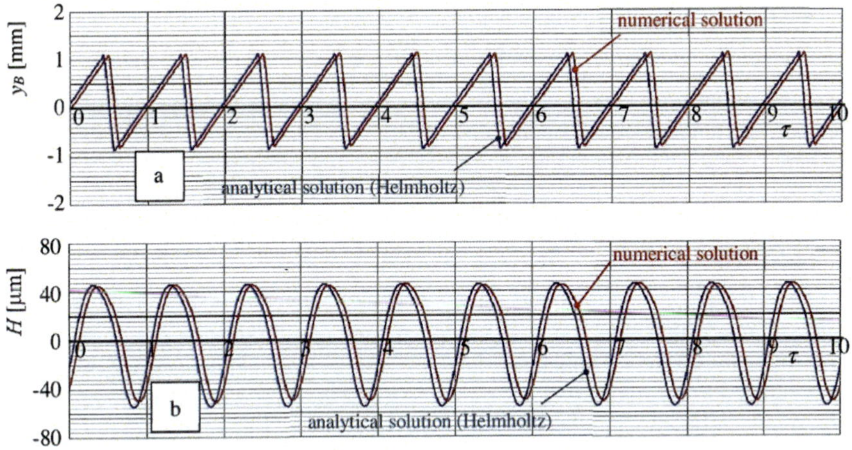

Fig. 2. String and soundbox response after a time $t_i \cong 0.1$ s. Time scale $\tau = (t - t_i)v_W/l_{sA4}$. (a) Stick-slip motion of bowed point B. (b) Harmonic table vibration, $H = \sum_J Q_J(t)E_J(X_{P_T}, Y_{P_T})$. Data: $n = 20$, $\mu_s S_s = 0.0008$ kg/m, $l_s = l_{sA4} = 325$ mm, $x_B = 285$ mm, $v_B = 1$ m/s, $T = 66$ N. Soundbox: frequencies = 275, 400, 450, 530, 620, 850, 980 [Hz]; quality factors = 50 (all modes).

In parallel, analytical approximations for the soundbox motion can be searched by expressing the $q_{i\sim}(t)$ by plausible functions, for example, referring to the Helmholtz motion and using the terms of its saw-tooth Fourier expansion,

$$q_{i\sim}(t) = \frac{2v_B}{\pi\omega a_i i^2} \times \frac{l_s}{(l_s - x_B)} \times \sin i\omega t \tag{11}$$

adding the static terms q_{i0}, replacing these quantities into the differential equations of the soundbox vibrations, Eqs. (4), and solving for the $Q_{J\sim}(t)$ and Q_{J0} by use of realistic bandwidths of the frequency response of the soundbox to calculate the Z_J.

Figure 2a, b shows the motion of the bowed point of the string and that of the reference soundbox point P_T, as they can be obtained by the numerical and the analytical procedure, for $l_s = 325$ mm (A4: 440 Hz). The former is quite similar to the Helmholtz motion (Fig. 2a) and points out that the very short run of the string endpoint does not affect so much the oscillations of the string itself. Due to the mutual incommensurability of the string and soundbox frequencies, the steady motions of the bowed point B and the soundbox point P_T, which occur after a transient period of one tenth of a second roughly, are not rigorously periodic but quasi-periodic with slight fluctuations of the amplitude and slow phase shifts. The analytical results show a good approximation anyway. Figure 3 shows results analogous to Fig. 2b when the string length is reduced by the violinist's finger to get the note C5 (523 Hz): there is a good agreement also in this case and a similar agreement may be found in many other test cases.

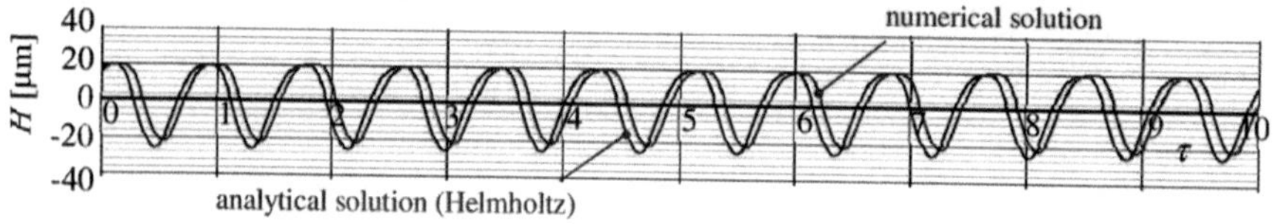

Fig. 3. Harmonic table vibration $H = \sum_J Q_J(t)E_J(X_{P_T}, Y_{P_T})$ after a time $t_i \cong 0.1$ s, as a function of the dimensionless time $\tau = (t - t_i)v_w/l_{sA4}$ (same time scale as in Fig. 2). Same data as in Fig. 2, except $l_s = l_{sC5} = 273$ mm (C5).

5 Conclusion

The present report proposes a simple analytical approach to describe the low-frequency behaviour of the bowed string instruments. It proves sufficiently consistent with the more accurate numerical results and might be completed by adding the higher frequency response obtainable by other methods described in the literature. The whole frequency spectra of the various individual instruments are certainly different from each other, as all luthiers are well aware and careful experimental tests should be carried out to characterize their tone colour. Yet, the present methodology may provide a useful tool to analyse the influence of possible structural changes of the soundbox parts on the global performances of the instruments in the low-frequency range.

References

1. Von Helmholtz, H.: On the Sensations of Tone. Longmans, Green and Co., London (1912). (Transl. by A. J. Ellis)
2. Raman, C.V.: On the mechanical theory of vibrations of bowed strings and of musical instruments of the violin family, with experimental verification of the results. Bull. Indian Assoc. Cultiv. Sci. **15**, 1–158 (1918)
3. Schelleng, J.C.: Bowed string and player. J. Acoust. Soc. Am. **53**, 26–41 (1973)
4. Hutchins, C.M.: A history of violin research. J. Acoust. Soc. Am. **73**, 1421–1440 (1983)
5. Woodhouse, J.: The acoustic of a violin: a review. Rep. Prog. Phys. **77**, 1–41 (2014)
6. Woodhouse, J.: On the "bridge hill" of the violin. Acta Acust. Acust. **91**, 155–165 (2005)

7. Bissinger, G.: Modal analysis of a violin octet. J. Acoust. Soc. Am. **113**, 2105–2113 (2003)
8. Gliga, V.G., Stanciu, M.D., Nastac, S.M., Campean, M.: Modal analysis of violin bodies with back plates made of different wood species. BioResources **15**(4), 7687–7713 (2020)
9. Gough, C.E.: The acoustics of thin-walled shallow boxes – a tale of coupled oscillators. Acoust. Today **12**(2), 22–30 (2016)

Study of the Influence of Wood Mechanical Properties Variability on the Sound Synthesis of a Simplified String Instrument

Guilherme O. Paiva[1](✉), Marcelo Queiroz[1], and Marcela R. Machado[2]

[1] Computer Science Department, University of São Paulo, São Paulo, Brazil
{gpaiva,mqz}@usp.br
[2] Department of Mechanical Engineering, University of Brasília, Brasília-DF, Brazil
marcelam@unb.br
https://compmus.ime.usp.br/

Abstract. The influence of wood choices on the sound produced by string instruments is a recurrent issue in musical acoustics. Moreover, the intrinsic variability of wood's mechanical properties requires considerable attention to any analysis attempt. This paper presents a numerical investigation of the wood mechanical properties variability influence on the sound synthesis of a simplified string instrument consisting of a single string coupled to a square soundboard. The adopted synthesis model is based on a hybrid modal approach that combines analytical expressions to describe the string vibration and numerical data obtained from a finite element analysis of the instrument soundboard. For the sound synthesis multiple sets of random wood mechanical properties are assumed. Samples are generated using Monte Carlo simulations and the synthesized sounds statistical moments are performed. Results show the influence of parametric uncertainty in the instrument sounds.

Keywords: String musical instruments · Sound synthesis · Wood variability

1 Introduction

String musical instruments in general produce sound through the interaction of several subsystems: the excitation mechanism, the strings, the instrument body and the air. Once the strings are excited, the most of the energy that will be converted into sound is transmitted to the instrument body, which in turns amplifies and radiates the sound to the air, acting as a 'filter' that colors the timbre.

Supported by the São Paulo Research Foundation (FAPESP), grant number 20/07978-0.

Z. Dimitrovová et al. (Eds.): WMVC 2022, MMS 125, pp. 890–899, 2023.
https://doi.org/10.1007/978-3-031-15758-5_92

Physical modelling for string instruments sound synthesis has attracted scientists, musicians and manufacturers interest in the last decades due to the increasing advent of computational processing. The prediction potential of such synthesis models has been extensively investigated in various research contexts [3,11,18,24]. Moreover, different aspects concerning the influence of wood choices on the produced sound have been addressed [5,17]. On the other hand, wood variability effects on the sound production is an issue that still requires considerable attention.

Uncertainties in material properties are intrinsic into real structures, and these structural parameters are spatially distributed by nature as demonstrated in [13,14]. Numerous authors have studied wood species for musical instruments (hereinafter called 'tonewoods') in terms of their properties variability [4,6,22]. Brémaud *et al.* [4] reviewed the literature data on tonewoods vibrational properties and investigated the dependence of their anisotropy variability on grain orientation. Carlier *et al.* [6] characterized physical and mechanical properties in a wide range of spruce and maple tonewoods according to several scales of wood variability and correlated with the criteria used by the instrument makers to choose their material. Viala *et al.* [22] measured the mechanical properties in different directions of 20 spruce samples and investigated the correlation between their variabilities and the grading attributed by the tonewood seller. Stochastic finite element approaches using variable wood mechanical properties have been also developed to study the influence of material uncertainties on the dynamic behaviour of violins [9,20] and guitars [19,20,23] in terms of their modal responses. Merchel *et al.* [15] investigated the influence of physical and mechanical properties variations on the tonal quality assessment of nine steel-string acoustical guitars of the same model. Preference listening tests were performed using short music sequences recorded from the nine instruments. Results showed significant preference for the guitars whose soundboards and bracewoods had relative low Young's modulus and low densities.

This paper addresses the implication of wood mechanical properties variability on the synthesized sound of a simplified string instrument. The instrument consists of a single string coupled to a square soundboard made of spruce tonewood, *Picea abies*. The adopted synthesis model is based on a hybrid modal approach introduced by Paiva *et al.* [16]. Such approach is adapted in the present work to combine analytical expressions describing the string vibration and numerical data obtained from a finite element analysis of the instrument plate. For the sake of simplicity, string and soundboard vibrations are assumed to be only planar and normal to the soundboard plane. From previous sensitivity analysis results for the spruce tonewood [22], the following wood mechanical properties are assumed to be random: the shear modulus in the LR plane, the elasticity modulus in longitudinal and radial directions to the wood fiber, $G_{LR}(\theta)$, $E_L(\theta)$, and $E_R(\theta)$, respectively. Random synthesized sound samples are generated using Monte Carlo simulations and the statistical moments in the spectral domain are presented. Results show the influence of parametric variability of wood mechanical properties in the instrument sounds. The use of a stochastic model is a way to predict variability's in the sounds quality.

2 Methods

2.1 A Brief Description on the Sound Synthesis Model

The sound synthesis model adopted in this study has been first presented in [16], where the interested reader can find the detailed model characterization. The model main features are summarized as follows.

Principle of the Model. The model uses a hybrid modal-based approach, what means that analytical expressions of mode shapes, natural frequencies and damping factors are used to obtain the modal basis of the uncoupled string while modal parameters of the uncoupled soundboard are extracted from a numerical modal analysis using the finite element method. String and soundboard are coupled by imposing displacement continuity at the coupling point. A temporal finite difference scheme is used to compute, at each instant, the string pluck response in terms of its physical velocity at the string/soundboard coupling point, $v(x_c, t)$. This transient signal is assumed to correspond to the instrument synthesized sound.

String and Soundboard with Random Properties Kinematics. A stiff string with small-amplitude vibration is assumed. The string is characterized by the following physical parameters: length L, mass per unit length μ, tension T, bending stiness B. The string is simply supported at one end located at $x = 0$ and attached to the soundboard at the other end at $x_c = L$. It is assumed that string and soundboard have only transverse displacements, normal to the soundboard plane, i.e., in the y direction. Axial and torsional motions of the string and its intrinsic geometric nonlinearities are neglected. The string displacement, $y^s(x, t)$, is described as the sum of N_s modes associated to the string with simply supported ends plus an interface mode contribution $y_0^s(x, t)$, which corresponds to the string static response when it is simply supported at $x = 0$ and loaded at $x_c = L$. Thus, the transverse string displacement at instant t and axial position x is given as

$$y^s(x,t) = \sum_{j=1}^{N_s} a_j(t) \sin\left(\frac{j\pi x}{L}\right) + a_0(t)\left(\frac{x}{L}\right), \tag{1}$$

where $a_j(t)$, with $j = \{0, 1, \ldots, N_s\}$, are the string modal amplitudes. The random soundboard transverse motion at the string coupling point $\mathbf{p_c} = (x_c, y_c, z_c)$ is expressed as the sum of N_{sb} modes so that

$$y^{sb}(\mathbf{p_c}, \theta, t) = \sum_{k=1}^{N_{sb}} b_k(t)\phi_k(\mathbf{p_c}, \theta), \tag{2}$$

where $b_k(t)$ and $\phi_k(\mathbf{p_c}, \theta)$, with $k = \{1, 2, \ldots, N_b\}$, are respectively the soundboard modal amplitudes and corresponding random mode shapes evaluated at the point $\mathbf{p_c}$.

String and Soundboard with Random Properties Dynamics. A modal framework is adopted for the string and soundboard dynamics description, where the rigidity soundboard's properties are assumed random. String and soundboard equation of motion for forced vibration are formulated as a set of secondary-order ordinary differential equations of the form as

$$\mathbf{M}^s\ddot{\mathbf{a}}(t) + \mathbf{C}^s\dot{\mathbf{a}}(t) + \mathbf{K}^s\mathbf{a}(t) = \mathbf{f}^e(t) - \mathbf{f}^c(t), \tag{3}$$

$$\mathbf{M}^{sb}\ddot{\mathbf{b}}(t) + \mathbf{C}^{sb}\dot{\mathbf{b}}(t) + \mathbf{K}^{sb}(\theta)\mathbf{b}(t) = \mathbf{f}^c(t), \tag{4}$$

where $\mathbf{a}(t)$ and $\mathbf{b}(t)$ are the string and soundboard modal amplitude vectors, respectively. The matrices $\mathbf{M}^l$, $\mathbf{C}^l$ and $\mathbf{K}^l$ are respectively the mass, damping and stiffness modal matrices associated to the string and soundboard, for $l = s$ and $l = sb$, respectively. The excitation force exerted by the pluck is expressed by the modal vector $\mathbf{f}^e$ in Eq. 3. String and soundboard modal equations are coupled through the string/body coupling force vector $\mathbf{f}^c$, which is initially not known. Such vector is computed using a temporal finite difference scheme described in [16], that finally permits the string pluck response calculation. All the string modal parameters required to build the matrices $\mathbf{M}^s$, $\mathbf{C}^s$ and $\mathbf{K}^s$, as well as the modal plucking force vector $\mathbf{f}^e(t)$, are calculated using analytical expressions described in [16]. The random soundboard mode shapes $\phi_k(\mathbf{p_c}, \theta)$ and all the parameters necessary to compute the matrices $\mathbf{M}^{sb}$ and $\mathbf{K}^{sb}(\theta)$ are obtained by means of FE modal analysis described in the Subsect. 2.2. As referred in the Subsect. 3.1, the soundboard modal damping factor values needed to compute the matrix $\mathbf{C}^{sb}$ are estimated from previous experimental measurements presented in [21].

2.2 Finite Element Analysis of the Soundboard with Random Properties

A finite element modal analysis of the unstrung soundboard was performed to obtain its modal parameters, that is, natural frequencies, modal masses, modal stiffnesses and modal shapes evaluated at the string/soundboard coupling point. The soundboard geometry model and respective mesh are built using the Salome-Meca [2] software and all the calculations were carried out using the Code-Aster solver [1]. Soundboard dimensions are $(0.5 \times 0.5 \times 0.003)$ m. The geometry model is meshed using quadratic triangular plate elements. A sufficiently refined mesh containing more than 20 elements per standing bending structural wave is obtained [12]. The FE model contains 9006 elements and 4630 nodes. Free boundary conditions are applied.

Based on previous sensitivity analysis results for the spruce tonewood presented by Viala *et al.* [22], the following spruce mechanical properties are assumed to be stochastic variables defined as uniform distributions between upper and lower values as given in Table 1: the elasticity moduli in longitudinal and radial directions to the wood fiber, and the shear modulus in the longitudinal-radial plane (LR plane), $E_L(\theta)$, $E_R(\theta)$ and $G_{LR}(\theta)$, respectively. Maximum and minimum values are given as $\pm 10\%$ of the respective nominal values. The nominal values are assigned to the remaining material properties, as shown in Table 1. The Monte Carlo simulation is used to perform sets of multiple numerical modal analyses, from which random samples of soundboard modal parameters are generated to feed the sound synthesis model. The random variations $E_L(\theta)$, $E_R(\theta)$ and $G_{LR}(\theta)$ are carried out individually, for one parameter at a time, and for all parameters simultaneously so that four sets of 300 simulations are obtained. Nominal values are considered while variations are not taken into account.

Table 1. Material properties of spruce tonewood used in the numerical modal analysis. The tangential elasticity modulus, E_T, and all Poisson's coefficients, ν_{LR}, ν_{RT}, ν_{TL}, are from [4,10]; remaining values from [22].

Parameter	Nominal value	Min. value	Max. value
E_L [MPa]	12800	12160	13440
E_R [MPa]	1010	959.5	1060.5
E_T [MPa]	3363	–	–
G_{LR} [MPa]	810	769.5	850.5
G_{RT} [MPa]	46	–	–
G_{TL} [MPa]	790	–	–
ν_{LR} [-]	0.37	–	–
ν_{RT} [-]	0.48	–	–
ν_{TL} [-]	0.02	–	–
Density [Kg/m]	440	–	–

3 Results and Discussions

This section first describes the parameters used in all simulations. Finally, spectral representations of synthesized sounds are presented and discussed.

3.1 Model Parameters

String Parameters. Table 2 shows the string parameters used in the sound synthesis. A G-guitar-string from the D'Addario Pro Arte Composite Hard Tension set is chosen. The values of tension and mass per unit length for the string are considered according to the manufacturer's specifications. The bending stiffness value is obtained from [24]. The string natural frequency, f_0, and respective overtones are computed up to 2 kHz taking into account the inharmonicity effects. String modal Q-factors are computed using the string damping model and respective parameter values presented in [24].

Table 2. G-string parameters used in the sound synthesis. [a]Reference system origin at the soundboard center.

String parameter name	Variable	Value
Fundamental frequency [Hz]	f_0	196.1
Length [m]	L	0.650
Number of modes [-]	N_s	10
Coupling point coordinates [m]	(x_c, y_c, z_c)	(0.05, 0.05, 0)[a]
Tension [N]	T	61.57
Mass/unit length [g/m]	μ	0.9
Bending stiffness [$\times 10^{-6}$ N.m^2]	B	310

Soundboard Parameters. All soundboard modal parameters, except damping factors, are obtained from the FE modal analysis described in Subsect. 2.2. Modes are extracted over the frequency range of 0–2 kHz. Since random variations of some spruce parameters are considered along the Monte Carlo simulations, the total number of modes, N_{sb}, varied from 122 to 124. The values of soundboard damping factors are assumed to be 1% for all modes. This choice is based on previous experimental results for spruce guitar soundboards [21].

General Parameters. All simulations are sampled at a frequency $f_s = 22050$ Hz. The string is plucked at 15 cm from the coupling point $\mathbf{p_c}$. The pluck is modelled as in [16]: a linear ramp force in the time domain of maximum amplitude $\text{F}_{pluck} = 1$ N is transversely applied to the string, in the y-direction, during an interval of 8 ms. All synthesized sounds have a duration of 8 s.

3.2 Synthesized Sounds

Deterministic Approach. Since the temporal velocity response at the coupling point, $v(x_c, t)$, is assumed to correspond the synthesized sound, its respective spectrum, $V(x_c, f)$, is given as $V(x_c, f) = \mathcal{F}\{v(x_c, t)\}$. Figure 1 shows the spectrum magnitude over the frequency range of 0–2 kHz of the entire sound when no random variations of soundboard mechanical parameters are applied. It can be clearly seen the strong contribution of the string modes which are represented by the harmonic comb of equally spaced peaks with a fundamental frequency 198 Hz. In addition to such resonating components, it is also visible a group of less prominent resonance peaks with distinct spacing and magnitudes representing the contribution due to the excitation of the soundboard modes. Mainly 250 Hz, individual resonance peaks of soundboard-mode contributions are increasingly less observable, tending to a smooth curve, since the modal overlap in plate-like systems increases considerably in the mid and high frequency domains [7]. The perceptual significance of soundboard-mode components has been reported on the early phase of classical guitar tones, as well as the strong dependence of their emergence to the pluck conditions [8].

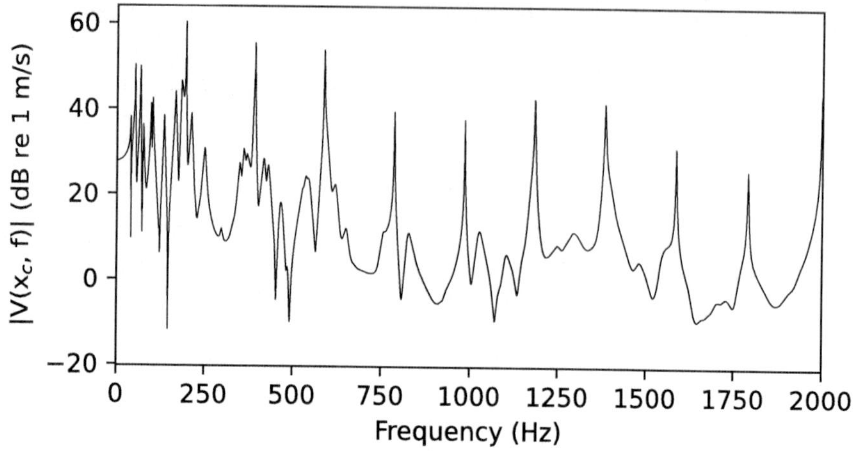

Fig. 1. Spectrum magnitude of the entire sound, with duration of 8 s, when no random variations of soundboard mechanical parameters are applied.

Stochastic Approach. Figure 2, from (a) to (d), represents respectively the dispersion of the sound spectrum magnitudes, between 0 250 Hz, for the individual random variations $E_L(\theta)$, $E_R(\theta)$, $G_{LR}(\theta)$ and all random variations simultaneously. Figure 2, from (e) to (h), displays the respective corresponding spectrum magnitudes in the overall analysed frequency range, between 0 and 2000 Hz Hz. Mean spectra are given by red curves. The blue areas are centered on these mean spectra, and their boundaries are two standard deviation below and above the mean spectra. For comparison purposes, black dashed curves represent the nominal spectrum magnitudes, which are obtained without any parameter variation.

From a qualitative comparison of such spectra, one can observe distinct uncertainty effects on the sound synthesis. For all variations applied, in general, dispersions are more accentuated in the frequency range of 0–250 Hz, as shown in Fig. 2, from (a) to (d). In Fig. 2, from (e) to (h), less prominent but significant dispersions are observed 250 Hz, mainly in the regions between the string-mode peaks, where individual soundboard-mode peaks are less and less observable. This suggests that dispersions in these regions result from the superposition of multiple soundboard-mode fluctuations. Figure 2, (d) and (h), presents in the overall frequency range the most pronounced dispersion, which is expected since all the three random variations are taken into account. Considering only the individual random variations, Fig. 2, (b) and (f), presents the most dispersed results in most of the frequency range, which means that for the studied instrument, in general, sound spectrum fluctuations are more sensitive to uncertainties around the radial elasticity modulus, E_R, than the other analysed mechanical properties. On the other hand, Fig. 2, (c) and (g), shows that the uncertainties around the shear modulus in the longitudinal-radial plane, G_{LR}, leads to the less dispersed results. In all the spectra presented, dispersions are observed mainly around soundboard-mode contributions, which is expected since random variations are applied only to soundboard mechanical parameters. Apparently, string-mode peaks remain without significant dispersion. At several frequency regions (see, for example, spectrum magnitude in Fig. 2(d)) nominal spectrum magnitudes surpass the standard deviation bounds, which evidences the disparity between the deterministic and stochastic sound synthesis approaches.

Although listening tests were not performed, some analyses concerning the perceptual differences in the generated sounds may be suggested. In Fig. 2(d), one can observe that the spectrum magnitude reaches an apparent dispersion of more than 10 dB. Moreover, some soundboard-mode components are only about 10 dB to 20 dB below the string fundamental frequency peak, at 196.1 Hz, which suggests that perceptually different sounds may be produced due to the spectral variability observed in this frequency range. On the other hand, as shown in Fig. 2(h), regions of significant spectral variability are identified about 40 dB or more below the string fundamental frequency, which suggests their perceptual irrelevance.

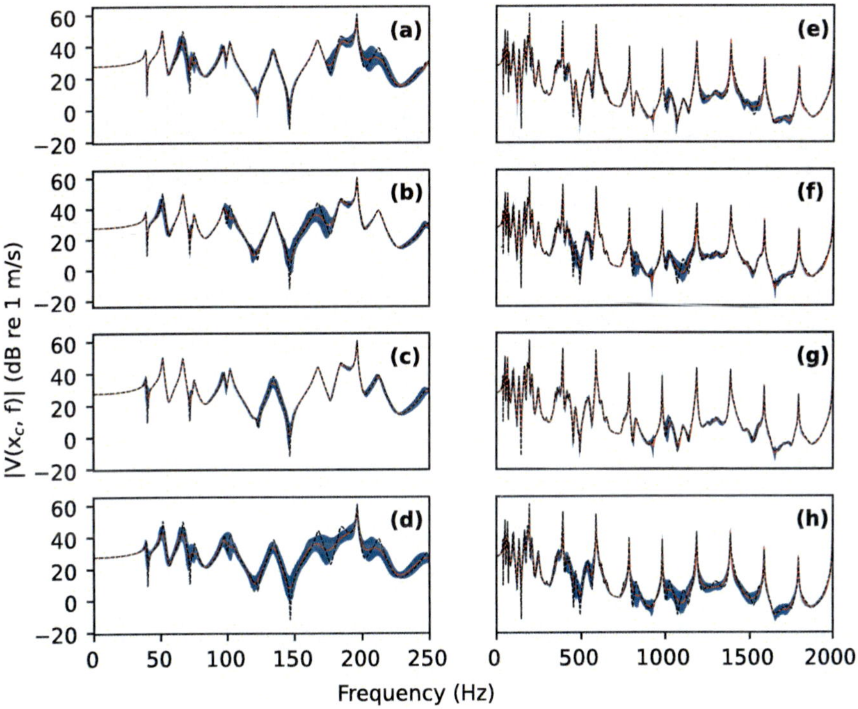

Fig. 2. From (a) to (d), respectively: sound spectrum magnitudes for the individual random variations $E_L(\theta)$, $E_R(\theta)$, $G_{LR}(\theta)$ and all random variations simultaneously. From (e) to (h): respective corresponding spectrum magnitude between 0 and 2000 Hz Hz. Mean spectra are given by red curves. The blue areas are centered on these mean spectra, and their boundaries are two standard deviation below and above the mean spectra. Black dashed curves represent the nominal spectrum magnitudes, which are obtained without any parameter variation.

4 Conclusions

An innovative framework for string musical instruments sound synthesis including material's variability has been presented. The effects of the variability of several spruce tonewood mechanical properties has been shown in the sound synthesis of a simplified instrument. In general, for the studied instrument, the radial elasticity modulus, E_R, has shown to be the most influential parameter in the sound spectrum magnitude generation, and the shear modulus in the longitudinal-radial plane, G_{LR}, has shown to be the less influential. Qualitative comparison of generated spectra allowed to conclude that the deterministic and stochastic sound synthesis approaches lead to sounds with different spectral content. Studies on the perceptual relevance of such differences is left as an important further work. Furthermore, the presented methodology should be applied in the investigation of material variability effects on the sound production of popular string musical instruments like classical guitars and violins.

References

1. Code_Aster – structures and thermomechanics analysis for studies and research. http://www.code-aster.org. Acessed 08 Mar 2022
2. Salome – the open source integration platform for numerical analysis. http://www.salome-platform.org/. Accessed 08 Mar 2022
3. Bilbao, S., et al.: The ness project: physical modeling, algorithms and sound synthesis. Comput. Music J. (2019)
4. Brémaud, I., Gril, J., Thibaut, B.: Anisotropy of wood vibrational properties: dependence on grain angle and review of literature data. Wood Sci. Technol. **45**(4), 735–754 (2011)
5. Carcagno, S., Bucknall, R., Woodhouse, J., Fritz, C., Plack, C.J.: Effect of back wood choice on the perceived quality of steel-string acoustic guitars. J. Acoust. Soc. Am. **144**(6), 3533–3547 (2018)
6. Carlier, C., Alkadri, A., Gril, J., Brémaud, I.: Revisiting the notion of "resonance wood" choice: a decompartementalised approach from violin makers' opinion and perception to characterization of material properties' variability (2019)
7. Elie, B., Gautier, F., David, B.: Estimation of mechanical properties of panels based on modal density and mean mobility measurements. Mech. Syst. Sig. Process. **40**(2), 628–644 (2013)
8. Fréour, V., Gautier, F., David, B., Curtit, M.: Extraction and analysis of body-induced partials of guitar tones. J. Acoust. Soc. Am. **138**(6), 3930–3940 (2015)
9. Gonzalez, S., Salvi, D., Baeza, D., Antonacci, F., Sarti, A.: A data-driven approach to violin making. Sci. Rep. **11**(1), 1–9 (2021)
10. Guitard, D., El Amri, F.: Modèles prévisionnels de comportement élastique tridimensionnel pour les bois feuillus et les bois résineux. In: Annales des sciences forestières, vol. 44, pp. 335–358. EDP Sciences (1987)
11. Issanchou, C., Le Carrou, J.L., Touzé, C., Fabre, B., Doaré, O.: String/frets contacts in the electric bass sound: Simulations and experiments. Appl. Acoust. **129**, 217–228 (2018)

12. Langer, P., Maeder, M., Guist, C., Krause, M., Marburg, S.: More than six elements per wavelength: the practical use of structural finite element models and their accuracy in comparison with experimental results. J. Comput. Acoust. **25**(04), 1750025 (2017)
13. Machado, M.R., Dos Santos, J.: Effect and identification of parametric distributed uncertainties in longitudinal wave propagation. Appl. Math. Model. **98**, 498–517 (2021)
14. Machado, M., Adhikari, S., Dos Santos, J., Arruda, J.: Estimation of beam material random field properties via sensitivity-based model updating using experimental frequency response functions. Mech. Syst. Sig. Process. **102**, 180–197 (2018)
15. Merchel, S., Altinsoy, M.E., Olson, D.: Perceptual evaluation of bracewood and soundboard wood variations on the preference of a steel-string acoustic guitar. J. Acoust. Soc. Am. **146**(4), 2608–2618 (2019)
16. Paiva, G.O., Ablitzer, F., Gautier, F., dos Santos, J.M.C.: Collisions in double string plucked instruments: physical modelling and sound synthesis of the viola Caipira. J. Sound Vib. **443**, 178–197 (2019)
17. Paté, A., Carrou, J.L.L., Navarret, B., Dubois, D., Fabre, B.: Influence of the electric guitar's fingerboard wood on guitarists' perception. Acta Acustica united Acustica **101**(2), 347–359 (2015)
18. Paté, A., Le Carrou, J.L., Fabre, B.: Predicting the decay time of solid body electric guitar tones. J. Acoust. Soc. Am. **135**(5), 3045–3055 (2014)
19. Shepherd, M.R., Hambric, S.A., Wess, D.B.: The effects of wood variability on the free vibration of an acoustic guitar top plate. J. Acoust. Soc. Am. **136**(5), EL357–EL361 (2014)
20. Viala, R.: Towards a model-based decision support tool for stringed musical instruments making: Outil d'aide à la décision pour la facture d'instruments de musique à cordes. Ph.D. thesis, Bourgogne Franche-Comté (2018)
21. Viala, R., Pérez, M.A., Placet, V., Manjon, A., Foltête, E., Cogan, S.: Towards model-based approaches for musical instruments making: validation of the model of a Spanish guitar soundboard and characterization features proposal. Appl. Acoust. **172**, 107591 (2021)
22. Viala, R., Placet, V., Cogan, S.: Simultaneous non-destructive identification of multiple elastic and damping properties of spruce tonewood to improve grading. J. Cult. Heritage **42**, 108–116 (2020)
23. Viala, R., Placet, V., Cogan, S.: Model-based evidence of the dominance of the guitar brace design over material and climatic variability for dynamic behaviors. Appl. Acoust. **182**, 108275 (2021)
24. Woodhouse, J.: On the synthesis of guitar plucks. Acta Acustica united Acustica **90**(5), 928–944 (2004)

VTE: Vibration Transmission and Energy Flow Analysis of Engineering Structures and Nonlinear Systems

A Modal Approach to Sound Propagation in Elliptical Ducts with Lined Walls

Paulo J. S. Gil(✉) and João M. G. S. Oliveira

LAETA, IDMEC, Instituto Superior Técnico, Universidade de Lisboa, Lisbon, Portugal
{paulo.gil,joliveira}@tecnico.ulisboa.pt

Abstract. Sound propagation in ducts with elliptical cross sections and lined walls is considered using a modal decomposition approach. The acoustic fields can be described in terms of Mathieu functions and Mathieu radial functions but the use of impedance boundary conditions leads to a coupled system of infinite algebraic equations. As a result, coupling of modes with different orders arise. Expressions for the determination of the eigenvalues and eigenfunctions are derived in the general case and an approximation is introduced for small eccentricity ($e < 0.3$) that leads to the uncoupling of the system of equations and of the modes. For small eccentricity the eigenmodes and eigenvalues for the axial wavenumber determined were similar to those of ducts with circular cross section, which can be considered as the limiting case as the eccentricity tends to zero. The attenuation of modes is always larger in elliptical ducts when compared to circular ducts and the imaginary part of the axial wavenumber can be more than 20% higher in the examples shown. The real part of the axial wavenumber for elliptical ducts, however, can be either smaller or larger than for circular ducts, depending on the frequency and mode order. For ducts with larger eccentricities the amount of mode coupling depends on the value of the eccentricity.

Keywords: Elliptical ducts · Sound propagation · Lined ducts · Modal · Mathieu functions

1 Introduction

The acoustics of ducts with elliptical cross sections has been less studied than that of ducts with other simple geometries, such as rectangular or circular cross sections. Despite their use in applications that include automotive and motorcycle engine exhausts and aircraft jet engines, ducts with elliptical cross sections have been studied mostly assuming rigid walls. However, acoustic liners are used to attenuate the noise radiated into the far field by the intake or exhaust ducts of jet engines. Therefore, an understanding of the effect of liners on the acoustics of ducts with elliptical cross-sections is important.

Analytical approaches have been used to study ducts with slowly varying elliptic cross-section with rigid walls, using multiple-scales methods [1], and the effect of flow on the acoustics of elliptic-cylinder waveguides and enclosures with rigid walls [2]. More recently, a numerical discontinuous Galerkin method was used to study the acoustic

Z. Dimitrovová et al. (Eds.): WMVC 2022, MMS 125, pp. 903–912, 2023.
https://doi.org/10.1007/978-3-031-15758-5_93

modes in flow ducts of several geometries, including ones with elliptical cross sections, with liners and rigid splices [3].

In this paper an analytical approach will be used to study the acoustics of a duct with elliptic cross section, possibly carrying a uniform flow, and with lined walls. Following [4], the solution of the propagation equations in terms of Mathieu functions will be discussed. The application of the impedance boundary conditions at the duct walls leads to mode coupling, and expressions for the determination of eigenvalues and modes will be derived in the general case and in the small eccentricity approximation. Finally, a numerical example will be presented for small eccentricity.

2 Sound Propagation in Elliptical Ducts

We consider the propagation of sound in an infinite duct with elliptical cross section. The duct walls can be rigid or acoustically lined, with a locally reacting liner with impedance Z. For a sound wave of angular frequency ω the acoustic pressure can be given as a superposition of modes

$$p(x, y, z, t) = \hat{p}(x, y)e^{-i(\omega t - k_z z)}, \tag{1}$$

where z is the axial coordinate and k_z is the axial wavenumber of the mode. The function $\hat{p}(x, y)$ is a solution of the transverse Helmholtz equation, and the transverse wavenumber $k_\perp$ is given by the dispersion relation

$$k_\perp^2 = \left(\frac{\omega}{c}\right)^2 - k_z^2. \tag{2}$$

The transverse Helmholtz equation can be solved in elliptical coordinates (ξ, η), which are defined by:

$$\begin{cases} x = f \cosh \xi \cos \eta, \\ y = f \sinh \xi \sin \eta, \end{cases} \tag{3}$$

with $0 \leq \xi$ and $0 \leq \eta < 2\pi$. For $\xi = \xi_0$, the coordinate curve is an ellipse with semi-major axis $a_e = f \cosh \xi_0$ and semi-minor axis $b_e = f \sinh \xi_0$. Therefore, the eccentricity of the ellipse is $e = 1/\cosh \xi_0$ and the focal distance is $f = a_e e$. As a result, the duct wall can be defined by either (f, ξ_0) or (a_e, e). Assuming that $\hat{p}(\xi, \eta) = \phi(\eta)\psi(\xi)$, the method of separation of variables leads to

$$\frac{d^2\phi(\eta)}{d\eta^2} + (a - 2q \cos 2\eta)\phi(\eta) = 0, \tag{4}$$

$$\frac{d^2\psi(\xi)}{d\xi^2} - (a - 2q \cosh 2\xi)\psi(\eta) = 0, \tag{5}$$

where a is a separation constant to be determined, and q is given by

$$q = \left(\frac{k_\perp f}{2}\right)^2 = \left(\frac{k_\perp a_e e}{2}\right)^2. \tag{6}$$

These are just the angular and radial Mathieu equations and its solutions can be given in terms of Mathieu functions. Solutions of Eq. (4) that are periodic in η, given q, exist only for certain values of a (called characteristic values) [5]. The periodic solutions are the cosine elliptic $ce_m(\eta; q)$ function (m is a non-negative integer), with a set of characteristic values given by $a = a_m(q)$; and the sine elliptic $se_n(\eta; q)$ function (n is a positive integer), with a set of characteristic values given by $a = b_n(q)$. $ce_m(\eta; q)$ is an even function of η and $se_n(\eta; q)$ is an odd function of η.

Since the set of characteristic values for the $ce_m(\eta; q)$ and $se_n(\eta; q)$ functions are different, each function is a solution of a different Mathieu angular equation Eq. (4). Therefore, given q, the periodic solution ϕ can be either $ce_m(\eta; q)$ or $se_n(\eta; q)$, but not a linear combination of both.

Solutions of Eq. (5) must be finite inside the duct. The characteristic values of Eqs. (4) and (5) are the same. Following [6], the radial Mathieu functions related to the even angular functions are denoted $Je_m(\xi, q)$ and the radial Mathieu functions related to the odd angular functions are denoted $Jo_m(\xi, q)$. Solutions of the transverse Helmholtz equation in elliptical coordinates can be either even modes, $\hat{p}_\perp(\xi, \eta) = ce_m(\eta; q)\, Je_m(\xi; q)$, with $a = a_m(q)$ and $m \geq 0$, or odd modes, $\hat{p}_\perp(\xi, \eta) = se_m(\eta; q)\, Jo_m(\xi; q)$ with $a = b_m(q)$, $m > 0$. These solutions are periodic in η and finite inside the duct.

3 Boundary Conditions and Solutions for Impedance Walls

In the absence of flow the boundary condition for lined walls at the ellipse defined by $\xi = \xi_0$ is

$$\frac{\partial \hat{p}}{\partial \xi} - i\,\beta\,\Omega\sqrt{\cosh^2\xi - \cos^2\eta}\;\hat{p} = 0, \tag{7}$$

where $\beta = \rho c/Z$ is the specific admittance and Ω is the reduced non dimensional frequency,

$$\Omega = \frac{\omega a_e}{c}. \tag{8}$$

For even modes the solutions can be written as a linear combination of products of even Mathieu functions and even radial Mathieu functions,

$$\hat{p}(\xi, \eta) = \sum\nolimits_{m=0}^{\infty} A_m\, ce_m(\eta, q)\, Je_m(\xi, q). \tag{9}$$

As shown in [4], substituting Eq. (9) in the boundary condition Eq. (7) eventually leads to an infinite homogeneous system of equations for the coefficient A_m that can be written as

$$\sum\nolimits_{m=0}^{\infty} M^e_{nm}(\xi_0, q)\, A_m = 0, \tag{10}$$

where the matrix elements M^e_{nm}, given by

$$M^e_{nm}(\xi_0, q) = N^e_m Je'_m(\xi_0, q)\delta_{nm} - i\,\beta\,\Omega\, I^e_{nm}(\xi_0, q)\, Je_m(\xi_0, q), \tag{11}$$

are written in terms of the integrals $I^e_{nm}(\xi_0, q)$ and $N^e_m(q)$:

$$I^e_{nm}(\xi_0, q) = \int_0^{2\pi} \sqrt{1 - e^2\cos^2\eta}\, ce_m(\eta, q)\, ce_n(\eta, q)\, d\eta, \tag{12}$$

$$N^e_m = \int_0^{2\pi} ce^2_m(\eta, q)\, d\eta. \tag{13}$$

The system defined by Eq. (10) can have non trivial solutions if and only if

$$\det M^e_{nm}(\xi_0, q) = 0. \tag{14}$$

The roots q^e_r of Eq. (14), with $r = 1, 2, \ldots, \infty$, define the allowed values of q. Each even mode r is then given by

$$\hat{p}_r(\xi, \eta, z) = \sum_{m=0}^{\infty} A_{r,m}\, ce_m\left(\eta, q^e_r\right) Je_m\left(\xi, q^e_r\right) \exp\left[ik_{z_r}(q^e_r)z\right], \tag{15}$$

where the coefficients $A_{r,m}$ are the solutions of Eq. (14) with $q = q^e_r$.

Similarly, for odd modes, the solution

$$\hat{p}(\xi, \eta) = \sum_{m=0}^{\infty} B_m\, se_m(\eta, q)\, Jo_m(\xi, q), \tag{16}$$

with the boundary condition Eq. (7) leads to the matrix equation

$$\sum_{m=0}^{\infty} M^o_{nm}(\xi_0, q)\, B_m = 0, \tag{17}$$

where the matrix elements M^o_{nm} and the integrals $I^o_{nm}(\xi_0, q)$ and N^o_m are

$$M^o_{nm}(\xi_0, q) = N^o_m Jo'_m(\xi_0, q)\delta_{nm} - i\,\beta\,\Omega I^o_{nm}(\xi_0, q)\, Jo_m(\xi_0, q) \tag{18}$$

$$I^o_{nm}(\xi_0, q) = \int_0^{2\pi} \sqrt{1 - e^2\cos^2\eta}\, se_m(\eta, q)\, se_n(\eta, q)\, d\eta, \tag{19}$$

$$N^o_m = \int_0^{2\pi} se^2_m(\eta, q)\, d\eta. \tag{20}$$

Equation (17) is a system of infinite homogeneous equations for the coefficients B_m which can have non trivial solution if and only if

$$\det M^o_{nm}(\xi_0, q) = 0. \tag{21}$$

The roots q^o_s of Eq. (21), with $s = 1, 2 \ldots, \infty$, define the allowed values of q. Each odd mode s is then given by

$$\hat{p}_s(\xi, \eta, z) = \sum_{m=0}^{\infty} B_{s,m}\, se_m\left(\eta, q^o_s\right) Je_m\left(\xi, q^o_s\right) \exp\left[ik_{z_s}(q^o_s)z\right], \tag{22}$$

where the coefficients $B_{s,m}$ are the solutions of (19) with $q = q^o_s$. Equations (15) and (22) show that the impedance boundary condition couples modes with different values of m.

Determining the roots q^e_r, q^o_s using the (truncated) determinant can be challenging because of the oscillatory nature of the integrating function in the integrals $N^{e,o}_m$ and $I^{e,o}_{nm}$. A simpler, approximate solution, valid for small eccentricity, will be derived next.

4 Approximate Solution for Small Eccentricity

For small eccentricity the condition $e^2 \ll 1$ holds. Using the approximation $\sqrt{1 - e^2\cos^2\eta} \approx 1 - \frac{1}{2}e^2\cos^2\eta + \cdots \approx 1$, the integrals I^e_{nm} and I^o_{nm} can be written as

$$I^e_{nm}(\xi_0, q) \approx \int_0^{2\pi} ce_m(\eta, q)ce_n(\eta, q)\, d\eta = N^e_m(q)\delta_{nm}, \tag{23}$$

$$I^o_{nm}(\xi_0, q) \approx \int_0^{2\pi} se_m(\eta, q)se_n(\eta, q)\, d\eta = N^o_m(q)\delta_{nm}, \tag{24}$$

where the orthogonality relations were used. Matrices M^e_{nm} and M^o_{nm} become diagonal:

$$M^e_{nm}(\xi_0, q) = N^e_m(q)\Big[Je'_m(\xi_0, q) - i\,\beta\,\Omega\, Je_m(\xi_0, q)\Big]\,\delta_{nm}, \tag{25}$$

$$M^o_{nm}(\xi_0, q) = N^o_m(q)\Big[Jo'_m(\xi_0, q) - i\,\beta\,\Omega\, Jo_m(\xi_0, q)\Big]\,\delta_{nm}, \tag{26}$$

and the coefficients A_m and B_m are all independent. Therefore, in the small eccentricity approximation there is no mode coupling.

The roots of Eqs. (14) and (21) in this approximation are obtained by setting each of the diagonal terms of Eqs. (25) and (26) equal to zero:

$$Je'_m(\xi_0, q) - i\,\beta\,\Omega\, Je_m(\xi_0, q), \tag{27}$$

$$Jo'_m(\xi_0, q) - i\,\beta\,\Omega\, Jo_m(\xi_0, q). \tag{28}$$

The solution no longer depends on the integrals $I^{e,o}_{nm}$ and $N^{e,o}_m$, which also simplifies its evaluation.

To effectively determine the solution it is necessary to find the roots q of Eq. (27) and Eq. (28) for concrete values of the admittance β, reduced frequency Ω, and eccentricity e.

5 Example of a Solution for the Case of Small Eccentricity

In the small eccentricity approximation, determining the roots q of Eq. (27) and Eq. (28) is equivalent, by Eq. (6), to determine the eigenvalues q of the Helmholtz equation. They were determined for ducts with elliptical cross-section of eccentricities $e = 0.1, 0.2, 0.3$ and several values of the reduced frequencies Ω, for both rigid walls and walls with admittance $\beta = 0.4 + 0.06\,i$. This value of the specific admittance corresponds to a specific impedance $\frac{Z}{\rho c} = R + i\chi = 2.44 - 0.37\,i$.

5.1 Location of Wavenumbers on Complex Plane

Mode attenuation depends on the axial wavenumbers k_z, which were calculated using Eqs. (2) and (6). The axial wavenumbers can be made dimensionless using the semi-major axis of the ellipse,

$$\mathrm{K_z} = k_z a_e. \tag{29}$$

The dimensionless axial wavenumber can be directly obtained from the dimensionless frequency and dimensionless transverse wavenumber:

$$\mathrm{K_z^2} = \Omega^2 - K_\perp^2. \tag{30}$$

The location of the dimensionless axial eigenvalues Eq. (29) in the complex plane are shown in Fig. 1 for circular and elliptical ducts with lined walls, for two values of eccentricity ($e = 0.1$, left, and $e = 0.3$, right) and three different values of the reduced frequencies ($\Omega = 1, 5, 10$, from top to bottom) for the $m = 0$ even modes. Only right-running modes (with positive real part) are shown. The symmetrical solutions of Eq. (2) that lie in the third quadrant and correspond to left-running modes are not shown. For circular ducts the dimensionless wavenumbers $\widetilde{K}_z^2 = k_z R$ were determined using

$$\widetilde{\mathrm{K}}_\mathrm{z}^2 = \widetilde{\Omega}^2 - \widetilde{K}_\perp^2, \tag{31}$$

and

$$\widetilde{K}_\perp J_n'(\widetilde{K}_\perp) - i\beta\,\widetilde{\Omega}\,J_n(\widetilde{K}_\perp) = 0, \tag{32}$$

with a radius $R = a_e$. Note that in the case of circular ducts the dimensionless wavenumber $\widetilde{K}_\perp = k_\perp R$, and the reduced frequency $\widetilde{\Omega} = \omega R/c$ were used.

For lined ducts, all modes are attenuated, but some lie near the real axis and are lightly attenuated, corresponding to the propagating modes in rigid ducts. Others lie near the imaginary axis and are strongly attenuated, corresponding to cut-off modes in rigid ducts. This is clearly seen in Fig. 1: for $\Omega = 1$ (top graphics) all modes except the first ($n = 1$) are strongly attenuated. As the frequency increases, more modes approach the real axis, and some modes are in a "transition region", moving from the vicinity of the imaginary axis to the vicinity of the real axis. The figure also shows that for the smaller eccentricity (graphics on the left), the axial wavenumbers for the elliptical ducts are similar to those of the circular duct. For $e = 0.3$ the imaginary part of the axial wavenumbers for elliptical ducts is greater than for circular ducts, which means that axisymmetric ($m = 0$) modes in elliptical duct have higher attenuation. The real parts are also different, especially on the "transition" modes, that is, the modes that are neither near the real axis (lightly attenuated modes) nor near the imaginary axis (highly attenuated modes). However, it is difficult to identify a definite trend.

The location in the complex plane of the axial eigenvalues for the $m = 1$ modes for circular and elliptical ducts with lined walls, for two values of eccentricity ($e = 0.1$, left, and $e = 0.3$, right) and three different values of the reduced frequencies ($\Omega = 1, 5, 10$, from top to bottom) is shown in Fig. 2 for even modes. The main features are the same as in the previous case. In both cases the modes for elliptical ducts have larger imaginary part and are therefore more attenuated than the corresponding modes for circular ducts, and this effect is stronger for larger eccentricity, as would be expected. The real part of the odd modes is smaller than for even modes and tends to be smaller than for circular ducts for the two larger frequencies.

6 Towards the Solution for All Values of the Eccentricity

The determinants Eq. (14) and Eq. (21) are infinite and to find their roots q_r^e, q_r^o they must be truncated. The larger the truncated determinant the better, as it is expected that the solutions of increasingly large truncated determinants will be increasingly close to the real solution, but also increasingly difficult to solve. It is therefore worthwhile to try to simplify the task.

The initial procedure to solve the problem will be illustrated with the determinant Eq. (21); for Eq. (14) the procedure is similar. We start by noticing that the constants $i\beta\Omega$ of the generic entry Eq. (18) of the matrix are always different from zero and appear multiplying all the entries of the matrix except the first term of each entry of the principal diagonal. We can therefore factor the constants out from all the lines of the determinant, making them appear only in the denominator of the mentioned first term. Also, the functions $Jo_m(\xi_0, q)$, $m = 1, 2, \ldots$ are the same for each line of the determinant. The parameter ξ_0 is constant for each considered value of the eccentricity. We start by determining the roots of $Jo_m(\xi_0, q)$. For these values of q the matrix $M_{nm}^o(\xi_0, q)$ is now diagonal and it is easy to check if they are also roots of the determinant in Eq. (21). To find the roots other than those of all the $Jo_m(\xi_0, q)$, $m = 1, 2, \ldots$ we can factor those terms out of the determinant of Eq. (21) in all lines, leading to

$$\det\left[M_{nm}^o(\xi_0, q)\right] = Jo_1(\xi_0, q) \times Jo_2(\xi_0, q) \times \cdots \times Jo_m(\xi_0, q) \times \ldots$$
$$\times \det\left[\frac{N_m^o(q)\, Jo_m^{'}(\xi_0, q)}{i\beta\Omega\, Jo_m(\xi_0, q)}\delta_{nm} - I_{nm}^o(\xi_0, q)\right] = 0, \text{ with } Jo_m(\xi_0, q) \neq 0, m = 1, 2, \ldots \tag{33}$$

or

$$\det\left[\frac{iN_m^o(q)\, Jo_m^{'}(\xi_0, q)}{\beta\Omega\, Jo_m(\xi_0, q)}\delta_{nm} + I_{nm}^o(\xi_0, q)\right] = 0, \text{ with } Jo_m(\xi_0, q) \neq 0, m = 1, 2, \ldots, \tag{34}$$

where the imaginary unit was passed to the numerator and the minus signs also factored out. The determinant Eq. (34) is simpler than Eq. (21) because now the constants and $Jo_m(\xi_0, q)$ only appear in the principal diagonal and, most importantly, since the integrals $I_{nm}^o(\xi_0, q)$ are symmetric in $\{m, n\}$, the matrix became symmetric.

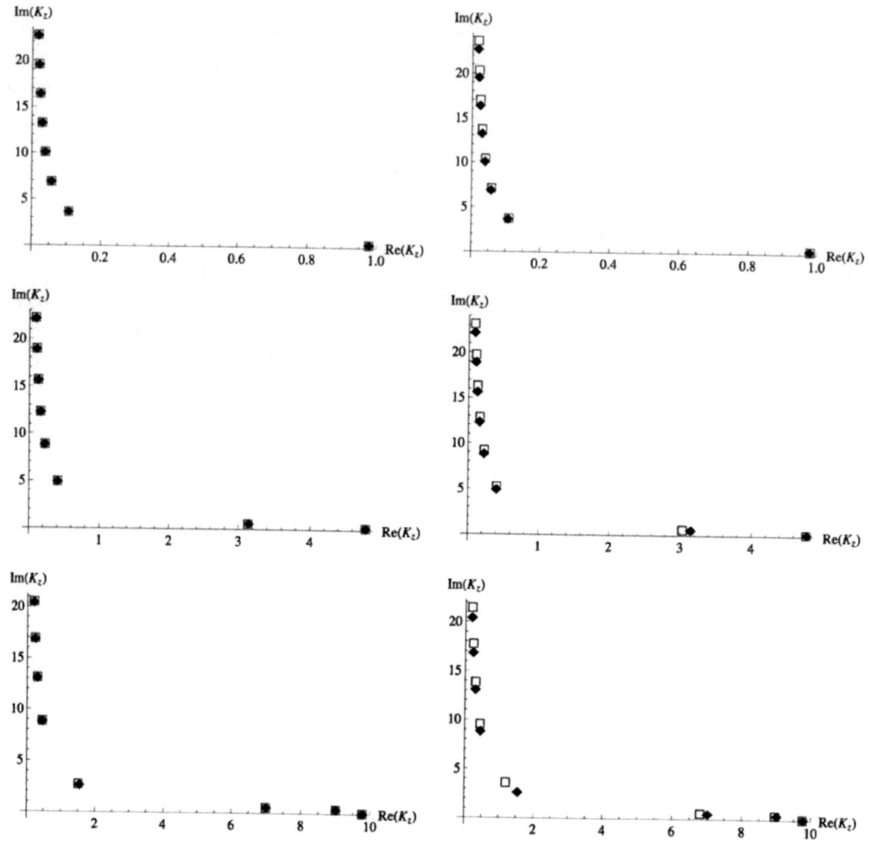

Fig. 1. Location of axial eigenvalues in complex plane for circular ducts (◆) and for even modes of elliptical ducts (□), $m = 0$ and $\Omega = 1$ (top), 5 (middle) and 10 (bottom), in the case of $e = 0.1$ (left) and $e = 0.3$ (right), for lined duct with wall admittance $\beta = 0.4 + 0.06\,i$.

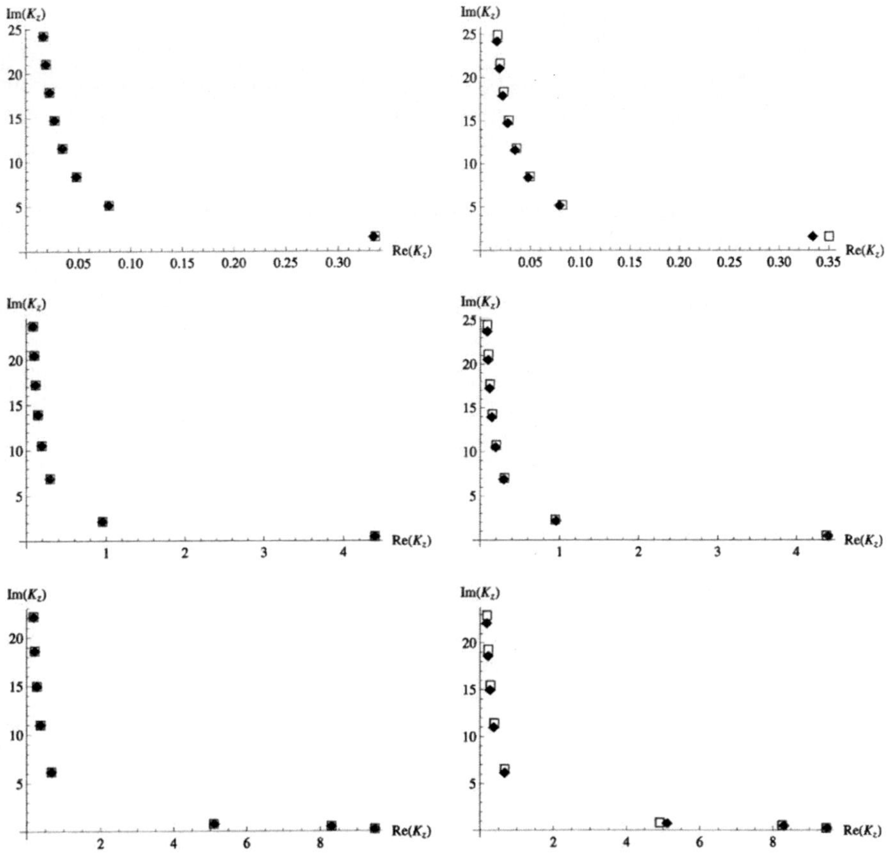

Fig. 2. Location of axial eigenvalues in complex plane for circular ducts (•) and for even modes of elliptical ducts (□), $m = 1$ and $\Omega = 1$ (top), 5 (middle) and 10 (bottom), in the case of $e = 0.1$ (left) and $e = 0.3$ (right), for lined duct with wall admittance $\beta = 0.4 + 0.06\,i$.

7 Conclusions

In this paper sound propagation in ducts of elliptical cross-section and lined with locally reacting liners was considered. The acoustic pressure field can be described in terms of Mathieu functions and radial Mathieu functions. The impedance boundary conditions lead to a system of an infinite number of algebraic equations, which results in the coupling of modes of different orders. The approximation for small eccentricity leads to the uncoupling of the system of equations and of the modes.

The eigenmodes and eigenvalues for the axial wavenumber that were determined for the case of small eccentricity were similar to those of ducts with circular cross-section, which can be considered as the limiting case as $e \to 0$. The attenuation of modes is always larger in elliptical ducts when compared to circular ducts and $\text{Im}(K_z)$ can be more than 20% higher. The real part of the axial wavenumber K_z for elliptical ducts, however, can be either smaller or larger than for circular ducts, depending on the frequency and mode order.

The cut-off frequency of rigid ducts was found to be an important parameter for the behaviour of the modes, at least for the example shown, in which a small admittance was used. The usual classification in evanescent and propagating modes is no longer valid. However, modes can be classified as: (i) strongly attenuated, for frequencies much smaller than the cut-off frequency of the mode; (ii) lightly attenuated (or "propagating") modes, for frequencies much larger than the cut-off frequency; and (iii) transition modes, for frequencies comparable to the cut-off frequency. This classification has been used before for lined circular ducts, and it remains valid for lined elliptical ducts of small eccentricity.

A difficulty in dealing with the general case of arbitrary eccentricity was identified that can suggest a line of action to find the general solution.

Acknowledgements. This work was supported by FCT, through IDMEC, under LAETA, project UIDB/50022/2020.

References

1. Peake, N., Cooper, A.J.: Acoustic propagation in ducts with slowly varying elliptic cross-section. J. Sound Vib. **243**, 381–401 (2001)
2. Willatzen, M., Lew Yan Voon, L.C.: Flow-acoustic properties of elliptic-cylinder waveguides and enclosures. J. Phys. Conf. Ser. **52**, 1–13 (2006)
3. Pascal, L., Piot, E., Casalis, G.: Discontinuous Galerkin method for the computation of acoustic modes in lined flow ducts with rigid splices. J. Sound Vib. **332**, 3270–3288 (2013)
4. Oliveira, J.M.G.S., Gil, P.J.S.: Sound propagation in acoustically lined elliptical ducts. J. Sound Vib. **333**, 3743–3758 (2014)
5. McLachlan, N.: Theory and Application of Mathieu Functions. Oxford University Press, Oxford (1947)
6. Gutiérrez-Veja, J.C., Rodrígues-Dagnino, R.M., Neneses-Nava, M.A., Chávez-Cerda, S.: Mathieu functions, a visual approach. Am. J. Phys. **71**, 233–242 (2003)

Design and Performance Analysis of a Novel Quasi-Zero Stiffness Vibration Isolator

Huang Mengting, Zhang Tao(✉), and Chen Cong

School of Naval Architecture and Ocean Engineering, Huazhong University of Science and Technology, Wuhan, China
zhangt7666@hust.edu.cn

Abstract. Quasi-zero stiffness (QZS) indicates the characteristics of high static stiffness and low dynamic stiffness. In present study, the QZS vibration isolator has been theoretically investigated. A novel QZS vibration isolator, which consists of a vertical string, and three pairs of horizontal spring-cam-roller mechanisms, is proposed. The differential equation of the cam profile is derived. The expression of the restoring force (RF) is obtained by introducing the Gaussian function, and the dynamic characteristics of the isolator are analyzed. The results indicate that the vibration isolator meets the QZS characteristics. With the increase of the external excitation frequency, the amplitude of the system increases to the peak value, and then drops rapidly, finally decreases slowly. The force transmission rate has a similar trend that the initial value is one, and the vibration isolation effect occurs at a certain frequency. Generally, the proposed design of the QZS vibration isolator is feasible for vibration isolation in the low-frequency range.

Keywords: Quasi-zero stiffness · Gaussian function · Dynamic characteristic

1 Introduction

Vibration isolation has been largely explored [1]. The quasi-zero stiffness (QZS) vibration isolator is a kind of nonlinear vibration isolator. Different from most the existing isolators (especially linear isolators), which have no obvious effect on low-frequency isolation, the QZS vibration isolator can achieve the characteristics of high static stiffness and low dynamic stiffness by adequately adjusting the parameters of the system. So it has been extensively studied at home and abroad in recent years.

In order to realize QZS characteristics, scholars usually combine the elements with negative and positive stiffness together. The key to this combination is the selection of negative stiffness elements. In the past decades, researchers have proposed several different negative stiffness elements such as oblique springs [2, 3], push rods [4, 5], link of horizontal springs [6], and so on [7, 8]. Yang et al. [9] proposed a nonlinear inertance mechanism, which enhanced vibration isolation performance. Dai et al. [10] presented a nonlinear vibration isolator with nonlinear elements created by the geometric nonlinearity of a linkage mechanism with embedded linear springs. Zhao et al. [11] studied the QZS isolator with one pair of oblique bars by considering pre-compression

Z. Dimitrovová et al. (Eds.): WMVC 2022, MMS 125, pp. 913–920, 2023.
https://doi.org/10.1007/978-3-031-15758-5_94

of horizontal springs. Shi et al. [12] developed displacement- and kinetic energy-based tuning methods for the design of the tuned inerter dampers. These are all the important works and recent research advances on nonlinear vibration isolators.

Among many types of QZS vibration isolators, the isolator, which consists of a vertical string and three pairs of horizontal spring-cam-roller mechanisms, is less studied. Different from other QZS isolators, the cam profile of the proposed QZS vibration isolator can be designed according to different parameters, which is also the difficulty in designing this kind of isolator. In order to fit the restoring force (RF) better, the Gaussian function is introduced, and on this basis, the differential equation of the cam profile is derived. The dynamic characteristics are also discussed. Design and static analysis of vibration isolator.

2 Design of Vibration Isolator and Static Analysis

2.1 Design of Vibration Isolator

This design has three pairs of horizontal spring-cam-roller mechanisms. Their resultant force, together with the force of the vertical spring, balances the external load. In this design, the vertical spring provides the positive stiffness, and the horizontal spring-cam-roller mechanisms provide the negative stiffness.

Figure 1 is the schematic diagram of the QZS vibration isolator. Two sets of coordinate systems (static and dynamic) are established. The cam profile can be obtained after determining the dynamic trajectory of *C*. Once the cam profile is properly designed, the QZS system can be established at the equilibrium position and realize the low-frequency isolation.

F_1 is the vertical spring force. F_2 is the horizontal spring force. F_3 is the force exerted by the cam on the roller. F_{r} is the RF. α is the angle between two coordinate systems. β is the angle between F_3 and Y-axis. θ is the angle between F_3 and V-axis. k_1 is the stiffness of vertical spring. k_2 is the stiffness of horizontal springs. d is the distance from the roller center to Y-axis. h is the initial height of the horizontal spring. b is the pre-compressed length of the horizontal spring. H is the initial height of the roller center. R is the radius of the roller.

Suppose the static coordinates of *C* are (x, y), and the dynamic coordinates are (u, v). Equations (1), (2) and (3) show the transformation between two sets of coordinates.

$$\begin{cases} x = u \cdot \cos\alpha - v \cdot \sin\alpha \\ y = u \cdot \sin\alpha + v \cdot \cos\alpha \end{cases} \tag{1}$$

$$x^2 + y^2 = u^2 + v^2 \tag{2}$$

$$x = d \tag{3}$$

Balance equations of the QZS system under design load can be obtained as:

$$F_1 + 3F_3 \cdot \cos\beta - G - F_r = 0 \tag{4}$$

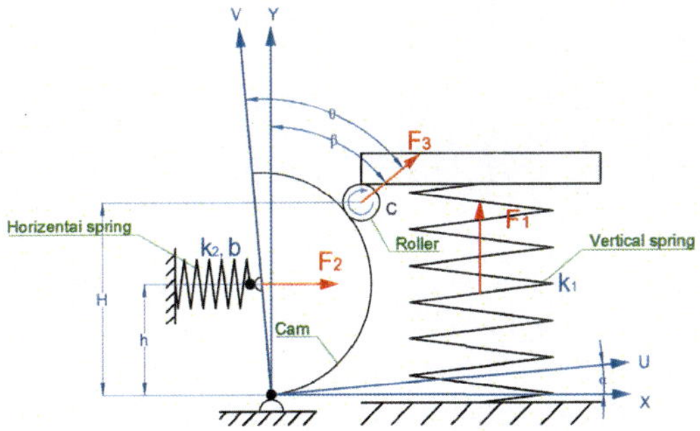

Fig. 1. Schematic diagram of the QZS vibration isolator.

In order to achieve the QZS characteristics, the relationship between the RF and vertical displacement must be constructed. In this paper, the Gaussian function is introduced to modify the RF of the system. With a series of transformations, the QZS characteristics shown can be achieved.

Therefore, the RF can be constructed as a function. The design load of the proposed isolator is *G*. Equation (5) is the constructed RF- dimensionless vertical displacement function of the QZS isolator.

$$F_r(\hat{y}) = \begin{cases} G(1 - e^{-10(n\hat{y}-m)^2}), & \hat{y} \geq m/n \\ 0, & |\hat{y}| \leq m/n \\ G(e^{-10(n\hat{y}+m)^2} - 1), & \hat{y} \leq -m/n \end{cases} \tag{5}$$

where *m*, *n* are two design parameters of the function, and they are called Gaussian parameters in this paper. $\hat{y}$ is the dimensionless vertical displacement of the isolated object shown in Eq. (6).

$$\hat{y} = 2(y - H)/L + 1 \qquad (-1 \leq \hat{y} \leq 1) \tag{6}$$

where *L* is the effective vertical stroke of the isolator.

Considering that the general bending moment of the cam in the quasi-static process is zero, the equilibrium equations can be obtained:

$$F_3 \cdot \cos\beta \cdot d - F_3 \cdot \sin\beta \cdot y + F_2 \cdot h \cdot \cos\alpha = 0 \tag{7}$$

$$F_2 = k_2 \cdot (b + h \cdot \tan\alpha) = k_2 \cdot (b + h \cdot \frac{y \cdot u - d \cdot v}{d \cdot u + y \cdot v}) \tag{8}$$

where $\alpha \leq 5°$, so cosα $\approx$1.

In dynamic coordinate system, the differential equation of the motion curve at C can be expressed:

$$\frac{dv}{du} = -\tan\theta \tag{9}$$

Equation (10) is obtained by solving Eqs. (5), (6), and (9):

$$\frac{dv}{du} = \frac{\left[\frac{d\cdot v - y\cdot u}{d\cdot u + y\cdot v}\right] - \left[\frac{3F_2\cdot h}{G - F_1 + F_r} + d\right]\cdot\frac{1}{y}}{1 + \left[\frac{d\cdot v - y\cdot u}{d\cdot u + y\cdot v}\right]\cdot\left[\frac{3F_2\cdot h}{G - F_1 + F_r} + d\right]\cdot\frac{1}{y}} \tag{10}$$

In Eq. (10), F_r is a function with y as its independent variable, while y, F_1, and F_2 are functions with u and v as their only independent variables. By solving the differential equation in Eq. (10), the dynamic trajectory of C can be obtained.

In the dynamic coordinate system, if the cam profile is T (u_1, v_1), Eq. (11) can be known from the geometric relationship:

$$\begin{cases} -\tan\theta = \frac{v - v_1}{u - u_1} \\ (u - u_1)^2 + (v - v_1)^2 = R^2 \end{cases} \tag{11}$$

Similarly, the profile curve can be obtained by discretizing Eq. (11).

2.2 Static Analysis

The specific values are substituted into the designed parameters. Assume $m = 0.03$, $n = 0.4$, $G = 1500$ N, $d = 5$ mm, $H = 80$ mm, $h = 40$ mm, $b = 5$ mm, $R = 5$ mm, $L = 7$ mm, $k_1 = 200$ N/mm, $k_2 = 100$ N/mm.

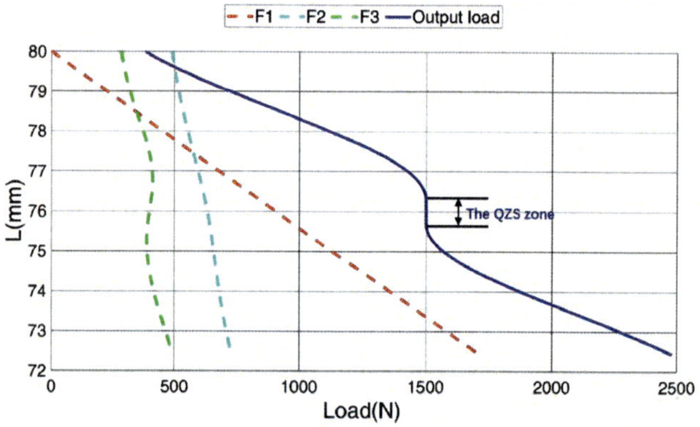

Fig. 2. Static analysis of the QZS vibration isolator.

By introducing the above parameters into Eqs. (10) and (11), the motion curve of the roller center and cam profile in the dynamic coordinate system can be obtained. Additionally, static analysis of the QZS vibration isolator are shown in Fig. 2. It is evident that the designed isolator meets the QZS characteristics.

3 Dynamic Analysis

3.1 Amplitude-frequency Response

It is known that damping has poor effect on static characteristics. However, in dynamic analysis, Damping is an important factor. Therefore, damping is considered in this part. A single-DOF system with damping is showed in Fig. 3.

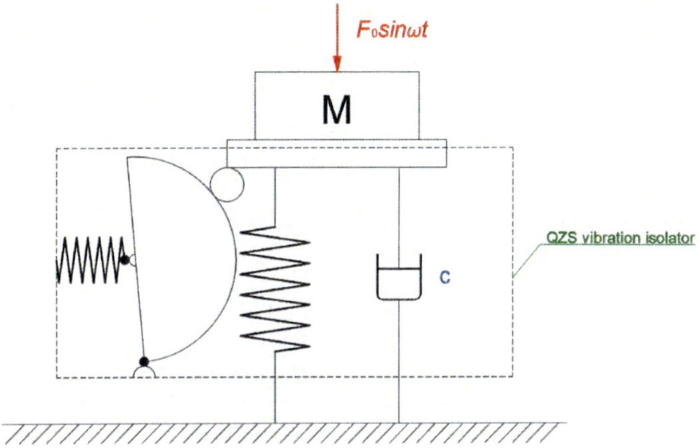

Fig. 3. A single-DOF system.

The RF is simplified into the Taylor formula in segments (retain first three terms):

$$f(u) = \begin{cases} p(u-q)^2 \cdots\cdots\cdots\cdots\cdots u \geqslant q \\ 0 \cdots\cdots\cdots\cdots\cdots\cdots\cdots |u| \leqslant q \\ -p(u+q)^2 \cdots\cdots\cdots\cdots u \leqslant -q \end{cases} \tag{12}$$

where $-1 \leq u \leq 1, p = 10Gn^2, q = \frac{m}{n}, G = Mg$.

When an object with mass M is installed and suffers a harmonic excitation load $F_0 \sin \omega t$, The differential equation of motion of the system can be set:

$$m \cdot \ddot{u} + c \cdot \dot{u} + f(\frac{u}{l}) = F_0 \sin \omega t \tag{13}$$

where c is the damping ratio, $l = L/2$, F_0 is the amplitude of the excitation load, and ω is the frequency of the excitation load.

Use the average method to solve the system dynamics equation. The amplitude-frequency response function of the nonlinear system can be obtained:

$$A^2\omega^2c^2 + [AM\omega^2 - 2AM\omega N(A)]^2 = F_0^2 \tag{14}$$

where A is the amplitude of the system, $N(A)$ is a nonlinear piecewise function:

$$N(A) = \begin{cases} 0 \cdots\cdots\cdots\cdots\cdots\cdots\cdots\cdots\cdots\cdots\cdots\cdots\cdots\cdots\cdots 0 \leq A \leq ql \\ \frac{4}{3}\frac{Ap}{\omega M\pi l^2}\sqrt{1-(\frac{ql}{A})^2} + \frac{2}{3}\frac{pq^2}{A\omega M\pi}\sqrt{1-(\frac{ql}{A})^2} - 2\frac{pq}{\omega Ml\pi}\arccos(\frac{ql}{A}) \cdot A \geq ql \end{cases} \tag{15}$$

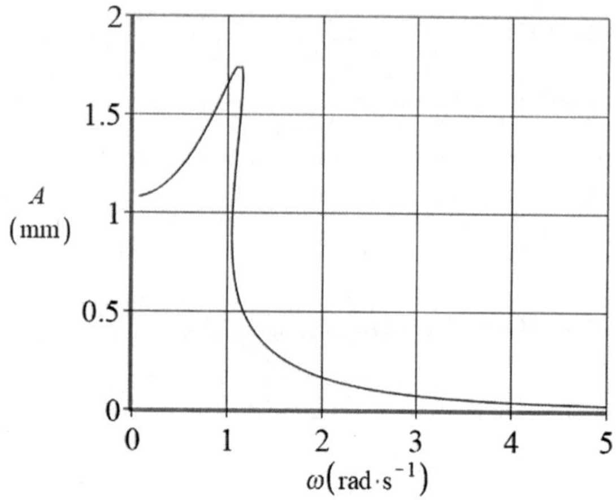

Fig. 4. Amplitude-frequency response.

Assume $m = 0.03$, $n = 0.4$, $M = 150$ kg, $c = 50$ N·s/mm, $F_0 = 100$ N, L = 7 mm. Discretizing the Eq. (14), the amplitude-frequency response can be obtained in Fig. 4.

It can be seen from Fig. 4 that jumping occurs in the amplitude-frequency of the QZS system. With the increase of the external excitation frequency, the corresponding amplitude of the system first gradually increases, then rapidly decreases after reaching the peak value, and carries on decreasing.

3.2 Force Transmissibility Analysis

In order to analyse the vibration isolation effect, the force transmissibility of the system is derived. Suppose the force transmitted to the foundation is F_T.

Equation (16) shows the F_t under the external harmonic excitation load $F_0 \sin \omega t$:

$$F_T = -cu' - f(u) = Mu'' - F_0 \sin \omega t \tag{16}$$

After derivation, the maximum value of F_T is obtained in Eq. (17).

$$F_{T\max} = \sqrt{3M^2A^2\omega^4 - 4M^2A^2\omega^3N(A) + F_0^2} \tag{17}$$

Therefore, the equation of force transmissibility T can be derived in Eq. (18).

$$T = \frac{F_{T\max}}{F_0} = \frac{\sqrt{3M^2A^2\omega^4 - 4M^2A^2\omega^3N(A) + F_0^2}}{F_0} \tag{18}$$

In order to analyse the influence of damping on transmissibility, different values are substitute in c. Assume $m = 0.03$, $n = 0.4$, $M = 150$ kg, $F_0 = 100$ N, L = 7 mm, $c_1 = 100$ N·s/mm, $c_2 = 50$ N·s/mm, $c_3 = 10$ N·s/mm. After discretizing the Eq. (18), the force transmissibility can be obtained in Fig. 5.

When the external excitation frequency is ultra-low, the force transmissibility of the system is greater than 1, and gradually increases with the excitation frequency; When

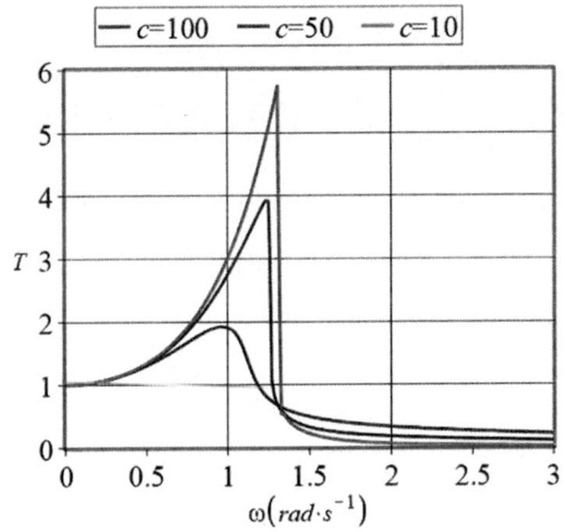

Fig. 5. Force transmissibility.

the frequency reaches a certain value, the system transmission rate reaches the peak value, and rapidly drops below 1, it is called initial vibration isolated frequency in this study; After this frequency, the vibration isolator shows the effect of vibration isolation. Additionally, with the increase of damping ratio, the peak of transmissibility decreases, and the initial isolated frequency of the system decreases, but the transmissibility after the initial isolated frequency is higher.

4 Conclusion

This paper studies the QZS characteristics of a cam-roller vibration isolator. The differential equation of the cam profile is deduced. The RF is described in a piecewise function. The result indicates that the designed vibration isolator has the QZS characteristics, so the system can achieve the low-frequency isolation performance at the equilibrium position.

References

1. Zhang, L., Zhang, T., Ouyang, H., Li, T.: Receptance-based antiresonant frequency assignment of an uncertain dynamic system using interval multiobjective optimization method. J. Sound Vib. **529**, 116944 (2022)
2. Zhao, F., Ji, J.C., Ye, K., Luo, Q.T.: An innovative quasi-zero stiffness isolator with three pairs of oblique. Int. J. Mech. Sci. **192**(15), 106093 (2021)
3. Lan, C.C., Yang, S.A., Wu, Y.S.: Design and experiment of a compact quasi-zero-stiffness isolator capable of a wide range of loads. J. Sound Vib. **333**(20), 4843–4858 (2014)
4. Yang, J., Xiong, Y.P., Xing, J.T.: Dynamics and power flow behaviour of a nonlinear vibration isolation system with a negative stiffness mechanism. J. Sound Vib. **332**(1), 167–183 (2013)
5. Huang, X., Chen, Y., Hua, H., Liu, X., Zhang, Z.: Shock isolation performance of a nonlinear isolator using Euler buckled beam as negative stiffness corrector: theoretical and experimental study. J. Sound Vib. **345**(9), 178–196 (2015)

6. Le, T.D., Ahn, K.K.: Experimental investigation of a vibration isolation system using negative stiffness structure. Int. J. Mech. Sci. **70**(5), 99–112 (2013)
7. Le, T.D., Ahn, K.K.: Active pneumatic vibration isolation system using negative stiffness structures for a vehicle seat. J. Sound Vib. **333**(5), 1245–1268 (2014)
8. Xu, J., Sun, X.T.: A multi-directional vibration isolator based on Quasi-Zero-Stiffness structure and time-delayed active control. Int. J. Mech. Sci. **100**, 126–135 (2015)
9. Yang, J., Jiang, J.Z., Neild, S.A.: Dynamic analysis and performance evaluation of nonlinear inerter-based vibration isolators. Nonlinear Dyn. **99**(3), 1823–1839 (2019). https://doi.org/10.1007/s11071-019-05391-x
10. Dai, W., Li, T., Yang, J.: Energy flow and performance of a nonlinear vibration isolator exploiting geometric nonlinearity by embedding springs in linkages. Acta Mech. **233**(4), 1663–1687 (2022)
11. Zhao, F., Cao, S., Luo, Q., Li, L., Ji, J.: Practical design of the QZS isolator with one pair of oblique bars by considering pre-compression and low-dynamic stiffness. Nonlinear Dyn. **108**(4), 3313–3330 (2022)
12. Shi, B., Yang, J., Jiang, J.Z.: Tuning methods for tuned inerter dampers coupled to nonlinear primary systems. Nonlinear Dyn. **107**(2), 1663–1685 (2022)

Experimental Investigation of Dynamic Response and Wave Dissipation of a Horizontal Plate Breakwater

Tengxiao Wang and Heng Jin(✉)

NingboTech University, Ningbo, China

Abstract. In recent years, floating breakwaters are considered for protecting the offshore engineering structures in the deep sea. Further, to expand the capabilities of the horizontal plate breakwater, an elastic supported horizontal plate (ESHP) breakwater is developed as an eco-friendly and high energy dissipation structure. In this study, the wave dissipation effect of an ESHP breakwater is investigated with experimental tests. Then the hydrodynamic coefficients and the wave force acting on the breakwater are analyzed with a variable stiffness of support spring and wave conditions. The experimental study shows that the interaction between the radiation wave of the heaving plate and scattered waves causes additional vortex flow, and the wave height is reduced rapidly at the lee side of the breakwater. Then the wave dissipation mechanism of ESHP breakwater for incident waves with different incident wave heights is discussed. At last, the wave force changes due to plate motion are also revealed to ensure the robustness of the structure.

Keywords: Breakwater · Heaving horizontal plate · Wave dissipation · Dynamic responses

1 Introduction

Different types of breakwaters for wave protection are mainly used to resist wave attacks on the nearshore structures such as coastline or harbors. Especially, the floating breakwater was recommended as an economical and efficient wave protection structure [1, 2]. Ji et al. conducted extensive experimental and analytical work on a new type of floating breakwaters [3, 4]. He and Huang [5] experimentally investigated the hydrodynamic performance of a pile-supported oscillating water column structure as a breakwater. Compared with floating breakwater, submerged horizontal plate (SHP) breakwater have the advantage of low construction costs and has been attracting increasing attention and there has been extensive research.

Recently, there has been extensive research on the wave dissipating performance of SHP breakwaters and the wave impact around the breakwaters. Patarapanich and Cheong [6] proved that the submerged horizontal plate has a great wave blocking performance when the radio of the plate width to wave length above the plate was in the rage of 0.5–0.7 and the ratio of plate submergence to water depth was within 0.05–0.15. The nonlinear transmission of waves over the SHP was analyzed by Liu and Huang et al. [7]

Z. Dimitrovová et al. (Eds.): WMVC 2022, MMS 125, pp. 921–929, 2023.
https://doi.org/10.1007/978-3-031-15758-5_95

by using a fully nonlinear numerical wave tank based on the desingularized boundary integral equation method. Porter discussed the scattering of incident waves and radiation of waves by forced motion by thin horizontal plates using a Galerkin method. In the dynamic characteristics studies of the SHP breakwater, Poupardin et al. [8] showed the vortex dynamics generated by the interaction of a submerged horizontal plate through experiments. Hayatdavoodi et al. [9] experimentally measured horizontal and vertical forces caused by Conidal waves on a submerged horizontal plate.

The elastic support structure is a widely used vibration damping device in mechanical engineering. Combining it with the horizontal plate breakwater, Koo and Kim [10] found that the movable floating horizontal plate has better wave dissipation effect than the fixed one and the heaving motion of the horizontal plate leads to an essential role in it. In the numerical studies of a heaving horizontal plate breakwater, it was found that a suitable spring stiffness could dissipate the vortex energy at the back wave surface of the structure under the resonant response of the floating plate [11–13]. In the dynamic characteristics studies of the elastic supported horizontal plate breakwater, the phenomenon of impact pressure forces induced by water waves was first examined by Wilde et al. [14]. Bing et al. [15] indicate that the uplift forces and acceleration of the elastically plates with small natural frequencies are different from those of the plates with a larger natural frequencies. However, investigations into the interaction between wave dissipating performance, wave force, and motion response of a submerged ESHP breakwater are not yet found in the published literature.

This paper proposes a submerged fixed plate breakwater and a submerged ESHP breakwater [16]. The addition of the spring support system is intended to improve the performance under incident waves. Motion response and wave force acting on the breakwater were examined using physical model tests in a two-dimensional wave flume. The design of the model test including incident wave condition, submergence ratio, etc. is discussed in the following section.

2 Experimental Setup

The experiments were conducted in a two-dimensional wave flume of the hydrodynamic Laboratory of Ningbo Institution of Technology. The wave flume is 11 m long, 0.6 m wide, and 1 m in height. The physical model of the heaving plate was made of acrylic and the dimensions of the model are the length (L) of 0.58 m, and the width (B) of 0.06 m. In the present experiment, the sides of the heaving plate were in contact with the flume, and the 0.01 m gap was filled by two universal wheels on both sides. In addition, a steel rod and two flanged ball bearings were installed on the horizontal plate which ensures the movement of the horizontal plate only in the heave direction.

Three wave probes were installed at front and rear of the breakwater (Fig. 1) to measure the wave height. 8 pressure sensors (YPS301-L) were fixed on the top and bottom sides of the plate (P1-8). In addition, a motion sensor (FASTRACK, POLHEMUS) was fixed on the plate surface for capturing the motion response.

In order to investigate the wave protection effect of the ESHP breakwater, the wave force and hydrodynamic characteristics of the fixed horizontal plate under wave actions were first tested. The water level in the wave flume is $d = 0.4$ m. The incident wave

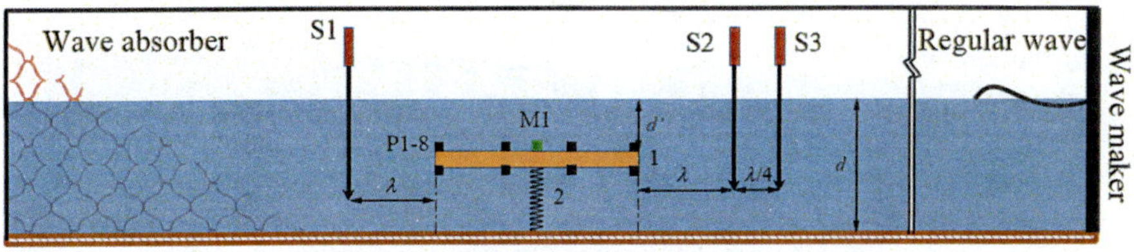

Fig. 1. The layout of the experimental wave flume and parameter setup; (1) horizontal plate, (2) spring, (P1-8) pressure sensors, (S1-3) wave probes. (M1) motion sensor, λ is the wavelength, d' is submergence depth, d is water depth in the water flume.

with a period of $T = 1$ s and the incident wave height (H) varied from 0.03m to 0.04m. The submergence depth d' was adjusted through a coupling and the submergence depth ratio d'/d ranged from 0.1 to 0.3.

The wave force and hydrodynamic characteristics of the ESHP breakwater under wave actions were then tested. A spring supporting system was considered in this condition. The stiffness coefficient of spring K varied from 154 kg/m to 537 kg/m. In the discussion, the K^* was nondimensionalized into $K/\rho d'H$ (from 21.9 to 76.5).

The duration of each test is set to 90 s and the stable intermediate data 30–80 s was intercepted for the following analysis. The motion response amplitude of each direction was obtained and the wave force was calculated by the pressure data obtained by the 8 pressure sensors.

3 Results and Discussion

3.1 Hydrodynamic Characteristics of ESHP and Fixed Plate Breakwater

Figure 2 exhibits the transmittance coefficient of two different horizontal plate breakwater under three wave height conditions. It implies that the wave dissipation effect of ESHP breakwater is better than that of the fixed breakwater, especially in deeper submergence depth ($d'/d = 0.02, 0.03$). This may be related to the added mass surrounding the plate [12]. From Fig. 2, the minimum transmittance coefficient of ESHP breakwater occurs at $h = 0.03$m and $d'/d = 0.2$. In this case, the transmitted wave height is 27% of the incident wave height. In addition, Fig. 3 shows the frequency domain of transmitted wave height behind the breakwater. The results indicate that the ESHP breakwater suppressed the wave amplitude at first and second-order wave frequencies.

Differing from the ESHP breakwater, the SHP breakwater is particularly effective at $d'/d = 0.1$. In this condition, the amplitude of the transmitted wave height is about half that of the other two cases. Moreover, the effect of the incident wave height (H) does not play a significant role here.

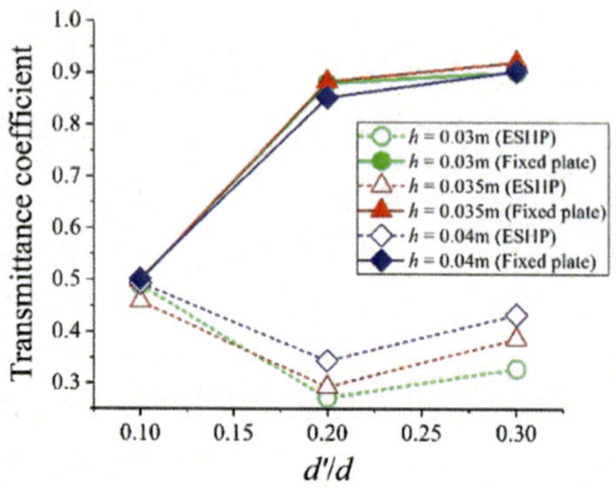

Fig. 2. The transmission coefficient of ESHP breakwater and fixed plate breakwater under different wave conditions.

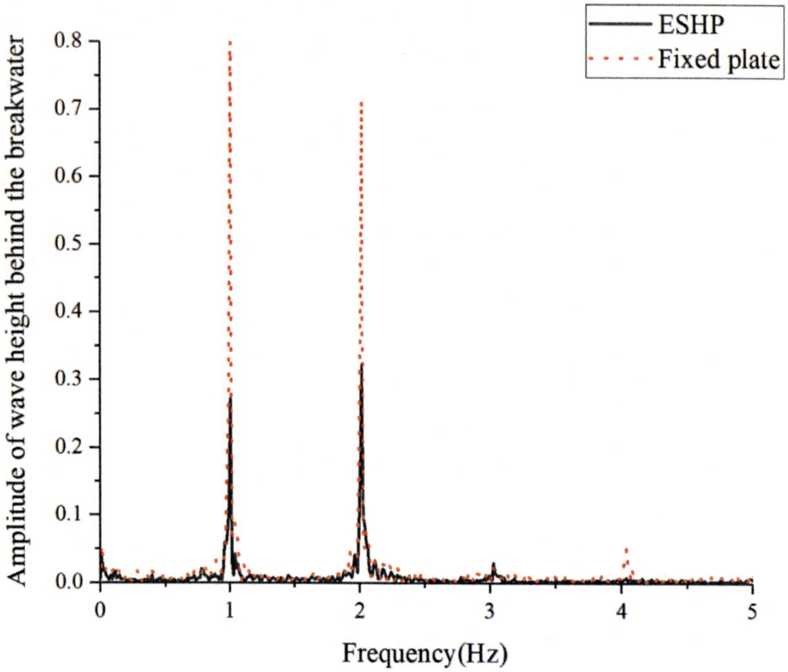

Fig. 3. The amplitude of wave behind the breakwater.

3.2 Spring Stiffness

In order to understand the wave dissipation effect of ESHP breakwater, the test condition with (H = 0.03, d'/d = 0.02) the best dissipating performance in Fig. 4 is chosen for further analysis. One of the most important factors was the stiffness of the spring which supported the horizontal plate. It offered the rigid force for the ESHP breakwater and meanwhile controlled the motion of the horizontal plate. It can be seen that a low spring stiffness in the ESHP breakwater system has a better dissipating performance. Meanwhile, the motion response of the breakwater is greater, with a portion of wave energy transformed into the kinetic of the breakwater.

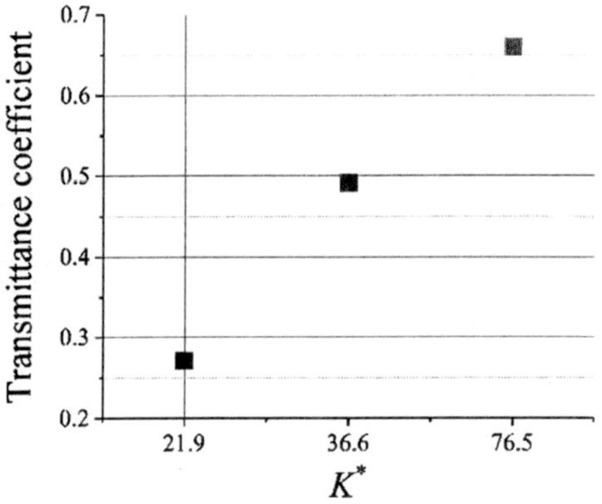

Fig. 4. Transmittance coefficient of ESHP breakwater ($H = 0.03$, $d'/d = 0.02$).

3.3 Wave Force and Motion Response of ESHP Breakwater and Fixed Plate Breakwater

The test conditions in this section are shown in Table 1. In this section, the K^* of spring of ESHP breakwater with the best wave dissipating effect (21.9) is considered, and the wave force was obtained by integrating the maximum value of the data from pressure sensors on the horizontal plates and shown in Fig. 5. It can be seen that the wave force increases as the incident wave height is gradually increases. This may be due to the fact that the higher value of wave height, has a greater wave energy and more wave force impact on the horizontal plate. Moreover, it is worth noting that the wave force significantly increases during the wave dissipating performance of the fixed plate breakwater gets better in Fig. 2. The remarkable upwards indicates that the wave force influences the wave dissipating performance of the fixed plate breakwater.

Table 1. Test condition of 9 cases in Fig. 5.

Cases	H(m)	d' (m)	d'/d
1	0.03	0.04	0.1
2	0.035	0.04	0.1
3	0.04	0.04	0.1
4	0.03	0.08	0.2
5	0.035	0.08	0.2
6	0.04	0.08	0.2
7	0.03	0.12	0.3
8	0.035	0.12	0.3
9	0.04	0.12	0.3

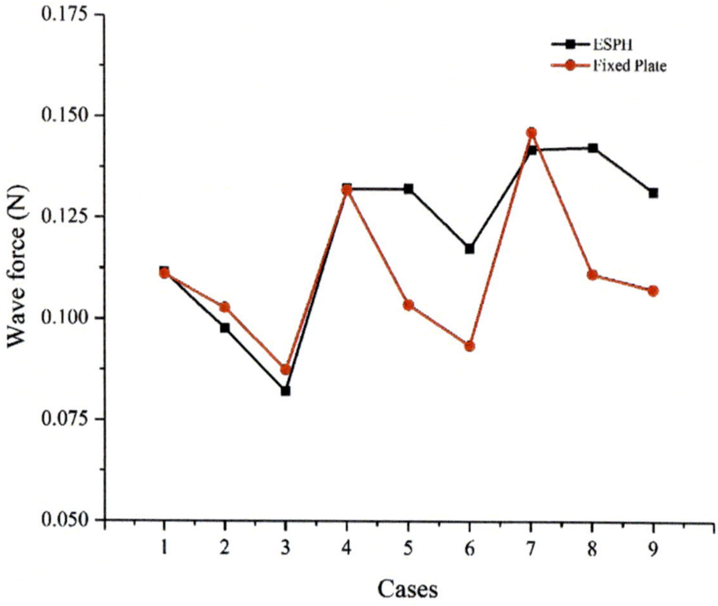

Fig. 5. Wave force of ESHP and fixed plate breakwater.

Figure 6 depicts the motion response, wave force and transmission coefficient of ESHP breakwater. It is clear that the wave dissipating performance is better than that of the smaller amplitude of motion response of the ESHP breakwater. The motion response of the ESHP breakwater is partly converted from the wave energy, which means that the wave would have an extra force on the structure.

In Fig. 7, as the spring stiffness enlarges, the amplitude of the motion response of the breakwater decreases and shows a worse wave dissipating performance. Meanwhile, the wave force increases with the increase of the supporting stiffness. This also strengthens the fact that the ESHP breakwater has a better wave dissipating performance than fixed plate breakwater.

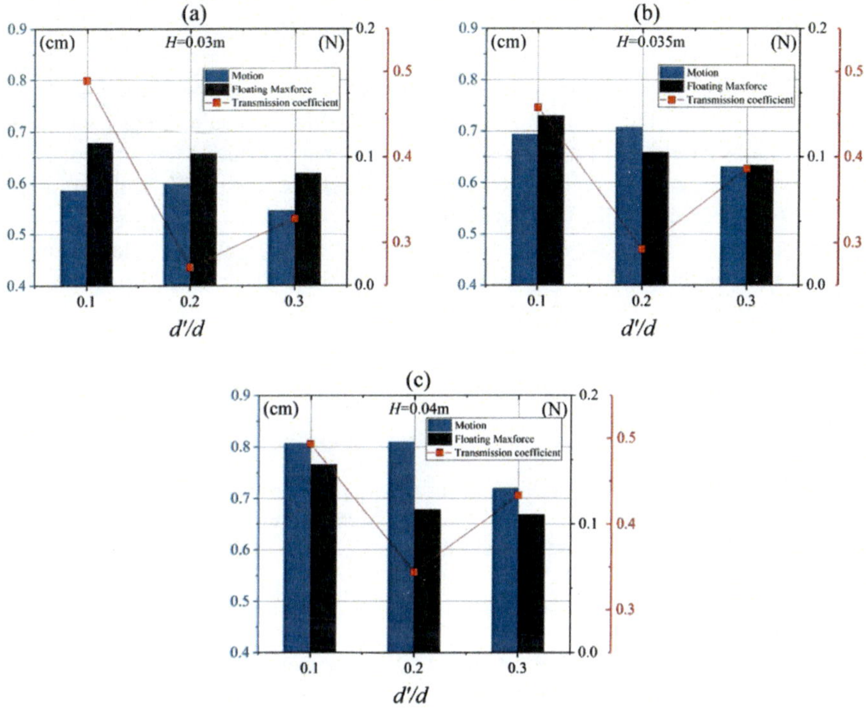

Fig. 6. Variation of motion response, wave force and transmission coefficient of ESHP breakwater.

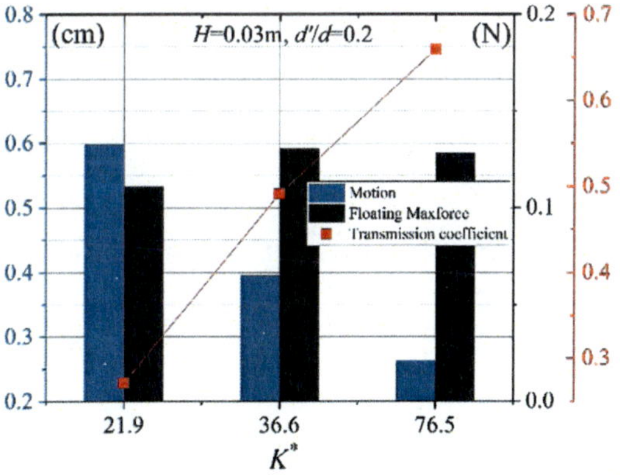

Fig. 7. Variation of motion response, wave force and transmission coefficient of ESHP breakwater.

4 Conclusions

The wave dissipation effect, wave force and the motion response of ESHP breakwater and fixed plate breakwater were investigated based on the experimental test in this research. The incident wave height, spring stiffness and submergence ratio of the plate were adjusted as variables. The main conclusions are listed as follows:

1. The wave dissipating performance of ESHP breakwater is better than that of fixed plate breakwater, the minimum transmission coefficient could reach 27%, which occurs at submergence ratio $(d'/d) = 0.2$. This may be due to the present elastic support system causing the additional energy dissipation and the larger the amplitude of motion response of the horizontal plate.
2. The wave height reduced rapidly at the leeside of the breakwater with elastic support.
3. Incident wave height has the minimal effect on wave dissipation, wave force and motion response of the breakwater.

References

1. Hales, L.Z.: Floating breakwaters: state-of-the-art literature review. US Army Engineer Waterways Experiment Station, Vicksburg, Mississippi, USA (1981)
2. Tong, C., Yan, Y.: Study on wave dissipation characteristics of floating breakwaters. Water Transp. Eng. **08**, 32–35 (2002)
3. Ji, C.-Y., Chen, X., Cui, J., Yuan, Z.-M., Incecik, A.: Experimental study of a new type of floating breakwater. Ocean Eng. **105**, 295–303 (2015)
4. Ji, C.-Y., Guo, Y.-C., Cui, J., Yuan, Z.-M., Ma, X.-J.: 3D experimental study on a cylindrical floating breakwater system. Ocean Eng. **125**, 38–50 (2016)
5. He, F., Huang, Z.: Hydrodynamic performance of pile-supported OWC-type structures as breakwaters: an experimental study. Ocean Eng. **88**, 618–626 (2014)
6. Patarapanich, M., Cheong, H.F.: Reflection and transmission characteristics of regular and random waves from a submerged horizontal plate. Coast. Eng. **13**(2), 161–182 (1989)
7. Liu, C., Huang, Z., Tan, S.K.: Nonlinear scattering of non-breaking waves by a submerged horizontal plate: experiments and simulations. Ocean Eng. **36**(17–18), 1332–1345 (2009)
8. Poupardin, A., Perret, G., Pinon, G., et al.: Vortex kinematic around a submerged plate under water waves. Part I: experimental analysis. Eur. J. Mech. B Fluids **34**, 47–55 (2012)
9. Hayatdavoodi, M., Ertekin, R.C.: Wave forces on a submerged horizontal plate-Part I: theory and modelling. J. Fluids Struct. **54**, 566–579 (2015)
10. Koo, W.C., Kim, D.H.: Numerical analysis of hydrodynamic performance of a movable submerged breakwater. J. Soc. Naval Archit. Korea **48**(1), 23–32 (2011)
11. Liu, C., Huang, Z., Chen, W.: A numerical study of a submerged horizontal heaving plate as a breakwater. J. Coast. Res. **33**(4), 917–930 (2017)
12. He, M., Xu, W., Gao, X., Ren, B.: SPH Simulation of wave scattering by a heaving submerged horizontal plate. Int. J. Ocean Coastal Eng. **1**(2), 1840004 (2018)
13. Wang, X., Zhao, X., Ding, F.: Numerical simulation of wave interaction with undulating horizontal plate breakwater. Mar. Eng. (China) **37**(3), 61–68 (2019)
14. Wilde, P., Szmidt, K., Sobierajski, E.: Phenomena in standing wave impact on a horizontal plate. Coast. Eng., 1489–1501 (1998)

15. Ren, B., Liu, M., Li, X.-L., Wang, Y.-X.: Experimental investigation of wave slamming on an open structure supported elastically. China Ocean Eng. **30**(6), 967–978 (2016). https://doi.org/10.1007/s13344-016-0063-1
16. Jin, H., Chen, J., Song, R., Liu, Y., Zhao, Z.: An elastic supported pontoon type breakwater, China, CN108643120B (2020)

Free Vibration Analysis of Laminated Composite Plate with a Cut-Out

Chen Zhou[1,2], Yingdan Zhu[2](✉), Xiaosu Yi[1], and Jian Yang[1,3](✉)

[1] Department of Mechanical, Materials and Manufacturing Engineering, The University of Nottingham Ningbo China, Ningbo 315100, China
jian.yang@nottingham.edu.cn

[2] Zhejiang Provincial Key Laboratory of Robotics and Intelligent Manufacturing Equipment Technology, Ningbo Institute of Materials Technology and Engineering, Chinese Academy of Sciences, Ningbo 315201, China
y.zhu@nimte.ac.cn

[3] Centre for Sustainable Energy Technologies, The University of Nottingham Ningbo China, Ningbo 315100, China

Abstract. This study investigates the free vibration behaviour of simply supported square laminated composite plates with a square cut-out and straight fibre orientation designs. The cut-out is placed at the centre of the plate. Different cases are considered with fibre angles changed from 0° to 45°, at an interval of 5°. The analytical method based on the first-order shear deformation laminate theory (FSDT) and the numerical method are both applied to obtain the free vibration properties in terms of the natural frequencies and mode shapes. The results obtained from these two methods are compared for verification. The effects of the cut-out size and fibre angles on the mode shapes are studied. It is found that the natural frequencies are influenced substantially by the size of cut-out and the fibre orientation, which can also be used to modify the mode shape and nodal line locations for the purpose of vibration suppression. With increasing the size of the cut-out of a square plate, the 3rd natural frequencies of the structure generally decrease. It is shown that the 2nd natural frequency increases with fibre angle increasing from 0° to 45°. The mode shape can also be tailored by the design of the fibre angle with the nodal line changing from being approximately the horizontal to the diagonal of the plate. The study provides some enhanced understanding on the free vibration properties of the laminated composite plate with a cut-out to achieve better dynamic designs.

Keywords: Composite plate · Free vibration · Cut-out · Natural frequency

1 Introduction

Composite laminates are increasingly used in aerospace, automotive, and civil engineering industries [1]. Laminated composite structures have a greater strength-to-weight ratio than common metal materials, as well as the ability to optimize design by controlling fiber orientation, stacking pattern, and fiber and matrix material selection [2]. Plates with

Z. Dimitrovová et al. (Eds.): WMVC 2022, MMS 125, pp. 930–938, 2023.
https://doi.org/10.1007/978-3-031-15758-5_96

cut-outs are widely required in engineering structures. Cut-outs are designed for many purposes, such as ventilation, inspection, weight reduction, access and quality control [3]. In the design of aerospace structures, cut-outs are mainly used for electrical and mechanical system entrance, weight saving, fuel lines and etc. [4]. In addition, many laminated composite plate structures will be subjected to dynamic loads during service, which will affect their working efficiency, especially at resonance [5]. Therefore, it is necessary to study the vibration properties of such plate structures.

For laminated plates with specific fibre angels, the classical laminated plate theory (CLPT) [6] and first order shear deformation theory (FSDT) [7, 8] can be used for analytical determination of natural frequencies and mode shapes. For laminated composite plates with complex fibre orientations and these with cut-outs, the finite element method can be used to obtain the vibration properties of the plate [9]. To analyze the vibration performance of simply supported rectangular plates with cut-outs, a semi-analytical technique based on the optimized Rayleigh–Ritz method was proposed [10]. The relationships between the position, the size of cut-outs and the natural frequencies of the plate with cut-outs were studied as well [11]. The natural frequencies of laminated plates with square holes were also determined analytically [12].

Although the vibration properties of the plates with cut-outs have attracted much attention in these years, many studies focused on metal plates, rather than composite laminated plates. The effects of the fibre orientation on the free vibration of composite plates with cut-outs need further exploration. In this paper, the free vibration properties of simply-supported laminated composite plates with cut-out and variable fibre orientation design are investigated. With the FSDT and the numerical methods, the effects of the cut-out size and fibre orientation on the natural frequencies and mode shapes of composite plates are explored.

2 Mathematical Modelling

Figure 1 shows a schematic illustration of a rectangular N-layered laminated composite plate with a central square cut-out. The plate is with length a in the direction of OX, width b in the direction of OY and thickness c in the direction of OZ. In addition, a square cut-out is placed in the centre with length L and width W. $OXYZ$ are a set of orthogonal axes defined in a global Cartesian coordinates system. The angle between the fibre direction and OX is θ_i for the i-th layer, measured from the positive OX axis. The plate is simply supported at its four edges.

In the local coordinate system, the constitutive equations for the k-th layer may be written as [13]:

$$\begin{Bmatrix} \sigma_1 \\ \sigma_2 \\ \tau_{12} \end{Bmatrix} = \begin{bmatrix} Q_{11} & Q_{12} & 0 \\ Q_{12} & Q_{22} & 0 \\ 0 & 0 & Q_{66} \end{bmatrix} \begin{Bmatrix} \varepsilon_1 \\ \varepsilon_2 \\ \gamma_{12} \end{Bmatrix}, \tag{1a}$$

$$\begin{Bmatrix} \tau_{23} \\ \tau_{13} \end{Bmatrix} = \begin{bmatrix} Q_{44} & 0 \\ 0 & Q_{55} \end{bmatrix} \begin{Bmatrix} \gamma_{23} \\ \gamma_{13} \end{Bmatrix}, \tag{1b}$$

where σ is the normal stress, τ is the shear stress, ε is the normal strain, γ is the shear strain, and Q_{ij} are the plane stress-reduced stiffnesses, and are related to the engineering constants by

$$Q_{11} = \frac{E_1}{(1 - \upsilon_{12}\upsilon_{21})}, \quad Q_{12} = \frac{\upsilon_{12}E_2}{(1 - \upsilon_{12}\upsilon_{21})}, \quad Q_{22} = \frac{E_2}{(1 - \upsilon_{12}\upsilon_{21})}, \tag{2}$$

$$Q_{66} = G_{12}, Q_{44} = G_{23}, Q_{55} = G_{13}. \tag{3}$$

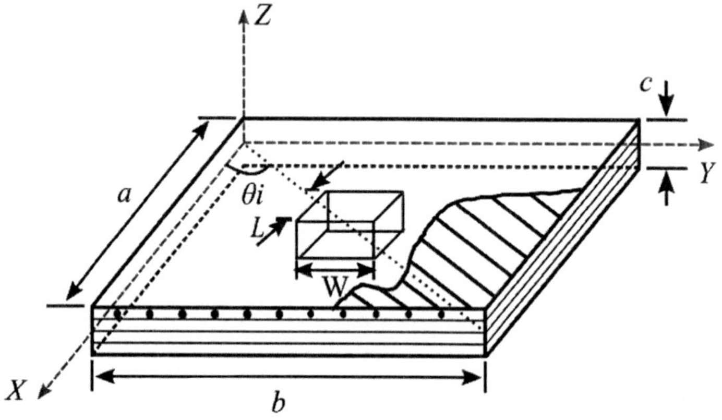

Fig. 1. Schematic illustration of a composite plate with a cut-out.

For an orthotropic material, the stress-strain relation is given by:

$$\begin{Bmatrix} \sigma_{xx} \\ \sigma_{yy} \\ \sigma_{xy} \end{Bmatrix} = \begin{bmatrix} \overline{Q}_{11} & \overline{Q}_{12} & 0 \\ \overline{Q}_{12} & \overline{Q}_{22} & 0 \\ 0 & 0 & \overline{Q}_{66} \end{bmatrix} \begin{Bmatrix} \varepsilon_{xx} \\ \varepsilon_{yy} \\ \gamma_{xy} \end{Bmatrix}, \tag{4}$$

where

$$\overline{Q}_{11} = Q_{11}\cos^4\theta + 2(Q_{12} + 2Q_{66})\sin^2\theta\cos^2\theta + Q_{22}\sin^4\theta, \tag{5a}$$

$$\overline{Q}_{12} = (Q_{11} + Q_{22} - 4Q_{66})\sin^2\theta\cos^2\theta + Q_{12}\left(\sin^4\theta + \cos^4\theta\right), \tag{5b}$$

$$\overline{Q}_{22} = Q_{11}\sin^4\theta + 2(Q_{12} + 2Q_{66})\sin^2\theta\cos^2\theta + Q_{22}\cos^4\theta, \tag{5c}$$

$$\overline{Q}_{66} = (Q_{11} + Q_{22} - 2Q_{12} - 2Q_{66})\sin^2\theta\cos^2\theta + Q_{66}\left(\sin^4\theta + \cos^4\theta\right). \tag{5d}$$

The displacements in the FSDT are designated by u, v and w, in the directions of OX, OY and OZ, respectively, and the relationships are

$$u(x, y, z, t) = u_0(x, y, t) + z\phi_x(x, y, t), \tag{6a}$$

$$v(x, t, z, t) = v_0(x, y, t) + z\phi_y(x, y, t), \tag{6b}$$

$$w(x, t, z, t) = w_0(x, y, t), \tag{6c}$$

where ϕ_x and $-\phi_y$ denote rotations about the OY and OX axes, respectively.

The governing equation of the plate is

$$D_{11}\frac{\partial^4 w_0}{\partial x^4} + 2(D_{12} + 2D_{66})\frac{\partial^4 w_0}{\partial x^2 \partial y^2} + D_{22}\frac{\partial^4 w_0}{\partial y^4} = q, \tag{7}$$

where D_{ij} is defined in terms of the lamina stiffnesses $\overline{Q}_{ij}^{(k)}$ as

$$D_{ij} = \int_{-\frac{h}{2}}^{\frac{h}{2}} \overline{Q}_{ij}\left(z^2\right)\mathrm{d}z = \sum_{k=1}^{N} \int_{z_k}^{z_{k+1}} \overline{Q}_{ij}^{(k)}\left(z^2\right)\mathrm{d}z. \tag{8}$$

With consideration of the boundary condition, the transverse deflection can be expressed as:

$$w_0(x, y) = \sum_{n=1}^{\infty}\sum_{m=1}^{\infty} W_{mn} \sin\alpha x \sin\beta y, \tag{9}$$

where $\alpha = m\pi/a$, $\beta = n\pi/b$, and W_{mn} is the coefficient to be determined.

3 Results and Discussion

The free vibration properties of a single layer square thin plate is investigated. All edges are simply supported, and the material parameters are set as $a = b = 10\,\mathrm{m}$, $c = 0.05\,\mathrm{m}$, $E = 200.0\,\mathrm{GPa}$, $\upsilon = 0.3$, and $\rho = 8000\,\mathrm{kg/m^3}$. The dimensionless natural frequencies of the plates are obtained with the equation $\overline{\omega}_n = \left(\omega_{11}{}^2\rho c b^4 / D_0\right)^{1/4}$, where $D_0 = Ec^3/\left[12\left(1 - \upsilon^2\right)\right]$. In the FE modelling process, the composite plate with a cut-out is discretized into a total number of 400 four-node ANSYS Shell 181 elements with 6 degrees of freedom for each node. The FSDT is used to set up the element in ANSYS. All calculations are completed on a PC running the Windows 10 operating system. The PC has Intel Core i7-8750 CPU with 16 GB RAM and the frequency of the CPU is 2.2 GHz. Table 1 shows the fundamental natural frequency of the plate obtained from previous research based on the edge-based smoothing technique [8], and from FSDT analytical and numerical methods used here. After comparison, results show good agreement. The table shows that both analytical and numerical methods can be used to obtain the free vibration properties of the composite plate accurately.

Table 1. Non-dimensional frequencies $\overline{\omega}_n$ of the square composite plate with various methods

Sources	Non-dimensional natural frequency		
	Mode 4	Mode 5	Mode 6
Ref. [1]	3.672	4.437	4.941
Analytical method	3.675	4.440	4.945
Numerical method	3.674	4.438	4.943

Table 2 and Fig. 2 show the effects of the size of the cut-out on the non-dimensional natural frequencies. Figure 3 shows the mode shape of the composite plates with a cut-out. The fibre angle θ of this four-layers plate is all the same as 45°. All edges of the considered plates are simply supported. Besides, the material properties herein are set as $a = b = 1\,\text{m}$, $c = 0.01\,\text{m}$, $E_{11} = 120\,\text{GPa}$, $E_2 = 7.9\,\text{GPa}$, $G_{12} = 5.5\,\text{GPa}$, $G_{23} = 1.58\,\text{GPa}$, $\upsilon_{12} = 0.33$, $\upsilon_{23} = 0.022$ and $\rho = 1580\,\text{kg/m}^3$. The dimensionless natural frequencies of the plates are obtained with the equation $\overline{\omega}_n = \omega_{11} b^2 \sqrt{\rho/(E_{22}c^2)}$. Table 2 shows that with the increase of the size of the cut-out, the natural frequencies in the whole first four modes generally decrease. Figure 2 shows that for the third mode, when the size of the cut-out enlarges from 0.01 m to 0.2 m, there is a reduction of 11.67% in frequency from 35.497 to 31.357. However, the changes in the 2nd and 4th natural frequencies are very small, despite of the changes in the cut-out size. Figure 3 shows the second mode shapes of composite plates with various sizes of the cut-out. The figure shows large displacements located at the top left and bottom right parts. The figure shows that the variations of the cut-out size does not lead to significant changes in the patterns of the mode shape.

Table 2. First four frequency parameters for laminated composite plate with various cut-out sizes

The length of the square cut-out L (m)	Mode 1	Mode 2	Mode 3	Mode 4
0.01	12.370	23.531	35.497	37.353
0.02	12.357	23.531	35.494	37.336
0.04	12.314	23.530	35.480	37.283
0.06	12.253	23.499	35.432	37.218
0.08	12.180	23.512	35.314	37.156
0.1	12.102	23.490	35.087	37.112
0.12	12.020	23.453	34.707	37.089
0.14	11.957	23.399	34.146	37.092
0.16	11.900	23.327	33.385	37.109
0.18	11.858	23.234	32.441	37.140
0.2	11.833	23.125	31.357	37.176

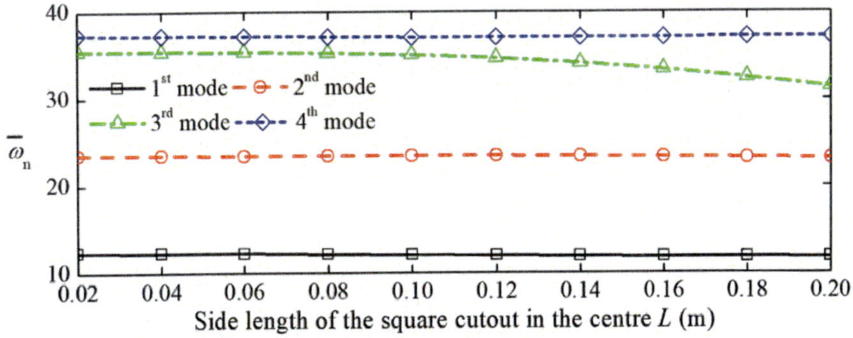

Fig. 2. The effects of the cut-out size on the non-dimensional natural frequency of square laminated composite plates with a cut-out.

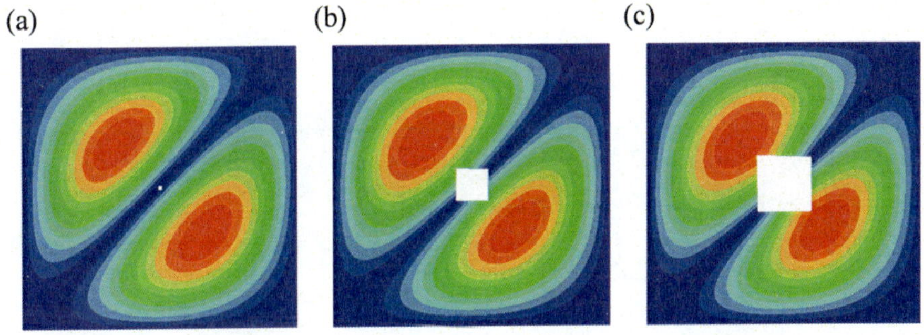

Fig. 3. The second mode shapes of square laminated composite plates with different cut-out sizes (a) $L = 0.02$ m, (b) $L = 0.1$ m, (c) $L = 0.2$ m.

The effects of fiber orientation on non-dimensional natural frequencies are shown in Table 3 and Fig. 4. The fibre angles for all the layers of this plate are set as the same. Different cases with various fibre angles from 0° to 45°, with an interval of 5° are studied. The length of the square cut-out is constant as 0.2 m. All edges are simply supported. The material properties are set as same as those used for Fig. 3. Table 3 shows that the natural frequencies of the plate with a cut-out are significantly influenced by the fibre orientation. Figure 4 shows that, as the fibre angle changes from 0° to 45°, the non-dimensional frequency $\overline{\omega}_n$ associated with the second mode increases 25.18% rapidly from 18.474 to 23.125. For the third mode, the natural frequency firstly increases with the fibre angle from 0° to 20°, and then decreases when θ changes from 20° to 45°. A peak of the 3rd resonant frequency is reached when the fibre angle $\theta = 20°$. In comparison, the opposite trend can be seen for the variations of the fourth natural frequency. With the increase of the fibre angle, the non-dimensional 4th natural frequency reduces to a trough value of 35.499 when $\theta = 20°$.

Table 3. First four frequency parameters for cut-out composite plates with various fibre orientations.

The fibre orientation θ	Mode 1	Mode 2	Mode 3	Mode 4
0	11.357	18.474	31.469	38.255
5	11.354	18.609	31.722	38.058
10	11.350	18.999	32.424	37.493
15	11.363	19.595	33.432	36.612
20	11.410	20.329	34.539	35.499
25	11.496	21.119	34.297	35.528
30	11.610	21.881	33.154	36.283
35	11.723	22.526	32.211	36.789
40	11.804	22.967	31.579	37.078
45	11.833	23.125	31.357	37.176

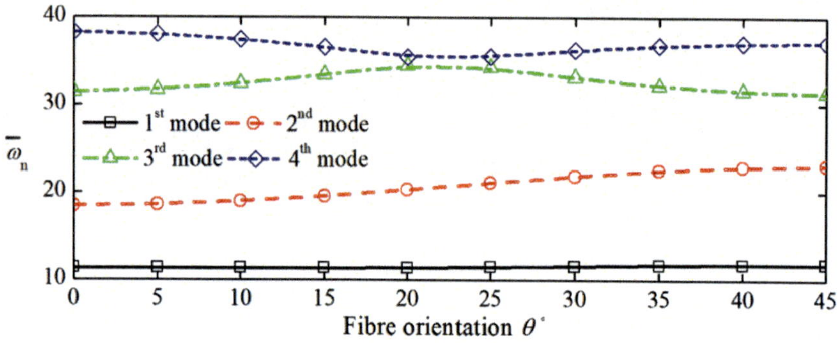

Fig. 4. The effect of the fibre orientation on the non-dimensional natural frequency of square laminated composite plates with a cut-out.

Figure 5 shows the changes in the mode shape of the composite plates with a cut-out. The changes in the second mode shape of the cut-out plate with various fibre angles are examined. The figure shows that the mode shape pattern changes little in plates with different fibre angles. When the fibre angle $\theta = 0°$, the largest deformations are found at the top and bottom middle parts. In contrast, with the increase of the fibre angle, the boundary between two deformations is rotated. When the fibre angle is set to be $\theta = 45°$, the nodal line is near to the diagonal of the plate, and points having the displacements are found at the top left and bottom right parts.

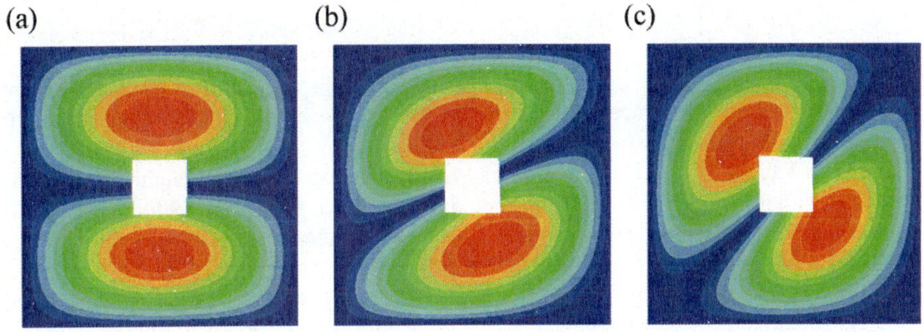

Fig. 5. The second mode shapes of square cut-out composite plates with different fibre orientations (a) $\theta = 0°$, (b)$\theta = 30°$, (c) $\theta = 45°$.

4 Conclusions

This study investigated the free vibration properties of laminated composite plates with a cut-out. The effects of the cut-out size and fibre angle on the non-dimensional natural frequencies and mode shapes have been explored. The results obtained by various approaches show a good agreement. The results reveal that with the changing of the cut-out size, the significant differences are found in the third mode. In addition, it is found that the natural vibration properties of the plate depend highly on the fibre orientation. With the fibre angle increasing from 0° to 45°, the 3rd natural frequency increases firstly, and then decreases, while the opposite trend is found in the fourth mode. It demonstrates that the free vibration characteristics of the composite plate with a cut-out can be tailored and optimized by proper design of fibre orientation.

Acknowledgements. This work was supported by National Natural Science Foundation of China under Grant number 12172185 and by Zhejiang Provincial National Science Foundation of China under Grant number LY22A020006.

References

1. Dai, K.Y., Liu, G.R., Lim, K.M., Chen, X.L.: A mesh-free method for static and free vibration analysis of shear deformable laminated composite plates. J. Sound Vib. **269**(3–5), 633–652 (2004)
2. Bhashyam, G.R., Gallagher, R.H.: A triangular shear-flexible finite element for moderately thick laminated composite plates. Comput. Methods Appl. Mech. Eng. **40**(3), 309–326 (1983)
3. Li, Y., Zhou, M., Li, M.: Analysis of the free vibration of thin rectangular plates with cut-outs using the discrete singular convolution method. Thin Walled Struct. **147**, 106529 (2020)
4. Sinha, L., Das, D., Nayak, A.N., Sahu, S.K.: Experimental and numerical study on free vibration characteristics of laminated composite plate with/without cut-out. Compos. Struct. **256**, 113051 (2021)
5. Shi, B., Yang, J.: Quantification of vibration force and power flow transmission between coupled nonlinear oscillators. Int. J. Dyn. Control **8**(2), 418–435 (2019). https://doi.org/10.1007/s40435-019-00560-7

6. Zhu, C., Yang, J.: Free and forced vibration analysis of composite laminated plates. In: The 26th International Congress on Sound and Vibration (2019)
7. Zhu, C., Yang, J., Rudd, C.: Vibration transmission and energy flow analysis of L-shaped laminated composite structure based on a substructure method. Thin Walled Struct. **169**, 108375 (2021)
8. Zhu, C., Yang, J., Rudd, C.: Vibration transmission and power flow of laminated composite plates with inerter-based suppression configurations. Int. J. Mech. Sci. **190**, 106012 (2021)
9. Boay, C.G.: Free vibration of laminated plates with a central circular hole. Compos. Struct. **35**, 357–368 (1996)
10. Laura, P.A.A., Romanelli, E., Rossi, R.E.: Transverse vibrations of simply supported rectangular plates with rectangular cutouts. J. Sound Vib. **202**, 275–283 (1997)
11. Merneedi, A., RaoNalluri, M., Rao, V.V.S.: Free vibration analysis of a thin rectangular plate with multiple circular and rectangular cut-outs. J. Mech. Sci. Technol. **31**(11), 5185–5202 (2017). https://doi.org/10.1007/s12206-017-1012-5
12. Rajamani, A., Prabhakaran, R.: Dynamic response of composite plates with cut-outs, Part I: simply-supported plates. J. Sound Vib. **54**(4), 549–564 (1977)
13. Reddy, J.N.: Mechanics of Laminated Composite Plates and Shells: Theory and Analysis. CRC Press, Boca Raton (2003)

The 2D Rectangular Tank Sloshing Response Under the Planar Tilt Settlement

Sunyu Jia, Heng Jin(✉), Mengfan Lou, and Tengxiao Wang

NingboTech University, Ningbo, China
J18855485261t@163.com

Abstract. The large rigid storage tanks are often used to store hazardous chemicals. It is very important to consider the non-linear load of strong wind or seismic excitation when designing the storage structure, because of the dangerous of petrochemical products. However, the large weight of the tank acted on the ground may cause foundation settlement over its service life. And above phenomenon is more serious in coastal areas. Meanwhile, when the planar tilt settlement of the tank is coupled with the external load such as ground motion, the local stress of the tank wall may exceed the allowable stress in related standard and lead to accident. Therefore, the sloshing phenomenon of storage tank under horizontal and tilting states is studied tentatively in this study. Three famous strong earthquake records are selected to test the dynamic response of storage tanks under different settlement conditions. And four different tilt angles are implemented. Through the overturning stability analysis, the maximum allowable tilt angle of the storage tank under dynamic load is given.

Keywords: Planar settlement · Tank sloshing · Critical angle · Dynamic response · Seismic excitation

1 Introduction

China's oil reserves are mostly stored in above-ground storage tanks, large storage tanks initially designed as well as placed, has taken into account its structural performance and other safety issues to avoid accidents, but with a series of problems such as large storage tanks service life and coastal rust, coupled with the weight of the storage tank itself, making the overall settlement of the tank. This effect has a great potential danger to make the structure of the storage tank change, resulting in its durability, seismic resistance decreases year by year, thus triggering accidents such as chucking, rupture, oil leakage.

In order to understand the settlement problem, according to API 653, it is divided into tank wall plate settlement and bottom plate settlement. Where the planar tilt belongs to the tank wall plate settlement. Palmer et al. [1] studied the deformation response of floating roof tanks under local uneven settlement, fixed tanks in the direction of their base diameter settlement difference can not exceed 0.007D (D is the internal diameter of the tank).

Z. Dimitrovová et al. (Eds.): WMVC 2022, MMS 125, pp. 939–947, 2023.
https://doi.org/10.1007/978-3-031-15758-5_97

For rigid foundations, the main settlement form is plane settlement. For the plane settlement in the settlement of the tank wall plate, Malik Z et al. [2, 3] derived a simple relationship between the radial displacement of the tank top and the uneven settlement of the tank bottom. Gong et al. [4] used ADINA software to calculate the modal state of a large oil storage tank. The results showed that: Under the plane settlement, the coupled vibration of the storage tank mainly includes multi-wave vibration, radial vibration and beam vibration. In addition to settlement problems occurring in storage tanks after long-term service, other problems may also exist, such as: spalling of concrete on the outside of the tank foundation ring beam, corrosion of steel reinforcement [5], and corrosion of the tank bottom plate [6], which are also not negligible and deserve attention. For the seismic study of vertical liquid storage tanks, the research object is mainly anchored tanks, and the more classic model in this regard is the Housner model [7], which can be used in the study of the vibration characteristics of liquid storage tanks considering the role of tank-liquid coupling Haroun- Housner simplified theoretical model [8], the model takes into account the convection pressure and pulsation pressure of the liquid inside the tank (including the flexible pulsation pressure of synchronous motion with the tank wall and the rigid pulsation pressure of motion with the ground). The domestic seismic standards for storage tanks are based on the Haroun-Housner model. In the early years, most of the storage tanks were anchored, and the seismic study of liquid storage tanks was carried out based on this, while with the gradual enlargement of the tanks, their connection with the foundation was gradually changed to non-anchored.

Kang et al. [9] selected 3 sets of representative earthquakes for numerical simulations and compared them with experimental results with a good agreement. Gurusamy et al. [10] studies the wave height and liquid level variation at shallow water for small, moderate, and large excitation amplitudes and compared them with numerical calculations. It is shown that with increasing amplitude, the hydraulic jump characteristics are evident at relatively heigh liquid depths.

For the case of coupled settlement and external loads, Sarkar et al. [11] explored the effect of inhomogeneous settlement on the seismic response of building structures and his results showed that different settlement types may be a key factor in determining the response of a structure. In his study, Bao et al. [12] analysed the difference between the seismic response of structures and the uneven settlement of foundations. On this basis, the relationship between the intensity of ground shaking and the damage probability of the structure was investigated by varying the magnitude of the peak seismic acceleration. Wu et al. [13] established a numerical analysis method for the dynamic response of storage tanks with uneven settlement under seismic loading to study the effect of uneven settlement on the seismic response of storage tanks.

Previous studies have focused on the response analysis of tanks under settlement or an external load, but the reality may be that the tanks will settle after a period of service and then be suddenly hit by an earthquake, and there has been little research on the dynamic response of tanks with uniformly tilted settlement under seismic loading. Therefore, this paper focuses on the seismic corresponding analysis of storage tanks under uniformly tilted settlement.

2 Experimental Methods

This research program is divided into two parts, the experimental study part and the numerical simulation part.

2.1 Experimental Study

As shown in Fig. 1 (a), the model carrier used for the experiment is a 2D rectangular tank made of acrylic sheets with a 0.01m thickness, which the internal dimensions are $L = 0.58$ m in length, $W = 0.1$ m in width and $H = 0.6$ m in height. In the experimental study, the filling level is 60%, that is, the filling level is $h = 0.36$ m, and staining the water in green is to better observe the movement of the liquid. Four pressure sensors are installed on the wall of the 2D rectangular tank, which named *P1*, *P2*, *P3* and *P4*. *P1* and *P3* are 0.3 m from the bottom of the tank, *P2* and *P4* are 0.08 m. Wall height sensors are named *W1* and *W2*, which placing them inside the 2D rectangular tank and 1 cm away from the wall on both sides respectively. The Six Degrees of Freedom platform is the movement carrier of the experiment, which is able to perform 6-DOF motions regularly or randomly according to an appropriate input of time histories. As shown in Fig. 1 (b), the experimental simulates the excitation by controlling the horizontal sway of the 6-DOF platform, and by controlling the tilt angle of the 6-DOF platform to simulate the angle of the planar tilt settlement. 3 different tilt angles ($\alpha = 0.83°$, $\beta = 1.2°$, $\gamma = 1.85°$) are implemented in experiment studies.

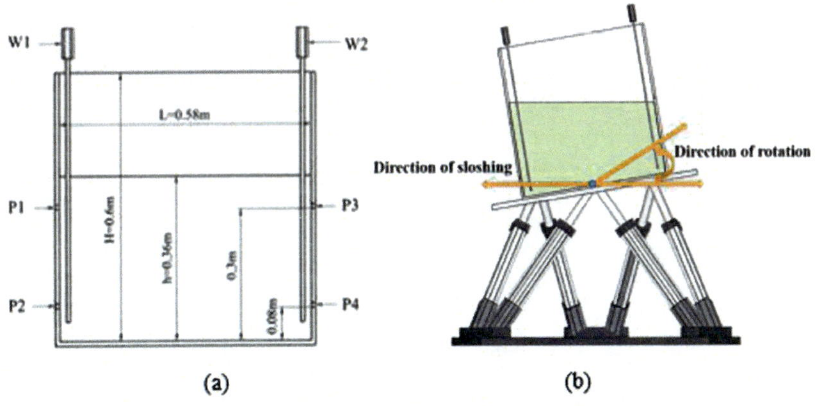

Fig. 1. The 2D rectangular tank and the 6-DOF platform. (a): The inner dimension of the 2D rectangular tank with the layout of four pressure sensors and two wave height sensors. (b): The movement direction of the 6-DOF platform.

There are two kinds of excitation applied in the experiment, simple harmonic excitation and nonlinear seismic excitation. The simple harmonic excitation equation is as

$$S = Asin\omega t \tag{1}$$

where A is the amplitude and $A = 2$ mm, ω is excitation frequency, t is the time and $t =$ 30 s.

When the external excitation frequency is equal to the tank natural frequency, at this time the tank resonates and the shakes most obviously. With the assumption of potential flow theory, the natural frequencies can be analytically determined as [14]

$$\omega = \sqrt{g\frac{\pi i}{L}tanh\left(\frac{\pi i}{L}h\right)} \quad i = 1, 2, 3 \tag{2}$$

where L is internal length of the tank, h is the filling level, g is the Gravity acceleration and i is the mode number. According to Eq. (2), the first-mode natural frequencies is 7.143 rad/s. In the experimental study, the external excitation frequencies include $\omega 1$ to $\omega 13$.

About nonlinear seismic excitation, three famous seismic excitations were selected as the Kobe, the Imperial Valley and the San Fernando [15]. As shown in Table 1, the seismic excitations are scaled to facilitate comparison of the tests. Figure 2 shows the displacement and acceleration after scaling for three different seismic excitations.

Table 1. The acceleration of simple harmonic excitation and nonlinear seismic excitation

Excitation type	Acceleration (g)
Simple harmonic excitations	0.087–0.114
Seismic excitations	0.095(Same acceleration as $\omega 5$)

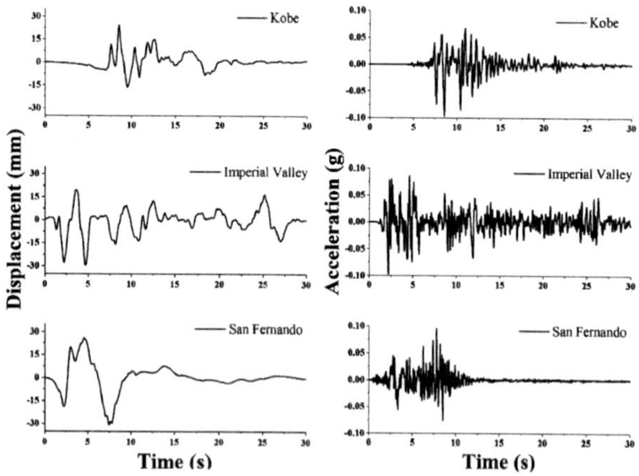

Fig. 2. Displacement time histories and acceleration time histories of three Seismic excitations.

The flow of the experimental study, through the computer output excitation signal to the 6-DOF platform, drive the 2D rectangular tank to produce the corresponding motion, pressure sensors and wave height sensors for data acquisition and data feedback to the computer, while the whole test process for the camera (4k 60bit).

2.2 Numerical Simulation

An experimental model based on OpenFOAM is built for experimental validation and evaluation of the tank. Firstly, the appropriate mesh is established in OpenFOAM, and set the sensor positions as the monitoring points in the numerical simulation, and the motion of the 6-DOF platform is simulated for numerical simulation (Monitoring point 1 is equivalent to *p1*, and so on). The monitoring points in the numerical simulations are compared with those in the experimental studies, while the maximum wave height change in the numerical simulation is compared with the experimental studies, and the Liquid surface level change is compared with the video comparison. Finally, Anti-Overturning stability and performance of the 2D rectangular tank are evaluated by both experimental studies and numerical simulations.

3 Experimental Validation

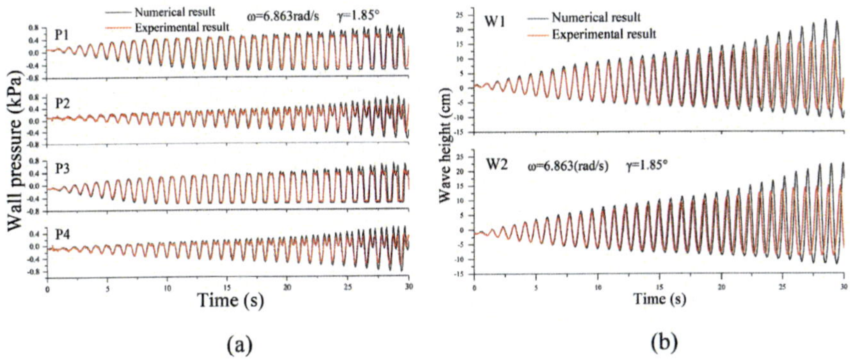

Fig. 3. Tilt angle of the tank is 1.85°, $\omega 4 = 6.863$ rad/s. (a): Pressure comparison between numerical simulations and experiments. (b): Wall height comparison between numerical simulations and experiments.

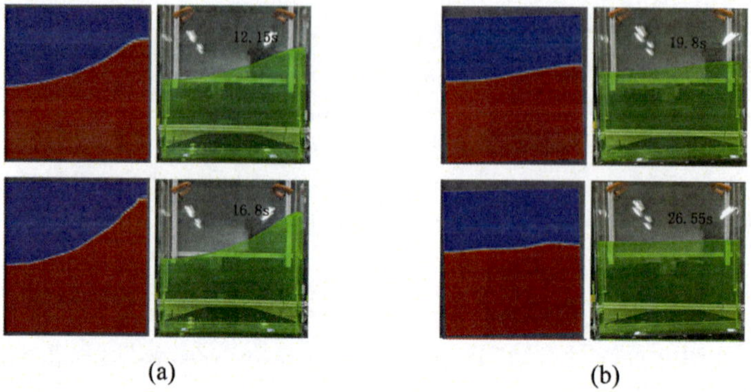

Fig. 4. Liquid surface comparison between numerical simulations and experiments. (a): The tank is placed horizontally, $\omega 6 = 6.918$ rad/s. (b): Tilt angle of the tank is 1.85°, $\omega 13 = 7.543$ rad/s.

Figure 3 (a) shows the variations of the wall pressure on the 2D rectangular tank when the tank is placed horizontally and the external excitation frequency is $\omega 4 = 6.863$ rad/s. From the initial time 0 point can be seen, in the planar tilt settlement, the pressure on the left wall of the tank is greater than that on the right. As shown in Fig. 3(b), wall height comparison between numerical simulation and experiment under tank is tilt angle of the tank is 1.85° ($\omega 4 = 6.863$ rad/s). With the increase of the time, it can be found that the experimental results and numerical results will be a certain phase difference in the latter seconds, and the numerical results are slightly larger than the experimental results, probably because it does not take into account the viscosity and compressibility of the fluid. Figure 4 Shows liquid surface comparison between numerical simulations and experiments, the numerical results agree with the experimental results.

The numerical results by excitation frequencies and tilt angles can be in good agreement with the experimental results. The numerical simulation of liquid sloshing established in this study can accurately predict the distribution of wall pressure, wave height, and liquid surface change.

4 Results and Discussion

Figure 5 shows the simple harmonic motion at different frequencies with the maximum wave heights on the left and right sides in the case of horizontal placement of the liquid chamber, and it can be seen that the wave height is not maximum at $\omega = 7.143$ rad/s. The maximum value of wave height occurs before $\omega = 7.143$ rad/s. In deep water sloshing, this phenomenon is known as the 'hard-spring' response. Its characteristic is a slow descent after a rapid rise to the maximum response value and maximum response frequency is less than natural frequency.

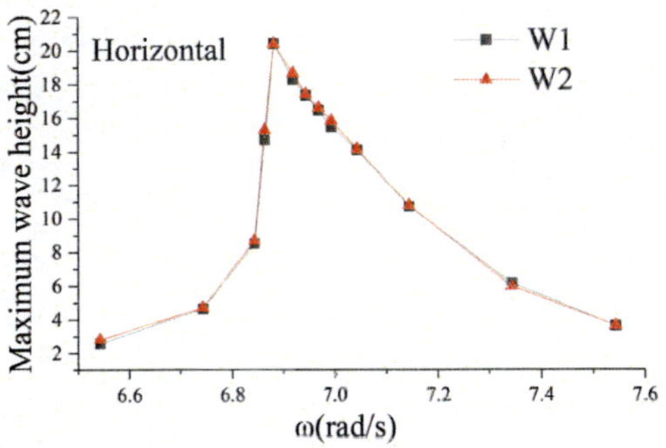

Fig. 5. The maximum wave heights on the left and right sides for different excitation frequencies under the horizontal tank placed.

The numerical simulation can simulate the sloshing of the liquid tank at different angles. Figure 6 shows the graph of the wave height variation of the sloshing of the liquid chamber with its Fourier amplitude spectra when simulating at the horizontal with the inclination angle of 1.2°. It can be seen that the wave height of the sloshing on the left side of the liquid tank will be larger than that when the liquid tank is placed horizontally when the liquid tank is in the planar tilt settlement. By Fourier amplitude spectra analysis, the first frequency plays a dominant role in the sloshing when the liquid chamber is subjected to simple harmonic excitations or seismic excitations.

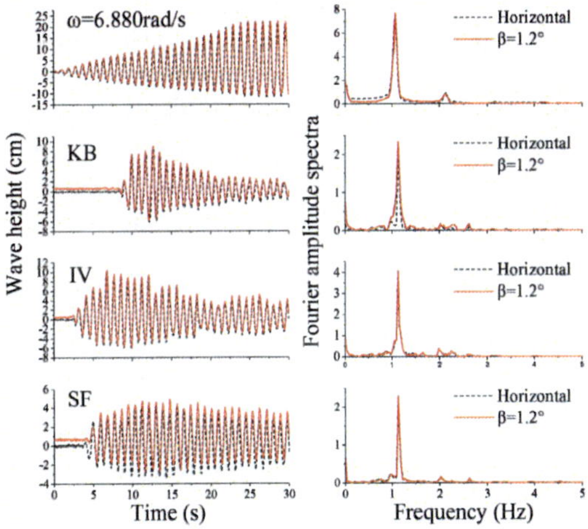

Fig. 6. Wave height variation and its Fourier amplitude spectra.

As shown in Fig. 7 numerical simulations are made to derive the liquid tank sloshing at different angles. By the position of the monitoring points, the overturning moment at the maximum pressure on the monitoring points of the wall is made. When the angle of plane settlement becomes larger, the overturning moment will also become larger and the pressure on the wall will also become larger, which will cause some impact on the liquid tank structure.

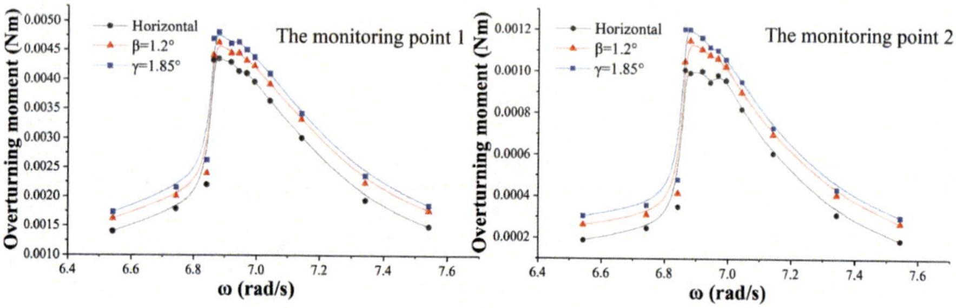

Fig. 7. The overturning moment at monitoring point 1 and monitoring point 2.

5 Conclusions

The following conclusions can be drawn from this study:

1. When the $h/L = 0.62 > 0.2$, this belongs to the category of deep liquid tank shaking. The resonance phenomenon occurs in advance.
2. By Fourier amplitude spectra analysis, the first frequency plays a dominant role in the sloshing when the liquid chamber is subjected to simple harmonic excitations or seismic excitations.
3. As the angle of the planar tilt settlement increases, the higher the overturning moment, the more unfavorable the seismic performance of the whole structure, and the more damage the structure will suffer after the shaking. In the case of the planar tilt settlement and external coupling, the liquid tank is subject to greater hazards.

References

1. Kamyab, H., Palmer, S.C.: Analysis of displacements and stresses in oil storage tanks caused by differential settlement. In: ARCHIVE Proceedings of the Institution of Mechanical Engineers Part C Journal of Mechanical Engineering Science, pp. 61–70 (1989)
2. Malik, Z., Morton, J., Ruiz, C.: Ovalization of cylindrical tanks as a result of foundation settlement. J. Strain Anal. **12**, 339–348 (1977)
3. Mouse, E.A., Ruiz, C.: Stresses in cylindrical tanks due to uneven circumferential settlement. J. Strain Anal. **15**, 7–9 (1979)
4. Gong, J., Tao, J.: Modal analysis of large oil storage tanks under ground plane settlement. In: Proceedings of the 7th Annual China CAE Engineering Analysis Technology Conference and the 2011 National Advanced Seminar on Computer Aided Engineering (CAE) Technology and Application, pp. 165–170 (2011)

5. Zhou Li, H.: Discussion on the rectification measures for the corrosion of the concrete spalling steel bar outside the ring beam of the atmospheric tank foundation. China Pet. Chem. Stand. Qual. **2013**(13), 243–244 (2013)
6. Jiang, L., Han, W., Wang, Z., Liu, H.: Analysis and suggestions on common corrosion problems in comprehensive inspection of crude oil storage tanks. Pet. Eng. Constr. **47**(03), 79–83 (2021)
7. Housner, G.W.: Dynamic pressures on accelerated fluid containers. Bull.seism.soc.am, 48 (1957)
8. Haroun Medhat, A.: Vibration studies and tests of liquid storage tanks. Earthq. Eng. Struct. Dyn. **11**, 179–206 (1983)
9. Kang, T.W., Yang, H.I., Jeon, J.S.: Earthquake-induced sloshing effects on the hydrodynamic pressure response of rigid cylindrical liquid storage tanks using CFD simulation. Eng. Struct. **197**, 109376 (2019)
10. Gurusamy, S., et al.: Sloshing dynamics of shallow water tanks: modal characteristics of hydraulic jumps. J. Fluids Struct. **104**, 103322 (2021)
11. Sarkar, R., Dutta, S.C., Saw, R., et al.: Effect of differential settlement on seismic response of building structure. Munic. Eng. **173**(3), 1–32 (2018)
12. Bao, C., Xu, F., Chen, G., et al.: Seismic vulnerability analysis of structure subjected to uneven foundation settlemen. Appl. Sci. **9**, 3507 (2019)
13. Wu, J.: Seismic dynamic response analysis of liquid storage tank under uneven settlement of foundation. Zhejiang University, p. 120 (2021)
14. Faltinsen, O.M., Timokha, A.N.: Sloshing. Cambridge University Press, Cambridge (2009)
15. Jin, H., Calabrese, A., Liu, Y.: Effects of different damping baffle configurations on the dynamic response of a liquid tank under seismic excitation. Eng. Struct. 229, 111652 (2021)

Vibration Analysis of Laminated Composite Panels with Various Fiber Angles

Chendi Zhu[1,2], Gang Li[1,2(✉)], and Jian Yang[3(✉)]

[1] Ningbo Institute of Dalian University of Technology, 26 Yucai Road, Ningbo 315016, People's Republic of China
ligang@dlut.edu.cn

[2] State Key Laboratory of Structural Analysis for Industrial Equipment, Department of Engineering Mechanics, Dalian University of Technology, Dalian 116023, People's Republic of China

[3] Faculty of Science and Engineering, University of Nottingham Ningbo China, 199 Taikang East Road, Ningbo 315100, People's Republic of China
jian.yang@nottingham.edu.cn

Abstract. Laminated composite structures have superior material properties such as high stiffness and high strength-to-weight ratio and have been increasingly used in advanced structures such as automobiles to replace conventional metal structures to reduce weight for enhanced energy efficiency. Composite structures in practical working condition are often subjected to external excitation force, which can result in severe vibration problems. This paper investigates the vibration characteristics of rectangular laminated composite panels with various fiber orientations subjected to harmonic loading. Both free and forced vibration analysis have been carried out to obtain natural frequencies and the steady-state dynamic responses. Both analytical and numerical FE methods based on the first-order shear deformation theory (FSDT) are used to carry out vibration analysis. The numerical FE analysis is employed to validate the accuracy of proposed analytical method. The vibration transmission behaviour of laminated composite panels is compared with that of steel panels. It is found that fiber orientations have significant influence on vibration characteristics of laminated composite panels. The results explicitly show that natural frequencies and dynamic responses could be altered by designing fiber orientation for vibration mitigation. The findings provide some improved understanding on the structural design of laminated composite structures for enhanced vibration transmission behaviour.

Keywords: Vibration analysis · Laminated composite panels · Fiber orientations · Natural frequencies · Mode shapes

1 Introduction

Laminated composite structures with superior properties such as high stiffness and high strength-to-weight ratio and light weight have been increasingly used in many advanced engineering structures [1]. In the automobile industry, the laminated composite structures

Z. Dimitrovová et al. (Eds.): WMVC 2022, MMS 125, pp. 948–956, 2023.
https://doi.org/10.1007/978-3-031-15758-5_98

have been increasingly used to replace the conventional metallic structures for saving weight and to increase the energy efficiency. The high-level vibrations can result in problems such as fatigue failure, destruction of mechanical system and high noise. There is a need for deeper understanding of vibration behaviour of laminated composite structures.

There have been a large number of studies on the free vibration behaviour of laminated composite panels based on several theories such as the classical laminate plate theory, first [2] and third-order shear deformation theory [3], and higher-order refined theory [4] to improve the accuracy of analytical methods. However, there have been much less studies reported on effects of the fiber orientations on vibration behaviour of the laminated composite panels subjected to complex dynamic loading. For instance, Dobyns [5] and Carvalho *et al.* [6] proposed a theoretical approach for vibration analysis of simply-supported rectangular laminated composite plate. Dynamic responses of laminated composite plates with respect to effects of various fiber orientations and stacking sequences were studied [7, 8]. The fibers could be tailored for vibration suppression designs [9, 10].

This paper presents an analytical approach and numerical FE method to examine the vibration behaviour of harmonically excited laminated composite panels with simply supported edges. Analytical method based on the first-order shear deformation theory (FSDT) can be used for special symmetric cross-ply laminated composite panel. For the laminated composite panels with various fiber orientations, the numerical ANSYS finite element (FE) method is used for investigation and validation of the analytical results. The dynamic responses are determined and compared with that of the steel panel. The effects of fiber orientations on the dynamic behaviour are examined.

2 Dynamic Analysis of Laminated Composite Panels

2.1 Model Description

Figure 1 shows the schematic illustration of a four-layered laminated composite rectangular panel with length a and width b and thickness h in the directions of OX, OY and OZ, respectively. Its four edges are simply supported and a harmonic external excitation force with amplitude $\tilde{f}_e$ and frequency ω is applied at the point A(x_e, y_e). Also a general point B(x_r, y_r) is labelled in the Fig. 1 for which the response is of interest. The angle θ_i that is measured from the positive direction of OX to the principal material axis is used to indicate the fiber orientation of the i-th layer.

The constitutive equations for orthotropic lamina can be written as [1]:

$$\begin{bmatrix} \sigma_1 \\ \sigma_2 \\ \tau_{12} \end{bmatrix} = \begin{bmatrix} Q_{11} & Q_{12} & 0 \\ Q_{12} & Q_{22} & 0 \\ 0 & 0 & Q_{66} \end{bmatrix} \begin{bmatrix} \varepsilon_1 \\ \varepsilon_2 \\ \gamma_{12} \end{bmatrix}, \begin{bmatrix} \tau_{23} \\ \tau_{13} \end{bmatrix} = \begin{bmatrix} Q_{44} & 0 \\ 0 & Q_{55} \end{bmatrix} \begin{bmatrix} \gamma_{23} \\ \gamma_{13} \end{bmatrix}, \tag{1a-1b}$$

where $Q_{ij}(i, j = 1, 2, 6)$ are the material constants in the material axes of the layer expressed by

$$Q_{11} = \frac{E_1}{1 - \nu_{12}\nu_{21}}, Q_{12} = \frac{\nu_{12}E_2}{1 - \nu_{12}\nu_{21}} = \frac{\nu_{12}E_1}{1 - \nu_{12}\nu_{21}}, Q_{22} = \frac{E_2}{1 - \nu_{12}\nu_{21}},$$

$$Q_{66} = G_{12}, Q_{44} = G_{23}, Q_{55} = G_{13},$$

where E_{11} and E_{22} are the in-plane moduli of elasticity in local coordinate system, G_{ij} are the shear moduli, while ν_{ij} are the Poisson's coefficients of the orthotropic materials. The stress-strain relations for the i-th orthotropic layer in the local coordinate system are

$$\begin{bmatrix} \sigma_x \\ \sigma_y \\ \tau_{xy} \end{bmatrix} = \begin{bmatrix} \bar{Q}_{11} & \bar{Q}_{12} & 0 \\ \bar{Q}_{12} & \bar{Q}_{22} & 0 \\ 0 & 0 & \bar{Q}_{66} \end{bmatrix} \begin{bmatrix} \varepsilon_x \\ \varepsilon_y \\ \gamma_{xy} \end{bmatrix}, \begin{bmatrix} \tau_{yz} \\ \tau_{xz} \end{bmatrix} = \begin{bmatrix} \bar{Q}_{44} & 0 \\ 0 & \bar{Q}_{55} \end{bmatrix} \begin{bmatrix} \gamma_{yz} \\ \gamma_{xz} \end{bmatrix}. \tag{2}$$

The constitutive relations can be transformed from the local coordinate system to the global coordinate system by using the following transformation matrix:

$$[T_1] = \begin{bmatrix} \cos^2\theta & \sin^2\theta & 2\sin\theta\cos\theta \\ \sin^2\theta & \cos^2\theta & -2\sin\theta\cos\theta \\ -\sin\theta\cos\theta & \sin\theta\cos\theta & \cos^2\theta - \sin^2\theta \end{bmatrix}, [T_2] = \begin{bmatrix} \cos\theta & -\sin\theta \\ \sin\theta & \cos\theta \end{bmatrix}. \tag{3}$$

The stiffness matrix for the i-th layer in global coordinate system can be obtained from

$$[\bar{Q}] = \left([T]^{(i)}\right)^{-1}[Q]^{(i)}[T]^{(i)}. \tag{4}$$

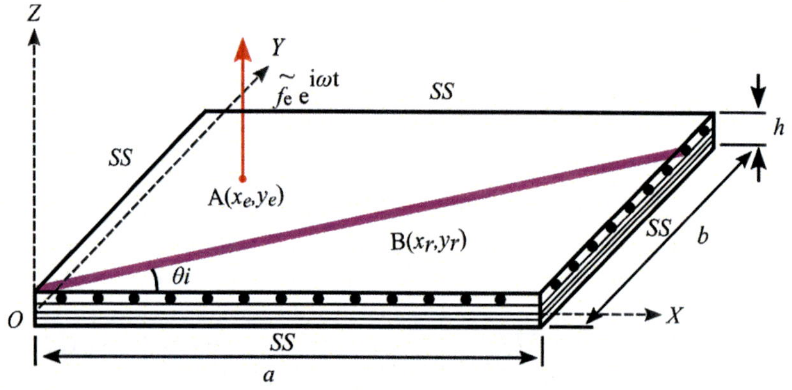

Fig. 1. Schematic illustrations of a laminated composite rectangular panel with simply supported (SS) edges.

2.2 Vibration Analysis of Laminated Composite Panels

For a symmetric cross-ply orthotropic plate, some of stress stiffness coefficients can be assumed as zero, $B_{ij} = 0, A_{16} = A_{26} = D_{16} = D_{26} = 0$. Hence, the equations of motion are

$$kA_{44}\left(w^0{}_{,yy} + \phi_{y,y}\right) + kA_{55}\left(w^0{}_{,xx} + \phi_{x,x}\right) + p_z = \rho h\ddot{w}^0, \tag{5a}$$

$$D_{11}\phi_{x,xx} + (D_{12} + D_{66})\phi_{y,xy} + D_{66}\phi_{x,yy} - kA_{55}\left(w^0{}_{,x} + \phi_x\right) = \frac{\rho h^3}{12}\ddot{\phi}_x, \quad (5b)$$

$$D_{66}\phi_{y,xx} + (D_{12} + D_{66})\phi_{x,xy} + D_{22}\phi_{y,yy} - kA_{44}\left(w^0{}_{,y} + \phi_y\right) = \frac{\rho h^3}{12}\ddot{\phi}_y, \quad (5c)$$

where h is the plate thickness, ρ is material density, w is the plate displacement in the OZ direction at the plate midplane, ϕ_x and ϕ_y are the shear rotations on the OY and OX directions. k is used to account for the effect of shear deformation, which is normally taken as 5/6.

With simply supported edges, we have

$$w = \phi_{x,x} = 0, \text{ at } x = 0, \text{ and } x = a.$$
$$w = \phi_{y,y} = 0, \text{ at } y = 0, \text{ and } y = b.$$

The solution of the dynamic response is based on expansions of the loads, displacement and rotations in double Fourier series. Each expression is the product of independent time coefficients by trigonometric functions, which satisfy the boundary condition:

$$\phi_x(x, y, t) = \sum_m \sum_n A_{mn}(t)\cos\alpha x \sin\beta y, \quad (6a)$$

$$\phi_y(x, y, t) = \sum_m \sum_n B_{mn}(t)\sin\alpha x \cos\beta y, \quad (6b)$$

$$w(x, y, t) = \sum_m \sum_n W_{mn}(t)\sin\alpha x \sin\beta y, \quad (6c)$$

where $\alpha = m\pi/a$, $\beta = n\pi/b$, m and n are integers, and $A_{mn}(t)$, $B_{mn}(t)$ and $W_{mn}(t)$ are the time-dependent coefficients. The load function is represented by

$$q(x, y, t) = \sum_m \sum_n Q_{mn}(t)\sin\alpha x \sin\beta y. \quad (7)$$

For a concentrated load located at the point A (x_e, y_e), we have

$$Q_{mn}(t) = \frac{4F(t)}{ab}\sin\alpha x_e \sin\beta y_e. \quad (8)$$

A substitution of Eq. (6) into Eq. (5) leads to

$$\begin{bmatrix} L_{11} & L_{12} & L_{13} \\ L_{12} & L_{22} & L_{23} \\ L_{13} & L_{23} & L_{33} \end{bmatrix} \begin{Bmatrix} A_{mn}(t) \\ B_{mn}(t) \\ W_{mn}(t) \end{Bmatrix} = \begin{Bmatrix} 0 \\ 0 \\ Q_{mn}(t) - \mu \ddot{W}_{mn}(t) \end{Bmatrix}, \quad (9)$$

where $\mu = \rho h$ is mass per unit area and the elements of the symmetric matrix L_{ij} are

$$L_{11} = D_{11}(\alpha)^2 + D_{66}(\beta)^2 + \kappa A_{55}, L_{12} = (D_{12} + D_{66})\alpha\beta, L_{13} = \kappa A_{55}\alpha,$$

$$L_{22} = D_{66}(\alpha)^2 + D_{22}(\beta)^2 + \kappa A_{44}, L_{33} = \kappa A_{55}(\alpha)^2 + \kappa A_{44}(\beta)^2, L_{23} = \kappa A_{44}\beta.$$

The Eq. (9) can be reduced to a single differential equation by the following transformation:

$$A_{mn}(t) = \frac{L_{12}L_{23} - L_{13}L_{22}}{L_{11}L_{22} - L_{12}^2} W_{mn}(t), \tag{10a}$$

$$B_{mn}(t) = \frac{L_{12}L_{13} - L_{11}L_{23}}{L_{11}L_{22} - L_{12}^2} W_{mn}(t). \tag{10b}$$

The equation of motion for a rectangular panel can be expressed as

$$\ddot{W}_{mn}(t) + \omega_{mn}^2 W_{mn}(t) = \frac{Q_{mn}(t)}{\mu}, \tag{11}$$

where the natural frequency can be obtained from:

$$\omega_{mn} = \sqrt{\frac{L_{13}K_A + L_{23}K_B + L_{33}}{\mu}}. \tag{12}$$

To solve the equation of motion, the modal displacement can be written as

$$W_{mn}(t) = X\mathrm{e}^{\mathrm{i}(\omega t - \theta)} = \left(X\mathrm{e}^{-\mathrm{i}\theta}\right)\mathrm{e}^{\mathrm{i}\omega t} = \overline{W_{mn}}\mathrm{e}^{\mathrm{i}\omega t}. \tag{13}$$

In a damped system, the damping effect can be included by multiplying the stiffness term by a factor of $(1 + \mathrm{i}\eta)$, where η denotes the structural damping. The displacement can be obtained and expressed as:

$$\tilde{w}(x, y, t) = \frac{4}{ab\mu} \sum_m \sum_n \sin \alpha x_e \sin \beta y_e \sin \alpha x \sin \beta y \times \frac{\tilde{f}_e}{\omega_{mn}^2(1 + \mathrm{i}\eta) - \omega^2} \mathrm{e}^{\mathrm{i}\omega t}. \tag{14}$$

3 Results and Discussions

Here a rectangular panel with dimensions of 1 m by 0.5 m with total thickness of 0.005 m is considered. The material used for composite panel is T300/934 CFRP and the material properties are set as $E_{11} = 120$ GPa, $E_{22} = E_{33} = 7.9$ GPa, $G_{12} = G_{13} = 5.5$ GPa, $G_{23} = 1.58$ GPa, $\nu_{12} = \nu_{13} = 0.33$, $\nu_{23} = 0.022$, $\rho = 1580\,\mathrm{kg/m^3}$. The material properties used for the counterpart steel panel are listed as: $E = 205.8$ GPa, $\rho = 7800\,\mathrm{kg/m^3}$ and $\nu = 0.3$. The structural damping coefficient η is 0.01. Both analytical and numerical ANSYS FE methods are employed for the free and forced vibration analysis of different types of panels. The free vibration analysis of a steel panel is carried out analytically based on the classical plate theory (CPT) and that of laminated composite panels with specific fiber orientation such as $[0°]_4$ and $[90°]_4$ can be conducted by analytical method based on FSDT. The numerical FE method using element Shell 281 based on FSDT is used when the analytical method for the laminated composite panels with various fiber orientation is not available.

Table 1 shows first five natural frequencies (in Hz) of different steel and laminated composite panels. The numerical and analytical results show good agreement, verifying both methods. It is found that first five natural frequencies of the steel panel are larger than those of the $[0°]_4$ laminated composite panel. It shows that the fundamental frequency of the steel panel is larger than that of the $[0°]_4$, $[30°]_4$ and $[45°]_4$ laminated composite panels but lower than that of the $[60°]_4$ and $[90°]_4$ laminated composite panels. An increase in the fiber angle θ leads to the increase of fundamental frequencies. It is found that the second natural frequencies decrease firstly and then increase when angle θ changes from 0° to 90°. The third, fourth and fifth natural frequencies are increased with fibre orientation with a maximum value being reached and then they are decreased.

Table 1. First five natural frequencies (in Hz) of different types of panels

Fibre angles	Method	Mode number				
		1	2	3	4	5
Steel	CPT	61.0413	97.6661	158.7074	207.5405	244.1653
	FE	61.1679	97.8526	159.3501	209.7384	246.2012
$[0°]_4$	FSDT	34.0765	90.0339	91.7675	136.1537	187.8359
	FE	34.1228	90.2194	92.6991	136.8970	188.8816
$[30°]_4$	FE	46.2024	83.9190	134.5403	140.2765	195.2814
$[45°]_4$	FE	55.9063	83.4566	123.5800	172.6839	196.4226
$[60°]_4$	FE	67.6608	85.1224	114.8423	156.3408	208.4675
$[90°]_4$	FSDT	81.5943	90.0339	107.4791	136.1537	176.6830
	FE	81.8147	90.2194	107.7478	136.8970	178.6634

Figure 2 shows that dynamic responses of the steel panel and the $[0°]_4$ panel at the forcing position (0.5 m, 0.25 m) in an examined frequency range of 0–500Hz. The solid and dashed lines represent the analytical results for steel and laminated composite panels, respectively and the corresponding numerical results from FE are denoted by circles and squares. It shows that the analytical results have a great agreement with FE results especially in the low-frequency range. It is found that at a lower excitation frequency, the laminated composite panel exhibits relatively high vibration responses. It is found that the peak amplitudes of laminated composite panel are higher than those of the considered steel panel. The steel panel with heavy weight has relatively lower dynamic responses when subjected to the excitation force.

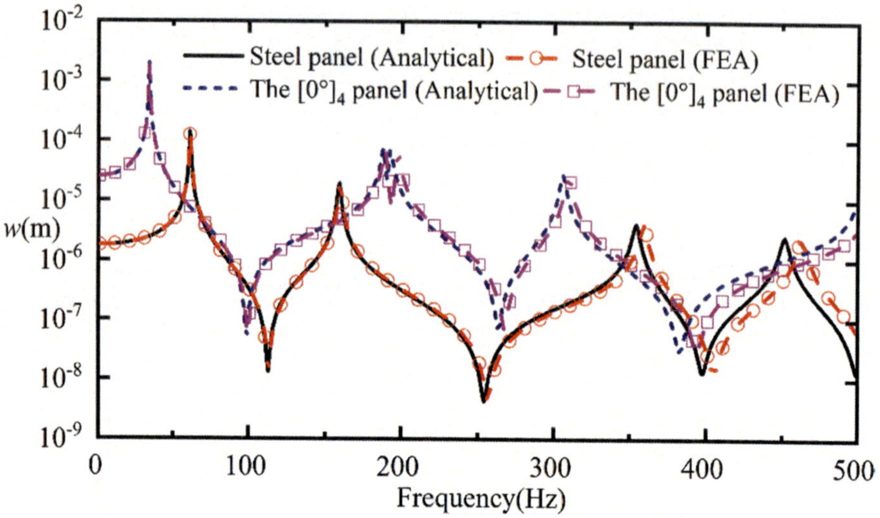

Fig. 2. Dynamic responses at the centre point (0.5 m, 0.25 m) on the steel panel and the laminated composite $[0°]_4$ panel with the excitation at the same point.

Figure 3 shows the variations of the dynamic responses at the forcing position with the excitation frequency. In this case, the effects of fiber orientations such as 0°, 30°, 45° and 60° on the dynamic responses are examined. It is found that with the increase of fibre angles from 0° to 60°, the first resonance peak is shifted to the higher frequency and the peak values are effectively reduced. The first resonance frequency of the $[60°]_4$ panel is larger than that of steel panel. The frequency range between the first and the second peaks of the laminated composite panels are narrowed with the increase of the fiber angle. The frequency range between the first and second peaks of the $[0°]_4$ panel is the widest. In the frequency band, the panel has relatively low dynamic response. In the frequency range approximately from 200 to 350 Hz, the vibration level of the steel panel is lower than that of laminated composite panels, whereas those of the laminated composite panels in the frequency range of 150 to 200 Hz are lowered. It demonstrates that reinforced composite laminates have an ability of reducing vibration levels in a described excitation frequency range and the peak values of dynamic response can be reduced by tailoring the fiber orientations.

In Fig. 4, the vibration transmission is investigated by examining the influences of fiber angles on the ratio R_t between the dynamic response amplitude of point B (0.75 m, 0.25 m) to that of the input point A (0.25 m, 0.25 m). In a lower frequency range of 0–55 Hz, it is found that the response amplitude ratio of the $[0°]_4$ panel is larger than in other cases. With the increase of fiber angle, the first peak of R_t is moved to larger frequencies. The first peak value of the $[60°]_4$ panel is the lowest, which indicates a low vibration transmission level. The first and second peak values of the $[0°]_4$ panel and the steel panel are close with the same vibration transmission behaviour. The frequency range between the first and second peak can be narrowed by increasing the fiber angle from 0° to 60°.

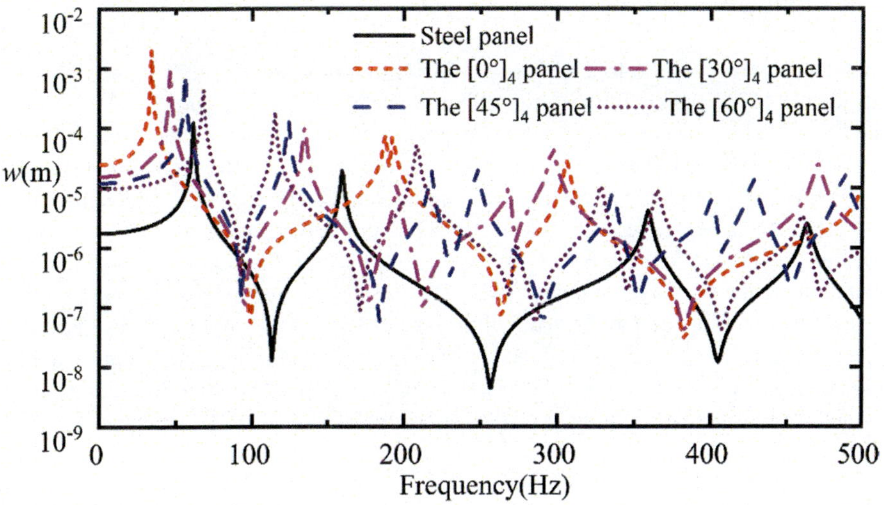

Fig. 3. Dynamic responses at the centre point (0.5 m, 0.25 m) on the steel panel and laminated composite panels with different fiber angles. —: the steel panel; ------: the $[0°]_4$ panel; -·-·-·-: the $[30°]_4$ panel; - - -: the $[45°]_4$ panel; ······: the $[60°]_4$ panel.

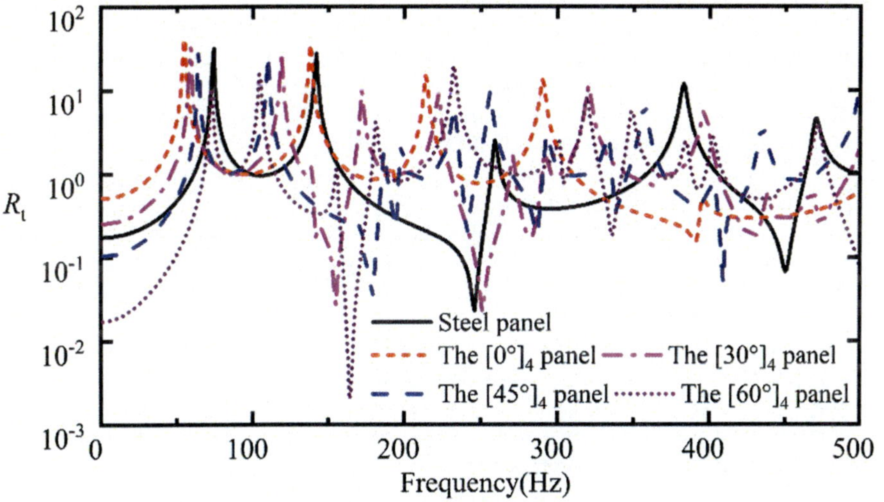

Fig. 4. Ratio between response amplitudes at Points B (0.75m, 0.25m) and A (0.25 m, 0.25 m) of the steel panel and laminated composite panels with different fiber angles. —: the steel panel; ------: the $[0°]_4$ panel; -·-·-·-: the $[30°]_4$ panel; - - -: the $[45°]_4$ panel; ······: the $[60°]_4$ panel.

4 Conclusions

This paper presented the free and forced vibration characteristics of rectangular laminated composite panels with various fiber orientations such as $[0°]_4$, $[30°]_4$, $[45°]_4$ and $[60°]_4$, with comparisons to vibration performance of a steel panel. The analytical method based on the FSDT has been shown to be accurate by verification with numerical ANSYS finite element results. This proposed method can be applied in investigation of forced vibration response at any designated positions. For panels with complex fiber orientations, the numerical FE method has been employed. The results showed that resonance peak values of dynamic responses of the steel panel are generally lower than that of laminated composite panels. The vibration responses and vibration transmission at a prescribed frequency can be reduced and designed by tailoring fiber orientation. The findings show the potential advantages of laminated composite panels with large design space to reduce the vibration level with reduced weight.

Acknowledgements. This work was supported by the National Key Research and Development Program under Grant number 2019YFA0706803, Zhejiang Provincial National Science Foundation of China under Grant number LY22A020006, and National Natural Science Foundation of China under Grant number 12172185.

References

1. Reddy, J.N.: Mechanics of Laminated Composite Plates and Shells: Theory and Analysis, 2nd edn. CRC Press, Washington (2003)
2. Thai, H.T., Choi, D.H.: A simple first order shear deformation theory for laminated composite plates. Compos. Struct. **106**, 754–763 (2013)
3. Aagaah, M.R., Mahinfalah, M., Jazar, G.N.: Natural frequencies of laminated composite plates using third order shear deformation theory. Compos. Struct. **72**, 273–279 (2006)
4. Kan, T., Swaminathan, K.: Analytical solution for free vibration of laminated composite and sandwich plates based on a higher-order refined theory. Compos. Struct. **53**, 73–85 (2001)
5. Dobyns, A.L.: Analysis of simply-supported orthotropic plates subject to static and dynamic loads. AIAA J. **19**, 642–650 (1981)
6. Carvalho, A., Soares, C.G.: Dynamic response of rectangular plates of composite materials subjected to impact loads. Compos. Struct. **34**, 55–63 (1996)
7. Zhu, C.D., Yang, J.: Free and forced vibration analysis of composite laminated plates. In: proceedings of the 26th International Congress on Sound and Vibration 2019 (ICSV26), IIAV, Montreal, Canada, 7–11 July (2019)
8. Zhu, C.D., Yang, J.: Vibration analysis of harmonically excited antisymmetric cross-play and angle-ply composite laminated plates. In: Proceedings of the 18th Asia Pacific Vibration Conference (APVC2019), Sydney, Australia, November 18–20, pp. 129–135 (2019)
9. Zhu, C.D., Yang, J., Rudd, C.: Vibration transmission and power flow of laminated composite plates with inerter-based suppression configurations. Int. J. Mech. Sci. **190**, 106012 (2021)
10. Zhu, C.D., Yang, J., Rudd, C.: Vibration transmission and energy flow analysis of L-shaped laminated composite structure based on a substructure method. Thin Walled Struct. **169**, 108375 (2021)

Vibration Power Dissipation in a Spring-Damper-Mass System Excited by Dry Friction

Cui Chao[1], Baiyang Shi[1], Jian Yang[1,2](✉), and Marian Wiercigroch[3]

[1] Department of Mechanical, Materials and Manufacturing Engineering, University of Nottingham Ningbo China, Ningbo 315100, People's Republic of China
jian.yang@nottingham.edu.cn

[2] Centre for Sustainable Energy Technologies (CSET), Faculty of Science and Engineering, University of Nottingham Ningbo China, Ningbo 315100, People's Republic of China

[3] Centre for Applied Dynamics Research, School of Engineering, Fraser Noble Building, King's College, University of Aberdeen, Aberdeen AB24 3UE, Scotland, UK

Abstract. This study investigates the vibration transmission and power dissipation behaviour of a mass-spring-damper system mounted on a conveyor belt. Coulomb friction exists between the mass and the belt moving at a constant velocity and acts as the external force for the mass. The steady-state power flow characteristics are obtained based on numerical integrations. The vibration energy dissipation at the contact interface and by the viscous damper is evaluated and quantified. The vibration transmission is measured by force transmissibility. For the system without the viscous damper, the instantaneous friction power can be positive or negative, depending on the motion characteristics of the mass. For the system with the viscous damper, in the steady-state motion, the vibration energy input caused by the friction can be dissipated by the viscous damper and also by the friction. Furthermore, effects of the magnitude of conveyor belt speed, damping ratio and friction force on the dynamic behaviour of systems are examined, and the power dissipation ratio of the system is analyzed. The results are expected to provide insights into the vibration transmission and suppression design of systems with friction.

Keywords: Vibration transmission · Power dissipation · Dry friction

1 Introduction

Friction is a very complex phenomenon and occurs at the interface contacting bodies. It is usually inevitable and plays a significant role in various engineering fields, such as seismology, mechanical engineering and civil engineering. A great number of published studies have revealed rich dynamic behavior of frictional systems [1, 2]. Popp *et al.* [3, 4] studied discrete and continuous models with stick-slip phenomena and observed abundant bifurcation and chaotic behaviors. Kruse *et al.* [5] studied the influence of joints on the stability and bifurcation behavior of a friction-induced flutter system. In Ref. [6],

Z. Dimitrovová et al. (Eds.): WMVC 2022, MMS 125, pp. 957–965, 2023.
https://doi.org/10.1007/978-3-031-15758-5_99

both the numerical simulation on low-degree-of-freedom models and the experimental validation on the real test rigs concerning the friction-induced vibration of systems were implemented.

For a system of a mass placed on conveyor belt, the dry friction has a significant effect on its dynamics. The analysis of the non-linear dynamics for the mass-on-belt system is a fundamental problem for many engineering fields. Many researchers carried out a number of investigations on the frictional non-linear dynamics [7–9]. However, few studies have considered the energy transmission and dissipation of such system [10]. The vibrational power flow analysis approach is a valuable tool to characterize the dynamic behaviour of complex systems [11]. Royston and Singh [12] examined the energy flow in a hydraulic engine mount system and showed that significant amount of vibration energy can be transmitted through a nonlinear path to a flexible base. Vakakis *et al.* [13] observed the phenomenon of energy transfer and noted that nonlinear attachment can be used to channel and dissipate the vibration energy of a main structure. Yang *et al.* [14, 15] developed power flow analysis (PFA) method for nonlinear dynamical systems, which reexamines typical nonlinear systems from a power flow perspective. Recently, the application of PFA to vibration control and vibration energy collection systems are investigated [16–21].

In this paper, the vibration energy flow transmission and dissipation characteristics of nonlinear non-smooth conveyor belt systems are investigated. The Runge-Kutta method is employed to investigate the vibration force transmission and power flow behaviour of systems with dry friction nonlinearity. Effects of the feeding speed, damping ratio and friction force on results are studied.

2 Single-Degree of Freedom (DOF) Coulomb Friction Models

Figure 1(a) shows the mass-on-belt frictional dynamic model, in which a mass block m is placed on the moving belt with a constant speed v_b. The mass is connected to a fixed wall through a viscous damper with damping coefficient c and a linear spring with stiffness coefficient k, connected in parallel. The Coulomb friction exists at the interface between the block and the belt. Figure 1(b) depicts the Karnopp model [22, 23] used to represent Coulomb friction.

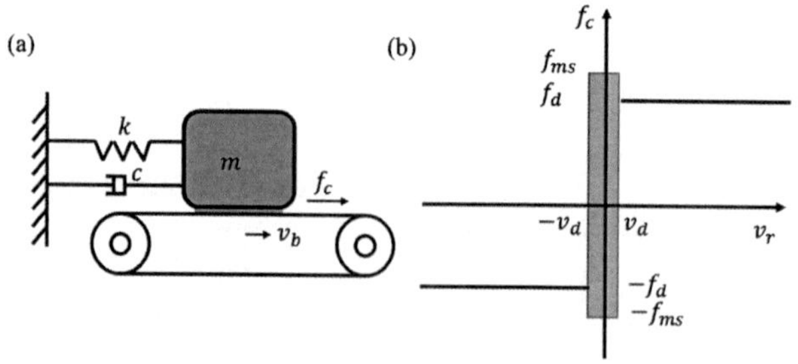

Fig. 1. (a) The spring–mass-damper system on the moving belt, and (b) the Karnopp model with magnitude of the dynamic friction force f_d and maximum static friction force f_{ms}. In (b), v_d is the limiting velocity of the assumed zeros velocity interval $[-v_d, v_d]$ for Karnopp model.

The mass block is subjected to the combined action of the friction, the spring and the damping forces. According to the Newton's second law, the equation of motion of the mass can be expressed as follows:

$$m\ddot{x} + c\dot{x} + kx = f_{ex} = f_c \tag{1}$$

where overdots denote derivative with respect to time t, f_{ex} is the external force and f_c is the nonlinear dry friction force acting as the excitation force to the mass. The friction f_c is described by the Karnopp model, written as:

$$f_c = \begin{cases} f_d \mathrm{sgn}(v_r), & \text{if } |v_r| > v_d, \\ f_{ms} \mathrm{sgn}(f_e), & \text{if } |v_r| \le v_d \text{ and } |f_e| \ge f_{ms}, \\ f_e, & \text{if } |v_r| \le v_d \text{ and } |f_e| < f_{ms}. \end{cases} \tag{2}$$

where f_d, f_{ms} and f_e are the dynamic friction force, the maximum static friction force and the resultant external force in tangential direction, respectively; v_d is the boundary velocity of the dead zone for Karnopp model. In this paper, it is assumed that $f_d = f_{ms}$ [24]. When the Karnopp friction model is used, we have $v_r = v_b - \dot{x}$ and $f_e = kx$ in Eq. (2). Following non-dimensional parameters are defined for parametric studies:

$$\begin{aligned} &\omega_0 = \sqrt{\frac{k}{m}}, \ \zeta = \frac{c}{2m\omega_0}, \ X = \frac{x}{l_0}, \ F_d = \frac{f_d}{kl_0}, \ V_b = \frac{v_b}{\omega_0 l_0}, \\ &V_d = \frac{v_d}{\omega_0 l_0}, \ V_r = V_b - X', \ \tau = \omega_0 t \end{aligned} \tag{3}$$

where ω_0 and ζ are the undamped natural frequency and the damping ratio of the system without friction consideration, l_0 and X arc the undeformed length of the linear spring and the non-dimensional displacement of the mass, F_d is the dimensionless magnitude of the dynamic dry friction force named magnitude of friction hereafter, V_b, V_d and V_r are the dimensionless velocity of the belt, limiting velocity of the assumed zeros velocity interval in the Karnopp model and relative velocity between the block and the mass, respectively, and τ is the dimensionless time. By using these dimensionless parameters and variables, Eq. (1) can be transformed into a dimensionless form:

$$X'' + 2\zeta X' + X = F_c \tag{4}$$

where primes denote derivative with respect to time τ, F_c is the non-dimensional friction force and is expressed by

$$F_c = \begin{cases} F_d \mathrm{sgn}(V_r), & \text{if } |V_r| > V_d, \\ F_d \mathrm{sgn}(F_e), & \text{if } |V_r| \le V_d \text{ and } |F_e| \ge F_d, \\ F_e, & \text{if } |V_r| \le V_d \text{ and } |F_e| < F_d. \end{cases} \tag{5}$$

where $F_e = X$ is the non-dimensional resultant force applied to the contacting interface in the tangential direction.

Two cases are considered in this paper: Case *I* considers the absence of the viscous damper (i.e., damping coefficient is set to zero), while Case *II* considers the presence of the viscous damper. The fourth-order Runge-Kutta method is used for the dynamic analysis to obtain the response and power flow variables.

3 Energy Flow and Force Transmissibility

The force transmissibility has been widely adopted as indicators for evaluating the vibration transmission between subsystems of an integrated linear or nonlinear structure [25]. For the current frictional SDOF system, the force transmissibility TR_B can be defined as the ratio between the maximum magnitude of the transmitted force to the wall and the amplitude of the external force:

$$TR_B = \frac{\max(|\Re(F_{tB})|)}{F_{ex}} \tag{6}$$

where F_{tB} represents the dimensionless transmitted force from mass to the wall. $F_{tB} = X$ is for the system without dampers (Case I), while $F_{tB} = X + 2\zeta X'$ is for the system with a damper (Case II). For the frictional system without external force excitation, the friction can be treated as the input force of the mass-spring-damper system, so that the F_{ex} in Eq. (6) can be replaced by F_d.

Pre-multiplying the governing Eq. (4) by the velocity X', the system's equation of power balance is obtained:

$$X'X'' + 2\zeta X'X' + X'X = X'F_c(\Delta(X')) \tag{7}$$

where $\Delta(X')$ is the relative velocity between the belt and the mass. Alternatively, Eq. (7) can be written in the following form:

$$\dot{K} + p_{dv} + \dot{U} = P_f \tag{8}$$

where $\dot{K} = X'X''$ and $\dot{U} = X'X$ are the non-dimensional rates of change of system kinetic and potential energies. $P_{dv} = 2\zeta X'X'$ and $P_f = X'F_c(\Delta(X'))$ are dimensionless instantaneous dissipated power and friction related power. In this paper, time-averaged behaviour of power flows is considered. Using an averaging time span of τ_p:

$$\bar{P}_{dv} = \frac{1}{\tau_p}\int_{\tau_i}^{\tau_i+\tau_p} P_{dv}\mathrm{d}\tau = \frac{1}{\tau_p}\int_{\tau_i}^{\tau_i+\tau_p} 2\zeta X'X'\mathrm{d}\tau, \tag{9a}$$

$$\bar{P}_{df} = \frac{1}{\tau_p}\int_{\tau_i}^{\tau_i+\tau_p} \mathrm{H}(-P_f)\mathrm{d}\tau = \frac{1}{\tau_p}\int_{\tau_i}^{\tau_i+\tau_p} \mathrm{H}(-X'F_c(\Delta(X')))\mathrm{d}\tau \tag{9b}$$

$$\bar{P}_{f\_in} = \frac{1}{\tau_p}\int_{\tau_i}^{\tau_i+\tau_p} \mathrm{H}(P_f)\mathrm{d}\tau = \frac{1}{\tau_p}\int_{\tau_i}^{\tau_i+\tau_p} \mathrm{H}(X'F_c(\Delta(X')))\mathrm{d}\tau \tag{9c}$$

where H() denotes the Heaviside step function, and τ_i is the starting time for averaging. $\bar{P}_{dv}$ denotes the time-averaged dissipated power by the damper, $\bar{P}_{df}$ is the power dissipated by the friction that is converted into heat, and $\bar{P}_{f\_in}$ is the input power by friction. For a periodic response, we have $\bar{P}_{dv} + \bar{P}_{df} = \bar{P}_{f\_in}$ with the averaging time set as one periodic cycle. $R_c = \frac{\bar{P}_{dv}}{\bar{P}_{f\_in}}$ and $R_f = \frac{\bar{P}_{df}}{\bar{P}_{f\_in}}$ are time-averaged power dissipation ratio by the damping and the friction.

4 Results and Discussions

This study focuses on the power flow characteristics of system with dry friction contact at the interface. It is assumed that initially at $t = 0$, the mass is placed on the belt such that their velocities are the same. The friction points to the right and forces the block to slide to the right, while the spring and the damping force act against the motion. The non-smoothness of the friction force is reflected by the conditional statement in the algorithm for numerical integrations of the governing equation. Another system parameter is fixed as $V_d = 10^{-3}$. The steady-state responses and the vibration dynamics of the system are of interest here and initial conditions can be set as $X(0) = F_d, X'(0) = V_b$.

For case *I* without the presence of the viscous damper, based on the given initial conditions, the steady-state instantaneous total power of the friction can be positive or negative depending on the direction of friction and velocity. Compared with the classical mass-spring-damper system without conveyor belt, the friction force in case *I* is the energy source when the mass slides on the conveyor, and it provides energy input into this system. By calculating, over a cycle of periodic oscillation, the total energy input by the friction equals to the energy dissipated by the friction, indicating that the work done by sliding friction on the block in case *I* is dissipated by itself.

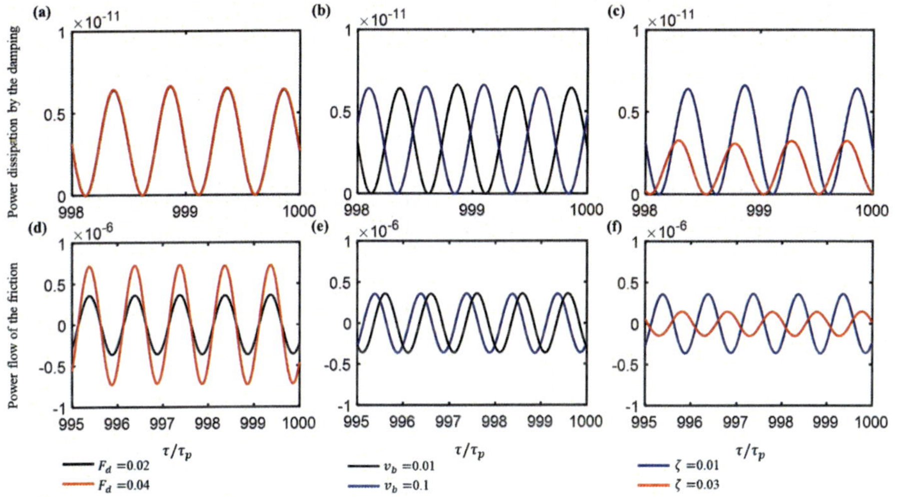

Fig. 2. Instantaneous power flow quantities of system on steady-state motion for case *II* (the effect of the viscous damper is considered). Initial conditions: $X(0) = F_d, X'(0) = V_b$.

Figure 2 presents the instantaneous power flow of the system considered in case *II* with the viscous damper. In Fig. 2(a) and (d), the black and red lines are characteristics for $F_d = 0.02$ and 0.04 with consistent parameters of $\zeta = 0.01$ and $V_b = 0.01$. In Fig. 2(b) and (e), the black and blue lines are characteristics for $V_b = 0.01$ and 0.1 with consistent parameters of $F_d = 0.02$ and $\zeta = 0.01$. In Fig. 2(c) and (f), the blue and red lines are characteristics for $\zeta = 0.01$ and 0.03 with consistent parameters of $F_d = 0.02$ and $V_b = 0.01$. Figure 2(a), (b) and (c) are instantaneous power dissipation characteristics by the viscous damping. Figure 2(d), (e) and (f) are power flow quantities of the friction force.

As depicted in Fig. 2(a) and (d), when the magnitude of friction F_d increases from 0.02 to 0.04, the power dissipation by the viscous damping is not affected while the amplitude of the friction power flow increases. As shown in Fig. 2(b) and (e), by changing the value of the conveyor belt speed V_b from 0.01 to 0.1, the figure shows only phase changes in the damping dissipated energy and the frictional power flows. From Fig. 2(c) and (f), as the damping ratio ζ increases from 0.01 to 0.03, amplitudes of the power flow magnitude of the viscous damper and the friction are further reduced. By a comparison of case *I* without the presence of the viscous damper, the friction force also provides power input into the system but the positive part of the frictional power flow is slightly larger than the negative part through calculation. The reason is that the value of power dissipation by the viscous damper is small being smaller than 10^{-11}. In contrast, the order of magnitude of frictional power flow is larger, 10^{-6} in Fig. 2(d), (e) and (f).

The energy dissipation is further investigated to reveal dynamics of the system. Figure 3(a) and (b) shows the effects of frictional contact on the power flow behavior of the system in the steady-state motion. The blue and green lines are characteristics for $F_d = 0.02$ and 0.04 with consistent parameters of $V_b = 0.01$, $\zeta = 0.01$. The pink and black lines are characteristics for $F_d = 0.02$ and 0.04 with consistent parameters of $V_b = 0.01$, $\zeta = 0.03$. The red symbol indicates the case for $V_b = 0.1$ of $\zeta = 0.01$ and $F_d = 0.02$. In Fig. 3(a), the power dissipation ratio by the viscous damper R_c, decreases significantly with the increase of the magnitude of the friction from 0.02 to 0.04, and increases slightly with the damping ratio from 0.01 to 0.03. When the feeding speed from the belt increases to 0.1, the figure shows little change in power flow quantities. In comparison, it is also found that most of the power is dissipated by the friction, about 99.99%. Which is consist with findings in Fig. 2 that the magnitude of the power dissipation by the damper is much smaller than the frictional dissipated power.

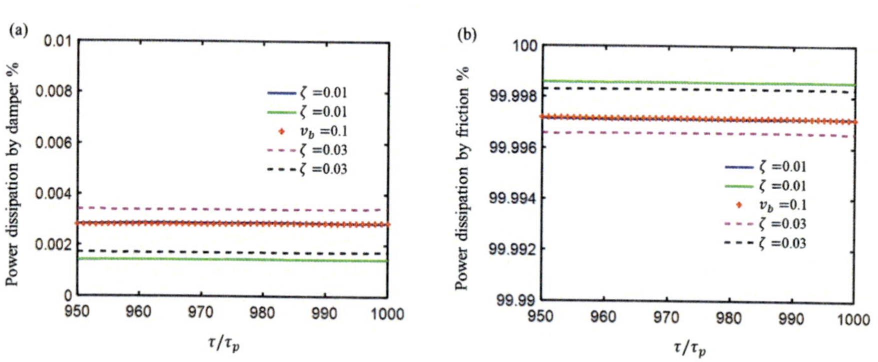

Fig. 3. Effects of the magnitude of friction F_d, damping coefficient ζ and conveyor belt speed V_b on the time-averaged power dissipation ratio (a) by the damper R_c and (b) by the friction R_f of the system. Initial conditions: $X(0) = F_d$, $X'(0) = V_b$.

In Fig. 4, as the damping ratio ζ of the viscous damper equals zero (i.e., case *I*), effects of the magnitude of dry friction F_d and conveyer belt speed V_b on the force transmissibility TR_B are obtained. The black and blue lines are the characteristics of $V_b = 0.01$ for $F_d = 0.02$ and 0.04. The red and green lines are the characteristics of $V_b = 0.1$ for $F_d = 0.02$ and 0.04. It can be found that TR_B decreases with friction force increase from 0.02 to 0.04 due to frictional resistance. On the other hand, TR_B increases as the belt speed V_b increases from 0.01 to 0.1. For case *II* considering the viscous damper, changes of the magnitude of friction force, damping ratio and feeding velocity can hardly affect the value of TR_B because of small change of the transmitted force F_{tB}. Figure 4 also shows that the force transmission to the wall has a downward trend with the increase of the damping ratio. Compared with the effects of the belt speed, variations in the level of dry friction and damping have a relatively large effect on the force transmissibility.

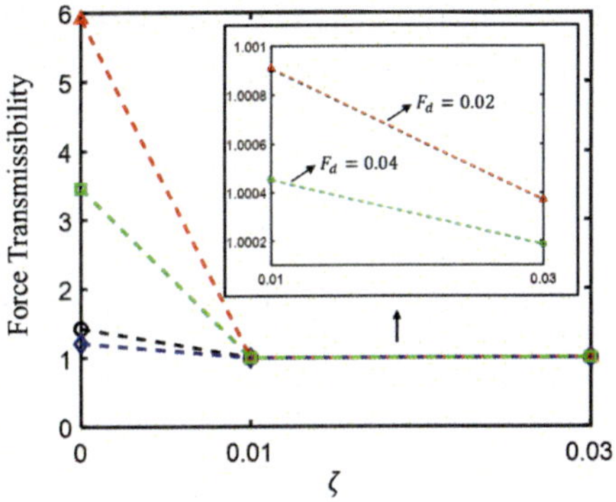

Fig. 4. Force transmissibility of system versus damping ratio ζ in steady-state motion. Initial conditions: $X(0) = F_d$, $X'(0) = V_b$.

5 Conclusions

This study investigated the power flow characteristic of a mass-spring-damper with a mass placed on a belt moving with constant velocity. Coulomb friction with Karnopp model is considered at the mass-belt interface. Time histories of the power flow quantities are obtained. The net power flow is zero when the viscous damping is ignored, that is, the work done by friction is dissipated by itself. When the damping is considered, the power flow magnitude of the friction is suppressed with the increase of the damping ratio ζ (which is the relative value between damping force and critical damping force). The magnitude of the friction has significant influence on frictional power flow but little effect on the damper's power flow. It is also found that the feeding speed of the system influences the phase of the steady-state response. By analyzing the time-averaged power dissipation ratio of the system, it is found that most of the power is dissipated by the friction not by the viscous damper. It is also shown that the force transmissibility of the system is mainly affected by the friction force and damping coefficient. These findings improve the understanding of the vibration transmission and suppression design of frictional systems.

Acknowledgement. This work was supported by National Natural Science Foundation of China under Grant number 12172185 and by the Zhejiang Provincial Natural Science Foundation under Grant number LY22A020006.

References

1. Ibrahim, R.A.: Friction-induced vibration, chatter, squeal, and chaos, Part II: dynamic and modeling. Appl. Mech. Rev. **47**(7), 227–253 (1994)
2. McMillan, A.J.: A non-linear friction model for self-excited vibrations. J. Sound Vib. **205**(3), 323–335 (1997)
3. Popp, K., Hinrichs, N., Oestreich, M.: Dynamical behaviour of a friction oscillator with simultaneous self and external excitation. Sadhana **20**(2), 627–654 (1995)
4. Popp, K., Stelter, P.: Stick-slip vibrations and chaos. Philos. Trans. Phys. Sci. Eng. **332**, 89–105 (1990)
5. Kruse, S., Tiedemann, M., Zeumer, B., Reuss, P., Hetzler, H., Hoffmann, N.: The influence of joints on friction induced vibration in brake squeal. J. Sound Vib. **340**, 239–252 (2015)
6. Wang, X.C., Huang, B., Wang, R.L., Mo, J.L., Ouyang, H.: Friction-induced stick-slip vibration and its experimental validation. Mech. Syst. Signal Process. **142**, 106705 (2020)
7. Saha, A., Wiercigroch, M., Jankowski, K., Wahi, P., Stefański, A.: Investigation of two different friction models from the perspective of friction-induced vibrations. Tribol. Int. **90**, 185–197 (2015)
8. Marques, F., Flores, P., Pimenta Claro, J.C., Lankarani, H.M.: A survey and comparison of several friction force models for dynamic analysis of multibody mechanical systems. Nonlinear Dyn. **86**(3), 1407–1443 (2016). https://doi.org/10.1007/s11071-016-2999-3
9. Saha, A., Wahi, P., Wiercigroch, M., Stefański, A.: A modified LuGre friction model for an accurate prediction of friction force in the pure sliding regime. Int. J. Non Linear Mech. **80**, 122–131 (2016)
10. Do, N., Ferri, A.A.: Energy transfer and dissipation in a three-degree-of-freedom system with stribeck friction. In: Proceedings of the ASME International Mechanical Engineering Congress and Exposition, Orlando, pp. 195–204. ASME (2005)
11. Goyder, H.G.D., White, R.G.: Vibrational power flow from machines into built-up structures, Part I: introduction and approximate analyses of beam and plate-like foundations. J. Sound Vib. **68**(1), 59–75 (1980)
12. Royston, T.J., Singh, R.: Vibratory power flow through a nonlinear path into a resonant receiver. J. Acoust. Soc. Am. **101**(4), 2059–2069 (1997)
13. Vakakis, A.F., Gendelman, O.V., Bergman, L.A., McFarland, D.M., Kerschen, G., Lee, Y.S.: Nonlinear Targeted Energy Transfer in Mechanical and Structural Systems. Springer, New York (2008). https://doi.org/10.1007/978-1-4020-9130-8
14. Yang, J.: Power flow analysis of nonlinear dynamical systems. University of Southampton (2013)
15. Yang, J., Xiong, Y.P., Xing, J.T.: Dynamics and power flow behaviour of a nonlinear vibration isolation system with a negative stiffness mechanism. J. Sound Vib. **332**(1), 167–183 (2013)
16. Yang, J., Shi, B., Rudd, C.: On vibration transmission between interactive oscillators with nonlinear coupling interface. Int. J. Mech. Sci. **137**, 238–251 (2018)
17. Yang, J., Jiang, J.Z., Neild, S.A.: Dynamic analysis and performance evaluation of nonlinear inerter-based vibration isolators. Nonlinear Dyn. **99**(3), 1823–1839 (2019). https://doi.org/10.1007/s11071-019-05391-x
18. Shi, B., Yang, J., Rudd, C.: On vibration transmission in oscillating systems incorporating bilinear stiffness and damping elements. Int. J. Mech. Sci. **150**, 458–470 (2019)

19. Dai, W., Yang, J., Shi, B.: Vibration transmission and power flow in impact oscillators with linear and nonlinear constraints. Int. J. Mech. Sci. **168**, 105234 (2020)
20. Dong, Z., Shi, B., Yang, J., Li, T.: Suppression of vibration transmission in coupled systems with an inerter-based nonlinear joint. Nonlinear Dyn. **107**(2), 1637–1662 (2021)
21. Dai, W., Yang, J.: Vibration transmission and energy flow of impact oscillators with nonlinear motion constraints created by diamond-shaped linkage mechanism. Int. J. Mech. Sci. **194**, 106212 (2021)
22. Karnopp, D.: Computer simulation of stick-slip friction in mechanical dynamic systems. J. Dyn. Syst. Meas. Contr. **107**(1), 100–103 (1985)
23. Olsson, H.M.: Control systems with friction. Department of Automatic Control, Lund Institute of Technology (LTH), Sweden (1997)
24. Dai, W., Yang, J., Wiercigroch, M.: Vibration energy flow transmission in systems with Coulomb friction. Int. J. Mech. Sci. **214**, 106932 (2022)
25. Xiong, Y.P., Xing, J.T., Price, W.G.: A general linear mathematical model of power flow analysis and control for integrated structure-control systems. J. Sound Vib. **267**(2), 301–334 (2003)

Vibration Power Flow and Wave Transmittance Analysis of Inerter-Based Dual-Resonator Acoustic Metamaterial

Yuhao Liu[1], Dimitrios Chronopoulos[2], and Jian Yang[1](✉)

[1] University of Nottingham Ningbo China, Ningbo 315100, People's Republic of China
jian.yang@nottingham.edu.cn
[2] KU Leuven, 9000 Leuven, Belgium

Abstract. This paper investigates the bandgap characteristics of inerter-based dual-resonator metamaterials and analyzes the low-frequency vibration suppression performance from the perspective of wave transmittance and vibration power flow. The studied metamaterial configuration is a one-dimensional mass-spring chain system with N identical lumped masses and each lumped mass has two resonators attached. Each unit cell of metamaterial is considered as a 1-DoF system with the effective mass varying with the excitation frequency, which can be negative in specific ranges of excitation frequencies. With dual resonators, dispersion relation diagrams show that there will be two separate bandgaps in which vibration transmission is suppressed, the frequency ranges for the bandgaps are similar to those for negative effective mass. Wave transmittance and power flow analysis also provides new perspectives to evaluate the dynamic characteristics. The results indicate that the wave transmittance is low and the energy is blocked within the bandgaps. The bandwidths of two bandgaps will not be influenced by cell position and the power transmittance of high excitation frequency is decreased as position number increases. The effect of inertance change on bandgap characteristics is examined and it shows that when the other parameters are the same, the two bandgaps are merged into one complete wide bandgap with identical inertance. These findings can provide a better understating of the dynamic behavior of dual-resonator metamaterials and their optimal design.

Keywords: Acoustic metamaterials · Dual-resonator · Inerter · Vibration power flow · Wave transmittance

1 Introduction

As one of the artificial composite materials, locally resonant acoustic metamaterials (LRAMs) have attracted extensive attention in the past two decades. LRAMs can achieve a variety of valuable physical properties such as wave absorption [1, 2] and reflection [3, 4], exhibiting excellent low-frequency vibrations suppression and noise control performance. Due to their extraordinary physical characteristics induced by proper internal

Z. Dimitrovová et al. (Eds.): WMVC 2022, MMS 125, pp. 966–975, 2023.
https://doi.org/10.1007/978-3-031-15758-5_100

mechanical structure design, LRAMs break through the restraints of traditional material properties, and they have a good application prospect in engineering fields such as vehicles, aviation, ships and precision instrument systems.

Liu *et al.* [5] firstly proposed a localized resonant material with effective negative elastic constants and a full-wave reflector within a certain frequency range, which led researchers to start investigating the application of metamaterials for locally resonant configuration design. Since the bandgap will be located around the natural frequency of the resonator [6], some researchers have tried to introduce the inerter [7] to the LRAM configuration design [8]. The inerter a two-terminal mechanical element that can provide large effective mass with small physical mass and reduce the natural frequencies of the vibration systems [9]. Kulkarni and Manimala [10] have investigated the wave propagation characteristics of various inerter based LRAM configurations using 1-D lattice models. The results indicate that the bandgap frequency and extent can be both up and down shifted depending on inerter configurations while maintaining the static mass of resonator. Dong *et al.* [11] studied the wave dissipation properties of inerter-based LRAM beam configuration with possible negative-stiffness elements. It presented that the response amplitude is greatly decreased within the bandgaps and anti-resonance frequencies are induced. Fang *et al.* [12] studied an inerter based metamaterial configuration to broaden bandgap width. It was found that an additional bandgap will be generated and the two bandgaps are possible to be merged into one wide bandgap. That is another potential research direction, inducing more band gaps and merging them together for wider band gaps. Li *et al.* [13] proposed a multi-resonator metamaterial for impact stress vibration suppression. The analysis presented that the multi-resonator metamaterial has wider bandgaps than those of a single-resonator metamaterial and it can provide a thin and light structure with a wider bandgap. Bao *et al.* [14] presented an enhanced dual-resonator LRAM and it showed that with the addition of cell number, mass and stiffness ratios, the bandgap is greatly influenced, showing good low-frequency wave attenuation performance. Some 3-D dual-resonator LRAM structures were also studied [15, 16], and dual band metamaterial was designed for K-band applications [17] and further improved for 5G sub-6 GHz applications [18].

In this study, the bandgap characteristics of an inerter-based dual-resonator locally resonant acoustic metamaterials configuration are investigated to gain a better understanding of LRAM study and design. In Sect. 2, the LRAM model is introduced and the effective mass for each unit cell and the dispersion relation for the infinite system are calculated. The wave transmittance and power flow analysis are also applied for new perspectives to evaluate the vibration suppression performance of the investigated LRAM. The influences of cell position and inertance on dynamic behavior are also studied. The conclusions are stated in Sect. 3.

2 Dual-Resonator Locally Resonant Metamaterials

2.1 Mathematical Modelling

For a dual-resonator LRAM configuration, it is known that there will be two bandgaps since the bandgaps are induced by local resonances. The investigated inerter based dual-resonator LRAM configuration is depicted in Fig. 1. It is a mass-spring structure with

N identical lumped masses, m_0, which are connected in series by dampers of damping coefficient c_0 and springs of stiffness coefficient k_0. Two resonator masses m_1 and m_2 are attached to each lumped mass by springs of stiffness coefficient k_1, k_2 and inerters of inertance b_1, b_2, respectively, forming a unit cell, and the initial distance between adjacent unit cells is L. The first lumped mass of the LRAM configuration is jointed to a harmonic moving base of displacement, $d_0 \cos \omega_f t$, which acts as the motion excitation.

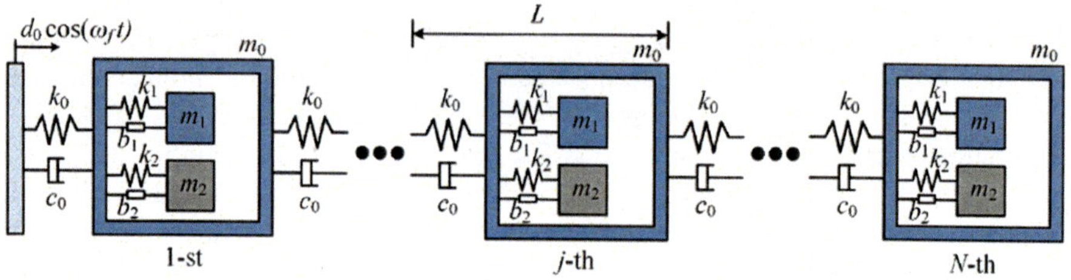

Fig. 1. The investigated inerter based dual-resonator LRAM configuration with excitation displacement.

The governing equations of motion of the dual-resonator LRAM are

$$m_0\ddot{x}_j + c_0\left(2\dot{x}_j - \dot{x}_{j-1} - \dot{x}_{j+1}\right) + k_0\left(2x_j - x_{j-1} - x_{j+1}\right) + m_1\ddot{y}_j + m_2\ddot{z}_j = 0, \tag{1a}$$

$$m_1\ddot{y}_j + k_1\left(y_j - x_j\right) + b_1\left(\ddot{y}_j - \ddot{x}_j\right) = 0, \tag{1b}$$

$$m_2\ddot{z}_j + k_2\left(z_j - x_j\right) + b_2\left(\ddot{z}_j - \ddot{x}_j\right) = 0, \tag{1c}$$

where x, y, z represent the displacements of lumped mass, m_0, resonator m_1 and resonator m_2, and the subscript j represents the cell positions.

The following parameters and variables are introduced

$$X_j = \frac{x_j}{L}, Y_j = \frac{y_j}{L}, Z_j = \frac{z_j}{L}, D_0 = \frac{d_0}{L}, \omega_0 = \sqrt{\frac{k_0}{m_0}}, \tau = \omega_0 t, \Omega = \frac{\omega_f}{\omega_0},$$
$$\mu_1 = \frac{m_1}{m_0}, \mu_2 = \frac{m_2}{m_0}, \beta_1 = \frac{k_1}{k_0}, \beta_2 = \frac{k_2}{k_0}, \zeta = \frac{c_0}{2m_0\omega_0}, \lambda_1 = \frac{b_1}{m_0},$$
$$\lambda_2 = \frac{b_2}{m_0}, \tag{2}$$

where X, Y, Z and D_0 represent the non-dimensional displacements, ω_0 is the natural frequencies of the unit cell without the resonators, τ and Ω are the non-dimensional time and excitation frequency, respectively, μ, β, ζ and λ are the mass ratio, stiffness ratio, damping ratio and the inertance-to-mass ratio, respectively. Using these variables and parameters, the motion excitation imposed on the first lumped mass is nondimensionalized as $D_0 \cos \Omega\tau$. The non-dimensional governing equations of motion of the system shown in Fig. 1 are

$$X_j'' + 2\zeta\left(2X_j' - X_{j-1}' - X_{j+1}'\right) + \left(2X_j - X_{j-1} - X_{j+1}\right)$$

$$+\mu_1 Y_j'' + \mu_2 Z_j'' = 0, \tag{3a}$$

$$\mu_1 Y_j'' + \beta_1 (Y_j - X_j) + \lambda_1 (Y_j'' - X_j'') = 0, \tag{3b}$$

$$\mu_2 Z_j'' + \beta_2 (Z_j - X_j) + \lambda_2 (Z_j'' - X_j'') = 0. \tag{3c}$$

2.2 Dispersion Relation and Wave Transmittance

To solve the governing equations, the displacements of the lumped mass and resonator masses in the j-th unit cell are expressed as

$$X_j = \hat{X}_j e^{i(jqL - \Omega\tau)},\ Y_j = \hat{Y}_j e^{i(jqL - \Omega\tau)},\ Z_j = \hat{Z}_j e^{i(jqL - \Omega\tau)}, \tag{4a-c}$$

where $\widehat{X}$, $\widehat{Y}$ and $\widehat{Z}$ are the complex amplitudes of the waves for lumped mass and two resonator masses, and q is the wave number.

According to Bloch's theorem [19], phase difference exists when two adjacent masses vibrate at identical amplitude. Then the motions of the adjacent $(j \pm 1)$-th masses are further expressed as

$$X_{j+1} = \widehat{X}_j e^{i((j+1)qL - \Omega\tau)} = \widehat{X}_j e^{i(jqL - \Omega\tau)} e^{iqL}, \tag{5a}$$

$$X_{j-1} = \widehat{X}_j e^{i((j-1)qL - \Omega\tau)} = \widehat{X}_j e^{i(jqL - \Omega\tau)} e^{-iqL}. \tag{5b}$$

Note that $e^{iqL} + e^{-iqL} = 2\cos(qL)$ and for dispersion relation analysis, the dampers are neglected since the damping effect will not influence the bandgap location. The equations for dispersion relation in Eq. (3) can be solved by Eqs. (4) and (5) and written in matrix form

$$\begin{vmatrix} 2(1-\cos(qL)) - \Omega^2 & -\mu_1\Omega^2 & -\mu_2\Omega^2 \\ -\beta_1 + \lambda_1\Omega^2 & -\mu_1\Omega^2 + \beta_1 - \lambda_1\Omega^2 & 0 \\ -\beta_2 + \lambda_2\Omega^2 & 0 & -\mu_2\Omega^2 + \beta_2 - \lambda_2\Omega^2 \end{vmatrix} = 0. \tag{6}$$

The 3-DOF unit cell can be considered as a SDOF system of effective mass M_{eff}, where the effective mass can be calculated by substituting Eq. (3b) and (3c) to Eq. (3a) to replace the Y_j and Z_j. The obtained effective mass is

$$M_{\text{eff}} = 1 + \mu_1 + \mu_2 + \frac{\mu_1^2\Omega^2}{\beta_1 - \lambda_1\Omega^2 - \mu_1\Omega^2} + \frac{\mu_2^2\Omega^2}{\beta_2 - \lambda_2\Omega^2 - \mu_2\Omega^2}. \tag{7}$$

As it is shown in Eq. (7), the effective mass of the lumped mass with two resonators depends on the excitation frequency when the material parameters are constant. The effective mass against excitation frequency diagram is depicted in Fig. 2. The terms related to materials parameters are set as $\mu_1 = \mu_2 = 0.2$, $\beta_1 = \beta_2 = 0.5$, $\lambda_1 = 0.1$ and $\lambda_2 = 0.2$. The black solid line represents the effective mass curve and the red dashed-dotted line is the zero effective mass reference line. Figure 2 shows two clear frequency

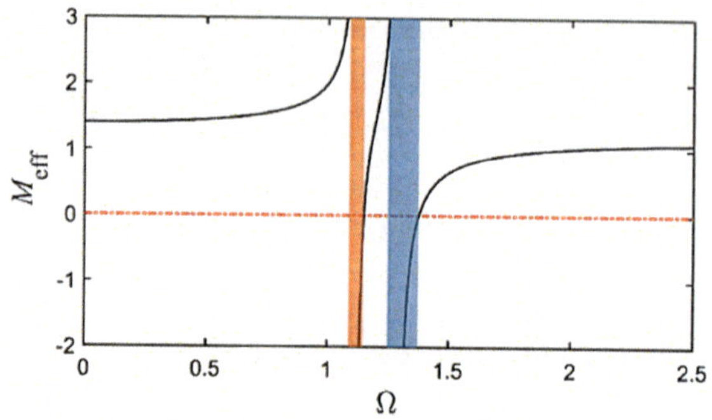

Fig. 2. Effective mass diagram of one dual-resonator unit cell.

gaps shadowed in orange and blue where the effective masses are negative. Based on the Newton's second law, the acceleration direction of the unit cell will be opposite to the applied force when effective mass is negative, suggesting this property can be used for the vibration suppression purpose.

Since the 3-DOF unit cell can be represented as a SDOF system by using effective mass, the dual-resonator LRAM configuration can be simplified to a system of effective masses interconnected by effective springs and dampers. Then the motion equation of the simplified finite LRAM configuration consisting of N unit cells is.

$$\mathbf{M}\mathbf{X}'' + \mathbf{C}\mathbf{X}' + \mathbf{K}\mathbf{X} = \mathbf{F}_e, \tag{8}$$

where

$$\mathbf{M} = M_{\text{eff}}\begin{bmatrix} 1 & 0 & \cdots & 0 & 0 \\ 0 & 1 & \ddots & \ddots & 0 \\ \vdots & \ddots & \ddots & \ddots & \vdots \\ 0 & \ddots & \ddots & 1 & 0 \\ 0 & 0 & \cdots & 0 & 1 \end{bmatrix}, \mathbf{C} = 2\zeta\begin{bmatrix} 2 & -1 & 0 & \cdots & 0 \\ -1 & 2 & -1 & \ddots & \vdots \\ 0 & \ddots & \ddots & \ddots & 0 \\ \vdots & \ddots & -1 & 2 & -1 \\ 0 & \cdots & 0 & -1 & 1 \end{bmatrix}$$
$$\mathbf{K} = K_{\text{eff}}\begin{bmatrix} 2 & -1 & 0 & \cdots & 0 \\ -1 & 2 & -1 & \ddots & \vdots \\ 0 & \ddots & \ddots & \ddots & 0 \\ \vdots & \ddots & -1 & 2 & -1 \\ 0 & \cdots & 0 & -1 & 1 \end{bmatrix}, \quad \mathbf{F}_e = \begin{bmatrix} D_0\cos\Omega\tau - 2\zeta D_0\sin\Omega\tau \\ 0 \\ \vdots \\ 0 \\ 0 \end{bmatrix}, \mathbf{X} = \begin{bmatrix} \hat{X}_1 \\ \hat{X}_2 \\ \vdots \\ \hat{X}_{N-1} \\ \hat{X}_N \end{bmatrix} \tag{9a-d}$$

represent the mass, damping and stiffness matrices, as well as the external force and displacement vectors, respectively. The displacement vector $\mathbf{X}$ can be derived

$$\mathbf{X} = \left[\mathbf{K} + i\Omega\mathbf{C} - \Omega^2\mathbf{M}\right]^{-1}\mathbf{F}_e, \tag{10}$$

In this study, the wave transmittance is expressed as the logarithm ratio of the last and first lumped mass displacement amplitudes.

$$T = 20\log_{10}(\hat{X}_N/\hat{X}_1). \tag{11}$$

The dispersion relation can be obtained by Eq. (6) and the wave transmittance can be calculated according to Eq. (11), which are presented in Fig. 3(a) and (b), respectively. The same parameters are selected as $\mu_1 = \mu_2 = 0.2$, $\beta_1 = \beta_2 = 0.5$, $\lambda_1 = 0.1$ and $\lambda_2 = 0.2$. The dispersion diagram shows that there are two regions with no real solutions for constant qL, which means the wave will be suppressed within these two frequency gaps called the bandgap. The two bandgaps cover the ranges between 1.0860 to 1.1486 and 1.2434 to 1.3766, which are very similar to the negative effective mass gaps in Fig. 2. Figure 3(b) depicts the wave transmittance figure of the finite metamaterial structure with 20 unit cells. It shows that there are also two gaps of low transmittance, defined as less than being −20dB. The low wave transmittance ranges are exactly the same as the bandgap ranges in the dispersion diagram.

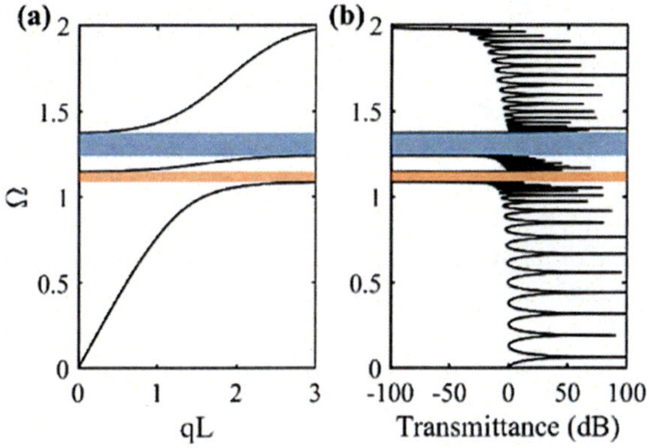

Fig. 3. (a) Dispersion figure and (b) wave transmittance figure of the investigated LRAM.

2.3 Power Flow Analysis

Power flow analysis (PFA) is a widely accepted method to investigate the dynamic performance of complex mechanical systems [20]. The instantaneous power to a system is defined as the product of the external force and the system velocity [21, 22].

The nondimensionalized power at time τ [23] is given by.

$$P(\tau) = \mathrm{Re}\{\overline{\mathbf{F}}_e(\tau)\}\mathrm{Re}\{\overline{\mathbf{X}}'(\tau)\}, \tag{12}$$

The time averaging input and output power for the LRAM can be further obtained:

$$\bar{P}_{in} = \frac{1}{\tau_s}\int_{\tau_0}^{\tau_0+\tau_s} -D_0\Omega\sin\Omega\tau(X_1 - D_0\cos\Omega\tau + 2\zeta_0 X_1{}' + 2\zeta_0 D_0\,\Omega\sin\Omega\tau)d\tau \tag{13a}$$

$$\bar{P}_{out} = \frac{1}{\tau_s}\int_{\tau_0}^{\tau_0+\tau_s} X_{N-1}{}'(X_{N-1} - X_N + 2\zeta_0 X_{N-1}{}' - 2\zeta_0 X_N{}')d\tau \tag{13b}$$

where τ_0 is the nondimensionalized start time and $\tau_s = 2\pi/\Omega$. The PFA is based on the numerical analysis of the input and output energy flow difference and ratio.

Based on Eq. (13), the time-averaged power flow diagrams can be obtained as shown in Fig. 4. The input power flow is plotted in red while the output power flow is plotted in blue as presented in Fig. 4(a). The power flow transmittance is defined as $\log_{10}(\overline{P}_{out}/\overline{P}_{in})$, and it is presented in Fig. 4(b). The input power flow is always higher than the output power flow and their difference is stable except in the two low power transmittance regions, which have same frequency ranges as the bandgaps. It indicates that the energy transmission is blocked, and the vibration is suppressed within the bandgaps.

The power transmission map of the investigated LRAM of 100-unit cell is drawn in Fig. 5. The power transmittance at unit position *j* refers to the logarithm ratio of the *j*-th unit cell and input power flow and the jet color map represents the power transmittance in decibel scale (dB). Based on the color map, the power transmittance is slightly reduced when the cell position number increases. There are two low-transmittance gaps in blue indicating that for all the cell positions, the energy is blocked within the bandgaps. The bandwidths of two bandgaps are not influenced by cell position when the position number is larger than 5. However, the power transmittance at high excitation frequency is decreased as position number increases while the lower frequency is slightly affected, which shows that it is harder to dissipate energy associated with the low-frequency components.

The material parameters will affect the bandgap characteristics and here we will focus on the influence of the inertance. The wave and power transmittance for the 100-unit dual-resonator LRAM are obtained and the results are depicted in Fig. 6 (a) and (b), respectively. The non-dimensional material parameters are selected as $\mu_1 = \mu_2 = 0.2$, $\beta_1 = \beta_2 = 0.5, \lambda_2 = 0.2$, and the inertance λ_1 varies from 0 to 0.5. Both diagrams exhibit two similar blue bandgaps. When λ_1 increases from 0 to 0.2, the lower bandgap keeps almost the same while the higher bandgap moves to a lower frequency with narrower bandwidth and the two bandgaps merged when $\lambda_1 = \lambda_2 = 0.2$. When λ_1 increases from 0.2 to 0.5, the higher bandgap remains constant while the lower bandgap moves to a lower frequency region and the bandwidth is reduced. When the other parameters are the same, identical inertance is preferred for dual-resonator LRAM, merging the two bandgaps into a wide single bandgap.

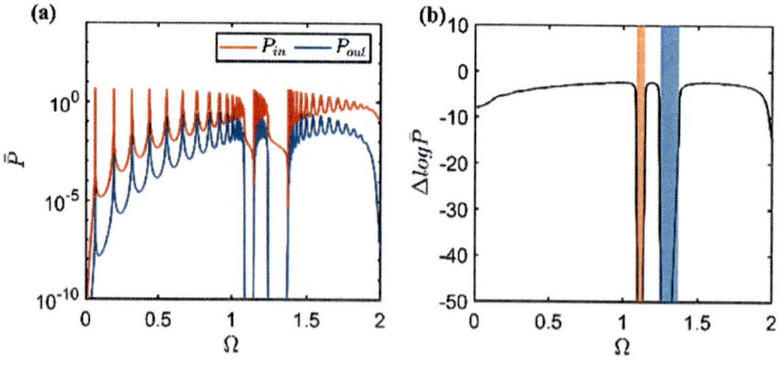

Fig. 4. (a) Time averaging power flow and (b) the power flow transmittance.

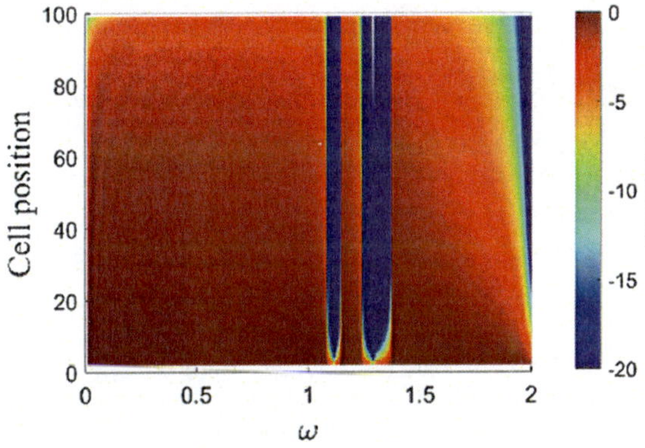

Fig. 5. The power flow transmittance map for a finite dual-resonator LRAM with 100 cells.

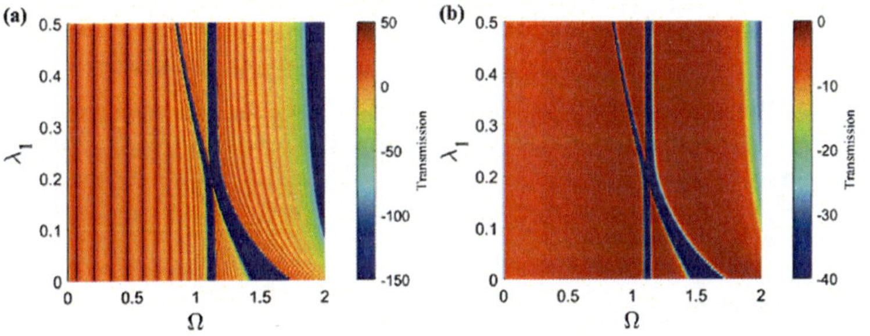

Fig. 6. (a) The wave transmittance map and (b) the power flow transmittance map for a finite 100-unit cell dual-resonator LRAM with λ_1 change from 0 to 0.5.

3 Conclusion

The dynamic behaviour of inerter-based dual-resonator locally resonant metamaterial configuration was investigated in this study. Wave transmittance and power flow analysis were carried out. It was shown that the use of inerters in LRAMs help reduce the bandgaps frequency for low-frequency vibration suppression. It has been found that:

- The effective mass of each unit cell showed that for a lumped mass with two resonators, there would be two negative effective mass regions which were related to specific excitation frequency ranges.
- There were dual bandgaps which were close to the negative mass regions. The power flow and wave transmittance figures had good agreement over the bandgap locations, indicating that the wave was attenuated and the energy transmission was blocked within the bandgaps.
- The bandgap location and width were shown not influenced by the unit cell position in the finite LRAM system.

- The two bandgaps could be merged if the material parameters on two resonators are identical and the bandgap would shift to lower frequency region with increasing inertance.

In conclusion, vibrtion power flow analysis of dual-resonator inerter-based LRAM was carried out and the effects of inertance on the bandgap characteristics were investigated. The results provided a new perspective for performance evaluation and design analysis of LRAMs.

Acknowledgement. This work was supported by the National Natural Science Foundation of China (NSFC) under Grant No. 12172185, the Zhejiang Provincial National Science Foundation of China under Grant No. LY22A020006, and the State Key Laboratory of Performance Monitoring and Protecting of Rail Transit Infrastructure of China under Grant No. HJGZ2021106.

References

1. Fang, X., Lun Tseng, M., Ou, J.-Y., MacDonald, K.F., Ping Tsai, D., Zheludev, N.I.: Ultrafast all-optical switching via coherent modulation of metamaterial absorption. Appl. Phys. Lett. **104**(14), 141102 (2014)
2. Peng, H., Pai, P.F.: Acoustic metamaterial plates for elastic wave absorption and structural vibration suppression. Int. J. Mech. Sci. **89**, 350–361 (2014)
3. Gu, S., Barrett, J., Hand, T., Popa, B.-I., Cummer, S.: A broadband low-reflection metamaterial absorber. J. Appl. Phys. **108**(6), 064913 (2010)
4. Plum, E., Fedotov, V., Zheludev, N.: Planar metamaterial with transmission and reflection that depend on the direction of incidence. Appl. Phys. Lett. **94**(13), 131901 (2009)
5. Liu, Z., et al.: Locally resonant sonic materials. Science **289**(5485), 1734–1736 (2000)
6. Hussein, M.I., Frazier, M.J.: Metadamping: an emergent phenomenon in dissipative metamaterials. J. Sound Vib. **332**(20), 4767–4774 (2013)
7. Smith, M.C.: Synthesis of mechanical networks: the inerter. IEEE Trans. Autom. Control **47**(10), 1648–1662 (2002)
8. Liu, Y., Yang, J., Yi, X., Chronopoulos, D.: Enhanced suppression of low-frequency vibration transmission in metamaterials with linear and nonlinear inerters. J. Appl. Phys. **131**(10), 105103 (2022)
9. Chen, M.Z., Hu, Y., Huang, L., Chen, G.: Influence of inerter on natural frequencies of vibration systems. J. Sound Vib. **333**(7), 1874–1887 (2014)
10. Kulkarni, P.P., Manimala, J.M.: Longitudinal elastic wave propagation characteristics of inertant acoustic metamaterials. J. Appl. Phys. **119**(24), 245101 (2016)
11. Dong, Z., Chronopoulos, D., Yang, J.: Enhancement of wave damping for metamaterial beam structures with embedded inerter-based configurations. Appl. Acoust. **178**, 108013 (2021)
12. Fang, X., Chuang, K.-C., Jin, X., Huang, Z.: Band-gap properties of elastic metamaterials with inerter-based dynamic vibration absorbers. J. Appl. Mech. **85**(7), 071010 (2018)
13. Li, Q., He, Z., Li, E., Cheng, A.: Design of a multi-resonator metamaterial for mitigating impact force. J. Appl. Phys. **125**(3), 035104 (2019)
14. Bao, H., Wu, C., Wang, K., Yan, B.: An enhanced dual-resonator metamaterial beam for low-frequency vibration suppression. J. Appl. Phys. **129**(9), 095106 (2021)
15. Zhang, Y., et al.: Dual band visible metamaterial absorbers based on four identical ring patches. Phys. E. **127**, 114526 (2021)

16. Deng, G., et al.: 3D rampart-based dual-band metamaterial absorber with wide-incident-angle stability. Appl. Phys. Express **14**(2), 022005 (2021)
17. Hakim, M.L., Alam, T., Almutairi, A.F., Mansor, M.F., Islam, M.T.: Polarization insensitivity characterization of dual-band perfect metamaterial absorber for K band sensing applications. Sci. Rep. **11**(1), 1–14 (2021)
18. Hasan, M., et al.: Polarization insensitive dual band metamaterial with absorptance for 5G sub-6 GHz applications. Sci. Rep. **12**(1), 1–20 (2022)
19. Brillouin, L.: Wave Propagation in Periodic Structures: Electric Filters and Crystal Lattices. Dover Publications, New York (1953)
20. Shi, B., Yang, J.: Quantification of vibration force and power flow transmission between coupled nonlinear oscillators. Int. J. Dyn. Control **8**(2), 418–435 (2019). https://doi.org/10.1007/s40435-019-00560-7
21. Al Ba'ba'a, H.B., Nouh, M.: Mechanics of longitudinal and flexural locally resonant elastic metamaterials using a structural power flow approach. Int. J. Mech. Sci. **122**, 341–354 (2017)
22. Shi, B., Yang, J., Rudd, C.: On vibration transmission in oscillating systems incorporating bilinear stiffness and damping elements. Int. J. Mech. Sci. **150**, 458–470 (2019)
23. Xing, J.T., Price, W.: A power–flow analysis based on continuum dynamics. Proc. R. Soc. Lond. Ser. A Math. Phys. Eng. Sci. **455**(1982), 401–436 (1999)

Vibration Suppression of Acoustic Black Hole Beam by Piezoelectric Shunt Damping with Different Positions

Zhiwei Wan[1,2], Xiang Zhu[1,2,3](✉), Tianyun Li[1,2,3], Sen Chen[1,2], and Junyong Fu[1,2]

[1] School of Naval Architecture and Ocean Engineering, Huazhong University of Science and Technology, Wuhan 430074, China
zhuxiang@hust.edu.cn

[2] Hubei Key Laboratory of Naval Architecture and Ocean Engineering Hydrodynamics, Wuhan 430074, China

[3] Collaborative Innovation Center for Advanced Ship and Deep-Sea Exploration, Shanghai 200240, China

Abstract. In this talk, a piezoelectric shunt damping is introduced to an acoustic black hole beam to form an acoustic black hole (ABH) piezoelectric composite structure, and its vibration characteristics are analyzed by a semi-analytical method. Based on the Hamilton principle, the Mexican hat wavelet is used as the shape function, and the energy method is used to solve the free and forced vibration of the acoustic black hole beam structure with PZT. The present results agree with those of the finite element method. To improve the effectiveness of the acoustic black hole, an external shunt circuit is connected to the PZT and shunt damping with the local resonance mechanism is introduced. The vibration characteristics of the beam with shunt damping and ordinary damping are compared and analyzed. The vibration suppression of ABH beam by piezoelectric shunt damping with different positions is discussed, and the optimal position is obtained. The designed acoustic black hole beam with shunt damping is significantly attenuated than the traditional damping layer acoustic black hole beam, which provides a new idea for the low-frequency vibration control of the acoustic black hole structure.

Keywords: Acoustic black hole · Piezoelectric shunt damping · Vibration suppression · Hamilton principle

1 Introduction

In a beam structure, the beam thickness becomes smaller in the form of a power law, and this wedge-shaped structure in the beam is called the acoustic black hole (ABH) structure [1]. The acoustic black hole structure has the characteristics of no additional mass and strong designability, etc., which has attracted widespread attention in the field of vibration engineering, such as suppression of vibration [2–5], energy accumulation [6–10], embedded in complex structures [11–15] and so on. When the bending wave propagates into the structure, the wave speed will gradually decrease as the thickness of

Z. Dimitrovová et al. (Eds.): WMVC 2022, MMS 125, pp. 976–983, 2023.
https://doi.org/10.1007/978-3-031-15758-5_101

the plate decreases. In the ideal case, the wave velocity will be reduced to zero, and the wave cannot propagate finally. However, due to the limitation of manufacturing accuracy, a truncation often exists in the central part of ABHs, which will affect the performance of ABHs [16]. Attaching a small amount of damping on the wedge surfaces can effectively dissipate the accumulated energy and significantly improve the vibration suppression ability of the ABH [17]. The disadvantage of ABHs is their poor performance below the cut-off frequency [18].

Shunt damping can produce local resonance [19–21], a strong advantage in low-frequency vibration control. Wan [22] has utilized shunt damping to suppress the multi-mode vibration of the ABH beam. This paper introduces the shunt damping composed of piezoelectric material shunted with a circuit to replace the traditional ordinary damping. The influence on the structural vibration characteristics was studied with the piezoelectric patch pasted at different positions through the energy method, and the best position was obtained. The first three-order vibration peaks of the ABH beam pasted with shunt damping were compared with ordinary damping. It is found that the ABH beam pasted with shunt damping has excellent vibration performance in low-frequency vibration.

2 Basic Theory

As shown in Fig. 1, the ABH beam with shunt damping consists of three parts: the ABH region with varying thickness $h(x) = x^2$ ($x_{b0} \leq x \leq x_{b1}$), the region with uniform thickness, and the shunt damping composed of piezoelectric material shunted with a circuit. The shunt circuit is composed of resistance element R and inductance element L in series. The right end of the beam is connected with a rotating spring and translational spring to simulate arbitrary boundary conditions. Table 1 shows the material parameters and dimension parameters of the model.

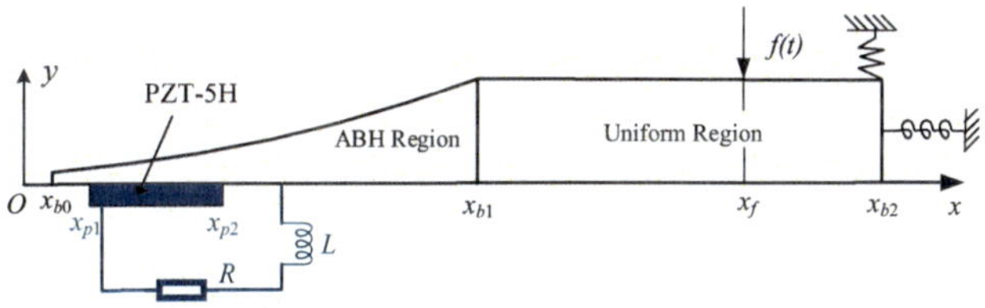

Fig. 1. Acoustic black hole beam structure with shunt damping

Table 1. Material parameters and size parameters

Geometrical parameters	Material parameters
$x_{b0} = 0.01$ m	$E_b = 210$ GPa
$x_{b1} = 0.05$ m	$\rho_b = 7800$ kg/m^3
$x_{b2} = 0.10$ m	$\eta_b = 0.005$

The equivalent Young's modulus can be calculated by the equivalent medium method for shunt damping. The shunt damping can be treated as an ordinary homogeneous material, and the characteristic is that its Young's modulus will change with the shunt circuit's impedance value. Detailed derivation can be found in the references [23, 24].

The piezoelectric material shunted with a circuit appears as a homogeneous material in the overall structure[24]. Its density is the one of piezoelectric material, and its Young's modulus is the equivalent one E_p:

$$E_p = \frac{h_p(1 + sZC_p)}{s_{11}^E h_p(1 + sZC_p) - sZd_{31}^2 A_s} \quad (1)$$

where s is the Laplace constant, h_p and A_s is the thickness and area of the piezoelectric patch, Z is the complex impedance of the shunt circuit, C_p is the capacitance of the piezoelectric patch, s_{11}^E is the short-circuit elastic compliance coefficient, and d_{31} is the piezoelectric strain constant.

Based on the Euler beam theory, the expression of the beam displacement field can be obtained. The Mexican hat-shaped wavelet function is selected as the shape function. Through Hamilton's principle, Lagrange equations related to system kinetic energy, potential energy and work done by external forces can be obtained. The matrix equation composed of mass matrix, stiffness matrix, response vector, and excitation vector can be obtained by substituting the displacement function expression into the Lagrange equation and simplifying it. The characteristic frequency and the corresponding system response can be obtained by solving this equation. The specific formula derivation can refer to the research work [5, 6]. In this method, the piezoelectric patch can be used as a part of the energy term so that the full coupling effect of the piezoelectric patch and the ABH beam can be considered in this coupled system. Besides, springs can be used to simulate different boundary conditions.

3 Method Verification

The ABH beam model is established in the finite element software COMSOL. The piezoelectric patch is made of PZT-5H, giving inherent anisotropic material properties. The position of the piezoelectric patch is $x_{p1} = 0.01$ m, $x_{p2} = 0.02$ m. The ABH beam is excited by $F = F_0 e^{i\omega t}$ ($F_0 = 1$N) at $x_f = 0.08$ m. The vibration acceleration level of $x = 0.06$ m is solved by the finite element method and the present method when the shunt circuit is open. The result is shown in Fig. 2. It can be seen that the first three-order forced vibration responses calculated by both methods are very consistent.

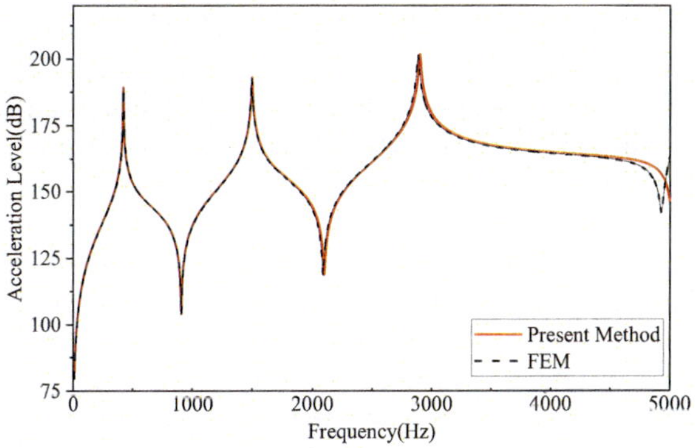

Fig. 2. Comparison of forced vibration responses between present method and FEM

4 The Effect of the Position of Piezoelectric Patch

The linear strain of the longitudinal fibers in the beam is proportional to the neutral axis's distance h.

$$\varepsilon = \frac{h}{\rho} \tag{2}$$

where ρ is the radius of curvature of the neutral layer. Suppose the perturbation of the bending deformation of the beam is w, and the mathematical analysis shows that the curvature ρ satisfies:

$$\frac{1}{\rho(x)} = \pm \frac{w''}{(1 + w'^2)^{3/2}} \approx \pm w'' \tag{3}$$

Therefore, the magnitude of the linear strain on the bottom surface of the beam is:

$$\varepsilon = hw'' \tag{4}$$

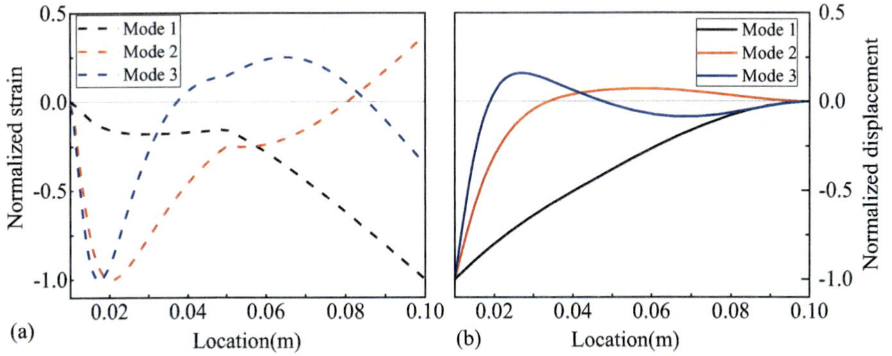

Fig. 3. (a) The first three bending strains (b) The first three modes

For the ABH beam without piezoelectric material attached, calculate the first three-order strain and modes of the ABH beam, as shown in Fig. 3 (a) and (b), respectively. The free end of an ABH produces the most significant level of displacement in free vibration. Since $x = 0.05$ m is the junction of the ABH part and the uniform thickness part, the first three strains all have abrupt changes. For the strain in the ABH region, the strain reaches the maximum near $x = 0.02$ m. The positive piezoelectric effect of the piezoelectric crystal is produced by deformation. Pasting the piezoelectric patch on the place where the strain of the beam is most significant can excite more electric charge and generate a larger current, which can better convert the bending strain energy into electrical energy.

The ABH beam's vibration characteristics are discussed when the piezoelectric plate is pasted on the part of the ABH. The shunt circuit connected to the piezoelectric patch cannot resonate locally, only to consider the influence of the position change on the structure vibration. The capacitance of the piezoelectric patch is 30.10nF, and the parameters of the shunt circuit are set to $L = 0.05$H, $R = 100\ \Omega$. The frequency of the electromagnetic oscillation generated by the shunt circuit is $f_e = \left(2\pi\sqrt{LC_p}\right)^{-1}$. So that the resonance frequency will occur near 4102 Hz, the shunt damping will not be coupled with the beam structure in the first three-order vibration frequency range. Paste the piezoelectric patch at different positions in the ABH area. The left end of the piezoelectric patch coordinates is $x_{p1} = 0.01 + 0.005 \times (p - 1)$ m, $(p = 1, 2 \cdots 7)$.

The output points are the nodes at different positions of the beam $x_{out} = 0.01 \times k$ m, $(k = 0, 1 \cdots 9)$, and the averaged vibration acceleration level of these nine nodes is calculated as the output index.

$$AL = 20\log_{10}\left(\omega^2\sqrt{\frac{1}{9}\sum_{i=1}^{9} A_i^2/a_p}\right) \tag{5}$$

A_i is the corresponding displacement amplitude of each output point at the angular frequency ω. a_p is the reference value, usually 1×10^{-6} m/s^2.

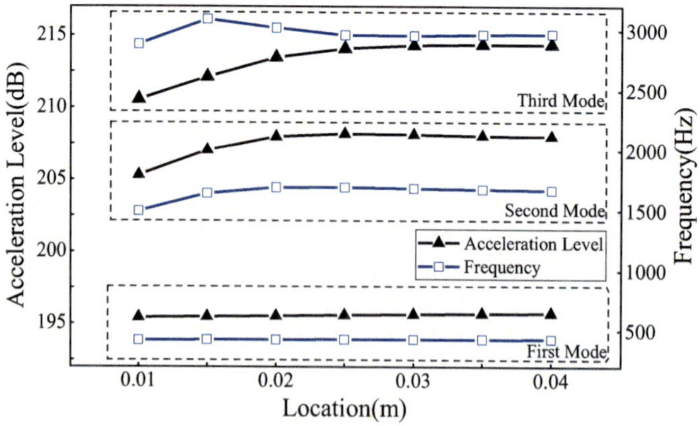

Fig. 4. The amplitude of output point and corresponding resonance frequency when the piezoelectric patch is pasted at different positions

Figure 4 shows the vibration acceleration level and frequency corresponding to the first three resonance peaks when the position of the piezoelectric patch changes. The abscissa of the figure is the position of the piezoelectric patch, the left ordinate is the vibration acceleration level, and the right ordinate is the frequency. The structure's first three vibration peaks and frequencies are the lowest when the piezoelectric patch is located at the root of the beam ($x_{p1} = 0.01$ m, $x_{p2} = 0.02$ m), where the bending strains and the deformation reach maximum. Due to its positive piezoelectric effect, the piezoelectric patch will generate charges on its surface when deformed, thereby converting the mechanical energy into electrical energy. The greater the deformation, the greater the electrical energy that the piezoelectric patch can store, and the smaller the mechanical energy of the beam structure. When the piezoelectric patch is located in the thicker part of the ABH, it has little effect on the first three orders' vibration. Because the ABH beam structure's strain is slight here, and the thickness of the piezoelectric patch is thinner than the beam structure here. The inherent characteristics of the structure have little effect. In summary, when the piezoelectric patch is located at the end of the ABH, the vibration responses and frequencies corresponding to the first three orders are lower. Therefore, for the ABH beam model studied in this paper, the optimal position for attaching the piezoelectric patch is $x_{p1} = 0.01$ m.

When the piezoelectric patch is pasted at $x_{p1} = 0.01$ m, the appropriate external circuit parameters are designed, and the suppression of a single resonance peak is studied. The resonance peak of the ABH structure with an ordinary damping layer is compared. The density of the ordinary damping layer material is 950 kg/m^3, Young's modulus is 5 GPa, and the material loss factor is 0.3. Figure 5 shows the suppression of the first-order resonance peak after the shunt damping is connected with the resistance $R_1 = 100\ \Omega$ and the inductance $L_1 = 5.35$H. The average vibration acceleration level of the output point $x_{out} = 0.01 \times k$, ($k = 0, 1 \cdots 9$ can be attenuated by 4.62 dB. Similarly, for the second resonance peak, the resonance peak can be attenuated by 20.12 dB when $R_2 = 250\ \Omega$, $L_2 = 0.425$H. For the third resonance peak, the resonance peak can be attenuated by 22.09 dB when $R_3 = 200\ \Omega$, $L_2 = 0.115$H.

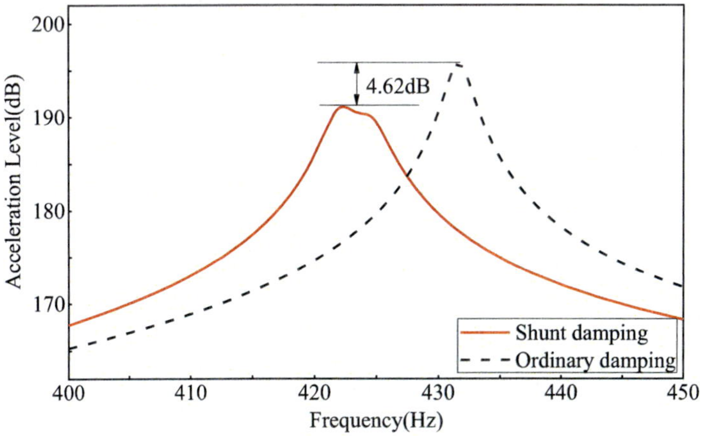

Fig. 5. First resonance peak by different method

The shunt piezoelectric damping could be considered as a multi-port impedance, whose dissipation comes from the relative motion of two system nodes [19]. Due to its positive piezoelectric effect, the piezoelectric patch generates electric charges on its surface when deformed, converting the mechanical energy of vibration into electrical energy. Part of the electric energy in the shunt circuit can be converted into heat by the resistance element and dissipated. The other part is electromagnetic resonance due to the interaction of the inductance and the inherent capacitance of the piezoelectric, forming a local oscillator. For different deformations of the base structure, the ability of the connection shunt damping to dissipate energy is also different. This difference can reflect the best position of the actual shunt damping. The above discussion shows that for this model, the effect of pasting the piezoelectric patch on the root of the ABH is the best. By designing different circuit parameters, compared with the traditional damping layer, the shunt damping can attenuate the first three-order resonance peaks by 4.62 dB, 20.12 dB, and 22.09 dB, respectively.

5 Conclusion

A piezoelectric shunt damping is introduced to the acoustic black hole beam, and the shunt damping is equivalent to a homogeneous material by the equivalent medium method. The energy method is used to solve the vibration of the ABH composite structure with shunt damping. The first three-order strain of the ABH structure is analyzed, and the influence of the position of the piezoelectric patch on the vibration characteristics of the structure is discussed. Finally, the first three-order resonance peaks of the ABH structure are suppressed. Compared with pasting the traditional damping layer, the shunt damping can attenuate the first three-order resonance peaks by 4.62 dB, 20.12 dB, and 22.09 dB, respectively. It provides ideas for improving the effectiveness of ABH in low-frequency ranges.

Acknowledgements. The authors wish to express their gratitude to the National Natural Science Foundation of China (Contract Nos. 51879113, 51839005 and 51479079) that have supported this work.

References

1. Krylov, V.V., Shuvalov, A.L.: Propagation of localised flexural vibrations along plate edges described by a power law. Institute of Acoustics, Milton Keynes, UK, pp. 263–70 (2000)
2. Deng, J., Zheng, L., Zeng, P.Y., Zuo, Y.F., Guasch, O.: Passive constrained viscoelastic layers to improve the efficiency of truncated acoustic black holes in beams. Mech. Syst. Signal Process. **118**, 461–476 (2019)
3. Gao, N., Wei, Z., Zhang, R., Hou, H.J.: Low-frequency elastic wave attenuation in a composite acoustic black hole beam. Appl. Acoust. **154**, 68–76 (2019)
4. Huang, W., Zhang, H., Inman, D.J., et al.: Low reflection effect by 3D printed functionally graded acoustic black holes. J. Sound Vib. **450**, 96–108 (2019)
5. Tang, L.L., Cheng, L., Ji, H.L., Qiu, J.H.: Characterization of acoustic black hole effect using a one-dimensional fully-coupled and wavelet-decomposed semi-analytical model. J. Sound Vib. **374**, 172–184 (2016)

6. Deng, J., Guasch, O., Zheng, L., Song, T., Cao, Y.: Vibration. Semi-analytical model of an acoustic black hole piezoelectric bimorph cantilever for energy harvesting. J. Sound Vib. **494**, 115790 (2020)
7. Ji, H.L., Liang, Y.K., Qiu, J.H., Cheng, L., Wu, Y.P.: Enhancement of vibration based energy harvesting using compound acoustic black holes. Mech. Syst. Signal Process. **132**, 441–456 (2019)
8. Zhao, L.X., Conlon, S.C., Semperlotti, F.: Broadband energy harvesting using acoustic black hole structural tailoring. Smart Mater. Struct. **23**, 065021 (2014)
9. Zhao, L.X., Conlon, S.C., Semperlotti, F.: An experimental study of vibration based energy harvesting in dynamically tailored structures with embedded acoustic black holes. Smart Mater. Struct. **24**(6), 065039 (2015)
10. Ning, L., Wang, Y.Z., Wang, Y.S.: Active control of a black hole or concentrator for flexural waves in an elastic metamaterial plate. Mech. Mater. **142**, 103300 (2020)
11. Deng, J., Guasch, O., Maxit, L., Zheng, L.: Vibration of cylindrical shells with embedded annular acoustic black holes using the Rayleigh-Ritz method with Gaussian basis functions. Mech. Syst. Signal Process. **150**, 107225 (2020)
12. Gao, N., Wei, Z., Hou, H., Krushynska, A.: Design and experimental investigation of V-folded beams with acoustic black hole indentations. J. Acoust. Soc. Am. **145**(1), 79–83 (2019)
13. Georgiev, V.B., Cuenca, J., Gautier, F., Simon, L., Krylov, V.V.: Damping of structural vibrations in beams and elliptical plates using the acoustic black hole effect. J. Sound Vib. **330**, 2497–2508 (2011)
14. Lee, J.Y., Jeon, W.: Wave-based analysis of the cut-on frequency of curved acoustic black holes. J. Sound Vib. **492**, 115731 (2021)
15. Lee, J.Y., Jeon, W.: Vibration damping using a spiral acoustic black hole. J. Acoust. Soc. Am. **141**(3), 1437 (2017)
16. Mironov, M.A.: Propagation of a flexural wave in a plate whose thickness decreases smoothly to zero in a finite interval. Sov. Phys. Acoust. **34,** 318–319 (1988)
17. Krylov, V.V., Tilman, F.: Acoustic 'black holes' for flexural waves as effective vibration dampers. J. Sound Vib. **274**(3–5), 605–619 (2004)
18. Conlon, S.C., Fahnline, J.B., Semperlotti, F.: Numerical analysis of the vibroacoustic properties of plates with embedded grids of acoustic black holes. J. Acoust. Soc. Am. **137**(1), 447–457 (2015)
19. Hagood, N.W., Flotow, A.V.: Damping of structural vibrations with piezoelectric materials and passive electrical networks. J. Sound Vib. **146**, 243–268 (1991)
20. Wu, S.: Piezoelectric shunts with a parallel R-L circuit for structural damping and vibration control. In: SPIE 2720, Smart Structures & Materials 1996: Passive Damping & Isolation, San Diego CA, pp. 259–269 (1996)
21. Wu, S.: Structural vibration damping experiments using improved piezoelectric shunts. In: SPIE 3045, Smart Structures & Materials 1997: Passive Damping & Isolation, San Diego CA, pp. 40–50 (1997)
22. Wan, Z.W., Zhu, X., Li, T.Y., Nie, R.: Low-frequency multimode vibration suppression of an acoustic black hole beam by shunt damping. J. Vib. Acoust. **144**(2), 021012 (2022)
23. Wang, G., Chen, S., Wen, J.: Low-frequency locally resonant band gaps induced by arrays of resonant shunts with Antoniou's circuit: experimental investigation on beams. Smart Mater. Struct. **20**, 015026 (2010)
24. Airoldi, L., Ruzzene, M.: Design of tunable acoustic metamaterials through periodic arrays of resonant shunted piezos. New J. Phys. **13**, 113010 (2011)

VWL: Vibrations and Waves Energy Transmission and Loss

Modelling the Effect of Introducing Flexible Coupling Between the SI Engine and Generator

Mohamed Brayek[1,2](✉) and Zied Driss[2]

[1] Architectural Engineering Department, College of Engineering, University of Prince Mugrin (UPM), Medina 42241, Saudi Arabia
m.brayek@upm.edu.sa

[2] Laboratory of Electro-Mechanic Systems (LASEM), National School of Engineers of Sfax (ENIS), University of Sfax (US), 3038 Sfax, Tunisia

Abstract. The current work involves testing the effect of using a flexible coupling between the engine and the generator on the rotational irregularity of an inverter generator Honda EU2200i. Simulation was carried on the generator's engine Honda GXR120 Engine of 123 cm^3 at 3000 rpm (6.3 Nm). An approach was developed and implemented in MATLAB environment. With reference to experimental setup, the calculated and simulated cylinder pressure curve shows a high degree of agreement. The key to higher electrical generator performance is the regularity of input speed. The obtained result of the present model shows that a higher degree of stability in the generator speed could be obtained with the introduction of a flexible coupling between the engine and the generator.

Keywords: Matlab · Rotational irregularity · Simulation · Spark ignition engine · Performance

1 Introduction

For the generation of electrical energy in desert, islands and isolated places, Honda EU2200i is one of the quietest, lightest, and most powerful generator in the 2,200-W range. It is currently primarily used in mobile and stationary operation. The high energy density of fossil fuels is converted into electrical energy by coupling an electricity generator with a combustion engine (usually SI or gas engines). Already in the design process of the plant, it is increasingly required to consider interactions between the system components and to identify an optimized system configuration and mode of operation. For example, the effects of dynamic load changes in the electrical grid and the reaction of the combustion engine have a direct impact on fluctuations in the grid frequency in stand-alone grids. Especially in applications with high electrical loads in isolated operation, such as cement and steel plants or ships with hybrid propulsion systems, considerable influences on the stability of the grid frequency can occur and cause negative effects on other grid components. For the design and operation of such isolated grids, it is of great importance to be able to reliably determine the expected interactions quantitatively. Up to now, such calculations have been carried out manually

Z. Dimitrovová et al. (Eds.): WMVC 2022, MMS 125, pp. 987–997, 2023.
https://doi.org/10.1007/978-3-031-15758-5_102

or in partial simulations. Uncertainties often lead to the selection of an oversized plant category with correspondingly high procurement costs.

In general, conventional simulations of such systems are calculated based on the engine speed [1]. However, the approach of the project includes the assumption that by developing a model based on rotational angle-resolved process calculation, a more precise statement on the system behavior can be made and thus the system configuration can be optimized regarding more significant parameters. As a result of the design process, a more economical plant can be selected, if necessary, which nevertheless enables stable system performance with low network fluctuations.

In addition to safe and best possible system performance, other economic and ecological aspects, such as efficiency and emissions, can also be defined as targets of a model-based system design.

In the simulation, two system models are to be merged at an interface. Thereby the models are a mechanical-thermodynamic engine model and a mechanical-electrical generator-network model. For the development of the model, a 121 cm^3 SI engine and a generator as an asynchronous machine, as they are commonly used in the middle east region, are to be modeled first.

This paper describes the simulation approach developed for the sub-model of the internal combustion engine and compares the simulation results obtained with measurement results from literature.

2 Mathematical Modeling

For the simulation of the SI engine, a calculation approach based on relevant 0D models and available engine data was selected.

Although several commercial tools for the calculation of drive systems are available, it was decided to develop a corresponding simulation model in-house, considering the research objectives. Reasons for this are:

- Exact knowledge of the structure of the models.
- Special adaptation of the models to the conditions of large engines and generators in single and multi-engine systems with mutual interaction.
- Free choice of the level of detail and user complexity of the sub-models and the overall system.
- Expandability of the model with respect to further interfaces, e.g. coupling to heat networks in case of networks for waste heat utilization.

The choice of the simulation environment falls on the program MATLAB. A flexible environment with several predefined differential equations solvers and offers the ability to build a model using mathematical and logical formulation. In this way physical relationships can be programmed in a comprehensible and adaptable way.

To simulate the combustion engine, a zero-dimensional model is used, which is based on the 1st law of thermodynamics, in which the combustion chamber is a transient open system [2, 3].

The energy balance [4], derived according to the crank angle φ, is generally described as follows:

$$\frac{dU}{d\varphi} = -p \cdot \frac{dV}{dt} + \frac{dQ_B}{d\varphi} - \frac{dQ_W}{d\varphi} + h_I \frac{dm_I}{d\varphi} - h_O \frac{dm_O}{d\varphi} \tag{1}$$

To solve the balance equation, a variable must be specified in the simulation, the heat input due to combustion from measured data or an empirically determined process. The internal energy in the combustion chamber is calculated via the caloric equation of state for ideal gases.

$$\frac{dU}{d\varphi} = mc_v \frac{dT}{d\varphi} \tag{2}$$

According to Brayek [3], in the simulation the Eq. (2) is derived according to the time, so the derivative of the constant reference temperature disappears, and one obtains:

$$\Delta \dot{U} = m \cdot c_v \cdot \dot{T} \tag{3}$$

Using this approach [2], the 1st law can be formulated as a 1st order differential equation as follows.

$$\dot{T} = \frac{-p \cdot \frac{dV(\varphi)}{dt} + \dot{Q}_B - \dot{Q}_w + \dot{H}_I - \dot{H}_O}{m \cdot c_v} \tag{4}$$

The specific heat capacity c_v is assumed to be constant during this study.

The cylinder pressure can be calculated using the ideal gas equation.

$$p = \frac{m \cdot R \cdot T}{V} \tag{5}$$

Besides the determination of pressure and temperature it is still necessary to calculate the supplied heat flux $\dot{Q}_B$ from the fuel. Combustion reaction is the release of the chemical heat in the fuel (the heat release) occurs following the burning rate [3].

The dynamic relations used in modeling are differential equations obtained from conservation of mass and energy laws.

To represent the combustion process in SI engine as realistically as possible, the Wieb function is used for the simulation. This function intends to describe the three typical phases of pre-combustion, main combustion and after burning [5].

The combustion curve is generated via the first part of the MATLAB program and passed to the second layer to conduct the flywheel study. Using Newton's approach, the dissipated heat output Q_W, can be calculated by heat transfer to cylinder wall and piston, describe.

$$\dot{Q_w} = \alpha A(\varphi) \cdot (T_{\text{Gas}} - T_{\text{Wall}}) \tag{6}$$

In Eq. (6), A is the heat-transferring surface area. It is made up of the cylinder head area, the piston crown area and the instantaneous surface area of the combustion chamber, depending on the piston position.

For the heat transfer coefficient α, Eq. (7) according to Woschni's correlation [6] was chosen in the simulation environment since it considers not only the high-pressure part (compression and expansion) but also the charge exchange (push-out and suction) is considered.

$$\alpha = 130 \cdot d^{-0,2} \cdot p^{0,8} \cdot T^{-0,53} \cdot (C_1 \cdot v)^{0,8} \tag{7}$$

The dimensionless constant C_1 is specified with the following values:
$C_1 = 2,28 + 0,308 \frac{v_u}{v_{mean}}$ for the high-pressure part.
$C_1 = 6,18 + 0,417 \frac{v_u}{v_{mean}}$ for the low-pressure part.
The velocity term v is given in Eq. (8) according to Wu et al. [7] as follows:

$$v = v_{mean}\left[1 + 2 \cdot \left(\frac{V_c}{V}\right)^2 \cdot p_i^{-0,2}\right] \tag{8}$$

Here, p_i is the indexed mean pressure calculated in the simulation environment as an integral quantity from the volume work of change, relative to the stroke volume of the cylinder.

The terms $\dot{H}_I$ and $\dot{H}_O$ are the enthalpy flows of the gas masses flowing in and out through the valves, which are formulated in the simulation as follows [4].

$$\dot{H}_I = \dot{m}_I \cdot (u_I + R_I \cdot T_I) \tag{9}$$

$$\dot{H}_O = \dot{m}_O \cdot (u_O + R_O \cdot T_O) \tag{10}$$

The index I denotes the state just before the inlet valve and, in an analogous way, the index O denotes the state after the outlet valve. The calculation of the specific internal energy is based on fuel data of Heywood [4], which are expressed in polynomial equations, as a function of temperature, for the respective air ratio. The following Eqs. (11), (12) are used to calculate the mass flow through the valves in the gas exchange [4].

$$\dot{m}_I = \mu \cdot A_{VI(\varphi)} \cdot \frac{p_I}{\sqrt{R \cdot T_I}} \sqrt{\frac{2k}{k-1}\left[\left(\frac{p_C}{p_I}\right)^{\frac{2}{k}} - \left(\frac{p_C}{p_I}\right)^{\frac{k+1}{k}}\right]} \tag{11}$$

$$\dot{m}_O = \mu \cdot A_{VO(\varphi)} \cdot \frac{p_C}{\sqrt{R \cdot T_C}} \sqrt{\frac{2k}{k-1}\left[\left(\frac{p_O}{p_C}\right)^{\frac{2}{k}} - \left(\frac{p_O}{p_C}\right)^{\frac{k+1}{k}}\right]} \tag{12}$$

Accordingly, the indices C stand for cylinder, I for inlet and O for outlet, each considered just before and after the valve, respectively.

A_v describes the area of the gap from the inlet or outlet valve, which changes as a function of the valve lift. The flow coefficient μ is a measure of the flow resistance at the valve channels, which considers losses due to friction. These can usually only be determined on flow test rigs or through costly CFD analyses [5]. For the simulation, flow coefficients were assumed as a function of the relative valve lift from [1, 5].

3 Modeling the Generator on MATLAB

3.1 Simulation Environment

The MATLAB environment offers the ability to build a model using the previous mathematical formulation presented in appropriate logical manner. MATLAB contain several differential equations solvers, in the present work the pre-defined ode23 solver was used to solve the pressure and temperature differential equations.

The engine model has been constructed using MATLAB. A graphical user interface GUI is created with a block of customizable parameters. It offers easier method to introduce different engine characteristics and to read and graphically present simulation results.

Figure 1 shows the Graphical user interface for the calculation program. The MATLAB GUI contain editable boxes where the engine geometry and operating parameters could be introduced as simulation input. This interface allows a simple interaction with different users where it allows to change engine geometry or operating condition if needed. Once the input cases are loaded, the GUI is fitted with a Start button that run the script and shows the results. The simulation starts by specifying an initial temperature, a starting speed and an air mass already present in the combustion chamber. The piston is at top dead center at 360° CA.

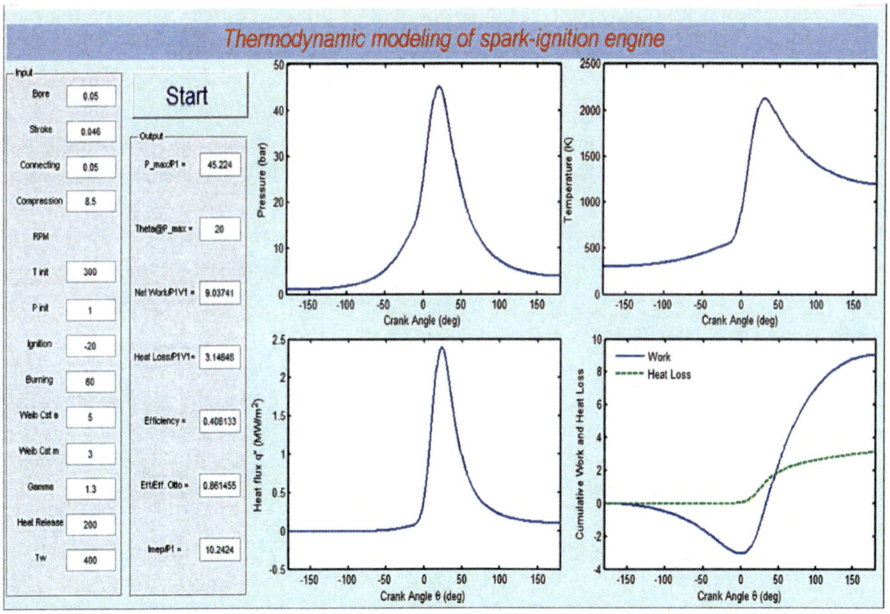

Fig. 1. GUI of the simulation program

In the present work the MATLAB routine was essentially following Ferguson investigation. This mathematical model allows the prediction of SI engine in cylinder variable as a function of crank angle. The present investigation focus on the effects of only the heat release introduced in the combustion chamber on engine performance. Calculations were made to simulate the performance of a selected engine with characteristics shown in Table 1. The engine speed was assumed to 3000 rpm corresponding to the maximum torque of 6.3 Nm.

Table 1. Engine specifications

Parameters	Value
Bore and stroke	60 mm × 43 mm
Displacement	123 cm^3
Compression ratio	8.6:1
Maximum torque	6.3 Nm at 3000 rpm

3.2 Model Validation

To verify the simulation model, the calculation results are compared with measured data by Dashti and Asghar [8].

Figure 2 shows a comparison of the cylinder pressure curve over a complete operating cycle at an engine speed of 4000 rpm. A combustion curve determined from the measurements was used as the input parameter for the simulation. A comparison of the determined cylinder pressure curves shows that simulation and calculation agree to a high degree. This is an important prerequisite for an accurate description of the temporal torque curve at the interface between the combustion engine and the connected generator.

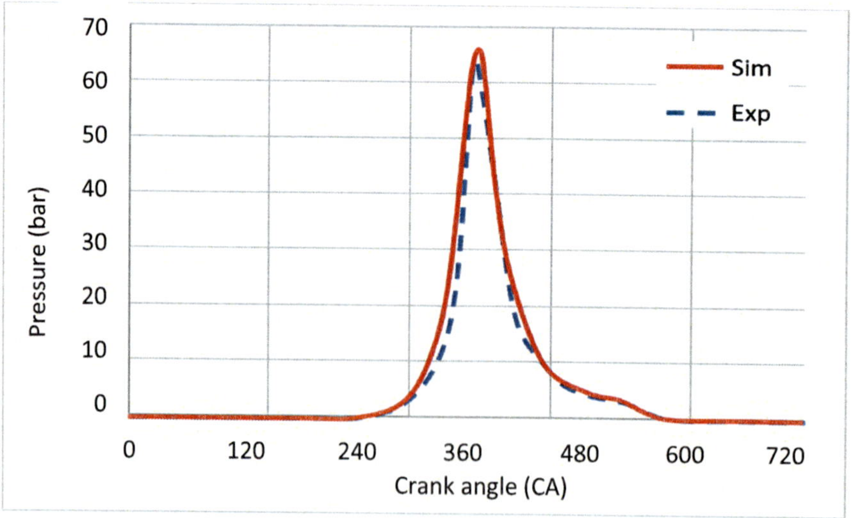

Fig. 2. Simulated (Sim) and experimental (Exp) Cylinder pressure values versus the crank angle

A comparison of the mean pressure as an integral variable for characterizing the engine load yields a deviation of less than 5% between simulation and measurement. Similar good results were also found at other operating points, which confirms the applicability of the simulation approach and the selected models, at least for the stationary operating points.

4 Results and Discussion

Figure 3 shows the simulated cylinder pressure curve and the variable cylinder volume over a working cycle (720° CA). As the piston moves upward, thc cylinder volume decreases and the compression phase begins, causing pressure and temperature to rise. At approx. 360° CA, combustion begins and the pressure rises. As the piston moves downward, the cylinder volume increases and the expansion phase follows, during which the pressure and temperature continue to decrease. Shortly before 480° KW, the exhaust valve opens and the exhaust gases begin to be expelled.

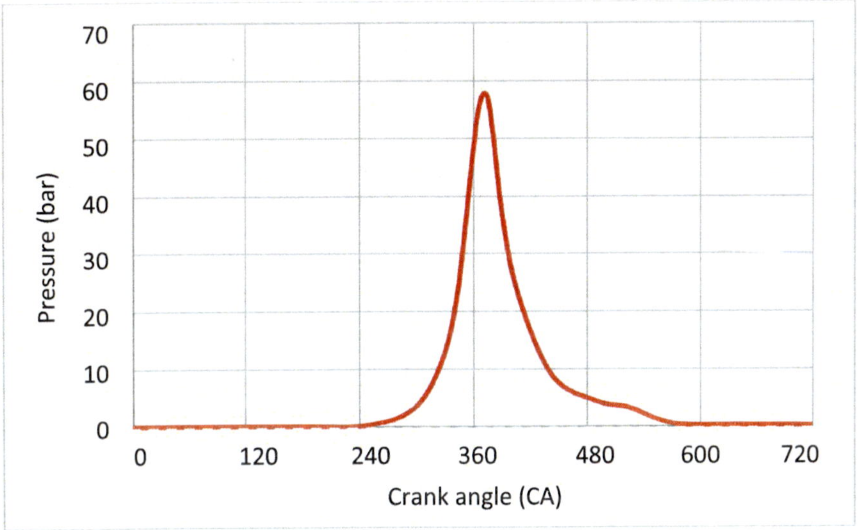

Fig. 3. Cylinder pressure and volume over 720° KW

For the calculation of the speed curve, the terms of the volume changes performed on the piston are used in each case.

The energy output $-pdV/d\varphi$, the dissipated useful power, the necessary power, etc.

The friction power and the power for the oscillating acceleration of the pistons are used. The determination of the friction medium pressure is done by a Polynomial approach of Chen and Flynn [9], which allows a good adaptability to the boundary and operating conditions.

A start speed must be defined in advance to be able to calculate the energy balance considering the oscillating mass torque and the friction torque. The load requirement in

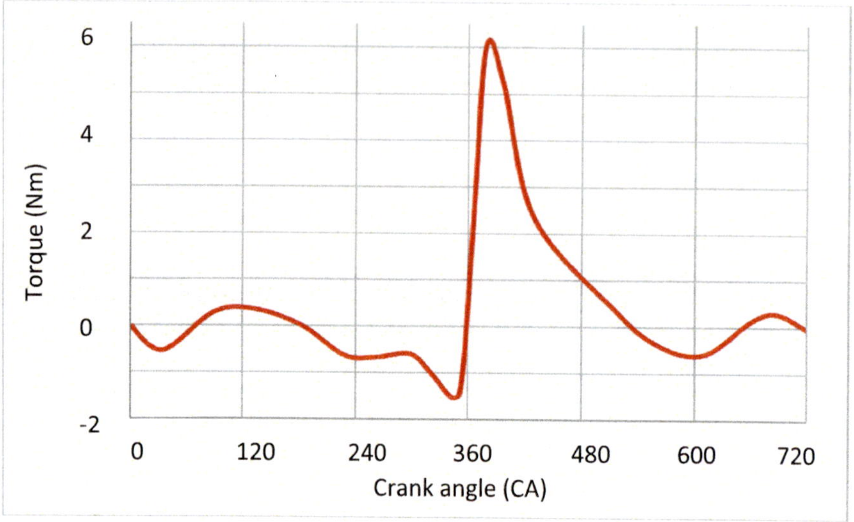

Fig. 4. Simulated torque curve

this system is stored as a constant, which can be increased or decreased depending on the simulation conditions. The energy state of the flywheel can be described with Eq. (13).

$$E_{Rot} = \frac{1}{2} \cdot J \cdot \omega^2 \tag{13}$$

Figure 4 shows a simulated torque curve for a cylinder, over a working cycle.

At the cumulative point, all inflowing energies are combined. By rearranging Eq. (13), the speed curve can be calculated as follows:

$$n = \sqrt{\frac{E_{Rot\,neu}}{2 \cdot J \cdot \pi^2}} \tag{14}$$

Since the rotational irregularity of the motor has a considerable influence on an electric machine coupled to it, the behavior of an elastic coupling element must also be simulated. According to the basic dynamic law for rotational motion, the following equations result for the rotating masses [2].

$$-T_m(\varphi_M - \varphi_G) - T_d(\dot{\varphi}_M - \dot{\varphi}_G) + M = J_M \cdot \ddot{\varphi}_M \tag{15}$$

$$T_m(\varphi_M - \varphi_G) + T_d(\dot{\varphi}_M - \dot{\varphi}_G) = J_G \cdot \ddot{\varphi}_G \tag{16}$$

In Eqs. (15), (16) $T_d(\dot{\varphi_M} - \dot{\varphi_G})$ is the damping torque and $T_m(\dot{\varphi_M} - \dot{\varphi_G})$ is the torque from the rotation of both rotating masses with respect to each other. M is the excitation torque introduced from the engine side.

Rearranged according to the highest derivative, the following Eq. (17) is obtained:

$$\begin{gathered} -T_m\left(\frac{1}{J_M} + \frac{1}{J_G}\right) \cdot (\varphi_M - \varphi_G) - T_d\left(\frac{1}{J_M} + \frac{1}{J_G}\right) \cdot (\dot{\varphi}_M - \dot{\varphi}_G) \\ + \frac{M}{J_M} = \ddot{\varphi}_M - \ddot{\varphi}_G \end{gathered} \tag{17}$$

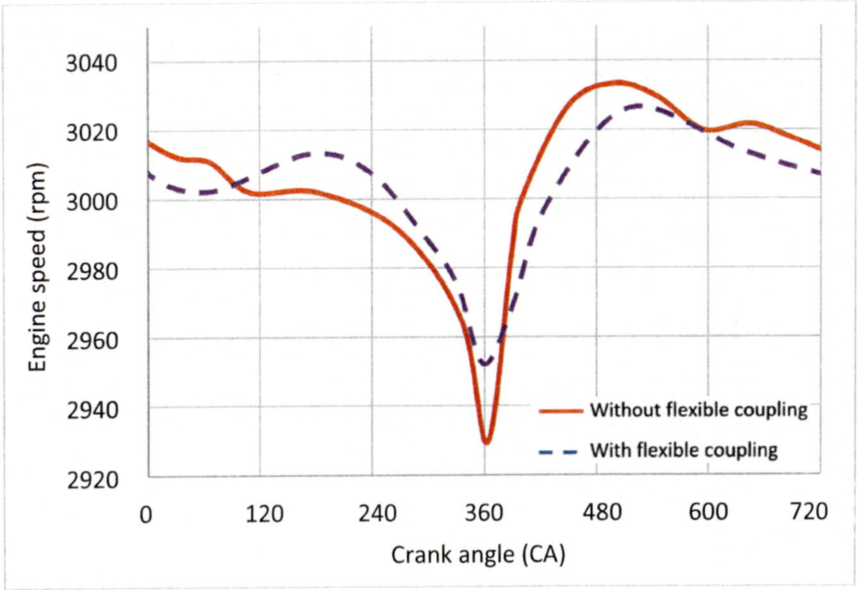

Fig. 5. Simulated rotational irregularity including coupling.

Figure 5 shows a simulated speed curve before and after the elastic coupling element. Simulation is done engine at a speed of 3000 rap corresponding to the maximum torque. It can be observed that the engine speed (rotational regularity) is not constant over two the four engine strokes it decreases during the compression stroke as the pressure is pushing the piston against the rotation direction. The combustion causes the pressure to increase rapidly and therefore increase the piston speed (engine speed). This is called the rotational irregularity of the engine.

When comparing results, it can be seen that the rotational irregularity of the internal combustion engine is transferred to the generator by the flexible coupling at a significantly reduced rate.

5 Conclusions

The operating behavior of a power generator with dynamic loads is to be mapped based on a modular simulation model. The model offers the possibility to vary different engine and flywheel configurations as well as different load scenarios.

Based on thermodynamic-mechanical correlations and individual sub models developed specifically for internal combustion engines, the overall model of a single cylinder, 4-stroke SI engine presented here was built in MATLAB. The results calculated using the model were compared with measurement results from an engine test bench at steady-state operating points. A comparison of the results calculated using the model shows very good agreement with measurement results from an engine test bench at steady-state operating points, which supports the applicability of the selected simulation approach. Simulated speed curve before and after the elastic coupling element. The obtained results shows that the use of a flexible coupling between to transfer power from the engine to the generator significantly reduces the rotational irregularity rate.

Nomenclature

U- internal energy [J]
p- pressure [Pa]
Q_B- energy release from fuel [J/s].
Q_W- wall heat flux [W]
φ- crank angle [°]
CA- crank angle [°]
h- specific enthalpy [J/kg]
m- mass [kg]
T- temperature [K]
v- *volume [m3]*
cv- *specific heat capacity [J/(kg.K)]*
$\dot{H}_{I,O}$- enthalpy current [J/s]
p_i- indexed mean pressure [bar]
μ- flow coefficient [-].
A_v- cylinder area [m^2]
J- moment of inertia [kg.m^2]
α- heat transfer coefficient [W/(m^2.K)]
J_M- moment of inertia motor [kg.m^2]
J_G- moment of inertia generator [kg.m^2]
T_m- the torque from the rotation of both rotating masses [Nm/ rad].
T_d- the damping torque [Nm/rad]
φ_M- torsional angle motor sideways [rad].
φ_G- angle of twist generator sideways [rad]
$\dot{\varphi}_M$- angular velocity motor sideways [rad/s].
$\dot{\varphi}_G$- angular velocity generator side [rad/s].
$\ddot{\varphi}_M$- angular acceleration motor sideways [rad/s^2]
$\ddot{\varphi}_G$- angular acceleration generator side [rad/s^2]
$\dot{m}$ mass flow [kg/s].
R- gas constant [J/(kg K)]
k- isentropic exponent [-]
ω angular velocity [s^{-1}]
A- heat transfer surface [m^2]
d- bore diameter [m]
v_u- swirl velocity [m/s].
v_{mean}- mean piston velocity [m/s]
u- specific internal energy [J/kg].

References

1. Wu, C.-M., Roberts, C.-E., Matthews, R.-D., Hall, M.-J.: Effects of engine speed on combustion in SI engines. Comparisons of Predictions of a Fractal Burning Model with Experimental Data journal of engines **102**, 2277–2291 (1993)
2. Ferguson, C., Kirkpatrick, A.: Internal Combustion Engines: Applied Thermosciences, 3rd edn. Wiley (2015)
3. Brayek, M., Jemni, M.A., Ibraim, A., Damak, A., Driss, Z., Abid, M.S.: Simulation of the effects of heat introduced during combustion on SI engine performance. In: Ben Amar, M., Bouguecha, A., Ghorbel, E., El Mahi, A., Chaari, F., Haddar, M. (eds.) A3M 2021. LNME, pp. 217–230. Springer, Cham (2022). https://doi.org/10.1007/978-3-030-84958-0_24
4. Heywood, J.-B.: Internal Combustion Engine Fundamentals. McGraw-Hill, New York (1988)

5. Mathur, M-L.: Internal combustion engines. Dhanpat Rai Publ (2005)
6. Woschni, G.: A Universally Applicable Equation for the Instantaneous Heat Transfer Coefficient in the Internal Combustion Engine. SAE International, Warrendale, PA (1967)
7. Wu, Y.-Y., Chen, B.-C., Hsieh, F.-C.: Heat transfer model for small-scale air-cooled spark-ignition four-stroke engines. Int. J. Heat Mass Transf. **49**, 3895–3905 (2006)
8. Dashti, M., Asghar, H.-A.: Thermodynamic model for prediction of performance and emission characteristics of SI engine fuelled by gasoline and natural gas with experimental verification. J. Mech. Sci. Technol. **26**(7), 2213–2225 (2012)
9. Chen, S., Flynn, P.: Development of a Single Cylinder Compression Ignition Research Engine. SAE Technical Paper 650733 (1965)

Transient Vibration of the Ship Power Train in Polar Conditions

Zeljan Lozina(✉), Damir Sedlar, and Andela Bartulovic

University of Split-FESB, Split, Croatia
{zeljan.lozina,damir.sedlar,abartulo}@fesb.hr
http://marjan.fesb.hr/∼lozina/

Abstract. The transient vibration of the ship power propulsion system in polar conditions is considered from numerical analysis point of view. Besides linear components the models of the ship power propulsion includes damper components that typically leads to the models that are non-classically damped. Additionally, some components have relatively high ratio of the stiffness and mass and that push higher frequencies to the very high level. That is why implementation of classification rules than impose very small time step which as consequence has low numerical efficacy. The direct integration and modal superposition are compared and implemented to the practical example of the ship power propulsion. The procedure for handling non-classical damping and load which depend on position in modal superposition approach is developed, implemented and compared with direct approach. It is demonstrated that nonlinearities due to non-classical damping and displacement/position dependent forces can be efficiently handled with proposed procedure. The paper includes a case study of the electrically powered propulsion system that includes coupling and gear reducer.

Keywords: Torsional vibration · Non-classical damping · Modal superposition · Transient vibration

1 Introduction

Transient vibration analysis of ship power propulsion system in polar conditions is regulated by classification rules, [1]. Due to specific design and rule requirements direct integration when applied involves very small time step and therefore inefficient numerical integration.

Modal analysis and numerical modal superposition are often favorable approach but typically limited to linear systems, [2]. Under some conditions these procedures limited to linear systems can be efficiently applied to nonlinear ones. The important special case of non-classical damped system excited by position dependent force is considered for the modal superposition procedure.

A mod decomposition in state space is proposed, [3,4], too. However, modal decomposition in state space has blurred physical meaning [5], and therefor here,

Z. Dimitrovová et al. (Eds.): WMVC 2022, MMS 125, pp. 998–1006, 2023.
https://doi.org/10.1007/978-3-031-15758-5_103

we prefer approach in the modal space. The implementation of the ice induced load of non-classically damped ship power train is selected for the practical implementation example.

2 Decoupling Techniques for Non-classically Damped Systems

The linear dynamical models are governed by the Eqs. (1):

$$\mathbf{M}\ddot{\mathbf{q}} + \mathbf{D}\dot{\mathbf{q}} + \mathbf{K}\mathbf{q} = \mathbf{Q}(\mathbf{q}, \mathbf{t}) \tag{1}$$

where:

$\mathbf{M}$ - n × n global mass matrix
$\mathbf{D}$ - n × n global damping matrix
$\mathbf{K}$ - n × n global stiffness matrix
$\mathbf{q}$ - n × 1 displacement vector that depends on time
$\mathbf{Q}(\mathbf{q},t)$ - n × 1 force vector as function of displacement and time

When the damping is the classical one, it can be presented as follows, (2):

$$\mathbf{D} = \alpha\mathbf{M} + \beta\mathbf{K} \tag{2}$$

where α and β - are scalar coefficients selected in order to fit damping matrix $\mathbf{D}$. Let $\boldsymbol{\Phi}$ and $\boldsymbol{\Lambda} = \lceil \omega^2 \rfloor$ be matrices of eigenpairs that correspond to eigenproblem of undamped system (3):

$$(\mathbf{K} - \lambda\mathbf{M})\phi = 0. \tag{3}$$

Additionally, let the eigenvectors be normalized to mass matrix $\mathbf{M}$, i.e.

$$\boldsymbol{\Phi}^T\mathbf{M}\boldsymbol{\Phi} = \mathbf{I}$$

where $\mathbf{I}$ is unitary matrix. In this case, in modal space, the equations of motion read, (4):

$$\ddot{r}_i + (\alpha + \beta\omega_i^2)\dot{r}_i + \omega_i^2 r_i = f_i(\mathbf{q}, t) \tag{4}$$

where:

r_i - are modal coordinates and
$f_i(\mathbf{q}, t)$ - are modal loads

defined with:

$$\mathbf{r}(t) = \boldsymbol{\Phi}^{-1}\mathbf{q}(t)$$
$$\mathbf{f}(\mathbf{q}, t) = \boldsymbol{\Phi}^T\mathbf{Q}(\mathbf{q}, t)$$

In this specific case differential equations of motion are decoupled and rigid body modes and elastic modes can be readily decoupled, presented separately and filtered out when necessary. Modal superposition is proved as efficient and reliable.

2.1 Decoupling Equations of Motion in Modal Space of Non-classically Damped System

Let the system be non-classically damped and let the damping matrix be decomposed in classical and non-classical part, i.e.: $\mathbf{D} = \mathbf{D}_L + \mathbf{D}_{NL}$. In this case, in modal space we have damping split in two matrices: $\mathbf{C} = \mathbf{C}_L + \mathbf{C}_{NL}$ (with $\mathbf{\Phi^T D_L \Phi} = \mathbf{C}_L = \lceil \alpha + \beta\omega_i^2 \rfloor$).

Now, equations of motion become, (5):

$$\mathbf{I\ddot{r}} + \mathbf{C}_L\mathbf{\dot{r}} + \mathbf{\Lambda r} = \mathbf{f(q,t)} - \mathbf{C}_{NL}\mathbf{\dot{r}} \tag{5}$$

where the elements of non-classical damping matrix are in $\mathbf{C}_{NL}$ which becomes part of the right-hand side of equation of motion. Physically, this is regarded as a part of the external force that depends on velocity. The equation can now be reshaped in $N \leq n$ decoupled equations (that are still coupled by nonlinear force component), (6):

$$\ddot{r}_i + 2\xi_i\omega_i\dot{r}_i + \omega_i^2 r_i = f_i(\mathbf{q}, t) - f_{NLi}(\mathbf{\dot{r}}), \quad i = 1, 2, ..., N \tag{6}$$

where proportional damping is assumed:

$$2\xi_i\omega_i = \alpha + \beta\omega_i^2 \tag{7}$$

The coefficients α and β can be calculated e.g. from known relative damping coefficients for two selected modes. Now we have nonlinear part of the non-classical damping matrix (8):

$$\mathbf{C}_{NL} = \mathbf{\Phi}^T(\mathbf{D} - \alpha\mathbf{M} - \beta\mathbf{K})\mathbf{\Phi} \tag{8}$$

and the problem can be solved in modal space. The solution in generalized coordinates can be found (9):

$$\mathbf{q} = \mathbf{\Phi r} \tag{9}$$

where $\mathbf{\Phi} = \mathbf{\Phi}_{N\times M}$, $\mathbf{r} = \mathbf{r}_{M\times 1}$ and M corresponds to number of modal equations involved in solution.

Algorithm 1. Direct algorithm: time dependent force

1: On input: $\mathbf{M}, \mathbf{D}, \mathbf{K}, \mathbf{q_0}, \mathbf{q_{t0}}, \mathbf{F}(t)$
2: Solution of eigenproblem: $(\mathbf{K} - \lambda\mathbf{M})\phi = 0, \quad \mathbf{\Lambda}, \mathbf{\Phi}$
3: Calculate α and β
  For the calculation strategy for α and β see paragraph 2.1
4: Calculate $\mathbf{D}_{NL} = \mathbf{D} - (\alpha\mathbf{M} + \beta\mathbf{K})$
5: Select modal equations
6: Perform modal superposition of transient problem with $\mathbf{f}(t) = \mathbf{\Phi}^T(\mathbf{F}(t) - \mathbf{D}_{NL}\mathbf{q}t)$
  6.1 Select convergence criteria
  6.2 Do nonlinear tightening iterations until convergence
7: On output: $\mathbf{q}(t), \mathbf{q}t(t), \mathbf{q}tt(t)$

2.2 Case Study of Ship Propulsion System

Practical nonlinear illustrative example of non-classically damped system with position dependent load is provided: transient torsional vibration of ship propulsion system in ice condition, [1]. Typical model sketch of the electrically powered system that consists of electromotor, elastic coupling, reducer gear, propeller shaft and propeller is given in Fig. 1

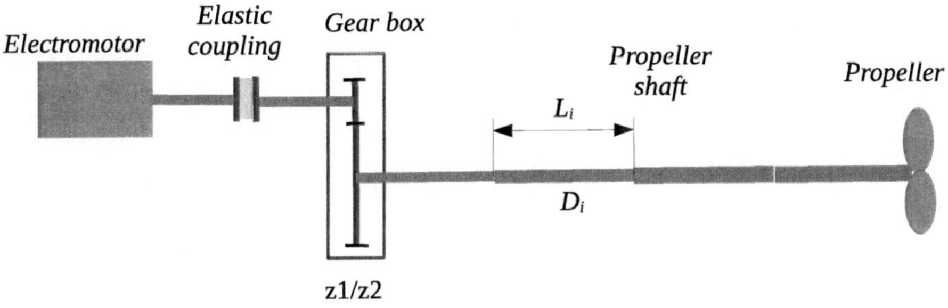

Fig. 1. The sketch of the ship power propulsion system powered with electrical motor

The model that corresponds to the sketch in Fig. 1 is given in Fig. 2.

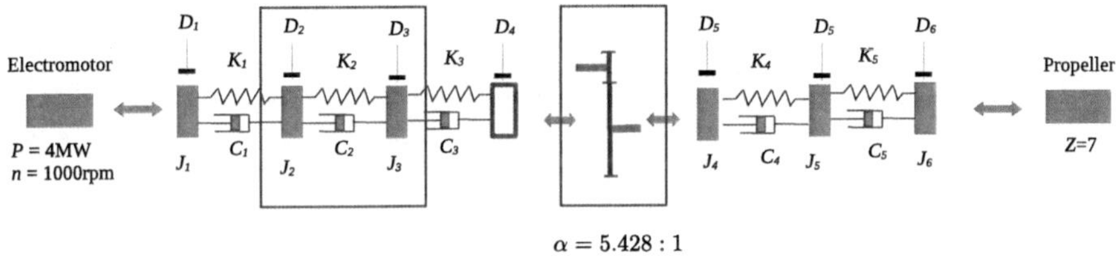

Fig. 2. The model of the considered ship propulsion

The considered model is characterized with electric motor (power 4 MW, MCR 1000 rpm, gear ratio $z_1/z_2 = 21/114$, FBP 7 blades propeller) with data according to Table 1. Transient problem non-classical damped model can be integrated directly or by proposed modified modal superposition. The implemented algorithm is supported with tight results for direct integration and modified modal superposition technique when same integration scheme is implemented, e.g. run up in Fig. 3 and 4.

Table 1. Propulsion model basic data of stiffness, damping and inertia

i		1	2	3	4	5	6
K	Nm	15.0e6	0.504e6	30.0e6	1500.0e6	8.44e6	
C	Nms	2105	1809	4210	21052	1185	
J	kgm^2	200	30	15	900	331.44	6060

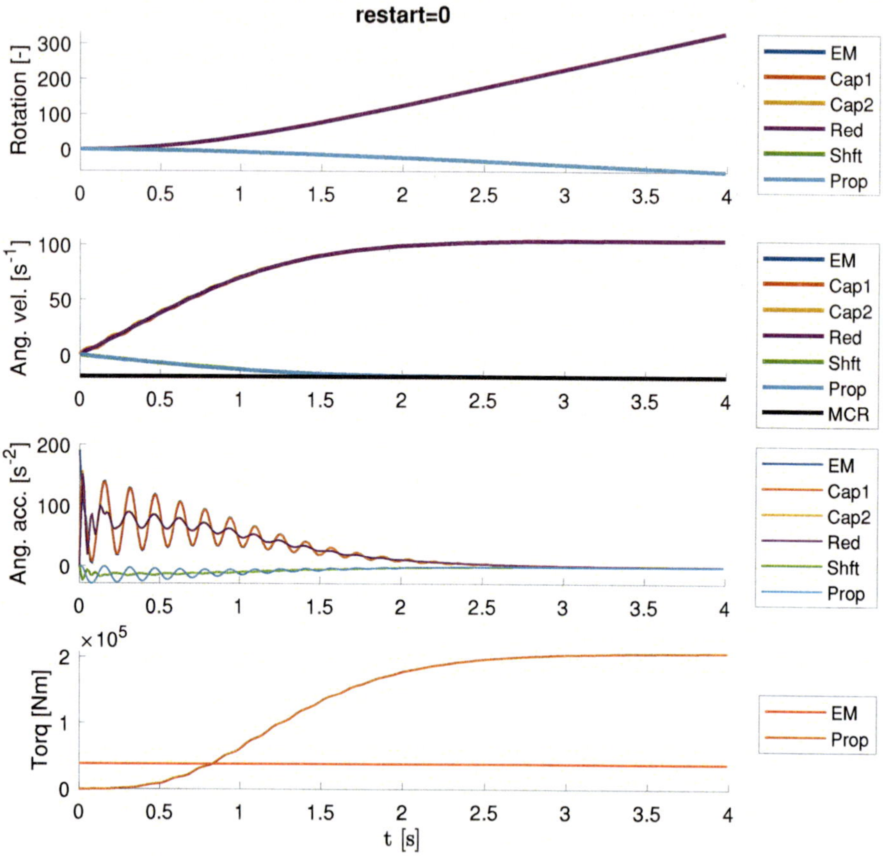

Fig. 3. Run up from zero speed to MCR (EM- electromotor, Cap- capling, Shft- propeller shaft, Prop- propeller, MCR- Maximum Continuous Rate)

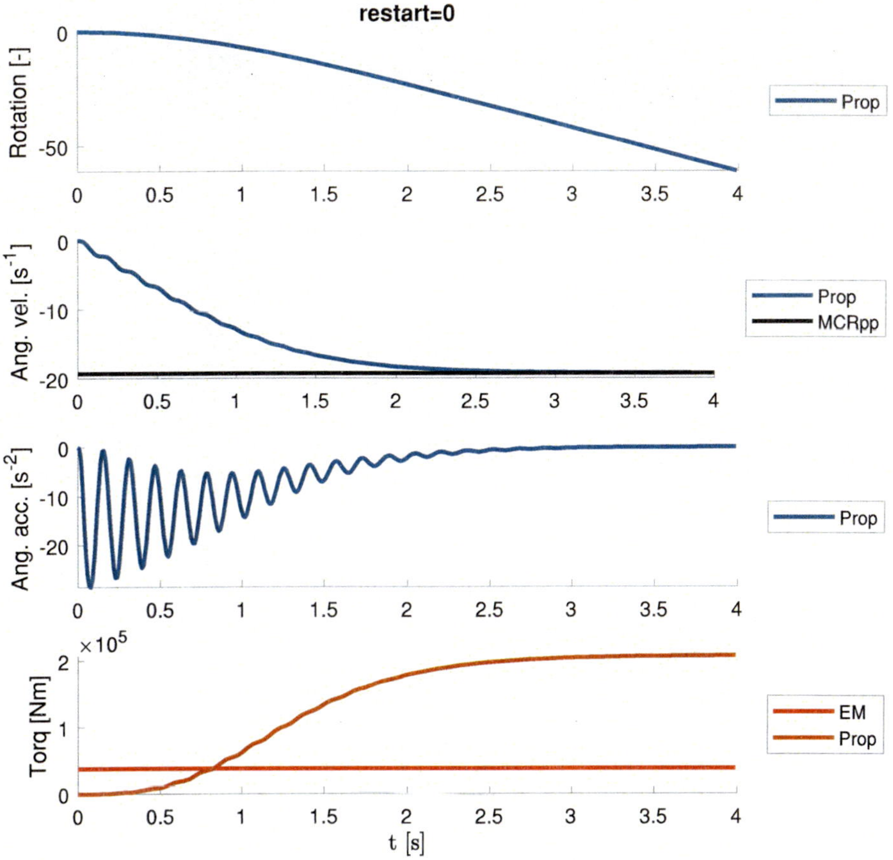

Fig. 4. Run up from zero speed to MCR, propeller shaft only

Modal superposition facilitate filtering of unwanted modes, e.g. rigid body modes, in solution phase and so improve efficiency and transparency of procedure, Fig. 5.

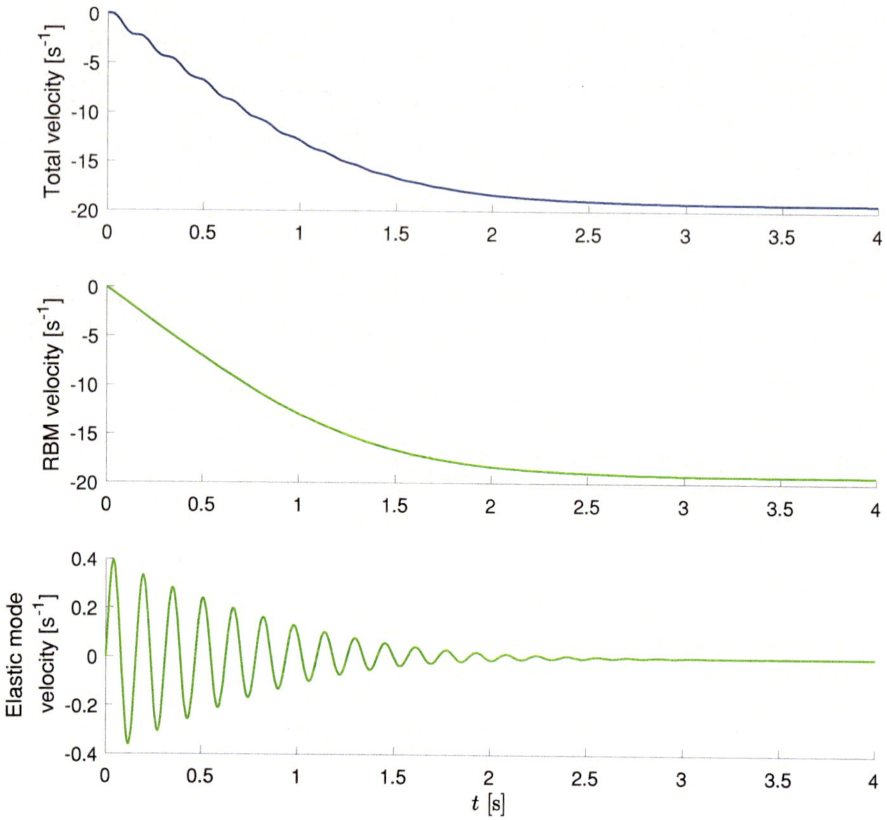

Fig. 5. Run up from zero to MCR: total velocity, rbm velocity, elastic mode velocity (filtered rbm)

The filtering of rigid body modes in solution phase in the case of position dependent forces like ice induced load is not possible because nonlinearity due to load dependence: angle of rotation is major influence parameter in ice load generation. However, decoupled presentation of the rigid body and elastic modes is possible, see Fig. 6.

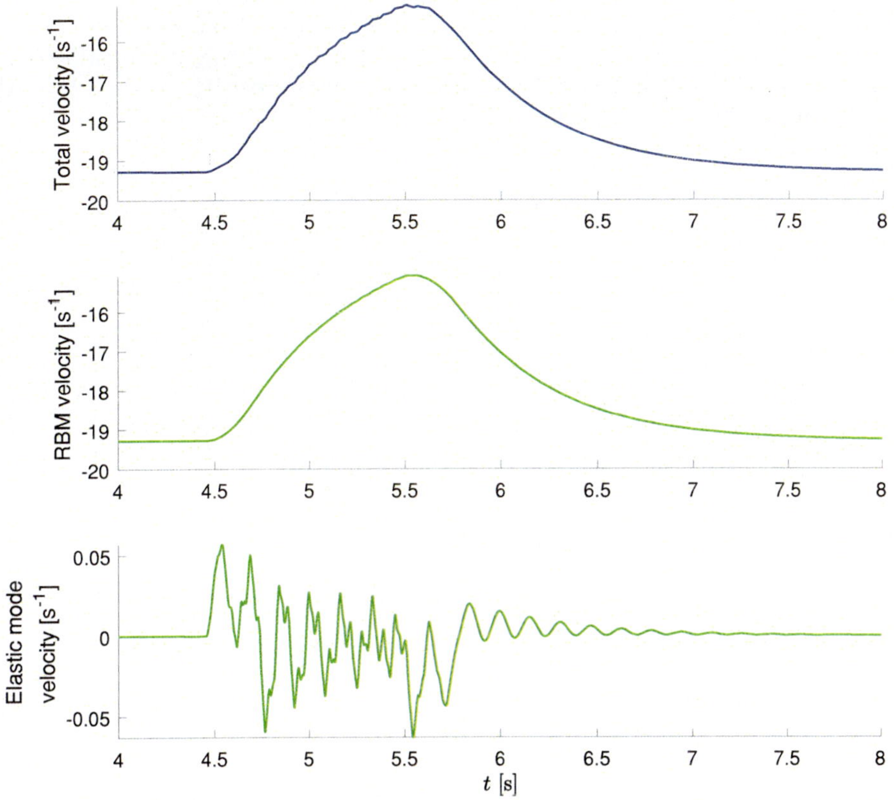

Fig. 6. Ice impacts, "Case 1", angular velocity of the propeller shaft, decoupled contributions of rbm and elastic modes

The requirements for accuracy according [1] enforce very small integration step. Therefore, direct integration is practically possible only in lumped models with low number $dofs$ where model is built keeping in mind that any subsystem with high stiffness and low mass has to be avoided. When modal analysis can be applied an simple and efficient mechanism is available to control complexity of the model and accuracy of the solution.

In the practical example of a ship power propulsion system in ice condition the capacity and limitations of proposed modal analysis of non-classical damped system with position dependent load is illustrated.

3 Conclusion

Modal decoupling of specific nonlinear systems that involves non-classical damping and position conditioned nonlinear external force are considered and solved directly and implementing modal superposition technique. The theoretical background and implementation issues are discussed in details. It is demonstrated that it is possible efficiently apply modal superposition technique in order to have advantages of the modal decomposition and still handle nonlinearities in considered case. Unfortunately, there is not a general procedure for any kind of nonlinear problem. Each case requires carefully tailored procedure. The practical implementation example is provided.

Acknowledgments. This research has been supported by the Croatian Science Foundation grant IP-2018-01-6774, Evolutionary Shape Synthesis with Integral and Partitioned 3D Phenotypes, Dynamic Parameterizations and Meshless Modeling

References

1. DNV-GL: Rules for classification - Ships, Part 6, Ch 6: Cold climate (2020)
2. Bathe, K.J.: Finite Element Procedures. Prentice-Hall, NJ (1996)
3. Caughey, T.K., Ma, F.: Complex modes and solvability of nonclassically damped linear systems. ASME J. Appl. Mech. **60**(5), 26–28 (1993)
4. Ma, F., Imam, A., Morzfeld, M.: The decoupling of damped linear systems in oscillatory free vibration. J. Sound Vibr. **324**(1–2), 408–428 (2009)
5. Ma, F., Morzfeld, M., Imam, A.: The decoupling of damped linear systems in free or forced vibration. J. Sound Vibr. **329**(15), 3182–3202 (2010)

WGA: 1-D and 2-D Waveguides and their Applications

Numerical Development of a Low Height Acoustic Barrier for Railway Noise Mitigation

João Lázaro[1](✉), Matheus Pereira[2], Pedro Alves Costa[1], and Luís Godinho[2]

[1] CONSTRUCT, Faculty of Engineering (FEUP), University of Porto, Rua Dr. Roberto Frias, 4200-465 Porto, Portugal
jlazaro@fe.up.pt

[2] ISISE, Department of Civil Engineering, University of Coimbra, Pólo II, Rua Luís Reis Santos, 3030-788 Coimbra, Portugal

Abstract. Although railway system is the most sustainable mode of transport, with the lowest energy consumption, the noise induced by rail traffic in urban regions is a significant drawback. Mitigation of railway noise can be performed by different solutions, namely by the implementation of acoustic barriers. Although they offer a significant reduction in noise levels, their height makes people feel enclosed. Therefore, in the case of the railway infrastructure, the solution to the problem may lie in the use of barriers with a lower height placed close to the track. The purpose of this paper is to illustrate the development of a barrier solution to be used in a railway context through numerical modelling with Boundary Elements Method. The solutions developed were placed close to the track and have a low height (approximately 0.8m above the rail head). The geometry was defined, as mentioned, using 2D BEM from the numerical simulation of a sound wave. Thus, it was possible to match the inner face of the barrier with the geometry of the wavefront, favouring the normal reflection of the sound waves and directing the energy back to the track to take advantage of the acoustic properties of the ballast. The addition of a porous granular material on the inner face of the barrier, through the numerical simulation of an equivalent fluid and corresponding coupling to the acoustic medium, allows the control of reflections between the vehicle body and the barrier, increasing its acoustic efficiency. The solutions presented show great efficiency in terms of energy loss in the receivers with and without the barrier, i.e., in terms of Insertion Loss, the solutions presented give losses greater than 10 dB over a wide range of frequencies. In the case of the barrier with the presence of porous granular material, the barrier efficiency is even more remarkable with an increase of at least 5 dB compared to the solution without porous granular material.

Keywords: First keyword · Second keyword · Another keyword

Z. Dimitrovová et al. (Eds.): WMVC 2022, MMS 125, pp. 1009–1018, 2023.
https://doi.org/10.1007/978-3-031-15758-5_104

1 Introduction

The objective of the following work is to demonstrate the development and numerical optimisation of an acoustic barrier performed using BEM. The type of barrier that is intended to be developed, of low height, presents itself as a barrier an alternative to traditional acoustic barriers, as proven by the studies of several authors authors [1–5]. Use this numerical model allows to study how the propagation of sound waves [4,6,7] is affected by the presence of an obstacle such as the acoustic barrier and thus understand which solution is best suited to mitigate the noise generated by rail traffic. Firstly, the choice of the frequency content analysed is an extremely important step, since it is a decisive factor. The frequency content was defined 200 Hz and 4000 Hz. In order to approximate as much as possible the numerical studies of the real conditions, the vehicle and track geometry were modelled according to data provided by the company Metro do Porto, SA. To evaluate the different geometries defined was use insertion loss, i.e., the difference in sound pressure levels between the case with and without the barrier. Although the insertion loss calculation is an effective way of systematising the acoustic performance of the barrier, the fact that the receptor mesh is large makes it difficult to analyse each receptor due to the variability of the results. As such, a compromise solution was chosen where the performance criterion involves calculating the area under the insertion loss line in the frequency range previously considered, i.e., 200 Hz and 4000 Hz, as shown in the Fig. 1. It should be noted that this criterion has no physical meaning however it allows for the definition of a single value of the performance criterion and as such facilitates comparison between the various solutions.

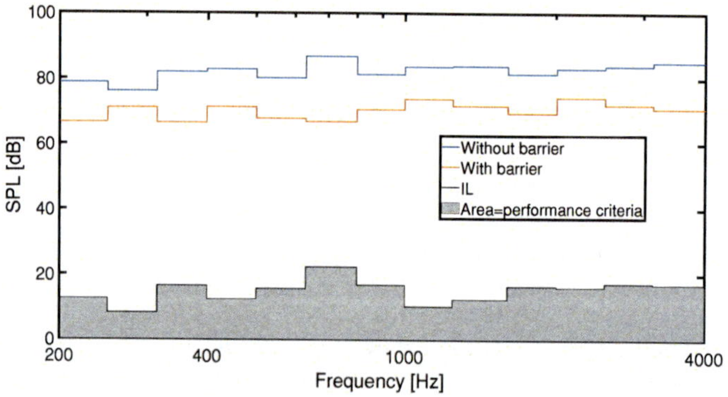

Fig. 1. Illustrative scheme of the evaluation criteria used.

2 Geometry and Positioning of the Barrier

Opting for low height barriers allows designing solutions which do not constitute a visual obstacle for those travelling on the train and for people living near the railway tracks, as well as reducing the noise levels in the vicinity of the source. It is therefore necessary to define the location in relation to other elements of the track where the barrier will be placed and to define its geometry. Taking into account information gathered from the company Metro do Porto, SA, it was decided that the inner face of the barrier had to be at least 1,225 m away from the rail. This dimension is precisely related to safety distances during maintenance operations. The aim of this work is to show the changes in the barrier's geometry in order to understand which parameters most influence its acoustic performance. The variable parameters are, the height, the depth of the barrier and the shape of the inner face of the barrier. The development of the barrier in a first phase involved fixing the height without prejudice to its primary objective, i.e. not to be an obstacle to the field of view. In a second phase, we tried to endow the barrier with geometric features that favour reflection normal to the contact surface and consequently increase the amount of energy that is reflected in the direction opposite to the propagation.In order to evaluate the barrier's behaviour, a grid of 6 receivers was defined (see Fig. 2), trying to represent the places where passers-by and buildings can be found.

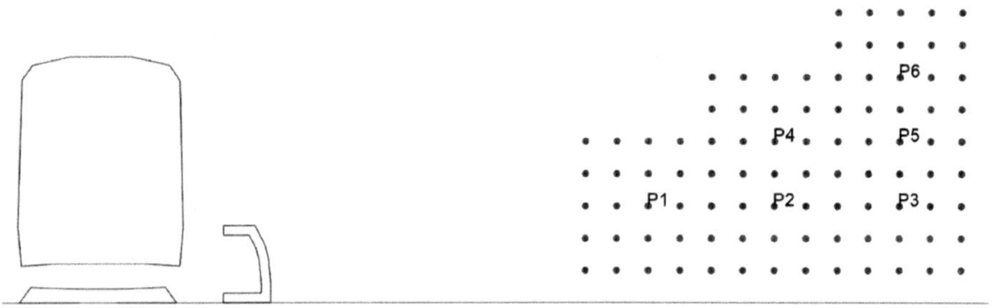

Fig. 2. Schematic representation of the boundaries and the mesh of receivers

2.1 Barrier Height

The studies were thus conducted on the premise, that is the most importance to have a low height, and it was concluded that the total height of the barrier should be about 1.20 m. The first tests were performed with a simple geometry barrier, a vertical barrier 0.15 m thick and whose height varied between 0.8 m and 1.20 m, respectively. The results were perennial and as expected we can observe a vertical displacement of the performance curve as a function of height increment, as represented in Fig. 3. From the point of view of acoustic performance it was proven, as expected, that the higher the barrier the greater is the increase of efficiency. On the other hand, this dimension does not compromise the objective of design barriers that do not constitute an obstacle in the visual field.

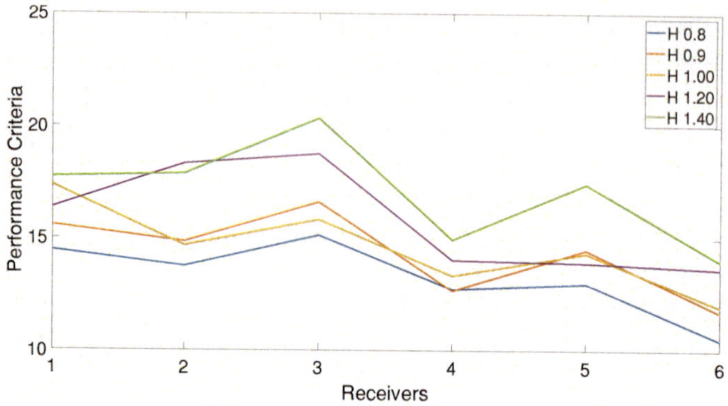

Fig. 3. Performance criterion for the height variation of a vertical barrier.

2.2 Barrier Depth

Taking the conclusions reached in point Sect. 2.1, the barrier depth has been subjected to several changes according to performance improvements. The objective of providing the barriers with depth, instead of being only straight surfaces, arises with the perception that increasing the air volume between the vehicle surface and the barrier promotes a higher energy dissipation. By placing the vertical face of the barrier in a more rearward position and the respective placement of a horizontal surface on top of the barrier, whose limit is at the distance defined above, reflections between the various faces of the barrier are promoted and, as such, the amount of energy transmitted to the receivers is reduced. Given the above, the methodology adopted was to find a compromise solution. In other words, once again the priority was to design a solution with the best possible performance and, on the other hand, not to turn the barrier into an obstacle that would affect the surrounding population. Thus, for a fixed height of 1.20 m, three solutions called U-barriers were tested varying their depth, namely: 0.40 m, 0.50 m and 0.6 m. From the results presented in Fig. 4 it can be seen that there is considerable variability between the various receivers, and it is not possible to observe a trend towards improvement in the performance criteria with increasing depth. This fact means that the following studies must take this phenomenon into account in order to understand how depth can influence the acoustic performance of the barrier.

2.3 Shape of the Inner Face of the Barrier

Given the variability of the results obtained in terms of the depth of the barrier, the following section seeks to combine the evolution of the geometry of the inner face of the barrier with the definition of the depth to be adopted. In the third and last phase of the barrier construction, the main idea is to enforce the properties of the sound waves to achieve a better acoustic performance. The main difference in relation to the previous case is the adaptation of the

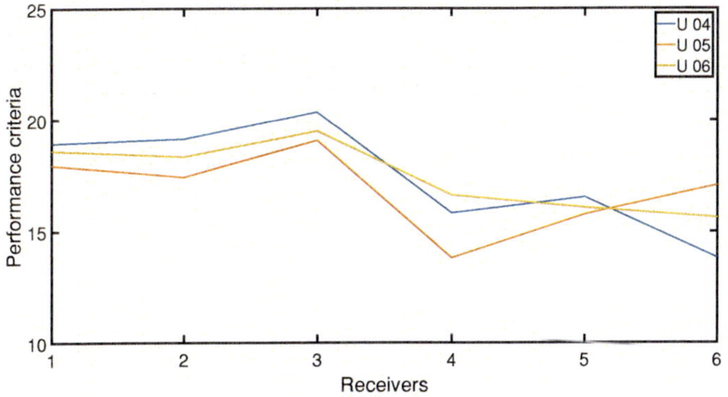

Fig. 4. Performance criteria for the depth variation of the acoustic barrier prototype.

inner face to the wavefront so that the reflection is carried out according to the normal to the barrier's surface, i.e., so that the reflected wave is equal to the incident wave. This phenomenon allows the energy carried by the sound wave to be sent back to its place of origin, where it can be partially absorbed by the track, depending on its acoustic properties. For the definition of the inner face geometry, the simulation of the wave propagation for the case without the barrier was necessary. As explained, the perpendicular reflection of sound waves has important advantages in terms of acoustic performance so it was intended to make the inner face of the barrier coincide with the wave front, as illustrated in Fig. 5.

Fig. 5. Schematic of the wave propagation that allowed the curved geometries of the barriers under analysis to be defined.

Within the two types of barriers tested, it is clear that the 0.6 m depth barriers have a substantially superior behaviour, when compared to the 0.4 m depth barriers. In the Fig. 6 shows that for each of the 6 receivers considered, the two 0.6 m barriers are superior to the 0.4 m barriers.

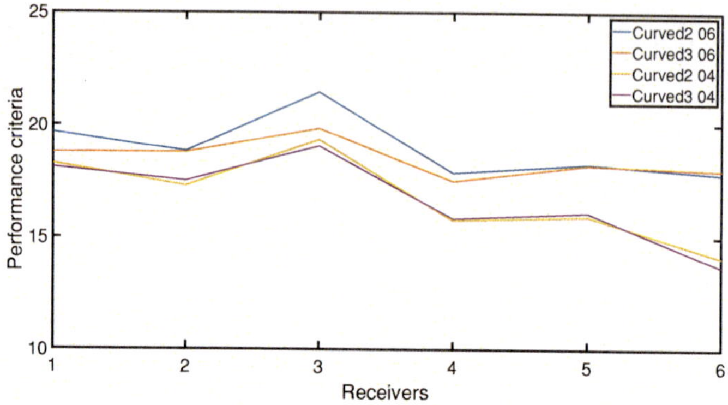

Fig. 6. Performance criteria for two curved barriers with two different depths

Finally, a comparative analysis of the curved barriers with the U-shaped barriers can be made. It can be observed in the Fig. 7 a good efficiency of all the barriers under analysis, however, the curved barriers present overall higher levels of reduction, standing out the higher receivers. Thus, for the two 0.6 m barriers, the one called Curve2 is acoustically more efficient than Curve3, particularly in the lower receptors. For this reason, the Curve2 barrier was selected to continue the study presented here.

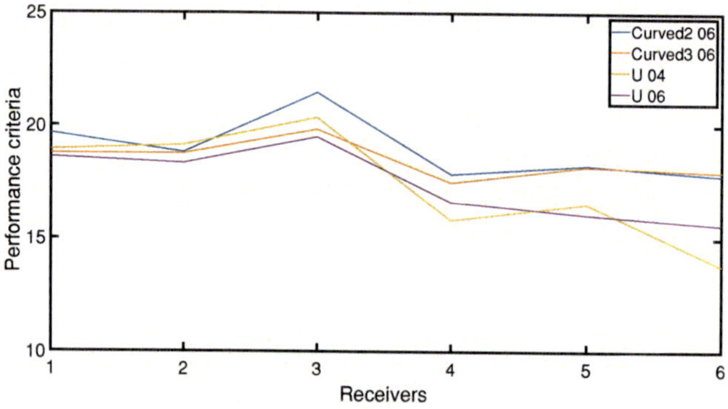

Fig. 7. Comparison between the performance criteria of curved barriers and U-barriers, with better performance.

3 Increase the Acoustic Efficiency with Porous Materials

The efficiency of the acoustic barrier prototype explained so far depends entirely on its geometry. Thus, to increase its efficiency and control reflections between the train and the barrier, the use of porous materials appears to be a very effective way of improving the acoustic behaviour of the barrier. This type of material is suitable for outdoor conditions due to its great durability and resistance to outdoor elements [5,10–18]. In this way, the study that follows refers to the use of porous materials, in this case concrete with expanded clay, as an absorbent material. Porous materials were introduced into the BEM model as an equivalent fluid, and to this end the model of Horoshenkov and Swift [9] was used to prescribe the behaviour of these materials. Given the intrinsic behaviour of porous material, several geometries of porous material were tested in order for it to be more efficient. Irregular geometries allow to improve over the whole frequency range the acoustic absorption behaviour of the porous material [8]. In Fig. 8 the different geometries of the tested porous material are presented.

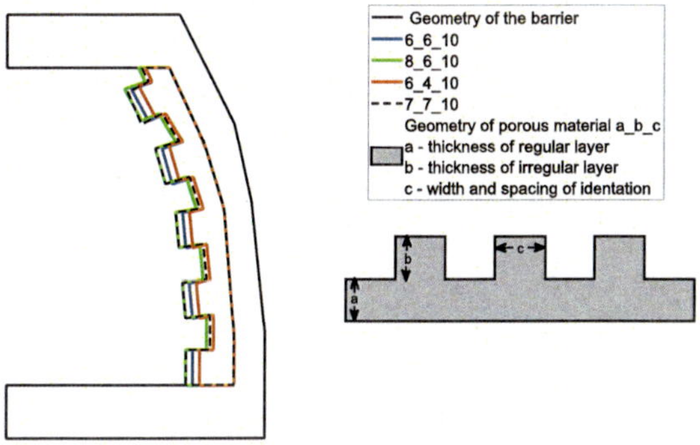

Fig. 8. Comparison between the geometry of the rigid barrier and the barriers with porous material; Details of the geometries of the studied porous material solutions.

Figure 9 shows the results of the performance criteria for each of the defined barriers. Firstly, it is observed that the porous material increases the efficiency of the barrier in a very considerable way, namely in the receptors closer to the ground. It is also possible to observe that increasing the thickness of both the regular layer and the indented part means that overall the barrier's behaviour is improved. Finally, a comparison between solutions 8_6_1 and 7_7_10, which present a total thickness of 0.14m of porous material, must be highlighted, although the former is more efficient than the latter. This assumption makes it clear that the study and design of each of these layers of porous material must be careful in order to obtain the best behaviour according to the scenarios under study.

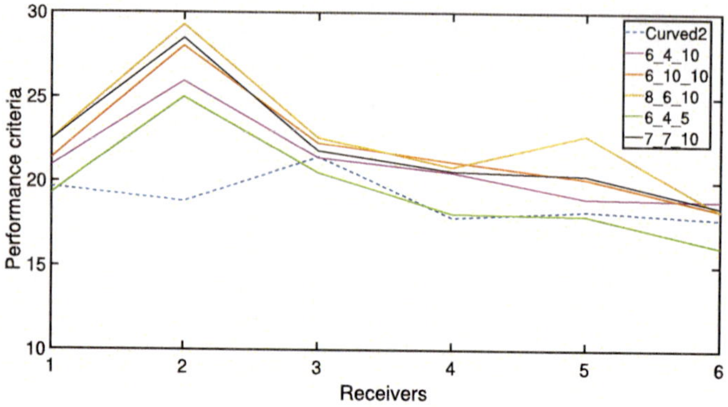

Fig. 9. Performance criteria for the curved barrier and the various porous material solutions studied.

Finally, Fig. 10 presents the comparison between the IL levels introduced by the two barriers with porous material 7_7_10 and 8_6_10 and the curved barrier without porous material, . For this purpose, the average insertion loss calculated on the mesh of receptors presented above was evaluated. Firstly, it is highlighted that the porous material allows increasing the IL by at least 5dB in some frequencies in which the barrier without porous material had a more deficient behaviour, namely 630 Hz, 1000 Hz and 2500 Hz. Among the two solutions with porous material, the solution with geometry 8_6_10 shows a substantial increase in efficiency, compared to the other solution under analysis, corroborating what was already observed in Fig. 9

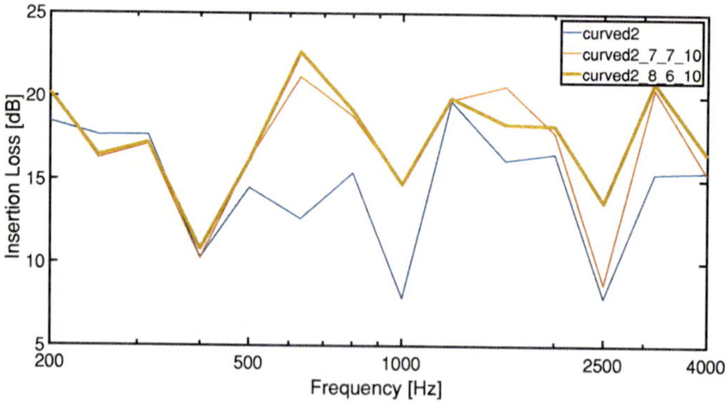

Fig. 10. Average IL for the rigid curved barrier and the two best performing porous material solutions: 7_7_10 e 8_6_10.

4 Conclusions

The presented study makes use of BEM to develop the geometry of an acoustic barrier to be used in the railway context. The low height acoustic barriers, such as the one presented, are solutions that allow mitigating the noise induced by rail traffic without constituting an obstacle in everyday life, both for passers-by and for those who circulate inside the trains. The final solution is a curved solution that takes advantage of the acoustic properties of the track, sending the energy back to the origin through its geometry, which favours the reflections normal to the inner face. Complementarily, in order to increase the acoustic efficiency of the prototype, the use of porous material was studied as a way to control the reflections between the train body and the barrier. It was proved that the use of this kind of material, with a well defined geometry, can increase the IL by 10 dB in certain frequencies where the rigid barrier has a poor performance.

References

1. Koussa, F., Defrance, J., Jean, P., Blanc-Benon, P.: Acoustic performance of gabions noise barriers: numerical and experimental approaches. Appl. Acoust. **74**, 189–197 (2013)
2. Jolibois, A., Defrance, J., Koreneff, H., Jean, P., Duhamel, D., Sparrow, V.W.: In situ measurement of the acoustic performance of a full scale tramway low height noise barrier prototype. Appl. Acoust. **94**, 57–68 (2015)
3. Nieuwenhuizen, E., Yntema, N.: The effect of close proximity , low height barriers on railway noise. In: Proceedings of the Euronoise, Heraklion, Crete, Greece 2018, pp. 1375–1380 (2018)
4. Kasess, C.H., Kreuzer, W., Waubke, H.: Deriving correction functions to model the efficiency of noise barriers with complex shapes using boundary element simulations. Appl. Acoust. **102**, 88–99 (2016)
5. Krezel, Z.A., McManus, K.: Recycled aggregate concrete sound barriers for urban freeways. In: Waste Management Series, vol. 1, pp. 884–892. Elsevier, Harrogate (2000)
6. Crocker, M.J.: Handbook of Noise and Vibration Control. Wiley, Hoboken (2007)
7. Kirkup, S.M.: The Boundary Element Method in Acoustics. Integrated Sound Software. University of Central Lancashire, Preston (2007)
8. Pereira, M., Mareze, P.H., Godinho, L., Amado-Mendes, P., Ramis, J.: Proposal of numerical models to predict the diffuse field sound absorption of finite sized porous materials - BEM and FEM approaches. Appl. Acoust. **180**, 80–92 (2021)
9. Horoshenkov, K.V., Swift, M.: The acoustic properties of granular materials with pore size distribution close to log-normal. J. Acoust. Soc. Am. **110**, 2371–2378 (2001)
10. Magrini, U.; Ricciardi, P. Surface sound acoustical absorption properties of multilayer panels. In: Proceedings of the INTER-NOISE and NOISE-CON Congress and Conference Proceedings, Nice, France, 27–31 August 2000, vol. 2000, pp. 2953–2959. Institute of Noise Control Engineering: Reston (2000)
11. Asdrubali, F., Horoshenkov, K.: The acoustic properties of expanded clay granulates. Build. Acoust. **9**, 85–98 (2002)

12. Vašina, M., Hughes, D., Horoshenkov, K., Lapčík Jr. L.: The acoustical properties of consolidated expanded clay granulates. Appl. Acoust. **67**, 787–796 (2006)
13. Bartolini, R., Filippozzi, S., Princi, E., Schenone, C., Vicini, S.: Acoustic and mechanical properties of expanded clay granulates consolidated by epoxy resin. Appl. Clay Sci. **48**, 460–465 (2010)
14. Kim, H.K., Lee, H.K.: Influence of cement flow and aggregate type on the mechanical and acoustic characteristics of porous concrete. Appl. Acoust. **71**, 607–615 (2010)
15. San Martín, J.C., Esquerdo-Lloret, T.V., Ramis-Soriano, J., Nadal-Gisbert, A.V., Denia, F.D.: Acoustic properties of porous concrete made from arlite and vermiculite lightweight aggregates. Cons. Super. Investig. CientíFicas **65**, e072 (2015)
16. Tie, T.S., Mo, K.H., Putra, A., Loo, S.C., Alengaram, U.J., Ling, T.C.: Sound absorption performance of modified concrete: a review. J. Build. Eng. **30**, 101219 (2020)
17. Pereira, A., Godinho, L., Morais, L.: The acoustic behavior of concrete resonators incorporating absorbing materials. Noise Control Eng. J. **58**, 27–34 (2010)
18. Lázaro, J., Pereira, M., Costa, P.A., Godinho, L.: Performance of low-height railway noise barriers with porous materials. Appl. Sci. **12**, 2960 (2022)

WPA: Wave Propagation in Pipes with Applications

A Simplified Model of the Ground Surface Vibration Arising from a Leaking Pipe

J. M. Muggleton[1(✉)], O. Scussel[1], E. Rustighi[2], M. J. Brennan[3], F. Almeida[4], M. Karimi[5], and P. F. Joseph[1]

[1] Institute of Sound and Vibration Research, University of Southampton, Highfield, Southampton SO17 1BJ, UK
jmm@isvr.soton.ac.uk

[2] Industrial Engineering Department, University of Trento, Via Sommarive, 9-38123 Povo, Trento, Italy

[3] Department of Mechanical Engineering, UNESP, Ilha Solteira, São Paulo 15385-000, Brazil

[4] Department of Mechanical Engineering, UNESP, Bauru, São Paulo 17602-496, Brazil

[5] Centre for Audio, Acoustics and Vibration, University of Technology Sydney, Ultimo, NSW 2007, Australia

Abstract. Acoustic techniques remain the bedrock of pipeline leak detection, particularly for the water industry. The correlation technique, in which leak noise measurements are made at accessible locations on the pipe, either side of the leak, is used world-wide. Unfortunately, especially in the case of plastic pipes, access points are often not spaced closely enough for effective leak detection to take place. An alternative to sensing on the pipe is to measure directly on the ground surface, using discrete sensors such as geophones or accelerometers. However, to do this, the vibrational field on the ground, produced by the leak, needs to be fully understood. The present author, alongside colleagues, has developed an analytical model to show how axisymmetric elastic waves propagating within the pipe radiate to the ground surface. The model, only valid directly above the pipe, shows that, dependent on the soil properties, both a conical shear wave and a conical compressional wave may radiate into the soil, and thence propagate to the ground surface. Moreover, the axial dependence of the ground surface response mirrors the axial dependence of the waves propagating within the pipe. Here, a simplified analytical model of the conical pipe-radiated waves, which encapsulates the essential phase-related features of the more complex development described previously, is presented. This then allows a relatively simple extension to predict the off-axis ground surface as well as that directly above the pipe. Numerical simulations and experimental investigations are also carried out to demonstrate the potentialities of the proposed model to reveal the underlying physics through a simple way.

Keywords: Leak detection · Elastic wave propagation · Vibration · Buried pipeline · Ground surface measurements

Z. Dimitrovová et al. (Eds.): WMVC 2022, MMS 125, pp. 1021–1030, 2023.
https://doi.org/10.1007/978-3-031-15758-5_105

1 Introduction

According to the World Bank, approximately 90 billion litres of water are lost due to leakage globally each day, this representing 30–50% of the world's pumped water [1]. The developed world is responsible for half this amount, with the UK's leakage alone accounting for approximately 3 billion litres per day, or over 3% of that lost globally [2]. Apart from the enormous loss of a treated resource, between 2% and 3% of the world's energy consumption is used to pump and treat water for urban residents and industry [3] so the energy wasted because of leakage corresponds to approximately 1% of the total global carbon footprint.

Whilst other methods are available, and have recently been extensively evaluated [3], acoustic techniques remain the bedrock of leak detection, particularly for the water industry [4, 5]. The correlation technique, in which leak noise measurements are made at accessible locations on the pipe (for example, at hydrants) either side of the leak, is used world-wide. Unfortunately, in the case of plastic pipes, hydrants are often not spaced closely enough for effective leak detection to take place [6].

Acoustic alternatives to sensing on a hydrant or at a convenient access point include: measuring inside the pipe with hydrophones [7]; measuring directly at the ground surface, using discrete sensors such as geophones or accelerometers; measuring below the ground surface using a continuous or distributed sensor, such as a fibre optic cable; or sensing above the ground surface using a non-contact sensor such as a laser vibrometer. The present authors have developed a vibration-based method for locating buried pipes from the ground surface [8], supported by analytical modelling [9, 10], but this has yet to be applied to the detection of leaks.

Leak noise from water (and, indeed, gas and oil) pipes tends to be concentrated at low frequencies [11, 12]. At these frequencies, well below the pipe ring frequency, four wave types are responsible for most of the energy transfer in fluid-filled pipes [13–15]: three axisymmetric waves ($n = 0$) and the $n = 1$ wave, related to beam bending. Of the $n = 0$ waves, the first, termed $s = 1$, is a predominantly fluid-borne wave; the second wave, $s = 2$, is predominantly a compressional wave in the shell; the third wave, $s = 0$, is a torsional wave uncoupled from the fluid. However, the authors' own work on buried plastic water pipes and others' work on a buried, gas-filled steel pipe [16] show that it is the fluid-dominated wave which dominates the radiated response from such structures. Previous work by the present authors, studying the radiation of the fluid-dominated wave into surrounding soil [9, 10] has shown that, depending on the soil properties, both a conical shear wave and a conical compressional wave may radiate into the soil, and thence propagate to the ground surface. Moreover, directly above the pipe, the axial dependence of the ground surface response mirrors the axial dependence of the waves propagating within the pipe. Broadly, if the wavespeed of the wave in the pipe exceeds the respective wavespeed in the soil, the conical wave will radiate. For a typical sandy-type soil, for example, for which both the compressional and shear wavespeeds are low, both waves radiate. For a typical clay-type soil, only the compressional wave radiates. This behaviour has been successfully demonstrated at test sites in Brazil and the UK [17]. However, in addition to the waves radiated from the pipe, the source applied to the pipe (in this case a leak) will also excite body waves directly in the ground. The body waves will radiate spherically from the source location. A portion of the energy of these waves will also reach the measurement points, alongside the waves radiated from the pipe. When measuring at the ground surface, potentially both magnitude and phase information will be useful, although to visualize the wave fields described above, phase is likely to be the more important measure. Lines of constant phase represent wave fronts, encapsulating the relevant time delay information as waves travel from the input (via the pipe or otherwise) to the measurement location. Moreover, unwrapped phase is extremely robust in the presence of noise [8].

Here we present a simplified analytical model of the conical pipe-radiated waves which encapsulates the essential phase-related features of the more complex development described in [9, 10]. This then allows a relatively simple extension to predict the off-axis ground surface as well as that directly above the pipe.

The paper is organized as follows. Followed by this introduction Sect. 2 presents the simplified model for the ground-surface response along with some theoretical and numerical investigations. Experimental investigations are then conducted in Sect. 3. Finally, the paper is closed with some conclusions and suggestions for future avenues of research in Sect. 4.

2 Simplified Model

A model of the expected ground surface phase response resulting from a leaking buried plastic water pipe is developed here. This model is based on the observed features of the response of analytical models developed previously [19–23]. We have established that there are body waves arising from excitation of the pipe due to a leak, which are radiated from the pipe towards the ground surface response as showed in Fig. 1.

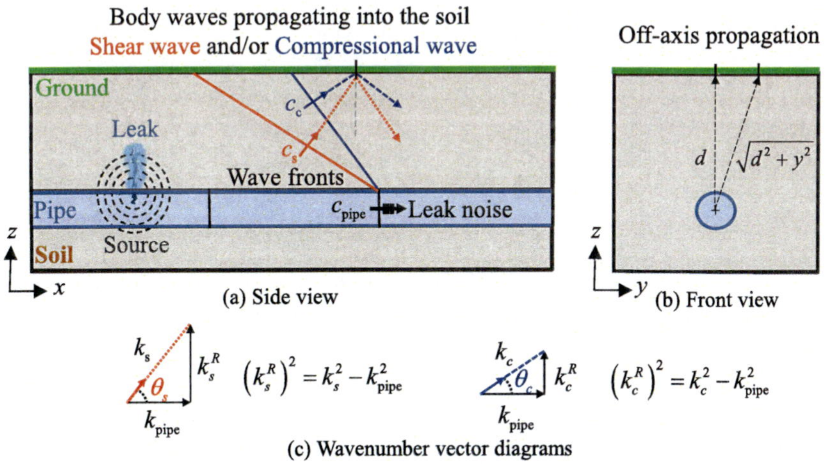

Fig. 1. Waves from a Leaking Pipe: (a) side view, (b) front view, (c) wavenumber diagrams. The terms c_S and c_C denote the wavespeed of shear and compressional wave respectively.

2.1 Radiated Conical Waves

In [9, 10] and [23], expressions are derived for the ground surface vibration resulting from the axisymmetric, fluid-dominated ($n = 0$, $s = 1$) propagating in a buried pipe. The in-axis horizontal and vertical ground surface displacements, U_x and U_z, directly above the pipe are given as [10]:

$$\begin{Bmatrix} U_x \\ U_z \end{Bmatrix} = \sqrt{\frac{2}{\pi k_c^R d}} e^{\mathrm{i}\pi/4} \mathbf{u}_c A_m \mathrm{e}^{-\mathrm{i}k_c^R d} e^{-ik_{\text{pipe}} x} + \mathrm{i}\sqrt{\frac{2k_s^R}{\pi d}} e^{\mathrm{i}\pi/4} \mathbf{u}_s B_m \mathrm{e}^{-\mathrm{i}k_s^R d} e^{-ik_{\text{pipe}} x} \quad (1)$$

where: d is the pipe depth; A_m and B_m are the compressional and shear wave potentials respectively; $\mathbf{u}_c$ and $\mathbf{u}_s$ are compressional and shear wave amplitude vectors incorporating the effects of reflection at the ground surface and functions of the axial and radial components of the shear and compressional wavenumbers in the ground only; and these surrounding medium compressional and shear radial wavenumbers k_c^R and k_s^R respectively are given by $\left(k_c^R\right)^2 = k_c^2 - k_{\text{pipe}}^2$ and $\left(k_s^R\right)^2 = k_s^2 - k_{\text{pipe}}^2$; k_c and k_s are the compressional (dilatational) and shear (rotational) wavenumbers in the surrounding soil, and k_{pipe} is the axial wavenumber in the pipe. Valid at frequencies for which $\left|k_c^R d\right| > 1$ and $\left|k_s^R d\right| > 1$, i.e. in the far field, a key observation regarding Eq. (1) may be made. Provided that the magnitude terms A_m and B_m, $\mathbf{u}_c$ and $\mathbf{u}_s$ vary slowly with frequency, the phase of the response at the ground surface is controlled by the terms $e^{-ik_c^R d}$ and $e^{-ik_s^R d}$, , i.e. the phase at the surface is largely the same as had the ground surface not been present. Closer examination of the forms of the magnitude terms (Eqs. (9), (18) and (20) in [9]), shows that they do indeed vary slowly with frequency compared with the phase terms above.

An important consequence of noting that the phase of the ground surface response can be determined on the basis of outgoing waves only is that the phase of the off-axis ground surface response can be estimated. This necessarily neglects the surface waves that can be generated at lateral distances from the pipe axis comparable or greater than the pipe depth [24, 25].

However, it is anticipated that, due to the high attenuation in soils, the effect of these will be small. By incorporating the axial and time dependences, and with reference to Fig. 1(b), at a lateral distance from the pipe axis (off-axis), the horizontal, perpendicular and vertical ground-surface displacements can now be re-expressed as:

$$U_x(x, y, d) = \left(u_{xc} e^{-ik_c^R \sqrt{y^2+d^2}} + u_{xs} e^{-ik_s^R \sqrt{y^2+d^2}}\right) \sqrt{\frac{a}{\sqrt{y^2+d^2}}} e^{-ik_{\text{pipe}} x} \quad (2a)$$

$$U_y(x, y, d) = \left(u_{yc} e^{-ik_c^R \sqrt{y^2+d^2}} + u_{ys} e^{-ik_s^R \sqrt{y^2+d^2}}\right) \sqrt{\frac{y^2 a}{\left(y^2+d^2\right)^{\frac{3}{2}}}} e^{-ik_{\text{pipe}} x} \quad (2b)$$

$$U_z(x, y, d) = \left(u_{zc} e^{-ik_c^R \sqrt{y^2+d^2}} + u_{zs} e^{-ik_s^R \sqrt{y^2+d^2}}\right) \sqrt{\frac{d^2 a}{\left(y^2+d^2\right)^{\frac{3}{2}}}} e^{-ik_{\text{pipe}} x} \quad (2c)$$

where a is the pipe radius and the u's are complex wave magnitudes (functions of the compressional and shear wavenumbers and the pipe radius). Wave attenuation is encapsulated in the exponential terms in which the wavenumbers comprise both real and imaginary components. This convenient form acknowledges both the cylindrically spreading nature of the waves and the dominant phase dependence. Moreover, it is a similar form to that previously derived in [15] when only one wave radiates (in this case a torsional wave). A 3D view of the conical waves is depicted in Fig. 2 by two cones describing the propagation of each wave type: shear wave (inner cone in red) and compressional wave (outer cone in blue). Depending on the soil properties, a conical shear wave and/or a conical compressional wave may propagate into the soil. If the speed of the leak noise wave in the pipe exceeds the respective wavespeed in the soil, the conical wave will then radiate towards the ground surface. The wave front of each conical wave has a certain angle of propagation which relates the wavespeed of the leak noise in the pipe with the wavespeed in the soil as depicted in Fig. 1(c) by the wavenumber diagrams. Attenuation, both in the pipe and in the soil, may be accounted for by introducing imaginary components into the wavenumbers k_c, k_s and k_{pipe}.

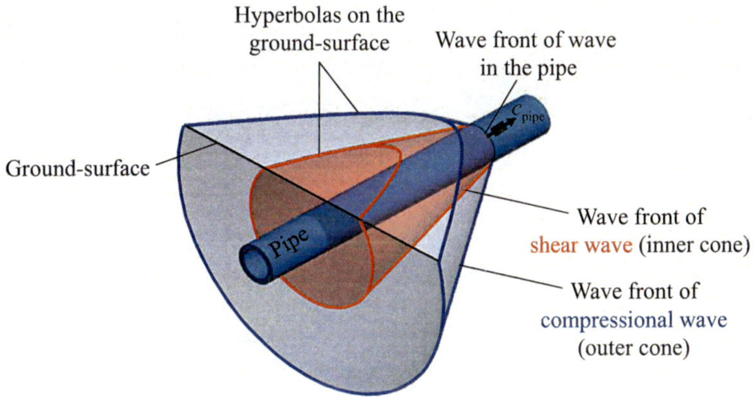

Fig. 2. Typical conical wave intersections with the surface of the ground.

2.2 Investigations Using Theoretical Model and Finite Element Model

In order to estimate the ground surface responses using both theoretical and numerical models, a rectangular grid of points was considered. The investigations were performed by simulating only responses in the vertical direction, given by Eq. 2(c), at various position over a grid of 32 × 7 points 0.5 m apart (axial and lateral distances) on the layer above a plastic water pipe which is buried at a depth of 1 m. Effects from reflected waves were neglected for simplification purposes.

The finite element model (FEM), implemented in COMSOL Multiphysics commercial software, describes a pipe system modelled as a four-part system comprising water, the pipe, the surrounding soil and a layer that corresponds to the ground. A monopole

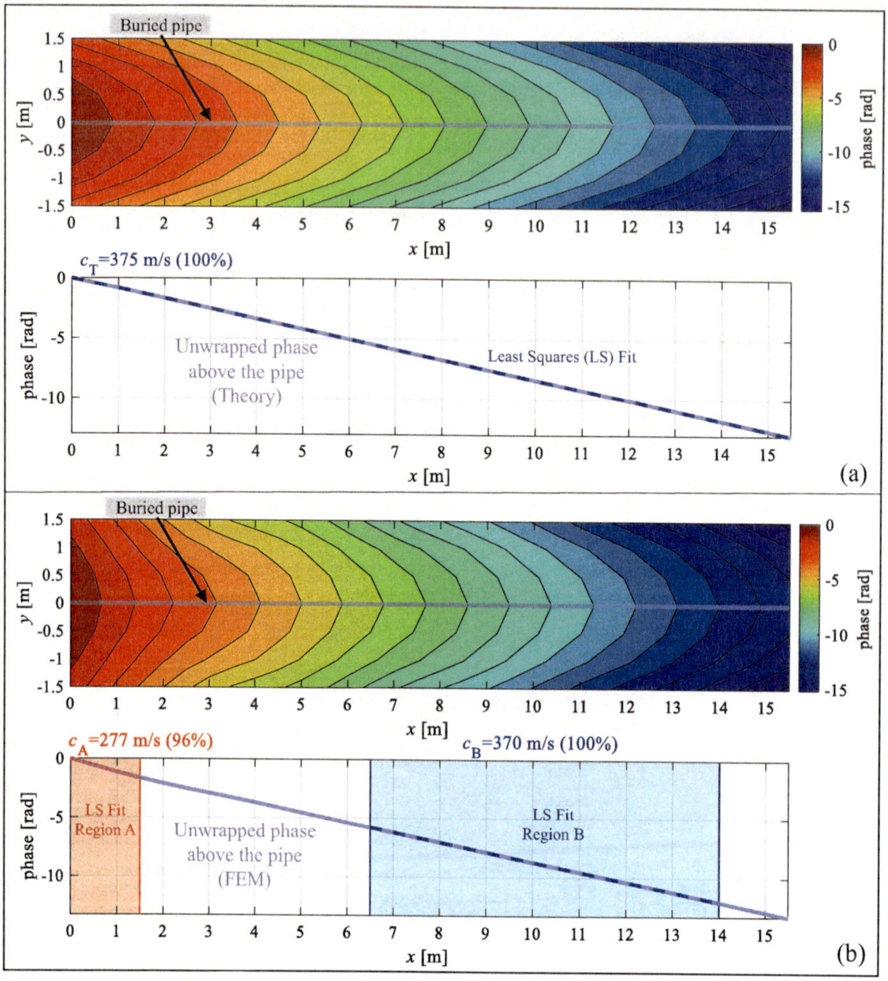

Fig. 3. Contour plots of spatially unwrapped phase for ground-surface response (top), phase gradients right above a plastic pipe (bottom) at 56 Hz: (a) Theoretical model; (b) FEM. The term C_T is the wavespeed predicted above the pipe using least squares fit.

pulsating source was placed 1 m away from the first row of sensor to excite the fluid-dominated wave (s = 1) in the pipe. A fine mesh of triangular elements was used in the discretisation of the model along with a perfectly matched layer (PML) applied on the boundary of the computational domain to simulate an infinite medium.

The phase analysis was performed by using a two-dimensional unwrapping algorithm applied over the entire grid for both theoretical and numerical ground surface responses. Overall, the contour plots represent wave fronts. Figure 3 shows the results for a particular frequency of 56 Hz. The wavespeed predicted is showed below the contour plot together the metric Value of Fit (VoF) which measures in % the accuracy of the results. For the FEM results, the analysis is divided into two regions: region A that has influence from the monopole source and region B where the prediction matches with the wavespeed in the pipe.

3 Experimental Results Using a Test Rig Located in the UK

A typical problem of mapping and locating a buried plastic pipe using ground vibration measurements is depicted in Fig. 4. The test site is located near Blithfield reservoir in Staffordshire in the UK and consists of a 120 m long medium-density polyethylene (MDPE) pipe buried at a depth of about 1 m. A leak was generated by opening the valve and the vibration signals were measured by accelerometers 1 and 2 to predict the wavespeed of $s = 1$ wave in the pipe, which was found to be around 375 m/s.

The ground surface is grass and the soil in which the pipe is buried, typically found in this region, is a mixture of gravel, sand and clay. Only main features of the test rig were included here, and more details about the test rig can be found in [18]. The cross power spectral density (CPSD) between the velocity measured by each geophone right above the pipe and the acceleration measured at the hydrant by the source was calculated for the entire grid of points as depicted in Fig. 4. The spatially unwrapped phases using the radial responses were then estimated. Results using the geophones that are positioned directly above the pipe (when $y = 0$) are shown below each contour plot as a function of axial distance along the pipe.

At a first glance, the run of the pipe can be detected and interaction between different wave types radiating into the surrounding medium are evident for distances up to 5 m away from the source as showed by the phase contour plots. In order to estimate the wavespeed using the ground responses, a least-squares fit was applied based on the linear interpolation of the measured phase gradients for region A (up to 1.5 m) as well as for the region B (from 6.5 m up to 14 m). The jumps observed in the phase plots arise due to several reasons such as external noise, discontinuities from wave interactions and effects of reflections. Moreover, some uncertainties present in the soil also affected the quality of the measured data. The results provided evidence that after some distance from the source, approximately 7 m, only the pipe wave dominates as verified in both theoretical and numerical examples. For region A (up to 2.5 m away from the source) there is a combination of waves from the source and body waves.

Overall, it has been found that, when calculating the CPSD between the vibrational velocity on the ground surface and the vibration at the source, the phase information is important and the contour plots of the spatially unwrapped phase on the ground surface can be an effective tool to reveal the location of the pipe. As can be observed in the phase contour plots, the run of the pipe using the experimental data is less evident and noisier compared to the theoretical and numerical results, since some effects mentioned earlier were neglected in both models.

Another marginal difference is the excitation mechanism adopted in each case. In Blithfield test rig, a standpipe was attached to a hydrant to reproduce a leak as depicted in Fig. 4, and their dynamical effects have not been included in both theoretical and numerical simulations. Furthermore, by examining the gradients of the spatially unwrapped phase it is possible to estimate the wavespeed on the ground-surface which is mostly dominated by the $s = 1$ wave, allowing to identify which type of wave propagating from the pipe within the soil is responsible for the ground vibration.

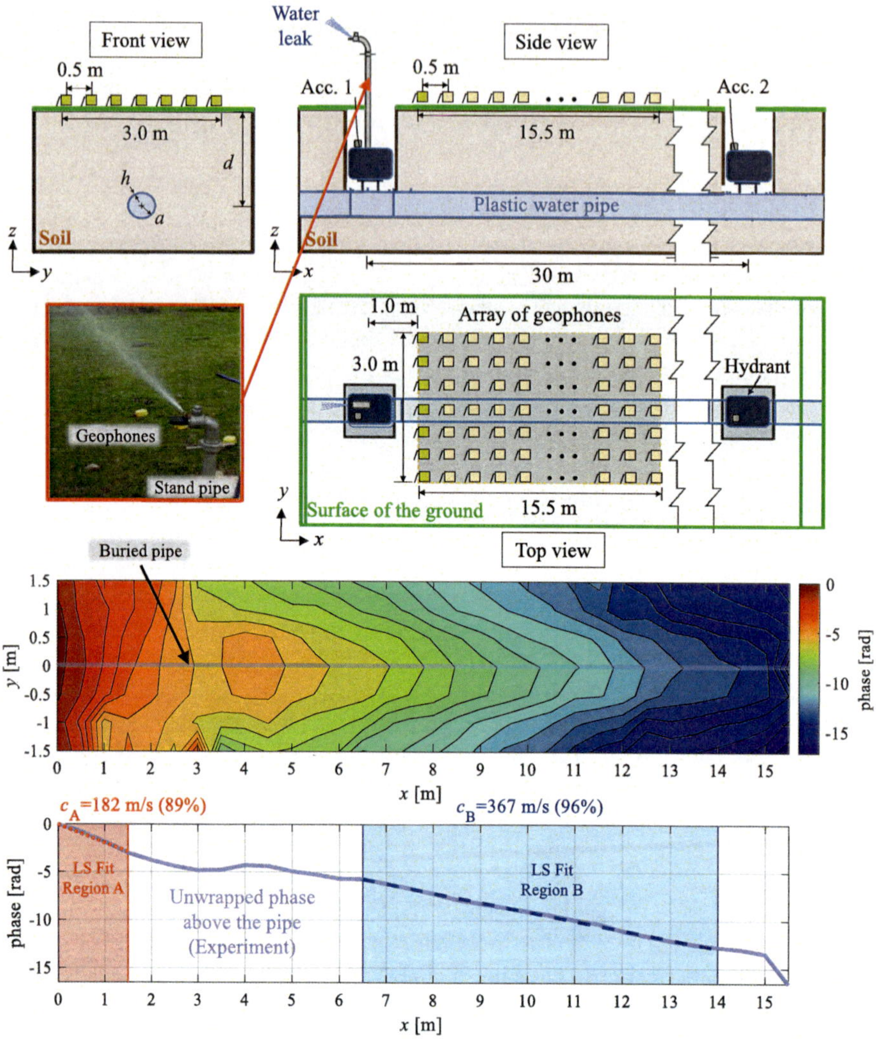

Fig. 4. Schematic diagram of the test rig in Blithfield-UK (not to scale) and the contour plots of spatially unwrapped phase for the grid of geophones at a frequency of 56 Hz together with the phase gradients using the geophones positioned right above the buried plastic pipe.

4 Conclusions

In this paper, a simplified approach for understanding the interaction of waves propagating from a plastic buried pipe towards the surface of the ground has been presented. This allows to exploit features from the leak noise wave radiation using ground vibration response phase data. Physical insight on the interaction of waves is given for a better understanding about how the soil properties are acting on the propagation characteristics of a leak noise wave travelling along the pipe-wall and propagating into the surrounding soil towards the ground surface. In addressing such a problem, there is always likely to be a trade-off between the need to provide a mathematical model which is sophisticated enough to encapsulate the key elements of the observed behaviour, yet simple enough to be of general applicability and amenable to straightforward interpretation. Moreover, it is important to limit both the number of unknown parameters and the overall size of the parameter space to reduce overall computational costs, but perhaps more crucially, to increase the likelihood of identifying globally optimal solutions, rather than local minima.

Acknowledgements. The authors gratefully acknowledge the financial support provided by AP Sensing, the EPSRC under RAINDROP (EP/V028111/1) and Fundação de Amparo à Pesquisa do Estado de São Paulo, process 2020/12251-1.

References

1. https://blogs.worldbank.org/water/what-non-revenue-water-how-can-we-reduce-it-better-water-service. Accessed 14 Mar 2022
2. www.discoverwater.co.uk/leaking-pipes. Accessed 14 Mar 2022
3. pdf.usaid.gov/pdf_docs/PNACT993.pdf. Accessed 14 Mar 2022
4. www.ukwir.org/eng/ukwir-big-question-zero-leakage. Accessed 14 Mar 2022
5. Fuchs, H.V., Riehle, R.: Ten years of experience with leak detection by acoustic signal analysis. Appl. Acoust. **33**, 1–19 (1991)
6. Hunaidi, O., Chu, W.T.: Acoustical characteristics of leak signals in plastic water distribution pipes. Appl. Acoust. **58**, 235–254 (1999)
7. Hunaidi, O.: Detecting leaks in water-distribution pipes. J. Am. Water Works Assoc. **92**, 82–94 (2000)
8. Khulief, Y.A., Khalifa, A., Mansour, R.B., Habib, M.A.: Acoustic detection of leaks in water pipelines using measurements inside pipe. J. Pipeline Syst. Eng. Practice **3**(2), 47–54 (2012)
9. Muggleton, J.M., Brennan, M.J., Gao, Y.: Determining the location of underground plastic water pipes from measurements of ground vibration. J. Appl. Geophys. **75**, 54–61 (2011)
10. Gao, Y., Muggleton, J.M., Liu, Y., Rustighi, E.: An analytical model of ground surface vibration due to axisymmetric wave motion in buried fluid-filled pipes. J. Sound Vib. **395**, 142–159 (2017)
11. Gao, Y., Muggleton, J.M., Liu, Y., Rustighi, E.: Corrigendum to an analytical model of ground surface vibration due to axisymmetric wave motion in buried fluid-filled pipes. J. Sound Vib. **423**, 520–525 (2018)
12. Meng, L., Yuxing, L., Wuchang, W., Juntao, F.: Experimental study on leak detection and location for gas pipeline based on acoustic method. J. Loss Prev. Process Ind. **25**(1), 90–102 (2012)
13. https://www.researchgate.net/publication/267598694_Monitoring_Acoustic_Noise_in_Steel_Pipelines. Accessed 14 Mar 2022
14. Pinnington, R.J., Briscoe, A.R.: Externally applied sensor for axisymmetric waves in a fluid filled pipe. J. Sound Vib. **173**(4), 503–516 (1994)
15. Fuller, C.R., Fahy, F.J.: Characteristics of wave propagation and energy distributions in cylindrical elastic shells filled with fluid. J. Sound Vib. **81**(4), 501–518 (1982)
16. Muggleton, J.M., Kalkowski, M.K., Gao, Y., Rustighi, E.: A theoretical study of the fundamental torsional wave in buried pipes for pipeline condition assessment and monitoring. J. Sound Vib. **374**, 155–171 (2016)

17. Jette, A.N., Parker, J.G.: Surface displacements accompanying the propagation of acoustic waves within an underground pipe. J. Sound Vib. **69**(2), 265–274 (1980)
18. Brennan, M.J., et al.: On the effects of soil properties on leak noise propagation in plastic water distribution pipes. J. Sound Vib. **427**, 120–133 (2018)
19. Muggleton, J.M., Rustighi, E.: Mapping the Underworld: recent developments in vibro-acoustic techniques to locate buried infrastructure. Géotechnique Lett. **3**(3), 137–141 (2013)
20. Muggleton, J.M., Brennan, M.J., Pinnington, R.J.: Wavenumber prediction of waves in buried pipes for water leak detection. J. Sound Vib. **249**(5), 939–954 (2002)
21. Muggleton, J.M., Brennan, M.J., Linford, P.W.: Axisymmetric wave propagation in fluid-filled pipes: wavenumber measurements in-vacuo and buried pipes. J. Sound Vib. **270**(1–2), 171–190 (2004)
22. Muggleton, J.M., Brennan, M.J.: Leak noise propagation and attenuation in submerged plastic water pipes. J. Sound Vib. **278**(3), 527–537 (2004)
23. Muggleton, J.M., Yan, J.: Wavenumber prediction and measurement of axisymmetric waves in buried fluid-filled pipes: inclusion of shear coupling at a lubricated pipe/soil interface. J. Sound Vib. **332**(5), 1216–1230 (2013)
24. Gao, Y., Sui, F., Muggleton, J.M., Yang, J.: Simplified dispersion relationships for fluid-dominated axisymmetric wave motion in buried fluid-filled pipes. J. Sound Vib. **375**, 386–402 (2016)
25. Jette, A.N., Parker, J.G.: Excitation of an elastic half-space by a buried line source of conical waves. J. Sound Vib. **67**(4), 523–531 (1979)
26. Ewing, W.M., Jardetzky, W.S., Press, F.: Elastic Waves in Layered Media. McGraw-Hill, New York (1957)

An Investigation into the Factors Affecting the Bandwidth of Measured Leak Noise in Buried Plastic Water Pipes

Oscar Scussel[1](✉), Michael J. Brennan[2], Fabricio C. L. de Almeida[3,4], Mauricio K. Iwanaga[2], Jennifer M. Muggleton[1], Phillip F. Joseph[1], and Yan Gao[5]

[1] Institute of Sound and Vibration Research, University of Southampton, Highfield, Southampton SO17 1BJ, UK
O.Scussel@soton.ac.uk

[2] Department of Mechanical Engineering, UNESP-FEIS, Ilha Solteira, São Paulo 15385-000, Brazil

[3] Faculty of Science and Engineering, UNESP-FCE, Tupã, São Paulo 17602-496, Brazil

[4] Department of Mechanical Engineering, UNESP-FEB, Bauru, São Paulo 17033-360, Brazil

[5] Key Laboratory of Noise and Vibration Research, Institute of Acoustics, Chinese Academy of Sciences, Beijing, China

Abstract. Inspection and preservation of buried water pipelines is of importance in the modern world. The wastage of water due to leaks is a global problem and existing technologies/methods to detect leaks in buried pipelines still face challenges, such as how to predict the bandwidth of measured leak noise using acoustic correlators, and what are the main factors affecting this frequency range. The leak noise bandwidth is useful information for operators to know before carrying out tests in the field, and currently there is no practical way of predicting this frequency range. This paper presents an approach to predict the bandwidth and investigates the main factors affecting it such as the distance between the sensors, wave speed and attenuation of the fluid-dominated wave, which is the main carrier of leak noise. To achieve this, a water-pipe-soil-sensor model is represented in terms of filters, allowing an investigation into the corresponding physical/geometric characteristics that affect the bandwidth of the measured leak noise. It is shown that the dominant factors are the material and geometry of the pipe, the properties of the surrounding soil and the type of transducer used.

Keywords: Leak noise bandwidth · Buried pipelines · Vibro-acoustic correlators · Water distribution systems

1 Introduction

The loss of water from distribution systems due to leakage and many other causes is a global problem, and is a major challenge for water utility companies [1]. In order to detect and locate such leaks several techniques have been developed throughout the last decades, each with its pros and cons [2]. This paper is concerned with one particular

Z. Dimitrovová et al. (Eds.): WMVC 2022, MMS 125, pp. 1031–1038, 2023.
https://doi.org/10.1007/978-3-031-15758-5_106

technique, that of locating leaks using a leak noise correlator. This device has been used for many years [2] by sensing acoustic pressure in the pipe or pipe vibration either side of a leak. The peak in the cross-correlation of the signals gives the difference in arrival times of the leak noise at the sensor positions, and this is combined with the distance between the sensors and the wave speed of the leak noise to determine the location of the leak [3]. The leak noise is filtered as it propagates along the pipe towards the sensors and the spectrum of the noise is further shaped by the frequency response characteristics of the sensors. The combination of the filtering effect of the pipe system and the sensor effectively acts as either a low-pass filter or a band-pass filter depending on the sensors used [4]. Since these articles have been published, more comprehensive models of the pipe system including the effects of the soil have been developed [5, 6] and investigated experimentally [7]. Moreover, a procedure for estimating soil properties using vibro-acoustic measurements in/on the pipe has been recently proposed in [8]. Most of the investigations conducted so far point towards the importance of estimating the bandwidth in which there is leak noise content, along with the main factors affecting it.

The bandwidth of leak noise varies widely depending upon the type of pipe, its location and the type of sensors used. From a practical standpoint, it is useful for operators to know approximately what this frequency range is before carrying out tests in the field. Currently, the only way to predict this is by performing measurements in the field, which can be time-consuming and cumbersome as it requires a dedicated experimental setup. This paper resolves this problem by using a simple analytical model of the water pipe system coupled with the frequency response of the sensors. The aim of this paper is to investigate the ways in which the physical parameters (pipe geometry/properties, soil properties and sensor type) of a submerged/buried plastic water pipe system affect the cross-power spectral density (CPSD) of measured signals, and hence how they affect the bandwidth of the leak noise which is then used to determine the location of a leak by an acoustic correlator.

This paper is organzied as follows. Following this introduction, Sect. 2 presents an approach for prediction of the bandwidth of measured leak noise together with discussions on the main factors affecting it. Some numerical examples related to typical measurements in the field are carried out using plastic pipes and different surrounding media: submerged in water and buried in different types of soil. In Sect. 3, three experimental data sets from different parts of the world are also analysed to determine the efficacy of the model in predicting the bandwidth of measured leak noise. The predictions are then compared with the bandwidth given by the coherence function between the measured data. The paper is then closed with some conclusions in Sect. 4.

2 Factors Affecting the Bandwidth of Measured Leak Noise

An alternative way of representing a typical water leak detection problem is to describe each component in terms of a filter so the entire system simply becomes a cascade of filters as shown in Fig. 1. Either side of the leak, three of the filters respectively relate to the water-filled massless in-air pipe (pipe stiffness), the pipe mass, and the external medium. One filter is also related to the time it takes for the leak noise to propagate from the leak position towards the measurement position (a pure time delay) and one is related to the dynamics of the transducer, which is shown as an accelerometer in Fig. 1. The filters that affect the bandwidth of measured leak noise are the ones related to the transducer and those related to the pipe as well as the surrounding medium. To illustrate the filtering effects of the various components of the system, the modulus of the CPSD related to the system in Fig. 1, not including accelerometers is considered. This is effectively related to the pressure measurements and is given by $e^{-\omega\beta d_i} = e^{-\omega(\beta_1+\beta_2+\beta_3)d_i}$, where $i = 1,2$.

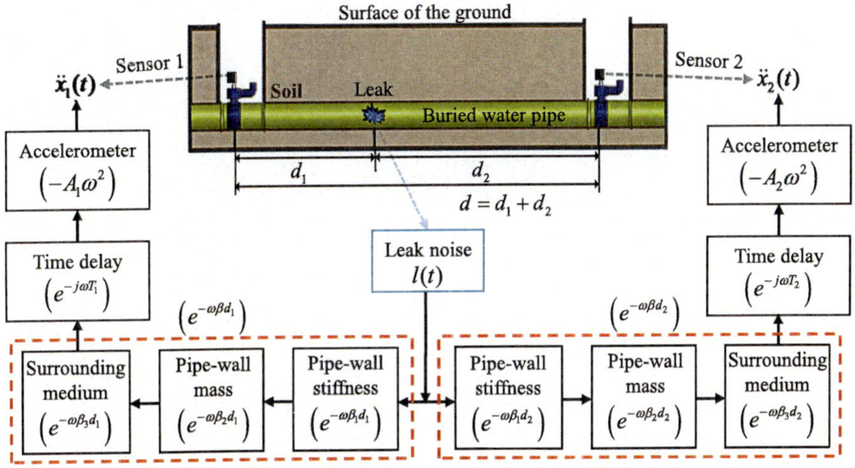

Fig. 1. The water-pipe-soil-sensor system modelled as a set of filters connected in series.

The term β is a measure of the attenuation of the leak noise as it propagates along the pipe and the contribution of each filter is given by.

$\beta_1 = -\mathrm{Im}\{k_{\beta_1}\}$, $\beta_2 = -\mathrm{Im}\{k_{\beta_2}\}$ and $\beta_3 = -\mathrm{Im}\{k_{\beta_3} - k_{\beta_2} - k_{\beta_1}\}$ (1a,b,c).

in which $k_{\beta_1} = k_{\mathrm{water}}\left(1 + \frac{K_{\mathrm{water}}}{K_{\mathrm{pipe\_massless}}}\right)^{\frac{1}{2}}$, $k_{\beta_2} = k_{\mathrm{water}}\left(1 + \frac{K_{\mathrm{water}}}{K_{\mathrm{pipe}}}\right)^{\frac{1}{2}}$ and $k_{\beta_3} = k_{\mathrm{water}}\left(1 + \frac{K_{\mathrm{water}}}{K_{\mathrm{pipe}}+K_{\mathrm{medium}}}\right)^{\frac{1}{2}}$, where k_{water} is the wavenumber of the water within the pipe, K_{water} is the dynamic stiffness of the water, $K_{\mathrm{pipe\_massless}}$ and K_{pipe} are the dynamic stiffnesses of the pipe without inertial term and including inertial term, respectively. The term K_{medium} corresponds to the dynamic stiffness of the surrounding medium [7]. In Figs. 2(ai-iii), the CPSDs for different combination of filters are plotted for an external medium of water, stiff clay soil and sandy soil respectively, with distance being arbitrarily set to 20 m. The chosen nominal values for the plastic pipe system can be found in [9] for the case in which the pipe is submerged in water, and in [10] for the cases where the pipe is buried in either stiff clay soil or sandy soil.

The first filter is related to the mass-less *in-vacuo* pipe, the second one involves the addition of mass, and the third one further involves the addition of the external medium. Note they are plotted on a dB scale so that the different low-pass behaviour in each case can be distinguished easily. It is clear from the figures, that the external medium is a dominant factor in each case. The system with sandy soil surrounding the pipe has the largest value of β, and thus has the most pro-found effect. It has been shown that

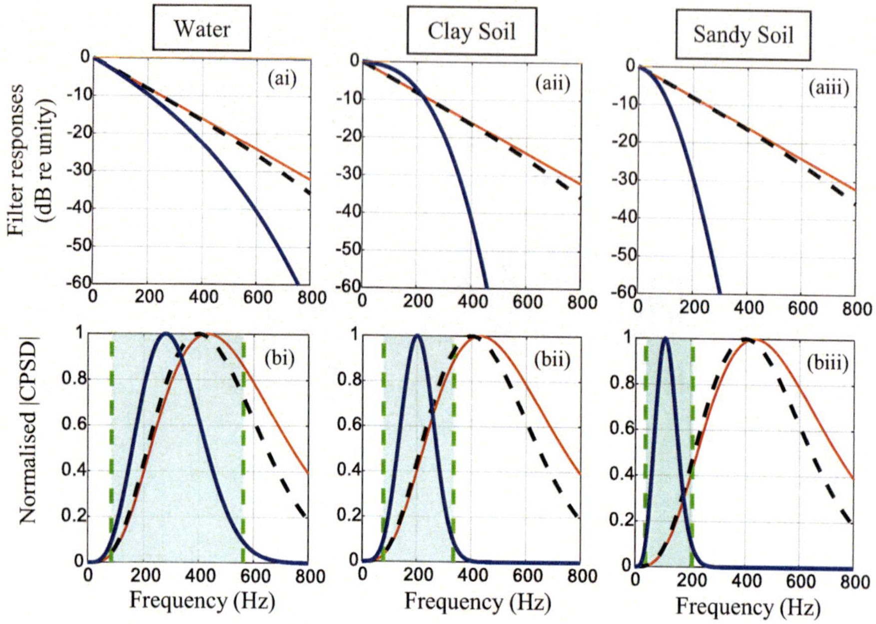

Fig. 2. Behaviour of the water pipe system modelled as a set of filters connected in series, with the measurement points 20 m apart. (a) modulus of the response of the massless *in-vacuo* pipe, (thin solid red line); modulus of the response of the *in-vacuo* pipe with inertia term, (thick dashed black line); modulus of the response of the pipe surrounded by the medium, (thick solid blue line). (b) modulus of the CPSDs corresponding to the systems in (a), including accelerometers at each measurement point. The green shaded region bounded by vertical dashed line denotes the bandwidth for the complete system (pipe, surrounding medium, accelerometers) in which the modulus of the CPSD is within 10 dB of the maximum. In [11], a webpage for frequency bandwidth estimation is given where it is possible to set parameters such as: type of sensor, distance between the sensors, pipe geometry/properties as well as type of surrounding medium: air or soil.

β generally increases due to the external medium, and hence its effects are to limit the measured leak noise to lower frequencies as showed in Figs. 2(ai-iii).

The CPSDs are then combined with the frequency response functions (FRFs) for the accelerometers and are plotted in a normalized form (by the maximum value) where the modulus of the CPSD is greater than one tenth of its maximum value (within 10 dB of the maximum). It can be seen in Figs. 2(bi-iii) that there is a gradual shifting to a lower frequency of this bandwidth, as the mass effect of the pipe and the external medium are accounted for. It is also clear that there is a progressive shift to a lower frequency bandwidth (and narrowing of the bandwidth) when the external medium changes from water to stiff clay soil and further to sandy soil. This occurs due to the following reasons: (i) Shear stiffness is introduced into the model by changing the surrounding medium from water to soil. (ii) Bulk and shear modulus of soil are much smaller in sandy soil than in clay soil, consequently both shear and dilatational waves radiate from the pipe creating a large radiation damping effect on the pipe.

3 Experimental Investigations

To investigate whether the bandwidth of leak noise can be predicted using the approach described in Sect. 2, some experimental data is compared with predictions given by the water-pipe-soil model. Moreover, to evaluate the robustness of the model, experimental data from three test rigs with very different conditions are considered. The test rigs are located in Blithfield-UK, Ottawa-Canada, and São Paulo-Brazil. Their schematic diagrams and photos depicting the leak noise mechanism along with the access points are showed in Fig. 3

.

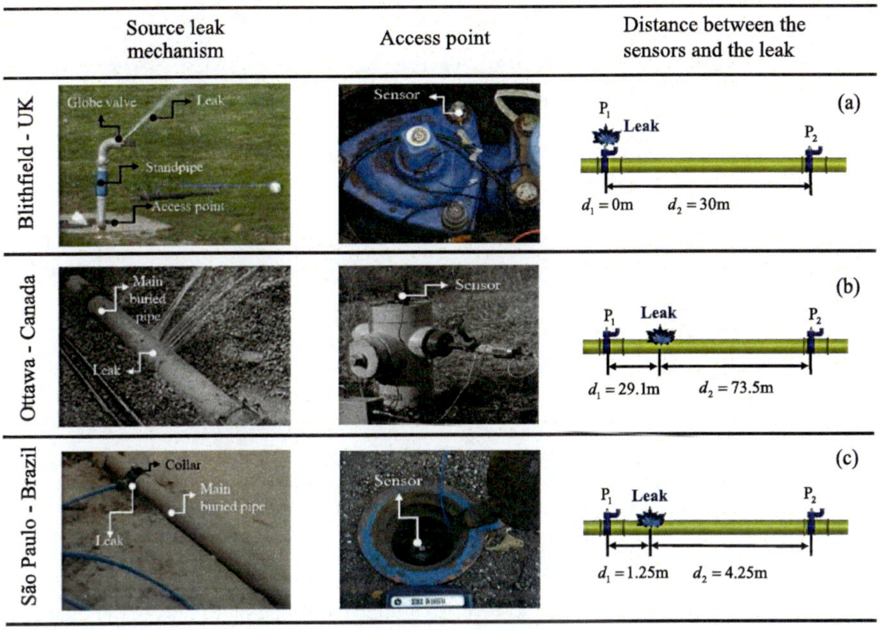

Fig. 3. Schematic diagrams of the three test-rigs (not to scale) showing the source leak mechanism, the access points and corresponding distance between the sensors: (a) Blithfield-UK (b) Ottawa-Canada, and (c) São Paulo-Brazil.

The processed experimental data is plotted in Fig. 4 for all three test rigs. Details of these test rigs have been reported previously (Blithfield [4, 10], Ottawa [3, 10], São Paulo [7, 8, 10]). For each test rig, there are 3 graphs corresponding to the modulus of the CPSD normalized by its maximum value, the phase of the CPSD, and the coherence. Also plotted in the CPSD graphs are the predicted quantities for the parameters given in [10]. The experimental data contains leak noise within certain frequency bands because of the band-pass filtering effects of the pipe-sensor systems and measurement noise.

The frequency bands for each case are determined approximately by examining the coherence function and are denoted as a shaded area in each case whose edges are marked with vertical dotted red lines. Also plotted on the CPSD graphs are the frequency

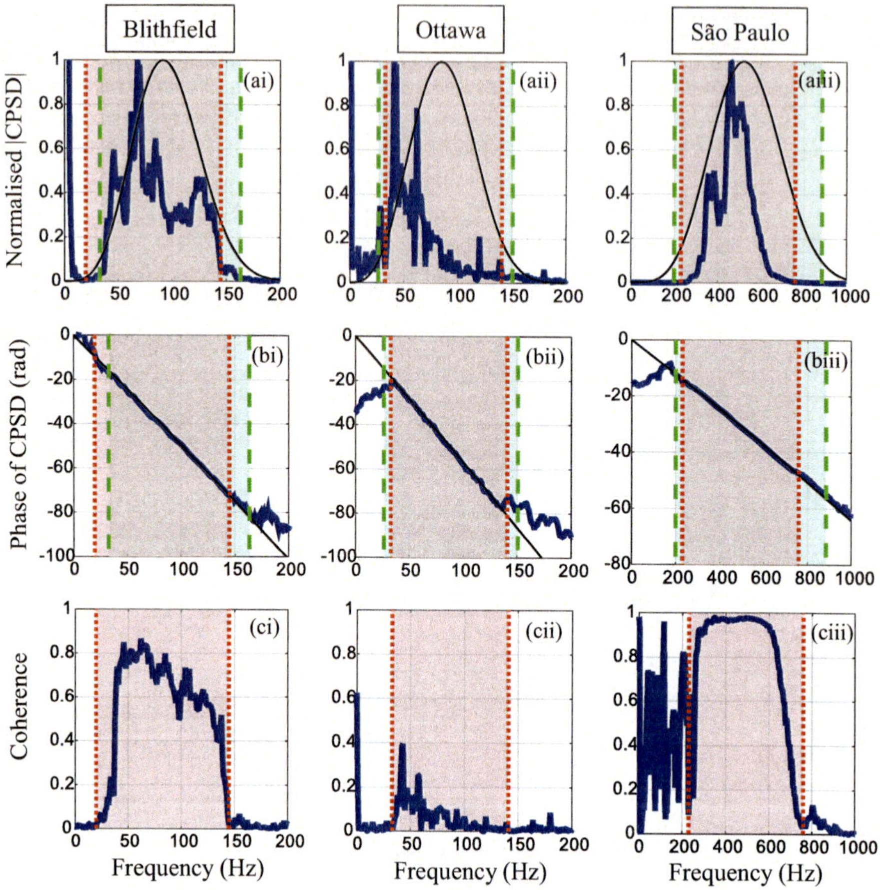

Fig. 4. Comparison of measurements made on the Blithfield, Ottawa and São Paulo sites and predictions made using the model with the parameters given in [10]. (a) normalized modulus of the CPSD, (b) phase of the CPSD, (c) coherence; Measurements (thick solid blue lines); Predictions using the water-pipe-soil-sensor model (thin solid black lines). The shaded region bounded by the thick dotted red lines denotes the bandwidth where there is good coherence; The shaded area bounded by the thick dashed green lines denotes the bandwidth predicted by the model.

ranges in which the moduli of the CPSDs are within 10 dB of their maximum values. These are also shaded areas whose edges are marked with vertical dashed green lines. Examining Figs. 4, it can be seen for all the sites that the differences in the modulus of CPSD between the experimental data and the model are due to measurement noise and unmodelled effects, such as the ground surface, inhomogeneity of the soil, uncertain interface between the pipe and soil, pipe connections/reflections, etc. However, the model still predicts the bandwidths over which there is leak noise adequately for practical applications. Furthermore, note that the model captures the phase behavior of leak noise content reasonably well within the predicted frequency ranges. The three main factors that govern such frequency ranges are the distance between the sensors, the wave-speed and the attenuation of the fluid dominated wave. The distance between the sensors affects the corresponding center frequency and does not affect the upper and lower frequencies of this range. Note that a large (small) distance between the sensors and a small (large) shear stiffness of the soil, which affect both wave-speed and wave attenuation, results in a low (high) centre frequency. Furthermore, it is important to note that the attenuation of the wave in the pipe due to energy leakage into the external medium has a profound effect on the bandwidth of the leak noise, both in terms of the centre frequency and the upper and lower frequencies of the frequency range. If a substantial amount of vibrational energy propagates into the external medium, then this significantly lowers the center frequency of the CPSD and consequently reduces the bandwidth.

4 Conclusions

This paper has shown how the physical parameters of the water-pipe-soil-sensor system affect the bandwidth of measured leak noise. To achieve this, a simple analytical model of the pipe system coupled with the frequency response of the sensor has been integrated into a model of the CPSD between the two sensors. Using accelerometers as sensors, the influence of the various physical parameters on the CPSD, and hence the frequency bandwidth for three types of surrounding medium, namely water, stiff clay soil and sandy soil have been investigated. It has been shown that the external media has a profound effect on the frequency range in which leak noise is measured. The approach to predict the bandwidth of the leak noise presented in this paper has been validated using experimental data from three different test sites and compared with results predicted by the coherence function between the measured signals. The main findings of this paper are relevant for enhancing existing techniques to locate a leak more accurately and could be potentially incorporated into acoustic correlators.

Acknowledgments. The authors are grateful for the financial support provided by the Engineering and Physical Sciences Research Council (EPSRC) under reference EP/V028111/1, entitled RAINDROP: tRansforming Acoustic SensINg for leak detection in trunk mains and water DistRibutiOn Pipelines. The authors also would like to thank the São Paulo Research Foundation (FAPESP) under Grant numbers 2018/25360–3 and 2019/00745–2. Oscar Scussel is grateful for the support from Coordination for the Improvement of Higher Education Personnel (CAPES) under Grant number 88887.374001/2019–00. The authors also thank the Brazilian water and waste management company (Sabesp) for part-funding this work and for providing one of the test rigs, and Osama Hunaidi from the National Research Council of Canada who provided the test data for the Ottawa test rig.

References

1. Hu, Z., Chen, B., Chen, W., Tan, D., Shen, D.: Review of model-based and data-driven approaches for leak detection and location in water distribution systems. Water Supply **21**(7), 3282–3306 (2021). https://doi.org/10.1016/j.jsv.2003.08.045
2. Hu, Z., Tariq, S., Zayed, T.: A comprehensive review of acoustic based leak localization method in pressurized pipelines. Mech. Syst. Signal Process. **161**, 107994 (2021). https://doi.org/10.1016/j.ymssp.2021.107994

3. Gao, Y., Brennan, M.J., Joseph, P.F., Muggleton, J.M., Hunaidi, O.: A model of the correlation of leak noise in buried plastic water pipes. J. Sound Vib. **277**, 133–148 (2004). https://doi.org/10.1016/j.jsv.2003.08.045
4. Almeida, F.C.L., Brennan, M.J., Joseph, P.F., Whitfield, S., Dray, S., Paschoalini, A.T.: On the acoustic filtering of the pipe and sensor in a buried plastic water pipe and its effect on leak detection: an experimental investigation. Sensors **14**, 5595–5610 (2014). https://doi.org/10.3390/s140305595
5. Muggleton, J.M., Yan, J.: Wavenumber prediction and measurement for buried fluid-filled pipes: inclusion of shear coupling at a lubricated pipe/soil interface. J. Sound Vib. **332**(5), 1216–1230 (2013). https://doi.org/10.1016/j.jsv.2012.10.024
6. Gao, Y., Sui, F., Muggleton, J.M., Yang, J.: Simplified dispersion relationships for fluid-dominated axisymmetric wave motion in buried fluid-filled pipes. J. Sound Vib. **375**, 386–402 (2016). https://doi.org/10.1016/j.jsv.2016.04.012
7. Brennan, M.J., et al.: On the effects of soil properties on leak noise propagation in plastic water distribution pipes. J. Sound Vib. **427**, 120–133 (2018). https://doi.org/10.1016/j.jsv.2018.03.027
8. Scussel, O., Brennan, M.J., Muggleton, J.M., de Almeida, F.C.L., Paschoalini, A.T.: Estimation of the bulk and shear moduli of soil surrounding a plastic water pipe using measurements of the predominantly fluid wave in the pipe. J. Appl. Geophys. **164**, 237–246 (2019). https://doi.org/10.1016/j.jappgeo.2019.01.010
9. Muggleton, J.M., Brennan, M.J.: Leak noise propagation and attenuation in submerged plastic water pipes. J. Sound Vib. **278**(3), 527–537 (2004). https://doi.org/10.1016/j.jsv.2003.10.052
10. Scussel, O., Brennan, M.J., Almeida, F.C.L., Muggleton, J.M., Rustighi, E., Joseph, P.F.: Estimating the spectrum of leak noise in buried plastic water distribution pipes using acoustic or vibration measurements remote from the leak. Mech. Syst. Signal Process. **147**, 107059 (2021). https://doi.org/10.1016/j.ymssp.2020.107059
11. Homepage - Frequency bandwidth estimation for leak noise correlators: http://correlacionador.tupa.unesp.br/, Accessed 23 Feb 2022

Pure Flexural Guided Wave Excitation Under Helical Tractions in Hollow Cylinders Based on the Normal Mode Expansion

Hao Dong and Wenjun Wu(✉)

Wuhan University of Technology, Wuhan 430070, China
wuwenjun@whut.edu.cn

Abstract. The flexural guided wave modes in hollow cylinders can be used as a supplement to the axisymmetric guided wave testing. Thus, the excitation of specific flexural modes is essential to the guided wave testing with flexural modes. Because the wave motion of flexural modes in hollow cylinders can be approximately considered as Lamb waves or SH waves in plates propagating in a spiral path relative to the cylinder axis, helical excitation can be used to excite specific flexural modes by setting the helix angle of helical excitation. The forced responses of hollow cylinders under helical loads are analyzed with the classic Normal Mode Expansion (cNME) method, and then verified by numerical simulations. Taking the testing of an AISI 316L steel pipe as an example, helical excitation with a helical angle of 12.8° corresponding to the flexural mode T(1,1) at the excitation frequency of 50 kHz is applied, the theoretical predictions agree well with the finite element simulation results, validating the theoretical derivations.

Keywords: Guided wave · Hollow cylinder · Nondestructive testing · Flexural mode · Helical excitation · Normal Mode Expansion

1 Introduction

Guided wave technology [1–4] has been widely used in the inspection of pipelines, plates, rails and etc., for its high efficiency and 100% cross section coverage. Although the wave motion of flexural modes is complicated and it is difficult to generate pure flexural modes in hollow cylinders, the flexural mode inspection can provide supplementary information about the integrity of the structure, such as the circumferential location of the defect, the oblique angle of cracks, and so on. It is of great interest to investigate the pure flexural mode excitation in hollow cylinders.

The excitation response of structures is usually analyzed with the Normal Mode Expansion (NME) method [5]. As for hollow cylinders, Ditri et al. [6] first derived the orthogonality relation of normal modes in hollow cylinders, and then analyzed the excitation response of the cylinder by expanding the fields in terms of orthogonal normal modes.

Z. Dimitrovová et al. (Eds.): WMVC 2022, MMS 125, pp. 1039–1048, 2023.
https://doi.org/10.1007/978-3-031-15758-5_107

Because the wave motion of flexural modes is complicated, directly using the flexural modes for inspections is rare. However, for hollow cylinders with a big diameter to thickness ratio, the wave modes in such cylinders can be approximated to be wave modes, including the lamb waves and SH waves, in an unwrapped isotropic plate. Wang et al. [7] proposed a new EMAT to generate multiple helical shear horizontal (mhsh) guided waves in view of this shortcoming that the propagation of the guided wave generated by present transducers is either concentrated with narrow beam directivity or omni-directional with disperse energy.

The helical load and time-delay phase control method are used to excite the flexrual mode in the pipeline. Ehsan et al. [8] proposed a new method called the ray plate theory to analyze the wave motion of flexrual guided waves in complex waveguides based on a phenomenological approach instead of a mathematical approach. Based on this understanding, helical loading is proposed for pure flexural mode excitation. At the same time, a beam control method is proposed. Linear phased array is used to guide guided waves in hollow cylinders.

The flexrual mode and helical propagated guided waves have the advantage of obtaining defect information, which has not been fully studied. In this paper, NME will be used to analyze the problem of flexrual guided waves in excited pipes under helical loads. Firstly, the Helix Angle dispersion curve of the pipe is given, and the flexrual guided wave mode with pure single mode excitation potential is selected according to the spiral Angle dispersion curve. Secondly, the optimal helix Angle of a flexrual mode is obtained theoretically. Finally, the previous theory is verified by finite element simulation experiment.

2 Guided Waves in Hollow Cylinders

2.1 Notation for the Guided Wave Modes in Cylinders

The axisymmetric guided wave is divided into longitudinal wave L(0,m), torsional wave T(0,m) and the flexural guided wave is represented by F(N,m) ($N \neq 0$).

2.2 Ray-Plate Theory

As shown in Fig. 1, the hollow cylinder can be approximated to be an unwrapped plate, and then the cylinder modes are approximated to be plate modes obliquely propagating at a helix angle with respect to the z-axis. The longitudinal flexural modes correspond to the Lamb waves and the torsional flexural modes correspond to the shear-horizontal (SH) waves.

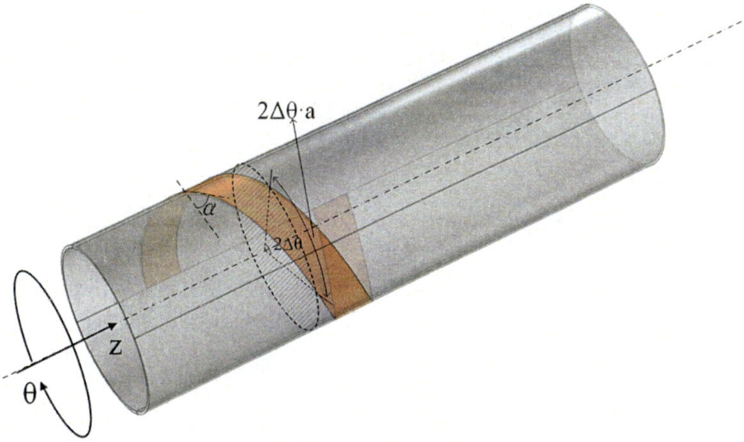

(a) The hollow cylinder

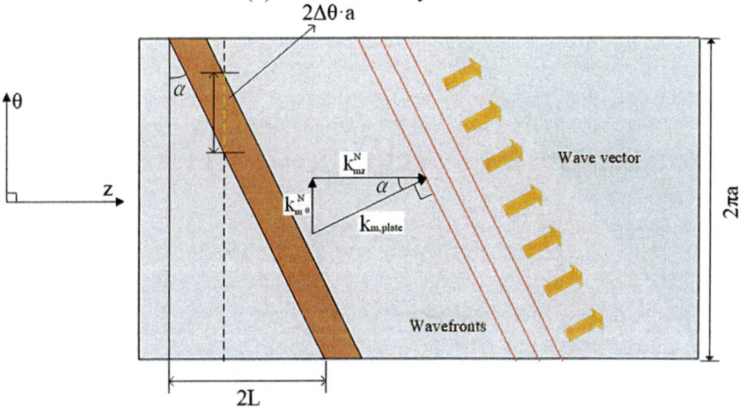

(b) The unwrapped plate

Fig. 1. System geometry

The displacement in the circumferential direction can be expressed as:

$$\begin{aligned}\mathbf{u}(r,\theta,z) &= \mathbf{U}(r)e^{i(k_{m,\text{plate}}\cos(\alpha)z + k_{m,\text{plate}}\sin(\alpha)a\theta - \omega t)} \\ &= \mathbf{U}(r)e^{i(k^N_{m,z}\cdot z + k^N_{m,\theta}\cdot a\theta - \omega t)}\end{aligned} \tag{1}$$

The guided wave displacement in the circumferential direction has a certain periodicity, we have: where a is the outer radius of the pipe, ω is the angular frequency, $k_{m,\text{plate}}$ stands for the wavenumber of the plate mode corresponding to cylinder modes of the m family, and $k^N_{m,z}$, $k^N_{m,\theta}$ are the circumferential and longitudinal component of $k_{m,\text{plate}}$. To be noted, $k^N_{m,z}$ is also the wavenumber of F(N,m), and this notation will be used in the rest of this paper.

$$\alpha_{\text{ray}} = \tan^{-1}\left(\frac{Ncp}{2\pi fa}\right), \tag{2}$$

where the subscript “ray” indicates that the helix angle α is derived by the ray-plate theory.

3 Application of Classical Normal Mode Expansion and Theoretical Analysis of Spiral Excited Flexrual Mode

3.1 Classical Normal Mode Expansion

The velocity field **v** and stress field **T** of some cylinder mode F(N,m) can be expressed as

$$\mathbf{v}_m^N(r,\theta)e^{i(\omega t-k_{m,z}^N z)} = \sum_{p=r,\theta,z} R_{mp}^N(r)\Theta_p^N(N\theta)e^{i(\omega t-k_{m,z}^N z)}\mathbf{e}_p = \begin{cases} R_{mr}^N(r)\cos(N\theta)e^{i(\omega t-k_{m,z}^N z)}\mathbf{e}_r \\ +R_{m\theta}^N(r)\sin(N\theta)e^{i(\omega t-k_{m,z}^N z)}\mathbf{e}_\theta \, , \\ +R_{mz}^N(r)\cos(N\theta)e^{i(\omega t-k_{m,z}^N z)}\mathbf{e}_z \end{cases} \tag{3}$$

$$\mathbf{T}_m^N(r,\theta)e^{i(\omega t-k_{m,z}^N z)} = \sum_p\sum_q R_{mpq}^N(r)\Theta_{pq}^N(N\theta)e^{i(\omega t-k_{m,,z}^N z)}\mathbf{e}_p\otimes\mathbf{e}_q, p,q\in\{r,\theta,z\}, \tag{4}$$

where the functions $R_{mp}^N(r)$ and $\Theta^N(N\theta)$ denote the radial and circumferential field distribution of the F(N,m) mode.

The amplitude of the generated F(N,m) mode by the classic NME method

$$A_m^N(z) = -\frac{e^{-ik_{m,z^2}^N}}{4P_{mm}^{NN}}\int_{-L}^{L} e^{ik_{m,zN}^N}\left(\oint_{\partial_1}\mathbf{v}_m^{N*}\cdot(\mathbf{T}\cdot\mathbf{n}_1)ds\right)dz, \tag{5}$$

As shown in Fig. 1, a tangential traction is applied on the colored helical strip in Fig. 1, and the traction direction is along the spiral axis of the strip, that is

$$\mathbf{T}\cdot\mathbf{n}_1 = \begin{cases} p\cdot\cos\alpha\cdot\mathbf{e}_z + p\cdot\sin\alpha\cdot\mathbf{e}_\theta, \ |z|\le L, r=R, |\theta-\theta_0|\le\Delta\theta \\ 0, \qquad\qquad otherwise, \end{cases}, \tag{6}$$

where p is the applied tangential traction, $2L$ is the axial length of the helical strip, $2\Delta\theta$ is the circumferential radian length, and

$$\theta_0 = \cot\alpha\cdot z/a. \tag{7}$$

Substituting Eq. (6) into Eq. (5) gives.

$$A_m^N(z) = -\frac{ape^{-ik_{m,z}^N}}{4P_{mm}^{NN}}\int_{-L}^{L} e^{ik_{m,z}^N}\left(\int_{\theta_0-\Delta\theta}^{\theta_0+\Delta\theta}\mathbf{v}_m^{N*}\cdot(\sin\alpha\cdot\mathbf{e_z}+\cos\alpha\cdot\mathbf{e_\theta})d\theta\right)dz. \tag{8}$$

Before substituting the velocity field (Eq. (3)) into Eq. (8), the original expression of the velocity field of Eq. (3) should be generalized to be

$$\mathbf{v}_m^N = \begin{cases} R_{m,r}^N(r)\cos\left(N\theta+\theta'\right)e^{i\left(\omega t-k_{m,z^N}^N\right)}\cdot\mathbf{e}_r \\ +R_{m,\theta}^N(r)\sin\left(N\theta+\theta'\right)e^{i\left(\omega t-k_{m,z}^N,z\right)}\cdot\mathbf{e}_\theta \\ +R_{m,z}^N(r)\cos\left(N\theta+\theta'\right)e^{i\left(\omega t-k_{m,z}^N,z^2\right)}\cdot\mathbf{e}_z \end{cases}$$
$$= \begin{cases} R_{m,r}^N(r)\cdot cos\theta'\cdot cos(N\theta)e^{i\left(\omega t-k_{m,z}^N z\right)}\cdot\mathbf{e}_r \\ +R_{m,\theta}^N(r)\cdot cos\theta'\cdot sin(N\theta)e^{i\left(\omega t-k_{m,z}^N z\right)}\cdot\mathbf{e}_\theta \\ +R_{m,z}^N(r)\cdot cos\theta'\cdot cos(N\theta)e^{i\left(\omega t-k_{m,z}^N z\right)}\cdot\mathbf{e}_z \end{cases} \begin{cases} -R_{m,r}^N(r)\cdot\sin\theta'\cdot\sin(N\theta)e^{i\left(\omega t-k_{m,z}^N z\right)}\cdot\mathbf{e}_r \\ +R_{m,\theta}^N(r)\cdot\sin\theta'\cdot\cos(N\theta)e^{i\left(\omega t-k_{m,z}^N z\right)}\cdot\mathbf{e}_\theta \\ -R_{m,z}^N(r)\cdot\sin\theta'\cdot\sin(N\theta)e^{i\left(\omega t-k_{m,z}^N z\right)}\cdot\mathbf{e}_z \end{cases} \tag{9}$$

where $\theta^{'}$ is the original phase of the circumferential distribution of the F(N,m) mode.

Without the loss of generality, the amplitude $A_m^N(z)$ of the F(N,m) mode can be calculated by computing the amplitudes these two orthogonal wave structures, respectively, represented by $A_m^{N,1}(z)$ and $A_m^{N,2}(z)$, and then combining $A_m^{N,1}(z)$ and $A_m^{N,2}(z)$ to obtain $A_m^N(z)$.

Substituting these two velocity fields into Eq. (7) gives

$$A_m^{N,1}(z) = -\frac{ap\sin(N\Delta\theta)}{2NP_{mm}^{NN}}\left(\begin{array}{c} -i\cdot R_{m,\theta}^{N*}(a)\cdot\cos\alpha\cdot\left(\dfrac{\sin B_1L}{B_1}-\dfrac{\sin B_2L}{B_2}\right) \\ +R_{m,z}^{N*}(a)\cdot\sin\alpha\cdot\left(\dfrac{\sin B_1L}{B_1}+\dfrac{\sin B_2L}{B_2}\right) \end{array}\right)\cdot e^{-ik_{m,z}^N z} \tag{10}$$

$$A_m^{N,2}(z) = -\frac{ap\sin(N\Delta\theta)}{2NP_{mm}^{NN}}\left(\begin{array}{c} R_{m,\theta}^{N*}(a)\cdot\cos\alpha\cdot\left(\frac{\sin B_1L}{B_1}+\frac{\sin B_2L}{B_2}\right) \\ +i\cdot R_{m,z}^{N*}(a)\cdot\sin\alpha\cdot\left(\frac{\sin B_1L}{B_1}-\frac{\sin B_2L}{B_2}\right) \end{array}\right)\cdot e^{-ik_{m,z^2}^N}, \tag{11}$$

where

$$B_1 = (N/a)\cdot\cot\alpha + k_{m,z}^N,\ B_2 = (N/a)\cdot\cot\alpha - k_{m,z}^N. \tag{12}$$

In fact, it is found in the following finite element simulation that the oblique angle of the F(N,m) mode excited by helical tractions always resembles the helix angle of the excitation trip, not the helix angle derived by the ray-plate theory of Eq. (2).

For simplifying the calculation of $A_m^N(z)$, $A_m^{N,1}(z)$ and $A_m^{N,2}(z)$ are approximated to be in phase, and then we have

$$\left|A_m^N(z)\right| = \sqrt{\left|A_m^{N,1}(z)\right|^2+\left|A_m^{N,2}(z)\right|^2}$$
$$= \frac{ap\sin(N\Delta\theta)}{2NP_{mm}^{NN}}\cdot\sqrt{\left(R_{m,\theta}^{N*}(a)^2\cdot\cos^2\alpha+R_{m,z}^{N*}(a)^2\cdot\sin^2\alpha\right)\cdot\left(\frac{\sin^2(B_1L)}{B_1^2}+\frac{\sin^2(B_2L)}{B_2^2}\right)}. \tag{13}$$

3.2 Pure Flexrual Mode Excitation

Pure flexural mode generation means maximizing the target flexural mode while suppressing the other modes simultaneously.

According to Eq. (13), the amplitudes of the generated modes depend on the helix angle α and their wave structures $\left(R_{m,\theta}^{N}(a), R_{m,z}^{N}(a)\right)$. Let

$$X = R_{m,\theta}^{N^*}(a)^2 \cdot \cos^2\alpha + R_{m,z}^{N^*}(a)^2 \cdot \sin^2\alpha, \tag{14}$$

$$Y = \frac{\sin^2(B_1 L)}{B_1^2} + \frac{\sin^2(B_2 L)}{B_2^2}. \tag{15}$$

Next, pure flexrual mode generation will be studied with respect to X and Y, respectively.

a. Pure flexrual mode generation with respect to X

Thus, the optimal α for maximizing X is

$$\alpha_{X,\mathrm{opt}} = \arctan\left(\frac{R_{m,z}^{N^*}(a)}{R_{m,\theta}^{N^*}(a)}\right). \tag{16}$$

Equation (14) also implies that the amplitude of X highly depends on the the wave structures of modes.

b. Pure flexrual mode generation with respect to Y

The axial length $2L$ of the helical excitation strip always takes one pitch, that is

$$2L = 2\pi a \tan\alpha. \tag{17}$$

Without the loss of generality, we let $k_{m,\theta}^{N} = 25$ and $k_{m,z}^{N} = 100$. Figure 2 gives the Y verses α curve. It can be seen from Fig. 3 that the optimal helix angle α_{ray} derived by the plate-ray theory is close to optimal helix angle $\alpha_{Y,\mathrm{opt}}$ for Y.

c. Guidances for pure flexural mode generation

(1) When surface tractions are applied, only the modes with dominant displacements in circumferential and axial directions can be effectively excited.

(2) The optimal helix angle α_{ray} derived by the plate-ray theory is a quasi-optimal helix angle, and can be used to generate flexural modes in most cases.

(3) The excitation frequency should be carefully chosen to improve the purity. The frequency at which the angle difference of α_{ray} for the target mode and the sub-largest mode is large enough should be chosen.

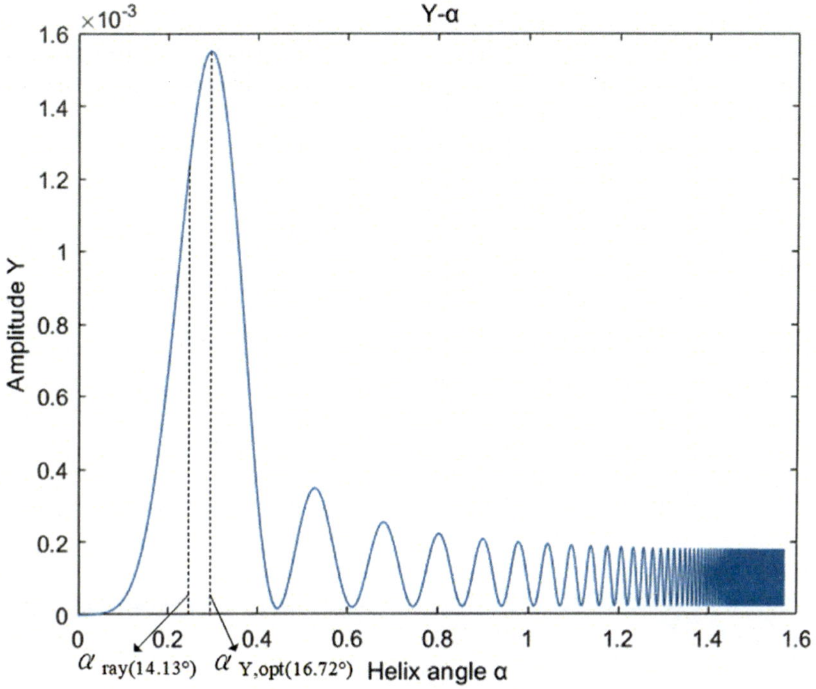

Fig. 2. *Y* verses helix angle α

4 Case study

Excitation of flexural torsional mode T(1,1) at frequency of 50 kHz in a steel pipe with parameters given in Table 1 is considered in this study. The setup of simulation experiment is shown in the Fig. 3. The helix angle at 50 kHz calculated from Eq. (2) is 12.8°. The screenshot of guided wave propagation in the simulation is shown in the Fig. 4, and by calculating the time difference Δt and comparing the phase speed dispersion curve, we can confirm that the echo in the Fig. 5 is the flexrual mode T(1,1).

Table 1. The parameters of pipe material.

Density ρ (kg/m^3)	Young's modulus E (GPa)	Poisson´s ratio μ
7930	206	0.27

The simulation results are shown in Fig. 7. Comparing the simulation results with the theoretical calculation results, it can be found that N = 0 accounts for about 5% in the theoretical calculation, which corresponds to the T (0,1) mode in the simulation results. N = 2 accounts for about 15% in the theoretical calculation, which is consistent with the L (2,2) mode in the simulation results. N = 3 accounts for about 5% in the theoretical calculation, which can also correspond to t (3,1) in the simulation results. To sum up, the results of the simulation experiment are in good agreement with the results of the theoretical calculation, which shows the correctness of this set of theoretical methods (Fig. 6).

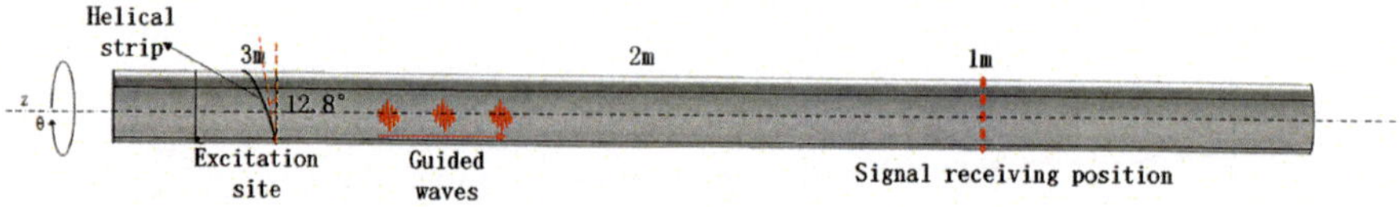

Fig. 3. Experiment setup for exciting flexural mode T(1,1).

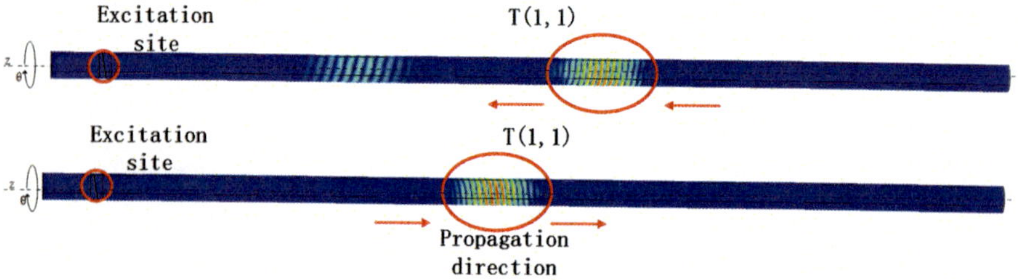

Fig. 4. The screenshot of guided wave propagation in simulation.

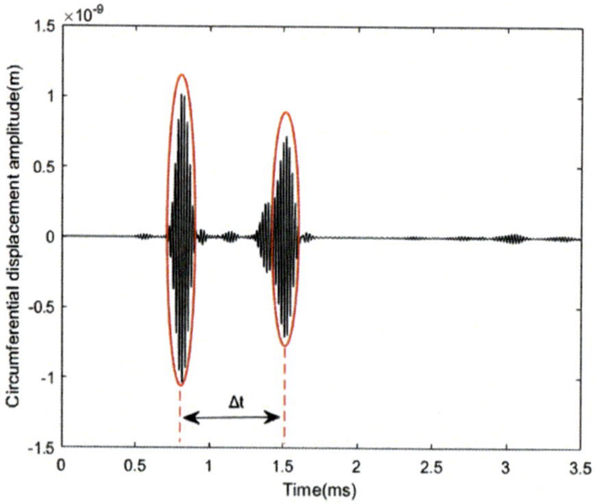

Fig. 5. Time domain diagram of N = 1 at 50 kHz

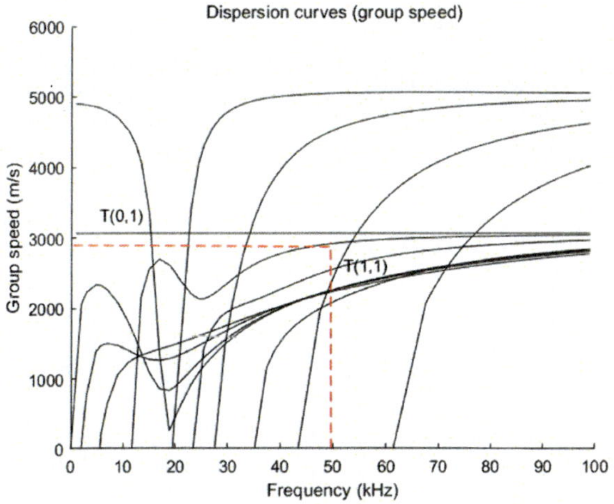

Fig. 6. Group speed dispersion curve

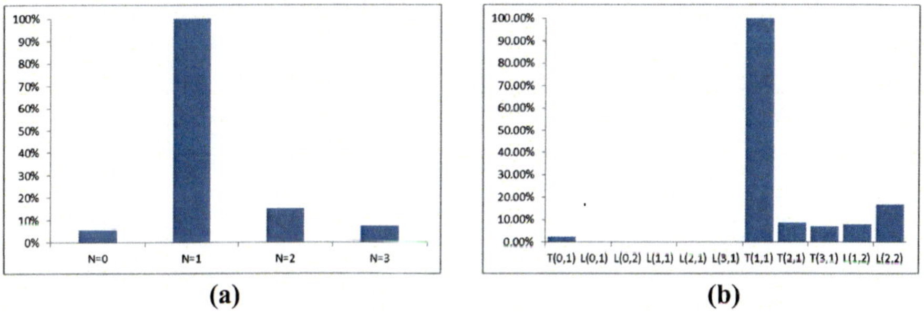

Fig. 7. Theoretical calculation and finite element simulation results. (a) Simulation experiment purity. (b) Theoretical calculation of the proportion of each mode.

5 Conclusion

The simulation results are in good agreement with the predicted results, which proves that this method can excite pure flexural modes under the conditions specified in the theory, which shows the correctness of this set of theoretical methods; At the same time, from the results of the simulation experiment, the pure target flexrual mode T (1,1) is excited, which has guiding significance for the application of flexrual mode and non-destructive testing research in the future.

References

1. Rose, J.L.: Ultrasonic Guided Waves in Solid Media. Cambridge University Press, Cambridge (2014)
2. Graff, K.F.: Wave Motion in Elastic Solids. Dover Publications (1975)

3. Drozdz, M.B.: Efficient finite element modeling of ultrasonic wave in elastic media (Ph.D. thesis), Imperial College London, London (2008)
4. Rose, J.L.: A baseline and vision of ultrasonic guided wave inspection potential. Press Vessel Technol. **124**(3), 273–82 (2002)
5. Auld, B.A., Kino, G.S.: Normal Mode Theory for Acoustic Waves and its Application to the Interdigital Transducer. IEEE Trans. Electron Dev. **18** (10), 898–907(1971)
6. Ditri, J.J., Rose, J.L.: Excitation of guided elastic wave modes in hollow cylinders by applied surface tractions. Appl. Phys. **72**(7), 2589–2597(1992)
7. Wang, Z., Huang, S., Wang, S., Wang, Q., Zhao, W.: Development of a new EMAT for multi-helical SH guided waves based on magnetostrictive effect. In: [IEEE 2018 IEEE International Instrumentation and Measurement Technology Conference (I2MTC) - Houston, TX, USA (2018.5.14–2018.5.17)] 2018 IEEE International Instrumentation and Measurement Technology Conference (I2MTC), pp. 1–6 (2018)
8. Khajeh, E.: Guided wave propagation in complex curved waveguides: Fleural guided waves and their application for defect classification in pipes. The Pennsylvania State University. The Graduate School (2013)

Flexural Vibration Analysis and Improvement of Wave Filtering Capability of Periodic Pipes

Mohd Iqbal(✉) and Anil Kumar

Department of Mechanical and Industrial Engineering, Indian Institute of Technology Roorkee, Roorkee 247667, India
miqbal@me.iitr.ac.in

Abstract. This article examines the propagation properties of flexural wave in periodic pipelines. In order to realize the effect of boundary condition on stopbands, a pipe with two types of periodic supports is investigated: (i) pipe on translational and rotational springs, and (ii) pipe on simple supports restraint with rotational springs. Initially, the dispersion relationships for the corresponding homogeneous pipe with different boundary conditions are derived in the context of Bloch-Floquet theory of periodic structures. Successively, the accuracy of resulting band structures is verified based on the vibration transmission spectrum computed by finite element models. In the examined frequency range, both Bragg and resonance type of stopbands are evolved in the pipe with first kind of supports while in second case only Bragg type stopbands are emerged. The waves corresponding to frequency range other than stopbands indicate passbands and they can freely propagate through the pipe. Hence, in order to control a specific passband, a single-degree-of-freedom resonator is employed at the center of each span of the pipe. The width and position of stopband due to resonator depend upon the mass and stiffness properties of the resonator. Therefore, the stopbands properties can be tuned by means of properly adjusting parameters of the resonator. The dispersion relations provided herein are promising to realize the characteristics of flexural waves and to design the resonator for the similar periodic structures with different boundary conditions.

Keywords: Periodic pipes · Phononic crystals · Wave propagation · Stopbands · Local resonance

1 Introduction

Fluid-conveying pipelines are very common in power and process industries. They are used to carry fluid between processing units and storage tanks in various industries including petrochemical plants, natural gas plants, etc. Vibrations in pipelines can occur by flow pulsation due to operation of valves, from machinery and support excitation which may results in loosening of connections, fatigue failure and fretting. Such a failure in pipelines can cause fire and explosion in different components of a plant which may lead to economical loss, safety and environmental issues. Therefore, it is extremely

Z. Dimitrovová et al. (Eds.): WMVC 2022, MMS 125, pp. 1049–1058, 2023.
https://doi.org/10.1007/978-3-031-15758-5_108

significant to examine the propagation of waves/vibrations in the pipes which eventually help in developing the efficient vibration control strategies [1].

A structure composed of periodic identical units is defined as periodic structure. The idea of phononic crystals (PCs) taken from solid-state physics opened new ways of vibration suppression in the periodic structures. Unlike to conventional periodic structures, PCs are artificial structures which carry material, geometric and boundary type spatial periodicity due to which they exhibit stopband properties wherein the waves get attenuated/stopped [2]. These stopbands in periodic structures are the results of Bragg scattering [3] or local resonance [4].

To date, extensive research has been carried out on the Bragg type stopbands in various engineering structures. Singh and Mallik [5] investigated the free harmonic waves propagation behavior and stopbands in infinite periodic pipe by means of wave approach. Domadiya et al. [6] studied the vibro-acoustic stopbands in one-dimensional infinite and finite periodic waveguides and concluded that such structures can be used as noise and vibration filters. Recently, the longitudinal and flexural vibration stopbands in bi-material periodic rod and beam were respectively investigated by Prasad and Sarkar [7].

On the other hand, PCs consisting of locally resonant (LR) units are defined as elastic/acoustic metamaterials. Along with Bragg stopbands, LR PCs yield additional stopbands produced by local resonances. In recent years, researchers exploit this concept to control the freely propagating waves in metamaterial pipes [8, 9], beams [10] and railway tracks [11]. Moreover, flexural vibration transmission characteristics of infinite and finite beams with periodically coupled vibration absorbers was examined by Xiao et al. [12]. Wu et al. [13] investigated the propagation behavior of flexural wave and vibration mitigation in pipe using periodically attached single-degree-of-freedom (SDoF) dynamic vibration absorbers.

In this paper, propagation behavior of flexural wave and vibration suppression in periodic pipelines is studied. Initially, wave propagation behavior in pipeline on two types of periodic supports is examined: (i) pipe kept on translational and rotational springs (i.e., Type #A support), and (ii) pipe kept on simple supports restraint with rotational springs (i.e., Type #B support). The dispersion relationships for both the cases are obtained by Floquet-Bloch theorem which are verified through numerical models. Figure 1(a) illustrates the layout of pipe without resonators kept on Type #A support. When stiffness of translational spring is infinitely large in Type #A support then it leads to the case of pipe with Type #B support. Although, pipe conveying fluid will show different dynamic characteristics due to hydrodynamic mass or internal pressure/gyroscopic effects as compared to an empty pipe. However, for simplicity, the latter case is considered herein.

Further, to control the freely propagating waves in the passband, a SDoF resonator is introduced at the center of each span/unit cell of the pipe. The configuration of pipe with resonators is shown in Fig. 1(b) for Type #A support. Moreover, the investigation of resonator parameters is carried out to understand their effects on the width and location of stopbands. For this, numerical models are developed to perform these studies and its accuracy is verified with the theoretical results. Similar models of pipe without and with resonators are employed for Type #B support.

The article is structured as follows: after introduction, Sect. 2 is concerned to the theoretical and numerical modelling of the periodic piping systems. Section 3 devotes to the results and discussion of pipe without and with resonators and the study of resonator parameters on stopbands. Finally, main results are concluded in Sect. 4.

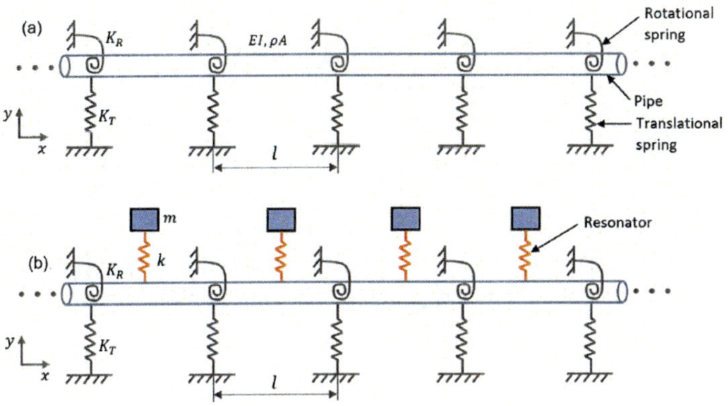

Fig. 1. Pipe with translational and rotational spring supports–Type #A –. (a) without resonators; (b) with resonators.

2 Methods

2.1 Theoretical Modelling

To study the propagation characteristics of flexural wave in pipe without resonators, the model depicted in Fig. 1(a) is utilized. Two spans of this illustrated in Fig. 2(a) are used to derive the dispersion relationship. The pipe herein is considered to be undamped and modelled based on Euler- Bernoulli beam theory, its motion equation is read as

$$\frac{\partial^2}{\partial x^2}\left[EI\frac{\partial^2 y(x,t)}{\partial x^2}\right] + \rho A\frac{\partial^2 y(x,t)}{\partial t^2} = 0 \tag{1}$$

where E, I, A and ρ respectively refer Young's modulus, sectional inertia moment about z axis, cross-sectional area and density.

The solution for Eq. (1) is assumed as

$$y(x,t) = Y(x)\exp(i\omega t) \tag{2}$$

Substitution of (2) in (1) entails

$$EIY^4(x) - \rho A\omega^2 Y(x) = 0 \tag{3}$$

here ω is radian frequency.

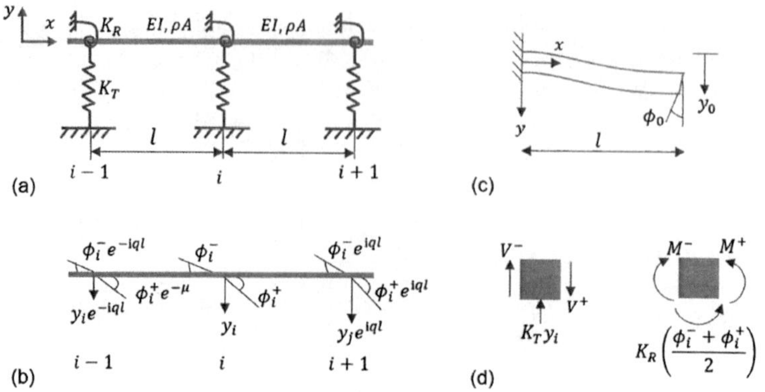

Fig. 2. Modelling of pipe: (a) two spans of pipe with Type #A support; (b) periodic Bloch-Floquet condition applied at the generic nodes; (c) simple beam with transverse displacement y_0 and rotation ϕ_0 at right end and clamped at left end; and (d) equilibrium of moments and forces on node i.

The solution of (3) provides displacement amplitude of the beam $Y(x)$ as,

$$Y(x) = \alpha_1 \cos(\beta x) + \alpha_2 \sin(\beta x) + \alpha_3 \cosh(\beta x) + \alpha_4 \sinh(\beta x) \tag{4}$$

where $\beta = \left(\frac{\rho A \omega^2}{EI}\right)^{\frac{1}{4}}$.

W.r.t Fig. 2(b), the transverse displacement of node $i-1$ and $i+1$ with node i are given as

$$y_{i-1} = y_i e^{-iql}, \quad y_{i+1} = y_i e^{iql} \tag{5}$$

where q is wave number and l is unit-cell length.

The constants α_1, α_2, α_3 and α_4 in (4) are determined utilizing the boundary conditions shown in Fig. 2(c), and are successively applied to find the shear force V and bending moment M at two sides of node i. The dynamic compliance coefficients [14] on $x = 0$ and $x = l$ for $y_0 = 1$ and $\phi_0 = 0$ are read as,

$$\begin{aligned} V_0' &= \frac{\beta^3 EI[\sin(\beta l) + \sinh(\beta l)]}{1 - \cos(\beta l)\cosh(\beta l)} \\ V_l' &= \frac{\beta^3 EI[\cosh(\beta l)\sin(\beta l) + \cos(\beta l)\sinh(\beta l)]}{1 - \cos(\beta l)\cosh(\beta l)} \\ M_0' &= \frac{\beta^2 EI[\cos(\beta l) - \cosh(\beta l)]}{1 - \cos(\beta l)\cosh(\beta l)} \\ M_l' &= \frac{\beta^2 EI[\sinh(\beta l)\sin(\beta l)]}{1 - \cos(\beta l)\cosh(\beta l)} \end{aligned} \tag{6}$$

For $y_0 = 0$ and $\phi_0 = 1$, the coefficients are expressed as

$$\begin{aligned} V_0'' &= \frac{-\beta^2 EI[\cosh(\beta l) - \cos(\beta l)]}{1 - \cos(\beta l)\cosh(\beta l)} \\ V_l'' &= \frac{-\beta^2 EI[\sinh(\beta l)\sin(\beta l)]}{1 - \cos(\beta l)\cosh(\beta l)} \\ M_0'' &= \frac{-\beta EI[\sin(\beta l) - \sinh(\beta l)]}{1 - \cos(\beta l)\cosh(\beta l)} \\ M_l'' &= \frac{-\beta EI[\cosh(\beta l)\sin(\beta l) - \cos(\beta l)\sinh(\beta l)]}{1 - \cos(\beta l)\cosh(\beta l)} \end{aligned} \tag{7}$$

W.r.t Fig. 2(b), the expressions of S and M at node i are obtained as

$$\begin{aligned} V^- &= -V_0' y_i e^{-iql} + V_l' y_i + V_0'' \phi_i^+ e^{-iql} + V_i'' \phi_i^- \\ V^+ &= V_0'' y_i e^{iql} - V_l' y_i + V_0'' \phi_i^- e^{iql} + V_l'' \phi_i^+ \\ M^- &= M_0' y_i e^{iql} + M_l' y_i - M_0'' \phi_i^+ e^{-iql} + M_i'' \phi_i^- \\ M^+ &= M_0' y_i e^{iql} + M_l' y_i + M_0'' \phi_i^+ e^{-iql} - M_i'' \phi_i^- \end{aligned} \tag{8}$$

The kinematic compatibility condition for angular displacement at node i is read as,

$$\phi_i^- = \phi_i^+ \tag{9}$$

and the equilibrium condition of moments and forces in Fig. 2(d) gives,

$$\begin{aligned} M^+ &= M^- - K_R\left(\frac{\phi_i^- + \phi_i^+}{2}\right) \\ V^+ &= V^- + K_T y_i \end{aligned} \tag{10}$$

where, K_T and K_R respectively denote the stiffness of translational and rotational springs.

Equations (8)–(10) yield a system of equations in terms of ϕ_i^-, ϕ_i^- and y_i as,

$$\begin{aligned} &\phi_i^- - \phi_i^+ = 0 \\ &(V_0'' e^{iql} - V_l'')\phi_i^- + (V_l'' - V_0'' e^{-iql})\phi_i^+ + [2V_0' \cos h(iql) - 2V_l' - k_y]y_i = 0 \\ &\left(M_0'' e^{iql} - M_l'' + \frac{K_R}{2}\right)\phi_i^- + \left(M_0'' e^{-iql} - M_l'' + \frac{K_R}{2}\right)\phi_i^+ + [-2iM_0' \sin(ql)]y_i \\ &= 0 \end{aligned} \tag{11}$$

A non-trivial solution of above set of equation yields the dispersion relation as,

$$\left[4M_0' V_0'' \sin^2(ql)\right] + \left[2V_0' \cos(ql) - 2V_l' - K_T\right]\left[2M_0'' \cos(ql) - 2M_l'' + K_R\right] = 0 \tag{12}$$

2.2 Numerical Modelling

To Validate the theoretical dispersion curves, numerical models of the pipe with Type #A and Type #B supports are developed as Euler-Bernoulli beam. If sufficient unit cells are composed by a finite structure, then it can replicate the same propagation/attenuation properties of waves as of infinite structure. Here, finite pipe with thirty spans is used. In order to compute the propagation behavior of flexural waves, a harmonic rotation $\phi_{i/p}e^{i2\pi ft}(f = \omega/2\pi)$ is prescribed at left end of the pipe and output rotation $\phi_{o/p}(f)$ is extracted at right end. The vibration transmission behavior $T(dB)$ is calculated as

$$T = 20\, log_{10}\left|\frac{\phi_{o/p}(f)}{\phi_{i/p}(f)}\right| \tag{13}$$

3 Results and Discussion

The flexural vibration stopband properties in pipe with Type #A and Type #B supports are initially examined and discussed. Subsequently, in order to control the freely propagating waves in the relevant passband, the SDoF resonator is attached to the center of each unit-cell of the pipe. Finally, the effect of resonator parameters on stopbands is investigated.

3.1 Stopband Properties of Pipe with Type #A and Type #B Supports

To investigate the propagation behavior of flexural wave in pipe without resonators with Type #A support, a model demonstrated in Fig. 1(a) is considered. In calculation, a pipe with following geometric and material properties is used:

Outer diameter $D_0 = 406.40$ mm, Inner diameter $D_i = 390.56$ mm, length $l = 6$ m, density $\rho = 7800\,\text{kg/m}^3$ and Young's modulus $E = 200$ GPa.

The stiffness values of translational and rotational springs are $K_T = 5E8$ N/m and $K_R = 2.63E7$ Nm/rad.

By substituting the above parameters in Eq. (12), the variation of imaginary part of ql with frequency $f = \omega/2\pi$ is calculated for pipe with Type #A support. The resulting dispersion curves are reported in Fig. 3(a) which are validated by numerical results (computed using Eq. (13)) shown in Fig. 3(b). In the frequency range [0 – 300] Hz, three stopbands with the range [0 – 40.25] Hz, [66.25 – 134] Hz and [159.75 – 263.25] Hz are observed and are depicted by shaded areas. It can be noted that the theoretical results demonstrate good correspondence with the numerical results. The first two stopbands are Bragg type stopbands which occurs by spatial periodicity while the third one is caused by resonance of pipe spring system.

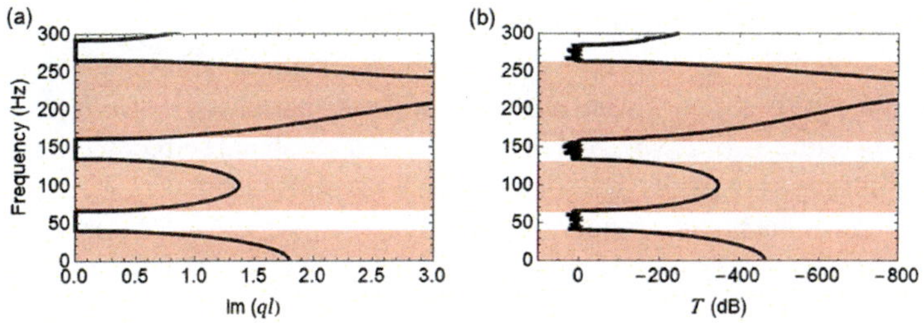

Fig. 3. Imaginary part of non-dimensional wave number and vibration transmission behavior in case of Type #A support: (a) Im (ql); and (b)T(dB).

When translational spring stiffness is infinitely large ($K_T \to \infty$), the configuration shown in Fig. 1(a) reduces to the pipe with Type #B support. Substitution of this condition in Eq. (12) yield the dispersion relation as

$$\left[2M_0'' \cos(ql) - 2M_l'' + K_R\right] = 0 \tag{14}$$

The stopband properties of pipe with Type #B support are calculated using Eq. (14). Similar to Type #A support, the dispersion curves are computed in the frequency range of [0 – 300] Hz. First three stopband frequency ranges are [0 – 41] Hz, [70.50 – 134] Hz and [192.50 – 285.50] Hz and are illustrated by shaded areas in Fig. 4. The theoretical results depicted in Fig. 4(a) show good agreement with the numerical results in Fig. 4(b). It can be noticed herein that only Bragg-type stopbands are emerged in the pipe with Type #B support.

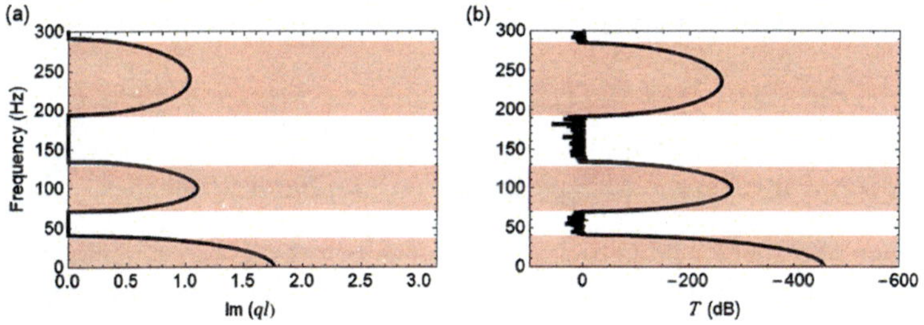

Fig. 4. Imaginary part of non-dimensional wave number and vibration transmission behavior in case of Type #B support: (a) Im (ql); and (b)T(dB).

3.2 Stopband Properties of Pipe Coupled with SDoF Resonators

The wave propagation properties in pipe without resonators have been investigated in Sect. 3.1. It is found in Figs. 3 and 4 that two passbands exist in the examined frequency range for pipe with Type #A and Type #B supports. Thereby, waves belong to these passbands can freely propagate through the pipe. Therefore, to control the particular passband, a SDoF resonator is endowed at the center of each unit cell of pipe.

To examine the vibration transmission behavior of the pipe endowed with resonators, model represented in Fig. 1(b) is used for Type #A support. Similar configuration is utilized for Type #B support. The mass and stiffness parameters of the resonator are assumed to be same as of m and k to control the first passband of pipe with two type of support conditions. If mass of each span of pipe is m_p and μ is mass ratio, then mass and stiffness values of the resonator are selected as, $m = \mu m_p,\ \mu = 15\%,\ m_p = \rho Al,\ k = 1E7$ N/m.

The vibration transmission behavior T of pipe with resonators for Type #A and Type #B supports is demonstrated in Figs. 5(a) and (b), respectively. It is observed that four stopbands are obtained in the frequency response curves for both type of supports. For Type #A, the respective stopbands frequency ranges are $[0 - 33]$ Hz, $[47 - 72]$ Hz, $[84 - 134]$ Hz and $[160 - 264]$ Hz. However, for Type #B, they are $[0 - 34]$ Hz, $[49 - 73]$ Hz, $[87 - 134]$ Hz and $[192.50 - 288]$ Hz. In both the cases, the second stopband from left (in blue) is by periodic resonators, while all others (in red) are due to the periodic pipe system. It can be observed that an additional stopband is created in the pipe with resonators. Thereby, the attenuation capability of the pipe is significantly enhanced.

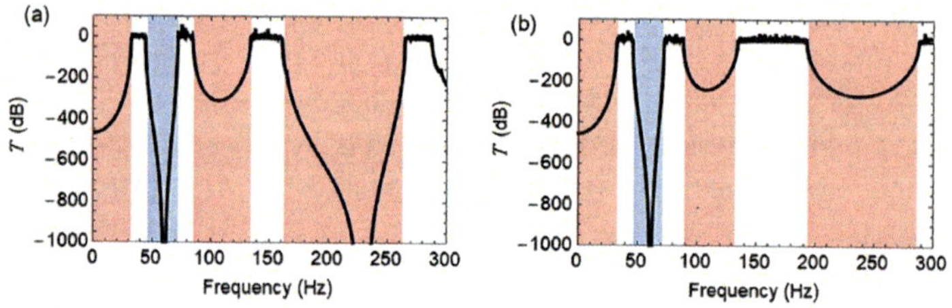

Fig. 5. T of pipe with resonators in case of: (a) Type #A support; and (b) Type #B support.

3.3 Influence of the Resonator Parameters on Stopband Properties

The influence of resonator properties on stopbands is investigated in depth for both the cases of pipe with Type #A and Type #B supports. To realize the effect of resonator parameters on stopbands, the different values of mass ratio (μ) are assumed to be 10%, 15% and 20%, while the values of stiffness (k) are considered as 1 $E7$ N/m, 2.25$E7$ N/m and 3.5$E7$ N/m. The same values of μ and k are considered when pipe with Type #B support is examined. In comparative study, one parameter is changed however other keeps constant as considered in Sect. 3.2.

Figures 6(a) and (b) respectively depict the influence of μ on stopband for Type #A and Type #B supports. In both the cases, it is observed that the second stopband shifts to low-frequency regime and its width gets widen with the increases of μ. The ending frequency of first stopband slightly moves to left hence its band width decreases. The starting frequency of third stopband also shifts to left while ending frequency remains identical, therefore, its width increases. However, the width and position of fourth stopband remains identical.

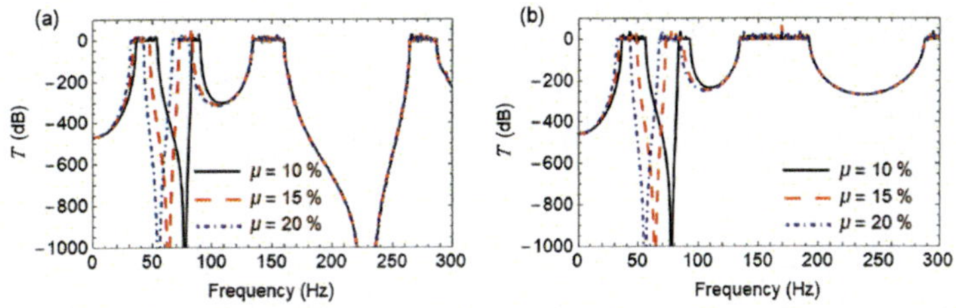

Fig. 6. Influence of resonator mass on stopband properties of pipe in case of: (a) Type #A support; and (b) Type #B support.

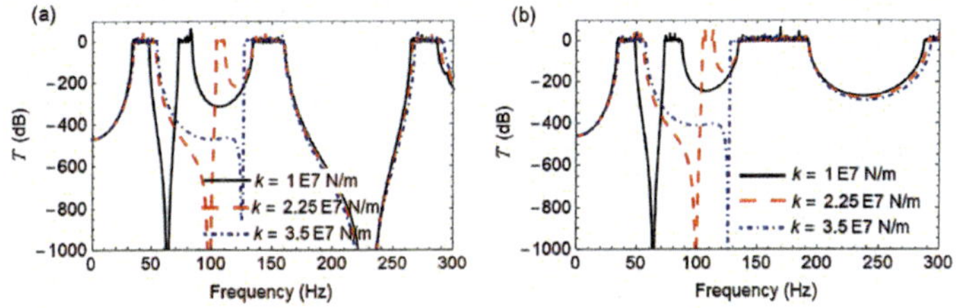

Fig. 7. Influence of resonator stiffness on stopband properties of pipe in case of: (a) Type #A support; and (b) Type #B support.

On the other hand, the variation of k significantly affects the width and location of stopbands which can be noticed in Figs. 7 (a) and (b) for Type #A and Type #B supports, respectively. It can be seen that the second stopband gets widen and moves to the high-frequency range with the increase of k. However, the starting frequency of third stopband shifts to right and the ending frequency stays at the same position, but, its bandwidth significantly decreases. For $k = 3.5E7$ N/m, both second and third stopbands gets merged and yields a wide coupled stopband with higher attenuation capacity which can be seen in respective figures. While in case of fourth stopband, the width is slightly increased and its position remains same.

4 Conclusion

The flexural vibration characteristics of pipe with Type #A and Type #B supports were examined theoretically and numerically. The dispersion relations for pipe with various support conditions were formulated by means of Floquet-Bloch theory which were successively verified by numerical Models. Pipe with Type #A support exhibit both Bragg and locally resonant stopbands while with Type #B it shows only Bragg type stopbands.

Further, to tuned the stopband properties, the SDoF resonator was coupled with each span of the pipe. A new stopband was produced in the pipe due to the periodically attached resonators. Thus, the waves attenuation strength of the pipe is substantially increased. Furthermore, the effect of variation of mass and stiffness of resonator was studied to realize their influence on the location and width of stopbands. It was found that the stopband properties of pipe significantly alter with the change of resonator parameters.

The realization of non-linear mechanisms in the modeling of supports and resonator deserves further research.

References

1. Iqbal, M., Jaya, M.M., Bursi, O.S., Kumar, A., Ceravolo, R.: Flexural band gaps and response attenuation of periodic piping systems enhanced with localized and distributed resonators. Sci. Rep. **10**, 85 (2020)
2. Mead, D.J.: Free wave propagation in periodically supported, infinite beams. J. Sound Vib. **11**, 181–197 (1970)
3. Brillouin, L.: Wave Propagation in Periodic Structures, 2nd edn. Dover Publications, New York (1953)
4. Liu, Z., et al.: Locally resonant sonic materials. Science **289**, 1734–1736 (2000)
5. Singh, K., Mallik, A.K.: Wave propagation and vibration response of a periodically supported pipe conveying fluid. J. Sound Vib. **54**, 55–66 (1977)
6. Domadiya, P.G., Manconi, E., Vanali, M., Andersen, L.V., Ricci, A.: Numerical and experimental investigation of stop-bands in finite and infinite periodic one-dimensional structures. Journal Vib. Control. **22**, 920–931 (2016)
7. Prasad, R., Sarkar, A.: Broadband vibration isolation for rods and beams using periodic structure theory. J. Appl. Mech. Trans. ASME. **86**, 1–10 (2019)
8. Iqbal, M., Kumar, A., and Bursi, O. S.: Vibration control of a periodic piping system employing metamaterial concept. In: 15th International Congress on Artificial Materials for Novel Wave Phenomena – Metamaterials, pp. 167–169. IEEE, NY, USA (2021)
9. Iqbal, M., Kumar, A., Bursi, O.S.: Lateral flexural vibration reduction in a periodic piping system enhanced with two-degrees-of-freedom resonators. Proc. Inst. Mech. Eng. Part L J. Mater. Des. Appl. **0**, 1–11 (2021)
10. Bao, H., Wu, C., Wang, K., Yan, B.: An enhanced dual-resonator metamaterial beam for low-frequency vibration suppression. J. Appl. Phys. **129**, 095106 (2021)
11. Iqbal, M., Kumar, A., Murugan Jaya, M., Bursi, O.S.: Flexural band gaps and vibration control of a periodic railway track. Sci. Rep. **11**, 1–13 (2021)
12. Xiao, Y., Wen, J., Yu, D., Wen, X.: Flexural wave propagation in beams with periodically attached vibration absorbers: band-gap behavior and band formation mechanisms. J. Sound Vib. **332**, 867–893 (2013)
13. Wu, J.H., Zhu, H.Z., Sun, Y.D., Yin, Z.Y., Su, M.Z.: Reduction of flexural vibration of a fluid-filled pipe with attached vibration absorbers. Int. J. Press. Vessel. Pip. **194**, 104525 (2021)
14. Clough, R.W., Penzien, J.: Dynamics of Structure, 3rd edn. Computers and Structures, USA (1975)

Focussing Acoustic Waves with Intent to Control Biofouling in Water Pipes

Austen Stone(✉), Timothy Waters, and Jennifer Muggleton

Institute of Sound and Vibration Research,University of Southampton, Highfield, Southampton SO17 1BJ, UK
as1u20@soton.ac.uk

Abstract. The colonisation of water pipes by macro-fouling organisms, such as barnacles and mussels, has presented a significant problem to industries drawing water from infested sources. Some of these creatures have been shown to be sensitive to low frequency sound and vibration, which have the potential to disrupt settlement and control population growth without the need for chemical interventions. The applicability of acoustic techniques to this problem is critically dependent on the achievable range of guided waves in the fluid or pipe wall which attenuate with distance from the actuation position due to mechanical losses.

In this paper, fluid waves are considered owing to their typically lower attenuation rates. A fluid-filled pipe is modelled analytically as a 2D rigid walled duct. Higher order acoustic waves, which are dispersive immediately above cut-on, are focussed at a target position using a transient excitation. The input waveform is obtained by filtering and time-reversing the impulse response so as to compensate for dispersion thereby compressing the signal in time and space. Simulations show that peak pressures can be obtained that are more than an order of magnitude higher than those achievable by harmonic excitation. Future work will model focussing of waves in a 3D pipe with fluid-structure coupling for which experimental validation will be sought.

Keywords: Fluid waves · Pipes · Biofouling · Energy focussing · Time reversal

1 Introduction

Bio-fouling is the unwanted accumulation of organisms on engineering structures, particularly those submerged in water. Aside from the build-up of smaller organisms, larger animals such as barnacles and mussels can firmly attach themselves to solid surfaces, causing a range of problems for affected structures. Perhaps the most notable example of this is the recent and ongoing issue of invasive freshwater mussel species colonising water pipes in industrial plants. In affected systems, a large accumulation of biomass can cause a reduction of flow and, if left untreated, organic matter will be carried farther into the piping system where

Z. Dimitrovová et al. (Eds.): WMVC 2022, MMS 125, pp. 1059–1068, 2023.
https://doi.org/10.1007/978-3-031-15758-5_109

it can interfere with downstream equipment. To date, most practical treatments for this problem involve chemicals or manual removal, neither of which are ideal. The use of chemicals is under increasing regulation out of environmental concern and manual removal is difficult, expensive, and not always possible.

There has, however, been research to suggest that sound and vibration can be used as an effective and environmentally friendly biofouling control technique [8]. For example, a study by Habibi et al. [5] demonstrated no fouling of a submerged metal plate after a month when subject to continuos structural vibration. Donskoy et al. [2] have shown that fluid borne sound can inhibit the settlement of zebra mussels on submerged surfaces. Sound treatment of bio-fouling affected piping systems could therefore prove an attractive control method for this type of bio-fouling.

Mechanical guided waves in water pipes can travel relatively far due to lack of geometric spreading. In this paper, a 2D acoustic waveguide is used as a pipe analogue to simplify analysis. For the purposes of bio-fouling control, it is of interest to find out exactly how far a given sound level can be maintained along the waveguide, or similarly, the maximum sound level which can be achieved at a given location in the waveguide. Both of these objectives can be realised through the principles of time reversal (TR) focussing. Time reversal is a process whereby the response of some distant excitation is measured by one or more transducers, reversed in time, and then retransmitted such that energy is refocussed back onto the source. Due to reciprocity, the source and receivers can be exchanged before retransmission for the same effect. Time reversal has been studied in a number of contexts, mostly for increasing signal to noise ratio by acting as a matched filter. In acoustic waveguides similar to the one studied in this paper, time reversal has been explored most notably for source location in the field of oceanography [3,7]. Roux and Fink also studied time reversal focussing in a 2D acoustic waveguide at high frequencies, using ray theory to explain the temporal and spatial focussing properties [11].

In the interest of obtaining a larger peak at the focus, several techniques for processing the impulse response before retransmission have been identified [12]. One notable technique involves digitising the impulse response over 1-bit, initially performed by Derode et al. [1], which leads to a much larger peak at the focal point. This is since the time reversal waveform now has a much higher signal power, whilst still preserving most of the phase information contained in the unprocessed impulse response. This technique has been used to great effect by several researchers [6,9,10].

In a practical setting, time reversal between two or more transducers cannot always be realised if parts of the system may be inaccessible. However, the principles of time reversal may be used more broadly to study energy focussing and to design focal waveforms which can be applied to real systems. With this in mind, the aim of this paper is to demonstrate the ability to use transient excitations to focus energy in a dispersive waveguide, and the relative advantages of doing so over a simple harmonic excitation.

Firstly, the system under investigation is described mathematically using the modal expansion technique. Transient excitations from a monopole source are then modelled and the time reversal technique applied to demonstrate the energy focussing capabilities. The relative advantages of this technique are then assessed in comparison to a harmonic excitation, with particular emphasis on peak response quantities. The dependence on excitation bandwidth is explored and it is shown that only a few of the lowest frequency modes are required for the focussing technique to yield the maximum response.

2 Modal Expansion in a 2D Waveguide

A two-dimensional acoustic waveguide (Fig. 1) is modelled analytically in the frequency domain using the modal expansion technique [4]. The pressure field of a single mode at any given frequency can be separated into axial and transverse components in the x and y direction respectively.

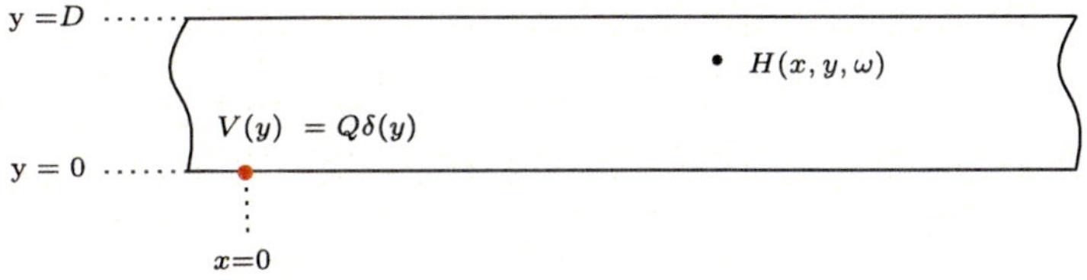

Fig. 1. Two-dimensional rigid walled acoustic waveguide. The system has infinite extent in the axial direction

$$P = \Phi(y)e^{-i\kappa x} \tag{1}$$

where the transverse mode function Φ satisfies the 2D Helmholtz equation.

$$\frac{d^2\Phi}{dy^2} + \mu^2\Phi = 0, \quad \mu^2 = k^2 - \kappa^2 \tag{2}$$

Here, $k = \omega/c_0$ is the free-space wavenumber and c_0 is the characteristic sound speed. This wavenumber is formed of the axial and transverse wavenumbers κ and μ. For perfectly rigid boundary conditions, i.e. zero normal velocity at $y = 0$ and $y = D$, the eigenfunctions of the Helmholtz equation are given by cosine functions representing transverse standing waves

$$\Phi_n(y) = B_n \cos \mu y, \quad \mu = \frac{n\pi}{D}, \quad n = 0, 1, 2, 3... \tag{3}$$

The normalisation coefficients are $B_n = 1$ for $n = 0$, otherwise $B_n = \sqrt{2}$. The free wave propagation in this system can be understood through the axial

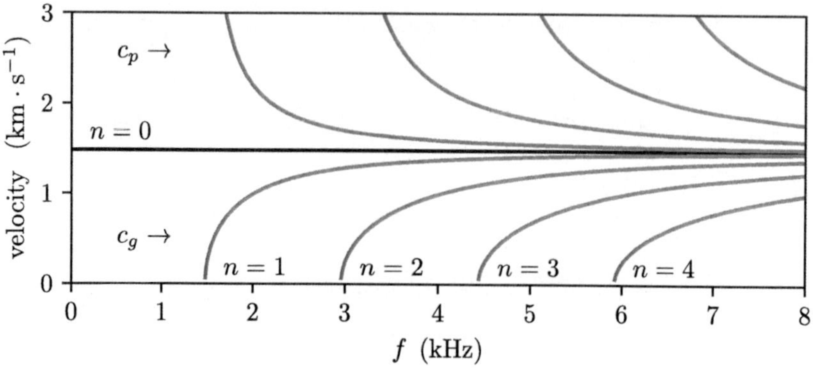

Fig. 2. Dispersion curves for a 2D rigid walled waveguide with system parameters in Table 1

dispersion relation $\kappa(\omega)$ which gives corresponding axial phase and group velocities

$$c_p = c_0 \left(1 - \left(\frac{n\pi c_0}{\omega D}\right)^2\right)^{-\frac{1}{2}} \qquad c_g = c_0 \left(1 - \left(\frac{n\pi c_0}{\omega D}\right)^2\right)^{\frac{1}{2}} \tag{4}$$

which are shown in Fig. 2 for the parameters listed in Table 1. The wave propagation consists of the $n = 0$ plane wave, which propagates at all frequencies, and an infinite number of higher order modes which only propagate above their cut-on frequency. The higher order modes can be physically understood as superimposing pairs of plane waves which travel down the waveguide via a series of oblique reflections at the boundaries. At cut-on, a given mode travels normal to the waveguide axis, hence the zero c_g and infinite c_p. As the frequency is increased above cut-on, the phase and group velocities tend towards c_0.

Monopole Source Expansion. For an acoustic monopole (point) source located at $(0, y_0)$, an explicit description of the pressure field can be found with a weighted summation of the modes [4].

$$P(x, y, \omega) = \rho_0 c \frac{\hat{Q}}{D} \sum_{n=0}^{\infty} \frac{k}{\kappa_n} \Phi_n(y_0)\, \Phi_n(y) e^{-i\kappa_n x} \tag{5}$$

where $\hat{Q}$ is the volume velocity and ρ_0 is the fluid density. For a given range of frequencies, this amounts to obtaining the transfer function $H(x, y, \omega)$ at a point in the waveguide. With this, the appropriate frequency-time transforms can be implemented to find the response of the system to a transient excitation. In practice the series in Eq. 5 must be truncated at an appropriate value of n, chosen here to be 3 times the maximum value of n for which κ_n is real.

It is noted that the presence of κ_n in the denominator of Eq. 5 leads to poles in the transfer function at the various cut-on frequencies. The unbounded response is dealt with by adding a small amount of hysteretic damping to the model. The bulk modulus K of the fluid is given an imaginary part with loss factor η resulting in a complex sound speed

$$c_0 = \sqrt{\frac{K}{\rho_0}} \qquad K = K'\,(1 + \eta i)\,. \tag{6}$$

While this damping model is not strictly causal in the time domain, small values of η will not have a significant effect on the causality of the response.

3 Time Reversal Focussing

Time reversal is now used to create a large peak pressure at a desired location in the waveguide. The objective is to transiently excite the monopole source in such a way that the frequency dependent wavespeed is exploited for maximum response at the focus. By exciting different frequencies with the appropriate delay, each spectral component of the wave can be made to converge at a focus simultaneously. Such an excitation can be built from the impulse response of the system from the source to the focus.

For this study, the monopole source was located at the lower boundary of the waveguide at $(0, 0)$, as in Fig. 1. The waveguide had parameters shown in Table 1. The volume velocity $\hat{Q}$ was set to unity, since the exact values are inconsequential in a linear system.

Table 1. Physical parameters of the acoustic waveguide.

Parameter	Value	Unit
D	0.5	m
K'	2.2	GPa
ρ_0	1000	$\mathrm{kg\,m^{-3}}$
η	0.001	

By evaluating Eq. 5 for 2^{17} frequencies from 0–40960 Hz, the transfer function at any point in the waveguide can be found $H(x, y, \omega)$ and with this, the corresponding band-limited impulse response function via the inverse Fourier transform.

$$h(x, y, t) = \mathcal{F}^{-1}\{H(x, y, \omega)\}. \tag{7}$$

According to the principles of reciprocity and time reversal[1], this impulse response may be flipped in time and transmitted from the monopole source, whereupon acoustic energy shall converge at the coordinate from which it was sampled. The waveform first, however, is normalised prior to transmission so

[1] Strictly speaking time reversal can only be achieved in a lossless system, however this does not negate its use as a focussing technique here.

that its maximum amplitude is constrained to ±1. For 'regular' time reversal, the focal waveform for a given location in the waveguide is then given by

$$f_{\text{reg}}(t) = \frac{h(-t)}{\max(|h(t)|)}. \tag{8}$$

In addition, the impulse response may be further modified before transmission in order to increase the maximum value at the focus, as discussed in Sect. 1. Clipping, a technique very similar to 1-bit digitisation [6,12], was used to significantly increase the response at the focus. For the present study, the clipping focal waveform took the form

$$f_{\text{clip}}(t) = \begin{cases} \text{sgn}(f_{\text{reg}}(t)), & \text{if } |f_{\text{reg}(t)}| \geq 0.01 \\ f_{\text{reg}}(t), & \text{otherwise} \end{cases} \tag{9}$$

where sgn represents the sign/signum function. To avoid time domain aliasing in the FFT, the clipped waveform was then filtered to remove the very highest frequency content in the signal. This low fidelity waveform will surprisingly perform far better than the unmodified impulse response when seeking a maximum pressure peak. As previously stated, this is in part because of the large increase in signal power which is achieved whilst retaining most of the phase information in the original signal.

To illustrate the focussing potential using time reversal, the impulse response at the coordinate $(20D, D/2)$ was used with Eqs. 8 & 9 to generate the focal waveforms shown in Fig. 3, which were then used as excitations to the monopole source at the origin. The spatial focussing in the waveguide is shown in Fig. 4 for both regular time reversal and that with the clipping technique applied. The spatio-temporal compression achieved with time reversal results in a large peak pressure amplitude around coordinate $(20D, D/2)$. Whilst the unaltered impulse response results in a peak which is better localised in time and space, the advantage of the clipping modification is seen in its much larger peak response. In this case, the peak pressure amplitude from the clipped waveform is 6 times greater.

3.1 Comparison to Harmonic Excitation

Having introduced regular and clipped time reversal based focussing, these techniques are now assessed through comparison to harmonic excitation. That is, the question of how much larger the pressure response is using time reversal focussing than a simple harmonic driving signal, which will serve as a benchmark here. The highest amplitude response for a tonal driving signal at a location in the waveguide is given by $\max(|H(x, y, \omega)|)$.

To evaluate the relative advantage of time reversal, the response along the bottom boundary was evaluated over a range of axial distances. At each distance, the peak response from the regular and clipped time reversal techniques was compared to the harmonic benchmark. The relative pressure gain over a harmonic excitation is shown in Fig. 5. It is clear that whilst regular time reversal offers no significant increase in pressure, the clipping technique can produce

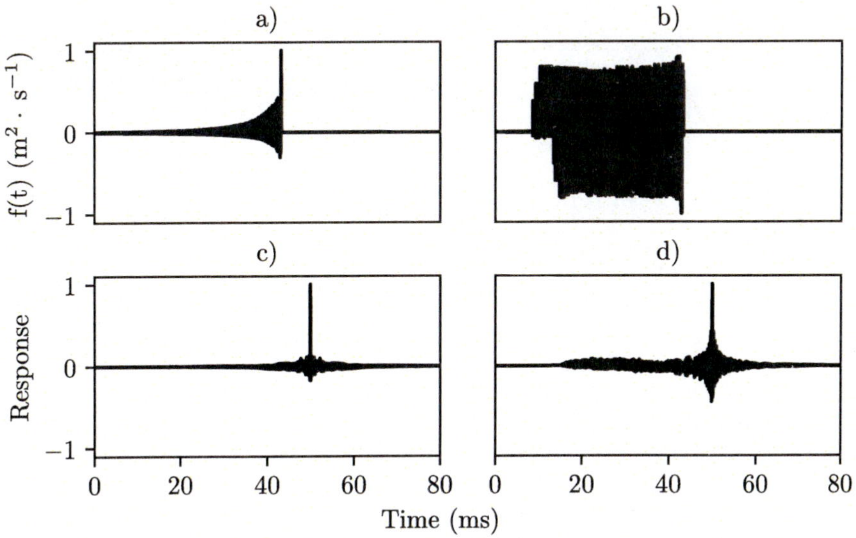

Fig. 3. Input waveforms for a) regular and b) clipped time reversal, along with their respective responses c) & d). The responses are amplitude normalised to show their general form. The clipped response is in fact several times larger.

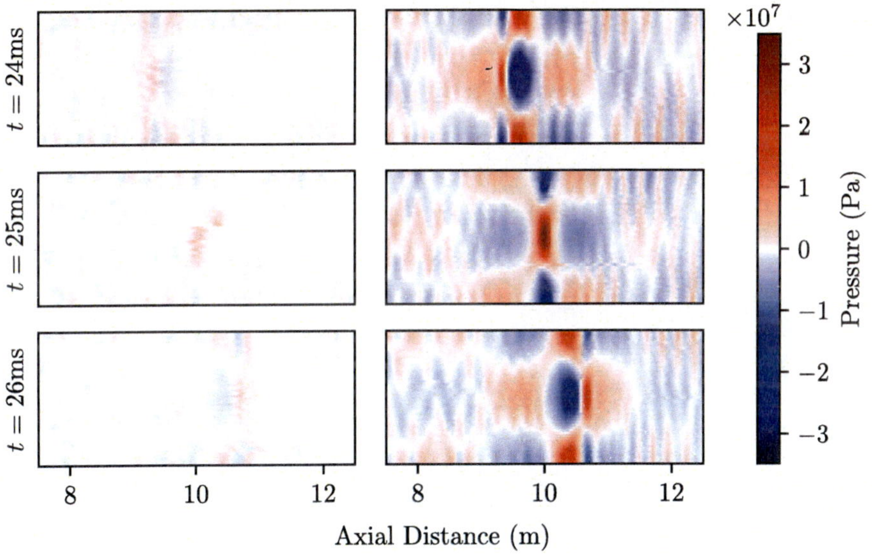

Fig. 4. Response to the time reversed impulse response sampled at $(20D, D/2)$ for regular (left) and clipped (right) time reversal. A focal peak appears at the location that the impulse response was originally sampled from.

peak responses in excess of 10 times that of a harmonic excitation in this simple model.

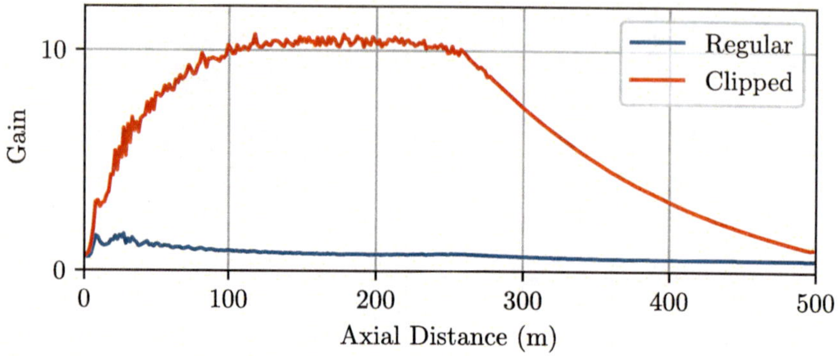

Fig. 5. The ratio of peak focussed pressure to peak harmonic pressure, evaluated along the bottom boundary of the waveguide.

The amplitude gain seen here also provides an increase to the range at which any given pressure can be achieved. With the same setup as before, Fig. 6 shows the maximum response along the waveguide boundary to a single focal waveform designed to converge at 100 m. Compared with a harmonic excitation, a level of pressure previously obtainable only at 7 m or less can now be achieved over 15 times farther from the point of excitation. However it must be noted that due to the 0.25 duration of the focal waveform, the pressure can now at best be delivered as a pulse train with a fundamental frequency of 4 Hz instead of the continuous sinusoid in the low kHz range for the harmonic case.

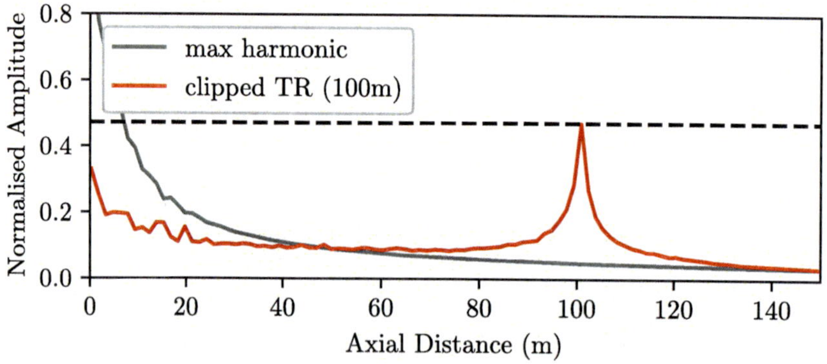

Fig. 6. Maximum response along the bottom boundary of waveguide to a waveform designed to converge at 100 m.

Effect of Bandwidth: Eq. 5 can be truncated to extract the impulse response of only the first few modes. It was found that the performance of clipped time reversal focussing is relatively insensitive to the number of modes included when generating the focal waveform. In fact, inclusion of only the first two modes is enough to generate a focal waveform capable of the tenfold pressure increase seen previously. The bandwidth required for focussing to work optimally is therefore roughly constrained by the dispersion curve of the $n = 1$ mode.

In a practical setting, this means focussing may be implemented by knowing the behaviour of only lowest order modes in the system, and without requiring high bandwidth transducers.

4 Conclusions

A simple model of a damped two-dimensional acoustic waveguide was used as a pipe analogue in order to study energy focussing capabilities of dispersive waves for biofouling applications. Through modal expansion, the responses to transient excitations from a single monopole source were modelled, and focussing at arbitrary locations was achieved by utilising the principles of time reversal. Modification of the time-reversed impulse response, in this case by clipping, was shown to significantly increase the peak pressure response at the focus. In the case of 0.1% structural damping, the peak response for clipped time reversal exceeded 10 times that of any harmonic excitation in the same bandwidth over a considerable distance from the source. In the context of biofouling control, this demonstrates the ability to extend the range of a given actuator or to concentrate energy at targeted areas in a dissipative environment.

References

1. Derode, A., Tourin, A., Fink, M.: Ultrasonic pulse compression with one-bit time reversal through multiple scattering. J. Appl. Phys. **85**(9), 6343–6352 (1999). https://doi.org/10.1063/1.370136
2. Donskoy, D.M., Ludyanskiy, M., Wright, D.A.: Effects of sound and ultrasound on Zebra Mussels. J. Acoust. Soc. Am. **99**(4), 2577–2603 (1996). https://doi.org/10.1121/1.415087
3. Feuillade, C., Clay, C.S.: Source imaging and sidelobe suppression using time-domain techniques in a shallow-water waveguide. J. Acoust. Soc. Am. **92**(4), 2165–2172 (1992). https://doi.org/10.1121/1.405228
4. Ginsberg, J.H.: Acoustics—A Textbook for Engineers and Physicists. Volume I: Fundamentals, Springer, Cham (2018). https://doi.org/10.1007/978-3-319-56844-7
5. Habibi, H., et al.: An acoustic antifouling study in sea environment for ship hulls using ultrasonic guided waves. J. Mar. Sci. Technol. Int. J. Eng. Technol. Manag. Res. **3** (2016). https://doi.org/10.29121/ijetmr.v3.i4.2016.59
6. Heaton, C., Anderson, B.E., Young, S.M.: Time reversal focusing of elastic waves in plates for an educational demonstration. J. Acoust. Soc. Am. **141**(2), 1084–1092 (2017). https://doi.org/10.1121/1.4976070
7. Hodgkiss, W.S., Song, H.C., Kuperman, W.A., Akal, T., Ferla, C., Jackson, D.R.: A long-range and variable focus phase-conjugation experiment in shallow water. J. Acoust. Soc. Am. **105**(3), 1597–1604 (1999). https://doi.org/10.1121/1.426740

8. Legg, M., Yücel, M., de Carellan, I.G., Kappatos, V., Selcuk, C., Gan, T.: Acoustic methods for biofouling control: a review. Ocean Eng. **103**, 237–247 (2015). https://doi.org/10.1016/j.oceaneng.2015.04.070
9. Montaldo, G., Roux, P., Derode, A., Negreira, C., Fink, M.: Ultrasound shock wave generator with one-bit time reversal in a dispersive medium, application to lithotripsy. Appl. Phys. Lett. **80**(5), 897–899 (2002). https://doi.org/10.1063/1.1446996
10. Montaldo, G., Roux, P., Derode, A., Negreira, C., Fink, M.: Generation of very high pressure pulses with 1-bit time reversal in a solid waveguide. J. Acoust. Soc. Am. **110**(6), 2849–2857 (2001). https://doi.org/10.1121/1.1413753
11. Roux, P., Fink, M.: Time-reversal in a waveguide. J. Acoust. Soc. Am. **110**, 2631 (2001). https://doi.org/10.1121/1.4776902
12. Willardson, M.L., Anderson, B.E., Young, S.M., Denison, M.H., Patchett, B.D.: Time reversal focusing of high amplitude sound in a reverberation chamber. J. Acoust. Soc. Am. **143**(2), 696–705 (2018). https://doi.org/10.1121/1.5023351

On the Pipe Localization Based on the Unwrapped Phase of Ground Surface Vibration Between a Roving Pair of Sensors

Mauricio Kiotsune Iwanaga[1(✉)], Michael John Brennan[1], Oscar Scussel[2], Fabrício César Lobato de Almeida[3,4], and Mahmoud Karimi[5]

[1] Department of Mechanical Engineering, UNESP-FEIS, Ilha Solteira, São Paulo 15385-000, Brazil
mkiwanaga23@gmail.com
[2] Institute of Sound and Vibration Research, University of Southampton, Highfield, Southampton SO17 1BJ, UK
[3] Department of Mechanical Engineering, UNESP-FEB, Bauru, São Paulo 17033-360, Brazil
[4] Faculty of Science and Engineering, UNESP-FCE, Tupã, São Paulo 17602-496, Brazil
[5] Centre for Audio, Acoustics and Vibration, University of Technology Sydney, Sydney, New South Wales, Australia

Abstract. Buried pipes are used worldwide to transport water. Although they are convenient, a large amount of water is wasted during the transportation. To minimize such a problem, water companies apply different technologies to locate leaks in their pipe networks, where the buried pipe is usually located first. Electromagnetic techniques can be used to locate buried pipes, but their performance is limited by the moisture content of the surrounding soil. Active vibro-acoustic techniques have also been investigated to locate buried pipes, in which vibro-acoustic energy is introduced into the soil by an excitation source. Although they are promising, their practical application can be expensive and complex due to the setup of an excitation mechanism. The aim of this paper is to present a passive vibro-acoustic localization technique for buried water pipes, in which a leak is the source of excitation. The localization technique is based on the calculation of an approximate slope for the unwrapped phase between a pair of sensors placed on the ground surface. The performance of the two-sensor technique is tested with two datasets, one extracted from a numerical model of the buried pipe system and the other extracted from an experiment carried out on a test rig. The results highlight the potential of the two-sensor technique in locating buried water pipes.

Keywords: Buried pipe localization · Ground surface vibration · Unwrapped phase · Pair of sensors

Z. Dimitrovová et al. (Eds.): WMVC 2022, MMS 125, pp. 1069–1076, 2023.
https://doi.org/10.1007/978-3-031-15758-5_110

1 Introduction

Water is commonly transported through buried pipes all over the world. Although they are convenient, a large amount of water is lost during the transportation [1]. Water companies apply different solutions to mitigate such a problem, in which one of them is to decrease the number of leaks in their pipe networks [2]. A wide range of technologies is available to locate leaks in buried water pipes [3], where the pipe is usually located first.

Sophisticated techniques have shown to be alternative solutions for the localization of buried pipes. Electromagnetic techniques, such as Ground Penetrating Radar (GPR), are well-established and are used to locate buried objects. However, the performance of such techniques depends on the moisture content of the surrounding soil as the electromagnetic energy is heavily attenuated by water [4]. Alternatively, vibro-acoustic techniques have also been investigated in the localization of shallow-buried objects [4, 5]. Most of the vibro-acoustic techniques are active, in which vibro-acoustic energy is introduced into the soil by an excitation source. There are several ways of exciting the soil and extracting information about the buried object from measurements of ground surface vibration. The technique proposed in [6], for example, consists of using a hammer to excite the soil (impact source) and the ground surface vibration is measured by a pair of sensors. The fluctuations in the resulting dispersion curves are used to locate the buried object. In [7, 8] an actuator excites the soil, and a set of sensors measures the ground surface vibration at different positions on the ground surface. An underground cross-section image is then generated to locate the buried object. Although active vibro-acoustic techniques are promising in locating buried pipes, their practical applicability can be expensive and complex as they require an excitation source to introduce vibro-acoustic energy into the soil.

The main goal of this paper is to present a passive vibro-acoustic localization technique for buried water pipes, in which a leak is the source of excitation. The technique is based on the calculation of an approximate slope for the unwrapped phase of a pair of sensors placed on the ground surface. This paper is organized as follows. A description for the two-sensor technique is presented in Sect. 2 to introduce the methodology in which the localization is performed. The application of the two-sensor technique is presented in Sect. 3, where the numerical and experimental setups used for testing are introduced together with the corresponding results. The conclusions for the application of the approach proposed in this paper are given in Sect. 4.

2 Two-Sensor Technique - Description

A leak in a buried water pipe generates noise in the form of waves that travel in the pipe, that radiate into the surrounding soil and towards the ground surface. The frequency regimes for the waves that propagate in a fluid-filled pipe are separated by the pipe ring frequency, which is the frequency at which a compressional wavelength is equal to the pipe circumference [9]. Well below the pipe ring frequency, the predominantly fluid-borne axisymmetric wave is mainly responsible for leak noise propagation in buried water pipes [9]. A schematic diagram for the wave propagation in a buried pipe system is illustrated in Fig. 1, where only one side of the pipe in relation to the leak is shown and the following simplifying assumptions are considered:

- The pipe and surrounding soil are of infinite extent in the axial direction, and the surrounding soil is also of infinite extent in the radial direction.
- The surrounding soil is a homogeneous, isotropic, elastic medium that can sustain both compressional and shear waves.
- The frequency range of interest is well below the pipe ring frequency, so that bending in the pipe-wall is neglected and the predominantly fluid-borne axisymmetric wave is the only wave propagating in the pipe.
- The frequency range of interest is such that an acoustic wavelength of water is much greater than the diameter of the pipe.

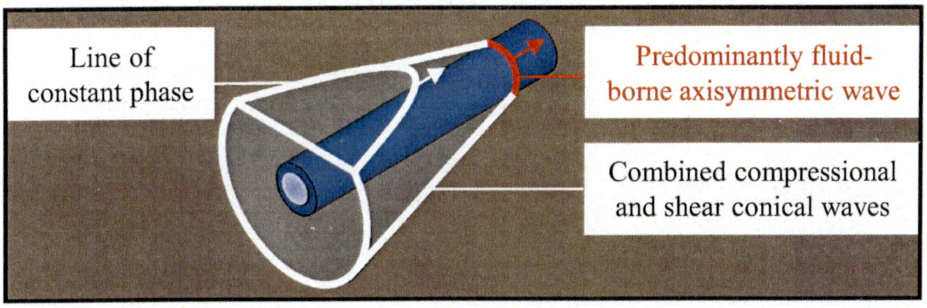

Fig. 1. Schematic diagram for the wave propagation in a buried pipe system.

The predominantly fluid-borne axisymmetric wave propagates in the pipe, and the combined compressional and shear conical waves radiate from the pipe into the surrounding soil [9]. All wavefronts illustrated in Fig. 1 are in-phase and the arrows indicate the corresponding direction of propagation. To idealize the lines of constant phase that travel on the ground surface due to a buried pipe with a leak, an illustration for an infinite plane parallel to the pipe and intersecting the combined conical waves is shown in Fig. 1, where the reflections are not considered for simplicity. Note that the intersection between the infinite plane and the combined conical waves results in an approximate hyperbola of constant phase, that travels at the same direction of the predominantly fluid-borne axisymmetric wave.

The shape and direction of propagation of the line of constant phase are used to introduce the two-sensor technique, which is based on the unwrapped phase measured for different positions on the ground surface by the sensors l and r. For this technique, the pipe direction is assumed to be known but its location is undetermined. An example of buried pipe localization with the two-sensor technique is illustrated in Fig. 2, where the pair of sensors is moved along the measurement line perpendicular to the buried pipe, as shown in Figs. 2(a.i-iii), and the unwrapped phase between the sensors $\phi_{l,r}(\omega) = -\omega\tau$ is determined for each measurement position, as shown in Figs. 2(b.i-iii), where ω is the angular frequency and τ is the time difference between the vibration from the pipe reaching each of the sensors.

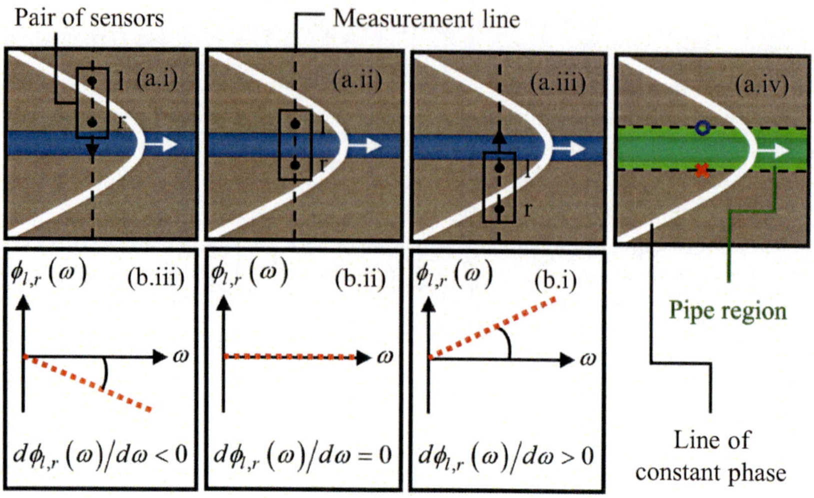

Fig. 2. Buried pipe localization with the two-sensor technique. (a.i) 1st measurement position, (a.ii) 2nd measurement position, (a.iii) 3rd measurement position and (a.iv) pipe region resulting from the two-sensor technique. (b.i) Unwrapped phase $\phi_{l,r}(\omega)$ for the 1st measurement position, (b.ii) unwrapped phase $\phi_{l,r}(\omega)$ for the 2nd measurement position and (b.iii) unwrapped phase $\phi_{l,r}(\omega)$ for the 3rd measurement position.

Note in Figs. 2(b.i-iii) that the slope $d\phi_{l,r}(\omega)/d\omega$ is negative, zero, or positive because it is governed by τ, and the arrival times for the line of constant phase depends on the position of the pair of sensors in relation to the pipe. The algorithm of the two-sensor technique checks the sign of $d\phi_{l,r}(\omega)/d\omega$ for each position along the measurement line. If the sign of $d\phi_{l,r}(\omega)/d\omega$ changes, the algorithm perceives that the pair of sensors crossed the buried pipe on the ground surface and marks the position where $d\phi_{l,r}(\omega)/d\omega$ is closer to zero as illustrated in Fig. 2(a.iv). Then, two lines parallel to the pipe direction are traced passing through the markers and the pipe region is circumscribed on the ground surface.

The algorithm of the two-sensor technique also indicates the direction for the pair of sensors to cross the buried pipe based on the sign of $d\phi_{l,r}(\omega)/d\omega$, i.e., if $d\phi_{l,r}(\omega)/d\omega < 0$, the algorithm points to the side of sensor r as shown by the black arrow in Fig. 2(a.i); if $d\phi_{l,r}(\omega)/d\omega = 0$, the algorithm does not point to any side as shown in Fig. 2(a.ii); and if $d\phi_{l,r}(\omega)/d\omega > 0$, the algorithm points to the side of sensor l as shown by the black arrow in Fig. 2(a.iii).

3 Application of the Two-Sensor Technique

The performance of the two-sensor technique described in the previous section is tested with two datasets: one generated from the numerical model of the buried pipe system shown in Fig. 3, and the other measured on a test rig situated in Blithfield, UK, shown in Fig. 4. For both figures, the distance parallel to pipe on the ground surface in relation to the leak is represented by x, and the distance perpendicular to pipe on the ground surface in relation to the leak is represented by y.

The numerical model is an axisymmetric Finite Element Model (FEM) developed in the software COMSOL Multiphysics, that describes a combined system composed by the water, the pipe, and the surrounding soil as shown in Fig. 3. The boundaries of the numerical model are Perfectly Matched Layers (PMLs) to simulate that the pipe has infinite length in the axial direction, and the surrounding soil has infinite length in both axial and radial directions. The leak is represented by a monopole source in the pipe positioned 15 m away from the measurement line, where the sensors are spaced 1 m apart from each other. The ground surface is represented by a plane parallel to the pipe direction and 1 m above the pipe, where the reflections of the incident waves are not considered. The velocity signals generated with the numerical model were sampled at a frequency of 1,024 Hz, and the frequency response for each pair of sensors along the measurement line was calculated with a 1,024-point fast Fourier transform and a Hanning window with 50% overlap, resulting in a frequency resolution of 1 Hz.

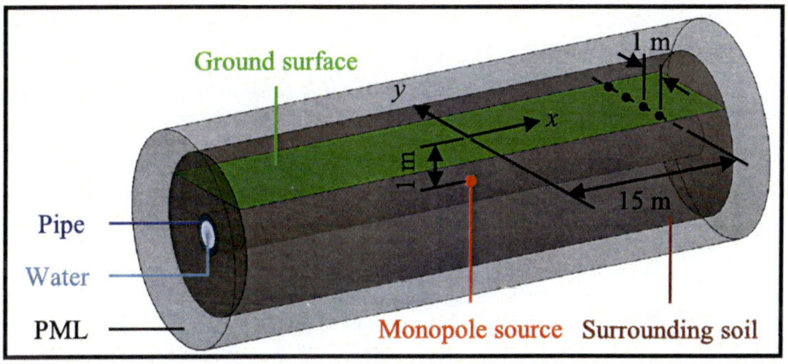

Fig. 3. Schematic diagram for the numerical model of the buried pipe system (not to scale).

The experiment carried out in the test rig consists of a buried pipe, whose depth is approximately 1 m, pressurized by the mains water distribution pipe as shown in Fig. 4. The leak was created through a standpipe connected to the buried pipe 15 m away from

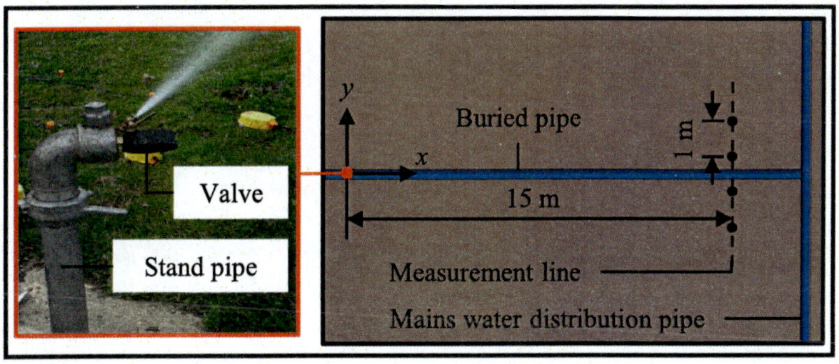

Fig. 4. Schematic diagram for the experiment carried out in the test rig (not to scale).

the measurement line on the ground surface, and a small valve was used to control the leak strength. Four I/O SM-24 geophones were placed 1 m apart from each other along the measurement line, and a Prosig DATS data acquisition system was used to acquire the experimental data. The velocity signals were sampled at a frequency of 1 kHz, and the cross-spectral density for each pair of sensors was calculated with a 1,000-point fast Fourier transform and a Hanning window with 50% overlap, resulting in a frequency resolution of 1 Hz.

For the practical application of the two-sensor technique, the algorithm approximates the line $\hat{\phi}_{l,r}(\omega)$ for the values of $\phi_{l,r}(\omega)$ via least squares method, to estimate the approximated slope $d\hat{\phi}_{l,r}(\omega)/d\omega$ for the unwrapped phase. The results for the application of the two-sensor technique on the numerical and experimental datasets are presented in Figs. 5 and 6, respectively, where the 1st, 2nd and 3rd positions for the pair of sensors on the measurement line are illustrated in (a.i-iii), and the unwrapped phase $\phi_{l,r}(\omega)$ with the corresponding approximated line $\hat{\phi}_{l,r}(\omega)$ are presented in (b.i-iii) in each figure. Note that $d\hat{\phi}_{l,r}(\omega)/d\omega < 0$, $d\hat{\phi}_{l,r}(\omega)/d\omega = 0$ and $d\hat{\phi}_{l,r}(\omega)/d\omega > 0$ for the 1st, 2nd and 3rd positions of the pair of sensors in Fig. 5, respectively, and $d\hat{\phi}_{l,r}(\omega)/d\omega < 0$, $d\hat{\phi}_{l,r}(\omega)/d\omega < 0$ and $d\hat{\phi}_{l,r}(\omega)/d\omega > 0$ for the 1st, 2nd and 3rd positions of the pair of sensors in Fig. 6, respectively. For both numerical and experimental datasets, the algorithm for the two-sensor technique detects that the sign of $d\hat{\phi}_{l,r}(\omega)/d\omega$ changes between the 1st and 3rd positions and marks the 2nd one, as shown in Figs. 5(a.iv) and 6(a.iv), since the corresponding $d\hat{\phi}_{l,r}(\omega)/d\omega$ is closer to zero. The pipe region is then marked on the ground surface by tracing two lines parallel to the pipe direction, passing through the markers of both results. Note that the pipe location and the direction to cross

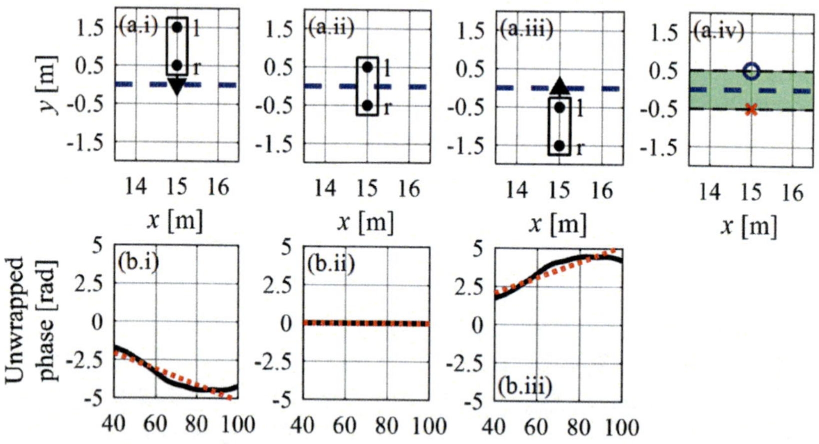

Fig. 5. Results for the application of the two-sensor technique applied on the numerical data. (a.i-iii) 1st, 2nd and 3rd positions of the pair of sensors, where the dashed blue line corresponds to the buried pipe. (b.i-iii) Unwrapped phase $\phi_{l,r}(\omega)$ and approximated line $\hat{\phi}_{l,r}(\omega)$ for the 1st, 2nd and 3rd positions, where the solid black line corresponds to $\phi_{l,r}(\omega)$ and the dashed red line corresponds to $\hat{\phi}_{l,r}(\omega)$.

the buried pipe are correct in both results, as the buried pipe is inside the estimated region and the arrows point to the pipe.

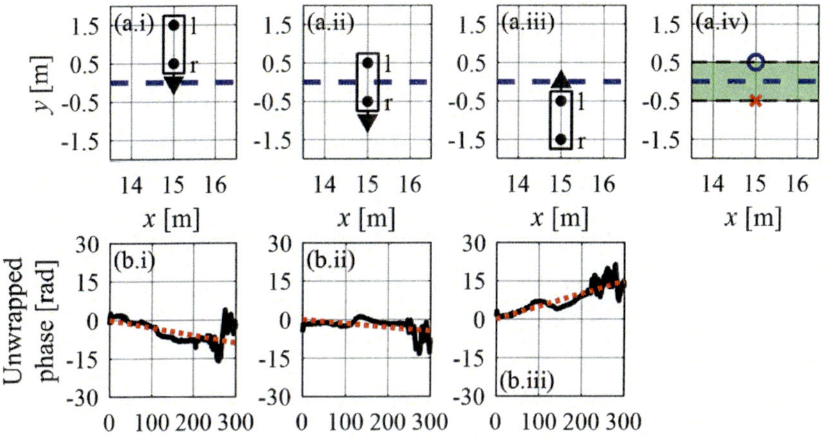

Fig. 6. Results for the application of the two-sensor technique applied on the experimental data. (a.i-iii) 1st, 2nd and 3rd positions of the pair of sensors, where the dashed blue line corresponds to the buried pipe. (b.i-iii) Unwrapped phase $\phi_{l,r}(\omega)$ and approximated line $\hat{\phi}_{l,r}(\omega)$ for the 1st, 2nd and 3rd positions, where the solid black line corresponds to $\phi_{l,r}(\omega)$ and the dashed red line corresponds to $\hat{\phi}_{l,r}(\omega)$.

4 Conclusions

The main purpose of this paper is to present a two-sensor passive vibro-acoustic localization technique for buried water pipes, in which a leak is the source of excitation. The localization procedure is based on the calculation of an approximate slope for the unwrapped phase measured by a pair of sensors for different positions on the ground surface, along the measurement line which is perpendicular to the pipe direction. The algorithm detects if the sign of the approximate slope changes over the measurement line and marks the position whose approximate slope is closer to zero. The performance of the technique is tested on two datasets, one extracted from a numerical model of the buried pipe system and the other extracted from an experiment performed in a test rig. For both applications, the two-sensor technique estimated the pipe location correctly emphasizing its potential in locating buried water pipes, particularly when measurements are made far from the leak position.

Acknowledgements. The authors are grateful for the financial support provided by the Brazilian water and waste management company (SABESP), and for the experimental data provided by Jennifer Muggleton from the University of Southampton. Oscar Scussel is grateful for the support from Coordination for the Improvement of Higher Education Personnel (CAPES) under Grant number 88887.374001/2019–00, and the EPSRC under RAINDROP (EP/V028111/1).

References

1. Liemberger, R., Wyatt, A.: Quantifying the global non-revenue water problem. Water Supply **19**(3), 831–837 (2019)
2. Frauendorfer, R., Liemberger, R.: The Issues and Challenges of Reducing Non-revenue Water. Asian Development Bank, Mandaluyong City (2010)
3. Hunaidi, O., Chu, W., Wang, A., Guan, W.: Detecting leaks in plastic pipes. Am. Water Works Assoc. **92**(2), 82–94 (2000)
4. Metje, N., et al.: Mapping the underworld – state-of-the-art review. Tunn. Undergr. Space Technol. **22**(5–6), 568–586 (2007)
5. Muggleton, J.M., Rustighi, E.: 'Mapping the underworld': recent developments in vibro-acoustic techniques to locate buried infrastructure. Géotechnique Lett. **3**(3), 137–141 (2013)
6. Ganji, V., Gucunski, N., Maher, A.: Detection of underground obstacles by SASW method – Numerical aspects. J. Geotech. Geoenviron. Eng. **123**(3), 212–219 (1997)
7. Papandreou, B., Brennan, M.J., Rustighi, E.: On the detection of objects buried at a shallow depth using seismic wave reflections. J. Acoust. Soc. Am. **129**(3), 1366–1374 (2011)
8. Muggleton, J.M., Papandreou, B.: A shear wave ground surface vibration technique for the detection of buried pipes. J. Appl. Geophys. **106**, 164–172 (2014)
9. Gao, Y., Sui, F., Muggleton, J.M., Yang, J.: Simplified dispersion relationships for fluid-dominated axisymmetric wave motion in buried fluid-filled pipes. J. Sound Vib. **375**, 386–402 (2016)

Passive Measurement of Pressure Wave Speed in Water Pipelines Using Ambient Noise

Zhao Li[1(✉)], Pedro Lee[1], Mathias Fink[2,4], and Ross Murch[3,4]

[1] Civil and Natural Resources Engineering, University of Canterbury, Christchurch, New Zealand
derek.li@canterbury.ac.nz
[2] Institut Langevin, ESPCI Paris, CNRS, PSL University, 1 rue Jussieu, 75005 Paris, France
[3] Department of Electronic and Computer Engineering, The Hong Kong University of Science and Technology, Hong Kong, China
[4] Institute for Advanced Study, The Hong Kong University of Science and Technology, Hong Kong, China

Abstract. The wave speed of pressure waves in water pipelines is sensitive to the reduction in pipe wall thickness and material strength, and it has been utilized as an indicator of water pipeline deterioration. These probing pressure waves are usually generated actively and is challenging to be incorporated into an automated senor network. This paper proposed a passive wave speed estimation method, which takes cross-correlation of the ambient noise in water pipeline networks measured by two synchronized pressure sensors to estimate the wave travel time. Field experiments were carried out in the operating water reticulation system at University of Canterbury campus for validation. In the experiments, pressure sensors were attached to fire hydrants to measure the ambient noise for 20 min. The experiment results indicate that pressure wave speed can be estimated using the proposed passive method, and the accuracy is at the same level compared with the conventional active method.

Keywords: Pressure wave speed · Ambient noise · Passive imaging · Cross-correlation · Water pipe deterioration

1 Introduction Section

The majority of water supply pipelines is deployed underground for decades [1]. During this period, chemical reactions from the water and soil have resulted in corrosion on pipe walls [2]. As the pipe wall deteriorates, it usually leads to

Supported by the University of Canterbury.

Z. Dimitrovová et al. (Eds.): WMVC 2022, MMS 125, pp. 1077–1084, 2023.
https://doi.org/10.1007/978-3-031-15758-5_111

decreased hydraulic efficiency, accumulation of pipe wall decay by-products, and development of leaks and large-scale bursts [3,4]. Therefore, it is critical to regularly assess the condition of the buried pipelines for proactive asset management.

Different techniques have been applied to assess pipeline physical conditions. Common commercial approach for pipe condition assessment involves physical inspection by excavating a section of pipe from a representative location in the network and taking CT scans and using crush test to measure the dimension and strength of the pipe sample [5]. In some cases, water mains are also inspected using CCTV, which requires the main to be depressurised and drained [6]. These approaches are destructive/intrusive, expensive, time-consuming and only yields information for a limited length of the total pipe work, where the condition of the test location may not be representative of the rest of the pipe. Other techniques using non-destructive methods, such as electromagnetic and ultrasonic testing [7]. However, these methods generally require access to the pipe wall and the signals are subject to significant attenuation, limiting their working range.

Alternatively, pressure waves, such as fluid transients or acoustic signals, are actively generated in pipeline systems and use wave speed variations to assess pipeline conditions remotely [8–13]. This method is based on the fact that pipeline deterioration reduces the pipe size and strength, resulting in a reduction in the wave speed. Numerous experimental studies have been carried out for verification [14–18]and it was found that wave speed is a sensitive indicator for pipe wall deterioration. However, to overcome the strong ambient noise (over 120 dB re 1 μPa/Hz [19]) and achieve long detection range, the pressure wave are usually generated with strong intensity using specially designed wave generator or rapid manual valve closure. These wave generations need to be repeated at a number of different locations to cover a regional water supply network, and it is not appropriate to be incorporated as an automated sensor network. In addition, the strong pressure waves may also pose extra stress on the network during detection, potentially exacerbating pipeline aging and deterioration.

These limitations can be accounted for by passively using the pressure waves that already exist in the water supply network. Acoustic emission methods use the hissing noise generated by water leaks to passively locate leaks [20], but this leak-induced noise does not exist when there is no leak in the system. Therefore, this paper investigate applying the passive imaging method [21,22] to use the ambient pressure noise in the system generate by turbulent flow, water usage activities, etc. as natural signals to probe the water pipeline and estimate the wave speed. Field experiments were carried out in an operating water supply system at the campus of University of Canterbury for methodology verification. The wave speeds estimated by this passive method are also compared with the conventional active wave speed measurement results for accurate assessment.

2 Passive Wave Speed Measurement Methodology

The framework of passive wave speed measurement is based on the cross-correlation analysis of the ambient noise. As shown in Fig. 1, a pipe is deployed along x axis and two sensors are at $x = h$ and $x = -h$, which are symmetric to the assumed origin of the 1D coordinate. The reason of using this 1D coordinate it that the ambient pressure noise in pipeline system is usually up 100 Hz [21], and it propagates as a plane wave in the pipeline systems within this low frequency range [23,24]. If the location of a noise source is y, the wave propagation in the pipeline can be modeled by the homogeneous wave function:

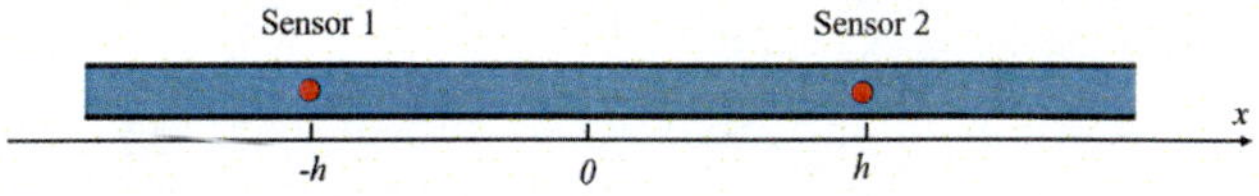

Fig. 1. Schematic of a pipeline coordinate system with two sensors.

$$\frac{\partial^2 p}{\partial x^2} + \frac{1}{c^2}\frac{\partial^2 p}{\partial t^2} = \delta(x - y) \tag{1}$$

where p is the pressure, t represents time and c is the wave speed to be measured. The frequency domain solution of Eq. 1 is the 1D Green's function $G(f, x, y)$:

$$G(f, x, y) = p = \frac{A}{2jk} e^{-\alpha|x-y|} e^{jk|x-y|} \tag{2}$$

where A is the wave amplitude, $k = 2\pi f/c$ is the wavenumber and f is the wave frequency. If the spectrum of the noise source is $N(f)$, the received noise at location x is $G(f, x, y) \cdot N(f)$. So the cross-correlation output of the noise signals measured by the two sensors can be expressed as

$$\begin{aligned} Corr(f) &= E[G(f, -h, y) \cdot N(f) \cdot G(f, h, y)^* \cdot N(f)^*] \\ &= \tfrac{A^2 E[N(f)^2]}{4k^2} e^{-\alpha(|-h-y|+|h-y|)} e^{jk(|-h-y|-|h-y|)} \end{aligned} \tag{3}$$

where $E[\cdot]$ presents the statistical expectation to account for the stochastic characteristics of the noise. Note that the term $E[N(f)^2]$ is the power spectrum of the noise source. In the following derivation, the sources are assumed as band-limited white noise signals for simplification. Colored noise sources can be whitened and the results will be the same.

If the noise sources are uniformly spread along the pipeline, such as noise generated by turbulent flow, the overall output of the cross-correlation is the integral of Eq. 3 over y, so in this scenario the overall cross-correlation output can be simplified as

$$Corr(f) = \begin{cases} \int_{-\infty}^{-h} \frac{A^2}{4k^2} e^{-2jkh} \cdot e^{2\alpha y} dy & y < -h \\ \int_{-h}^{h} \frac{A^2}{4k^2} e^{-2jky} \cdot e^{-2\alpha h} dy & -h < y < h \\ \int_{h}^{+\infty} \frac{A^2}{4k^2} e^{2jkh} \cdot e^{-2\alpha y} dy & y > h \end{cases} \tag{4}$$

Using inverse Fourier transform, Eq. 4 can be presented more intuitively in time domain as

$$Corr(\tau) = \begin{cases} \frac{1}{2\alpha} F(\tau) \cdot e^{-2\alpha h} * \delta(\tau - \frac{2h}{c}) & y < -h \\ \frac{1}{2} F(\tau) \cdot e^{-2\alpha h} * S(\frac{2h}{c}) & -h < y < h \\ \frac{1}{2\alpha} F(\tau) \cdot e^{-2\alpha h} * \delta(\tau + \frac{2h}{c}) & y > h \end{cases} \tag{5}$$

where $S(2h/c)$ is a step function that $S(2h/c) = 1(-2h/c < \tau < 2h/c)$ and $S(2h/c) = 0$ elsewhere. $F(\tau)$ is the inverse Fourier transform of $A^2/4k^2$ term in Eq. 4. If we ignore the $F(\tau)$ term, Eq. 5 indicates that for noise sources at the left side of sensor 1, the cross-correlation output will be a large pulse at $\tau = 2h/c$ since typically $\alpha \ll 1$. Similarly, for noise sources at the right side of sensor 2, the correlation output will be a large pulse at $\tau = -2h/c$. For noise sources between the two sensors, the output will be a step function with a small value between the two large pulses.

This prediction can be confirmed from the simulation in Fig. 2. In the simulation, the wave speed is set as 1000 m/s and two sensors are deployed in a pipeline with $h = 100$ m. There is no junction, branch or loop in the pipeline and noise sources are uniformly distributed within $[-250, 250]$ m. In the cross-correlation output, we can see there are two dominant peaks at $\tau = \pm 2h/c = \pm 0.2$ s. Therefore, if we can measure the ambient coherent noise in water pipeline using two sensors and take cross-correlation of the two sensor measurement, we will find two red spikes in the cross-correlation output, whose time instants are linked to wave speed as $\tau = \pm 2h/c$. If we know the distance between the two sensors, $2h$, the wave speed can be directly measured by $c = 2h/\tau$.

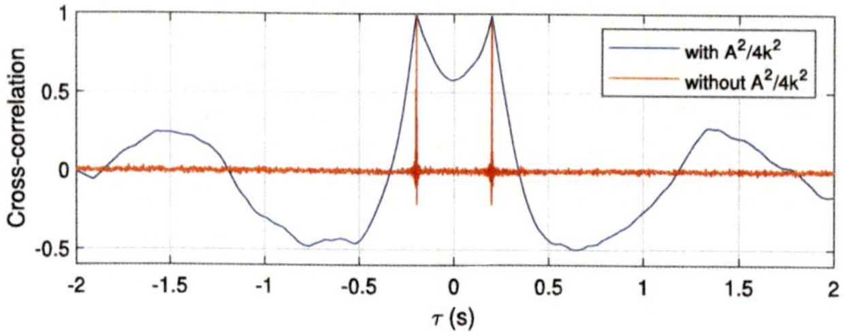

Fig. 2. Simulation results of the cross-correlation output of the ambient noise with and without considering the $A^2/4k^2$ term.

The blue curve in Fig. 2 shows the simulation results of the cross-correlation output when $F(\tau)$ term in Eq. 5 is considered, and it is found that the two dominant peaks are replaced by two triangles. This is because the Fourier transform of $A^2/4k^2$ term has a triangle function component. But the peak of each triangle is still the same as the red curve, meaning it still can be used for wave speed estimation. But in case of false peak identification, we can multiply k over the measured noise frequency spectrum $N(f)$ at each sensor then take the cross-correlation. This compensates the $A^2/4k^2$ term and the cross-correlation output is converted back into the red curve.

3 Field Experiment Procedure and Results

To verify the feasibility of the proposed passive wave speed measurement method, field experiments were carried out in the operating water supply network at the campus of University of Canterbury. These experiments also aimed to test the accuracy of passive wave speed measurement by comparing with the measurement results of conventional active method using hydraulic transients generated by rapid manual valve closures.

3.1 Experiment Site, Setup and Procedure

In the field experiment, two fire hydrant were selected to access a buried water supply pipeline. The distance between the two hydrants is 62 m. The pipe between the hydrants is straight without any junction or branch, and is made of asbestos cement with a 150 mm diameter.

Two recording systems were used to passively measure the ambient pressure noise. These recording system were made up of a piezoelectric pressure sensor (type 102A07), a National Instrument (NI) data acquisition board (type USB-6212), a GPS module, and the user interface on a laptop. The sensor only capture the dynamic pressure variations, which is suitable for noise measurement. The NI board was controlled by the interface software programmed by LabVIEW and the acquired pressure data were saved to the laptop hard disk. The sampling rate was set as 2 kHz. A trigger from the GPS module was then used to ensure both sensors were synchronized.

The sensor was mounted on top of a standpipe and it was installed at each fire hydrant. Each standpipe had a bleed valve to flush air out of the system before measurement started. The ambient noise in the water supply network were recorded by these systems for 20 min. Another separate experiment was also carried out, where the rapid closure of the bleed valve was used to generate transient pressure waves at one of the hydrant and received at the other hydrant to estimate pressure wave speed using the conventional active method. This active results were used as benchmarks to test the accuracy of the wave speed measured by the passive method.

3.2 Experiment Results

The power spectrum for the measure pressure noise is calculated using the Welch's method (90 s window length, 50% overlapping) and is shown in Fig. 3: At 0.1 Hz, the power is 175 dB, and this decreases to 97 dB 100 Hz. A100 Hz, the power spectrum flattens out. Therefore, frequency components between [0.1, 100] Hz were selected as the noise data to be further for processed.

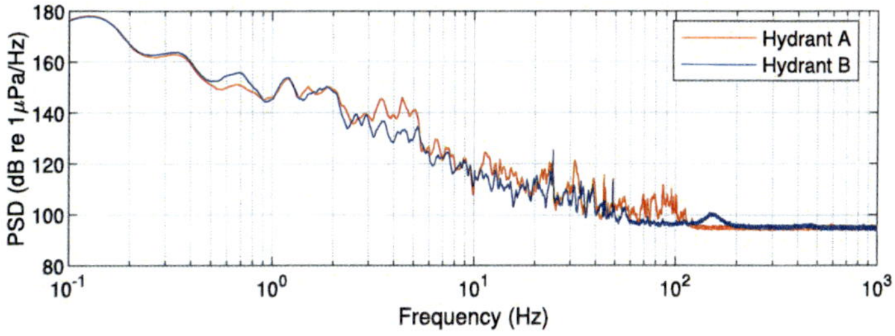

Fig. 3. The power spectra of the ambient noise measured at Hydrant A and B.

Figure 4a shows the cross-correlation output of the 20 min pressure data measured by the two sensors, and its shape is consistent with the theoretical derivation and numerical simulation, with two dominant peaks at symmetric time instants. Reading off this graph, the two peaks are at $\tau = \pm 0.0573$ s, which corresponds to a wave speed of 1082.66 m/s. Next, we divide the 20 min data into 5 groups in time order and process each data set separately using cross-correlation and the results are listed in Table 1. The standard deviation of passive measurement results is 13.64 m/s. Furthermore, the 20 min data was divided into 4 groups with 5 min duration and went through the same cross-correlation processing. The results show that longer time window is helpful in reducing the measured wave speed uncertainty, but 4 min time window is sufficient to obtain consistent wave speed measurement.

Table 1. Pressure wave speed (m/s) measured by different methods

Method	#1	#2	#3	#4	#5	Mean	Std
Passive (4 min)	1082.02	1083.92	1065.29	1103.20	1078.26	1082.26	13.64
Passive (5 min)	1082.02	1076.39	1097.35	1080.14		1083.98	9.22
Active	1068.97	1085.81	1072.66	1070.81	1074.52	1074.56	6.63

For the active method, one of pressure waves recorded at the two hydrants are shown in Fig. 4b, and the wave speed can be found by comparing the time delay of the peaks of the two signals. Such active measurement were repeated 5 times in the experiment and the estimated wave speeds are listed in Table 1. It is obvious that the wave speed measured by active and passive method are consistent: Comparing the averaged active results and passive results of 20 min time window, the difference in wave speed is 7.9 m/s, corresponding to a relative error of 0.7%.

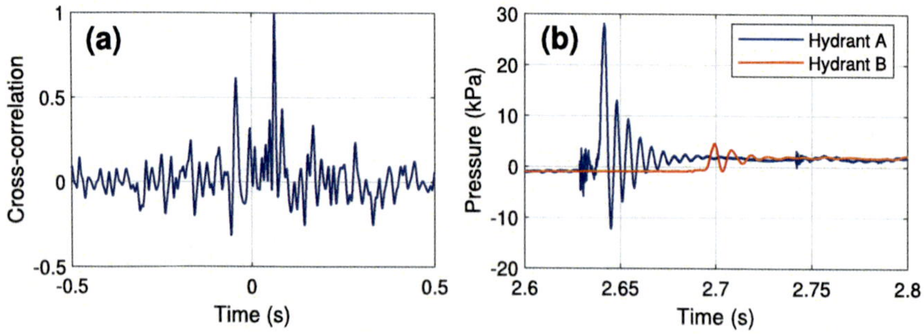

Fig. 4. (a) Cross-correlation output of the ambient noise. (b) Pressure transient measured at Hydrant A and B.

4 Conclusion

This paper proposed a passive wave speed estimation method, which is based on the cross-correlation processing of the ambient noise in water pipeline networks measured by two pressure sensors. Field experiment results show that the wave speed can be estimated using the proposed passive method, and the accuracy is at the same level as the conventional active method. It is also found that longer measurement and processing time window is helpful in reducing the uncertainty of the passive method, and 4 min time window is sufficient to obtain consistent passive wave speed measurement.

References

1. Water New Zealand: National performance review 2018–2019. Technical report, The New Zealand Water and Wastes Association (2019)
2. Rodhe, H., Herrera, R.: Acidification in Tropical Countries. Wiley, Hoboken (1988)
3. Vreeburg, I.J., Boxall, J.B.: Discolouration in potable water distribution systems: a review. Water Res. **41**, 519–529 (2007)
4. Cole, I.S., Marney, D.: The science of pipe corrosion: a review of the literature on the corrosion of ferrous metals in soils. Corros. Sci. **56**, 5–16 (2012)
5. Kumar, S., Taheri, F.: Neuro-fuzzy approaches for pipeline condition assessment. Nondestr. Test. Eval. **22**(1), 35–60 (2007)
6. Jo, B.Y., Laven, K., Jacob, B.: Advances in CCTV technology for in-service water mains. In: Pipelines 2010: Climbing New Peaks to Infrastructure Reliability: Renew, Rehab, and Reinvest, pp. 538–547 (2010)
7. Liu, Z., Kleiner, Y.: State of the art review of inspection technologies for condition assessment of water pipes. Measurement **46**, 1–15 (2012)
8. Li, Z., Lee, P., Murch, R.: Compressive-sensing based super-resolution detection for leakage and uniform blockage in water pipelines. Mech. Syst. Sig. Process. **158**, 107686 (2021)
9. Wang, W., Li, Z., Jing, L., Lee, P., Murch, R.: A straightforward method for estimating the size of leaks in water pipelines using acoustic transients. J. Acoust. Soc. Am. **144**(5), EL404–EL409 (2018)
10. Gong, J., Simpson, A.R., Lambert, M.F., Zecchin, A.C., Kim, Y.-I., Tijsseling, A.S.: Detection of distributed deterioration in single pipes using transient reflections. J. Pipeline Syst. Eng. Pract. **4**(1), 32–40 (2013)
11. Jing, L., et al.: An approximate inverse scattering technique for reconstructing blockage profiles in water pipelines using acoustic transients. J. Acoust. Soc. Am. **143**(5), EL322–EL327 (2018)
12. Alexander, J., et al.: Experimental investigation of the interaction of fluid transients with an in-line air pocket. J. Hydraul. Eng. **146**(3), 04019067 (2020)
13. Alexander, J., Li, Z., Lee, P.J., Davidson, M., Duan, H.-F.: Comparison of numerical models for the interaction of a fluid transient with an offline air pocket. J. Hydraul. Eng. **147**(5), 04021014 (2021)
14. Stephens, M.L.: Transient response analysis for fault detection and pipeline wall condition assessment in field water transmission and distribution pipelines and networks. Ph.D. thesis, University of Adelaide (2008)

15. Gong, J., Lambert, M.F., Simpson, A.R., Zecchin, A.C., et al.: Distributed deterioration detection in single pipes using the impulse response function. In: 14th Water Distribution Systems Analysis Conference (2012)
16. Tuck, J., Lee, P.: Inverse transient analysis for classification of wall thickness variations in pipelines. Sensors **13**(12), 17057–17066 (2013)
17. Gong, J., Lambert, M.F., Simpson, A.R., Zecchin, A.C.: Detection of localized deterioration distributed along single pipelines by reconstructive MOC analysis. J. Hydraul. Eng. **140**(2), 190–198 (2014)
18. Gong, J., Stephens, M.L., Arbon, N.S., Zecchin, A.C., Lambert, M.F., Simpson, A.R.: On-site non-invasive condition assessment for cement mortar-lined metallic pipelines by time-domain fluid transient analysis. Struct. Health Monit. **14**(5), 426–438 (2015)
19. Dubey, A., Li, Z., Lee, P., Murch, R.: Measurement and characterization of acoustic noise in water pipeline channels. IEEE Access **7**, 56890–56903 (2019)
20. Ahadi, M., Bakhtiar, M.S.: Leak detection in water-filled plastic pipes through the application of tuned wavelet transforms to acoustic emission signals. Appl. Acoust. **71**(7), 634–639 (2010)
21. Wang, W., Li, Z., Dubey, A., Lee, P., Fink, M., Murch, R.: Passive imaging of water pipelines using ambient turbulence noise. Mech. Syst. Sig. Process. **160**, 107882 (2021)
22. Garnier, J., Papanicolaou, G.: Passive Imaging with Ambient Noise. Cambridge University Press, Cambridge (2016)
23. Li, Z., Jing, L., Murch, R.: Propagation of monopole source excited acoustic waves in a cylindrical high-density polyethylene pipeline. J. Acoust. Soc. Am. **142**(6), 3564–3579 (2017)
24. Li, Z., Jing, L., Wang, W., Lee, P., Murch, R.: The influence of pipeline thickness and radius on guided wave attenuation in water-filled steel pipelines: theoretical analysis and experimental measurement. J. Acoust. Soc. Am. **145**(1), 361–371 (2019)

Preliminary Numerical Simulation for the Development of a Seismic Camera

Pedro Matos Casado[1(✉)], Emiliano Rustighi[1], Vigilio Fontanari[1], and Jennifer Muggleton[2]

[1] Industrial Engineering Department, University of Trento, Trento, Italy
{ph.demelocasadomatos,emiliano.rustighi,vigilio.fontanari}@unitn.it
[2] Institute of Sound and Vibration Research, University of Southampton, Southampton, UK
jmm@isvr.soton.ac.uk

Abstract. The acoustic camera is an established and highly effective device for localising acoustic sources by the use of a number of simultaneously acquired signals from an array of pressure sensors (microphones). The acoustic camera essentially provides for a highly directional sensor in which the signals arriving from the noise source in the steered main beam of the array are highly amplified relative to the background noise, which arrives at the camera from all directions outside the main beam and is therefore suppressed. The underlying principle of the acoustic camera is the beamforming data processing method which is widely applied in sensor array configurations and acts as a spatial filtering operation. The long term vision of the team is to develop an analogous device, termed here seismic camera, which allow to locate the direction of the noise sources generated from water leaks. This is an array of 3-axis geophones distributed on the ground in the vicinity of the suspected leak to localise and quantify water leaks with significantly greater accuracy and reliability than conventional methods that use just two sensors either side of the leak. The seismic camera differs from the acoustic camera since the array of data is vectorial (three axis geophones provide velocities instead of a scalar pressure field), two or more wave types (compressional, shear and surface waves) propagate simultaneously and the soil properties varies greatly with location, type and condition. In this preliminary feasibility study a time-domain solution calculated from the analytical elasticity equations is considered to generate the numerical data. The wave field is composed by spherical compressional waves radiating directly from the leak which is modelled here as a spherical cavity of radius a. The obtained numerical data is elaborated in order to look at the implementation of the Delay-and-Sum beamforming algorithm for the detection of the leak. Finally, the effects of wave reflection caused by a free surface and the sensor direction of measurement are discussed and it is shown that the beamforming algorithm works better and more precisely in infinity medium models, although the half-space model still presents satisfactory result.

Z. Dimitrovová et al. (Eds.): WMVC 2022, MMS 125, pp. 1085–1095, 2023.
https://doi.org/10.1007/978-3-031-15758-5_112

Keywords: Water leak source detection · Buried pipelines · Seismic camera

1 Introduction

The acoustic camera, often referred to as the acoustic beamforming methods, is a standard method for localising sources of sound waves in free space and has been widely used in different acoustic applications for the past hundred years [1,2]. The technique is often performed in engineering applications that involves moving objects localisation such as air-crafts, submarines and vehicles and is commonly applied in signals that describes scalar wave fields from pressure fluctuations measurements through microphones. The beamforming algorithms can be interpreted as a spatial filtering operation to map the distribution of sound sources at a certain distance of an array of sensors [2]. However, the technique is yet to be analysed in applications that involves detection of buried sound sources through the use of vector wave field measurements performed by state-measuring sensors such as geophones or accelerometers. Through this approach, the feasibility of the seismic camera and its usage for the localisation of buried leaks in pipes is investigated. The actual state of the art in leak detection is the cross-correlation techniques that involves estimating the leak position [3,4].

The application of the beamforming method requires prior information on the free-field wave velocity, although this information can either be calculated from theoretical models and ground properties measurements or estimated in-situ with other types of experimental techniques [5]. Zheng et al. [6] applied the beamforming algorithm to estimate the wave speed and then detect leak position along a gas pipeline for different number of sensors. The measurements were made directly on the pipe wall by accelerometers in a no-buried condition. The technique showed good estimates for both the wave velocity and the leak position for all tested cases, although the algorithm required a high computational cost to iterate through the estimates of wave speed and leak position with good resolution. However, no investigation on the beamforming method performance to detect leaks in buried pipes through an array of directional sensors has been made.

The aim of this paper is to investigate the beamforming algorithm performance in detecting a buried noise radiator spatially measured by a geophones array. The analytical solution for the displacements of a spherical wave propagating in an infinite medium is calculated to generate the numerical data. The propagating wave is radiated as an impulse by a symmetrical spherical source of constant radius. As spherical symmetry is assumed, the spatial solution is given only in terms of the radial coordinate which is decomposed in two perpendicular directions to further investigate the effect of the direction of measurement of the wave fields (perpendicular and parallel to the ground surface) and to investigate the method applicability in situations limited by 1D-axis sensors. The infinite medium solution is obtained from the theory of elasticity equations by applying a time pulse in the boundary condition for the radial stress around the spherical source. The half-space medium solution is obtained by considering reflection coefficients and wave mode conversion effects that are calculated from the no-stress boundary conditions at the free-surface. The array of sensors is considered to be a one-dimensional horizontal line of evenly distributed sensors. A spherical source is placed in any position in relation to any one of the sensors, and the numerical signals are generated from the analytical solutions. At last, the generated data is performed through the Delay-and-Sum beamforming algorithm to generate the signal amplitude map for different trial points in the array surroundings.

The paper is organised as follows. Section 2 presents the beamforming algorithm and review some of its applications. Section 3 presents the propagating spherical wave solution radiated by an impulsive spherical cavity. Then, the formulation that gives the wave reflection coefficients is presented and the expressions for the sensor displacements under reflecting conditions are calculated. Numerical results are presented in Sect. 4 and the effects of the measurement direction and the reflection on the surface are highlighted. At last, Sect. 5 presents the concluding remarks of the preliminary study and suggests future analysis.

2 The Beamforming Algorithm

The beamforming methods refers to a family of array signal processing algorithms that act as a spatial filtering process of the spatially measured signals within an array of sensors. Different variations and strategies of the beamforming algorithm may be found in the literature [2,7]. Differences between the wide variety of beamforming algorithm strategies include the sensor distribution in the arrays aperture and the deconvolution, inverse and acoustic imaging methods [8]. For a initial feasibility study, the time-domain beamforming Delay-and-Sum algorithm is chosen for the numerical data processing.

The Delay-and-Sum is a simple yet robust beamforming algorithm for source localisation. The algorithm consists in applying a delay to the measured signals of an array and then stacking up the phased signals to obtain an equivalent signal which is amplified in the directions where a source is present. The technique works either if the array is in the far-field or in the near-field with respect to the source, although for the far-field case only the source direction is obtained [2]. Through the delaying process, a main beam of the array is formed and can be mathematically steered acting as if the aperture of the array of sensors was an antenna amplifying signals arriving from sources and suppressing background noise arriving from other directions.

The phase centre is defined as a mean geometric position of the array of sensors, and the time delay applied in each sensors is done in such a way that all the signals would be received there. The distance between the phase-centre and the source trial position is called the main beam of the array. The equivalent signal measured by the phase-centre is mathematically described as [7]

$$z(t) = \Sigma_{m=0}^{M-1} w_m y_m(t - \Delta_m) \tag{1}$$

where M is the number of sensors in the array, w_m is a weight function that is also interpreted as a spatial windowing coefficient useful to relieve spatial aliasing and the leakage effects common to all digital signals, $y_m(t - \Delta_m)$ is the signal measurement of the m-th sensor which is phased by a time delay of Δ_m. The time delay that phases each sensor depends on the propagating wave, its wave speed and its propagation path. For the case of a spherical wave front, the delay for each sensor can be calculated using the expression

$$\Delta_m = \frac{r^0 - r_m^0}{c_l} \tag{2}$$

where r^0 is the distance between the phase centre and the source trial position, r_m^0 is the distance between the m-th sensor and the source trial position and c_l is the longitudinal wave velocity in the medium. When the source trial position coincides with the source actual position, the signal is amplified to its maximum, pinpointing the noise source location.

3 Elastic Wave Propagation in Free Space

This section presents the solution for the displacements field in an infinite elastic medium given a radial step-pressure stress excitation around a spherical cavity. Moreover, longitudinal incident wave reflection and wave mode conversion on a free-surface are discussed and the amplitude ratio between reflected and incident waves are derived from a no-stress boundary condition at the free-surface.

The analytical solution that generates the numerical data is based on the theory of elasticity which are derived from the stress-strain relationship in a continuous solid and may be rewritten in terms of potentials through the Helmholtz decomposition of a vector quantity. The main assumptions of the formulation are that the solid is elastic, isotropic and continuous and the Lamé constants are the only material properties considered [9]. From this assumptions, it can be shown that two basic wave types can exist in the elastic medium, namely the longitudinal and shear waves. Such waves propagate independently from each other with their own specific velocities that in turn depends only on the medium elastic properties.

The stress equation in terms of the spherical scalar potential ϕ in an infinite medium is given as [9]

$$\tau_{rr} = \lambda \nabla^2 \phi + 2\mu \frac{\partial^2 \phi}{\partial r^2}, \tag{3}$$

where τ_{rr} is the normal stress in the radial direction in N/m^2, λ and μ are the first and second Lamé parameters which are functions of the shear and bulk moduli of the surrounding medium and ∇ is the space differential vector operator reduced to $\partial/\partial r$ due to the spherical symmetry.

In practical applications the arriving wave fronts detected by the sensors in the array often carry a wide-band frequency energy, and thus, impulse stress boundary condition is considered to obtain the analytical solution that generates the numerical data. The analytical solution for a longitudinal spherical impulse radiated by a spherical cavity of radius a is obtained by considering the following stress condition around the cavity's surface

$$\tau_{rr}(a, \theta, t) = -p_0 \delta(t), \tag{4}$$

where p_0 is the pressure amplitude in $\mathrm{N/m^2}$ and $\delta(t)$ is the Dirac Delta function. The initial conditions are considered to be homogeneous and are written as $u(r,0) = \dot{u}(r,0) = 0$. Also, the spherical symmetry of the problem is exploited so that the displacements are described only in terms of the radial component of the coordinate system. The radial component of the displacement will be further decomposed in two perpendicular directions to simulate the sensor direction of measurement. By applying the stress condition around the cavity in Eq. (4) in Eq. (3), the displacement field for the medium is obtained as

$$u_r(r, \tau) = \frac{1}{r} \frac{a^2}{\rho c_l} \frac{(1-\nu)}{\sqrt{1-2\nu}} \exp(\xi\tau) \left[\left(\frac{1}{r} + \frac{a}{c_l} \right) \sin \omega\tau + \cos \omega\tau \right] \tag{5}$$

where r is the radial distance from the spherical source in metres, τ corresponds to the difference between a reference time and the flight duration of the radiated wave and is related to the position of the wave front for all time. For a pulsating sphere of radius a, the time delay may be written as $\tau = t - (r-a)/c_l$, where c_l is the longitudinal wave speed in m/s. The frequency parameters ξ and ω are function of the medium Poisson's ratio and the source radius and are expressed as $\xi = (1-2\nu)c_l/(1-\nu)a$ and $\omega = \xi(1-2\nu)^{-1/2}$.

Figure 1 illustrates the described model showing a spherical longitudinal wave propagating with velocity c_l. The wave is radiated by a spherical pulsating source and the medium properties are known. When a free-surface is added to the model, both wave reflection and mode conversion may occur. Upon arrival onto the free-surface, the incident wave energy may be distributed between a reflected shear and/or reflected longitudinal wave whose energy distribution depends on the boundary condition and incident angle θ_1.

To calculate the energy distribution between the reflected waves in terms of the reflection coefficients, the free-surface is set to have a no-stress boundary condition. Then the general D'Alambert solution of the wave equation is assumed for the potentials. This is consistent with the impulse formulation since the system is observed to be non-dispersive and thus no distortion of the wave shape occurs. The boundary condition on the free surface is modelled as the no-stress condition and may be expressed as

$$\tau_{yy} = \tau_{yx} = \tau_{yz} = 0. \tag{6}$$

The amplitude ratio between the incident and the reflected waves may be obtained by considering the D'Alambert solution for the scalar and the vector potentials as [9]

$$\phi = A_1 f(k_p(x \sin\theta_1 + y\cos\theta_1 - c_p t)) + A_2 f(k_p(x\sin\theta_1 - y\cos\theta_1 - c_p t)), \tag{7}$$

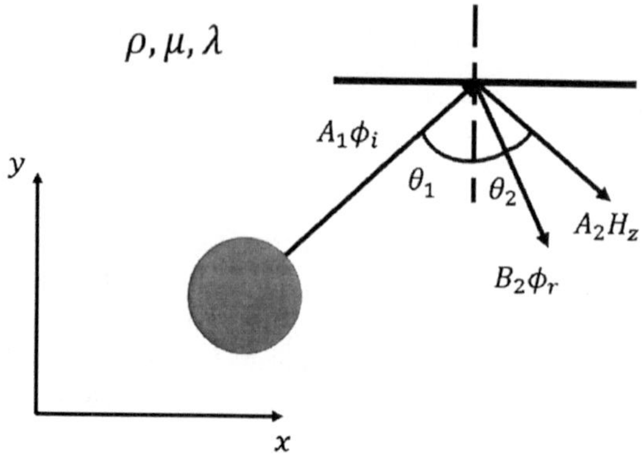

Fig. 1. A compressional wave with amplitude A_1 is emitted by a spherical cavity of radius a and then is reflected upon arrival in the free surface with incidence angle θ_1. Because of wave mode conversion, a shear wave of amplitude B_2 is reflected with reflection angle θ_2 in conjunction with a compressional wave of amplitude A_2 with reflection angle θ_1.

$$H_z = B_2 f(k_s(x\sin(\theta_2) - y\cos(\theta_2) - c_s t)), \tag{8}$$

where k_p and k_s are the compressional and shear wavenumbers respectively in unit of rad/m and θ_1 is the incident and reflection angle for the compressional wave whereas θ_2 is the reflection angle for the shear wave. The angles θ_1 and θ_2 are defined as the angles between the surface normal direction, defined in Fig. 1 as y, and the wave direction of propagation. For the present paper, it is considered that only compressional waves impinge in the surface and thus only the reflected shear waves exist in the solution. Consequently, only the shear amplitude B_2 is considered in Eq. (8). By substituting Eqs. (7)–(8) in Eq. (6), the amplitude ratios between the incident and reflected compressional waves are obtained as

$$\frac{A_2}{A_1} = \frac{\sin 2\theta_1 \sin 2\theta_2 - k^2 \cos^2 2\theta_2}{\sin 2\theta_1 \sin 2\theta_2 + k^2 \cos^2 2\theta_2}, \tag{9}$$

and for shear wave, the amplitude ratio is defined by

$$\frac{B_2}{A_1} = \frac{2\sin 2\theta_1 \cos 2\theta_2}{\sin 2\theta_1 \sin 2\theta_2 + k^2 \cos^2 2\theta_2}. \tag{10}$$

Note that, since the reflection coefficients and purely real, they may be used directly in the displacement equation described in Eq. (5). Therefore, Eqs. (9) and (10) describes how the energy of the incident wave is distributed between the reflected longitudinal and shear waves, respectively. Note that the energy distribution is a function of both the angle of incidence θ_1 and of the medium material properties that appear in the equation as k^2 in Eqs (9)–(10) and represents the ratio between the longitudinal and shear wave velocities, i.e., $k = c_l/c_s$.

Through the Helmholtz decomposition, the total displacement in the x and y directions are written in terms of the scalar potential field ϕ which is related to the compressional waves and the vector potential field H_z which is related to the shear waves. The relationship between the potential fields and the displacements may be written as

$$u_x = \frac{\partial \phi}{\partial x} + \frac{\partial H_z}{\partial y}, \qquad u_y = \frac{\partial \phi}{\partial y} - \frac{\partial H_z}{\partial x}. \tag{11}$$

Note that, since shear waves direction of propagation is perpendicular to the particles motion, the horizontal and vertical contributions of the shear wave are inverted in relation to those of the longitudinal wave. This behaviour may also be observed in Eq. (11) in the partial derivatives of the vector potential H_z. Finally, the x and y components of the displacement in the surface of a half-medium are obtained by substituting the D'Alambert solutions in Eqs. (7)–(8) in Eq. (11) as

$$u_x = \sin\theta_1(A_1 u_i + A_2 u_r) + B_2 \cos\theta_2 u_s, \tag{12}$$

$$u_y = \cos\theta_1(A_1 u_i - A_2 u_r) + B_2 \sin\theta_2 u_s, \tag{13}$$

where u_i, u_r and u_s are the radial displacement component of the incident, reflected and converted waves respectively, as shown in Fig. 1.

4 Results and Discussion

The numerical data is generated by the analytical solution of the pressure impulse discussed in Sect. 3. The array is considered to contain four sensors that measure the displacements in the horizontal and the vertical directions. The vector component along the radial direction is calculated through vector decomposition and summation of the orthogonal measurements. The sensors are distributed in the array one meter apart from each other and a spherical source (magenta circle) with radius of 5 cm is placed in the point $(-0.75, -2)$ of the coordinate system as represented in Fig. 2. The phase-centre is represented as the green dot and the sensors as red dots in Fig. 3. The pressure amplitude of the impulse is of the order of 1 MPa. The soil has properties of 2000 kg/m^3 density and a bulk and shear moduli of 0.053 GPa and 0.02 GPa respectively which results in longitudinal and shear wave speeds of 200 m/s and 100 m/s respectively. The soil properties were chosen based on the typical sand soil properties described in the work presented in [10]. The sensor distribution through the array and the source position were chosen arbitrarily. The source localisation occur by mapping an arbitrary parameter of the algorithm output and for this analysis the maximum absolute value of the output signal was selected to avoid sign problems with the signal phase. At this step of the algorithm implementation, it is necessary to define a grid of possible source positions from which the delay in Eq. 2 is calculated for each sensor. For the present work, a square grid of a 29×29 trial points distant 0.25 m apart is chosen to calculate the beamforming delay for the sensors. Note that the choice of the distance between the trial points only limits the source location error, thus a 0.25 m precision is satisfactory for the purpose of this paper. Figure 3 presents the mapping for the

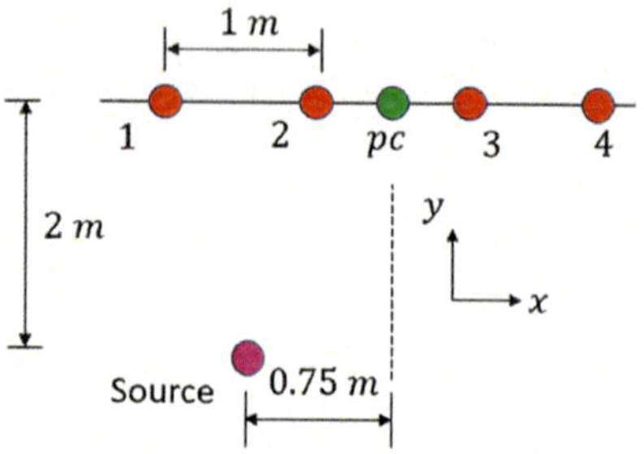

Fig. 2. Schematic representation of the numerical analysis. An array of four sensors (red dot) with the phase centre (green dot) at the origin measures the signal radiated by a spherical source (magenta dot).

source location for the infinite space (left) and the half space (right) cases for the radial (top row), vertical (middle row) and horizontal (bottom row) components.

A preliminary analysis of Fig. 3 indicates that the beamforming algorithm performance is sensible to the direction of measurement of the signal. Specifically, the horizontal component result in Figs. 3(e) and (f) presents a continuous beam of possible positions for the source location, i.e., the algorithm is only capable of estimating the direction of propagation of the noise. This behaviour is due to the reference frame of the sensors pointing all to the positive x direction of the phase-centre reference point. The reference frame of the sensor defines the sign of the measured signal, thus the number of sensors either side of the source defines the maximum amplitude of the beamforming output. If the number of sensors were equal either side of the leak then the output would be approximately zero and the horizontal coordinate of the source would be defined by a vertical line representing a local minimum of the map. On the other hand, the components r and y were able to pinpoint the source location by the superposition of all the beams formed by the algorithm. As the vertical components of the wave and of the sensors reference frame points to the same direction, no interference occurs and the beamforming output is maximised when the assumed source location coincides with its actual location. Also, it is possible to observe that for the half space results the beam associated with the horizontal direction measurement is reinforced in all directions of measurement along with other minor changes. This effect may be explained by the destructive wave interference caused between the incident and the reflected longitudinal waves, consequently reducing the relative amplitude between the vertical and the horizontal beamforming outputs. In addition, the half-space model considers reflection of shear waves that also contribute for the sensors displacement in both orthogonal directions, thus, depending on the angle of incidence and on the soil properties, the shear displacement contribution in each direction may alter the algorithm mapping.

The results suggest that the infinite medium x-component beamforming simulations may be representative of the half-space results. This possibility may

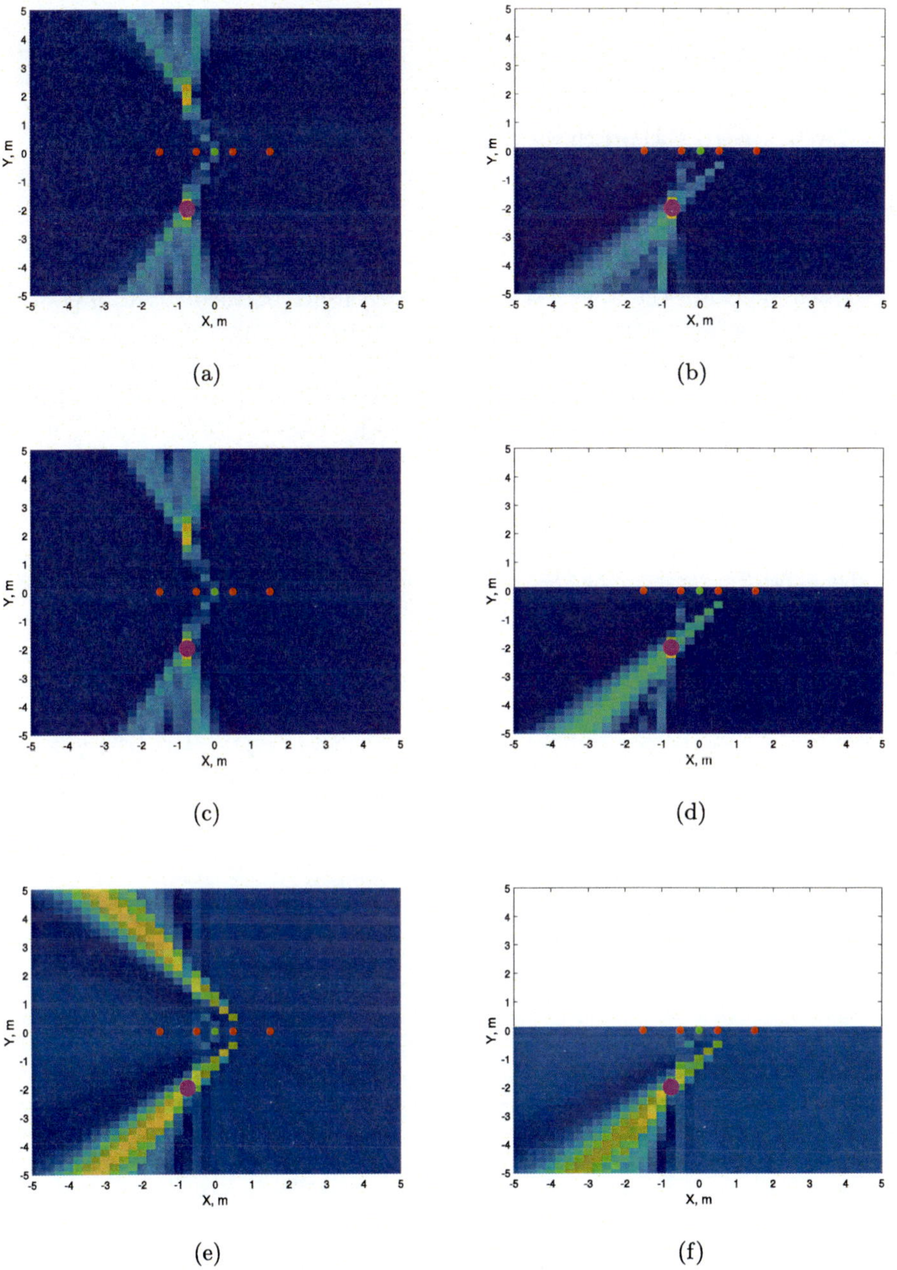

Fig. 3. Source localisation mapping obtained through the application of the beamforming. The left column are the results for the infinite medium and the right column are the results for the half medium. The results are divided by the direction of the components used in the algorithm, where the first row gives the results for the radial component, the middle row for the vertical component and the bottom row for the horizontal component.

greatly simplify the modelling requirements for more realistic cases and it will also be investigated in future research.

5 Concluding Remarks

The delay-and-sum beamforming technique has been proposed for locating underground sources through an array of directional sensors. A numerical investigation was carried out to analyse the effects of the direction of measurement of the sensor and of the wave reflection on the surface on the algorithm performance. The results for the horizontal direction presented a particular behaviour of estimating only the source direction rather than its approximate location. This behaviour was attributed to the signal cancellation on the x components of the displacements due to the sensors reference frame pointing all to the same direction. In contrast, the results for the vertical and the radial components resulted in the pinpointing of the source location due to the fact that no signal cancelling occurs to signals measured on those directions since the source is located beneath the array. The wave reflection effect did not have much influence on the algorithm performance on locating the source but some output anomalies were observed and correlated to the different contributions of the incident and reflected waves on the signal. In conclusion, the numerical investigation results suggests that the delay-and-sum beamforming algorithm may be a potential alternative technique to remotely locate leaks in buried pipes. Further investigation regarding the beamforming algorithm variants and optimisations are to be carried out in the future, considering more sensors for the array, a more precise model for the wave propagation solution and the effect of the wave velocity estimation on the algorithm performance as well as the soil properties and pressure amplitude effects.

References

1. Michel, U.: History of acoustic beamforming. In: 1st Berlin Beamforming Conference, p. 18 (2006)
2. Chiariotti, P., Martarelli, M., Castellini, P.: Acoustic beamforming for noise source localization - reviews, methodology and applications. Mech. Syst. Sig. Process. **120**, 422–448 (2019)
3. Puust, R., Kapelan, Z., Savic, D.A., Koppel, T.: A review of methods for leakage management in pipe networks. Urban Water J. **7**(1), 25–45 (2010)
4. Fuchs, H.V., Riehle, R.: Ten years of experience with leak detection by acoustic signal analysis. Appl. Acoust. **33**(1), 1–19 (1991)
5. Almeida, F.C.L., Brennan, M.J., de Lima, F.K., Iwanaga, M.K., Scussel, O.: Using a geophone as an actuator to estimate the velocity of leak noise propagation in buried water pipes. Appl. Acoust. **184**, 108251 (2021)
6. Zheng, X., Wang, Q., Xue, S., Zheng, C.: A beamforming-based joint estimation method for gas pipeline leak localization. Measurement **177**, 109264 (2021)
7. Johnson, D.H., Dudgeon, D.E.: Array Signal Processing: Concepts and Techniques. Prentice Hall (1992)

8. Leclère, Q., Pereira, A., Bailly, C., Antoni, J., Picard, C.: A unified formalism for acoustic imaging based on microphone array measurements. Int. J. Aeroacoustics **16**(4–5), 431–456 (2017)
9. Graff, K.F.: Wave Motion in Elastic Solids. Dover Publications (2012)
10. Gao, Y., Liu, Y., Muggleton, J.M.: Axisymmetric fluid-dominated wave in fluid-filled plastic pipes: loading effects of surrounding elastic medium. Appl. Acoust. **116**, 43–49 (2017)

Pressure Pulsation and Pipeline Vibration Damping with the Use of 3D Printed Nozzles

Przemysław Młynarczyk(✉), Damian Brewczyński, Joanna Krajewska-Śpiewak, Paweł Lempa, Jarosław Błądek, and Kamil Chmielarczyk

Faculty of Mechanical Engineering, Cracow University of Technology, 31-155 Cracow, Poland
pmlynarczyk@pk.edu.pl

Abstract. Manifold vibrations in volumetric compressors systems are often the result of pressure pulsations and flow characteristics as well as compressor-generated mechanical vibrations. Standard methods of damping pressure pulsations in compressor installations using Helmholtz resonators focus on damping the main pulsation frequency resulting from compressor operation. Currently, in both refrigeration and industrial compressors, continuous operation with variable rotational speed of the compressor drive shaft is becoming a standard. As a result, the compressor, while working, generates pressure pulsations with different frequency values, which cannot be effectively suppressed using the standard method. In the works published in recent years, there are informations that shaped nozzles, introduced into the discharge manifold of compressor, can significantly affect the damping of pressure pulsations in a wide range of their frequencies. The production of damping nozzles using 3D printing techniques allows us to design unprecedented shapes. The article presents the effect of nozzles with complex shapes (impossible or difficult to produce using conventional methods) on the damping of pressure pulsations and pipeline vibrations.

Keywords: Pressure pulsations · 3D printing · Vibration · Damping · Compressors

1 Introduction

Pressure pulsations which occur in installations of displacement compressors are one of the main negative phenomena generated by the operation of the compressor. One of the effects of the occurrence of pressure pulsation is pipeline vibration. Due to the safety requirements (in industrial gas installations) and parameters related to the comfort of their use in cooling installations, this pipeline vibration should be limited.

The most popular method of damping pulsations are mounting Helmholtz resonators. The tuned mass dampers are used often to dampen pipeline vibrations. These technologies only work on the designed frequencies. The publications [1–3] proved the effectiveness of pulsation damping in a wide frequency range by shaped nozzles. However, in this paper the influence of elements with complex shapes (impossible or difficult to produce with the use of classical production techniques) is presented. In order to produce such

Z. Dimitrovová et al. (Eds.): WMVC 2022, MMS 125, pp. 1096–1104, 2023.
https://doi.org/10.1007/978-3-031-15758-5_113

elements, 3D printing technologies were used. Currently, this type of approach seems to be more common in this type of research. In several papers different approach of 3D printing and usage of that elements was described. An unexpected acoustic absorption phenomenon in 3D printed Helmholtz resonators is described in paper [4]. The possibility to use 3D printed materials in high temperatures was described in paper [5]. Influence of different geometry in Tuned Sloshing Dampers on wave energy dissipation is presented in [6]. This paper shows that 3D printing can be used in a wide range of applications related to damping. From the very beginning of the use of displacement compressors the problem of pressure pulsations phenomena is very well known. Nevertheless, the current research topic is still widely described in scientific publications. The authors in paper [7] present the pressure pulsations in piping system and describe the damping characteristics. The influence of introducing flat orifice plate on the pressure pulsations is also presented in this paper.

Nonlinear vibrations of thin plate in pulsating flow are described and the theoretical approach is presented in [8]. The installation of multi-dimensional vibration isolation and mitigation device is described in Xu et al. [9]. The influence of the orifice on the decrease in the pressure pulsation level with the reduction of the orifice cross-sectional area was described by Rydlewicz et al. [10]. The usability of this element in piping systems to attenuate pressure pulsations was indicated. Effectiveness in damping pressure pulsations when applied to a resonant piping system by mechanism of a Herschel-Quincke (HQ) device was presented by Lato et al. [11]. In this paper, experimental investigations of the passive pressure pulsation and pipeline vibration damping with the use of 3D printed elements are described.

2 Damping Elements

During the research 12 different elements, made with 3D printing techniques were tested. Changes in static fluid pressure, flow velocity, and vortex characteristics are different for obstacles with different shape characteristics. Also the forces acting on different nozzles affect the pipeline differently and may amplify or dampen its vibrations. For this reason, the tests were carried out for nozzles of various shapes, classified into four groups. Elements of hyperboloidal shapes belong to the first group presented in Fig. 1. The effectiveness of these elements, in damping pressure pulsations, was presented in papers [1–3]. In presented research, different variations of hyperboloidal nozzle were examined. Shape A.1. is the simply hyperboloidal nozzle which allows to compare the effect of other shapes on the damping of pulsations and vibrations. Nozzles A.2., A.3. And A.5. Are twin hyperboloidal shapes, where A.5. is also shape used in previous investigation. Shapes A.2. And A.3. Had cylindrical holes in the material. In the case of A.2. The holes are larger and their arrangement is less frequent than in the A.3. Shape A.4. is a single hyperboloid shape with a void space between the outer diameter and a thin wall with a hyperboloidal profile. Additionally, the wall has openings with similar dimensions and arrangement as in the shape A.3.

The shapes of the second group are shapes that narrow the flow cross-section along the length of the nozzle similar as in the case of hyperboloid nozzles. These shapes are shown in Fig. 2. Shape B.1. is based on the shape of the mouthpiece for brass instruments. In the case of the instrument, the sound wave spreads evenly over the instrument, which may have a positive effect on the damping of pressure pulsations and maintaining flow consistency.

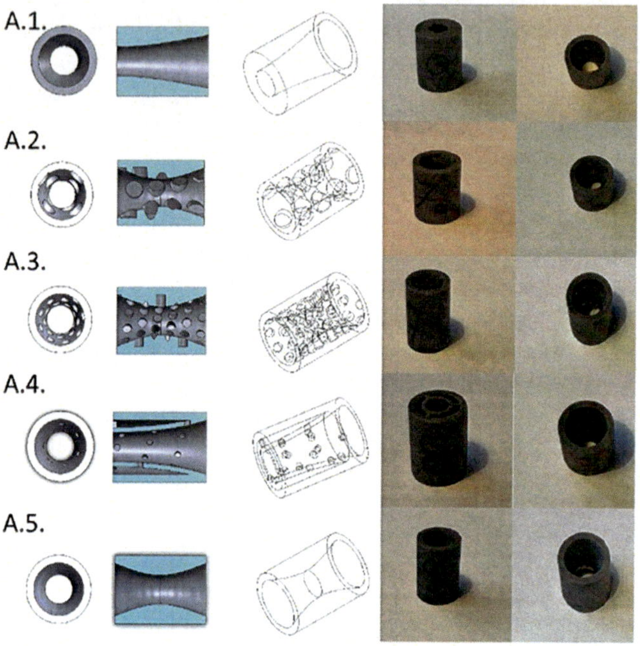

Fig. 1. Group A – hyperboloidal-type shapes

The concept B.2. is derived from a silencer for firearms, however, there is a narrowing of the inner diameter. The shape B.3. Consists of many helical ribs which have the effect of breaking up the flow around the circumference of the nozzle. Due to its complicated shape it is impossible to produce with subtractive methods.

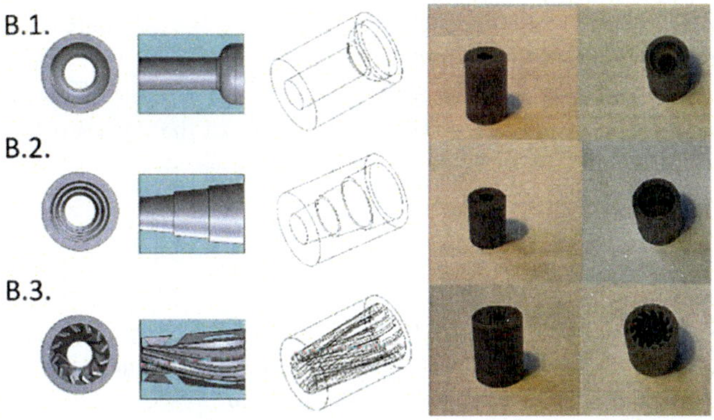

Fig. 2. Group B – narrowing shapes

Group three are non-standard nozzles with an irregular constriction. In Fig. 3 it can be seen that due to their shape, the flow through these elements is more irregular than in the previous groups.

A turbine-like element, shown in Fig. 4, was examined as the last one.

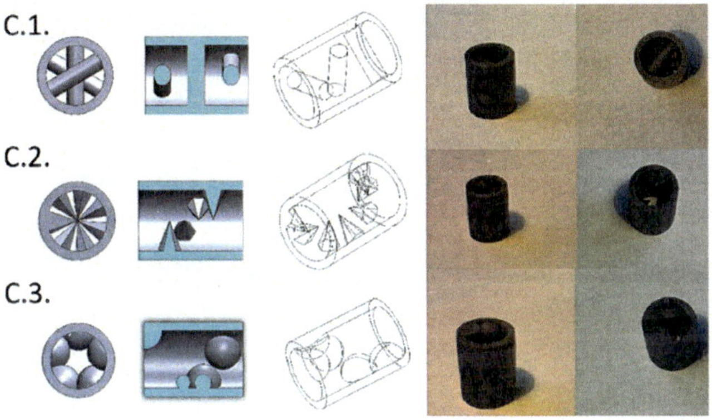

Fig. 3. Group C – irregular shapes

All elements were designed to meet a condition in which the change of the gas flow surface area in the passage is equal to 70% of the flow area of the pipe in which the elements are assembled.

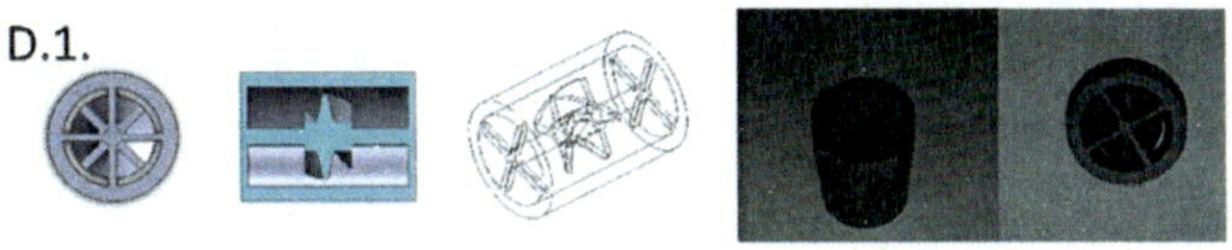

Fig. 4. Group D – turbine-like element

The point of reference for the results obtained for the presented shapes was the installation with an element maintaining the diameter of the pipeline, later described as an empty pipe. On the basis of previous tests, the large value of the narrowing used in the tests was selected, to emphasize the significant impact of the elements on the damping of pressure pulsations and on pipeline vibration.

3 Test Stand

The experimental investigations were performed in the discharge manifold of the two-cylinder reciprocating air compressor SAF-23 driven by an electric motor which is controlled by inverter. Test stand system is presented in Fig. 5.

The LabView/SignalExpress software was used to measure the pressure pulsation, while the DASYLab was used to measure the vibration signal.

For all measurements the pressure was set with a reducing valve at 1.5 bar. The temperature for the highest rotational speed stabilized each time at around 104 °C. While waiting for the compressor to reach a stable operating temperature, the temperature in the hall and the atmospheric pressure were recorded. After the parameters stabilized, measurements were taken. The duration of data recording for each measurement was approximately 5 s long. This amount of time was sufficient to catch the appropriate signal range. Due to the tracking the signal in real time, it was possible to detect incorrect readings caused by detaching the sensors during the measurement. In order to avoid random measurement errors, these measurements were repeated at 5 min intervals. After a series of measurements of the tested element, the compressor and all related devices were turned off.

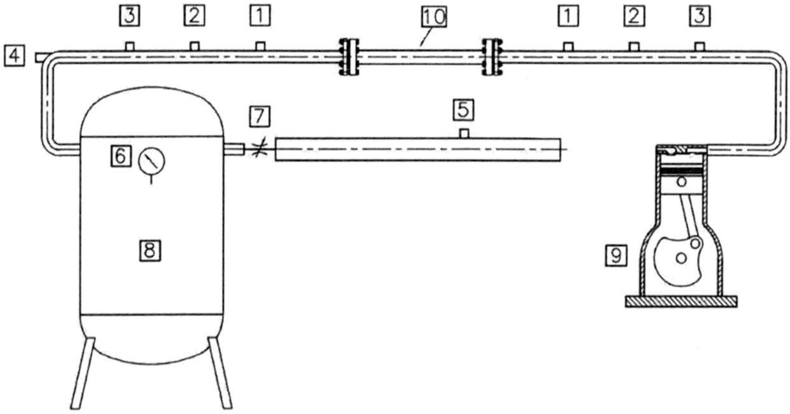

Fig. 5. The scheme of the test stand. 1 – Differential pressure measuring points, 2 – Dynamic pressure sensors, 3 – Thermocouples, 4 – Vibration sensor, 5 – Thermoanemometer, 6 – Manometer, 7 – Throttle valve, 8 – Compressed air tank, 9 – Compressor, 10 – Damping element mounted in the assembly part

4 Results

A series of experimental tests made it possible to determine the peak-to-peak value of the pressure pulsation amplitude for different values of the shaft rotational speed, and thus for different values of the pulsation frequency. The results obtained by the sensor behind nozzle mounting point for all investigated shapes for all pulsation frequencies are presented in the Fig. 6.

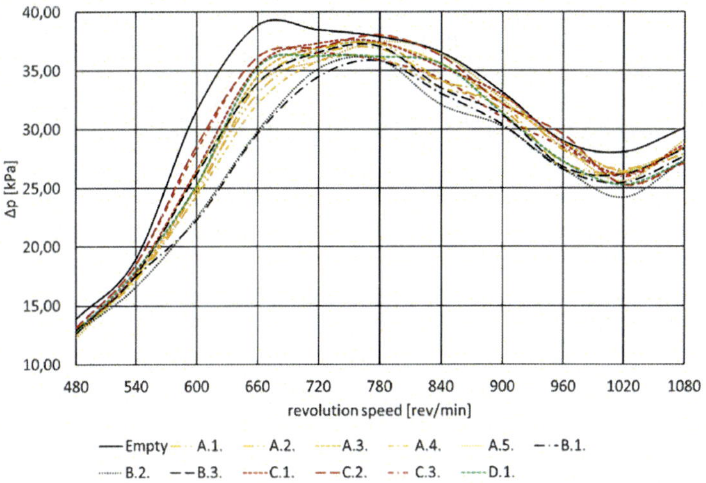

Fig. 6. Pressure pulsations peak-to-peak values for different shapes and different shaft revolution speed – pressure pulsation sensor located behind the nozzle mounting point

For all values the average pressure pulsation damping, for all frequencies, was calculated as a percentage damping related to empty installation as it is described in the Eq. (1):

$$\psi_{nozzle.avg} = 100\% - \frac{\Delta p_{nozzle.avg}}{\Delta p_{empty.avg}} \cdot 100\% \tag{1}$$

where:

$\psi_{nozzle.avg}$ – average nozzle pressure pulsation damping for wide range of frequencies [%],

$\Delta p_{nozzle.avg}$ – average pressure peak-to-peak value [kPa] with nozzle in the installation,

$\Delta p_{empty.avg}$ – average pressure peak-to-peak value [kPa] for straight pipe in the installation.

The calculated values of the average pulsation damping, measured in front and behind the nozzle for wide range of frequencies, are presented in Table 1.

Table 1. Pressure pulsation damping by different shapes.

No.		A.1.	A.2.	A.3.	A.4.	A.5.	B.1.	B.2.	B.3.	C.1.	C.2.	C.3.	D.1.
$\psi_{nozzle.avg}$[%]	*In front*	3,0	4,5	3,8	5,0	4,9	2,6	3,8	4,5	3,8	3,5	2,9	3,0
	Behind	7,9	8,0	7,2	9,5	7,1	12,0	12,9	8,4	6,0	4,8	6,3	8,0

The average damping of pressure pulsations behind the damping element is the highest for elements from group B as it can be seen in the Table 1. Damping in front and behind the nozzles have different nature for different shapes. In the A group the A.4. Shape stands out as it suppresses the pulsation value the most, both in front of and behind the nozzle. For the shapes B.1. And B.2. Pulsation damping in front of the nozzle is very low – that suggests a significant wave reflection on the orifice.

As during the tests the level of longitudinal vibrations of the installation was also measured, Table 2 shows the average vibration damping of the first harmonic for the tested frequency range. For all values the average vibration acceleration damping for all frequencies was calculated as a percentage damping related to empty installation, as it is described in the Eq. (2):

$$\Theta_{nozzle.avg} = 100\% - \frac{\Delta a_{nozzle.avg}}{\Delta a_{empty.avg}} \cdot 100\% \tag{2}$$

where:

$\Theta_{\text{nozzle.avg}}$ – average nozzle acceleration damping for wide range of frequencies [%],

$\Delta a_{\text{nozzle.avg}}$ – average acceleration peak-to-peak value [$\text{m}\cdot\text{s}^{-2}$] with nozzle in the installation,

$\Delta a_{\text{empty.avg}}$ – average acceleration peak-to-peak value [$\text{m}\cdot\text{s}^{-2}$] for straight pipe in the installation.

Table 2. Longitudinal vibration damping by different shapes.

No.	A.1.	A.2.	A.3.	A.4.	A.5.	B.1.	B.2.	B.3.	C.1.	C.2.	C.3.	D.1.
$\boldsymbol{\Theta}_{nozzle.avg}$[%]	10,4	5,3	5,1	17,9	8,7	11,3	16,3	−2,0	0,4	3,3	−18,5	−4,2

As can be seen, the influence of different shapes on vibration damping is highly heterogeneous in different groups. It can be seen that for shapes A.2. and A.3., which are the same hyperboloidal shapes, but with different hole diameters, the influence on the vibration damping is similar. The same effect is for hyperboloidal shapes without holes. In this group, for the shape with openings and free cavity between the hyperboloid profile and the pipe vibrations damping reach highest value. Vibrations are not damped or are actually amplified in the nozzles, in which the pulsating flow can cause higher values of radial and oblique forces than axial ones. These shapes are distinguished by the fact that they do not constrict the flow in an axially symmetric manner.

5 Conclusions

Performed investigations of the influence of 3D printed elements on the pressure pulsation and pipeline vibration damping indicate that shape is a key factor and have big influence on the described phenomena. The article presents how different shapes influence these important phenomena and shows a linear relationship between pressure pulsation damping and vibration damping for shapes narrowing along the length of the nozzle which can be described as axially symmetrical (excluding the holes contained therein). What is very important, it has been proven that even very different shapes placed inside the pipeline suppress pulsations in a wide frequency range. Thanks to the conducted experimental research, it is possible to choose the further direction of shape development in this type of damper. The results of the conducted research suggest that the most effective may be shapes that narrow along the length of the nozzle. Shape A.4 turned out to be particularly effective in simultaneous damping of pressure pulsations and vibrations of the pipeline, as it combines an internal hyperboloidal profile with an empty void into which gas flows through holes made in the hyperboloidal profile. Further research will be carried out, to determine the influence of a given shape on the flow resistance. Future research will focus on optimization of the shapes of this type of damping inserts.

Acknowledgements. This research was funded by Polish National Centre for Research and Development, grant number LIDER/40/0140/L-11/19/NCBR/2020.

References

1. Młynarczyk, P., Cyklis, P.: The application of nozzles for the attenuation of volumetric compressor pressure pulsation. Int. J. Refrig. **90**, 108–118 (2018)
2. Młynarczyk, P., Cyklis, P.: The estimation of the pressure pulsation damping coefficient of a nozzle. Int. J. Sound Vibr. **464** (2020)
3. Młynarczyk, P., Cyklis, P., Ryncarz, T.: Innovative hybrid device for the simultaneous damping of pressure pulsations and vibrations in positive displacement compressor manifolds. Int. J. Refrig. **132**, 109–118 (2021)
4. Costa-Baptista, J., Fosting, E.R., Therriault, D., Ross, A.: Impact of 3D printed thin wall on Helmholtz resonator acoustic response: experimental investigations. In: Proceedings of the 26th International Congress on Sound and Vibration, Montreal Bridges (2019)
5. Park, S.J., Lee, J.E., Park, J., Lee, N.-K., Son, Y., Park, S.-H.: High-temperature 3D printing of polyetheretherketone products: Perspective on industrial manufacturing applications of super engineering plastics. Mater. Des. **211**, 110163 (2021)
6. Cavalagli, N., Agresta, A., Biscarini, C., Ubertini, F., Ubertini, S.: Enhanced energy dissipation through 3D printed bottom geometry in Tuned Sloshing Dampers. J. Fluids Struct. **106**, 103377 (2021)
7. Hayashi, I., Kaneko, S.: Pressure pulsations in piping system excited by a centrifugal turbomachinery taking the damping characteristics into consideration. J. Fluids Struct. **45**, 216–234 (2014)
8. Tubaldi, E., Amabili, M., Alijani, F.: Nonlinear vibrations of plates in axial pulsating flow. J. Fluids Struct. **56**, 33–55 (2015)

9. Zhao-Dong, X., Ge, T., Miao, A.: Experimental and theoretical study on a novel multi-dimensional vibration isolation and mitigation device for large-scale pipeline structure. Mech. Syst. Signal Process. **129**, 546–567 (2019)
10. Rydlewicz, W., Rydlewicz, M., Palczynski, T.: Experimental investigation of the influence of an orifice plate on the pressure pulsation amplitude in the pulsating flow in a straight pipe. Mech. Syst. Signal Process. **117**, 634–652 (2019)
11. Lato, T., Mohany, A.: Passive damping of pressure pulsations in pipelines using Herschel-Quincke tubes. J. Sound Vibr. **448**, 160–177 (2019)

Simulation and Experimental Estimation of the Free Wavenumbers for Helically Grooved Tubes

Milena Watanabe Bavaresco[1](✉), Neil Ferguson[1], Claus Hessler Ibsen[2], and Atul Bhaskar[1]

[1] University of Southampton, Southampton SO17 1BJ, UK
mwb1f19@soton.ac.uk
[2] Vestas Aircoil, Smed Hansens Vej 13, 6940 Lem, Denmark

Abstract. The dynamic characterisation of elastic tubes is relevant to many fields in the industry. They are a key component for many processes involving fluid flow, heating, or cooling, etc., where grooves are frequently introduced to alter flow, and thus enhance heat transfer. Thin-walled tubes with helical features have been previously studied numerically and analytically, but experimental observation is still missing. In this paper, the wave propagation characteristics of a cylindrical tube with helicoidal grooves are investigated experimentally. The Wave Finite Element method is applied to a 3D model of this structure and a correlation method is used on a physical tube to experimentally observe the splitting of the degenerate dispersion branch into two. As opposed to a tube without grooves, both numerical and experimental methods show that the tubes with helicoidal grooves possess two distinctive but closely valued bending wave branches in the dispersion curves. The effect of the grooves is also visible in the mode shapes of the tubes of finite length, which will also be presented, where bending modes become unrestricted to lying in a single preferential direction.

Keywords: Pipes · Helicoidal grooves · Wave propagation · Experimental estimation of wavenumbers

1 Introduction

The characterisation of wave properties of mechanical structures has gained increased interest for topics such as damage detection, wave focusing and general understanding of the dynamic behavior of structures, to name a few. In this context, long slender structures with helical aspects have been analysed to identify existing propagating waves ([1–4]).

A characteristic effect of the helical geometry in the wave properties of these structures is that two distinctive flexural branches are visible in their dispersion curves. This contrasts to the two identical flexural branches related to degenerate modes occurring in orthogonal planes in an equivalent non-helical structure. This effect has been reported for helical waveguides in the form of beams through an analytical approach by Frikha [5], rods and wires through the Semi-Analytical Finite Element (SAFE) method in

Z. Dimitrovová et al. (Eds.): WMVC 2022, MMS 125, pp. 1105–1113, 2023.
https://doi.org/10.1007/978-3-031-15758-5_114

Tresseyde's work [3], through the Scaled Boundary Finite Element Method (SBFEM) in Liu's et al. work [4], and for helical springs through the Wave Finite Element (WFE) method in Renno and Mace's work [6]. A similar phenomenon was observed for a non-helical but curved pipe in work of Demma et al. [7], in which a mode that is described by one dispersion curve for a straight pipe splits into two dispersion curves in the curved pipe. Having observed this effect through these analytical and numerical investigations, this work is motivated by the need to experimentally observe the two distinctive flexural branches in a structure with helical features.

Among the existing experimental methods for the estimation of wavenumbers in one-dimensional waveguides, this paper focuses on the method used for the estimation of free wavenumbers introduced by Ferguson et al. [8]. It is based on a correlation between measurements in the form of transfer functions and a wave considering trial wavenumbers, where the maxima indicate the wavenumber values present in the response. It was used by Souza et al. [9] to estimate the wavenumber of propagating waves in metamaterial beams and by El Masri et al. [10] in reinforced concrete beams. A similar correlation-based method was also presented by Berthaut et al. [11] where complex-valued trial wavenumbers were introduced along with directional flexibility and a scaling to restrict the correlation results to lie within the unity range. This method has been used in many works for the estimation of wavenumbers for varied types of structures ([12]).

In the current work, a numerical investigation using WFE is described in Sect. 2. A unit cell of the real pipe is modelled and discretised to obtain the mass and stiffness matrices necessary for the calculation of wavenumbers. These results serve as reference and are supported by experimental observations, which are obtained through the correlation technique. The test setup and methodology are described in Sect. 3, followed by implementation details for the experiments. Finally, Sect. 4 contains a visual representation of how these two bending branches manifest in the mode shapes of the structure through the results of experimental and numerical modal analyses.

2 Numerical Analysis

A 1D WFE approach is proposed to estimate the propagating wavenumbers of the helically grooved pipe. A periodic cell is modelled as a 11 mm long tube with cross-sectional dimensions extracted from a microscopic picture of a segment of the real structure. The 3D CAD model was meshed in Ansys Workbench 18.1 with different types of elements as with several levels of refinement until mesh convergence was achieved with a solid shell type of element with three solid shell elements through the thickness. The material properties were selected as the nominal values for stainless steel. The mesh of the model is shown in Fig. 1.

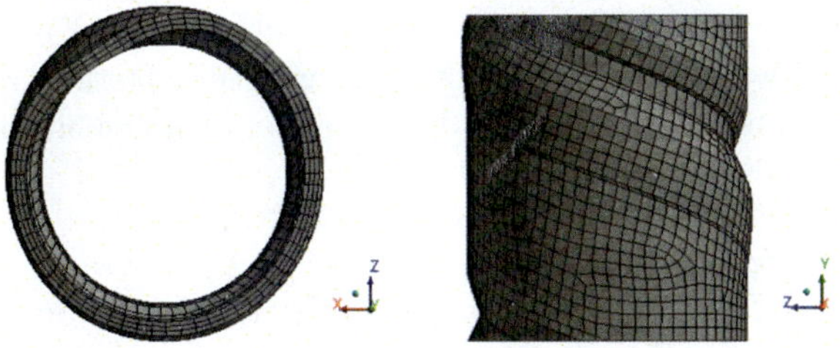

Fig. 1. Meshed finite element model of a periodic cell of the helically grooved tube.

The FE model was used to extract the mass and stiffness matrices, $\boldsymbol{M}$ and $\boldsymbol{K}$, which compose the dynamic stiffness matrix $\boldsymbol{D} = \boldsymbol{K} - \omega^2\boldsymbol{M}$. Damping was considered negligible. The equation of motion is expressed as

$$\boldsymbol{M}\ddot{\boldsymbol{q}} + \boldsymbol{K}\dot{\boldsymbol{q}} = \boldsymbol{f} \tag{1}$$

where $\boldsymbol{f}$ and $\boldsymbol{q}$ are $(2n \times 1)$ vectors representing the loads and nodal displacement at the n DOFs on each side of the cell. It can be written in terms of the dynamic stiffness matrix $\boldsymbol{D}$ as

$$\boldsymbol{D}\boldsymbol{q} = \boldsymbol{f} \tag{2}$$

Writing Eq. (2) in matrix form to reveal the association of the right or left nodes of the segment gives

$$\begin{bmatrix} \boldsymbol{D}_{LL} & \boldsymbol{D}_{LR} \\ \boldsymbol{D}_{RL} & \boldsymbol{D}_{RR} \end{bmatrix} \begin{Bmatrix} \boldsymbol{q}_L \\ \boldsymbol{q}_R \end{Bmatrix} = \begin{Bmatrix} \boldsymbol{f}_{\mathrm{L}} \\ \boldsymbol{f}_{\mathrm{R}} \end{Bmatrix}' \tag{3}$$

where the subscripts L and R discriminate between the left and right sides of the periodic cell (in this case, the top and bottom cross-section of the model shown in Fig. 1). The corresponding nodes from left and right sides can also be related through periodic conditions by defining a propagation constant λ, where the displacements and loadings are related through the expressions

$$\boldsymbol{q}_{\mathrm{R}} = \boldsymbol{\lambda}\boldsymbol{q}_{\mathrm{L}}, \ \boldsymbol{f}_{\mathrm{R}} = -\boldsymbol{\lambda}\boldsymbol{f}_{\mathrm{L}} \tag{4}$$

With

$$\lambda_j = e^{-jk_j\Delta} \tag{5}$$

where k is the unknown wavenumber, of which there are n solutions and Δ is the length of the periodic cell. The internal nodes of the periodic cell were dealt with by performing a dynamic condensation.

Among the existing formulations for the WFE eigenproblem, the one used in this work can be reached by rewriting Eq. (3) with the substitutions from Eq. (4), and projecting the equations of motion onto the left-hand DOFs, leading to the eigenvalue problem ([13, 14]):

$$\left[\begin{bmatrix} 0 & \boldsymbol{D}_{RL} \\ -\boldsymbol{D}_{RL} & -(\boldsymbol{D}_{LL} + \boldsymbol{D}_{RR}) \end{bmatrix} - \lambda \begin{bmatrix} \boldsymbol{D}_{RL} & 0 \\ 0 & \boldsymbol{D}_{LR} \end{bmatrix}\right] \begin{Bmatrix} \boldsymbol{q}_L \\ \lambda \boldsymbol{q}_L \end{Bmatrix} = 0 \tag{6}$$

This eigenproblem has been reportedly useful for the estimation of free propagating waves in undamped waveguides ([14]). Numerical errors in the form of FE discretization error, round-off errors in the dynamic stiffness matrix and conditioning issues were verified in accordance to the guidelines from [14].

The handling of the stiffness and mass matrices into the eigenproblem of Eq. (6) and its solution were performed through a MATLAB script. The wavenumbers related to the free propagating waves were selected as the real-valued solutions from $k = -\ln\lambda/(i\Delta)$ and are also shown in Fig. 2. For reference, the wavenumbers resulting from the same analysis applied to a pipe without helical grooves but with otherwise identical geometry and material properties are shown in Fig. 2. Only the positive wavenumber axis is shown for simplicity. The main difference observed in this frequency range is the existence of two distinct flexural wavenumber branches, labelled $F(1,1)^+$ and $F(1,1)^-$ for the helically grooved pipe. For the non-grooved tube, only one branch is visible, labelled $F(1,1)$ – or in fact two identical-valued branches representing the bending waveguide modes of degeneracy two in orthogonal directions of the pipe.

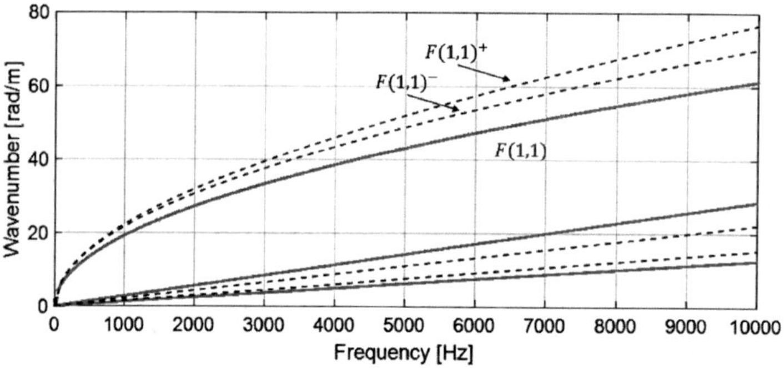

Fig. 2. Predicted wavenumbers calculated through the WFE for the helically grooved tube (black dashed line) and for the non-grooved tube (grey continuous line).

The additional branches containing lower values of wavenumbers presented in Fig. 2 are for the torsional and longitudinal propagating waves and are not investigated experimentally in this work. In the following section of this paper, the experimental observations of the distinctive flexural branches in the dispersion curves of the helically grooved pipe are described.

3 Experimental Investigation

The experimental approach to estimate wavenumbers considered herein is the correlation technique introduced by Ferguson [8]. It consists in finding, for each frequency, the wavenumber value that produces the maximum absolute value of the complex correlation function:

$$W\left(k_{tx}, k_{ty}, \omega\right) = \int_{-\infty}^{\infty}\int_{-\infty}^{\infty} w(x, y, \omega)e^{-jk_{tx}x}e^{-jk_{ty}y}dxdy \quad (7)$$

where the correlation of the measurements $w(x, y, \omega)$ with the wavefield $e^{-ik_{tx}x}e^{-ik_{ty}y}$ [for a 2D-plate] is evaluated. This can be assessed for any $(k_{tx}\ k_{ty})$ pair and these are designated as the trial wavenumbers. For a 1D waveguide with discrete measurement locations, Eq. (7) simplifies to:

$$\hat{W}(k_{tx}, \omega) \approx \left(\frac{l_x}{N}\right) \sum_{i=1}^{N} w(x_i, \omega) e^{-jk_{tx}x_i}. \tag{8}$$

The correlation technique is an approach to convert the measured data from the spatial domain to the wavenumber domain, with the same purpose and in a fashion similar to that of a Spatial Discrete Fourier Transform (SDFT). In relation to the SDFT, the correlation technique has the advantage of having the wavenumber resolution determined arbitrarily by the user, which improves the estimation of wavenumbers for a given dimension of the measurement grid ([8]). However, the energy leakage that exists in the SDFT is also present in the wavenumber spectrum generated by the correlation technique. It can affect the estimation of wavenumbers when they are closely spaced at a given frequency.

The measurements $w(x, \omega)$ are mobility functions measured along the length of the pipe with a roving hammer technique. The pipe was placed on a soft foam to simulate free boundary conditions. A laser Doppler vibrometer Polytec PDV100 was placed in a single location while an instrumented hammer PCB 086E80 excited 35 points (impact rover test), equally separated at a distance of 22 mm. The data was acquired and processed through the Data Physics Quattro data acquisition module along with its SignalCalc ACE software for signal processing.

Measurements for two pipes were made, one with and one without helical groove. Taking the non-grooved pipe as a reference and applying the correlation technique with the measured data, with trial wavenumbers ranging between ±100 rad/m and scaling each frequency step so that the maximum correlation amplitude is unitary led to the plot shown in Fig. 3a. Figure 3b contains the plot with the dominant wavenumbers extracted from the colormap of Fig. 3a.

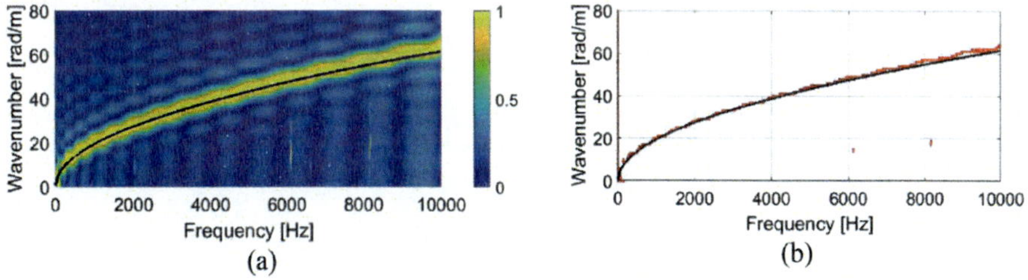

Fig. 3. Predicted wavenumber with WFE (in black) versus estimated wavenumbers (in color) of the non-grooved pipe in (a) colormap of correlation and (b) extracted dominant wavenumbers.

The plots in Fig. 3 show that the predictions using WFE and the experimentally estimated wavenumbers are in good agreement and only one branch of the flexural mode is visible in the measured direction. The same procedure was applied to the helically grooved pipe, yielding the results shown in the plots of Fig. 4.

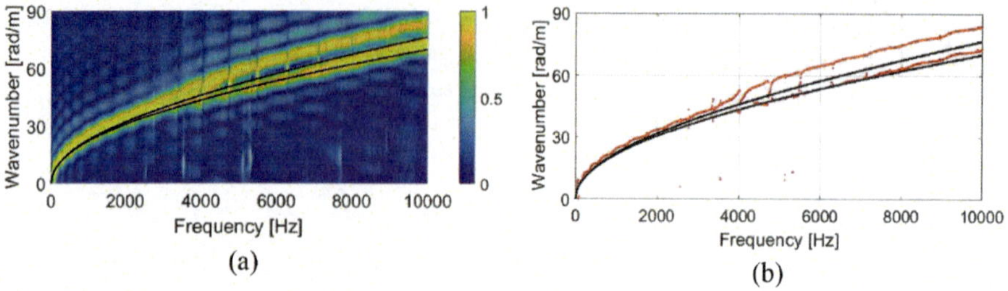

Fig. 4. Predicted wavenumber with WFE (in black) versus estimated wavenumbers (in color) of the helically grooved pipe in (a) colormap of correlation and (b) extracted dominant wavenumbers.

From the numerical and experimental results, it is possible to observe two distinct branches in Fig. 4, though some quantitative disagreement exists. At frequencies below 4 kHz, the correlation technique is unable to distinguish two separate wavenumbers due to the leakage effect of converting the spatial samples into the *k*-space. Above 4 kHz, although two separate branches are discernible in the experimental results, there is some deviation in the results which could arise from differences in the material or geometrical properties used in the FE model.

Empirically, and inspired by the work of Manconi et al. [1], it was observed that decreasing the Young's modulus of the grooved section of the pipe by a factor of 3 leads to a better fit between the results. This is shown in Fig. 5. Further investigation of the material properties in this structure would be necessary to make the numerical model more accurate, although this is not envisaged in this work.

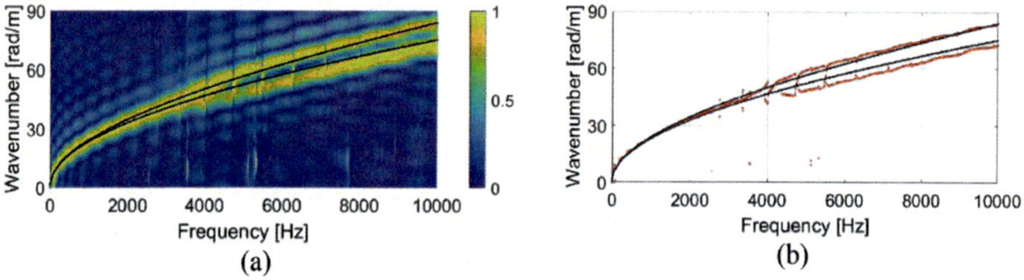

Fig. 5. Predicted wavenumber with WFE (in black) versus estimated wavenumbers (in color) of the helically grooved pipe in (a) colormap of correlation and (b) extracted dominant wavenumbers.

4 The Effect on the Mode Shapes Due to the Helical Groove

The presence of the two distinctive flexural wave branches indicates potentially different mode shapes exist and this was analysed through an experimental modal analysis. The mobilities obtained previously were used and the experimental modal analysis performed along one direction of the pipes yielding the selected mode shapes shown in Fig. 6.

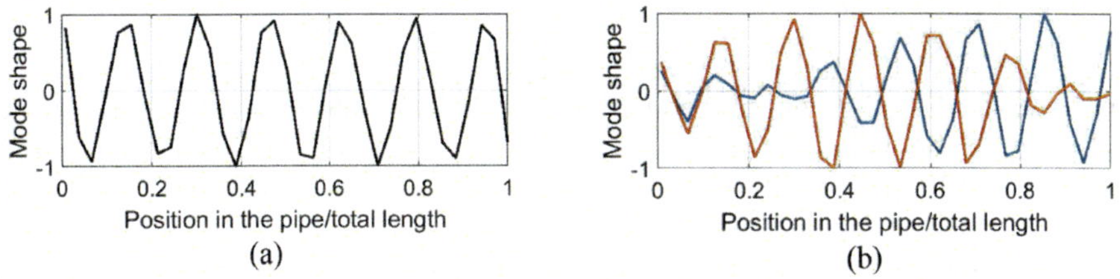

Fig. 6. Mode shapes of the 12th order obtained experimentally for (a) the non-grooved pipe (at 5785 Hz and $k = 47.5$ rad/m) and (b) the helically grooved pipe: two modes with varying magnitude at distinct and slightly different frequencies (4022 and 4042 Hz and $k \approx 47$ and 50.3 rad/m).

The mode shapes of the helically grooved pipe distinguish themselves from the ones of the non-grooved pipe through the change in magnitude at certain locations of each mode shape and through the existence of two mode shapes at slightly different frequencies for each mode order. Due to the one-directional view of the experimental modal analysis, a numerical modal analysis performed in Ansys Workbench for a pipe with simplified helical groove is used to visualize the three-dimensional behaviour of the pipe.

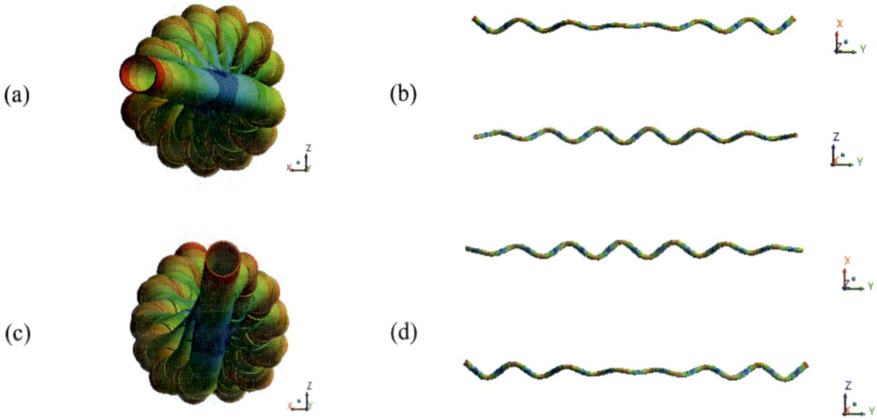

Fig. 7. Two mode shapes of the helically grooved tube representing the 12th mode order: Mode 12-1 from it's (a) front view and (b) side views, Mode 12-2 (c) front view and (d) side views.

The mode shapes shown in Fig. 7 provide a clearer insight into the bending behaviour of the helically grooved tube. The fluctuations in magnitude that can be seen in Fig. 6b are explained in the side views of Fig. 7b and Fig. 7d as deflections that take place in a different direction than the one that was measured. The end result is a bending mode that is not restricted to lying in one direction only but occurs in a continuously rotating direction, as shown in Fig. 7a and c.

5 Conclusions

In this paper, the effect of a helical pattern on breaking degeneracy of the bending modes of a pipe has been shown experimentally. Both wave and modal approaches were investigated. The wavenumbers were initially estimated numerically through the WFE method and a real structure was tested with the correlation technique to estimate the wavenumbers experimentally. The excellent qualitative and quantitative agreement between the experimental and the numerical results corroborates the existence of two distinctive flexural branches. These contrast with the two identical branches existent on degenerate structures and are brought by the break of symmetry arising due to the helical pattern along the pipe. Mode shapes obtained experimentally and numerically evidence the effect of these distinctive propagating waves on the modal behaviour of the helically grooved pipe. The bending motion is seen not to be restricted to one fixed direction of the pipe for each mode (as it occurs for an axisymmetric pipe), but to a direction that continuously rotates along the length of the pipe.

Improvements can still be made on the material properties of the FE model for a more accurate representation of the effect of the manufacturing technique on the real structure.

Acknowledgements. This paper is supported by the European Union's Horizon 2020 research and innovation programme under the Marie Skłodowska-Curie grant agreement No 765636, project InDEStruct (Integrated Design of Engineering Structures).

References

1. Manconi, E., Sorokin, S., Garziera, R., Søe-Knudsen, A.: Wave motion and stop-bands in pipes with helical characteristics using wave finite element analysis. J. Appl. Comput. Mech. **4**, 420–428 (2018)
2. Treyssède, F., Laguerre, L.: Investigation of elastic modes propagating in multi-wire helical waveguides. J. Sound Vib. **329**, 1702–1716 (2010)
3. Treyssède, F.: Numerical investigation of elastic modes of propagation in helical waveguides. J. Acoust. Soc. Am. **121**(6), 3398–3408 (2007)
4. Liu, Y., Han, Q., Li, C., Huang, H.: Numerical investigation of dispersion relations for helical waveguides using the Scaled Boundary Finite Element method. J. Sound Vib. **333**, 1991–2002 (2014)
5. Frikha, A., Treyssède, F., Cartraud, P.: Effect of axial load on the propagation of elastic waves in helical beams. Wave Motion **48**, 83–92 (2011)
6. Renno, J.M., Mace, B.R.: Vibration modelling of helical springs with non-uniform ends. J. Sound Vib. **331**(12), 2809–2823 (2012)
7. Demma, A., Cawley, P., Lowe, M., Pavlakovic, B.: The effect of bends on the propagation of guided waves in pipes. J. Pressure Vessel Technol. **127**(3), 328–335 (2005)
8. Ferguson, N.S., Halkyard, C.R., Mace, B.R., Heron, K.H.: The estimation of wavenumbers in two-dimensional structures. In: Proceedings of ISMA 2002: International Conference on Noise and Vibration Engineering, vols. 1–5, pp. 799–806 (2002)
9. Souza, M.R., Beli, D., Ferguson, N.S., de Arruda, J.R., Fabro, A.T.: A Bayesian approach for wavenumber identification of metamaterial beams possessing variability. Mech Syst. Signal Process. **135**, 106437 (2020)

10. El Masri, E., Ferguson, N., Waters, T.: Wave propagation and scattering in reinforced concrete beams. J. Acoust. Soc. Am. **146**, 3283–3294 (2019)
11. Berthaut, J., Ichchou, M., Jézéquel, L.: Wavenumbers identification in two-dimensional structures: comparison between two new methods. In: Proceedings of the Tenth International Congress on Sound and Vibration, pp. 1155–1162 (2003)
12. Ichchou, M.N., Berthaut, J., Collet, M.: Multi-mode wave propagation in ribbed plates: Part I. Wavenumber-space characteristics. Int. J. Solids Struct. **45**(5), 1179–1195 (2008)
13. Cicirello, A., Mace, B.R., Kingan, M.J., Yang, Y.: Sensitivity analysis of generalised eigenproblems and application to wave and finite element models. J. Sound Vib. **478**, 115345 (2020)
14. Waki, Y., Mace, B., Brennan, M.: Numerical issues concerning the wave and finite element method for free and forced vibrations of waveguides. J. Sound Vib. **327**, 92–108 (2009)

Waves Propagating in Water-Filled Plastic Pipes Due to Leak Noise Excitation: A Numerical and Experimental Investigation

Matheus Mikael Quartaroli[1](✉), Fabricio Cesar Lobato de Almeida[1], Michael Brennan[1], Mahmoud Karimi[2], and Bruno Cavenaghi Campos[1]

[1] UNESP, Bauru, São Paulo 17033-360, Brazil
matheus.quartaroli@unesp.br
[2] Centre for Audio, Acoustics and Vibration, University of Technology Sydney, Sydney, Australia

Abstract. Water loss has been one of the main concerns of water distribution companies in many countries for many years. Among the techniques used to detect and locate leaks in buried water pipes, leak-noise correlators are routinely employed. This method uses the measurements of pipe vibration due to leak noise that propagates along the pipe. Studies suggest that the predominantly fluid-borne wave carries most of the leak energy. In this paper, however, it is shown experimentally that two wave-types can propagate in an in-air plastic pipe when excited by a leak source. One wave-type is related to the predominantly fluid-borne wave, characterized by a low-frequency content. The other wave-type is predominant in the pipe wall. This wave is not well coupled to the fluid and has high-frequency content. A bespoke test rig is used to perform measurements and to collect pipe vibration data related to leak-noise under controlled conditions. The experimental data is used to show the presence of these two wave-types. Numerical results using the finite element method are also presented and validated against experimental data. Further, a discussion of the physics behind the phenomena is presented using analytical and numerical models.

Keywords: Plastic pipe · Leak excitation · Fluid borne wave · Structural wave-type · Finite element method

1 Introduction

One of the major concerns of water companies is to reduce non-revenue water in urban distribution systems, as it increases operational costs and can lead to structural disruptions during system repair [1]. This water loss is generally divided into apparent and real losses. The former relates the volume of water that is consumed by the customer, but, for some reason, is not billed. This can be caused by hydrometer measurement errors, fraud, and clandestine connections. The latter, however, is associated with water supply via distribution systems that do not reach consumers. Hence, such water loss is generally related to leaks in the distribution network [2].

Z. Dimitrovová et al. (Eds.): WMVC 2022, MMS 125, pp. 1114–1121, 2023.
https://doi.org/10.1007/978-3-031-15758-5_115

According to reports from the National Sanitation Information System (SNIS), from studies conducted in 2019, it is estimated that about 39.2% of water is wasted in Brazil's distribution system. This rate is much higher than that found in developed countries such as Australia, Japan, and New Zealand, where the rate of water wastage is about 10% [2]. Faced with these water losses, water companies have been using techniques to detect and locate leaks in the early stages, such as vibro-acoustic ones [3]. These techniques are passive methods that use the characteristic of the leak excitation and the associated distribution pipe to aid in its location [4, 5]. Leak noise energy is mostly carried in the wave where the fluid and pipe-wall are well-coupled, namely the $s = 1$ wave. However, recent measurements in a bespoke test rig for leak simulations show the presence of a second wave type which is believed to be the structural wave namely the $s = 2$ wave [6]. The aim of this paper is to investigate the presence of this wave in measured leak vibration data and to support the experimental observations with a finite element model of the system and the classical wave approach using an analytical model.

2 Experimental Work

The test rig, shown in Fig. 1-a, is located in the city of Tupã in Brazil and belongs to the water company (Sabesp). It is an unburied plastic pipe (PVC – Polyvinyl Chloride) connected to the water mains where the system can be filled and pressurized via two control valves. Moreover, a drainage valve is also provided to remove any air bubbles present in the system which can affect the coupling between the pipe-wall and water (fluid). The plastic pipe has an outside diameter of 60 mm and a wall thickness of 3.3 mm. These measurements were conducted on the straight pipe section with a length of approximately 45 m as depicted in Fig. 1-a. The leak noise response for the system excitation (vibration source) was collected at the leak position and at two other measurement positions, P1 and P2, at distances of 3 and 6 m from the leak, respectively. Figure 1-b shows photograph of the test rig, with the unburied pipe section and the measurement positions highlighted accordingly. All tests were carried out in a controlled environment, so that the leak was manually activated by removing a plug from a small hole in the pipe.

Three PCB PIEZOTRONICS triaxial sensors each with a sensitivity of 500 mV/g were used for three simultaneous measurements in 3 orthogonal directions. However, for the results presented in the work, radial direction measurements were used because this is the one generally taken for leak detection problems. They were glued to the pipe at the measurement points described previously. The signals were acquired using a Siemens LMS SCADAS XS at a sampling frequency of 12.8 kHz for a time duration of 60 s.

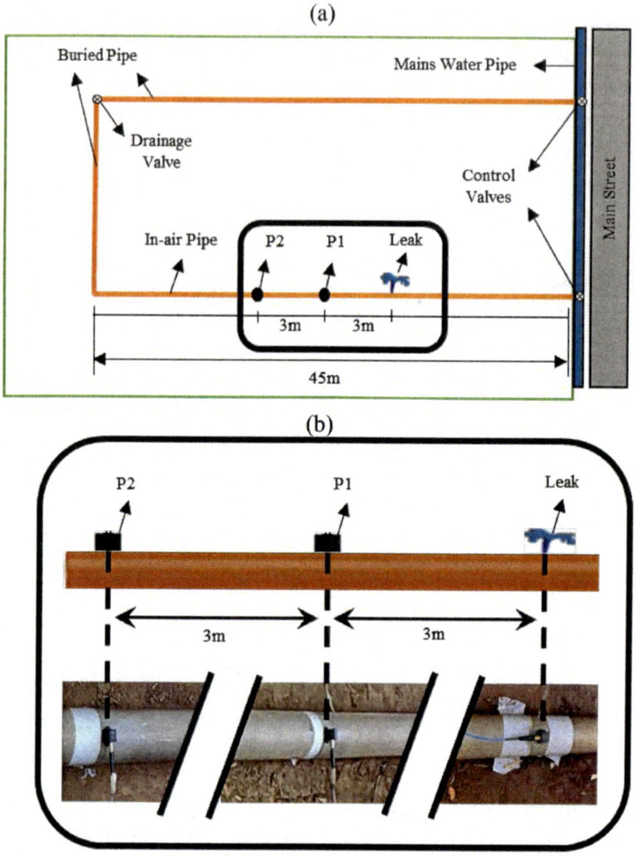

Fig. 1. The test rig used in the experiments. (a) Schematic of the test rig and (b) Measurement positions used to fix accelerometers and collect the leak response.

The data (time histories) acquired using an LMS Scada acquisition system were processed to estimate the frequency response function (FRF) between the pipe vibration at the leak position and the pipe response at P1 and P2. The FRFs together with coherence were calculated using a Hanning window, frequency resolution of 10 Hz and an overlap of 50%. Figures 2(a), (b), and (c) show the estimated modulus of the transfer function, the phase of the transfer function, and the coherence, respectively. The labels "(i)" and "(ii)" are used for the measurements between the leak and P1 and and P2, respectively.

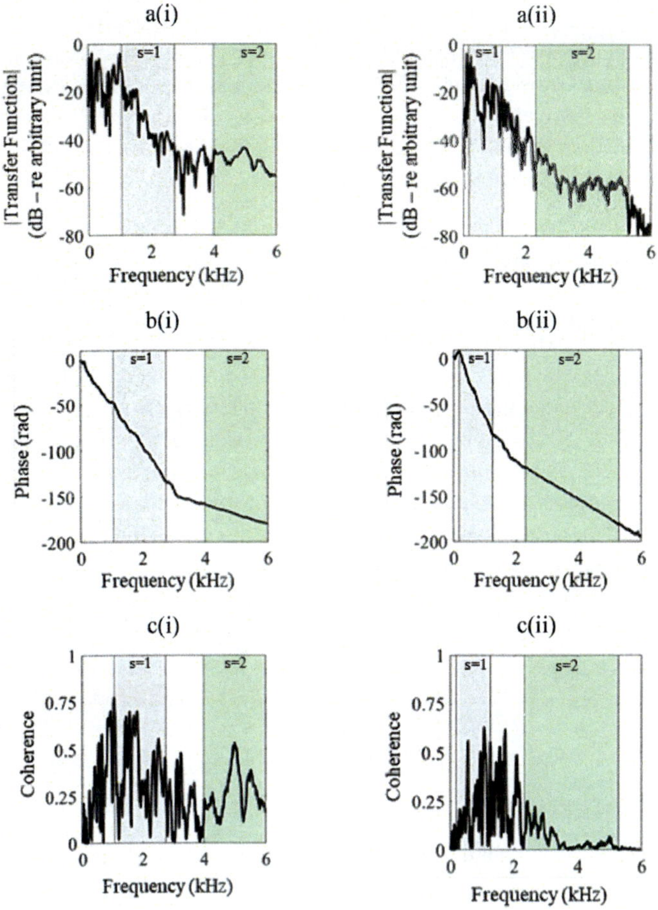

Fig. 2. The transfer function modulus (a), phase (b), and coherence (c) obtained in the test field for the measurement points P1 (i) and P2 (ii).

Examining Fig. 2, it can be seen that two frequency bands are marked, which are related to the frequencies regions where the phases have broad distinct behaviour, i.e., they have different, but constant gradients. These bands can also be observed in the modulus of the FRF and the coherence via the amplitude difference, but it is not as clear as in the phase behavior. This pattern suggests that there are different dominant waves in the two frequency bands with two different velocities propagating along the pipe, which can be calculated through the signal delay For leak-P1 measurements, for example, it seems that the predominant wave $s = 1$ is located within the lower frequency range at frequencies up to about 2.75 kHz with velocity of 362 m/s. However, for leak-P2 the $s = 1$ wave seems to be dominant up to 1.25 kHz, with velocity of 367 m/s, which is similar to that calculated from the leak-P1 data. There is a region where the interaction between the two waves can be observed specially via the deviations in the straight-line behaviour of the phase. This region is in between the two frequency ranges depicted in

Fig. 2. The $s = 2$ wave, appears to be dominant in the high frequency range. This wave has a much greater speed of propagation than the $s = 2$ wave, and was estimated from the measurements to be 1669 m/s and 1786m/s for the leak-P1 and leak-P2 measurements respectively. This motivates the investigation of such characteristic using two models: a numerical one modeled via the finite element (FE) method and a model based on the classical wave approach. These models are described in the next section.

3 Pipe System Modeling

3.1 Finite Element Model

This model is developed using a commercial FE software using COMSOL Multiphysics 5.3. The FE model takes the advantage of the problem symmetry to reduce the computational cost via carrying out the one quarter model, as shown in Fig. 3. The leak was simulated by using a monopole source in the pipe, and the radial pipe vibration was measured at three points corresponding to the leak position and positions P1 and P2. The FRFs were calculated directly from the model. The acoustic-solid interaction module of COMSOL was used in the modeling of this system.

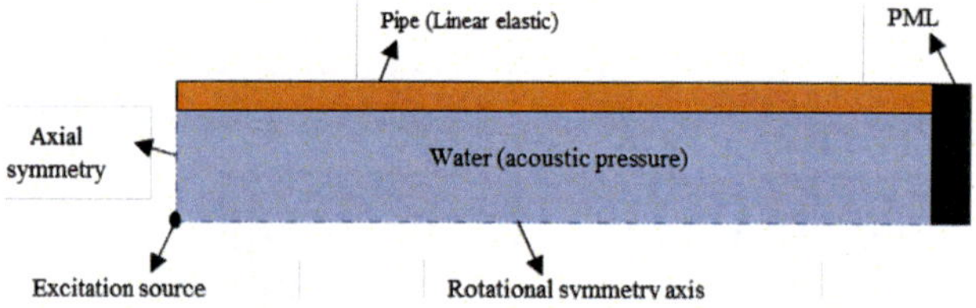

Fig. 3. A quarter of the finite element model of the water-pipe system.

When selecting the acoustic-solid interaction module, the program presents some standard numerical methods for solving the problem. The water is modeled as a lossless fluid, using the Acoustic Pressure module, and the pipe is modeled as a linear elastic medium, using the Solid Mechanics module. To connect the acoustic pressure model (water) with the structural component (pipe), the Acoustics-Structure boundary condition is used. PML (Perfectly Matched Layer) is applied at the computational limit to simulate an infinite pipe system, to avoiding reflections.

3.2 Analytical Model: Wave Approach

This model is based on the classic plane wave approach. Figure 4 shows the schematic of the leak noise propagating along a plastic pipe. In the model developed by Pinnington and Briscoe [6], there are three types of axisymmetric waves that are responsible for most of the energy transfer in leaking pipes. It was found that the highest leakage energy in pipelines is predominantly fluid-borne $s = 1$ wave [7, 8]; the second wave, $s = 2$, is a shell-dominated structural wave; the third wave, s = 0, is a torsion wave decoupled from the fluid and has no significant radial motion and is not of interest in this work.

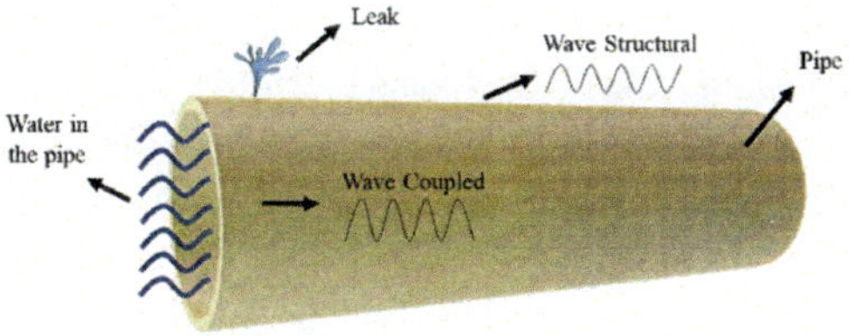

Fig. 4. Leak wave propagation scheme along the unburied pipe.

For the wave $s = 1$, the wavenumber is given by $k_1 = \omega/c_1(1 - j\eta_1/2)$, where $j = \sqrt{-1}$, ω is the frequency in rad/s, c_1 is the speed at which the predominantly fluid-borne wave propagates in m/s, and η_1 is the loss factor [8]. For the wave $s = 2$, the wavenumber is given by $k_2 = \omega/c_2(1 - j\eta_1/2)$, where c_2 is the speed at which the predominantly structural wave propagates. The radial displacement of the pipe wall is considered to be the sum of these waves, generated by the leak, and is given by

$$W(x) = W_1 e^{-jk_1 x} + W_2 e^{-jk_2 x}, \tag{1}$$

where W_1 and W_2 are the amplitude of the fluid-borne wave and structural, respectively, x is the distance from the leak to the measurement position.

4 Results and Discussions

The models were developed using the geometric characteristics of the test rig located in Tupã, described previously. However, the material properties were selected from standard tables. Table 1 shows the main parameters used in the models.

Table 1. Parameters used in the models.

Parameters	Value
Poisson's Ratio	0.4
Outer Diameter	60 mm
Pipe Thickness	3.3 mm
Pipe Young's modulus	3.5e9 Nm^{-2}
Pipe Density	1350 kgm^{-3}
Pipe Loss Factor	0.07
Amplitude Ratio (W_1/W_2) at point P1	0.02
Amplitude Ratio (W_1/W_2) at point P2	0.01
Wavespeed s = 1	365 m/s
Wavespeed s = 2	1700 m/s

Figures 5(a) and (b) show the modulus and phase of the FRF calculated using the leak data collected in the test rig (grey solid line) together with the results given by the numerical model (black line) and the wave approach (red line). The labels "i" and "ii" are for the analysis between leak-P1 and leak-P2, respectively. An amplitude ratio of 0.02 and 0.01 between wave $s = 1$ and $s = 2$ was used in the analytical model, for points P1 and P2, respectively. This ratio was chosen to match the frequency at which the phase slope changes, as the loss factor is considered the same for both waves. The difference in the amplitude ratio for points P1 and P2 is perhaps due to the fact that the simple model does not take into account the mass component of the pipe wall, as observed in reference [6, 7]. It is observed that there is a very good match for the phase, but a good trend for the amplitude. Although the phase of the $s = 1$ wave for the leak-P2 measurement does not overlay quite well with the simulated ones due to an extra phase shift next to 2 kHz, the slope has a very similar behavior though. This emphasizes that the extra wave presented in the actual leak data is related to the structural $s = 2$ wave.

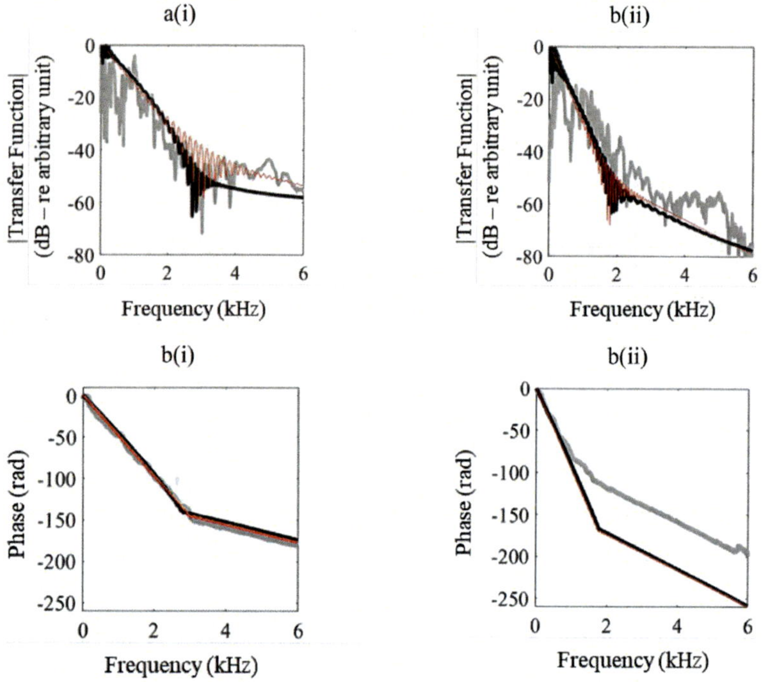

Fig. 5. Comparison of transfer function modulus (a) and phase (b) for experimental (gray line), numerical (black line), and analytical (red line) models with an amplitude ratio of 0.02 in the point P1 and 0.01 in the point P2. Leak-P1 (i) and Leak-P2 (ii).

5 Conclusions

This work has shown experimentally that two types of waves can propagate in an in-air water plastic pipe when excited by a leak source. One wave type is related to the predominantly fluid-borne wave $s = 1$ where the fluid is well-coupled with the pipe-wall, and the other one is related to the predominantly structural wave type $s = 2$. Numerical results using the finite element method and analytical results using the simplified wave propagation model are presented and used to support the presence of these two waves in different frequency bands. It has been shown that there is a low-frequency wave that is related to the predominantly fluid-borne wave $s = 1$ type with a wave speed of about 365 m/s, and the other wave-type which appears to be the predominantly structural $s = 2$ wave which is not well coupled to the fluid and has a high-frequency content with wave speed of about 1700 m/s. The results numerical and analytical have a good match with the actual leak data, especially for the phase suggesting that the waves observed are those predicted by the models. For the case of a buried pipe, the surrounding medium (soil) will considerably increase the wave attenuation, therefore, the s = 2 wave may not be clearly observed, as it has a lower energy compared to the s = 1 wave.

Acknowledgements. The authors would like to thank the financial support provided by the São Paulo Research Foundation (FAPESP) under grant numbers 2020/12251-1, CNPQ under grant 432272/2018-6 and by the Coordenação de Aperfeiçoamento de Pessoal de Nível Superior – Brasil (CAPES) inder grand 88882.461711/2019-01. The authors would also like to thank the Brazilian company (Sabesp) for funding this work and for providing the test rig.

References

1. AL-Washali, T., Sharma, S., Kennedy, M.: Methods of assessment of water losses in water supply systems: a review. Water Resour. Manage **30**(14), 4985–5001 (2016). https://doi.org/10.1007/s11269-016-1503-7
2. Brasil: Ministério do Desenvolvimento Regional. Secretaria Nacional de Saneamento – SNS. Sistema Nacional de Informações sobre Saneamento: 25° Diagnóstico dos Serviços de Água e Esgotos – 2019. Brasília (2020)
3. Ociepa, E., Mrowiec, M., Deska, I.: Analysis of Water Losses and Assessment of Initiatives Aimed at Their Reduction in Selected Water Supply Systems. Water (2019)
4. de Almeida, F.C.L., Brennan, M.J., Joseph, P.F., Dray, S., Whitfield, S., Paschoalini, A.T.: Measurement of wave attenuation in buried plastic water distribution pipes. J. Mech. Eng. **60**(5), 298–306 (2014)
5. Brennan, M.J., et al.: On the effects of soil properties on leak noise propagation in plastic water distribution pipes. Journal of Sound and Vibration **427**, 120–133 (2018)
6. Pinnington, R.J., Briscoe, A.R.: Externally applied sensor for axisymmetric waves in a fluid filled pipe. Journal of Sound and Vibration **173**(4) (1994)
7. Muggleton, J.M., Brennan, M.J., Pinnington, R.J.: Wavenumber prediction of waves in buried pipes for water leak detection. J. Sound Vib. **249**(5), 939–954 (2002)
8. Gao, Y., Brennan, M.J., Muggleton, J.M., Hunaidi, O.: A model of the correlation function of leak noise in buried plastic pipes. J. Sound Vib. **277**, 133–148 (2004)

WSI: Wave Mechanics for Structural Interfaces

Dynamic Pile Response During Vibratory Driving and Modal-Based Strain Field Mapping

Sergio S. Gómez(✉), Athanasios Tsetas, Apostolos Tsouvalas, and Andrei V. Metrikine

Faculty of Civil Engineering and Geosciences, Delft University of Technology, Stevinweg 1, 2628 CN Delft, The Netherlands
{s.sanchezgomez-1,a.tsetas,a.tsouvalas,a.metrikine}@tudelft.nl

Abstract. For offshore wind turbines (OWTs), the monopile comprises the most common type of foundation and vibratory driving is one of the main techniques for monopile installation (and decommissioning). In practice, prior to pile installation, a pile driving analysis is performed to select the appropriate installation device and the relevant settings. However, pile penetration results from a complicated vibrator-pile-soil interaction and better understanding of the latter is necessary for an efficient installation process. During the course of installation, the interface and boundary conditions of the pile continuously alter due to the soil layering and the non-linearity of the soil reaction. In this paper, a set of experimental data from an onshore experimental campaign are employed in a numerical scheme to identify the pile strain field based on *in vacuo* modes of simpler yet related systems. By mapping the pile strain field onto physically-based shape functions, the evolution of the soil reaction during pile installation can be studied, in order to facilitate the back-analysis of driving records and, by extension, improve pile drivability and vibro-acoustics predictions.

Keywords: Pile driving · Vibrations of shells · Mode-matching · Soil-structure interaction

1 Introduction

The installation of monopile foundations for offshore wind turbines (OWTs) is one of the most critical phases during the construction of an offshore wind farm [6]. Customarily, the monopile installation is performed via impact hammering, albeit this method has raised significant environmental concerns in the past decade due to underwater noise emissions [5]. In response to these concerns, alternative installation methods are examined with vibratory pile driving being one of the most suitable candidates. This method poses a number of open

Z. Dimitrovová et al. (Eds.): WMVC 2022, MMS 125, pp. 1125–1134, 2023.
https://doi.org/10.1007/978-3-031-15758-5_116

research questions stemming from the complex vibrator-pile-soil interaction during installation. To that end, inflow of new field data and development of suitable numerical approaches for analysis and prediction purposes are necessary to improve our understanding of vibratory pile driving.

In line with this work, an experimental campaign has been performed, which included installation tests via different driving methods, i.e. axial vibratory driving (VH) and the novel 'Gentle Driving of Piles' (GDP) [3,4]. The latter method introduces a high-frequency torsional excitation in addition to the classical axial vibration to improve the installation performance and reduce noise emissions. The collected data of the instrumented piles that were installed by means of the already mentioned pile driving techniques are used.

In this paper, a set of experimental data from the GDP campaign are employed to study the pile response during installation. The pile is modelled as a thin cylindrical shell and the modes *in vacuo* are found for two limit case configurations, i.e. for both ends being i) free and ii) fixed. These two modal sets are combined and the recorded axial strain is mapped onto them, as they can address stresses at the pile ends, as well as admit rigid body motion which is essential for pile penetration. The mode-matching process is performed by means of a least-squares optimization and the obtained field is in great agreement with the field measurements. Conclusively, this approach can be valuable for back-analysis of field measurements and aid in the identification of the dynamic soil reaction during driving, which is one of the most uncertain parameters in vibratory pile installation.

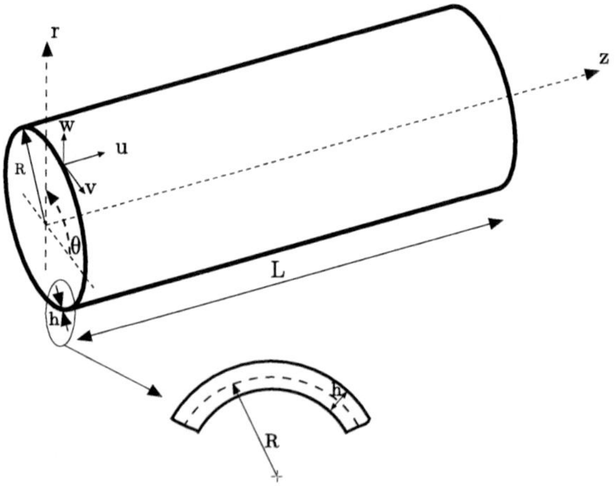

Fig. 1. Geometry and coordinate system of a cylindrical shell.

2 Governing Equations

A linear elastic cylindrical shell is considered, with finite length L, radius R, and wall thickness h, as shown in Fig. 1.The equations of motion of a cylindrical shell can be written in the following general form:

$$(\mathbf{L}+\mathbf{I})\,\mathbf{u}=\mathbf{0} \tag{1}$$

where $\mathbf{L}$ and $\mathbf{I}$ are the stiffness and inertial operators, respectively. The vector $\mathbf{u}=\begin{bmatrix} u & v & w \end{bmatrix}^{\mathrm{T}}$ describes the displacement of the mid-surface of the shell in the cylindrical coordinate system (r, θ, z). In this work, the shell motion is considered axisymmetric (i.e. $\partial(\cdot)/\partial\theta = 0$) and the membrane shell theory is employed to describe the pile [1]. Therefore, the non-zero elements L_{ij} and I_{ij} of the stiffness and inertial operators are defined as [2]:

$$L_{11}=\frac{\partial^2}{\partial z^2},\ L_{13}=\frac{\nu}{R}\frac{\partial}{\partial z},\ L_{22}=\frac{1-\nu}{2}\frac{\partial^2}{\partial z^2},\ L_{31}-\frac{\nu}{R}\frac{\partial}{\partial z},\ L_{33}=-\frac{1}{R^2},$$
$$I_{11}=I_{22}=I_{33}=-\frac{1}{c^2}\frac{\partial^2}{\partial t^2} \tag{2}$$

where $c=\sqrt{E/\left(\rho(1-\nu^2)\right)}$ is the thin plate wave velocity, E is the Young's modulus, ρ is the mass density and ν is the Poisson ratio of the shell material. From Eq. (2), it can be noted that the circumferential motion (v) is uncoupled from the longitudinal (u) and radial (w) motions of the shell, due to the axisymmetric conditions. For the coupled axial-radial motion of the shell, the solution to the axisymmetric free vibration problem reads:

$$u(z,t)=\sum_{i=1}^{\infty} U_i(z)q_i(t)$$
$$w(z,t)=\sum_{i=1}^{\infty} W_i(z)q_i(t) \tag{3}$$

where $U_i(z)$ and $W_i(z)$ are the axial and radial modal displacements of the i-th mode, respectively, and $q_i(t)$ is the corresponding modal coordinate. For the case of a shell that is clamped at both ends, the boundary conditions read:

$$u(z=0,t)=u(z=L,t)=0 \tag{4}$$

while in the case of a shell with traction-free ends, the boundary conditions read:

$$N_{zz}(z=0,t)=N_{zz}(z=L,t)=0 \tag{5}$$

where N_{zz} is the so-called axial force resultant. By substitution of Eq. (3) into Eq. (1), the free vibration problem can be formulated and the axial wavenumbers and natural frequencies of the cylindrical shell can be found. The relevant treatment is well-known in the topic of shell vibrations [2] and will be omitted here for brevity. The free vibration solutions are given for the case of clamped-clamped (C-C):

$$\begin{aligned} u_{\mathrm{c}}(z,t) &= \sum_{i=1}^{\infty} \sin(\gamma_i z) q_i(t) \\ w_{\mathrm{c}}(z,t) &= \sum_{i=1}^{\infty} \frac{\nu R \gamma_i c^2}{R^2\omega_i^2 - c^2} \cos(\gamma_i z) q_i(t) \end{aligned} \tag{6}$$

and free-free (F-F) boundary conditions:

$$\begin{aligned} u_{\mathrm{f}}(z,t) &= \sum_{i=0}^{\infty} \cos(\gamma_i z) q_i(t) \\ w_{\mathrm{f}}(z,t) &= \sum_{i=0}^{\infty} \frac{-\nu R \gamma_i c^2}{R^2\omega_i^2 - c^2} \sin(\gamma_i z) q_i(t) \end{aligned} \tag{7}$$

where ω_i and $\gamma_i = \omega_i \sqrt{(R^2\omega_i^2 - c^2)/(c^2(R^2\omega_i^2 - c^2(1-\nu^2)))}$ denote the i-th natural frequency and the i-th axial wavenumber of the cylindrical shell. As can be seen, the index of the F-F shell modes begins from $i = 0$, which corresponds to the rigid body mode of axial translation. The two chosen configurations will be used in Sect. 5 in an optimization scheme for the analysis of the experimental data, as they correspond to two limit cases of the pile during installation.

3 Experimental Tests

The experimental data presented and analysed in this paper were obtained during the GDP field tests campaign executed at the Maaskvlakte II in the Port of Rotterdam (the Netherlands) [3,4]. During the GDP tests, a number of piles were installed by means of different driving methods. The present work focuses on two different driving methods, namely the axial vibratory driving (VH) and the GDP method. The geometrical characteristics of the test piles are described in Table 1.

As regards to the pile instrumentation, the types and locations of the pile sensors are shown in Fig. 2. It is remarked that two vertical series of sensors were installed at diametrically opposite locations along the pile circumference (i.e. $\theta = 0$ and $\theta = \pi$). The sampling frequency was set equal to 1 kHz for all sensors.

The same installation protocol was followed for all tested piles. Specifically, in the first 3 m of penetration, the pile was laterally supported to eliminate any inclination that may develop during installation in shallow depth. Therefore, all the results presented in this paper correspond to final 5 m of pile penetration (from 3 m to 8 m), in which the pile was unrestrained and penetrated into the soil under the shaker excitation and its self-weight.

Table 1. Geometrical characteristics of the piles.

Length	Embedded length	Outer diameter	Thickness
L	L_e	D_o	h
10 m	8 m	0.762 m	0.0159 mm

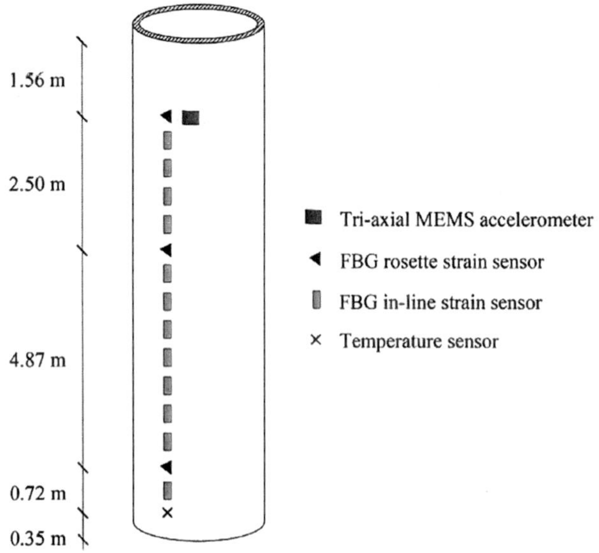

Fig. 2. Instrumentation of test piles.

4 Mapping of the Strain Field Based on Mode-Matching

In order to gain insight into the structural response of the pile during installation, the axial strains (ε_{zz}) along the pile length are studied. Furthermore, a process to map the recorded strains onto appropriate shape functions for post-processing purposes is presented. Specifically, the free vibration modes *in vacuo* of the cylindrical shell (see Sect. 2) are employed in this process. The two shell configurations, outlined in Sect. 2 (i.e. C-C and F-F), comprise two limit cases for the pile during installation, as both of its ends have tractions (C-C case) and rigid body motion is admissible (F-F case). Therefore, an optimization process is set up to obtain the multipliers of each mode from the C-C and F-F sets, such that the axial strain data and the mode-matching based strains agree during the whole installation process. The strain error $R_\varepsilon(z_j, t_k)$ at each sensor location z_j and time moment t_k is used in a least-squares optimization process to obtain the $q_i(t_k)$:

$$R_\varepsilon(z_j, t_k) = \hat{\varepsilon}_{zz}(z_j, t_k) - \varepsilon_{zz}(z_j, t_k) = \frac{\partial \hat{u}(z_j, t_k)}{\partial z} - \sum_{i=1}^{N} \frac{\partial U_i(z_j)}{\partial z} q_i(t_k) \tag{8}$$

where $\hat{\varepsilon}_{zz}(z_j, t_k)$ and $\varepsilon_{zz}(z_j, t_k)$ denote the measured and computed axial strain, respectively. By N the total amount of modes from both C-C and F-F sets is denoted and accordingly these functions are arranged in the following matrix form:

$$\mathbf{U}_{\mathrm{m}} = \left[\mathbf{U}_{\mathrm{c}}\ \mathbf{U}_{\mathrm{f}}\right] = \begin{bmatrix} U_{\mathrm{c},1}(z_1) & \dots & U_{\mathrm{c},N_{\mathrm{c}}}(z_1) & U_{\mathrm{f},0}(z_1) & \dots & U_{\mathrm{f},N_{\mathrm{f}}-1}(z_1) \\ \vdots & \ddots & \vdots & \vdots & \ddots & \\ U_{\mathrm{c},1}(z_M) & \dots & U_{\mathrm{c},N_{\mathrm{c}}}(z_M) & U_{\mathrm{f},0}(z_M) & \dots & U_{\mathrm{f},N_{\mathrm{f}}-1}(z_M) \end{bmatrix} \tag{9}$$

where the matrices $\mathbf{U}_{\mathrm{c}}$ and $\mathbf{U}_{\mathrm{f}}$ encapsulate the C-C and F-F modal sets at the sensor location points z_j. The respective number of modes included from each set (i.e. N_{c} and N_{f}) can vary such that greater agreement between the measured data and the model computation can be acquired.

The data collected by the diametrically opposite FBG sensors (at the same height level) are averaged and the offset is removed via high-pass filtering. Given the total amount of averaged signals ($M = 12$), we adopt the following constraint in the mode-matching optimization process: $N_{\mathrm{c}} + N_{\mathrm{f}} \leq M$, such that the number of unknowns $q_i(t_k)$ does not exceed the number of equations at hand.

5 Results

In this section, the structural response of two piles installed by means of two distinct pile driving techniques (i.e. VH and GDP) is presented. First, in Fig. 3 the time-frequency analysis of the longitudinal acceleration ($\partial^2 u/\partial t^2$) of the two piles during installation is shown in the form of a spectrogram. As can be seen, the axial driving frequency was approximately 16 Hz for the GDP pile and 25 Hz for the VH pile. In the spectrograms of both piles, super-harmonics of the driving frequency with appreciable magnitude are present. In principle, the driving process can be considered almost stationary as the frequency components and their respective magnitudes are approximately constant during the pile installation.

In Fig. 3, information of the frequency content of the pile motion and its temporal evolution can be extracted. As regards to the spatial distribution of the pile deformation, the axial strains (see Fig. 2) along the pile length are employed. The mode-matching process presented in Sect. 4 is applied to the axial strain data and the results are shown in Figs. 4 and 5. For the problem at hand, $N_{\mathrm{c}} = N_{\mathrm{f}} = 6$ is chosen in the analysis.

The results of the optimization process shown in Fig. 4 indicate that the mode-matching based fit is in great agreement with the experimental data for three distinct penetration depths(4 m, 6 m and 8 m). As can be seen, the strains

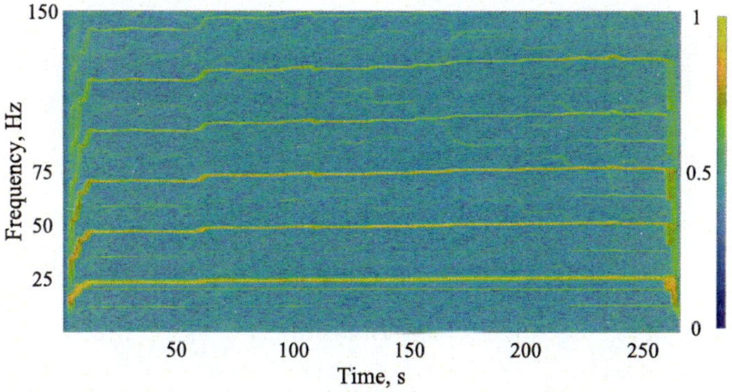

(a) VH Pile

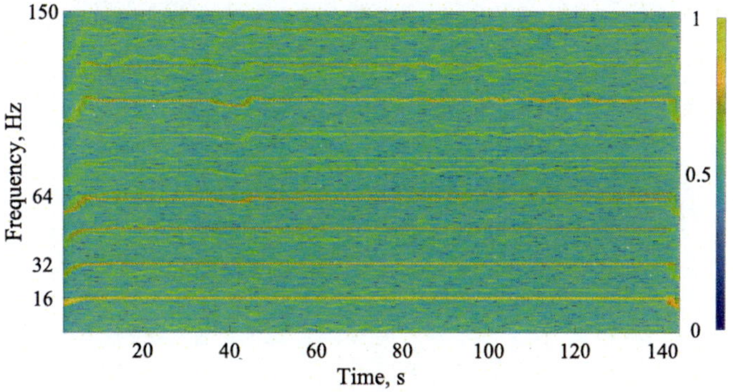

(b) GDP pile

Fig. 3. Normalized time-frequency spectrum of the pile acceleration in the longitudinal direction ($\partial^2 u/\partial t^2$) for (a) the VH pile and (b) the GDP pile.

are relatively uniform along the pile for any time moment, with decreasing amplitude from the top to the pile tip resulting from the soil reaction at the pile shaft.

In the case of pile GDP, the mode-matching results appear to be in agreement with the data as well (see Fig. 5). The spatial distribution of the axial strains slightly deviated from the respective patterns observed in the case of the VH pile. By comparing Fig. 3(a) and 3(b), one can note that the number of frequency components with appreciable amplitude are many more in the case of the GDP pile. This is a result of the potential interaction between the axial and torsional motion and results in the richer vibration pattern observed compared to the VH pile, albeit the main trend of diminishing axial strain amplitude with depth is present in both piles.

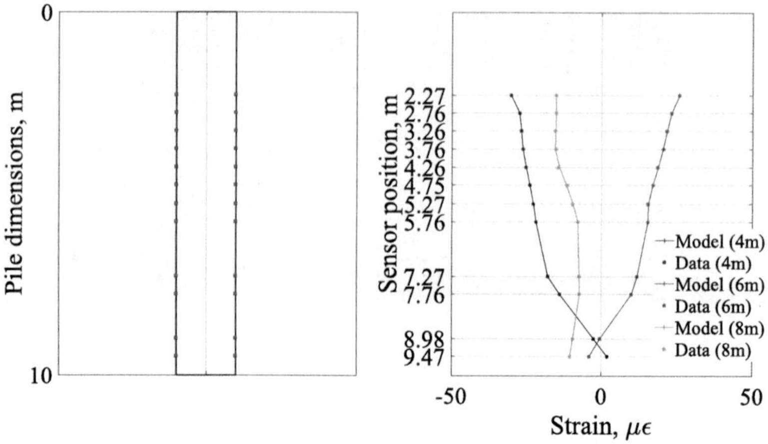

Fig. 4. Mode-matching based fit of the axial strain data of the VH pile at three different penetration depths (4 m, 6 m and 8 m).

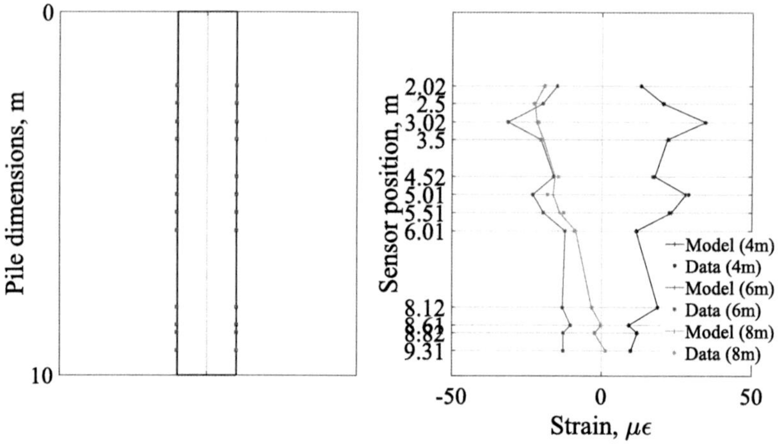

Fig. 5. Mode-matching based fit of the axial strain data of the GDP pile at three different penetration depths (4 m, 6 m and 8 m).

In Figs. 6 and 7 the axial strain shapes at different time moments are presented, during one period of vibration and around 6 m of penetration depth. The axial strain shown in Figs. 6 and 7 corresponds to the sensor at the top of the pile in both cases. The observations from Figs. 4 and 5 are further supported, and the response of the GDP pile also resembles the periodic pattern seen in the VH pile.

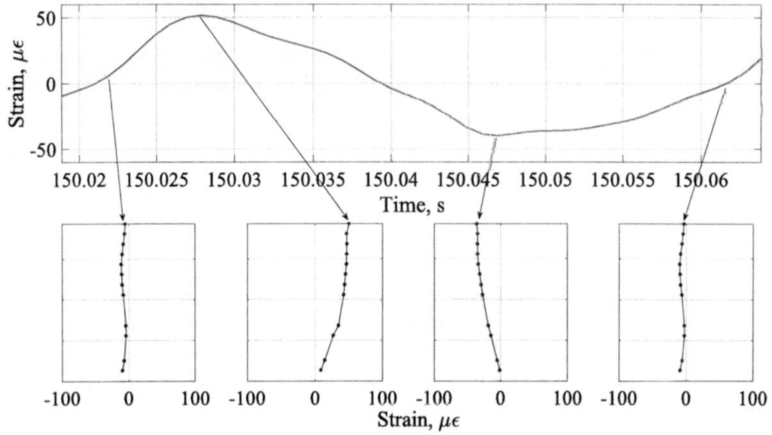

Fig. 6. Spatial distribution of axial strains for the VH pile at different time moments around 6 m penetration depth.

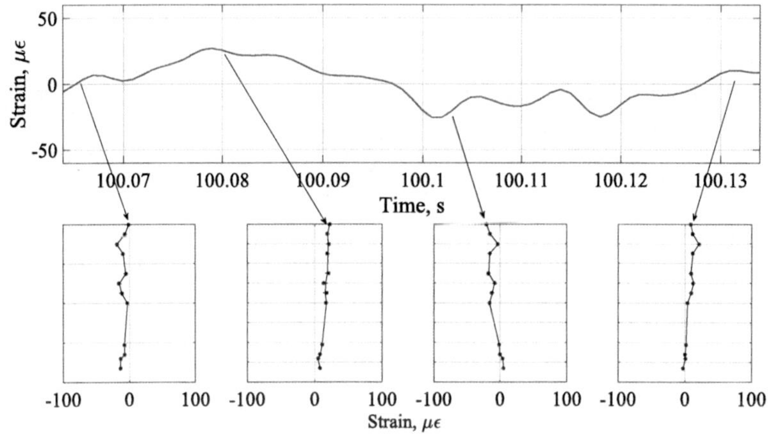

Fig. 7. Spatial distribution of axial strains for the GDP pile at different time moments around 6 m penetration depth.

6 Conclusions

In this paper, a set of field data from the GDP field campaign are analysed with a focus on the pile structural response during installation. The pile vibrations indicate that the vibratory driving is an almost stationary process, as the frequency components and their respective magnitudes are approximately constant during the pile installation. Furthermore, a mode-matching procedure is developed and applied to two test piles, with a view to map the strain field onto two distinct sets of *in vacuo* pile modes. The mode-matching derived strain field is in great agreement with the strain field data for all the cases examined.

The proposed mode-matching approach is considered to be a valuable tool for analysing and processing such datasets in the absence of extensive/dense instrumentation. The utilization of the present approach can aid in both research and engineering practice works such as the identification of the dynamic soil reaction during installation, the back-analysis of penetration records and the source identification/description for vibro-acoustics modelling.

Acknowledgments. This research is associated with the GDP project in the framework of the GROW joint research program. Funding from "Top sector Energie subsidie van het Ministerie van Econsomische Zaken" under grant number TE-HE117100 and financial/technical support from the following partners is gratefully acknowledged: Royal Boskalis Westminster N.V., CAPE Holland B.V., Deltares, Delft Offshore Turbine B.V., Delft University of Technology, ECN, Eneco Wind B.V., IHC IQIP B.V., RWE Offshore Wind Netherlands B.V., SHL Offshore Contractors B.V., Shell Global Solutions International B.V., Sif Netherlands B.V., TNO and Van Oord Offshore Wind Projects B.V.

References

1. Graff, K.: Wave Motion in Elastic Solids. Oxford University Press (1975)
2. Leissa, A.W.: Vibration of shells, vol. 288. Scientific and Technical Information Office, National Aeronautics and Space Administration (1973)
3. Metrikine, A.V., et al.: GDP: a new technology for gentle driving of (mono)piles. In: Proceedings of the 4th International Symposium on Frontiers in Offshore Geotechnics, Austin, TX, USA, 16–19 August 2020 (2020)
4. Tsetas, A., et al.: Experimental identification of the dynamic behaviour of pile-soil system installed by means of three different pile-driving techniques. In: Proceedings of the XI International Conference on Structural Dynamics, EURODYN 2020. vol. II, pp. 3005–3015. European Association for Structural Dynamics (2020)
5. Tsouvalas, A.: Underwater noise emission due to offshore pile installation: a review. Energies **13**(12), 3037 (2020)
6. Wu, X., et al.: Foundations of offshore wind turbines: a review. Renew. Sustain. Energy Rev. **104**, 379–393 (2019)

Unidirectional Interfacial Waves in Gyroscopic Elastic Systems

Giorgio Carta[1(✉)], Michael J. Nieves[2], and Michele Brun[1]

[1] Department of Mechanical, Chemical and Materials Engineering, University of Cagliari, 09123 Cagliari, Italy
giorgio_carta@unica.it

[2] School of Computing and Mathematics, Keele University, Keele ST5 5BG, UK

Abstract. Wave propagation in an elastic triangular lattice connected to a system of gyroscopic spinners is studied. The attention is focussed on waves travelling at the interface between two half-planes, where the spinners rotate in opposite directions. The relation describing the dispersive properties of these interfacial waves is given in closed form. These interfacial waves exist within a narrow frequency interval in the internal stop-band for the bulk medium, and are characterised by very low group velocity. The results of the dispersion analysis are confirmed by independent finite element simulations, which show that time-harmonic forces applied in proximity of the interface generate waves that are localised at the interface and do not propagate inside the bulk of the medium. In addition, these waves mainly propagate in one specific direction, that can be predicted from the dispersion curves and that can be reversed by changing the direction of rotation of the spinners. The model described here is an alternative to other elastic structures proposed in the literature and, considering its tuning properties, may be useful for different engineering applications including wave guiding, vibration isolation and energy harvesting.

Keywords: Structural interfaces · Gyroscopic elastic system · Unidirectional interfacial waves

1 Introduction

The interest of the scientific community in the design of structural interfaces for optimal control of guided waves has significantly grown in recent years. The possibility to ensure wave propagation despite the presence of imperfections along the wave path has been allowed for by creating topologically non-trivial band-gaps. The latter are intervals of frequencies where waves can propagate along edges or interfaces of a structure without being back-scattered by geometrical or material defects, in analogy with the Quantum Hall Effect [1].

In order to generate topologically protected waves, two different approaches have been adopted in the literature. The first one consists in designing passive

Z. Dimitrovová et al. (Eds.): WMVC 2022, MMS 125, pp. 1135–1142, 2023.
https://doi.org/10.1007/978-3-031-15758-5_117

systems possessing specific classes of symmetries, that are perturbed with the aim of creating band-gaps in correspondence of broken Dirac cones. Mechanical systems belonging to this category include lattices [2,3], plates [4,5], origami and kirigami structures [6], granular crystals [7] and chiral continuous systems [8]. The second approach is based on the use of active elements to break time-reversal symmetry. For elastic structures, one of the most successful methods involves the exploitation of gyroscopic elements, and it has been applied both to elastic lattices [9–16] and plates [17,18].

In this work, we study the simplest model of a gyroscopic elastic discrete system, represented by a triangular lattice whose nodes are connected to gyroscopic spinners. We examine a structure divided into two sub-domains, where gyroscopic systems rotate in opposite directions. The dispersion relation of waves localised at the interface between the two sub-domains is obtained in closed form. The associated dispersion curves are non-symmetric for positive and negative values of the wave number. As a consequence, interfacial waves exhibit a strong preferential directionality.

The paper is organised as follows. In Sect. 2, the gyroscopic discrete system is described in detail, and the expressions for the dispersion surfaces of the bulk waves are provided. In Sect. 3, the procedure for deriving the dispersion properties of the interfacial waves is outlined, and the dispersion relation is presented. In Sect. 4, an independent finite element simulation is performed, and a highly localised unidirectional waveform is illustrated. Finally, in Sect. 5, concluding remarks are provided.

2 Description of the Model

We consider a triangular lattice, consisting of particles with mass m connected by elastic springs of stiffness γ, length L and negligible mass with respect to m (see Fig. 1(a)). Each particle is also attached to a gyroscopic spinner, whose motion is described by the time evolution of the Euler angles θ, ϕ and ψ (see Fig. 1(b)). We assume that the nutation angle θ is small, hence the displacement of the particle connected to the gyroscopic spinner is also small and the equations of motion can be linearised.

Each gyroscopic spinner is a solid of revolution, having length l, moments of inertia I_1 and I_0 about the axis of revolution and the two perpendicular axes respectively, and gyricity Ω. The latter is an independent constant parameter, given by the sum of the initial precession and spin rates [19,20]: $\Omega = \dot{\phi} + \dot{\psi} =$ Const. The gyroscopic spinner couples the displacement components along the x_1- and x_2-directions of the lattice's particle to which it is attached. In addition, the connection between the gyroscopic spinner and the particle is such that the twisting motion of the spinner is not transmitted to the particle. We introduce the *effective mass* m^* and the *effective gyricity* Ω^*, given by [14]

$$m^* = m\left(1 + \frac{I_0}{ml^2}\right) \tag{1}$$

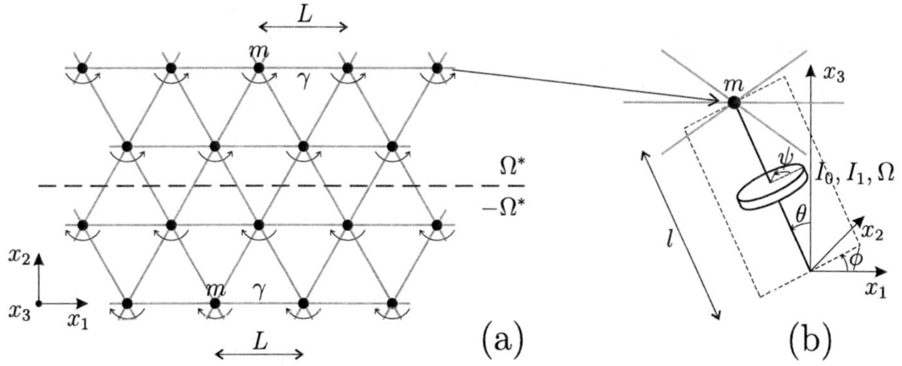

Fig. 1. (a) Triangular lattice made of masses m, linked by elastic rods with stiffness γ, length L and negligible inertia, and connected to gyroscopic spinners. The lattice consists of two sub-domains, where the gyroscopic spinners are characterised by equal and opposite values of the effective gyricity Ω^*. (b) Gyroscopic spinner, pinned at the base and attached to a lattice particle at the tip; θ, ϕ and ψ represent the nutation, precession and spin angles, respectively.

and

$$\Omega^* = \Omega \frac{I_1}{I_0 + ml^2}, \tag{2}$$

respectively.

The triangular lattice is divided into two sub-domains, separated by an interface, represented by the dashed line in Fig. 1(a). In the two sub-domains, the properties of the structure are all the same, except for the value of the effective gyricity: in the upper part it is positive, while in the lower part it is negative and has the same absolute value.

We consider time-harmonic conditions. Accordingly, the displacement vector is expressed by

$$\boldsymbol{u}(\boldsymbol{x}) = \boldsymbol{U}(\boldsymbol{x})\mathrm{e}^{-\mathrm{i}\omega t}, \tag{3}$$

where t is time, ω is the radian frequency, $\boldsymbol{x}$ is the position vector and $\boldsymbol{U}(\boldsymbol{x})$ is the complex amplitude to be determined. Further, we denote by $\boldsymbol{k}$ the wave vector, that will be used when describing the periodicity of the lattice.

2.1 Normalisation

We introduce the following normalisations:

$$\begin{aligned} &\boldsymbol{x} = \tilde{\boldsymbol{x}}L, \quad \boldsymbol{u} = \tilde{\boldsymbol{u}}L, \quad \boldsymbol{U} = \tilde{\boldsymbol{U}}L, \quad \boldsymbol{k} = \tilde{\boldsymbol{k}}/L, \\ &t = \tilde{t}\sqrt{m^*/\gamma}, \quad \omega = \tilde{\omega}\sqrt{\gamma/m^*}, \quad \Omega^* = \tilde{\Omega}^*\sqrt{\gamma/m^*}, \end{aligned} \tag{4}$$

where the symbol tilde indicates a dimensionless quantity.

The symbol tilde will be omitted in the following for ease of notation, and all the variables will be assumed as dimensionless.

2.2 Dispersion Properties of the Infinite Lattice with Uniform Effective Gyricity

In each half-plane of Fig. 1(a), the effective gyricity Ω^* is the same everywhere. In order to determine the frequencies at which waves can propagate in the bulk of each half-plane, we can use the dispersion relation of the infinite periodic system with uniform Ω^*, given by [14]

$$\omega^4 - \left[\mathrm{tr}(\boldsymbol{C}) + (\Omega^*)^2\right]\omega^2 + \det(\boldsymbol{C}) = 0\,, \tag{5}$$

which yields the two dispersion surfaces defined by

$$\omega_{1,2} = \sqrt{\frac{\mathrm{tr}(\boldsymbol{C}) + (\Omega^*)^2 \mp \sqrt{\left[\mathrm{tr}(\boldsymbol{C}) + (\Omega^*)^2\right] - 4\det(\boldsymbol{C})}}{2}}\,. \tag{6}$$

In (5) and (6), the matrix $\boldsymbol{C}$ is expressed by

$$\boldsymbol{C} = \begin{pmatrix} 3 - 2\cos(\zeta+\xi) - [\cos(\zeta)+\cos(\xi)]/2 & \sqrt{3}[\cos(\zeta)-\cos(\xi)]/2 \\ \sqrt{3}[\cos(\zeta)-\cos(\xi)]/2 & 3 - 3[\cos(\zeta)+\cos(\xi)]/2 \end{pmatrix}, \tag{7}$$

where $\zeta = (k_1 + \sqrt{3}k_2)/2$ and $\xi = (k_1 - \sqrt{3}k_2)/2$, being k_1 and k_2 the wave vector components along the x_1- and x_2-direction, respectively (see Fig. 1(a)).

We note that the dispersion relation (5) does not depend on the sign of Ω^*. Consequently, the dispersion diagrams for the two sub-domains with equal and opposite values of the effective gyricity are identical.

The projections of the dispersion surfaces, given by (6) with $\Omega^* = 2.5$, in the $\omega - k_1$ plane are shown in Fig. 2 by solid lines, corresponding to different values of k_2. The thick horizontal lines represent the limits of the two pass-bands.

3 Dispersive Properties of Interfacial Waves

We seek for waves localised at the interface between the two semi-infinite planes in Fig. 1(a), where the interface is parallel to the x_1-direction. The amplitudes of these waves should decay exponentially as the distance from the interface is increased. Therefore, we can use the same solution for the bulk waves determined in relation with the Rayleigh waves problem, analysed in [15,16].

In the present case, however, the boundary conditions are different. In the Rayleigh waves problem, the nodes on the boundary of the half-plane are connected only to the nodes of the adjacent row belonging to the same half-plane. Here, the nodes belonging to the rows adjacent to the interface (with opposite effective gyricity) are mutually connected; in addition, they are also linked to the nodes of the neighbouring rows in the same half-plane (with the same effective gyricity).

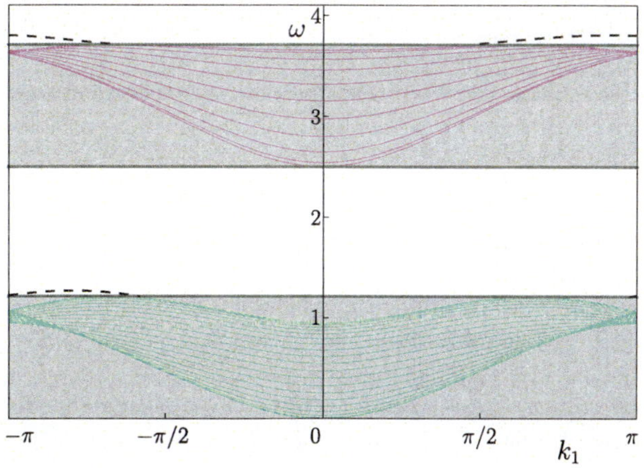

Fig. 2. Dispersion diagram in the $\omega - k_1$ plane for an effective gyricity $\Omega^* = 2.5$. The solid lines are the projections of the dispersion surfaces for the infinite lattice with uniform effective gyricity; the thick horizontal lines indicate the limits of their pass-bands. The dashed lines represent the dispersion curves associated with the interfacial waves, here shown in the stop-bands of the infinite system with uniform effective gyricity.

Distinguishing between symmetric and antisymmetric modes, we can derive the dispersion relation in closed form, that is given by

$$h_1 \left(1 \pm \frac{1}{\lambda_1}\right)\left(1 \mp \frac{1}{\lambda_2}\right) - h_2 \left(1 \mp \frac{1}{\lambda_1}\right)\left(1 \pm \frac{1}{\lambda_2}\right) = 0\,, \tag{8}$$

where

$$h_j(\lambda_j) = -\mathrm{i}\frac{2\Omega^*\omega - \sqrt{3}(\lambda_j - 1/\lambda_j)\sin(k_1/2)}{6 - 2\omega^2 - 3(\lambda_j + 1/\lambda_j)\cos(k_1/2)} \quad (j = 1, 2) \tag{9}$$

and

$$\lambda_1 = \frac{\sqrt{g^+ + 1} - \sqrt{g^+ - 1}}{\sqrt{g^+ + 1} + \sqrt{g^+ - 1}}, \quad \lambda_2 = \frac{\sqrt{g^- + 1} - \sqrt{g^- - 1}}{\sqrt{g^- + 1} + \sqrt{g^- - 1}}. \tag{10}$$

In the formulae above,

$$g^\pm = -C \pm \sqrt{C^2 - D}\,, \tag{11}$$

where

$$C = \cos\left(\frac{k_1}{2}\right)\left[\frac{2\omega^2}{3} - 1 - 2\sin^2\left(\frac{k_1}{2}\right)\right], \tag{12a}$$

$$D = \frac{\omega^2}{3}\left\{\omega^2 - (\Omega^*)^2 - 4\left[1 + \sin^2\left(\frac{k_1}{2}\right)\right]\right\} + 1 + 3\sin^2\left(\frac{k_1}{2}\right). \tag{12b}$$

The dispersion curves, obtained from (8) and calculated in the stop-bands of the infinite periodic system with uniform effective gyricity, are shown in Fig. 2 by dashed lines, taking $\Omega^* = 2.5$. It is apparent that these dispersion curves are not symmetric with respect to the axis $k_1 = 0$. In particular, the branch within the internal stop-band of the infinite system only exists for negative values of the wave number k_1. Hence, if the lattice is excited by a time-harmonic force having a frequency within that branch, the generated waves will be localised at the interface and mainly propagate to the left of the force. This will be demonstrated in the next section.

4 Forced Problem

In this section, we determine the response of the lattice to an external excitation, represented by a time-harmonic displacement. As in Fig. 1(a), the discrete system is partitioned into two sub-domains with equal and opposite values of effective gyricity (taken again as ± 2.5). The excitation is applied at the interface between the two sub-domains. The problem is solved numerically with a finite element model built in *Comsol Multiphysics* (version 5.6). In order to avoid the disturbance of reflected waves, Adaptive Absorbing Layers (AAL) are inserted near the boundaries of the finite computational domain.

In the present simulation, the imposed time-harmonic displacement has unit amplitude and radian frequency $\omega = 1.255$. This frequency falls inside the branch of the dispersion curves associated with interfacial waves, located inside the internal stop-band of the infinite system and existing only for negative values of k_1 (see Fig. 2).

The amplitude of the total displacement in the lattice is plotted in Fig. 3. We observe a wave localised at the interface, which propagates to the left of the external excitation. This is in accordance with the dispersion analysis discussed in Sect. 3.

5 Conclusions

In this paper, we have shown that gyroscopic spinners connected to the nodes of an elastic lattice can create unidirectional interfacial waves. The lattice's particles can be arranged in the simplest way, namely in a triangular pattern, differently from previous works where honeycomb configurations were considered [11,14]. In addition, the presence of Dirac cones in the dispersion diagram is not required to achieve preferential directionality.

The relation describing the dispersive behaviour of interfacial waves is given in closed analytical form here for the first time. For specific values of the effective gyricity, this relation leads to the appearance of dispersion curves in the stop-bands of the dispersion diagram associated with bulk waves. Since these dispersion curves are not symmetric, waves are characterised by a preferential directionality.

We also emphasise that the direction of the interfacial wave can be reversed by changing the signs of the effective gyricities in the two sub-domains of the medium. In addition, we point out that neither the type of excitation (namely, force or displacement) nor the inclination of the excitation in the lattice's plane affect the robustness of the phenomenon.

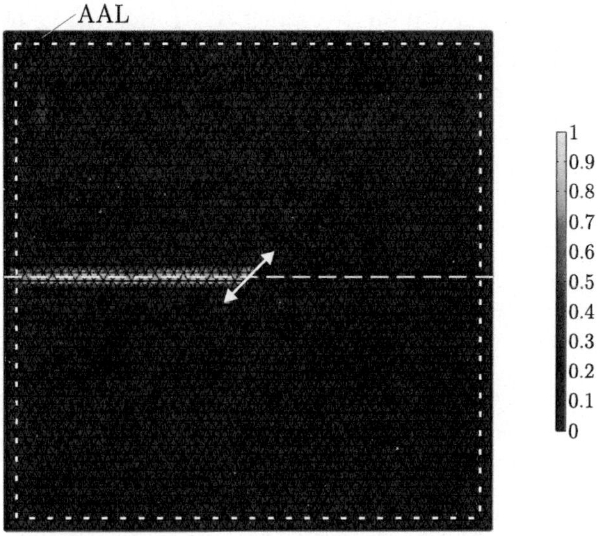

Fig. 3. Strongly localised unidirectional waveform, produced by a time-harmonic excitation (whose direction of application is indicated by the arrow) of frequency $\omega = 1.255$.

Acknowledgements. M.J.N gratefully acknowledges the support of the EU H2020 grant MSCA-RISE-2020-101008140-EffectFact. The work by G.C. and M.B. was performed under the auspices of GNFM-INDAM.

References

1. Haldane, F.D.M.: Model for a quantum Hall effect without Landau levels: condensed-matter realization of the parity anomaly. Phys. Rev. Lett. **61**, 2015 (1988)
2. Pal, R.K., Schaeffer, M., Ruzzene, M.: Helical edge states and topological phase transitions in phononic systems using bi-layered lattices. J. Appl. Phys. **119**, 084305 (2016)
3. Vila, J., Pal, R.K., Ruzzene, M.: Observation of topological valley modes in an elastic hexagonal lattice. Phys. Rev. B **96**, 134307 (2017)
4. Pal, R.K., Ruzzene, M.: Edge waves in plates with resonators: an elastic analogue of the quantum valley Hall effect. New J. Phys. **19**, 025001 (2017)
5. Miniaci, M., Pal, R.K., Manna, R., Ruzzene, M.: Valley-based splitting of topologically protected helical waves in elastic plates. Phys. Rev. B **100**, 024304 (2019)
6. Chen, R.G., et al.: Topological mechanics of origami and kirigami. Phys. Rev. Lett. **116**, 135501 (2016)
7. Zheng, L.Y., Teocharis, G., Tournat, V., Gusev, V.: Quasitopological rotational waves in mechanical granular graphene. Phys. Rev. B **97**, 060101(R) (2018)
8. Li, S., Kim, I., Iwamoto, S., Zang, J., Yang, J.: Valley anisotropy in elastic metamaterials. Phys. Rev. B **100**, 195102 (2019)
9. Wang, P., Lu, L., Bertoldi, K.: Topological phononic crystals with one-way elastic edge waves. Phys. Rev. Lett. **115**, 104302 (2015)

10. Nash, L.M., Kleckner, D., Read, A., Vitelli, V., Turner, A.M., Irvine, W.T.M.: Topological mechanics of gyroscopic metamaterials. Proc. Natl. Acad. Sci. **112**(47), 14495–14500 (2015)
11. Garau, M., Carta, G., Nieves, M.J., Jones, I.S., Movchan, N.V., Movchan, A.B.: Interfacial waveforms in chiral lattices with gyroscopic spinners. Proc. R. Soc. A **474**, 20180132 (2018)
12. Lee, C.H., Li, G., Jin, G., Liu, Y., Zhang, X.: Topological dynamics of gyroscopic and Floquet lattices from Newton's laws. Phys. Rev. B **97**, 085110 (2018)
13. Mitchell, N.P., Nash, L.M., Irvine, W.T.M.: Tunable band topology in gyroscopic lattices. Phys. Rev. B **98**, 174301 (2018)
14. Garau, M., Nieves, M.J., Carta, G., Brun, M.: Transient response of a gyro-elastic structured medium: unidirectional waveforms and cloaking. Int. J. Eng. Sci. **143**, 115–141 (2019)
15. Nieves, M.J., Carta, G., Pagneux, V., Brun, M.: Rayleigh waves in micro-structured elastic systems: non-reciprocity and energy symmetry breaking. Int. J. Eng. Sci. **156**, 103365 (2020)
16. Nieves, M.J., Carta, G., Pagneux, V., Brun, M.: Directional control of Rayleigh wave propagation in an elastic lattice by gyroscopic effects. Front. Mater. **7**, 602960 (2021)
17. Carta, G., Colquitt, D.J., Movchan, A.B., Movchan, N.V., Jones, I.S.: One-way interfacial waves in a flexural plate with chiral double resonators. Philoso. Trans. R. Soc. A **378**(2162), 20190350 (2019)
18. Carta, G., Colquitt, D.J., Movchan, A.B., Movchan, N.V., Jones, I.S.: Chiral flexural waves in structured plates: directional localisation and control. J. Mech. Phys. Solids **137**, 103866 (2020)
19. Carta, G., Nieves, M.J., Jones, I.S., Movchan, N.V., Movchan, A.B.: Elastic chiral waveguides with gyro-hinges. Q. J. Mech. Appl. Math. **71**, 157–185 (2018)
20. Nieves, M.J., Carta, G., Jones, I.S., Movchan, N.V., Movchan, A.B.: Vibrations and elastic waves in chiral multi-structures. J. Mech. Phys. Solids **121**, 387–408 (2018)

WVS: Waves and Vibration in Nonsmooth Systems and Structures

Effect of an Impulsive Source on Shear Wave Propagation in a Piezo-electro-magnetic Structure Mounted Over a Heterogeneous Isotropic Substrate Carrying a Viscoelastic Thin Film

Arindam Nath(✉) and Sudarshan Dhua

National Institute of Technology Andhra Pradesh, Tadepalligudem, India
arindamnath.math@gmail.com

Abstract. The present paper studies the propagation behavior of shear wave through piezo-electro-magnetic material (PEMM) due to the impact of an impulsive point source at the lower boundary of PEMM. The PEMM lies between an isotropic viscoelastic layer and an isotropic heterogeneous half-space. Green's function technique along with Fourier's transformation has been employed to obtain the dispersion equation. For some particular cases, dispersion equation has been obtained and is validated and matched with pre-established standard results. The effects of viscoelasticity and functional gradient parameters on the phase velocity of propagating wave are shown graphically through numerical computations.

Keywords: Green's function · Point source · Viscoelasticity · Piezo-electro-magnetic medium · Heterogeneity · Shear wave

1 Introduction

Piezo-electro-magnetic materials (PEMM), which are composition of a piezoelectric and piezomagnetic phase, not only have original piezoelectric and piezomagnetic properties but also show a remarkable magnetoelectric coupling effect that is not present in the constituents. As a result, they are extensively used in several engineering fields such as magnetic field probes, electronic packaging, hydrophones, strain gauges, medical ultrasonic imaging, and in general, as magnetic/ electric transducers, sensors, and actuators [1]. In general, magneto-electro-elastic materials are brittle, and as a result, cracks inevitably form during the manufacturing process and subsequent handling. Therefore, to provide damping to the composite, a thin viscoelastic film is considered on the top of PEMM layer. Viscoelastic materials such as polymers are characterized by their time and frequency-dependent mechanical properties. That means the functionality and applicability of polymeric products can change considerably after a certain period. These are currently critical structural materials, particularly in the aerospace and automotive industries. Some notable works on wave propagation in viscoelastic medium are performed by

Z. Dimitrovová et al. (Eds.): WMVC 2022, MMS 125, pp. 1145–1157, 2023.
https://doi.org/10.1007/978-3-031-15758-5_118

Cui et al. [2], Chen et al. [3]. More recently, Nath and Dhua have studied the propagation of SH-wave in multi-layered magneto-visco-elastic orthotropic multi-layered structure [4].

For the present study, the heterogeneity taken is caused by variation in rigidity, density, and internal friction. Functionally graded materials have been known to have good performances in many applications. Different authors have taken various forms of variation, like harmonic, linear, quadratic, *etc.*, for simulating the variation in density and other geological parameters inside the Earth, such as Dhua [5]. Sari and Salk [6] took the variation in sediment density with depth as a hyperbolic function.

The external force of disturbances originating due to point source is one of the reasons for wave motions in a different medium. This action force or impulse is usually described by the Dirac-delta function. The solutions of equations of motion represent the elastic displacement due to a unit impulse force in space-time coordinates. Green's function plays a vital role in solving these boundary value problems. Many authors have studied wave propagation due to point source in various media, like Chattopadhyay [7].

This article analyzes the shear wave propagation generated due to a point source impulse in a piezo-electro-magnetic layer situated between viscoelastic film and heterogeneous half-space. Green's function technique and Fourier's transformation are employed to solve the problem analytically. Based on the governing equations and appropriate boundary conditions, the closed form of the dispersion equation is obtained for shear wave propagation. Moreover, special cases of the dispersion equation have been deduced and are validated with the classical Love wave equation. Numerical computation and graphical interpretation have been carried out to delineate the effects of viscoelastic parameter, heterogeneity parameter, magnification factor related to the heterogeneity, piezoelectric constant, dielectric constant, magnetic permeability, piezomagnetic and piezo-electro-magnetic constant.

2 Mathematical Formulation of the Problem

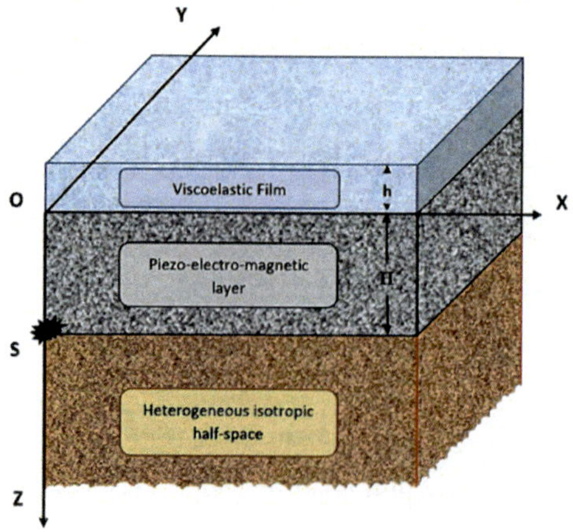

Fig. 1. Geometry of the problem

Consider a model consisting of PEMM layer situated between an isotropic viscoelastic thin film and an isotropic heterogeneous half-space. The viscoelastic film and PEMM layer thickness are h and H, respectively. We choose the cartesian coordinate system so that the x-axis is taken along the wave propagation, and the z-axis points vertically downwards. The origin O is taken at the interface of viscoelastic and PEMM layers. An impulsive point source S of disturbance is located at the common surface of the PEMM layer and half-space (Fig. 1).

Constitutive Relations:
The stress-strain relation for isotropic viscoelastic upper layer is given by

$$\tau_{ij} = \lambda_1 \delta_{ij} s_{kk} + 2\mu_1 s_{ij} \tag{1}$$

The constitutive relations for transversely isotropic PEMM layer are given by [8]

$$\left.\begin{aligned} \sigma_{ij} &= c_{ijkl} S_{kl} - e_{kij} E_k - d_{kij} H_k, \\ D_k &= e_{kij} S_{ij} + \varepsilon_{ki} E_i + g_{ki} H_i, \\ B_k &= d_{kij} S_{ij} + g_{ki} E_i + \mu_{ki} H_i \end{aligned}\right\} \tag{2}$$

where $c_{ijkl}, e_{kij}, d_{kij}, \varepsilon_{ki}, g_{ki}, \mu_{ki}$ are elastic constants, PE constants, PM constants, dielectric constants, electromagnetic constants, and magnetic permeability, respectively. D_k, B_k are electric displacement and magnetic flux. $E_i(= -\phi_{,i}), H_i(= -\psi_{,i})$ are electric field and magnetic field. ϕ, ψ are electric potential and magnetic potential.

The stress-strain relation for isotropic heterogeneous lower half-space is

$$\pi_{ij} = \lambda_3 \delta_{ij} a_{kk} + 2\mu_3 a_{ij} \tag{3}$$

Here $\tau_{ij}, \sigma_{ij}, \pi_{ij}$ are stress components and s_{ij}, S_{ij}, a_{ij} are strain components of upper, intermediate and lower half-space mediums, respectively. λ_n, μ_n are elastic constants ($n = 1,3$).

In Cartesian coordinate system, let (u_n, v_n, w_n) be the displacement components in (x,y,z) direction for upper, intermediate and half-space, respectively ($n = 1, 2, 3$). For the propagation of shear wave in the x-direction and causing displacement in the y-direction, it may be assumed

$$v_n = v_n(x, z, t), \quad u_n = w_n = 0 \tag{4}$$

where t denotes time.

Dynamics of Layer 1:
The non-vanishing equation of motion without body force for the upper viscoelastic layer is given by

$$\frac{\partial}{\partial x}\left(\mu_1 \frac{\partial v_1}{\partial x}\right) + \frac{\partial}{\partial z}\left(\mu_1 \frac{\partial v_1}{\partial z}\right) = \rho_1 \frac{\partial^2 v_1}{\partial t^2} \tag{5}$$

where $\mu_1 = \underset{1}{\frown{\mu}} + \underset{1}{\frown{\eta}} \frac{\partial}{\partial t}$ is elastic constant, $\underset{1}{\frown{\eta}}$ is the parameter due to the internal friction of viscoelastic medium, ρ_1 is the density of the medium.

Dynamics of Layer 2:
The governing equations for the PEMM intermediate layer are:

$$\sigma_{ij,j} + F_i = \rho_2 \ddot{u}_i, \;\; D_{i,i} = 0, \;\; B_{i,i} = 0 \tag{6}$$

Therefore, the equations of motion for the PEMM intermediate layer with force density distribution $f(r, t)$ due to point source existing at the interface of the PEMM layer and half-space is obtained as

$$c_{44}\frac{\partial^2 v_2}{\partial x^2} + c_{44}\frac{\partial^2 v_2}{\partial z^2} + e_{15}\frac{\partial^2 v_2}{\partial x^2} + e_{15}\frac{\partial^2 v_2}{\partial z^2} - \rho_2\frac{\partial^2 v_2}{\partial t^2} = 4\pi f(r, t) \tag{7}$$

$$e_{15}\frac{\partial^2 v_2}{\partial x^2} + e_{15}\frac{\partial^2 v_2}{\partial z^2} - \varepsilon_{11}\frac{\partial^2 \phi}{\partial x^2} - \varepsilon_{11}\frac{\partial^2 \phi}{\partial z^2} - g_{11}\frac{\partial^2 \psi}{\partial x^2} - g_{11}\frac{\partial^2 \psi}{\partial z^2} = 0 \tag{8}$$

$$d_{15}\frac{\partial^2 v_2}{\partial x^2} + d_{15}\frac{\partial^2 v_2}{\partial z^2} - g_{11}\frac{\partial^2 \phi}{\partial x^2} - g_{11}\frac{\partial^2 \phi}{\partial z^2} - \mu_{11}\frac{\partial^2 \psi}{\partial x^2} - \mu_{11}\frac{\partial^2 \psi}{\partial z^2} = 0 \tag{9}$$

where r is the position of the point source impluse and ρ_2 is the density of the medium.
Substituting eqs. (8), (9) in (7), we get

$$\frac{\partial^2 v_2}{\partial x^2} + \frac{\partial^2 v_2}{\partial z^2} - \frac{\rho_2}{\tilde{c}_{44}}\frac{\partial^2 v_2}{\partial t^2} = \frac{4\pi f(r, t)}{\tilde{c}_{44}} \tag{10}$$

where $\tilde{c}_{44} = c_{44} + \left(2e_{15}d_{15}g_{11} - e_{15}^2\mu_{11} - d_{15}^2\varepsilon_{11}\right)/\left(g_{11}^2 - \varepsilon_{11}\mu_{11}\right)$

Dynamics of Layer 3:
For the isotropic half-space, vertical heterogeneity in elastic constants and density is considered as follows:

$$\left.\begin{aligned} \mu_3 &= \mu_3^{(0)} + \varepsilon \sinh b(z - H) \\ \rho_3 &= \rho_3^{(0)} + \varepsilon^* \sinh b(z - H) \end{aligned}\right\} \tag{11}$$

where b is the heterogeneity parameter associated with the hyperbolic type of heterogeneity in the isotropic half-space and ε, ε^* are the magnification factors of hyperbolic kind of heterogeneity in shear modulus and density, respectively and ρ_3 is the density of the medium.

The equation of motion of the isotropic heterogeneous substrate is given by

$$\frac{\partial}{\partial x}\left(\mu_3\frac{\partial v_3}{\partial x}\right) + \frac{\partial}{\partial z}\left(\mu_3\frac{\partial v_3}{\partial z}\right) = \rho_3\frac{\partial^2 v_3}{\partial t^2} \tag{12}$$

Now, for harmonic solution of Love type wave propagation, the displacement and force density term be of the form

$$\left.\begin{aligned} v_n(x, z, t) &= \tilde{v}_n(x, z)e^{i\omega t} \\ f(r, t) &= f(r)e^{i\omega t} \end{aligned}\right\} (n = 1, 2, 3) \tag{13}$$

where $\omega(=ck)$ is circular frequency, k is wavenumber, and c is shear wave velocity.

The disturbances caused by the impulsive force $f(r)$ may be represented in terms of Dirac-δ function at the source point as

$$f(r) = \delta(x)\delta(z-H) \tag{14}$$

Using (13) in (5), (10), (12) the equations of motion can be written as

$$\frac{\partial^2 \tilde{v}_1}{\partial x^2} + \frac{\partial^2 \tilde{v}_1}{\partial z^2} + \frac{\omega^2}{\beta_1^2}\tilde{v}_1 = 0\left(\beta_1 = \sqrt{\mu_1/\rho_1}\right) \tag{15}$$

$$\frac{\partial^2 \tilde{v}_2}{\partial x^2} + \frac{\partial^2 \tilde{v}_2}{\partial z^2} + \frac{\omega^2}{\beta_2^2}\tilde{v}_2 = \frac{4\pi\delta(x)\delta(z-H)}{\tilde{c}_{44}}\left(\beta_2 = \sqrt{\tilde{c}_{44}/\rho_2}\right) \tag{16}$$

$$\mu_3\frac{\partial^2 \tilde{v}_3}{\partial x^2} + \frac{d\mu_3}{dz}\frac{\partial \tilde{v}_3}{\partial z} + \mu_3\frac{\partial^2 \tilde{v}_3}{\partial z^2} + \rho_3\omega^2\tilde{v}_3 = 0\left(\beta_2 = \sqrt{\mu_3^{(0)}/\rho_3}\right) \tag{17}$$

Now applying Fourier transform on the eqs. (15), (16) and (17), we get

$$\frac{d^2 V_1}{dz^2} - \alpha_1^2 V_1 = 0 \tag{18}$$

$$\frac{d^2 V_2}{dz^2} - \alpha_2^2 V_2 = \frac{2\delta(z-H)}{\tilde{c}_{44}} \tag{19}$$

$$\frac{d^2 V_3}{dz^2} - \alpha_3^2 V_3 = 4\pi g(z) \tag{20}$$

where $V_n(\xi, z)$ is the Fourier transform of $\tilde{v}_n(x, z)$, ξ is the transform parameter.

$$4\pi g(z) = -\frac{\varepsilon}{\mu_3^{(0)}}\left[\begin{array}{l} \sinh b(z-H)\dfrac{d^2V_3}{dz^2} + b\cosh b(z-H)\dfrac{dV_3}{dz} \\ -\left(\xi^2\sinh b(z-H) - \dfrac{\varepsilon^*}{\varepsilon}\omega^2\sinh b(z-H)\right)V_3 \end{array}\right] \tag{21}$$

3 Boundary Conditions

The boundary conditions can be written as follows:

i. The upper surface of the viscoelastic layer is stress free, i.e., at $z = -h$

$$\tau_{yz} = 0 \tag{22}$$

ii. The stress and displacement components are continuous at the common interface of the viscoelastic layer and PEMM layer, i.e., at $z = 0$

$$a)\ \tau_{yz} = \sigma_{zy} \tag{23}$$

$$b)\ V_1 = V_2 \tag{24}$$

The electrical and magnetic boundary conditions are

$$c)\ \phi = 0 \tag{25}$$

$$d)\ \psi = 0 \tag{26}$$

iii. The stress and displacement components are continuous at the common interface of the PEMM layer and heterogeneous half-space, i.e., at $z = H$

$$a)\ \sigma_{yz} = \pi_{yz} \tag{27}$$

$$b)\ V_2 = V_3 \tag{28}$$

Also, the electrical and magnetic boundary conditions are

$$c)\ \phi = 0 \tag{29}$$

$$d)\ \psi = 0 \tag{30}$$

4 Solution to the Problem

Using the boundary condition (22), we may derive $\frac{dV_1}{dz} = 0$.

Let $G_1(z, z_0)$ be Green's function for the viscoelastic upper layer satisfying the condition $\frac{dG_1(z,z_0)}{dz} = 0$ at $z = -h$ and $z = 0$.

Therefore, $G_1(z, z_0)$ will satisfy the equation

$$\frac{d^2G_1(z, z_0)}{dz^2} - \alpha_1^2 G_1(z, z_0) = \delta(z - z_0) \tag{31}$$

where z_0 is any arbitrary point in the top layer.

Now, multiplying Eq. (18) by $G_1(z, z_0)$ and Eq. (31) by $V_1(z)$, then subtracting and integrating with respect to z from $z = -h$ to $z = 0$ results in the following expression after employing the symmetric property of Green's function

$$V_1(z) = -G_1(z, 0)\left[\frac{dV_1(z)}{dz}\right]_{z=0} \tag{32}$$

Again, let $G_2(z, z_0)$ be Green's function for the intermediate PEMM layer satisfying the condition $\frac{dG_2(z,z_0)}{dz} = 0$ at z = 0 and z = H.

Then, $G_2(z, z_0)$ is the solution to the equation

$$\frac{d^2G_2(z, z_0)}{dz^2} - \alpha_2^2 G_2(z, z_0) = \delta(z - z_0) \tag{33}$$

Multiplying eqs. (19) and (33) by $G_2(z, z_0)$ and $V_2(z)$, respectively, then subtracting and integrating with respect to z from $z = 0$ to $z = H$, using the symmetric property of Green's function, we get

$$V_2(z) = \frac{2}{\tilde{c}_{44}} G_2(z, H) - G_2(z, H)\left[\frac{dV_2(z)}{dz}\right]_{z=H} + G_2(z, 0)\left[\frac{dV_2(z)}{dz}\right]_{z=0} \tag{34}$$

Similarly, let $G_3(z, z_0)$ be Green's function for lower heterogeneous isotropic half-space satisfying the condition $\frac{dG_3(z,z_0)}{dz} = 0$ at $z = H$ and tends to zero as $z \to \infty$.

Then, $G_3(z, z_0)$ is the solution to the equation

$$\frac{d^2 G_3(z, z_0)}{dz^2} - \alpha_3^2 G_3(z, z_0) = \delta(z - z_0) \tag{35}$$

Again, proceeding in a similar manner, the value of $V_3(z)$ at any point z in the half-space can be obtained as

$$V_3(z) = G_3(z, H)\left[\frac{dV_3(z)}{dz}\right]_{z=H} + \int_H^\infty 4\pi g(z_0) G_3(z, z_0) dz_0 \tag{36}$$

Using boundary conditions (23) and (24), eqs. (32) and (34), we obtain

$$\left[\frac{dV_1(z)}{dz}\right]_{z=0} = \frac{1}{A}\left(G_2(0, H)\left[\frac{dV_2(z)}{dz}\right]_{z=H} - \frac{2}{\tilde{c}_{44}} G_2(0, H)\right) \tag{37}$$

where $A = G_1(0, 0) + \frac{\mu_1}{\tilde{c}_{44}} G_2(0, 0)$

Given the boundary condition (27) and (28), from eqs. (34) and (36), we derive

$$\left[\frac{dV_2(z)}{dz}\right]_{z=H} = \frac{1}{B}\left[\frac{2}{\tilde{c}_{44}}\left(G_2(H, H) - \frac{\mu_1}{c_{44}} \frac{G_2(H, 0) G_2(0, H)}{A}\right) - \int_H^\infty 4\pi g(z_0) G_3(H, z_0) dz_0\right] \tag{38}$$

where $B = G_2(H, H) + \frac{c_{44}}{\mu_3^{(0)}} G_3(H, H) - \frac{\mu_1}{c_{44}} \frac{G_2(H,0) G_2(0,H)}{A}$

Using eqs. (37) and (38), we deduce

$$\left[\frac{dV_1(z)}{dz}\right]_{z=0} = \frac{1}{A}\left(\frac{G_2(0, H)}{B}\left[\begin{array}{l}\frac{2}{\tilde{c}_{44}}\left(G_2(H, H) - \frac{\mu_1}{c_{44}} \frac{G_2(H, 0) G_2(0, H)}{A}\right) \\ - \int_H^\infty 4\pi g(z_0) G_3(H, z_0) dz_0\end{array}\right] - \frac{2}{\tilde{c}_{44}} G_2(0, H)\right) \tag{39}$$

In the view of Eq. (39), Eq. (32) results in

$$V_1(z) = -\frac{G_1(z, 0)}{A}\left(\frac{G_2(0, H)}{B}\left[\begin{array}{l}\frac{2}{\tilde{c}_{44}}\left(G_2(H, H) - \frac{\mu_1}{c_{44}} \frac{G_2(H, 0) G_2(0, H)}{A}\right) \\ - \int_H^\infty 4\pi g(z_0) G_3(H, z_0) dz_0\end{array}\right] - \frac{2}{\tilde{c}_{44}} G_2(0, H)\right) \tag{40}$$

Furthermore, employing the expression of (21), (27), (36), and (38), we deduce

$$V_3(z) = \frac{G_3(z,H)}{B}\frac{c_{44}}{\mu_3^{(0)}}[\frac{2}{\tilde{c}_{44}}\left(G_2(H,H) - \frac{\mu_1}{c_{44}}\frac{G_2(H,0)G_2(0,H)}{A}\right)$$
$$+\int_H^\infty \left\{\begin{array}{l} -\frac{\varepsilon}{\mu_3^{(0)}}[\sinh b(z-H)\frac{d^2V_3}{dz^2} + b\cosh b(z-H)\frac{dV_3}{dz} \\ -\left(\xi^2 \sinh b(z-H) - \frac{\varepsilon^*}{\varepsilon}\omega^2 \sinh b(z-H)\right)V_3] \end{array}\right\} G_3(H,z_0)dz_0]$$
$$-\int_H^\infty \left\{\frac{\varepsilon}{\mu_3^{(0)}}\left[\begin{array}{l} \sinh b(z-H)\frac{d^2V_3}{dz^2} + b\cosh b(z-H)\frac{dV_3}{dz} \\ -\left(\xi^2 \sinh b(z-H) - \frac{\varepsilon^*}{\varepsilon}\omega^2 \sinh b(z-H)\right)V_3 \end{array}\right]\right\} G_3(z,z_0)dz_0 \tag{41}$$

The value $V_3(z)$ may be obtained from Eq. (41) by the method of successive approximation. On neglecting the higher powers of ε in the Eq. (41), we obtain the first-order approximation of $V_3(z)$ as

$$V_3(z) = \frac{G_3(z,H)}{B}\frac{c_{44}}{\mu_3^{(0)}}\frac{2}{\tilde{c}_{44}}\left\{G_2(H,H) - \frac{\mu_1}{c_{44}}\frac{G_2(H,0)G_2(0,H)}{A}\right\} \tag{42}$$

On using eqs. (21), (40) and (42), the expression of $V_1(z)$ after simplification as

$$V_1(z) = -\frac{G_1(z,0)}{A}[\frac{G_2(0,H)}{B}\{\frac{2}{\tilde{c}_{44}}\left(G_2(H,H) - \frac{\mu_1}{c_{44}}\frac{G_2(H,0)G_2(0,H)}{A}\right)$$
$$+\frac{2\varepsilon c_{44}}{B(\mu_3^{(0)})^2\tilde{c}_{44}}\left(G_2(H,H) - \frac{\mu_1}{c_{44}}\frac{G_2(H,0)G_2(0,H)}{A}\right) \times \int_H^\infty (\sinh b(z_0-H)$$
$$\frac{d^2G_3(z_0,H)}{dz^2} + b\cosh b(z_0-H)\frac{dG_3(z_0,H)}{dz}$$
$$-\left(\xi^2 \sinh b(z_0-H) - \frac{\varepsilon^*}{\varepsilon}\omega^2 \sinh b(z_0-H)\right)G_3(z_0,H))G_3(H,z_0)dz_0\} - \frac{2}{\tilde{c}_{44}}G_2(0,H)] \tag{43}$$

Constructing $G_1(z,z_0)$, $G_2(z,z_0)$ and $G_3(z,z_0)$ from eqs. (31), (33) and (35) respectively, we have

$$G_1(z,z_0) = -\frac{1}{2\alpha_1}\left[e^{-\alpha_1|z-z_0|} + \frac{e^{\alpha_1 z}(e^{\alpha_1(h+z_0)} - e^{-\alpha_1(h+z_0)})}{e^{\alpha_1 h} - e^{-\alpha_1 h}} + \frac{e^{\alpha_1 z}(e^{\alpha_1(h+z_0)} - e^{-\alpha_1(h+z_0)})}{e^{\alpha_1 h} - e^{-\alpha_1 h}}\right] \tag{44}$$

$$G_2(z,z_0) = -\frac{1}{2\alpha_2}\left[e^{-\alpha_2|z-z_0|} + \frac{e^{\alpha_2 z}(e^{\alpha_2(H+z_0)} - e^{-\alpha_2(H+z_0)})}{e^{\alpha_2 H} - e^{-\alpha_2 H}} + \frac{e^{\alpha_2 z}(e^{\alpha_2(H+z_0)} - e^{-\alpha_2(H+z_0)})}{e^{\alpha_2 H} - e^{-\alpha_2 H}}\right] \tag{45}$$

$$G_3(z,z_0) = -\frac{1}{2\alpha_3}\left[e^{-\alpha_3|z-z_0|} + e^{-\alpha_3(z+z_0-2H)}\right] \tag{46}$$

Substituting the values of $G_1(z,z_0)G_2(z,z_0)$ and $G_3(z,z_0)$ in Eq. (43), we deduce

$$V_1(z) = -\frac{2\cosh(\alpha_1 z)c_{44}}{\alpha_1\alpha_2\alpha_3\tanh(\alpha_1 h)\tanh(\alpha_2 H)\bar{c}_{44}\mu_3^{(0)}AB}\left[1-\frac{b\left(B\alpha_3\mu_3^{(0)}+c_{44}\right)\left(\varepsilon\xi^2-\varepsilon^*\omega^2+\varepsilon\alpha_3^2\right)}{B(\mu_3^{(0)})^2\alpha_3^2(4\alpha_3^2-b^2)}\right] \tag{47}$$

Eliminating the terms containing higher powers of ε the corresponding displacement at any point in the uppermost layer may be obtained from the above Eq. (47) by taking inverse Fourier transformation as

$$\tilde{v}_1(x,z) = -\int_{-\infty}^{\infty}\frac{2\cosh(\alpha_1 z)c_{44}e^{-i\xi x}}{\alpha_1\alpha_2\alpha_3\tanh(\alpha_1 h)\tanh(\alpha_2 H)\bar{c}_{44}\mu_3^{(0)}AB\left[1+\frac{b\left(B\alpha_3\mu_3^{(0)}+c_{44}\right)\left(\varepsilon\xi^2-\varepsilon^*\omega^2+\varepsilon\alpha_3^2\right)}{B(\mu_3^{(0)})^2\alpha_3^2(4\alpha_3^2-b^2)}\right]}d\xi \tag{48}$$

The value of the integral (48) entirely depends on the poles of the integrand, which are located at the zeros of the equation

$$B(\mu_3^{(0)})^2\alpha_3^2(4\alpha_3^2-b^2)+b\left(B\alpha_3\mu_3^{(0)}+c_{44}\right)\left(\varepsilon\xi^2-\varepsilon^*\omega^2+\varepsilon\alpha_3^2\right)=0 \tag{49}$$

Replacing α_1 by $i\alpha_{10}$, α_2 by $i\alpha_{20}$ and ξ to k, Eq. (49) may be reduced to

$$\tan(\alpha_{02}H) = \frac{c_{44}\alpha_{02}\left\{\mu_3^{(0)}\alpha_3\left(4\alpha_3^2-b^2\right)+Lb\right\}-c_{44}\mu_1\alpha_{02}\alpha_{01}\tan(\alpha_{01}h)\left(4\alpha_3^2-b^2\right)}{\left\{\mu_3^{(0)}\alpha_3\left(4\alpha_3^2-b^2\right)+Lb\right\}\mu_1\alpha_{01}\tan(\alpha_{01}h)+c_{44}^2\alpha_{02}^2\left(4\alpha_3^2-b^2\right)} \tag{50}$$

where $L=\left(\varepsilon k^2-\varepsilon^*c^2k^2+\varepsilon\alpha_3^2\right)$, $\alpha_{01}=k\sqrt{c^2/\beta_1^2-1}$, $\alpha_{02}=k\sqrt{c^2/\beta_2^2-1}$

which is the dispersion equation of the shear wave propagation in a piezo-electro-magnetic layer between viscoelastic film and a heterogeneous half-space, influenced by a point source of disturbance.

5 Special Cases

a) In the absence of piezo-electro-magnetic effect and material gradient in the half-space, i.e., $e_{15}=0, q_{15}=0, \varepsilon_{15}=0, \alpha_{15}=0, \mu_{15}=0, b=0, \varepsilon^*=0, \varepsilon=0$, the dispersion equation is

$$\tan(\alpha'_{02}H) = \frac{c_{44}\alpha'_{02}\mu_3^{(0)}\alpha_3 - c_{44}\mu_1\alpha'_{02}\alpha_{01}\tan(\alpha_{01}h)}{\mu_3^{(0)}\mu_1\alpha_3\alpha_{01}\tan(\alpha_{01}h)+c_{44}^2\alpha'^2_{02}} \tag{51}$$

b) When the viscoelastic film is absent, $h\to 0$

$$\tan(\alpha'_{02}H) = \frac{\mu_3^{(0)}\alpha_3}{c_{44}\alpha'_{02}}\left(\alpha'_{02}=\sqrt{\omega^2\rho_2/c_{44}-k^2}\right) \tag{52}$$

The dispersion eq. (52) exactly matches the classical Love wave equation [9].

6 Numerical Results and Discussions

Numerical simulation and graphical illustration have been carried out to show the effect of different parameters on the phase velocity of the shear wave. The numerical results in this section have been rendered with the following data [10, 11]:

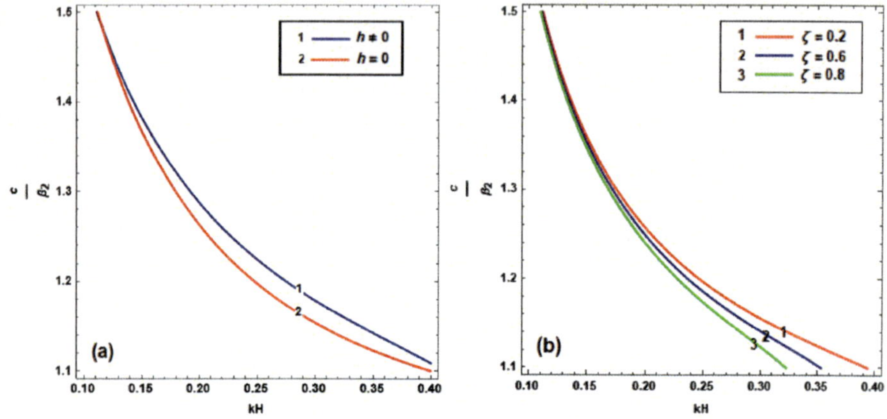

Fig. 2. Variation of dimensionless phase velocity against dimensionless wavenumber **(a)** in presence/absence of viscoelastic thin film and **(b)** viscoelastic parameter ζ of the top film.

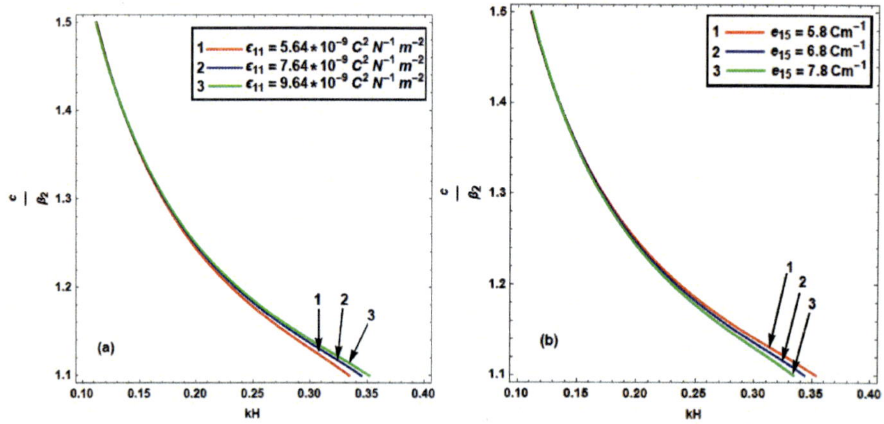

Fig. 3. Variation of dimensionless phase velocity against dimensionless wavenumber for different values of **(a)** dielectric constant ε_{11} and **(b)** piezoelectric constant e_{15} associated with the PEMM layer.

Figure 2(a) illustrates the variation of dimensionless phase (c/β_2) velocity of shear wave against dimensionless wavenumber (kH) in the presence and absence of top viscoelastic thin film, and Fig. 2(b) portrays the effect of the viscoelastic parameter $\zeta = \left(\omega \,\widehat{\eta}_1/\widehat{\mu}_1\right)$ associated with it. Figures 3 and 4 illustrate the variation of dimensionless phase velocity for different values of dielectric constant, piezoelectric constant, magnetic permeability and piezomagnetic constants of the intermediate PEMM layer.

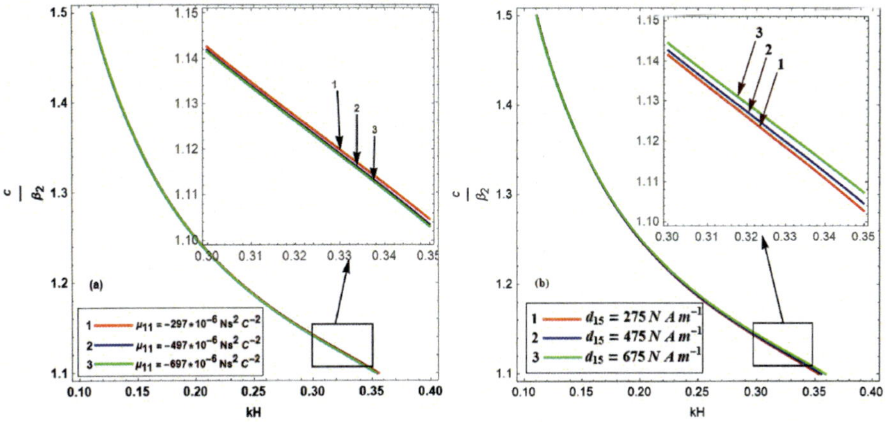

Fig. 4. Variation of dimensionless phase velocity against dimensionless wavenumber for different values of **(a)** magnetic permeability constant μ_{11} and **(b)** piezomagnetic constant d_{15} associated with the PEMM layer.

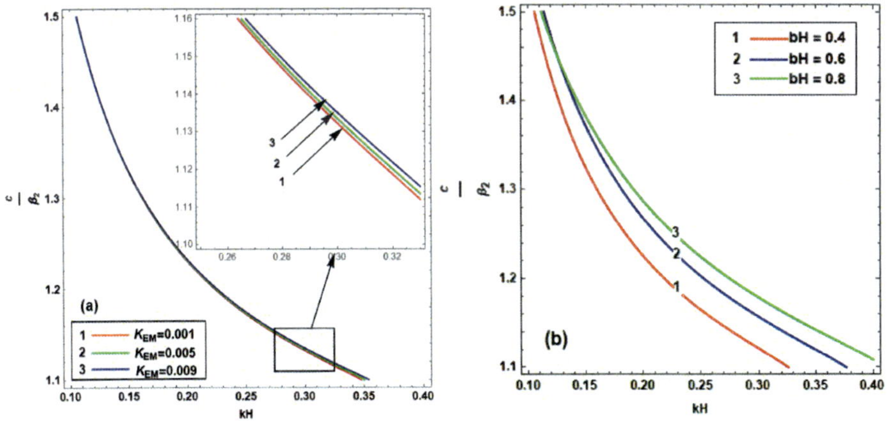

Fig. 5. Variation of dimensionless phase velocity against dimensionless wavenumber for different values of **(a)** piezo-electro-magnetic coupling parameter K_{EM} of the PEMM layer and **(b)** dimensionless heterogeneity parameter bH of the half-space.

Figure 5(a) maps the graph of dimensionless phase velocity against dimensionless wavenumber for different values of piezo-electro-magnetic coupling parameter $K_{EM} = (g_{11}e_{15}d_{15}/\mu_{11}\varepsilon_{11}c_{44})$ of the intermediate PEMM layer. Figure 5(b) enlightens the influence of dimensionless heterogeneity parameter (bH) of the half-space on the phase velocity of shear wave.

7 Conclusion

The critical points of this study may be concluded as follows:

i. The phase velocity is increased when the top viscoelastic thin film is present and diminished when it is absent.
ii. The viscoelastic film disfavors the phase velocity of the shear wave.
iii. The dielectric constant parameter ε_{11} related to the PEMM layer enhances the phase velocity of the shear wave. Whereas the piezoelectric constant parameter e_{15} lowers the phase velocity.
iv. The magnetic permeability constant μ_{11} associated with the PEMM layer reduces the phase velocity of the shear wave. However, the piezomagnetic constant d_{11} related to the PEMM layer increases the phase velocity.
v. The piezo-electro-magnetic coupling parameter enhances the phase velocity of the shear wave.
vi. The heterogeneity parameter *(bH)* related to shear modulus and density of the isotropic half-space has a favoring effect on phase velocity.

It is well known that smart materials such as piezo-electro-magnetic materials are inherently capable of actuating and sensing. Since the sensor surface lies on the interface in the design of Love wave sensors, the interface response with or without viscoelastic film due to point source impulse plays a crucial role in better designing such sensors in the long run. Therefore, the analysis of shear wave characteristics in the considered geometrical model may be applicable in developing high-end Love wave sensors.

References

1. Achenbach, J.D.: Quantitative nondestructive evaluation. Int. J. Solids Struct. **37**, 13–27 (2000)
2. Cui, J., Du, J.K., Ma, T.F., Wang, J.: SH-SAW propagation in a piezoelectric structure with viscoelastic layer. In: Proceedings of the 2012 Symposium on Piezoelectricity, Acoustic Waves and Device Applications, Spawda 2012 (2012)
3. Chen, K., Wu, Z., Jin, Y., Hu, J., Du, J., Zhang, M.: Love wave propagation in piezoelectric structures bonded with conductive polymer films. Ultrasonics. p. 118 (2021)
4. Nath, A., Dhua, S.: Influence of initial stress on shear wave scattering in a functionally graded magneto-visco-elastic orthotropic multi-layered structure. Waves in Random and Complex Media (2022)
5. Dhua, S., Chattopadhyay, A.: Torsional wave in an initially stressed layer lying between two inhomogeneous media. Meccanica **50**(7), 1775–1789 (2015). https://doi.org/10.1007/s11012-015-0119-y

6. Sari, C., Şalk, M.: Analysis of gravity anomalies with hyperbolic density contrast: An application to the gravity data of Western Anatolia. J. Balk. Geophys. Soc. 5 (2002)
7. Chattopadhyay, A., Gupta, S., Singh, A.K., Sahu, S.A.: Effect of point source, self-reinforcement and heterogeneity on the propagation of magnetoelastic shear wave. Appl. Math. 02 (2011)
8. Singh, S., Singh, A.K.: Anti-plane surface and interfacial waves influenced by layer reinforcement in Piezo-Electro-Magnetic structures with surface energy. The European Physical Journal Plus **136**(3), 1–20 (2021). https://doi.org/10.1140/epjp/s13360-021-01288-0
9. Ewing, W.M., Jardetzky, W.S., Press, F.: Elastic Waves in Layered Media. GFF. 80 (1958)
10. Gubbins, D.: Seismology and plate tectonics. Seismol. plate tectonics (1990)
11. Ke, L.L., Wang, Y.S., Yang, J., Kitipornchai, S.: The size-dependent vibration of embedded magneto-electro-elastic cylindrical nanoshells. Smart Mater. Struct. 23 (2014)

Nonlinear Normal Modes of an N Degrees of Freedom Cyclically Symmetric Piecewise Linear System

Mohit Kumar(✉) and Abhijit Sarkar

Department of Mechanical Engineering, Indian Institute of Technology Madras, Chennai 600036, India
mohitkumar.2k1@gmail.com, asarkar@iitm.ac.in

Abstract. Nonlinear normal modes (NNMs) provide a useful framework for understanding nonlinear vibrations. This article discusses the NNMs of cyclically symmetric piecewise linear systems possessing N degrees of freedom. Such structures find practical applications, for example, in modeling turbomachine blade assemblies. We compute the system's NNMs via a shooting-based method. In contrast to standard applications of this technique, our nonlinear system does not possess a clear underlying linear system, which is required to initiate the numerical method. To overcome this, we propose a linear system that incorporates cyclic symmetry and tends to the nonlinear system in the limiting case. This methodology is used to compute the NNMs of systems with 3 and 7 degrees of freedom. These examples show that the underlying linear system provides significant qualitative inferences about the NNMs, such as their mode shapes. Further, the NNMs split into groups with different frequencies based on their nodal diameters. These observations are generalized to the N degree of freedom case.

Keywords: Nonlinear vibrations · Nonlinear normal modes · Contact nonlinearity · Cyclic symmetry

1 Introduction

Nonlinear vibrations have been analyzed using the concept of nonlinear normal modes (NNMs). Since NNMs were introduced by Rosenberg [7], they have been thoroughly studied and applied in many mechanical systems [1,6].

Cyclically symmetric (CS) structures with nonlinearities are of significant theoretical and practical interest. The NNMs of these systems have been studied analytically [4,10] and numerically [2,3]. However, these works deal with smooth nonlinearities. To the best of the authors' knowledge, CS systems with contact nonlinearities have not been studied. Such systems are practically relevant: they can model a turbomachine's shrouded blade assembly, which naturally possesses a contact nonlinearity.

Z. Dimitrovová et al. (Eds.): WMVC 2022, MMS 125, pp. 1158–1165, 2023.
https://doi.org/10.1007/978-3-031-15758-5_119

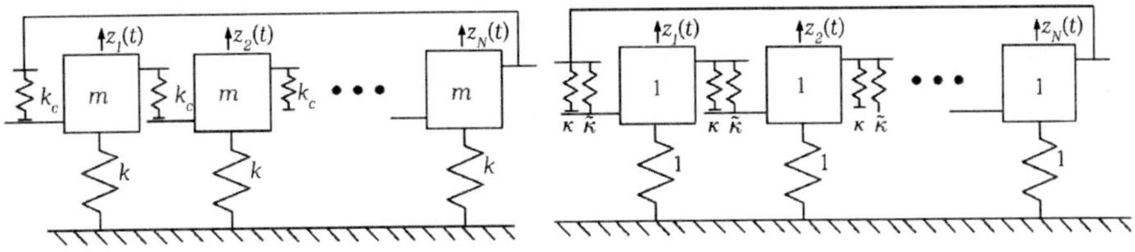

Fig. 1. A physical representation of: (a) the system under consideration, (b) a proposed underlying linear system in nondimensional form. Notice the additional linear springs of stiffness $\tilde{\kappa}$.

This work examines the NNMs of a CS N degrees of freedom nonlinear system, as shown in Fig. 1. The nonlinearity results from the bilinear contact springs between the adjacent masses. The objectives of this study are two-fold. Firstly, we describe a systematic implementation of the shooting method to compute the NNMs of the above class of dynamical systems. Secondly, using these computations, we describe general trends observed in the NNMs.

2 System Description

We consider a CS lumped oscillator model illustrated in Fig. 1a. The system represents a simplistic model of a turbine blade assembly: each of the N blades is modeled as a lumped mass m and grounded spring of stiffness k. Let z_i denote the displacement of the ith mass. As a convention, we will treat $i = 0$ as equivalent to $i = N$ and $i = N + 1$ as $i = 1$. Each mass may interact with its nearest neighbor (when their relative displacement decreases) by a contact spring of stiffness k_c. Consequently, the stiffness of the contact spring connecting the ith and $(i-1)$th masses may be mathematically modeled as $k_c\mathrm{H}(z_i - z_{i-1})$, where $\mathrm{H}(\cdot)$ represents the Heaviside function. With these considerations, the governing equations of motion are

$$m\ddot{z}_i + kz_i + k_c(z_i - z_{i-1})\mathrm{H}(z_i - z_{i-1}) + k_c(z_i - z_{i+1})\mathrm{H}(z_{i+1} - z_i) = 0\,, \quad (1)$$

where $i = 1, 2, \ldots, N$. We non-dimensionalize the above equation by introducing $y_i = z_i/\tilde{z}$ ($\tilde{z}$ is some reference displacement), $\tau = t\sqrt{k/m}$, and $\kappa = k_c/k$:

$$\ddot{y}_i + y_i + \kappa(y_i - y_{i-1})\mathrm{H}(y_i - y_{i-1}) + \kappa(y_i - y_{i+1})\mathrm{H}(y_{i+1} - y_i) = 0\,. \quad (2)$$

As noted by Zuo and Curnier [11], scaling y_i ($i = 1, \ldots, N$) by a positive constant leaves Eq. (2) invariant. This is reminiscent of the scaling property of linear systems. The significance of this property will be seen in our numerical examples.

3 Numerical Computation of the NNMs

There are several numerical techniques to compute the NNMs of mechanical systems [6]. The current study employs a shooting method owing to its simplicity, low computational costs, and suitability for nonsmooth systems [5,8]. We implemented this algorithm on MATLAB using its nonlinear equation solver *fsolve* and numerical integrator *ode45* with event detection.

The numerical method proposed in the present work is similar to the shooting-based continuation scheme used to compute the NNMs of a nonlinear system [5]. However, we do not require any continuation in the type of systems under consideration (Eq. (2)). This simplification results from the positive-scaling property of the system. In particular, suppose an NNM solution is found at a given energy. At any other energy, this NNM merely manifests as a scaled copy of itself.

The shooting-based continuation method has been widely used in the literature for computing the NNMs of smooth nonlinear dynamical systems [5]. Usually, these systems possess a clear underlying linear system, the linear normal modes (LNMs) of which are used as low-energy initial guesses for the NNM computation. A direct application of this approach is not possible for the present system since no reasonable underlying linear system exists upon ignoring the contact springs. In the following section, we artificially introduce an underlying linear system to overcome this challenge.

3.1 Linear Approximations of the NNMs

We recognize that (i) when $\kappa \ll 1$, the NNMs must closely resemble a linear response, and (ii) the NNMs of the system are intimately linked to the system's cyclic symmetry. Motivated by these ideas, we construct a new system by connecting additional linear springs of (nondimensional) stiffness $\tilde{\kappa}$ adjacent to the contact springs, as illustrated in Fig. 1. Under the condition $\kappa \ll \tilde{\kappa} \ll 1$, one expects that the LNMs of this system will satisfactorily approximate the NNMs of our original system, and the shooting-based procedure can be initiated.

The LNMs of this proposed underlying linear system are well-understood in the literature [9,10]. Most LNMs occur in pairs and are known as 'double modes'; they have repeated natural frequencies and span two-dimensional eigenspaces. Consequently, any linear combination of two double modes is also a linear mode. The other LNMs are referred to as 'single modes.' Each of these families of LNMs may be characterized by the number of nodal diameters p in the mode shape. The analytical expressions for the LNMs' mode shapes are [10]:

$$\begin{aligned}
&\text{Single modes:} \quad & \mathbf{u}_0 &= [1, 1, \ldots, 1] & & \\
& & \mathbf{u}_{N/2} &= [1, -1, 1, -1, \ldots, 1], & &\text{if } N \text{ is even} \\
&\text{Double modes:} \quad & \mathbf{v}_p &= [v_{p,1}, v_{p,2}, \ldots, v_{p,N}], & v_{p,i} &= \cos(i\phi_p) \\
& & \mathbf{w}_p &= [w_{p,1}, w_{p,2}, \ldots, w_{p,N}], & w_{p,i} &= \sin(i\phi_p)
\end{aligned} \tag{3}$$

where $\phi_p = 2\pi p/N$, p is an integer with $p \in \{1, \ldots, (N-1)/2\}$ for N odd or $p \in \{1, \ldots, (N-2)/2\}$ for N even.

Each of the LNMs of Eq. (3) gives rise to an NNM of our original system. For example, the LNM with zero nodal diameters (i.e., when all masses oscillate identically) is trivially an NNM of the nonlinear system by virtue of the system's symmetry. Since no contact springs are activated in such a motion, the time period is independent of the contact spring and equals 2π.

However, NNMs generated from the underlying LNMs are not the only NNMs present. In CS systems, Vakakis showed that additional NNMs result from the interactions between a pair of double modes, representing a type of internal resonance [10]. These NNMs are called 'traveling wave NNMs,' owing to their peculiar form: the masses undergo identical oscillatory motions spaced by a delay. In the configuration space, traveling wave NNMs manifest as deformed ellipses. By definition of the traveling wave NNM, we may make an educated guess of a harmonic approximation for these NNMs as follows:

$$\mathbf{q}_p(t) = [\cos(t-\phi_p), \cos(t-2\phi_p), \ldots, \cos(t-i\phi_p), \ldots, \cos(t)] \ , \tag{4}$$

where $\phi_p = 2\pi p/N$, p is an integer with $p \in \{1, \ldots, (N-1)/2\}$ for N odd or $p \in \{1, \ldots (N-2)/2\}$ for N even. We present a more formal derivation in the appendix and show that $\mathbf{q}_p(t)$ is a linear combination of the double modes.

We use the linear vibrations defined by Eqs. (3) and (4) as initial guesses to initiate the shooting procedure for some $\kappa_0 \ll 1$. After this, we find the initial guesses for other values of κ by a sequential continuation procedure as follows: the NNM solution at κ_0 is specified as the initial guess for the solution at $\kappa_1 = \kappa_0 + \Delta\kappa$, where $\Delta\kappa$ is a small increment. This procedure is repeated to compute the NNM responses for any value of κ.

4 Results

We applied the methodology detailed in the previous section to compute the NNMs of two systems: (i) $N = 3$, $\kappa = 1$ and (ii) $N = 7$, $\kappa = 2$. The case $N = 3$ highlights some of the basic properties of the NNMs. $N = 7$ allows us to generalize our findings to systems having larger degrees of freedom. To concisely emphasize the main properties of the NNMs, we use the frequency-energy plot (FEP) representation of the NNMs and present their time series.

4.1 3DOF System

The LNMs and traveling wave vibration of the proposed underlying linear system with $N = 3$ are given by Eqs. (3) and (4) as follows:

$$\begin{aligned} &\text{Single modes:} && \mathbf{u}_0 = [1, 1, 1] \\ &\text{Double modes:} && \mathbf{v}_1 = [-1/2, -1/2, 1],\ \mathbf{w}_1 = [\sqrt{3}/2, -\sqrt{3}/2, 0] \ , \\ &\text{Traveling wave:} && \mathbf{q}_1(t) = [\cos(t - 2\pi/3), \cos(t + 2\pi/3), \cos(t)] \ . \end{aligned} \tag{5}$$

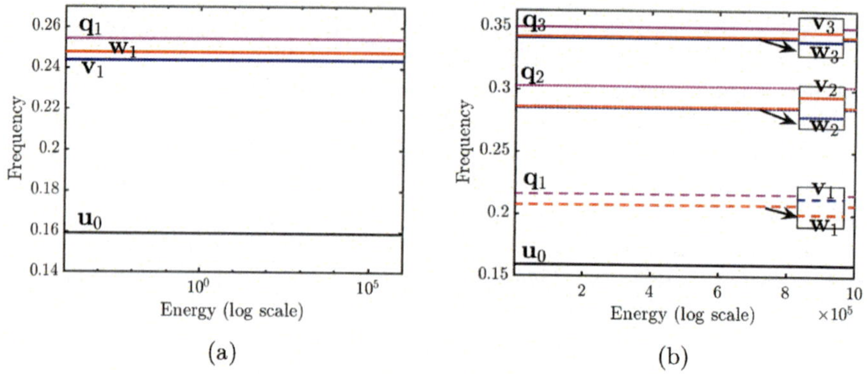

Fig. 2. Frequency-energy plot (FEP) for (a) $N = 3$, $\kappa = 1$ and (b) $N = 7$, $\kappa = 2$. The NNMs are marked according to their parent linear vibrations. The insets in panel (b) zoom into the double modes.

These linear vibrations are used to find the four corresponding NNMs of the system. The resulting FEP is presented in Fig. 2a. (Note, the 'frequency' referred to here is the inverse of the time period.) Remarkably, all the 'curves' on the plot are horizontal lines, reminiscent of LNMs. This is because of the positive-scaling property highlighted earlier: an NNM solution only scales with changes in system energy. The lowermost line corresponds to the 2π-periodic NNM, where all masses oscillate identically and do not trigger the contact springs. Other NNMs have shorter time periods since contact springs are activated in their motion.

The time series corresponding to each NNM is shown in Fig. 3. Notice that $\mathbf{u}_0$ trivially extends to the nonlinear system, as seen in Fig. 3a. The LNM $\mathbf{v}_1$ manifests as the NNM in Fig. 3b. Analogous to its linear approximation $\mathbf{w}_1$, only two masses oscillate in the NNM shown in Fig. 3c. Finally, the traveling wave NNM (Fig. 3d) is a distorted form of the three equally spaced cosine waves in $\mathbf{q}_1(t)$. These plots show that the NNMs are similar to their parent linear vibrations $\mathbf{u}_0$, $\mathbf{v}_1$, $\mathbf{w}_1$, and $\mathbf{q}_1(t)$. In this manner, the information on the mode shapes gained from the proposed linear system is valuable in qualitatively understanding the system's NNMs.

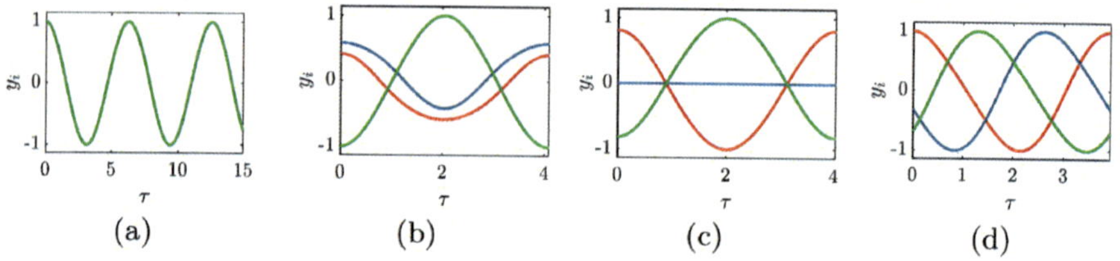

Fig. 3. Time series plots of the NNMs ($N = 3$, $\kappa = 1$) corresponding to (a) $\mathbf{u}_0$ (b) $\mathbf{v}_1$ (c) $\mathbf{w}_1$ (d) $\mathbf{q}_1$. The maximum displacements have been normalized to 1.

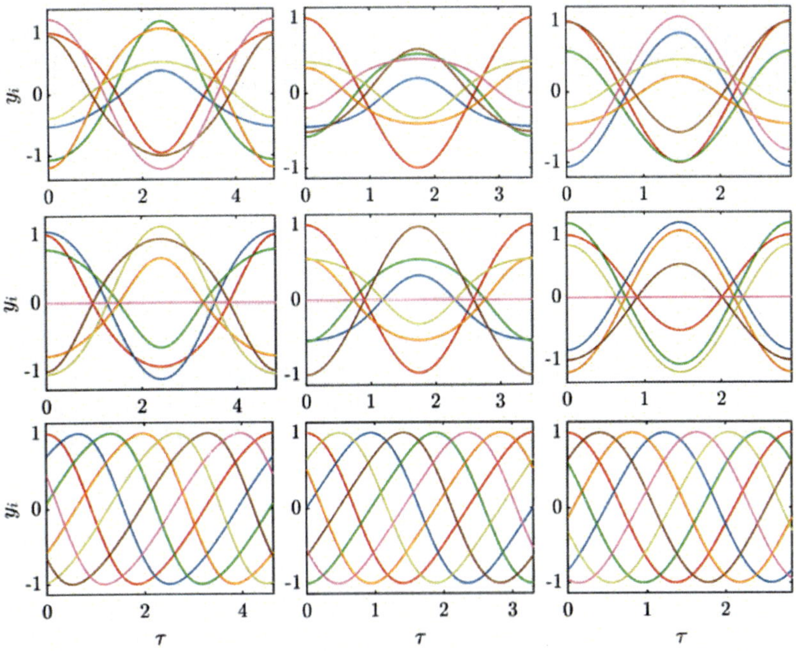

Fig. 4. Time series plots of the NNMs ($N = 7$, $\kappa = 2$). The time series in column i has i nodal diameters. Row 1 corresponds to NNMs arising from $\mathbf{v}_p$, row 2 provides the same for $\mathbf{w}_p$, and row 3 for $\mathbf{q}_p(t)$ (c.f. Eqs. (3) and (4)).

4.2 7DOF System

The LNMs and traveling wave linear vibrations of the proposed underlying linear system with $N = 7$ are computed using Eqs. (3) and (4) to yield:

$$\begin{aligned}
&\text{Single modes:} && \mathbf{u}_0 = [1, 1, 1, 1, 1, 1, 1]\\
&\text{Double modes:} && \mathbf{v}_1 = [0.6235, -0.2225, -0.9010, -0.9010, -0.2225, 0.6235, 1],\\
& && \mathbf{w}_1 = [0.7818, 0.9749, 0.4339, -0.4339, -0.9749, -0.7818, 0],\\
& && \mathbf{v}_2 = [-0.2225, -0.9010, 0.6235, 0.6235, -0.9010, -0.2225, 1],\\
& && \mathbf{w}_2 = [0.9749, -0.4339, -0.7818, 0.7818, 0.4339, -0.9749, 0],\\
& && \mathbf{v}_3 = [-0.9010, 0.6235, -0.2225, -0.2225, 0.6235, -0.9010, 1],\\
& && \mathbf{w}_3 = [0.4339, -0.7818, 0.9749, -0.9749, 0.7818, -0.4339, 0],\\
&\text{Traveling waves:} && \mathbf{q}_1(t) = [\cos(t - 2\pi/7), \dots, \cos(t - 2i\pi/7), \dots, \cos(t)],\\
& && \mathbf{q}_2(t) = [\cos(t - 4\pi/7), \dots, \cos(t - 4i\pi/7), \dots, \cos(t)],\\
& && \mathbf{q}_3(t) = [\cos(t - 6\pi/7), \dots, \cos(t - 6i\pi/7), \dots, \cos(t)]\,.
\end{aligned} \tag{6}$$

The NNMs on the FEP are shown in Fig. 2b. The time series plots corresponding to the 9 NNMs (excepting the NNM with 0 nodal diameters) are shown in Fig. 4. The mode shapes resemble their underlying linear modes given in Eq. (6).

Interestingly, the NNMs split into different bands on the FEP depending on the number of nodal diameters in the associated LNM. As we move upwards on the FEP, the number of nodal diameters increases. The NNMs arising from single modes lie alone while the other NNMs form groups of three. In other words, for a given number of nodal diameters, say p, the NNMs corresponding to $\mathbf{v}_p$, $\mathbf{w}_p$, and $\mathbf{q}_p$ are closely spaced on the FEP. This observation further highlights the information provided by the underlying linear system.

The clustering of the NNMs in the FEP is explained in the following manner. All the masses straddled between two nodal radii possess qualitatively similar motions: they rise and fall together. Consequently, the contact springs between such masses do not offer high resistance and do not appreciably contribute to increasing the natural frequency of that normal mode. The situation is exactly the opposite for two adjacent masses on different sides of a nodal radius. The masses move roughly antiphase with each other, and the contact spring between them strongly resists their motion, thereby increasing the frequency of oscillations. As the number of nodal diameters increases, the number of springs strongly engaged, and therefore the natural frequency of the mode, increases. This also explains why NNMs with the same number of nodal diameters have similar natural frequencies.

This property of the NNM frequencies splitting into bands shall hold for systems with larger degrees of freedom. Finally, we mention that the observations are similar for a system with even degrees of freedom. The only significant difference is an additional NNM (resulting from $\mathbf{u}_{N/2}$ in Eq. (3)) with $N/2$ nodal diameters, which resides above all others on the FEP.

5 Conclusions

This article examined the NNMs of a class of CS piecewise linear systems. We employed a shooting method to compute the NNMs. To initiate the numerical procedure, the LNMs of an artificially constructed underlying linear system were used. Two representative examples for systems with 3 and 7 degrees of freedom were presented. These examples provided insight into some general characteristics of the NNMs: (i) they do not exhibit frequency-energy dependence, (ii) their mode shapes resemble their parent linear vibrations, and (iii) their frequencies segregate depending on the number of nodal diameters. We believe that these general findings shall help design and analyze CS piecewise linear systems.

A Traveling Wave NNM

We present a systematic method to find a linear vibration that gives rise to the traveling wave NNM. To this end, we construct $\mathbf{q}_p(t)$ as a general linear combination of the double modes of the underlying linear system (Fig. 1) possessing p nodal diameters and enforce all masses' amplitudes to be equal. Consider

$\mathbf{q}_p(t) = \mathbf{v}_p \cos t + \alpha \mathbf{w}_p \cos(t - \theta)$. Using Eq (3), the amplitude of the ith mass A_i can be written as

$$A_i^2 = \cos^2(i\phi_p) + \alpha^2 \sin^2(i\phi_p) + 2\alpha \sin(i\phi_p) \cos(i\phi_p) \cos\theta \,. \quad (7)$$

Setting the amplitudes of all masses to be identical is equivalent to making A_i in Eq. (7) independent of i. This is done by setting $\alpha = 1$ and $\theta = \pi/2$. Substituting this result into $\mathbf{q}_p(t)$ and using Eq. (3), we arrive at

$$\mathbf{q}_p(t) = [\cos(t - \phi_p), \cos(t - 2\phi_p), \ldots, \cos(t - i\phi_p), \ldots, \cos(t)] \,. \quad (8)$$

References

1. Avramov, K.V., Mikhlin, Y.V.: Review of applications of nonlinear normal modes for vibrating mechanical systems. Appl. Mech. Rev. **65**(2) (2013)
2. Georgiades, F., Peeters, M., Kerschen, G., Golinval, J.C., Ruzzene, M.: Modal analysis of a nonlinear periodic structure with cyclic symmetry. AIAA J. **47**(4), 1014–1025 (2009)
3. Grolet, A., Thouverez, F.: Vibration analysis of a nonlinear system with cyclic symmetry. J. Eng. Gas Turbines Power **133**(2) (2011)
4. King, M.E., Vakakis, A.F.: A very complicated structure of resonances in a nonlinear system with cyclic symmetry: nonlinear forced localization. Nonlinear Dyn. **7**(1), 85–104 (1995). https://doi.org/10.1007/BF00045127
5. Peeters, M., Viguié, R., Sérandour, G., Kerschen, G., Golinval, J.C.: Nonlinear normal modes, part ii: toward a practical computation using numerical continuation techniques. Mech. Syst. Signal Process. **23**(1), 195–216 (2009)
6. Renson, L., Kerschen, G., Cochelin, B.: Numerical computation of nonlinear normal modes in mechanical engineering. J. Sound Vib. **364**, 177–206 (2016)
7. Rosenberg, R.: On nonlinear vibrations of systems with many degrees of freedom. In: Advances in Applied Mechanics, vol. 9, pp. 155–242. Elsevier (1966)
8. Seydel, R.: Practical Bifurcation and Stability Analysis, vol. 5. Springer, New York (2009). https://doi.org/10.1007/978-1-4419-1740-9
9. Thomas, D.: Standing waves in rotationally periodic structures. J. Sound Vib. **37**(2), 288–290 (1974)
10. Vakakis, A.: Dynamics of a nonlinear periodic structure with cyclic symmetry. Acta Mech. **95**(1), 197–226 (1992). https://doi.org/10.1007/BF01170813
11. Zuo, L., Curnier, A.: Non-linear real and complex modes of conewise linear systems. J. Sound Vib. **174**(3), 289–313 (1994)

General Conference Topic

Antennas Based on Metamaterials and Their Application in Modern Communication Technologies

Marie Richterova[1(✉)], Miroslav Popela[2], Jana Olivova[1], and David Fritz[1]

[1] Department of Communication Technologies, Electronic Warfare and Radiolocation, University of Defence, Brno, Czech Republic
{marie.richterova,jana.olivova2}@unob.cz, david.fritz@seznam.cz
[2] Department of Radio Electronics, Brno University of Technology Brno, Brno, Czech Republic
miroslav.popela@unob.cz

Abstract. This paper presents the analysis and simulation of the properties of microstrip antennas based on metamaterial. The Matlab and Antenna Toolbox software environments were used for the simulations. CuFlon metamaterial was chosen for the simulations and subsequent fabrication of microstrip antennas. The following antennas and antenna arrays were selected: dipole, crossed dipole, microstrip antenna array with crossed dipoles, microstrip phased array for the frequency band from 1.5 GHz to 4 GHz were selected for simulations and subsequent practical experiments with CuFlon metamaterial. Problems of electromagnetic wave polarization are solved very often in practical applications, so it was necessary to take this aspect into account to the antenna design. Based on the results obtained in the simulations, the optimization of the designed antennas was performed and subsequently the antennas were fabricated. The power radiation patterns of the antennas were measured in an anechoic chamber, and the impedance and standing wave ratio were measured by a vector analyzer. The measured results were compared with the simulation results.

Keywords: Metamaterials · Microstrip patch antenna · Antenna radiation pattern · Radiocommunication technology · Anechoic chamber

1 Introduction

Many fields of science are concerned with the manipulation of electromagnetic waves, where composite materials, otherwise known as metamaterials (MMs), have become a great asset for the development of new technologies. MMs are three-dimensional artificial composite nanostructures with unusual properties for guiding and controlling the flow of electromagnetic waves that are not achievable with natural materials [1–3]. Metamaterials exhibit unusual properties when the internal structure is suitably designed, such as negative refractive index, reverse Doppler effect or negative permittivity [1–3].

Metamaterials are a newly introduced structure for microwave and optical technologies. With these properties, existing microwave and optical devices can be improved or completely new concepts of these devices can be designed.

Z. Dimitrovová et al. (Eds.): WMVC 2022, MMS 125, pp. 1169–1176, 2023.
https://doi.org/10.1007/978-3-031-15758-5_120

This paper presents the analysis and simulation of the properties of microstrip antennas based on metamaterial for applications in modern communication technologies. The Matlab and Antenna Toolbox software environments were used for the simulations. Problems of electromagnetic wave polarization are solved very often in practical applications, so it was necessary to take this aspect into account to the antenna design.

2 Antenna Design Based on Metamaterial

An antenna is a device that is capable of transmitting and receiving radio waves at different frequencies [6]. For modern radio communication technology, not only the small size of the device, the low weight of the device, but also the higher frequency band above 1.5 GHz is important. The demands of the times have also brought significant changes to antenna designs for frequencies above 1.5 GHz. The MMs appear to be a suitable solution through their very low weight and suitable properties in terms of electromagnetic wave radiation.

There are a number of MMs, but not all are suitable for microstrip antenna design in the frequency band from 1.5 to 4 GHz. An analysis of the properties of MMs was carried out and it was found that CuFlon material with Polytetrafluoroethylene (PTFE) dielectric is used as an ideal substrate with very low losses for high frequency circuits especially for frequencies above 5 GHz [4]. The basic properties of CuFlon metamaterial are listed in Table 1.

Table 1. Basic properties of CuFlon metamaterial.

Name	CuFlon
Type	CF-A-10–7-7
Dielectric type	PTFE
Dielectric thickness	254 ± 25.4 (μm)
Relative permittivity ε_r	2.1

Before the actual design of the microstrip antenna based on CuFlon metamaterial, it was necessary to verify that the relative permittivity value corresponded to the value that was given in the data sheet (see Table 1). To verify the relative permittivity value, Eq. (1) for calculating the dielectric capacitance was chosen

$$C = \varepsilon_0 \cdot \varepsilon_r \cdot \frac{S}{d}, \tag{1}$$

where C denotes the capacitance, ε_0 the vacuum permittivity, ε_r the relative permittivity, S the surface area content and d the thickness of the metamaterial under test.

After modifying Eq. (1) to calculate the relative permittivity, the following Eq. (2) is obtained

$$\varepsilon_r = \frac{C \cdot d}{\varepsilon_0 \cdot S} \tag{2}$$

From Eq. (2) it follows that to calculate the relative permittivity, it is necessary to know the value of the capacitance of the chosen material. For this purpose, a 2x2 cm square was cut out from the MMs CuFlon, on which the capacitance was repeatedly measured with an impedance analyzer Analog Discovery 2, manufactured by Digilent. The capacitance measurements were performed ten times in total and the average value was calculated from the measured values. The resulting capacitance value is 30.034 pF. Subsequently, the relative permittivity ε_r was calculated and the value is 2.154.

On the basis of the results obtained from the measurements and taking into account the measurement error and the deviation in averaging the capacitance values, it can be concluded that the value of the relative permittivity ε_r for the MMs CuFlon, as given in the manufacturer's data sheet, agrees with the measured value of this parameter obtained by real measurement and subsequent mathematical calculation.

3 Simulation of Antenna Properties Based on CuFlon Metamaterial

The Matlab and Antenna Toolbox software environments were used for the simulations. The following antennas and antenna arrays were selected: dipole, crossed dipole, microstrip antenna array with crossed dipoles, microstrip phased array for the frequency band from 1.5 GHz to 4 GHz were selected for simulations and subsequent practical experiments with CuFlon metamaterial.

3.1 Microstrip Dipole Antenna

The frequency of the microstrip dipole antenna was designed at 2 GHz to meet the microwave range condition and to allow the antenna properties to be measured in an anechoic chamber. The chosen frequency determines the dimensions of the dipole itself, which were calculated from Eq. (3)

$$l = \frac{\lambda}{2}, \tag{3}$$

where l is the dipole arm length and λ the wavelength.

The proposed microstrip dipole antenna is shown in Fig. 1. The dipole consists of a rectangle with dimensions 75 × 2 mm. The dipole is placed on a dielectric of 77 × 15 mm and a ground plane of the same dimension. The antenna feed is placed in the centre of the dipole.

An important parameter of antennas is the radiation pattern. In this and the following simulations described, the antenna power radiation patterns have been solved. This was decided on the basis of the characteristics of the available anechoic chamber in which the antenna power radiation pattern can be measured. The power radiation pattern of a microstrip dipole antenna at resonant frequency is shown in Fig. 2. The power radiation pattern has two dominant lobes for a resonant frequency of 3.09 GHz, the antenna achieves a maximum radiated power of −50.2 dB. The radiated power by this dipole is therefore very small. The minimum radiated power is between the described lobes and reaches a level of −80 dB.

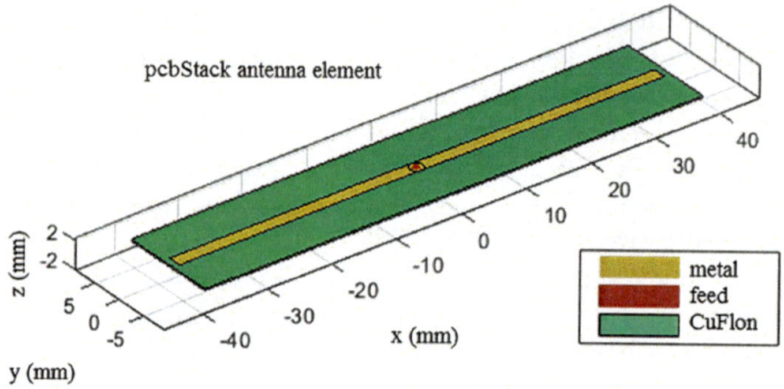

Fig. 1. The microstrip dipole antenna based on CuFlon.

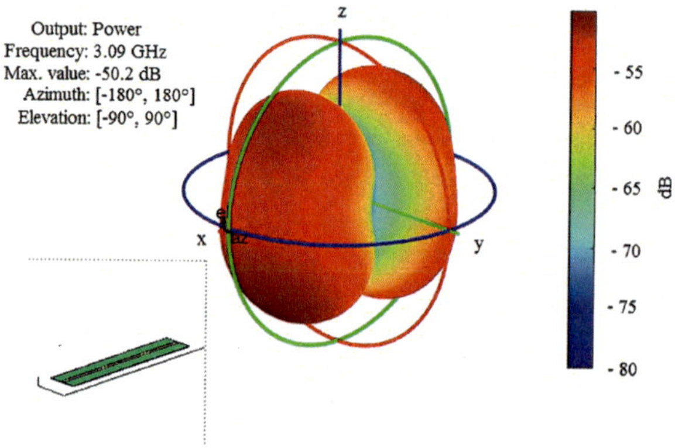

Fig. 2. The 3D power radiation pattern of microstrip dipole antenna based on CuFlon.

The proposed microstrip dipole can be used for vertical or horizontal polarization of the radiated electromagnetic wave. In practical applications, it is often required that the antenna be capable of receiving and transmitting both a vertically polarized electromagnetic wave and a horizontally polarized electromagnetic wave. This requirement has been accepted in the selection of other types of microstrip antennas, which are described in the following subsections.

3.2 Microstrip Crossed Dipole Antenna

After a successful simulation with a microstrip dipole, a more complex antenna structure was chosen, namely a microstrip cross dipole antenna. Structures with a cross-shaped elementary cell are often used for metasurfaces [5], and therefore this type of antenna was chosen for further experiments. The frequency chosen was 2 GHz, hence the dimensions of the two perpendicular dipoles of which the antenna is composed, did not change. The Eq. (3) was used to calculate the antenna dimensions. The proposed antenna is shown in Fig. 3. The arms of the cross antenna are symmetrical and the feed of the antenna was placed in the centre of the cross. The total dimension of the dielectric and ground plane is 77 × 77 mm.

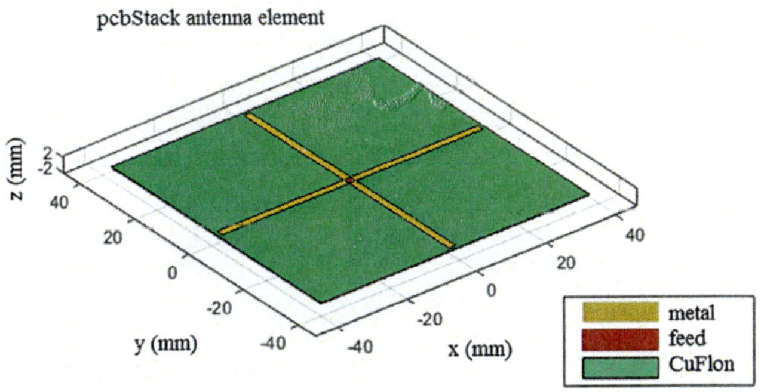

Fig. 3. The microstrip crossed dipole antenna based on CuFlon.

The Fig. 4 shows the 3D power radiation pattern of a microstrip crossed dipole antenna based on CuFlon. Four main lobes can be observed in the radiation pattern. Thus, the proposed antenna is capable of receiving both vertically and horizontally polarized electromagnetic wave. The maximum radiated power is −44.1 dB. It can therefore be concluded that the use of a microstrip crossed dipole antenna instead of a microstrip dipole will improve the maximum radiated power by 6 dB, which is four times more power in Watts.

3.3 Crossed Dipole Antenna Array

After performing simulations of the properties of the microstrip dipole and microstrip crossed dipole and based on the results obtained, it was decided to focus on a more complex antenna array. The simplicity of the elementary cell was preferred when choosing the structure of the antenna array. Considering this requirement, a cross-shaped elementary cell was chosen.

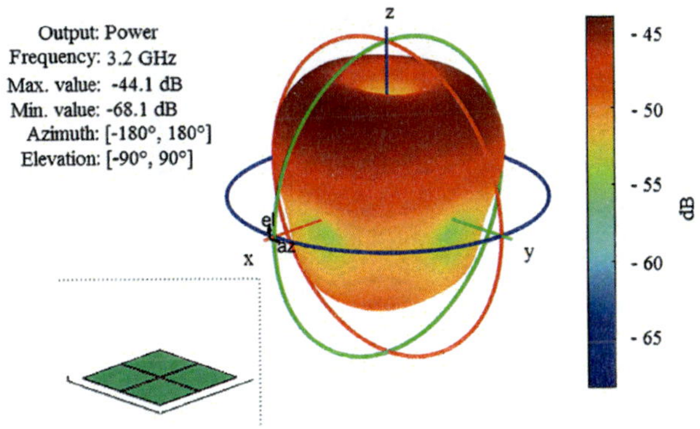

Fig. 4. The 3D power radiation pattern of microstrip crossed dipole antenna based on CuFlon.

A similar structure is addressed in [5]. The dimensions of the elementary cell were related to the size requirement of the whole antenna, which was set to 100 × 100 mm. Another criterion was the number of repetitions of the elementary cell relative to the given antenna array dimensions. Due to the computational capabilities of Matlab and the computing technology itself, it was determined that the antenna array would have 4x4 elementary elements. From these data, the dimensions of the elementary cell were chosen, namely, the length of the cross dipole arm was 23 mm and the width of the dipole was 5 mm. Given these dimensions, the overall antenna size and the number of elementary cells, the distance between adjacent antenna elements is 2 mm. The feed of each element of the antenna array has been moved off-centre. The overall view of the proposed antenna array is shown in Fig. 5.

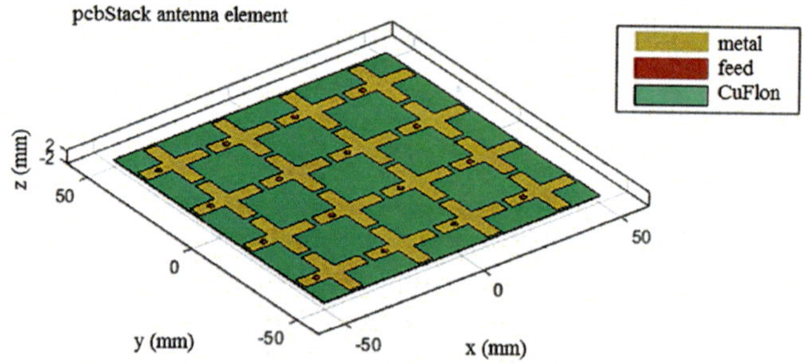

Fig. 5. The microstrip crossed dipole antenna array based on CuFlon.

The power radiation pattern of the 4x4 microstrip crossed dipole antenna array is shown in Fig. 6. The calculated power radiation pattern for the frequency 5.76 GHz proves the validity of the theoretical assumptions presented in [6].

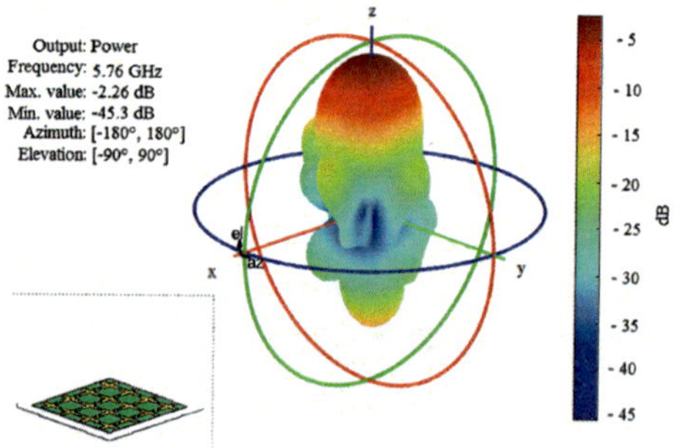

Fig. 6. The 3 D power radiation pattern of microstrip crossed dipole antenna array based on CuFlon.

4 Implementation of Antennas and Achieved Results

The proposed microstrip antennas based on CuFlon metamaterial were fabricated (see Fig. 7) and their power radiation patterns were measured in an anechoic chamber (see Fig. 8).

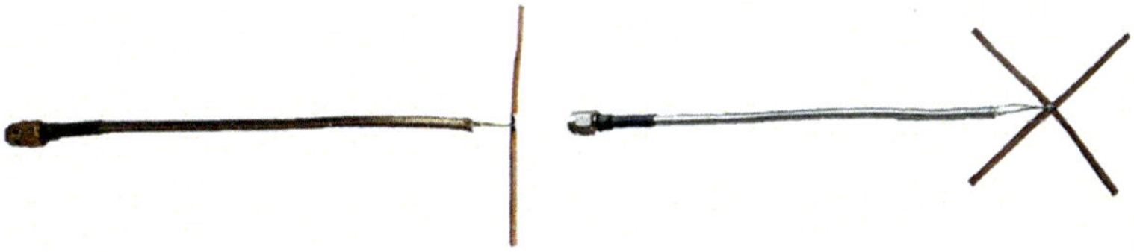

Fig. 7. The microstrip dipole antenna and microstrip crossed dipole antenna based on CuFlon.

The impedances of the designed microstrip antennas and the standing wave ratio were measured using a vector analyzer manufactured by Rohde & Schwarz. The results showed that the resonant frequency of the microstrip dipole antenna is 2.93 GHz and the resonant frequency of the microstrip crossed dipole antenna is 2.90 GHz (see Fig. 9).

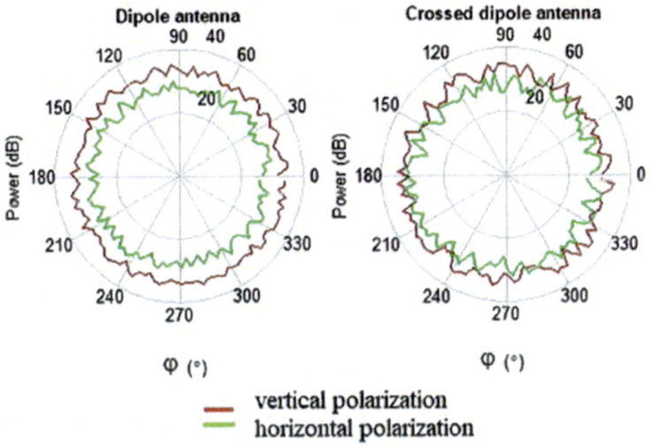

Fig. 8. The power radiation pattern measurement of microstrip antennas.

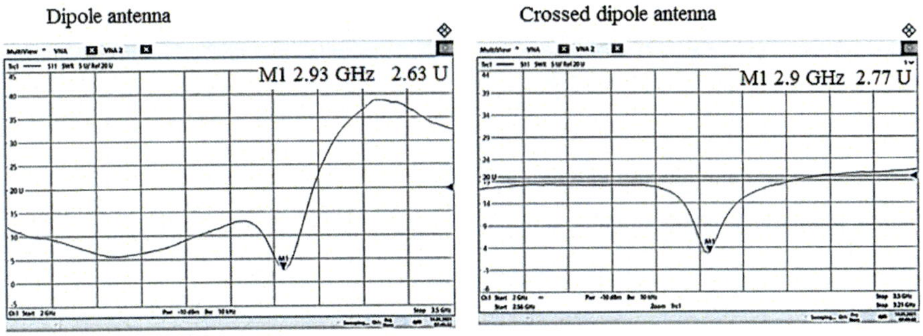

Fig. 9. The stand wave ratio of microstrip antennas.

5 Conclusion

The measured results were compared with the simulation results. There is a difference between the results obtained by simulations in Matlab and the results obtained by measurements on the vector analyzer and in the anechoic chamber. If the specific procedure for the fabrication of metamaterial-based microstrip antennas is followed, the measurement results can be expected to be close to the simulated values.

The obtained results demonstrate that crossed antennas and antenna array based on metamaterial are suitable for applications where miniature size and low weight of the device are required. The designed and implemented antennas can be used for military applications as well as for civil applications, e.g. for small satellites and for 5G.

The work presented in this article was supported by specific research of the University of Defence entitled "Modern Antenna Systems and Methods of their Management", and by the Czech Republic Ministry of Defence – the University of Defence development program "AIROPS".

References

1. Elefteriades, G.V., Engheta, N.: Metamaterials: fundamentals and applications in the microwave and OpticalRegimes. In: Proceedings of the IEEE [online], vol. 99(10), pp. 1618–1621. IEEE (2011)
2. Glybovski, S.B., Tretyakov, S.A., Belov, P.A., Kivshar, Y.S., Simovski, C.R.: Metasurfaces: from microwaves to visible. Physics Reports [online] **2016**(634), 1–72 (2016)
3. Tretyakov, S.A.: Metamaterials with wideband negative permittivity and permeability. Microwave and optical technology letters [online] **31**(3), 163–165 (2001)
4. Designing Microstrip Circuits with low-loss CuFlon Substrates. In: CRANE Areospace & Electronic [online]. Lynnwood: Crane Co. https://www.craneae.com Last accessed 20 February 2022
5. Badawe, M.E., Almoneff, T.S., Ramani, O.M.: A True Metasurface Antenna. Scientific Reports [online] **6**(1), 1–8 (2016)
6. Balanis, C.A.: Antenna theory analysis and design, 3rd edn. John Wiley, New Jersey (2005)

Influence of Multiple Potential Wells, Excitation Intensities and Electro-Mechanical Parameters on Vibratory Energy Harvesting from Nonlinear Oscillators

Pankaj Kumar[1(✉)] and S. Narayanan[2]

[1] Senior Manager, Vibration Analysis Group, Bharat Heavy Electricals Limited, Nagpur 440001, India
pankajiit1@yahoo.co.in

[2] Professor (Retired), Mechanical Engineering, Indian Institute of Technology Madras, Chennai 600036, India
narayans@iitm.ac.in

Abstract. Numerical investigation of multi-stable vibrational energy harvesting under Gaussian white noise excitation is presented by solution of the corresponding Fokker-Planck (FP) equation. The effects of the system parameters, potential well function and noise intensity on the mean square voltage are studied. Under random excitation, for lower intensity of noise the energy harvester with single potential well outperforms the harvester with multiple deep potential wells. Beyond a threshold value of noise intensity, energy harvester with multiple well potentials outperforms as the jump from one potential well to another becomes more frequent with inter-well dynamics functioning. It is observed that the energy harvested can be optimized by suitable choice of the potential function, coupling parameters and the noise intensity.

Keywords: Piezoelectric energy harvesting · Multiple well potentials · Optimal energy harvesting parameters

1 Introduction

Vibrational energy harvesting has become an important topic of research in the last decade and half. It finds applications to power low energy devices such as charging batteries of electronic devices, wireless sensors, safety monitoring devices, biomedical and microelectro mechanical system applications [1–4]. The linear vibratory energy harvester, represented as a linear spring-mass-damper system coupled with piezoelectric or electromagnetic transducers [1], operates based on the principle of resonance limiting its effectiveness only to a narrow frequency range close to the resonance frequency. The vibration energy harvester

Z. Dimitrovová et al. (Eds.): WMVC 2022, MMS 125, pp. 1177–1186, 2023.
https://doi.org/10.1007/978-3-031-15758-5_121

(VEH) can be made more effective over a broader frequency range with nonlinear oscillators and considering random excitations [2–5]. Thus the analysis of nonlinear vibration energy harvester subjected to random excitations have been investigated widely in the literature. [3,4,6,7].

Nonlinear oscillators with monostable [8], bistable [3,4] and tristable [6] potential wells have been investigated in the literature. They are found to enhance the effectiveness of the vibration energy harvesters (VEH) over a wider frequency range. Daqaq [8] theoretically investigated a nonlinear monostable VEH and observed that it does not outperform the linear counterpart under broadband random excitation. A significant number of theoretical as well as experimental investigations on bistable energy harvester under Gaussian white noise excitation [3,4] showed direct correlation between the power output and the noise intensity. Recently, a new class of energy harvesters with triple potential wells has been proposed to improve the broadband performance of randomly excited VEH [6]. The influence of potential well depth for a tristable energy harvesting was experimentally investigated in [6]. It was observed that the VEH with shallower potential enhanced the broadband performance.

In this paper, the effectiveness of single, double and triple well potential wells on energy harvesting is investigated. In the case of random excitation, the finite element(FE) method is used to solve the corresponding Fokker-Planck (FP) equation giving the joint probability density functions (JPDF) of the response and the voltage generated. The electromechanical coupling coefficient, damping ratio, intensity of noise and time constant and the shape of the potential wells have significant effects on the harvested energy. For very low noise levels, the energy harvester with single potential well outperforms the harvester with multiple deep potential wells. Energy harvesters with multiple well potentials perform better with higher noise levels as the response increases and jumps from one potential well to another with inter-well dynamics operating. Hence the depth and distributed positions of potential wells are factors which directly influence the performance of multistable VEH. The conflicting interplay of the oscillator parameters including the shape of the potential wells leads to an optimization problem for maximum energy harvesting, which is also considered in this paper.

2 Mathematical Model of Multistable Energy Harvester

VEH with multi potential wells can be constructed by coupling the nonlinear oscillator to an electrical circuit as shown in Fig. 1. The ferromagnetic cantilever beam [2] is supported between two symmetric permanent magnets with the variable inclination angles, near the free end and subjected to support random Gaussian white noise excitation. The tip of the cantilever too has a magnet which acts as proof mass and its interaction with the other two magnets produce a magnetic force. In addition to the monostable state, the multi stable configuration such as bistable and tristable states can be obtained by adjusting the design parameters of magneto-elastic interaction. Piezoelectric layers are bonded to the cantilever

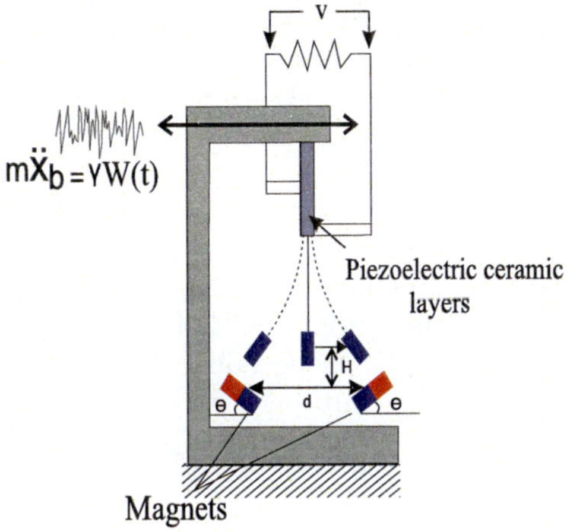

Fig. 1. Piezomagnetoelastic cantilever beam subjected to Random excitation.

beam near the root and connected to an electrical load.The voltage generated from the piezoelectric layers across the load due to the random excitation contributes to the energy harvesting.

The coupled nonlinear electro-mechanical equations of the motion of tristable VEH, governing the system first natural mode and electric voltage generated can be written in the following form [3]

$$m\ddot{X} + c\dot{X} + U^{'}(X) - \theta V = -m\ddot{X}_b, \tag{1}$$

$$C_p\dot{V} + \frac{V}{R_l} + \theta\dot{X} = 0, \tag{2}$$

where X represents the relative transverse displacement of the beam tip with respect to support motion X_b, V is the voltage output across the load resistance R_l, c is a linear viscous damping coefficient, θ is a linear coupling piezoelectric coupling coefficient in the two equations, C_p is the capacitance of the piezoelectric material, $\ddot{X}_b$ is the support acceleration, $U(X)$ is the potential energy function of the mechanical system which depends on the nonlinearity present in the harvester and ($'$) denotes differentiation with respect to X. In this work potential energy function is chosen in the following general form

$$U(X) = \frac{1}{2}k_1X^2 + \frac{1}{4}k_2X^4 + \frac{1}{6}k_3X^6, \tag{3}$$

where k_1, k_2 and k_3 are respectively the linear, cubic and quintuple nonlinear stiffness coefficients. We assume $m\ddot{X}_b = \gamma W(t)$ in Eq. 1, where $W(t)$ is a stationary, zero mean unit Gaussian white noise process with $E[W(t)W(t+\tau)] = \delta(\tau)$ and γ is the intensity of excitation.

With the appropriate change of variables, Eqs. 1 and 2 can be rewritten in the following form as

$$\ddot{X} + 2\zeta\omega_n \dot{X} + \kappa X + \alpha X^3 + \mu X^5 - \chi V = \sigma W(t), \tag{4}$$

$$\dot{V} + \lambda V + \beta \dot{X} = 0, \tag{5}$$

where $\zeta = c/2m\omega_n$, $\kappa = k_1/m$, $\alpha = k_2/m$, $\mu = k_5/m$, $\sigma = \gamma/m$, $\lambda = 1/R_l C_p$, $\beta = \theta/C_p$, $\chi = \theta^2/k_1 C_p$. Regardless of the coupling mechanism, chosen potential energy function permits Eqs. 4 and 5 to characterize the dynamics of nonlinear energy harvesting system into following four categories (i) linear $(\kappa > 0, \alpha = \mu = 0)$,(ii) nonlinear monostable $(\kappa \geq 0, \alpha \neq 0, \mu = 0)$,(iii) nonlinear bistable $(\kappa < 0, \alpha > 0, \mu = 0)$ and (iv) nonlinear tristable $(\kappa > 0, \alpha < -2\sqrt{\mu}, \mu > 0)$.

3 Fokker-Planck Equation of VEH

Introducing the variables $X_1 = X$, $X_2 = \dot{X}$ and $X_3 = V$, Eqs. (4) and (5) can be expressed in terms of a set of first order stochastic differential equation(SDE) form as

$$\begin{Bmatrix} \mathrm{d}X_1(t) \\ \mathrm{d}X_2(t) \\ \mathrm{d}X_3(t) \end{Bmatrix} = \begin{Bmatrix} X_2 \\ -\kappa X_1 - \alpha X_1^3 - \mu X_1^5 - 2\zeta\omega_n X_2 + \chi X_3 \\ -\beta X_2 - \lambda X_3 \end{Bmatrix} \mathrm{d}t + \begin{Bmatrix} 0 \\ \sigma \\ 0 \end{Bmatrix} \mathrm{d}B(t), \tag{6}$$

where, $\mathrm{d}B(t) = B(t_{n+1}) - B(t_n) \sim \sqrt{t_{n+1} - t_n} N(0,1)$ is the unit Wiener process increment and $N(0,1)$ is the unit normal random variate. For Gaussian white noise excitation, the response state vector $\mathbf{X} = [X_1, X_2, X_3]^T$ corresponding to Eq. (6) of the energy harvesting system is Markovian and consequently the time evolution of the JPDF $p \equiv p(\mathbf{X}, t|\mathbf{X}_0, t_0)$ will be governed by the following FP equation [9]

$$\begin{aligned} \frac{\partial p}{\partial t} = & -X_2 \frac{\partial p}{\partial X_1} + (2\zeta\omega_n X_2 + \kappa X_1 + \alpha X_1^3 + \mu X_1^5 - \chi X_3)\frac{\partial p}{\partial X_2} + \\ & (\beta X_2 + \lambda X_3)\frac{\partial p}{\partial X_3} + \frac{\sigma^2}{2}\frac{\partial^2 p}{\partial X_2^2} + (2\zeta\omega_n + \lambda)p, \end{aligned} \tag{7}$$

where p is the JPDF of the state variables $\mathbf{X}$. The stationary joint probability density function (PDF) of the response variables of the corresponding FP equation $(\partial p(\mathbf{X}, t|\mathbf{X}_0, t_0)/\partial t = 0)$ of VEH is numerically obtained using the FE method [4]. The response statistics, the mean square voltage and the mean square of the displacement of the VEH are obtained from the respective joint PDFs and also verified using Monte Carlo simulation (MCS) studies.

4 Numerical Results

The dynamics of the piezomagnetoelastic nonlinear energy harvester represented by Eqs. (4) and (5) is numerically investigated. The following system parameters $\omega_n^2 = 1$, $\kappa = 1$, $\lambda = 0.05$, $\beta = 0.5$ and $\chi = 0.05$ for the oscillator are kept fixed in most of the investigations. The coefficients of the nonlinear terms α and μ, electromechanical coupling parameter λ, piezoelectric coupling coefficients χ, the damping ratio ζ and noise intensity σ are varied.

Figure 2 shows the variation of mean square voltage as a function of the damping ratio ζ for four different values of the noise intensity $\sigma = 0.05$, 0.1, 0.15 and 0.2. The other parameters adopted for the oscillator are $\alpha = -2.3$ and $\mu = 1$. For a given value of the noise intensity as the damping ratio ζ increases the mean square voltage decreases. Figure 3 shows the JPDF of X_1 and X_2 for three values of ζ and for $\sigma = 0.1$. The three separate peaks in the JPDF tend to come closer as the value of ζ decreases and merges for $\zeta = 0.0125$. This signifies that there is larger probability of the system response to move between the potential wells for

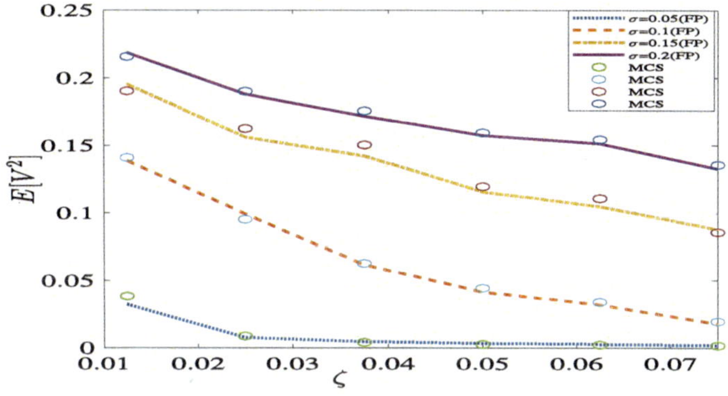

Fig. 2. Variation of $E[V^2]$ with damping coefficient ζ for different value of noise intensity; FP solution (_____); MCS (∘)

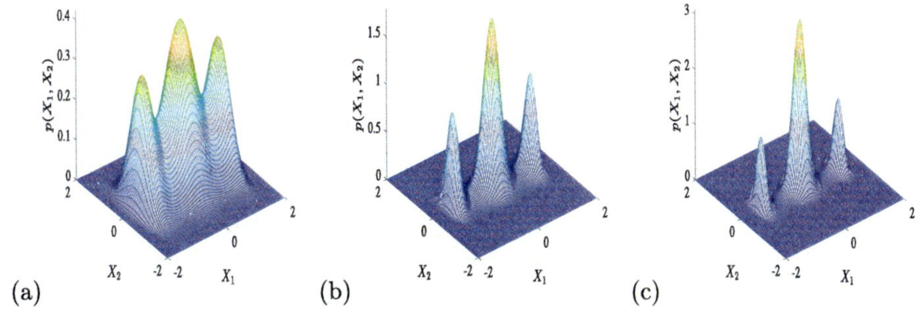

Fig. 3. Joint PDF (a) $\zeta = 0.0125$, (b) $\zeta = 0.05$, (c) $\zeta = 0.075$.

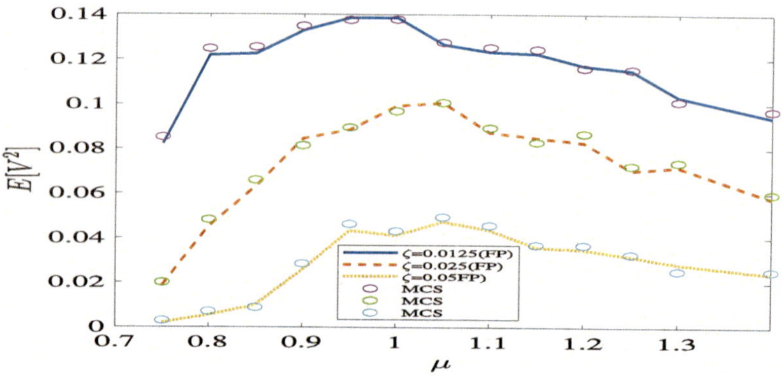

Fig. 4. Variation of $E[V^2]$ with μ for different value of c; FP solution (____); MCS (○)

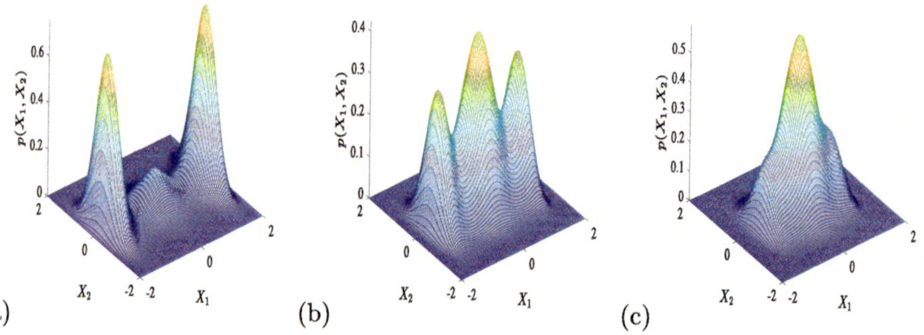

(a) (b) (c)

Fig. 5. Joint PDF $c = 0.025$ and (a) $\mu = 0.75$, (b) $\mu = 1$, (c) $\mu = 1.4$.

lower values of ζ. For higher damping there is higher probability of the response staying in the neighborhood of the stable equilibrium points of the deterministic system and the probability of the response to move between potential wells is very negligible.

Figure 4 shows the variation of mean square voltage as a function of the nonlinear stiffness parameter μ for three damping coefficients $\zeta = 0.0125, 0.025$ and 0.05. The other parameters adopted for the oscillator are $\alpha = -2.3$ and $\sigma = 0.1$. The mean square voltage initially increases with increase in μ, reaches a maximum for a particular μ and decreases with further increase in μ. Thus it may be concluded that for a given value of the damping ratio ζ, there is an optimum value of μ for which the mean square voltage is a maximum. As seen from Fig. 5 for the given damping coefficient, at lower value of μ separation distance between the potential wells and depth of the potential wells increase and larger intensity of noise is required to enable the inter-well motion. As μ increases the three peaks in the JPDF merge. This signifies that there is larger probability of the system response moving from one potential well to the other. This also signifies

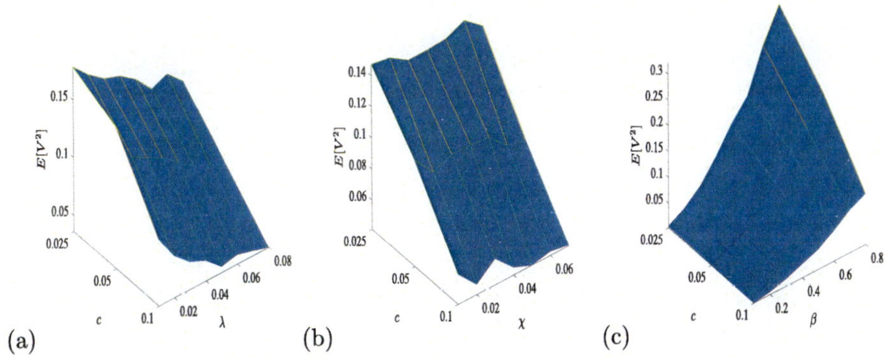

Fig. 6. Variation of $E[V^2]$ with (a) λ;(b) χ; (c) β for different value of damping coefficient c.

that there is increase in the power output. With further increase in μ the JPDF tends to become unimodal with the JPDF peaking around the zero equilibrium point with consequent decrease in the mean square voltage. This is the reason for the mean square voltage to attain a maximum value for a particular value of μ.

Figure 6 demonstrates the effect of the system parameters λ, the electromechanical coupling coefficient in the mechanical system χ and the coupling coefficient in the electrical equation β, on the mean square voltage for three different values of the damping ratio ζ for the tristable oscillator. The other parameters adopted for the oscillator are $\alpha = -2.3$, $\mu = 1$ and $\sigma = 0.1$. The mean square voltage decreases with increase in λ, decreases slightly with increase in χ, but increases considerably with increase in β. Hence the energy harvesting capability of the nonlinear oscillator can be enhanced by proper tuning of the parameters of the system in addition with an optimum choice of the noise intensity.

Figure 7 presents a comparison of the mean square voltage for different types of potential wells as a function of noise intensity. Upto some threshold value of σ the mean square voltage is very low, but it increases steeply with σ beyond this value. This behaviour is irrespective of the shape of the potential wells.It is observed that the mean square voltage and hence the harvested energy is significantly larger with bistable shallow potential well and tristable potential well than with linear, monostable potential well and bistable deep potential well. Moreover, for low excitation intensity as shown in Fig. 7 (b) the linear VEH performs better than the nonlinear oscillators with monostable potential well and bistable deep potential well. Time waveforms of displacement and voltage for nonlinear monostable, bistable with shallow potential and tristable VEH at $\sigma = 0.1$ corresponding to Fig. 7 are shown in Figs. 8 and 9 respectively. It can be seen that for the given noise intensity both bistable and tristable VEH undergo frequent jumps between potential wells and produce large amplitude of response, thus generating a high output voltage, whereas for nonlinear monostable VEH the response is restricted to a single well and the output voltage is relatively low.

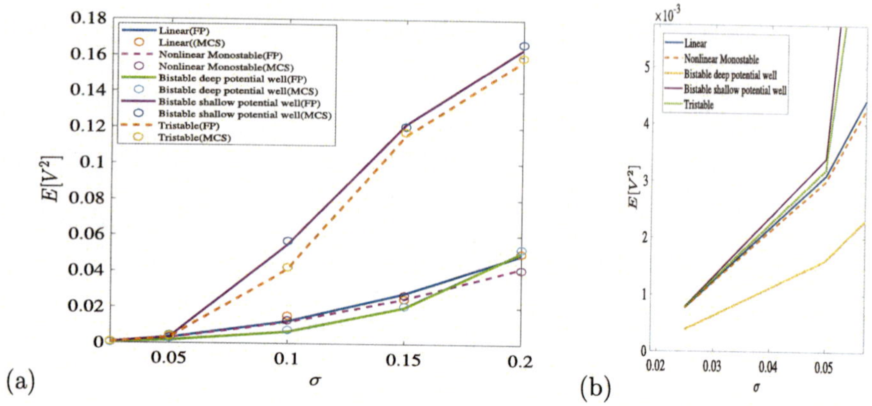

Fig. 7. A relative comparison of $E[V^2]$ (a) for different type of potential function (b) magnified view for low intensity; FP solution (——); MCS (○).

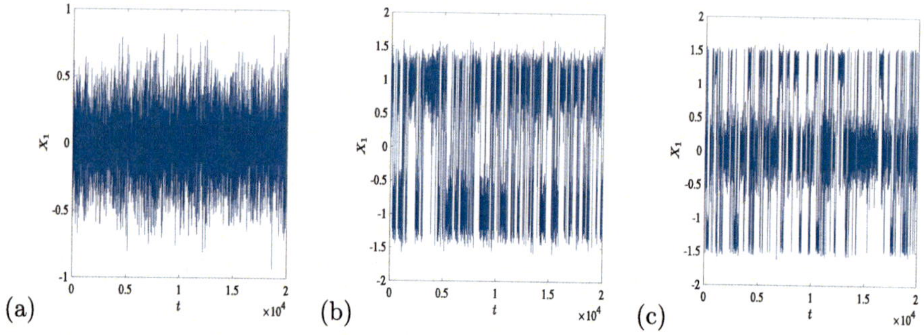

Fig. 8. Time waveform of displacement at $\sigma = 0.1$ for Fig. 7 for (a) nonlinear monostable, (b) bistable with shallow potential, (c) tristable.

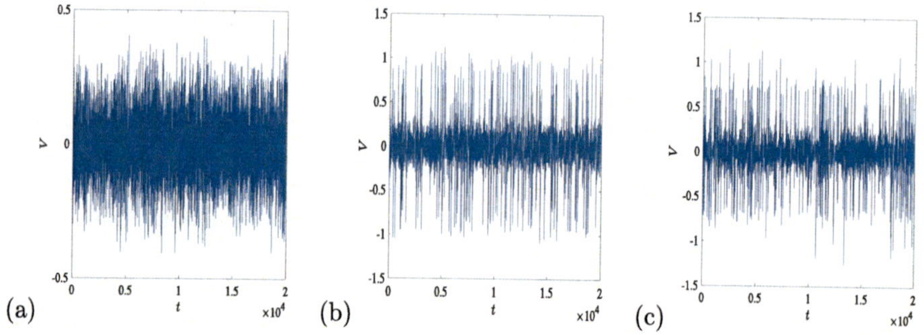

Fig. 9. Time waveform of voltage at $\sigma = 0.1$ for Fig. 7 for (a) nonlinear monostable, (b) bistable with shallow potential, (c) tristable.

In Figs. 2, 4 and 7, the mean square voltage obtained by Monte Carlo simulation (MCS) is also shown alongside by the symbol ∘ which show very good agreement with FE results of FP equation solutions.

5 Conclusions

The relative performance of piezomagnetoelastic nonlinear energy harvester with different types of potential wells under Gaussian white noise excitation is investigated by FE solution of the corresponding FP equation. The influence of various types of potential well functions, coupling parameters, damping coefficient and the intensity of white noise on the mean square voltage is investigated.The linear VEH outperforms a nonlinear VEH with monostable potential well and bistable deep potential well especially for low noise intensities. This may be attributed to the low probability of inter well movement with low intensity of noise. The nonlinear VEH with bistable shallow potential well and tristable potential well outperforms the linear VEH and the nonlinear VEH with monostable potential well and bistable deep potential well significantly. In all the cases the mean square voltage increases slowly with increase in noise intensity initially but rapidly after a threshold value of noise intensity. The higher mean square voltage of nonlinear VEH with bistable shallow potential well and tristable potential well may be due to the higher probabilities of inter well motions. The mean square voltage also varies with respect to other system parameters such as the damping coefficient c, electromagnetic coupling coefficient β and χ and the load resistance capacitance parameter λ in a complex and conflicting way which leads to an optimization problem for maximum energy harvested in terms of mean square voltage with respect to the parameters as well as the noise intensity and shape of potential wells which can be investigated further.

References

1. Sodano, H., Inman, D., Park, G.: A review of power harvesting from vibration using piezoelectric materials. Shock Vibr. Digest **36**(3), 197–205 (2004)
2. Erturk, A., Hoffmann, J., Inman, D.J.: A piezomagnetoelastic structure for broadband vibration energy harvesting. Appl. Phys. Lett. **94**(25), 254102 (2009)
3. Litak, G., Friswell, M.I., Adhikari, S.: Magnetopiezoelastic energy harvesting driven by random excitation. Appl. Phys. Lett. **96**, 214103 (2010)
4. Kumar, P., Narayanan, S., Adhikari, S., Friswell, M.I.: Fokker–Planck equation analysis of randomly excited nonlinear energy harvester. J. Sound Vib. **333**, 2040–2053 (2014)
5. Xiao, S., Jin, Y.: Response analysis of the piezoelectric energy harvester under correlated white noise. Nonlinear Dyn. **90**(3), 2069–2082 (2017). https://doi.org/10.1007/s11071-017-3784-7
6. Cao, J., Zhou, S., Wang, W., Lin, J.: Influence of potential well depth on nonlinear tristable energy harvesting. Appl. Phys. Lett. **106**, 173903 (2015)
7. Daqaq, M.F.: Transduction of a bistable inductive generator driven by white exponentially correlated Gaussian noise. J. Sound Vib. **330**, 2254–2664 (2011)

8. Daqaq, M.F.: Response of unimodal Duffing type harvesters to random forced excitations. J. Sound Vib. **329**, 3621–3631 (2010)
9. Risken, H.: The Fokker-Planck Equation: Methods of Solution and Applications. Springer, Heidelberg, New York (1996). https://doi.org/10.1007/978-3-642-61544-3

Optimal Design of Magnetorheological Damper for Prosthetic Ankle

Sachin Kumar, Sujatha Chandramohan(✉), and Sujatha Srinivasan(✉)

Machine Design Section, Department of Mechanical Engineering,
Indian Institute of Technology Madras, Chennai 600036, India
me19d414@smail.iitm.ac.in, {sujatha,sujsree}@iitm.ac.in

Abstract. MR fluid is being used, of late, in several applications such as vibration isolation, prosthetics and haptic devices. Application of a magnetic field controls the rheological properties of the MR fluid enabling pressure drop, high shear stress, and reversibility of these properties. The present work aims at designing a prosthetic ankle and utilizes a pressure drop formulation using the Bingham model. Subsequently, magnetostatic analysis is done for solving an approximate parametric magnetic model (PMM) assuming linear magnetic systems. The usage of PMM reduces the computation time. Finite element magnetostatic (FEMS) validation of PMM for fluid parameters has been done. Finally, the design parameters of the MR valve are optimized with a multi-objective genetic algorithm (MOGA). The selection of design parameters has been done considering anthropometric constraint of a below-knee amputee. The objective functions chosen for the optimization are to: (a) maximize pressure drop and (b) minimize mass of the damper valve, thus reducing the mass of the prosthetic ankle. The optimization results indicate effective trade-off solutions between pressure drop and mass of the damper. The most optimal design is arrived at by awarding higher priority to the mass of the MR damper.

Keywords: Magnetorheological fluid · Multi-objective optimization · Prosthetic ankle

1 Introduction

Human gait is a complicated process involving a number of muscles around the hip, knee and ankle joints functioning in a coordinated manner. Lower limb amputation necessitates a change in biomechanical motion and load patterns, which can lead to overload injuries [1]. According to the 2011 census of India, out of a population of 121 crore people, 2.21% people are specially abled. 20% of the impaired in this 2.21% handicapped population had a mobility disability. [2].

A prosthesis (artificial limb) with a knee or ankle joint allows a person with a transtibial or below-knee amputation to walk again by allowing them to control their ankle. MR dampers have recently been offered as a satisfactory semi-active device in prosthetics [3, 4]. Actuator design is a vital part of the design to get the desired moment during toe-off phase.

Z. Dimitrovová et al. (Eds.): WMVC 2022, MMS 125, pp. 1187–1195, 2023.
https://doi.org/10.1007/978-3-031-15758-5_122

In literature, the shape of the MR valve of semi-active dampers has been modelled and optimized; fluid flow has been modelled using Bingham plastic model [5–8]. Equation (1) represents the Bingham model.

$$\tau = \tau_0 + \mu \times (\dot{\gamma}) \tag{1}$$

Here, τ and τ_0 are shear and yield stresses respectively, μ is viscosity and $\dot{\gamma}$ is shear rate. Authors have conventionally used a constant value of relative magnetic permeability of the MR fluid [8] for calculating the magnetic flux density in the damper. A lot of research work has been reported by several researchers on structural optimization of an MR valve constrained in a volume [4, 8, 9].

The current study focuses on the investigation of an MR valve in a prosthetic device. The work utilizes a magnetostatic analysis for solving an approximate parametric magnetic model (PMM). The use of linear magnetic systems and PMM reduces the computation time. Using the Bingham fluid model between parallel plates, the flow of MR fluid is first investigated. This yields the pressure drop along the duct. Subsequently, the MR valve configuration is determined, and the magnetic circuit design is examined. The results are confirmed using commercial MR fluid characteristics and ANSYS-APDL software package's finite element (FE) modelling. Finally, an optimization approach is developed to determine the best geometric parameters of the MR valve. The objective functions chosen for the optimization are to (a) maximize pressure drop, (b) minimize mass of the MR valve and (c) reduce the mass of the prosthetic device. The optimization results indicate effective trade-off between pressure drop and mass.

2 Pressure Drop and Force Analyses of MR Damper

The MR damper consists of a steel core, coil and MR fluid as shown in Fig. 1. After applying current in the coil, magnetic field lines are generated and they enhance viscosity and so yield stress. As a result, damping force is generated due to the large pressure drop; this pressure drop is of two types: (i) viscous pressure drop in the passive region and (ii) field induced pressure (Δp_τ) drop in the active region. The pressure drop using Bingham model [6] in the active region is given by Eq. 2 where Δp_τ is the pressure drop of a single annular MR valve. The velocity profile of the movement of the MR fluid determines the coefficient c.

$$\Delta p_\tau = \frac{2cL_p\tau_y}{t_d} sgn(\dot{x}_p) \tag{2}$$

$$c = 2.07 + \frac{12Q\eta}{12Q\eta + 0.8\pi R_d t_d^2 \tau_y} \tag{3}$$

Here, L_p is the magnetic pole length, τ_y is the induced yield stress, η is post-yield viscosity of the MR fluid when no magnetic field is applied, Q is the volume flow rate in the duct, t_d is the gap length of the approximated parallel plate and R_d is the average radius of the annular duct given by $R_d = R - t_h - 0.5t_d$.

The yield stress τ_y (in kPa) in terms of magnetic flux density B_{mr} in Tesla is.

$$\tau_y = 52.962\, B_{mr}^4 - 176.51\, B_{mr}^3 + 158.79\, B_{mr}^2 + 13.708\, B_{mr} + 0.144 \tag{4}$$

Now the total pressure drop $(\Delta p)_{total}$ in the MR valve is.

$$(\Delta p)_{total} = \Delta p_n + \Delta p_\tau \tag{5}$$

where the two terms on the right hand side denote the pressure drops in the passive region and active regions respectively. The pressure drop in the passive region is.

$$\Delta p_\eta = \frac{6\eta L}{\pi t_d^3 \left(R - t_h - \frac{t_d}{2}\right)} Q \tag{6}$$

where L is the the effective length. After calculating the pressure drop, the damping force F_{d} can be estimated as.

$$F_d = (\Delta p)_{total} \times A_p \tag{7}$$

where A_{p} is the effective cross sectional area of the piston.

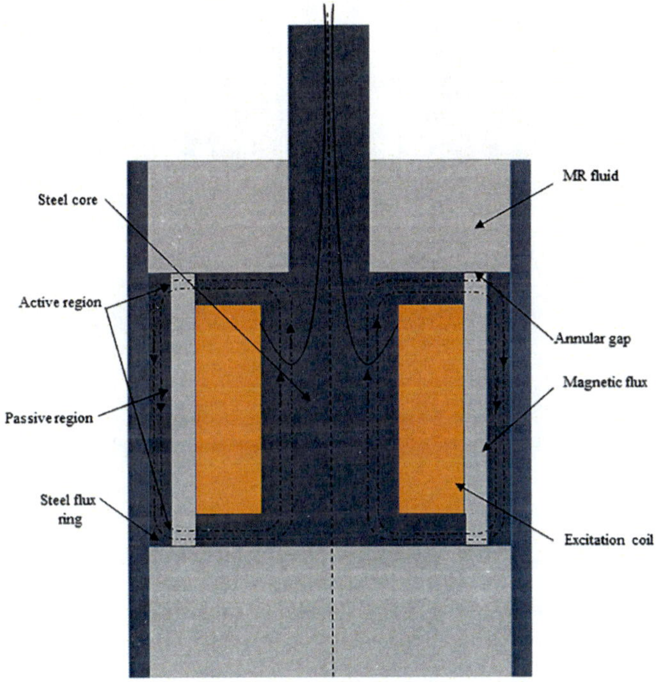

Fig. 1. Schematic configuration of annular channel of a single-coil MR damper.

3 MR Valve and Damper Configuration and Modeling

A geometric illustration of the MR valve structure with a simplified magnetic circuit is depicted in Fig. 2. Here, the MR fluid, the valve coil and the valve core iron are shown in blue, orange and white respectively. The MR fluid travels from the inlet to the outlet via annular ducts between the core and the housing. Between the magnetic poles, a magnetic field is applied to the annular channel containing MR fluid when the power to the coil is turned on. The MR fluid in the active region changes state to semi-liquid or solid and ceases to flow as a result. The fluid can only flow through the valve again when the intake pressure is large enough to compensate for the active MR fluid's yield stress.

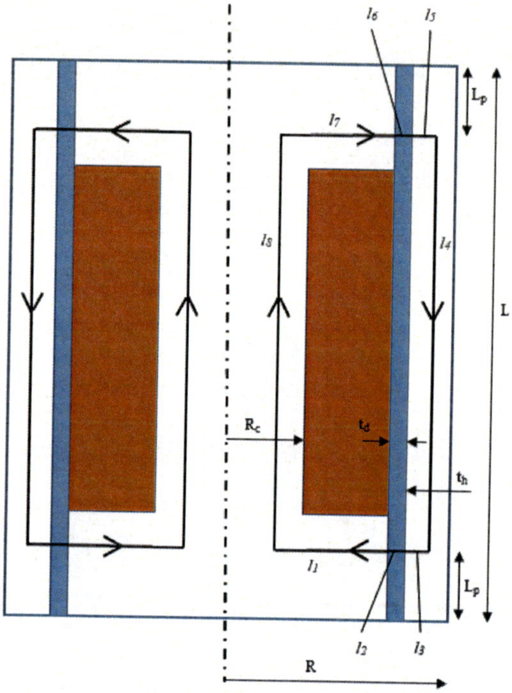

Fig. 2. Approximate magnetic circuit developed in MR valve structure.

3.1 Lumped Parameter Formulation

The magnetic flux density in the active zone can be determined [10] using Kirchhoff's law of magnetics by solving the magnetic circuit formed around the coil (Fig. 2) as.

$$\sum H_k l_k = N_c I \tag{8}$$

Here H_k and l_k are the magnetic field intensity in the kth link of the circuit and effective length of the link respectively. The MR valve's magnetic circuit is divided into eight links in this work, as illustrated in Fig. 2. The applied current is I, and N_c is the total number of turns in the coil. The magnetic flux described below is the result of magnetic flux conservation in magnetic circuits.

$$\phi = B_K A_k \tag{9}$$

Here the magnetic flux density of the kth link is B_k and its cross-sectional area is A_k. The magnetic flux density grows linearly with magnetic field intensity for low magnetic fields as follows:

$$B_k = \mu_0 \mu_k H_k \tag{10}$$

where μ_k is the relative magnetic permeability of the material of the kth link in the electromagnetic circuit and μ_0 is the magnetic permeability of empty space ($\mu_0 = 4\pi * 10^{-7}$ TmA^{-1}). The length and cross-sectional area of the magnetic link can be expressed using the geometric characteristics shown in Fig. 2 [7]. The magnetic flux density across the active areas of the flow channel can then be approximated using Eqs. (10), (11) and (12) for low magnetic fields as shown below.

$$B_{mr} = B_2 = \frac{\mu_0 \mu_{mr} N_c I}{2t_d + \frac{2l_1 \mu_{mr} A_2}{\mu A_1} + \frac{2l_3 \mu_{mr} A_2}{\mu A_3} + \frac{2l_4 \mu_{mr} A_2}{\mu A_1} + \frac{2l_8 \mu_{mr} A_2}{\mu A_8}} \tag{11}$$

μ_{mr} is the relative permeability of the MR fluid and μ that of the piston valve material. The magnetic flux density in the valve core material, i.e., the 8th link is.

$$B_8 = \frac{B_2 A_2}{A_8} \tag{12}$$

The FE model of the MR valve that has been modelled as an axisymmetric structure is shown in Fig. 3. The FE model was studied using the ANSYS APDL software. Figure 4 shows the magnetic flux density in the active region obtained using Eq. (11); a comparison with the result of FE modelling is also shown.

Fig. 3. Axisymmetric finite element.

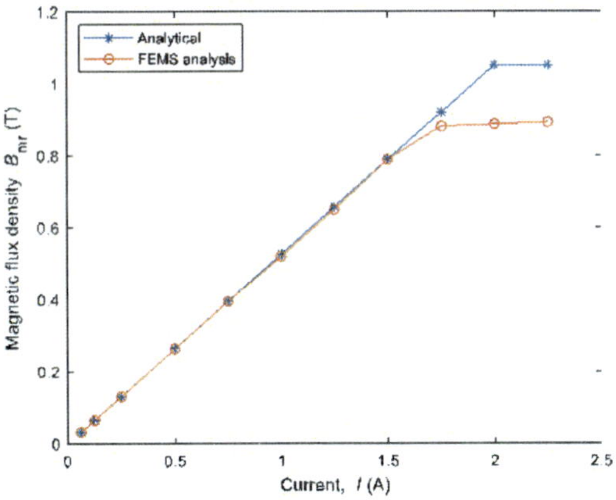

Fig. 4. Magnetic flux density vs. current in model the active area.

Here, flux densities obtained from Eq. (11) match well with FE results for small currents. However, for larger currents, FE results reveal saturation of magnetic flux, although analytical results do not. Analytical values continue to increase until saturation is reached for a larger current. The geometric details of the valve used for the design are shown in Table 1. The magnetic characteristics of various valve components are shown in Table 2.

Table 1. Initial dimensions of MR valve.

Parameters	Value (mm)
Outer radius of damper housing (R)	22.2
Housing thickness (t_h)	2.5
Annular duct gap (t_d)	1.1
Piston length (L)	26.1
Magnetic pole length (L_p)	9.6
Piston core radius (R_c)	11

Table 2. Piston material and MR fluid magnetic characteristics

Component	Material	μ	*Bsat*
Valve core	Silicon steel	B-H Curve [6]	1.5 T
Coil	Copper	1	--
MR fluid	--	B-H curve [10]	1.7 T

4 Geometric Optimization of the MR Valve Structure

The damping force of the damper for powering prosthesis can be increased by enhancing the pressure drop. Furthermore, the weight of the prosthesis is a significant consideration. The lighter a prosthesis is, the more comfortable it is. Based on these considerations, MOGA is used to optimize the geometrical dimensions of the MR actuator in order to achieve a damper that is both energy efficient and light.

4.1 Pressure Drop

For a specified flow rate and current, the pressure drop is provided by.

$$(\Delta p)_{total} = \frac{6\eta L}{\pi t_d^3\left(R - t_h - \frac{t_d}{2}\right)}Q + \frac{2cL_p\tau_y}{t_d}sgn(\dot{x}_p) \tag{13}$$

As the current increases, the pressure falls; as the current increases, so does the density of magnetic flux in valve material, increasing up to the valve material's saturation limit. There is a limit to how much the current may be increased due to the current rating of the wire utilized. As a result, when the valve material approaches its magnetic saturation limit, the maximum pressure loss occurs.

4.2 Mass of Damper

Although mass falls under the category of performance criteria, it can have an impact on material costs. The first goal in the optimization problem in this study is mass minimization.

$$m_d = m_h + m_{mr} + m_w \tag{14}$$

where m_h is the mass of the damper housing, m_{mr} represents the mass of the MR fluid, and m_w the mass of the coil of wire.

5 Design Optimization

For the optimization problem, the performance indices discussed in the previous section have been used as objective functions. The optimization problem is stated as follows:

$$Tominimize : \begin{cases} Objective1 : -\Delta P \\ Objective2 : m_d \end{cases} \tag{15}$$

subjected to B_{mr} <1.7 T with lower and upper bounds (LB and UB) on design variables (DV) listed in Table 3. The optimal solutions are shown in Table 4.

Table 3. DV parameters

Parameters	LB (mm)	UB (mm)
Radius of damper housing (R)	13	25
Housing thickness (t_h)	2.5	5
Annular duct gap (td)	0.4	1.2
Piston length (L)	22	35
Magnetic pole length (L_p)	4	10
Piston core radius (R_c)	7	15

The design variable bounds are decided by considering anthropometric and manufacturing constraints, considering prosthetic ankle requirement. Each time the optimization problem is iterated, the design variable is changed in MATLAB software and τ_y is calculated using Eq. (4); after that the objective function is calculated. The current is increased until the magnetic flux density in the valve core approaches magnetic saturation. As stated in Table 4, the method created 10 optimal solutions that fulfil the given goal functions, and the pareto front is shown in Fig. 5.

Table 4. Optimal objective functions for different chosen design variable bounds.

Sl. No.	Radius of damper housing (mm)	Piston core radius (mm)	Housing thickness (mm)	Annular duct gap (mm)	Piston length (mm)	Magnetic pole length (mm)	Pressure drop (Pa)	Mass (kg)
1	13.6	8.7	4.359	0.434	28.7	9.56	5207173	0.139
2	21.5	11.3	4.408	0.432	31.9	9.89	4739443	0.042
3	16.3	9.9	4.345	0.433	31.3	9.89	5141685	0.058
4	13.5	8.5	4.336	0.449	28.3	9.58	5240700	0.202
5	13.6	8.6	4.333	0.434	29.8	9.86	5182322	0.096
6	15.2	10.1	4.387	0.434	30.1	9.88	3389621	0.038
7	20.2	11	4.380	0.433	31.4	9.89	4046467	0.039
8	13.5	8.4	4.336	0.451	28.3	9.28	4859164	0.045
9	13.5	8.5	4.337	0.452	28.2	8.96	4353736	0.049
10	21.5	11.3	4.408	0.432	31.9	9.89	4338097	0.038

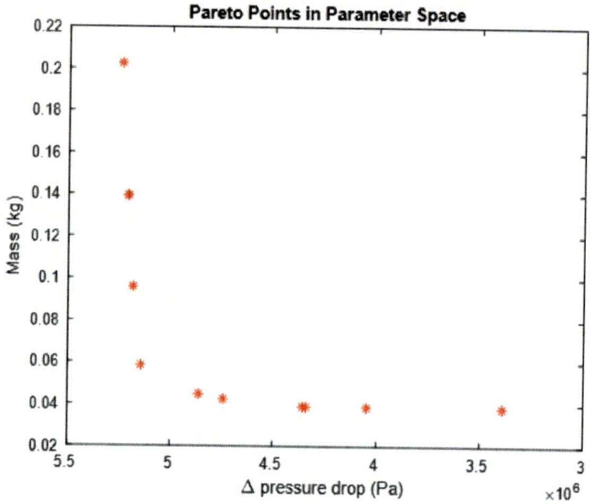

Fig. 5. Pareto optimal solutions.

6 Conclusions

The current study focused on the investigation of an MR valve in a prosthetic ankle. In this study, pressure drop formulation has been done using Bingham model. The governing parameters for magnetic flux generated in the MR fluid region were determined using an analytical formulation and validation was done using finite element magnetostatic analysis. The optimum dimensions of the valve were determined with the objective of maximization of pressure drop and minimization of mass of the valve using multi-objective genetic algorithm optimization technique. Optimal solutions for different values of design variables were obtained and the best one was chosen considering the trade-off between valve weight and pressure drop.

References

1. Espinoza, M.J., García, S.D.: Amputation level in lower limbs: impact on the patient's future performance. Sciencedirect, p. 1. https://doi.org/10.1016/s0716-8640(14)70038-0 (2014)
2. Verma, D., et al.: Disabled persons in India: a statistical profile 2016. Ministry of Statistics and Programme Implementation, Government of India (2016)
3. Gao, F., Liu, Y.-N., Liao, W.-H.: Optimal design of a magnetorheological damper used in smart prosthetic knees. Smart Mater. Struct. **26**(3) (2017). https://doi.org/10.1088/1361-665X/aa5494
4. Seid, S., Chandramohan, S., Sujatha, S.: Optimal design of an MR damper valve for prosthetic knee application. J. Mech. Sci. Technol. **32**(6), 2959–2965 (2018). https://doi.org/10.1007/s12206-018-0552-7
5. Nguyen, Q.H., Choi, S.B.: Optimal design of MR shock absorber and application to vehicle suspension. Smart Mater. Struct. **18**(3), 035012 (2009). https://doi.org/10.1088/0964-1726/18/3/035012
6. Parlak, Z., Engin, T., Çalli, I.: Optimal design of MR damper via finite element analyses of fluid dynamic and magnetic field. Mechatronics **22**(6), 890–903 (2012). https://doi.org/10.1016/j.mechatronics.2012.05.007
7. Gołdasz, J., Sapiński, B.: Insight into Magnetorheological Shock Absorbers. https://doi.org/10.1007/978-3-319-13233-4. Springer (2015)
8. Nguyen, Q.H., Choi, S.B., Kim, K.S.: Geometric optimal design of MR damper considering damping force, control energy and time constant. J. Phys: Conf. Ser. **149**, 012076 (2009). https://doi.org/10.1088/1742-6596/149/1/012076
9. Nguyen, Q.H., Choi, S.B., Lee, Y.S., Han, M.S.: An analytical method for optimal design of MR valve structures. Smart Mater. Struct. **18**(9) (2009). https://doi.org/10.1088/0964-1726/18/9/095032
10. http://www.lordmrstore.com/_literature_231215/Data_Sheet_MRF-132DG_Magneto-Rheological_fluid

Optimization of the Wiebe Function Parameters and a New Function for the Filling Coefficient for Dual-Fuel Engines

Mohamed Brayek[1,2](✉), Zied Driss[2], and Mohamed Ali Jemni[2]

[1] Architectural Engineering Department, College of Engineering, University of Prince Mugrin (UPM), Medina 42241, Saudi Arabia
m.brayek@upm.edu.sa

[2] Laboratory of Electro-Mechanic Systems (LASEM), National School of Engineers of Sfax (ENIS), University of Sfax (US), 3038 Sfax, Tunisia

Abstract. In this paper, an approach was developed and implemented in MATLAB environment to define the effects of hydrogen addition to the combustion characteristics in SI engine. The current work involves predicting the combustion duration, the pressure and temperature and the mass of air entering the cylinder. Firstly, the combustion duration of air-fuel mixtures for different operating conditions is determined by simulation. High accuracy of the chosen model results was observer when it was compared with experimental results. Then, the Wiebe adjustable parameters were adopted to correspond to experimental data. To ensure the reliability of the modelling results, experimental results from the literature are used for comparison. Finally, a new method to predict the filling coefficient of cylinders for different hydrogen fraction is proposed and validated. Results show that these parameters are adopted effectively especially for low hydrogen fractions.

Keywords: Filling coefficient · Hydrogen · Matlab · Simulation · Spark ignition engine · Performance · Wiebe function

1 Introduction

This paper is the continuation of previous experimental and theoretical studies. The effects of adding hydrogen to gasoline on the performance of a SI engine were experimentally investigated [1, 3]. A model was built in MATLAB platform to simulate the effects of heat introduced during combustion on SI engine performance [4]. The aim of the present work is to study analytically the effects of hydrogen addition on the combustion characteristics of the SI engine.

The addition of hydrogen results on the one hand in a strong decrease of the combustion duration and on the other hand in an increase of the amount of heat introduced into the cylinder. To get closer to reality in the simulation, these two parameters must be coupled.

The objective is to make a model compatible to simulate the performances of engines fuelled with mixture of hydrogen–gasoline, while keeping a good compromise between

Z. Dimitrovová et al. (Eds.): WMVC 2022, MMS 125, pp. 1196–1206, 2023.
https://doi.org/10.1007/978-3-031-15758-5_123

accuracy and complexity of the model. The solution proposed in this paper consists in defining a relation linking the variations of the quantity of hydrogen introduced in the cylinder to the Wiebe constants and to the filling coefficient of the cylinder. To give accuracy and reliability to the modelling results, experimental results from the literature are used to compare with the simulation results.

2 Simulation Environment

The MATLAB environment offers the ability to build a model using the previous mathematical formulation presented in appropriate logical manner. MATLAB contain several differential equations solvers, in the present work the pre-defined ode23 solver was used to solve the pressure and temperature differential equations.

The engine model has been constructed using MATLAB. A graphical user interface GUI is created with a block of customizable parameters. It offers easier method to introduce different engine characteristics and to read and graphically present simulation results.

Figure 1 shows the Graphical user interface for the calculation program.

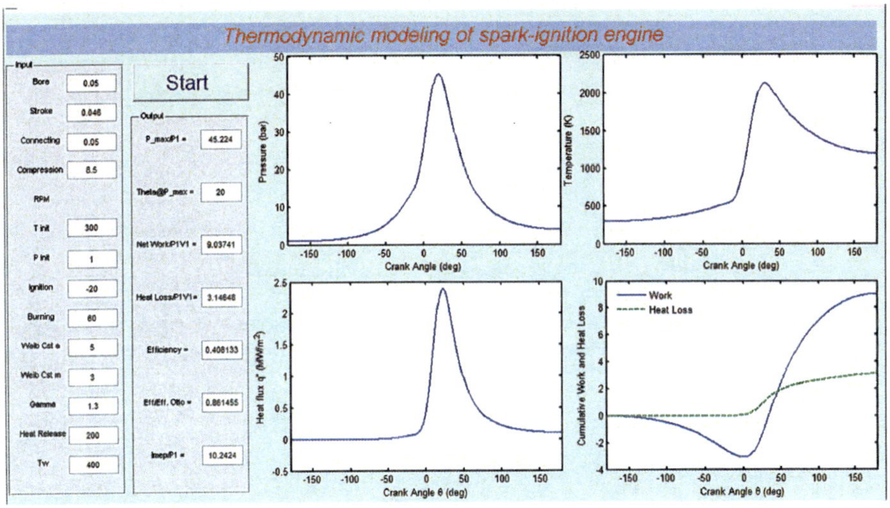

Fig. 1. GUI of the simulation program.

In the present work the MATLAB routine was essentially following Ferguson investigation. This mathematical model allows the prediction of SI engine in cylinder variable as a function of crank angle. The present investigation focusses on the effects of adding hydrogen on the burning duration, the Wiebe parameters, pressure and temperature inside the combustion chamber and the mass flow of air entering the cylinder.

3 The Combustion Duration of the Air Flow Mixture

The total mass burned and the amount of heat released during the combustion process are independent of the combustion duration ($\Delta\theta$). However, if the same heat is released at a short combustion duration, the heat release rate and pressure peak will be higher than for a long combustion duration [5]. The following equation is used to estimate the combustion duration:

$$\Delta\theta(r, N, \varphi, \theta_0) = f_1(r) f_2(N) f_3(\varphi) f_4(\theta_0) \Delta\theta_1, \tag{1}$$

with

$$\begin{aligned} f_1(r) &= 3.2989 - 3.3612\left(\tfrac{r}{r_1}\right) + 1.0800\left(\tfrac{r}{r_1}\right)^2, \\ f_2(N) &= 1.222 + 0.9717\left(\tfrac{N}{N_1}\right) + 5.0510 \times 10^{-2}\left(\tfrac{N}{N_1}\right)^2, \\ f_3(\varphi) &= 4.3111 - 5.6393\left(\tfrac{\varphi}{\varphi_1}\right) + 2.3040\left(\tfrac{\varphi}{\varphi_1}\right)^2, \\ f_4(\theta_0) &= 1.0685 - 0.2902\left(\tfrac{\theta_0}{\theta_1}\right) + 0.2545\left(\tfrac{\theta_0}{\theta_1}\right)^2. \end{aligned} \tag{2}$$

Thus, the combustion duration for all operating conditions for a particular engine can be estimated from the previous equation. The reference parameters r_1,N_1, θ_1 and $\Delta\theta_1$ can be accounted and refined by referring to observations of the experimental results.

Firstly, a comparison with experimental Changwei's [6] results is conducted. They used a spark ignition engine whose characteristics are shown in Table 1. The engine idle speed and ignition advance were kept at initial values of 790 rpm and top dead center, respectively.

Table 1. Engine specifications used by Ji and Wang [6].

Parameters	Value
Engine type	4 cylinders in line
Manufacturer	Beijing hyundai motors
Stroke (mm)	77.4
Bore (mm)	85.0
Compression ratio	10:1
Torque (N.m - rpm)	143.28–4500
Power (kW - rpm)	82.32–6000
Displacement (l)	1.599

Tests were performed at stoichiometric conditions with three volume fractions of hydrogen 0%, 3.23% and 6.52%.

The engine operates in stoichiometric condition. Table 2 shows a comparison of the results of the test and those obtained by the simulation model.

Table 2. Comparison of the experimentally measured combustion durations with those obtained analytically with a mixture equivalence ratio of 1 and a rotation frequency of 790 rpm.

Hydrogen fraction	Engine speed (rpm)	Equivalence ratio	Ignition advance	Combustion duration (experimental)	Combustion duration (analytical)
0	790	1	0	89.5	89.2
3.23	790	1	0	75.1	74.55
6.52	790	1	0	62.4	61.52

Secondly, we kept the same advance and tested the validity of the model to calculate the Combustion duration of the mixture. The engine was tested at 1400 rpm with an equivalence ratio 0.76. As can be seen in the Table 3 there is a good agreement between simulation and experimental data for the combustion duration at different operating conditions.

Table 3. Comparison of the experimentally measured combustion times with those obtained analytically with a mixture equivalence ratio of 0.76 and a rotation frequency of 1400 rpm.

Hydrogen fraction	Engine speed (rpm)	Equivalence ratio	Ignition advance	Combustion duration (experimental)	Combustion duration (analytical)
0	1400	0.76	0	105	104.6
3	1400	0.76	0	73.5	73.6

A third comparison was conducted between simulation and experimental results at an engine speed 1400 rpm and absolute pressure in the intake manifold at 61.5 kPa, which could represent one of the typical operating conditions for city driving, and an equivalence ratio of 0.83. The volumetric fraction of hydrogen in the total intake gas (H_2 – air) is 0% and 3% respectively. The ignition timing is varied from $-20°$ to $-36°$ with a step of 2°. Table 4 shows a comparison of the experimental and simulation results. With the good adjustment of the parameters a good agreement between the simulation results and the reference ones can be observed. The agreement is clearly observed with the addition of hydrogen when comparing with the initial case with pure gasoline.

4 Calibration and Validation of the Wiebe Model with Fuel Mixtures

The mass fraction of fuel burned (x_b) in an engine can be expressed as a function of the crank angle using the Wiebe function [7].

$$x_b(\theta) = 1 - exp\left[-a_w\left(\frac{\theta - \theta_1}{\Delta\theta}\right)^n\right], \tag{3}$$

The parameters a_W and n are two adjustable parameters used to fit the experimental data. At the star of combustion $\theta = \theta_1$, therefore $x_b(\theta_1) = 0$ there is no energy released. Since the cumulative heat release curve asymptotically approaches 1. The end of combustion is defined by a chosen limit of 99% complete combustion.

Table 4. Comparison of the experimentally measured combustion times with those obtained analytically for different ignition advances.

Ignition advance	Percentage of hydrogen by volume: 0		Percentage of hydrogen by volume: 3	
	Combustion duration (experimental)	Combustion duration (analytical)	Combustion duration (experimental)	Combustion duration (analytical)
−36	54.2	55.7	40.5	39.1
−34	54.2	55.8	41	39.2
−32	53.7	56.1	40	39.4
−30	54.7	56.4	40.2	39.7
−28	54.6	56.9	40.7	40
−26	55.4	57.6	40.6	40.5
−24	56.3	58.7	41.5	41.2
−22	58.3	60.1	42.4	42.2
−20	63	62.1	43.4	43.6

With the engine described in Table 1 and the combustion duration calculated by the model, the chosen model will be calibrated with different volume fractions of hydrogen: 0, 3.23 and 6.52. Figure 2. Shows a comparison of the burned mass fractions as a function of the crankshaft angle. With the increase of the percentage of hydrogen the burning duration is reduced. A good agreement and superposition between the curves of the burned mass fraction obtained experimentally by Wang et al. [8] and those obtained by simulation.

Ceviz and Yüksel [9] measured the mass fraction burned of a Ford MVH-418 (SI engine). The test engine is operated at 2000 rpm, and measurements are made over 500 consecutive engine cycles at $\lambda = 1.3$ for 2.14%, 5.28% and 7.74% hydrogen additions, respectively. As can be seen in Fig. 3, the curve obtained by calculation represents an average value of the presented cycles.

Table 5 summarizes the parameters of the Wiebe function.

Table 5. Calibration of the parameters of the Wiebe function.

Percentage by volume of hydrogen	Efficiency factor of the Wiebe function (a_W)	Form factor of the Wiebe function (n)
0	5	5
2.14	4	4
5.28	3.8	3.8
7.74	3	3

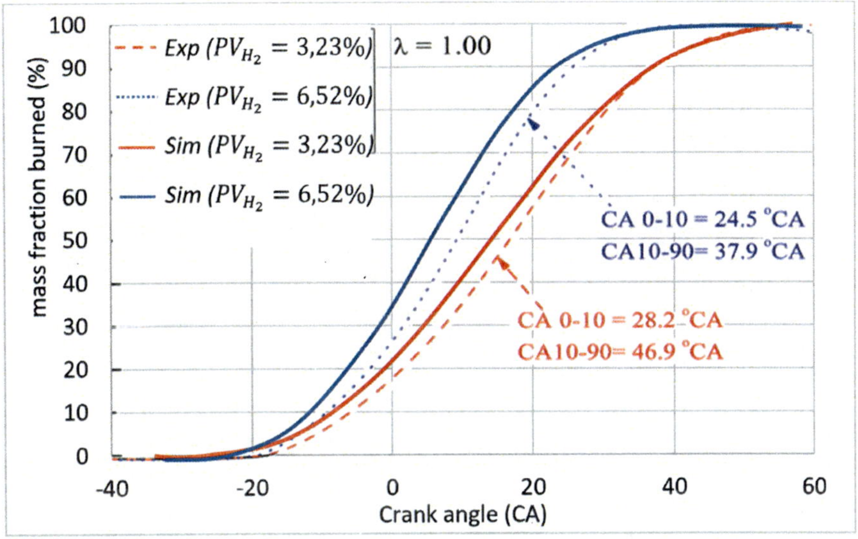

Fig. 2. Comparison of experimentally measured and analytically calculated burnt mass fraction curves for different hydrogen fractions.

Experimental measurements were used to develop the following equations for the Wiebe function:

$$n = 5 - 0.8106.PV_{H_2} + 0.1945.PV_{H_2}^2 - 0.0159.PV_{H_2}^3, \tag{4}$$

$$a_w = 3 - 0.1378.PV_{H_2} + 0.0244.PV_{H_2}^2 - 0.0017.PV_{H_2}^3, \tag{5}$$

With PV_{H2} the volumetric percentage of H_2 in the mixture.

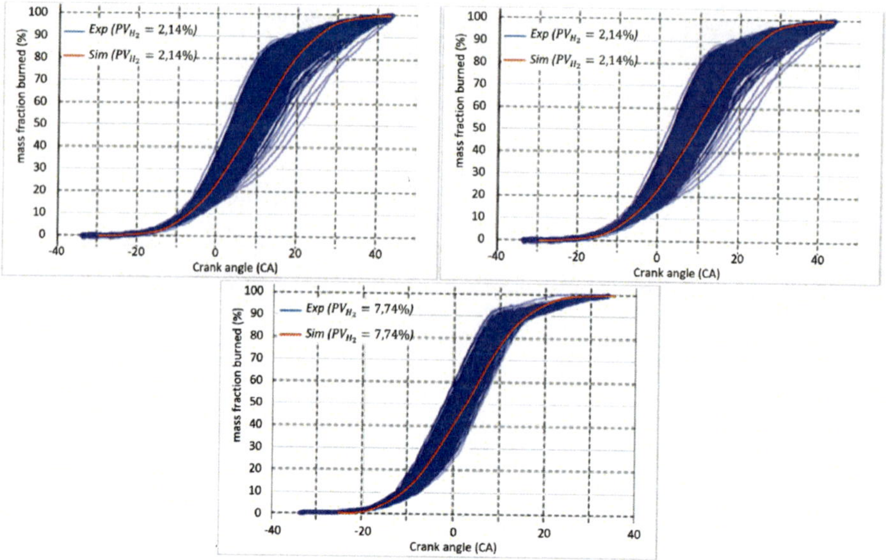

Fig. 3. Comparison of experimentally plotted 500 cycle burned mass fraction curve and analytically calculated curve for different hydrogen fractions.

It can be concluded that this model can be used for different air-fuel equivalence ratio, so it can be used with pure fuels as well as with hydrogen enriched mixtures. This gives credibility to the chosen burnt mass fraction calculation model.

5 Pressure and Temperature

The relationship between pressure and temperature versus the crankshaft angle in the combustion chamber are defined as follow:

$$\frac{dP}{d\theta} = \left(\frac{\gamma - 1}{V}\right)\left[Q_{ch}\frac{dx_b}{d\theta} - \frac{h(\theta)A(\theta)}{\omega}(T(\theta) - T_w)\right] - \gamma\frac{P}{V}\frac{dV}{d\theta} \tag{6}$$

$$\frac{dT}{d\theta} = \frac{1}{mc_v}\left(\frac{\delta Q_{ch}}{d\theta} - \frac{h(\theta)A(\theta)(T(\theta) - T_w)}{2\pi N} - P\frac{dV}{d\theta}\right) \tag{7}$$

It can be observed in Fig. 4 that a good agreement between the pressure evolution curves with the use of a low hydrogen percentage the agreement was not affected by the equivalence ratio of the mixtures. Comparison of the simulation results with those of Wang et al. [8] shows that the values of the maximum pressure obtained by simulation are slightly higher than those obtained experimentally.

This is an important prerequisite for an accurate description of the temporal torque curve at the interface between the combustion engine and the connected generator. A comparison of the mean pressure as an integral variable for characterizing the engine load yields a deviation of less than 5% between simulation and measurement. Similar good results were also found at other operating points, which confirms the applicability of the simulation approach and the selected models, at least for the stationary operating points.

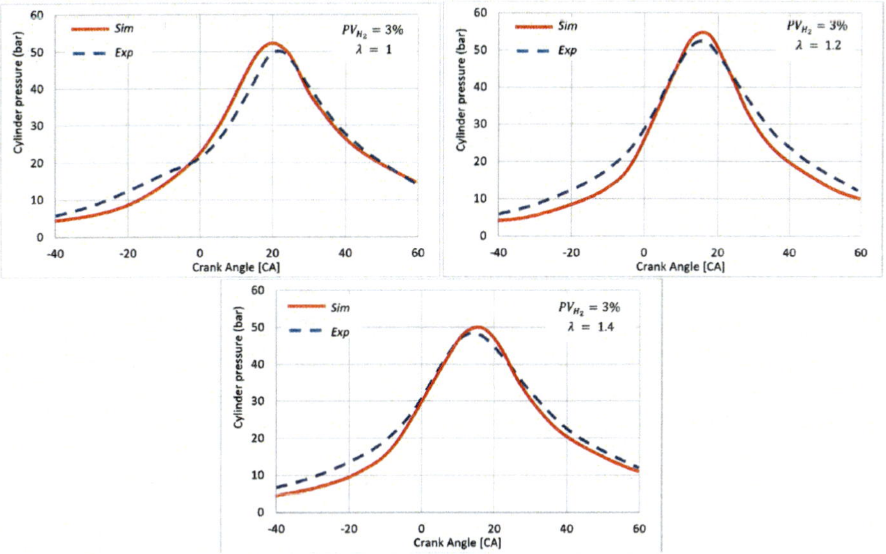

Fig. 4. Comparison of the evolution profiles of the pressure in the cylinder obtained by simulation with that obtained experimentally (gasoline-hydrogen).

A comparison of two curves of temperature evolution in the combustion chamber obtained by simulation and experimentally is presented in Fig. 5. It is observed that the results obtained confirm the validity of the chosen model. The difference between the hypotheses chosen to make the calculations and the real phenomena taking place during combustion may be the origin of the differences between the simulation results and the results obtained experimentally.

The studies conducted can help to affirm the validation of the selected models. It should also be mentioned that the proposed models require the determination of several experimental data (initial pressure and temperature in the combustion chamber and ignition advance, etc.…) which are subject to error.

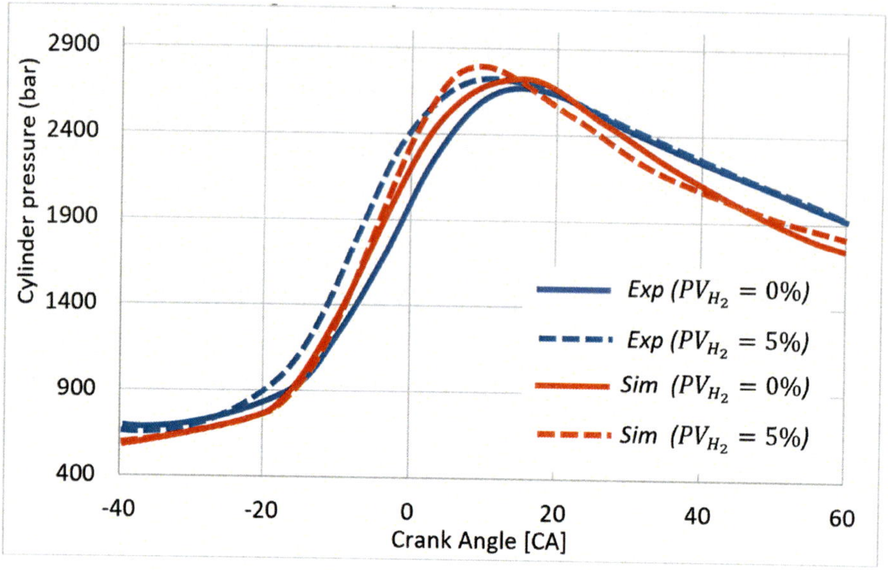

Fig. 5. Comparison of the temperature evolution profiles in the cylinder obtained by simulation with that obtained experimentally.

6 Air Mass Flow Rate

Referring to the experimentally measured results, the following laws are interpreted for the filling coefficient

$$\eta_{rem} = 0.7761 - 0.0962.PV_{H_2} + 0.0345.PV_{H_2}^2 - 0.0046.PV_{H_2}^3 \tag{8}$$

and

$$\eta_{rem} = 0,776 + 0,0229.PE_{H_2} + 0,0017.PE_{H_2}^2 - 5.10^{-5}.PE_{H_2}^3 \tag{9}$$

with PV_{H2} and PE_{H_2} PE_{H2} the volume and energy percentages of H_2 in the mixture.

The increase in the amount of hydrogen in the intake replaces more intake air. In the following the ratio between the volumetric and energetic ratio of hydrogen to air entering the engine is used.

The engine with the characteristics shown in Table 6 is maintained at 790 rpm in stoichiometric conditions. The air mass flow rate obtained analytically by the developed model will be compared with the experimental results. Figure 6 shows the variation of the air mass flow rate with the increase of hydrogen energy fraction at stoichiometric conditions. There is a good agreement between the obtained results and experimental ones at low energy (or volumetric) fraction of hydrogen. The proposed equations were not examined at high hydrogen fractions exceeding 10% in energy and 5% in volume. Several other parameters must be considered when exceeding these limits such as the automatic adjustment of the engine idle valve after the hydrogen addition, which results in a decrease of the intake air flow.

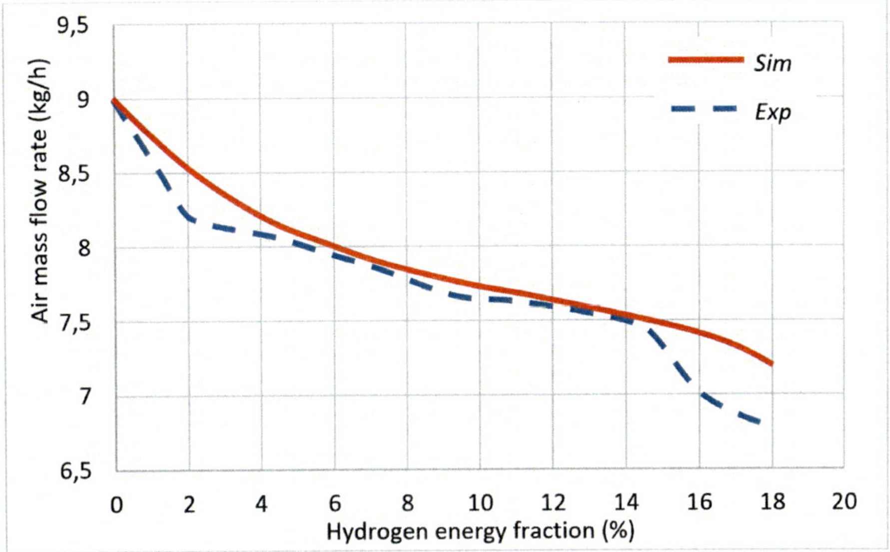

Fig. 6. Air mass flow versus hydrogen energy fraction.

7 Conclusion

The combustion behaviour of an SI mapped based on a modular simulation model. The model offers the possibility to vary different mixture configurations as well as different hydrogen fractions.

Based on thermodynamic-mechanical correlations and individual sub models developed specifically for the Wiebe function and the filling coefficient the overall model of a single cylinder 4-stroke SI engine presented built in MATLAB could be more accurate and reliable. The results calculated using the model were compared with experimental measurement from different references. The comparison of the obtained results using the model shows very good agreement with experimental measurements, which supports the applicability of the developed simulation approach. The obtained results shows that the use of a hydrogen could significantly reduce the combustion duration which has positive effect on the combustion propriety however this arrives with a coast regarding the filling coefficient.

Nomenclature

A	heat transfer surface [m^2]
a_w	the efficiency factor of the Wiebe function [--]
c_v	specific heat capacity [J/(kg.K)]
h	heat transfer coefficient [W/(m^2.K)]
m	fuel mass [kg]
n	the form factor of the Wiebe function [--]
N	engine speed [rpm]

P	in cylinder pressure [bar]
PE_{H2}	the energy percentage of H_2 in the mixture
PV_{H2}	the volume percentage of H_2 in the mixture
Q_{ch}	chemical heat release from fuel combustion [J]
r	compression ratio [--]
T_w	cylinder walls temperature [°C]
V	cylinder volume [m^3]
x_b	mass fraction burned [--]
θ	crank angle [°]
θ_0	the crank angle for start of combustion [°]
θ_d	combustion duration [°]
$\Delta\theta$	combustion duration [°]
φ	the equivalence ratio [--]
γ	ratio of the specific heats [--]

References

1. Brayek, M., Jemni, M.-A., Kantchev, G., Abid, M.-S.: Effect of hydrogen–oxygen mixture addition on exhaust emissions and performance of a spark ignition engine. Arab. J. Sci. Eng. **41**(11), 4635–4642 (2016)
2. Brayek, M., Jemni, M.-A., Kantchev, G., Abid, M.-S.: Study of spark-ignition engine fueled with hydrogen produced by the reaction between aluminum and water in presence of KOH. Arab. J. Sci. Eng. **44**(2), 695–705 (2019)
3. Brayek, M., Jemni, M.-A., Kantchev, G., Abid, M.-S.: Effect of different hydrogen-oxygen flow rates on the performance and emissions of SI engine. Int. J. Oil, Gas Coal Technol. **15**(1), 94–106 (2017)
4. Brayek, M., Jemni, M.A., Ibraim, A., Damak, A., Driss, Z., Abid, M.S.: Simulation of the effects of heat introduced during combustion on SI engine performance. In: Ben Amar, M., Bouguecha, A., Ghorbel, E., El Mahi, A., Chaari, F., Haddar, M. (eds.) A3M 2021. LNME, pp. 217–230. Springer, Cham (2022). https://doi.org/10.1007/978-3-030-84958-0_24
5. Liu, Z., Karim, G.-A.: Simulation of combustion processes in gas-fuelled diesel engines. Proceedings of the Institution of Mechanical Engineers, Part A: Journal of Power and Energy **211**(2), 159–169 (1997)
6. Ji, C., Wang, S.: Effect of hydrogen addition on combustion and emissions performance of a spark ignition gasoline engine at lean conditions. Int. J. Hydrogen Energy **34**(18), 7823–7834 (2009)
7. Ferguson, C, Kirkpatrick, A.: Internal Combustion Engines: Applied Thermosciences, 3rd edn. Wiley (2015)
8. Wang, S., Ji, C., Zhang, B., Zhou, X.: Analysis on combustion of a hydrogen-blended gasoline engine at high loads and lean conditions. Energy Procedia **61**, 323–326 (2014)
9. Ceviz, M.-A., Yüksel, F.: Engine performance, exhaust emissions, and cyclic variations in a lean-burn SI engine fueled by gasoline hydrogen blends engine. Appl. Therm. Eng. **2012**(36), 314–324 (2006)

Parametric Dependence of Spectral Properties of Elastic Vibrations in Composite Thin Films

S. Cojocaru(✉)

Horia Hulubei National Institute for Physics and Nuclear Engineering, Magurele, Romania
scojocaru@theory.nipne.ro

Abstract. Solution for the long-wavelength region of the elastic vibration spectrum is obtained in an explicit form for a bi-layered plate based on recently proposed approach, which significantly simplifies both numeric and analytic studies of layered structures. We explore the behavior in the full range of elastic moduli, mass densities, and thicknesses of the layers on account of properly defined scaled parameters. The analysis reveals some non-trivial properties of the spectrum in composite free plates. In particular, it is shown that dependence of acoustic properties on material parameters of the added layer can be, rather unexpectedly, strongly non-monotonous. Thus, for a number of natural modes with the increase of a thickness ratio of a stiffer material the eigenfrequency may, e.g., decrease before it starts increasing, or have a sharp rise followed by a significant drop and only then resumes a steady growth. A quasi-oscillatory behavior is also possible. The property of avoided crossing between spectral branches allows to establish a continuous one-to-one correspondence between surface or interface modes and volume acoustic waves at shorter wavelengths, on one side, and fundamental or gapped modes at long wavelengths, on the other side. For composites with non-symmetric stacking the existence of a spectral property reminiscent of the A and S mode alternation of the Lamb solutions is demonstrated. We also discuss the implications of the above results to the electron-phonon heat exchange in composite nanostructures.

Keywords: Rayleigh-Lamb waves · Layered structures · Vibration spectrum

1 Introduction

One of the challenges in engineering the required vibration properties in composites is the complexity of their dependence on constituent materials even for the simplest structures. On the other side, understanding this dependence extends multiple application possibilities at the macro-scale to emerging nanoscale technologies. For instance, in electronic on-chip micro-coolers or micro-bolometers

Z. Dimitrovová et al. (Eds.): WMVC 2022, MMS 125, pp. 1207–1215, 2023.
https://doi.org/10.1007/978-3-031-15758-5_124

composed of metal and insulator layers [1] it is essential to consider the quantum nature of acoustic waves [2]. Then it turns out that a 20% decrease in wave velocity can almost triple the heat power transferred from hot electrons to vibration quanta (phonons) [3]. As the number of independent parameters characterizing constituent materials is large, their choice is usually a matter of empirical consideration and numerical analysis for specific materials. In particular, when there are two homogeneous layers ($\gamma = a, b$) the parametric space is spanned by two values of the thicknesses (h_γ), two values of the mass densities (ρ_γ) and four values of the Lamé elastic constants (μ_γ and λ_γ). It may also be convenient to use the transverse and longitudinal sound velocities, $s_\gamma = \sqrt{\mu_\gamma/\rho_\gamma}$ and $\ell_\gamma = \sqrt{(\lambda_\gamma + 2\mu_\gamma)/\rho_\gamma}$, or Poisson ratios ν_γ, etc. In any case, the whole set of dispersion curves of a composite system corresponds to a single point in an eight-dimensional space which can hardly be explored by direct variation of separate material parameters. An alternative way is suggested by the analysis in [4] of the Rayleigh surface wave velocities, v_R, in a layer-over-substrate system where one switches to relative quantities and the substrate serves as a reference point. Importantly, the reference velocity of the Rayleigh mode in a substrate is a monotonic function of material parameters, as can be seen directly from the explicit analytic expressions in [5,6], e.g., it is an increasing function of ν_γ or a decreasing function of $J_\gamma = s_\gamma^2/\ell_\gamma^2$. It has then been shown that a larger thickness of the layer generally results in softening ("loading") or stiffening of the spectrum depending on whether the transverse velocities ratio "layer/substrate" is, respectively, smaller or larger than unity. This correlation between material characteristics and wave properties has been observed experimentally and can be exploited in various applications. For instance, as has been demonstrated in the microwave detection of corrosion in metallic plates [7,8], the measured resonance frequency varies monotonously with moist exposure time. For acoustic waves the rust thickness is effectively quantified by tracking the frequency shift of Lamb waves [9] based on the concept of frequency compensation of cutoff values [10]. However, as exemplified in [4] for a few material combinations near the parametric threshold, $s_a/s_b \approx 1$, the dependence of v_R on the layer thickness may become non-monotonic. In the following we will present a general overview of the behavior which captures the full range of parameters of a bilayered plate based on the approach proposed in [11]. We consider the long-wavelength part of the spectrum where explicit solutions can be obtained.

2 Solution by Surface Adapted Partial Waves

The main point of the approach is that exterior boundary conditions (BC) are accounted for by adapting the form of the basis set functions used in the superposition of partial waves. These BC are satisfied *exactly* by construction since the equations defining respective surfaces (planes in this case) are present in the arguments of the basis functions. In the context of a bilayered system we choose the horizontal plane $z = 0$ at the interface, so that the equations $z - d_\gamma = 0$ with $d_\gamma = -h_a, h_b$ correspond to the boundaries. An example of such a basis for the thickness (z) dependent displacement amplitudes in the sagittal plane (xz) for the Rayleigh-Lamb waves in a multilayer system is given below

$$u_{x,\gamma} = A_\gamma w_\gamma \sinh((z-d_\gamma)kw_\gamma) + X_\gamma v_\gamma \sinh\left((z-d_\gamma)kv_\gamma\right) + Y_\gamma \cosh((z-d_\gamma)kw_\gamma) + B_\gamma \cosh\left((z-d_\gamma)kv_\gamma\right),$$

$$iu_{z,\gamma} = Y_\gamma w_\gamma \sinh((z-d_\gamma)kw_\gamma) + B_\gamma \frac{\sinh\left((z-d_\gamma)kv_\gamma\right)}{v_\gamma} + A_\gamma w_\gamma^2 \cosh((z-d_\gamma)kw_\gamma) + X_\gamma \cosh\left((z-d_\gamma)kv_\gamma\right), \quad (1)$$

Here k is the in-plane (x, y) propagation wavenumber defining the direction of the $x-$axis and A, B, X, Y stand for the eight superposition coefficients. The phase velocity $c = \omega/k$ is contained in

$$w_\gamma = \sqrt{1-(c/\ell_\gamma)^2}, \quad v_\gamma = \sqrt{1-(c/s_\gamma)^2}. \quad (2)$$

It can be easily verified that the amplitudes (1) satisfy the wave equations in a layered medium (see, e.g., [12] Ch. 12.2). Solving for eigenfrequencies is significantly simplified because application of conditions on exterior boundaries reduces to simple linear expressions for a number of unknown coefficients (e.g., $\{A_\gamma, B_\gamma\}$ in terms of the remaining ones $\{X_\gamma, Y_\gamma\}$). Thus, for the bilayered plate with traction-free boundaries we obtain for the relevant components of the stress tensor

$$\sigma_{xz}(z)/k\mu_\gamma = X_\gamma\left(1+v_\gamma^2\right)\left[\cosh\left((z-d_\gamma)kv_\gamma\right) - \cosh\left((z-d_\gamma)kw_\gamma\right)\right] - Y_\gamma\left[\frac{\sinh\left((z-d_\gamma)kv_\gamma\right)}{2v_\gamma} - 2w_\gamma \sinh\left((z-d_\gamma)kw_\gamma\right)\right],$$

$$i\sigma_{zz}(z)/k\mu_\gamma = X_\gamma\left[2v_\gamma \sinh\left((z-d_\gamma)kv_\gamma\right) - \left(1+v_\gamma^2\right)^2 \frac{\sinh\left((z-d_\gamma)kw_\gamma\right)}{2w_\gamma}\right] - Y_\gamma\left(1+v_\gamma^2\right)\left[\cosh\left((z-d_\gamma)kv_\gamma\right) - \cosh\left((z-d_\gamma)kw_\gamma\right)\right]. \quad (3)$$

These expressions demonstrate the main point of our approach as introduced above. As a result we get a 4×4 secular matrix determined solely by conditions on the interface at $z = 0$. Also note that in this new representation the elements of the matrix are differentiable functions of all the parameters and therefore no singular terms and no spurious solutions may appear. It can also be shown that this matrix complies with the theorem proven in [13] on the non-crossing of the spectral lines or absence of degeneracy for layered structures without stacking symmetry. This allows to classify the spectral lines by introducing a band index which can be used, for instance, to establish a one-to-one correspondence between spectra at long and short waves (e.g., the fundamental flexural mode, FFM, and the lowest Rayleigh surface wave). As it will be clear from the analysis of the dependence on layer thickness ratio, the spectrum in the long-wavelength domain is continuously related to that of symmetric and antisymmetric Lamb waves in a homogeneous plate. At long waves, in addition to FFM, the spectrum consists of fundamental dilatational, FDM, and a series of branches originating from cutoff modes. We mention that when generalized to include the parametric degrees of freedom the theorem mentioned above implies that the spectral lines unfold into the corresponding non-crossing spectral surfaces.

The FFM and FDM velocities are considered relative to the known expressions for the corresponding velocities of the A_0 and S_0 modes in a free plate of material "a" with thickness d. The variables defining the parametric space are chosen according to [4] and are denoted as follows

$$\delta = h_b/h_a, \quad \rho = \rho_b/\rho_a, \quad s = s_b/s_a, \quad j = (1 - J_b)\,/\,(1 - J_a)\,. \tag{4}$$

We point out that j is a differentiable monotonic function for the physically allowed interval of J_γ , so that $1/4 < j < 4$. We compare velocities at the same wavenumber and at the fixed total thickness $d = h_b + h_a$ with the values $c_{S_0}(a) = 2s_a\sqrt{1 - J_a}$ and $c_{A_0}(a) = dks_a\sqrt{(1 - J_a)/3}$ of the respective velocities in a single layer. It is then easy to deduce that variation of the FDM velocity $(C_D = c_{FDM}/c_{S_0})$ is monotonic in all the coordinates

$$C_D = \frac{1 + \delta \times \rho \times j \times s^2}{1 + \delta \times \rho}. \tag{5}$$

The two regions where the function (5) strictly increases or decreases with δ and ρ are separated by the surface $j \times s^2 = 1$ where $c_{FDM} = c_{S_0}$ for any value of $\rho \times \delta$. This behavior is perfectly consistent with expectation as explained earlier, in particular, the above surface being determined by the condition $c_{S_0}(a) = c_{S_0}(b)$.

In contrast, expression for the flexural mode $(C_F = c_{FFM}/c_{A_0})$

$$C_F = \frac{\left(1 - \rho \times j \times s^2 \times \delta^2\right)^2 + 4\rho \times j \times s^2 \times \delta \times (1 + \delta)^2}{(1 + \delta)^2\,(1 + \rho \times j \times s^2 \times \delta)\,(1 + \delta \times \rho)}, \tag{6}$$

reveals the existence of a large region of parametric space where variation of velocity with δ (or ρ) may change its sense. To capture the full parametric space we first note that this phenomenon is signaled by the vanishing of the first derivative in δ. Its explicit form is omitted here since the result is best represented in a graphic form in the Fig. 1, where j has been fixed at 3/4 to pick an example. Using Eq. (6) it can be shown that C_F increases monotonically in the j and s coordinates for the physically allowed range of values. One can also note that changing the value of the $j-$coordinate amounts to rescaling of the s^2 coordinate, so that the surface in Fig. 1 preserves its overall shape. The completely monotonic dependence of the FFM velocity on thickness ratio can only take place if a ray parallel to the δ axis does not cross the surface. It can then be seen that there are two disjoint regions in the (j, ρ, s) space where this is possible. The Fig. 2 gives a "snapshot" view of the separation surface in this space for a given value of δ to illustrate the point that for any thickness ratio there exist values of other parameters where the derivative vanishes.

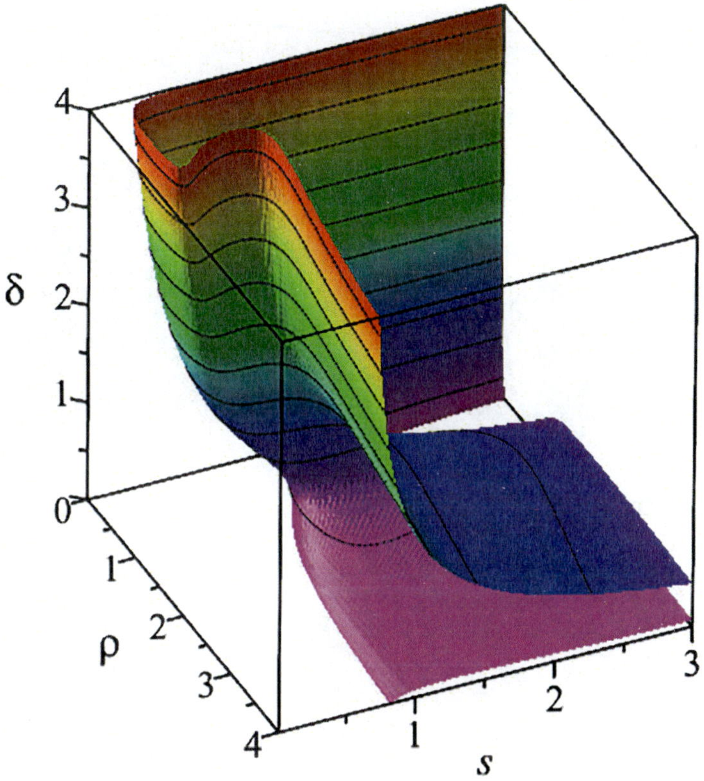

Fig. 1. The surface corresponding to the condition on the scaled phase velocity of the flexural mode, $\partial C_F/\partial\delta = 0$, which separates regions in the (δ, ρ, s) parametric space where C_F either increases or decreases with thickness ratio, $\delta = h_b/h_a$. Here $j = 3/4$, definitions are given in the text.

The large complementary region, corresponding to alternating variation with δ extends over the whole range of bulk ratios s. By construction, crossing of the separation surface means switching the variation direction of C_F. Although the shape of this surface is quite non-trivial, some simple estimates can be extracted in the limiting cases. For instance, in the limit of thin coating ($\delta \ll 1$) we find the estimate $\rho \simeq 2/\left(3js^2 - 1\right)$ which sets a lower bound in this region to $s > 1/\sqrt{3j}$ and is a good approximation for the lowest foil in the Fig. 2.

It turns out that behavior of the higher frequency spectral surfaces is even more non-trivial. This can be better understood by considering the plate resonances or the cutoff modes due to their simple polarization pattern, "thickness-stretch" and "thickness-shear", similar to case of a single layer plate (see, e.g., Ref. [14] Ch.8). The resonance with the latter polarization type coincide with the $k = 0$ limit of the shear horizontal modes, so that their governing equation

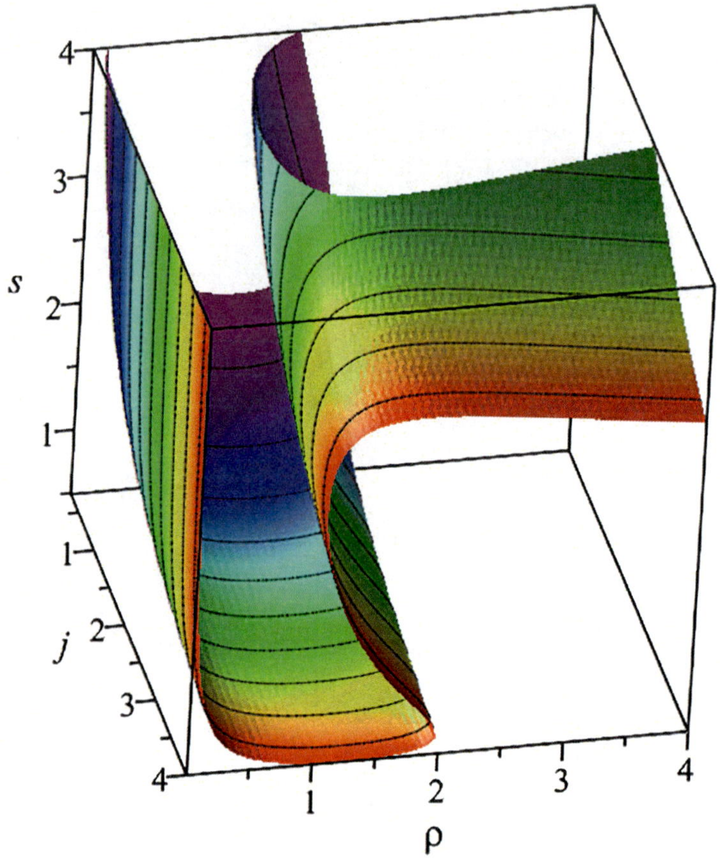

Fig. 2. The surface corresponds to the condition $\partial C_F/\partial\delta = 0$ in the (ρ, j, s) parametric space at a fixed value of the thickness ratio, $\delta = 0.2$.

takes the following form

$$\sin\left(\frac{\Omega}{1+\delta}\right)\cos\left(\frac{\Omega\delta}{1+\delta}\frac{1}{s}\right) + \rho s \sin\left(\frac{\Omega\delta}{1+\delta}\frac{1}{s}\right)\cos\left(\frac{\Omega}{1+\delta}\right) = 0. \quad (7)$$

Here we have introduced the scaled eigenfrequency $\Omega = \omega d/s_a$ which quantifies the deviation from the frequencies of the corresponding thickness resonances $\omega_n = \pi n s_a/d$ in a single layer, "a", indexed by the integer n. Then, to emphasize the key features of the solution, we show in the Fig. 3 a representative region of the (δ, s) space with a few lowest spectral surfaces for some fixed mass density $\rho > 1$. It can be expected that a higher (lower) bulk velocity would induce a rise (fall) in frequency. However, we find that for small thickness ratios (thin coating) the frequencies decrease with δ irrespective of the bulk velocity s_b. This may appear surprising because adding a layer of an even much acoustically faster material $(s \gg 1)$ should increase the resonance frequency of the composite. Such behavior is indeed observed as soon as the layer thickness becomes sufficiently large. The same transition is observed for the inverted density ratio $\rho < 1$. Namely, in the thin coating regime the eigenfrequency of the bilayered structure keeps growing, even if the appended layer is of a much slower material ($s \ll 1$), until the expected tendency sets in at larger thickness ratios. Still, the expected monotonic correlation of the $\delta-$dependence on the bulk velocities is not totally fulfilled in this region as well. Indeed, it can be shown that if parameters satisfy the condition $(1-\rho)\left(1-\rho s^2\right) > 0$, then the dependence $\Omega(\delta)$ for the cutoff modes is non-monotonic and is actually quasi-oscillatory. Otherwise, the dependence is strictly monotonic, as can be seen in the last figure. There exist a large number of material combinations which fall on both sides of this condition. In particular, for the $Cu - Si_3N_4$ layered structure used in the electronic micro-refrigeration devices already mentioned [1,3] the above inequality is not satisfied, while for $Cu - SiO_2$ it is safely satisfied. This implies that increasing the thickness of the Cu layer deposited on the supporting insulating membrane for the latter will result in a strongly non-monotonic variation of the vibration spectrum and, consequently, of the heat flux carried by phonons.

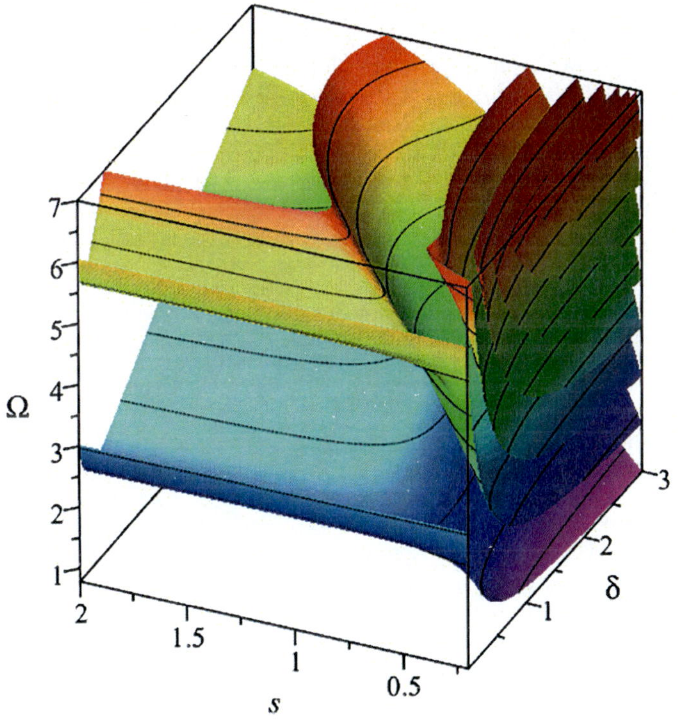

Fig. 3. Surfaces represent the eigenfrequency evolution of the few lowest shear plate resonance modes with variation of the bulk velocity ratio $s = s_b/s_a$ and layer thickness ratio $\delta = h_b/h_a$ for a density ratio $\rho = \rho_b/\rho_a = 5$.

3 Conclusions

Thus, the vibration spectrum in a layered structure demonstrates an unexpectedly complex dependence on the characteristics of constituent materials. The perfectly monotonic behavior in the entire parametric space is found for the only mode which shares linear frequency-vs-wavenumber dispersion with the bulk acoustic waves, i.e., for the dilatational fundamental mode. At both lower and higher frequencies, the rest of the spectrum follows a strongly non-monotonic variation upon changing the relative thicknesses of the layers in a broad range of values for the other parameters. For the flexural mode, for instance, one can observe this phenomenon for practically any value of the bulk shear velocity ratio, given the density and Poisson ratios are chosen appropriately. Still, there exists a large volume of the parametric space, limited by the surfaces discussed in the present work, where monotonic dependence persists. Both the outlined approach and general results can be useful for the design of devices exploiting the vibration properties of layered structures.

References

1. Saira, O.P., Matheny, M.H., Wang, L., Pekola, J., Roukes, M.: Modification of electron-phonon coupling by micromachining and suspension. J. App. Phys. **127**, 024307-1–024307-9 (2020)
2. Stroscio, M.A., Dutta, M.: Phonons in Nanostructures. Cambridge University Press, Cambridge (2003)
3. Cojocaru, S., Anghel, D.V.: Low-temperature electron-phonon heat transfer in metal films. Phys. Rev. B **93**, 115405 (2016)
4. Farnell, G.W., Adler, E.L.: Elastic wave propagation in thin layers. In: Mason, W.P., Thurston, R.N. (eds.) Physical Acoustics, vol. 9, pp.35-127. Academic Press, New York (1972)
5. Malischewsky, P.G.: Comparison of approximated solutions for the phase velocity of Rayleigh waves. Nanotechnology **16**, 995 (2005)
6. Malischewsky, P.G.: Surface waves. In: Altenbach, H., Öchsner, A. (eds.) Encyclopedia of Continuum Mechanics, pp. 1–9. Springer, Heidelberg (2018). https://doi.org/10.1007/978-3-662-53605-6_232-2
7. Zhang, H., He, Y., Gao, B., Tian, G.Y., Xu, L., Wu, R.: Evaluation of atmospheric corrosion on coated steel using K-band sweep frequency microwave imaging. IEEE Sens. J. **16**(9), 3025–3033 (2016)
8. Marindra, A.M.J., Tian, G.Y.: Chipless RFID sensor for corrosion characterization based on frequency selective surface and feature fusion. Smart Mater. Struct. **29**(12), 125010 (2020)
9. Cao, X., Zeng, L., Lin, J., Hua, J.: A correlation-based approach to corrosion detection with lamb wave mode cutoff. J. Nondestruct. Eval. **38**(87), 8716 (2019). https://doi.org/10.1007/s10921-019-0629-y
10. Zhu, W., Rose, J.L., Barshinger, J.N., Agarwala, V.S.: Ultrasonic guided wave NDT for hidden corrosion detection. Res. Nondestr. Eval. **10**, 205–225 (1998). https://doi.org/10.1007/PL00003908
11. Cojocaru, S.: Surface adapted partial waves for the description of elastic vibrations in bilayered plates. Wave Motion **92**, 102430 (2020)

12. Rose, J.L.: Ultrasonic Guided Waves in Solid Media. Cambridge University Press, Cambridge (2014)
13. Kausel, E., Malischewsky, P., Barbosa, J.: Osculations of spectral lines in a layered medium. Wave Motion **56**, 22–42 (2015)
14. Graff, K.F.: Wave Motion in Elastic Solids. Dover Publications, New York (1991)

Research of Dynamic Process in Water Cistern of Fire Automobile During Its Moving Along Rough Woodland

Serhii Pozdieiev[1](✉), Olexandr Tarasenko[2], Kamran Almazov[2], and Valeriia Nekora[1]

[1] Cherkasy Institute of Fire Safety Named After Chernobyl Heroes of National University of Civil Defence of Ukraine, Cherkasy, Ukraine
svp_chipbbk@ukr.net
[2] National University of Civil Defence of Ukraine, Kharkiv, Ukraine

Abstract. This work is dedicated to the research of the dynamic process in the fire automobile's water cistern with different levels of filling. The work's aim is to obtain data on the fluctuations of the dynamic system's mass center that combines water and its tank during a fire automobile moving through a rough woodland at different speed. Determined data allow to predict the danger of fire automobile overturning under such conditions. During the solution of this task, the geometric configuration of the fire automobile cistern with water was modeled using the universal program complex LS-DYNA for computer modeling of dynamic systems. Time relations of cistern's turn angles were set to recreate the dynamic impact on the fire automobile water cistern from the side of the rough woodland relief. The obtained relations were used as boundary conditions for recreating the dynamic impact of the rough woodland relief. The turning angles for the water cistern were determined depending on the relief bumps. The geometric form and parameters of these bumps were calculated using a pseudorandom number generator. Upon the explicit method and smoothed particles hydrodynamics method realized in LS-DYNA program code, the center mass fluctuating regularity for the water cistern of a fire automobile was determined in dependence of filling level and movement speed. This regularity can be used as a basis for predicting the danger of a fire automobile overturning as it moves through rough woodland.

Keywords: Water cistern of fire automobile · Rough woodland · Mass center fluctuation · Smoothed particles hydrodynamics method · Explicit method · Danger of overturning

1 Introduction

Statistical analysis of the fires' current state shows that the number and the forest fires' scale tend to increase, due to modern challenges of natural, environmental and man-made nature [1]. The main vehicles that deliver fire extinguishers to fires and extinguish fires are fire trucks, including fire tank trucks with water cistern [2]. When operating

Z. Dimitrovová et al. (Eds.): WMVC 2022, MMS 125, pp. 1216–1223, 2023.
https://doi.org/10.1007/978-3-031-15758-5_125

a fire tank truck that is not completely filled with water, there is an increased risk of overturning when it crosses rough forested terrain due to fluctuations in its masses center. Therefore, when operating fire tanks, it is recommended for fire departments to avoid traffic on rough forested terrain with an incompletely filled tank, which is equivalent to a ban. Such recommendations have to be violated every time, since a strict ban on the operation of fire tanks is economically and practically unjustified. The main data of the recommendation should be revised taking into account the scientific research results. Considering the foregoing, we can conclude that the fluctuation patterns' study of the fire tank trucks' masses center, caused by the dynamic influence of inequalities when they move over rough forested terrain, is an urgent scientific and technical task.

2 Analysis of Literary Sources and Problem Statement

In terms of the danger level associated with the distribution scale, socio-economic losses and environmental consequences, landscape fires are among the most dangerous wildfires[1]. To create the preconditions for successful control of these fires, intensive research is conducted in almost all areas of the industry [1]. The increase in the means' mobility of transporting fire extinguishing agents to fire cells while maintaining safe working conditions for fire departments remains an important aspect in extinguishing landscape fires. Among the main means of extinguishing landscape and forest fires are special fire tank trucks, for which an essential aspect of increasing efficiency is to prevent their overturning. The tipping hazard is especially pronounced when fire trucks are not completely full. As a result, dangerous waves can form there, which can cause large fluctuations in the masses center and, as a result, overturning of the tank. Providing scientifically based recommendations on the parameters of the fire trucks' movement across rough forest terrain would create prerequisites for improving their safe operation and would increase the efficiency of their use. The works [2–4] are devoted to the dynamic processes occurring in the fire trucks' tanks. In these papers, different approaches to predicting the dynamics of these systems are formulated. In [2], an approach that represents the water mass in a tank in the form of a pendulum dynamic system is proposed. This approach allows to determine the fluctuations of the masses center of the dynamic system, however, the disadvantage of this approach is the limitation when taking into account the influence of the internal system of breakwaters and partitions. These articles [3] are dedicated to the study of dynamic processes in tanks using a general theoretical approach based on the use of the Navier-Stokes equation system. This approach quite accurately reproduces the dynamic processes of waves' formation and spreading in tanks with baffles, however, in this case, difficulties arise in determining the masses center's fluctuations of this dynamic system. The smoothed particles hydrodynamics (SPH) method [4, 5] is found to be advantageous for modeling dynamic processes during the fire truck's movement, according to which water in a tank can be simulated using particles, this method allows you to simulate the mutual viscous interaction between particles and the tank fencing. Under such conditions, it is convenient to study the fluctuation of the masses center of the formed dynamical system. Given the complexity of mathematical models, their implementation is possible only with the use of special software. In the work [6] in order to solve scientific and technical problems of fire safety a universal software package LS-DYNA, developed by Livermore Software Technology Corporation (LSTC) for the study of dynamic systems was effectively used. The LS-DYNA computing software code implements the approach to modeling the dynamics of liquids based on the SPH method in combination with the explicit method of integrating differential equations of dynamics with their approximation by the finite element method, which is a prerequisite for its mathematical modeling of dynamic processes in a fire tank truck during its movement through rough forest terrain. Considering the above information, the aim and objectives of the research were set.

The Aim of the Study. The aim of this study was to identify patterns of fluctuations in the masses center of fire trucks caused by the dynamic impact of irregularities in their movement through rough forest terrain as a scientific basis for recommendations to fire departments in the operation of fire tank trucks to improve their efficiency and safety.

Main Research Objectives. To achieve the aim set, the following main objectives were formulated:

- to develop a geometric and water tank's mathematical model of one of the most common fire trucks using the LS-DYNA software package capabilities;
- to develop a mathematical model of the rough forest terrain for its reproduction of its influence when applying boundary conditions to the dynamic system of the fire truck's water tank when it moves through rough woodland;
- using the developed mathematical model of the tank to investigate the dynamic processes in the water tank and to identify fluctuations' patterns of the masses center of fire trucks caused by the dynamic influence of irregularities in their movement through forest cross-country terrain;
- to provide recommendations on the proportions between the level of water filling the tank of the fire truck and its speed movement through rugged forest areas on the basis of the identified patterns.

3 Mathematical Modeling of the Surface Relief Influence on Dynamic Processes in the Tank of a Fire Truck During Its Movement Through Rough Forest Terrain

3.1 Design Features of the Fire Truck's Tank

To study the impact of surface relief on the dynamic processes occurring in the tank of a fire truck during its movement through forest terrain, one of the most common options was considered, fire truck AC-4,5–60 (TGM 12.240) -364, which can be used for extinguishing landscape fires with water and air-mechanical foam. The appearance of the truck together with the located water tank is shown in Fig. 1. The internal structure of the truck's tank is given here:

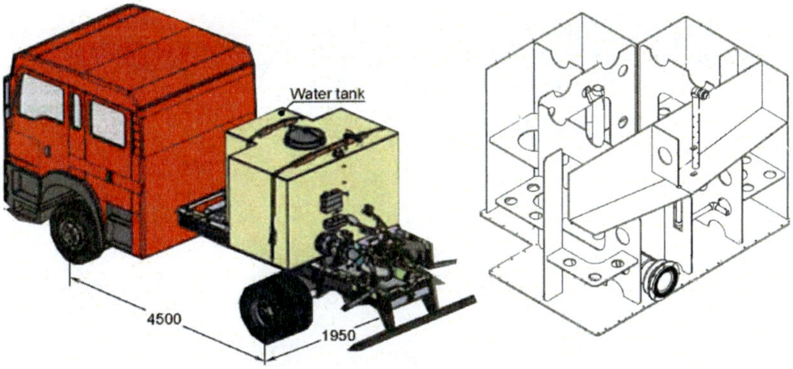

Fig. 1. Location of the water tank in the fire truck.

3.2 Mathematical Model of the Influence of the Surface Relief

The main parameters that determine the movement of the tank when overcoming the forest cross-country terrain, taking into account the influence of local relief are rotations around horizontal axes and vertical movement, as shown in the diagram shown in Fig. 2.

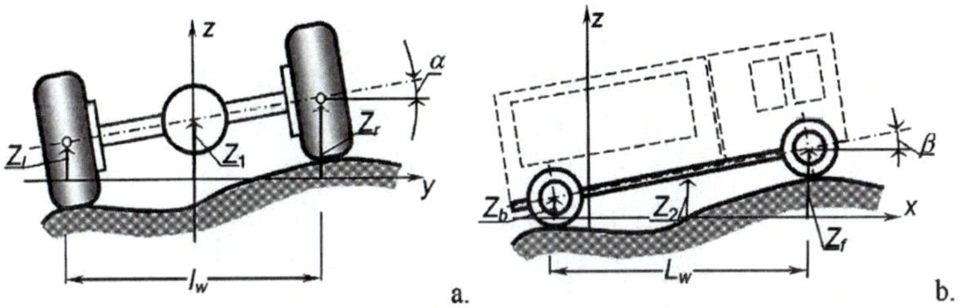

Fig. 2. Calculation schemes for determining the tilt (a) and the yaw (b) of a fire truck.

According to the technical characteristics of fire trucks and according to the recommendations [7], the maximum angular displacements and geometric characteristics of the reference track were adopted. When determining the rotation angles of the local relief, a separate sequence of surface irregularities is set for the right and left wheels of the rear axle using a pseudo-random number generator. To reproduce the possible options, three cases of tank movement with speeds of 6 km/h, 4 km/h and 2 km/h are considered. Thus, 3 sets of displacement laws for a fire truck tank were obtained. Figure 3, as an example, represents the laws of tank truck's movement for a speed of 6 km/h.

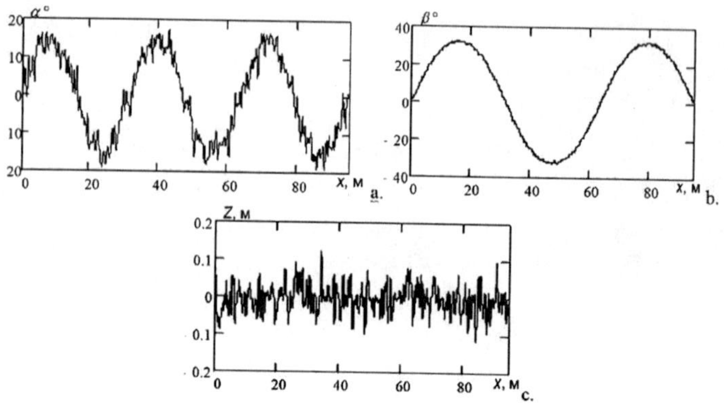

Fig. 3. Laws of tank movements when a fire truck moves through forest terrain at a speed of 6 km/h: a - tilt angle; b - yaw angle; in - vertical movement.

3.3 Mathematical Model of the Tank

Using the accepted assumptions about the tank of the fire truck, the finite-element scheme presented in Fig. 4.

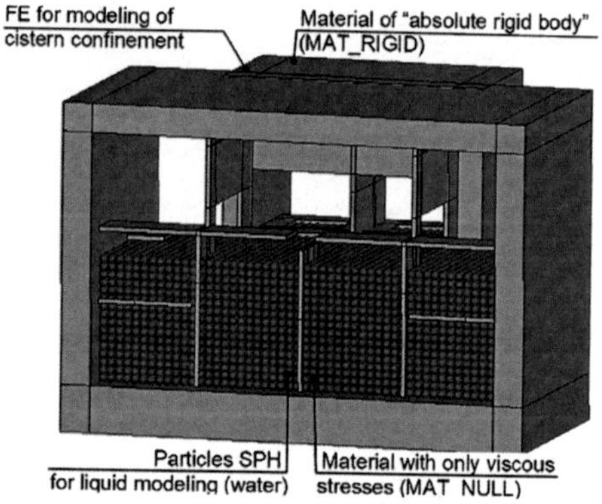

Fig. 4. Finite element scheme of the fire truck's tank.

The main components of the mathematical model of dynamic processes in the fire truck's tank during its movement through the rough forest terrain are given in Table 1.

Table 1. The main components of the mathematical model and methods of its implementation.

Component (method)	Way of implementation	Source
Basic system of equations	System of differential equations based on conservation laws	[5]

(*continued*)

Table 1. (*continued*)

Component (method)	Way of implementation	Source
Mathematical model of a liquid	Smoothed particles hydrodynamics method	[4, 5]
Numerical implementation method	Explicit method with finite element method approximation	[5]
Computing tool	Software complex LS-DYNA	[5]
Contact interaction	Penalty Method	[5]

4 The Results Obtained During the Research

The position of the water tank with the level of filling with 25% for different moments of time during its movement at a speed of 6 km/h by the reference track of rough forest terrain is shown in Fig. 5.

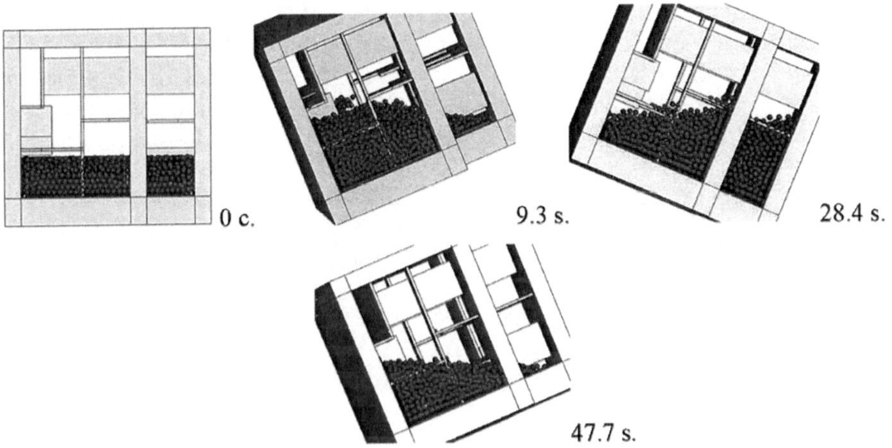

Fig. 5. Position of the tank with 25% level of water filling of the fire truck's tank for different moments of time at its movement at a speed of 6 km/h rough forest area.

In order to trace the position of the tank's masses center with different levels of its filling with water during the movement of the fire truck a trajectory of its movement at a speed of 6 km/h was built. Figure 6 presents the curves that reproduce these trajectories.

To identify the fluctuations' regularities of the masses center of the fire truck, a regression dependence was made by the method of full factorial experiment in accordance with [8]. The resulting regression looks like this:

$$y = 192.75 + 1.23 \cdot V - 21.625 \cdot v - 0.105 \cdot V \cdot v \tag{1}$$

where V is the level of filling the fire truck's tank (%), v is the speed of movement (km/h).

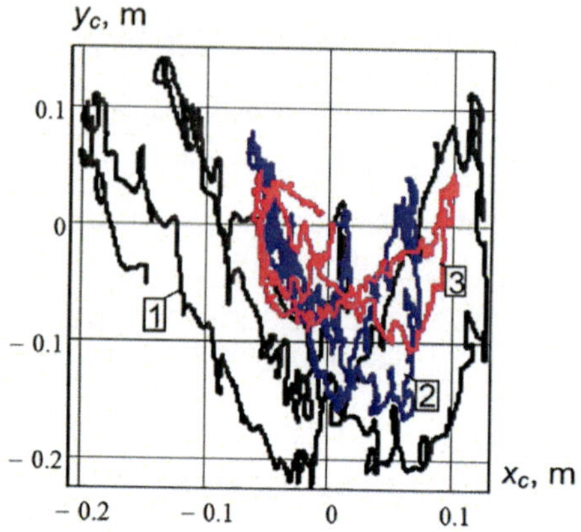

Fig. 6. The trajectories of the masses center position of the tank at a speed of the fire truck 6 km/h through rough forest terrain with different levels of water filling: 1–25%; 2–50%; 3–75%.

Assessing the influence of the inertia transverse force, the corresponding calculations of the difference between the moments of the resulting forces in the vertical and transverse directions were made according to the scheme in Fig. 7.

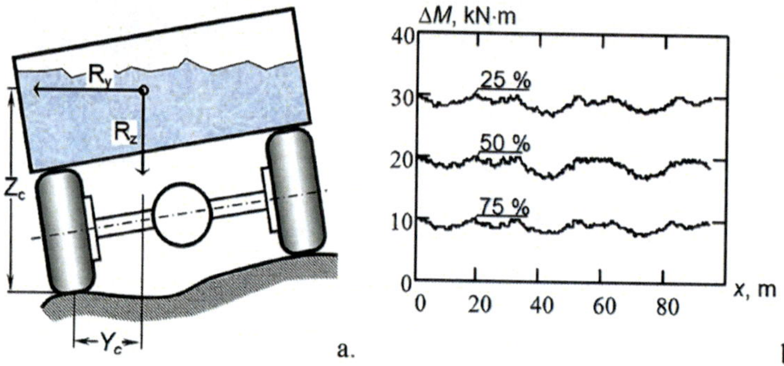

Fig. 7. Diagram for determining the moments' difference of equivalent forces in a fire truck (a) and moments' curves of fire truck's equivalent forces depending on time for different levels of the tank's filling (b).

The curves (look Fig. 7 (b)) of equivalent forces' moments in the tank depending on time show that the forces of inertia acting in the transverse direction do not significantly affect the possibility of overturning, as the difference between the vertical and horizontal components of the equivalent force have only positive values that are large enough. This can be explained by the effect of extinguishing large splashes and waves with the existing breakwaters, which are part of the tank design.

5 Conclusions

The conducted research allows to draw the following conclusions:

- the mathematical model on research of fluctuations of the tank's masses center with water depending on speed of the fire tank truck at its movement through rough forest terrain was created and the corresponding calculations were made;
- it was found that the fluctuations of the masses center of the fire truck's tank are most affected when it moves through rough forest terrain at a speed of 6 km/h when filling the tank with water by 25%;
- revealed fluctuations' regularities of the fire truck's masses center, on which the dependence's mathematical model of the maximum displacement of the masses center of the tank with water depending on speed and tank's filling level in the form of regression was obtained;
- it was proven that the inertia forces in the transverse direction do not significantly affect the possibility of overturning.

References

1. Nyzhnyk, V.V., Tarasenko, O.A., Kyrychenko, O.V., Kosiarum, S.O., Pozdieiev, S.V.: The criteria of estimating risks of spreading fire to adjacent building facilities. In: IOP Conference Series: Materials Science and Engineering, vol. 708, pp. 99–110. Kharkiv (2019)
2. Vikovich, I.A., Lavrivskiy, M.Z., Zinko, R.V.: Teoriya Adaptuvannya ta Zastosuvannya Pozhezhnyh Avtomobiliv Dlya Likvidacii Nadzvychaynyh Sytuatsiy. Lviv (2020)
3. Nicolici, S., Bilegan, R.M.: Fluid structure interaction modeling of liquid sloshing phenomena in flexible tanks. Nucl. Eng. Des. **258**, 51–56 (2013)
4. Monaghan, J.J., Gingold, R.A.: Shock Simulation by the particle method of SPH. J. Comput. Phys. **52** (1983)
5. Hallquist, J.O.: LS-DYNA Theory Manual. Livermore Software Technology Corporation, California, USA (2005)
6. Pozdieiev, S., Nizhnyk, V., Pidhoretskiy, Y.: Research of disclosure of relief venting structures with polycarbonate fencing in conditions of explosion. IOP Conf. Ser.: Mater Sci. Eng. **1021**, 012025 (2021)
7. TGL 12.240–220 4 × 2 BL (KI) EURO 6c: MAN Truck & Bus France. Autres Caractéristiques. In: Editor, F., Editor, S. (eds.) Conference 2016, LNCS, vol. 9999, pp. 1–13. Springer, Heidelberg (2016)
8. Johnson, N., Leone, F.: Statistics and experimental design in engineering and the physical sciences, vol. 1. New York (1977)

Study of Seismic Effect on Reinforced Concrete Building Due to Swimming Pool on Roof Top

Anish Bhujel[1], Mahesh Raj Bhatt[1](✉), and Prachand Man Pradhan[1,2]

[1] Kathmandu University, Dhulikhel, Nepal
mahesh.bhatt@ku.edu.np
[2] Manmohan Technical University, Morang, Nepal

Abstract. This study evaluates the dimensional parameters (area and depth) of the swimming pool that yields to minimum roof displacement of the Reinforced Concrete (RC) building under seismic loading and also compares the roof displacement of the building considering the water as static dead load and dynamic load (considering sloshing of water). The frame model considered for this study are 10 storied Reinforced Concrete models with swimming pool centrally located at the roof. The area ratio (ratio of the area of the pool to the top floor area of the building) is varied from 0.2 to 0.8 with an increment of 0.1 and the depths of the pool considered are 1.22 m (4 feet), 1.35 m (4.5 feet), and 1.52 m (5 feet). The Linear Time History Analysis (LTHA) of the building is performed considering the three recorded earthquake time histories and eight different sinusoidal loadings. Two different approaches (IS 1893(Part 2): 2014 code method and Yu et al. 1999 Model) are used to represent fluid-structure interaction. The roof displacements are obtained minimum for the case when the area ratio is 0.4 and the depth is 4 feet (1.22 m). Furthermore, the results show that responses are different while considering water as static load and dynamic load.

Keywords: Spring mass model · RC building · Area ratio · Time history analysis · Swimming pool · Water sloshing

1 Introduction

The trend of construction of building with terrace swimming pools is increasing in Nepal. The sloshing of water distinguishes such structures from normal behaviors. During the application of any load additionally with the strength of structure, the feeling of its occupants also matters. Consequently, while designing such a structure the limitation of maximum displacement or vibration should be considered. Now a day's tuned liquid damper (TLD) has been adopted to reduce vibrations of tall buildings. A tuned liquid damper is a passive damper, which relies on shallow liquid sloshing in a rigid tank to suppress the horizontal structural vibration induced by wind or other dynamic loads (Sun et al. 1992). Several studies were performed to reduce structural response using TLDs (Sun et al. 1992; Sun et al. 1995; Fujino et al. 1992; Banerji et al. 2000; Banerji et al. 2010). It will be beneficial if a swimming pool could be used to control excess vibration of

Z. Dimitrovová et al. (Eds.): WMVC 2022, MMS 125, pp. 1224–1234, 2023.
https://doi.org/10.1007/978-3-031-15758-5_126

buildings just like by a tuned liquid damper. TLD is found to be more effective when the fundamental frequency of liquid is tuned to the fundamental frequency of the structure (Banerji et al. 2000). The shallow liquid is often used to obtain high damping of liquid sloshing in TLD (Sun et al. 1992). Most of the researchers have restricted the depth ratio (ratio of the depth of water to a length of the tank) to less than 0.1 to maintain shallow depth and increase the effectiveness of the liquid damper (Banerji et al. 2000). However, the primary purpose of the swimming pool is for swimming so, it is not always possible to tune its frequency to the frequency of the structure and maintain the depth ratio to 0.1. So, the objective of this study is to evaluate the depth of the pool and area ratio that yields to least roof displacement under seismic loads. Similarly, another objective is to evaluate and compare the peak roof displacement (Maximum displacement occurring during Time History Analysis) of RC building considering the water as static dead load and dynamic load.

2 Literature Review

The most well-known model to represent fluid-structure interaction is Housner's Model. According to Housner (1963), if a tank is partially filled then the total mass is divided into two masses i.e., impulsive mass (m_o) and convective mass (m_c). The mass that moves along with the reservoir is called impulsive mass (m_o) and another mass that represents the sloshing behaviors is called convective mass (m_c). Similarly, K_c is the stiffness coefficient for convective mass. The mechanical model by Housner is shown in "Fig. 1".

IS 1893 (Part 2): 2014 and ACI 350.3 (2001) also provide the spring-mass model's parameters for the analysis of liquid retaining structures, but they are slightly different from those given by Housner (1963). The expressions for parameters of the spring-mass model for a rectangular tank in ACI 350.3 (2001) which are similar to parameters of the spring-mass model in IS 1893 (Part 2): 2014 are as follows:

$$\frac{m_i}{m} = \frac{\tanh\left(0.866\frac{L}{h}\right)}{0.866\frac{L}{h}} \tag{1}$$

$$\frac{h_i}{h} = 0.375 \quad \text{for } h/L \leq 0.75 \tag{2}$$

$$= 0.5 - \frac{0.09375}{h/L} \text{ for } h/L > 0.75 \tag{3}$$

$$\frac{m_c}{m} = 0.264\frac{\tanh\left(3.16\frac{h}{L}\right)}{h/L} \tag{4}$$

$$\frac{h_c}{h} = 1 - \frac{\cosh\left(3.16\frac{h}{L}\right) - 1.0}{3.16\frac{h}{L}\sinh\left(3.16\frac{h}{L}\right)} \tag{5}$$

$$K_C = 0.833\frac{mg}{h}\tanh^2\left(3.16\frac{h}{L}\right) \tag{6}$$

where m is total mass of water, m_i is impulsive mass, m_c is convective mass, h is the depth of water in a tank, h_i is height of impulsive mass, h_c is height of convective mass, g is the acceleration due to gravity, and K_c is the stiffness coefficient for convective mass.

Yu et al. (1999) addressed the nonlinearities of the dynamic behavior of water in a TLD through non-linear stiffness and damping coefficient by representing the TLD as a TMD (Tuned Mass Damper) as shown in "Fig. 2". Ultimately, the following equations define the stiffness and damping coefficients of the liquid (Malekghasemi et al. 2015).

$$k_d = \begin{cases} 1.075\left(\frac{A}{L}\right)^{0.07} M(2\pi f)^2, \text{ if } \left(\frac{A}{L}\right) \le 0.03 \\ 2.52\left(\frac{A}{L}\right)^{0.25} M(2\pi f)^2, \text{ if } \left(\frac{A}{L}\right) > 0.03 \end{cases} \quad (7)$$

$$c_d = 1.04\sqrt{Mk_d}\left(\frac{A}{L}\right)^{0.35} \quad (8)$$

$$f = \left(\frac{1}{2\pi}\right)\sqrt{\frac{\pi g}{L}\tanh\left(\frac{\pi h}{L}\right)} \quad (9)$$

where k_d is the nonlinear damper stiffness coefficient, c_d is the nonlinear damping coefficient, M is the total water mass equal to m_w, A is the amplitude of excitation, g is the acceleration due to gravity, L is the length of the tank, h is the depth of water in tank and f is the fundamental frequency of the fluid that depends on the surface waves.

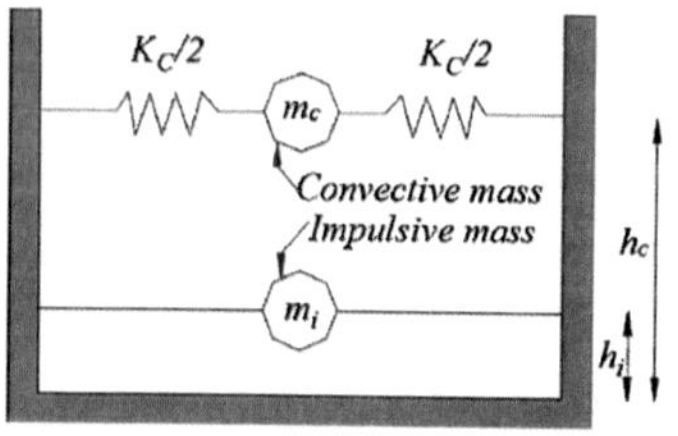

Fig. 1. Housner's model for water inside the tank.

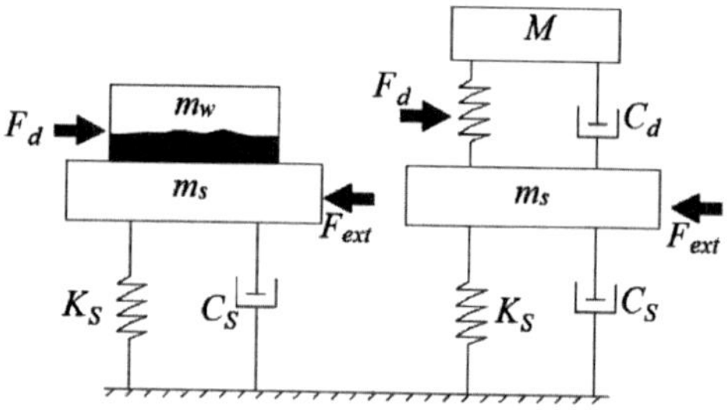

Fig. 2. Representation of a TLD. (Yu et al. 1999)

3 Methodology

For this study, 10 storied building with 3 bays in each direction with a spacing of 4 m and floor height of 3 m is considered. The area ratio is varied from 0.2 to 0.8 with an increment of 0.1 and the depths of the pool considered are 1.22 m (4 feet), 1.35 m (4.5 feet), and 1.52 m (5 feet). Three-Dimensional Finite Element Analysis of the building model is done in ETABS V 17.0.1. The analysis considerations are shown in "Table 1".

Three different ground motion time histories are selected where two ground motions satisfy the strong ground motion criteria by (Maniatakis et al. 2008). The ground motion is then matched with the target spectrum (IS 1893(Part 1): 2016) in the Time domain. Similarly, eight different sinusoidal loadings with 0.1g acceleration amplitude are selected and frequencies are taken such that the frequency ratio (Ratio of frequency of structure to the frequency of excitation) varies from 0.5 to 1.2.

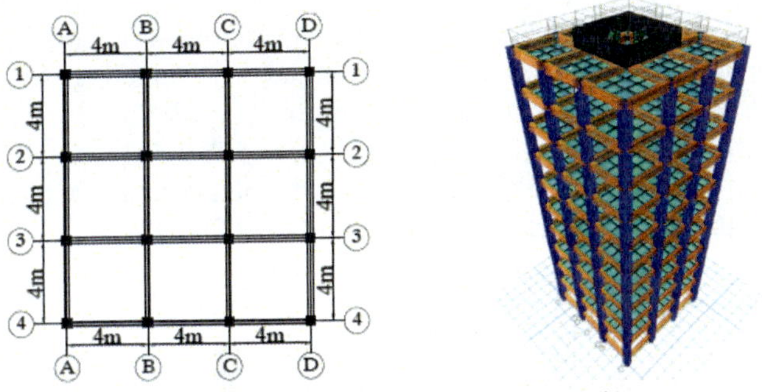

Fig. 3. (a) Typical Plan; and (b) 3D Finite Element Model (ETABS).

Table 1. Analysis and Design Consideration (IS 1893 (Part 1): 2016)

Particulars	Details	Particular	Details
seismic zone	Zone-V	Damping ratio	5%
Zone factor, Z	0.36	Grade of concrete	M25
Building frame system	SMRF	Grade of reinforcement	Fe-500
Response reduction factor. R	5	beam size	300 mm × 550 mm
Importance factor, I	1	Column size	550 mm × 550 mm
Rock/Soil type	Medium Soil	Slab thickness	125 mm

Spring mass model in IS 1893 (Part 2): 2014 and Yu et al. (1999) are used to represent fluid-structure interaction. Spring mass model in IS 1893(Part 2): 2014 provides the impulsive mass, convective mass, the height of impulsive and convective masses, and the spring stiffness neglecting damping factor. The spring stiffness provided is constant throughout the analysis of a particular structural model. Consequently, the modeling of the spring-mass model in ETABS is possible using linear links and the joint masses. However, Yu et al. (1999) model considered non-linear stiffness and non-linear damping dependent on the amplitude of excitation. The stiffness coefficient and damping coefficient need to be updated as per the amplitude of roof displacement so MATLAB coding is used to solve the Multi-Degree of Freedom (MDOF) system using Newmark's method. In 1959, N.M. Newmark developed a time-stepping method based on "Eqs. (10) and

(11)" to solve the equation of motion for the case where the excitation applied force or ground motion varies arbitrarily with time or if the system is non-linear (Chopra 2008):

$$\dot{u}_{i+1} = \dot{u}_i + \left[(1-\gamma)\Delta t\right]\ddot{u}_i + (\gamma\ \Delta t)\dot{u}_{i+1} \tag{10}$$

$$u_{i+1} = u_i + (\Delta t)\dot{u}_i + \left[(0.5-\beta)(\Delta t)^2\right]\ddot{u}_i + \left[\beta(\Delta t)^2\right]\ddot{u}_{i+1} \tag{11}$$

where $\ddot{u}_i$, $\dot{u}_i$, and u_i are the acceleration, velocity, and displacement of the respective masses. γ and β are constants that define the acceleration variation within each time interval (Δt) and determine the stability and accuracy of the method. So, in this study algorithm based on Newmark's method is used to solve MDOF for Yu et al. (1999) spring-mass model.

Similarly, in water as static load case, the load is calculated and applied as pressure intensity using simple hydrostatic formula as shown in "Fig. 4".

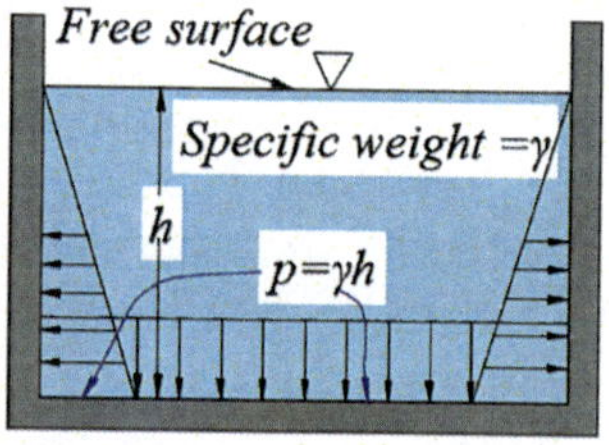

Fig.4. Pressure Distribution on the wall and bottom of a pool (Static Load). (Bhujel 2022)

The parameters (types of building model, varieties in ground motion) considered for this study are limited and the site consideration of building is only for zone V as per IS 1893 (Part 1): 2016, the conclusions may not be general however this study describes the effect of the geometric configuration of the pool on the response of building and feasibilities of swimming pool to reduce the response of building under considered circumstances using analytical approaches.

4 Results and Discussions

4.1 Comparison of Roof Displacement Considering Water Load as Dynamic Load (Sloshing of Water is Considered)

Spring mass model in IS 1893 (Part 2): 2014 is used to represent the fluid-structure interaction. The Linear Time History Analysis method using three recorded ground motions and eight sinusoidal loadings with different excitation frequency ratios are used.

Linear Time History Method using Recorded Ground Motion

"Figure 5a, 5b, and 5c" show the comparison of peak roof displacement for 1.22 m (4 Feet), 1.35 m (4.5 feet), and 1.52 m (5 feet) depth pool respectively. The Peak roof displacements of building without swimming pools are 159.33 mm, 153.66 mm, and 147.71 mm for the Kobe Japan earthquake, Northridge earthquake, and El Centro earthquake respectively.

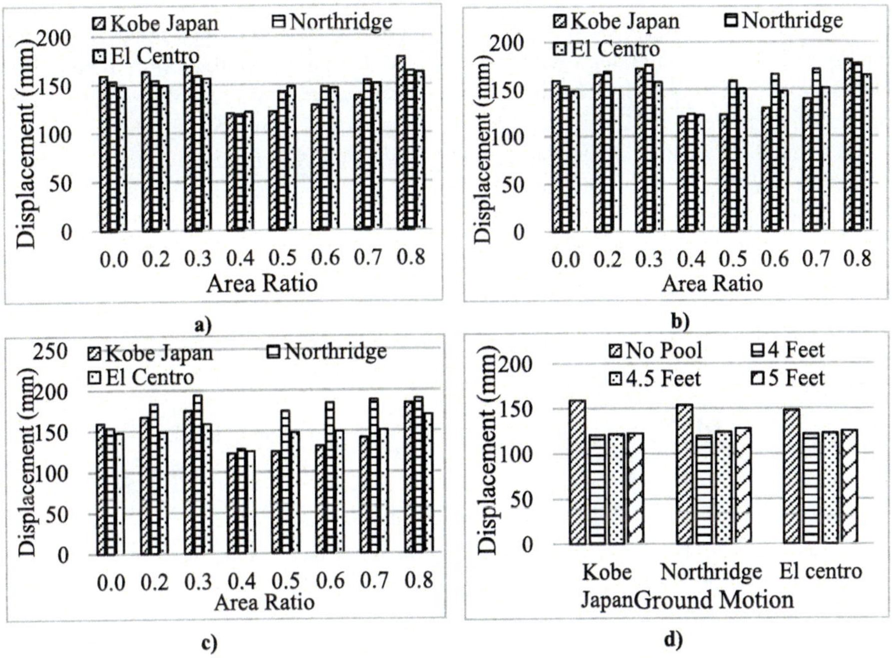

Fig. 5. Comparison of Peak Roof Displacement for **a)** 1.22 m (4 Feet) Depth Pool, **b)** 1.35 m (4.5 Feet) Depth Pool, **c)** 1.52 m (5 Feet) Depth Pool and **d)** 0.4 Area Ratio and different pool Depth.

"Figure 5a, 5b, and 5c" show that for each depth, the area ratio that yields to minimum peak roof displacement is 0.4 and "Fig. 5d". Shows that at area ratio of 0.4, the maximum reduction in peak roof displacement is at 1.22 m (4 feet) depth pool. Ultimately, the area ratio and depth corresponding to the maximum reduction in peak roof displacement is found to be 0.4 and 1.22 m (4 feet). The roof displacement for area ratio of 0.4 and depth 1.22 m (4 feet) are 120.59 mm, 119.49 mm, and 121.51 mm for Kobe Japan earthquake, Northridge earthquake, and El Centro earthquake respectively.

Consequently, the percentage reductions are 24.31%, 22.23%, and 17.74% for the Kobe Japan earthquake, Northridge earthquake, and El Centro earthquake respectively.

Linear Time History Method with Sinusoidal Loading
"Figure 6a, 6b, and 6c" show the comparison of peak roof displacement for 1.22 m (4 Feet), 1.35 m (4.5 feet), and 1.52 m (5 feet) depth pool respectively. The peak roof displacements of a building without a swimming pool are 44.20 mm, 63.95 mm, 81.98 mm, 109.75 mm, 159.80 mm, 278.50 mm, 182.33 mm, and 147.16 mm for eight sinusoidal loading with different frequency ratios.

"Figure 6a, 6b, and 6c" show that for each depth, the area ratio that yields to minimum peak roof displacement is about 0.4 and "Fig. 6d" shows that at area ratio 0.4, the maximum reduction in peak roof displacement is at 1.22 m (4 feet) depth pool. Ultimately, the area ratio and depth corresponding to the maximum reduction in peak roof displacement is found to be 0.4 and 1.22 m (4 feet). The peak roof displacement for area ratio of 0.4 and depth 1.22 m (4 feet) are 42.98 mm, 52.13 mm, 56.65 mm, 97.04 mm, 134.78 mm, 121.67 mm, 159.63 mm, and 139.41 mm. Consequently, the percentage reductions are 2.76%, 18.48%, 30.89%, 11.58%, 15.65%, 56.31%, 12.45%, and 5.26% for eight sinusoidal loading with different time period. It is also observed that for an area ratio of 0.4 and at each depth, the maximum reduction in roof displacement is observed when the frequency ratio is unity.

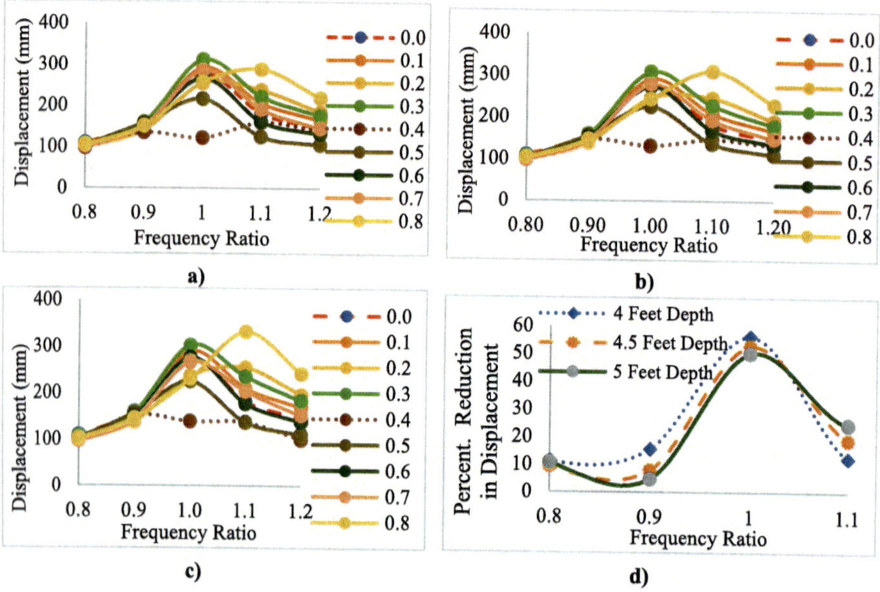

Fig. 6. Comparison of Peak Roof Displacement for **a)** 1.22 m (4 Feet) Depth Pool, **b)** 1.35 m (4.5 Feet) Depth Pool, **c)** 1.52 m (5 Feet) Depth Pool and **d)** 0.4 Area Ratio and different pool Depth.

4.2 Results from YU'S Et.Al Model (Modal Analysis in MATLAB Using NEWMARk's Method)

The Linear Time History Analysis method for Yu et al (1999) model is performed using Newmark's Method in MATLAB. The results of peak roof displacement from time history analysis using three recorded ground motions are described here. "Fig. 7a, 7b, and 7c" show the comparison of peak roof displacement for 1.22 m (4 Feet), 1.35 m (4.5 feet), and 1.52 m (5 feet) depth pool respectively.

"Figure 7a, 7b, and 7c" show that for each depth, the area ratio that yields to minimum peak roof displacement is 0.4 and "Fig. 7d" shows that at area ratio of 0.4, the maximum reduction in peak roof displacement is at 1.22 m (4 feet) depth pool. Ultimately, the area ratio and depth corresponding to the maximum reduction in peak roof displacement is found to be 0.4 and 1.22 m (4 feet).

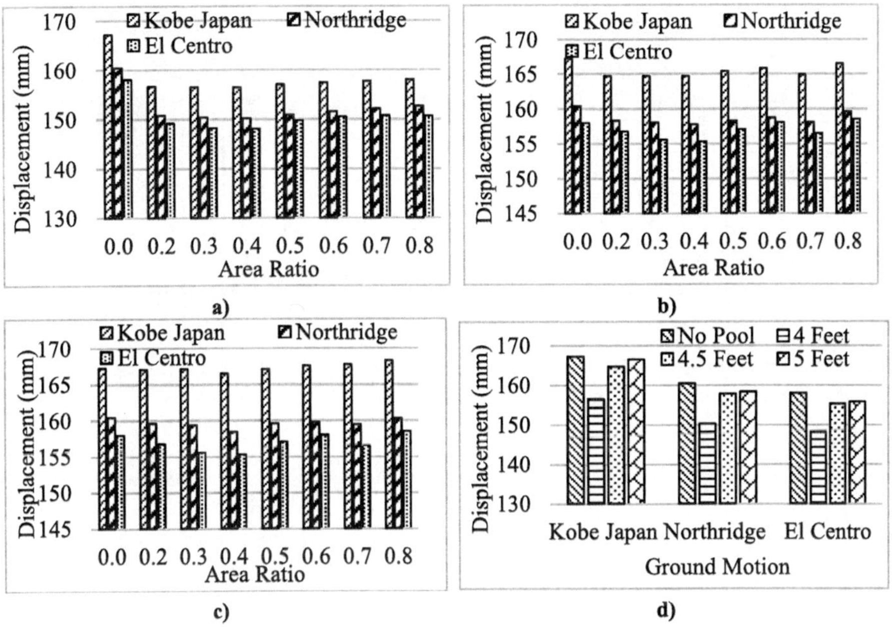

Fig. 7. Comparison of Peak Roof Displacement (mm) for **a)** 1.22 m (4 Feet) Depth Pool, **b)** 1.35 m (4.5 Feet) Depth Pool, **c)** 1.52 m (5 Feet) Depth Pool and **d)** for 0.4 Area Ratio and different pool Depth.

From each method of analysis and different loading cases, it is observed that the area ratio and the depth corresponding to minimum peak roof displacement are 0.4 and 1.22 m (4 feet) respectively. Similarly, the percentage reductions in roof displacement are higher for the ground motion satisfying the strong ground motion criteria. The damping increases with increase in sloshing of water masses. For more sloshing, strong excitation force, more water mass, and area ratio are required however the increase in water mass and the area also increases the inertial force so there exists an area ratio and depth that contribute to damping more than to inertial force. Consequently, there exists a particular depth and an area of pool which yield to the minimum roof displacement.

4.3 Comparison of Roof Displacement Considering Water Load as Dead Load and Dynamic Load

The Peak roof displacement obtained from two approaches 1) water as static dead load where the water load is applied simply as uniform varying load and 2) water as dynamic load, where the spring-mass model (IS codal method) is used to represent dynamic load, is compared. The Linear Time History method using three recorded ground motions is used. The results obtained from the three recorded ground motions are of a similar pattern so, the results for Kobe Japan and for 1.22 m (4 feet) and 1.52 m (5 feet) depth are only shown here.

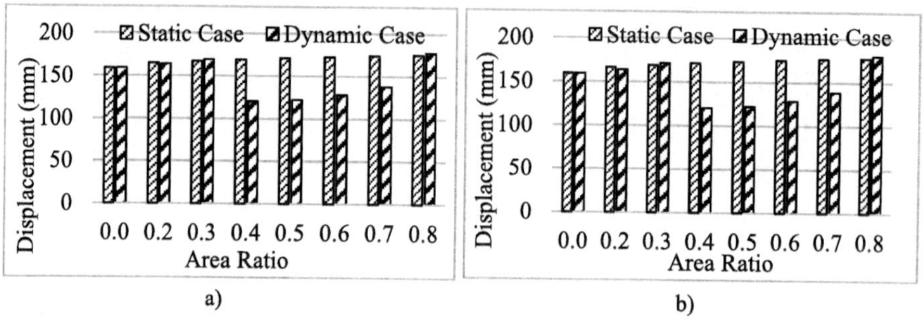

Fig. 8. Comparison of Peak Roof Displacement (mm) for water load as Static Case and Dynamic Case for **a)** 4 Feet Depth Pool, **b)** 5 Feet Depth Pool.

The peak roof displacements observed are different for both cases (water load as static dead and dynamic load) in each area ratio and depth of pool as shown in "Fig. 8a and 8b" however there is significantly higher difference in peak roof displacement for area ratio 0.4. The percentage differences in peak roof displacement at 0.4 area ratio are 40.57% and 41.28% for 1.22 m (4 feet) and 1.52 m (5 feet) depth respectively. It is also observed that the roof displacements are increasing with an increase in area ratio when water load is considered as static dead load. In the case of water as a static dead load, an increase in area ratio and depth increases the dead load at the roof without contributing to any damping. However, in the case of water as dynamic load where sloshing phenomena occur, water also contributes to damping through the sloshing phenomena along with the increase in load. So, the responses observed are different for the two cases.

5 Summary and Conclusions

This study presents the evaluation of the swimming pool existing in the building to reduce the response (peak roof displacement) and also compares the difference in response of building while considering water as static load and dynamic load. The Linear Time History Method is used to obtain the roof displacement for three recorded ground motions and eight sinusoidal loadings. Following conclusions can be made from the study:

1. The roof displacement of buildings while considering water as static load and dynamic load are different. The maximum difference in roof displacement occurs at

area ratio and depth where excessive reduction in roof displacement occurs in the case of water as dynamic load (i.e., at Area ratio 0.4 and depth 4 feet).

2. In water as static load cases, the roof displacement increases with an increase in area ratio and depth, which is due to increase in water mass as a static dead load at top of the building.
3. In water as dynamic load cases, the roof displacements were obtained minimum when the area ratio is 0.4 and the depth of 4 feet (1.22 m).
4. In water as dynamic load cases, reduction in roof displacement is higher when the frequency ratio is equal to 1(which is observed from the Linear Time History Analysis using sinusoidal loading time history). Thus, the roof displacement also depends upon the frequency of the loading.
5. Reduction in roof displacement is comparatively more for the ground motion satisfying the strong ground motion criteria (Kobe Japan and Northridge).

References

1. ACI 350.3: Seismic Design of Liquid Containing Concrete Structures. American Concrete Institute, Michigan (2001)
2. Banerji, P., Murudi, M., Shah, A., Popplewell, N.: Tuned liquid dampers for controlling earthquake responses of structure. Earthquake Eng. Struct. Dynam. **29**(5), 587–602 (2000). https://doi.org/10.1002/(SICI)1096-9845(200005)29:5%3c587::AID-EQE926%3e3.0.CO;2-I
3. Banerji, P., Samanta, A., Chavan, S.: Earthquake vibration control of structures using tuned liquid dampers: experimental studies. Int. J. Adv. Struct. Eng. **2**(2), 133–152 (2010). https://www.sid.ir/en/journal/ViewPaper.aspx?id=209216
4. Bhujel, A.: Seismic Performance of RCC Building with Swimming Pool on the Terrace. Unpublished Master's thesis, Kathmandu University, Nepal (2022)
5. Chopra, A.K.: Dynamics of Structures, Theory and Applications to Earthquake Engineering, 3rd edn. Pearson Education, Inc., Englewood Cliffs (2008)
6. Fujino, Y., Sun, L.M., Pacheco, B.M., Chaiseri, P.: Tuned Liquid Damper (TLD) for suppressing horizontal Motion of Structures. J. Eng. Mech. **118**(10), 2017–2030 (1992). https://doi.org/10.1061/(ASCE)0733-9399(1992)118:10(2017)
7. Housner, G.W.: The dynamic behaviour of water tanks. Bull. Seismol. Soc. Am. **53**(2), 381–387 (1963). https://doi.org/10.1785/BSSA0530020381
8. IS 1893 (Part 1): Criteria for Earthquake Resistant Design of Structures. Bureau of Indian Standards, Delhi, India (2016)
9. IS 1893 (Part 2): Criteria for Earthquake Resistant Design of Structures. Bureau of Indian Standards, Delhi, India (2014)
10. Malekghasemi, H., Ashasi-Sorkhabi, A., Ghaemmaghami, A.R., Mercan, O.: Experimental and numerical investigations of the dynamic interaction of tuned liquid damper–structure systems. J. Vib. Control **21**(14), 2707–2720 (2013). https://doi.org/10.1177/1077546313514759
11. Maniatakis, C.A., Taflambas, I.M., Spyrakos, C.C.: Identification of Near-Fault earthquake record characteristics. In The 14th World Conference on Earthquake Engineering. Beijing, China (2008)
12. Sun, L.M, Fujino, Y., Chaiseri, P., Pacheco, B.M: The properties of Tuned Liquid Dampers Using a TMD Analogy. Earthquake Engineering & Structural Dynamics 24(7), 967–976. (1995). https://doi.org/10.1002/eqe.4290240704

13. Sun, L.M., Fujino, Y., Pacheco, B.M., Chaiseri, P.: Modeling of tuned liquid damper. J. Wind Eng. Ind. Aerodyn. **43**(1–3), 1883–1894 (1992). https://doi.org/10.1016/0167-6105(92)90609-E
14. Yu, J.-K., Toshihiro, W., A. Reed, D.: A non-linear numerical model of the tuned liquid damper. Earthq. Eng. Struct. Dyn. **28**(6), 671–686 (1999). https://doi.org/10.1002/(SICI)1096-9845(199906)28:6<671::AID-EQE835>3.0.CO;2-X